Osvaldo Gervasi   Marina L. Gavrilova
Vipin Kumar   Antonio Laganà
Heow Pueh Lee   Youngsong Mun
David Taniar   Chih Jeng Kenneth Tan (Eds.)

Computational Science and Its Applications – ICCSA 2005

International Conference
Singapore, May 9-12, 2005
Proceedings, Part I

Volume Editors

Osvaldo Gervasi
University of Perugia
E-mail: ogervasi@computer.org

Marina L. Gavrilova
University of Calgary
E-mail: marina@cpsc.ucalgary.ca

Vipin Kumar
University of Minnesota
E-mail: kumar@cs.umn.edu

Antonio Laganà
University of Perugia
E-mail: lag@dyn.unipg.it

Heow Pueh Lee
Institute of High Performance Computing, IHPC
E-mail: hplee@ihpc.a-star.edu.sg

Youngsong Mun
Soongsil University
E-mail: mu@computing.soongsil.ac.kr

David Taniar
Monash University
E-mail: David.Taniar@infotech.monash.edu.au

Chih Jeng Kenneth Tan
Queen's University Belfast
E-mail: cjtan@optimanumerics.com

Library of Congress Control Number: Applied for

CR Subject Classification (1998): D, F, G, H, I, J, C.2.3

ISSN	0302-9743
ISBN-10	3-540-25860-4 Springer Berlin Heidelberg New York
ISBN-13	978-3-540-25860-5 Springer Berlin Heidelberg New York

This work is subject to copyright. All rights are reserved, whether the whole or part of the material is concerned, specifically the rights of translation, reprinting, re-use of illustrations, recitation, broadcasting, reproduction on microfilms or in any other way, and storage in data banks. Duplication of this publication or parts thereof is permitted only under the provisions of the German Copyright Law of September 9, 1965, in its current version, and permission for use must always be obtained from Springer. Violations are liable to prosecution under the German Copyright Law.

Springer is a part of Springer Science+Business Media

springeronline.com

© Springer-Verlag Berlin Heidelberg 2005
Printed in Germany

Typesetting: Camera-ready by author, data conversion by Scientific Publishing Services, Chennai, India
Printed on acid-free paper   SPIN: 11424758   06/3142   5 4 3 2 1 0

Preface

The four-volume set assembled following *the 2005 International Conference on Computational Science and Its Applications*, ICCSA 2005, held in Suntec International Convention and Exhibition Centre, Singapore from 9 May 2005 till 12 May 2005, represents the fine collection of 540 refereed papers selected from nearly 2700 submissions.

Computational science has firmly established itself as a vital part of many scientific investigations, affecting researchers and practitioners in areas ranging from applications such as aerospace and automotive, to emerging technologies such as bioinformatics and nanotechnologies, to core disciplines such as mathematics, physics, and chemistry. Due to the sheer size of many challenges in computational science, the use of supercomputing, parallel processing, and sophisticated algorithms is inevitable and becomes a part of fundamental theoretical research as well as endeavors in emerging fields. Together, these far-reaching scientific areas contribute to shape this conference in the realms of state-of-the-art computational science research and applications, encompassing the facilitating theoretical foundations and the innovative applications of such results in other areas.

The topics of the refereed papers span all the traditional as well as emerging computational science realms, and are structured according to six main conference themes:
- Computational Methods and Applications
- High-Performance Computing, Networks and Optimization
- Information Systems and Information Technologies
- Scientific Visualization, Graphics and Image Processing
- Computational Science Education
- Advanced and Emerging Applications

In addition, papers from 27 workshops and technical sessions on specific topics of interest, including information security, mobile communication, grid computing, modeling, optimization, computational geometry, virtual reality, symbolic computations, molecular structures, Web systems and intelligence, spatial analysis, bioinformatics and geocomputation, to name a few, complete this comprehensive collection.

The warm response of the great number of researchers to the offer to present high-quality papers in ICCSA 2005 took the conference to record heights. The continuous support of computational science researchers has helped build ICCSA into a firmly established forum in this area. We look forward to building on this symbiotic relationship together to grow ICCSA further.

We recognize the contribution of the International Steering Committee and we deeply thank the International Program Committee for their tremendous support in putting this conference together, nearly 900 referees for their diligent work, and the Institute of High Performance Computing, Singapore for its generous assistance in hosting the event.

We also thank our sponsors for their continuous support without which this conference would not have been possible.

Finally, we thank all authors for their submissions and all invited speakers and conference attendees for making the ICCSA conference truly one of the premium events in the scientific community, facilitating the exchange of ideas, fostering new collaborations, and shaping the future of computational science.

May 2005

Marina L. Gavrilova
Osvaldo Gervasi

on behalf of the co-editors

Vipin Kumar
Antonio Laganà
Heow Pueh Lee
Youngsong Mun
David Taniar
Chih Jeng Kenneth Tan

Organization

ICCSA 2005 was organized by the Institute of High Performance Computing (Singapore), the University of Minnesota (Minneapolis, MN, USA), the University of Calgary (Canada) and the University of Perugia (Italy).

Conference Chairs

Vipin Kumar (Army High Performance Computing Center and University of Minnesota, USA), Honorary Chair
Marina L. Gavrilova (University of Calgary, Canada), Conference Co-chair, Scientific
Osvaldo Gervasi (University of Perugia, Italy), Conference Co-chair, Program
Jerry Lim (Institute of High Performance Computing, Singapore), Conference Co-chair, Organization

International Steering Committee

Alexander V. Bogdanov (Institute for High Performance Computing and Information Systems, Russia)
Marina L. Gavrilova (University of Calgary, Canada)
Osvaldo Gervasi (University of Perugia, Italy)
Kurichi Kumar (Institute of High Performance Computing, Singapore)
Vipin Kumar (Army High Performance Computing Center and University of Minnesota, USA)
Andres Iglesias (University de Cantabria, Spain)
Antonio Laganà (University of Perugia, Italy)
Heow Pueh Lee (Institute of High Performance Computing, Singapore)
Youngsong Mun (Soongsil University, Korea)
Chih Jeng Kenneth Tan (OptimaNumerics Ltd., and Queen's University Belfast, UK)
David Taniar (Monash University, Australia)

Local Organizing Committee

Kurichi Kumar (Institute of High Performance Computing, Singapore)
Heow Pueh Lee (Institute of High Performance Computing, Singapore)

Workshop Organizers

Approaches or Methods of Security Engineering
Haeng Kon Kim (Catholic University of Daegu, Korea)
Tai-hoon Kim (Korea Information Security Agency, Korea)

Authentication, Authorization and Accounting
Eui-Nam John Huh (Seoul Women's University, Korea)

Component-Based Software Engineering and Software Process Models
Haeng Kon Kim (Catholic University of Daegu, Korea)

Computational Geometry and Applications (CGA 2005)
Marina Gavrilova (University of Calgary, Canada)

Computer Graphics and Geometric Modeling (TSCG 2005)
Andres Iglesias (University of Cantabria, Spain)
Deok-Soo Kim (Hanyang University, Korea)

Computer Graphics and Rendering
Jiawan Zhang (Tianjin University, China)

Data Mining and Bioinformatics
Xiaohua Hu (Drexel University, USA)
David Taniar (Monash University, Australia)

Digital Device for Ubiquitous Computing
Hong Joo Lee (Daewoo Electronics Corp, Korea)

Grid Computing and Peer-to-Peer (P2P) Systems
Jemal H. Abawajy (Deakin University, Australia)
Maria S. Perez (Universitad Politecnica de Madrid, Spain)

Information and Communication Technology (ICT) Education
Woochun Jun (Seoul National University, Korea)

Information Security and Hiding, ISH 2005
Raphael C.W. Phan (Swinburne University of Technology, Sarawak, Malaysia)

Intelligent Multimedia Services and Synchronization in Mobile Multimedia Networks

Dong Chun Lee (Howon University, Korea)
Kuinam J. Kim (Kyonggi University, Korea)

Information Systems Information Technologies (ISIT)

Youngsong Mun (Soongsil University, Korea)

Internet Comunications Security (WICS)

Josè Sierra-Camara (University Carlos III of Madrid, Spain)
Julio Hernandez-Castro (University Carlos III of Madrid, Spain)
Antonio Izquierdo (University Carlos III of Madrid, Spain)
Joaquin Torres (University Carlos III of Madrid, Spain)

Methodology of Information Engineering

Sangkyun Kim (Somansa Ltd., Korea)

Mobile Communications

Hyunseung Choo (Sungkyunkwan University, Korea)

Modeling Complex Systems

Heather J. Ruskin (Dublin City University, Ireland)
Ruili Wang (Massey University, New Zealand)

Modeling of Location Management in Mobile Information Systems

Dong Chun Lee (Howon University, Korea)

Molecular Structures and Processes

Antonio Laganà (University of Perugia, Italy)

Optimization: Theories and Applications (OTA 2005)

In-Jae Jeong (Hanyang University, Korea)
Dong-Ho Lee (Hanyang University, Korea)
Deok-Soo Kim (Hanyang University, Korea)

Parallel and Distributed Computing

Jiawan Zhang (Tianjin University, China)

Pattern Recognition and Ubiquitous Computing

Woongjae Lee (Seoul Women's University, Korea)

Spatial Analysis and GIS: Local or Global?

Stefania Bertazzon (University of Calgary, Canada)
Borruso Giuseppe (University of Trieste, Italy)
Falk Huettmann (Institute of Arctic Biology, USA)

Specific Aspects of Computational Physics for Modeling Suddenly Emerging Phenomena

Paul E. Sterian (Politehnica University, Romania)
Cristian Toma (Titu Maiorescu University, Romania)

Symbolic Computation (SC 2005)

Andres Iglesias (University of Cantabria, Spain)
Akemi Galvez (University of Cantabria, Spain)

Ubiquitous Web Systems and Intelligence

David Taniar (Monash University, Australia)
Wenny Rahayu (La Trobe University, Australia)

Virtual Reality in Scientific Applications and Learning (VRSAL 2005)

Osvaldo Gervasi (University of Perugia, Italy)
Antonio Riganelli (University of Perugia, Italy)

Program Committee

Jemal Abawajy (Deakin University, Australia)
Kenny Adamson (EZ-DSP, UK)
Srinivas Aluru (Iowa State University, USA)
Frank Baetke (Hewlett-Packard, USA)
Mark Baker (Portsmouth University, UK)
Young-Cheol Bang (Korea Polytechnic University, Korea)
David Bell (Queen's University Belfast, UK)
Stefania Bertazzon (University of Calgary, Canada)
Sergei Bespamyatnikh (Duke University, USA)
J.A. Rod Blais (University of Calgary, Canada)
Alexander V. Bogdanov (Institute for High Performance Computing and Information Systems, Russia)
Richard P. Brent (University of Oxford, UK)
Peter Brezany (University of Vienna, Austria)
Herve Bronnimann (Polytechnic University, NY, USA)
John Brooke (University of Manchester, UK)
Martin Buecker (Aachen University, Germany)
Rajkumar Buyya (University of Melbourne, Australia)
YoungSik Choi (University of Missouri, USA)

Hyunseung Choo (Sungkyunkwan University, Korea)
Bastien Chopard (University of Geneva, Switzerland)
Min Young Chung (Sungkyunkwan University, Korea)
Toni Cortes (Universidad de Catalunya, Spain)
Yiannis Cotronis (University of Athens, Greece)
Danny Crookes (Queen's University Belfast, UK)
Josè C. Cunha (New University of Lisbon, Portugal)
Brian J. d'Auriol (University of Texas at El Paso, USA)
Alexander Degtyarev (Institute for High Performance Computing and Data Bases, Russia)
Frédéric Desprez (INRIA, France)
Tom Dhaene (University of Antwerp, Belgium)
Beniamino Di Martino (Second University of Naples, Italy)
Hassan Diab (American University of Beirut, Lebanon)
Ivan Dimov (Bulgarian Academy of Sciences, Bulgaria)
Iain Duff (Rutherford Appleton Laboratory, UK and CERFACS, France)
Thom Dunning (NCSA, USA)
Fabrizio Gagliardi (CERN, Switzerland)
Marina L. Gavrilova (University of Calgary, Canada)
Michael Gerndt (Technical University of Munich, Germany)
Osvaldo Gervasi (University of Perugia, Italy)
Bob Gingold (Australian National University, Australia)
James Glimm (SUNY Stony Brook, USA)
Christopher Gold (Hong Kong Polytechnic University, China)
Yuriy Gorbachev (Institute of High Performance Computing and Information Systems, Russia)
Andrzej Goscinski (Deakin University, Australia)
Jin Hai (Huazhong University of Science and Technology, China)
Ladislav Hlucky (Slovak Academy of Science, Slovakia)
Shen Hong (Japan Advanced Institute of Science and Technology, Japan)
Paul Hovland (Argonne National Laboratory, USA)
Xiaohua Hu (Drexel University, USA)
Eui-Nam John Huh (Seoul Women's University, Korea)
Terence Hung (Institute of High Performance Computing, Singapore)
Andres Iglesias (University of Cantabria, Spain)
In-Jae Jeong (Hanyang University, Korea)
Elisabeth Jessup (University of Colorado, USA)
Peter K. Jimack (University of Leeds, UK)
Christopher Johnson (University of Utah, USA)
Benjoe A. Juliano (California State University at Chico, USA)
Peter Kacsuk (MTA SZTAKI Research Institute, Hungary)
Kyung Woo Kang (KAIST, Korea)
Carl Kesselman (University of Southern California, USA)
Daniel Kidger (Quadrics, UK)
Deok-Soo Kim (Hanyang University, Korea)

Haeng Kon Kim (Catholic University of Daegu, Korea)
Jin Suk Kim (KAIST, Korea)
Tai-hoon Kim (Korea Information Security Agency, Korea)
Yoonhee Kim (Syracuse University, USA)
Mike Kirby (University of Utah, USA)
Jacek Kitowski (AGH University of Science and Technology, Poland)
Dieter Kranzlmueller (Johannes Kepler University Linz, Austria)
Kurichi Kumar (Institute of High Performance Computing, Singapore)
Vipin Kumar (Army High Performance Computing Center and University of
    Minnesota, USA)
Domenico Laforenza (Italian National Research Council, Italy)
Antonio Laganà (University of Perugia, Italy)
Joseph Landman (Scalable Informatics LLC, USA)
Francis Lau (University of Hong Kong, Hong Kong, China)
Bong Hwan Lee (Texas A&M University, USA)
Dong Chun Lee (Howon University, Korea)
Dong-Ho Lee (Hanyang University, Korea)
Heow Pueh Lee (Institute of High Performance Computing, Singapore)
Sang Yoon Lee (Georgia Institute of Technology, USA)
Tae Jin Lee (Sungkyunkwan University, Korea)
Bogdan Lesyng (ICM Warszawa, Poland)
Zhongze Li (Chinese Academy of Sciences, China)
Laurence Liew (Scalable Systems Pte., Singapore)
David Lombard (Intel Corporation, USA)
Emilio Luque (Universitat Autonoma of Barcelona, Spain)
Michael Mascagni (Florida State University, USA)
Graham Megson (University of Reading, UK)
John G. Michopoulos (US Naval Research Laboratory, USA)
Edward Moreno (Euripides Foundation of Marilia, Brazil)
Youngsong Mun (Soongsil University, Korea)
Jiri Nedoma (Academy of Sciences of the Czech Republic, Czech Republic)
Genri Norman (Russian Academy of Sciences, Russia)
Stephan Olariu (Old Dominion University, USA)
Salvatore Orlando (University of Venice, Italy)
Robert Panoff (Shodor Education Foundation, USA)
Marcin Paprzycki (Oklahoma State University, USA)
Gyung-Leen Park (University of Texas, USA)
Ron Perrott (Queen's University Belfast, UK)
Dimitri Plemenos (University of Limoges, France)
Richard Ramaroson (ONERA, France)
Rosemary Renaut (Arizona State University, USA)
Alexey S. Rodionov (Russian Academy of Science, Russia)
Paul Roe (Queensland University of Technology, Australia)
Reneé S. Renner (California State University at Chico, USA)
Heather J. Ruskin (Dublin City University, Ireland)

Ole Saastad (Scali, Norway)
Muhammad Sarfraz (King Fahd University of Petroleum and Minerals, Saudi Arabia)
Edward Seidel (Louisiana State University, USA and Albert Einstein Institute, Germany)
Josè Sierra-Camara (University Carlos III of Madrid, Spain)
Dale Shires (US Army Research Laboratory, USA)
Vaclav Skala (University of West Bohemia, Czech Republic)
Burton Smith (Cray, USA)
Masha Sosonkina (University of Minnesota, USA)
Alexei Sourin (Nanyang Technological University, Singapore)
Elena Stankova (Institute for High Performance Computing and Data Bases, Russia)
Gunther Stuer (University of Antwerp, Belgium)
Kokichi Sugihara (University of Tokyo, Japan)
Boleslaw Szymanski (Rensselaer Polytechnic Institute, USA)
Ryszard Tadeusiewicz (AGH University of Science and Technology, Poland)
Chih Jeng Kenneth Tan (OptimaNumerics, UK and Queen's University Belfast, UK)
David Taniar (Monash University, Australia)
John Taylor (Quadrics, UK)
Ruppa K. Thulasiram (University of Manitoba, Canada)
Pavel Tvrdik (Czech Technical University, Czech Republic)
Putchong Uthayopas (Kasetsart University, Thailand)
Mario Valle (Visualization Group, Swiss National Supercomputing Centre, Switzerland)
Marco Vanneschi (University of Pisa, Italy)
Piero Giorgio Verdini (University of Pisa and Istituto Nazionale di Fisica Nucleare, Italy)
Jesus Vigo-Aguiar (University of Salamanca, Spain)
Jens Volkert (University of Linz, Austria)
Koichi Wada (University of Tsukuba, Japan)
Kevin Wadleigh (Hewlett-Packard, USA)
Jerzy Wasniewski (Technical University of Denmark, Denmark)
Paul Watson (University of Newcastle upon Tyne)
Jan Weglarz (Poznan University of Technology, Poland)
Tim Wilkens (Advanced Micro Devices, USA)
Roman Wyrzykowski (Technical University of Czestochowa, Poland)
Jinchao Xu (Pennsylvania State University, USA)
Chee Yap (New York University, USA)
Osman Yasar (SUNY at Brockport, USA)
George Yee (National Research Council and Carleton University, Canada)
Yong Xue (Chinese Academy of Sciences, China)
Igor Zacharov (SGI Europe, Switzerland)
Xiaodong Zhang (College of William and Mary, USA)

Aledander Zhmakin (SoftImpact, Russia)
Krzysztof Zielinski (ICS UST/CYFRONET, Poland)
Albert Zomaya (University of Sydney, Australia)

Sponsoring Organizations

The Institute of High Performance Computing, Singapore
University of Perugia, Perugia, Italy
University of Calgary, Calgary, Canada
University of Minnesota, Minneapolis, USA
Queen's University Belfast, UK
Society for Industrial and Applied Mathematics, USA
The Institution of Electrical Engineers, UK
OptimaNumerics Ltd., UK
MASTER-UP, Italy

Table of Contents – Part I

Information Systems and Information Technologies (ISIT) Workshop

The Technique of Test Case Design Based on the UML Sequence Diagram for the Development of Web Applications
  Yongsun Cho, Woojin Lee, Kiwon Chong 1

Flexible Background-Texture Analysis for Coronary Artery Extraction Based on Digital Subtraction Angiography
  Sung-Ho Park, Jeong-Hee Cha, Joong-Jae Lee, Gye-Young Kim 11

New Size-Reduced Visual Secret Sharing Schemes with Half Reduction of Shadow Size
  Ching-Nung Yang, Tse-Shih Chen 19

An Automatic Resource Selection Scheme for Grid Computing Systems
  Kyung-Woo Kang, Gyun Woo 29

Matching Colors with KANSEI Vocabulary Using Similarity Measure Based on WordNet
  Sunkyoung Baek, Miyoung Cho, Pankoo Kim 37

A Systematic Process to Design Product Line Architecture
  Soo Dong Kim, Soo Ho Chang, Hyun Jung La 46

Variability Design and Customization Mechanisms for COTS Components
  Soo Dong Kim, Hyun Gi Min, Sung Yul Rhew 57

A Fast Lossless Multi-resolution Motion Estimation Algorithm Using Selective Matching Units
  Jong-Nam Kim 67

Developing an XML Document Retrieval System for a Digital Museum
  Jae-Woo Chang 77

WiCTP: A Token-Based Access Control Protocol for Wireless Networks
  Raal Goff, Amitava Datta 87

An Optimized Internetworking Strategy of MANET and WLAN
  Hyewon K. Lee, Youngsong Mun 97

An Internetworking Scheme for UMTS/WLAN Mobile Networks
Sangjoon Park, Youngchul Kim, Jongchan Lee 107

A Handover Scheme Based on HMIPv6 for B3G Networks
*Eunjoo Jeong, Sangjoon Park, Hyewon K. Lee, Kwan-Joong Kim,
Youngsong Mun, Byunggi Kim* 118

Collaborative Filtering for Recommendation Using Neural Networks
Myung Won Kim, Eun Ju Kim, Joung Woo Ryu 127

Dynamic Access Control Scheme for Service-Based Multi-netted Asymmetric Virtual LAN
Wonwoo Choi, Hyuncheol Kim, Seongjin Ahn, Jinwook Chung 137

New Binding Update Method Using GDMHA in Hierarchical Mobile IPv6
*Jong-Hyouk Lee, Young-Ju Han, Hyung-Jin Lim,
Tai-Myung Chung* 146

Security in Sensor Networks for Medical Systems Torso Architecture
Chaitanya Penubarthi, Myuhng-Joo Kim, Insup Lee 156

Multimedia: An SIMD – Based Efficient 4x4 2 D Transform Method
*Sang-Jun Yu, Chae-Bong Sohn, Seoung-Jun Oh,
Chang-Beom Ahn* 166

A Real-Time Cooperative Swim-Lane Business Process Modeler
Kwang-Hoon Kim, Jung-Hoon Lee, Chang-Min Kim 176

A Focused Crawling for the Web Resource Discovery Using a Modified Proximal Support Vector Machines
YoungSik Choi, KiJoo Kim, MunSu Kang 186

A Performance Improvement Scheme of Stream Control Transmission Protocol over Wireless Networks
*Kiwon Hong, Kugsang Jeong, Deokjai Choi,
Choongseon Hong* 195

Cache Management Protocols Based on Re-ordering for Distributed Systems
SungHo Cho, Kyoung Yul Bae 204

DRC-BK: Mining Classification Rules by Using Boolean Kernels
Yang Zhang, Zhanhuai Li, Kebin Cui 214

General-Purpose Text Entry Rules for Devices with 4x3 Configurations of Buttons
  Jaewoo Ahn, Myung Ho Kim 223

Dynamic Load Redistribution Approach Using Genetic Information in Distributed Computing
  Seonghoon Lee, Dongwoo Lee, Donghee Shim, Dongyoung Cho 232

A Guided Search Method for Real Time Transcoding a MPEG2 P Frame into H.263 P Frame in a Compressed Domain
  Euisun Kang, Maria Hong, Younghwan Lim, Youngsong Mun, Seongjin Ahn 242

Cooperative Security Management Enhancing Survivability Against DDoS Attacks
  Sung Ki Kim, Byoung Joon Min, Jin Chul Jung, Seung Hwan Yoo 252

Marking Mechanism for Enhanced End-to-End QoS Guarantees in Multiple DiffServ Environment
  Woojin Park, Kyuho Han, Sinam Woo, Sunshin An 261

An Efficient Handoff Mechanism with Web Proxy MAP in Hierarchical Mobile IPv6
  Jonghyoun Choi, Youngsong Mun 271

A New Carried-Dependence Self-scheduling Algorithm
  Hyun Cheol Kim 281

Improved Location Management Scheme Based on Autoconfigured Logical Topology in HMIPv6
  Jongpil Jeong, Hyunsang Youn, Hyunseung Choo, Eunseok Lee 291

Ontological Model of Event for Integration of Inter-organization Applications
  Wang Wenjun, Luo Yingwei, Liu Xinpeng, Wang Xiaolin, Xu Zhuoqun 301

Secure XML Aware Network Design and Performance Analysis
  Eui-Nam Huh, Jong-Youl Jeong, Young-Shin Kim, Ki-Young Mun 311

A Probe Detection Model Using the Analysis of the Fuzzy Cognitive Maps
  Se-Yul Lee, Yong-Soo Kim, Bong-Hwan Lee, Suk-Hoon Kang, Chan-Hyun Youn 320

Mobile Communications (Mobicomm) Workshop

QoS Provisioning in an Enhanced FMIPv6 Architecture
  Zheng Wan, Xuezeng Pan, Lingdi Ping 329

A Novel Hierarchical Routing Protocol for Wireless Sensor Networks
  Trong Thua Huynh, Choong Seon Hong 339

A Vertical Handoff Algorithm Based on Context Information in CDMA-WLAN Integrated Networks
  Jang-Sub Kim, Min-Young Chung, Dong-Ryeol Shin 348

Scalable Hash Chain Traversal for Mobile Device
  Sung-Ryul Kim 359

A Rate Separation Mechanism for Performance Improvements of Multi-rate WLANs
  Chae-Tae Im, Dong-Hee Kwon, Young-Joo Suh 368

Improved Handoff Scheme for Supporting Network Mobility in Nested Mobile Networks
  Han-Kyu Ryu, Do-Hyeon Kim, You-Ze Cho, Kang-Won Lee, Hee-Dong Park 378

A *Prompt Retransmit* Technique to Improve TCP Performance for Mobile Ad Hoc Networks
  Dongkyun Kim, Hanseok Bae 388

Enhanced Fast Handover for Mobile IPv6 Based on IEEE 802.11 Network
  Seonggeun Ryu, Younghwan Lim, Seongjin Ahn, Youngsong Mun 398

An Efficient Macro Mobility Scheme Supporting Fast Handover in Hierarchical Mobile IPv6
  Kyunghye Lee, Youngsong Mun 408

Study on the Advanced MAC Scheduling Algorithm for the Infrared Dedicated Short Range Communication
  Sujin Kwag, Jesang Park, Sangsun Lee 418

Design and Evaluation of a New Micro-mobility Protocol in Large Mobile and Wireless Networks
  Young-Chul Shim, Hyun-Ah Kim, Ju-Il Lee 427

Performance Analysis of Transmission Probability Control Scheme in Slotted ALOHA CDMA Networks
  In-Taek Lim 438

RWA Based on Approximated Path Conflict Graphs in Optical Networks
  *Zhanna Olmes, Kun Myon Choi, Min Young Chung, Tae-Jin Lee,
  Hyunseung Choo* 448

Secure Routing in Sensor Networks: Security Problem Analysis and Countermeasures
  Youngsong Mun, Chungsoo Shin 459

Policy Based Handoff in MIPv6 Networks
  *Jong-Hyouk Lee, Byungchul Park, Hyunseung Choo,
  Tai-Myung Chung* 468

An Effective Location Management Strategy for Cellular Mobile Networks
  In-Hye Shin, Gyung-Leen Park, Kang Soo Tae 478

Authentication Authorization Accounting (AAA) Workshop

On the Rila-Mitchell Security Protocols for Biometrics-Based Cardholder Authentication in Smartcards
  Raphael C.-W. Phan, Bok-Min Goi 488

An Efficient Dynamic Group Key Agreement for Low-Power Mobile Devices
  Seokhyang Cho, Junghyun Nam, Seungjoo Kim, Dongho Won 498

Compact Linear Systolic Arrays for Multiplication Using a Trinomial Basis in $GF(2^m)$ for High Speed Cryptographic Processors
  Soonhak Kwon, Chang Hoon Kim, Chun Pyo Hong 508

A Secure User Authentication Protocol Based on One-Time-Password for Home Network
  Hea Suk Jo, Hee Yong Youn 519

On AAA with Extended IDK in Mobile IP Networks
  Hoseong Jeon, Min Young Chung, Hyunseung Choo 529

Secure Forwarding Scheme Based on Session Key Reuse Mechanism in HMIPv6 with AAA
  Kwang Chul Jeong, Hyunseung Choo, Sungchang Lee 540

A Hierarchical Authentication Scheme for MIPv6 Node with Local
Movement Property
 Miyoung Kim, Misun Kim, Youngsong Mun 550

An Effective Authentication Scheme for Mobile Node with Fast
Roaming Property
 Miyoung Kim, Misun Kim, Youngsong Mun 559

A Study on the Performance Improvement to AAA Authentication in
Mobile IPv4 Using Low Latency Handoff
 Youngsong Mun, Sehoon Jang 569

Authenticated Key Agreement Without Subgroup Element Verification
 Taekyoung Kwon 577

Multi-modal Biometrics with PKIs for Border Control Applications
 Taekyoung Kwon, Hyeonjoon Moon 584

A Scalable Mutual Authentication and Key Distribution Mechanism in
a NEMO Environment
 Mihui Kim, Eunah Kim, Kijoon Chae 591

Service-Oriented Home Network Middleware Based on OGSA
 Tae Dong Lee, Chang-Sung Jeong 601

Implementation of Streamlining PKI System for Web Services
 *Namje Park, Kiyoung Moon, Jongsu Jang, Sungwon Sohn,
Dongho Won* 609

Efficient Authentication for Low-Cost RFID Systems
 Su Mi Lee, Young Ju Hwang, Dong Hoon Lee, Jong In Lim 619

An Efficient Performance Enhancement Scheme for Fast Mobility
Service in MIPv6
 Seung-Yeon Lee, Eui-Nam Huh, Sang-Bok Kim, Young-Song Mun ... 628

Face Recognition by the LDA-Based Algorithm for a Video Surveillance
System on DSP
 Jin Ok Kim, Jin Soo Kim, Chin Hyun Chung 638

Computational Geometry and Applications (CGA'05) Workshop

Weakly Cooperative Guards in Grids
 Michal Małafiejski, Paweł Żyliński 647

Mesh Generation for Symmetrical Geometries
  Krister Åhlander 657

A Certified Delaunay Graph Conflict Locator for Semi-algebraic Sets
  François Anton 669

The Offset to an Algebraic Curve and an Application to Conics
  *François Anton, Ioannis Emiris, Bernard Mourrain,
  Monique Teillaud* 683

Computing the Least Median of Squares Estimator in Time $O(n^d)$
  Thorsten Bernholt 697

Pocket Recognition on a Protein Using Euclidean Voronoi Diagram of Atoms
  *Deok-Soo Kim, Cheol-Hyung Cho, Youngsong Cho, Chung In Won,
  Dounguk Kim* 707

Region Expansion by Flipping Edges for Euclidean Voronoi Diagrams of 3D Spheres Based on a Radial Data Structure
  Donguk Kim, Youngsong Cho, Deok-Soo Kim 716

Analysis of the Nicholl-Lee-Nicholl Algorithm
  Frank Dévai 726

Flipping to Robustly Delete a Vertex in a Delaunay Tetrahedralization
  Hugo Ledoux, Christopher M. Gold, George Baciu 737

A Novel Topology-Based Matching Algorithm for Fingerprint Recognition in the Presence of Elastic Distortions
  Chengfeng Wang, Marina L. Gavrilova 748

Bilateral Estimation of Vertex Normal for Point-Sampled Models
  Guofei Hu, Jie Xu, Lanfang Miao, Qunsheng Peng 758

A Point Inclusion Test Algorithm for Simple Polygons
  Weishi Li, Eng Teo Ong, Shuhong Xu, Terence Hung 769

A Modified Nielson's Side-Vertex Triangular Mesh Interpolation Scheme
  Zhihong Mao, Lizhuang Ma, Wuzheng Tan 776

An Acceleration Technique for the Computation of Voronoi Diagrams Using Graphics Hardware
  Osami Yamamoto 786

On the Rectangular Subset Closure of Point Sets
Stefan Porschen 796

Computing Optimized Curves with NURBS Using Evolutionary Intelligence
Muhammad Sarfraz, Syed Arshad Raza, M. Humayun Baig 806

A Novel Delaunay Simplex Technique for Detection of Crystalline Nuclei in Dense Packings of Spheres
A.V. Anikeenko, M.L. Gavrilova, N.N. Medvedev 816

Recognition of Minimum Width Color-Spanning Corridor and Minimum Area Color-Spanning Rectangle
Sandip Das, Partha P. Goswami, Subhas C. Nandy 827

Volumetric Reconstruction of Unorganized Set of Points with Implicit Surfaces
Vincent Bénédet, Loïc Lamarque, Dominique Faudot 838

Virtual Reality in Scientific Applications and Learning (VRSAL 2005) Workshop

Guided Navigation Techniques for 3D Virtual Environment Based on Topic Map
Hak-Keun Kim, Teuk-Seob Song, Yoon-Chu Choy, Soon-Bum Lim 847

Image Sequence Augmentation Using Planar Structures
Juwan Kim, Dongkeun Kim 857

MultiPro: A Platform for PC Cluster Based Active Stereo Display System
Qingshu Yuan, Dongming Lu, Weidong Chen, Yunhe Pan 865

Two-Level 2D Projection Maps Based Horizontal Collision Detection Scheme for Avatar in Collaborative Virtual Environment
Yu Chunyan, Ye Dongyi, Wu Minghui, Pan Yunhe 875

A Molecular Modeling System Based on Dynamic Gestures
Sungjun Park, Jun Lee, Jee-In Kim 886

Face Modeling Using Grid Light and Feature Point Extraction
Lei Shi, Xin Yang, Hailang Pan 896

Virtual Chemical Laboratories and Their Management on the Web
  *Antonio Riganelli, Osvaldo Gervasi, Antonio Laganà,
  Johannes Froehlich* 905

Tangible Tele-meeting System with DV-ARPN (Augmented Reality
Peripheral Network)
  Yong-Moo Kwon, Jin-Woo Park 913

Integrating Learning and Assessment Using the Semantic Web
  *Osvaldo Gervasi, Riccardo Catanzani, Antonio Riganelli,
  Antonio Laganà* 921

The Implementation of Web-Based Score Processing System for WBI
  Young-Jun Seo, Hwa-Young Jeong, Young-Jae Song 928

ELCHEM: A Metalaboratory to Develop Grid e-Learning Technologies
and Services for Chemistry
  *A. Laganà, A. Riganelli, O. Gervasi, P. Yates, K. Wahala,
  R. Salzer, E. Varella, J. Froeklich* 938

Client Allocation for Enhancing Interactivity in Distributed Virtual
Environments
  Duong Nguyen Binh Ta, Suiping Zhou 947

IMNET: An Experimental Testbed for Extensible Multi-user Virtual
Environment Systems
  Tsai-Yen Li, Mao-Yung Liao, Pai-Cheng Tao 957

Application of MPEG-4 in Distributed Virtual Environment
  Qiong Zhang, Taiyi Chen, Jianzhong Mo 967

A New Approach to Area of Interest Management with
Layered-Structures in 2D Grid
  Yu Chunyan, Ye Dongyi, Wu Minghui, Pan Yunhe 974

Awareness Scheduling and Algorithm Implementation for Collaborative
Virtual Environment
  Yu Sheng, Dongming Lu, Yifeng Hu, Qingshu Yuan 985

M of N Features vs. Intrusion Detection
  Zhuowei Li, Amitabha Das 994

Molecular Structures and Processes Workshop

High-Level Quantum Chemical Methods for the Study of Photochemical Processes
   Hans Lischka, Adélia J.A. Aquino, Mario Barbatti, Mohammad Solimannejad 1004

Study of Predictive Abilities of the Kinetic Models of Multistep Chemical Reactions by the Method of Value Analysis
   Levon A. Tavadyan, Avet A. Khachoyan, Gagik A. Martoyan, Seyran H. Minasyan 1012

Lateral Interactions in O/Pt(111): Density-Functional Theory and Kinetic Monte Carlo
   A.P.J. Jansen, W.K. Offermans 1020

Intelligent Predictive Control with Locally Linear Based Model Identification and Evolutionary Programming Optimization with Application to Fossil Power Plants
   Mahdi Jalili-Kharaajoo 1030

Determination of Methanol and Ethanol Synchronously in Ternary Mixture by NIRS and PLS Regression
   Q.F. Meng, L.R. Teng, J.H. Lu, C.J. Jiang, C.H. Gao, T.B. Du, C.G. Wu, X.C. Guo, Y.C. Liang 1040

Ab Initio and Empirical Atom Bond Formulation of the Interaction of the Dimethylether-Ar System
   Alessandro Costantini, Antonio Laganà, Fernando Pirani, Assimo Maris, Walther Caminati 1046

A Parallel Framework for the Simulation of Emission, Transport, Transformation and Deposition of Atmospheric Mercury on a Regional Scale
   Giuseppe A. Trunfio, Ian M. Hedgecock, Nicola Pirrone 1054

A Cognitive Perspective for Choosing Groupware Tools and Elicitation Techniques in Virtual Teams
   Gabriela N. Aranda, Aurora Vizcaíno, Alejandra Cechich, Mario Piattini 1064

A Fast Method for Determination of Solvent-Exposed Atoms and Its Possible Applications for Implicit Solvent Models
   Anna Shumilina 1075

Thermal Rate Coefficients for the $N + N_2$ Reaction: Quasiclassical, Semiclassical and Quantum Calculations
Noelia Faginas Lago, Antonio Laganà, Ernesto Garcia, X. Gimenez 1083

A Molecular Dynamics Study of Ion Permeability Through Molecular Pores
Leonardo Arteconi, Antonio Laganà 1093

Theoretical Investigations of Atmospheric Species Relevant for the Search of High-Energy Density Materials
Marzio Rosi 1101

Pattern Recognition and Ubiquitous Computing Workshop

ID Face Detection Robust to Color Degradation and Facial Veiling
Dae Sung Kim, Nam Chul Kim 1111

Detection of Multiple Vehicles in Image Sequences for Driving Assistance System
SangHoon Han, EunYoung Ahn, NoYoon Kwak 1122

A Computational Model of Korean Mental Lexicon
Heui Seok Lim, Kichun Nam, Yumi Hwang 1129

A Realistic Human Face Modeling from Photographs by Use of Skin Color and Model Deformation
Kyongpil Min, Junchul Chun 1135

An Optimal and Dynamic Monitoring Interval for Grid Resource Information System
Angela Song-Ie Noh, Eui-Nam Huh, Ji-Yeun Sung, Pill-Woo Lee 1144

Real Time Face Detection and Recognition System Using Haar-Like Feature/HMM in Ubiquitous Network Environments
Kicheon Hong, Jihong Min, Wonchan Lee, Jungchul Kim 1154

A Hybrid Network Model for Intrusion Detection Based on Session Patterns and Rate of False Errors
Se-Yul Lee, Yong-Soo Kim, Woongjae Lee 1162

Energy-Efficiency Method for Cluster-Based Sensor Networks
Kyung-Won Nam, Jun Hwang, Cheol-Min Park, Young-Chan Kim 1170

A Study on an Efficient Sign Recognition Algorithm for a Ubiquitous
Traffic System on DSP
   Jong Woo Kim, Kwang Hoon Jung, Chung Chin Hyun 1177

Real-Time Implementation of Face Detection for a Ubiquitous
Computing
   Jin Ok Kim, Jin Soo Kim 1187

On Optimizing Feature Vectors from Efficient Iris Region Normalization
for a Ubiquitous Computing
   Bong Jo Joung, Woongjae Lee 1196

On the Face Detection with Adaptive Template Matching and Cascaded
Object Detection for Ubiquitous Computing Environment
   Chun Young Chang, Jun Hwang 1204

On Improvement for Normalizing Iris Region for a Ubiquitous
Computing
   *Bong Jo Joung, Chin Hyun Chung, Key Seo Lee, Wha Young Yim,
Sang Hyo Lee* 1213

Author Index 1221

Table of Contents – Part II

Approaches or Methods of Security Engineering Workshop

Implementation of Short Message Service System to Be Based Mobile Wireless Internet
Hae-Sool Yang, Jung-Hun Hong, Seok-Hyung Hwang, Haeng-Kon Kim 1

Fuzzy Clustering for Documents Based on Optimization of Classifier Using the Genetic Algorithm
Ju-In Youn, He-Jue Eun, Yong-Sung Kim 10

P2P Protocol Analysis and Blocking Algorithm
Sun-Myung Hwang 21

Object Modeling of RDF Schema for Converting UML Class Diagram
Jin-Sung Kim, Chun-Sik Yoo, Mi-Kyung Lee, Yong-Sung Kim 31

A Framework for Security Assurance in Component Based Development
Gu-Beom Jeong, Guk-Boh Kim 42

Security Framework to Verify the Low Level Implementation Codes
Haeng-Kon Kim, Hae-Sool Yang 52

A Study on Evaluation of Component Metric Suites
Haeng-Kon Kim 62

The K-Means Clustering Architecture in the Multi-stage Data Mining Process
Bobby D. Gerardo, Jae-Wan Lee, Yeon-Sung Choi, Malrey Lee 71

A Privacy Protection Model in ID Management Using Access Control
Hyang-Chang Choi, Yong-Hoon Yi, Jae-Hyun Seo, Bong-Nam Noh, Hyung-Hyo Lee 82

A Time-Variant Risk Analysis and Damage Estimation for Large-Scale Network Systems
InJung Kim, YoonJung Chung, YoungGyo Lee, Dongho Won 92

Efficient Multi-bit Shifting Algorithm in Multiplicative Inversion Problems
Injoo Jang, Hyeong Seon Yoo 102

Modified Token-Update Scheme for Site Authentication
Joungho Lee, Injoo Jang, Hyeong Seon Yoo 111

A Study on Secure SDP of RFID Using Bluetooth Communication
Dae-Hee Seo, Im-Yeong Lee, Hee-Un Park 117

The Semantic Web Approach in Location Based Services
Jong-Woo Kim, Ju-Yeon Kim, Hyun-Suk Hwang, Sung-Seok Park, Chang-Soo Kim, Sung-gi Park 127

SCTE: Software Component Testing Environments
Haeng-Kon Kim, Oh-Hyun Kwon 137

Computer Security Management Model Using MAUT and SNMP
Jongwoo Chae, Jungkyu Kwon, Mokdong Chung 147

Session and Connection Management for QoS-Guaranteed Multimedia Service Provisioning on IP/MPLS Networks
Young-Tak Kim, Hae-Sun Kim, Hyun-Ho Shin 157

A GQS-Based Adaptive Mobility Management Scheme Considering the Gravity of Locality in Ad-Hoc Networks
Ihn-Han Bae, Sun-Jin Oh 169

A Study on the E-Cash System with Anonymity and Divisibility
Seo-Il Kang, Im-Yeong Lee 177

An Authenticated Key Exchange Mechanism Using One-Time Shared Key
Yonghwan Lee, Eunmi Choi, Dugki Min 187

Creation of Soccer Video Highlight Using the Caption Information
Oh-Hyung Kang, Seong-Yoon Shin 195

The Information Search System Using Neural Network and Fuzzy Clustering Based on Mobile Agent
Jaeseon Ko, Bobby D. Gerardo, Jaewan Lee, Jae-Jeong Hwang 205

A Security Evaluation and Testing Methodology for Open Source Software Embedded Information Security System
Sung-ja Choi, Yeon-hee Kang, Gang-soo Lee 215

An Effective Method for Analyzing Intrusion Situation Through
IP-Based Classification
  *Minsoo Kim, Jae-Hyun Seo, Seung-Yong Lee, Bong-Nam Noh,
  Jung-Taek Seo, Eung-Ki Park, Choon-Sik Park* 225

A New Stream Cipher Using Two Nonlinear Functions
  Mi-Og Park, Dea-Woo Park 235

New Key Management Systems for Multilevel Security
  *Hwankoo Kim, Bongjoo Park, JaeCheol Ha, Byoungcheon Lee,
  DongGook Park* 245

Neural Network Techniques for Host Anomaly Intrusion Detection
Using Fixed Pattern Transformation
  ByungRae Cha, KyungWoo Park, JaeHyun Seo 254

The Role of Secret Sharing in the Distributed MARE Protocols
  Kyeongmo Park 264

Security Risk Vector for Quantitative Asset Assessment
  *Yoon Jung Chung, Injung Kim, NamHoon Lee, Taek Lee,
  Hoh Peter In* 274

A Remote Video Study Evaluation System Using a User Profile
  Seong-Yoon Shin, Oh-Hyung Kang 284

Performance Enhancement of Wireless LAN Based on Infrared
Communications Using Multiple-Subcarrier Modulation
  Hae Geun Kim 295

Modeling Virtual Network Collaboration in Supply Chain Management
  Ha Jin Hwang 304

SPA-Resistant Simultaneous Scalar Multiplication
  Mun-Kyu Lee 314

HSEP Design Using F2mHECC and ThreeB Symmetric Key Under
e-Commerce Environment
  Byung-kwan Lee, Am-Sok Oh, Eun-Hee Jeong 322

A Fault Distance Estimation Method Based on an Adaptive Data
Window for Power Network Security
  *Chang-Dae Yoon, Seung-Yeon Lee, Myong-Chul Shin,
  Ho-Sung Jung, Jae-Sang Cha* 332

Distribution Data Security System Based on Web Based Active
Database
  *Sang-Yule Choi, Myong-Chul Shin, Nam-Young Hur,
  Jong-Boo Kim, Tai-hoon Kim, Jae-Sang Cha* 341

Efficient DoS Resistant Multicast Authentication Schemes
  JaeYong Jeong, Yongsu Park, Yookun Cho 353

Development System Security Process of ISO/IEC TR 15504 and
Security Considerations for Software Process Improvement
  Eun-ser Lee, Malrey Lee 363

Flexible ZCD-UWB with High QoS or High Capacity Using Variable
ZCD Factor Code Sets
  *Jaesang Cha, Kyungsup Kwak, Changdae Yoon,
  Chonghyun Lee* 373

Fine Grained Control of Security Capability and Forward Security in a
Pairing Based Signature Scheme
  *Hak Soo Ju, Dae Youb Kim, Dong Hoon Lee, Jongin Lim,
  Kilsoo Chun* 381

The Large Scale Electronic Voting Scheme Based on Undeniable
Multi-signature Scheme
  Sung-Hyun Yun, Hyung-Woo Lee 391

IPv6/IPsec Conformance Test Management System with Formal
Description Technique
  *Hyung-Woo Lee, Sung-Hyun Yun, Jae-Sung Kim, Nam-Ho Oh,
  Do-Hyung Kim* 401

Interference Cancellation Algorithm Development and Implementation
for Digital Television
  Chong Hyun Lee, Jae Sang Cha 411

Algorithm for ABR Traffic Control and Formation Feedback Information
  *Malrey Lee, Dong-Ju Im, Young Keun Lee, Jae-deuk Lee,
  Suwon Lee, Keun Kwang Lee, HeeJo Kang* 420

Interference-Free ZCD-UWB for Wireless Home Network Applications
  *Jaesang Cha, Kyungsup Kwak, Sangyule Choi, Taihoon Kim,
  Changdae Yoon, Chonghyun Lee* 429

Safe Authentication Method for Security Communication in Ubiquitous
  Hoon Ko, Bangyong Sohn, Hayoung Park, Yongtae Shin 442

Pre/Post Rake Receiver Design for Maximum SINR in MIMO
Communication System
  Chong Hyun Lee, Jae Sang Cha 449

SRS-Tool: A Security Functional Requirement Specification
Development Tool for Application Information System of Organization
  Sang-soo Choi, Soo-young Chae, Gang-soo Lee 458

Design Procedure of IT Systems Security Countermeasures
  Tai-hoon Kim, Seung-youn Lee 468

Similarity Retrieval Based on Self-organizing Maps
  *Dong-Ju Im, Malrey Lee, Young Keun Lee, Tae-Eun Kim,
  SuWon Lee, Jaewan Lee, Keun Kwang Lee, Kyung Dal Cho* 474

An Expert System Development for Operating Procedure Monitoring
of PWR Plants
  Malrey Lee, Eun-ser Lee, HeeJo Kang, HeeSook Kim 483

Security Evaluation Targets for Enhancement of IT Systems Assurance
  Tai-hoon Kim, Seung-youn Lee 491

Protection Profile for Software Development Site
  Seung-youn Lee, Myong-chul Shin 499

Information Security and Hiding (ISH 2005) Workshop

Improved RS Method for Detection of LSB Steganography
  Xiangyang Luo, Bin Liu, Fenlin Liu 508

Robust Undetectable Interference Watermarks
  *Ryszard Grząślewicz, Jarosław Kutyłowski, Mirosław Kutyłowski,
  Wojciech Pietkiewicz* 517

Equidistant Binary Fingerprinting Codes. Existence and Identification
Algorithms
  Marcel Fernandez, Miguel Soriano, Josep Cotrina 527

Color Cube Analysis for Detection of LSB Steganography in RGB
Color Images
  Kwangsoo Lee, Changho Jung, Sangjin Lee, Jongin Lim 537

Compact and Robust Image Hashing
  Sheng Tang, Jin-Tao Li, Yong-Dong Zhang 547

Watermarking for 3D Mesh Model Using Patch CEGIs
  Suk-Hwan Lee, Ki-Ryong Kwon 557

Related-Key and Meet-in-the-Middle Attacks on Triple-DES and
DES-EXE
  *Jaemin Choi, Jongsung Kim, Jaechul Sung, Sangjin Lee,
  Jongin Lim* 567

Fault Attack on the DVB Common Scrambling Algorithm
  Kai Wirt 577

HSEP Design Using F2mHECC and ThreeB Symmetric Key Under
e-Commerce Envrionment
  Byung-kwan Lee, Am-Sok Oh, Eun-Hee Jeong 585

Perturbed Hidden Matrix Cryptosystems
  Zhiping Wu, Jintai Ding, Jason E. Gower, Dingfeng Ye 595

Identity-Based Identification Without Random Oracles
  Kaoru Kurosawa, Swee-Huay Heng 603

Linkable Ring Signatures: Security Models and New Schemes
  Joseph K. Liu, Duncan S. Wong 614

Practical Scenarios for the Van Trung-Martirosyan Codes
  Marcel Fernandez, Miguel Soriano, Josep Cotrina 624

Obtaining True-Random Binary Numbers from a Weak Radioactive
Source
  Ammar Alkassar, Thomas Nicolay, Markus Rohe 634

Modified Sequential Normal Basis Multipliers for Type II Optimal
Normal Bases
  *Dong Jin Yang, Chang Han Kim, Youngho Park, Yongtae Kim,
  Jongin Lim* 647

A New Method of Building More Non-supersingular Elliptic Curves
  Shi Cui, Pu Duan, Choong Wah Chan 657

Accelerating AES Using Instruction Set Extensions for Elliptic Curve
Cryptography
  Stefan Tillich, Johann Großschädl 665

Modeling of Location Management in Mobile Information Systems Workshop

Access Control Capable Integrated Network Management System for TCP/IP Networks
  *Hyuncheol Kim, Seongjin Ahn, Younghwan Lim,
  Youngsong Mun* 676

A Directional-Antenna Based MAC Protocol for Wireless Sensor Networks
  Shen Zhang, Amitava Datta 686

An Extended Framework for Proportional Differentiation: Performance Metrics and Evaluation Considerations
  Jahwan Koo, Seongjin Ahn 696

QoS Provisioning in an Enhanced FMIPv6 Architecture
  Zheng Wan, Xuezeng Pan, Lingdi Ping 704

Delay of the Slotted ALOHA Protocol with Binary Exponential Backoff Algorithm
  Sun Hur, Jeong Kee Kim, Dong Chun Lee 714

Design and Implementation of Frequency Offset Estimation, Symbol Timing and Sampling Clock Offset Control for an IEEE 802.11a Physical Layer
  *Kwang-ho Chun, Seung-hyun Min, Myoung-ho Seong,
  Myoung-seob Lim* 723

Automatic Subtraction Radiography Algorithm for Detection of Periodontal Disease in Internet Environment
  Yonghak Ahn, Oksam Chae 732

Improved Authentication Scheme in W-CDMA Networks
  *Dong Chun Lee, Hyo Young Shin, Joung Chul Ahn,
  Jae Young Koh* 741

Memory Reused Multiplication Implementation for Cryptography System
  Gi Yean Hwang, Jia Hou, Kwang Ho Chun, Moon Ho Lee 749

Scheme for the Information Sharing Between IDSs Using JXTA
  Jin Soh, Sung Man Jang, Geuk Lee 754

Workflow System Modeling in the Mobile Healthcare B2B Using
Semantic Information
  *Sang-Young Lee, Yung-Hyeon Lee, Jeom-Goo Kim,
  Dong Chun Lee* 762

Detecting Water Area During Flood Event from SAR Image
  Hong-Gyoo Sohn, Yeong-Sun Song, Gi-Hong Kim 771

Position Based Handover Control Method
  Jong chan Lee, Sok-Pal Cho, Hong-jin Kim 781

Improving Yellow Time Method of Left-Turning Traffic Flow at
Signalized Intersection Networks by ITS
  Hyung Jin Kim, Bongsoo Son, Soobeom Lee, Joowon Park 789

Intelligent Multimedia Services and Synchronization in Mobile Multimedia Networks Workshop

A Multimedia Database System Using Dependence Weight Values for a
Mobile Environment
  Kwang Hyoung Lee, Hee Sook Kim, Keun Wang Lee 798

A General Framework for Analyzing the Optimal Call Admission
Control in DS-CDMA Cellular Network
  Wen Chen, Feiyu Lei, Weinong Wang 806

Heuristic Algorithm for Traffic Condition Classification with Loop
Detector Data
  Sangsoo Lee, Sei-Chang Oh, Bongsoo Son 816

Spatial Data Channel in a Mobile Navigation System
  Yingwei Luo, Guomin Xiong, Xiaolin Wang, Zhuoqun Xu 822

A Video Retrieval System for Electrical Safety Education Based on a
Mobile Agent
  Hyeon Seob Cho, Keun Wang Lee 832

Fuzzy Multi-criteria Decision Making-Based Mobile Tracking
  Gi-Sung Lee 839

Evaluation of Network Blocking Algorithm based on ARP Spoofing
and Its Application
  Jahwan Koo, Seongjin Ahn, Younghwan Lim, Youngsong Mun 848

Design and Implementation of Mobile-Learning System for Environment Education
Keun Wang Lee, Jong Hee Lee 856

A Simulation Model of Congested Traffic in the Waiting Line
Bongsoo Son, Taewan Kim, Yongjae Lee 863

Core Technology Analysis and Development for the Virus and Hacking Prevention
Seung-Jae Yoo 870

Development of Traffic Accidents Prediction Model with Intelligent System Theory
SooBeom Lee, TaiSik Lee, Hyung Jin Kim, YoungKyun Lee 880

Prefetching Scheme Considering Mobile User's Preference in Mobile Networks
Jin Ah Yoo, In Seon Choi, Dong Chun Lee 889

System Development of Security Vulnerability Diagnosis in Wireless Internet Networks
Byoung-Muk Min, Sok-Pal Cho, Hong-jin Kim, Dong Chun Lee 896

An Active Node Management System for Secure Active Networks
Jin-Mook Kim, In-sung Han, Hwang-bin Ryou 904

Ubiquitous Web Systems and Intelligence Workshop

A Systematic Design Approach for XML-View Driven Web Document Warehouses
Vicky Nassis, Rajugan R., Tharam S. Dillon, Wenny Rahayu 914

Clustering and Retrieval of XML Documents by Structure
Jeong Hee Hwang, Keun Ho Ryu 925

A New Method for Mining Association Rules from a Collection of XML Documents
Juryon Paik, Hee Yong Youn, Ungmo Kim 936

Content-Based Recommendation in E-Commerce
Bing Xu, Mingmin Zhang, Zhigeng Pan, Hongwei Yang 946

A Personalized Multilingual Web Content Miner: PMWebMiner
Rowena Chau, Chung-Hsing Yeh, Kate A. Smith 956

Context-Based Recommendation Service in Ubiquitous Commerce
 Jeong Hee Hwang, Mi Sug Gu, Keun Ho Ryu 966

A New Continuous Nearest Neighbor Technique for Query Processing
on Mobile Environments
 Jeong Hee Chi, Sang Ho Kim, Keun Ho Ryu 977

Semantic Web Enabled Information Systems: Personalized Views on
Web Data
 *Robert Baumgartner, Christian Enzi, Nicola Henze, Marc Herrlich,
 Marcus Herzog, Matthias Kriesell, Kai Tomaschewski* 988

Design of Vehicle Information Management System for Effective
Retrieving of Vehicle Location
 Eung Jae Lee, Keun Ho Ryu 998

Context-Aware Workflow Language Based on Web Services for
Ubiquitous Computing
 Joohyun Han, Yongyun Cho, Jaeyoung Choi 1008

A Ubiquitous Approach for Visualizing Back Pain Data
 T. Serif, G. Ghinea, A.O. Frank 1018

Prototype Design of Mobile Emergency Telemedicine System
 *Sun K. Yoo, S.M. Jung, B.S. Kim, H.Y. Yun, S.R. Kim,
 D.K. Kim* 1028

An Intermediate Target for Quick-Relay of Remote Storage to Mobile
Devices
 Daegeun Kim, MinHwan Ok, Myong-soon Park 1035

Reflective Middleware for Location-Aware Application Adaptation
 *Uzair Ahmad, S.Y. Lee, Mahrin Iqbal, Uzma Nasir, A. Ali,
 Mudeem Iqbal* 1045

Efficient Approach for Interactively Mining Web Traversal Patterns
 Yue-Shi Lee, Min-Chi Hsieh, Show-Jane Yen 1055

Query Decomposition Using the XML Declarative Description Language
 Le Thi Thu Thuy, Doan Dai Duong 1066

On URL Normalization
 Sang Ho Lee, Sung Jin Kim, Seok Hoo Hong 1076

Clustering-Based Schema Matching of Web Data for Constructing Digital Library
Hui Song, Fanyuan Ma, Chen Wang 1086

Bringing Handhelds to the Grid Resourcefully: A Surrogate Middleware Approach
Maria Riaz, Saad Liaquat Kiani, Anjum Shehzad, Sungyoung Lee 1096

Mobile Mini-payment Scheme Using SMS-Credit
Simon Fong, Edison Lai 1106

Context Summarization and Garbage Collecting Context
Faraz Rasheed, Yong-Koo Lee, Sungyoung Lee 1115

EXtensible Web (xWeb): An XML-View Based Web Engineering Methodology
Rajugan R., William Gardner, Elizabeth Chang, Tharam S. Dillon 1125

A Web Services Framework for Integrated Geospatial Coverage Data
Eunkyu Lee, Minsoo Kim, Mijeong Kim, Inhak Joo 1136

Open Location-Based Service Using Secure Middleware Infrastructure in Web Services
Namje Park, Howon Kim, Seungjoo Kim, Dongho Won............ 1146

Ubiquitous Systems and Petri Nets
David de Frutos Escrig, Olga Marroquín Alonso, Fernando Rosa Velardo 1156

Virtual Lab Dashboard: Ubiquitous Monitoring and Control in a Smart Bio-laboratory
XiaoMing Bao, See-Kiong Ng, Eng-Huat Chua, Wei-Khing For 1167

On Discovering Concept Entities from Web Sites
Ming Yin, Dion Hoe-Lian Goh, Ee-Peng Lim 1177

Modelling Complex Systems Workshop

Towards a Realistic Microscopic Traffic Simulation at an Unsignalised Interscetion
Mingzhe Liu, Ruili Wang, Ray Kemp........................... 1187

Complex Systems: Particles, Chains, and Sheets
  R.B Pandey 1197

Discretization of Delayed Multi-input Nonlinear System via Taylor
Series and Scaling and Squaring Technique
  Yuanliang Zhang, Hyung Jo Choi, Kil To Chong 1207

On the Scale-Free Intersection Graphs
  Xin Yao, Changshui Zhang, Jinwen Chen, Yanda Li 1217

A Stochastic Viewpoint on the Generation of Spatiotemporal Datasets
  MoonBae Song, KwangJin Park, Ki-Sik Kong, SangKeun Lee 1225

A Formal Approach to the Design of Distributed Data Warehouses
  Jane Zhao 1235

A Mathematical Model for Genetic Regulation of the Lactose Operon
  Tianhai Tian, Kevin Burrage 1245

Network Emergence in Immune System Shape Space
  Heather J. Ruskin, John Burns 1254

A Multi-agent System for Modelling Carbohydrate Oxidation in Cell
  Flavio Corradini, Emanuela Merelli, Marco Vita 1264

Characterizing Complex Behavior in (Self-organizing) Multi-agent
Systems
  Bingcheng Hu, Jiming Liu 1274

Protein Structure Abstraction and Automatic Clustering Using
Secondary Structure Element Sequences
  *Sung Hee Park, Chan Yong Park, Dae Hee Kim, Seon Hee Park,
  Jeong Seop Sim* 1284

A Neural Network Method for Induction Machine Fault Detection with
Vibration Signal
  Hua Su, Kil To Chong, A.G. Parlos 1293

Author Index 1303

Table of Contents – Part III

Grid Computing and Peer-to-Peer (P2P) Systems Workshop

Resource and Service Discovery in the iGrid Information Service
  *Giovanni Aloisio, Massimo Cafaro, Italo Epicoco, Sandro Fiore,
  Daniele Lezzi, Maria Mirto, Silvia Mocavero* 1

A Comparison of Spread Methods in Unstructured P2P Networks
  Zhaoqing Jia, Bingzhen Pei, Minglu Li, Jinyuan You 10

A New Service Discovery Scheme Adapting to User Behavior for Ubiquitous Computing
  Yeo Bong Yoon, Hee Yong Youn 19

The Design and Prototype of RUDA, a Distributed Grid Accounting System
  M.L. Chen, A. Geist, D.E. Bernholdt, K. Chanchio, D.L. Million .. 29

An Adaptive Routing Mechanism for Efficient Resource Discovery in Unstructured P2P Networks
  Luca Gatani, Giuseppe Lo Re, Salvatore Gaglio 39

Enhancing UDDI for Grid Service Discovery by Using Dynamic Parameters
  Brett Sinclair, Andrzej Goscinski, Robert Dew 49

A New Approach for Efficiently Achieving High Availability in Mobile Computing
  M. Mat Deris, J.H. Abawajy, M. Omar 60

A Flexible Communication Scheme to Support Grid Service Emergence
  Lei Gao, Yongsheng Ding 69

A Kernel-Level RTP for Efficient Support of Multimedia Service on Embedded Systems
  Dong Guk Sun, Sung Jo Kim 79

Group-Based Scheduling Scheme for Result Checking in Global Computing Systems
  *HongSoo Kim, SungJin Choi, MaengSoon Baik, KwonWoo Yang,
  HeonChang Yu, Chong-Sun Hwang* 89

Service Discovery Supporting Open Scalability Using FIPA-Compliant Agent Platform for Ubiquitous Networks
  Kee-Hyun Choi, Ho-Jin Shin, Dong-Ryeol Shin 99

A Mathematical Predictive Model for an Autonomic System to Grid Environments
  Alberto Sánchez, María S. Pérez 109

Spatial Analysis and GIS: Local or Global? Workshop

Spatial Analysis: Science or Art?
  Stefania Bertazzon 118

Network Density Estimation: Analysis of Point Patterns over a Network
  Giuseppe Borruso 126

Linking Global Climate Grid Surfaces with Local Long-Term Migration Monitoring Data: Spatial Computations for the Pied Flycatcher to Assess Climate-Related Population Dynamics on a Continental Scale
  Nikita Chernetsov, Falk Huettmann 133

Classifying Internet Traffic Using Linear Regression
  Troy D. Mackay, Robert G.V. Baker 143

Modeling Sage Grouse: Progressive Computational Methods for Linking a Complex Set of Local, Digital Biodiversity and Habitat Data Towards Global Conservation Statements and Decision-Making Systems
  Anthonia Onyeahialam, Falk Huettmann, Stefania Bertazzon 152

Local Analysis of Spatial Relationships: A Comparison of GWR and the Expansion Method
  Antonio Páez 162

Middleware Development for Remote Sensing Data Sharing and Image Processing on HIT-SIP System
  Jianqin Wang, Yong Xue, Chaolin Wu, Yanguang Wang, Yincui Hu, Ying Luo, Yanning Guan, Shaobo Zhong, Jiakui Tang, Guoyin Cai 173

A New and Efficient K-Medoid Algorithm for Spatial Clustering
  Qiaoping Zhang, Isabelle Couloigner 181

Computer Graphics and Rendering Workshop

Security Management for Internet-Based Virtual Presentation of Home Textile Product
  Lie Shi, Mingmin Zhang, Li Li, Lu Ye, Zhigeng Pan 190

An Efficient Approach for Surface Creation
  L.H. You, Jian J. Zhang 197

Interactive Visualization for OLAP
  Kesaraporn Techapichetvanich, Amitava Datta 206

Interactive 3D Editing on Tiled Display Wall
  Xiuhui Wang, Wei Hua, Hujun Bao 215

A Toolkit for Automatically Modeling and Simulating 3D Multi-articulation Entity in Distributed Virtual Environment
  Xiaohui Liang, Chuanpeng Wang, Yinghui Che, Jiangying Yu, Na Qu 225

Footprint Analysis and Motion Synthesis
  Qinping Zhao, Xiaoyan Hu 235

An Adaptive and Efficient Algorithm for Polygonization of Implicit Surfaces
  Mingyong Pang, Zhigeng Pan, Mingmin Zhang, Fuyan Zhang 245

A Framework of Web GIS Based Unified Public Health Information Visualization Platform
  Xiaolin Lu 256

An Improved Colored-Marker Based Registration Method for AR Applications
  Xiaowei Li, Yue Liu, Yongtian Wang, Dayuan Yan, Dongdong Weng, Tao Yang 266

Non-photorealistic Tour into Panorama
  Yang Zhao, Ya-Ping Zhang, Dan Xu 274

Image Space Silhouette Extraction Using Graphics Hardware
  Jiening Wang, Jizhou Sun, Ming Che, Qi Zhai, Weifang Nie 284

Adaptive Fuzzy Weighted Average Filter for Synthesized Image
  Qing Xu, Liang Ma, Weifang Nie, Peng Li, Jiawan Zhang, Jizhou Sun 292

Data Mining and Bioinformatics Workshop

The Binary Multi-SVM Voting System for Protein Subcellular
Localization Prediction
  *Bo Jin, Yuchun Tang, Yan-Qing Zhang, Chung-Dar Lu,
  Irene Weber* 299

Gene Network Prediction from Microarray Data by Association Rule
and Dynamic Bayesian Network
  Hei-Chia Wang, Yi-Shiun Lee 309

Protein Interaction Prediction Using Inferred Domain Interactions and
Biologically-Significant Negative Dataset
  Xiao-Li Li, Soon-Heng Tan, See-Kiong Ng 318

Semantic Annotation of Biomedical Literature Using Google
  *Rune Sætre, Amund Tveit, Tonje Stroemmen Steigedal,
  Astrid Lægreid* 327

Fast Parallel Algorithms for the Longest Common Subsequence
Problem Using an Optical Bus
  Xiaohua Xu, Ling Chen, Yi Pan, Ping He 338

Estimating Gene Networks from Expression Data and Binding Location
Data via Boolean Networks
  *Osamu Hirose, Naoki Nariai, Yoshinori Tamada, Hideo Bannai,
  Seiya Imoto, Satoru Miyano* 349

Efficient Matching and Retrieval of Gene Expression Time Series Data
Based on Spectral Information
  Hong Yan 357

SVM Classification to Predict Two Stranded Anti-parallel Coiled Coils
Based on Protein Sequence Data
  Zhong Huang, Yun Li, Xiaohua Hu 374

Estimating Gene Networks with cDNA Microarray Data Using
State-Space Models
  Rui Yamaguchi, Satoru Yamashita, Tomoyuki Higuchi 381

A Penalized Likelihood Estimation on Transcriptional Module-Based
Clustering
  Ryo Yoshida, Seiya Imoto, Tomoyuki Higuchi 389

Conceptual Modeling of Genetic Studies and Pharmacogenetics
  Xiaohua Zhou, Il-Yeol Song 402

Parallel and Distributed Computing Workshop

A Dynamic Parallel Volume Rendering Computation Mode Based on Cluster
Weifang Nie, Jizhou Sun, Jing Jin, Xiaotu Li, Jie Yang, Jiawan Zhang 416

Dynamic Replication of Web Servers Using Rent-a-Servers
Young-Chul Shim, Jun-Won Lee, Hyun-Ah Kim 426

Survey of Parallel and Distributed Volume Rendering: Revisited
Jiawan Zhang, Jizhou Sun, Zhou Jin, Yi Zhang, Qi Zhai 435

Scheduling Pipelined Multiprocessor Tasks: An Experimental Study with Vision Architecture
M. Fikret Ercan 445

Universal Properties Verification of Parameterized Parallel Systems
Cecilia E. Nugraheni 453

Symbolic Computation, SC 2005 Workshop

2d Polynomial Interpolation: A Symbolic Approach with Mathematica
Ali Yazici, Irfan Altas, Tanil Ergenc 463

Analyzing the Synchronization of Chaotic Dynamical Systems with Mathematica: Part I
Andres Iglesias, Akemi Gálvez 472

Analyzing the Synchronization of Chaotic Dynamical Systems with Mathematica: Part II
Andres Iglesias, Akemi Gálvez 482

A Mathematica Package for Computing and Visualizing the Gauss Map of Surfaces
Ruben Ipanaqué, Andres Iglesias 492

Numerical-Symbolic *Matlab* Toolbox for Computer Graphics and Differential Geometry
Akemi Gálvez, Andrés Iglesias 502

A LiE Subroutine for Computing Prehomogeneous Spaces Associated with Real Nilpotent Orbits
Steven Glenn Jackson, Alfred G. Noël 512

Applications of Graph Coloring
  Ünal Ufuktepe, Goksen Bacak 522

Mathematica Applications on Time Scales
  Ahmet Yantır, Ünal Ufuktepe 529

A Discrete Mathematics Package for Computer Science and Engineering Students
  Mustafa Murat Inceoglu 538

Circle Inversion of Two-Dimensional Objects with Mathematica
  Ruben T. Urbina, Andres Iglesias 547

Specific Aspects of Computational Physics for Modeling Suddenly-Emerging Phenomena Workshop

Specific Aspects of Training IT Students for Modeling Pulses in Physics
  Adrian Podoleanu, Cristian Toma, Cristian Morarescu,
  Alexandru Toma, Theodora Toma 556

Filtering Aspects of Practical Test-Functions and the Ergodic Hypothesis
  Flavia Doboga, Ghiocel Toma, Stefan Pusca, Mihaela Ghelmez,
  Cristian Morarescu 563

Definition of Wave-Corpuscle Interaction Suitable for Simulating Sequences of Physical Pulses
  Minas Simeonidis, Stefan Pusca, Ghiocel Toma, Alexandru Toma,
  Theodora Toma 569

Practical Test-Functions Generated by Computer Algorithms
  Ghiocel Toma 576

Possibilities for Obtaining the Derivative of a Received Signal Using Computer-Driven Second Order Oscillators
  Andreea Sterian, Ghiocel Toma 585

Simulating Laser Pulses by Practical Test Functions and Progressive Waves
  Rodica Sterian, Cristian Toma 592

Statistical Aspects of Acausal Pulses in Physics and Wavelets Applications
  Cristian Toma, Rodica Sterian 598

Wavelet Analysis of Solitary Wave Equation
Carlo Cattani 604

Numerical Analysis of Some Typical Finite Differences Simulations of
the Waves Propagation Through Different Media
Dan Iordache, Stefan Pusca, Ghiocel Toma........................ 614

B–Splines and Nonorthogonal Wavelets
Nikolay Strelkov 621

Optimal Wavelets
Nikolay Strelkov, Vladimir Dol'nikov 628

Dynamics of a Two-Level Medium Under the Action of Short Optical Pulses
Valerică Ninulescu, Andreea-Rodica Sterian 635

Nonlinear Phenomena in Erbium-Doped Lasers
Andreea Sterian, Valerică Ninulescu............................ 643

Internet Communications Security (WICS) Workshop

An e-Lottery Scheme Using Verifiable Random Function
Sherman S.M. Chow, Lucas C.K. Hui, S.M. Yiu, K.P. Chow 651

Related-Mode Attacks on Block Cipher Modes of Operation
Raphael C.-W. Phan, Mohammad Umar Siddiqi 661

A Digital Cash Protocol Based on Additive Zero Knowledge
Amitabh Saxena, Ben Soh, Dimitri Zantidis 672

On the Security of Wireless Sensor Networks
Rodrigo Roman, Jianying Zhou, Javier Lopez..................... 681

Dependable Transaction for Electronic Commerce
Hao Wang, Heqing Guo, Manshan Lin, Jianfei Yin, Qi He, Jun Zhang 691

On the Security of a Certified E-Mail Scheme with Temporal Authentication
Min-Hua Shao, Jianying Zhou, Guilin Wang 701

Security Flaws in Several Group Signatures Proposed by Popescu
Guilin Wang, Sihan Qing 711

A Simple Acceptance/Rejection Criterium for Sequence Generators in
Symmetric Cryptography
Amparo Fúster-Sabater, Pino Caballero-Gil 719

Secure Electronic Payments in Heterogeneous Networking: New
Authentication Protocols Approach
*Joaquin Torres, Antonio Izquierdo, Arturo Ribagorda,
Almudena Alcaide* 729

Component Based Software Engineering and Software Process Model Workshop

Software Reliability Measurement Use Software Reliability Growth
Model in Testing
Hye-Jung Jung, Hae-Sool Yang 739

Thesaurus Construction Using Class Inheritance
Gui-Jung Kim, Jung-Soo Han 748

An Object Structure Extraction Technique for Object Reusability
Improvement Based on Legacy System Interface
Chang-Mog Lee, Cheol-Jung Yoo, Ok-Bae Chang 758

Automatic Translation Form Requirements Model into Use Cases
Modeling on UML
Haeng-Kon Kim, Youn-Ky Chung 769

A Component Identification Technique from Object-Oriented Model
Mi-Sook Choi, Eun-Sook Cho 778

Retrieving and Exploring Ontology-Based Human Motion Sequences
*Hyun-Sook Chung, Jung-Min Kim, Yung-Cheol Byun,
Sang-Yong Byun* 788

An Integrated Data Mining Model for Customer Credit Evaluation
Kap Sik Kim, Ha Jin Hwang 798

A Study on the Component Based Architecture for Workflow Rule
Engine and Tool
Ho-Jun Shin, Kwang-Ki Kim, Bo-Yeon Shim 806

A Fragment-Driven Process Modeling Methodology
Kwang-Hoon Kim, Jae-Kang Won, Chang-Min Kim 817

A FCA-Based Ontology Construction for the Design of Class Hierarchy
  Suk-Hyung Hwang, Hong-Gee Kim, Hae-Sool Yang 827

Component Contract-Based Formal Specification Technique
  Ji-Hyun Lee, Hye-Min Noh, Cheol-Jung Yoo, Ok-Bae Chang 836

A Business Component Approach for Supporting the Variability of the Business Strategies and Rules
  Jeong Ah Kim, YoungTaek Jin, SunMyung Hwang 846

A CBD Application Integration Framework for High Productivity and Maintainability
  Yonghwan Lee, Eunmi Choi, Dugki Min 858

Integrated Meta-model Approach for Reengineering from Legacy into CBD
  Eun Sook Cho 868

Behavior Modeling Technique Based on EFSM for Interoperability Testing
  Hye-Min Noh, Ji-Hyen Lee, Cheol-Jung Yoo, Ok-Bae Chang 878

Automatic Connector Creation for Component Assembly
  Jung-Soo Han, Gui-Jung Kim, Young-Jae Song 886

MaRMI-RE: Systematic Componentization Process for Reengineering Legacy System
  Jung-Eun Cha, Chul-Hong Kim 896

A Study on the Mechanism for Mobile Embedded Agent Development Based on Product Line
  Haeng-Kon Kim 906

Frameworks for Model-Driven Software Architecture
  Soung Won Kim, Myoung Soo Kim, Haeng Kon Kim 916

Parallel and Distributed Components with Java
  Chang-Moon Hyun 927

CEB: Class Quality Evaluator for BlueJ
  Yu-Kyung Kang, Suk-Hyung Hwang, Hae-Sool Yang, Jung-Bae Lee, Hee-Chul Choi, Hyun-Wook Wee, Dong-Soon Kim 938

Workflow Modeling Based on Extended Activity Diagram Using ASM Semantics
Eun-Jung Ko, Sang-Young Lee, Hye-Min Noh, Cheol-Jung Yoo, Ok-Bae Chang 945

Unification of XML DTD for XML Documents with Similar Structure
Chun-Sik Yoo, Seon-Mi Woo, Yong-Sung Kim 954

Secure Payment Protocol for Healthcare Using USIM in Ubiquitous
Jang-Mi Baek, In-Sik Hong 964

Verification of UML-Based Security Policy Model
Sachoun Park, Gihwon Kwon 973

Computer Graphics and Geometric Modeling (TSCG 2005) Workshop

From a Small Formula to Cyberworlds
Alexei Sourin 983

Visualization and Analysis of Protein Structures Using Euclidean Voronoi Diagram of Atoms
Deok-Soo Kim, Donguk Kim, Youngsong Cho, Joonghyun Ryu, Cheol-Hyung Cho, Joon Young Park, Hyun Chan Lee 993

C^2 Continuous Spline Surfaces over Catmull-Clark Meshes
Jin Jin Zheng, Jian J. Zhang, Hong Jun Zhou, Lianguan G. Shen .. 1003

Constructing Detailed Solid and Smooth Surfaces from Voxel Data for Neurosurgical Simulation
Mayumi Shimizu, Yasuaki Nakamura 1013

Curvature Estimation of Point-Sampled Surfaces and Its Applications
Yongwei Miao, Jieqing Feng, Qunsheng Peng 1023

The Delaunay Triangulation by Grid Subdivision
Si Hyung Park, Seoung Soo Lee, Jong Hwa Kim 1033

Feature-Based Texture Synthesis
Tong-Yee Lee, Chung-Ren Yan 1043

A Fast 2D Shape Interpolation Technique
Ping-Hsien Lin, Tong-Yee Lee 1050

Triangular Prism Generation Algorithm for Polyhedron Decomposition
Jaeho Lee, JoonYoung Park, Deok-Soo Kim, HyunChan Lee 1060

Tweek: A Framework for Cross-Display Graphical User Interfaces
Patrick Hartling, Carolina Cruz-Neira 1070

Surface Simplification with Semantic Features Using Texture and Curvature Maps
Soo-Kyun Kim, Jung Lee, Cheol-Su Lim, Chang-Hun Kim 1080

Development of a Machining Simulation System Using the Octree Algorithm
Y.H. Kim, S.L. Ko 1089

A Spherical Point Location Algorithm Based on Barycentric Coordinates
Yong Wu, Yuanjun He, Haishan Tian 1099

Realistic Skeleton Driven Skin Deformation
X.S. Yang, Jian J. Zhang 1109

Implementing Immersive Clustering with VR Juggler
Aron Bierbaum, Patrick Hartling, Pedro Morillo, Carolina Cruz-Neira 1119

Adaptive Space Carving with Texture Mapping
Yoo-Kil Yang, Jung Lee, Soo-Kyun Kim, Chang-Hun Kim 1129

User-Guided 3D Su-Muk Painting
Jung Lee, Joon-Yong Ji, Soo-Kyun Kim, Chang-Hun Kim 1139

Sports Equipment Based Motion Deformation
Jong-In Choi, Chang-Hun Kim, Cheol-Su Lim 1148

Designing an Action Selection Engine for Behavioral Animation of Intelligent Virtual Agents
Francisco Luengo, Andres Iglesias 1157

Interactive Transmission of Highly Detailed Surfaces
Junfeng Ji, Sheng Li, Enhua Wu, Xuehui Liu 1167

Contour-Based Terrain Model Reconstruction Using Distance Information
Byeong-Seok Shin, Hoe Sang Jung 1177

An Efficient Point Rendering Using Octree and Texture Lookup
Yun-Mo Koo, Byeong-Seok Shin 1187

Faces Alive: Reconstruction of Animated 3D Human Faces
Yu Zhang, Terence Sim, Chew Lim Tan 1197

Quasi-interpolants Based Multilevel B-Spline Surface Reconstruction from Scattered Data
Byung-Gook Lee, Joon-Jae Lee, Ki-Ryoung Kwon 1209

Methodology of Information Engineering Workshop

Efficient Mapping Rule of IDEF for UMM Application
Kitae Shin, Chankwon Park, Hyoung-Gon Lee, Jinwoo Park 1219

A Case Study on the Development of Employee Internet Management System
Sangkyun Kim, Ilhoon Choi 1229

Cost-Benefit Analysis of Security Investments: Methodology and Case Study
Sangkyun Kim, Hong Joo Lee 1239

A Modeling Framework of Business Transactions for Enterprise Integration
Minsoo Kim, Dongsoo Kim, Yong Gu Ji, Hoontae Kim 1249

Process-Oriented Development of Job Manual System
Seung-Hyun Rhee, Hoseong Song, Hyung Jun Won, Jaeyoung Ju, Minsoo Kim, Hyerim Bae 1259

An Information System Approach and Methodology for Enterprise Credit Rating
Hakjoo Lee, Choon Seong Leem, Kyungup Cha 1269

Privacy Engineering in ubiComp
Tae Joong Kim, Sang Won Lee, Eung Young Lee 1279

Development of a BSC-Based Evaluation Framework for e-Manufacturing Project
Yongju Cho, Wooju Kim, Choon Seong Leem, Honzong Choi 1289

Design of a BPR-Based Information Strategy Planning (ISP) Framework
Chiwoon Cho, Nam Wook Cho 1297

An Integrated Evaluation System for Personal Informatization Levels
and Their Maturity Measurement: Korean Motors Company Case
   Eun Jung Yu, Choon Seong Leem, Seoung Kyu Park,
   Byung Wan Kim 1306

Critical Attributes of Organizational Culture Promoting Successful KM
Implementation
   Heejun Park 1316

Author Index.. 1327

Table of Contents – Part IV

Information and Communication Technology (ICT) Education Workshop

Exploring Constructivist Learning Theory and Course Visualization on Computer Graphics
  Yiming Zhao, Mingming Zhang, Shu Wang, Yefang Chen 1

A Program Plagiarism Evaluation System
  Young-Chul Kim, Jaeyoung Choi 10

Integrated Development Environment for Digital Image Computing and Configuration Management
  Jeongheon Lee, YoungTak Cho, Hoon Heo, Oksam Chae 20

E-Learning Environment Based on Intelligent Synthetic Characters
  Lu Ye, Jiejie Zhu, Mingming Zhang, Ruth Aylett, Lifeng Ren, Guilin Xu 30

SCO Control Net for the Process-Driven SCORM Content Aggregation Model
  Kwang-Hoon Kim, Hyun-Ah Kim, Chang-Min Kim 38

Design and Implementation of a Web-Based Information Communication Ethics Education System for the Gifted Students in Computer
  Woochun Jun, Sung-Keun Cho, Byeong Heui Kwak 48

International Standards Based Information Technology Courses: A Case Study from Turkey
  Mustafa Murat Inceoglu 56

Design and Implementation of the KORI: Intelligent Teachable Agent and Its Application to Education
  Sung-il Kim, Sung-Hyun Yun, Mi-sun Yoon, Yeon-hee So, Won-sik Kim, Myung-jin Lee, Dong-seong Choi, Hyung-Woo Lee ... 62

Digital Device for Ubiquitous Computing Workshop

A Space-Efficient Flash Memory Software for Mobile Devices
  Yeonseung Ryu, Tae-sun Chung, Myungho Lee 72

Security Threats and Their Countermeasures of Mobile Portable
Computing Devices in Ubiquitous Computing Environments
  Sang ho Kim, Choon Seong Leem 79

A Business Model (BM) Development Methodology in Ubiquitous
Computing Environment
  *Choon Seong Leem, Nam Joo Jeon, Jong Hwa Choi,
  Hyoun Gyu Shin* 86

Developing Business Models in Ubiquitous Era: Exploring
Contradictions in Demand and Supply Perspectives
  Jungwoo Lee, Sunghwan Lee 96

Semantic Web Based Intelligent Product and Service Search Framework
for Location-Based Services
  Wooju Kim, SungKyu Lee, DeaWoo Choi 103

A Study on Value Chain in a Ubiquitous Computing Environment
  Hong Joo Lee, Choon Seong Leem 113

A Study on Authentication Mechanism Using Robot Vacuum Cleaner
  Hong Joo Lee, Hee Jun Park, Sangkyun Kim 122

Design of Inside Information Leakage Prevention System in Ubiquitous
Computing Environment
  Hangbae Chang, Kyung-kyu Kim 128

Design and Implementation of Home Media Server Using TV-Anytime
for Personalized Broadcasting Service
  Changho Hong, Jongtae Lim 138

Optimization: Theories and Applications (OTA) 2005 Workshop

Optimal Signal Control Using Adaptive Dynamic Programming
  Chang Ouk Kim, Yunsun Park, Jun-Geol Baek 148

Inverse Constrained Bottleneck Problems on Networks
  Xiucui Guan, Jianzhong Zhang 161

Dynamic Scheduling Problem of Batch Processing Machine in
Semiconductor Burn-in Operations
  Pei-Chann Chang, Yun-Shiow Chen, Hui-Mei Wang 172

Polynomial Algorithm for Parallel Machine Mean Flow Time Scheduling
Problem with Release Dates
  Peter Brucker, Svetlana A. Kravchenko 182

Differential Approximation of MIN SAT, MAX SAT and Related Problems
  Bruno Escoffier, Vangelis Th. Paschos 192

Probabilistic Coloring of Bipartite and Split Graphs
  *Federico Della Croce, Bruno Escoffier, Cécile Murat,
  Vangelis Th. Paschos* 202

Design Optimization Modeling for Customer-Driven Concurrent
Tolerance Allocation
  *Young Jin Kim, Byung Rae Cho, Min Koo Lee,
  Hyuck Moo Kwon* 212

Application of Data Mining for Improving Yield in Wafer Fabrication
System
  Dong-Hyun Baek, In-Jae Jeong, Chang-Hee Han 222

Determination of Optimum Target Values for a Production Process
Based on Two Surrogate Variables
  Min Koo Lee, Hyuck Moo Kwon, Young Jin Kim, Jongho Bae 232

An Evolution Algorithm for the Rectilinear Steiner Tree Problem
  Byounghak Yang 241

A Two-Stage Recourse Model for Production Planning with Stochastic
Demand
  K.K. Lai, Stephen C.H. Leung, Yue Wu 250

A Hybrid Primal-Dual Algorithm with Application to the Dual
Transportation Problems
  Gyunghyun Choi, Chulyeon Kim 261

Real-Coded Genetic Algorithms for Optimal Static Load Balancing in
Distributed Computing System with Communication Delays
  Venkataraman Mani, Sundaram Suresh, HyoungJoong Kim 269

Heterogeneity in and Determinants of Technical Efficiency in the Use
of Polluting Inputs
  Taeho Kim, Jae-Gon Kim 280

A Continuation Method for the Linear Second-Order Cone
Complementarity Problem
  Yu Xia, Jiming Peng 290

Fuzzy Multi-criteria Decision Making Approach for Transport Projects
Evaluation in Istanbul
  E. Ertugrul Karsak, S. Sebnem Ahiska 301

An Improved Group Setup Strategy for PCB Assembly
  V. Jorge Leon, In-Jae Jeong 312

A Mixed Integer Programming Model for Modifying a Block Layout to
Facilitate Smooth Material Flows
  Jae-Gon Kim, Marc Goetschalckx 322

An Economic Capacity Planning Model Considering Inventory and
Capital Time Value
  S.M. Wang, K.J. Wang, H.M. Wee, J.C. Chen 333

A Quantity-Time-Based Dispatching Policy for a VMI System
  Wai-Ki Ching, Allen H. Tai 342

An Exact Algorithm for Multi Depot and Multi Period Vehicle
Scheduling Problem
  Kyung Hwan Kang, Young Hoon Lee, Byung Ki Lee 350

Determining Multiple Attribute Weights Consistent with Pairwise
Preference Orders
  Byeong Seok Ahn, Chang Hee Han 360

A Pricing Model for a Service Inventory System When Demand Is Price
and Waiting Time Sensitive
  Peng-Sheng You 368

A Bi-population Based Genetic Algorithm for the Resource-Constrained
Project Scheduling Problem
  Dieter Debels, Mario Vanhoucke 378

Optimizing Product Mix in a Multi-bottleneck Environment Using
Group Decision-Making Approach
  Alireza Rashidi Komijan, Seyed Jafar Sadjadi 388

Using Bipartite and Multidimensional Matching to Select the Roots of
a System of Polynomial Equations
  Henk Bekker, Eelco P. Braad, Boris Goldengorin 397

Principles, Models, Methods, and Algorithms for the Structure
Dynamics Control in Complex Technical Systems
  B.V. Sokolov, R.M. Yusupov, E.M. Zaychik 407

Applying a Hybrid Ant Colony System to the Vehicle Routing Problem
Chia-Ho Chen, Ching-Jung Ting, Pei-Chann Chang 417

A Coevolutionary Approach to Optimize Class Boundaries for
Multidimensional Classification Problems
Ki-Kwang Lee 427

Analytical Modeling of Closed-Loop Conveyors with Load Recirculation
Ying-Jiun Hsieh, Yavuz A. Bozer 437

A Multi-items Ordering Model with Mixed Parts Transportation
Problem in a Supply Chain
Beumjun Ahn, Kwang-Kyu Seo 448

Artificial Neural Network Based Life Cycle Assessment Model for
Product Concepts Using Product Classification Method
Kwang-Kyu Seo, Sung-Hwan Min, Hun-Woo Yoo 458

New Heuristics for No-Wait Flowshop Scheduling with Precedence
Constraints and Sequence Dependent Setup Time
Young Hae Lee, Jung Woo Jung 467

Efficient Dual Methods for Nonlinearly Constrained Networks
Eugenio Mijangos 477

A First-Order ε-Approximation Algorithm for Linear Programs
and a Second-Order Implementation
*Ana Maria A.C. Rocha, Edite M.G.P. Fernandes,
João L.C. Soares* 488

Inventory Allocation with Multi-echelon Service Level Considerations
Jenn-Rong Lin, Linda K. Nozick, Mark A. Turnquist 499

A Queueing Model for Multi-product Production System
Ho Woo Lee, Tae Hoon Kim 509

Discretization Approach and Nonparametric Modeling for Long-Term
HIV Dynamic Model
Jianwei Chen, Jin-Ting Zhang, Hulin Wu 519

Performance Analysis and Optimization of an Improved Dynamic
Movement-Based Location Update Scheme in Mobile Cellular Networks
Jang Hyun Baek, Jae Young Seo, Douglas C. Sicker 528

Capacitated Disassembly Scheduling: Minimizing the Number of
Products Disassembled
    Jun-Gyu Kim, Hyong-Bae Jeon, Hwa-Joong Kim, Dong-Ho Lee,
    Paul Xirouchakis 538

Ascent Phase Trajectory Optimization for a Hypersonic Vehicle Using
Nonlinear Programming
    H.M. Prasanna, Debasish Ghose, M.S. Bhat,
    Chiranjib Bhattacharyya, J. Umakant 548

Estimating Parameters in Repairable Systems Under Accelerated Stress
    Won Young Yun, Eun Suk Kim.................................. 558

Optimization Model for Remanufacturing System at Strategic and
Operational Level
    Kibum Kim, Bongju Jeong, Seung-Ju Jeong 566

A Novel Procedure to Identify the Minimized Overlap Boundary of
Two Groups by DEA Model
    Dong Shang Chang, Yi Chun Kuo 577

A Parallel Tabu Search Algorithm for Optimizing Multiobjective VLSI
Placement
    Mahmood R. Minhas, Sadiq M. Sait 587

A Coupled Gradient Network Approach for the Multi-machine Earliness
and Tardiness Scheduling Problem
    Derya Eren Akyol, G. Mirac Bayhan 596

An Analytic Model for Correlated Traffics in Computer-Communication
Networks
    Si-Yeong Lim, Sun Hur.. 606

Product Mix Decisions in the Process Industry
    Seung J. Noh, Suk-Chul Rim 615

On the Optimal Workloads Allocation of an FMS with Finite In-process
Buffers
    Soo-Tae Kwon... 624

NEOS Server Usage in Wastewater Treatment Cost Minimization
    Isabel A.C.P. Espoírito-Santo, Edite M.G.P Fernandes,
    Madalena M. Araújo, Eugenio C. Ferreira 632

Branch and Price Algorithm for Content Allocation Problem in VOD Network
  Jungman Hong, Seungkil Lim 642

Regrouping Service Sites: A Genetic Approach Using a Voronoi Diagram
  Jeong-Yeon Seo, Sang-Min Park, Seoung Soo Lee, Deok-Soo Kim ... 652

Profile Association Rule Mining Using Tests of Hypotheses Without Support Threshold
  Kwang-Il Ahn, Jae-Yearn Kim 662

The Capacitated max-k-cut Problem
  Daya Ram Gaur, Ramesh Krishnamurti 670

A Cooperative Multi-Colony Ant Optimization Based Approach to Efficiently Allocate Customers to Multiple Distribution Centers in a Supply Chain Network
  Srinivas, Yogesh Dashora, Alok Kumar Choudhary, Jenny A. Harding, Manoj Kumar Tiwari 680

Experimentation System for Efficient Job Performing in Veterinary Medicine Area
  Leszek Koszalka, Piotr Skworcow 692

An Anti-collision Algorithm Using Two-Functioned Estimation for RFID Tags
  Jia Zhai, Gi-Nam Wang 702

A Proximal Solution for a Class of Extended Minimax Location Problem
  Oscar Cornejo, Christian Michelot 712

A Lagrangean Relaxation Approach for Capacitated Disassembly Scheduling
  Hwa-Joong Kim, Dong-Ho Lee, Paul Xirouchakis 722

General Tracks

DNA-Based Algorithm for 0-1 Planning Problem
  Lei Wang, Zhiping P. Chen, Xinhua H. Jiang 733

Clustering for Image Retrieval via Improved Fuzzy-ART
  Sang-Sung Park, Hun-Woo Yoo, Man-Hee Lee, Jae-Yeon Kim, Dong-Sik Jang 743

Mining Schemas in Semi-structured Data Using Fuzzy Decision Trees
Sun Wei, Liu Da-xin 753

Parallel Seismic Propagation Simulation in Anisotropic Media by
Irregular Grids Finite Difference Method on PC Cluster
Weitao Sun, Jiwu Shu, Weimin Zheng 762

The Web Replica Allocation and Topology Assignment Problem in
Wide Area Networks: Algorithms and Computational Results
Marcin Markowski, Andrzej Kasprzak 772

Optimal Walking Pattern Generation for a Quadruped Robot Using
Genetic-Fuzzy Algorithm
Bo-Hee Lee, Jung-Shik Kong, Jin-Geol Kim 782

Modelling of Process of Electronic Signature with Petri Nets and
(Max, Plus) Algebra
Ahmed Nait-Sidi-Moh, Maxime Wack 792

Evolutionary Algorithm for Congestion Problem in Connection-Oriented
Networks
Michał Przewoźniczek, Krzysztof Walkowiak 802

Design and Development of File System for Storage Area Networks
Gyoung-Bae Kim, Myung-Joon Kim, Hae-Young Bae 812

Transaction Reordering for Epidemic Quorum in Replicated Databases
Huaizhong Lin, Zengwei Zheng, Chun Chen 826

Automatic Boundary Tumor Segmentation of a Liver
Kyung-Sik Seo, Tae-Woong Chung 836

Fast Algorithms for l1 Norm/Mixed l1 and l2 Norms for Image
Restoration
*Haoying Fu, Michael Kwok Ng, Mila Nikolova, Jesse Barlow,
Wai-Ki Ching* 843

Intelligent Semantic Information Retrieval in Medical Pattern Cognitive
Analysis
Marek R. Ogiela, Ryszard Tadeusiewicz, Lidia Ogiela 852

FSPN-Based Genetically Optimized Fuzzy Polynomial Neural Networks
Sung-Kwun Oh, Seok-Beom Rob, Daehee Park, Yong-Kah Kim 858

Unsupervised Color Image Segmentation Using Mean Shift and
Deterministic Annealing EM
   Wanhyun Cho, Jonghyun Park, Myungeun Lee, Soonyoung Park 867

Identity-Based Key Agreement Protocols in a Multiple PKG
Environment
   Hoonjung Lee, Donghyun Kim, Sangjin Kim, Heekuck Oh 877

Evolutionally Optimized Fuzzy Neural Networks Based on Evolutionary
Fuzzy Granulation
   *Sung-Kwun Oh, Byoung-Jun Park, Witold Pedrycz,
Hyun-Ki Kim* 887

Multi-stage Detailed Placement Algorithm for Large-Scale Mixed-Mode
Layout Design
   Lijuan Luo, Qiang Zhou, Xianlong Hong, Hanbin Zhou 896

Adaptive Mesh Smoothing for Feature Preservation
   Weishi Li, Li Ping Goh, Terence Hung, Shuhong Xu 906

A Fuzzy Grouping-Based Load Balancing for Distributed Object
Computing Systems
   Hyo Cheol Ahn, Hee Yong Youn 916

DSP-Based ADI-PML Formulations for Truncating Linear Debye and
Lorentz Dispersive FDTD Domains
   Omar Ramadan 926

Mobile Agent Based Adaptive Scheduling Mechanism in Peer to Peer
Grid Computing
   *SungJin Choi, MaengSoon Baik, ChongSun Hwang, JoonMin Gil,
HeonChang Yu* 936

Comparison of Global Optimization Methods for Drag Reduction in
the Automotive Industry
   Laurent Dumas, Vincent Herbert, Frédérique Muyl 948

Multiple Intervals Versus Smoothing of Boundaries in the Discretization
of Performance Indicators Used for Diagnosis in Cellular Networks
   Raquel Barco, Pedro Lázaro, Luis Díez, Volker Wille 958

Visual Interactive Clustering and Querying of Spatio-Temporal Data
   Olga Sourina, Dongquan Liu 968

Breakdown-Free ML(k)BiCGStab Algorithm for Non-Hermitian Linear Systems
   Kentaro Moriya, Takashi Nodera 978

On Algorithm for Efficiently Combining Two Independent Measures in Routing Paths
   Moonseong Kim, Young-Cheol Bang, Hyunseung Choo 989

Real Time Hand Tracking Based on Active Contour Model
   Jae Sik Chang, Eun Yi Kim, KeeChul Jung, Hang Joon Kim 999

Hardware Accelerator for Vector Quantization by Using Pruned Look-Up Table
   Pi-Chung Wang, Chun-Liang Lee, Hung-Yi Chang, Tung-Shou Chen 1007

Optimizations of Data Distribution Localities in Cluster Grid Environments
   Ching-Hsien Hsu, Shih-Chang Chen, Chao-Tung Yang, Kuan-Ching Li 1017

Abuse-Free Item Exchange
   Hao Wang, Heqing Guo, Jianfei Yin, Qi He, Manshan Lin, Jun Zhang 1028

Transcoding Pattern Generation for Adaptation of Digital Items Containing Multiple Media Streams in Ubiquitous Environment
   Maria Hong, DaeHyuck Park, YoungHwan Lim, YoungSong Mun, Seongjin Ahn 1036

Identity-Based Aggregate and Verifiably Encrypted Signatures from Bilinear Pairing
   Xiangguo Cheng, Jingmei Liu, Xinmei Wang 1046

Element-Size Independent Analysis of Elasto-Plastic Damage Behaviors of Framed Structures
   Yutaka Toi, Jeoung-Gwen Lee 1055

On the Rila-Mitchell Security Protocols for Biometrics-Based Cardholder Authentication in Smartcards
   Raphael C.-W. Phan, Bok-Min Goi 1065

On-line Fabric-Defects Detection Based on Wavelet Analysis
   Sungshin Kim, Hyeon Bae, Seong-Pyo Cheon, Kwang-Baek Kim 1075

Application of Time-Series Data Mining for Fault Diagnosis of
Induction Motors
  Hyeon Bae, Sungshin Kim, Yon Tae Kim, Sang-Hyuk Lee 1085

Distortion Measure for Binary Document Image Using Distance and
Stroke
  Guiyue Jin, Ki Dong Lee 1095

Region and Shape Prior Based Geodesic Active Contour and
Application in Cardiac Valve Segmentation
  Yanfeng Shang, Xin Yang, Ming Zhu, Biao Jin, Ming Liu 1102

Interactive Fluid Animation Using Particle Dynamics Simulation and
Pre-integrated Volume Rendering
  Jeongjin Lee, Helen Hong, Yeong Gil Shin 1111

Performance of Linear Algebra Code: Intel Xeon EM64T and ItaniumII
Case Examples
  Terry Moreland, Chih Jeng Kenneth Tan 1120

Dataset Filtering Based Association Rule Updating in Small-Sized
Temporal Databases
  Jason J. Jung, Geun-Sik Jo 1131

A Comparison of Model Selection Methods for Multi-class Support
Vector Machines
  Huaqing Li, Feihu Qi, Shaoyu Wang 1140

Fuzzy Category and Fuzzy Interest for Web User Understanding
  SiHun Lee, Jee-Hyong Lee, Keon-Myung Lee, Hee Yong Youn 1149

Automatic License Plate Recognition System Based on Color Image
Processing
  Xifan Shi, Weizhong Zhao, Yonghang Shen 1159

Exploiting Locality Characteristics for Reducing Signaling Load in
Hierarchical Mobile IPv6 Networks
  Ki-Sik Kong, Sung-Ju Roh, Chong-Sun Hwang 1169

Parallel Feature-Preserving Mesh Smoothing
  Xiangmin Jiao, Phillip J. Alexander 1180

On Multiparametric Sensitivity Analysis in Minimum Cost Network
Flow Problem
  Sanjeet Singh, Pankaj Gupta, Davinder Bhatia 1190

Mining Patterns of Mobile Users Through Mobile Devices and the
Musics They Listen
  John Goh, David Taniar 1203

Scheduling the Interactions of Multiple Parallel Jobs and Sequential
Jobs on a Non-dedicated Cluster
  Adel Ben Mnaouer 1212

Feature-Correlation Based Multi-view Detection
  Kuo Zhang, Jie Tang, JuanZi Li, KeHong Wang 1222

BEST: Buffer-Driven Efficient Streaming Protocol
  *Sunhun Lee, Jungmin Lee, Kwangsue Chung, WoongChul Choi,
  Seung Hyong Rhee* 1231

A New Neuro-Dominance Rule for Single Machine Tardiness Problem
  Tarık Çakar 1241

Sinogram Denoising of Cryo-Electron Microscopy Images
  Taneli Mielikäinen, Janne Ravantti 1251

Study of a Cluster-Based Parallel System Through Analytical Modeling
and Simulation
  Bahman Javadi, Siavash Khorsandi, Mohammad K. Akbari 1262

Robust Parallel Job Scheduling Infrastructure for Service-Oriented
Grid Computing Systems
  J.H. Abawajy 1272

SLA Management in a Service Oriented Architecture
  James Padgett, Mohammed Haji, Karim Djemame 1282

Attacks on Port Knocking Authentication Mechanism
  *Antonio Izquierdo Manzanares, Joaquín Torres Márquez,
  Juan M. Estevez-Tapiador, Julio César Hernández Castro* 1292

Marketing on Internet Communications Security for Online Bank
Transactions
  José M. Sierra, Julio C. Hernández, Eva Ponce, Jaime Manera ... 1301

A Formal Analysis of Fairness and Non-repudiation in the RSA-CEGD
Protocol
  *Almudena Alcaide, Juan M. Estévez-Tapiador, Antonio Izquierdo,
  José M. Sierra* 1309

Distribution Data Security System Based on Web Based Active
Database
  *Sang-Yule Choi, Myong-Chul Shin, Nam-Young Hur,
  Jong-Boo Kim, Tai-Hoon Kim, Jae-Sang Cha* 1319

Data Protection Based on Physical Separation: Concepts and
Application Scenarios
  Stefan Lindskog, Karl-Johan Grinnemo, Anna Brunstrom 1331

Some Results on a Class of Optimization Spaces
  K.C. Sivakumar, J. Mercy Swarna 1341

Author Index 1349

The Technique of Test Case Design Based on the UML Sequence Diagram for the Development of Web Applications*

Yongsun Cho, Woojin Lee, and Kiwon Chong

Department of Computing, Soongsil University, Seoul, Korea
yongsuns@hanafos.com
bluewj@empal.com
chong@comp.ssu.ac.kr

Abstract. The systematic testing is frequently regretted in recent web applications because of time and cost pressure. Moreover developers have difficulties with applying the traditional testing techniques. A technique for generating test cases from the UML sequence diagrams of a web application is proposed for the rapid and effective testing. A test of the web applications is composed of a single web page test, a mutual web page test and an integrated web page test. The test cases for a single web page test are generated from self-call messages and the test cases for a mutual web page test are generated from the messages between web pages. The test cases for an integrated web page test are generated from the messages which are sent to the system by an actor and received back from the system.

1 Introduction

Recent business environments have been changing into Internet business environments and web applications have been developed continuously for various fields such as advertisement, sale of goods and customer support in Internet business environments [1]. Moreover, accurate and rapid development of web applications and preoccupation of market are required according as businesses and services become various and companies compete with each other.

The accuracy of web applications is emphasized in these environments. If the web application does not operate correctly or it discontinues because of malfunctions, it leads to corporate losses and disrepute. To prevent these situations before they occur, it is necessary to test the reliability of web applications. Although many techniques for testing web applications have been studied, these are not enough to deal with this problem. Most of the early techniques for testing web applications checked syntax and the context of html, jsp and asp files or the correctness of the links among them. Furthermore most of the recent techniques used for testing web applications check the operations of a single web page or the call-relations among web pages. The clustering

---

* This work was supported by the Soongsil University Research Fund.

test for testing the collaboration of web pages is also required for the reliability of web applications.

Therefore, this paper proposes a test case design technique based on the UML [2] sequence diagram for insurancing accuracy of web applications.

2 Related Works

RUP provides guidelines about the technique of extracting test cases from use case for functional test of system [3]. As flow of event in each use case, there are a basic flow and several alternate flows. Scenarios of the use case are made from compounding these scenarios. Variables relative these scenarios are extracted to test data and test cases are extracted by adding the test data and adding necessary conditions. RUP mentions the level of test such as unit test, integration test and system test, but RUP does not provide the technique of extracting test cases according to the level of test. RUP uses use case for functional test of system and uses supplementary specification for non-functional tests. However, RUP does not provide the technique of extracting test cases for non-functional tests and test cases are extracted by heuristic manner. RUP provides only guidelines for extracting test cases.

Ye Wu and Offutt propose the technique of test by extracting test cases from flow of sections of a server program [4]. The kinds of section are atomic section which is an elementary physical unit to identify a part of server program and composite section which is a set of atomic section.

An atomic section is a static HTML file or a section of a server program that prints HTML. An atomic section has an "all-or-nothing property", that is, either the entire section is sent to clients or none of the section is sent.

Possible execution flows are formally expressed by analyzing program codes and each expression is used to test as a test case in this technique. However, it is difficult to represent test conditions or value of test data with only the expressions. Moreover, innumerable test cases can be extracted if server program is complex and very complex expressions are derived in the case of applying integration test of several pages and system test because the technique considers all execution flows based on white box testing. The technique is difficult to apply in real system so that the practical use of the expressions for testing web applications is remained future work in this study.

Filippo Ricca and Tonella propose two techniques of static verification and dynamic verification for testing web applications [5]. Static verification is a technique to scan the HTML pages in a web site and detect possible faults and anomalies. The syntax of HTML pages or links to other pages is examined in this technique. These examinations are performed in many published tools.

Dynamic verification is a technique to extract test cases by analyzing the relationship of web pages. First, the graph for expressing relationships among web pages is made by ReWeb tool. Moreover tests are executed with TestWeb tool.

This technique has the weak point in that so many candidates of test cases can be extracted and the inside of a web page is not fully tested.

3 Test Case Design

Testing of web applications is achieved, from the smallest unit test to a whole system test. If a submodule operates incorrectly, the test result of the module is not reliable. Therefore it is necessary to test an application level-by-level.

A test of the web applications is composed of a single web page test, a mutual web page test and an integrated web page test. Each level of the test is performed iteratively because detected failures from the test should be corrected and the regression test must be performed in the same environment again to confirm the correctness of the result. Furthermore, related parts of the web application should be tested again because changes after testing can affect other parts of the application.

The transmission of messages among web pages can be expressed using the notations of the UML sequence diagram. The test cases are generated from the sequence diagram of web pages. The test cases for a single web page test are generated from self-call messages and the test cases for a mutual web page test are generated from the messages between web pages. Furthermore, the test cases for an integrated web page test are generated from the messages which are sent to the system by an actor and received back from the system. The technique for generating test cases is as follows:

1. Generating test cases for a single web page test: The self-call messages of a web page are used to call script functions of the page or the page is re-executed by itself. The test cases for a single web page test are generated from these messages.
2. Generating test cases for a mutual web page test: The test cases for a mutual web page test are generated from the messages transmitted between web pages.

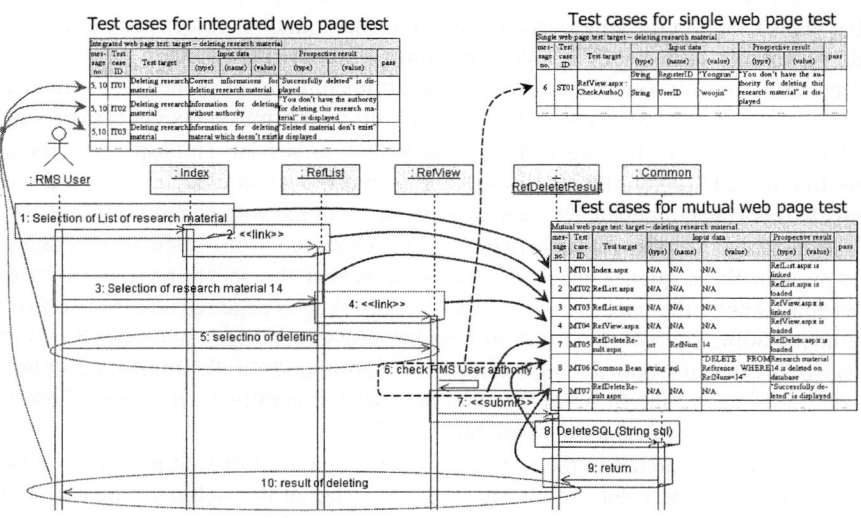

Fig. 1. Test case generation from sequence diagram

3. Generating test cases for an integrated web page test: The test cases for an integrated web page test are generated from the messages transmitted to the system by an actor and the response messages received from the system.

The technique for generating test cases of the system from the sequence diagrams which are developed in the use case analysis step is shown in Figure 1.

The technique for generating test cases from the factors of the sequence diagrams which are developed in the use case analysis step and targets of the test are described in Table 1.

Table 1. The technique for generating test cases of web pages from the sequence diagram

Factors of the sequence diagram		Test case	Target of test
Self-call message	→	Test case for a single web page test	* Script function * Re-execution of web page
Message between web pages	→	Test case for a mutual web page test	* Logic of web page * Transmission of message between web pages
Message related to actor	→	Test case for an integrated web page test	* Achievement of function through several web pages

3.1 Extracting Test Cases for Single Web Page Test

First of all, single web page test is performed for testing of web applications. Each web page is tested in the single web page test. Context and resources are examined for static pages like html pages. Context examination confirms whether the page made out according to syntax and resources examination confirms existence of the linked or called URL. These are easily tested with html editors or html syntax checking tools.

The main target for test are dynamic pages such as servlet, jsp, asp, aspx and php pages that include logics and classes such as bean classes that associated with the web pages. In this stage, independent execution modules in the web pages are tested.

The test cases for single web page test are extracted from self-call messages of each page in the sequence diagram. This case is that the server page calls its own script functions or the page is reexecuted by itself. A web page reexecutes itself in the case of including several functions. For example, a server page reexecutes by itself if the page has the function for registering information and the function for displaying the registeration result. In this case, the web page generally selects one of the two functions according to value of a variable.

The test data are added to the test case if script functions or web pages needs input values. Moreover, the values of objects in the form tag of the web page are considered as test data if the script functions reference the values of objects in the form tag using document object [6]. The related web pages, script functions and variables with these test cases are referenced from the page diagram [7].

Table 2 is an example of extracting test cases from sequence diagram of Figure 1 for single web page test. ST01 is a test case for testing the script function that confirms the authority of user for deleting a research material. In this example, it is tested if the action of system is correct in the case that an inputed user id is different from registered id of the research material. Although one test case extracted from self-call message of sequence diagram is represented in this example, valid input values and invalid input values should be tested.

Table 2. An example of extracting test cases for single web page test

Single web page test: scope – deleting research material								
message no.	Test case ID	Test target	Input data			Prospective result		pass
			(type)	(name)	(value)	(type)	(value)	
6	ST01	RefView.aspx:: CheckAuthor()	String	RegisterID	"yongsun"	"You do not have the authority for deleting this research material" is displayed		
			String	UserID	"woojin"			
...	...	...	...	...	...	...	...	...

3.2 Extracting Test Cases for Mutual Web Page Test

Independent execution modules such as java script functions in the web page and reexecution of the page are tested in the single web page test. The mutual web page test which examines if the pages are correctly performed in their mutual relation is performed after single web page test. The purpose of the mutual web page test is to examine if a page is linked to another page without loss of information, incorrect transfer of information or error.

The test cases for mutual web page test are extracted from the messages of each page that receives from actors or other pages. The messages of number 1, 2, 3, 4, 7, 8 and 9 in Figure 1 are extracted in this case.

The additional test data are necessary in test cases. The test data are different according to purpose of test and many branchs may occur according to the test data. The typical techniques for determining test data are equivalence partitioning and boundary value analysis [8]. The area of test data is classified for efficient testing and the test data of all class should be tested in the equivalence partitioning technique. It is basis of boundary value analysis technique the fact that many failures occurr around the boundary of input area rather than center. The boundary values are used to test data in this technique. The proper test data for target and purpose of the test should be included in the test cases based on these principles.

Table 3 is an example of extracting test cases for mutual web page test from sequence diagram of Figure 1. MT01, MT02, MT03, MT04 and MT05 are test cases to examine whether each page is correctly loaded on the browser through link. MT06 is a test case to examine if Common Bean class correctly deletes the information from database using inputed SQL statement. MT07 is a test case to examine if the system correctly displays the result of deleting the information for user.

Table 3. An example of extracting test cases for mutual web page test

message no.	Test case ID	Test target	Input data (type)	(name)	(value)	Prospective result (type)	(value)	pass
1	MT01	Index.aspx	N/A	N/A	N/A		RefList.aspx is linked	
2	MT02	RefList.aspx	N/A	N/A	N/A		RefList.aspx is loaded	
3	MT03	RefList.aspx	N/A	N/A	N/A		RefView.aspx is linked	
4	MT04	RefView.aspx	N/A	N/A	N/A		RefView.aspx is loaded	
7	MT05	RefDelete-Result.aspx	int	RefNum	14		RefDelete.aspx is loaded	
8	MT06	Common Bean	string	sql	"DELETE FROM Reference WHERE RefNum=14"		Research material 14 is deleted on database	
9	MT07	RefDelete-Result.aspx	N/A	N/A	N/A		"Successfully deleted" is displayed	
...	...	...	...	...	...		...	...

Mutual web page test: scope – deleting research material

3.3 Extracting Test Cases for Integrated Web Page Test

The integrated web page test is performed after single and mutual web page test. The purpose of integrated web page is to examine if the prospective results come through several web pages according to the request of an actor. The test cases are extracted from the messages transferred to the system by an actor and the result of the test cases is extracted from the response messages received from the system.

Table 4. An example of extracting test cases for integrated web page test

message no.	Test case ID	Test target	Input data (type)	(name)	(value)	Prospective result (type)	(value)	pass
5, 10	IT01	Deleting research material	Correct informations for deleting research material				"Successfully deleted" is displayed and selected material is deleted in database.	
5, 10	IT02	Deleting research material	Information for deleting without authority				"You do not have the authority for deleting this research material" is displayed and selected material is not deleted in database.	
5, 10	IT03	Deleting research material	Information for deleting materal which does not exist				"Seleted material do not exist" is displayed	
...	...	...	...				...	...

Integrated web page test: scope – deleting research material

The message 5 is a test case and the message 10 is a result of the test case in Figure 1. Table 4 is an example of extracting test cases for integrated web page test from sequence diagram in Figure 1. IT01 is a test case to test the function that deletes the specified research material and IT02 is to examine if the system prevents the action when an unauthorized user is going to delete a material. IT03 is a test case to examine if the system announces error to user when a user is going to delete a nonexistent material.

4 Testing with OnlineTestWeb

In this section, OnlineTestWeb is proposed. OnlineTestWeb is a tool for testing web applications. OnlineTestWeb on-line executes web applications on web server or application server with extracted test case and display the result of execution. It archive and manage sequent test result to analyze and test web applications more efficiently.

This tool is made using Microsoft Visual Basic 6. Figure 2 and figure 3 are pictures of OnlineTestWeb. The left side of user interface of OnlineTestWeb is for setting test case. The address of server page which is test target is set on "Test Web Page" item and the names and values for testing is inputted in turn. The inputted names and values are displayed on spread sheet to offer simple view.

When the "Test Execution" button is pushed and executing test case has finished, the result of execution is displayed on the right two mini browsers. The upper browser is for displaying the information of browser and the lower browser is for displaying the result of execution of test case. The result of test is reviewed and conclusion of test is inputted on "Conclusion" item on the lower left corner of user interface of OnlineTestWeb.

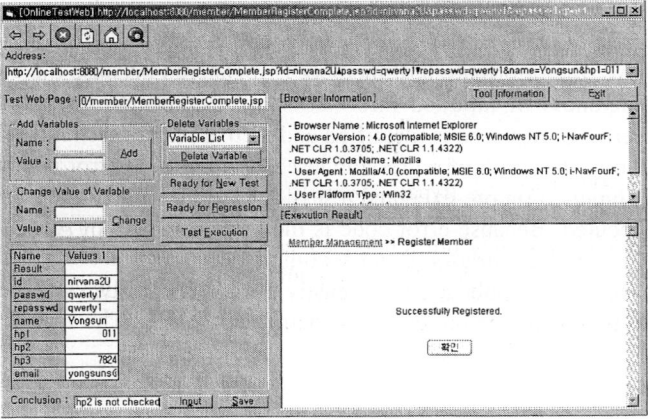

Fig. 2. OnlineTestWeb – Check Error

The "Ready for New Test" button is for preparing new test. If "Ready for New Test" button is pushed, the pre-executed test values are changed to gray color and the new column for new values is created.

If an error is discovered after testing, test target should be changed to correct the error and regression test should be executed with same environment. The "Ready for Regression" button is for preparing regression test. It the "Ready for Regression" button is pushed, the pre-executed test values are changed to gray color new column with same value is created on the right in spread sheet. The title of new column for regression test is created with the sign "-R" which indicates regression test.

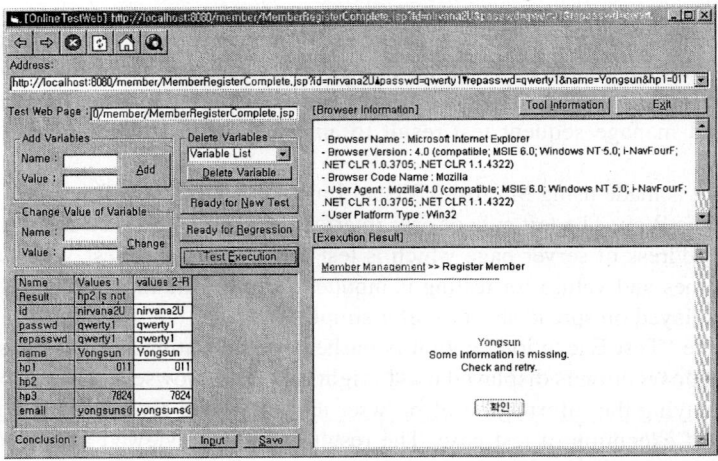

Fig. 3. OnlineTestWeb – Regression Test

Figure 2 is an example of executing a mutual web page test. In this case, hp2 which is a value for the middle number of cellular phone is omitted so error message is displayed. However the information of member is registered because the code for checking the completeness of inputted information in "MemberRegisterComplete.jsp" is omitted. The error of test case is identified and web application should be modified.

Figure 3 shows the result of regression test after identified error is modified and the data which is inputted on pre-executed test is deleted on database. The same test with figure 2 is executed. Because error code is modified, missing of some information is noticed and incorrect registration is not accomplished.

The test of web application is easily and efficiently accomplished using OnlineTestWeb and the technique of extracting test cases. The archived test results helped the test.

5 Comparison and Evaluation with Other Techniques

The test level is classified into three levels of single, mutual and integrated web page test, and the technique for the generation of test cases for each level is proposed in this paper. However test level is not classified for the generation of test cases in other techniques of related works. Test cases are generated by grouping flows of events of use cases for integrated web page test in RUP [3]. Test cases are generated by grouping control flows of source code and flows of web pages for all level tests in the technique of Wu, Offutt [4], so it is very complex to test web applications using the technique. Test cases are generated based on the relation model of web pages for integrated web page test in the technique of Ricca, Tonella [5].

The technique for testing script functions which are atomic units for performing logic in web applications is proposed in this paper, while it is not proposed in the technique of RUP and Ricca, Tonella.

Test data for both of normal and abnormal cases are used in this paper and RUP, while it is difficult to present test data because only the flows of logic and web pages are presented in the technique of Wu, Offutt, and test cases can be greatly increased according to the number of test data in the technique of Ricca, Tonella.

The number of test cases is the number of messages of sequence diagrams in this paper, while the number of test cases can be greatly increased because the combination of flows of events, control flows of source code or flows of web pages is used to generate test cases in other technique.

Table 5 shows the result of comparison of the techniques for the generation of test cases.

Table 5. Comparison of the techniques for the generation of test cases

Technique \ Item	This paper	RUP	Wu, Offutt	Ricca, Tonella
Source of test cases	Sequence diagram	Use case model	Source code	Relation model of web pages
Test level	Single, mutual, integrated level	Integrated level	Integration of three levels	Integrated level
Test of script functions	O	X	O	X
The number of test cases	Few (The number of messages of sequence diagrams)	Many (The number of combinations of flows of events)	Many (The number of combinations of control flows of source code)	Medium (The number of combinations of flows of web pages)

O: supported   X: not supported

6 Conclusion and Future Work

A test design technique based on the UML sequence diagram for developing web applications is proposed. A test of the web applications is composed of a single web page test, a mutual web page test and an integrated web page test. The test cases for a single web page test are generated from self-call messages and the test cases for a mutual web page test are generated from the messages between web pages. The test cases for an integrated web page test are generated from the messages which are sent to the system by an actor and received back from the system.

The flow of logic between web pages is easily understood, so test cases are easily generated because a sequence diagram shows a number of objects and the messages that are passed between these objects according to the time ordering of messages [9]. The notation of UML is frequently used in the analysis and design of web applications. Therefore, the technique of this paper will be more useful.

We plan to connect analysis and design with testing of web applications in order to test efficiently web applications.

References

[1] Abhijit Chaudhury, et al., *Web channels in E-Commerce*, Communications of the ACM, Jan. 2001.
[2] *Unified Modeling Language Specification Version 1.4*, OMG, September 2001.
[3] *The Rational Unified Process*, Rational Software Corporation (a wholly owned subsidiary of IBM), 2003.
[4] Ye Wu, Jeff Offutt, *Modeling and Testing Web-based Applications*, GMU ISE Technical ISE-TR-02-08, Nov. 2002.
[5] Filippo Ricca, Paolo Tonella, *Analysis and testing of Web applications*, Proceedings of the 23rd International Conference on Software Engineering, 2001.
[6] Mashito Hamba, Ryuichi Okakura, *HTML & Java Script Dictionary*, Youngjin.com, 2000.
[7] Jim Conallen, *Building Web Application with UML Second Edition*, Addison Wesley Longman, Inc., 2002.
[8] Roger S. Pressman, *Software Engineering: A Practitioner's Approach (5th Edition)*, McGraw-Hill, 2000.
[9] Grady Booch, et al., *The Unified Modeling Language User Guide*, Addison Wesley Longman, Inc., 1999.

Flexible Background-Texture Analysis for Coronary Artery Extraction Based on Digital Subtraction Angiography

Sung-Ho Park[1], Jeong-Hee Cha[1], Joong-Jae Lee[1], and Gye-Young Kim[2]

[1] School of Computing, Soong-Sil University, Korea
{kboar, arbitlee, pelly}@vision.ssu.ac.kr
[2] Department of Computing, Soong-Sil University, Korea
gykim@computing.ssu.ac.kr

Abstract. Image subtraction makes it possible to enhance the visibility of the difference between two images. This technique has been used in digital subtraction angiography (DSA) that is a well established modality for the visualization of coronary arteries. DSA involves the subtraction of a mask image – an image of the heart before injection of contrast medium – from live image. A serious disadvantage of this method, inherent to the subtraction operation, is sensitivity to distortion and variance of background gray-level intensity. Among the causes of these distortions are mean gray-level shift, and motions of heart and lung. In this paper, by choosing the image that has the minimum distortion through similarity analysis of background texture, we suggest the way to solve fundamental problems caused by background distortion, and then how to segment vessel from colony angiography through local gray-level correction of selected image.

Index Terms: Digital subtraction angiography (DSA), motion correction, texture analysis.

1 Introduction

Digital Subtraction Angiography (DSA) makes it possible to enhance the visibility of the blood vessel in the human body. With this method, a sequence of two-dimensional X-ray projection images are acquired to show the passage of a bolus of injected contrast medium through the vessel of interest. Resulting of DSA shows opacity vessel called live image, background structures are largely removed by subtracting an image acquired prior to injection called mask.

In the resulting subtraction, it works well if the images are perfectly aligned and no change in mean gray level. But, in practice, artifact of distortions is occurred as a result of changes in mean gray level and motion of object in the image. The changes in gray level are due to fluctuations in the power of X-ray source or noise in the image intensifier and the diffusion of the contrast material into ancillaries within parts of the heart outside the coronary arteries. Motion artifacts are due to cardiac motion and respiration. Clinical evaluations of DSA, following its introduction in the early 1980's, revealed that this is not the case for a substantial number of examinations

In order to cope with this problem, various techniques have been developed in coronary DSA. An overview of DSA and related works can be found in [1]. Other methods [2]-[5] have also been developed to deal with the problem. Respiration and cardiac motion may cause misregistration artifacts in images of the thoracic and abdominal regions [2]. Yanagisawa *et al.* first finds the global geometric translation and rotation based on estimated local misregistration of manually selected windows, performs the global correction and then corrects the residual local misregistration [3]. This method requires manual selection windows based on prior knowledge of the operator. The finding of the global image translation and rotation based on local misregistration may be difficult since local misregistration vectors can have arbitrary direction. Besides, the matching is done only through shifting of window pixels. Even though Van Tran *et al.* [4] introduce exclusion template, and sub-pixel precision, the exclusion template simply derived from mean and variance makes it difficult to divide the template into two part, vessel area and background. M. Thorsten *.et al.* [5], by using entropy of histogram of intensity, shows how to segment vessel area, but the histogram obtained from coronary angiography produces illegible shape which is difficult to get feature value. Thus, it is hard to apply it in this case immediately

In this paper, we present the way how to minimize error cause by motion artifacts.

We solve structural problems between before and after injection by selecting the best image close to the background similarity in the comparison with the subtraction of a mask - an image of the heart before injection of contrast medium - from a live image. And then we can segment coronary artery through local gray-level correction.

2 Methodology

Overall system configuration is as shown in Fig.1. Coronary angiography has the characteristic that image is geometrically deformed with cardiac motion and pulmonary air flow. And this cycle is periodically repeated. Accordingly, after approximate

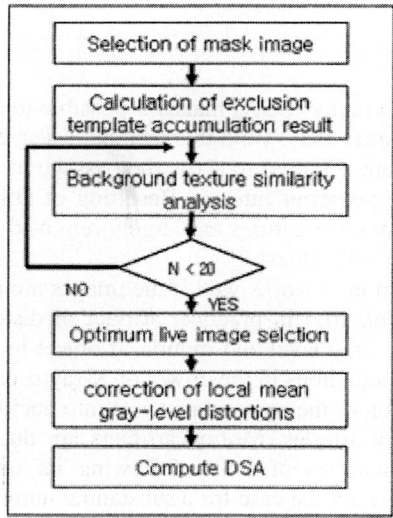

Fig. 1. Overall System Configuration

background area is obtained with use of exclusion template to detect vessel area, similarity of background texture is measured between mask and live image. And then, the most similar image is selected. After searching an image that is structurally most similar to mask image through determination of background texture similarity, subtraction is performed through correction of local variation in mean gray level, and then, only vessel area of actual coronary artery may rapidly and accurately be extracted.

Prior to obtain background similarity between two images, it is important to get approximate background area which is excluded from similarity measurement. As the area with contrast medium diffusion is relatively dark in comparison to the pixels of image with no diffusion, the ones containing contrast material may be obtained with use of exclusion template. Equation-(1) shows definition of exclusion template.

$$I_E(x, y) = I_M(x, y) - I_L(x, y)$$
$$I_E(x, y) = \begin{cases} 0 & I_E(x, y) \leq 0 \\ I_E(x, y) & , otherwise \end{cases} \quad (1)$$

Where, $I_E(x, y)$ is Exclusion template image, $I_M(x, y)$ is mask image, and $I_L(x, y)$ is live image. Exclusion template in Equation-(1) is the method used in distinguishing approximate vessel area and contrast background. When $I_E(x, y)$, the result of Equation-(1), can be binary value using of threshold-value, approximate vessel area and background area may be divided. And using of appropriate threshold value can eliminates noise. The range of threshold to be used is from 1 to 5.

Fig. 2 shows the distribution ratio of areas, which is obtained by Exclusion template calculation in Equation-(1) for five kinds of image streams displaying the diffusion process of contrast medium. This is consistent with the process that contrast medium is gradually appeared to maximum level and then disappeared.

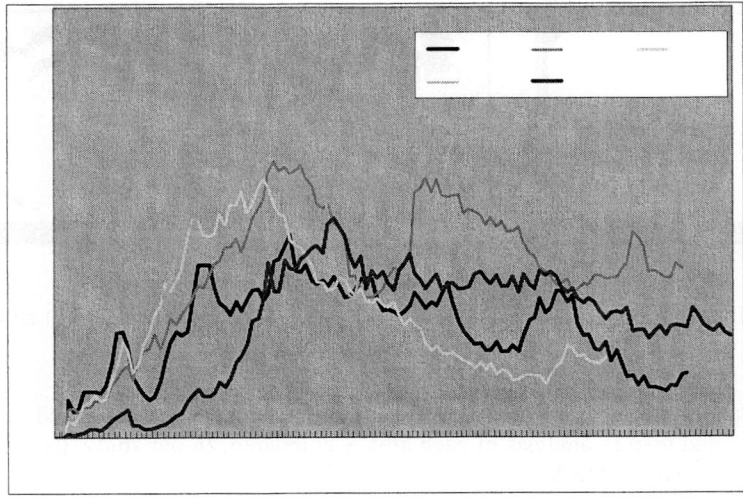

Fig. 2. Exclusion template area distribution

In Fig. 3-(a) and (b) show 43rd image that has, in terms of vascular area distribution and in total image sequences, the largest distribution among total 120 or more frames showing diffusion of contrast medium.

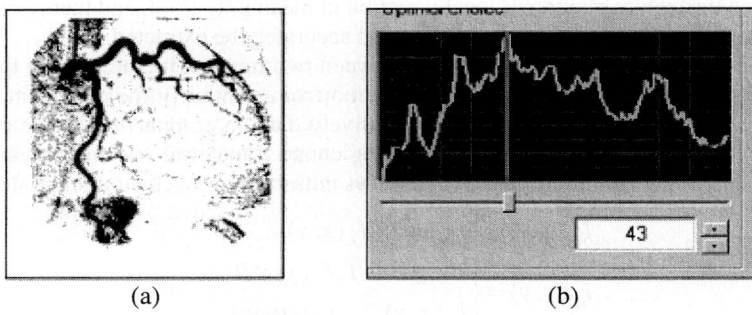

Fig. 3. The selected image and area distribution after exclusion template

Although the image with the highest value of area is the case that diffusion of contrast medium is the largest, it has lots of noise on the image. Therefore, twenty frames are selected starting form the image with the largest area distribution, and similarity with mask image is compared to choose optimum image. As one same background area should be defined for the all images, by accumulating exclusion template for continuous twenty images, similarity of same background area may be measured.

Fig. 4 shows the process to obtain common background of twenty images to determine the similarity of background texture. Fig. 4-(a) shows the frame border of image. Fig. 4-b shows accumulated result of twenty images after exclusion template process. Common background region may be obtained by overlapping these two images.

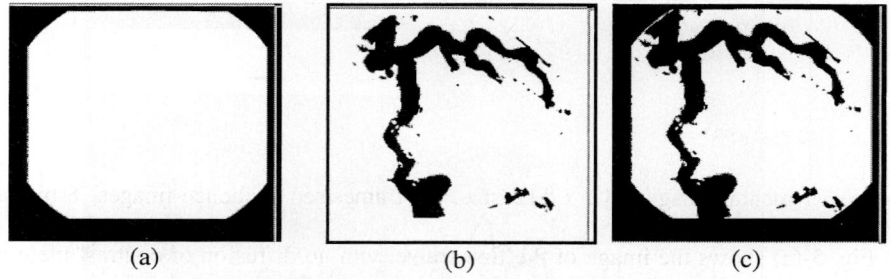

Fig. 4. Determination of background area : (a) detection of frame border (b) accumulated result of exclusion template on twenty images (c) final result area

To improve the accuracy of similarity measure, the image was separated in terms of sub block and texture analysis of each area is compared. In our study, the image was divided into 6x6 sub block.

Equation (2) shows the entropy of background texture pattern for comparison of backgrounds similarities.

$$H = -\sum_i \sum_j P[i,j]\log(P[i,j]) \qquad (2)$$

Matrix P is gray-level co-occurrence matrix and the element p[i,j] is defined by first specifying a displacement vector and counting all pairs of pixels separated by displacement vector having gray levels i and j. As entropy measurement method is, among texture analysis methods, less dependent to intensity variation of image and can be used in measurement of background similarity between two images. When the entropy obtained from each image and the entropy obtained from mask image are compared and the image with highest similarity of background is selected, an image that has the most approximate statistical value on blood vessels and background may be obtained. If correction of local mean gray-level distortions after that, it's possible to obtain final vessel area only.

Equation (3) and (4) show the correction of local variation in mean gray level. After creation of image D of which the size is same to mask image, the image is divided into windows n x n pixels. At this time, window size should be adjusted so as to be overlapped by one pixel in every direction. In our experiments, n is chosen to be 21.

$$D(i,j) = (1/P) \Sigma I(i,j) - M(i,j),$$
$$\text{where } (i,j) \notin \text{Exclusion template} \qquad (3)$$

$$M(i,j) = M(i,j) + D(i,j) \qquad (4)$$

Where, $I(i,j)$ and $M(i,j)$ are live image and mask image, respectively, and P is the number of pixels included in the calculation. At this time, all the calculation is applied to the areas that do not belong to exclusion template. Now our experiments find absolute value of the pixel-by-pixel difference between the mask image and the final corrected live.

3 Experiment

As experimental image, 512 x 512 size, 15 frames/sec sequence-images, 8-bitmap images were saved.

Fig. 5-(a) shows the image of the first frame with no diffusion of contrast medium and Fig. 5-(b) shows the image, among the images showing good diffusion during diffusion process of contrast medium, which was selected because it had the highest background similarity with mask image.

Fig. 6 shows the comparison of finally obtained area of the vessels with the result of previous study. 6-(a) is a live image obtained through analysis of background texture; 6-(b) is the image showing ideal vessel area of 6-(a) image; 6-(c) is the result of pixel-to-pixel subtraction; and 6-(d) shows differencing result on flexible mask subtraction [4]. In this case, the live image with largest area after exclusion template

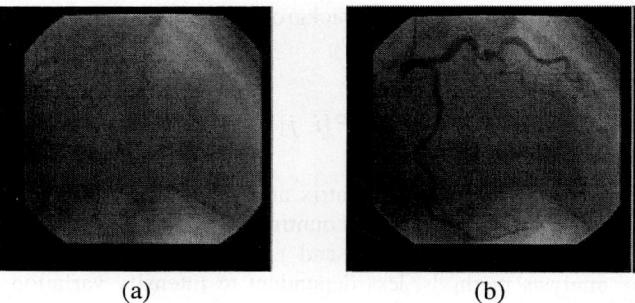

Fig. 5. Mask image and selected live image

process was selected as mask image. Fig. 6-(e) is the final area of the vessel obtained by selecting an image that has the highest similarity background-texture between mask and live images, be correcting local mean gray-level distortions, and by labeling of obtained image, according to the method suggested in our study. In the result, terminal branching area of blood vessels has insufficient contrast of the contrast material and some no-detection areas are established.

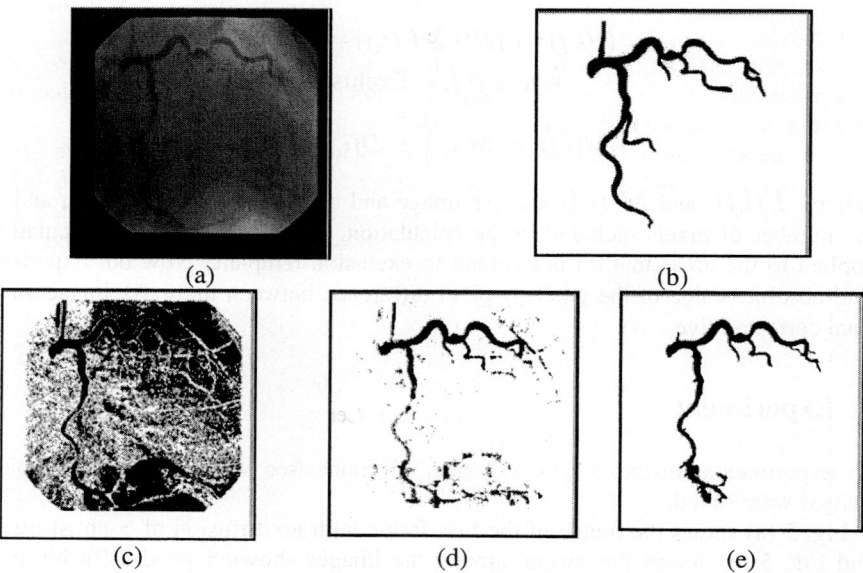

Fig. 6. Comparison of finally obtained results: (a) live image (b) ideal vessel area (c) pixel-to-pixel subtraction (d) flexible mask subtraction (e) proposed method

Table 1 shows comparison of accuracy, no-detection rates, and false-detection rates that background area is recognized as area of the vessels. Accuracy is the percentage of matched pixels in total pixels of the image. No-detection rate is the percentage of not-detected pixels for the pixels in ideal vessel area, and false-detection rate is the percentage of false-detected pixels in the pixels of background area. In the method suggested, accuracy and false-detection rate showed satisfactory results, however, no-detection rate is somewhat high due to no recognition of the blood vessels with insufficient contrast.

Table 1. Comparison of accuracy, no-detection rates, and false-detection rates

	Accuracy (%)	No-detection rate (%)	False-detection rate (%)
Pixel to pixel subtraction	65.67	14.61	34.50
Flexible mask subtraction	79.53	32.11	19.33
Suggested method	**97.13**	**20.73**	**0.76**

4 Conclusion

This paper presents the way how to automatically segment vessel area without human intervention. We select the best image close to the mask image through background texture analysis, which minimizes error caused by motion artifacts. Generally, motion artifacts arise from patient motion. However, in case of coronary angiography, it is more affected by cardiac motion than patient one. Due to repeated motion, this motion characteristic can be solved by selecting the best image close to mask image through background texture analysis. And then, correction of local mean gray-level distortions after that, it's possible to obtain final vessel area only

For more accurate vessel segmentation, more-than-one mask images can be used. In case of one mask image, it sometimes doesn't match live image because of variance between the mask image and next frames. Thus, selecting appropriate mask image before diffusion of contrast medium can be better way to accurate vessel extraction.

Acknowledgements

This work was supported by grant No. (R01-2002-000-00561-0(2004)) from the Basic Research Program of the Korea Science & Engineering Foundation.

References

1. W. R. Brory, "Digital subtraction angiography", IEEE Trans Nucl Sci, vol. NS-29, pp. 1176-1180, june 1982
2. L. M. Boxt, "Intravenous digital subtraction angiography of the thoracic and abdominal aorta", CardioVascular Interventional Radiology, vol. 6, pp. 205-213, 1983
3. M. Yanagisiwa, S. Shigemitsu, and T. Akatsuka, "Registration of locally distorted images by multiwindow pattern matching and displacement interpolation: The proposal of an algorithm and its appication to digital subtraction angiography", in Proc. IEEE seventh Inf. Conf. Pattern Recognition 1984, pp. 1188-1291

4. Luong Van Tran, Jack Sklansky, "Flexible Mask Subtraction for Digital Angiography", IEEE Trans. Medical Imaging. vol. 11. NO. 3. September, 1992
5. T. M. Buzug and J. Weese, "Similarity measures for subtraction methods in medical imaging.", in Proc 18th Ann. Int. Conf. IEEE Engineering Medicine Biology Society, 1996, pp. 140-14
6. Erik H. W. Meijering, Wiro J. Niessen, "Retrospective Motion Correction in Digital Subtraction Angiography: A Review", IEEE Trans. Medical Imaging, Vol. 18, No. 1. January 1999

New Size-Reduced Visual Secret Sharing Schemes with Half Reduction of Shadow Size

Ching-Nung Yang and Tse-Shih Chen

Department of Computer Science and Information Engineering,
National Dong Hwa University,
#1, Da Hsueh Rd, Sec. 2, Hualien, 974-01, Taiwan
Tel: +886-3-8634025   Fax: +886-3-8634010
cnyang@mail.ndhu.edu.tw

Abstract. Visual Secret Sharing (VSS) scheme proposed by Naor and Shamir is a perfect secure scheme to share the secret image. By using m subpixels to represent one pixel, the secret image is divided into several shadow images whose size is m times than the secret image. The value of m is known as the *pixel expansion*. In this paper, we propose the new size-reduced VSS schemes and dramatically decrease the *pixel expansion* to a half and meantime the contrast is not compromised.

1 Introduction

For a VSS scheme with the general access structure (Γ_{Qual}; Γ_{Forb}), each pixel is divided into m black and white sub pixels in n shadows given to n participants in a set $\mathcal{P}=\{1, 2, \ldots, n\}$, where $\Gamma_{Qual} \subseteq 2^{\mathcal{P}}$, $\Gamma_{Forb} \subseteq 2^{\mathcal{P}}$, and $\Gamma_{Qual} \cap \Gamma_{Forb} = \phi$ [1], [2], [3], [4]. VSS Schemes can be described by $n \times m$ Boolean matrix $S = [s_{ij}]$, where $s_{ij} = 1$ if and only if the jth sub pixel in the ith shadow is black, otherwise $s_{ij} = 0$. When shadows $i_1, i_2, \ldots, i_r$ in a set $X(\subseteq \mathcal{P}) \in \Gamma_{Qual}$ are stacked, the rows $i_1, i_2, \ldots i_r$ in S are OR-ed. The gray level of the recovered image is proportional to the Hamming weight of this OR-ed m-vector V. If $H(V) \geq d_B$, this gray level is interpreted by the user's visual system as *Black* and if $H(V) \leq d_W$ the result is interpreted as *White*, where $d_B > d_W$. Thus, finally one can "see" the secret image by human visual system directly without the help of computing device and algorithm. But if the stacked shadows are in the forbidden set Γ_{Forb}, one can get no information. Other image secret sharing schemes with perfect quality secret image are shown in [5], [6], [7]; however they need computations.

Most recent papers about VSS schemes are dedicated to the reduction of pixel expansion. Recently, in [8], Kuwakado and Tanaka proposed size-reduced VSS schemes by deleting some columns from the black and white sets in the original VSS schemes. In this paper, we proposed new size-reduced (k, n) VSS schemes to further reduce the pixel expansion by processing two-pixeled block each time; meantime our size-reduced (k, n) VSS schemes have the same contrast as the conventional VSS schemes.

The rest of this paper is organized as follows. Section 2 describes the basic (k, n) VSS schemes. In Section 3, the constructions for the proposed size-reduced (k, n)

VSS schemes are proposed; also, analyses of the contrast and the recognition of the edge are given. Comparison with the previous size-reduced scheme and experimental results are shown in Section 4. Finally, conclusions are presented in Section 5.

2 k-Out-of-n VSS Schemes with the Pixel Expansion m

A popular (k, n) VSS scheme has the qualified set Γ_{Qual} including all the sets of k or more participants and the forbidden set $\Gamma_{\text{Forb}} = 2^P - \Gamma_{\text{Qual}}$. For example, a (2, 2) VSS scheme, the qualified and forbidden sets are $\Gamma_{\text{Qual}} = \{\{1,2\}\}$ and $\Gamma_{\text{Forb}} = \{\{1\},\{2\}\}$. We use (k, n, m) VSS scheme to denote a (k, n) VSS scheme with the pixel expansion m. Suppose that two B_1 and B_0 are the black and white base $n \times m$ matrices for (k, n, m) VSS schemes, and C_1 and C_0 are their corresponding black and white sets including all matrices obtained by permuting the columns of B_1 and B_0, respectively. When one black (resp. white) pixel is shared, we randomly select one matrix from the collection set C_1 (resp. C_0) and then choose one row of this matrix to a relative shadow.

Next, we formally list two conditions, *contrast* and *security* conditions, where a (k, n) VSS Scheme needs to satisfy [1].

C1 (Ccontrast Condition):
For any S of the $n \times m$ matrix in C_1 (resp. C_0), the OR-ed V of rows $i_1, i_2, \ldots, i_r$, for $r \geq k$, satisfies $H(V) \geq d_B$ (resp. $H(V) \leq d_W$).

C2 (Security Condition):
For $r < k$, the two collections of $r \times m$ matrices obtained from C_1 and C_0 are same in the sense that they contain the same matrices with the same frequencies.

More pixel expansion increases the shadow size and makes VSS schemes impractical for real application. Thus, how to further reduce the pixel expansion is the important issue for VSS schemes.

3 The Proposed Size-Reduced (n, k, m) VSS Schemes Using (n, k, m') VSS Schemes

3.1 Design Concept

We now describe the design concept our size-reduced (n, k, m) VSS schemes, where $m = m'/2$, from (k, n, m') VSS schemes, i.e. the shadow size is half reduced.

The concept of our proposed scheme is to process a two-pixeled block each time, i.e. ■■, □□, ■□, □■, by using the four corresponding sets C_{11}, C_{00}, C_{10} and C_{01} of $n \times m'$ Boolean matrices, respectively. Thus, finally the pixel expansion is

$$m = m'/2 \tag{1}$$

The first two sets, C_{11} and C_{00}, contribute the contrast of the recovered image, while the last two sets, C_{10} and C_{01}, determine the clearness of edges between black and white areas.

For our new two-pixel-based VSS schemes, we rewrite the condition *C1* to *C1'-1* (contrast condition) and *C1'-2* (precision of edge condition). Also, the condition *C2* is rewritten to *C2'* (security condition).

C1'-1 (Ccontrast Condition):
For any S of the matrix in C_{11} (resp. C_{00}), the OR-ed V of rows $i_1, i_2, \ldots, i_r$ for $r \geq k$, satisfies $H(V) \geq d_B$ (resp. $H(V) \leq d_W$), where $d_B > d_W$.

C1'-2 (Precision of Edge Condition):
For any S of the matrix in C_{10} (resp. C_{01}), the OR-ed V of rows $i_1, i_2, \ldots, i_r$ for $r \geq k$, satisfies the precision of edge $V_n/|C_{10}| \geq P_{BW}$ (resp. $V_n/|C_{01}| \geq P_{WB}$), where V_n is the maximum number of same V and $0 < P_{BW}, P_{WB} \leq 1$.

C2' (Security Condition):
For $r < k$, the four collections of $r \times m$ matrices obtained from C_{11}, C_{00}, C_{11} and C_{00} are same in the sense that they contain the same matrices with the same frequencies.

We herein use Naor-Shamir contrast $C_{NS} = ((m - d_w) - (m - d_B))/m$ [1] and the average contrast definition in [8] to define the contrast of our two-pixeled-basd VSS scheme as

$$C_{YC} = \left(\frac{1}{|C_{11}|} \sum_{\text{All } S \text{ in } C_{11}} H(V) - \frac{1}{|C_{00}|} \sum_{\text{All } S \text{ in } C_{00}} H(V) \right) / m' \quad (2)$$

Also, we define the average precision of edge P_{MEAN}, the distinction of black line in white area D_{BIW} (i.e. the distinction between C_{10} and C_{00}, and the distinction between C_{01} and C_{00}), the distinction of white line in black area D_{BIW} (i.e. the distinction between C_{10} and C_{11}, and the distinction between C_{01} and C_{11}) and the average distinction D_{MEAN} as follows:

$$P_{MEAN} = (P_{BW} + P_{WB})/2 \quad (3)$$

and

$$\begin{aligned} D_{BIW} &= (D_{C_{10}-C_{00}} + D_{C_{01}-C_{00}})/2, \\ D_{WIB} &= (D_{C_{10}-C_{11}} + D_{C_{01}-C_{11}})/2, \text{ where } D_{C_{ij}-C_{kl}} = |C_{ij} - (C_{ij} \cap C_{kl})|/|C_{ij}| \\ D_{MEAN} &= (D_{BIW} + D_{WIB})/2. \end{aligned} \quad (4)$$

The value C_{YC} shows the contrast between black and white areas. The average precision of edge P_{MEAN} determines the clearness of the edge (Notice that if the stacked patterns V are not same then it will lack consistency and the edge is irregular.); the distinction D_{MEAN} is the value to measure whether we can distinguish thin lines in the recovered images. In the following constructions, we discuss how to design (n, k, m) VSS schemes to achieve the high C_{YC}, P_{MEAN} and D_{MEAN}.

3.2 Constructions for Size-Reduced (n, k, m) VSS Schemes from (n, k, m') VSS Schemes with $m = m'/2$

Construction 1 shows five construction methods for the size-reduced $(n, k, m'/2)$ VSS schemes using (n, k, m') VSS schemes based on two-pixel-based structure.

Construction 1: Let C_1 and C_0 be the black and white sets for (n, k, m') VSS schemes. Then the four sets, C_{11}, C_{00}, C_{10} and C_{01} for our two-pixeled-based (n, k, m) VSS schemes can be constructed as the following five methods:

1. $C_{11} = C_1, C_{00} = C_0$, and $C_{10} = C_{01} = C_1 \cup C_0$;
2. $C_{11} = C_1, C_{00} = C_0$, and $C_{10} = C_{01} = C_1$;
3. $C_{11} = C_1, C_{00} = C_0$, and $C_{10} = C_{01} = C_0$;   (5)
4. $C_{11} = C_1, C_{00} = C_0, C_{10} = C_1$ and $C_{01} = C_0$;
5. $C_{11} = C_1, C_{00} = C_0, C_{10} = C_0$ and $C_{01} = C_1$.

Theorem 1: The scheme from *Construction 1* is a $(n, k, m'/2)$ VSS scheme.

Proof: To prove *Construction 1-1*, let the values d_B and d_W be defined in *C1* conditions for (n, k, m') VSS schemes. From (5), since $C_{11} = C_1$ and $C_{00} = C_0$, then d'_B and d'_W in the size-reduced VSS scheme are $d'_B = d_B$ and $d'_W = d_W$; since $C_{10} = C_{01} = C_1 \cup C_0$, so $P_{BW} = P_{WB} = Max\{1/2C_{d_B}^{m'}, 1/2C_{d_W}^{m'}\}$. Thus, condition *C1'* is met. To prove the security condition, it is observed that every row of these four sets is same to the row in C_1 and C_0. Thus, the scheme satisfies *C2'* condition. From (1), the pixel expansion is $m = m'/2$.

The proof of *Construction 1-2, 1-3, 1-4* and *1-5* are similar. □

The values C_{YC}, P_{MEAN} and D_{MEAN} for *Construction 1-1* are calculated as follows:

$$C_{YC} = \left(\frac{1}{|C_{11}|}\sum_{\text{All } S \text{ in } C_{11}} H(V) - \frac{1}{|C_{00}|}\sum_{\text{All } S \text{ in } C_{00}} H(V)\right)\Big/m'$$

$$= \left(\frac{1}{|C_1|}\sum_{\text{All } S \text{ in } C_1} H(V) - \frac{1}{|C_0|}\sum_{\text{All } S \text{ in } C_0} H(V)\right)\Big/m'$$

$$= (d_B - d_W)/m'$$

$$= C_{NS};$$

$$P_{MEAN} = \left(Max\{1/2C_{d_B}^{m'}, 1/2C_{d_W}^{m'}\} + Max\{1/2C_{d_B}^{m'}, 1/2C_{d_W}^{m'}\}\right)\Big/2$$

$$= Max\{1/2C_{d_B}^{m'}, 1/2C_{d_W}^{m'}\};$$

$$D_{C_{10}-C_{00}} = D_{C_{01}-C_{00}} = D_{C_{10}-C_{11}} = D_{C_{01}-C_{11}} = m'/2m'! = 1/2,$$

$$D_{BIW} = (1/2 + 1/2)/2 = 1/2, \qquad D_{WIB} = (1/2 + 1/2)/2 = 1/2,$$

$$D_{MEAN} = (1/2 + 1/2)/2 = 1/2.$$

Table 1. The size-reduced (k, n, m) VSS schemes from (k, n, m') VSS schemes

	P_{BW}	P_{WB}	P_{MEAN}	D_{BIW}	D_{WIB}	D_{MEAN}	m	C_{YC}
Construction 1-1	$\text{Max}\begin{cases}1/2C_{d_B}^{m'}, \\ 1/2C_{d_W}^{m'}\end{cases}$	$\text{Max}\begin{cases}1/2C_{d_B}^{m'}, \\ 1/2C_{d_W}^{m'}\end{cases}$	$\text{Max}\begin{cases}1/2C_{d_B}^{m'}, \\ 1/2C_{d_W}^{m'}\end{cases}$	1/2	1/2	1/2	$m'/2$	C_{NS}
Construction 1-2	$1/C_{d_B}^{m'}$	$1/C_{d_B}^{m'}$	$1/C_{d_B}^{m'}$	1	0	1/2	$m'/2$	C_{NS}
Construction 1-3	$1/C_{d_W}^{m'}$	$1/C_{d_W}^{m'}$	$1/C_{d_W}^{m'}$	0	1	1/2	$m'/2$	C_{NS}
Construction 1-4	$1/C_{d_B}^{m'}$	$1/C_{d_W}^{m'}$	$\left(1/C_{d_B}^{m'}+1/C_{d_W}^{m'}\right)/2$	1/2	1/2	1/2	$m'/2$	C_{NS}
Construction 1-5	$1/C_{d_W}^{m'}$	$1/C_{d_B}^{m'}$	$\left(1/C_{d_W}^{m'}+1/C_{d_B}^{m'}\right)/2$	1/2	1/2	1/2	$m'/2$	C_{NS}

Table 2. (2, 2, 1) VSS schemes

	P_{BW}	P_{WB}	P_{MEAN}	D_{BIW}	D_{WIB}	D_{MEAN}	m	C_{YC}, C_{NS}
Construction 1-1	1/2	1/2	1/2	1/2	1/2	1/2	1	1/2
Construction 1-2	1	1	1	1	0	1/2	1	1/2
Construction 1-3	1/2	1/2	1/2	0	1	1/2	1	1/2
Construction 1-4	1	1/2	3/4	1/2	1/2	1/2	1	1/2
Construction 1-5	1/2	1	3/4	1/2	1/2	1/2	1	1/2

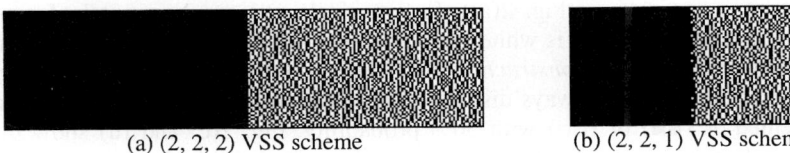

(a) (2, 2, 2) VSS scheme   (b) (2, 2, 1) VSS scheme

Fig. 1. The contrasts for (2, 2, 2) and (2, 2, 1) VSS schemes

All the values C_{YC}, P_{BW}, P_{WB}, P_{MEAN} and D_{MEAN} for *Construction 1* are listed in Table 1.

Construct the size-reduced (2, 2, 1) VSS scheme from Naor-Shamir (2, 2, 2) VSS scheme, by using *Construction 1*. The values in Table 1 are calculated and shown in the following table.

To confirm the definition of contrast in (2), from Fig. 1, it is observed that the following recovered images (the left side is the black region and right side is white region) using the conventional (2, 2, 2) VSS scheme and the size-reduced (2, 2, 1) VSS scheme (by *Construction 1-1*) have almost the same contrast; however the edge between black and white areas is irregular in Fig. 1(b).

From Table 2, we successfully reduce the pixel expansion of (2, 2, 2) VSS scheme to 1 with the same contrast, $C_{YC}=C_{NS}=1/2$. $P_{MEAN}=1/2$ (*Construction 1-1* and *1-3*) means that the edge of black and white areas will be in the precision with 50%; the value $D_{MEAN}=1/2$ shows that we can distinguish the thin lines (line width less than 3 pixels) with 50%.

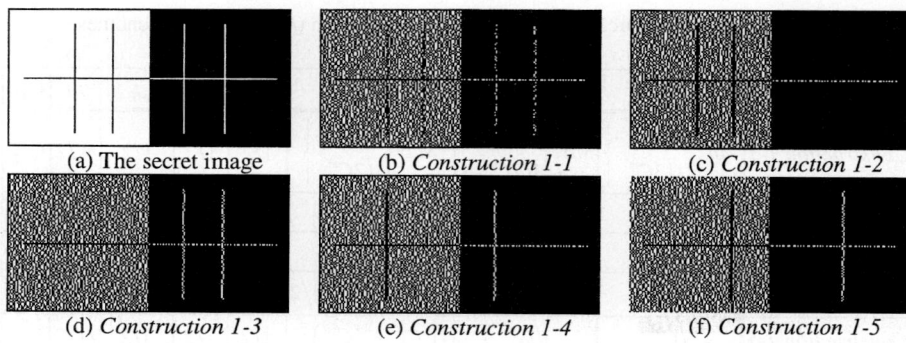

Fig. 2. The recovered images for five (2, 2, 1) VSS schemes

Some constructions either improve the precision of the edge or the distinction of the thin lines. *Construction 1-2*, *1-4* and *1-5* improved the average precision of edge (large than 50% precision), i.e. the recovered images will have the more regular edges. Moreover, *Construction 1-2* has the better recognition of the black line on the white area since $D_{BIW}=1$ and $D_{WIB}=0$; *Construction 1-3* has the better recognition of the white line on the black area since $D_{BIW}=0$ and $D_{WIB}=1$; *Construction 1-1*, *1-4* and *1-5* have half and half distinction on both black and white areas. All of these five constructions have average distinction 50%.

We use a test pattern (Fig. 2(a)), four verticals and one horizontal of one-pixel width; the left background is white region and the right is black. Fig. 2(b)~(f) are the recovered images using *Construction 1-1* ~ *Construction 1-5*, respectively. Fig. 2(b) shows that the verticals always display because the two-pixeled blocks, (■□;□■), are represented as (■■) or (□□) with 50% probability. Fig. 2(c) and (d) show that the while and black verticals disappear since two-pixeled blocks, (■□;□■), are all represented as (■■) for *Construction 1-2* and (□□) for *Construction 1-3*, respectively. In Fig. 2(e), the black (resp. white) verticals will disappear when we only just process (□■) (resp. (■□)) and it will remain when processing (■□) (resp. (□■)). The same analysis can be applied to *Construction 1-5* and shown in Fig. 2(f).

For observing the edge irregularity, Fig. 2 (c) has the best regularity because the average precision of edge $P_{MEAN}=1$.

Notice that Fig. 2 is consistent with the definitions of D_{BIW}, D_{WIB} and P_{MEAN}. From Fig. (2), we know that some constructions may have the line disappearance problem except *Construction 1-1*; however it is observed that the edge of recovered image will have irregularity using *Construction 1-1*.

4 Experimental Results and Comparison

4.1 Experimental Results

In this section, we use (2, 2, 1) and (2, 2, 1.5) VSS schemes to observe the clearness of the edge in the recovered image. The secret image is Fig. 2(a) but with the line width 2. Experimental results are shown in Fig. 3. Fig. 3(b) and (c) show the possible disappearance of two-pixeld lines; however Fig. 3(d) and (e) always display the two-pixeled lines and have clear line rather than Fig. 3(a). It is obvious that when the line width is no less than two, all constructions will show the line. The recovered images of all constructions will be "seen" clear except *Construction 1-1*.

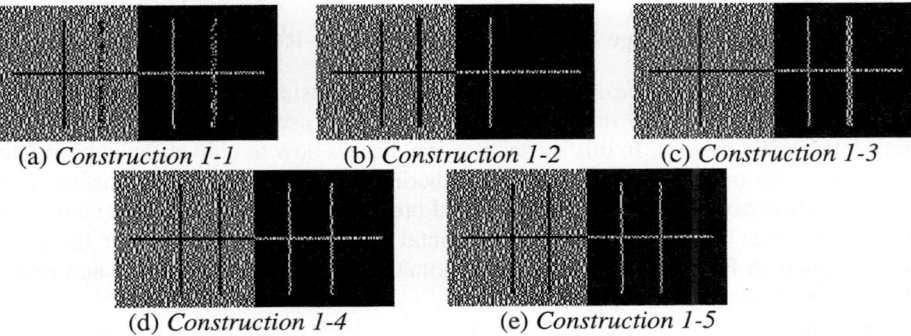

(a) *Construction 1-1* (b) *Construction 1-2* (c) *Construction 1-3*
(d) *Construction 1-4* (e) *Construction 1-5*

Fig. 3. Recovered images of the (2, 2, 1) VSS schemes for the Fig. 2(a) with 2-pixeled line

More experimental results are shown in Fig. 4. One standard image *House* (from USC-SIPI image database) is used to test the effects of thin lines and image qualities. From Fig. 4, it is observed that *Construction 1-5* has the best image quality of *House*. There is no definite answer which construction is the best way to share the secret image. It depends on the pattern of secret image.

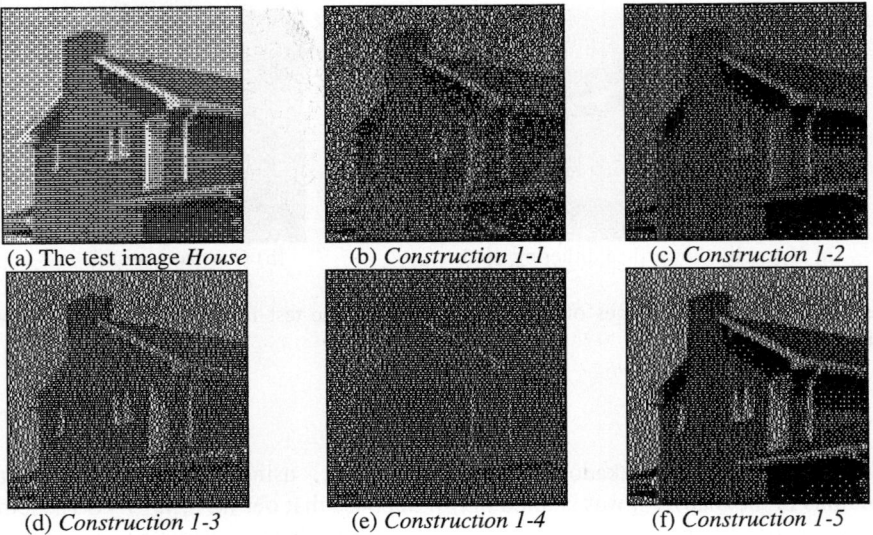

(a) The test image *House* (b) *Construction 1-1* (c) *Construction 1-2*
(d) *Construction 1-3* (e) *Construction 1-4* (f) *Construction 1-5*

Fig. 4. The recovered images of the (2, 2, 1) VSS schemes for the test image *House*

In the next subsection, we experiment on the conversion of a gray secret image to a black and white one for our constructions. Because the sets C_{10} and C_{01} in *Construction 1-1 ~ Construction 1-5* are constructed by C_1 or C_0 or their union, thus the performance of the recovered image is compromised when the processed patterns are (■□) and (□■). Therefore, conversion of a gray secret image to a black and white one needs more carefulness.

4.2 Convert a Gray Image Appropriate for Our Size-Reduced Constructions

As we know *dithering* is a kind of technology to transfer a gray level image into halftone (black and white) image [9], [10], and it was used in [11] to construct the gray-level VSS schemes. In this subsection, we discuss how to convert an appropriate gray image for our constructions. When dithering, every gray pixel is transferred to black or white pixel by a dithering mask and produces the different representation of black and white image. We use the horizontal dithering mask to transfer the gray image *House* to Fig. 4(a) that has more continuous black and white pixels and make the recovered image with a good contrast.

By using another dithering mask, the vertical dithering mask, which is orthogonal form the previous mask used in Fig. 4(a). The image *House* is transferred to Fig. 5(a) that has the same image quality as Fig. 4(a). The major difference is the more serial vertical homogeneous pixels increases. Since the new dithering mask is vertical, so our processed patterns will have more two-pixeled blocks, (■□), (□■), and this will compromise the recovered image. Fig. 5(b) is the recovered image of *Construction 1-2* when using Fig. 5(a) as the secret image. When compared Fig. 4(c) and Fig. 5(b); although the secret image *House* (Fig. 4(a) and Fig. 5(a)) have the same image quality, Fig. 5(b) is worse than Fig. 4(c).

(a) *House* by the vertical dithering mask   (b) *Construction 1-2*

Fig. 5. The recovered images of *Construction 1-2* for the test image *House* by the vertical dithering mask

4.3 Comparison

The size-reduced Kuwakado-Tanaka VSS scheme, using the deletion of some columns in the matrices, was studied in [8]. Suppose that our (k, n, m_{YC}) VSS schemes and (k, n, m_{KT}) Kuwakado-Tanaka VSS schemes are both constructed by (k, n, m') VSS schemes; thus, they have the same contrasts $C_{YC} = C_{KT} = C_{NS}$. The comparison is shown in Table 4 and our scheme is better than Kuwakado-Tanaka scheme, i.e. the pixel expansion of our scheme is no large than that in Kuwakado-Tanaka VSS scheme. Moreover, our scheme can be constructed from any (k, n, m') VSS schemes to achieve further reduction, but the size-reduced Kuwakado-Tanaka VSS scheme does not work for reducing the shadow size of (k, k, m') VSS schemes.

Table 3. Comparison of the proposed VSS schemes with the size-reduced VSS schemes in [8] for n=3, 4 and 5

		m'	m_{KT}	$m_{YC} = m'/2$	C_{YC}, C_{KT}, C_{NS}
n=3	k=2	3	2	1.5	1/3
n=4	k=2	6	3	3	1/3
	k=3	6	3	3	1/6
n=5	k=2	5	4	2.5	1/5
		10	5	5	3/10
	k=3	8	5	4	1/8
	k=4	15	8	7.5	1/15

Fig. 6 shows the recovered images for our proposed (2, 3, 1.5) VSS scheme and (2, 3, 2) Kuwakado-Tanaka VSS scheme for the secret image *House* (Fig. 4(a)). Both of them are constructed based on the (2, 3, 3) VSS scheme.

We use *Construction 1-5* in our (2, 3, 1.5) VSS scheme which has the best image qualities for this case. Although the authors declare that they achieve the same contrast $C_{KT} = C_{NS} = 1/3$ (see the first row in Table 4), in fact it is observed that our (2, 3, 1.5) VSS scheme has the better visual image quality than (2, 3, 2) Kuwakado-Tanaka VSS scheme. Moreover, our pixel expansion m=1.5 is less than m=2 in (2, 3, 2) Kuwakado- Tanaka VSS scheme.

(a) The proposed (2, 3, 1.5) scheme

(b) Kuwakado-Tanaka (2, 3, 2) scheme

Fig. 6. The recovered images for the proposed size-reduced VSS scheme and Kuwakado-Tanaka VSS scheme for the test image *House*

5 Concluding Remarks

In this paper, the new size-reduced (k, n, m) VSS schemes from the (k, n, m') VSS schemes with $m = m'/2$ are proposed. When compared to the existing size-reduced

VSS scheme, Kuwakado-Tanaka scheme [8], our size-reduced schemes work better either on the pixel expansion or the image quality. It is shown that the edges in our recovered images are a little irregular but the recovered images have the same contrast. It is indefinite that which construction has the best performance, i.e. the clear edge and the high contrast, because it depends on the arrangement of black and white pixels in the secret image. So how to dither a gray image to a black and white secret image for our constructions is also discussed here. The two-pixel-based construction method may be extended to three-pixel-based construction for further reduction of shadow size. But, we need to design the sets C_{110}, C_{101}, C_{011}, C_{001}, C_{010} and C_{100} appropriately such that the edge clearness is assured. To find a generalized multi-pixel-based construction is interesting and needs further studying.

References

[1] M. Naor and A. Shamir, "Visual cryptography," *Advances in Cryptology - EUROCRYPT'94, Lecture Notes in Computer Science*, No. 950, pp. 1-12, 1995.
[2] G. Ateniese, C. Blundo, A. De Santis, and D.R. Stinson, "Visual cryptography for general access structures," *ECCC, Electronic Colloquium on Computational Complexity* (TR96-012), 1996.
[3] E. R. Verheul and H.C.A. Van Tilborg, "Constructions and properties of k out of n visual secret sharing schemes," *Designs, Codes and Cryptography*, Vol. 11, No. 2, pp. 179-196, May, 1997.
[4] P.A. Eisen and D.R. Stinson, "Threshold visual cryptography schemes with specified whiteness," *Designs, Codes and Cryptography*, Vol. 25, No. 1, pp.15-61, Jan., 2002.
[5] R. Lukac and K.N. Plataniotis, "Colour image secret sharing," *IEE Electronics Letters*, Vol. 40, No. 9, pp. 529-530, April 2004.
[6] R. Lukac and K.N. Plataniotis, "Bit-level based secret sharing for iImage encryption," *Pattern Recognition*, Vol. 38, No. 5, pp. 767-772, May 2005.
[7] R. Lukac and K.N. Plataniotis, "A color image secret sharing scheme satisfying the perfect reconstruction property, " *Proceedings of MMSP'04*, pp. 351-354, Italy, Sep. 29 - Oct. 1, 2004.
[8] H. Kuwakado and H. Tanaka, "Size-Reduced Visual Secret Sharing Scheme," *IEICE trans. Fundamentals*, Vol. E87-A, No. 5, pp.1193-1197, May 2004.
[9] R. Ulichney, Digital Halftoning, MIT Press, Cambridge, MA, 1987.
[10] Y. Zhang, "Space-filling curve ordered dither," *Computers and Graphics* Vol. 22, No. 4, pp. 559–563, Aug., 1998.
[11] C. C. Lin and W.H. Tsai, "Visual cryptography for gray-level images by dithering techniques," *Pattern Recognition Letters*, 24, pp.349-358, 2003.

An Automatic Resource Selection Scheme for Grid Computing Systems

Kyung-Woo Kang[1] and Gyun Woo[2]

[1] Department of Computer and Communication Engineering,
Cheonan University, 115, Anseo-dong, Cheonan 330-704,
Choongnam, Republic of Korea
kwkang@cheonan.ac.kr
[2] Department of Computer Science and Engineering,
Pusan National University,
Busan 609-735, Republic of Korea
woogyun@pusan.ac.kr

Abstract. Our grid system *META* allows CFD users to access computing resources across the network [1]. There are many research issues involved in the grid computing, including fault-tolerance issues, computing resource selection issues, and user-interface design issues. In this paper, we propose an automatic resource selection scheme for executing the parallel SPMD application written in MPI. *META* uses the network latency time and the speed of the test-run of the kernel loop for the efficient management of resources. The network latency time highly influences the performance of the CFD application on the grid environment. The kernel loop is a loop which consumes the most of the computation time of a CFD program.

Keywords: CFD, grid computing, resource selection.

1 Introduction

As the number of supercomputers is increased and the speed of the network becomes faster, the necessity of the grid system of supercomputers is being raised. A grid system is an assembly of distinct computers or processors which are designed to work together to tackle a single task or a set of problems. The supercomputers may be located in the same building, or separated by thousands of miles. However, in order to make the grid system easy to use, it should be made to show its capabilities as one system on the user's PC screen [2, 3, 4, 5].

The goal of *META* is to help the user to run any CFD (Computational Fluid Dynamics) analysis program on the grid environment efficiently [1]. The main technical challenge in implementing scientific applications on the grid lies in accommodating the long latencies incurred in distributed computations on geographically-separated machines [2, 6, 7]. The structure of a CFD solver is a repetition structure of same flow work. A CFD is a method for finding an approximate solution of the equations like the Navier-Stokes equation using computers.

The kernel loop of a CFD program is the loop that performs the most dominant repeated work of the program. On the grid environment, our experiments show that the CPU-time for each iteration of a kernel loop is uniformly distributed.

This paper is organized as follows. Section 2 provides the properties of CFD analysis program and its example in order to describe the scope of this research. *META* should choose a set of supercomputers that might generate the result of CFD solver for shortest CPU time. Section 2 also describes the voting mechanism of *META* for selecting the fastest set of supercomputers. Section 3 presents the results of this approach for CFD solvers. We conclude in Section 4.

2 The Research Target and Approaches

2.1 CFD Analysis Program

The fluid dynamics could be represented by the Navier-Stokes equation. CFD is a method for finding an approximation of the Navier-Stokes equation using computer [6]. Because the Navier-Stokes equation is the system of partial differential equations, it is practically impossible to find an analytic solution of the equation. With the increasing speed of the computer, there have been many efforts to find approximately the numerical solution of the equation. At present, the supercomputers are used for finding the solution.

CFD transforms the Navier-Stokes equation under some assumption into the simultaneous algebraic equations. Then, the supercomputer could find the analytic solution of the simultaneous algebraic equations using a repeated computation. We can say that CFD analysis program is one that finds the solution of the transformed simultaneous algebraic equations. Because these algebraic equations are nonlinear, the analysis of the equations should be done on whole computation space for each time step. The concept of the analysis is shown in Fig. 1.

As Fig. 1 shows, the analysis program computes the solution of the simultaneous equation at every grid point (I, J) for each time step N. At each grid point (I, J), the inverse matrix of a $M \times M$ matrix should be computed because

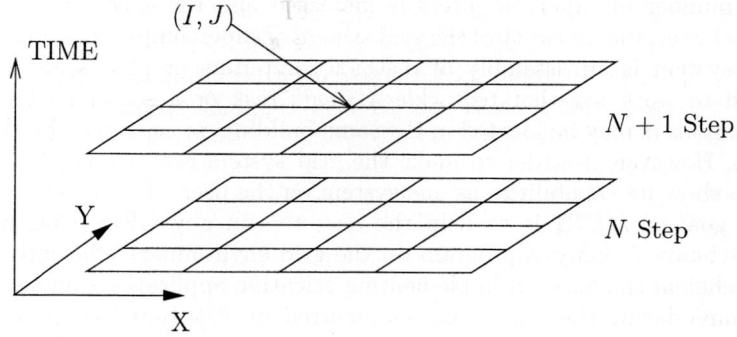

Fig. 1. The concept of a CFD analysis program

```
       program NS2D              procedure SOLVER
          . . . .                do n = 1, Nstep_Max
       C Reading Data            do I = 1, I_Max
       C Solving Phase           do J = 1, J_Max
                                 do n1 = 1, M
          Call SOLVER            do n2 = 1, M
       C Writing Data               . . . .
          stop                   x(n2) = A(n1,n2) * b(n2)
          end                       . . . .
                                 enddo
                                 enddo
                                 enddo
                                 enddo
                                 enddo
                                    . . . .
                                 end
```

Fig. 2. The structure of a typical CFD analysis program

the simultaneous equation is composed of M equations ($M \geq 4$). The structure of a typical CFD analysis program can be described as in Fig. 2.

The program read the data that is the initial values of solver. The solver routine might be a subroutine or not. In Fig. 2, the general structure of the solver shows that the index of the outermost loop is the time-step index. The elapsed CPU time of the solver is about $ELAPSED\_CPU\_TIME \times Nstep\_Max$ where $ELAPSED\_CPU\_TIME$ is the elapsed time of any time-step index.

2.2 Finding the Working Set Candidates

The speed of the network lines of the grid system has much influence on the overall performance of the execution of a CFD program because the parallel tasks of the CFD program should communicate one another for every time-step. Therefore, when we distribute the tasks on the grid environment, the first priority should be given to the network speed rather than the speed of the processing elements on which the tasks will be performed. In this section, we apply to the graph theory in order to select the least set of processing elements from the grid which will perform the CFD program in the fastest speed.

The machine selection algorithm of *META* is described in terms of the *latency graph* of the grid system, which is defined as follows:

Definition 1. *The* latency graph *of the grid system is an undirected weighted graph*

$$G = (M, E)$$

where M is the set of machines and E is the set of conceptual network lines between two different machines. Assuming that the grid system contains n machines,

$$M = \{m_1, m_2, \ldots m_n\}$$
$$E = \{e \mid e \subset M, |e| = 2\}$$

The latency graph is defined along with two functions P_t ($P_t : M \to \mathbb{N}$, $\mathbb{N}$ is the set of natural numbers) and L_t ($L_t : E \to \mathbb{R}$, $\mathbb{R}$ is the set of real numbers), where $P_t(m)$ returns the number of available processing elements of the machine m at time t and $L_t(\{m_1, m_2\})$ returns the communication latency time between m_1 and m_2 at time t. If the time t is assumed to the current time, the subscript t can be omitted and just $P(m)$ and $L(\{m_1, m_2\})$ can be used. Furthermore, the functions P and L can naturally be extended to accept the set of nodes M and the set of edges E, respectively: $P(M) = \sum_{m \in M} P(m)$ and $L(E) = \sum_{e \in E} L(e)$.

Note that an edge $e \in E$ is a set of two different nodes. Therefore, the number of edges $|E|$ is $n(n-1)/2$ (where $n = |M|$). This implies that the edge set E is empty ($|E| = 0$) when the number of machines in the grid system is less than two ($|M| < 2$).

With these definitions are assumed, the goal of *META* is to find a complete subgraph of the latency graph of the grid system, which corresponds the set of machines on which the CFD program will be performed. We will call the nodes of this kind of subgraph a *working set* of the task. *META* searches some candidates of the working set and the selection criteria is the communication latency time. The selection algorithm to find the candidates of the working set is described as Algorithm 1.

Algorithm 1: Finding the Candidates of Working Set

Input : Latency Graph $G = (M_G, E_G)$;
$NumberOfPENeeded$ = the number of processing elements needed;
$NumberOfCandidates$ = the maximum number of candidates to find
Output : The Set of Working Set W, the element of which is a subset of M_G

1 $M_G :=$ the set of nodes of G;
2 $C := \{M \mid M \subseteq M_G, P(M) \geq NumberOfPENeeded\}$;
3 $W := \emptyset$;
4 **while** $|W| < NumberOfCandidates$ and $C \neq \emptyset$ **do**
5 $\quad$ select $M \in C$ with minimum $L(E)$
$\quad\quad$ where E is the set of edges of M, i.e. $E = \{e \mid e \subset M, |e| = 2\}$;
6 $\quad$ $C := C - \{M\}$;
7 $\quad$ $W := W \cup \{M\}$;
8 **return** W;

Line 2 of Algorithm 1 computes the set of valid candidates C, each of which contains the enough number of processing elements for the task. If the whole grid system G does not have enough processing elements for the task, C becomes empty and so does W, for the while loop at line 4 will be exit early. For most cases, G contains a large set of machines enough to process the task and the algorithm selects some candidates of working sets. In this case, the number of candidates is limited to the given argument *NumberOfCandidates*. In any case, the algorithm terminates with a finite set of candidate working sets.

2.3 The Kernel Loop Model

The performance of CFD analysis programs is highly influenced by the network latency time. Therefore, we selected the candidate working sets for the application considering the network latency time. The performance of the application is secondly influenced by the performance of processing elements. In this section, our scheme select the fastest working set for the CFD application from the candidate working sets. A CFD analysis program contains a kernel loop whose execution time occupies the majority of total execution time [6]. In this section, we propose a scheme for selecting the fastest working set by performing pre-computation of several iterations of the kernel loop.

Definition 2. *The* kernel loop *is the loop that exists in a CFD analysis program whose CPU-time occupies the majority of the whole execution time. Some properties of a kernel loop are follows:*

1. *The number of iterations of a kernel loop is between hundreds and thousands.*
2. *The execution-time of one iteration of a kernel loop is about some seconds.*
3. *The execution-time of one iteration of a kernel loop is almost same regardless of the iteration index.*

These properties implies that if a particular working set is faster than others in executing one iteration of a kernel loop, the same is true in executing the whole kernel loop.

META transforms a source program (a CFD analysis program) into another program that executes only some iterations of the kernel loop of the program. The transformed program is compiled and executed on the selected working sets. The elapsed-time of each execution is collected by the *META* server and used to select the fastest working set.

When the whole grid system is small, the original requirement for the number processing elements can be relaxed. The *NumberOfPENeeded* of the Algorithm 1 may be intentionally set to a smaller value than actually needed. In this case, the automatic selection and distribution steps (line 8–9) of the Algorithm 2 should do some more works. Assume that the number of domains of the job is N and that we have n supercomputers of the selected working set ($M = \{m_1, m_2, ..., m_n\}$). As a result of the test run of the stub code, the computation time T_i and the communication time T_i' of the execution of the stub code in m_i is collected into the *META* server. *META* computes the utility ratio U_i for each M_i: $U_i = (T_i + T_i')/P(m_i)$. Finally, *META* assigns Q_i processing elements for each m_i where $Q_i = N \times U_i / \sum_{i=1}^{n} U_i$.

Algorithm 2: Selecting the Fastest Working Set Using the Kernel Loop

Input : A CFD analysis program;
The data files for the execution of the source program;
The compilation command and execution method;
The working set candidates

Output : The execution of the source program on the fastest working set

1 *META* searches the source file that contains the kernel loop;
2 The source file is transformed into an AST (abstract syntax tree) in order to search the begin-statement and the end-statement of the kernel loop;
3 The kernel loop is deleted and *META* inserts a stub code into the source file. The stub measures the elapsed-time that it takes to execute one iteration of the kernel loop;
4 All the source files and the data files are transferred to every computer of the working set;
5 Every computer compiles the source;
6 *META* initiates the executables of all the computers;
7 As a result of each execution, the elapsed-time of one iteration of the kernel loop is sent to *META*;
8 *META* decides the fastest working set based on the elapsed-time information of the test-run;
9 The CFD analysis program and appropriate portions of the data files are distributed to the available nodes of the selected working set in order;

3 Implementation and Experiments

3.1 Implementation

The structure of the current *META* implementation is shown in Fig. 3. *META* is composed of three components: the Test Program Generator, the Resource Selector, and the Executor. The Test Program Generator detects the kernel loop and generates the test program P_t from the original CFD solver P. The Resource Selector performs the test run on each supercomputers and selects faster supercomputers based on the statistics of the test run. The Executor does the final job, executing the CFD solver program P using the selected resources.

META is implemented on top of PVM (Parallel Virtual Machine) and PVMmake [2]. The PVM is a software package that allows the utilization of a heterogeneous network of parallel and serial computers as a single computational resource. The PVM provides facilities for spawning processes, communicating between processes, and synchronizing the processes over a network of heterogeneous machines. The PVMmake is a tool for transferring the solver files and the data files to each computing machine and for compiling the source program.

3.2 Experiments

For verification of *META*, experiments have been conducted by using some CFD models designed for the solution to the Navier-Stokes equations. These models

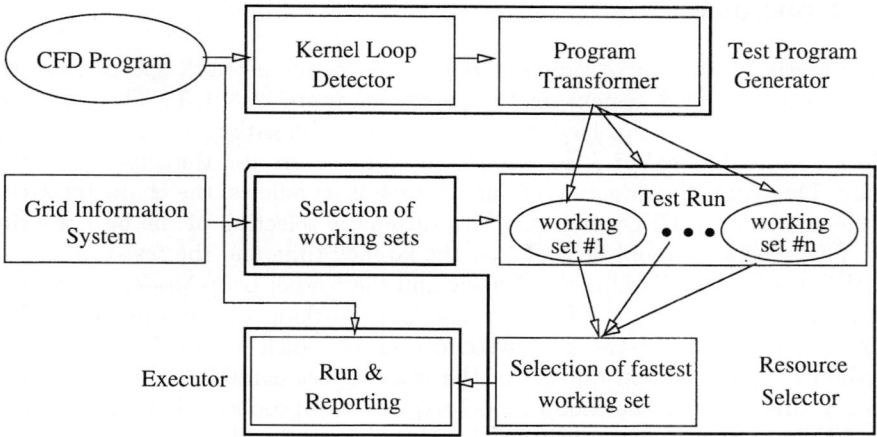

Fig. 3. The system architecture of *META*

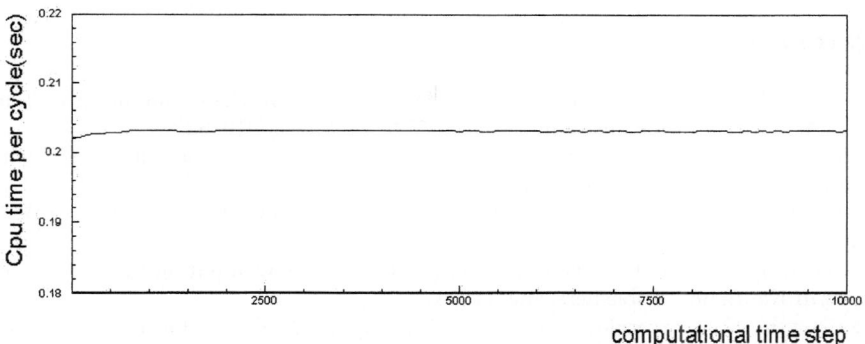

Fig. 4. The CPU time per a unit cycle of computation

are the parallel versions based on domain decomposition and explicit message passing; MPI, PVM. All the models are formulated by finite difference (volume) method on a structured grid system. Consequently, the basic frames of the numerical models are similar. The assumptions that were made for design of resource selection models in the previous section, are also valid. However, it is worth pointing out that these Navier-Stokes solvers were designed to tackle different problems. The code structures and the employed algorithms are different.

It is worth recalling the basic assumptions that the CFD simulations are of recreative nature and the computing time per unit repetition does not vary with other conditions. In order to ascertain this basic constraint, the elapsed time per a unit repetition is depicted in Fig. 4. Test runs on HPC320 and GS320 ascertain the validity of the above assumptions.

4 Conclusion

The increasing number of supercomputers and faster network speeds raise the necessity of the grid system on the supercomputers [2, 3, 4, 5]. There may be many research issues involved in the grid system according to the goal of the grid system. The goal of *META* is to help CFD users use the supercomputers easily. Therefore, the direction of our research is as follows: the characterization of the structure of CFD programs, the automatic selection of the best working set. The selection should be based on the latency time and the test-run.

We propose the Working Set Scheme and the Kernel Loop Model as a general modelling technique for CFD programs. The Working Set Scheme finds some complete subgraphs of the grid system, each of which is called a walking set candidate. The selection criteria of the working set candidates is the network latency time. The Kernel Loop Model exploits the property that the structure of a CFD program is a repetition of the same flow work. The Kernel Loop Model is a scheme that entrusts one of the repeated works to the selected working sets and that votes on the best one.

References

1. Kang, K.W., Woo, G.: A resource selection scheme for Grid computing system *META*. Lecture Notes in Computer Science **3251** (2004) 919–922
2. Sunderam, V.S.: PVM: A framework for parallel distributed computing. Concurrency: Practice and Experience **2** (1990) 315–340
3. Foster, I., Kesselman, C.: Globus: A metacomputing infrastructure toolkit. Intl. J. Supercomputer Applications **11** (1997) 115–128
4. Foster, I., Kesselman, C.: The Grid: Blueprint for a new Computing Infrastructure. Morgan Kaufmann Publishers, Inc. (1998)
5. Czajkowski, K., Fitzgerald, S., Foster, I., Kesselman, C.: Grid Information Services for Distributed Resource Sharing. In: Proceedings of the Tenth IEEE International Symposium on High-Performance Distributed Computing (HPDC-10), IEEE Press (2001) 181–184
6. Hoffmann, K.A.: Computational Fluid Dynamics for Engineers. Morgan Kaufmann Publishers, Inc. (1993)
7. Yang, X., Hayes, M.: Application of Grid techniques in the CFD field. Integrating CFD and Experiments in Aerodynamics (2003) Glasgow, UK.

Matching Colors with KANSEI Vocabulary Using Similarity Measure Based on WordNet

Sunkyoung Baek[1], Miyoung Cho[1], and Pankoo Kim[2,*]

[1] Dept. of Computer Science,
Chosun University, Gwangju 501-759 Korea
{zamilla100, irune80}@chosun.ac.kr
[2] Dept. of CSE, Chosun University, Korea
pkkim@chosun.ac.kr

Abstract. Recently, the image retrieval based on content is capable of understanding the semantics of visual information. However, it is hard to represent emotion or feeling of human. To approach more intelligent content-based retrieval, we focus on KANSEI information. This paper presents a method of matching color, which is part of visual information associated with KANSEI-vocabulary relation. We use WordNet that is a kind of lexical ontology by relations between words. We define relation for matching between color and KANSEI vocabulary using the meaning of color table. We propose the similarity measure between Color-KANSEI vocabulary and query. After experiment we can find the best pertinent color using Lesk algorithm. The significance of our study is finding semantically pertinent color according to various queries based on WordNet. This is the approach as computing vocabulary to show KANSEI of Human.

1 Introduction

With the increasing use of image data, sophisticated techniques have become necessary to enable this information to be accessed based on its content [1]. The existing system is capable of understanding the semantics of visual information based on generic features such as color, size, texture and shape, is well within the technical realm of possibility. Specially, color information is widely used in image retrieval. We focus on retrieval based on KANSEI more intelligent than that based on content.

KANSEI in Japanese means by sensibility that is to sense, recall, desire and think of the beauty in objects [2]. KANSEI is expressed usually with emotional words for example: beautiful, romantic, fantastic, comfortable, etc [3]. The concept of KANSEI is strongly tied to the concept of personality and sensibility. KANSEI is an ability that allows humans to solve problems and process information in a faster and more personal way. In addition, the color is the most expressive feature among the visual information. Also it is easy to communicate with the KANSEI and associate with meaning [4].

---

* Corresponding author.

In this paper, we propose a method to match color with vocabulary. First, we define relation between color and KANSEI vocabulary. We give name to "Color-KANSEI" as a presentation the meaning of color. The similarity measure is to match using the relation of adjectives on WordNet by the adapting Lesk algorithm. After finding Color-KANSEI similar to user's query, it finds the best matching color by Color-KANSEI. And it matches query and color that means Color-KANSEI.

The significance of our study is finding semantically pertinent color according to various queries based on WordNet. This is approach as computing vocabulary to show KANSEI of Human. It is possible to apply image retrieval by KANSEI vocabulary.

The rest of the paper is organized as follows: Section 2 introduces the current related work about color and KANSEI adjective. In Section 3, we suggest more details about creation of related Color-KANSEI vocabulary in WordNet. In section 4, we measure similarity between Color-KANSEI vocabulary and Query using WordNet, then use represent matching of color using Similarity measure. The final conclusion is given in Section 5.

2 Related Works

2.1 KANSEI Vocabulary and Adjective

KANSEI of Human is the psychological state through the five senses, which have the distinctive differences between sensitive and experience made by individual. To specify it in details, sensitive is indicated as pleasure, sorrow, angry, happy, love and hate. It's the Knowledge using by all of the people. On the other hand, KANSEI is Knowledge based on individual experience and differentiates each person. For example, sensitive is the good feeling of healthy and best conditions but the feeing of looking out of the window and realize it's snowing is different from the individual experience and knowledge [5].

KANSEI is the reflex and intuitive reaction. However, it possesses a lot of inconstant characteristics that are difficult to make it objective and typical. Therefore, we use the natural language for the representation of KANSEI because that is including the image structure of Human's idea that we can't observe it. In the natural language, adjective is representation of it.

2.2 Color and KANSEI Vocabulary

The image what human thinks recognizes KANSEI in the color, because each person gets a feeling in the same color. This means they get through the Knowledge and experience.

The color can be used as a communication tool, because of a symbol, the color image is being treated as a objective method. Kobayashi made a relation between color and language at the researching of color image standardization and Haruyoshi expresses a language which a lot of color included the image at the questionnaire in Japan [6][7]. Color Wheel Pro explains the meaning 9 based color including a Red and local color. And then, Hewlett-Packard defines the color meaning at 20 colors in the USA. In the Republic of Korea, IRI develop the I.R.I adjective image scale at a visual and symbol of Korean's KANSEI [8]. In conclusion, the changing from KANSEI's meaning of color to adjective is under researching.

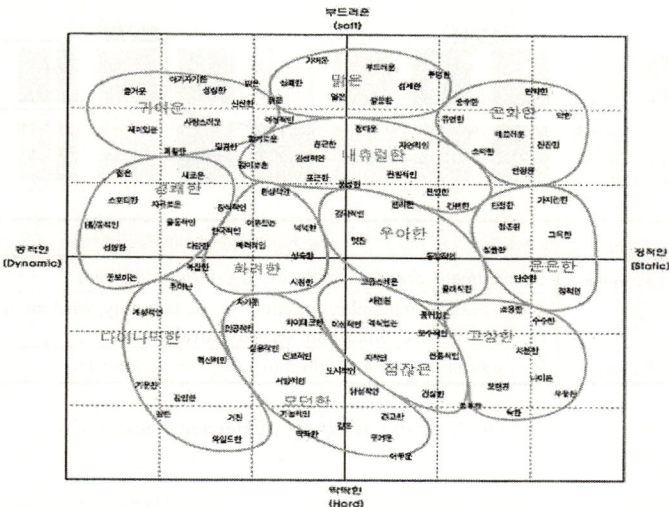

Fig. 1. I.R.I Adjective Image Scale

3 The Creation of Color-KANSEI Vocabulary Relation

This section creates relation between color and KANSEI vocabulary. It uses 20 colors that are defined "The meaning of color" by Hewlett-Packard. We extend various vocabularies as query based on finite KANSEI vocabulary that is defined [9]. In this paper, we propose a method that refers to WordNet for extension of semantic vocabulary. The following parts explain more clearly.

3.1 KANSEI Vocabulary Associated with Color

First of all, we need KANSEI information for color. Because the definition of color by human is subjective and ambiguous, most of the researchers used SD (Semantic Differential) technique through either replication or statistics. In this paper, we use 20 colors that are defined "The meaning of color" by HP as KANSEI Vocabulary. The table lists vocabularies that express KANSEI about each of the color. We define Color-KANSEI about color using this table. It is described in figure 2.

The meaning defined in figure 2 is an expression of various parts-of-speech. We unify from various part-of-speech to adjective. Because adjectives are vocabularies that represent KANSEI, we give a name to "Color-KANSEI" as adjective.

3.2 Adjective in WordNet

WordNet is the product of a research project at Princeton University which has attempted to model the lexical knowledge of a native speaker of English. Information in

WordNet is organized around logical groupings called synsets. Each synset consists of a list of synonymous word forms and semantic pointers that describe relationships between the current synset and other synset. WordNet stores information about words

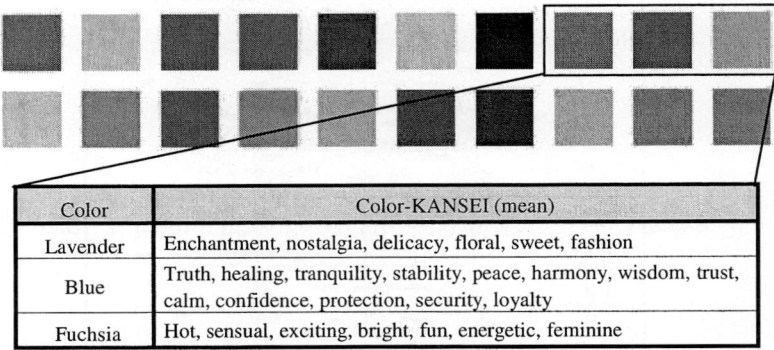

Color	Color-KANSEI (mean)
Lavender	Enchantment, nostalgia, delicacy, floral, sweet, fashion
Blue	Truth, healing, tranquility, stability, peace, harmony, wisdom, trust, calm, confidence, protection, security, loyalty
Fuchsia	Hot, sensual, exciting, bright, fun, energetic, feminine

Fig. 2. "The Meaning of Color" of Hewlett-Packard Co

that belong to four parts–of–speech: nouns, verbs, adjectives and adverbs. WordNet version 1.7 has 107,930 nouns arranged in 74,448 synsets, 10,860 verbs in 12,754 synsets, 21,365 adjective in 18,523 synsets, and 4,583 adverbs in 3,612 synsets. Prepositions and conjunctions do not belong to any synset. In this paper, we suggest to use Adjective in WordNet [10].

WordNet divides adjectives into two major classes: descriptive and relational. Descriptive adjectives ascribe to their head nouns values of (typically) bipolar attributes and consequently are organized in terms of binary oppositions (antonymy) and similarity of meaning (synonymy) [11]. Descriptive adjectives that do not have direct antonyms are said to have indirect antonyms by virtue of their semantic similarity to adjectives that do have direct antonyms.

WordNet contains pointers between descriptive adjectives expressing a value of an attribute and the noun by which that attribute is lexicalized. Reference-modifying adjectives have special syntactic properties that distinguish them from other descriptive adjectives. Relational adjectives are assumed to be stylistic variants of modifying nouns and so are cross-referenced to the noun files. Chromatic color adjectives are regarded as a special case.

3.3 The Relation Between Adjectives in WordNet

Adjectives and adverbs in WordNet are fewer in number than nouns and verbs, and adverbs have far fewer relations defined for them compared to the rest of the parts of speech. The most frequent relation defined for adjectives is that of similar to. This is a semantic relationship that links two adjective synsets that are similar in meaning, but not close enough to be put together in the same synset. WordNet defines a similar to relation between the first two synsets and another similar to relation between the third and fourth synsets. As with the also–see relationships, we are unsure of the criteria for deciding similarity but believe that human judgment was used to decide if two synsets should be linked through this relation [12].

As mentioned previously, the attribute relationship links noun and adjective synsets together if the adjective synset is a value of the noun synset. This is a symmetric relationship implying that for each attribute relation that links a noun synset to an adjective synset, there is a corresponding attribute relation that connects the adjective synset with the noun synset.

Table 1. Various relations for Adjective

Relationship	Adjectives
Antonym	4,132
Attribute	650
Also see	2,714
Similar to	22,492
Participle of	120
Pertainym of	4,433

It only uses 4 relations (Attribute, also see, similar to, pertainym of) for similarity measure between Color-KANSEI and query on table 1.

4 Matching Color with KANSEI Vocabulary

In section 4, we propose the method of similarity measure between query and Color-KANSEI using relation of adjective. We use the adaptive Lesk algorithm by Banerjee and Pedersen for measure of similarity [13]. That is the unique method that can measure similarity between adjective on WordNet.

4.1 Method for Similarity Measure Between Color-KANSEI and Query

Banerjee and Pedersen began this line of research by adapting the Lesk algorithm for word sense disambiguation to WordNet. Lesk's algorithm disambiguates a target word by selecting the sense whose dictionary gloss shares the largest number of words with the glosses of neighboring words.

The original Lesk Algorithm compares the glosses of a pair of concepts and computes a score by counting the number of words that are shared between them. This scoring mechanism does not differentiate between single word and phrasal overlaps and effectively treats each gloss as a bag of words. For example, it assigns a score of 3 to $bank_2$: (sloping land especially beside a body of water) and lake: (body of water surrounded by land), since there are 3 overlapping words: land, body, water. Note that stop words are removed, so of is not considered an overlap.

However, there is a Zipfian relationship between the lengths of phrases and their frequencies in a large corpus of text. The longer the phrase, the less likely it is to occur multiple times in a given corpus. A phrasal n-word overlap is a much rarer occurrence than a single word overlap. Therefore, we assign an n word overlap the score of n^2. This gives an n-word overlap a score that is greater than the sum of the scores assigned to those n words if they had occurred in two or more phrases, each less than n words long. This is true since the square of a sum of positive integers is strictly greater than the sum of their squares. That is, $(a_0 + a_1 + ... + a_n)^2 > a_0^2 + a_1^2 + ... + a_n^2$, where a_i is a positive integer. For the above gloss pair, we assign the overlap land a score of 1 and body of water a score of 9, leading to a total score of 10.

4.2 Color Matching Using Similarity Measure

Table 2 shows tracing result of similarity measure between KANSEI vocabulary and query using the adapting Lesk algorithm.

Table 2. Tracing similarity measure between *friendly#a#1* and *peaceful#a#1*

Synset 1: *friendly#a#1*
Synset 2: *peaceful#a#1*
Functions: also glos - also glos : 47
Overlaps: 1 × "of" 1 × "characterized by" 1 × "characterized by friendship and good will" 2 × "to" 4 × "or"
Functions: also glos - glos : 3
Overlaps: 1 × "by" 2 × "or"
Functions: also glos - sim glos : 20
Overlaps: 1 × "of" 1 x "conducive to" 1 × "and" 1 × "characterized by" 1 × "nature" 2 × "to" 1 × "disposed to" 1 × "inclined" 2 × "or"
…
Overlaps: 1 × "of" 2 × "the" 1 × "by" 1 × "and" 1 × "not" 3 × "or"
Functions: sim glos - glos : 4
Overlaps: 1 × "by" 1 × "not" 2 × "or"
Functions: sim glos - sim glos : 10
Overlaps: 1 × "of" 1 × "by" 1 × "and" 3 × "a" 1 × "disposed" 1 × "not" 2 × "or"

Table 2 shows similarity measure using semantic relation between *friendly#a#1* and *peaceful#a#1*. Here, *also* indicates **also see** relation between senses of word and then *sim* is **similar** to relation *attr* is **attribute**. For example, the overlap value is calculated using **also see** relation among grosses of synset 1 and synset 2. It is sum of square of overlapped words using the adapting Lesk algorithm. It is calculated 1 × "of", 1 × "characterized by", 1 × "characterized by friendship and good will", 2 × "to", 4 × "or". Result is $1^2+2^2+6^2+2+4=47$. Table 3 lists similarity measure between Color-KANSEI and query.

When we calculate, we consider the words whose values less than 100 are insignificant words such as article, adjunction and so on. So we just select more than 100 values from calculated values. The following table 4 indicates that how is the color matching with KANSEI vocabulary by using result of table 3. And we know that selected Color-KANSEI is the best matching with query. So, table 4 shows the result of matching adaptive colors to vocabulary impression according to user's query.

Table 3. Similarity measure

Color-KANSEI	Similarity according to Query					
	warm	energetic	alterative	fortunate	interest	peaceful
Optimistic	28	38	7	21	26	27
Dynamic	49	175	20	49	55	50
Energizing	25	99	11	6	37	26
Excite	68	25	11	18	54	26
Sexy	54	64	24	43	57	60
Intense	58	52	14	43	83	65
Aggressive	59	57	18	29	52	61
Powerful	69	23	15	46	57	38
Energetic	43	673	13	32	39	62
Vigorous	19	122	12	11	20	28
Elegant	45	63	21	35	44	61
Rich	24	25	10	21	17	26
Mature	198	17	12	28	57	34
Expensive	31	13	7	15	18	15
Truthful	10	10	6	12	14	15
Healing	23	11	251	9	32	23
Peaceful	49	62	22	31	57	712
Harmonious	20	19	10	18	27	29
Wise	43	44	11	37	42	43
Calm	27	20	14	14	24	139
Confidential	14	13	9	8	18	15
Protective	49	39	27	35	66	44
Secure	47	40	15	51	111	56
Loyal	18	22	9	22	32	24
Natural	39	34	16	31	49	46
Enviable	12	7	3	8	5	6
Fertile	44	20	15	15	32	32
Lucky	25	12	14	80	46	17
Hopeful	30	23	13	133	16	23
Stabile	21	10	8	10	19	20
Successful	65	37	18	142	71	48
Generous	23	20	15	8	29	30
Lovely	13	8	5	12	16	12
Romantic	45	14	10	9	33	23
Soft	63	52	14	46	63	59
Delicate	51	40	23	24	54	48
Sweet	52	33	10	38	67	40
Friendly	92	62	23	58	67	103
Tender	137	44	21	53	198	84
Hot	1284	80	33	77	115	88
Sensual	21	16	10	16	21	18
Bright	52	41	12	39	55	39
Fun	29	21	14	15	74	34

Table 4. Result of matching Colors

Query	Similarity	Color-KANSEI	Color
Warm	1284	Hot	Fuchsia
	198	Mature	Burgundy
	137	Tender	Light pink
Energetic	673	Energetic	Bright red
	673	Energetic	Fuchsia
	175	Dynamic	Bright red
Alterative	251	Healing	Blue
	251	Healing	Green
Fortunate	142	Successful	Green
	133	Fortunate	Green
Interest	198	Tender	Light pink
	115	Hot	Fuchsia
	111	Secure	Blue
Peaceful	712	Peaceful	Blue
	139	Calm	Blue
	120	Tranquil	Blue

5 Conclusions

In this paper, we introduced the definition of relation between color and KANSEI vocabulary. Furthermore we have shown result of color-matching according to query using WordNet and the adapting Lesk algorithm. Our future work will be focused on construct Color Retrieval System using query by KANSEI Vocabulary. And also we will study techniques for KANSEI-based retrieval, which can be extended to include multi-modal queries such as query by KANSEI and texture or query by KANSEI and shape. Therefore we will develop KANSEI-Ontology based on relationship of visual information and KANSEI vocabulary

Acknowledgement. This work was supported by the Korea Research Foundation Grant. (KRF-2004-042-D00171)

References

1. A.Ono, M. Amano, M. Hakaridani, T. Satou, M. Sakauchi, "A fiexible Content-based Image Retrival System with Combined Scene Description Keyword," Proceeding of Multimedia 96, pp.201-208, 1996
2. Hideki Yamazaki, Kunio Kondo, "A Method of Changing a Color Scheme with KANSEI Scales," Journal for Geometry and Graphics, vol. 3, no. 1, pp.77-84, 1999
3. Shunji Murai, Kunihiko Ono and Naoyuki Tanaka, "KANSEI-based Color Design for City Map," ARSRIN 2001, vol. 1, no. 3, 2001
4. Sunghee Park, Hyunjin Kim, Jaehun Choi, Myounggil Jang, "Color word-to-Color Matching or Natural Language Query in Multimedia Information Retrieval," HCI2001
5. Young-Joon Nam, "The Construction of Sensibility Thesaurus Based on Color," Journal of Information Management, vol. 34, no. 4, pp. 43-61, 2003.
6. Kobayshi Singenobu, "Color Image Scale," Kodansha America, 1990

7. Haruyoshi Nagumo, "Color Image Chart," Chohyung Publishing Co., 2000
8. I.R.I, http://www.iricolor.com/04_colorinfo/colorsystem.html, "Adjective Image Scale"
9. Hewlett-Packard, http://www.hp.com/united-states/public/color/meaning.html, "Color Printing Center-The Meaning of Color"
10. George A. Miller, "WordNet: a lexical database for English," Communications on the ACM, 1995
11. C. Fellbaum, D. Gross, K. Miller, http://www.cogsci.princeton.edu/~wn, "Adjectives in WordNet"
12. S. Banerjee, T. Pedersen, "An adapted Lesk algorithm for word sense disambiguation using WordNet," In Proceedings of the Third International Conference on Intelligent Text Processing and Computational Linguistics, Mexico City, pp. 136–145, 2002
13. S. Banerjee, T. Pedersen, "Extended gloss overlaps as a measure of semantic relatedness," In Proceedings of the Eighteenth International Joint Conference on Artificial Intelligence, Acapulco, pp. 805–810, 2003

A Systematic Process to Design Product Line Architecture*

Soo Dong Kim, Soo Ho Chang, and Hyun Jung La

Department of Computer Science, Soongsil University,
1-1 Sangdo-Dong, Dongjak-Ku, Seoul, Korea 156-743
sdkim@comp.ssu.ac.kr, {shchang, hjla}@otlab.ssu.ac.kr

Abstract. Product Line Engineering is being accepted as a representative software reuse methodology by using core assets and product line architecture is known as a key element of core assets. However, current research on product line engineering has room to provide specific and detailed guidelines of designing product line architectures and reflecting variability in the architecture. In this paper, we present a reference model and a process to design the architecture with detailed instructions. Especially architectural variability is codified by describing decision model representing variation.

1 Introduction

Product Line Engineering (PLE) has been widely accepted as a representative software reuse methodology using core assets. Architecture plays a key role in scoping applications and it defines overall structures for applications. A core asset in PLE provides a framework for developing various products in the product line. As a key element of core assets, product line architecture (PLA) should also be generic to be applied to various products.

Although processes or methods to design PLA have been suggested in various research works, there is a large room for improvement, providing specific and detailed process of defining PLA and reflecting architectural variability. Especially, how the essential elements of architecture design such as driver, view, and styles can be applied to PLA should be specified in detail.

In this paper, we first present a reference model of PLA and propose a systematic process to design PLA. Each activity of the process is elaborated with detailed instructions. In addition, architectural variability is codified by describing decision model representing variation points.

2 Related Works

Bosch proposes a design method for system family software architectures [1]. When designing family architecture, architects design architecture based on archetype that is

---

* This work was supported by Korea Research Foundation Grant. (KRF-2004-005-D00172)

core abstractions of the system, assess architecture for quality requirements using scenarios, and then transform quality requirements to functionality to improve the quality attributes of architecture. For transformation, system family architecture may require achieving variable requirements, optionality of parts of the architecture, and conflicts between components as well as imposing architectural style, architectural pattern, and design pattern.

QADA is a method standing for Quality-Driven Architecture Design Analysis method[2]. The method consists of activities; *Requirements engineering, Conceptual architecture design and analysis, Concrete architecture design and analysis.* In conceptual architecture analysis, analysis and representation of variability is focused using variation point description and product-line pattern.

Ceron and his colleagues propose processes for developing reference architecture and deriving the architecture as well as architectural meta-model[3]. The meta-model appends architectural variability to P1471[4]. The process for developing reference architecture consists of three activities; *Scoping, Choosing architectural style, Providing variability.* Applying functional and non-functional features into architecture, this work also points out conflict problems of variability, a stability of common requirements and architectural style in reference architecture as well.

Thiel suggests a process framework that supports the design of high-quality product family architectures, called QUASAR [5]. QUASAR is organized with three workflows; *Preparation, Modeling* and *Evaluation* that analyze the achievement of architectural qualities. To integrate variability with PLA design, this work gives guidelines for documenting variability about where variation points are in architectural views, how to instantiate them, and resolution rules.

These related works define more or less implicitly what is included in PLA and suggest overall process for designing PLA. Especially, they mention needs to represent and document variability in designing PLA. Hence, we can make more practical design process by supplementing detail instructions. To enhance importance of designing variability, we can classify types of the architectural variability and represent variability more concretely by using architectural decision model.

3 Meta-model of Product Line Architecture

Based on our survey, we now present our meta-model of PLA by taking the common elements of the related works and refining them as in **Fig. 1**.

Elements of PLA are distinguished into abstract elements and concrete elements. The abstract one is conceptual elements to which PLA should conform or refer, whereas the concrete one is elements which constitute PLA as physical parts. From the meta-model, we specify each element as followings.

Architectural View: Based on the requirements and PL analysis model derived from the requirements, we choose perspectives to illuminate PL, called as a *view* shown in the figure [6]. Although many kinds of view types are proposed, there is no standard on architectural view [7]. However, we choose the three kinds of view, Module View, C&C View and Allocation View, as specialized view types of PL architectural view since they are generally accepted [2][3][6].

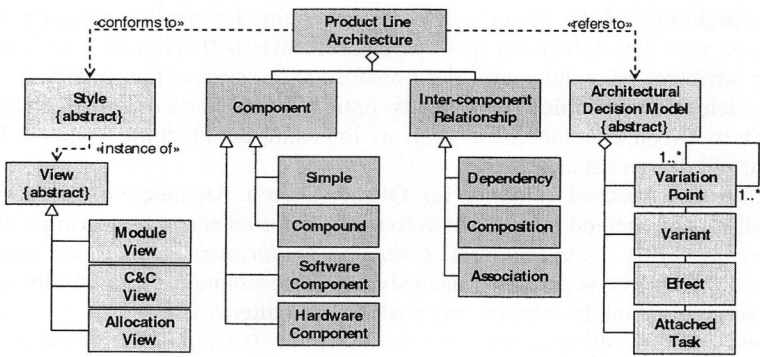

Fig. 1. Meta-model of PLA

Architectural Style: An architectural *style* is a specialization of elements and relation types [6] and helps simplify the architecting process [8]. From the notion, architecture design begins with choosing most appropriate architectural styles and components and inter-component relationships in architecture are more or less directly derived from the selected styles. Therefore, it is fair to state that the components and relationships effectively implement functional and nonfunctional requirements within architectural styles. Architectural styles can be defined as abstract elements to which architecture conforms rather than constituents of PLA. A style is a partial instance of a view, and a set of several architecture styles can realize a view [6][9].

Component and Inter-component Relationship: PLA consists of *Components* and *Inter-component Relationships*. Components implement functional and non-functional requirements. While functional requirements are directly designed, non-functional requirements may derive additional functional requirements which realize the quality attributes and are implemented into the components. A component is specialized into simple and compound components as their composition relationships. A component is also specialized into software and hardware component.

Inter-component relationship may have several stereotypes depending on architectural views by which the relationships are represented. In the meta-model, the relationship is specified as dependency, composition, or association. Generally a dependency is for message passing between components, composition is for relationships between simple and compound components, and an association is for persistent relationships between hardware components.

Architectural Decision Model (ADM): Since decision model is specification of variability in PL, it is not a design element only for PLA but a reference from which variability is designed into PLA. Hence, the relationship between PLA and ADM is shown as *refers to,* as in **Fig. 1**. We call the decision model capturing architectural variability *Architectural Decision Model (ADM). Variation Point*, *Variant*, *Effect*, and *Attached Task* are constituent to the architectural decision model [10][11].

To elaborate architectural variability, we propose following candidate variants types for architectural variation points;

- **Architectural Style:** Since a style represents part of architecture, architecture may contain a set of styles. Besides, architecture should be stable [3]. Therefore, a variation point of an architectural style set may have a few style variations. That is, a few styles in the set may vary from product to product such as optional or alternative.
- **Component:** Variations of components in architecture can be classified to optional and alternative. Optional variability is for the case in which a component is used or not. Alternative is for the case in which another component can be substituted for one component. Note that variations may be occurred in components such as logics, workflows, or data [12]. However it is not architectural variability but intra-component variability. Therefore descriptions of the intra-component variability are out of the scope of this paper.
- **Relationship:** Within the one style, message passing among components may be changed by applications. We define that variation occurs in inter-component relationship. Note that this variation point should be distinguished from architectural style variability.

4 Process and Instructions

We now present a process to design PLA in **Fig. 2**. The process consists of five activities and each activity has its detailed steps.

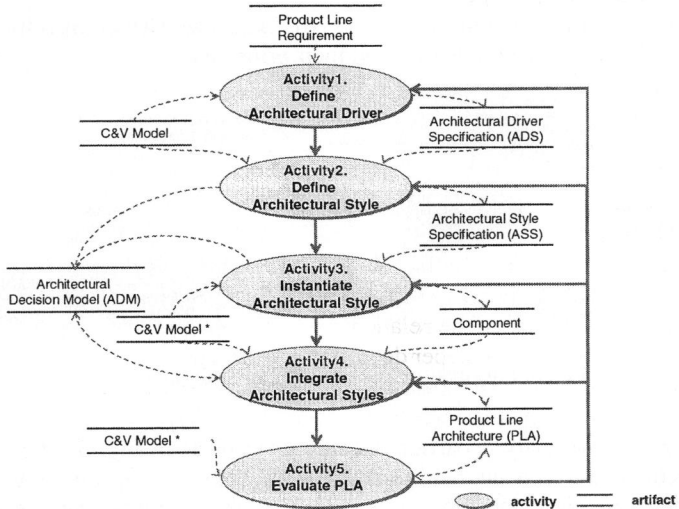

Fig. 2. Process and Artifacts for PLA Design

Since these activities are included in the phase for PLA design, domain analysis is preceded and component design is followed. PLA design phase begins with analysis model delivered from domain analysis phase and carries over a PLA to component design phase.

The five activities are Define Architectural Driver, Define Architectural Style, Instantiate Architectural Style, Integrate Architectural Styles and Evaluate Architecture. Each activity is decomposed by steps and provides instructions and templates with the steps.

4.1 Activity 1. Define PL Architectural Driver

Overview: This activity is to derive a set of PL architectural drivers for a product line. An architectural driver is a requirement which has influence on the design of architecture [1][13]. Therefore, acquiring right architectural drivers is an essential prerequisite for well-designed architecture. In this activity, both common and variable drivers are identified since there can be variation on product architectures.

Input and Output: As shown in **Fig. 3**, the input to this activity is both product line requirement specifications (PRS) and C&V model. A PRS specifies functional and nonfunctional requirement. Each type of requirement for variation can be mandatory, alternative or optional. C&V model is an analysis model of PRS, and the model specifies both common and variable requirements in systematic way. Examples of C&V model are feature analysis [14].

We assume that PRS is available before this process is applied. That is functional and nonfunctional requirements are separately defined and the PRS is relatively complete and consistent. If not, techniques of requirement engineering [15] can be applied to derive a high quality PRS. We also assume that C&V modeling has been completed before this process.

The output artifact of this activity is Architectural Driver Specification (ADS) which specifies architectural drivers and their priorities.

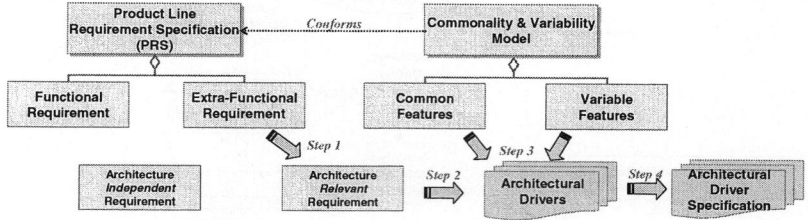

Fig. 3. Information Flow from PL Requirement to PLA

Instruction: This activity is carried out in four steps as in **Fig. 3**. The first step is to extract architecture-relevant requirements from PRS. As shown in **Fig. 3**, not all nonfunctional requirements are architecture-relevant. Hence, we need to extract nonfunctional requirements that have some impact on the design of PLA. In general, quality attributes and constrains are architecture-relevant. For example, reliability as a quality attribute may be applied into architecture as data mirroring.

The second step is to define architectural drivers from the architecture-relevant requirements. That is, architecture-relevant requirements are analyzed and itemized as architectural drivers. Each architecture driver is given a name for further references in subsequent activities.

The third step is to classify architectural drivers according to the common and variable features from C&V model. An architectural driver itself can be variable in two forms; alternative and optional.

The fourth step is to prioritize drivers according to the commonality and significance. Architectural drivers are the main source for choosing architectural styles and therefore different drivers may yield different styles. When a product line has several drivers, resulting architectural styles may be complicated. As a logical way to resolve the possible complications of styles, we suggest to prioritize drivers using product line dependent criteria. Each driver is given with its *priority, name, description* and *variability type* which can be mandatory, optional, or alternative.

4.2 Activity 2. Define Architectural Styles

Overview: This activity is to derive architectural styles. By using architectural style that eases the design process by providing routine solution for recurring problems, we can reuse design and code, easily understand a system's organization, and gain insight into style-specific analysis of solution characteristics [16]. In this activity, architectural styles which satisfy with architectural drivers and make PLA effective are defined.

Input and Output: To define an appropriate architectural style, this activity requires ADS. One output is an *Architectural Style Specification (ASS)* which addresses architectural views, styles, and rationales that are shared by PL applications. The other output is a part of an ADM which specifies architectural variation on PL architectural style set. ADM describes all architectural variability and only part of ADM, *Style Set Variability*, is defined in this activity.

Instruction: This activity is to choose appropriate architectural styles according to derived architectural drivers and consists of three steps as followings.

The first step is to select architectural views by which PLA is illustrated. At least one view should be chosen from the three views [6]. Note that a view may focus on several architectural drivers and an architectural driver may also be realized in one or more views. With this step, a view list with architectural drivers is listed and ranked as priority of the drivers.

The second step is to choose architectural style for each view. We firstly explore and list candidate architectural styles for each architectural driver and then, decide architectural style as our strategies, project policies, or constraints. The rank of architectural drivers can affect resolving conflicts among architectural styles.

The third step is to specify ADM for variation on the architectural style set. Style set variability which is identified and specified in this step is stemmed from variations on architectural drivers. Different architectural drivers drive different architectural styles covering each driver as shown **Fig. 4**.

For one architectural view, several architectural drivers have their styles. Especially an architectural driver which has variation has one or more style depending on variation type. From the **Fig. 4**, we can extract architectural style set {style a, style b, …, *style i-1|style i-b*, ..}. The style set has variation on *style i*.

According to variation type of architectural driver in ADS, variants of style set variability are defined as variable driver and its style. Variation type is equally transformed from variation type of ADS. Effect and task can be specified in this step, and further refined in *refining overlapped area step in activity 4*.

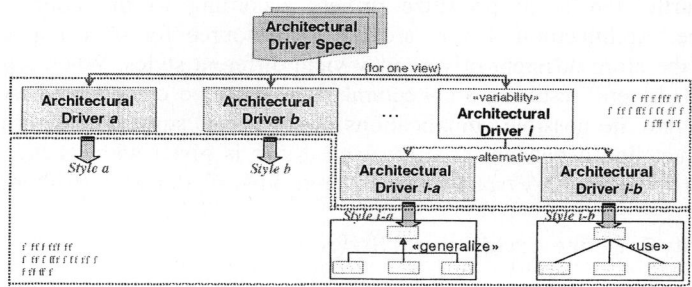

Fig. 4. Architectural Style Set Variability of PLA

4.3 Activity 3. Instantiate Architectural Style

Overview: This activity is to realize ASS and ADM. Architectural styles are represented by architectural units and their relationships. In this activity, a specification of architectural style set is transformed into concrete parts of architecture. During style instantiation, variation on a style is also applied into the instantiated styles.

Input and Output: The input to this activity is both an ASS and an ADM which are defined in activity 2. The output artifact of this activity is an embodied architectural style set which are represented by actual components and their relationships. ADM is also refined as appending *Architectural Style Variability*.

Instruction: This activity is carried out in three steps. The first step is to extract component. Types of components in architecture can be divided into types; software and hardware components. Software components are applied into logical view such as module view and process view such as C&C view. To extract the component, we may use a clustering method in [17]. Hardware components are represented in physical view such as allocation view and the component may be server, DBMS, and other hardware units. To extract these physical components, we may use strategic constraints.

The second step is to apply the extracted components into architectural style. For one architectural style, we arrange the extracted components and then elicit relationships among arranged components. Depending on types of components, dependency, composition, or association can be applied with specific stereo-types.

The third step is to append architectural style variability to ADM. Architectural style variability is discovered in component or inter-component relationship. **Fig. 5** shows an example of architectural style variability on a share-date style of Sale Domain.

During instantiating architectural style, one component may not be used or replaced with other component by one application. In addition one relationship between components may be omitted or changed depending on applications. These variations may be represented in ADM.

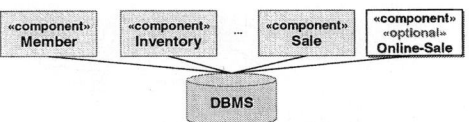

Fig. 5. Architectural Style Variability of PLA

4.4 Activity 4. Integrate Architectural Styles

Overview: This activity is to finalize PLA by combining instantiated architectural styles. Individual instantiated architectural style is a part of PLA, so it should be populated into whole PLA. During arranging several styles, overlapped area among styles should also be recognized and resolved in this activity. In addition, ADM is more refined by *Effects* and *Tasks* describing propagation of architectural variability.

Input and Output: The input to this activity is an instantiated architectural style set where styles may have architectural variability. The output artifact of this activity is PLA in which the whole range of architectural elements is represented in terms of components and their relationships on chosen views.

Instruction: This activity is carried out in two steps. The first step is to gather the instantiated styles on a same view. Since components in different styles may have different granularity, it is needed to normalize components in different styles into same-grained components in this step.

The second step is to link styles and refine overlapped areas among styles. Some components may be included in several instantiated styles and other components may be embedded in compound components which are included in other styles. From these cases, we define overlapped area which contains some components and their relations included in two or more styles. The overlapped area is distinguished two types; one is area having architectural variation and the other is area not having variation. In the case of overlapped area having variation, architectural variability should be handled more carefully. By resolving overlapped area and applying variability into the area, overlapped component may be refined and relations may be modified. **Fig. 6** shows an example of overlapped area during integrating instantiated styles.

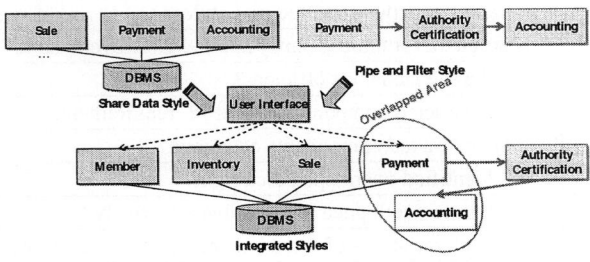

Fig. 6. An Example of Integrated Styles and Overlapped Area

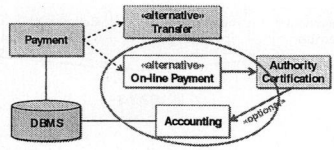

Fig. 7. An Example of Overlapped Area having Variability

Fig. 7 represents a resolution of overlapped area. In this case, *payment component* is refined *Transfer* and *On-line payment* components as alternatives. For its propagation, the relationship between *Authority Certification* component and *Accounting* component is refined as optional relationship. The refinement is also appended to ADM in this step

4.5 Activity 5. Validate PLA

Overview: This activity is to evaluate PLA with several check lists and decide whether PLA should be refined or finalized. As criteria of items in the check list, the process is returned to prior activities or closed. In this activity, we propose a check list which support to validate PLA and instruction using the check list.

Input and Output: The input to this activity is PLA defined through activity 1 to 4, C&V model, and a predefined check list. The output of this activity is evaluated PLA and result indicating process direction.

Instruction: First of all, we define a check list to evaluate PLA for this activity as shown **Table 1**.

Table 1. Check List to Evaluate PLA

Artifact	Check Point
Architectural Driver	Do the architectural drivers meet non-function requirement?
	Are derived architectural drivers used for designing PLA?
	Is the priority of driver right?
	Are variable architectural drivers defined based on adequate criteria?
Architectural Style	Are selected views satisfied with all architectural drivers?
	Are all of drivers applied into styles?
	Isn't a driver unnecessarily applied into several styles?
	Is the selected style efficient?
Instantiated Style	Do extracted components cover PL requirements?
	Are extracted components economic?
PLA	Is identified overlapped area right?
	Does the overlapped area resolve effectively?
Architectural Decision Model	Are all variations of drivers delegated to adequate variation point?
	Is all variants necessary?
	Are all variability propagations in overlapped area detected?
	Are all variation points and variants consistent through the artifacts?

Check Points in the check list are classified with artifact of each activity. Depending on activity goal, check points emphasize completeness, accuracy, efficiency, or conciseness. In architectural driver, check point focuses on completeness, accuracy of architectural driver for PL requirements. Check points of architectural style are efficiency of extracted styles for architectural driver, the point of instantiated style is completeness for quality attribute and functional requirements, and the point of PLA is its suitability for integrated style set. For ADM, completeness and efficiency are indicated. Based on the list, we may decide whether PLA design should be refined or finalized.

5 Concluding Remarks

The architecture of core asset should be generic to be applied to various products. Therefore, it is an essential element of core assets. We presented a reference model of PLA and proposed a systematic process having 5 activities. Each activity of the process was elaborated with detailed instructions and artifact templates. We also identified how the architectural variability is traced to elements of decision model.

We showed how architectural elements such as driver, views and styles can be applied to PLA. Using the proposed process, one can design a high quality PLA supporting architectural variability as well as architectural commonality.

References

[1] Bosch, J. Design and Use of Software Architectures, Addison-Wesley, 2000.
[2] Matinlassi, M., Niemela, E., and Dobrica, L., "Quality-driven architecture design and quality analysis method : A revolutionary initiation approach to a product line architecture," VTT Technical Research Center of Finland, ESPOO2002, 2002.
[3] Ceron, R., et. al., "Architectural Modeling in Product Family Context," *proceeding of EWAS, LNCS 3047*, Springer-Verlag Berlin Heidelberg, 2004.
[4] IEEE Recommended Practice for Architectural Description of Software-Intensive Systems (IEEE Standard P1471); IEEE Architecture Working Group (AWG); 2000.
[5] Thiel, S., and Hein, A., "Systematic Integration of Variability into Product Line Architecture Design," *proceeding of SPLC2, LNCS 2379*, Springer-Verlag Berlin Heidelberg, 2002.
[6] Clements, P., et al., Documenting Software Architectures Views and Beyond, Addison-Wesley, 2003.
[7] Woods, E., "Experiences Using Viewpoints for Information Systems Architecture: An Industrial Experience Report," *proceeding of EWSA 2004, LNCS 3047*, Springer-Verlag Berlin Heidelberg, 2004.
[8] Heineman, G., and council, W., *Component-Based Software Engineering*, Addison Wesley, 2001.
[9] Bass, L., Clements, P., Kazman, R., *Software Architecture in Practice*, Addison-Wesley, 2003.
[10] Sinnema, M., et al., "COVAMOF: A framework for Modeling Variability in Software Product Family," *proceeding of SPLC 2004, LNCS 3154*, Springer-Verlag Berlin Heidelberg, 2004.

[11] Kim, S., Chang, S., and Chang, C., "A Systematic Method to Instantiate Core Assets in Product Line Engineering," *Proceedings of APSEC 2004*, Nov. 2004.
[12] Kim, S., Her, J., and Chang, S., "A theoretical foundation of variability, in component-based development," Journal of Systems and Software, To Appear.
[13] America, P., et al., "Scenario-Based Decision Making for Architectural Variability in Product Families," *proceeding of SPLC 2004, LNCS 3154*, Springer-Verlag Berlin Heidelberg, 2004.
[14] Choi, S., Chang, S, and Kim, S., "A Systematic Methodology for Developing Component Frameworks," *LNCS 2984, Proceedings of the 7th FASE*, 2004.
[15] Lauesen, S., Software Requirements Styles and Techniques, Addison-Wesley, 2002.
[16] Garlan, D., Allen, R., Ockerbloom, J., "Exploiting Style in Architectural Design Environments," *Proceedings of SIGSOFT'94*, Foundations of Software Engineering, pp. 175-188, 1994.
[17] Kim, S., Chang, S., "A Systematic Method to Identify Component," *Proceedings of APSEC 2004*, Nov. 2004.

Variability Design and Customization Mechanisms for COTS Components[*]

Soo Dong Kim, Hyun Gi Min, and Sung Yul Rhew

Department of Computer Science,
Soongsil University,
1-1 Sangdo-Dong, Dongjak-Ku, Seoul, Korea 156-743
{sdkim, syrhew}@comp.ssu.ac.kr, hgmin@otlab.ssu.ac.kr

Abstract. Component-Based Development (CBD) is gaining popularity as an effective reuse technology. Components in CBD are mainly for inter-organizational reuse, rather than intra-organizational reuse [1]. One of the common forms of reusing commercial-off-the-shelf (COTS) components is to acquire and customize them for each application. Therefore, components must be developed with consideration of commonality and variability in a domain in order to increase the reusability and applicability [2]. One effective factor in determining the quality of components is how precisely the variability is modeled and how effective customization mechanisms are provided. COTS components often come in binary and blackbox form, therefore modifying the source code or re-linking object code with library are forbidden. However, much of current approaches to component customization are directed towards tailoring whitebox components, i.e. source code is modified. In this paper, we present a comprehensive set of techniques to realize variability into blackbox components and to provide effective interface-based customization mechanisms. Maintanbility, applicability and reusability can be enhanced by using the mechanism.

1 Introduction

CBD is gaining popularity in both industry and academia as an effective reuse technology. Components in CBD are mainly for inter-organizational reuse, rather than intra-organizational reuse. One of the common forms of reusing COTS components is to acquire and customize them for each application. Therefore, components must be developed with consideration of commonality and variability in a domain in order to increase the reusability and applicability. One effective factor in determining the quality of components is how precisely the variability is modeled and how effective the customization mechanisms are.

COTS components often come in binary and blackbox form to minimize the coupling between components and applications and to protect intellectual design assets. For blackbox components, modifying the source code or re-linking object code with library are forbidden. However, much of current approaches to component

---

[*] This work was supported by Korea Research Foundation Grant. (KRF-2004-005-D00172).

customization are directed towards tailoring whitebox components. Tailoring whitebox components involves understanding the internal design of the component, modifying source code, and rebuilding the component with any necessary library. While tailoring whitebox component is more effort and time consuming, tailoring blackbox component only involves invoking the customization interface to set appropriate variants and so it is more efficient and less time consuming.

However, it is challenging to be able to customize blackbox components without accessing their internal design and source code. In this paper, we present a comprehensive set of techniques to design components with variability so that blackbox form of components can be customized effectively only through interfaces. We focus on practical applicability of customization techniques which can be implemented in popular CBD platforms.

We survey representative works on variability design in section 2, and present a foundation on variability types and interfaces. The main customization techniques proposed are presented in section 4, 5 and 6. The proposed work is compared with other works in section 7.

2 Related Work

In this section, we present a survey of representative customization methods for whitebox and blackbox components.

Keepence and Mannion's Work suggests three patterns for variability design; *single adapter*, *multiple adapter* and *options* patterns [3] as in figure 1. In *single adapter*, generic features are modeled in a base class and specific features are modeled in subclasses. Only one subclass can be instantiated in any single system. *Multiple adapters* are similar to single adapter, but more than one subclass can be instantiated in any single system. In *options* pattern, two associated peer classes are created to realize a variation. Keepence's work suggests three types of variability mechanism. However, this research specifies range of variant. It doesn't include detailed implementation techniques.

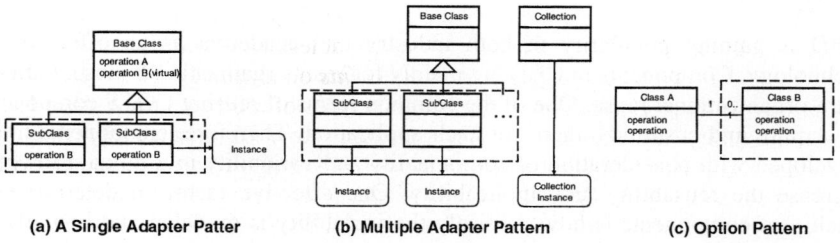

(a) A Single Adapter Patter   (b) Multiple Adapter Pattern   (c) Option Pattern

Fig. 1. Keepence's Patterns for Variability

Anastasopoulos and Gacek's Work identifies several customization methods; aggregation/delegation, AOP, conditional compilation, dynamic class loading, dynamic link libraries(DLL's), frames, inheritance, overloading, parameterization, properties and static libraries [4]. *Aggregation/delegation* method enables objects to delegate requests to other objects which provide a customized behavior. *Conditional compilation* method enables control over the multiple code segments; including or excluding selected code segment from a program compilation. *Dynamic class loading* is a feature in Java where all classes are loaded into memory as soon as they are needed at runtime. *Frame* is used to specify adaptable behavior. *Parameterization* is used to pass variants. *Static library* contains a set of external functions that can be linked to an application after it has been compiled. Among the proposed techniques, only aggregation/delegation, dynamic link library and parameterization methods can be effectively applied to blackbox components.

Svahnberg and Bosch's Work suggests five customization techniques; inheritance, extensions, parameterizations, configuration and generation of derived components [5]. *Extensions* mechanism is used when parts of a component can be extended with additional behavior, selected from a set of variations. *Configuration* enables selection of source code segment and files from a code repository to form a customized product. *Generation* of derived components is hard coded to a particular set of parameters. However, inheritance and generation methods cannot be used for customizing blackbox component.

3 Foundation

In this section, we summarize fundamental concepts and terms used for presenting our methods. We first define terms related to variability; *Variation Point (VP)*, *Variant*, and *Variability*. *Variation Point* is a place in software where a minor difference occurs among family members [6]. It is possible for a function to have more than one variation point. *Variant* is a value or instance that can validly fill in variation points, i.e. a variant resolves a variation point. A variation point typically has more than one variant. *Variability* is characterized by various variations within common requirement, and it consists of variation points and a set of their valid variants. Therefore, variability is a comprehensive description of variations occurring in a family.

There are four types of variability in CBD; variability on *Attribute*, *Logic*, *Workflow*, or *Persistence* [6]. Attribute is defined as an abstract storage to store values, and it is realized as constants, variables, or data structures. *Attribute variability* denotes occurrences of *variation points* on attributes. The typical forms of variations are the different number and/or data types of attributes for a given function. Logic describes an algorithm or a procedural flow of a relatively fine-grained function. *Logic variability* denotes occurrences of *variation points* on the algorithm or logical procedure. Workflow describes a sequence of method invocations to carry out a coarse-grained function. *Workflow variability* denotes occurrences of *variation points* on the sequence of method invocations. Persistency is maintained by storing attribute values of a component in a permanent storage so that the state of the component can alive over system sessions. Typically a component contains several classes, and classes contain persistent attributes. These attributes must be mapped to a representation on a secondary storage such as files on hard disk and relational database tables. *Persistency variability* denotes occurrences of *variation points* on the physical schema or representation of the persistent attributes on a secondary storage.

4 Selection Technique

The selection technique is to define classes and an customize interface for clients to select one of the variants realized inside the components. Once a variant is selected, the value is stored and remembered so that further invocations can refer to the selected variant. The selection mechanism works in four steps.

4.1 Step 1. Defining Variable Functions

This step is to define functions to handle the selection process for the classes which have variation points. As shown in figure 2, variation points are realized as functions which include a *switch* statement, and variants are specified as *case* clauses within the switch statement.

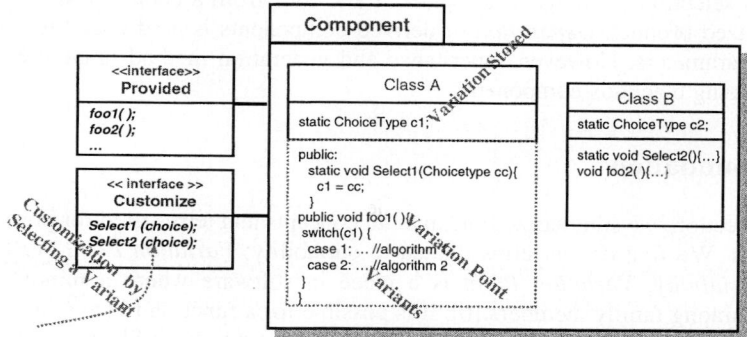

Fig. 2. Mechanism of Selection Technique

In the figure 2, the function *foo1 ( )* of *class A* realizes a variation point and each *case* clause of *foo1( )* realizes an associated variant. If *foo1( )* is a function computing a temperature, the unit of temperature can be a variation point with two variants; Centigrade and Fahrenheit. Consequently, one *case* clause is an algorithm using Centigrade and the other *case* clause can be an algorithm using Fahrenheit.

4.2 Step 2. Defining Static Attributes and Operations for Customization

This step is, for each class in a component, to define attributes for storing selected variants and to define *customize* operations that read selections made by clients and store the selected variants in these attributes. Once the selections are (possibly persistently) stored in these attributes, further invocations of *foo1( )* operation will refer to the values stored in the attributes.

In this way, one-time customization has a persistent effect on the variability. In order to keep the value of *c1* for a long period of time or persistently, the value must be stored in a secondary storage such as a file or a database. If the value isn't stored, it should be customized at a re-installation time such as Web Application Server (WAS) rebooting. The customization activity ends at this moment.

4.3 Step 3. Defining Customize Interface

This step is to collect *customize* operations in various classes in a component into a single *customize interface*. In figure 2, *Select1 (Choicetype cc)* and *Select2 (Choicetype cc)* are *customize* operations; Therefore, they are included in the *customize interface* of the component. The component consumer sets the variants using this interface to customize components.

4.4 Step 4. Setting Variants

This step is to customize components using *customize* operations. For example, a component consumer invokes a customize operation, *Select1(choice)*, with a parameter setting on an appropriate variant. The actual argument of the method is now stored in a static attribute *c1* by the assignment statement of *Select1( )* method. As a result, further invocation on *foo( )* will refer to this attribute. Now, customization on components is completed, and the operations in the *Provided* interface are invoked.

4.5 Remarks for Realizing Attribute Variability

Selection technique can be applied to various types of variability; attribute, logic, workflow, and persistence variability. The mechanism requires special attention when realizing attribute variability.

Selection mechanism for realizing attribute variability utilizes the mechanism of generic classes, which capture common behavior and instantiate concrete classes of specific data types. For example, the *template* construct of C++ provides the mechanism of generic classes.

The variation points are realized as functions which include a *switch* statement, and variation on data type is specified as *case* clauses within the switch statement. The objects which have different data type are created by the case clauses. Each *case* clause includes a statement to create an object that has attributes of an appropriate data type.

As shown in figure 3, the component includes two classes; class *A* and class *Account* which have variation points, and *accountID* and *createAccount( )*, of the attribute variability type. Class A has a attribute variability on a variation point *createAccount( );* this variation point has already implemented possible types of *accountID* using a switch-case statement. Furthermore, class *Account* has attribute variability on a variation point *accountID;* this variation point has been implemented using a parameterized type concept.

A static int attribute *c* in *Class A* stores variants. A *createAccount( )* method in class A performs two things according to variation storage *c*. The first thing is to instantiate template class Account conforming to *c*, the second thing is to create a new account object with particular type of *accountID*. If it can be an integer type or string type, the attribute is instantiated by "new Account<int> (12932)" and "new Account<string> ("AZ129")" statement.

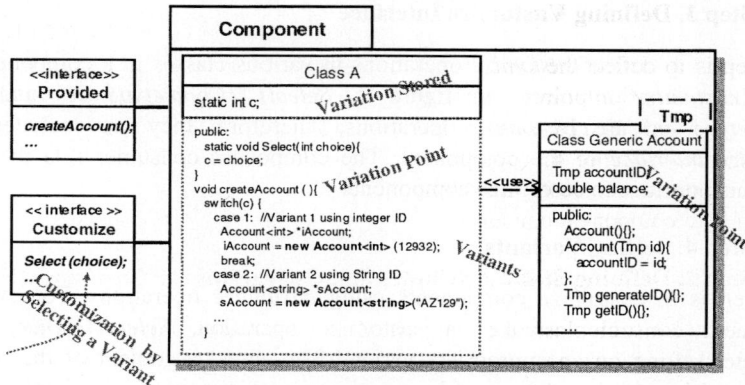

Fig. 3. A Mechanism of Attribute Variability using Selection Technique

5   Plug-in Technique

Plug-in technique is used to assign an external variant to a variation point of a component through a *customize* interface. By passing references of objects to components and setting these functions are objects invoked inside components, application-specific functionality is defined and supplied to the components. In this way, components can be customized for each application.

The effects of customizing components should be persistently stored in and around the components. With the plug-in technique, this is done by persistently maintaining the references, functions or objects passes as parameters.

The *plug-in* technique can be applied to various types of variability; attribute, logic, workflow, and persistence variability.

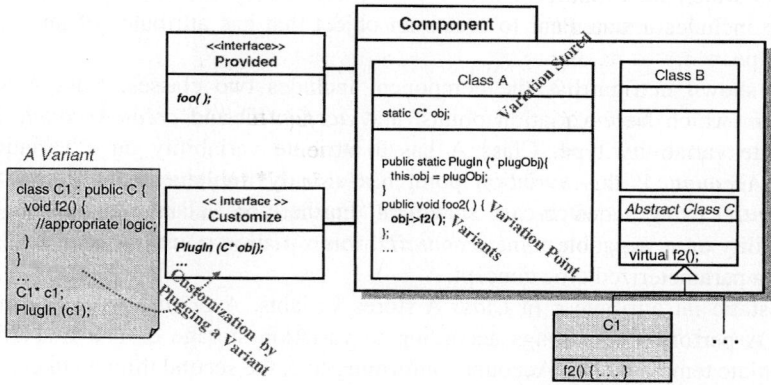

Fig. 4. Mechanism of Passing Pluggable Objects

5.1 Step 1. Defining Variable Functions

This step is to define the functions which handle the plug-in process for the classes which have variation points. The functions have hot spots for unknown variants. The hot spots will be filled by external functions and objects.

As shown in figure 4, an object as a variant can be plugged into a component. Variation points include object pointers that are hot spot. The object pointers will be plugged by a component consumer.

5.2 Step 2. Defining Static Attributes and Operations for Customization

This step is, for each class in a component, to define attributes for storing references of plugged functions and objects, and to define *customize* operations that read functions or objects made by clients and store the plugged functions or objects in these attributes. Once the references of functions and objects are (possibly persistently) stored in these attributes, further invocations of *foo1( )* operation will refer to the external functions. It is shown how pluggable functions and objects can be passed on to a component as a component customization technique.

To show an example of pluggable objects, In figure 4, the method *foo2( )* has a logic variability and the variation point is the method *f2( )* inside *foo2( )*. Hence, each family member may supply its own implementation of *f2( )*.

5.3 Step 3. Defining Customize Interface

This step is to collect *customize* operations in various classes in a component into a single *customize interface*. A component has a *customize* interface which takes the value or references of external elements and assigns it to its corresponding variation point.

In figure 4, an external object *c1* must be plugged into *obj* inside the component. To instantiate this variation point with an appropriate object, a customize interface is defined which contains *PlugIn (Classname* Obj)* method. Now, a family member can pass a reference to its own object that is extended by abstract class C. Class A stores the variant object that has the appropriate logic. The component client invokes the *foo( )* function. The *foo2( )* function invokes the *f2( )* method in a variant object by the dynamic binding of the object-oriented technique.

The method of *PlugIn(void (*fn)())* and *PlugIn (C* Obj);* are *customize* operations; thus, they are included in the *customize interface* of the component. Component consumer sets pluggable functions and objects using this interface to customize components.

5.4 Step 4. Setting Variants

This step is to customize components using *customize* operations. A component consumer makes an appropriate object that should be inherited from abstract class. The consumer invokes *PlugIn (C* Obj)* with the object. Therefore, further invocation on *foo1( )* and *foo2( )* will run the plugged parts.

6 External Profile Technique

External Profile technique is used to assign an external variant to a variation point of a component through an external profile such as a XML file. The external profile describes variants for customization. This can be done by storing the values of external profile. If a variant is changed, the component consumer only modifies the profile.

6.1 Step 1. Defining Variable Functions

The step for customization is similar to the technique of *passing pluggable object*. Variation points are realized as functions which include a sentence to read the variant. They will be read by XML profiles.

How variants in external profile can be passed on to a component as a component customization technique are exhibited. Figure 5 shows a component which contains class A. The class has the method *foo()* that is a variation point which is resolved by an external profile.

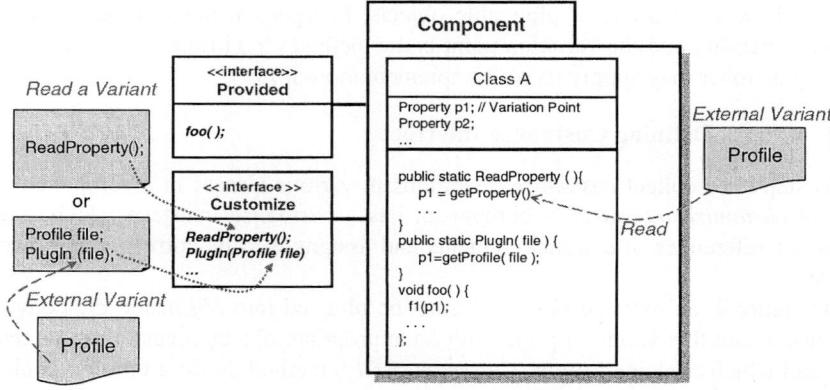

Fig. 5. Mechanism of 'Read the External Profile'

6.2 Step 2. Defining Static Attributes and Operations for Customization

The variable function reads variants from an external profile that includes variants for a family member. The component consumer invokes the *ReadProperty( )* in the *Customize* interface, the *ReadProperty( )* in *Class A* reads and analyses the external profile that should have a fixed location. The method *ReadProperty()* knows the position of the external profile.

In a different other mechanism, a component consumer invokes the *PlugIn (profileFile)* in *Class A*. The function reads and analyses the external profile. The profile does not have a fixed location. This technique is used to assign an external profile to a variation point of a component through a *customize* interface. These customize operations read variants from the external profile and store variants to the attributes. Therefore, the customize operations read XML files. Using Simple API for XML (SAX) rather than Document Object Model (DOM) is preferred.

However, the customize mechanism may read a XML profile once when a component consumer tailors components. The whole document is not needed to be fed into memory using a tree structure. There is no control over the order. The customize mechanism prefers to SAX rather than DOM. The *External Profile* technique can be implemented using Java and SAX API.

6.3 Step 3. Defining Customize Interface

This step for customization is similar to the technique of *Selection* and *Plug-In*. This step is to collect *customize* operations in various classes in a component into a single *customize interface*.

6.4 Step 4. Setting Variants

This step is to customize components using *customize* operations. The *External Profile* can be described by a XML file. For example, variability attributes are grouped thus; *Attr*, *Logic,* and *Workflow* in figure 6. If the requirement of the target system is changed, then the component consumer only modifies the XML profile.

```
<?xml version="1.0">
<Variability>
  <AttrType variantType = "var1" >…</AttrType>
  <AttrType variantType = "var2" >… </AttrType>
  <LogicType> … </LogicType>
  <WorkflowType> ...</WorkflowType>
  . . .
</Variability>
```

Fig. 6. Example of Variant Profile using XML

7 Assessment

In this paper, we propose variability design and customization mechanisms for COTS components. Our mechanism addresses variation types and scopes for variability mechanism. It is to increase component reusability and maintanability. We now compare our work to other representative works.

The catalysis presents two types of variability mechanism using inheritance. Keepence's work suggests three types of variability mechanism. However, these researches do not include detailed implement technique. Anastasopoulos' paper presents variability and feature type. However, the research is used to whitebox component. Svahnberg's research suggests five types of variability type. However, the research does not include inner detailed mechanism.

Our mechanisms are challenging to be able to customize blackbox components without accessing their source code. If new requirements that were covered by customization mechanism are discovered, we only select or plugin each variant by customizing interfaces to maintain components. Therefore, maintanbility can be enhanced by using the mechanism.

8 Concluding Remarks

As components are more for inter-organizational reuse, we need to model variability as well as commonality. One of the common forms of reusing COTS components is to acquire and customize them for each application. Therefore, components must be developed with consideration of commonality and variability in a domain.

In this paper, we present a comprehensive set of techniques to design components with variability so that blackbox form of components can be customized effectively only through interfaces. The COTS component is usually blackbox component. We focus practical applicability of customization techniques which can be implemented in popular CBD platforms.

The four types of component variability are covered by our customization mechanism. We proposed three techniques for variability implementation; *Selection*, *Plug-In* and *External Profile* technique. The techniques were presented more detailed customization methods. Through the three customization mechanism, we believe that the applicability, reusability, and maintainability of components can be greatly increased.

References

[1] Kim, S., "Lesson Learned from a Nationwide CBD Promotion Project," *Communications of the ACM, Vol. 45, Issue. 10, Oct.*, 2002.
[2] Kim, S., and Park, J., "C-QM: A Practical Quality Model for Evaluating COTS Components," *Proceedings of IASTED International Conference on Software Engineering*, Innsbruck, Austria, Feb., 2003.
[3] Aanastasopoulos, M., and Gacek, C., "Implementing Product Line Variabilities," *Proceedings of the 2001 symposium on Software reusability: putting software reuse in context*, Toronto, Cananda, May, 2001.
[4] Keepence, B., and Mannion, M., "Using patterns to model variability in product families," IEEE Software, Vol. 16, Issue. 4, July-Aug., 1999.
[5] Svanhnberg, M., and Bosch, J., "Issues Concerning Variability in Software Product Lines," *Lecture Notes in Computer Science 1951, Proceedings of the Third International Workshop on Software Architectures for Product Families*, 2000.
[6] Choi, S., Chang, S., and Kim, S.,"A Systematic Methodology for Developing Component Frameworks," *Lecture Notes in Computer Science 2984, Proceedings of 7th Fundamental Approaches to Software Engineering (FASE'04) Conference*, 2004.

A Fast Lossless Multi-resolution Motion Estimation Algorithm Using Selective Matching Units

Jong-Nam Kim

Dept. of Computer Engineering, PKNU (Pukyung National University)
jongnam@pknu.ac.kr

Abstract. We propose a new and fast multi-resolution motion estimation (MRME) algorithm using optimal matching units and scans for video coding, to significantly reduce the amount of computation in motion estimation. Our proposed algorithm has no any degradation of prediction quality compared to the original MRME algorithm. The computational reduction of our algorithm comes from fast elimination of unlikely candidate vectors. We fast eliminate inappropriate motion vectors using gradient magnitude in image data. The experimental results show that our algorithm reduces the computational amount of just 98~95% compared to full search algorithm and 55~80% compared to the original MRME algorithm. Our algorithm will be useful to real-time video coding applications such as MPEG-2 and MPEG-4 AVC.

1 Introduction

Motion estimation is defined as getting the best motion vector, which is the displacement of the coordinate of the best similar block in a previous frame for the block in a current frame. Of many approaches for motion estimation, the block-matching algorithm (BMA) is very popular in the framework of generic coding [1]-[2]. In last decades, so many BMA-based fast motion estimation algorithms have been published as follows [2]: unimodal error surface assumption (UESA) techniques[3]-[6], multi-resolution motion estimation (MRME) techniques [7]-[13], variable search range techniques with spatial/temporal correlation of the motion vectors, half-stop techniques using threshold of matching distortion, integral projection technique of matching block, low bit resolution techniques, sub-sampling techniques of matching block, successive elimination algorithm (SEA) [15]-[16], partial distortion elimination (PDE) [17], and so on.

In these fast algorithms, MRME is popular in video coding applications because of reduced computation, good prediction quality, its simplicity and easy hardware implementation. So many modified MRME algorithms of motion estimation have been reported in the last decades to further improve the original MRME algorithm. However, most of modified fast MRME algorithms can result in poor prediction quality for some cases, which can be a serious problem in actual applications [7]-[13].

In this paper, we remove only unnecessary computation in calculating block matching errors without any degradation of prediction quality compared to the original MRME algorithm. Our proposed algorithm employs multi-resolution approach, the

PDE, spiral search, and the adaptive matching scan from image complexity of the reference block. The MRME scheme has good prediction quality with significant computational reduction at same time. The PDE algorithm reduce only computational amount without affecting prediction quality, and the spiral search algorithm increases the probability to detect the optimal motion vectors in the search range as soon as possible because most of optimal motion vectors distribute about center area of the search range. And our proposed adaptive matching scan algorithm eliminates impossible candidates fast by first calculating block matching error for complex area of the reference block. From the combined algorithms, we further reduce only unnecessary computation without any degradation of prediction quality in calculating block-matching errors.

This paper is organized as follows. In section 2, we describe our proposed algorithm, which is based on MRME scheme, PDE, spiral search algorithm, and adaptive matching scan. In section 3, experimental results for various sequences and several fast algorithms and discussions of the results are given. Conclusion is followed in section 4.

2 Proposed Algorithm Using Spiral PDE and Adaptive Matching Scan Algorithm in MRME Scheme

In this section, we propose MRME based fast block-matching algorithm applicable to the current international video coding standards. Our algorithm reduces only unnecessary computation without any degradation of prediction quality compared to the original MRME algorithm. To do that, we use multi-resolution scheme, PDE (partial distortion elimination), and adaptive matching scan, which have different orders of calculations of matching errors in the matching blocks. With the concept, we further reduce the computations compared to the conventional MRME algorithm with the same prediction quality as that of the conventional MRME algorithm. From these concepts, we can remove impossible candidates faster, resulting in further reduced computations compared to the conventional MRME algorithm. We describe adaptive matching scan algorithm with PDE and spiral search based on image complexity.

2.1 Partial Distortion Elimination (PDE) and Spiral Search Algorithm

Before describing our algorithms, we will introduce the conventional PDE and spiral search algorithm. An efficient algorithm to reduce the computational complexity efficiently is the partial distortion elimination (PDE) method [1]-[2]. It uses the partial sum of matching distortion to eliminate impossible candidates before complete calculation of matching distortion in a matching block. That is, if an intermediate sum of matching error is larger than the minimum value of matching error at that time, the remaining computation for matching error is abandoned. In the partial distortion elimination (PDE), the *kth* partial sum of absolute difference (SAD) to check during the matching is as follows:

$$\sum_{i=1}^{k} \sum_{j=1}^{N} \left| f_t(i,j) - f_{t-1}(i+x, j+y) \right| \quad k = 1, 2, \dots N \tag{1}$$

N represents matching block size in the Eq. (1). If k is smaller than N from the partial sum of absolute difference (SAD) exceed the current SADmin, then we can quit the remaining summation of matching error calculation and kick out the impossible candidate motion vector (x,y). The PDE technique has been widely used to reduce the computational load in the full search algorithm. The reduction of calculation in obtaining motion vectors with the PDE algorithm depends on how fast global minimum of matching distortion is detected. If we find the global minimum of distortion in calculation of matching error faster, computational amount for matching error in a block is further reduced and k in the Eq. (1) is determined faster.

Ability to reject impossible candidates in the PDE algorithm depends on the search strategy which makes minimum matching error be detected faster. For the purpose, spiral search method shown in Fig. 1 (a) can be used. So, combined PDE algorithm uses Eq. (1) and spiral search. Fig. 1(b) shows top-to-bottom sequential search algorithm. In general, most of motion vectors are distributed about (0,0), which is the center of the search range. Thus, we can get higher probability for obtaining minimum matching error by using the spiral search scheme. If we get less matching error faster, we can ultimately remove more computation. Therefore, we employ the spiral search and PDE algorithm in our proposed algorithms.

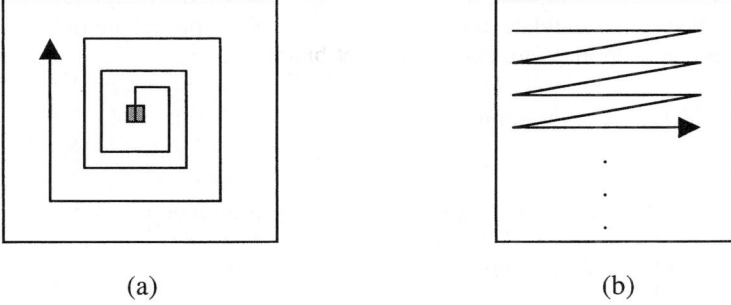

(a)   (b)

Fig. 1. Conventional spiral search direction and top-to-bottom sequential search: (a) spiral search direction and (b) top-to-bottom sequential search

As shown in Fig. 1(b), simple PDE algorithm means top-to-bottom matching scan based on Eq. (1) with top-to-bottom search in a given search range. Ability to reject impossible candidates in the PDE algorithm depends on the search strategy, which makes minimum matching error be detected faster. For the purpose, spiral search method shown in Fig. 1(a) is very efficient. So, combined PDE algorithm uses Eq. (1) and spiral search. As shown in the experimental results, the combined PDE algorithm rejects more the impossible candidates than simple PDE. Therefore, we employ the spiral search in the proposing matching scan algorithms.

2.2 Adaptive Matching Scan Algorithm Based on Image Complexity

2.2.1 Physical Meaning of the Relationship Between Matching Error and Image Complexity

Important thing in the PDE algorithm is that how fast impossible candidates are detected by removing unnecessary computation. To do so, we use the relationship between block matching error and image complexity of reference block using representative pixels. In this paper, we use the fact that the block-matching error is proportional to the complexity of the reference block with Taylor series expansion. The motivation of the proposing algorithm is using image complexity to find the impossible candidates faster. That is, we use image complexity and matching error for fast motion estimation. Thus, we measure image complexity and use sum of absolute difference (SAD) as matching criterion.

Usually, the matching distortion error, $d_t(p)$, is defined as the first expression in Eq. (2). Here, $f_t(p)$ represents pixel values at the block position p which represents a matching block position in the t_{th} frame, $cmv=(cmvx, cmvy)$ represents the position of the candidate motion vectors in the given search range. Additionally, the matching distortion can be approximated with the second expression of Eq. (2). It means that the matching block of the position p in the t_{th} frame can be approximated with the block of the $t-1_{th}$ frame. It has the displacement of optimal motion vector $mv=(mvx, mvy)$ of the given search range in the $t-1_{th}$ frame from the corresponding position of the t_{th} frame. Using the Taylor series expansion and taking only the first term of the series can obtain the third expression in Eq. (2). The higher order terms from the Taylor series expansion can be ignored in Eq. (2). Then, the t_{th} frame block can be exchanged approximately, with the $t-1_{th}$ frame block as the first approximation. As shown in Eq. (2), the matching distortion is expressed by the image gradient and the difference between candidate motion vectors and the optimal motion vector. The gradient and its magnitude are expressed as shown in Eq. (3).

$$\begin{aligned} d_t(p) &= |f_t(p) - f_{t-1}(p+cmv)| \\ &\approx |f_{t-1}(p+mv) - f_{t-1}(p+cmv)| \\ &\approx \left| \tfrac{\partial f_{t-1}(p+mv)}{\partial x}(cmvx-mvx) + \tfrac{\partial f_{t-1}(p+mv)}{\partial y}(cmvy-mvy) \right| \\ &\approx \left| \tfrac{\partial f_t(p)}{\partial x}(cmvx-mvx) \right| + \left| \tfrac{\partial f_t(p)}{\partial y}(cmvy-mvy) \right| \end{aligned} \qquad (2)$$

$$\begin{aligned} |G[f(x,y)]| &= \left\| \begin{bmatrix} Gx \\ Gy \end{bmatrix} \right\| = \left\| \begin{bmatrix} \tfrac{\partial f}{\partial x} \\ \tfrac{\partial f}{\partial y} \end{bmatrix} \right\| = \sqrt{G^2_x + G^2_y} \\ &\approx |Gx| + |Gy| \\ &\approx |f(x,y) - f(x+1,y)| + |f(x,y) - f(x,y+1)| \end{aligned} \qquad (3)$$

2.2.2 Fast Adaptive Matching Scan Algorithms Using Image Complexity of Reference Block

By localizing the image complexity well, we can get more reduction of unnecessary computation. In general, image complexity is well localized in pieces rather than the whole span of the image or of some fixed blocks. Our proposed algorithm using adaptive matching scan based on image complexity is validated by further reduced computation. That is, if we use localization of image complexity with small square subblocks then, we can get a faster elimination of impossible candidates than that of the previous works. This is realized with the first calculations of matching error for the area of larger distortion. Our adaptive matching scan algorithm from complexity order with square subblocks can be represented as shown in Eq. (4). Local complexity of subblock is defined as spatial complexity of image data for each subblock and measured with gradient magnitude as shown in Eq. (3). The size of matching unit of square subblocks and row vectors is the same for the fair comparison. We just change the shape of the matching unit from row vectors to square subblocks to localize image complexity well.

$$\text{Partial Distortion} = \sum_{u=1}^{k} \sum_{i=1}^{N/s} \sum_{j=1}^{N/s} \begin{vmatrix} f_t(i + q_u * (N/s), j + r_u * (N/s)) \\ -f_{t-1}(i + x + q_u * (N/s), j \\ + y + r_u * (N/s)) \end{vmatrix} \quad (4)$$

$$k = 1, 2, \ldots N, \quad q_u = floor((m_u - 1)/(N/s)),$$
$$r_u = (m_u - 1)\%(N/s),$$
$$m_u \in local\_complexity\_order[u]$$

In Eq. (4), s is the size of a small sub-block, N is the size of matching block, and the *local_complexity_order[]* is the matching scan order according to the local complexity from the square subblocks, *(x,y)* is a candidate vector in the search range, and m_u is the index of the subblock, which is determined by gradient magnitude and arranged in the array of *local_complexity_order[u]*. The operation, *floor(alpha)*, rounds off the alpha. The operation, *(alpha)%(beta)*, calculates remainder after division of *(alpha/beta)*. k is variable because of PDE scheme.

Our proposed algorithm employs the MRME scheme, the PDE, the spiral search, and the adaptive matching scan from the image complexity of the reference block. As described previously, the MRME scheme has good prediction quality with significant computational reduction at same time. The PDE algorithm reduce only computational amount without affecting prediction quality, and the spiral search algorithm increases the probability to detect the optimal motion vectors in the search range as soon as possible because most of optimal motion vectors distributes about center area of the search range. And our proposed adaptive matching scan algorithm eliminates impossible candidates fast by first calculating block matching error for complex area of the reference block. From the combined algorithms, we further reduce only computation without any degradation of prediction quality in calculating block-matching errors.

3 Experimental Results

To compare the performance of the proposed algorithm with the conventional algorithms, we use 100 frames of 'foreman', 'car phone', 'trevor', and 'clair' image sequences. In these sequences, 'foreman', and 'car phone' have big motions compared with other image sequences, while 'clair' is almost inactive sequences compared with first three sequences. 'trevor' sequence has intermediate motions.

We presented experimental results with the original full search (FS), the three-step search (TSS) [2], the new three-step search (NTSS) [3], the original multiresolution

resolution motion estimation (MRME) [1]-[2]. We further reduced computational complexity of MRME scheme as described previously with the partial distortion elimination (PDE) algorithm and adaptive matching scan algorithm which is our proposed algorithm.

The matching block size is 16 x 16 pixels and the search window is 31 x 31 pixels. Image format is QCIF (176 x 144) for each sequence and only forward prediction is used. Sum of absolute difference (SAD) as error criterion for finding motion vector is employed. The simulation results are shown in terms of average number of checking rows with the reference of that of full search without any fast operation and peak signal-to-noise ratio (PSNR). The average checking rows is used because the comparison for the partial distortion and the minimum distortion in the conventional PDE algorithm is performed in the unit of line-by-line of matching blocks. Here, the experiment were carried out by skipping zero and two frames. Therefore, the resulting frame rate is 30, and 10 frames per second (fps). All figures for the average checking rows of our proposed algorithm in the tables were considered with overhead computations for complexity measure.

We measured the image complexity with three directions of three neighboring points. The three directions are the right point, the below point, and the diagonal right bottom point from the corresponding point. Of course, we can extend to eight directions instead of three directions. In our experiments, the three directions were the most appropriate considering the trade-off between the computational amounts and the reduction of checking rows. The reduction of checking rows from more than three directions ones were close to that of three directions with only increased overhead computation. Also, all the matching scan algorithms employed spiral search scheme to make use of the distribution of motion vectors.

Table 1 present experimental results for average PSNR of several algorithms with 30 fps (frames per second). The average PSNR in these tables means the prediction quality for each algorithm. In these tables, we can see that the MRME schemes produce the prediction quality close to the full search (FS) algorithm. Our proposed MRME algorithms have the same prediction quality as that of the original MRME algorithm because our algorithm reduces only unnecessary computations without affecting the prediction quality. In general, it is reported MRME schemes has better prediction quality than that of TSS and NTSS in actual video sequences [1]-[2]. Generally, NTSS algorithm shows good prediction quality in inactive image sequences because it is operated with center-biased search method. Our test image sequences are QCIF format, thus motions in the sequences are not large.

Table 1. Experimental results for average PSNR of several algorithms with 30 fps

	Foreman	Car phone	Trevor	Clair
Original FS	34.43	33.44	33.28	41.29
Original MRME	33.89	33.27	33.21	41.29
TSS	33.84	33.24	33.21	41.29
NTSS	34.33	33.40	33.26	41.29

Table 2. Computational reduction ratio of several algorithms with 10 fps

	Foreman	Car phone	Trevor	Clair
Original FS	100 %	100 %	100 %	100 %
Original MRME	9%	9%	9%	9%
Fixed Direction	4.56%	4.20%	3.97%	2.11%
Adaptive Direction	4.04%	3.86%	3.69%	1.98%
TSS	11.11%	11.11%	11.11%	11.11%
NTSS	10.52%	9.65%	7.96%	7.68%

Table 3. Computational reduction ratio of several algorithms with 30 fps

	Foreman	Car phone	Trevor	Clair
Original FS	100 %	100 %	100 %	100 %
Original MRME	9%	9%	9%	9%
Fixed Direction	3.91%	3.62%	2.92%	1.75%
Adaptive Direction	3.37%	3.29%	2.74%	1.70%
TSS	11.11%	11.11%	11.11%	11.11%
NTSS	8.67%	8.70%	7.98%	7.61%

Table 2 ~ 3 present the computational reduction ratio for several algorithms for 10 fps and 30 fps. We can see that the MRME schemes obtain significant computational reduction compared to other TSS and NTSS algorithms. Additionally, the MRME schemes get good performance about computational reduction and prediction quality. Especially, our proposed algorithm, spiral search based adaptive matching MRME, obtains the most computational reduction ratio in MRME schemes, which include the original MRME and the spiral search based fixed matching MRME. There are three MRME schemes in the tables. The original MRME doesn't employ any fast method, and the MRME with fixed direction uses the spiral search method and the PDE method with the sequential fixed top-to-bottom matching direction. And our proposed algorithm, MRME with adaptive direction, uses the spiral search and the PDE with adaptive matching direction. To set the reference of computational reduction, we put the computation of the FS as 100%.

Fig. 2 ~ 3 represent average checking rows of high-resolution for "foreman" and "clair" sequences with 30 fps. We can find obvious difference in the computational reduction between the sequential matching scan MRME and our adaptive matching scan MRME. Our proposed MRME algorithm with adaptive matching scan shows better performance than the sequential matching scan algorithm for all the frames and all the frame rates of each sequence in these figures.

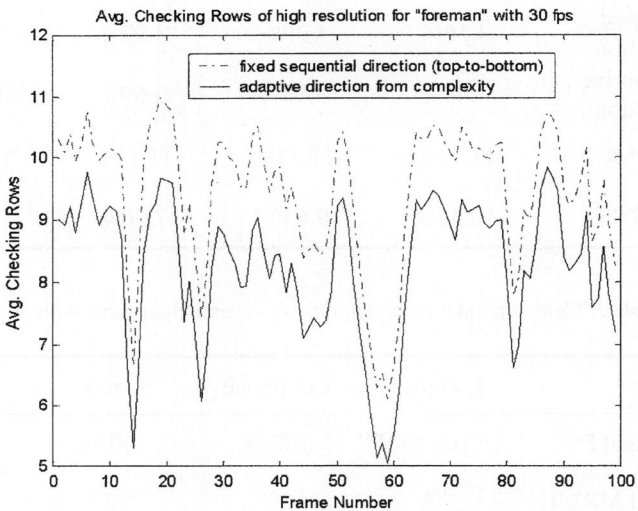

Fig. 2. Average checking rows of high resolution for "foreman" sequence with 30 fps

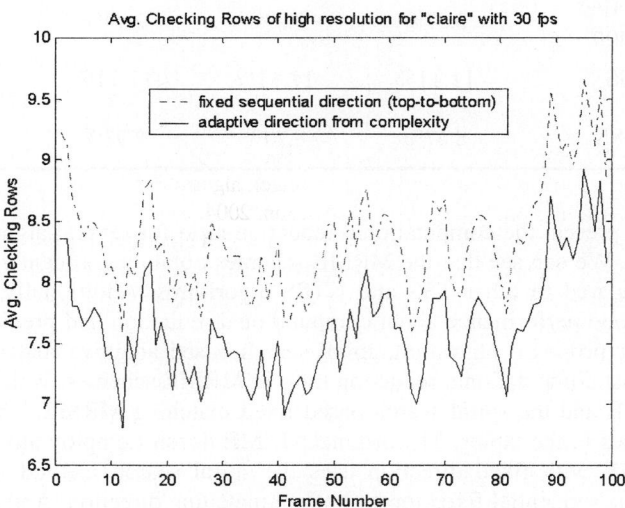

Fig. 3. Average checking rows of high resolution for "clair" sequence with 30 fps

From the above experimental results, we can conclude that the MRME scheme is excellent in terms of the prediction quality and the computational reduction compared to other fast motion estimation algorithms, and our proposed adaptive matching scan MRME algorithm obtains more computational reduction than other MRME schemes, meanwhile keeping the same prediction quality compared to the original MRME scheme.

4 Conclusions

In this thesis, we proposed MRME based fast block-matching algorithms applicable to the current international video coding standards. In our algorithm, we used image complexity, multi-resolution motion estimation, partial distortion elimination (PDE), and adaptive matching scan, which has different orders for matching error calculation in the matching blocks. The experimental results showed that our algorithm reduces the computational amount of just 98~95% compared to full search and 55~80% compared to the original MRME. Our algorithm can be applicable to real-time video coding applications such as MPEG-2 and MPEG-4 AVC encoders.

Acknowledgement

This work has been partially supported by "Research Center for Future Logistics Information Technology" and "New Professor Project" hosted by the Ministry of Education, "RIS" by KOTEF, and "New Professor Project" by PKNU in Korea.

References

1. M. Tecalp, *Digital Video Processing*, Prentice Hall, PP.72-129, 1995.
2. J.N. Kim, "A study on fast block matching algorithms of motion estimation for video compression," Ph. D Thesis in Gwang-Ju Institute of Science and Technology, 2001.
3. R. Li, B. Aeng and M. L. Liou, "A new three-step search algorithm for block motion estimation," *IEEE Trans. Circuits Syst. Video Technol.*, vol. 4, pp. 438-442, Aug. 1994.
4. X. Jing, L.P. Chau, "An efficient three-step search algorithm for block motion estimation," *IEEE Trans.Multimedia,* vol. 6, pp. 435-438, Jun. 2004.
5. C.H. Cheung, L.M. Po, "A novel cross-diamond search algorithm for fast block motion estimation, " *IEEE Trans. Circuits Syst. Video Technol.*, vol. 12, pp. 1168-1177, Dec. 2002.
6. C. zhu, X. Lin, L. Char, L.M. Po, "Enhanced hexagonal search for fast block motion estimation," *IEEE Trans. Circuits Syst. Video Technol.*, vol. 14, pp. 1210-1214, Oct. 2004.
7. Y. Q. Shi and X. Xia, "A thresholding multiresolution block matching algorithm," *IEEE Trans. Circuits Syst. Video Technol.*, vol. 7, pp. 437-440, Feb. 1997.
8. K.M. Uz, M. Vetterli, and D. LeGalll, "Interpolative multiresolution coding of advanced television with compatible subchannels," *IEEE Trans. Circuits Syst. Video Technol.*, vol. 1, pp. 86-99, Mar. 1991.
9. S.N. Kim, S.H. Rhee, J.G. Jeon, and K.T. Park, "Interframe coding using two-stage variable block-size multiresolution motion estimation and wavelet decomposition," *IEEE Trans. Circuits Syst. Video Technol.*, vol. 8, pp. 399-410, Aug. 1998.

10. K.W. Lim and J.B. Ra, "Improved hierarchical search block matching algorithm by using multiple motion vector candidates," *IEE Elect. Letters*, vol. 33, pp. 1771-1772, Oct. 1997.
11. J. Chalidabhongse and C.C.J. Kuo, "Fast motion vector estimation using multiresolution-spatio-temporal correlations," *IEEE Trans. Circuits Syst. Video Technol.*, vol. 7, pp. 477-488, Jun. 1997.
12. G.B. Rath and A. Makur, "Subblock matching-based conditional motion estimation with automatic threshold selection for video compression," *IEEE Trans. Circuits Syst. Video Technol.*, vol. 13, pp. 914-924, Sept. 2003.
13. B.C. Song and K.W. Chun, "Multi-resolution block matching algorithm and its VLSI architecture for fast motion estimation in an MPEG-2 video encoder,", *IEEE Trans. Circuits Syst. Video Technol.*, vol. 14, pp. 1119-1137, Sept. 2004.
14. J. Zan, M.O. Ahmad and M.N. Swamy, "New techniques for multi-resolution motion estimation," *IEEE Trans. Circuits Syst. Video Technol.*, vol. 12, pp. 793-802, Sept. 2002.
15. J.N. Kim, D.K. Kang, and S.C. Byun, "A fast full search motion estimation algorithm using sequential rejection of candidates from hierarchical decision structure," *IEEE Trans.Broadcasting,* vol. 48, pp. 43-46, Mar. 2002.
16. X.Q. Gao, C.J. Duanmu, and C.R. Zou, "A multilevel successive elimination algorithm for block matching motion estimation," *IEEE Trans. Image Processing*, vol. 9, pp. 501-504, Mar. 2000.
17. J.N. Kim, S.C. Byun, Y.H. Kim, and B.H. Ahn, "Fast full search motion estimation algorithm using early detection of impossible candidate vectors," *IEEE Trans. Signal Processing,* vol. 50, pp. 2355-2365, Sept. 2002.

Developing an XML Document Retrieval System for a Digital Museum

Jae-Woo Chang

Dept. of Computer Engineering, Research Center for Advanced LBS Technology,
Chonbuk National University, Chonju,
Chonbuk 561-756, South Korea
jwchang@chonbuk.ac.kr

Abstract. In this paper, we develop an XML document retrieval system for a digital museum. It can support unified retrieval on XML documents based on both document structure and image content. To achieve it, we perform the indexing of XML documents describing Korean porcelains used for a digital museum, based on not only their basic unit of element but also their image color and shape features. In addition, we provide a similarity measure for a unified retrieval to a composite query, based on both document structure and image content. Finally, we implement our XML document retrieval system designed for a digital museum and analyze its performance in terms of retrieval time, insertion time, storage overhead, as well as recall and precision measure.

1 Introduction

Recently, it has been very common for users to meet a variety of XML documents through a Web browser since the XML (eXtensible Markup Language) has become a standard markup language to represent Web documents [1]. To develop a digital information retrieval system which provides services in the Web, it is necessary to efficiently retrieve XML documents required by users. An XML document not only has a logical and hierarchical structure, but also contains its multimedia data, such as image and video. As a result, it is essential to develop an XML document retrieval system that can support both the retrieval based on both document structure and image content.

In this paper, we develop an XML document retrieval system used for a digital museum. It can support unified retrieval on XML documents based on both document structure and image content. In order to support retrieval based on document structure, we perform the indexing of XML documents describing Korean porcelains used for a digital museum, based on their basic unit of elements. Using this, we design four index structures, i.e., keyword, structure, element and attribute index structure. For supporting retrieval based on image content, we also do the indexing of the documents describing Korean porcelains, based on color and shape features of their images. This results in the design of a high-dimensional index structure using the CBF method. Finally, we provide a similarity measure for a unified retrieval to a composite query, based on both document structure and image content.

This paper is organized as follows. In Section 2, we introduce related work in the area of retrieval based on document structure and image content. In Section 3, we design an XML document retrieval system for a digital museum. In Section 4, we present the implementation of our XML document retrieval system designed for a digital museum and analyze its performance. Finally, we draw conclusions and provide future work in Section 5.

2 Related Work

Because an element is a basic unit that constitutes a structured (i.e., SGML or XML) document, it is essential to support not only retrieval based on element units but also retrieval based on logical inclusion relationships among elements. First, RMIT in Australia proposed a *subtree model* which indexes all the elements in a document and stores all the terms which appear in the elements [2] so as to support five query types for structure-based retrieval in SGML documents. Secondly, SERI in South Korea proposed a *K-ary Complete Tree Structure* which represents a SGML document as a K-ary complete tree [3]. In this method, a relationship between two elements can be acquired by calculation because each element corresponds to a node in a K-ary tree. Thirdly, Univ. of Wisconsin in Madison proposed a new technique to use the position and depth of a tree node for indexing each occurrence of XML elements [4]. For this, the inverted index was used to enable ancestor queries to be answered in constant time. Fourthly, IBM T.J. Watson research center in Hawthorne proposed ViST, a novel index structure for searching XML documents [5]. The ViST made use of tree structures as the basic unit of query to avoid expensive join operations and provided a unified index on both text content and structure of XML documents. However, these four indexing techniques were supposed to handle tree data. Finally, Univ. of Singapore proposed D(k)-Index, a structural summary for general graph structured documents [6]. The D(k) index possesses the adaptive ability to adjust its structure according to the current query load, thus facilitating efficient update algorithms

There have been a lot of studies on content-based retrieval techniques for multimedia or XML documents. First, the *QBIC(Query By Image Content) project* of IBM Almaden research center studied content-based image retrieval on a large on-line multimedia database [7]. The study supported various query types based on the visual image features such as color, texture, and shape. Secondly, the VisualSEEk project of Colombia University in the USA developed a system for content-based retrieval and browsing [8]. Its purpose was an implementation of CBVQ(Content-Based Visual Query) that combines spatial locations of image objects and their colors. Thirdly, the Chonbuk National Univ. in South Korea developed an XML document retrieval system that can support a unified retrieval based on both image content and document structure [9]. Fourthly, the Pennsylvania State Univ. presented a comprehensive survey on the use of pattern recognition methods for content-based retrieval on image and video information [10]. Finally, the Chinese Univ. of Hong Kong presented a multi-lingual digital video content management system, called iVIEW, for intelligent searching and access of English and Chinese video contents [11]. The iVIEW system allows full content indexing and retrieval of multi-lingual text, audio and video materials in XML documents.

3 Design of XML Document Retrieval System

We design an XML document retrieval system for a digital museum, which is mainly consists of five parts, such as a preprocessing part, an indexing part, a storage manager part, a retrieval part and a user interface part. Figure 1 shows the system architecture of our XML document retrieval system for a digital museum. When an XML document is given, we parse it and perform image segmentation from it through the indexing part, so that we can index its document structure consisting of element units and can obtain the index information of color and shape features of its image. The index information for document structure and that for image content are separately stored into its structure-based and content-based index structures, respectively. Using the index information extracted from a set of XML documents, some documents are retrieved by the retrieval part in order to obtain a unified result to answer user queries. Finally, the unified document result is given to users through a user interface part using a Web browser.

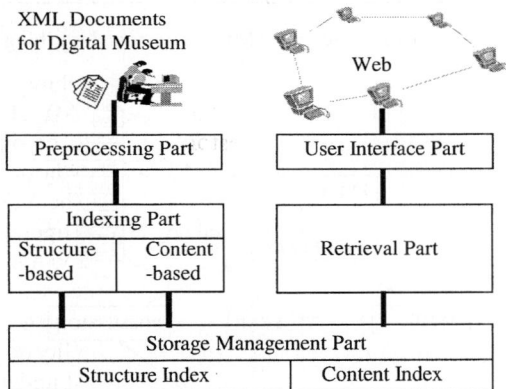

Fig. 1. XML document retrieval system architecture for a digital museum

3.1 Indexing

Because an element is a basic unit for retrieving an XML document, it is necessary to support a query based on a logical inclusion between elements and based on the characteristic value of elements. To achieve it, we construct a document structure tree for XML documents describing Korean porcelains used for a digital museum, after analyzing XML documents based on DTD. Figure 2 depicts a DTD grammar for representing XML documents describing Korean porcelains and an XML document instance following the DTD. The XML document instance contains not only a document structure between elements, but also attribute information. That is, the *porcelain* element has child elements, i.e., *name, year, description, and image* element. The *porcelain* element has an attribute 'TYPE' with a value 'Chung-ja'. To build a document structure tree for XML documents describing Korean porcelains, we parse the XML documents by using sp-1.3 parser [12]. Finally the storage manager extracts document structure information and image content information from the tree and stores them into a database. Figure 3 shows the document structure tree built from the XML DTD grammar in Figure 2.

```
<! ELEMENT relic(porcelain)*>
<!ELEMENT porcelain
 (name,year,possession,classification,description,image)>
    <!ATTLIST porcelain TYPE CDATA #REQUIRED>
    <!ELEMENT name (#PCDATA)>
    <!ELEMENT year (#PCDATA)>
    <!ELEMENT possession(museum|university|personal)>
        <!ELEMENT museum (#PCDATA)>
        <!ELEMENT university (#PCDATA)>
        <!ELEMENT personal (#PCDATA)>
               •
<! ELEMENT image(align)*>
<! ATTLIST image SRC CDATA#REQUIRED>
    <! ELEMENT align (#PCDATA)>
               •
```

```
<relic>
 <porcelain TYPE = "Chung-ja">
    <name>
       Chung-ja kettle
    </name>
    <year>
       A.D 1170
    </year>
       :
    <description>
       Chung-ja kettle is like a shape ...
    </description>
    <image SRC = "hc_1">
    </image>
 </porcelain>
</relic>
```

Fig. 2. DTD grammar and its instance for XML documents describing Korean porcelains

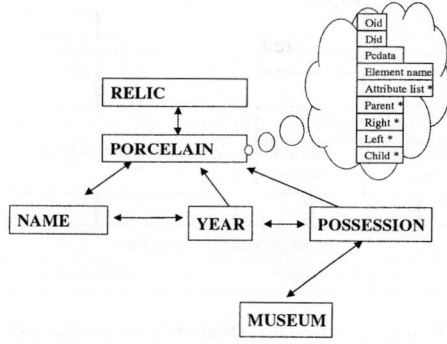

Fig. 3. Document structure tree built from XML DTD grammar

For content-based indexing of images contained in XML documents, we extract images from XML documents and analyze them for obtaining image feature vectors. To achieve it, it is necessary to separate a salient object, i.e., a Korean porcelain, from the background of an image. To extract only a region for the salient object, we use the fuzzy c-mean (FCM) algorithm that is a generally famous clustering one to divide object regions from color images [13]. When we divide an image into two clusters, the FCM algorithm calculates the distance of a pixel from the center point of each cluster and assigns it to a cluster with shorter distance. It has an advantage that the separation of a salient object of an image from its background can be performed well when an image has little noise, as shown in our porcelain images.

In order to obtain an image feature vector for shape, we obtain a salient object from an image by the preprocessing part and generate a 24-dimensional feature vector. An algorithm for generating a shape feature vector is as follows. First, we calculate the central point of a salient object using its maximum and minimum value. Secondly, we increase 15 degrees at the central point, starting from the X-axis, and select 24 points met at the edge. Thirdly, we compute the distance between the central point and the 24 edge points. Finally, we generate a 24-dimensional feature vector by normalizing the 24 distances by dividing them by the maximum distance. It can be very efficient for digital museums where an image has only a couple of salient objects. Since the proximity among colors in the RGB color space doesn't mean their similarity among colors, we use HSV color space model. The HSV model provides a uniform distribution of colors and makes color transformation easier. In this model, H means an aggregate of color, ranging from 0 to 360 degree. S means the saturation of color, and V means the brightness of color. An algorithm for generating a color feature vector is as follows. First, we transform all color pixels of an image object in the RGB color space into those in the HSV color space. Secondly, we generate a color histogram by using color histogram generation algorithm. Finally, we generate a 22-dimensional feature vector by normalizing the color histogram by dividing it by the number of the entire pixel.

3.2 Storage Management

The index information for document structure and that for image content are separately stored into structure and content index structures, respectively. The index structures for structure-based retrieval are constructed by indexing XML documents based on an element unit and consist of keyword, structure, element, and attribute index structures. First, the keyword index consists of three files, i.e., keyword index file being composed of keywords extracted from data token element (e.g., PCDATA, CDATA) of XML documents, posting file including the IDs of document and element where keywords appear, and location file containing the location of keyword appearance in elements. Secondly, because the structure index is used for searching an inclusion relationship among elements, it should represent the logical structure of a document and guarantee good performance on both retrieval time and storage overhead. To achieve it, we propose an element unit parse tree structure where an element contains the location of its parent, its left sibling, its right sibling, and its first left child. We can find an inclusion relationship among elements easily because the tree structure represents the hierarchical structure of a document well. The element index structure contains element information and the identifier of a document that an element belongs. Thirdly, the element index structure contains some element information and the identifier of a document that an element belongs. Finally, the attribute index structure contains some attribute information and the identifier of an element that an attribute belongs.

The index structure for content-based retrieval is a high-dimensional index structure based on the CBF method [14], so as to store and retrieve both color and shape feature vectors efficiently. A main focus on managing a large number of XML documents is retrieval performance. As the number of dimensions of feature vectors is increasing, the retrieval performance of the traditional index structures is exponentially increasing. However, the CBF method can achieve good retrieval performance even though the dimension of feature vectors is high. In addition, our CBF-based index structure can support a variety of types of queries, like point, range, and K-NN search.

3.3 Retrieval

Using the stored index information extracted from a set of XML documents, some documents are retrieved by the retrieval part in order to obtain a unified result to answer user queries. There is little research on retrieval models for integrating structure- and content-based information retrieval. To answer a query for retrieval based on document structure, a similarity measure (S_w) between two elements, say q and t, is computed as the similarity between the term vector of node q and that of node t by using a cosine measure [15]. Supposed that a document can be represented as D = { $E_0, E_1, ..., E_{n-1}$ } where E_i is an element i in a document D. Thus, a similarity measure (D_w) between an element q and a document D is computed as follows.

$$D_w = \text{MAX} \{ COSINE\ (NODE_q, NODE_{E_i}), 0 \leq i \leq n-1 \}$$

To answer a query for retrieval based on image content, we first extract color or shape feature vectors from a given query image. Next, we compute Euclidean distances between a query color (or shape) feature vector and the stored image color (or shape) vectors by searching the content index structure. A similarity measure, $C_W(q, t)$, between a query image q and a target image t in the database is calculated as the following equation. Here Distc(q, t) and Dists(q, t) are a color vectors distance and a shape vector distance between q and t, respectively. Nc and Ns are the maximum color and the maximum shape distances for normalization, respectively.

$$C_w = \begin{cases} 1 - \dfrac{Distc(q,t)}{Nc}, & \text{if a query contains only a color feature.} \\ 1 - \dfrac{Dists(q,t)}{Ns}, & \text{if a query contains only a shape feature.} \\ (1 - \dfrac{Distc(q,t)}{Nc}) \times (1 - \dfrac{Dists(q,t)}{Ns}), & \text{if a query contains both color and shape feature.} \end{cases}$$

Finally, when α is the relative weight of retrieval based on document structure over that based on image content, a similarity measure (T_w) for a composite query is calculated as follows. If the weight of the former retrieval is equal to that of the latter retrieval, α equals 0.5.

$$T_w = \begin{cases} C_w \times \alpha + D_w \times (1-\alpha), & \text{if results are document for user query} \\ C_w \times \alpha + S_w \times (1-\alpha), & \text{if results are element for user query} \end{cases}$$

4 Implementation and Performance Analysis

We implement our XML document retrieval system for a digital museum, under SUN SPARCstation 20 with GNU CCv2.7 compiler. For this, we make use of O_2-Store v4.6 [15] as a storage system and Sp-1.3 as an XML parser. For constructing a prototype digital museum, we make use of 630 XML documents describing Korean porcelains with a museum XML DTD, as shown in Table 1. They are extracted from several Korean porcelain books published by Korean National Central Museum.

Table 1. XML documents used for a digital museum

Document content	Korean porcelains
The number of documents	630 documents
Document average size	1.20K(text)+42K(image)
The number of elements	10754
The number of keyword	37807

For retrieving XML documents, we can classify user queries into two types, i.e., simple and composite queries. The simple queries can be divided into keyword-, structure-, attribute-, and content-based query. Their examples are as follows.

- Keyword-based: Find documents which contain 'Buddhist image' term.
- Structure-based: Find all children elements of [Porcelain] element.
- Attribute-based: Find documents whose attribute type is 'Chong-ja'.
- Content-based: Find documents whose image contains a specific color or shape.

The composite query is the combination of simple queries, i.e., structure + keyword, structure + attribute, keyword + color and structure + shape query. Figure 4 shows a digital museum interface for retrieving XML documents describing Korean porcelains. By using this interface, a user, for example, generates a composite query to retrieve documents that contain a keyword 'Buddhist image' in all children element of [porcelain] element whose TYPE attribute value is 'chong-ja' and that include an image whose color is 'blue' and whose shape is like 'kettle'. Figure 4 also shows a result for the composite query and its similarity with images in the database.

To evaluate the efficiency of our XML document retrieval system for a digital museum, we measure insertion time, retrieval time, and storage overhead. Our system requires about 6 seconds on the average to insert an XML document into keyword, attribute, structure, and element indexes. It also requires less than 1 second on the average to insert an image content into color and shape index. Figure 5 shows retrieval times for simple queries. The retrieval time for the structure-based query is 6.5 seconds. But the retrieval times for the other queries are less than 2 seconds, respectively. It is shown form the result that the structure-based query requires the largest amounts of time to answer it. We also measure retrieval times for the combinations of two simple queries, such as structure + keyword, structure + attribute, keyword + color, and structure + shape. Figure 6 shows retrieval times for composite queries. The retrieval time for a keyword + color query is less than 3 seconds. However, the retrieval times for structure + keyword, structure + attribute, and structure + shape are 8.7, 8.1, and 7.3 seconds, respectively. This shows that a composite query containing structure-based retrieval requires large amounts of time to answer it. Finally, we measure storage overhead as a ratio of the total size of our index files over that of the original XML documents. Our XML document retrieval system requires about 50% storage overhead.

To evaluate the retrieval effectiveness of our XML retrieval system, we measure recall and precision [16] by a test group of computer engineering graduate students from Chonbuk National University, South Korea. Table 2 shows the precision and recall measure of our XML document retrieval system when we choose K porcelains as

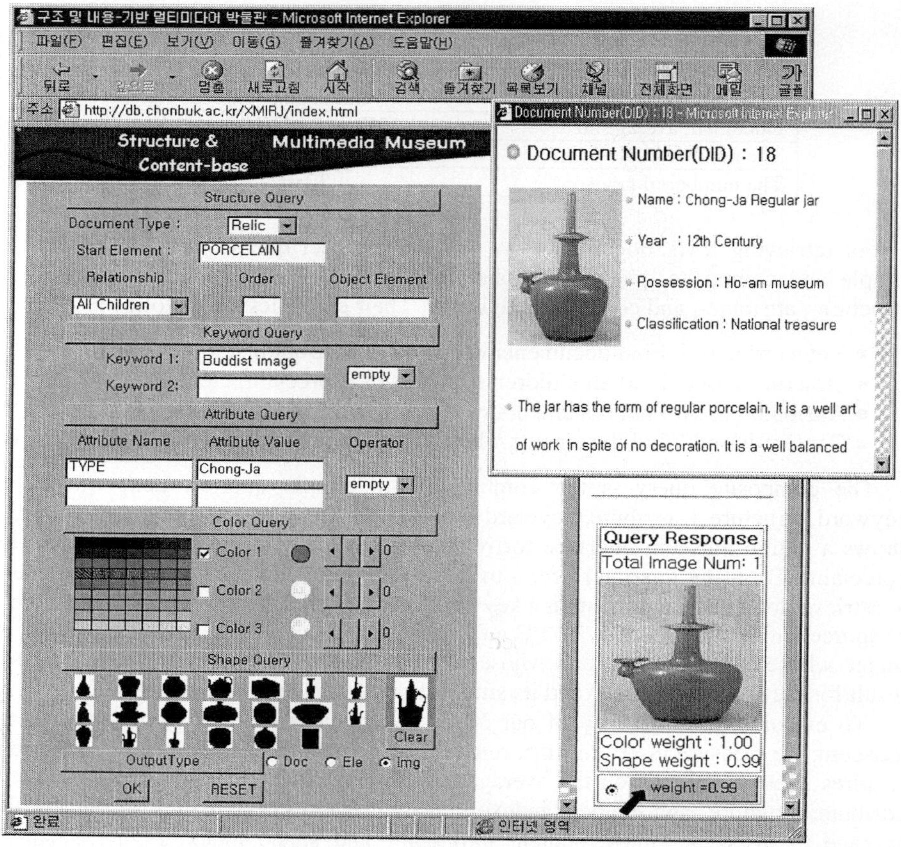

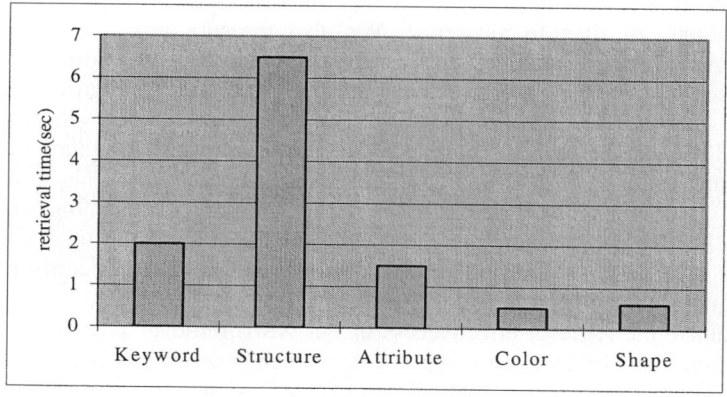

Fig. 4. A digital museum interface

Fig. 5. Retrieval times for simple queries

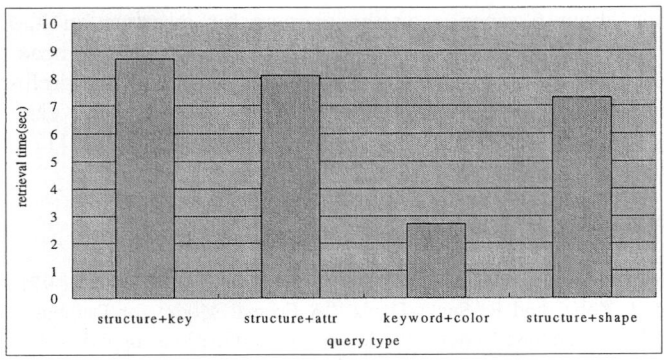

Fig. 6. Retrieval time for complex queries

the most similar ones. As K increases, the precision decreases and the recall increases. When K=10, the precision is about 0.6 and the recall is about 0.4, for both the color- and shape-based query. It is shown from our performance analysis that retrieval based on document structure has a great significance because a structure-based query requires much more time to answer it. Therefore, we compare our XML document retrieval system with the k-ary complete tree system [3] which is known as a promising system for structure-based queries. Our system requires 6.5 seconds for answering a structure-based query and the k-ary tree system requires 6.3 seconds. This shows nearly the same retrieval performance for structure-based queries.

Table 2. Precision and recall

	Color		Shape	
	Precision	Recall	Precision	Recall
K = 7	0.64	0.33	0.57	0.23
K = 10	0.60	0.37	0.55	0.32

5 Conclusions and Future Work

In this paper, we developed an XML document retrieval system for a digital museum. It can support efficient retrieval on XML documents for both structure- and content-based queries. In order to support structure-based queries, we performed the indexing of XML documents describing Korean porcelains for a digital museum, based on their basic unit of element. This resulted in designing four efficient index structures, i.e., keyword, structure, element, and attribute. In order to support content-based queries, we also performed the indexing of the XML documents based on both color and shape features of their images. This resulted in designing a high-dimensional index structure based on the CBF method. We also provided a similarity measure for a unified retrieval to a composite query, based on both document structure and image content. Our system for a digital museum requires about 6 seconds for answering a structure-based query and requires less than 2 seconds for the remaining queries. Our system spends about 6 seconds on the average for inserting an XML document describing a Korean porcelain and requires about 50 % storage overhead. Future work can be studied on new information retrieval models for integrating preliminary results acquired from both structure- and content-based queries. This can be achieved ultimately by trying to handle MPEG-7 compliant XML documents [17].

References

[1] eXtensible Markup Language(XML), http :// www.w3.org/TR/PR-xml-971208.
[2] B. Lowe, J. Zobel and R. Sacks-Davis, "A Formal Model for Databases of Structured Text," In Proc. Database Systems for Advanced Applications, pp 449-456, 1995.
[3] S.G. Han, et. al., "Design and Implementation of a Structured Information Retrieval System for SGML Documents," In Proc. Database Systems for Advanced Applications, pp 81-88, 1999.
[4] C. Zhang, J. Naughton, D. DeWitt, Q. Luo, and G. Lohman, "On Supporting Containment Queries in Relational Database Management Systems," In Proc. ACM SIGMOD. pp 425-436, 2001.
[5] H. Wang, S. Park, W. Fan, and P.S. Yu, "ViST: A Dynamic Index Method for Querying XML Data by Tree Structures," In Proc. ACM SIGMOD. pp 110-121, 2003.
[6] Q. Chen, A. Lim, and K.W. Ong, "D(k)-Index: An Adaptive Structural Summary for Graph-structured Data," In Proc. ACM SIGMOD. pp 134-144, 2003.
[7] M. Flickner, et. al., "Query by Image and Video Content: The QBIC System," IEEE Computer, Vol. 28, No.9, pp. 23-32, 1995.
[8] J. R. Smith and S. F. Chang, "VisualSEEk: a Fully Automated Content-Based Image Query System," In Proc. ACM Int'l Conf. on Multimedia, pp 87-98, 1996.
[9] K. Jin and J. Chang, "An Efficient Storage Manager for Content-based Multimedia Information Retrieval in NoD Applications," In Proc. the 3rd Int'l Conf. of Asia Digital Library, pp 275-281, 2000.
[10] S. Antani, R. Kasturi, and R. Jain, "A Survey on the Use of Pattern Recognition Methods for Abstraction, Indexing and Retrieval of Images and Video," Pattern Recognition. Vol. 35, No. 4, pp 945-965. (2002).
[11] M.R. Lyu, E. Yau, and S. Sze, "A Multilingual, Multimodal Digital Video Library System," In Proc. ACM/IEEE-CS Joint Conf. on Digital Libraries, pp 145-153, 2002.
[12] http://www.jclark.com/sp.
[13] J.C. Bezdek and M.M. Triedi, "Low Level Segmentation of Aerial Image with Fuzzy Clustering," IEEE Trans. on SMC, Vol. 16, pp 589-598, 1986.
[14] S.G. Han and J.W. Chang, "A New High-Dimensional Index Structure using a Cell-based Filtering Technique," Lecture Notes in Computer Science, Vol., pp 79-92, 2000.
[15] O. Deux et al. "The O_2 System," Communication of the ACM, Vol. 34, No. 10, pp 34-48, 1991.
[16] G. Salton and M. McGill, "An Introduction to Modern Information Retrieval," McGraw-Hill, 1983.
[17] U. Westermann and W. Klas, "An Analysis of XML Database Solutions for the Management of MPEG-7 Media Descriptions," ACM Computing Surveys. Vol. 35, No. 4, pp 331-373, 2003.

WiCTP: A Token-Based Access Control Protocol for Wireless Networks

Raal Goff and Amitava Datta

School of Computer Science & Software Engineering,
University of Western Australia, 35 Stirling Highway,
Perth, WA 6009, Australia

Abstract. The introduction and growth of large, metropolitan-sized wireless networks has identified a significant problem within the current 802.11 protocol specification, hidden nodes. Hidden nodes dramatically decrease performance and throughput of networks where they are present, and the current solution, RTS/CTS leaves much room for improvement. This paper proposes a new, token-based protocol called WiCTP that can be implemented on top of standard CSMA/CA, but which enables any higher level protocol to be used with it. WiCTP successfully improves network stability and throughput in wireless networks where hidden nodes may be present. The overhead of the WiCTP protocol in terms of the overall bandwidth used is small. We discuss an implementation of the WiCTP protocol as a Linux kernel module and experimental results from its implementation in a metropolitan-sized wireless network.

1 Introduction

Since the introduction of the Wireless Fidelity (WiFi) standard in 1997, the use of wireless networks has increased dramatically. While initially costly and having a somewhat slow transfer rate of 2mb/sec, later improvements of 802.11b (the WiFi standard) increased the transfer rate to 11mbps. Further improvements were made with the introduction of 802.11a and 802.11g, increasing the transfer rate again to 54mbps. Greater market penetration and demand also saw a drop in consumer and enterprise level equipment for it to become a viable alternative for wired Ethernet.

The use of a wireless network instead of a wired network results in a greater flexibility for the placement of workstations in offices and the home, since no cabling is needed to connect a computer to the network. Wireless cards have also been incorporated into laptops in order to give the user greater flexibility in where they can connect to the network from. These, and many other uses of the wireless standard are exactly what the protocol was designed for, allowing a greater flexibility in the way computers connect to a local network.

1.1 The Hidden Node Problem

The *hidden node* problem is present due to the collision detection scheme implemented in the WiFi standards. WiFi uses Collision Sense Multiple Access with Collision Avoidance (CSMA/CA) as a means to control access to the medium. The hidden node problem arises when one or more nodes on the network cannot determine if another node is currently transferring, because it is outside of its transmission range. These nodes are considered to be hidden from each other, and are termed hidden nodes.

In standard WiFi set-ups, all nodes are usually within transmission range of each other. There may be a few outlying nodes that are hidden from the rest of the network, but the majority of nodes are able to detect whether the medium is in use by other nodes. In these situations, the few hidden nodes that are present, do not impact the performance of the network considerably. However, when the range of a WiFi network is extended beyond its initial design, such as the case with public Freenets, and long-range connections, it is the common case for many nodes to be hidden from one another.

It is easy to see that with this process continuing, the number of collisions on the network increases dramatically, since each node continually believes that the medium is free, and will transfer their packet after each wait period. If each node continues to transfer, more collisions occur and so on, until the network degrades to a point where most protocols will fail.

Khurana *et al.* [3] have presented results detailing exactly how the presence of hidden nodes affects wireless networks. When approximately 10% of nodes are within the transmission range of one another the wireless LAN can still offer reasonable performance, with a throughput of around 65%. However, when the number of hidden nodes climbs above 10%, the network performance degrades dramatically. When the number of hidden nodes climbs to 30%, the network performance drops to only 22% throughput.

1.2 Current Solutions

Currently, the only widespread method of combating this problem is the Request to Send/Clear to Send (RTS/CTS) protocol. The RTS/CTS protocol allows nodes to reserve the medium for a short amount of time so that they may send their data collision-free. The RTS/CTS protocol works well in situations where most nodes are close to each other, and can receive any RTS/CTS packet sent. If only one or two nodes are outside of the main group of nodes, RTS/CTS works well to combat the hidden node problem in this situation [3], but, just like CSMA/CA, the protocol breaks down when most, if not all, nodes on a wireless network are hidden from each other, as is the case in Freenets. RTS/CTS still uses CSMA/CA to determine if the medium is free, and if a node can transfer data.

The rest of the paper is organized as follows. We discuss some previous work on token-based solutions for avoiding the hidden node problem in Section 2. We discuss our WiCTP protocol and its implementation in Section 3. The experimental results are presented in Section 4.

2 Previous Token-Based Solutions

In this paper, we propose a token-based access control mechanism to avoid collisions. The idea of a token-based access control protocol is not new. Other protocols that implement a similar access control mechanism include the open source projects Frottle [4] and the Wireless Central Coordinated Protocol (WiCCP) [1].

2.1 Frottle

Frottle is a Linux-only implementation of a token-based access control mechanism. It currently relies on the Linux kernels iptables' packet filtering abilities to control access to the network, and uses the TCP/IP stack to communicate between the master and clients. It runs as a user space application and utilizes the iptables QUEUE rule set to queue packets for transfer.

The Frottle package communicates between master and node via TCP/IP port 999. This is a major disadvantage of the package, since its reliance on both TCP and iptables limits its portability enormously. Without iptables, Frottle has no effective way of controlling packet queuing, and without a TCP/IP stack, it has no way of communicating between nodes. For any embedded system wanting to run Frottle, they will need both the TCP/IP stack and the iptables modules compiled into the kernel, increasing the size dramatically. While TCP/IP would probably be compiled into the system anyway, iptables can increase the required size by many kilobytes, a serious concern for any embedded system.

The Frottle package also has a weakness in that it allows unrestricted access on port 999. If another server runs a service on port 999, a user who is part of the Frottle ring can circumvent the token access control and directly access the service. This will result in a dramatic drop in performance and the re-appearance of the hidden node problem in the network if the service is heavily used. Frottle also has no auto-detection of master nodes.

2.2 WiCCP

An alternative to Frottle is the Wireless Central Co-ordinated Protocol (WiCCP). This package, unlike Frottle, is a kernel-level implementation of a cyclic token passing protocol. WiCCP is implemented as a Linux Kernel Module (LKM) and thus has no reliance on a TCP/IP stack or iptables. WiCCP addresses most of the problems associated with Frottle. It has no reliance on TCP/IP or iptables and automatically discovers master nodes and associates with them. WiCCP also encapsulates higher-level protocols in its own custom protocol, ensuring that any higher-level protocol can be used with it. Since it uses its own low-level protocol to communicate, masters and clients need not be on the same network subnet. Nodes only need to be connected to the same access point in order to receive the benefits of the WiCCP access control. This allows a far greater level of flexibility in network set-ups than any Frottle set up could achieve.

However, anecdotal evidence suggests that WiCCP may be much slower than Frottle, and have a higher level of packet loss in low quality connections. Another inefficiency of WiCCP is how new nodes associate with a master. The WiCCP specifies that at given intervals (the specification suggests 5 seconds), the master sends a Free Contention Period (FCP) packet. This packet is a signal for any un-associated nodes to send their association packet to the master. Thus, any node wishing to associate with a master has to wait for the FCP to arrive before being able to join the network proper.

3 Our WiCTP Protocol

The new protocol implements a token-passing access control mechanism, much like the mechanism implemented by both Frottle and WiCCP, called WiCTP. WiCTP works by passing a token around to each node on the network. Only the node that is in possession of the token may transfer data, thus only one node may transfer at any one time, eliminating collisions entirely. Each node on the network is classed as either a master or a slave. The master node has knowledge about every other node on the network, but each slave only knows about the master. The master node controls all access control to the network, since it handles the distribution of the token to each node.

This allows the master node a great amount of flexibility in its control of the network. It can re-order the passing of the token to better utilize the network, or to give preferential treatment to certain nodes. This allows the master to ensure that nodes that normally would be relegated to low performance will get a fair share of network bandwidth. While not implemented in this paper, the new protocol supports the integration of different Quality of Service (QoS) algorithms.

When a node connects to the wireless network, it will send a packet announcing its presence to the network. When the master receives this packet, it replies to the announce packet with data about the master to which it has been associated, making the node a slave. The master adds the new slave's information to its list of current slaves and places it in the queue for receiving the token. When the slave receives the token, it checks to see if the token is from the master it has been associated with. If it is, it will check if it has any data that needs to be sent. If there is data to be sent, it will examine the token to see how many data packets it is allowed to send in its time slot. The slave will then work through its queue of data packets, sending them to the recipient. When the slave has no more data packets to send, or it has reached the maximum number of packets it may send, it sends the token back to the master. Since only the slave that has the packet can transmit data, the hidden node problem is eliminated. No two nodes can transfer at the same time, so collisions are eliminated entirely.

3.1 Slave Operation

Nodes operating under the new protocol are classified as either a master or a slave. Each network has only 1 master, which controls the token packet for the network. The implementation of the new protocol for this paper has been built as a Linux Kernel Module (LKM) and resides between the Data Link Layer and the Network Layer. The LKM created for this project creates a new interface that attaches to an already existing interface. This new interface is called wethx, where x is the interface number (e.g, attaching to eth2 would create an interface called weth2).

The new interface operates exactly the same way as any other interface, except any packets sent out across the new interface are encapsulated in the new protocol. The new interface can be associated with an IPv4, IPv6, or any other protocol address just like an ordinary interface. Aliases and other such functions will also work with the new interface.

Whether the new interface is used to send data or not is usually determined by the routing table of the computer, but there may be instances where the user-level program explicitly specifies the new interface as the network interface that is used. As far as any user-space program is concerned, the new interface is simply another Ethernet interface that is available for use.

The LKM also registers several packet handlers for the following packets: PACKET_ANNOUNCE, PACKET_ACK, PACKET_TOKEN, PACKET_DATA.

The PACKET_ANNOUNCE packet is simply a broadcast packet containing the slave node's MAC address. The master only needs the slave's MAC address initially, so there is no reason to add any additional information. When it receives a PACKET_ANNOUNCE, the master will add the slave's MAC address to its list of slaves, if it is not already there. The master does not send any kind of acknowledgment to the slave indicating whether its announcement has been successful. A slave only knows that it has become part of the token ring when it receives its first token.

The *data transfer amount* value holds the number of packets a slave is allowed to send in this period. The slave can transfer until this value is reached, or it runs out of data to send. The *number of nodes* value is the current number of nodes in the ring. This enables the slave to estimate how long it will be until it receives the token again. This time value can be used to put the wireless card into energy saving mode until it is due to receive the token again, and can also be used to determine if the master node has crashed. If the slave node does not receive the token again within the calculated time, it starts announcing again to associate with a new master.

Once a slave has received a token, it checks the ID against the last ID it received. If it has changed, the node knows that the previous token was lost somewhere, and that the network could be temporarily unstable. After checking the ID, the slave checks its data queue to see if there are any packets that need to be delivered. If there are packets to deliver, the slave sends back a PACKET_ACK to the master, and then starts transferring the packets.

The acknowledgment packet signals to the master node that the slave has data to transfer, and that the master should alter its timeout values to wait for it to send all its data and return the token. If the slave has no data in its queue, has finished transferring all its data, or has reached the maximum number of data packets to send, it will send the token back to the master node in a specified format to indicate this.

3.2 Master Node Operation

The master node's operation is similar to a slave node with some notable exceptions. The master node LKM creates a new interface like the slave LKM, but it registers more packet handlers, namely, PACKET_ANNOUNCE, PACKET_ACK, PACKET_TOKEN and PACKET_DATA.

The master node registers PACKET_ACK and PACKET_ANNOUNCE in addition to the two registered with a slave. Once a master has initialized its new network interface and its packet handlers, it will initialize the ring ID and create an empty list of nodes to send the token to. It will then add itself to the list as a node so that it may transfer through the interface immediately. When a master receives a PACKET_ANNOUNCE packet, it will extract the source MAC address from the data packet and add the node to its list. It also initializes some additional data fields such as if the node has packets waiting to be sent, and the node's status.

When a master sends the token to a slave node, it first initializes some timers so that it may recover if the token is lost somehow. The master initializes two timers, a token timer and an acknowledgment timer. The master will initialize both the token timer and the acknowledgment timer, with the acknowledgment timer having a smaller timeout value than the token timer, and then send the token to the slave node with the initialized ring ID. If the slave node replies with either a PACKET_ACK or a PACKET_TOKEN, the acknowledgment timer is disabled. The token timer is only disabled if a PACKET_TOKEN is received.

If either of the timers expires the master will take steps to recover from losing the token. If the acknowledgment timer expires, this indicates to the master that the slave node to which it was addressed has died, or otherwise become unable to send acknowledgments to the master. To recover, the master will mark the slave node as dead in its node list and increment the ring ID. The master will then ignore this node as it moves through the node list. If the token timer expires, this indicates to the master that the slave node has either died, or been somehow delayed in its processing of data packets. In this case the master will mark the node as potentially dead in its node list and increment the ring ID. It will continue to send this potentially dead node tokens since another timeout will mark the node as either dead, if it fails to return the acknowledgment, or alive if it returns the token. If the node returns an acknowledgment packet, and then does not return the token for a second time, causing another token timeout, the node is marked as dead.

The ring ID is incremented in the case of a timeout so that it will not become confused if the slave node resumes normal operations. The master could have sent the token to a slave node, and the slave node then got delayed in its processing of the packet, causing timeouts on the master node. The master would then continue and send a new token to the next node in the list. If the initial slave node then processed its token packet, it would send its data and then attempt to return the token to the master node. Incrementing the ring ID enables the master to determine that the slave token was simply delayed instead of down, and can alter the node list accordingly. It will also be able to determine that the token returned was old, and thus should not be forwarded again to other nodes.

When a master node receives a token back from a slave node, it will check the source MAC address to make sure it came from one of its known nodes. If it is, and the ring ID contained in the packet matches the current ring ID, the master will disable both the acknowledgment and the token timers. After extracting the number of packets queued on the slave node it will store it in the node information. The master will then examine its node list to determine where to send the token next, initialize timers again, and send it.

4 Experimental Results

In order to test how well the new protocol performs against the standard CSMA/CA access control method, it was loaded onto several computers that were part of LWN. LWN is a community wireless network situated in Leeming, Perth, Australia, with client nodes connecting from the surrounding suburbs. LWN consists of approximately six wireless nodes connected to a central access point that is connected to a 14dB

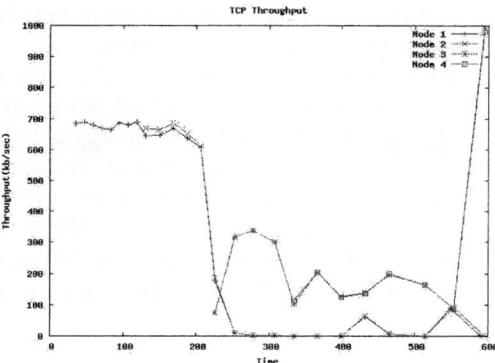

Fig. 1. Combined TCP Throughput using CSMA

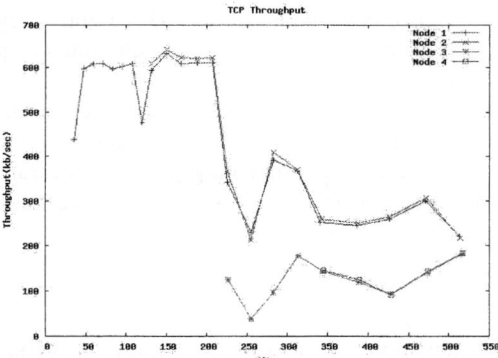

Fig. 2. Combined TCP Throughput using WiCTP

omni-directional antenna situated on top of a 10-metre mast. The client nodes each connect using a 24dB directional antenna and a client wireless card. The network uses the 802.11b WiFi standard to communicate, giving the network a maximum theoretical throughput of 11Mbps and a realistic maximum throughput of approximately 500KBps. Each node's card has a maximum transmission power of 15dBm (31.62mW). Each node on the network is anywhere between 500m to 2km from the access point, hence all nodes are hidden from each other, making the network ideal for testing the new protocol. In order for each node to maximize its transfer rates, the network benchmark tool netperf was run on each node.

The main set of results gathered by the tests are shown in Fig. 1 and 2. These figures show the sum of both RX (receive) and TX (transmit) transfer rates of each node. While the graphs indicate that 2 nodes are transferring at 700k/sec each, a figure far above the theoretical maximum of the network, nodes are actually transferring between one another. Thus Node 1 may be sending data to Node 2 at 350k/sec, and Node 2 may be sending data to node 1 at 350k/sec. Hence when the sum of both RX and TX rates are calculated, both nodes have a total transfer rate of 700k/sec.

The nodes used for the tests displayed in Fig. 1 and 2 were not all hidden from each other. Unfortunately some of the members of LWN moved away or otherwise could not participate in the experiments, thus, the set-up was as follows, node 1 was the central access point, nodes 2-4 were three laptops. This set-up used netsync to gradually increase the network load and return statistics on the nodes' performance. In this test Node 1 started to transfer to Node 2 until approximately 100 seconds, where Node 2 would start transferring back to Node 1. At Approximately 200 seconds, Node 3 would start transferring to Node 4, and then at 400 seconds, Node 4 would transfer back to node 3.

4.1 TCP Throughput

Nodes 2 and 4 were not hidden form each other, but this does not seriously undermine the results. Fig. 1 demonstrates how CSMA/CA performs in a 4 Node network. Nodes 1 and 2 sustain a high level of throughput, until Nodes 3 and 4 begin to transfer. Once nodes 3 and 4 begin to transfer, the throughput of each node drops significantly, indicating how badly hidden nodes affect the network.

Fig. 2 shows how WiCTP performed under the same set-up as Fig. 1. WiCTP clearly gives a much better level of stability. Nodes 1 and 2 clearly sustain maximum transfer rates until Nodes 3 and 4 begin transferring, where the transfer rate drops off to allow the new nodes to begin transferring. The bandwidth is efficiently shared between all nodes.

4.2 Ping Times

Fig. 3 and 4 show the ICMP Ping reply time for each node as the network load gradually increased. Fig. 3 shows the ping times for the 4 nodes under WiCTP. The ping time gradually increases as more and more data is transferred. Fig. 4 shows the ping

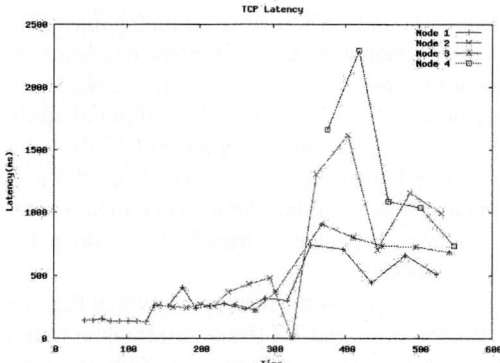

Fig. 3. Ping time for WiCTP

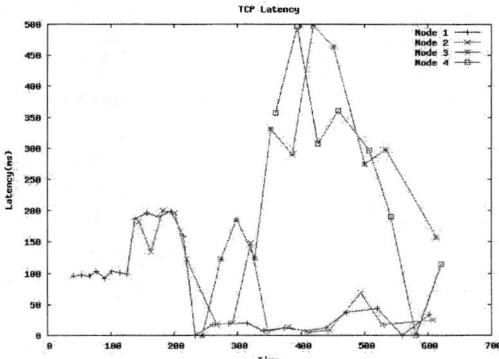

Fig. 4. Ping time for CSMA/CA

times under CSMA/CA, which also increase as more data is transferred. The WiCTP clearly has greater ping times than CSMA/CA with times in excess of 2000ms under the greatest load. Even the lowest ping time for WiCTP, approximately 100ms, is very high, however the CSMA/CA test shows that it does not have much lower ping times either. Under higher load however, the CSMA/CA protocol clearly has a better ping time, with a maximum of 500ms, compared to WiCTP's 2000ms.

4.3 Error Rates

Fig. 5 and 6 show the error rates of each node during the tests. Fig. 5 shows the error rates during the experiment for the nodes when they are running WiCTP and Fig. 6 shows the error rate when the nodes are using CSMA/CA. Each of the two scenarios shows an increase in errors as more nodes start to transfer at once, however it's clear that under WiCTP, the nodes have significantly less transfer errors than under CSMA/CA.

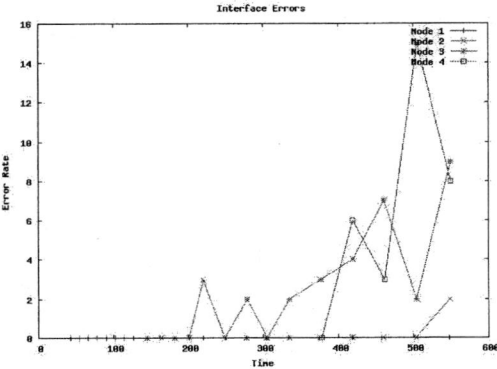

Fig. 5. Error Rates for WiCTP

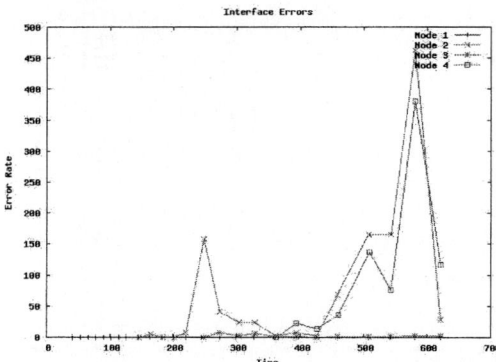

Fig. 6. Error Rates for CSMA/CA

As more nodes begin to transfer the error rate jumps from negligible amounts of fewer than 50 errors per sample to close to 500 errors per sample under CSMA/CA. Under WiCTP the error rate is kept at negligible levels, and has a maximum of fewer than 16 errors per sample even under the highest load.

References

1. Wireless central coordinated protocol. http://www.patraswireless.net/software.html, 2003.
2. Jason Jones. e3 :: blogging the wireless freenet. http://www.e3.com.au.
3. S. Khurana, A. Kahol, and A. P. Jayasumana. Effect of hidden terminals on the performance of ieee 802.11 mac protocol. In *Local Computer Networks, 1998. LCN '98. Proceedings., 23rd Annual Conference on*, pages 12–20, 1998.
4. Chris King. Frottle: Packet scheduling and qos for wireless networks. http://frottle.sf.net/.

An Optimized Internetworking Strategy of MANET and WLAN*

Hyewon K. Lee and Youngsong Mun

School of Computing, Soongsil University, Seoul, Korea
kerenlee@sunny.ssu.ac.kr, mun@computing.ssu.ac.kr

Abstract. In this paper, we propose internetworking strategy of WLAN and MANET. Both WLAN and MANET are based on IEEE 802.11 for MAC layer protocol while they adopt different mode, such as infrastructure and ad-hoc. To support internetworking between different IEEE 802.11 modes, switching algorithm on mobile node is proposed. This algorithm is mainly affected by user's preference, not only by signal strength. For MANET routing protocol, OLSR is employed, which is a proactive protocol and has optimally reduced signal broadcasting overhead. The effect of distances between a mobile node and MANET Gateway which is a link to WLAN is proved when the mobile node resides on MANET. Also, latencies between handoff from WLAN to MANET and handoff from MANET to WLAN are compared.

1 Introduction

The IEEE 802.11 specifies two different modes, such as infrastructure mode and ad-hoc mode [1]. MIPv4 and MIPv6 have been proposed to provide mobility service for infrastructure mode while mobile ad-hoc network (MANET) supports a multi-hop wireless network without any prepared base station (BS). In this paper, we propose internetworking strategy of WLAN and MANET. To interconnect two different modes, mode switching algorithm is implemented on mobile node. Proposed switching algorithm is strongly affected by user's preference when two different MAC frames with valid signal strength are received. Among many routing protocols in MANET, OLSR (optimized link state routing) is employed. OLSR is proactive routing protocol and reduces broadcasting overhead using MPRs (multipoint relays). We have proved the effect of distances between mobile node and MANET Gateway which is a link to WLAN when a mobile node is away from home link and resides on MANET. Also, latencies between handoff from WLAN to MANET and handoff from MANET WLAN are compared.

2 Related Works

2.1 IEEE 802.11

Networks in IEEE 802.11 are classified into two structures, such as infrastructure and ad hoc [2]. In infrastructured structure, access point (AP) is easily found in a network,

---
* This work was supported by the Korea Research Foundation Grant. (KRF-2004-005-D00147)

as shown Fig. 1. Basic unit for infrastructured structure is Basic Service Set (BSS), where station group communicates each other with no help from AP or others. The Extended Service Set (ESS) is the set of BSS. The ad-hoc network is Independent Basic Service Set (IBSS), where no AP is found, and mobile stations communicate each other as peer-to-peer.

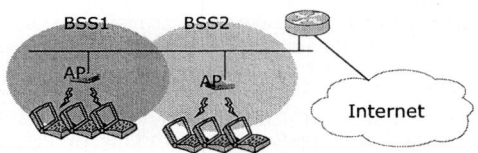

Fig. 1. Simplified structure for IEEE 802.11

2.1 OLSR (Optimized Link State Routing)

OLSR is optimized link state algorithm for mobile ad hoc networks. OLSR belongs to a proactive routing protocol [3] and helps nodes to keep all routing information in network. The unique point of OLSR is the set of selected nodes to forward broadcast control message, MPRs. A node in network selects MPRs among its one hop neighbors, and only MPRs flood control traffic. OSLR employs such hierarchical model to configure a network, which minimizes the overhead from flooding of control traffic to all nodes in the network. Basic requirement for MPR selection is whether there is bi-directional link between a given node and MPR, which avoids problems from data transfer over uni-directional link. To build the route from a given node to any destination in the network, the MPRs are also used [4].

In OLSR, three message types are defined; HNA, Hello and Topology Control (TC) messages. Main functions of OLSR are summarized as neighbor discovery and topology dissemination. For neighbor discovery, each node in network exchanges Hello message periodically. This Hello message contains the list of neighbors. The Hello message is only processed within all one-hop neighbors, not forwarded to others. The Hello message enables every node to discover one-hop and two-hop neighbors. MANET Gateway periodically broadcasts HNA message which contains prefix information for address configuration and subnet information where the Gateway belongs. This message will notify other nodes in network that the Gateway is the door to public network.

2.2 Mobile IPv6 Handoff

The mobile IPv6 protocol provides mobility service to mobile nodes in Internet [5]. When a mobile node exists on home link, communications with other nodes are performed via common routing mechanism, but when a mobile node moves into a foreign network, handover between different networks should be followed for seamless connection with address registration. The care-of address identifies mobile node's current location, and it is obtained via normal IPv6 automatic configuration mechanism. Generally, L2 and L3 handoffs are considered in MIPv6. L2 handoff is a process which a mobile node changes its physical link layer connection, AP, to another. The change of AP followed by change of access router (AR) leads to L3 handoff.

2.3 Integrating WLAN and MANETs to the IPv6 Based Internet

In [6], mode detection and switching algorithm to integrate MANET to public WLAN is proposed. This algorithm automatically enables a mobile node to handoff from WLAN to MANET and vice versa. OLSR and MIPv6 protocols are employed for MANET and WLAN, respectively. The weak point of the proposed algorithm in [6] is that it only considers signal strength for mode switching. If a mobile node boots or reboots, and it sensors both IEEE 802.11 MAC frames, it may prefer one mode to the other. In this case, the algorithm just selects one mode depending on signal strength, not on user's preference. Besides, [6] assumes that home agent and MANET Gateway reside on the same subnet. We need to consider the other situation; such home agent is several-hop distant from MANET Gateway, as well as 1-hop.

3 Proposed Architecture

In this paper, we propose integrated architecture of MANET and MIPv6. When a mobile node moves and enters into dead spot, where it sensors only ad-hoc network, not public network, connection to public network is supported by MANET Gateway. The mobile node is able to listen to any IEEE 802.11 frame format, but no standards for handoff between ad-hoc and infrastructure are present while handoffs between ad-hoc modes or between infrastructure modes are present. For internetworking between MANET and MIPv6, handoff mechanism between different modes is required.

MIPv6 protocol is employed for mobile devices within public network, and OLSR is employed in MANET. The OLSR protocol holds all networks routing information, and optimization is performed via MPRs, which is very good for large and dense networks. OLSR accepts original IP packet format, and no change is required, so OLSR seems to be very adequate for integration of MANET and MIPv6.

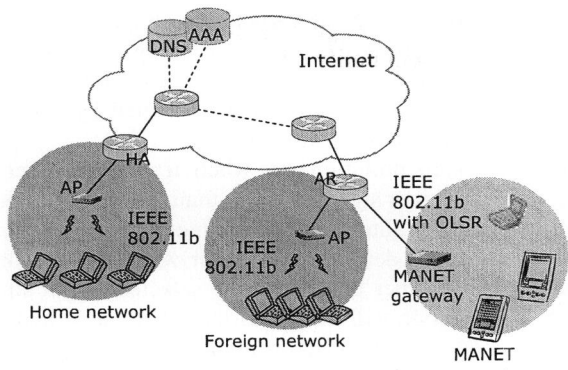

Fig. 2. Internetworking between WLAN and MANET

Proposed architecture is depicted in Fig. 2. The network is divided into three parts; WLANs based on MIPv6, MANET and Internet, mainly wired public networks. As shown in Fig. 2, WLAN is directly connected to public networks, while MANET is connected to public networks via MANET Gateway. The MANET Gateway is a link between MANET and WLAN. In OLSR-based MANET, two new node types are defined [3], as follows:

- MANET Gateway: A static node provides connection to nodes in MANET into WLAN. The MANET Gateway periodically broadcasts HNA message into network. Its main function is to notify foreign nodes in MANET of outside subnet information where it is attached and prefix information.

- MANET Router: General OLSR-enabled nodes are simply called MANET Router. MANET Router may be assigned with more than one address.

From MIPv6's angle, MANET is regarded as foreign network. Once a mobile node enters pollution area, where it receives different kinds of IEEE 802.11 MAC frames with higher signal strength than threshold, it may keep its current network connection or change into the different mode, depending on the switching algorithm specified below. As the mobile node determines to change mode, new care-of address is required. A periodical HNA message from MANET Gateway informs the mobile node of prefix information and its subnet routing information. From this message, the mobile node knows the MANET Gateway is the door to Internet. Now, the mobile node builds care-of address, and starts registration process with home agent. Besides, this mobile node will join in MANET. The mobile node exchanges Hello message with its neighbored nodes. When the mobile node keeps moving and goes away from the MANET Gateway by several hops, packets originated from the mobile node and destined to some node out of MANET will be delivered to MANET Gateway using OLSR, and the Gateway will forward it to AR using MIPv6 protocol.

3.1 Mode Switching

The handoff between different IEEE 802.11 modes is actively under researches. Many alternatives have been proposed, such as implementation on hardware or software. In this paper, we propose enhanced mode switching algorithm between different IEEE 802.11 modes, originated from [6]. The mode switching is strongly affected by user's own decision, which prevents frequent mode change between WLAN and MANET and reflects user preference, while [6] is simply affected by signal strength when a mobile node gets different mode MAC frames.

When a mobile node enters pollution area where it receives different kinds of IEEE 802.11 MAC frames with higher signal strength than threshold, it may keep its current network connection or change into the different mode. The switching between infrastructure and ad-hoc modes is determined by user whether he wants to keep current connection or handoff to another one. Once the selected signal strength drops from the threshold, the mobile node will switch to the other mode and wait for adequate beacon message. The mode switching algorithm is shown in Fig. 3.

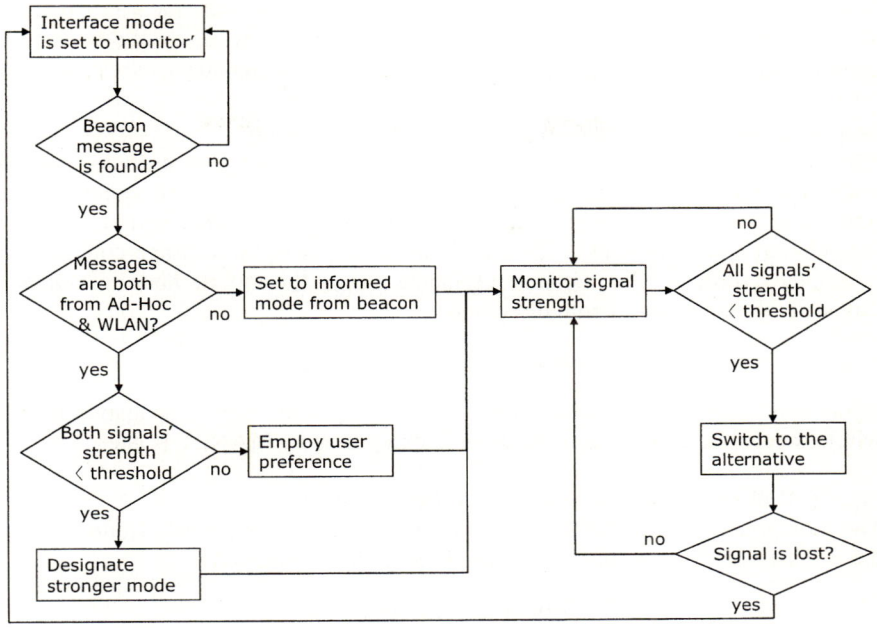

Fig. 3. Algorithm for mode switching

3.2 Handoff Operation from WLAN to MANET

As a mobile node enters into MANET network, it sensors different IEEE 802.11 mode and switches into IEEE 802.11 ad-hoc mode depending on algorithm shown in Fig. 3. Once the mobile node decides to handoff to MANET, the node should be the member of MANET, at first. At this stage, Hello messages are exchanged between one-hop distanced nodes, and MPR selection and Topology Dissemination are followed. Then, routing table of MANET is computed by each node in MANET.

After joining the MANET, the mobile node will start address configuration using HNA message from MANET Gateway. This HNA message also contains routing information about the subnet where the MANET Gateway is attached. Once the mobile builds new care-of address, and it starts home registration with home agent to notify node's new current address.

Total handoff time is expressed with the sum of time to handle signaling message for handoff by each intermediate nodes or end nodes, as shown in (1), where T_{iMj} is required time to handle message, M_j, at each i step on handoff. Retransmission due to link error is not considered [8].

$$T_s = \sum \sum T_{iM_j}. \tag{1}$$

Step 1: Mode Detection and Switching
A mobile node moving in WLAN decides to handoff to MANET and starts L1/L2 handoff.

$$T_1 = \gamma_{MN} + T_{wlan}, \qquad (2)$$

where γ_{MN} is processing time of mobile node, and T_{wlan} is channel access latency [9].

Step 2: Hello Message Exchange for Neighbor Discovery
After L1/L2 handoff, the mobile node is required to configure itself as a MANET node. The mobile node starts to send empty Hello message to its neighbor to learn link states of neighbors. This Hello message is only forwarded within 1-hop node. Neighbored nodes respond to the mobile node with a Hello message containing mobile node's address. This message exchange allows the mobile node to validate the link state of neighbors, in both directions.

$$T_2 = 2(\alpha + \beta_{wireless}) + \gamma_{MN} + \gamma_{MR}, \qquad (3)$$

where α is message transmission time [10], $\beta_{wireless}$ is message propagation time over wireless network, and γ_{MR} is message processing time by MANET Router.

Step 3: Hello Message Exchange for MPR Selection
The mobile node selects MPRs among neighbors' replies and sends Hello message to them. These MPRs send back Hello message to the mobile node.

$$T_3 = 2(\alpha + \beta_{wireless}) + \gamma_{MN} + \gamma_{MR}. \qquad (4)$$

Step 4: Topology Dissemination and Routing Table Computation
The MPRs broadcast TC message to all nodes in MANET. This message notifies other nodes that the MPRs are the final hop in a route to the mobile node.

$$T_4 = \alpha + \beta_{wireless} + \gamma_{MN}. \qquad (5)$$

Step 5: HNA Message Reception and Address Configuration
The MANET Gateway sends HNA message periodically, which contains prefix information and routing information about its attached subnet.

$$T_5 = n(\alpha + \beta_{wireless}) + (n-1)\gamma_{MR} + \gamma_{MN}, \qquad (6)$$

where n is the number of hop between MANET Gateway and the mobile node

Step 6: Home Registration
To inform mobile node's new location of home agent, Binding Update (BU) and Binding Acknowledgement (BA) messages are exchanged between mobile node and home agent via MANET Gateway.

$$T_6 = 2\{(n+m)\alpha + n\beta_{wireless} + m\beta_{wired} + (n-1)\gamma_{MR} + (m-1)\gamma_{AR} + \gamma_{MG}\} + \gamma_{HA} + \gamma_{MN}, \qquad (7)$$

where m is the number of hop between MANET Gateway and home agent, β_{wired} is message propagation time over wired network, γ_{AR} is message processing time by Access Router, and γ_{HA} is message processing time by home agent.

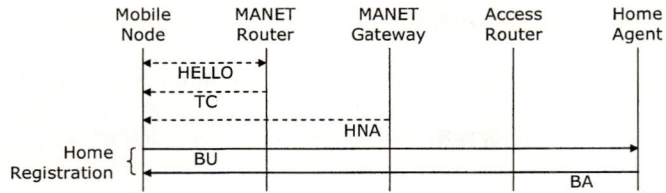

Fig. 4. Mobile node handoff scenario from WLAN to MANET

3.3 Handoff Operation from MANET to WLAN

When a mobile node comes back to WLAN, it sensors different IEEE 802.11 mode and switches into IEEE 802.11 infrastructure mode depending on algorithm shown in Fig. 3. Once the mobile node decides to handoff to WLAN, new care-of address acquisition and registration processes are required. Specific handoff procedure is as follows:

Step 1: Mode Detection and Switching
A mobile node decides to handoff to WLAN, and starts L1/L2 handoff. The latency is as same as the first stage of handoff from WLAN to MANET.

Step 2: Router Discovery
When the mobile node comes into WLAN, it may receive periodical Router Advertisement (RA) from AP, or it sends Router Solicitation (RS) to trigger RA. If the mobile node solicits RA message, the delay at this stage is as follows:

$$T_2 = 2(\alpha + \beta_{wireless}) + \gamma_{AP} + \gamma_{MN}, \tag{8}$$

where γ_{AP} is message processing time by AP.

Step 3: Home Registration
To inform mobile node's new location information of home agent, Binding Update (BU) and Binding Acknowledgement (BA) are exchanged between mobile node and home agent.

$$T_3 = 2\{(m+1)\alpha + \beta_{wireless} + \gamma_{AP} + m\beta_{wired} + (m-1)\gamma_{AR}\} + \gamma_{HA} + \gamma_{MN}. \tag{9}$$

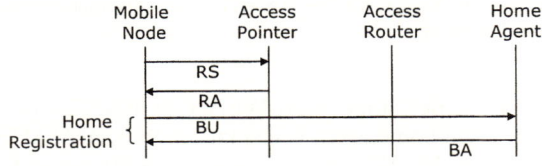

Fig. 5. Mobile node handoff scenario from MANET to WLAN

4 Performance Evaluation

Total handoff time is yielded in the former section through numerical analysis. System parameters specified in Table 1 is employed to get T_i and T_S. Table 1 specifies system parameters based on IEEE 802.11b for the experiment of proposed model.

Table 1. System parameters, based on IEEE 802.11b

Input Parameter	Values	Input Parameter	Values
Traffic type	UDP	Message Size	100 bytes
Link Parameters			
Wireless Link Data Rate	2Mbps	Transmission time (α)	0.4ms
Propagation time (β)			
wired link (β_{wired})	0.12ms	wireless link ($\beta_{wireless}$)	1μs
Processing Time(γ)			
$\gamma_{MN}, \gamma_{MR}, \gamma_{MG}, \gamma_{AP}$	15ms	γ_{AR}, γ_{HA}	5ms

In Fig. 6, the effect of distances between mobile node and MANET Gateway, and MANET Gateway and home agent is depicted when a mobile node is away from home link and resides on MANET. Distance between MANET Gateway and mobile node is fixed as n, and distance between MANET Gateway and home agent varies from 1 to 8. As shown in Fig. 6, handoff time is not strongly affected from m, but affected from n.

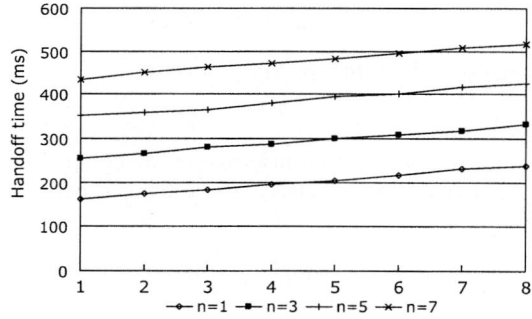

Fig. 6. Handoff latency from WLAN to MANET when distance between mobile node and MANET Gateway is fixed as n

Fig. 7 shows that handoff time from WLAN to MANET is very different from MANET to WLAN when home agent is distant 3-hop from MANET Gateway. When handoff from WLAN to MANET occurs, join to OLSR network is required before address registration with home agent, which mainly causes high handoff time.

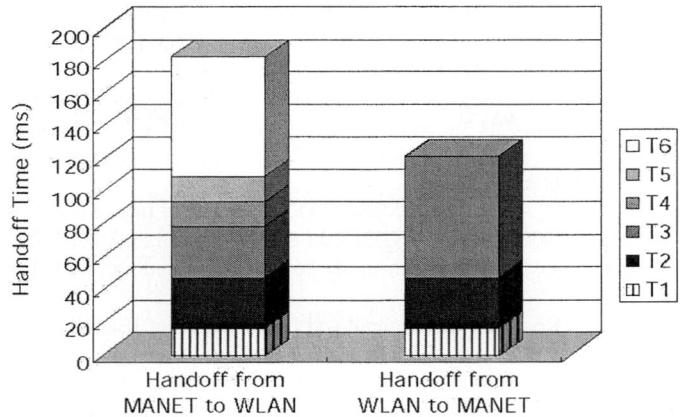

Fig. 7. Handoff latency from MANET to WLAN and from WLAN to MANET, respectively, when home agent is distant 3-hop from MANET Gateway. In case of handoff from WLAN to MANET, mobile node is assumed to be one-hop distant from MANET Gateway.

5 Conclusions

In this paper, we propose internetworking strategy of WLAN and MANET, which adopt a different IEEE 802.11 mode, respectively. For internetworking between them, enhanced switching algorithm is proposed. This switching algorithm is strongly affected by user's preference when two different MAC frames with valid signal strength are received. OLSR protocol is employed for MANET routing protocol, which elegantly cooperates with current MIPv6 protocol. We have proved that distance between MANET Gateway and mobile node strongly affects to handoff time rather than distance between MANET Gateway and home agent. Also, latencies between handoff from WLAN to MANET and handoff from MANET to WLAN are compared, where latency on handoff from WLAN to MANET is remarkably higher than the other due to joining process to MANET.

References

1. Information technology-Telecommunications and information exchange between systems-Local and metropolitan area networks-Specific requirement- Part 11: IEEE Wireless LAN Medium Access Control (MAC) and Physical Layer (PHY) Specifications, ANS/IEEE Std 802.11, 1999 Edition
2. M. Gast, "802.11 Wireless Networks: The Definitive Guide," O'Reilly, April 2002
3. T. Clausen, P. Jacquet, "Optimized Link State Routing Protocol (OLSR)," RFC 3626, October 2003
4. Y. Ge, "Quality-of-Service Routing in Ad-Hoc Networks Using OLSR," master thesis, December 2002
5. D. Johnson, C. Perkins and J. Arkko, "Mobility Support in IPv6," RFC 3775, June 2004

6. L. Lamont, M. Wang, L. Villasenor, T. Randhawa, S. Hardy, "Integrating WLANs & MANETs to the IPv6 based Internet," Communications, 2003. ICC'03. IEEE International Conference on , Volume: 2, May 2003
7. L. Lamont, M. Wang, L. Villasenor, T. Randhawa, R. Hardy, P. McConnel, "An IPv6 and OLSR based architecture for integrating WLANs and MANETs to the Internet," Wireless Personal Multimedia Communications, 2002. The 5th International Symposium on , Volume: 2 , 27-30 Oct. 2002
8. Janise McNair, Ian F. Akyildiz and Michael D. Bender, "An Inter-System Handoff Technique for the IMT-2000 System," INFOCOM 2000. Nineteenth Annual Joint Conference of the IEEE Computer and Communications Societies. Proceedings, Vol. 1, pp. 208 - 216, 26-30 March 2000
9. Yang Xino and Jon Rosdahl, "Throughput and Delay Limits of IEEE 802.11," IEEE Communications Letters, Vol. 6, pp. 355-357, August 2002
10. N. Efthymiou, Y. Fun Hu and Ray E. Sheriff, "Performance of Intersegment Handover Protocols in an Integrated Space/Terrestrial -UMTS Environment," IEEE Transactions on Vehicular Technology, Vol. 47, pp. 1179 - 1199, November 1998

An Internetworking Scheme for UMTS/WLAN Mobile Networks*

Sangjoon Park[1], Youngchul Kim[1], and Jongchan Lee[2]

[1] Information & Media Technology Institute, Soongsil University
[2] Department of Computer Information Science, Kunsan National University
lubimia@archi.ssu.ac.kr

Abstract. In this paper, we study a strategy of network architecture to the integration of Universal Mobile Telecommunications System (UMTS) and Wireless Local Area Network (WLAN) for the seamless vertical handover between these heterogeneous systems. For the internetworking strategy, we consider Hierarchical Mobile IPv6 (HMIP) networks based on Mobility Anchor Points (MAP) that the latency of the vertical handover can be decreased to provide fast handover to mobile nodes. Based on the integration strategy of HMIP, we also propose a vertical handover scheme which avoids the call blocking (new calls or hard handover calls). In integrated UMTS and WLAN networks, mobile nodes having dual system modes can use heterogeneous network services.

1 Introduction

UMTS and WLAN internetworking provide the network utilization increment by supporting inter-complementary characteristics so that a mobile node is able to use more flexible network services [1]. Seamless vertical handover between UMTS and WLAN networks should be implemented to provide stable network services. Hence, effective integration strategy of UMTS and WLAN should be provided to support rapid vertical handover. For implementing the vertical handover, it has been classified under two integration scenarios of UMTS and WLAN [2], [3]: tight coupling internetworking and loose coupling internetworking. In tightly coupled internetworking, WLAN is directly connected to the UMTS core network (CN). Signaling messages for the vertical handover to the heterogeneous networks will be exchanged between the WLAN gateway and the Serving GPRS Support Node (SGSN) via the Gateway GPRS Support Node (GGSN) in the UMTS CN. In this architecture, based on the Iu interface, same mobility management, quality of service (QoS), and security operations are provided to WLAN and UMTS networks. However, the WLAN system must implement UMTS modules such like authentication schemes based on UMTS Subscriber Identity Module (USIM) or Removable User Identity Module (R-UIM). Furthermore, if a mobile node in WLAN communicates with a correspondent node in Mobile-IP networks, it must use the

---

* This work was supported by the Korea Research Foundation Grant (KRF-2004-005-D00147).

UMTS CN for the connection. In loosely coupled internetworking, a WLAN gateway has a connection link to Mobile-IP networks so that WLAN networks cannot connect to a UMTS CN directly. The advantage of this scenario is that the internetworking cost is lower than the cost of first scenario because WLAN is deployed as an independent operator connected to the Internet. Also, a mobile node in WLAN is able to connect its correspondent node in Mobile-IP networks with no support from the UMTS CN.

In this paper, we study a loose coupling internetworking scenario to provide fast vertical handover for the node mobility in integrated UMTS/WLAN networks. Also, we study an inter system handover scheme that implements the vertical handover into a different network system if a mobile node cannot properly receive available channel from current network system because of the full state of current network system. Also, if the previous network has available resources, the mobile node in the overlaid network can return to the previous network by the vertical handover.

The rest of the paper is organized as follows. In Section 2, we propose an internetworking strategy for UMTS/WLAN networks. Section 3 presents the proposed handover scheme and its mathematical modeling. Section 4 shows the numerical results for the proposed scheme, and conclusions are given in Section 5.

2 Internetworking Architecture

Fig. 1 shows new integration architecture of UMTS and WLAN based on the loose coupling internetworking. The WLAN gateway connects to the direct MAP which is connected to the UMTS CN. When a mobile node moves into the WLAN area, the mobile node sends the registration message (BU – Binding Update) to a direct MAP rather than Home Agent (HA) because the HA is usually away from the mobile node [4]. Hence, the signaling delay of vertical handover can be reduced by using the directed MAP. Fig. 1 shows the signaling flow passed during the vertical handover between UMTS and WLAN. In the figure, the upper signaling part is about the mobility from UMTS to WLAN.

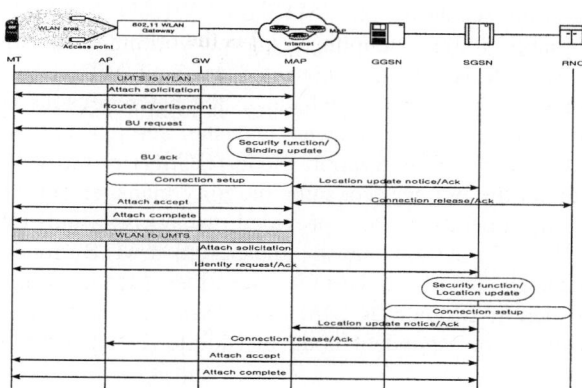

Fig. 1. Vertical handover procedures

If a mobile node starts in a UMTS cell, the mobile node or the UMTS system determines the vertical handover into a WLAN. The mobile node sends an Attach Solicitation message to a local MAP and gets on-link Care of Address (LCoA) and Regional CoA (RCoA) on the MAP subnet. Then, the mobile node sends a BU to the local MAP. The local MAP implements the Authentication, Authorization and Accounting (AAA) services and the BU function. Once, after the BU function, the connection setup will be implemented between the local MAP and the WLAN. The local MAP informs the mobile node that current location change into the SGSN in the UMTS CN, and the previous connection will be released. When the local MAP receives a connection release acknowledge message, it informs to the mobile node of the attach completion. Reverse vertical handover procedure from WLAN to UMTS is similar to the vertical handover from UMTS to WLAN. If a mobile node sends an Attach Solicitation message to a SGSN, the SGSN request the Identity of the mobile node. After the security function and the location update are implemented, a new connection is determined between the Node B and a corresponding node. From above vertical handover procedures, we can obtain the vertical handover delay. Let T_{UtoW} be the signaling time of vertical handover when the mobile node hands off into WLAN from UMTS. T_{UtoW} is divided into two phase signaling time: binding update (T_{MTBU}) and connection complete (T_{ConCom}). Hence, T_{UtoW} is given as follows:

$$T_{UtoW} = T_{MTBU} + T_{ConCom} \tag{1}$$

$$T_{MTBU} = T_{at\_sol} + T_{RA} + T_{BU\_r} + T_{sec\,andBU} + T_{BU\_ack} \tag{2}$$

$$T_{ConCom} = T_{Conset} + T_{Loupnt/ack} + T_{Conrel/ack} + T_{at\_done} \tag{3}$$

where T_{at_sol} is the attach solicitation delay, T_{RA} is the router advertisement delay, T_{BU_r} is the binding update request delay, $T_{sec\,andBU}$ is the implementation delay of security function and binding update, T_{BU_ack} is the binding acknowledge delay to MT after completing the binding update, T_{Conset} is the implementation time for connection setup, $T_{Loupnt/ack}$ is the delay of location update notice and the acknowledgement, $T_{Conrel/ack}$ is the connection release and acknowledgement delay to the UMTS CN, and $T_{at\_done} (= T_{at\_acc} + T_{at\_com})$ is the signaling delay of attach accept and complete.

T_{WtoU} is the signaling time of vertical handover from WLAN to UMTS. T_{WtoU} is also divided into two phase: location update (T_{LocUp}) and connection complete (T_{ConCom}). Therefore, T_{WtoU} is given by

$$T_{WtoU} = T_{LocUp} + T_{ConCom} \tag{4}$$

$$T_{LocUp} = T_{at\_sol} + T_{iden\_r/ack} + T_{sec\,andLocup} \tag{5}$$

$$T_{ConCom} = T_{Conset} + T_{Locupnt/ack} + T_{Conrel/ack} + T_{at\_done} \tag{6}$$

where $T_{iden\_r/ack}$ is the signaling delay of identity request and acknowledgement, and $T_{sec\,andLoup}$ is the implementation delay of security function and location update.

In this paper, we assume that IEEE 802.11 is adopted for the MAC layer. Hence, if a mobile node in WLAN sends a signal message to an AP, the mean signaling time ($T_{Wsignal}$) is given by

$$T_{Wsignal} = T_{Wtrans} + P_{propa} + T_{DIFS} + CW *$$
$$T_{Wtrans} = T_{pream} + T_{PLCP} + T_{payload} \quad (7)$$

where T_{Wtrans} is a message transmission time, P_{propa} is a message propagation time, T_{DIFS} is a distributed inter frame space (DIFS) time, $CW *$ is a contention delay time, T_{pream} is a physical preamble duration time, T_{PLCP} is a PLCP header duration time, and $T_{payload}$ is a transmission time of payload.

From (7), if a mobile node sends a signaling message to designated MAP via a gateway system, we can get the signaling delay from the mobile node to the designated MAP as follows:

$$T_{sig(MT->MAP)} = n \cdot T_{Wsignal(MT->AP)} + T_{msg\_delay(AP->GW)}$$
$$+ (T_{msg\_delay(GW->MAP)} + m \cdot T_{msg\_delay(MAP->MAP)}) \quad (8)$$

where n is the number of collision, and m is the relay number of MAP to approach the designated MAP.

We define the random variable T_U that denotes the signaling delay of the vertical handover from UMTS to WLAN. The random variable T_U is assumed to be exponentially distributed with mean value T_{UtoW}. Let $f_{T_U}(t)$ be the probability density function to the random variable T_U. $f_{T_U}(t)$ is given by

$$f_{T_U}(t) = e^{-t/T_{UtoW}}/T_{UtoW} \quad (t \geq 0) \quad (9)$$

Let T_W be the random variable that denotes the signaling delay of the vertical handover from WLAN to UMTS. We assume that T_W is exponentially distributed with mean value T_{WtoU} as follows:

$$f_{T_W}(t) = e^{-t/T_{WtoU}}/T_{WtoU} \quad (t \geq 0) \quad (10)$$

When a mobile node moves into an UMTS cell from a WLAN cell, the blocking probability of a vertical handover that the mobile node moves away from the handover area before the requested signaling time, T_{mt}, is given by

$$P_{WtoU} = \int_0^\infty [1 - F_{T_W}(t)] \cdot f_{T_{mt}}(t) dt \quad (11)$$

We assume that the WLAN cell is located in the UMTS cell. Hence, we don't consider the blocking probability of a vertical handover from UMTS to WLAN because a mobile node can receive the network service from UMTS even though the signaling time of the vertical handover is delayed. Therefore, we only consider the blocking probability of a vertical handover from WLAN to UMTS. In this paper, we consider two distribution functions about the residence time of mobile node: Exponential and Gamma distribution. Firstly, if the residence time (T_{mt}) of mobile node in the vertical handover area has the Exponential distribution, the probability density function (pdf) is given by

$$f_{T_{mt\_EXP}}(t) = \mu e^{-t\mu} \quad (t \geq 0) \qquad (12)$$

where $1/\mu$ is the average residence time.

Secondly, if T_{mt} has the generalized Gamma distribution, the pdf of T_{mt} is as follows:

$$f_{T_{mt\_gamma}}(t) = \eta e^{-\eta t}(\eta t)^{\omega-1}/\Gamma(\omega) \quad (t \geq 0) \qquad (13)$$

where ω is the shape parameter, η is the scale parameter represented as $\eta = \sigma\omega$, and the quantity $\Gamma(\omega)$ is the Gamma function defined by $\Gamma(\omega) = \int_0^\infty e^{-t}(t^{\omega-1})\,dt$

Note that the Gamma distribution becomes the Erlang distribution, if ω is an integer. Hence, the blocking probabilities of the vertical handover (P_{WtoU}) to $f_{T_{mt\_EXP}}(t)$ and $f_{T_{mt\_gamma}}(t)$ are respectively represented as follows:

$$P_{WtoU(\exp)} = 1 - \Pr\{T_W < T_{mt}\} = \frac{\mu T_{WtoU}}{1 + \mu T_{WtoU}} \qquad (14)$$

and

$$P_{WtoU(gamma)} = \int_0^\infty e^{-t/T_{WtoU}} \cdot \frac{\eta e^{-\eta t}(\eta t)^{\omega-1}}{\Gamma(\omega)}\,dt \qquad (15)$$

3 Dual Handover

We propose an inter-system handover, called dual handover that in overlaid UMTS/WLAN networks a mobile node use different overlaid network system if current network system has not network resources for the mobile node. In this paper, we consider that the dual handover occurs from a UMTS network to a WLAN. That is, if a UMTS system cannot provide network resources for a mobile node, an overlaid WLAN temporarily processes the network service of the mobile node. In general, a mobile node in a UMTS cell continuously measures radio quality and reports the measurement result to the Serving Radio Network Controller (SRNC). If a mobile node moves into next UMTS cell and implements a hard handover call, the SRNC selects a new radio link (RL) setup between the SRNC and the Node B of next cell. Also, when a new call to a mobile node occurs, the mobile node requests a new RL setup to the Node B. If Iub and Iur bearer setup such like Dedicated Channel (DCH) allocation are established, uplink and down link synchronizations are implemented. However, if the Node B is in the full state and cannot offer available channels to the mobile node, the UMTS network sends a warning message to the mobile node and recommends the vertical handover. Hence, the mobile node implements the vertical handover from UMTS to WLAN in the integrated UMTS/WLAN networks as mentioned Section 2 above. The mobile node sends the attach solicitation message to the Access Point (AP) of WLAN located in UMTS handover area. After an attach solicitation, WLAN implements the registration operation and the security function by AAA services. Hence, once WLAN has new connection setup for the mobile node, it informs the attach completion of the mobile node. Here, note that the UMTS network may temporarily maintain the mobile node information [e.g., such like Mobility Management (MM) context buffered in Serving GPRS support Node (SGSN)] for the mobile node returning. If radio resources in the Node B are available, the mobile node can return to the Node B by the vertical handover. We describe the analytic model to evaluate the performance of the proposed scheme. In UMTS networks, we assume that new call arrival and intra hard handover call arrival to the Node B have Poisson process with mean rates of λ_n and λ_{ih}, respectively. The service time to each arrival call is exponentially distributed with means of $1/\mu_n$ and $1/\mu_{ih}$. Also, we assume that K channels are available and will be allocated to mobile nodes. Hence, the basic call processing system is modeled as a $M/M/K/K$ queuing system. By using the Markov chain balance equation, the stationary probability is given by

$$\pi_k = \prod_{i=1}^{k}(\lambda_n + \lambda_{ih}/\mu_n + \mu_{ih}) \bigg/ \left[\sum_{k=1}^{K_m}(\lambda_n + \lambda_{ih}/\mu_n + \mu_{ih})^k \cdot (1/k!)\right] \quad k \leq K_m. \quad (16)$$

A mobile node can normally implement the vertical handover call to a WLAN network that the arrival process is assumed to be a Poisson process with a mean rate of λ_{vh}. The service time is exponentially distributed with a mean of $1/\mu_{vh}$. Here, we assume that the WLAN area is overlaid on handover area between UMTS cells. In WLAN networks, new calls and intra handover calls are generated with Poisson process with mean rates of λ'_n and λ'_{ih}, and each service time is exponentially distributed with means of $1/\mu'_n$ and $1/\mu'_{ih}$. In addition, the vertical handover call from WLAN to UMTS forms a Poisson process with a mean rate of λ'_{vh}. The service time of the vertical handover call is assumed to be exponentially distributed with a mean of $1/\mu'_{vh}$. We also assume that N traffic channels and M traffic sessions are available in a UMTS cell and a WLAN cell. Here, we can see that M is the agreeable maximum user number to the channel utilization for using network services in a WLAN cell [8]. Fig. 2 shows the state transition diagram when $\lambda_\alpha = \lambda_n + \lambda_{ih}$, $\mu_\alpha = \mu_n + \mu_{ih}$, $\lambda'_\alpha = \lambda'_n + \lambda'_{ih}$, $\mu'_\alpha = \mu'_n + \mu'_{ih}$, $\mu_\beta = \mu_n + \mu_{ih} + \mu'_{vh}$, $\mu'_\beta = \mu'_n + \mu'_{ih} + \mu_{vh}$, $\lambda_\gamma = \lambda_\alpha + \lambda'_\alpha$ and $\mu_\gamma = \mu_\alpha + \mu'_\alpha$. Here, if a Node B of UMTS network is in the full state, λ_α becomes the dual handover rate to the overlaid WLAN.

The system state can be represented by (i, j), where i $(0 \leq i \leq N)$ is the number of traffic channel used in a UMTS cell, and j $(0 \leq i \leq M)$ is the number of traffic channel used in a WLAN cell. Let $\pi_{i,j}$ denote the stationary probabilities of states (i, j). From the station transition diagram, we can derive the balance equations as follows: in states (i, j), $(0 \leq i \leq N-1, \ 0 \leq j \leq M-1)$, a Node B and a overlaid

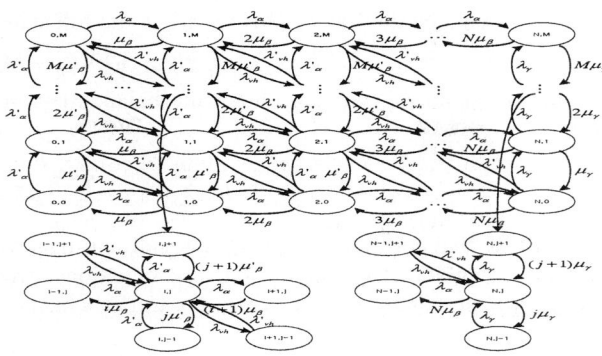

Fig. 2. State transition diagram

WLAN provide network services for mobile nodes so that global balance equations are given by

$$\pi_{i,0} = \frac{1}{(\lambda_\alpha + \lambda'_\alpha + \lambda'_{vh} + i\mu_\beta)} \cdot \left[\lambda_\alpha \pi_{i-1,0} + \lambda_{vh} \pi_{i-1,1} + (i+1)\mu_\beta \pi_{i+1,0} + \mu'_\beta \pi_{i,1}\right] \quad for \ 1 \leq i \leq N-1 \quad (17)$$

$$\pi_{0,j} = \frac{1}{(\lambda_\alpha + \lambda'_\alpha + \lambda_{vh} + j\mu'_\beta)} \cdot \left[\lambda'_\alpha \pi_{0,j-1} + \lambda'_{vh} \pi_{1,j} + (j+1)\mu'_\beta \pi_{0,j+1} + \mu_\beta \pi_{1,j}\right] \quad (18)$$
$$for \ 1 \leq j \leq M-1$$

$$\pi_{i,j} = \frac{1}{(\lambda_\alpha + \lambda'_\alpha + \lambda_{vh} + \lambda'_{vh} + i\mu_\beta + j\mu'_\beta)} \cdot \left[\lambda_\alpha \pi_{i-1,j} + \lambda'_\alpha \pi_{i,j-1} + \lambda_{vh} \pi_{i+1,j} + \lambda'_{vh} \pi_{i+1,j-1}\right. \quad (19)$$
$$\left. + (i+1)\mu_\beta \pi_{i+1,j} + (j+1)\mu'_\beta \pi_{i,j+1}\right] \quad for \ 1 \leq i \leq N-1, \ 1 \leq j \leq M-1$$

Also, $\pi_{N,0}$ and $\pi_{0,M}$ are given by

$$\pi_{N,0} = \frac{\lambda_\alpha \pi_{N-1,0} + \lambda_{vh} \pi_{N-1,1} + \mu_\gamma \pi_{N,1}}{\lambda_\gamma + \lambda'_{vh} + N\mu_\beta} \quad (20)$$

$$\pi_{0,M} = \frac{\lambda'_\alpha \pi_{0,M-1} + \mu_\beta \pi_{1,M}}{\lambda_\alpha + M\mu'_\beta}. \quad (21)$$

In states (i, M), $(0 \leq i \leq N-1, \ j = M)$, only a WLAN can provide network services that the balance equation is given by

$$\pi_{i,M} = \frac{1}{(\lambda_\alpha + \lambda_{vh} + i\mu_\beta + M\mu'_\beta)} \cdot \left[\lambda_\alpha \pi_{i-1,M} + \lambda'_\alpha \pi_{i,M-1} + \lambda'_{vh} \pi_{i+1,M-1} + (i+1)\mu_\beta \pi_{N,j+1}\right]. (22)$$
$$for \ 1 \leq i \leq N-1$$

The balance equation to states (N, j), $(i = N, \ 1 \leq j \leq M-1)$ is given by

$$\pi_{N,j} = \frac{1}{(\lambda_\gamma + \lambda_{vh} + N\mu_\beta + j\mu_r)} \left[\lambda_\alpha \pi_{N-1,j} + \lambda_\gamma \pi_{N,j-1} + \lambda'_{vh} \pi_{N,M-1} + (j+1)\mu_\gamma \pi_{N,j+1}\right] \quad (23)$$
$$for \ 1 \leq j \leq M-1$$

The state (N, M) is that a Node B and a WLAN are in the full state.

$$\pi_{N,M} = (\lambda_\alpha \pi_{N-1,M} + \lambda_\gamma \pi_{N,M-1})/(N\mu_\beta + M\mu_\gamma) \qquad (24)$$

From above balance equations, we can obtain the balance equation for the state $(0,0)$ as follows:

$$\pi_{0,0} = 1 - \left(\sum_{i=1}^{N} \pi_{i,0} + \sum_{j=1}^{M} \pi_{0,j} + \sum_{i=1}^{N} \sum_{j=1}^{M} \pi_{i,j} \right) \qquad (25)$$

In a Node B, if all of channels are occupied, new call and handover calls will be moved into overlaid WLAN cell. From the balance equations, we can obtain the probability to the vertical handover call when no traffic channel of Node B is available.

$$P_d = \sum_{i=1}^{N-1} \pi_{i,M} \qquad (26)$$

If UMTS and WLAN networks have not available channel, new calls and intra hard handover calls will be blocked. Therefore, the blocking probability is given by

$$P_b = \pi_{N,M} \qquad (27)$$

4 Numerical Results and Performance Comparisons

We assume that the MAC layer protocol used in the numeric analysis is the IEEE 802.11b, and the transmission rate is 2 Mbps. A UMTS cell includes WLAN cells so that the overlaid networks are structured. Table 1 shows the system parameters [5], [6]. Firstly, let us show the signaling delay of the proposed internetworking strategy. As above mentioned, if ω is an integer, the Gamma distribution becomes the Erlang distribution. When $\omega = 1$ and $\omega = 2$, the equation (15) can be represented by $(\sigma T_{WtoU}/1+\sigma T_{WtoU})$ and $(2\sigma T_{WtoU}/1+2\sigma T_{WtoU})^2$, respectively. Hence, we can obtain T_{WtoU} and the blocking probabilities $P_{WtoU\,(exp)}$ and $P_{WtoU\,(gamma)}$ from (14), (29) and (30) by using system parameters in Table 1. Fig. 3 shows the performance comparison to the HA/FA-based [7] and the proposed (MAP-based) schemes. Fig. 4 indicates that the MAP-based scheme has better performances than those of the HA/FA-based scheme. As an example, when the mean residence time is 60s, the blocking probability of the proposed scheme to Gamma distribution ($\omega = 1$) is 0.019, but the blocking probability to the HA-based scheme with same conditions is 0.024. The figure also compares $P_{WtoU\,(exp)}$ and $P_{WtoU\,(gamma)}$. By changing ω, we observe that ω does greatly affect $P_{WtoU\,(gamma)}$. When $\omega = 1$, $P_{WtoU\,(gamma)}$ is higher than $P_{WtoU\,(exp)}$. However, $P_{WtoU\,(gamma)}$ decreases dynamically as ω increases. When $\omega = 2$, $P_{WtoU\,(gamma)}$ becomes much lower than $P_{WtoU\,(exp)}$. From applying $\sum\sum \pi_{i,j} = 1$, $(0 \le i \le N, 0 \le j \le M)$ and above balance equations (in Section 3.2), we can obtain the performance evaluations for the comparison of the scheme with and without the dual handover implementation (dual handover scheme and non-dual handover scheme). In UMTS networks, mobile nodes can use DCH that carries dedicated traffics. As shown Table 1, the total number of DCH N is assumed to be

40, 30 and 20 traffic channels. Also, we assume that the channel utilization to WLAN is up to 0.3, and M is 20, 15 and 10 traffic sessions, respectively [8]. Each service average time is assumed to be 60s and 120s. Fig. 4 shows the performance comparisons with and without the dual handover. In Fig. 4(a), we assume that λ_{vh} is 0.01, $\lambda'_\alpha (=\lambda'_n + \lambda'_{ih})$ is 0.03, and λ'_{vh} is 0.02. Each service average time is assumed to be 60s. The figure shows that the blocking probability is lower in the dual handover scheme than in non-dual handover scheme. Also, in dual handover scheme, the blocking probability is dynamically decreased as N increase. In Fig. 4(b), each service average time is assumed to be 120s, and $\lambda_{vh} = 0.02$, $\lambda'_\alpha = 0.04$ and $\lambda'_{vh} = 0.02$. Fig. 4(b) shows that the blocking probability of dual handover is higher than the blocking probability in Fig. 4(a) because λ_{vh} and λ'_α increase. From numeric results, we know that the stable range to the blocking probability of the dual handover increases as the channel number increases and the service average time decreases. For example, in Fig. 4(a), we can see that the blocking probability of the dual handover scheme stays stable between 0.1 and 0.4 when UMTS channels = 30 and WLAN traffic sessions = 15.

Table 1. System parameters

WLAN MAC parameters		UMTS frame delay	10 ms	MAP	5 ms
T_{pream}	144 μs	Propagation time		The number of channel	
T_{PLCP}	48 μs	Wired line	50 μs	N	40, 30, 20
$T_{payload}$	Payload/data rate	Wireless line	0.12 ms	M	20, 15, 10
T_{DIFS}	50 μs	System processing time		Message size	
CW	Min=32, Max=1024	UMTS component	5 ms	WLAN	50 bytes
Slot time	20 μs	WLAN component	5 ms	UMTS	50 bytes

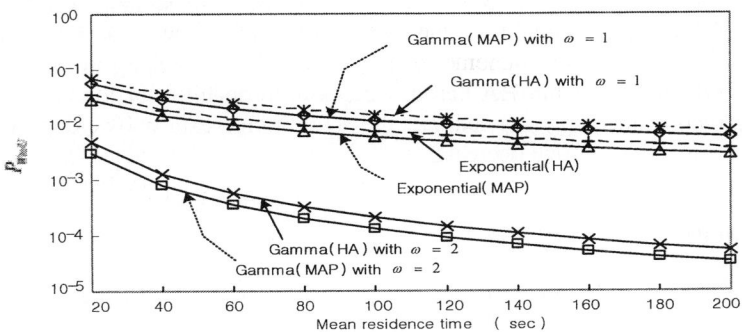

Fig. 3. Blocking probabilities versus mean residence time

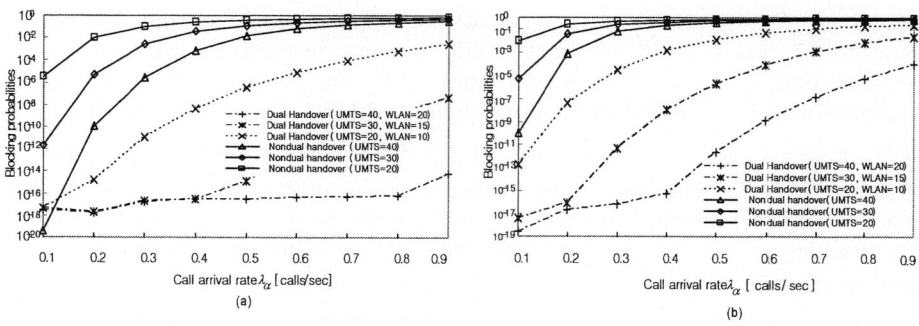

Fig. 4. Performance comparisons (blocking probabilities versus call arrival rates)

(a) $\lambda_{vh} = 0.01$, $\lambda'_{\alpha} = 0.03$, $\lambda'_{vh} = 0.02$, and each service average time is 60s.

(b) $\lambda_{vh} = 0.02$, $\lambda'_{\alpha} = 0.04$, $\lambda'_{vh} = 0.02$, and each service time is 120s.

However, when UMTS = 40 and WLAN traffic sessions = 20, the stable range increases from 0.1 to 0.8. Also, the stable range is greater in Fig. 4(a) than in Fig. 4(b).

5 Conclusions

In this paper, we study the integration strategy of UMTS/WLAN networks which can provide the seamless network service in the heterogeneous environments. We adapt the MAP of HMIPv6 that it can reduce the vertical handover delay in the loosely coupled internetworking. Based on the proposed internetworking strategy, we consider the dual handover scheme that it implements the vertical handover to use complementary network systems when a part network in UMTS/WLAN networks cannot provide available network resource to mobile nodes. We analyze the proposed scheme and provide the performance evaluation of the proposed scheme compared with the basic schemes. The MAP-based internetworking strategy can reduce the signaling delay to the vertical handover between UMTS and WLAN. Furthermore, by the proposed handover scheme, the results show that the blocking probability of the proposed scheme is much lower than the blocking probability of the basic handover scheme. Also, in the proposed scheme, the stable range to low blocking probability increases as the service time decreases and the channel number increases.

References

1. S.-L. Tsao, and C.-C Lin, "Design and Evaluation of UMTS-WLAN Internetworking Strategies," IEEE VTC'02, pp.777-781, September, 2002.
2. M. Buddhikot, G. Chandranmenon, S. Han, Y. W. Lee, S. Miller, and L. Salgarelli, "Integration of 802.11 and Third-Generation Wireless Data Networks," IEEE INFOCOM'03, pp. 502-512, April, 2003.

3. V. K. Varma, S. Ramesh, K. D. Wong, and J. A. Friedhoffer, "Mobility Management in Integrated UMTS/WLAN Networks," IEEE ICC'03, pp. 1048-1053, May, 2003.
4. Masaki Bandai, and Iwao Sasase, "A Load Balancing Mobility Management for Multilevel Hierarchical Mobile IPv6 Networks," IEEE PIMRC'03, pp.460-464, September, 2003.
5. F. Khan, and D. Zeghlache, "Effects of cell residence time distribution on the performance of cellular mobile networks," IEEE VTC'97, pp. 949-953, May, 1997.
6. Y. Xiao, and J. Rosdahl, "Throughput and Delay Limits of IEEE 802.11," IEEE Communications Letters, vol.6, no.8, pp.355-357, August, 2002.
7. 3GPP TR 22.934, V2.0.0 Feasibility study on 3GPP system to Wireless Local Area Network (WLAN) internetworking (Release 6).
8. R. Bruno, M. Conti and E. Gregori, "IEEE 802.11 Optimal Performances: RTS/CTS mechanism vs. Basic Access," IEEE PIMRC'02, pp. 1747-1751, September, 2002.

A Handover Scheme Based on HMIPv6 for B3G Networks*

Eunjoo Jeong[1], Sangjoon Park[2], Hyewon K. Lee[1], Kwan-Joong Kim[3], Youngsong Mun[1], and Byunggi Kim[1]

[1] School of Computing, Soongsil University, Seoul, Korea
desire79@archi.ssu.ac.kr, kerenlee@sunny.ssu.ac.kr,
{mun, bgkim}@comp.ssu.ac.kr
[2] Information Media Research Technology Institute,
Soongsil University, Seoul, Korea
lubimia@hanmail.net
[3] Dept. of Computer and Information, Hanseo University, Korea
kimkj@hanseo.ac.kr

Abstract. By complementary integration of UMTS and WLAN, a beyond third generation (B3G) mobile network has been proposed to establish better global roaming environments. The integrations of UMTS and WLAN are classified into two groups: loosely-coupled and tightly-coupled. A tightly-coupled network demands lots of investment and considerable amount of time to construct. On the other hand, a loosely-coupled network is more scalable and easier to implement than a tightly-coupled one while it has critical drawbacks of packet loss and blocking of services due to handover delay. To alleviate these drawbacks, this paper proposes a handover scheme between UMTS and WLAN, which is based on HMIPv6. The performance of the proposed scheme is evaluated adding the handover time of each step, and the blocking probability is computed in each scheme. The proposed internetworking scheme's performance based on HMIPv6 is two times more likely than that based on MIP.

1 Introduction

The ubiquitous age drew near, researches in mobile communications dramatically have increased allowing mobile access anywhere and anytime. Therefore, mobile technology should be needed and then Universal Mobile Telecommunication Systems (UMTS) [1] and IMT-2000 is revealed in the third generation (3G). Nevertheless developing 3G technology, supporting better services and internetworking with different systems are raised an important part. In Beyond Third Generation (B3G), Internetworking is possible with other different systems.

This work places focus on internetworking between IEEE 802.11 WLAN [2] and ETSI UMTS. While WLAN covers only small service area but supports high transmission speed and low cost, UMTS has wide area coverage with low speed and high cost. Accordingly, if WLAN and UMTS are combined together by completing

---

* This work was supported by the Korean Research Foundation Grant (KRF-2004-005-D00147).

O. Gervasi et al. (Eds.): ICCSA 2005, LNCS 3480, pp. 118–126, 2005.
© Springer-Verlag Berlin Heidelberg 2005

strength and weakness, the noticeable result is expected. In order to inter-network between UMTS and WLAN, this work propose the internetworking based on Hierarchical Mobile IP version 6 (HMIPv6) [3] instead of MIP, so that the handover delay will be reduced.

2 Related Works

2.1 The Background: Internetworking Mechanisms

Internetworking mechanism between WLAN and UMTS is classified into two groups; tightly-coupled and loosely-coupled mechanisms [4]. Tightly-coupled internetworking directly connects UMTS Core Network (CN) and WLAN via Inter Working Unit (IWU). In tightly-coupled based networks, WLAN cell is regarded as UTRAN cell, so users from different networks communicate with each other with no modification, and further any service guaranteed from UMTS CN, such as security, quality of service (QoS), mobility and billing, is also offered to WLAN cells. The weak point of this mechanism is a long period and high expenses to implement it because changes in WLAN devices, such as loading UMTS module, are mandatory.

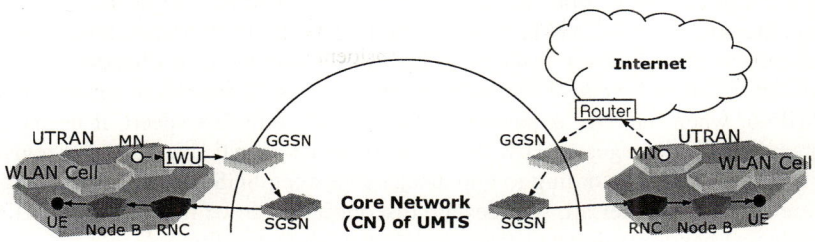

Fig. 1. Internetworking schemes where tightly-coupled internetworking and loosely-coupled internetworking is specified in the left and the right, respectively

In loosely-coupled internetworking scheme, no changes are required in WLAN devices. Nodes in different networks communicate with each other via Internet, as shown in the right of Fig. 1. In loosely-coupled based networks, aforementioned services for each different network should be provided separately. Interconnection of WLAN and UMTS is performed over Internet, so it has rather lower data rate than tightly-coupled based networks. However, this mechanism is preferred to the other because of high extensibility and scalability. Besides, no changes are necessary in WLAN for internetworking, and implementation will be accomplished within a shorter period with lower expenses.

2.2 Internetworking Between WLAN and UMTS Using MIPv4

For internetworking between WLAN and UMTS, [4] has proposed the loosely-coupled internetworking based on MIPv4. A Mobile Node (MN) operates the dual mode, which allows the MN to listen and handle any MAC frame whenever it roams. Each network has agents, such as home and foreign agents (HA/FA), for mobility service, and a WLAN is assumed home network. When a MN moves into UMTS, it starts with a L1 and L2 handover and then sends Attach Request and Active PDP/MM Context messages to SGSN for communication with UMTS. After that, A MN sends an Agent Solicitation message to a local FA and it receives an Agent Advertisement message, builds a CoA and sends a Registration Request message to HA for the binding update. After updating, HA sends a received packet to the foreign network.

Internetworking between UMTS and WLAN based on MIPv4 [4] has some problems, such as un-optimized routing paths, exhaustion of IP resources, weakness on security, disconnection due to roaming to ingress filtering employed networks, and may more. MIPv6 [5] is proposed to solve aforementioned problems and further provides robust mobility service. The foreign agent obsoletes in MIPv6, and instead access routers in foreign networks replace its role. By modifying [4], this work proposes the internetworking model based on HMIPv6 instead of MIPv4.

3 The Proposed Scheme: Internetworking Between WLAN and UMTS Using HMIPv6

When a MN communicates with correspondent node, it may enter the pollution area where it senses signals both from WLAN cell and UMTS cell at the same time. If signal strength from WLAN is closer to the threshold, the MN may lose connectivity or attempt to do handover to UMTS. If the MN decides to do handover, it reports handover to AP and Node B [6], switches its mode and handoffs to foreign networks. In MIPv6, when the MN is passed by lots of cell with fast speed, it needs to send frequently BU messages to HA. At the time, HA's signal overhead for registration and handover delay according to hop distance between SGSN (AR) and HA via Internet are increased. After all, services are disconnected so we'll apply the HMIPv6 [3] for reducing handover delay instead of MIPv6 as shown in Fig. 2.

In HMIPv6, three types of address are defined to identify MN's current location; home address, on-link care-of address (LCoA) and regional care-of address (RCoA). The MAP in HMIPv6 acts as an intermediate HA. When a MN moves from home to foreign network, MAP and home registrations are performed. While the MN roams within the same MAP domain, only local address (LCoA) will be changed, which triggers MAP registration. Only inter-MAP movement is notified to HA, which reduces signal message exchange between MN and HA.

This work assumes a MN is in WLAN as a home network, a MAP is selected one of GGSNs (ARs). That is, a MAP manages some GGSNs (ARs) for hierarchical structure as shown in Fig. 2. The detail explanation of proposed model, which the handover algorithm and MN's mode switching is shown by Fig. 3, is as follows.

When a MN handoffs from WLAN to UMTS for the first time, it detects its movement and performs L1 and L2 handovers. The MN sends a UMTS Attach message and active Packet Date Unit/Mobility Management (PDP/MM) Contexts [7] for sending or receiving packet data and for providing mobility information of it. Once the MN has valid session with UMTS, the UMTS Attach message exchange is not required. For router discovery, the MN may send Router Solicitation (RS) message, or it may receive unsolicited Router Advertisement (RA) message from access router on the visited network. From the RA message, a MN is able to build new addresses

(LCoA, RCoA) as specified [3] to identify its current location. Once a MN builds addresses, MAP and home registrations are followed.

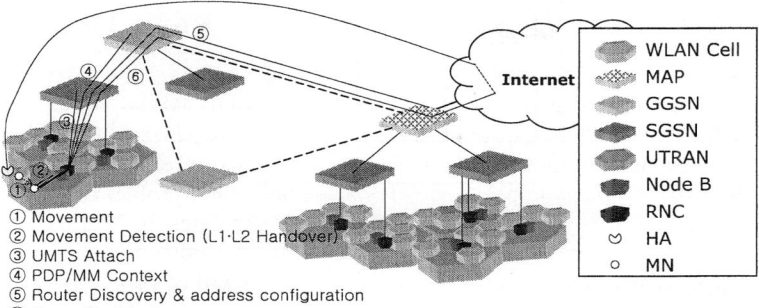

Fig. 2. Internetworking between WLAN and UMTS using HMIPv6

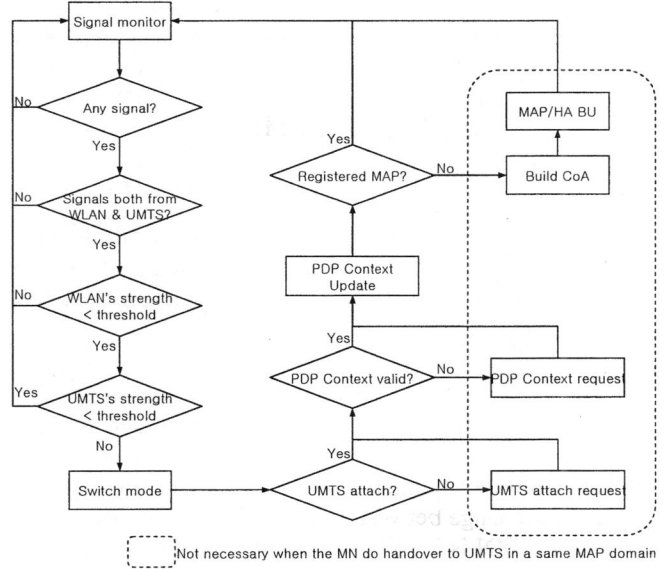

Fig. 3. Algorithm for Handover from WLAN to UMTS and mode switching

When a MN moves from WLAN into UMTS area for the first time, handover is processed as above. When the MN handoffs from WLAN to UMTS where it has visited before and it still has valid PDP/MM context, handover process becomes a different story. Simply, Routing Area Update message exchange between MN and SGSN operating as AR and MAP registration are performed for handover, as shown in Fig. 4.

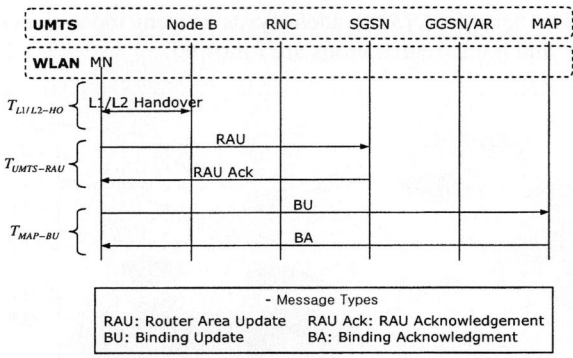

Fig. 4. Handover Signaling from WLAN to UMTS using HMIPv6 (When MN moves within the same MAP domain)

Handover latency is expressed with sum of $T_{L1/L2-HO}$, $T_{UMTS-RAU}$ and T_{MAP-BU}, as shown in (1), where $T_{L1/L2-HO}$ is L1/L2 Handover, $T_{UMTS-RAU}$ is Active PDP/MM Context, and T_{MAP-BU} is Binding Update to a MAP. The handover procedure is specified as follows.

$$T_{HMIPv6} = T_{L1/L2-HO} + T_{UMTS-RAU} + T_{MAP-BU} \tag{1}$$

- **L1/L2 Handover ($T_{L1/L2-HO}$)**

Changing of MN's mode and the access mechanism to wireless interface of foreign network is involved this step. L1 and L2 handovers delay time is control message processing time γ_{MN} of MN and access time delay T_{UMTS} between MN and Node B. Therefore, the processing time occurs in (2).

$$T_{L1/L2-HO} = \gamma_{MN} + T_{UMTS} \tag{2}$$

- **Active PDP/MM Context ($T_{UMTS-RAU}$)**

The MN has maintained a valid PDP/MM Context with UMTS, so it simply sends a Routing Area Update message. That is, a MN should use a previous original PDP/MM Context for the communication.

- **Binding Update to MAP (T_{MAP-BU})**

If a MN moves within a registered MAP domain, it sends a Binding Update message to the MAP instead of sending to HA.

4 The Numerical Result or Performance Analysis

For comparing and analyzing with the existing method and the proposed one, the system parameter is defined as show in Table 1.

Table 1. System Parameters

Input Parameter	Values	Input Parameter	Values
Traffic type	UDP	Message size	100 bytes
Link Parameters			
Wireless Link Data Rate	2 $Mbps$	UMTS link data rate	384 $kbps$
Transmission time (α)			
WLAN (α_{WLAN})	0.4 ms	WLAN (α_{UMTS})	2 ms
Propagation time (β)			
Wired link (β_{wire})	0.12 ms	Wireless link (β_{UMTS})	0.05 ms
Processing Time(γ)			
$\gamma_{MN}, \gamma_{NodeB}, \gamma_{AP}$	15 ms	$\gamma_{HA}, \gamma_{SGSN/AR}, \gamma_{RNC}, \gamma_{MAP}$	5 ms

4.1 Total Handover Delay

By applying system parameters from Table 1, the total handover delay is measured. The hop distance with HA and GGSN or with MAP and GGSN must influence the handover delay time because our internetworking scheme is attained via Internet.

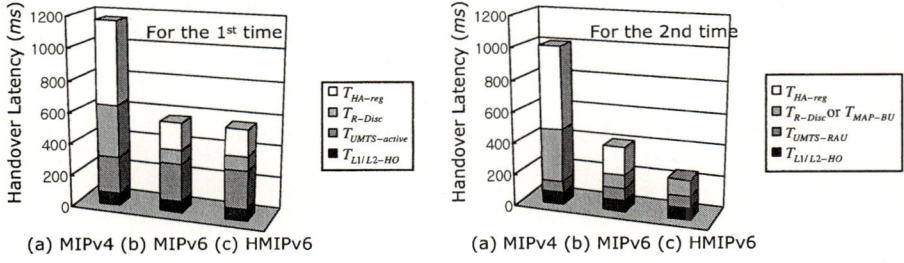

(a) MIPv4 (b) MIPv6 (c) HMIPv6   (a) MIPv4 (b) MIPv6 (c) HMIPv6

Fig. 5. Comparison of handover delay where the MN entered to UMTS for the first time and it has a valid session with UMTS is specified in the left and the right, respectively

When 5 hop distance between GGSN and HA, and 1 hop distance between GGSN and MAP are assumed, the total handover delay is as shown in Fig. 5. Fig. 5 shows that comparison of handover delay where the MN entered to UMTS for the first time and it has a valid session with UMTS is specified in the left and the right, respectively. When the MN has a valid PDP/MM Context, the handover latency using HMIPv6 is about 5 times more likely than using MIPv4 and is about 2 times more likely than using MIPv6.

4.2 Blocking Probability

When a MN moves from cell to other cells, if the cell residence time [10] is less than the total handover time, packets are lost and the network service is forcefully terminated by the link loss. This is called the blocking probability by the formula as below.

The total handover time, which is expressed as the exponential distribution function $f_{T_{HO}}$ with a mean value T_{HMIPv6}, if $T_{HMIPv6} > 0$, is given by (4).

$$f_{T_{HO}}(t) = \begin{cases} \dfrac{e^{-\frac{t}{T_{HMIPv6}}}}{T_{HMIPv6}}, & t \geq 0 \\ 0 & t < 0, \end{cases} \quad (4)$$

where T_{HO} is the random variable of signaling delay to the handover and T_{HMIPv6} is mean value of the total handover latency using HMIPv6. The dwell time of MN, which is presented by the exponential and gamma distributions, is given by (5) and (6).

$$f_{T_{cell-dwell}\exp}(t) = \begin{cases} \lambda e^{-\lambda t}, & t \geq 0 \\ 0, & t < 0, \end{cases} \quad (5)$$

$$f_{T_{cell-dwell}gam}(t) = \begin{cases} \dfrac{\rho e^{-\rho t}(\rho t)^{\omega-1}}{\Gamma(\omega)}, & t \geq 0 \\ 0, & t < 0, \end{cases} \quad (6)$$

where $T_{cell-dwell}$ is a residence time of MN in boundary cell, $\dfrac{1}{\lambda}$ is a mean dwell time of MN in the cell, ω is a shape parameter, σ is a mean value, and ρ is a scale parameter ($\rho = \sigma\omega$). The blocking probability of handover from WLAN to UMTS is given by (7) and (8).

$$\begin{aligned} P_{B_{\exp}} &= 1 - \Pr\{T_{HO} < T_{cell-dwell}\} = \Pr\{T_{HO} > T_{cell-dwell}\} \\ &= \int_0^\infty e^{\frac{-t}{T_{HMIPv6}}} \cdot f_{T_{cell-dwell}\exp}(t) dt \\ &= \int_0^\infty e^{\frac{-t}{T_{HMIPv6}}} \cdot \lambda e^{-\lambda t} dt \\ &= \frac{\lambda T_{HMIPv6}}{1 + \lambda T_{HMIPv6}}, \end{aligned} \quad (7)$$

$$\begin{aligned} P_{B_{gam}} &= 1 - \Pr\{T_{HO} < T_{cell-dwell}\} = \Pr\{T_{HO} > T_{cell-dwell}\} \\ &= \int_0^\infty e^{\frac{-t}{T_{HMIPv6}}} \cdot f_{T_{cell\_dwell}gam}(t) dt \\ &= \int_0^\infty e^{\frac{-t}{T_{HMIPv6}}} \cdot \frac{\rho e^{-\rho t}(\rho t)^{\omega-1}}{\Gamma(\omega)} dt \end{aligned} \quad (8)$$

In Fig. 6, the blocking probability is shown by exponential and gamma distributions according to the dwell time of MN from (7) and (8). When a MN moves from a WLAN cell into a UMTS cell, the blocking probability using MIPv6 is about two times more likely than that of using HMIPv6. Therefore, when the handover occurs, the proposed scheme reduces signal message exchanges so that the seamless handover can be safely provided.

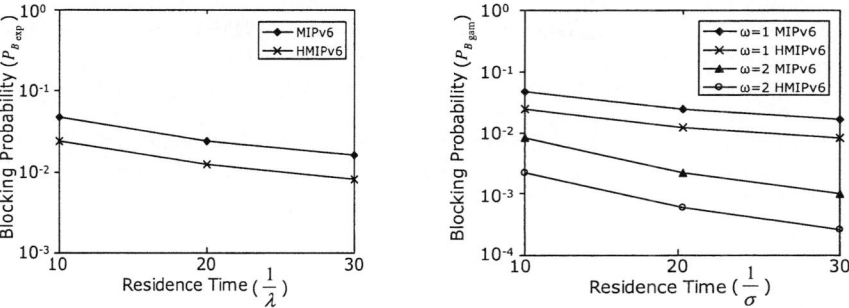

Fig. 6. Blocking probabilities versus residence time

5 Conclusions

Internetworking with other systems is considered as an important part in mobile communications. The loosely-coupled mechanism, which is important part of research on internetworking with UMTS and WLAN, is explained using MIPv4 and MIPv6. Internetworking with UMTS and WLAN becomes possible by using MIP. However, the MN has to register to HA via Internet when the MN handoffs to other network, so the handover latency takes no small times. Therefore, the packet loss and the non-guaranteed seamless service are occurred. In order to improve this problem, the internetworking with UMTS and WLAN based on HMIPv6 is proposed in this paper.

For evaluating the proposed scheme, the system parameter is defined. After that, the handover latency time and the blocking probability are enumerated. Hence, the results of proposed scheme where is the handover delay and the blocking probability is reduced twice more than the existing one, so it can provide the seamless service.

References

1. Heikki Kaaranen, Siamäk Naghian, Lauri Laitinen, Ari Ahtiainen and Valtteri Niemi: UMTS Networks: Architecture, Mobility and Services, John Wiley & Sons, (2001)
2. Mattbew S. Gast: 802.11 Wireless Networks: The Definitive Guide, O'Reilly & Associates, (2002)
3. Soliman, H., Catelluccia, C., Malki, K. and Bellier, L., "Hierarchical Mobile IPv6 Mobility Management," work in progress, (2004)
4. Shiao-Li Tsao and Chia-Ching Lin, "Design and Evaluation of UMTS-WLAN Interworking Strategies", Proceeding of IEEE Vehicular Technology Conference 2002(VTC 2002) 56th, Vol. 2, (2002) 777 - 781
5. Pete Loshin: IPv6;theory, protocol, and practice. 2nd edn. Morgan Kaufmann (2004)
6. Janise McNair, Ian F. Akyildiz and Michael D. Bender, "An Inter-System Handoff Technique for the IMT-2000 System", INFOCOM 2000. Nineteenth Annual Joint Conference of the IEEE Computer and Communications Societies. Proceedings, Vol. 1, (2000) 208 - 216

7. Yi-Bing Lin, Yieh-Ran Haung, Yuan-Kai Chen and Imrich Chlamtac, "Mobility management: from GPRS to UMTS," Wireless Communications and Mobile Computing, vol. 1 (2001) 339-359
8. Nektaria Efthymiou, Yim Fun Hu and Ray E. Sheriff, "Performance of Intersegment Handover Protocols in an Integrated Space/Terrestrial -UMTS Environment", IEEE Transactions on Vehicular Technology, Vol. 47, (1998) 1179 - 1199
9. 3rd generation Partnership Project. "Delay Budjet within the Access Stratum", TR 25.853, v.4.0.0, http://www.3gpp.org
10. F. Khan and D. Zeghlache, "Effect of cell residence time distribution on the performance of cellular mobile networks", Proceedings of IEEE Vehicular Technology Conference 1997 47th, Vol. 2 (1997) 949-953

Collaborative Filtering for Recommendation Using Neural Networks

Myung Won Kim, Eun Ju Kim, and Joung Woo Ryu

School of Computing, Soongsil University, 1-1, Sangdo 5-Dong, Dongjak-Gu, Seoul, Korea
mkim@comp.ssu.ac.kr, blue7786@ssu.ac.kr, ryu0914@orgio.net

Abstract. Recommendation is to offer information which fits user's interests and tastes to provide better services and to reduce information overload. It recently draws attention upon Internet users and information providers. Collaborative filtering is one of the widely used methods for recommendation. It recommends an item to a user based on the reference users' preferences for the target item or the target user's preferences for the reference items. In this paper, we propose a neural network based collaborative filtering method. Our method builds a model by learning correlation between users or items using a multi-layer perceptron. We also investigate integration of diverse information to solve the sparsity problem and selecting the reference users or items based on similarity to improve performance. We finally demonstrate that our method outperforms the existing methods through experiments using the EachMovie data.

1 Introduction

Recommendation is to offer information which fits user's interests and tastes to provide better service and to reduce information overload. For information filtering there are generally three different methods depending on the kinds of information used to filtering: content-based filtering, demographic filtering, and collaborative filtering. Content-based filtering estimates the preference for an item based on its similarity in some properties with the items with known preferences. Demographic filtering estimates a user's preference based on the user's demographic information such as gender, age, hobbies, and job. Collaborative filtering estimates a user's preference for an item based on other users' preferences for the item. Different filtering methods can be best applied to different application problems and they have their strengths and weaknesses. For example, content-based filtering is good for recommending books and documents for which contents can be clearly specified. However, it may not be efficient for recommending items for which contents cannot be clearly defined. It is also difficult to recommend items not similar in contents but closely related to some given items. Collaborative filtering can be best applied to problems for which items are recommended to users based on their correlated preference behaviors without the need of knowing the contents of items or demographic information of the users. Collaborative filtering can be considered more general than content-based and demographic filtering. It has been reported that content-based filtering and demographic filtering lack flexibility in recommendation and their performances are generally low compared with collaborative filtering [1], [2], [3].

There have been two typical methods for collaborative filtering: the *k*-NN method and the association rule method [4], [5]. In spite of some advantages, their performances are generally low. The low performances of the methods are mainly due to the limitation of their underlying models. The *k*-NN method assumes that the attributes of data are independent, however, it is usually not the case. Association rules are limited in representing complex relationships among data [6], [7].

In this paper, we propose collaborative filtering based on neural network. In general, it has several advantages over the conventional models. Some important features of a neural network include: (1) it can learn a complex relationship between input and output values; (2) it can easily integrate diverse information; (3) it can handle various data types; (4) it can handle incomplete information efficiently. These distinguishing features can well fit to problems such as collaborative filtering. In our method, a multi-layer perceptron is adopted as the basic neural network architecture. A multi-layer perceptron is trained to learn a correlation among preferences of the target user and the reference users. The resulting model is called a user model. The same principle can be applied to build an item model.

Our neural network model has several advantages over the existing methods. First of all, it is expected that its performance is improved because it can learn a complex relationship of preferences among users or items. In addition, the hidden nodes of the model represent latent concepts for recommendation and it causes performance improvement. Next, it is easy to handle diverse data types including continuous numeric data, binary or logical data, and categorical data. Finally, it is easy to integrate diverse kinds of information such as contents and demographic information for efficient filtering. We investigate integration of contents information into the user model to examine the possibility of solving the sparsity problem. In training a neural network it is important to use good features as input data. For this we also investigate selection of the reference users or items based on the similarity between the target user and the reference users or the target item and the reference items.

This paper is organized as following. In Section 2 we describe our neural network based collaborative filtering and some methods to improve the performance. In Section 3 we briefly review the existing collaborative filtering methods. In Section 4 we describe our experiments with our method and compare the performance of our method with those of the existing methods. Finally, in Section 5 we make a conclusion and propose future research.

2   Collaborative Filtering Based on Neural Network

Among many different types of neural networks we adopt as the basic neural network model the multi-layer perceptron(MLP) [8], which is commonly used in various applications. In this section we describe our method of building collaborative recommendation models based on the MLP. We also describe integration of additional information and selection of the reference data to improve the performance of our model.

2.1   Neural Network Models

There are two neural network models for collaborative filtering (CFNN): a user model called U-CFNN and an item model called I-CFNN. In the U-CFNN model the input nodes correspond to the reference users' preferences and the output node corresponds to the target user's preference for the target item. In the I-CFNN model the input nodes correspond to the target user's preferences for the reference items and the output node corresponds to his preference for the target item. CFNN can be designed to include more than one output node for multiple users or items. In this case it needs to train the whole network for incremental learning for even a small number of users or items. Furthermore, it is difficult to utilize user or item specific information such as the selected reference users or items. Thus we do not consider such models in this paper.

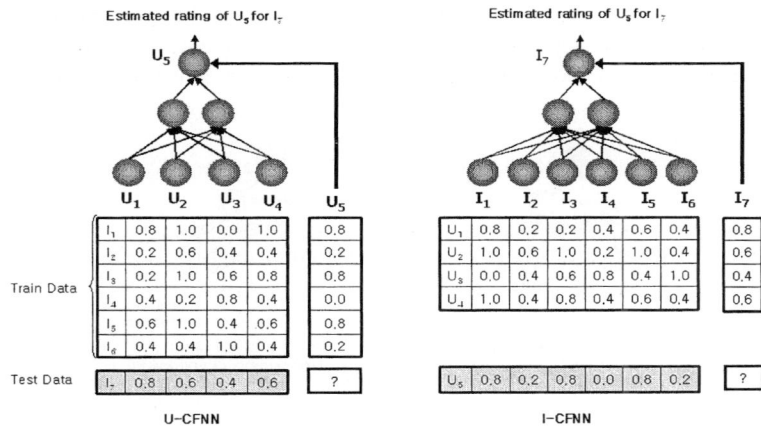

Fig. 1. Example of Training CFNN

The U-CFNN model is produced by learning the target user's (U_t) correlation with the reference users (U_1, U_2, ... U_{n-1}, U_n). When an item is given for recommendation the model outputs an estimation of the target user's preference for the item based on other users' preferences for the item. For example, suppose we are given the user-item preference ratings and we need to predict user U_5's rating for item I_7 by referencing other users' ratings. We associate the input nodes with users U_1, U_2, U_3 and U_4 as the reference users and the output node with the target user U_5. We train a U-CFNN model using the ratings of users for each item. The resulting model represents a general behavior of the target user's preference as associated with the reference users' ratings. Now, for a new item, for example I_7, the U-CFNN model can estimates the target user's (U_5) rating for the item based on the reference users' ratings for the item.

An I-CFNN model can be built in a similar way to that described in the above and it is basically of the same structure as that of the U-CFNN model. It differs from the U-CFNN model in that it represents preference association among items. Given an item for recommendation its rating is estimated based on ratings for other items rated by the target user.

2.2 Information Integration

Collaborative filtering often suffers the sparsity problem in which the performance is low when the rating information is not sufficient in the data. For a new item, only a limited number of users rate their preferences for the item. In this case collaborative filtering has difficulty to estimate the preference for the item correctly. To solve the sparsity problem and to improve the performance, [10] used the singular value decomposition (SVD) to use latent structure in the data. [1] and [11] proposed integration of collaborative filtering and content-based filtering. In [11] the content-based predictor estimates ratings of users for the target item and that rating information is fed into the collaborative filtering system to estimate the rating for the target item.

In this paper we investigate integration of additional information into CFNN as shown in Fig. 2, to solve the sparsity problem. We consider integrating content information of items into the U-CFNN model. For example, we can integrate content information such as the genre of movies into the U-CFNN model. Also we can integrate into the I-CFNN model user's demographic information such as age, gender, job, and hobbies.

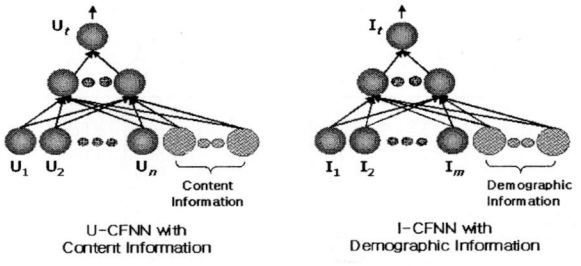

Fig. 2. CFNN with Information Integration

An MLP for integrating additional information can be built simply by adding new input nodes corresponding to the additional information and connecting them to hidden nodes as shown in Fig. 2. Our method takes advantage that it is very easy to integrate different kinds of information using neural networks. However, we describe later an experiment with integrating the genre information into the U-CFNN model. We do not consider integrating the demographic information into the I-CFNN model because of lacking useful demographic information in the data.

2.3 Selection of Reference Data

We investigate using similarity in selecting the reference users or items. In the CFNN model there is no restriction on selecting the reference users or items. However, For the U-CFNN model if we select those users who are similar to the target user in preference, the model can learn stronger correlation among users than random selection of the reference users and consequently, the performance can be improved. Fig. 3 illustrates a U-CFNN model with the selected reference users as input users. In this paper we adopt Pearson's correlation coefficient as similarity measure. We select k users with the highest Pearson's correlation coefficients.

Fig. 4 illustrates building a U-CFNNS (U-CFNN with the selected reference users). It computes the similarities of preferences between the target user (U_5) and other users (U_1, U_2, U_3, and U_4). If k is three, it produces a model by selecting three users (U_1, U_2,

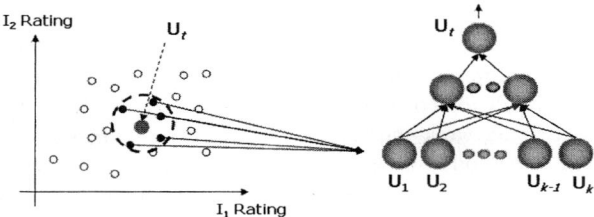

Fig. 3. U-CFNNS

and U_4) as the reference users, who have the highest similarities with the target user. The selected reference users correspond to the input nodes. Similarly, for the I-CFNNS model three items (I_1, I_4, and I_5) are selected as similar to the target item (I_7).

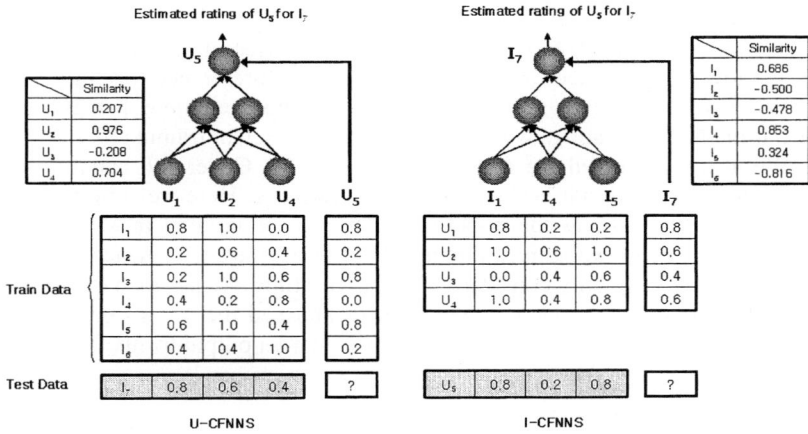

Fig. 4. Example of CFNNS

3 Related Works

The k-NN method, which was used in GroupLens for the first time [9], is a memory-based collaborative filtering. In the k-NN method, a subset of users are chosen based on their similarities to the target user in preference and a weighted combination of their ratings is used to produce a prediction for the target user. The k-NN method is simple and easy to use, however, its performance is generally low. The reason for this is that the k-NN algorithm assumes that the attributes are independent. Consequently, it may not be efficient if independence among attributes is not guaranteed. It also has the scalability problem that computation time increases as the size of data increases [4], [6].

Association rules represent the relations between properties of data in the form of 'IF A Then B.' In association rules we can consider that the correlations (weights) between users or items are represented in terms of support and confidence of rules.

However, more detail correlation may not be represented by support or confidence. In addition the performance is low because they rely on a simple statistics of co-occurrences of data patterns without considering intrinsic semantic structure in the data. It is also difficult to represent complex (not logical) relationships among data only in rules of attribute-value pairs.

[10] proposed a method, which is based on dimensionality reduction through the SVD of an initial matrix of user ratings. It exploits latent structure to essentially eliminate the need for users to rate common items and it solves the sparsity problem to some degree. [12] investigated combining the user model and the item model for performance improvement.

4 Experiments

We experimented with our method over the domain of movie recommendation. We used the EachMovie dataset [13]. It consists of 72,916 users, 1,628 movies, and 2,811,983 numeric ratings. Movies are rated by six different levels between 0.0 and 1.0 for the lowest preference (or "dislike"), the highest preference (or "like"), respectively. In our research we chose the first 1000 users who rated more than 100 movies.

For our method, the ratings were transformed for efficient training as shown in Table 2. We also represented the missing ratings by 0.0. Generally, handling missing values is one of the difficult problems in data processing. Here we simply considered missing ratings the same as the in-between rating to be ignored in MLP processing and in our experiments we noticed that it did not affect the performance significantly. In training the MLP we only considered the target ratings greater than 0.7 or less than 0.3 and other ratings were ignored. We also transformed the ratings greater than 0.7 into 1.0 and the ratings less than 0.3 into 0.0 as the target output for the MLP.

Table 1. Rating Encoding

User Rating	0.0	0.2	0.4	0.6	0.8	1.0	missing
CFNN Encoding	-1.0	-0.6	-0.2	0.2	0.6	1.0	0.0

For its generalization power we also fixed the number of hidden nodes to be small, such as five. In our experiments we used the learning rate 0.05 and the momentum constant 0.0.

We use four different measures for performance evaluation: accuracy, precision, recall, and F-measure as defined in [16]. For performance evaluation using those measures we quantized the output ratings of our model greater than or equal to 0.5 to "like" and less than 0.5 to "dislike", respectively.

In this section, we analyzed the performance of our method for recommendation and compared it with the existing methods in various aspects. First we analyzed and compared CFNN and CFNNS, and then evaluated the performance for integrating the genre information into the U-CFNNS model. Finally, we compared CFNN with the k-NN and the association rule methods.

4.1 Comparison of CFNN and CFNNS

We built 20 models for 10 users and 10 movies. We particularly selected 10 users whose preferences are unbiased in that they rated almost equal number of items as "like" and "dislike". In this case it is generally difficult to predict user's preferences correctly. Each model is evaluated by 4-fold cross validation. In our research for efficiency we limit the number of the reference users or movies to 100. In CFNN the reference users or movies are selected randomly while in CFNNS users whose preferences are similar to that of the target user are selected. Table 2 compares CFNN and CFNNS in performance. As shown in the table CFNNS clearly outperforms CFNN. The performance improvement can be accounted for by the difference of average similarities between random selection and similarity based selection of the reference users or items. We notice that the performance improvement for the item model is greater than for the user model. In this case the difference is 0.32 for the item model while it is 0.27 for the user model.

Table 2. Performance Comparison of CFNN and CFNNS (%)

	User Model		Item Model	
	U-CFNN	U-CFNNS	I-CFNN	I-CFNNS
Accuracy	82.1	83.6	77.7	81.5
Precision	81.7	82.6	77.9	82.0
Recall	84.1	85.6	75.8	79.4
F-Measure	82.9	84.1	76.2	80.7

4.2 Information Integration

It is often the case that integrating diverse relevant information or combining multiple methods yields better performance. This is also true with recommendation. [1] and [12] demonstrated performance improvement by integrating additional information and combining different filtering methods. Especially, such integration can solve the sparsity problem in collaborative filtering. In this paper we describe integration of diverse information for performance improvement and solving the sparsity problem using a neural network. Demographic information in the EachMovie data includes gender, age, residential area, and job. However, we decided to ignore it. It is because such demographic information is of little use: residential area and job are little correlated with movie preference while gender and age appear to be significantly biased in the data.

Instead, we integrated the genre of movies into the U-CFNN model as well correlated to user's preference. The EachMovie dataset has 10 different genres of movies and each movie can have more than one genre specified. Although each genre is represented in binary, when we trained the model we transformed it by multiplying the target user's rating for the movie and used the result as input to the neural network model as shown in Fig. 2. We experimented with the 10 users under the same conditions as described previously.

Table 3. Performance Comparison of U-CFNNS and U-CFNNS with Genre (%)

	No. of Reference Users (U-CFNNS)					
	10		50		100	
	User	User/Genre	User	User/Genre	User	User/Genre
Accuracy	77.3	82.8	80.5	84.2	83.6	84.1
Precision	75.2	81.4	80.3	83.4	82.6	84.0
Recall	85.9	87.8	82.8	87.0	85.6	85.2
F-measure	79.6	83.4	81.4	84.4	84.1	83.7

Table 3 compares the performances of U-CFNNS with and without the genre information. In the experiment we examined the effect of the genre information by varying the number of the reference users. As we can see in the table when the number of the reference users is smaller, the performance improvement is larger. Integration of diverse relevant information will help recommending items correctly to a new user who only has a limited number of reference users. This result demonstrates that integration of relevant information can solve the sparsity problem in collaborative filtering and our neural network model proves to be easy to integrate such information.

4.3 Performance Comparison of CFNN and the Existing Methods

In this section, we compare our collaborative filtering method with the existing methods. For comparison we used the first 1000 users in the EachMovie dataset, who rated more than 100 movies as the training data, and the first 100 users whose user IDs are greater than 70,000 and who rated more than 100 movies as the test data. Using the data we built 30 user and item models. Especially we selected 30 target users randomly among those who have user ID over 70,000. Two thirds of 30 models are trained using data of unbiased preferences and the rest are trained using data of a little biased preferences.

Tables 4 and 5 compare in performance our method with the existing methods such as k-NN and association rule methods. As shown in the tables the CFNN and CFNNS models show a significant improvement of performance compared with the existing methods.

Table 4. Perfomance Comparison of the User model and the Existing Methods (%)

	k-NN	Assoc. Rule	U-CFNN		U-CFNNS	
			User	User/Genre	User	User/ Genre
Accuracy	67.8	72.0	81.6	81.4	87.2	88.1
Precision	60.3	75.1	77.4	78.0	86.6	88.5
Recall	55.7	58.4	69.6	65.7	82.8	88.3
F-measure	57.9	65.7	73.3	71.3	83.3	85.7

Table 5. Performance Comparison of the Movie Model and the Existing Mehtods (%)

	k-NN	Assoc. Rule	I-CFNN	I-CFNNS
Accuracy	64.7	61.1	76.2	79.1
Precision	67.8	75.4	76.1	74.4
Recall	59.8	22.6	72.1	76.6
F-measure	62.4	34.	72.7	74.6

5 Conclusions

In this paper, we propose a collaborative filtering method based on neural network called CFNN. We also propose some methods to improve the performance including integration of additional information and selection of the reference users or items based on similarity. Our model utilizes the advantages of a neural network over other methods. It is powerful to learn a complex relationship among data and it is easy to integrate diverse information. The experiment results prove that our method show a significant improvement of performance compared with the existing methods. One of the weaknesses of our method is that the neural network model is not comprehensible, however, in recommendation the comprehensibility of a model should not be important.

Acknowledgement. This work was supported by grant no. 00013078 from the Intelligent and Interactive Module Development of the Ministry of Commerce, Industry and Energy(IIM).

References

1. Pazzani, M.J.: A Framework for Collaborative, Content-Based and Demographic Filtering. Artificial Intelligence Review. 13(5-6). (1999) 393-408
2. Sarwar, B.M., Karypis, G., Konstan, J.A. and Ried, J.: Item-based Collaborative Filtering Recommender Algorithms. Accepted for publication at the WWW10 Conference (2001)
3. Cheung, K.W., Kwok, J.T., Law, M.H., Tsui, K.C.: Mining Customer Product Ratings for Personalized Marketing. Decision Support Systems. Volume 35. Issue 2. (2003) 231-243
4. Sarwar, B.M., Karypis, G., Konstan, J.A., and Ried, J.: Analysis of Recommendation Algorithms for E-Commerce. In Proceedings of the ACM EC'00 Conference. Minneapolis. MN. (2000) 158-167
5. Lin, W., Ruiz, C., and Alverez, S.A.: Collaborative Recommendation via Adaptive Association Rule Mining. International Workshop on Web Mining for E-Commerce(WEBKDD2000) (2000)
6. Herlocker, J.L., Konstan, J.A., Borchers, A. and Riedl, J.: An Algorithmic Framework for Performing Collaborative Filtering. In Proceedings on the 22nd annual international ACM SIGIR conference on research and development in information retrieval. Berkeley. CA. (1999) 230-237
7. Claypool, M., Gokhale, A., Miranda, T., Murnikov, P., Netes, D., and Sartin, M.: Combining Content-Based and Collavorative Filters in an Online Newspaper. In Proceedings of ACM SIGIR'99 Workshop in Recommender Systems: Algorithms and Evaluation. Univ. of California. Berkely. (1999)

8. Haykin, S.: Neural Network : A Comprehensive Foundation, 2nd edition. Prentice-Hall. Inc. (1999)
9. Konstan, J., Miller, B., Maltz, D., Herlocker, J., Gordon, L. And Riedl, J.: GroupLens: Applying Collaborative Filtering to Usenet News. Communications of the ACM, 40(3). (1997) 77-87
10. Ungar, L.H., Foster, D.P.: Clustering Methods For Collaborative Filtering. Proceedings of the Workshop on Recommendation Systems. AAAI Press. (1998)
11. Billsus, D. and Pazzani, M.J.: Learning Collaborative Information Filters. In Proceedings of the Fifteenth International Conference on Machine Learing. (1998) pp. 46-53
12. Press, W. H., Teukolsky, S.A., Vetterling W.T., Flannery, B.P.: Numerical Recipes in C++. 2nd edition. Cambridge University Press. (2002)
13. Do, Y.A., Kim, J.S., Ryu, J.W. and Kim, M.W.: Efficient Method for Combining User and Article Models for Collaborative Recommendation. Journal of Kiss : Software and Application. Vol. 30. No. 56. (2003)
14. Melville, P., Mooney, R.J., and Nagarajan, R.: Content-Boosted Collaborative Filtering. Proceeding of the SIGIR-2001 Workshop on Recommender Systems. New Orleans. LA. (2001)
15. McJones, P.: Eachmovie Collaborative Filtering Data Set. http://www.rearchdigital.com/SRC/eachmovie. DEC Systems Research Center. (1997)
16. Myung Won Kim, Eun Ju Kim and Joung Woo Ryu : A Collaborative Recommendation Based on Neural Networks, Springer, LNCS 2973 (2004) pp. 425-430

Dynamic Access Control Scheme for Service-Based Multi-netted Asymmetric Virtual LAN

Wonwoo Choi[1], Hyuncheol Kim[1], Seongjin Ahn[2], and Jinwook Chung[1]

[1] Dept. of Electrical and Computer Engineering,
Sungkyunkwan University, 300 Chunchun-Dong,
Jangan-Gu,Suwon, Korea, 440-746
{wwchoi, hckim, jwchung}@songgang.skku.ac.kr
[2] Dept. of Computer Education, Sungkyunkwan University,
53 Myungryun-Dong, Jongro-Gu, Seoul, Korea, 110-745
sjahn@comedu.skku.ac.kr

Abstract. Virtual Local Area Network (VLAN) is a logical grouping of end stations such that end stations in the VLAN appear to be on the same physical LAN segment even though they may be geographically separated. Contrary to its primary expectations, server centralization, enterprise-wide collaborative applications trends raise various network resources need to be made available to users regardless of their VLAN membership. Unfortunately these trends also increase network security threats. It is general that the primary threat to network security is not caused by external users but the come from individuals inside and organization. Although network access is opened for every user in VLAN, it must be restricted to some degree. In this paper, we propose a new asymmetric VLAN management scheme in which users belonging to multiple VLANs to access another VLAN end station while both end stations are VLAN-unaware. In our scheme, an end station can communicate another end station belonging to different VLAN only after authentication. We also propose a novel VLAN access control scheme that allows only authorized users to access the multi-netted asymmetric VLAN.

1 Introduction

A VLAN is a logical grouping of end stations such that all end stations in the VLAN appear to be on the same physical LAN segment even though they may be geographically separated [1]. End stations are not constrained by their physical location and can communicate as if they were on a common LAN. Because VLAN permit hosts connected to a LAN switch to be grouped into logical groups as a function of some administration strategy, it is very flexible for user/host management, bandwidth allocation and resource optimization. IEEE 802.1Q defines a new format called tagged frame with VID (VLAN ID) and the mechanisms to support VLAN in Bridged LAN environment. The Filtering Database (FDB) contains the addressing information and frame filtering information in the form of MAC (Media Access Control) address entries and VLAN entries [2][3][4].

Contrary to its primary expectations, server centralization, enterprise-wide collaborative applications trends raise various network resources need to be made available to users regardless of their VLAN membership. Some end stations such as centralized servers need to communicate with multiple VLANs, in which the end stations simultaneously belong to more than one VLAN. But most end stations now are VLAN-unaware devices, all frames sent by one station can only be inserted with the same PVID (port VID) by the switch, thus an end station can't belong to multiple VLANs, and as a result, it can't be accessed by end stations in other VLANs without routers [3]. Although annex B of IEEE802.1 Q introduces an asymmetric VLAN, which can be used by users belonging to different VLANs to access a shared server while both users and the server are VLAN-unaware, it does not detail more about algorithms and processing for implementation.

On the other hand these trends in VLAN led to unexpected network security threats. Network security and survivable network have now become one of the most important and urgent aspects in VLANs. It is general that the primary threat to network security is not caused by external users but the come from individuals inside and organization. Hence more emphasis should be placed on internal control mechanisms, such as dynamic access control and host blocking [6][7][8]. However, it is difficult to prevent unauthorized network access at the VLAN connection points, and it is more difficult to protect against network attacks that are based on IP (Internet Protocol) and MAC (Media Access Control) address spoofing.

In this paper, we propose a new asymmetric VLAN management scheme in which users belonging to multiple VLANs to access another VLAN end station while both end stations are VLAN-unaware. In our scheme, an end station can communicate another end station belonging to different VLAN only after authentication. We also propose a novel VLAN access control scheme that allows only authorized users to access the multi-netted asymmetric VLAN.

The reminders of this paper are organized as follows. We discuss background relevant for diversity and dynamic access control problems on previous researches in section 2. The details of proposed asymmetric VLAN management and access control schemes are described in section 3 and section 4 respectively. The comparative numerical analysis is also explained in section 4. Finally, we make a conclusion.

2 Background

The IEEE 802.1Q specification establishes a standard method for tagging Ethernet frames with VLAN membership information. It also defines the operation of VLAN Bridges that permit the definition, operation and administration of Virtual LAN topologies within a Bridged LAN infrastructure. The 802.1Q standard is intended to address the problem of how to break large networks into smaller

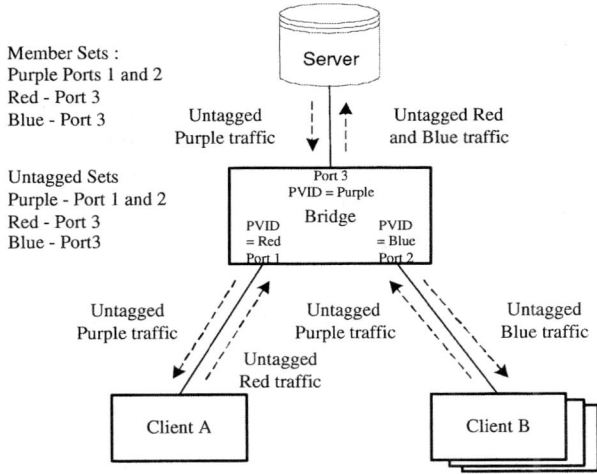

Fig. 1. Asymmetric VLAN Configuration

parts so broadcast and multicast traffic would not grab more bandwidth than necessary. The standard also helps provide a higher level of security between segments of internal networks.

One of the virtual workgroup may run into simple problem that users sometimes access to printers and servers in different VLAN and some end stations such as centralized servers need to communicate with multiple VLANs, in which the end stations simultaneously belong to more than one VLAN. It is most effective that these accesses should be provided without pass through a router. For this case, annex B of IEEE802.1 Q introduces an asymmetric VLAN, which can be used by users belonging to different VLANs to access a shared server. There are also some circumstances in which it is convenient to make use of two distinct VLANs, one used to transmit frames, and the other used to send information.

However the operation of this model is not complete and the annex does not detail more about algorithms and processing for implementation. To describe the detailed operational procedure and configuration, Sun Limin et al. [3], as shown in Fig. 1, suggested an asymmetric VLAN management protocol as an optional solution and implemented the protocol in switch with distributed architecture. But [3] controls VLAN membership not by host but by port unit, all hosts belonging to the VLAN can access server without restrictions. There are also some cases in which a server does not want to communicate some hosts belonging to same VLAN.

3 Service-Based Multi-netted VLAN

A host can be a member of the different VLAN simultaneously and temporarily in a multi-netted VLAN. Although multi-netted asymmetric VLAN offers

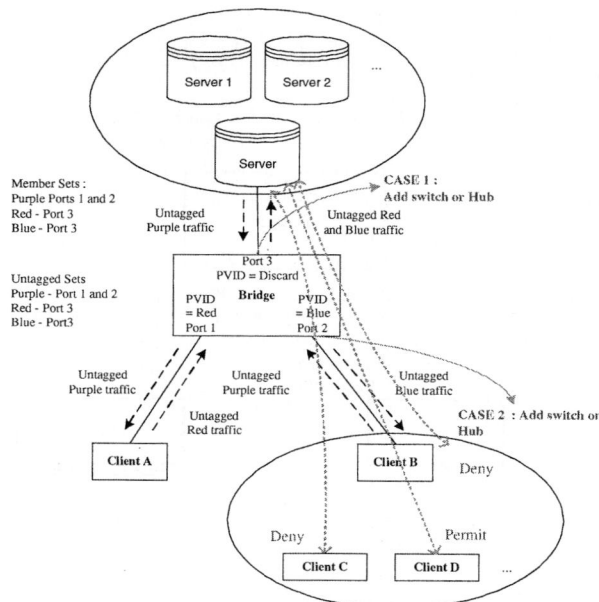

Fig. 2. Proposed Asymmetric VLAN Configuration

dynamic inter-VLAN connections, VLAN management scheme must have ability to block the host by way of a member host breaks the specific rule, such as illegal behavior and abnormal traffic. It also VLAN manager's responsibility to maintain the balance between supply and demand in the network, possibly preventing the setup of inter-VLAN connections. Moreover such control schemes must be reactive, that is, that the multi-netted VLAN will react in real-time to new requests for inter-VLAN connection setups.

The proposed service-based multi-netted asymmetric VLAN scheme is based on IEEE 802.1Q using MAC-based VLAN as default, which is not a simple but efficient way for VLAN grouping. Fig. 2 shows a simplified VLAN configuration for the proposed multi-netted asymmetric VLAN. Under normal circumstances, a pair of devices communicating in a VLAN environment will both send and receive using the same VLAN, However, as mentioned above, there are some circumstances in which it may be convenient to make use of two distinct VLANs.

The follow are described the process of the proposed multi-netted asymmetric VLAN operation.

- Step 1: Configure PVID for each switch port so that whenever a frame enters the switch, it is associated with a VID. In Fig. 2 there are three kinds of VID (Purple, Red , Blue). An administrator additionally configure the PVID (Purple, Port 1 and 2); (Red, Port2); (Blue, Port3).
- Step 2: If new address is found, asymmetric VLAN management has the chance to handle the new address message. According to the source and

Table 1. Initial Forwarding DB

Port #	PVID	MAC Address
1	Red, Purple	A_{MAC}
2	Blue, Purple	B_{MAC}
3	Red, Purple, Purple	S_{MAC}

Table 2. Forwarding DB: Case1

Port #	PVID	MAC Address
1	Red, Purple	A_{MAC}
2	Blue, Purple	B_{MAC}
3	Red, Purple, Purple	$S_{MAC} + (S1_{MAC}, S2_{MAC} \ldots)$

Table 3. Forwarding DB: Case2

Port #	PVID	MAC Address
1	Red, Purple	$A_{MAC} + (B_{MAC}, C_{MAC} \ldots)$
2	Bleu, Purple	B_{MAC}
3	Red, Purple, Purple	S_{MAC}

destination MAC address, all two addresses must satisfy certification, VID are added the entry of FDB.
- Step 3: Search for the mapping table defined in FDB, and discover the set of PVID that can communicate with the VID provided in the new address message, combine each VID in the PVID set with the new MAC address, source port etc.

Table 1 illustrates FDB fields based on Annex B of 802.1Q standard. There are some limitations, such as port based classification only, VLAN-unaware devise.

The several servers are added on Port2 in CASE 1. The FDB is the same of the following Table 2 in this case. Although a client having the right to access only one specific server by means of asymmetric VLAN, it can be approach both S1 and S2. Because a switch forwards frames by judging destination MAC and PVID, all servers of members in Port 3 can be accessed. That could be a serious security problem.

The clients in port 2 are added in CASE 2. The following Table 3 shows the FDB in this case. When only the one client communicates S, the others could have ability to reach S at the same way in CSAE1. That also could be a security problem.

As you can see the problems in CASE1 and CASE2, we have to solve the restricts for using asymmetric VLAN. At the same time, FDB are also asked for

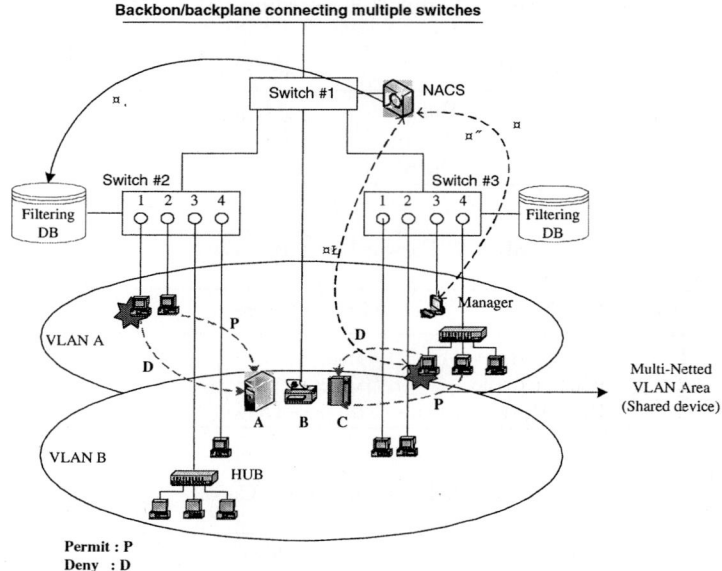

Fig. 3. Dynamic VLAN Access Control Scheme

making up the dynamic access control for multiple membership including FTDB (Filtering DB). The extra common access control scheme has a restriction that agent system is needed in each LAN segment or bridge. But if we have only one system, we can apply access control policy to cover broadcast domain without router in VLAN environment.

4 VLAN Access Control Scheme

VACS (VLAN Access Control Scheme) is defined to control the communication session between the devices in different asymmetric VLAN. This scheme helps to communicate a device as if they are same VLAN on condition that a client has a access right. In addition, the authority levels are decided by network administrator. To begin with doing this, Filtering DB keeps a tuple of Destination MAC, Source MAC, VLAN ID information for each device.

The function of VACS are described as below

- The Filtering DB sent the rank information set by manager(administrator) decides the authority to access specific devices when extending to asymmetric VLAN (ⓐ, ⓑ)
- Deny access the network originally. (①, ②)

The first function can solve the problem that original asymmetric VLAN forward the frames according to Destination MAC, PVID by storing Destination MAC, Source MAC, PVID tuple in filtering DB.

ⓐ : Any client, which want to be a member of a different VLAN, requests the right to admission of approach the shared devices.
ⓑ : The manager (Administrator) updates the current VLAN information. If an unauthorized client tries to connect A (shared device), switch #2 will not forward the frames to A. There is no information about an unauthorized client in FTDB.

The second function is directly implemented using VACS.

① : The manager who has monitored VLAN information commands a policy to block a illegal client.
② : VACS will send blocking frames added client's VID. There are two cases to block a client. One thing is to block the client being already a member of network. The other is to block the client using network in VLAN the access network originally.

VACS automatically collect network information to establish Filtering DB and block illegal host in real time. The collected information about network resource, such as VLAN ID, MAC address, etc, is sent to manager. The manager system gives a right to communicate a device in different VLAN and a protection to reach an important server using information received from VACS.

VACS component applied with dynamic access control scheme is shown in Fig. 4. There are three types of VACS component modules. The essential function of VACS is monitoring VLAN frame, VLAN frame creation and Authentication. Also, VACS must be a VLAN-aware device to analysis entire VLAN frame. Additionally VACS can have ability to block a client regardless of DHCP or Static environment. If unauthorized client try to be a member of network, VACS can protect network resources.

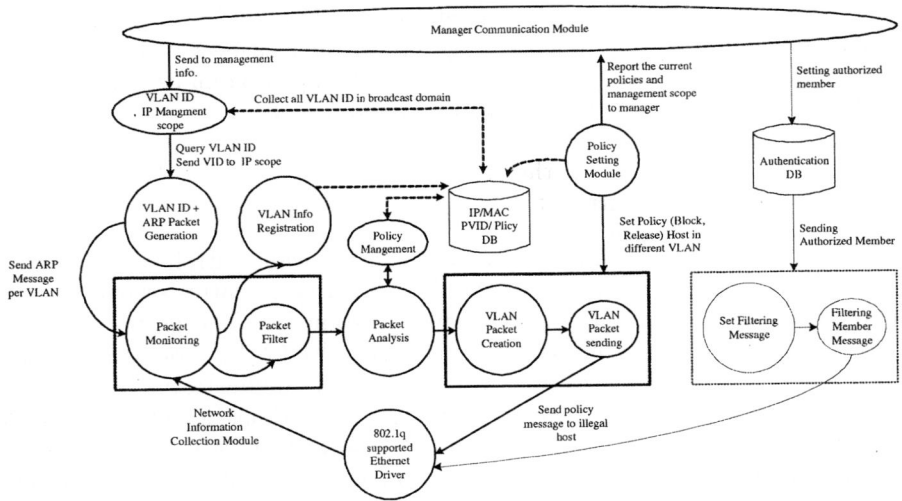

Fig. 4. VACS Components

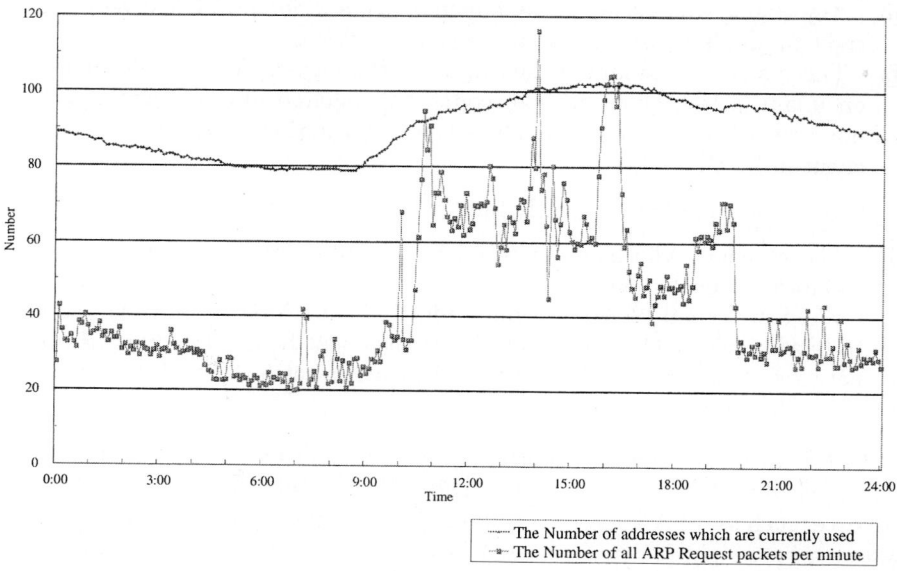

Fig. 5. Proposed Asymmetric VLAN Configuration

Because access control techniques using ARP (Address Resolution Protocol) is possible by just one packet, VACS must analyze all ARP packets. We analyzed the number of engaged IP address and ARP request packet for 7 days in C class network. This test examined IP address that is working by sending ARP request packet about all IP addresses every 5 minutes, and inspected the number of all ARP request packet except packet that is transmitted. As shown in Fig. 5, 90 IP address was used on an average and the number of ARP request packet measured by average 45 at 1 second. Because number of request packet is few and the number of packet that is required for IP address administration is less, it will not become an issue in performance aspects.

5 Conclusions

In this paper, a new algorithm for establishing a access controlled asymmetric VLAN is proposed. We also propse the dynamic access control scheme for the enhancement of asymmetric VLAN in 802.1Q standard. The proposed VACS address a way to solve the problem in asymmetric VLAN by processing the authentication for a different VLAN. Also, a network administrator having only one VACS can control entire members of VLAN in real time. In fact, most clients don't care about network architecture. So they want to be a member of different VLAN. But point of view in network administrators VACS is very efficient method to secure, manage VLAN. Because VACS can automatically learning VLAN information by capturing frames, it is applied both port-based membership and MAC-based membership, too. The proposed VACS can be applied dynamic access of VLAN membership to protect unauthorized users by adding VID on ethernet frame. VACS can insert a VLAN ID according to administration DB, administrator setting. The VACS mechanism can be also applied to the DHCP by adding function to check unauthorized user's MAC.

References

1. D. Ruffen, T. Len, J. Yanacek: Cabletrons SecureFast VLAN Operational Model Version 1.8, IETF RFC 2643, Aug. (1999)
2. IEEE Standards for Local and Metropolitan Area Networks: Virtual Bridged Local Area Networks, IEEE Standard 802.1Q, (1998)
3. Sun Limin, Zhang Hao, Wang Zaifang, Wu Zhimei: Asymmetric VLAN Management Protocol for Distributed Architecture, International Conference on Algorithms and Architectures for Parallel Processing (ICA3PP'02), (2002)
4. Xiaoying Wang, Hai Zhao, Mo Guan, Chengguang Guo, Jiyong Wang: Research and Implementation of VLAN Based on Service, IEEE Globcom'03, (2003) 2932-2936
5. Enrique J. Hernandez-Valencia, Pramod Koppol, Wing Cheong Lau: Managed Virtual Private LAN Services, Bell Labs Technical Journal , Vol. 7, No. 4, (2003) 61-76
6. Carter and Catz, Computer Crime: An emerging challenge for law enforcement, FBI Law Enforcement Bulleting, Dec. (1996) 1-8
7. J. PIKOULAS, W.BUCHANAN, M.MANNION, K.TRIANTAFYLLOPOULOS: An Intelligent Agent Security Intrusion System, International Conference and Workshop on the Engineering of Computer-Based Systems (ECBS.02), (2002) 1-6
8. T. Alj, J-Ch. Gregoire: Admission Control and Bandwidth Management for VLANs, IEEE Workshop on High Performance Switching and Routing, (2001) 130-134

New Binding Update Method Using GDMHA in Hierarchical Mobile IPv6

Jong-Hyouk Lee, Young-Ju Han, Hyung-Jin Lim, and Tai-Myung Chung

Cemi: Center for Emergency Medical Informatics,
School of Information and Communication Engineering,
Sungkyunkwan University,
Chunchun-dong 300, Jangan-gu, Suwon, Kyunggi-do,
Republic of Korea
{jhlee, yjhan, hjlim, tmchung}@imtl.skku.ac.kr

Abstract. Mobile IPv6 is a protocol that guarantees mobility of mobile node within the IPv6 environment. However current Mobile IPv6 supports simple the mobility of the mobile nodes and does not offer any special mechanism to reduce the handoff delay. For the purpose of reducing the handoff delay, Hierarchical Mobile IPv6 has been studied. However Hierarchical Mobile IPv6 supports the micro mobility only in the area managed by mobility anchor point. Therefore, the handoff delay problem still has been unsolved when mobile node moves to another mobility anchor point. In this paper, we propose GDMHA (Geographically Distributed Multiple HAs) mechanism to provide macro mobility between the mobility anchor points. Using this mechanism, a mobile node performs Binding Update with the nearest Home Agent, so that the delay of Binding Update process can be reduced. It also reduces the handoff delay.

1 Introduction

The user demand for mobile service has increased today. However, current IP protocol has a fatal problem that the MN (Mobile Node) can not receive IP packets on its new point when the MN moves to another network without changing its IP address. To solve this problem, IETF (Internet Engineering Task Force) has proposed a new protocol entitled 'Mobile IP'.

MIPv6 (Mobile IPv6) is a protocol that guarantees Mobility of MN within the IPv6 environment. Especially, it basically provides the route optimization and doesn't need FA (Foreign Agent), which is used for MIPv4 (Mobile IPv4). In MIPv6 world, a MN is distinguished by its home address. When it moves to another network, it gets a CoA (Care of Address), which provides information about MN's current location. The MN registers its newly given CoA with its HA (Home Agent). After that, it can directly communicate with CN (Correspondent Node) [1], [2]. The basic operation of MIPv6 is shown in Fig. 1.

MIPv6 supports just the mobility of MN and does not offer any special mechanism to reduce the handoff delay. For the purpose of reducing the handoff delay, Fast MIPv6 (Fast Handoff for Mobile IPv6) and HMIPv6 (Hierarchical Mobile IPv6) have been studied [3], [4].

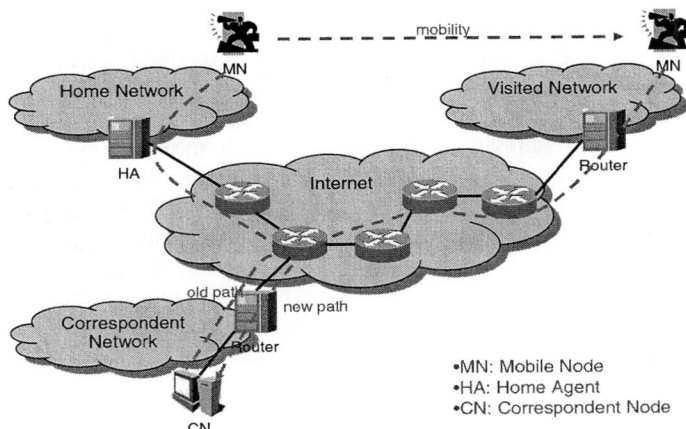

Fig. 1. The basic operation of MIPv6

Fast MIPv6 mechanism uses layer 2 triggers. HMIPv6 mechanism introduces MAP (Mobility Anchor Point) to manage the movement of MN locally. HMIPv6 supports the micro mobility only in the area managed by MAP. In other words, HMIPv6 does not care about its outside world. Therefore, the handoff delay problem still exists when MN moves to another MAP.

In this paper, we propose using geographically distributed multiple HAs to provide macro mobility within the HMIPv6 environment. When MN moves from a MAP to another MAP, it performs BU (Binding Update) with the nearest HA, so that the delay of BU process can be reduced. Since those HAs have the same Anycast Address, the nearest HA can be chosen easily.

In Chapter 2, we will see how the handoff process takes place section by section when a MN moves to a new network. Then, we will analyze the delay caused during BU. Also, we'll see the technique involved in regional MN management by HMIPv6 utilizing the MAP. At the end of this chapter, we will see about the GIA (Global IP Anycast), which is used for BU between multiple HAs and MN. Chapter 3 describes the GDMHA (Geographically Distributed Multiple HAs), which introduces GIA into HMIPv6, and Chapter 4 compares and analyzes the existing mechanisms that import HMIPv6 and GDMHA mechanism. Chapter 5 discusses this research's conclusion and the consequent technology.

2 Related Works

2.1 Mobile IPv6 Handoff

When a MN detects an L3 handoff, it performs DAD (Duplicated Address Detection) on its link-local address and selects a new default router in result of Router Discovery, and then performs Prefix Discovery with that new router to form new CoA. It registers its new primary CoA with its HA. After updating its home registration, the MN updates associated mobility bindings in CNs which are performing route optimization.

The time is needed to detect MN's movement detection, to configure a new CoA in the visiting network, and to resister with the HA or CN together make up the overall handoff delay.

Movement detection can be achieved by receiving RA (Router Advertisement) message from the new access router. The network prefix information in the RA message can be used to determine L3 handoff by comparing with the one received previously. After movement detection, the MN starts address configuration which makes topologically correct addresses in a visited network and verifies that these addresses are not already in use by another node on the link by performing DAD (Duplicated Address Detection). Then the MN can assign this new CoA to its interface. This CoA must be registered with the HA for mobility management and CN for route optimization [5], [6]. Fig. 2 illustrates the handoff process occurring as the MN is transferred to a new network.

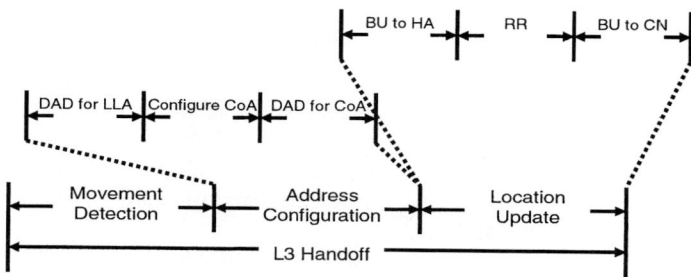

Fig. 2. MIPv6 Handoff

Fig. 2 illustrates the MN carrying out BU to HA of the new network during the Location Update process. Such a process increases the delay required within the BU process as the MN moves further away from HA and incurs the final result of increasing the Location Update delay. HMIPv6, whose research purpose is to reduce handoff within Layer 3, can carry on the role of an assumed HA by utilizing the MAP and can reduce Location Update.

2.2 Hierarchical Mobile IPv6

HMIPv6 introduces a special conceptual entity called MAP and regionally manages MN. The MAP acts as the local HA for the MN and it is a router that manages a binding between itself and MN currently visiting its domain. When a MN attaches itself to a new network, it registers to the MAP serving that domain. The Fig. 3 illustrates architecture and operation of HMIPv6.

MAP intercepts all the packets destined for the MN within its service area and tunnels them to the corresponding on-link CoA of the MN. If the MN moves into a new MAP domain, it needs to configure a regional address and an on-link address. After forming these addresses, the MN sends a binding update to the MAP, which will bind the MN's regional address to its on-link address. In addition to the binding at the MAP, the MN must also register its new regional address with HA and CNs by sending another binding updates that specify the binding between its home address and the regional address. If the MN changes its current address within a local MAP domain, it only needs to register the new on-link address with the MAP [3], [7], [8].

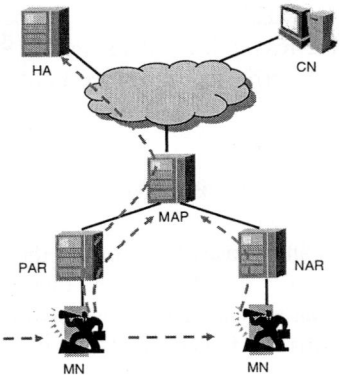

Fig. 3. Architecture and operation of HMIPv6

Since the location update delay mainly depends on transmission delay between MN and HA or CNs, this hierarchical architecture can reduce the location update delay drastically. This process is no concern of supporting the micro mobility, therefore it cannot reduce the delay occurring during the MN transfer between MAPs.

2.3 Global IP Anycast

Anycast routing arrives onto the node that's nearest to the Anycast's destination address. The nodes with same address can be made to function as Web mirrors and can gain the function of dispersing the node traffic function and can be chosen and utilized as the nearest server by the user. For example, this process can be considered when a user wishes to download a file from a Web page. In order for the user to download a file from the nearest file server, the user doesn't choose a file server but it can be automatically directed to the nearest file server to download the file via the operation of Anycast [9], [10].

The file download speed via Anycast not only has an improved speed but can reduce the overall network traffic. Fig. 4 illustrates the general operation process of Anycast.

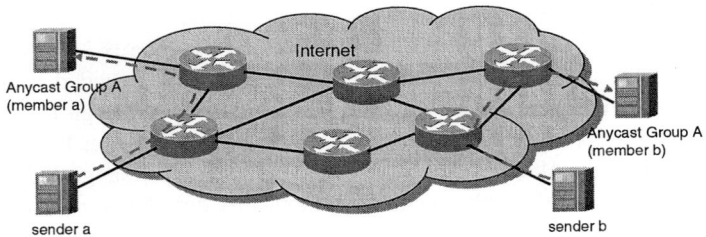

Fig. 4. Operation of Anycast

Although the existing Anycast operates within the nodes of the identical prefix, GIA supports global routing [10]. Because of that, a node that exists within a different network can be included in Anycast.

This research provides GDMHA mechanism to support macro mobility between the MAPs. When the MN carries out BU outside its MAP administrative domain, the geographically distributed HA carries out the nearest HA and BU and reduces the BU delay via GIA.

3 Mobility Management Between MAPs

We proposed GDMHA mechanism uses GIA to find the nearest HAs among several HAs. The HAs are geographically distributed. And these are belong to the same anycast group. A packet sent to an anycast address is delivered to the closest member in the GIA group. A MN performs BU to nearest HA by the GIA mechanism. According to the GIA, the BU sent to an anycast address is routed to the nearest HA.

3.1 The Mobility Management Using GDMHA

HMIPv6 manages the MN's movement within the MAP administrative domain. Because of that, when MN moves to another MAP, it must carry out BU with its HA within HMIPv6 environment as shown in Fig. 5.

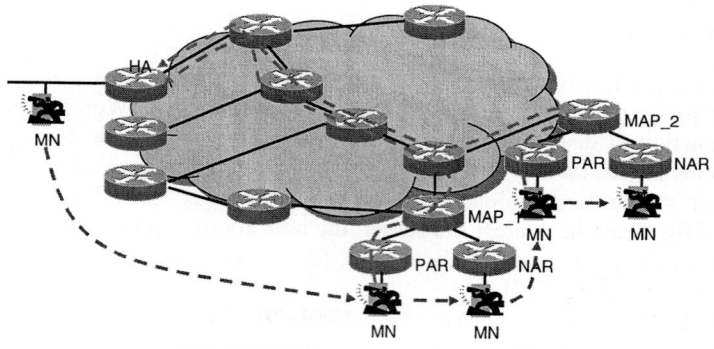

Fig. 5. BU operation within the HMIPv6 environment

At this time, if the distance of MN and HA is far away, the BU delay will be increased; this is the aspect of macro mobility that cannot be provided by HMIPv6's micro mobility utilizing the MAP. However, if we distribute through HAs belonging to an identical Anycast group, this problem should be manageable. Fig. 6 illustrates the geographically distributed HAs and MN's BU operation suggested by this research within the HMIPv6 environment. Geographically distributed multiple HAs will have the identical Anycast address.

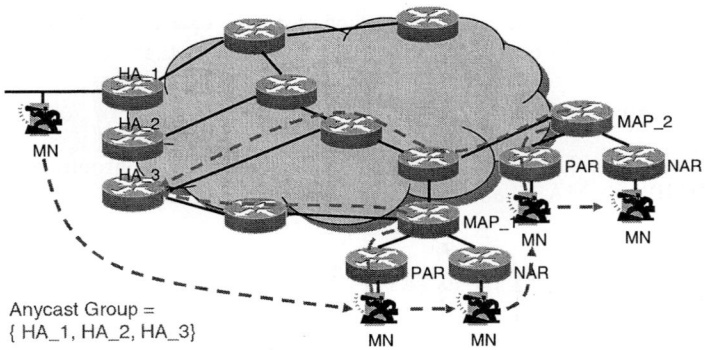

Fig. 6. BU operation within GDMHA environment

Fig. 6 illustrates the BU delay reduction when BU is carried out with a HA, which is the geographically closest one to the MN, provide that the MN moves to a different MAP for the HA belonging to the same Anycast routing group.

3.2 Binding Cache Router Advertisement

Since HA is geographically distributed, it is needed to synchronize the Binding Cache information that each HA. As shown in Fig. 7, for the purpose of synchronization, HA utilizes the BCRA (Binding Cache Router Advertisement) message, which is a modified RA message.

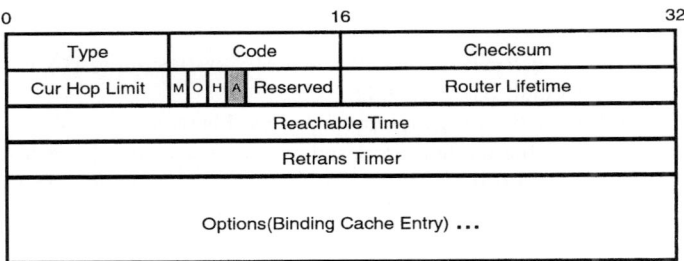

Fig. 7. Binding Cache Router Advertisement message

The destination address of the new BCRA message becomes the HA's unicast address belonging to the HA group. The basic value of the current Hop Limit field is assigned to 255 and made to reach geographically distributed HAs. The newly added A bit signifies the BCRA message. For the Option section, the BCRA message conveying HA's binding cache information is sent with the HA's Binding Cache information. The BCRA message is transmitted only when HA receives BU from the MN and must not exceed the maximum IPv6 MTU.

The HA that receives the BCRA message instantly update the Binding cache information for addition of new MN.

3.3 BU Operation Procedure

Fig. 8 illustrates the HA's BU operation in the GDMHA environment, when MN moved to HMIPv6's MAP then moved again to another MAP.

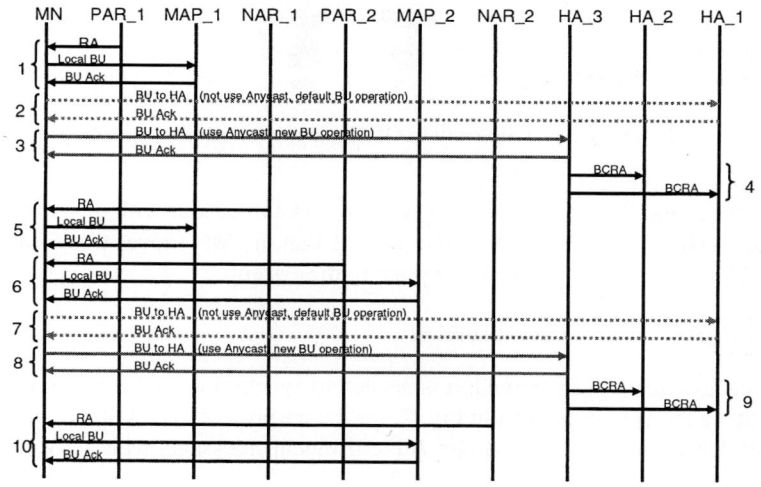

Fig. 8. MN's BU Operation Procedure

As illustrated in Fig. 9, the BU procedure that MN carries out to HA is as follows:
Procedures 2 and 7: Default BU operation
Procedures 3 and 8: New BU operation proposed by this paper
If we were to employ the method proposed by this paper, like the procedures 2 and 7, BU only carried out with the nearest HA. Therefore, BU delay can be extremely reduced.

4 Analysis of BU Delay

Like Fig. 2 during Location Update, proceeding after the Address Configuration, BU to HA and RR (Return Routability), and BU to CN process occur using a newly given CoA. All messages are not retransferred but obtained once; and when T_{work} represents work process being carried out, and $D_{x \cdot y}$ represents the message transfer delay between x and y, Location Update can be expressed by the following formula:

$$T_{LocationUpdate} = T_{BUtoHA} + T_{RR} + T_{BUtoCN} \qquad (1)$$

Location Update process divided up in stages can be expressed as follows:

$$T_{BUtoHA} = D_{MN \cdot HA} + D_{HA \cdot MN} \qquad (2)$$

$$T_{RR} = \max[(D_{MN \cdot HA} + D_{HA \cdot CN}), D_{MN \cdot CN}] + \max[(D_{CN \cdot HA} + D_{HA \cdot MN}), D_{CN \cdot MN}] \qquad (3)$$

$$T_{BUtoCN} = D_{MN \cdot CN} + D_{CN \cdot MN} \qquad (4)$$

Table 1. illustrates below, in hop units, BU operation between MN and HA_1, which is original HA of MN Fig. 5, BU operation between MN and the nearest HA_3 determined by using the geographically distributed multiple HA group's anycast operation.

Table 1. BU Delay Comparison in Existing BU and in Multiple HA Environment

	BU delay in current method		BU delay in proposed method	
	MAP_1	MAP_2	MAP_1	MAP_2
T_{BUtoHA}	14	14	8	10
$D_{MN \cdot HA}$	7	7	4	5
$D_{HA \cdot MN}$	7	7	4	5

As we see in Table 1, It is obvious that using GIA and carrying out BU with the nearest HA is more efficient then current way.

Fig. 9 and Fig. 10 show that there are noticeable decreases in delay with increase of the number of HAs.

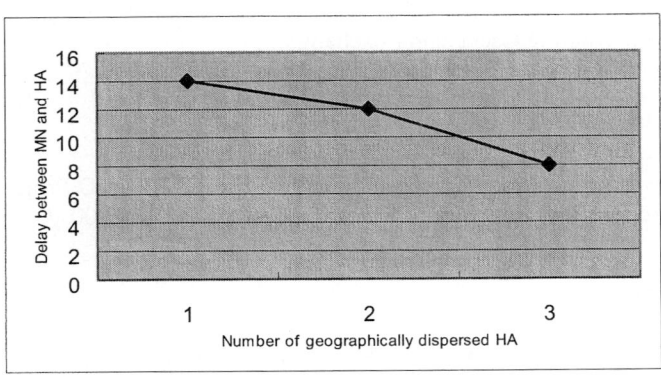

Fig. 9. $D_{MN \cdot HA}$ in MAP_1

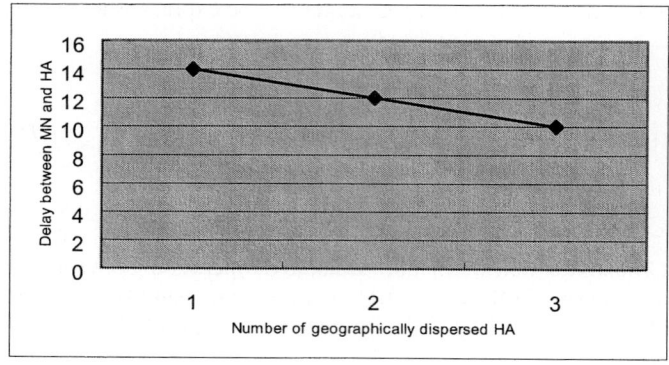

Fig. 10. $D_{MN \cdot HA}$ in MAP_2

5 Conclusions and Further Work

In order to reduce the MN's BU delay regarding HA, this paper has proposed the GDMHA (Geographically Distributed Multiple HAs). This mechanism resolved the BU delay problem of the existing HMIPv6's problem of BU delay between MAPs transfer. It is also effective when MN moves far away from the home network because BU delay can be reduced as HA is widely distributed over a large geographical area. Moreover, it would be a good example of and function well in anycast mobile environment, one of IPv6's address modes.

This mechanism can be effectively utilized within a single administrative domain, such as an ISP or a university campus. However, HA administrative information must be maintained and HA function needs to be augmented in order to manage the BCRA message for the purpose of Binding Cache entry exchange between HAs.

Further research activities proposed by this paper is research into the administrative information that HA, within GIA, needs to maintain and reducing the load generated during the Binding Cache entry exchange.

Acknowledgements

This study was supported by a grant of the Korea Health 21 R&D Project, Ministry of Health & Welfare, and Republic of Korea (02-PJ3-PG6-EV08-0001).

References

1. D. Johnson, C. Perkins, J. Arkko: Mobility Support in IPv6, RFC 3775, June 2004.
2. C. Perkins: Mobility Support in IPv4, RFC 3344, August 2002.
3. R. Koodli: Fast Handovers for Mobile IPv6, draft-ietf-mobileip-fast-mipv6-08.txt, October 2003.

4. H. Soliman, C. Castelluccia, K. El-Malki, and L. Bellier: Hierarchical Mobile IPv6 mobility management (HMIPv6), draft-ietf-mobileip-hmipv6-08.txt, Jun 2003.
5. R. Hsieh, Z.G. Zhou, A. Seneviratne: S-MIP: a seamless handoff architecture for mobile IP, INFOCOM 2003, Vol. 3, pp. 1173 - 1185, April 2003.
6. S. Sharma, Ningning Zhu, Tzi-cker Chiueh: Low-latency mobile IP handoff for infrastructure-mode wireless LANs, Selected Areas in Communications, IEEE Journal, Vol. 22, Issue 4, pp. 643 - 652, May 2004.
7. I. Vivaldi, M.H. Habaebi, B.M. Ali, V. Prakesh:Fast handover algorithm for hierarchical mobile IPv6 macro-mobility management, APCC 2003, Vol. 2, pp. 630 - 634, Sept. 2003.
8. I. Vivaldi, B.M. Ali, H. Habaebi, V. Prakash, A. Sali: Routing scheme for macro mobility handover in hierarchical mobile IPv6 network, Telecommunication Technology, 2003. NCTT 2003 Proceedings. 4th National Conference on , pp. 88 - 92, Jan. 2003.
9. S. Weber, Liang Cheng: A survey of anycast in IPv6 networks, IEEE Communications Magazine, Vol. 42, Issue 1, pp. 127 - 132, Jan. 2004.
10. D. Katabi, J. Wroclawski: A Framework for Scalable Global IP-Anycast (GIA), ACM SIGCOMM Computer Communication Review, Vol. 30, Issue 4, pp. 315, Oct 2000.

Security in Sensor Networks for Medical Systems Torso Architecture

Chaitanya Penubarthi[1], Myuhng-Joo Kim[2], and Insup Lee[1]

[1] Department of Computer and Information Science, University of Pennsylvania,
Philadelphia 19104-6389, USA
cpenubar@saul.cis.upenn.edu, lee@cis.upenn.edu
[2] Department of Information Security, Seoul Women's University,
Seoul 139-774, Korea
mjkim@swu.ac.kr

Abstract. Wireless sensor networks have become a ubiquitous part of the computing community, with applications ranging from health care to warfare, to pollution control, to the Mars rover. Key aspects of these sensing devices are the constrained computational and energy resources, and the communication security aspects. A secure communication between these sensing devices is a desired characteristic when it comes to medical applications, military and other mission critical applications. A trusted third party based architecture under which certificate authorities store the public-key certificates of participating hospitals and medical practitioners is a solution for security in tele-medicine, but this does not hold for resource constrained sensing devices. In this paper, we present a new Torso architecture distributed sensor network approach of monitoring patients as a more feasible, secure, and efficient sensor network mechanism in medical systems.

1 Introduction

The preparation Wireless Sensor Networks (WSNs) have been a subject of extensive research with their use being advocated for a wide variety of applications. WSNs applied to medical technologies have recently emerged as an important application, with fusion of wireless secure communication and medical techniques with sensing devices [1], [2], [3]. Biomedical sensors are being developed for retinal prosthesis to aid the visually impaired. Our research forms the basis for another medical application where the sensors form a distributed network over the patients body. The medical application is a new concept wherein we use the wireless network technology to monitor patient's vital functions and provide instantaneous medical feedback. There has been a lot of research into medical applications but all were aimed at making the network mobile. The security aspect of the communication was not considered as security in sensors is believed to be expensive. In this paper, we propose the Torso Architecture, which is a distributed layered approach to monitoring patients, and also explain the security protocol embedded in this architecture which makes the communication secure and efficient.

The envisioned Torso distributed sensor network for patient monitoring and care has a leaf node layer which follows a ring architecture consisting of patient's sensors which are self organizing, called the SENSOR LEVEL. The intermediate layer consists of a super node, which acts as a supervisor to the leaf nodes and also resides with the patient, called the SUPER-NODE LEVEL. The final layer is the root node or the central base station. This concept provides an individual the flexibility to roam around freely without having to wait at the treatment centers and thus giving the individuals higher QOL (Quality of Life) [4]. The concept of wireless medical treatment is achieved by the patient carrying a sensor network that communicates with the root, which is the doctor's access point.

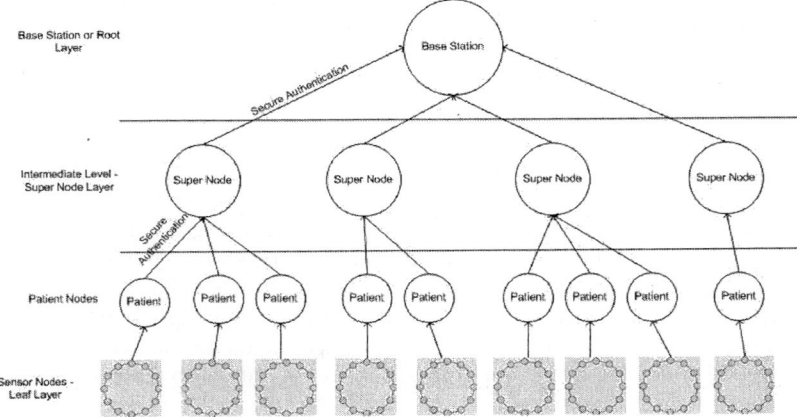

Fig. 1. The Torus Architecture

In this paper, we present a distributed sensor network architecture which is scalable, secure and efficient in monitoring patients. The architecture is termed as *the Torso Architecture* and we define a security protocol that utilizes the bandwidth and also provides secure communication using a key setup protocol that does not consume a lot of energy resources. The following sections are organized as follows. Next section presents the background required into distributed sensor networks. Section III presents the Torso Architecture in detail. Section IV describes the Communication Protocol. Section V gives the implementation details. We also present of an analysis of the protocol in this section. Section VI presents the conclusion.

2 Background

Sensor Networks are data-centric rather than address-centric. Queries are directed towards a cluster of sensor nodes rather than specific sensor addresses. The medical applications of sensor networks have long been a research area focusing on patient monitoring. However, in many cases the bottleneck of bandwidth limits the usage of these applications. In this paper, we present a way to monitor patients using distributed sensor networks to form a sensing ring architecture over the human body to monitor patient's vital information and provide instantaneous medical feedback. We present a layered approach to monitor the patient's health condition and propose a schema for better utilization of the bandwidth called *the Torso Architecture*. We define a *criticality quotient* for every patient that determines the amount of bandwidth allotted to them.

The Torso Architecture provides the capability independent of wired monitoring and diagnosis by way of a ring layered architecture spread over the patient's body. The patient communicates with the supervisor that will communicate in turn with the base station which is the treatment center. The leaf layer or the sensor layer is sensors that sense the information and transmit it to the next level in the hierarchy, which is the patient layer. The patient layer is followed by the intermediate node layer and then on top of the hierarchy is the root.

3 Layers of the Torso Architecture

3.1 Leaf Layer

The leaf layer is a collection of sensor nodes forming a ring architecture across the human body which communicate over a wireless network. The individual sensors do not have any processing power but are mere sensing devices. The leaf nodes form a ring architecture for reliability, efficiency and more accurate communication with the higher levels in the hierarchy. The leaf layer is also called the Sensor Node Layer or Sensor Node Level. The formation of the leaf layer is critical to the entire network as they form a ring architecture and based on this formation the nodes sense the vital information and send the information up the Torso architecture to the root which is the treatment center. The leaf layer sensors have no additional capability then mere sensing and passing the information one layer up the hierarchy.

Formation of the Ring Architecture. *Self-organization* refers to the ability of the system to achieve the necessary organizational structures without requiring human intervention, particularly by specially trained installers and operators. Self-organization is a critical attribute needed to achieve the wide use and applicability of distributed sensor networks.

Advantages of the Ring Architecture. The ring architecture formation over the human body gives the sensor network the all important stability with respect to the body movements. The communication protocol is simple and straightforward. The importance of any sensor or a leaf node is decided on the patient's health condition. For example, if a patient is suffering from a heart attack, the most important sensor node could be the heart beat sensor while for a patient suffering from asthma, the all important sensor node could be something totally different.

3.2 Patient Node Layer

This node layer represents the patient itself. They are two ways to present this node. The node can itself reside on the patient. The patient node is a supervisor of all the leaf nodes of that patient and gathers information from all the sensors and sends the information to the intermediate super node that it interacts with. The patient node collects all the information from the leaf nodes and does the processing of information and sends it to the intermediate layer. The patient node is responsible for communicating with all of the leaf nodes. The patient node receives feedback from all the sensor nodes and sends them to the intermediate node. The patient node is directly responsible to the above super node that it belongs to. Once authenticated the patient node is now responsible for that super node and sends the information to the intermediate node which in turn sends the information to the root.

3.3 Intermediate Layer – Super Node Layer

The super node layer is the next layer to the central root. This node is responsible for the up and down communications with the patient layer and the root. This layer receives information from all the patient nodes that are under this layer. The distribution of the patient nodes under a particular super node is done on the basis of proximity, geographical location. Depending on the current location of the patient, the patient node is controlled by a different super node. The patient node authenticates to the node before the node can send information to the intermediate node. The super node communicates with the root and sends all the information. The link from the patient node up to the intermediate node and the link between the intermediate node and the root node are both band-limited due to the limited wireless link capacity. Both the links are prone to malicious attacks and hence require a secure and efficient communication up and down the Torso architecture. Each node in this layer will be in contact with its geographical neighbors.

4 Communication Protocol

All the communication between the layers of the Torso architecture is wireless and as the result is bandwidth limited and is also prone to various attacks. The Denial of service and eavesdropping of crucial information have always been a threat to wireless networks. Eavesdropping of information may not be externally harmful but in case of medical applications it is not desirable to send information in the open as the data being sent may contain some confidential information. Trusted third party based architectures are impractical for sensor networks because of the resource constraints that sensing devices have. A unique light weight key exchange mechanism is required for secure communication. The communication bandwidth also forms a bottleneck because of the limited bandwidth of wireless networks, a huge number of remote patients cannot transmit information at the same time which is not desirable in medical applications as the number of patients in a particular region cannot be predetermined. In this paper, we address the three most important bottlenecks for sensor network communication.

- The bandwidth bottleneck of the wireless link from the patient node to the intermediate layer and from the intermediate layer up to the root node.
- The security, confidentiality and privacy of the information being transferred.
- The limitation of battery power of the patient nodes.

Bandwidth has always been a bottleneck in wireless communication. We propose a better utilization of the available bandwidth depending on the criticality of the patient. *Criticality* is the term used to describe the importance of the doctor's advice. We describe this value as the basis for determining the bandwidth allotted to that particular patient. The value can range anywhere from 0 to 100, where each of these values has its own meaning. The value of 100 meaning the patient requires attention and a value of 0 requires no attention and hence no bandwidth. Therefore, this scheme of better utilization of the bandwidth works really well in very constrained systems. The information exchange between the sensors is vital and should therefore be required to maintain confidentiality. The secure mechanism provides the required confidentiality at a very low cost computational power.

4.1 Bandwidth Bottleneck

The link between the root node and the intermediate layer is bandwidth constrained. The root node maintains a table for different intermediate nodes with the number of patients each of them are addressing. Based on this number, the root node determines the amount of bandwidth to be allotted to each of the intermediate nodes. Each patient is also associated with a criticality quotient, which determines the limit of bandwidth that the patient node gets allotted. The tabular column present at the root node looks somewhat like this.

Table 1. Example of a Bandwidth Allocation Table (BAT) on a root

Node ID	Cardinality	Bandwidth Allotted
12	5	50
13	3	30
14	1	10
16	1	10
20	0	0

The root node determines and calculates the Bandwidth Allocation Table based on the input from its intermediate children and generates the BAT table as shown and calculates the allocation bandwidth percentage. The *Cardinality* of the intermediate nodes is the defined as the number of patients it is supervising. We will discuss more about the cardinality as we go through the paper.

The root node determines bandwidth allocation based on the bandwidth allocation table at the particular instant. The bandwidth at this instant is allocated 50% based on the BAT to an intermediate node with the Node Id 12. Thus, the root node best utilizes the available bandwidth and allows the patient with more importance of treatment process more information, thus acting as a life saver. The Intermediate node calculates the bandwidth required directly based on the criticality of the patient and distributes the allotted bandwidth among all of its patient nodes. The intermediate node 12 in the above example, distributes the bandwidth allotted to it based on the criticality of the patient. Assuming the intermediate node has just one patient node, all the bandwidth allotted to that intermediate node is utilized by the single patient that it serves and receives information from. The bandwidth utilization is therefore best utilized.

4.2 Secure Communication

The focus of recent research on WSNs has been on extending their lives using energy-conserving communication protocols [5],[6]. Security is of utmost importance especially in the field of medical applications where strategic decisions are expected to be taken based on the information received from the sensor nodes [7]. We propose a lightweight key setup protocol that caters best to the energy conservation while providing the required confidentiality and privacy of information exchange.

4.3 Challenges for Security Implementation in SN

In this section we identify some of the unique characteristics that make security in SN different from usual networks.

Scalability: The patients authenticating with different intermediate nodes should be very efficient such that it will not add overhead to the communication of the network. Contributing key establishment protocols might not be most efficient in these networks where having such a large number of network nodes might actually slow down this process. The protocol should be scalable in terms of the patient nodes. We propose a mechanism where in the scalability of the overall network could be increased by using efficient communication mechanism there by making the network more reliable when a patient requires attention and also providing better utilization of bandwidth which adds to the scalability.

Resource Constraints: One of the most important aspects of WSNs is the limited energy constraint. Depending on their role within the network, some of these nodes have some power recharging mechanisms. In order to ensure longer life for the nodes, energy efficient mechanisms and power conserving methodologies should be adapted at every level in the network. Pottie et al. have established that the energy cost of transmitting 1Kb over a 100 m distance is the same as the energy required by a general-purpose 100 MIPS/W processor to execute 3 million instructions. The protocol implemented for security should minimize the exchange of security related setup messages. Also the cryptographic metric selected for encryption should be small enough to capture the resource constraints. Our protocol implementation takes advantage of the above aspects of sensor networks.

- Data aggregation is less expensive then transmitting data.
- Not all the data that has been transmitted need to be encrypted.

5 Implementation Issues

In this section, we describe the basic implementation issues of the security protocol. We detail key exchange between the patient node and the intermediate node for transmitting data and the communication between the intermediate and root node.

5.1 Assumption

Before describing the protocol, we define the assumptions underlying the model. We assume that the radio model is symmetric *i.e.*, given a signal-to-noise ratio, the energy required to transmit an m bit message from node A to node B is the same as the energy required to transmit the same m bit message from node B to A. We also assume that the root is more resourceful than the regular sensor node. We also assume that the intermediate nodes are more resourceful than the regular sensor nodes but not in the order of the root. The root with all its resources can store all the keys and access them directly without any overhead. We assume that each sensor node is created with a unique Device Identifier (DId) which is known only by that particular node. We also assume that the root has, built into it all the DIds for all the sensor nodes that have been dispersed into the network. We also assume that all the sensors have an in-built system clock.

5.2 Notation

We will use the following notation to represent our cryptographic operations.

- A message M encrypted with a symmetric key K is represented as $E_K(M)$.
- $E_{PUBLIC\ KEY}(M)$ is a message encrypted using the public key of root.
- $MAC(K,C)$ is the message authentication code computed over the counter C using key K.
- $A \rightarrow B - E_A(M)$ is a message from node A to node B and the message is M encrypted using key A.
- $A \Rightarrow E_{PUB}(M)$ is a *multicast* message to all the nodes in the network.

We define some related messages such as JOIN-ROOT, AUTHENTICATE-NEIGHBORS, JOIN-INTERMEDIATE NODE.

5.3 The Design Issues

We define two phases in the protocol, the first being the intermediate nodes joining the network and then each of the intermediate nodes authenticating with its neighbors. The other phase is the patient nodes authenticating when leaving or joining an intermediate node. The first part of the protocol implementation is key setup process. The key setup process is used to authenticate the nodes as part of the network and have the node assigned a key. We use the DId that has been determined for each of the sensor node and is stored in the root. Initially, each intermediate sends a JOIN-ROOT message to the root, by encrypting the Device Identifier along with the current Time stamp, using the public key of the root. When root receives the message, it decrypts using its private key and compares the Device Identifier with that of the database that it stores. If it matches, it authenticates the node as a part of the network and sends back an authentication message. The authentication message contains a unique temporary node identifier, which is used to communicate henceforth along with a randomly generated number, encrypted using a key that is the MAC of the DId and the time stamp. This is the encryption key for the sensor nodes. The symmetric key is also computed at the sensor node using the MAC of its DId and the time stamp and is decrypted.

$$A \to ROOT - E_{PUBLICKEY}(A_{DId}, TimeStamp)$$
$$KEY_A(M) = MAC(A_{DId}, TimeStamp)$$
$$ROOT \to A - E_{KEYA}(NodeId_A, R_A)$$

Once all the intermediate nodes are authenticated, we have the first two layers of the network setup. The same process is executed with the patient nodes and the intermediate nodes. The Communication between the sensor nodes and the patient nodes is assumed to be secure as they are part of the human body.

The patient nodes initially broadcast the JOIN-INTERMEDIATE NODE message that is encrypted using a predetermined symmetric key which is known to all intermediate nodes when manufacturing. We determine the Time To Live (TTL) to be the time for the message to travel from one end point to the other of the range of the intermediate node. We make sure initially that each of the intermediate node is at such a distance that there are no intersections in the areas covered nor there are any places that do not come under any of the intermediate node.

Initially, every patient node broadcasts a JOIN-INTERMEDIATE NODE message, which is encrypted using the symmetric key provided for communication with the intermediate nodes. The patient node broadcasts its DId and TS. All the intermediate nodes receive the message, decrypt it using their copy of the symmetric key and do any kind of action only if Time taken by the message, is less then the predetermined Time To Live (TTL). Our assumption ensures that only one intermediate node receives the message within the predetermined TTL and so that node is the parent of the patient node. The intermediate node, then communicates with the root once, for confirmation of the DId, that the Device actually belongs to the network and once confirmed, and it receives a Node Id and a random number, forwards the message to the patient node along with its Node Id and Key for further communication.

Time = System Time - Time Stamp
Time $\leq$ TTL for Node A

$$PN \Rightarrow E_{symkey}(PN_{DId}, TimeStamp)$$
$$A \to BS - E_A(PN_{DId}, TimeStamp)$$
$$BS \to A - E_A(NodeId_{PN}, R_{PN})$$
$$A \to PN - E_{PN}(NodeId_{PN}, R_{PN}, NodeId_{PN}, N_{DId}, E_{A-PN})$$

5.4 Node Joining and Leaving

The patient nodes are mobile and can leave and join different intermediate nodes based on proximity of the patient node to the intermediate node. When the patient node needs to send information, it sends the information along with a time stamp. Each intermediate node receives the message but only that node that receives the message within the specified TTL will reply to the message and hereafter the node sends the information to only that intermediate node. Once the patient node joins an intermediate node sends all the information to the intermediate node. But patient nodes are mobile and keep moving along with the patient. Every message to be sent from the patient node has a time stamp attached to it and if the time taken by the message to reach the intermediate node is more than the TTL then the intermediate node sends an invalid message to the patient node and the patient node broadcasts its Device identifier as done initially to know its new parent.

We compare the efficiency of Torso against some other common key setup protocols in terms of their corresponding energy costs. One of the simplest key setup protocols is pre-deployment of keys before the sensor nodes are put into active operation [5]. Once deployed, the nodes already share the cryptographic keys, and therefore the protocol only requires node authentication using a challenge-response scheme. Although this protocol has a minimum overhead, it raises scalability and security concerns especially for changing mission configurations. The security protocol is different for other usual security mechanisms as it takes into consideration the energy resources and is light weight. The key size of 64 bits should be sufficient to get the required security of information. The other usual protocol is the *Kerberos*, but Kerberos requires that the server share a long-term explicit master key with every sensor node which is a potential drawback, especially for large networks. Torso architecture doesn't make any such assumptions. Torso Security mechanism also makes sure the base station assigns all the node ids. We also reduce the number of communication messages to establish authentication as just a single JOIN message can serve as both join and authenticating is done in a single message.

6 Conclusion

The Torso architecture provides scalable, efficient and secure communication in Distributed Sensor Networks by reducing the flow of information over the wireless medium and increasing local processing at the host node so as to compensate the constraints of the resources. This protocol allows self-configurable operations in an autonomous network with minimum user intervention, which is suited for a high risk wireless sensor network with dynamically changing topology. The confidentiality is obtained by using the symmetric key and a unique device identifier to actually get the symmetric key. We also take advantage of the level of confidentiality required and make a good analysis of which messages should be encrypted. Torso architecture provides suitable bandwidth consumption and better utilization of the available bandwidth yet does not compromise the criticality of the patient. The importance of patient's condition actually drives the allotment of bandwidth for that particular patient. Depending upon the level of assurance required, suitable mechanisms can also be put in place to periodically refresh these keys in order to safeguard against brute-force attacks.

References

1. R. Colin Johnson, Companies test prototype wireless-sensor nets, EE Times (2003)
2. L. Schweibert, S. K.S. Gupta, J. Weinmann, Research Challenges in Wireless Network of Biomedical Sensors, Proc. Of the 7th Annual International Conference on Mobile Computing and Networking (ACM SIGMOBILE) (2001) 151-165
3. B.Woodward, M.F.A. Rasid, Wireless Telemedicine: The Next Step, Proc. Of the 4th Annual IEEE Conference on Information Technology Applications in Biomedicine (2003) 43-46

4. P.Bauer, M.Sichitiu, R.istepanian, K.Premaratne, the Mobile Patient: Wireless Distributed Sensor Networks for Patient Monitoring and Care, Proc. Of the First Annual IEEE Conference on Information Technology Applications in Biomedicine (2000) 17-21
5. D.W. Carman, P.S. Kruss, and B.J. Matt. Constraints and Approaches for Distributed Sensor Network Security. In NAI Labs Technical Report 00-010 (2000)
6. J.Stankovic, T. Abdelzaher, C. Lu, L. Sha, J. Hou, Real-time communication and coordination in embedded sensor networks, Proceeding of the IEEE, volume 91 (2003) 1002-1022
7. D. W. Carman, P. S. Kruus, B. J. Matt, Constraints and approaches for distributed sensor network security, NAI Lans TR #00-010 (2000)

Multimedia: An SIMD – Based Efficient 4x4 2 D Transform Method

Sang-Jun Yu, Chae-Bong Sohn, Seoung-Jun Oh, and Chang-Beom Ahn

VIA-Multimedia Center, Kwangwoon University, 447-1, Wolgye-Dong,
Nowon-Gu, 139-701, Seoul Korea
{yusj1004, bongbong, sjoh, cbahn}@viame.re.kr
http://www.viame.re.kr

Abstract. In this paper, we present an efficient scheme for the computation of 4x4 integer transform using SIMD instructions, which can be applied to discrete cosine transform (DCT) as well as Hadamard transform (HT) in MPEG-4 AVC/H.264, a video compression scheme for DMB. Even though it is designed for 64-bits SIMD operations, our method can easily be extended to 128-bits SIMD operations. On a 2.4G (B) Intel Pentium IV system, the proposed method can obtain 4.34x and 2.6x better performances for DCT and HT, respectively, than a 4x4 integer transform technique in an H.264 reference codec using 64-bits SIMD operations. We can still have 6.77x and 3.98x better performances using 128-bits SIMD operations, respectively.

1 Introduction

Transform coding is at the heart of the majority of video coding systems and standards such as H.26x and MPEG. Spatial image data can be transformed into a different representation, the transform domain since spatial image data are inherently difficult to compress without adversely affecting image quality: neighboring samples are highly correlated and the energy tends to be evenly distributed across the image. There are several desirable properties of a transform for compressions. It should compact the energy in the image into a small number of significant values, decorrelate the data and be suitable for practical implementation in software and hardware. The forward and inverse transforms are commonly used in 1D (Dimension) or 2D forms for image and video compression. The 1D version transforms a 1D array of samples into an 1D array of coefficients, whereas the 2D version transforms a 2D array (block) of samples into a block of coefficients [1][2].

Most modern microprocessors have multimedia instructions to facilitate multimedia applications. For example, the single-instruction-multiple-data (SIMD) execution model was introduced in Intel architectures. MMX (Multimedia Extension), SSE (Streaming SIMD Extension) and SSE2 technologies can execute several computations in parallel with a single instruction. These instructions can make parallel processing of several data at the same time. In general, better performance can be achieved if the data pre-arranged for SIMD computation. However, this may not always be possible. Even if, referencing unaligned SIMD register data can incur a performance penalty due to accesses to physical memory [3]. Also, when applying an SIMD operation, it may be happened that data packing/unpacking operation takes more time than arithmetic operations.

In this paper, we present an optimized SIMD method to carry out 2D DCT with an integer transform for a block of 4x4 in the processor supporting SIMD operation. The proposed method is multiplier free and can improve the speed by reducing the number of operations such as store, move, and load. Those operations dominate in 4x4 transform relatively to packing/unpacking process.

2 Transform Coding in H.264

Instead of using DCT/IDCT (Inverse DCT), MPEG-4 AVC/H.264 adopted as a video data compression technology in DMB (Digital Multimedia Broadcasting) service uses a 4x4 integer transform to convert spatial-domain signals into frequency-domain and vice versa. The "baseline" profile of H.264 uses three transforms depending on the type of residual data that is to bi coded: a transform for the 4x4 array of luma(luminance) DC(Direct Current) coefficients in intra macroblocks (predicted in 16x16 mode), a transform for the 2x2 array of chroma(chrominance) DC coefficients(in any macroblock) and a transform for all other 4x4 blocks in the residual data. Data within a macroblock are transmitted in the order shown in Fig.1. If the macroblock is coded in 16x16 Intra mode, then the block labeled "-1" is transmitted first, containing the DC coefficient of each 4x4 luma block. Next, the luma residual blocks "0~15" are transmitted in the order shown (with the DC coefficient set to zero in a 16x16 Intra macroblock). Blocks "16" and "17" contain a 2x2 array of DC coefficients from the Cb and Cr chroma components respectively. Finally, Chroma residual blocks "18~25" (with zero DC coefficients) are sent [4]-[6].

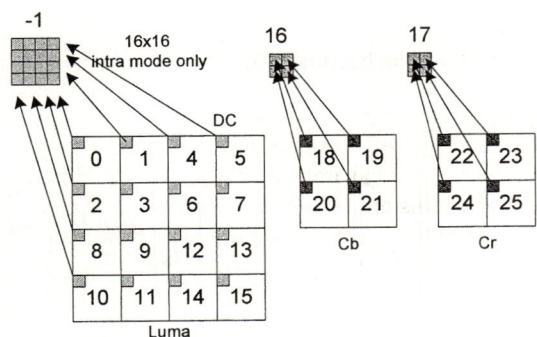

Fig. 1. Scanning order of residual blocks within macroblock

There are two types of 4x4 transforms for the residual coding in H.264. The one is DCT, and the other is HT [4].

2.1 Development from the 4x4 DCT

The 4x4 DCT of an input array is given by (1):

$$Y = AXA^T$$

$$= \begin{bmatrix} a & a & a & a \\ b & c & -c & -b \\ a & -a & -a & a \\ c & -b & b & -c \end{bmatrix} \begin{bmatrix} x_{00} & x_{01} & x_{02} & x_{03} \\ x_{10} & x_{11} & x_{12} & x_{13} \\ x_{20} & x_{21} & x_{22} & x_{23} \\ x_{30} & x_{31} & x_{32} & x_{33} \end{bmatrix} \begin{bmatrix} a & b & a & c \\ a & c & -a & -b \\ a & -c & -a & b \\ a & -b & a & -c \end{bmatrix} \quad (1)$$

where, $a = \frac{1}{2}$, $b = \sqrt{\frac{1}{2}}\cos(\pi/8)$, $c = \sqrt{\frac{1}{2}}\cos(3\pi/8)$.

This matrix multiplication can be factorized to the following equivalent form (2):

$$Y = (CXC^T) \otimes E$$

$$= \left(\begin{bmatrix} 1 & 1 & 1 & 1 \\ 1 & d & -d & -1 \\ 1 & -1 & -1 & 1 \\ d & -1 & 1 & -d \end{bmatrix} X \begin{bmatrix} 1 & 1 & 1 & d \\ 1 & d & -1 & -1 \\ 1 & -d & -1 & 1 \\ 1 & -1 & 1 & -d \end{bmatrix} \right) \otimes \begin{bmatrix} a^2 & ab & a^2 & ab \\ ab & b^2 & ab & b^2 \\ a^2 & ab & a^2 & ab \\ ab & b^2 & ab & b^2 \end{bmatrix} \quad (2)$$

where CXC^T is a core 2D transform. E is a matrix of scaling factors and the symbol $\otimes$ indicates that each element of CXC^T is multiplied by the scaling factor in the same position in matrix E. The constant d is $c/b \approx 0.414$.

To simplify the implementation of the transform d is approximated by 0.5. To ensure that the transform remains orthogonal, b also needs to be modified, so that $a = 1/2$, $b = \sqrt{2/5}$, $d = 1/2$.

The final forward transform becomes (3):

$$Y = (CXC^T) \otimes E$$

$$= \left(\begin{bmatrix} 1 & 1 & 1 & 1 \\ 2 & 1 & -1 & -2 \\ 1 & -1 & -1 & 1 \\ 1 & -2 & 2 & -1 \end{bmatrix} X \begin{bmatrix} 1 & 2 & 1 & 1 \\ 1 & 1 & -1 & -2 \\ 1 & -1 & -1 & 2 \\ 1 & -2 & 1 & -1 \end{bmatrix} \right) \otimes \begin{bmatrix} a^2 & ab/2 & a^2 & ab/2 \\ ab/2 & b^2/4 & ab/2 & b^2/4 \\ a^2 & ab/2 & a^2 & ab/2 \\ ab/2 & b^2/4 & ab/2 & b^2/4 \end{bmatrix} \quad (3)$$

which is an approximation to the 4x4 DCT. Because of the change to factors d and b, the output of the new transform will not be identical to the 4x4 DCT [2]. Equation (3) shows the mathematical form of the forward integer transform in H.264 standard, where CXC^T is a core 2-D transform which can be calculated using matrix multiplications. Since both the first and the third matrices have only constant coefficients of "±1" and "±2", it is possible to implement the calculation by using only additions, subtractions and shifts. This "multiply-free" method is quit efficient and, thus, has been implemented in the H.264 reference codec [6].

2.2 4 x 4 Luma DC Coefficients Transform (HT)

In H.264, Hadamard transform is applied to the luminance DC terms in 16 x16 intra prediction mode. The transform matrix is shown in (4) and its fast implementation is shown in Fig. 5(a). HT seems to be a simplified version of (3) with replacing the coefficient 2 by 1. The inverse HT has the same form as (4) since the transform matrix for HT is orthogonal and symmetric [7][8].

$$Y_D = \left(\begin{bmatrix} 1 & 1 & 1 & 1 \\ 1 & 1 & -1 & -1 \\ 1 & -1 & -1 & 1 \\ 1 & -1 & 1 & -1 \end{bmatrix} \begin{bmatrix} W_D \end{bmatrix} \begin{bmatrix} 1 & 1 & 1 & 1 \\ 1 & 1 & -1 & -1 \\ 1 & -1 & -1 & 1 \\ 1 & -1 & 1 & -1 \end{bmatrix} \right) \quad (4)$$

where W_D is the block of 4x4 DC coefficients and Y_D is the block after transformation. Each output coefficient $Y_{D(i,j)}$ is divided by 2 with rounding.

3 Representative SIMD-Based Schemes

One way for speed-up is to implement a parallel method on a parallel architecture such as an SIMD machine.

Figure 2 shows a typical SIMD computation. Two sets of four packed data elements (X1, X2, X3, X4, and Y1, Y2, Y3, Y4) are operated in parallel, with the same operation being performed on each corresponding pair of data elements (X1 and Y1, X2 and Y2, X3 and Y3, X4 and Y4). The results of four parallel computations are sorted as a set of four packed data elements. Thus, we can achieve speed-up on the SIMD machine in the areas of graphics, speech recognition, image/video processing, and other scientific applications [9].

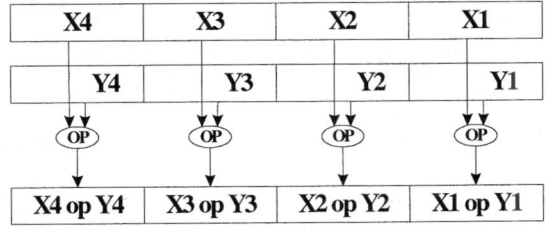

Fig. 2. Typical SIMD Operation

There exist three types of integer transform methods with SIMD operation: a 2D matrix multiplication method, a butterfly expression method and a parallel processing with four matrix [10]-[12].

In the 2D matrix multiplication method separability, which is an important property of the 2-D DCT, is used. Using that property 2-D DCT can be obtained by first performing 1-D DCTs of the rows of x_{ij} followed by 1-D DCTs of the columns of, where x_{ij} and c_{ij} are the (i,j)th elements in matrix X and C in (5), respectively [7].

$$XC = \begin{bmatrix} x_{00} & x_{01} & x_{02} & x_{03} \\ x_{10} & x_{11} & x_{12} & x_{13} \\ x_{20} & x_{21} & x_{22} & x_{23} \\ x_{30} & x_{31} & x_{32} & x_{33} \end{bmatrix} \begin{bmatrix} 1 & 2 & 1 & 1 \\ 1 & 1 & -1 & -2 \\ 1 & -1 & -1 & 2 \\ 1 & -2 & 1 & -1 \end{bmatrix} \quad (5)$$

If there is no multiplier in a processor when the 2D matrix multiplication method is used, it takes too much time on computation for a transform since a multiplication operation must be replaced with other several operations.

The second method is code rescheduling to exploit parallelism in the 2D DCT. Parallelism in the 2D DCT approached from two directions as shown in Fig. 3: within a single 4x4 DCT and among four 4x4 DCTs. While data are accessed by rows within the matrix in the former, data from the four matrices are interleaved to enable efficient use of the MMX instructions in the latter. The single DCT approach has some disadvantages; 1) The input matrix must be transposed in order to operate on several rows in parallel, and 2) packed shift instructions must be used to prevent overflow. The four DCTs approach has also some problems as following; 1) The input data from the four matrices must be interleaved before computing transform, and 2) more temporal use of memory is needed since it requires more register usage [8].

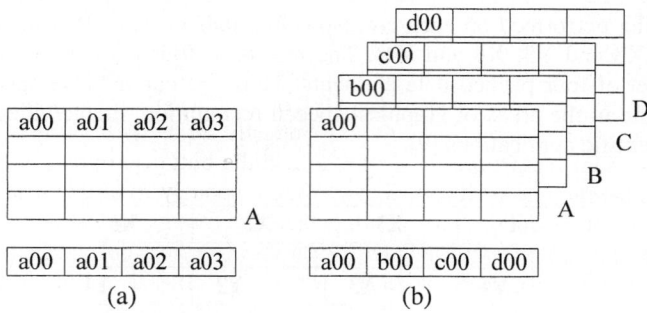

Fig. 3. Single DCT vs. Four DCTs: parallelism (a) within single matrix and (b) among four matrices [11]

The third method is to use the butterfly expression shown in Fig. 4. Even though it is easy to implement, it is hard to further optimize parallel processing in SIMD instructions because the operations on each operand are different.

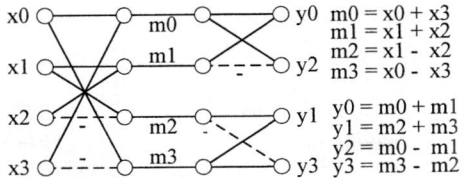

(a) Butterfly Expression of the Hadamard Transform

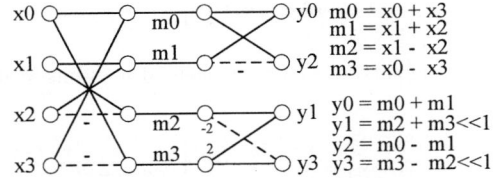

(b) Butterfly Expression of the Integer Transform

Fig. 4. Butterfly expression of the Transform in H.264

4 An SIMD-Based 4x4 2D Transform Method

As we mention before, each element of the DCT matrix in H.264 takes one of four values: ± 1 and ± 2 [6]. Therefore, we can compute DCT coefficients without any multiplication, that is, need only addition, subtraction and shift commands to calculate them. Zhou and his colleagues proposed a new method [3]. They showed that the original form of the integer transform was the 4x4 matrix multiplications, which could be implemented by using SIMD instructions, providing better performance than the multiplication-free method implemented in the H.264 reference code.

Equation (5) shows a typical 4x4 matrix multiplication. On the other hand, the butterfly expression has the multiplication-free method implemented in the H.264 reference code. Equation (5) shows a typical 4x4 matrix multiplication. On the other hand, the butterfly expression has the multiplication-free property. However, that expression may be hard to be implemented in parallel form using the SIMD instructions as mentioned before because the operations required in the butterfly are not identical.

To solve the problems in Zhou's and the butterfly expression, a transpose process is taken for 4x4 input matrix before 1D DCT, and the butterfly operation is applied for four input elements, that is, a row vector, simultaneously to utilize a series of SIMD instructions. From this point we assume that the 64-bits operations are used. The transpose process followed by butterfly operations in parallel applied in horizontal and vertical directions makes 2D DCT for the 4x4 input as schematically shown in Fig. 5.

The proposed method consists of four steps: the transpose step using the SIMD instructions shown in Fig. 6, the butterfly step for the row-direction 1D 4x4 integer transform using the series of SIMD operations shown in Fig. 8, the transpose step for the 1D output with the SIMD instructions, and the butterfly step for the column-direction 1D 4x4 integer transform using the series of SIMD operations.

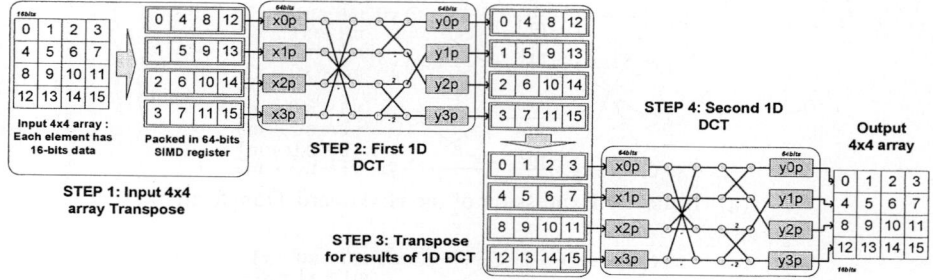

Fig. 5. Block diagram of the proposed method at 64-bits SIMD operation

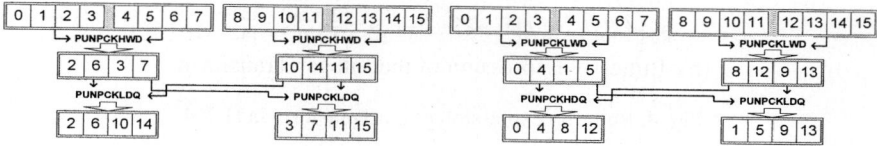

Fig. 6. 4x4 block transpose with SIMD instructions

With the SIMD instruction, PUNPCKx can be used to optimize the transpose process. The PUNPCKx instructions perform an interleave merge of the data elements of the destination and source operations into the destination register. The following example merges the two operations into the destination registers with interleaving. For example, PUNPCKLWD instruction interleaves the two low-order words of the *source1* operand (1_1 and 1_0) and the two low-order words of the *source2* operand (2_1 and 2_0) and writes them to the destination operand. PUNPCKHWD instruction interleaves the two high-order words of the *source1* operand (1_3 and 1_2) and the two high-order words of the *source2* operand (2_3 and 2_2) and writes them to the destination operand. One of the destination registers will have the combination illustrated in Fig. 7 [13].

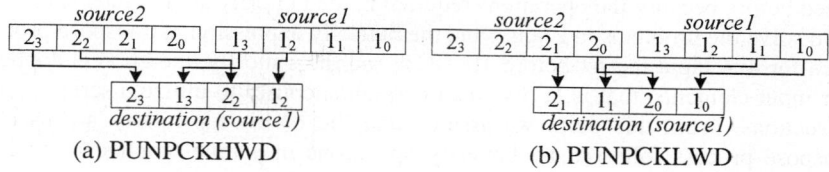

(a) PUNPCKHWD   (b) PUNPCKLWD

Fig. 7. Operation of the PUNPCKx instruction

We explain the butterfly step for the 1D 4x4 integer transform using the series of SIMD operations shown in Fig. 8. Let four elements in the first row be x00, x10, x20 and x30. They are loaded into x0p which is a 64-bits register. By the same token, other three rows can be loaded into registers x1p, x2p and x3p. The 1D DCT integer transform process is followed using the butterfly structure in Fig. 4(b).

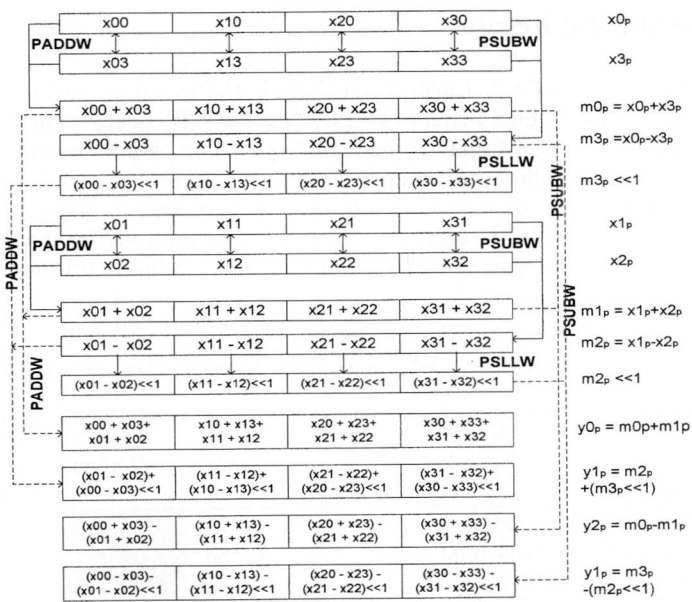

Fig. 8. Example of the 4x4 integer transform "core" implementation

The operations of addition, subtraction and shift in Fig. 4(b) are replaced with PADDW, PSUBW and PSLLW, respectively, supported by IA-32 Intel architecture [13]. Up to now we explain the SIMD-based 4x4 integer transform method on the 64-bits machine. That method can be expended to be executed on 128-bits machine. After expansion two consecutive 4x4 blocks can be simultaneously transformed. Figure 9 shows an additional process for expansion.

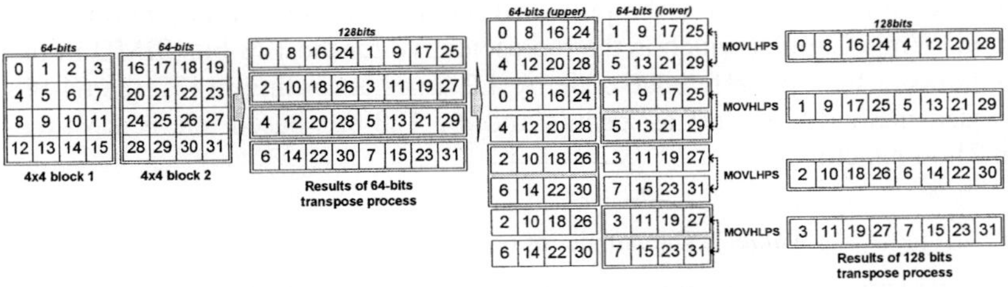

Fig. 9. 128-bits SIMD instructions for transpose

For the additional process two SIMD instructions, MOVHLPS and MOVLHPS, are used. Using the instruction of MOVHLPS (MOVLHPS) the upper (lower) 64-bits of the *source2* are loaded into the lower (upper) 64-bits of the 128-bit *destination*, and the upper(lower) 64-bits of *source1* are unchanged.

The proposed method can be applied to compute the 4x4 HT in H.264 with removing the shift instruction of PSLLW in Fig. 8.

5 Experimental Results

The reference code for the 4x4 integer transform in H.264, Zhou's and our methods are executed on a 2.4G(B) Intel Pentium IV PC. The 2.4G (B) Intel Pentium IV processor supports both 64-bits MMX and 128-bits SSE2 operations. A performance analysis tool Vtune [14] is used to collect the detailed characteristics of those methods.

Tables 1 and 2 show both execution times and a comparison of four different 4x4 integer transforms and HTs, respectively. Each execution time for the 64-bits MMX and the 128-bits implementations is measured after inputting 2 million and 1 million data, respectively.

Table 1. Comparison of different methods for the 4x4 integer transforms

Speedup methods	Times (sec)	H.264 reference code	Proposed method (64-bits)
H.264 reference code	0.817	1.000	0.230
Proposed method (64-bits)	0.188	4.341	1.000
Proposed method (128-bits)	0.121	6.776	1.561

Table 2. Comparison of different methods for the 4x4 Hadamard transform

Speedup methods	Times (sec)	H.264 reference code	Proposed method (64-bits)
H.264 reference code	0.458	1.000	0.384
Proposed method (64-bits)	0.176	2.602	1.000
Proposed method (128-bits)	0.115	3.983	1.530

As shown in the Table 3, the proposed methods do not use any multiplication operation. While Zhou's, which is the representative for the 128-bits SIMD implementation, is faster than the reference code by 4.2 times according to [3], the 64-bits based proposed method is as fast as Zhou's and the 128-bits based one is 1.5 times faster than Zhou's method.

Table 3. The number of operation in different methods

methods operations	Reference code	Proposed (64-bits)	Proposed (128-bits)
mov	522	114	71
add, sub, shift	102	26	14
multiplication	0	0	0
Else	76	25	22
total clocktic	700	165	107
Speedup	1.000	4.242	6.542

6 Conclusion

In this paper, we proposed the SIMD-based fast methods for the 4x4 integer transform as well as HT in MPEG-4 Part10 AVC/H.264. The experimental results showed that the proposed 64-bits and 128-bits SIMD-based 4x4 integer transform methods were 4.3 and 6.7 times faster than the H.264 reference code, respectively. Furthermore, the proposed 64-bits and 128-bits SIMD-based 4x4 HT methods were 2.6 and 3.98 times faster than the H.264 reference code, respectively

Acknowledgment

This work was supported by grant from IT-SoC Association, grant No. R01-2002-000-00179-0 from the Basic Research Program of the Korea Science & Engineering Foundation and Research Grant of Kwangwoon University in 2004

References

1. Iain E. G. Richardson, *Video Codec Design-Developing Image and Video Compression Systems*, John Wiley & Sons Ltd, England, 2002
2. Iain E.G Richardson, *H.264 and MPEG-4 VIDEO COMPRESSION*, WILEY, 2003
3. X. Zhou, Eric Q. Li and Yen-Kuang Chen, "Implementation of H.264 Decoder on General-Purpose Processors with Media Instructions," *Proceeding of SPIE conference on Image and Video Communication and Processing*, Vol. 5022, Jan. 2003, pp. 224-235
4. T. Wiegand, G. J Sullivan, G. Bjontegaard and A.Lutha, "Overview of the H.264/AVC Video Coding Standard," *IEEE Trans. on CSVT*, Vol. 13, No. 7, July 2003 pp. 560-576.
5. Henrique S. Malvar, Antti Hallapuro, Marta Karczewicz, and Louis Kerofsky, " Low-Complexity Transform and Quantization in H.264/AVC", *IEEE Transaction on Circuits and Systems for Video Technology*, Vol. 13, No.7, July 2003, pp. 598-603
6. Iain E., G Richardson, "H.264 White paper - White Papers describing aspects of the new H.264/MPEG-4 Part 10 standard", May 12, 2004
7. Vasudev B. and Konstanstinos k., *Image and Video Compression Standards*, Kluwer Academic Publishers, Norwell, Massachusetts 02061 USA, 2000
8. Tu-Chih Wang. Yu-Wen Hwang, Hung-Chi Fang, and Liang-Gee Chen, "Parallel 4x4 2D Transform and Inverse Transform Architecture for MPEG-4 AVC/H.264," *Circuits and Systems, 2003. ISCAS '03. Proceedings of the 2003 International Symposium* on , Vol. 2 , 25-28 May 2003 p:II-800 - II-803
9. Intel Corp.,"IA-32 Intel® Architecture Optimization Reference Manual," AP - 248966-010
10. Intel Corp., "Using Streaming SIMD Extensions in a Fast DCT Algorithm for MPEG Encoding," version 1.2, Jan. 1999
11. Intel Corp., "Using Streaming SIMD Extensions 2(SSE2) to Implement an Inverse Discrete Cosine Transform," version 2.0, July 2000
12. Intel Corp., "Streaming SIMD Extensions - Matrix Multiplication," AP-930, June 1999
13. Intel Corp., "Intel Pentium 4 and Intel Xeon Processor Optimization - Reference Manual," Order Number: 248966-05, 2002
14. Intel Corp., Intel® VTune TM Performance Analyzer, Version 7.0, 2003.

A Real-Time Cooperative Swim-Lane Business Process Modeler

Kwang-Hoon Kim[1], Jung-Hoon Lee[1], and Chang-Min Kim[2]

[1] Collaboration Technology Research Lab.,
Department of Computer Science, KYONGGI UNIVERSITY,
San 94-6 Yiuidong Youngtonggu Suwonsi Kyonggido,
442-760, South Korea
{kwang, jhlee}@kyonggi.ac.kr
[2] Division of Computer Science, SUNGKYUL UNIVERSITY,
147-2 Anyang8dong, Manangu, Anyangsi,
Kyonggido, 430-742, South Korea
kimcm@sungkyul.ac.kr

Abstract. In this paper[1], we propose an advanced business process modeling system, which is called a real-time cooperative swim-lane business process modeler, that enables several real actors/workers to cooperatively define a business process model in a real-time collaborative fashion. Through implementing the modeler, we are able to accomplish the goals - maximizing efficiency as well as involvement of real workers in the business process modeling activities. In the traditional approaches, the modeling work is done by a single designer who has to know all about the detailed and complex knowledge for business processes, such as relevant data, organizational data, roles, activities, application programs, scripts, etc. However, when we take into account the recent trend that a business process has become more complicated and large-scaled, it is hard to say that the approach is feasible, anymore. Therefore, we propose a more realistic approach that enables several real actors/workers to cooperatively define a business process at the same time.

The real-time cooperative swim-lane business process modeler is implemented by using Java programming language and EJB framework approach so as to be deployed on various heterogeneous platforms without any further technical consideration. And its major components might be EJB-based DB component, graphical swim-lane workflow modeler, organizational dependency analysis algorithm, organizations, relevant data, invoked applications, and repositories management components.

Keywords: Information Control Net, Real-time Cooperative Swim-lane Model, B2B e-Commerce, Cross-organizational Business Process Modeling System

---

[1] This work was supported by Korea Research Foundation Grant. (KRF-2002-003-D00247)

1 Introduction

In the workflow and BPM literature, we can see two evidences - Merging real-time groupware into business process and workflow issue and Cross-organizational workflow issue. The first issue is related with the complexity problem of workflow. A business process has been becoming gradually complex more and more in terms of the structural aspect of the workflow as well as the behavioral aspect of the workflow. The structural complexity is concerning about how to efficiently model and define a large-scale business process consisting of a massively parallel and large number of activities. Therefore, by using the real-time groupware technology, a group of people (business process and/or workflow designers) is able to cooperatively model and define the large-scale business process in a real-time collaborative fashion. The behavioral complexity is related with the run-time components of workflow enactment. If a certain activity's implementation (application program, for example) has to be collaboratively done by a group of actors, then in order to support the situation, the real-time groupware technology should be merged into the workflow enactment components. In this paper, we are looking for a method for merging groupware into the business process modeling work.

The second issue is concerning about the interoperability problem of workflow. The interoperability problem addresses the following two issues - the build-time (construction) issue and the runtime (enactment) issue of a cross-organizational workflow or business process. The runtime issue has been treated and standardized effectively by the WfMC and other workflow-related standard organizations, as one knows well. Also, the workflow description languages aspect of the build-time issue has been well formatted and standardized by the standard organizations, such as WfMC's WPDL/XPDL, BPMI's BPML/BPMN/BPQL, ebXML's BPSS, RosettaNet's PIP, and so on. However, the modeling methodologies and systems aspect of the build-time issue hasn't been well done yet, because almost all workflow and business process management systems have their own modeling methodology and system, it is very hard to draw up a single standard. Additionally, it confronts with the dilemma stated in [2] that is the independence (or security) of organizations versus the efficiency of construction. In constructing a cross-organizational business process, the methodology and the system in one place have advantages on the efficiency of construction. However, the independence of each organization must be abandoned because they have to open their own internal workflow information. In contrast to the situation, each organization wants to be more secured and has the ability or the right to decide things on its own, so they have to give up the efficiency of construction. In this paper, we would also propose a method for solving the dilemma.

Conclusively, in this paper, we would like to seek a feasible solution for resolving the previous two methods at once. It is called cooperative swim-lane business process modeler hat is used for constructing a cross-organizational business process in a way of not only that a group of actors can be engaged in the modeling work at anywhere and anytime, and but also that it avoids the dilemma. We implement a cooperative swim-lane business process modeling concept with respect to ICN (Information Control Net) and by embedding the real-time groupware functionality, as well. Therefore, a group of designers or actors are able to open a session, join to the session, and cooperatively construct a cross-organizational business process through the system. We describe about the system after introducing related works and backgrounds in the next section. Finally, we explain about the use and extension of the system.

2 Backgrounds and Related Works

Recently, electronic commerce and its related technologies have been swiftly adopted and hot-issued in the real world. This atmosphere booming e-commerce is becoming a catalyst for triggering explosion of the electronic logistics and supply chain management technologies and markets as well. This also means that organizations are much closer each other and need to support the inter-organizational cooperative activities with great efficiencies. Hence, the workflow and BPM technology does also become a core platform for technologies bringing organizations much closer and make them much more tightly coupled by supporting inter-organizational activities. At the same time, the traditional workflow systems are now fairly setting up as a core platform for automating intra-organizational business process. Therefore, in terms of the conceptual points of view, the workflow and BPM technology has to be required with supporting not only cooperative people works but also cooperative organization works, which can be dealt with the intra-organizational workflow and the inter-organizational workflow, respectively.

Particularly, in terms of the inter-organizational workflow modeling aspect, there are several related works in the literature. According to the degree of complexity in collaboration among organizations (or inter-organizations), there might be two types of collaborative organizations - loosely coupled collaborative organizations and tightly coupled collaborative organizations. The Interworkflow project [2] conducted at the Kanagawa Institute of Technology, Japan, is one of typical frontiers pioneering inter-organizational workflow modeling methodology and system for the loosely coupled collaborative organizations. They focus on the definition of a global-workflow model (which they call interworkflow) for a cross-organizational business process. This interworkflow model defines the basic interaction between the associated parties, and then it is transferred into the workflow management systems of the parties. Within the system, the (local) processes are modified to be suitable for the needs of the individual enterprises. On this basis, the interworkflow definition tool is used to define the interworkflow process. After this definition work, the translators automatically convert the interworkflow process definition data into the workflow engines used in each organization. While on the other, as a system supporting the tightly coupled collaborative organizations, we implement the cooperative swim-lane business process modeling approach. The detailed idea and concept of the system is described in the next section. Also we explain what are differences between the [2]'s approach and ours based upon the issues, too.

3 The White-Box Approach

As stated in the previous section, we would propose a system for modeling a business process in a completely different way from the traditional modeling approaches. Our approach reflects the basic philosophies of the previous two factors as much as possible. That is, we look for a way not only that makes people closer and more cooperative but also that makes inter-related organizations more collaborative in modeling a global business process. According for the business process to be more complicated and to be engaged with many organizations, our approach will be more worthy and more effective. It comes from a simple idea, in which it should be not reasonable for only a single designer to define a whole business process. Then, how can we make as many designers as possible to be engaged in modeling a business process? In this section, we conceptually and operationally illustrate the idea of the approach, as an answer for the question.

From the point of operational view, we would name the [2]'s approach "black box approach," in which the internal process in each organization is treated as a black box, because it makes only the linkage with others visible. While on the other, our approach would be named "white box approach," because all of the internal processes associated with a global business process can be precisely described in one place where the modeler is located, and broadcasted into the other coordinators. As we described in the previous, our methodology is pursuing the closed e-business framework, not the open e-business framework. That is, while the black box approach might provide a reasonable solution for constructing the open e-business framework, the white box approach, our approach, should be appropriate for constructing the closed e-business framework. In this section, we explain how the white box approach, our modeling methodology, defines a global business process and generates its collaborative local business process, each of which belongs to an organization.

In summary, we have introduced our white-box modeling methodology, so far. The methodology might fit very well into the system that provides tightly coupled interactions among a group of people as well as a group of organizations in collaboration. So, we have a basis for implementing a workflow modeling system that is operable based upon the methodology, which avoids the dilemma [2] and provides a great simplicity and efficiency in business process modeling work that should be perfectly applicable for not only intra-organizational workflows but also cross-organizational workflows. Also, it can be embedded into the closed or process-centric e-business framework. The methodology's advantages can be summarized as the followings:

1. It is a white box approach, because each activity can be clearly assigned into the corresponding organization that has the responsibility of execution. So, the construction unit of a global workflow model becomes the activity, which is exactly same with the conventional business process modeling methodology.
2. A group of cooperative designers are always aware of collaborative linkages, in real time, existing among internal processes (or local workflow/BP) of a

cross-organizational business process model, because the cooperative swim-lane modeling tool provides WYSIWIS functionality to the designers logged on a session, and it is able to automatically generate a cross-organizational BP model by combining and arranging the local BPs.
3. It needs not to develop a set of translators converting the description language of a global business process model into the language of each local BP model, because the description language of the global BP model collaboratively defined by a group of designers is WPDL/XPDL, and the corresponding local BPs are stored in the registry/repository component that provides interfaces to the runtime components of each organization.

4 The Cooperative Swim-Lane Business Process Modeler

In this section, we describe the implementation of a cooperative swim-lane business process modeler that reflects the white-box modeling approach. The CTRL research group of the Kyonggi University has been developing especially for the e-Logistics and e-Commerce framework, and is funded by the Korea Research Foundation Grant (KRF-2002-003-D00247). The system is based on the graphical notation of the information control net (ICN) [8] that is the most famous workflow model. Also we borrow the concept of swim-lane workflow model [9], too, which is very useful for composing the windows and user interfaces of the tool, because a swim-lane on the windows may represent an actor, a role, or an organization (and its local workflows). Also, the swim-lane becomes a boundary for access control. Therefore, we combine these two concepts, ICN and Swim-lane, to implement the approach. The system is completely operable now.

4.1 Architectural Components of the System

We realize the white-box modeling approach by implementing a business process modeling system that is based on the notations of the global information control net and the swim-lane workflow model as well. This system also pertains to the key principles of real-time groupware, such as concurrent work, WYSIWIS , group awareness, and supporting second-level language. The concurrent work is implemented through the event sharing mechanism, the group awareness is graphically represented by cursors with user's id, and the second-level language is supported through a chatting function. The functional architecture of the BP modeling system is presented in Fig. 1. The modeler has three major components - The cooperative Swim-lane Business Process modeler that is used to define a global BP through its graphical user interface, the cooperative Swim-lane BP modeling server that takes the roles of session control and event control, and the registry manager that has a repository for XML-based global BP models and provides APIs to a set of different BP enactment engines. We briefly describe about each of the architectural components of the business process modeler.

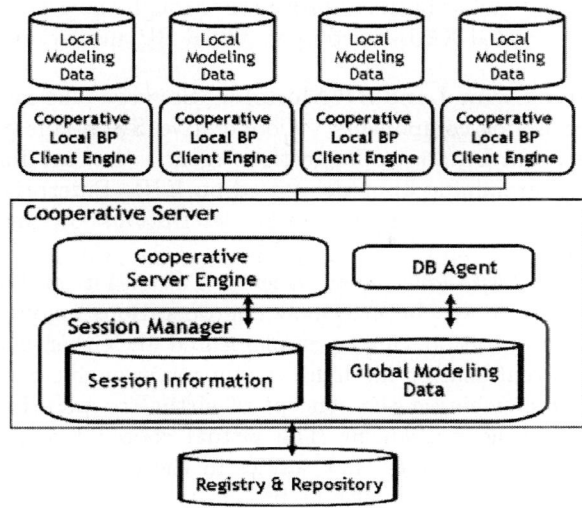

Fig. 1. Overall Architecture of the Cooperative Swim-lane BP Modeler

The Cooperative Swim-Lane BP Modeler. The modeler provides a set of functions and capabilities for the BP modeling work cooperatively performed by a group of users (or designers/builders). As a front-end component of the system, it has three major components - graphical user interface manager, cooperative fragment client engine, and local modeling data manager - that provide the group BP modeling functionalities, such as user management, session control, event handling and synchronization, floor control, space sharing, access control, message handling, modeling data management, model verification, and so on.

The graphical user interface manager enables a group of cooperative designers to define, edit, verify, and simulate a collaborative global business process on a shared space. The shared space is divided into the number of designers logged on the current session, each of whom is assigned into one swim-lane. Then, the owner of each swim-lane takes the responsibility for editing his/her workflow fragment. Of course, each of designers can't edit the others' local BPs, but they can only see and read the others' modeling works. As a result, the swim-lane means, in this modeler, the access control boundary (read/write permission for the owner, read-only for the others).

Finally, a group of designers is able to cooperatively draw and edit a global BP model through the graphical icons that are exactly identical to the graphical notations of the information control net. The major information dealt with the modeler consists of four dimensions - global BP process models, global organizational information, relevant data structures, and the invoked application programs. After modeling the global business process, those related information are stored on the registry/repository's database through the JDBC operations of the database connection module. Moreover, the modeler provides a feature for verifying the defined global business processes through the graphical animation approach, which is called XML-based open global BP animation tool.

The Cooperative Swim-Lane BP Modeling Server. The cooperative server consists of three major components - Cooperative Swim-lane Server Engine, Session Manager, and Database Connection Agent - that are implemented by the framework programming approach based on EJB (Enterprise Java Beans) infrastructure. From now, we give brief descriptions of the three major components.

The Cooperative Swim-lane Server Engine. The major responsibility of the cooperative server is to synchronize modeling operations among the modeling clients. The synchronization can be implemented by either a screen-sharing mechanism or an event-sharing mechanism. Our server engine performs the synchronization by implementing the concept of virtual cooperative server based upon the event-sharing mechanism. The virtual cooperative server is an instance of the cooperative server that resides in the client of the modeler and handles all events from the modeler to broadcast to the others. The types of events are listed such as insert, remove, move, awareness, pull, chat, and update.

The Session Manager. The session management functionality is the most crucial part of the server. It maintains opening and closing sessions, joining and dropping users on the sessions, and assigning access rights to the users. Especially, in terms of the access control mechanism, we newly implement the double-level floor control mechanism. The first-level floor control is for inter-swim-lanes of the modeler, and the second-level floor control is done within a swim-lane. That is, the users, each of whom is logged in a different swim-lane, are able to perform her/his own modeling work concurrently without any interference from others. In the contrast to this, the users who are logged in the same swim-lane can't perform their own modeling works at the same time. Through the two-level floor control mechanism, we accomplish that the concurrency level of inter-organizations' modeling work can be maximized, the concurrency level of intra-organization's modeling work can be minimized. From the access control policy, just like this, we can expect that the cooperative swim-lane BP modeling system can be primarily used for modeling cross-organizational business processes. Of course, the system can support the BP modeling work within an organization. In this case, the notion of swim-lane has to be replaced by a role not an organization. Then we can expend the system's usability without any further modifications.

The Database Connection Module. The module (agent) provides a set of JDBC-based APIs that are used for the server to access its database as well as the repository. That is, the module is located between the cooperative swim-lane server engine and the registry/repository system. Also it is deployed on EJB sever as a component. Through the EJB technology, we are able to guarantee a stable database agent that can be characterized by the distributed object management, reliable recovery mechanism from system failures, reliable large-scale transaction management, and the security functions.

The Registry/Repository Manager. The registry manages the model information of cooperative swim-lane BP models transformed in XML. It consists of two major modules, the APIs for workflow enactment engines and the database connection module. The former takes in charge of maintain the business processes, which is formatted in the XML format (WPDL/XPDL). The latter has the database access functionality that is exactly identical with the database connection module in the cooperative server. Moreover, it carries out connections with the cooperative swim-lane server engines that are operable on the EJB computing environment.

4.2 The Operational Examples

In this section, we would present a couple of captured screens of the modeling system. We would simply prove, through the selected screen, that the methodology is completely workable on the real environment. Fig. 2 is a screen of the modeler, which simply shows a global business process definition containing ten activities, one or-split and one and-split. As you see, the window has three vertical partitions - the left-hand side is for the global business processes with tree structures, the middle is reserved for a set of icons that represents, from the top, the start node, end node, activity, nested activity, cooperative activity, AND-split node, OR-split node, loop activity, control flow, data flow, and finally the relevant data, respectively, and the right-hand side is the shared space for the cross-organizational BP modeling work, on where eight workflow designers are cooperatively performing the modeling works. The pencils labeled with numbers on the window represent the designers, which mean the group awareness that is one of the real-time groupware's key features. As we described in the previous, the right-hand side is, again, horizontally partitioned into eight swim-lanes, each of which represents an actor or an organization. And the graphical notation of the modeler follows the rules of information control net.

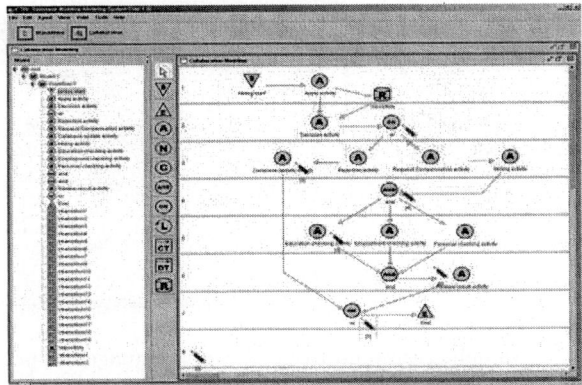

Fig. 2. Graphical User Interface of the Cooperative Swim-lane Modeling System

As mentioned in the previous sections, the system's design goal is based upon the fundamental principles of real-time groupware systems - Concurrent work, WYSIWIS, group awareness, double-level languages. So, the following is to summarize how the principles are implemented on the system:

- Concurrent Work: The Double-Level Floor Control Mechanism
- WYSIWIS: The Event Sharing Mechanism
- Group Awareness: Pencil labeled by a number
- Double-Level Languages Support: The First-level Language(Swim-lane and ICN), The Second-level Language(hatting Function)

Where, the double-level language support has a special meaning in a groupware system. Generally speaking, the language is a communication tool supporting a group of people to exchange knowledge each other. However, in a groupware system, a group of users communicates each other through its shared objects. That's why we call it the first-level language of the groupware system. But, the only single-level of language might not be enough for a group of users to communicate each other. So we need one more language during a work group session, such as chatting, email, telephone, and so on. And one of these languages becomes the second-level language of a groupware system. Fig. 3 shows the double-level languages support in the cooperative swim-lane BP modeler - The first-level language is swim-lane and ICN, themselves, and the second-level language is provided through a chatting function. We would not show other screens here that are used for defining activity information, performers, organizations, relevant data, invoked applications, and so on.

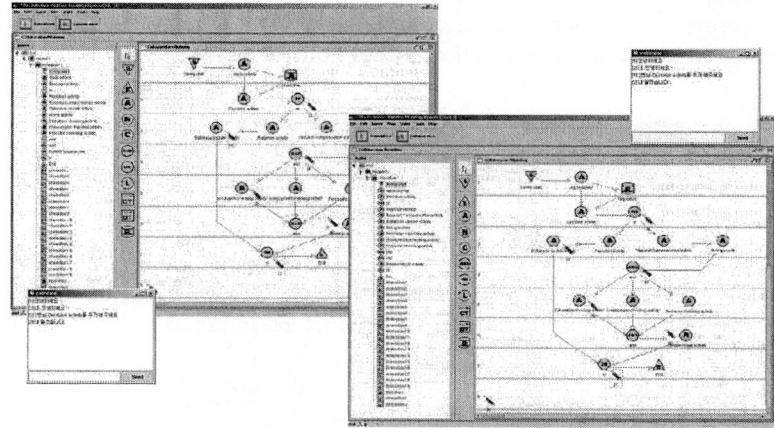

Fig. 3. Collaborative Modeling through Double-level Language Support in the System

5 Conclusion

So far, we have proposed the cooperative swim-lane business process modeling approach (the white-box modeling approach) and proved that the approach is a feasible and realistic by developing the cooperative swim-lane business process modeler. We strongly believe that the modeler fortifies the modeling approach, and vice versa. Also we assure that the modeler fits very well into the tightly-coupled framework for the workflow-driven e-Commerce domains. Especially, the approach might have a strong advantage in terms of the high degree of modeling ability for not only complex and maximally parallelized business processes but also collaborative global business processes. At the same time, it should be very applicable to the type of collaborative global business process models that is defined and characterized, by us, with tightly coupled global BPs in collaborative organizations.

Recently, BPM and its related technological fields, including process-centric e-Business and e-Logistics, are catching great attentions from the society of information science and database management fields. So, there are a lot of challenges to develop and commercialize e-business solutions. This paper should be one of those active attempts for pioneering global business process modeling methodologies and systems toward supporting cross-organizational BPs in collaboration.

Acknowledgement. This work was supported by Korea Research Foundation Grant (KRF-2002-003-D00247).

References

1. Edward A. Stohr, J. Leon Zhao, "Workflow Automation: Overview and Research Issues", Information Systems Frontiers, Volume 3, Issue 3, 2001
2. Haruo Hayami, Masashi Katsumata, Ken-ichi Okada, "Interworkflow: A Challenge for Business-to-Business Electronic Commerce", Workflow Handbook 2001, WfMC, October 2000
3. Sun Microsystems, "The JAVATM 2 Enterprise Edition Developer's Guide", Version 1.2.1, May 2000
4. Sun Microsystems, "Enterprise JavaBeans Programming", Revision A.1, May 2000
5. Kwang-Hoon Kim, Clarence A. Ellis, "A Framework for Workflow Architectures", University of Colorado, Department of Computer Science, Technical Reports, CU-CS-847-97, December 1997
6. Dong-Keun Oh, Kwang-Hoon Kim, "An EJB-Based Database Agent for Workflow Definition", Journal of Korean Society for Internet Information, Vol.4, No.5, December 2001
7. Kwang-Hoon Kim, Clarence A. Ellis, "Performance Analytic Models and Analyses for Workflow Architectures", Information Systems Frontiers, Vol. 3, No. 3, pp. 339-355, 2001
8. Clarence A. Ellis, "Formal and Informal Models of Office Activity", Proceedings of the 1983 Would Computer Congress, Paris, France, April 1983
9. Alec Sharp and Patrick McDermott, "Workflow Modeling - Tools for Process Improvement and Application Development", Artech Housw, Inc., 2001

A Focused Crawling for the Web Resource Discovery Using a Modified Proximal Support Vector Machines

YoungSik Choi, KiJoo Kim, and MunSu Kang

Department of Computer Engineering, Hankuk Aviation University,
200-1 Hwajun-Dong Dukyang-Gu
Kyounggi-Province, Koyang-City, Korea
{choimail, zgeniez, mskang}@hau.ac.kr
http://data-mining.hau.ac.kr

Abstract. With the rapid growth of the World Wide Web, a focused crawling has been increasingly of importance. The goal of the focused crawling is to seek out and collect the pages that are relevant to a predefined set of topics. The determination of the relevance of a page to a specific topic has been addressed as a classification problem. However, when training the classifiers, one can often encounter some difficulties in selecting negative samples. Such difficulties come from the fact that collecting a set of pages relevant to a specific topic is not a classification process by nature.

In this paper, we propose a novel focused crawling method using only positive samples to represent a given topic as a form of hyperplane, where we can obtain such representation from a modified Proximal Support Vector Machines. The distance from a page to the hyperplane is used to prioritize the visit order of the page. We demonstrated the performance of the proposed method over the WebKB data set and the Web. The promising results suggest that our proposed method be more effective to the focused crawling problem than the traditional approaches.

1 Introduction

The World Wide Web continues to grow rapidly at a rate of a million pages per day [1]. Searching a document from such enormous number of Web pages has brought about the need for a new ranking scheme beyond the traditional information retrieval schemes. Hyperlink analysis has been proposed to overcome such problems and showed some promising results [6][9]. Now, hyperlink analysis related technologies are somehow embedded into modern commercialized search engines, notably Google [9][11]. Although hyperlink based methods appear to be an effective approach showing reasonable retrieval results, there are still many problems yet to be solved. Such problems are mainly caused by the simple fact that there are too many pages out there to be crawled. Even the largest crawlers cover only 30 ~ 40% of the Web and the refreshes take up to a month [1][9]. Many ideas have been proposed in recent years to handle such problems; among them a focused crawling has gained much attention [1-5]. A focused crawling is to seek out and collect the pages that are relevant to specific topics so that it may be able to cover topical portions of the Web quickly without having to explore a whole Web [2][3]. One can view the focused crawling as a resource discovery process from the Web. For instance, the focused crawling has been used for constructing Knowledge Base from the Web [2].

The focused crawling is a process of selecting relevant pages from the Web. The selection criterion on relevant pages is generally based on the assumption of "topic locality" that child pages are likely to contain same topic to that of their parent pages [2][9]. Starting from seed pages, usually from a few representative pages on a given topic, the focused crawler explores pages that are relevant to the topic according to the predefined relevance measurement. The measurement of relevance affects the whole performance of the focused crawler. In general, the judgment of relevance has been interpreted as a classification process [1][9] where a classifier should be trained by using positive and negative examples. In such approach, we often encounter difficulties choosing negative samples. This is because selecting pages related to a specific topic is not a classification process by nature. Moreover, in a classification paradigm, there ought to be a hard decision making on the class membership, which causes sometimes to lose important pages. This can be alleviated by making soft decision [9] but still having to maintain only two classes, relative or irrelative. Considering the diversity of the Web content, one can hardly imagine that one classifier may be able to bisect the entire Web. Instead, it is more natural to represent a specific topic as a one-class and to measure relevance for that class.

The Proximal SVMs are well known for sustaining all the good features of Proximal SVMs including a powerful generalization capability out of training samples [7][8]. In this paper we present a modified Proximal SVM to seek out the proximal hyperplane that best represents the underlying distribution of a given set of positive samples. That is, our modified Proximal SVM only uses positive samples from a given topic so that one can use the distance from a page to the proximal hyperplane as a measurement of a page's relevance to a topic. This approach can be viewed as a traditional LS (Least Squares) Regression with regularization and therefore enjoys both a representation power from the LS Regression and a generalization power from the SVM [10]. In a training phase, we extract the proximal hyperplane over positive samples by using the modified Proximal SVMs. During the crawling, the distance to the proximal hyperplane from each crawled page is computed and quantized. According to the quantized distance, the offspring pages from each crawled page are put into one of the priority bins. Then, the focused crawler selects one page out of a non-empty highest priority bin for the next crawl.

This paper is organized as follows. In the next section, we present the general framework of the focused crawling using a modified Proximal SVM in Section 2. Section 3 discusses experimental results in detail. We make our conclusions and summaries in Section 4.

2 Framework for the Proposed Focused Crawling

2.1 Proximal Support Vector Machines

The linear PSVM (Proximal Support Vector Machine) seeks to find the "proximal planes" that best represent training samples while maximizing the distance between the two proximal planes. The obtained two proximal planes comprise the optimal separating hyperplane [7].

The training samples are represented as a matrix $\mathbf{A} \in \Re^{m \times n}$, where m denotes the number of the training samples and n represents the dimensionality. Matrix D denotes an $m \times m$ diagonal matrix with the values $\{-1, +1\}$ of each element and y represents a column vector whose element represents an error. Matrix $\mathbf{I}$ denotes an identity matrix.

In linearly separable cases, the optimal separating hyperplane can be represented as $\mathbf{x}^T \mathbf{w} - \gamma = 0$, where $\mathbf{x} \in \Re^n$. The objective function for the Proximal SVM can be stated as

$$\text{minimize } v\frac{1}{2}\|y\|^2 + \frac{1}{2}(\mathbf{w}^T\mathbf{w} + \gamma^2) \tag{1}$$

$$\text{subject to } D(\mathbf{A}\mathbf{w} - \mathbf{e}\gamma) + y = \mathbf{e}$$

Note that v is a non-zero constant and e is the column vector of ones. The objective function in equation (1) can be restated as the "regularized least squares." Figure 1 shows that the positive proximal plane ($xw - \gamma = 1$) and the negative proximal plane ($xw - \gamma = -1$) are located in the middle of positive and negative training samples, respectively.

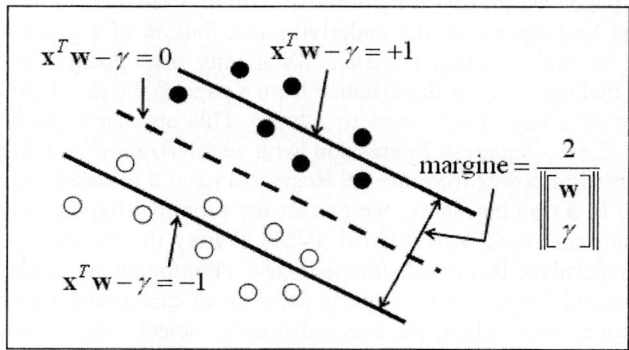

Fig. 1. Proximal Support Vector Machines: positive and negative proximal planes

One can solve the optimization problem in equation (1) using the Lagrangian multipliers $\boldsymbol{\alpha} \in \Re^m$ as follows.

$$L(\mathbf{w}, \gamma, y, \boldsymbol{\alpha}) = v\frac{1}{2}\|y\|^2 + \frac{1}{2}\left\|\begin{bmatrix}\mathbf{w} \\ \gamma\end{bmatrix}\right\|^2 - \boldsymbol{\alpha}^T(D(\mathbf{A}\mathbf{w} - \mathbf{e}\gamma) + y - \mathbf{e}) \tag{2}$$

Taking the derivative of equation (2) with respect to $\mathbf{w}$, γ, y and setting them to zero, and do some algebra, the following equations can be obtained.

$$\boldsymbol{\alpha} = \left(\mathbf{I}/v + D(\mathbf{A}\mathbf{A}^T + \mathbf{e}\mathbf{e}^T)D\right)^{-1}\mathbf{e} = \left(\mathbf{I}/v + \mathbf{H}\mathbf{H}^T\right)\mathbf{e} \tag{3}$$

$$H = D[A \quad -e]$$

The linear decision function can be obtained as follows.

$$x^T w - \gamma = x^T A^T D\alpha - \gamma = 0 \qquad (4)$$

One can obtain the kernel version of Proximal SVM by replacing A by K, where K is an abridged form of matrix $K(A, A^T)$ whose elements are dot products of two samples in a given kernel space.

$$\upsilon = (I/\nu + D(KK^T + ee^T)D)^{-1} e = (I/\nu + GG^T)^{-1} e, \qquad (5)$$
$$G = D[K \quad -e]$$

$$f(x) = (K(x^T, A^T)K(A, A^T)^T + e^T)D\upsilon = 0 \qquad (6)$$

In [10], it was shown that the performance of the PSVM should be comparable with that of the SVM with fast computational time.

2.2 A Modified Proximal Support Vector Machine

If we set the diagonal matrix D to identity matrix in equation (1), the corresponding equation becomes the following.

$$\text{minimize } \nu \frac{1}{2}\|y\|^2 + \frac{1}{2}(w^T w + \gamma^2) \qquad (7)$$
$$\text{subject to } Aw - e\gamma + y = e$$

In equation (2), we want to minimize the geometric distances from all samples to a hyperplane $wx-\gamma 1$ that is corresponding to a positive proximal hyperplane in Figure 1. Therefore the optimal hyperplane that satisfies equation (7) can be considered as a proximal hyperplane that best represents the underlying positive samples. We can obtain such a proximal hyperplane using Lagrangian multipliers as in Section 2.1.

$$L(w, \xi, \alpha) = \frac{1}{2}\|w\|^2 + \frac{1}{2}\gamma^2 + \frac{\nu}{2}\|y\|^2 - \alpha^T(Aw - \gamma + y - e) \qquad (8)$$

Starting from equation (7), one can easily show the following formula for the modified Proximal SVM.

$$\alpha = (AA^T + ee^T + I/\nu)^{-1} e \qquad (9)$$

$$f(x) = x^T w - \gamma = x^T A^T (AA^T + ee^T + I/\nu)^{-1} e + e^T \alpha \qquad (10)$$

From equation (10), the distance form proximal hyperplane is defined as follows.

$$\text{Distance} = |f(x) - 1| \qquad (11)$$

The kernel version of a modified Proximal SVM is defined as following, and the distance can be obtained using equation (11).

$$\alpha = (K(A, A^T) + ee^T + I/\nu)^{-1} e \qquad (12)$$

$$f(x) = K(x^T, A^T)K(A, A^T) + ee^T + I/\nu)^{-1} e + e^T \alpha \qquad (13)$$

2.3 Focused Crawling Using Proximal Hyperplane

The proximal hyperplane represents a distribution of data samples. In other words the proximal hyperplane is a maximal margin hyperplane where most training data samples reside [8]. Therefore, one can use the distance to a proximal hyperplane as a dissimilarity measurement for a class membership. We apply this idea to the focused crawling.

Algorithm for the Focused Crawling Using Proximal Hyperplane

A. Training Phase
1. Determine the Proximal Hyperplane over positive samples using equation (11)
2. Compute a mean value of the distances to Hyperplane from positive samples.

B. Crawling Phase
1. Set the number of Priority Bins to k
2. Get one URL from a non-empty highest Priority Bin
3. Download a corresponding page from the Web
   3.1 Compute distance from the downloaded page using equation (11)
   3.2 Quantize the distance as following
      If distance $\leq$ mean, then Priority = 1

      Else if mean < distance $\leq$ 0.5, then Priority = $2 + \left\lfloor (k-2) \dfrac{\text{distance} - \text{mean}}{0.5 - \text{mean}} \right\rfloor$

      Otherwise, Priority = k
   3.3 Extract all the URLs in this page and put them into a Priority Bin
4. Go to step 2.

Fig. 2. Proposed Focused Crawling Algorithm Using a Proximal Hyperplane

First, we determine the proximal hyperplane from a given positive samples. We also compute a mean value of the distances from the samples to the proximal plane. Then, we start to crawl with a few seed URLs. The seed URLs may be a set of representative URLs in a given topic, or can be a set of target hosts in which the crawler is confined to explore. After fetching a page of a selected URL, the crawler transforms the page into a vector, and extracts all the hyperlinks, URLs from the fetched page. Now, the crawler computes the distance from a transformed vector to the predetermined proximal hyperplane, and quantizes the distance so that the URLs extracted from the fetched page can be put into a corresponding priority bin. There are k priority bins and each bin corresponds to one quantized distance level. The focused crawler chooses one URL at a time from a non-empty highest priority bin so that it can crawl favorably pages closer to a proximal hyperplane. The overall focused crawling algorithm is shown in Fig. 2.

3 Results

3.1 Experimental Setup

The crawler that we used in this paper was implemented using Java, and the experiments shown here were conducted in Pentium IV with main memory of 1GBytes. The details of implementation are out of scope in this paper, and you can refer to [6] and [11] for the various implementation issues. For the preprocessing of the HTML documents, we did not execute the stemming, and we filtered the stop words. To make transformation of a page to a vector, we only used TF (Term Frequency) and normalized vectors of unit length.

The experiments were conducted in two steps. First, in Section 3.1, we tested our Modified Proximal Support Vector Machine using the WebKB data set [2]. Then, in Section 3.2, we applied the proposed crawling method on the Web with a specific topic. In order to evaluate the performance of focused crawlers, a harvest rate has often been used [9]. The harvest rate is an average relevance of the fetched pages at any given time. However, a 'true' relevance cannot be obtained as long as millions of Web pages are concerned. Therefore the relevance measure used in the focused crawling is usually used as a relevance measure in the evaluation [9]. In this paper, we used a cosine similarity as a relevance measure in order to compare our focused crawling and a traditional approach using a Naive Bayesian Classifier [9].

3.2 Results from the Modified Proximal SVM

In order to see how our modified Proximal SVM, we experimented over WebKB data set. In this experimental setup, we only used 4 classes, "course", "faculty", "project", and "student". We trained each proximal hyperplane using 10 pages randomly taken out of each class. Then we computed the distances from test pages out of 4 classes to the proximal hyperplane. We quantized the distances as described in Section 2.3 and we set k to 20. We conducted this process for each class.

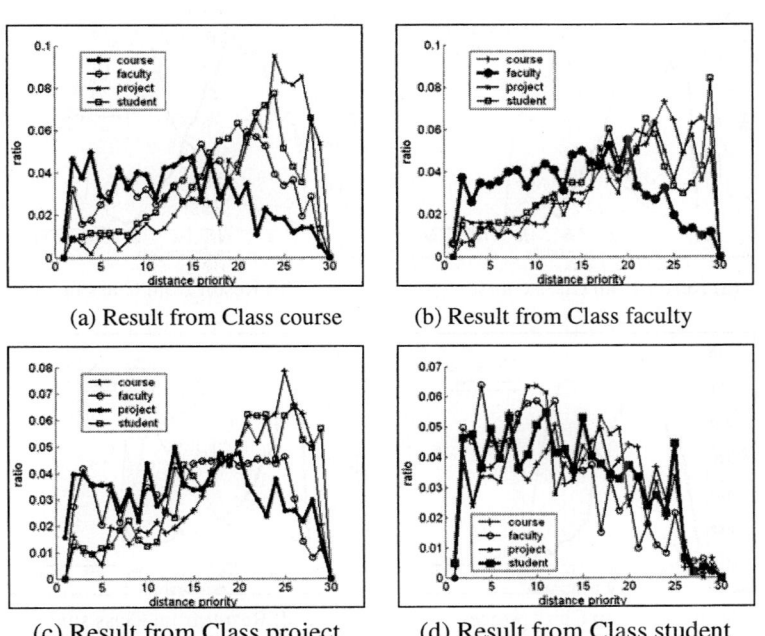

(a) Result from Class course (b) Result from Class faculty

(c) Result from Class project (d) Result from Class student

Fig. 3. Ratio of true positive pages and false positive pages over the Priority Bins for each class: The bold lines in (a)-(d) indicate a ratio of positive pages to total number pages in each class

We also tested the modified Proximal SVM with yet another class "others". Class others is a collection of pages that do not belong to a specific class defined in the WebKB project [2].

Figure 4 shows the results against Class others. As we can see in this experiment, all four Proximal Hyperplanes were able to separate themselves well from Class others. These two experiments indicate that the proposed Proximal Hyperplane approach should be effective to the focused crawling where negative examples are not well defined and sometimes only positive examples are available.

3.3 Results from the Proposed Focused Crawling

We tested the proposed focused crawling algorithm over the Web and compared the performance with Naïve Bayesian Classifier based focused crawling [9]. We selected "Computer Science" as a topic of interest and chose 5 positive URLs from the Web. We also chose 5 negative web pages from "Law and Medical" pages for the training of Naïve Bayesian Classifier. Note that the number of samples used in our experiments is much less than that used in others [1-5][9].

We experimented with four different sets of seed URLs. Figure 5 shows the result with http://www.harvard.edu and http://www.eecs.harvard.edu/index/cs/ as Seed URLs. Figure 6 shows the results with http://www.yahoo.com and some URLs in Yahoo! Category "Computer Science" as Seed URLs. In all cases, we confined the crawling range to the same domains as seed URLs.

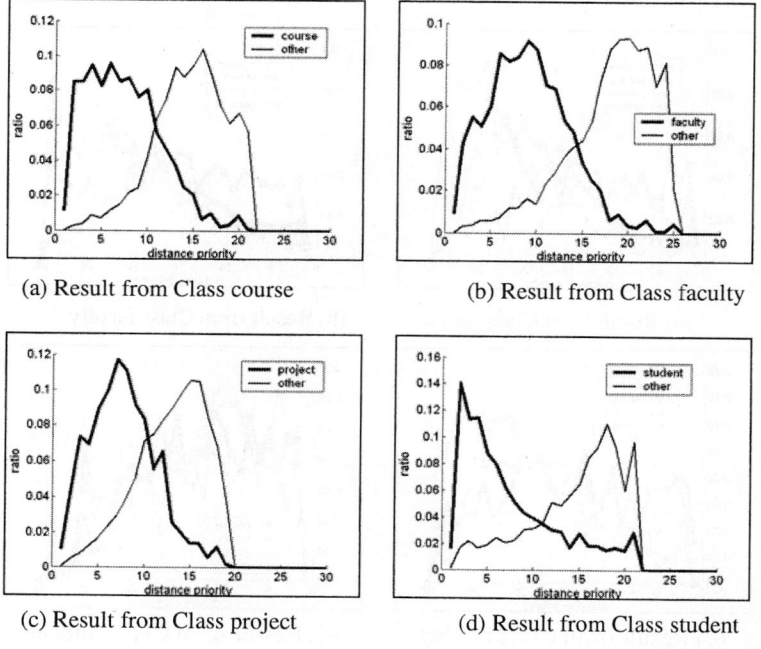

(a) Result from Class course (b) Result from Class faculty

(c) Result from Class project (d) Result from Class student

Fig. 4. Ratio of true positive pages and false positive pages over the Priority Bin for each class

As shown in Figure 5 and 6, our proposed method outperformed the Bayesian based approach over a whole crawling duration. It is interesting that the Bayesian approach degenerated into the breadth first crawler as a certain amount of time goes by. The experimental results indicate that the overall performance of the proposed focused crawling was much better than the traditional approach using Naïve Bayesian Classifier. Note also that the harvest ratio from the breadth first crawling goes almost constant across the crawling time as we expected.

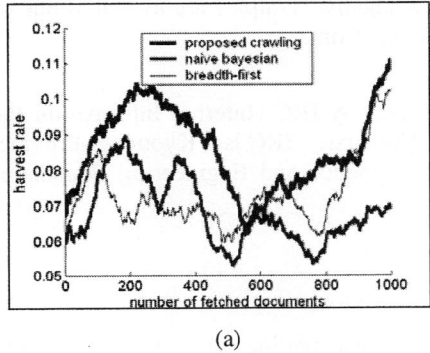
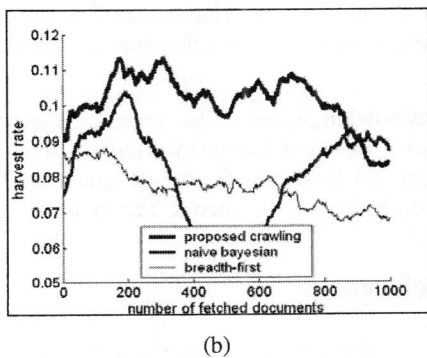

Fig. 5. The results from the proposed crawling, Naive Bayesian based Crawling, and the crawling without focusing starting from (a) http://www.harvard.edu (b) http://www.eecs.harvard.edu/index/cs/

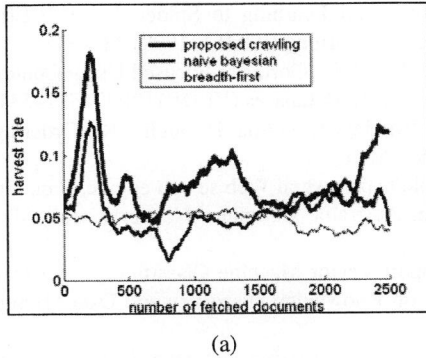
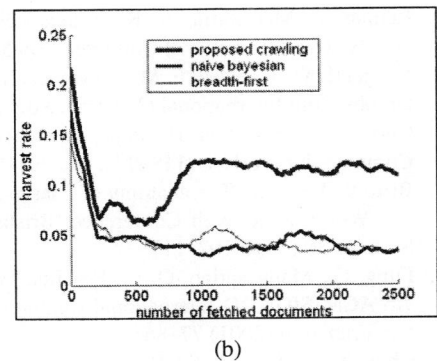

Fig. 6. The results from the proposed crawling, Naive Bayesian based Crawling, and the crawling without focusing starting from (a) http://www.yahoo.com (b) http://dir.yahoo.com/science/computer_science

4 Conclusions

The proposed crawling method in this paper starts from the idea that a focused crawling for a specific topic can be better formulated as a one-class problem than the two class classification problem. In order to model a specific topic out of a few positive samples, we modified the Proximal SVMs so that the obtained proximal hyperplane can represent a distribution of positive samples. We have also presented a focused crawling method using the distances to the proximal hyperplanes.

We tested our proposed method over various data sets and also compared the performance with a traditional Bayesian based crawling. The experimental results were very promising and encouraging to do more researches related to this approach. Among them are how to quantize the distances into priority bins properly. It is also worth investigating into the case where some negative samples are available but we are not sure how many different classes they come from.

Acknowledgement. This research was supported by IRC (Internet Information Retrieval Research Center) in Hankuk Aviation University. IRC is a Kyounggi-Province Regional Research Center designed by Korea Science and Engineering Foundation and Ministry of Science & Technology.

References

1. Chakrabarti, S., van den Berg, M., Dom, B.: Focused crawling: a new approach to topic-specific Web resource discovery. 8th International World Wide Web Conference, Toronto (1999) 1623–1640
2. Aggarwal, C. C., Al-Garawi, F., Yu, P. S.: Intelligent Crawling on the World Wide Web with Arbitrary Predicates. 10th International World Wide Web Conference, Hong Kong (2001) 96–105
3. Rennie, J., McCallum, A. K.: Using Reinforcement Learning to Spider the Web Efficiently. 16th International Conference on Machine Learning (ICML) (1999) 335–343
4. Diligenti, M., Coetzee, F. M., Lawrence, S., Giles, C. L., Gori, M.: Focused Using Context Graphs. 26th International Conference on Very Large Databases (VLDB) (2000) 527–534
5. Cho, J., Garcia-Mlina, H., Page, Lawrence.: Efficient Crawling Through URL Ordering. Computer Networks and ISDN Systems (1998) 161–172
6. Brin, S., Page, L.: The anatomy of a large-scale hypertextual Web search engine. Proc. 7th Int. World Wide Web Conference, Brisbane, Australia, Computer Networks and ISDN Systems 30 (1998)107–117.
7. Fung, G., Mangasarian, O. L.: Proximal Support Vector Machine Classifiers. KDD2001: 7th ACM SIGKDD International Conference on Knowledge Discovery and Data Mining, San Francisco (2001) 77–86
8. Choi, Y. S., Noh, J. S.: Relevance Feedback for Content-Based Image Retrieval Using Proximal Support Vector Machine. International Conference on Computational Science and Its Applications (ICCSA), Vol. 2. Assisi, Italy (2004) 942–951
9. Charkrabarti, S.: mining the web Discovering Knowledge from Hypertext Data. Morgan Kaufmann Publishers (2003)
10. Cristianini, N., Shawe-Taylor, J.: An Introduction to Support Vector Machines and other kernel-based learning methods. Cambridge University Press (2000)
11. Najork, M., Heydon, A.: High-performance Web crawling. Tech. Rep. Research Report 173, Compaq SRC(2001)

A Performance Improvement Scheme of Stream Control Transmission Protocol over Wireless Networks

Kiwon Hong[1], Kugsang Jeong[1], Deokjai Choi[1], and Choongseon Hong[2]

[1] Computer Science Department, Chonnam National University
300 Yongbong-dong, Buk-gu, Gwangju, 500-757, Korea
{kiwon77,handeum}@iat.chonnam.ac.kr, dchoi@chonnam.ac.kr
[2] Department of Computer Engineering, Kyung Hee University
1 Seocheon, Giheung, Yongil, Gyeonggl, 449-710, Korea
cshong@khu.ac.kr

Abstract. Today's computer network is shifting from wired networks to wireless networks. Several attempts have been made to assess the performance of TCP over wireless networks. Also, several solutions have been proposed to improve its performance. Many people believe SCTP would be replacement of TCP. However, the performance enhancement research for SCTP over wireless network is in beginning stage yet. In this paper we have measured performance characteristics of SCTP, and compared with ones of TCP and TCP-snoop over wireless network. After this experiment, we found the performance of SCTP is better than TCP, but worse than TCP-snoop. To improve the performance of SCTP over wireless network, we modified SCTP code by adopting TCP-snoop approach. We named it SCTP-snoop. With this SCTP-snoop, we experimented again, and found that it showed the best performance over wireless network among TCP, TCP-snoop, and SCTP.

1 Introduction

For the past 20 years, most Internet applications are implemented using TCP or UDP. But, those protocols are not good enough for multimedia traffics that are getting popular these days. In such background, SCTP was proposed by Internet Engineering Task Force and it was published as RFC 2960 in October 2000 as a Proposed Standard [1].

Like TCP, SCTP offers a point-to-point, connection-oriented, reliable delivery transport service for applications communicating over an IP network. It inherits many of the functions developed for TCP over the past two decades, including powerful congestion control and packet loss recovery functions. Also multiple stream mechanism of SCTP is designed to solve the head-of-the.line blocking problem of TCP, and association mechanism (the exchange of at least four SCTP packets) of SCTP is designed to solve the classic SYN flooding-type of denial-of-service attack of TCP [2].

Today Internet users are increasing rapidly. Also, various service extensions of mobile communication providers and increasing of mobile communication users require not only voice transmission, but also data transmission for various services. A communication service market will be changed from wired networks to wireless networks and to the consolidation trend of wired and wireless networks. So, data communication demands are increasing rapidly.

An original TCP was optimized in wired networks. It is not customized to the wireless networks, and it has many defects for wireless networks such as frequent cutoff and high error rate. So, if TCP is used over wireless networks, efficient transmission will not be guaranteed. Since wireless networks are less stable than wired networks, it experiences frequent packet loss. It affects RTT time, and TCP understands the network is in congestion state. According to the TCP retransmission policy, it will reduce the traffic and get slow down retransmission time too. It will affect the general performance of network [3].

To overcome that problem of TCP, researchers have proposed modified protocols actively such as Indirect-TCP and TCP-Snoop. But, in case of SCTP that is believed to be a next generation transport protocol, there has not been substantial effort to improve performance over wireless networks.

The Snoop protocol is a simple and efficient protocol for end-to-end communication [4]. Therefore we modified Snoop protocol for SCTP over wireless networks. The Snoop Agent is implemented on a base station (BS) that connects wired networks and wireless networks. We evaluate performance of SCTP-snoop in terms of required time to transmit certain amount of data, and the experiment shows that the performance of SCTP-snoop is better than TCP or SCTP over wireless networks.

This paper is organized as follows: In section 2, we present differences between SCTP and TCP congestion control. Section 3 describes a snoop protocol. In section 4, our simulation results and analysis are presented. Section 5 discusses future work and provides some conclusions.

2 Differences Between SCTP and TCP Congestion Control

The congestion control of SCTP is based on TCP's principles and it uses the SACK extension of TCP. It also includes slow start, congestion avoidance, and fast retransmission. There are subtle differences between the congestion control mechanisms of TCP and SCTP. The congestion control properties of SCTP that are different from those of TCP are as follows [5]:

1. The congestion window (cwnd) is increased according to the number of bytes acknowledged, not the number of acknowledgements received. Similarly, the flight-size variable, that represents how much data has been sent but not acknowledged on a particular destination address, is decreased by the number of bytes acknowledged. While in TCP, it is controlled by the number of new acknowledgement received.

2. The initial congestion window is suggested to be 2*MTU in SCTP, which is usually one MTU in TCP.

3. SCTP performs congestion avoidance when cwnd >ssthresh(slow start threshold). It is required to be in slow start phase when the ssthresh is equal to the cwnd. It is optional in TCP to be either in the ssthresh or in the congestion avoidance phase when the ssthresh is equal to the cwnd.

4. SCTP's Fast Retransmit algorithm is slightly different from TCP's. SCTP has no explicit fast recovery algorithm that is used in TCP. In SCTP, the parameter

Max.Burst is used after the fast retransmit to avoid flooding the network. Max.Burst limits the number of SCTP packets that may be sent after processing the SACK, which acknowledges the data chunk that has been fast retransmitted.

5. An unlimited number of GAP ACK blocks are allowed in SCTP. TCP allows a maximum of three SACK blocks

3 Snoop Protocol

The snoop is a TCP aware link layer protocol. Snoop was designed so that the wired infrastructure of the network would need no changes. Since Snoop protocol does not require changing wired network, it is good candidate for our system [4].

So, we modified snoop protocol for SCTP which means it supports multi-homing and multi-stream.

To support the multi-homing and the multi-streaming of SCTP, SCTP-Snoop agent executes followings: SCTP interchanges INIT Chunk and INIT-ACK Chunk, which are needed information for multi-homing and multi-streaming, during association establishment. In the process of exchange, SCTP-Snoop agent gets and stores addresses of Sender and receiver included INIT Chunk and INIT-ACK Chunk. If the packet losses have occurred by receiver, SCTP-Snoop agent will judge a problem by a transmission path. So, SCTP Snoop transfers lost packets by selecting one of transmission paths. Also, SCTP Snoop checks lost chunks through Gap Ack Block field of SCTP SACK-Chunk, and retransmits lost chunks stored in buffer.

The snoop module has two linked procedures, snoop_data() and snoop_ack(). Snoop_data() processes and caches packets intended for the mobile host(MH) while snoop_ack() processes acknowledgments (ACKs) coming from the MH and drives local retransmissions from the base station to the mobile host. The flowcharts summarizing the algorithms for snoop_data() and snoop_ack() are shown in Figure 1 and Figure 2 , and their working details are described in brief below.

In figure 1, when the data chunk packet is received by base station with snoop agent, if it is not the new data chunk which is adjudged through the Transmission Sequence Number (TSN) of SCTP, the agent forwards it to the receiver without storing to buffer.

If it is the new data chunk and an out-of-sequence, the agent forwards it to the receiver after marking packet loss by congestion. If it is the new SCTP packet and in order, it copies into buffer and forwards it to the receiver. At that time, if the new SCTP packet is INIT-Chunk or INIT-ACK chunk, Snoop agent could store addresses into buffers after verification address parameters. And Snoop agent stores INIT-Chunk and INIT-ACK chunk into buffers and forwards to MH.

In figure 2, when the SCTP sack chunk is received by base station, if it is the new sack chunk, the agent removes it from the buffer and modifies retransmission timer by measuring RTT to reflect new RTT, while if it is the first duplicate sack chunk, it means packet loss in wireless networks. So, Snoop agent retransmits lost chunks after verification Gap Ack block of SCTP SACK-Chunk and dumps the duplicate sack chunk. However, if it is not the first duplicate sack chunk but the duplicate sack chunk, the agent dumps the duplicate sack chunk.

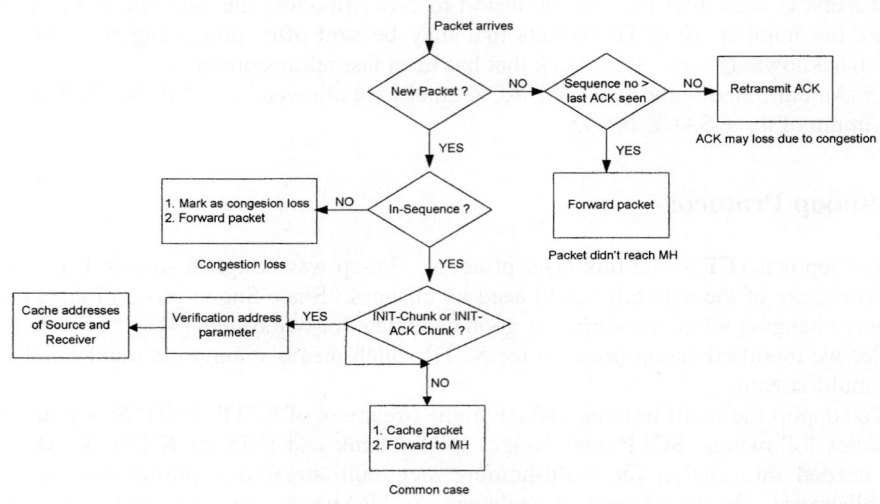

Fig. 1. Snoop_data()

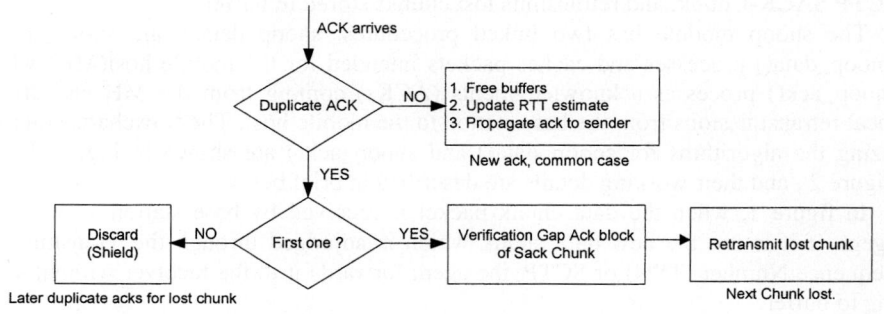

Fig. 2. Snoop_ack()

4 Performance Evaluation and Analysis

We estimated SCTP performance through comparison with original TCP which currently have been used as an transfer protocol for reliable transmission.

4.1 Simulation

All of the simulation results presented in this paper were obtained from an implementation of SCTP for the ns-2 simulation environment, which was developed by UC

Berkeley [6]. The SCTP module for ns-2 was developed by the Protocol Engineering Lab at the University of Delaware and is available as third party module [7].

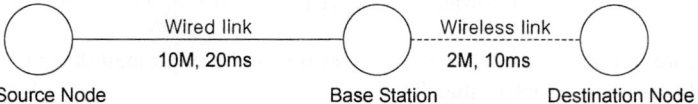

Fig. 3. Network Topology

Figure 3 illustrates the simulated network. Two end nodes are connected with each other via the base station that monitors every packet that passes through the connection in both directions. The network model consists of a 10 Mbps, 20ms delay wired channel and a 2Mbps wireless channel with 10ms delay.

In experimentation of this paper, TCP or SCTP agent exists at each edge nodes. We assumed that all packet losses are occurred in wireless network. So, error model of simulation is simplified and only considered transmission error of wireless networks. In all the simulation runs, there are 1000byte data segments transferred (excluding headers) for TCP and 1000byte data chunks (excluding headers) for SCTP.

We used the total time to transfer data for performance evaluation. We measured the performance using ftp application which transfers data from 1Mbyte to 10Mbyte.

4.2 SCTP vs TCP

We first ran a test to compare the performance of original SCTP with that of original TCP under same loss over wireless networks.

Figure 4 shows that the total time of original SCTP to transfer files are almost half of one of original TCP.

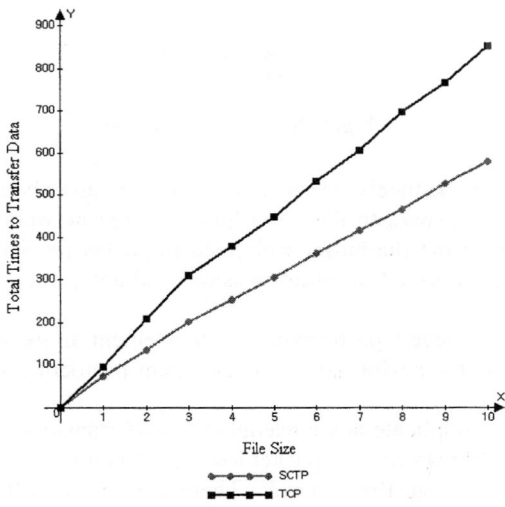

Fig. 4. SCTP vs TCP

Both original TCP and SCTP transfer data through wireless network with packet loss. If packet losses occurred in wireless networks, source node could regard it as congestion by packet loss of wireless network. So, both TCP and SCTP have executed congestion control in order to recover packet losses. This is affected by congestion control.

Therefore, the result of this experimentation can be explained that SCTP uses more enhanced congestion control than TCP.

4.3 SCTP vs TCP Snoop

In the second simulation, we have compared original SCTP with TCP Snoop. As shown in the figure 5, the total time of SCTP to transfer data has been delayed more than TCP Snoop.

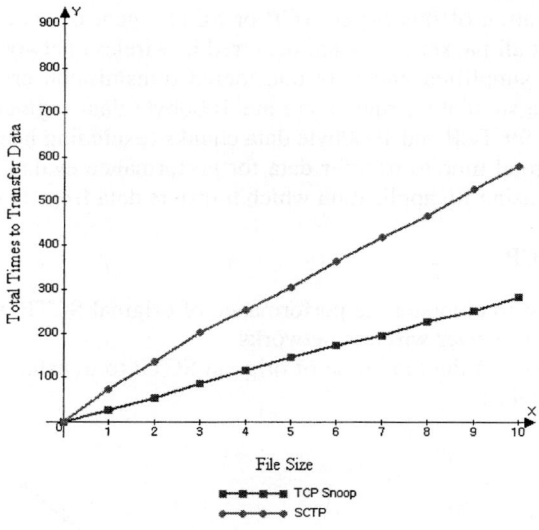

Fig. 5. SCTP vs TCP Snoop

TCP Snoop agent immediately doesn't deliver information about packet loss which occurred in wireless networks to the fixed host at wired networks, but Snoop agent processes packet loss using the buffer which stores packet received from the sender, thereby avoiding unnecessary fast retransmissions and congestion control invocations by the sender.

So, TCP Snoop achieved performance enhancement more than original TCP. But SCTP doesn't use the performance enhancement protocol such as Snoop Protocol of TCP.

If SCTP receives a duplicate acknowledgement and transmission error by retransmission timeout, SCTP regards it as packet loss. At this time, the performance of the network has fallen by congestion control. Therefore, SCTP will need SCTP Snoop like TCP Snoop for performance enhancement in wireless networks.

4.4 SCTP Snoop vs Original SCTP and TCP Snoop

In this paper, we used modified Snoop Protocol for SCTP because Snoop was TCP-aware protocol. Figure 6 shows the total time to transfer data of SCTP Snoop and TCP Snoop. As shown in the figure, the total time of SCTP Snoop is the shortest among TCP Snoop and Original SCTP.

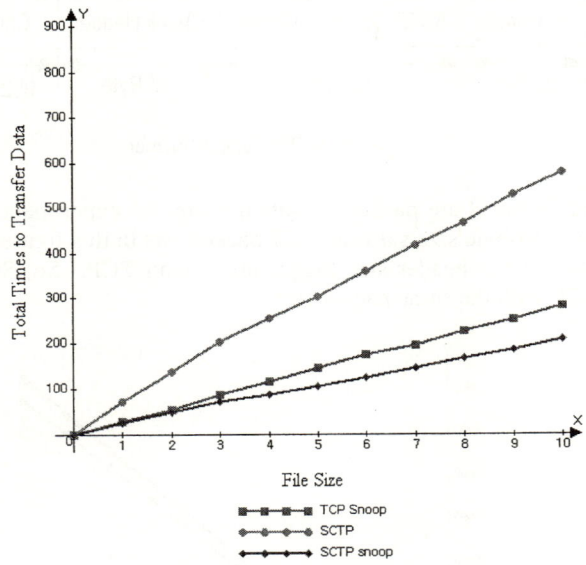

Fig. 6. SCTP-Snoop vs SCTP,TCP Snoop

Like TCP-Snoop, SCTP-Snoop hides the packet loss from the fixed host, thereby preventing unnecessary congestion control mechanism invocations.

The result of this experimentation can be explained that SCTP Snoop has achieved performance enhancement more than original SCTP just as TCP Snoop has achieved performance enhancement of TCP in wireless networks.

4.5 The Number of Packets

One of the main reasons that the preceding chunk-based format was chosen for SCTP was its extensibility.

SCTP and TCP header are slightly different [1]. SCTP packets are made up of an SCTP common header and specific building blocks called chunks. The SCTP common header provides SCTP with specific validation and associative properties. Chunks provide SCTP with the basic structure needed to carry information. Also, each chunk has header of its own in order to provide various information such as chunk type, TSN.

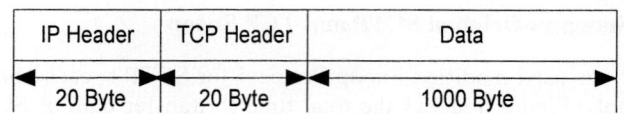

Fig. 7. TCP Packet Format

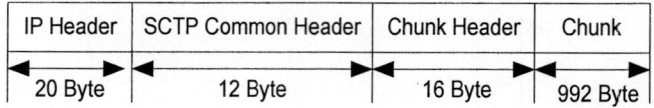

Fig. 8. SCTP Packet Format

Figure 7 and Figure 8 are packet formats used in our experimentations. We used SCTP packet of 1040byte size same as TCP packet. As in this figures, we ascertained the facts that SCTP has header size bigger 8byte than TCP. So, SCTP sends more packets than TCP with the same packet size.

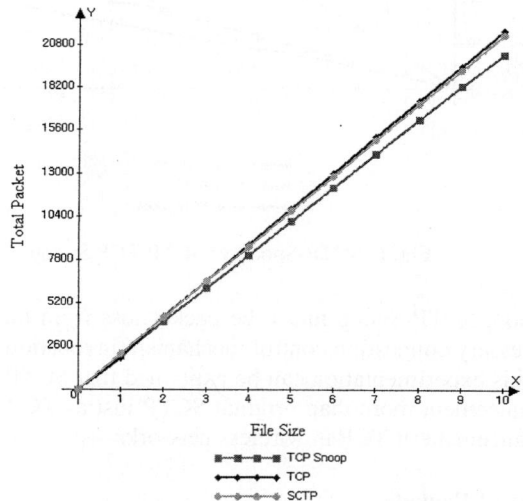

Fig. 9. The number of Packets comparison of SCTP and TCP

SCTP has a multi-homing feature. To support it, SCTP uses the HEARBEAT chunk type of 56byte. It is used to periodically probe reachability of the destination addresses and update the RTT of a destination address. So, the transmission of the HEARBEAT chunk is one of the reasons increasing total packet counts of SCTP.

Through figure 9, we show that SCTP interchanges more packets than TCP Snoop during the data transmission. The increase of packet counts not only cause more congestion control by packet loss in wireless networks, but also degrade the performance due to overload when many users use wireless networks with low bandwidth and high delay.

5 Conclusion and Future Work

We carried out 3 experimentations over wireless networks with packet loss as following: at first experimentation, we compared the performance of TCP with SCTP by sending 10 different file size. It showed the total time of SCTP to transfer files is almost half of one of TCP. We believe that original SCTP uses more enhanced congestion control than original TCP in wireless networks. Since our experiment environment has some packet loss, its congestion control scheme is effective. At second experimentations, we compared the performance of TCP-snoop with SCTP. It showed that original SCTP has less performance than TCP Snoop in wireless networks. The reason for this would be that TCP-snoop helps TCP engine to avoid unnecessary congestion control process which severely affects delay. At third experimentations, the performance of the proposed SCTP-Snoop was compared with original SCTP. We found that the performance of SCTP-Snoop in wireless network was improved over original SCTP and TCP-Snoop.

We have shown that SCTP sent more packets than TCP. It can be one of reasons of SCTP performance degradation because the increasing of packet counts have influence on the more congestion control execution and it can occur overload by sharing in wireless networks by many users.

In this research, we modified SCTP by including Snoop mechanism properly. We show that SCTP-snoop approach is effective over wireless networks.

For the future work, we want to find the sources of performance degradation thru continuous experimentations, and to improve SCTP protocol based on intrinsic SCTP features.

References

[1] R. Stewart, Q. Xie, K. Morneault, C. Sharp, H. Schwarzbauer, T. Taylor, I. Rytina, M. Kalla, L. Zhang and V. Paxson. "Stream Control Transmission Protocol" Proposed standard, RFC 2960, Internet Engineering Task Force(IETF),October 2000.
[2] R. Stewart, Qiaobing Xie, "Stream Control Transmission Protocol:a reference guide", Addison-Wesley 2001.
[3] G. Xylomenos, G. C. Polyzos, P. Mahonen and M.Saaranen, "TCP Performance Issues over Wireless Link ", IEEE Communications Vol. 39 No.4 pp.52-58, April 2001.
[4] Hari Balakrishnan, Srinivasan Seshan, Elan Amir, Randy H. Katz. "Improving TCP/IP Performance over Wireless Networks ", Proc. 1st ACM Conf. on Mobile Computing and Networking, Berkeley, CA, November 1995.
[5] R. Brennan and T. Curran, "SCTP congestion control : Initial simulation studies", International Teletraffic Congress(ITC 17), Brazil, 2001
[6] UC Bekely, LBL, USC/ISI, and Xerox Parc. Ns-2 documentation and software, Version 2.7, January 2004. http://www.isi. edu/nsnam/ns/index.html
[7] A. Caro and J. Iyengar. Ns-2 SCTP module, Version 3.4, August 2003. http://pel.cis.udel.edu

Cache Management Protocols Based on Re-ordering for Distributed Systems

SungHo Cho[1] and Kyoung Yul Bae[2]

[1] Dept. of Information Science & Telecommunication,
Hanshin University, Korea
zoch@hs.ac.kr
[2] Division of Computer Software, SangMyung University, Korea
jbae@smu.ac.kr

Abstract. Database systems have used a client-server computing model to support shared data in distributed systems such as Web systems. To reduce server bottlenecks, each client may have its own cache for later reuse. This paper suggests an efficient cache consistency protocol based on a optimistic approach. The main characteristic of our scheme is that some transactions that read stale data items can not be aborted, because it adopts a re-ordering mechanism to enhance the performance. This paper presents a simulation-based analysis on the performance of our scheme with other well-known protocols. The analysis was executed under the *Zipf* workload which represents the popularity distribution on the Web. The simulation experiments show that our scheme performs as well as or better than other schemes with low overhead.

1 Introduction

A potential weakness of the the client-server model is server bottlenecks that can arise due to the volume of data requested by clients[1,2]. To reduce this problem, each client may have its own cache to maintain some portion of data for later reuse [3,4]. When client caching is used, there should be a transactional cache consistency protocol between client and server to ensure that the client cache remains consistent with the shared data by use of transaction semantics [5,6].

In the literature, many transactional cache consistency algorithms have been proposed. The previous studies[6–8] indicate that Optimistic Two-Phase Locking(O2PL) performs as well as or better than the other approaches for most workloads, because it exploits client caching well and also has relatively lower network bandwidth requirements.

In this paper, we suggests an optimistic protocol called RCP(Re-orderable Cache Protocol). Compared with O2PL, the main advantage of RCP is to reduce

---

* This work was supported by Ministry of Education and Human Resources Development through Embedded Software Open Resource Center(ESORC) at SangMyung University.

unnecessary operations based on re-ordering. In addition, even if some schemes use multiple-versions to reduces unnecessary operations[1], RCP stores only a single version of each data item for re-ordering. finally, our scheme does not require global deadlock detection which tends to be complex and has frequently been shown to be incorrect.

This paper presents a simulation-based analysis on the performance of some schemes with the *Zipf* distribution which is governed by *Zipf*'s law[9]. So far, a number of groups have shown that, by examining the access logs for different Web servers, the popularity distribution for files on the Web is skewed following *Zipf*'s law. In [10], the authors showed that the *Zipf*'s distribution is strongly applied even to WWW documents serviced by Web servers. Our simulation experiments show that RCP performs as well as or better than other schemes with low overhead.

2 Re-orderable Caching Consistency Protocol

The RCP protocol is one of the avoidance-based algorithms, and it defers write intention declarations until the end of a transaction's execution phase. The difference between RCP and O2PL is that RCP uses a preemptive policy. If possible, RCP tries to re-order transactions which accessed read-write conflicting data items.

Basically, RCP uses the *current version check* for validation. For this, the server maintains read time-stamp $D.T^r$ and write time-stamp $D.T^w$ for each persistent data item. The time-stamps are the time-stamps of the youngest (i.e., lastest in time) committed transaction. When a transaction wants to read a data D_i, the server sends the data with current write time-stamp $D_i.T^w$ if data D_i is not in the local cache of the client. Clients also maintain data and their current time-stamps in the local cache. Hence, a transaction views each data item as a (name, version) pair. Note that we leave the granularity of logical data unspecified in this paper; in practice, they may be pages, objects, etc.

The client of transaction X maintains the following information for the transaction.

- **Set** S_X^R - Set of the data read by transaction X
- **Set** S_X^W - Set of the data written by transaction X
- **Set** S_X^I - Set of the data invalidated by the server during transaction X is running.
- **Time-stamp** T_X^L - The maximum time-stamp among the time-stamps of the data read by transaction X
- **Time-stamp** T_X^U - The minimum time-stamp among the time-stamps of the committing transactions that are conflicted with transaction X (during transaction X is running)

Consider the example in Section 1. If transaction Y is back-shifted, what is the valid time-stamp interval of transaction Y for re-ordering? Intuitively, a back-shifted time-stamp of transaction Y has to be larger than the maximum time-stamp among time-stamps of read data and has to be smaller than the minimum time-stamp among the time-stamps of conflicting and committing transactions. Based on the property, time-stamps T^L and T^U denote the lower-bound and upper-bound of valid interval for re-ordering, respectively. The initial values of time-stamps T^L and T^U are the smallest time-stamp in the system. Sets S^R and S^W are maintained for validation as in optimistic concurrency control. Set S^I is maintained to reduce unnecessary operations.

When transaction X is ready to enter its commit phase, the client sends to the server a message containing sets S_X^R, S_X^W, time-stamps T_X^L and T_X^U. When the server receives the message, it assigns a unique committing time-stamp T_X^C that is equal to the certification time. After that, the server sends a message to each client that has cached copies of any of the updated data items in set S_X^W. The message contains committing time-stamp T_X^C, sets S_X^R and S_X^W. When a remote client gets the invalidation message, it evicts local copies of the data updated by transaction X, and sends an acknowledgment(ACK) to the server. Once all ACKs have been received, the server sets read time-stamp $D_j.T^r$ for each data D_j in set S_X^R and write time-stamp $D_k.T^w$ for each data D_k in set S_X^W to committing time-stamp T_X^C.

Now, we describe how to update the lower-bound time-stamp T^L. Whenever transaction X reads data D_i, time-stamp T_X^L is compared with the current time-stamp $D_i.T^w$. If time-stamp T_X^L is lower than the time-stamp $D_i.T^w$, then T_X^L is set to $D_i.T^w$.

Next, consider the upper-bound time-stamp T^U. When the client of transaction Y gets the invalidation message issued by transaction X, the client updates time-stamp T_Y^U with the following rules;

- For each data D_i in set S_X^W, if data D_i also is in set S_Y^R, then time-stamp T_Y^U is updated based on the following procedures. Otherwise, the following procedures are ignored.
- For each data D_j in set S_X^R, if data D_j also is in set S_Y^W, the client aborts transaction Y, because transaction Y can not be re-ordered.
- If transaction Y is not yet aborted, time-stamp T_Y^U is compared to time-stamp T_X^C. If time-stamp T_Y^U is the initial value, then it is set to time-stamp T_X^C. Otherwise, time-stamp T_Y^U is set to time-stamp T_X^C if it is larger than time-stamp T_X^C.
- Each data D_k in sets S_X^W and S_Y^R is inserted into set S_Y^I to prevent unnecessary operations.

Whenever time-stamps T_Y^L or T_Y^U is changed, transaction Y is aborted if time-stamp T_Y^U is not the initial value and time-stamp T_Y^L is larger than or equal to time-stamp T_Y^U, because the server can not find any back-shifting time-stamp. In addition, whenever transaction Y tries to write the data in set S_Y^I, transaction Y is also aborted to prevent write-write conflicts.

When transaction Y which has accessed invalidated data is ready to enter its commit phase, the client sends to the server a message containing sets S_Y^R, S_Y^W, time-stamps T_Y^L and T_Y^U. When the server receives the message, if time-stamp T_Y^U is not the initial value, it sets committing time-stamp T_Y^C to T_Y^U - δ (δ is an infinitesimal quantity) for re-ordering instead of assigning a new time-stamp. Note that, instead of a specific value, we use an infinitesimal quantity for the value of δ. This approach has an advantage, because it reserves sufficient interval between time-stamps of committing transaction for accepting re-ordered transactions.

After setting T_Y^C to T_Y^U - δ, the server checks whether the back-shifted time-stamp T_Y^C is always larger than time-stamp $D_i.T^r$ for each data D_i in set S_Y^W (Indirect Conflict Check). If time-stamp T_Y^C does not satisfy the rule, transaction Y is aborted. Otherwise, transaction Y is committed with the back-shifted time-stamp T_Y^C.

In commit processing, the server sends time-stamp T_Y^C, sets S_Y^R and S_Y^W to each client that has cached copies of any of the updated data items by transaction Y. After all ACKs are obtained, for each data D_j in set S_Y^R, the server sets time-stamp $D_j.T^r$ to time-stamp T_Y^C if $D_j.T^r$ is less than T_Y^C. In addition, for each data D_k in set S_Y^W, it sets time-stamp $D_k.T^w$ to time-stamp T_Y^C if $D_k.T^w$ is less than T_Y^C.

3 Simulation Study

3.1 Simulation Model

In this section, we compare the proposed scheme(RCP) with O2PL-Invalidate (O2PL) and Call Back Locking(CBL). Our simulation shows the relative performance and characteristics of these approaches. Our study concentrates here mainly on performance aspects, since we are primarily interested in the relative suitability of the cache protocols. Table 1 describes the parameters used to specify the system resources and overhead.

Our simulation model consists of components that model diskless client workstations and a server machine that are connected over a simple network. A client or server is modeled as a simple processor with a microsecond granularity clock. This clock advances as event running as the processor makes "charges" against it. Charges in this model are specified using instruction count.

The number of clients are assumed to be parameter *No_Client*. This study ran experiments using 1 - 25 clients. Each client consists of a *Buffer Manager* that uses an LRU page replacement policy and a *Client Manager* that coordinates the execution of transactions. A *Resource Manager* provides CPU service and accesses to the network. Each client also has a *Transaction Source* which initiates transaction one-at-a-time at the client site.

Compared with client workstations, the server machine has the following differences. The server's *Resource Manager* manages a disk as well as a CPU, and *Concurrency Control Manager* has the ability to store information about the location of page copies in the system and also manages locks (for O2PL and CBL). Since all transactions originate at client workstations, there is no *Transaction Source* module at the server.

Table 1. *System and Overhead Parameter Setting*

Parameter	Meaning	Setting
$Page\_Size$	Size of a page	4Kbyte
DB_Size	Size of DB in pages	1250
No_Client	No. of clients	1 to 25
No_Tr	No. of transactions per client	1000
Tr_Size	Size of each transaction	20 page
$Write\_Prob$	Write probability	20%
Ex_Tr	Mean time between transactions	0 Sec.
Ex_Op	Mean time between operations	0 Sec.
$Client\_CPU$	Client CPU power	15 MIPS
$Server\_CPU$	Server CPU power	30 MIPS.
$Client\_Buf$	Per client buffer size	5%, 25% of DB
$Server\_Buf$	Server buffer size	50% of DB
Ave_Disk	Average disk access time.	20 millisecond
$Net\_Bandwidth$	Network bandwidth	8Mbps
$Page\_Inst$	Per page instruction	30K inst.
Fix_Msg_Inst	Fixed no. of inst. per msg	20K inst.
Add_Msg_Inst	No. of added inst. per msg	10K inst. per 4Kb
$Control\_Msg$	Size of a control msg.	256 byte
$Lock\_Inst$	Inst. per lock/unlock	0.3K inst.
$Disk\_Overhead$	CPU overhead to perform I/O	5K inst.
$Dead\_Lock$	Deadlock detection frequency	1 Sec.

Upon completion of one transaction, the *Transaction Source* module submits the next transaction. If a transaction aborts, it is re-submitted. It then begins making all of the same pages accesses over again. Eventually, the transaction may complete. The number of transactions in a client is assumed to be the parameter No_Tr. For a precise result, each client executes 1000 transactions in this study. The parameter Tr_Size denotes the mean number of operations accessed per transaction. The Ex_Tr parameter is the mean think time between client transactions, and the Ex_Op parameter is the mean think time between operations in a transaction. To make a high degree of data contention, we set both parameters to 0.

Pages are randomly chosen without replacement from among all of the pages according to the workload model described later. The number of pages in the database is assumed to be parameter DB_Size. The parameter $Page\_Size$ denotes the size of each page. A page access cost ($Page\_Inst$) is modeled as the fixed number of instructions. The probability that a page read by a transaction will also be written is determined by the parameter $Write\_Prob$.

The parameters $Client\_Buf$ and $Server\_Buf$ denote the client buffer size and the server buffer size, respectively. In this study, we assume that each client has a small cache (5% of the active database size) or a large cache (25% of the active database size). The CPU service time corresponds to the CPU MIPS rating and the specific instruction lengths given in Table 1. The simulated CPUs of the system are managed using a two-level priority scheme. System CPU requests, such as those for message and disk handling, are given higher priority than user requests. System CPU requests are handled using FIFO queuing discipline, while a processor-sharing discipline is employed for user requests. The disk has a FIFO queue of requests. The average disk access time is specified as the parameter *Ave_Disk*. The parameter *Disk_Overhead* denotes CPU overhead to access disk.

A simple network model is used in the simulator's *Network Manager* component. The network is modeled as a FIFO server with a specified bandwidth (*Net_Bandwidth*). The network bandwidth is set to 8 Mbits/sec which was chosen to approximate the speed of an Ethernet, reduced slightly to account for bandwidth lost to collisions, etc. The CPU cost for managing the protocol to send or receive a message is modeled as a fixed number of instructions per message(*Fix_Msg_Inst*) plus an additional charge per message byte(*Add_Msg_Inst*). The parameter *Control_Msg* denotes the size of a control message.

Our simulation was executed under the *Zipf* workload are governed by *Zipf*'s law. *Zipf*'s law states that if data items are ordered from most popular to least popular, then the number of references to a data tends to be inverse proportional to its rank. in the *Zipf* workload, the probability of choosing data item D_i ($i = 1$ to DB_Size is proportional to $1 / i$. Since the first few items in database are much more likely to be chosen than the last few items, the workload has a high degree of locality per client and very high degree of sharing and data contention among clients.

This study uses total system throughput in committed transactions per second (TPS) as our main metric in order to compare the performance of all schemes. Since we use a closed simulation model, throughput and latency are inversely related: the scheme that has better throughput also has low average latency.

3.2 Experiments and Results

In this section, we present the results from performance experiments. This study uses total system throughput in committed transactions per second (TPS) as our main metric in order to compare the performance of all schemes. Since we use a closed simulation model, throughput and latency are inversely related: the scheme that has better throughput also has low average latency.

Fig. 1 shows the total system throughput with a small cache(5% of DB). There is an extremely high degree of data contention. In this experiment, the proposed algorithm (RCP) performs the best with O2PL performing at a somewhat lower level. CBL has the lowest performance throughout the entire range of client population.

In the range from 1 to 5 clients, the performance of all schemes increases. However, beyond 5 clients, all protocols exhibits a "thrashing" behavior in which the aggregate throughput decreases significantly as clients are added to the system. This phenomenon is related with the abort rate as can be seen in Fig. 3.

Generally, the abort rate of locking system such as CBL is lower than that of optimistic based scheme. However, as Fig. 3 shows, the number of aborts of

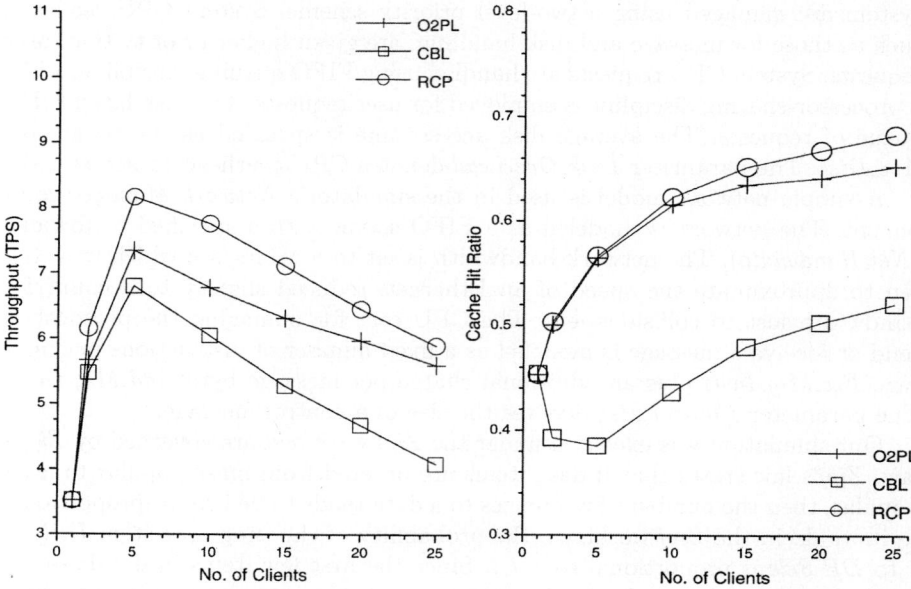

Fig. 1. Throughput (a small cache)

Fig. 2. Cache Hit Ratio (a small cache)

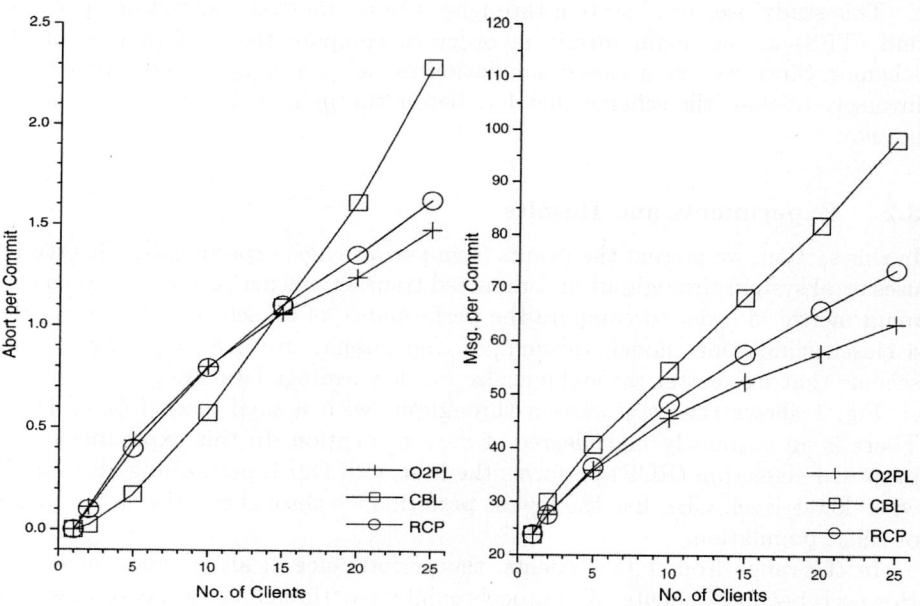

Fig. 3. Number of Aborts (a small cache)

Fig. 4. Message per Commit (a small cache)

CBL exceeds that of other schemes. While write-lock acquisitions are delayed until the end of execution phase in O2PL, write lock declarations occur during the transaction's execution phase in CBL. Hence, in the high client population, CBL generates more aborts than O2PL because its retaining time of write lock is larger than that of O2PL.

Even if the abort rate of RCP is a little higher than that of O2PL, RCP provides the best performance, because it increases data availability by using commit processing without waiting and reduces unnecessary operations by aborting write-write conflicting transactions in their execution phase. In addition, even though RCP does not use any locking method, RCP reduces abort rate significantly compared with CBL, because it re-orders read-write conflicting transactions. Eventually, as Fig. 2 shows, these phenomenons result that the cache hit ratio of RCP is higher than other schemes.

Even if O2PL reduces the abort rates with a lock method, there may be a sudden reduction in the number of active transactions due to transaction blocking. Such blocked transactions eventually leads to a severe degradation in performance, because other transactions which want to access the exclusively locked data are delayed accordingly. In addition, not all the blocked transactions can be committed in O2PL because of deadlock. It makes unnecessary operations.

Re-execution of a transaction is more efficient than its execution in the first phase, because pages that were accessed by the transaction are already available at the client of transaction execution. Note that CBL can not abort conflicting transactions in their execution phase. In contrast, RCP aborts write-

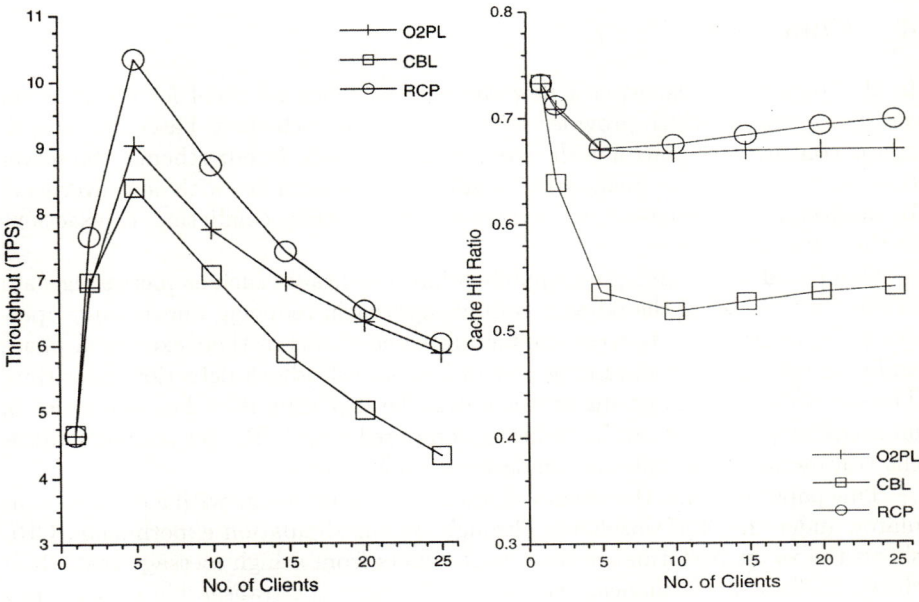

Fig. 5. Throughput (a large cache) **Fig. 6.** Cache Hit Ratio (a large cache)

write conflicting transactions in their execution phase. By these reasons, the abort rate has only a small impact on the performance of RCP compared with CBL.

As Fig. 4 represents, CBL sends significantly more messages per commit than other protocols throughout the entire range of client population because of the highest abort rate. Due high message cost and high aborts rate, CBL has the lowest performance throughout in the *Zipf* workload.

Fig. 5 shows the total system throughput with a large (25% of DB) client cache. Compared with Fig. 1, the performance of all protocols slightly increases because of extending buffer size. However, relatively large cache size does not significantly affect the performance of the schemes. It denotes that extremely high number of aborts leads the network and server to bottlenecks.

Fig. 6 represents the cache hit ratio. The ratio of CBL is extremely lower than other schemes. The reason is that the actions that remove the invalidated data items in caches under O2PL and RCP occur only after the server decides a transaction's commit. However, preemptive pages does not guarantee the transaction's commit under CBL. Compared with Fig. 2, Fig. 6 also shows that cache hit ratio is less affected by the restart-induced buffer hits compared with the case of the small cache buffer. The restart-induced buffer hits means that most of all data items needed by re-started transactions were available in the client buffer. This was further borne out by the client hit rate. Hence, it also explains why the cache hit ratio of RCP is higher than that of O2PL.

4 Conclusion

In this paper, we suggested a new cache consistency protocol for client-server database systems which provides serializability. Our scheme is based on an optimistic concurrency control with re-ordering approach. In our scheme, the server tries to re-order some read-write conflicting transactions with low overhead. In addition, the suggested scheme aborts write-write conflicting transactions in their execution phase.

Compared with O2PL, our approach has advantages such as increasing data availability by use of the no-wait commit approach, reducing unnecessary operations by aborting write-write conflicting transactions in their execution phase and eliminating the maintaining cost of lock and deadlock detection algorithm. The deadlock freedom of our protocol considerably simplifies the complexity of an actual implementation. In addition, compared with CBL, our scheme reduces the transaction abort rate and unnecessary operations.

This paper presents the results of simulation experiments with a detailed simulator under the *Zipf* workload. Throughout our simulation experiments, CBL shows the worst performance because it suffers from a high message cost. RCP shows the best performance, because O2PL limits the transaction concurrency level. By the experimental results, we show that our scheme performs as well as or better than the other approaches with low overhead.

References

1. E. Pitoura and P. K. Chrysanthis, "Multiversion Data Broadcast," *IEEE Transacions on Computers*, Vol.51, No.10, pp 1224–1230, 2002.
2. Daniel Barbara, "Mobile Computing and Database - a Survey," *IEEE Transactions on Knowledge and Data Engineering*, Vol.11, No.1, pp.108–117, 1999.
3. J. Jing, A. Elmagarmid, A. Helal and R. Alonso, "Bit-Sequences: An Adaptive Cache Invalidation Method in Mobile Client/Server Environments", *ACM/Baltzer Mobile Networks and Applications*, Vol.2, No.2, 1997.
4. C. F. Fong, C. S. Lui and M. H. Wong, "Quantifying Complexity and Performance Gains of Distributed Caching in a Wireless Network Environment", *Proceedings of the 13th International Conference on Data Engineering*, pp.104–113, April 1997.
5. V. Gottemukkala, E. Omiecinski and U. Ramachandran, "Relaxed Consistency for a Client-Server Database," *Proc. of International Conference on Data Engineering*, February, 1996.
6. A. Adya, R. Gruber, B. Liskov, and U. Maheshwari, "Efficient optimistic concurrency control using loosely synchronized clocks," *In Proc. of the ACM SIGMOD Conf. on Management of Data*, pp. 23–34, 1995.
7. M. J. Carey, M. J. Franklin, M. Livny, and Shekita, "Data caching tradeoffs in client-server DBMS architectures," *In Proc. of the ACM SIGMOD Conf. on Management of Data*, pp. 357–366, 1991.
8. M. J. Franklin, M. J. Carey, and M. Livny, "Local disk caching in client-server database systems," *In Proc. of the Conf. on Very Large Data Bases (VLDB)*, pp. 543–554, 1993.
9. G.K. Zipf, *Human Behavior and the Principles of Least Effort*, Reading, Mass., Addison Wesley, 1949.
10. V. Almeida, A. Bestavros, M. Crovella, and A. D. Oliveira, "Characterizing reference locality in the WWW," *Proceedings of the 1996 International Conference on Parallel and Distributed Information Systems (PDIS '96)*, pp. 92-103, 1996.

DRC-BK: Mining Classification Rules by Using Boolean Kernels

Yang Zhang, Zhanhuai Li, and Kebin Cui

Dept. of Computer Science & Software,
Northwestern Polytechnical University, China
zhangy@co-think.com, lzh@co-think.com,
ckebin@mail.nwpu.edu.cn

Abstract. An understandable classification models is very useful to human experts. Currently, SVM classifiers have good classification performance; however, their classification model is non-understandable. In this paper, we build *DRC-BK*, a decision rule classifier, which is based on structural risk minimization theory. Experiment results on UCI dataset and Reuters21578 dataset show that *DRC-BK* has excellent classification performance and excellent scalability, and that when applied with MPDNF kernel, *DRC-BK* performances the best.

Keywords: Boolean Kernel, SVM, Decision Rule Classifier.

1 Introduction

In this paper, we are trying to create a classifier with the following properties:

- It is based on structural risk minimization theory, and hence it has high classification accuracy.
- It is a decision rule classifier, with understandable classification model.

We use SVM, which is applied with a Boolean kernel, as a learning engine, and mine decision rules from the hyper-plane constructed by SVM, so as to build *DRC-BK* (Decision Rule Classifier based on Boolean Kernel), a decision rule classifier. In order to study the classification performance and scalability of *DRC-BK*, we made experiments on 16 binary dataset on UCI dataset and Reuters21578 dataset, and the experiment results are inspiring.

2 Related Works

[1~4] are devoted to the study of Boolean kernels for classification. [3] pointed out that the classification performance of Boolean kernels presented in [2] and [4] decreases rapidly with the increasing of dimensions in the input space. In this paper, we use the Boolean kernels presented by us in [1] for mining classification rules. Our pilot study about mining classification rules with help of Boolean kernels is reported in [5].

3 Boolean Kernel

In this paper, we write $\overset{n}{x}_i$ for vector i in a matrix; and write x_i for element i of vector x.

In this paper, we focus on classification tasks when sample data only have Boolean attributes. A sample dataset with d samples could be represented as $\{\overset{n}{x}_i, y_i\}$, $i=1, 2,\ldots,d$, with $\overset{n}{x}_i \in \{0,1\}^n$, representing a sample data, and $y_i \in \{-1,+1\}$, representing the class type of this sample.

Proposition 1: Suppose $U \in \{0,1\}^n$, $V \in \{0,1\}^n$, $\sigma > 0$, $p \in N$, I is the unit vector, then,

$$K_{MDNF}(U,V) = -1 + \prod_{i=1}^{n}(\sigma U_i V_i + 1) \tag{1}$$

$$K_{DNF}(U,V) = -1 + \prod_{i=1}^{n}(\sigma U_i V_i + \sigma(I - U_i)(I - V_i) + 1) \tag{2}$$

$$K_{MPDNF}(U,V) = -1 + (\sigma \langle U,V \rangle + 1)^p \tag{3}$$

$$K_{PDNF}(U,V) = -1 + (\sigma \langle U,V \rangle + \sigma \langle I - U, I - V \rangle + 1)^p \tag{4}$$

are Boolean kernels. For more detail of these Boolean kernels, please refer to [1].

4 DRC-BK Classifier

Here, we present *DRC-BK* classifier, which mine decision rules from the knowledge learned by non-linear SVM, and the detailed steps are: 1, Constructing the classification hyper-plane by non-linear SVM that is applied with MDNF, DNF, MPDNF, or PDNF kernel. 2, Mining the decision rules from this hyper-plane. 3, Classifying the sample data using these decision rules.

Because of lacking of space, here we only introduce the important rule mining algorithm, please refer to [5] for classification rule mining algorithm and classification algorithm.

4.1 MDNF and DNF Kernel

Definition 1: A rule is of the form: $r = \langle z, w_z \rangle$, where z is conjunction with j Boolean literals, and w_z is the weight of the rule.

Definition 2: The length of the rule $r = \langle z, w_z \rangle$ is defined as the number of Boolean literals in z.

Here, we write SVP (SVN) as the set of positive (negative) support vectors. We can get $w_{MDNF,z,SVP} = \sum_{i \in SVP} \alpha_i y_i \ddot{X}_{i,s1} \ddot{X}_{i,s2} ... \ddot{X}_{i,sj} \sigma^j$ and $w_{MDNF,z,SVN} = \sum_{i \in SVN} \alpha_i y_i \ddot{X}_{i,s1} \ddot{X}_{i,s2} ... \ddot{X}_{i,sj} \sigma^j$.

Definition 3: For a rule $r = \langle z, w_z \rangle$, if it satisfies $|w_{z,SVP}| \geq MinWeight$ or $|w_{z,SVN}| \geq MinWeight$, then we say it is an important rule. Here $MinWeight$ is a parameter denoting the minimum weight.

Let's consider the j-length conjunction $z = X_{s1} X_{s2} ... X_{sj}$. The ability of an arbitrary $j+1$-length conjunction, which consists of z, say, $z' = X_{s1} X_{s2} ... X_{sj} X_{sj+1}$, to identify positive and negative samples is:

$$w_{MDNF,z',SVP} = \sum_{i \in SVP} \alpha_i y_i \ddot{X}_{i,s1} \ddot{X}_{i,s2} ... \ddot{X}_{i,sj} \ddot{X}_{i,sj+1} \sigma^{j+1} \quad (5)$$

$$w_{MDNF,z',SVN} = \sum_{i \in SVN} \alpha_i y_i \ddot{X}_{i,s1} \ddot{X}_{i,s2} ... \ddot{X}_{i,sj} \ddot{X}_{i,sj+1} \sigma^{j+1} \quad (6)$$

It is shown in [1] that σ should satisfy $\sigma < 1$. So, we have:

$$|w_{MDNF,z',SVP}| \leq |w_{MDNF,z,SVP}| \quad (7)$$

$$|w_{MDNF,z',SVN}| \leq |w_{MDNF,z,SVN}| \quad (8)$$

These formulas means that for all the conjunctions, which consists of z, the ability to identify positive and negative samples can never exceed the ability of z.

Following the above discussion, we could mine the important rules for MDNF kernel in this way: Firstly, 1-length rules are mined out from SVP or SVN; then, 2-length rules are mined out based on the 1-length rules; ...; this step is repeated until we cannot mine out any $n+1$-length rules. Then, all the k-length rules (k=1, 2, ..., n) create the set of important rules for MDNF kernel.

Fig. 1 gives the important rule mining algorithm for MDNF kernel. Here, function $BF(R)$ is used to get the set of conjunctions $\{z\}$ from $R = \{\langle z, w_z \rangle\}$.

For DNF kernel, a classification task, which is defined on n-dimension input space, and use both positive and negative Boolean literals for classification, could be changed into another classification task, which is defined on $2n$-dimension input space, with n dimensions in the new input space representing n Boolean literals in the original input space, and n dimensions in the new input space representing the negative of n Boolean literals in the original input space. So, we can mine important rules for DNF kernel by an algorithm similar to algorithm 1.

Algorithm 1:	Algorithm for Mining Important Rules for MDNF Kernel
Input:	the set of support vectors SV ($SV \in \{SVP, SVN\}$); the weight vector α ; the minimum weight $MinWeight$;
Output:	the set of important rules

1. $R_1 = \{\langle z, w_{MDNF,z,SV}\rangle \mid \|w_{MDNF,z,SV}\| \geq MinWeight \}$;

2. $R_{all} = \Phi$; $n = 2$;

3. $R_n = \left\{ \langle zz', w_{MDNF,zz',SV}\rangle \; \middle| \; \begin{array}{l} z \in BF(R_{n-1}) \\ \wedge \; z' \in BF(R_1) \\ \wedge \; \|w_{MDNF,zz',SV}\| \geq MinWeight \end{array} \right\}$;

4. $R_{all} = R_{all} \cup R_n$;

5. If $R_n = \Phi$ goto 6; else $n = n + 1$, goto 3;

6. Output R_{all} ;

Fig. 1. Algorithm for Mining Important Rules for MDNF Kernel

4.2 MPDNF and PDNF Kernel

Here, we will first discuss mining important rules for MPDNF kernel. Suppose:

$$g(X) = \sum_{i=1}^{l} \alpha_i y_i K_{MPDNF}(\overset{\hbar}{X}_i, X) + b \qquad (9)$$

then, we have:

$$g(X) = \sum_{s_1,s_2,\ldots,s_{n+1}} (\sum_{i=1}^{l} \alpha_i y_i \frac{p!}{s_1! s_2! \ldots s_n! s_{n+1}!} \overset{\hbar}{X}_{i,1}^{s_1} \overset{\hbar}{X}_{i,2}^{s_2} \ldots \overset{\hbar}{X}_{i,n}^{s_n} X_1^{s_1} X_2^{s_2} \ldots X_n^{s_n} \sigma^{s_1+s_2+\ldots+s_n}) \qquad (10)$$
$$+ b$$

Here, $s_i \in \{0\} \cup N$, $i = 1, 2, \ldots, n, n+1$, $\sum_{i=1}^{n} s_i > 0$, $\sum_{i=1}^{n+1} s_i = p$, and $X_1^{s_1} X_2^{s_2} \ldots X_n^{s_n}$ ($\sum_{i=1}^{n} s_i = j$) here means a conjunctions with j ($j \leq p$) Boolean literals, with $X_1^{s_1}$ representing the conjunction of X_1 for s_1 times.

For MPDNF kernel, the weight of the j-length rule, with conjunction $z = X_1^{s_1} X_2^{s_2} ... X_n^{s_n}$, could be calculated by:

$$w_{MPDNF,z} = \sum_{i=1}^{l} \alpha_i y_i \frac{p!}{s_1!s_2!...s_n!s_{n+1}!} X_{i,1}^{\hbar s_1} X_{i,2}^{\hbar s_2} ... X_{i,n}^{\hbar s_n} \sigma^j \quad (11)$$

For MPDNF kernel, $w_{MPDNF,z,SVP}$ and $w_{MPDNF,z,SVN}$ could be calculated by

$$w_{MPDNF,z,SVP} = \frac{p!}{s_1!s_2!...s_n!s_{n+1}!} w_{MDNF,z,SVP} \text{ and } w_{MPDNF,z,SVN} = \frac{p!}{s_1!s_2!...s_n!s_{n+1}!} w_{MDNF,z,SVN}.$$

Let's consider the set of support vectors SV ($SV \in \{SVP, SVN\}$), 1-length rule $r_1 = \langle z_1, w_{MPDNF,z_1,SV} \rangle$ and j-length rule $r_j = \langle z_j, w_{MPDNF,z_j,SV} \rangle$:

If we have $\left|w_{MDNF,z_1,SV}\right| < \frac{MinWeight}{p!}$, then, we can get:

$$\left|w_{MPDNF,z_1z_j,SV}\right| =\leq \left|p! w_{MDNF,z_1,SV}\right| < p! \frac{MinWeight}{p!} < MinWeight \quad (12)$$

This formula means that all the rules which consists of z_1 are not important rules.

Similarly, if we have $\left|w_{MDNF,z_j,SV}\right| < \frac{MinWeight}{p!}$, then we get: $\left|w_{MPDNF,z_1z_j,SV}\right| < MinWeight$.

And this formula means that all the rules which consists of z_j are not important rules.

Therefore, all the $j+1$-length important rules are made up of z_1 that satisfying $\left|w_{MDNF,z_1,SV}\right| \geq \frac{MinWeight}{p!}$ and z_j that satisfying $\left|w_{MDNF,z_j,SV}\right| \geq \frac{MinWeight}{p!}$.

Definition 4: The conjunction that satisfies $\left|w_{MDNF,z,SV}\right| \geq \frac{MinWeight}{p!}$ is named as important conjunction.

Following this conclusion, we can mine important rules for MPDNF kernel in this way: Firstly, all the *1*-length important conjunctions and all *1*-length rules are mined out; then, all the *2*-length important conjunctions and all *2*-length rules are mined out;...; this step is repeated until we cannot mine out *n*-length important conjunctions, which means that we cannot mine out n+1-length rules. Then, all the k-length rules (k=1, 2, ..., n) create the set of important rules for MPDNF kernel. The detailed algorithm is shown in Fig. 2.

> **Algorithm 2:** Algorithm for Mining Important Rules for MPDNF Kernel
> **Input:** the set of support vectors SV ($SV \in \{SVP, SVN\}$);
> the weight vector α; the minimum weight $MinWeight$;
> **Output:** the set of important rules
>
> 1. $R'_1 = \left\{ \langle z, w_{MDNF,z,SV} \rangle \ \middle| \ |w_{MDNF,z,SV}| \geq \dfrac{MinWeight}{p!} \right\};$
>
> 2. $R_1 = \left\{ \langle z, w_{MPDNF,z,SV} \rangle \ \middle| \ |w_{MPDNF,z,SV}| \geq MinWeight \right\};$
>
> 3. $R_{all} = \Phi; \ n = 2;$
>
> 4. $R'_n = \left\{ \langle zz', w_{MDNF,zz',SV} \rangle \ \middle| \ z \in BF(R'_{n-1}) \wedge z' \in BF(R'_1) \wedge |w_{MDNF,zz',SV}| \geq \dfrac{MinWeight}{p!} \right\};$
>
> 5. $R_n = \left\{ \langle zz', w_{MPDNF,zz',SV} \rangle \ \middle| \ z \in BF(R'_{n-1}) \wedge z' \in BF(R'_1) \wedge |w_{MPDNF,zz',SV}| \geq MinWeight \right\};$
>
> 6. $R_{all} = R_{all} \cup R_n;$
>
> 7. If $R'_n = \Phi$ goto 8; else $n = n+1$, goto 4;
>
> 8. Combining the rules with the same conjunctions in R_{all};
>
> 9. Output R_{all};

Fig. 2. Algorithm for Mining Important Rules for MPDNF Kernel

A similar strategy as the one used for DNF kernel to mine important rules (section 4.1) is used for PDNF kernel to mine important rules, and the algorithm is similar to the algorithm shown in Fig. 2.

5 Experiment

In order to measure the classification performance and scalability of *DRC-BK*, we made experiment on UCI and Reuters21578 dataset. In the reminder of this section, DRC-MDNF, DRC-DNF, DRC-MPDNF, and DRC-PDNF represents *DRC-BK* that is applied with MDNF, DNF, MPDNF, and PDNF kernels, respectively.

5.1 UCI Dataset

Table 1 lists the comparison of classification performance between *DRC-BK* and some other 7 classifiers. Column 1 lists the name of 16 datasets used in our experiment.

Column 2, 3, and 4 give the classification accuracy of *C4.5*, *CBA*[6], and *CMAR*[7], respectively. These experiment results are copied from [7]. Column 5 and 6 gives the classification accuracy of *DEep*[8], *LB*[9], respectively. These experiment results are copied from [8]. Column 7 and 8 gives the classification accuracy of *CAEP*[10] and linear SVM, respectively. Column 9~12 gives the classification accuracy of DRC-MDNF, DRC-DNF, DRC-MPDNF, DRC-PDNF, respectively. In table 1, - means the experiment result is unseen from the literature.

For MDNF, DNF, MPDNF, and PDNF kernel, $MinWeight$ is set to $0.05*b$, $0.005*b$, $0.01*b$, and $0.01*b$; σ is set to 0.1, 0.05, 0.1, and 0.05, respectively. The hyper-parameter C is set to 0.5, and the hyper-parameter p is set to 2 for MPDNF and PDNF kernel.

Table 1. Comparison of Classification Performance of *DRC-BK* and 7 Other Classifiers

Dataset	C4.5	CBA	CMAR	DEep	LB	CAEP	SVM	MDNF	DNF	MPDNF	PDNF
AUSTRA	84.7	84.9	86.1	84.78	85.65	86.21	84.49	85.36	85.23	84.64	84.78
DIABETES	74.2	74.5	75.8	76.82	76.69	–	77.73	79.04	78.00	78.13	77.99
GERMAN	72.3	73.4	74.9	74.4	74.8	72.50	74.90	75.20	73.33	75.70	75.40
HEART	80.8	81.9	82.2	81.11	82.22	83.70	81.48	83.33	83.33	82.59	82.59
IONO	90	92.3	91.5	86.23	–	90.04	90.03	90.60	92.88	91.74	90.88
PIMA	75.5	72.9	75.1	76.82	75.77	75.00	77.21	77.86	77.47	76.82	77.08
SONAR	70.2	77.5	79.4	84.16	–	–	87.02	85.58	85.58	87.98	88.46
TIC-TAC	99.4	99.6	99.2	99.06	–	99.06	98.33	99.79	98.33	98.33	98.33
BREAST	95	96.3	96.4	96.42	96.86	97.28	96.42	96.85	96.71	96.71	96.71
CLEVE	78.2	82.8	82.2	87.17	82.19	83.25	83.17	83.17	83.83	84.16	84.16
CRX	84.9	84.7	84.9	84.18	–	–	86.24	85.51	85.80	85.80	85.80
HEPATIC	80.6	81.8	80.5	81.18	84.5	83.03	85.81	86.45	85.81	86.45	86.45
HORSE	82.6	82.1	82.6	84.21	–	–	82.34	84.51	83.42	83.97	83.15
HYPO	99.2	98.9	98.4	97.19	–	–	99.34	97.50	95.23	99.24	99.24
LABOR	79.3	86.3	89.7	87.67	–	–	92.98	92.98	94.74	94.74	94.74
SICK	98.5	97	97.5	94.03	–	–	97.29	97.32	96.14	97.25	97.25
Average	84.1	85.4	86.03	85.96	82.34	85.56	87.17	87.57	87.24	**87.77**	87.69

From table 1, it is obviously that *DRC-BK* has better classification performance than the other 7 classifiers on these datasets, with DRC-MPDNF performing the best.

5.2 Reuters21578 Dataset

We also made experiment on Reuters21578 dataset. In the experiments, we set *MinWeight* to $0.01*b$; for MPDNF kernel, σ is set to 0.08; and for MDNF kernel, σ is set to 0.1.

For DRC-MPDNF, Fig. 3 gives the experiment results measured in micro F1. Here, the horizontal axis represents the number of attributes, and the vertical axis represents micro F1 measure. $p=2$ represents the experiment result when the hyper-parameter p is set to 2; others are similar. From Fig. 3 we can see that, with the increasing of p, the classification performance of DRC-MPDNF kernel is decreasing.

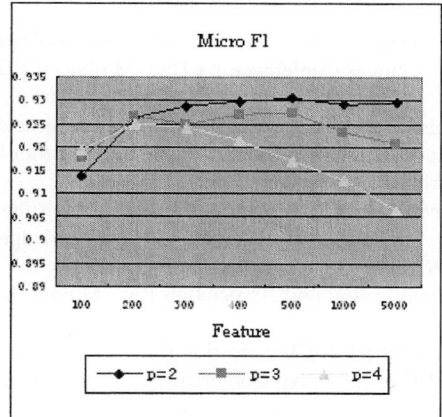
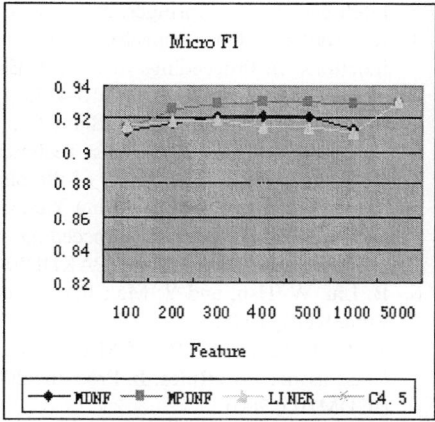

Fig. 3. Classification Performance of DRC-MPDNF

Fig. 4. Comparison of Classification Performance

Fig. 4 gives the comparison of classification performance. Here, the horizontal axis represents the number of attributes, and the vertical axis represents the micro F1 measure. *MDNF*, *MPDNF*, *LINEAR*, and *C4.5* represents the classification performance of DRC-MDNF, DRC-MPDNF kernel, linear SVM, and C4.5, respectively. For DRC-MPDNF, p is set to 2. From Fig. 4, it is obviously that DRC-MPDNF has the best performance. So, we can conclude that *DRC-BK* still has good classification performance and good scalability when it is used as a text classifier.

6 Conclusion and Future Work

In this paper, we present a novel decision rule classifier, *DRC-BK*, which is applied with a Boolean kernel, and mines decision rule from the classification hyper-plane that is constructed by SVM. We present the important rule mining algorithm for MDNF, DNF, MPDNF and PDNF Boolean kernel, respectively. From the experiment results on UCI dataset and Reuters21578 dataset we can see that *DRC-BK* has excellent classification performance and scalability.

In the future, we schedule to find out Boolean kernels with better structures, so as to improve the classification performance of SVM that is applied with Boolean kernels.

References

1. Zhang Yang, Li Zhanhuai, Kang Muning, Yan Jianfeng.: Improving the Classification Performance of Boolean Kernels by Applying Occam's Razor. In The 2nd International Conference on Computational Intelligence, Robotics and Autonomous Systems (CIRAS'03), 2003.
2. K. Sadohara.: Learning of Boolean functions using support vector machines. In Proceedings of the International Conference on Algorithmic Learning Theory, Lecture Notes in Artificial Intelligence 2225, Springer, 2001, pp.106-118.
3. K. Sadohara.: On a capacity control using Boolean kernels for the learning of Boolean functions. In Proceedings of 2002 IEEE International Conference on Data Mining, IEEE Computer Society, 2002, pp.410-417.
4. R. Khardon, D. Roth, and R. Servedio.: Efficiency versus convergence of Boolean kernels for on-line learning algorithms. Technical Report UIUCDCS-R-2001-2233, Department of Computer Science, University of Illinois at Urbana-Champaign, 2001
5. Zhang Yang, Li Zhanhuai, Tang Yan, Cui Kebin.: DRC-BK: Mining Classification Rules with Help of SVM, In the Proceedings of the 8th Pacific-Asia Conference on Knowledge Discovery and Data Mining (PAKDD'04), LNAI 3056, Springer-Verlag Press, 2004.
6. B. Liu, W. Hsu, and Y. Ma.: Intergrating Classification and Association Rule Mining. In Proc. KDD, 1998.
7. W. Li, J. Han, and J. Pei.: CMAR: Accurate and Efficient Classification Based on Multiple Class-association Rules. In Proc. the 2001 IEEE International Conference on Data Mining (ICDM'01), 2001.
8. Jinyan Li, Guozhu Dong, and Kotagiri Ramamohanarao.: Instance-based classification by emerging patterns. In PKDD-2000, 4th European Conference of Principles and Practice of Knowledge Discovery in Databases, September 13-16, 2000, Lyon, France
9. D. Meretakis, B. Wuthrich.: Extending Naïve Bayes Classifiers Using Long Itemsets. In Proceedings of the Fifth ACM SIGKDD, San Diego, pp.165-174, 1999.
10. Dong, G., Zhang, X., Wong, L., Li, J.: CAEP: Classification by aggregating emerging patterns. In Second International Conference on Discovery Science, 1999.

General-Purpose Text Entry Rules for Devices with 4x3 Configurations of Buttons

Jaewoo Ahn[1] and Myung Ho Kim[2]

[1] Mobience, Inc.,
924 Sigmapark, 276-1 Seohyun Bundang, Seongnam,
Gyeonggi 463-824, South Korea
jaewoo@mobience.com
http://www.mobience.com/
[2] School of Computing, Soongsil University,
Seoul 156-743, South Korea
kmh@comp.ssu.ac.kr

Abstract. General-purpose rules for entering texts on devices with 4x3 configurations of buttons are suggested. The rules support simple, consistent and intuitive entering of alphabets as well as numbers and symbols found on most computer keyboards. These rules together with appropriate layouts of alphabets and symbols may provide a common, consistent and powerful framework to devising text-input methods across diverse devices and languages.

1 Introduction

There are, and there will be many kinds of devices supporting text-input functions. Currently the most widely used one is the mobile handset. There have been devised many text-input methods to apply to these devices according to their hardware gadgets as the means for users to manipulate to make the input. These methods typically convert or interpret events which users generate to texts.

1.1 Keyboard-Like Methods vs. Predictive Methods

Input methods are largely categorized into two: keyboard-like ones and predictive ones. Methods in the former convert events in character-wide fashion with some direct simple conversion rules [1], and methods in the latter interpret events in word-wide fashion with indirect predictive rules and built-in databases of words [2]. In the former, users are required to acquire the rules by heart so that they may improve their skills and input speeds by developing so-called muscle or nervous memories as they practice. On the other hand, in the latter, devices need some processing powers with memories so that they may suggest appropriate words from their built-in databases of words with usage statistics. As users select words to make input, they may adapt their databases to reflect the patterns of words for the convenience of users. Users need only to acquire minimal rules to follow.

Keyboard-like methods are a lot easier to be supported on simple devices than predictive ones. They may require users to get used to, and even memorize, their layouts of alphabets for each language supported, if they are different from those used in predictive methods, which is surely the most troublesome task for users. They may also require familiarizing with their input rules for conversion or composition as well. Nevertheless users may prefer to these methods over predictive ones because of the inescapable characteristics of restrictions, alterations and dependencies of the latter despite their merits. Conversion or composition rules, once familiarized with, are also easy to get transferred across languages and devices of the same or similar structure. We may need a universal and powerful keyboard-like text-input method for every appropriate small electronic device in the end that overcomes restrictions and varieties of diverse devices while retaining efficiency and learnability.

1.2 Configurations of Buttons

Specific configuration of buttons, or button-like means, is necessary for a device to materialize a text-input function. Some of the conceivable ones are depicted in Fig. 1. If there are only a small number of buttons, then it might not be easy to find an efficient rule. And if there are a large number of buttons, then it might be difficult for users to press one of them without physical difficulty. For small devices, in particular, buttons may get too small and close with one another.

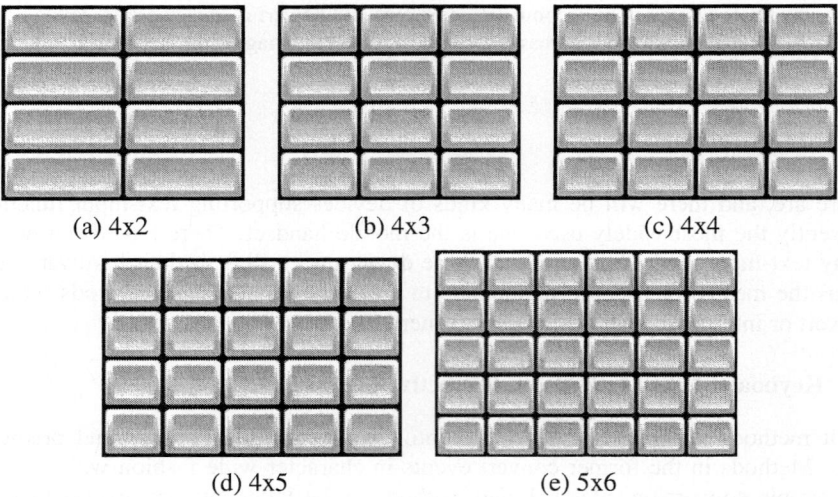

(a) 4x2   (b) 4x3   (c) 4x4

(d) 4x5   (e) 5x6

Fig. 1. Configurations of buttons. 4x3 is experientially ideal for small handheld devices that conveniently facilitates one-handed and two-handed input

The 4-row by 3-column, or 4x3, configuration of buttons, as depicted in Fig. 1(b), is one that is widely adopted particularly on telephones and handsets, and users are familiar with it. It is easy to be supported on small handheld devices with buttons of enough size arranged regularly without being too close. Moreover, it facilitates one-handed input with one to three fingers, as well as two-handed input with two thumbs or fingers. For these reasons we consider only this layout of buttons in this paper, and it becomes evident that this layout is general enough to support a keyboard-like text-input method.

1.3 Text Entry with Alphabets and Symbols

Texts in general are composed mainly of alphabets of a language as well as numbers and symbols. Numbers are rather conceptually and psychologically distinctive from alphabets when thinking out a sentence; one may pause a while without getting a feeling of being interrupted when entering numbers for a sentence in progress.

Symbols, especially punctuation marks, however, are just essential part of a sentence; pausing a while to use a menu to select a symbol appropriate to the context of a sentence being composed makes it not only inconvenient and inefficient but also disturbing and desultory. We may consider symbols found on standard computer keyboards are just those that are commonly used and text-input methods that feature intuitive entries of at least these symbols as well as alphabets are preferred.

Entering a symbol just like an alphabet in a keyboard-like text-input method can be naturally solved by introducing a layout of symbols just like those of alphabets. Some symbols are after all considered a worldwide common alphabet included in each language.

2 Example Layouts

Explaining text-input rules is not easy without resorting to specific example layouts of alphabets, numbers and symbols even if they are not strictly dependent on specific layouts. Fig. 2 shows a few of possible layouts [3] that meet the assumptions of the

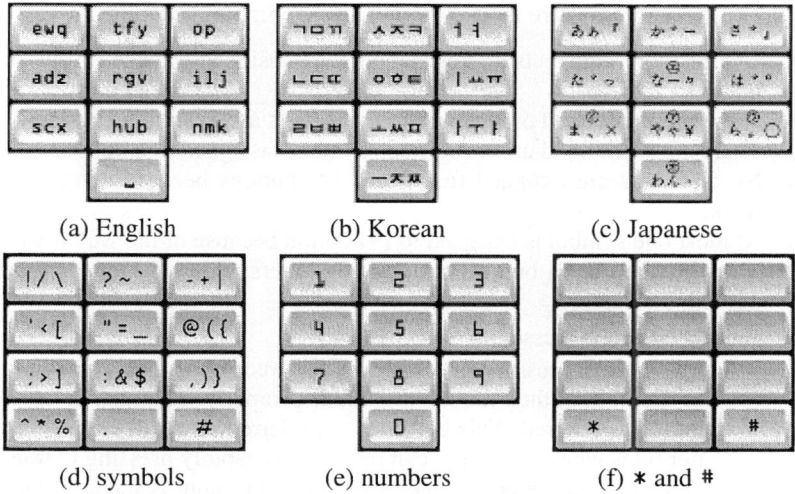

Fig. 2. Example layouts of numbers and symbols together with layouts of alphabets for English, Korean and Japanese. [*] and [#] buttons are specially treated

rules as dictated in the next section. Incidentally, these layouts for alphabets support 26 alphabets and space for English (Fig. 2(a)), 19 consonants and 10 vowels for Korean to compose 11,172 possible letters (Fig. 2(b)), and 10 basic letters and 5 shifts to get 80 variations of letters along with 11 common symbols for Japanese (Fig. 2(c)). Layouts for other languages could also be well devised and supported in a similar way.

Symbols in Fig. 2(d) are those found on a standard computer keyboard. Some symbols may be removed from this layout and/or additional symbols may be added to this layout, of course, to adjust to languages and computer keyboards publicly used in diverse countries.

Standard telephone keypad layout is shown in Fig. 2(e) and Fig. 2(f). [*] and [#] buttons are shown separate from numbers to emphasize their roles in our text-entry rules.

3 General-Purpose Text Entry Rules

Text-input rules with a 4x3 configuration of buttons are introduced. They support entering of alphabets as well as numbers and symbols found on most computer keyboards. These rules together with appropriate layouts of alphabets and symbols, as in Fig. 2, may provide a common, consistent and powerful framework to devising text-input methods across diverse devices and languages.

The positions of [*] and [#] buttons as well as those of numbers as in the figure are assumed by common sense and experience. In particular, [*] is easy to press with the left hand, and [#] with the right hand, when using both thumbs, for example.

3.1 Assumptions

The following assumptions are made as to layouts and implementations:

    a. One or more alphabets and/or symbols are assigned to each of one or more buttons.
    b. There are predefined orders between alphabets assigned to a button.
    c. There are predefined orders between symbols assigned to a button.
    d. No alphabets are assigned to [*] and [#] buttons because of the way they work.
    e. At most one symbol is assigned to [#] button because of the way it works.
    f. There may be extra buttons for deleting letters, changing input modes, and moving the cursor.
    g. When a button is pressed:
        i. If it is not necessary to distinguish between short press and long press of the button, then the input is made promptly. The later release of the button is ignored. This behavior is preferred to avoid confusion for input rules where fast input can be made by rapidly pressing buttons.
        ii. Otherwise, an internal timer starts, and the input is made according to which comes first between the release of the button (short press) and the expiration of the timer (long press). In the former case the timer is removed, and in the latter case the later release of the button is ignored.

h. The minimum durations for long button presses may be set according to users' preferences or skills gained after practices. Those for [*] and [#] buttons are preferred to be somewhat longer than those for the other buttons.

3.2 Input Modes

These input modes are supported which are either regular or temporary:

a. Alphabet mode for each language supported.
b. Number mode either regular or temporary.
c. Temporary symbol mode.

It would be desirable that there are extra buttons for specifying regular input modes. For example, the extra buttons in Fig. 3 are for those in Fig. 2.

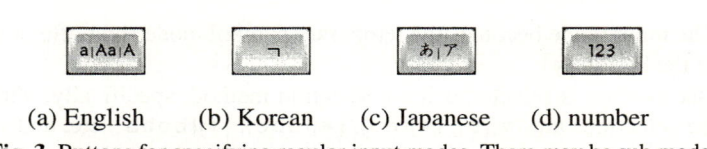

   (a) English   (b) Korean   (c) Japanese   (d) number

Fig. 3. Buttons for specifying regular input modes. There may be sub modes

3.3 Selection

Two methods to select one among a number of alphabets or symbols in a button are described; one is the well-known *multitap*, and the other is the *directap*.

Multitap:
a. If there is only one alphabet or one symbol assigned to a button, then the input is completed right after each press of the button.
b. Otherwise, each press of the button selects an alphabet or a symbol in the predefined order one after another and alters the current input letter with the selected. The input is completed when another button is pressed, or there is no button presses within a fixed period. In case button [#] is pressed, the input is just completed without further action, but any other button is processed normally after completing the current input.
c. We may return to the first alphabet or the first symbol after the last one, or the input may be completed right after the last one is selected.
d. The input is denoted according to the selection as [b], [bb], [bbb], [bbbb], etc.

Directap:
a. An alphabet or a symbol may be selected directly without sequential alterations.
b. There can be made a selection among maximum four alphabets or four symbols of a button; the first one is from a short press of the button, the second from a long press, the third from a short press with a preceding short press of

button [#], and the fourth from a long press with a preceding short press of button [#].
c. The input is denoted according to the selection as [b], [b~], [#b], and [#b~].

For the layouts of English and Korean in Fig. 2, the first alphabets of each button correspond to about three quarters, second ones about a fifth, and the third one about a twentieth according to the usage statistics of the languages. Thus the average number of button presses can be kept minimal using multitap and directap in particular.

3.4 Temporary Symbol Mode

Symbols, especially punctuation marks, are essential part of a sentence. Temporary symbol mode supports intuitive entry of each of these symbols:

a. The input mode becomes the temporary symbol mode right after a short press of the button [*].
b. One symbol is selected using a selection method; specifically, with multitap the selections are [*][b], [*][bb], [*][bbb], [*][bbbb], etc. and with directap the selections are [*][b], [*][b~], [*][#b], and [*][#b~].
c. If there is a symbol assigned to the button [#], this is selected as [*][#] in multitap, and [*][#~] in directap. In case where there is no button with more than two symbols assigned, then the symbol in the button [#], if any, can be selected using [*][#] also in directap, since symbols in the other buttons can be selected without using [*][#b] or [*][#b~].
d. The input mode returns to the previous input mode right after the input of a symbol is completed.

3.5 To Enter a Number in Alphabet Mode

Entering a single-digit number in alphabet mode is supported for convenience:

a. If a symbol is selected with multitap in the temporary symbol mode, the number can be input with [b~] or [*][b~]. Using the latter is preferred to the former to avoid discrimination between the short press and the long press of a button for prompt entries of alphabets (see the section 3.1 g i).
b. Or, if a symbol is selected with directap in the temporary symbol mode:
   i. If there is no button with more than three alphabets assigned, then the number can be input with [#][b~], considering it as the fourth alphabet of the button.
   ii. Or, if there is no button with more than zero, one, two, or three symbols assigned, then the number can be input with [*][b], [*][b~], [*][#b], or [*][#b~], considering it as the first, the second, the third, and the fourth symbol, respectively, of the button.
   iii. Otherwise, the number can be input after the input mode is changed into the number mode either regular or temporary.

c. In case it is desired to suppress any long presses of buttons at all, each number may be considered as the next alphabet or the next symbol after the last one in the corresponding button. For example, using multitap selection with a button of three alphabets and three symbols, numbers are input with either [bbbb] or [*][bbbb].

3.6 Temporary Number Mode

Consecutive entering of more than one single-digit numbers in alphabet mode is supported for convenience in this temporary number mode:

a. The input mode becomes the temporary number mode right after a long press of the button [*], or [*~].
b. Every short press of each number button enters its corresponding number ([b], b = 0 ~ 9).
c. The input mode returns to the previous input mode right after the press of any other button. In case button [#] is pressed, there is no further action other than the mode change, but any other button is processed normally after the mode change.

3.7 Number Mode

Entering numbers mixed with spaces and symbols can be done conveniently in the number mode:

a. Each number is input with a short press of its corresponding button [b], b = 0 ~ 9.
b. The rules for entering a number in alphabet mode and for changing to temporary number mode are not applied in this mode; and thus, for prompt input relevant buttons do not need to distinguish their short and long presses.

3.8 Space and Enter

A space, or a blank, in any of the regular input modes is entered thus:

a. In case where multitap selection is used, each short press of the button [#] enters a space if the button is not pressed to complete a selection in multitap (see the section 3.3 Multitap a), or to end the temporary number mode (see the section 3.6 c).
b. In case where directap selection is used:
   i. If there is no button with more than two letters assigned for the current input mode, each short press of the button [#] enters a space.
   ii. Otherwise, every two short presses of it enter a space ([#][#]).

Entering a new-line character, in any of the regular input modes, is supported thus:

a. [#~]

3.9 Sub Modes for English

There are three sub modes in the English mode. They are the lowercase mode (a), the uppercase mode (A), and the sentence mode (Aa). In the sentence mode, the first input is made with an uppercase alphabet, and then the input mode changes to the lowercase mode automatically. Sub modes are changed thus:

a. The default sub mode of the English mode is the lowercase mode.
b. Each press of the English mode button changes the sub mode successively (a → Aa → A → a → ...). In particular, when starting a sentence in the lowercase mode, it suffices to press the English mode button once at the outset.
c. Or, each short press of the English mode button toggles between the lowercase and the uppercase sub modes, and a long press of the English mode button changes the sub mode to the sentence mode. Incidentally, a short press of the mode button in the sentence mode changes to the lowercase mode.
d. Spaces are entered frequently in English, thus a separate button is preferred for entering the space only, as in Fig. 2. And in this case, since the sentence mode is necessary mostly right after a space is entered, two consecutive presses of the space button enters a space and changes to the sentence mode. Any supplementary presses of the space button enter spaces in the sentence mode without changing the mode. Note also that spaces can be entered with the button [#] without changing input modes.

4 Remarks

The rules introduced in this paper may seem complicated; however, they are naturally derived from the necessities for entering alphabets, symbols, and numbers with and without mode changes using the familiar layout of 4x3 buttons, and thus are considered to be easily learned with some practice. The rules do not depend on particular language, and thus can be applicable to any languages.

Allocating all the 12 buttons to alphabets may improve the situation for entering alphabets, but additional button(s) for symbols and numbers may be necessary as well as a button with the role of button [#]. The overall rules can be applied even in this case with appropriate modifications on the button [*] and [#].

5 Conclusion

General-purpose rules for entering texts on devices with 4x3 configurations of buttons have been presented. The rules support simple, consistent and intuitive entering of alphabets as well as numbers and symbols. These rules together with appropriate layouts of alphabets and symbols may provide a common, consistent and powerful framework to devising text-input methods across diverse devices and languages in either one-handed or two-handed environments.

References

1. MacKenzie, I.S., Soukoreff, R.W.: Text entry for mobile computing: Models and methods, theory and practice. Human-Computer Interaction, 17 (2002) 147-198.
2. MacKenzie, I.S., Kober, H., Smith, D., Jones, T., Skepner, E.: LetterWise: Prefix-based disambiguation for mobile text input. Proceedings of the ACM Symposium on User Interface Software and Technology – UIST 2001, 111-120, New Work (2001).
3. Mobience, Inc. "Character arrangements, input methods and input device," International Application #PCT/KR2004/000577, published March 17, 2004: patent pending.

Dynamic Load Redistribution Approach Using Genetic Information in Distributed Computing

Seonghoon Lee[1], Dongwoo Lee[2], Donghee Shim[3], and Dongyoung Cho[3]

[1] Department of Computer Science, Cheonan University
115, Anseo-dong, Cheonan, Choongnam, Korea, 330-180
shlee@cheonan.ac.kr
[2] Department of Computer Science, Woosong University
17-2 Jayang-dong dong-ku Daejon, Korea 300-718
dwlee@woosong.ac.kr
[3] School of Information Technology & Engineering, Jeonju University
1200, 3rd Street Hyoja-dong Wansan-Koo, Jeon-Ju, Chonbuk, Republic of Korea
{dhshim, chody}@jj.ac.kr

Abstract. Under sender-initiated load redistribution algorithms, the sender continues to send unnecessary request messages for load transfer until a receiver is found while the system load is heavy. Because of these unnecessary request messages it results in inefficient communications, low cpu utilization, and low system throughput. To solve these problems, we propose a genetic algorithm approach for improved sender-initiated load redistribution in distributed systems, and define a suitable fitness function. This algorithm decreases response time and increases acceptance rate.

1 Introduction

An objective of load redistribution in distributed systems is to allocate tasks among the processors to maximize the utilization of processors and to minimize the mean response time. Load redistribution algorithms can be largely classified into three classes: static, dynamic, adaptive. Our approach is based on the dynamic load redistribution algorithm. In dynamic scheme, an overloaded processor(sender) sends excess tasks to an underloaded processor(receiver) during execution.

Dynamic load redistribution algorithms are specialized into three methods: sender-initiated, receiver-initiated, symmetrically-initiated. Basically our approach is a sender-initiated algorithm.

Under sender-initiated algorithms, load redistribution activity is initiated by a sender trying to send a task to a receiver[1, 2]. In sender-initiated algorithm, decision of task transfer is made in each processor independently. A request message for the task transfer is initially issued from a sender to an another processor randomly selected. If the selected processor is receiver, it returns an accept message. And the receiver is ready for receiving an additional task from sender. Otherwise, it returns a reject message, and the sender tries for others until receiving an accept message. If all the request messages are rejected, no task transfer takes place. While distributed systems remain to light system load, a sender-initiated algorithm performs well. But when a distributed system becomes to heavy system load, it is difficult to find a suitable receiver because most processors have additional tasks to send. So, many request and reject messages are repeatedly sent back and forth, and a lot of time is consumed before execution. Therefore, much of the task processing time is consumed, and causes low system throughput, low cpu utilization

To solve these problems in sender-initiated algorithm, we use a new genetic algorithm. A new genetic algorithm evolves strategy for determining a destination processor to receive a task in sender-initiated algorithm. In this scheme, a number of request messages issued before accepting a task are determined by proposed genetic algorithm. The proposed genetic algorithm applies to a population of binary strings. Each gene in the string stands for a number of processors which request messages should be sent off.

The rest of the paper is organized as follows. Section 2 presents the Genetic Algorithm-based sender-initiated approach. Section 3 presents several experiments to compare with conventional method. Finally the conclusions are presented in Section 4.

2 Genetic Algorithm-Based Approach

In this section, we describe various factors to be needed for GA-based load redistribution. That is, load measure, representation method, fitness function and algorithm.

2.1 Load Measure

We employ the CPU queue length as a suitable load index because this measure is known the most suitable index[5]. This measure means a number of tasks in CPU queue residing in a processor.

We use a 3-level scheme to represent a load state on its own CPU queue length of a processor. Table 1 shows the 3-level load measurement scheme. T_{up} and T_{low} are algorithm design parameters and are called *upper* and *lower thresholds* respectively.

Table 1. 3-level load measurement scheme

Load state	Meaning	Criteria
L-load	light-load	$CQL \leq T_{low}$
N-load	normal-load	$T_{low} < CQL \leq T_{up}$
H-load	heavy-load	$CQL > T_{up}$

(CQL : CPU Queue Length)

The transfer policy use the threshold policy that makes decisions based on the CPU queue length. The transfer policy is triggered when a task arrives. A node identifies as a *sender* if a new task originating at the node makes the CPU queue length exceed T_{up}. A node identifies itself as a suitable *receiver* for a task acquisition if the node's CPU queue length will not cause to exceed T_{low}.

2.2 Representation

Each processor in distributed systems has its own population which genetic operators are applied to. There are many encoding methods; Binary encoding, Character and real-valued encoding and tree encoding[12]. We use binary encoding method in this paper. So, a string in population can be defined as a binary-coded vector $<v_0,v_1,...,v_{n-1}>$ which indicates a set of processors to which the request messages are sent off. If the request message is transferred to the processor P_i(where $0 \leq i \leq n-1$, n is the total number of processors), then $v_i=1$, otherwise $v_i=0$. Each string has its own fitness value. We select a string by a probability proportional to its fitness value, and transfer the request messages to the processors indicated by the string. When ten processors exist in distributed system, the representation is displayed as fig 1.

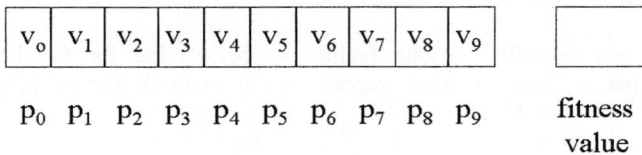

Fig. 1. Representation for processors

2.3 Sender-Based Load Redistribution Approach

2.3.1 Overview

In sender-based load redistribution approach using genetic algorithm, Processors received the request message from the sender send accept message or reject message depending on its own CPU queue length. In the case of more than two accept messages returned, one is selected at random.

Suppose that there are 10 processors in distributed systems, and the processor P_0 is a sender. Then, genetic algorithm is performed to decide a suitable receiver. It is selected a string by a probability proportional to its fitness value. Suppose a selected string is <-, 1, 0, 1, 0, 0, 1, 1, 0, 0>, then the sender P_0 sends request messages to the processors (P_1, P_3, P_6, P_7). After each processor(P_1, P_3, P_6, P_7) receives a request message from the processor P_0, each processor checks its load state. If the processor P_3 is a light load state, the processor P_3 sends back an accept message to the processor P_0. Then the processor P_0 transfers a task to the processor P_3.

2.3.2 Fitness Function

Each string included in a population is evaluated by the fitness function using following formula in sender-initiated approach.

$$F_i = \left(\frac{1}{\alpha \times TMP + \beta \times TMT + \gamma \times TTP} \right)$$

a, β, γ used above formula mean the weights for parameters such as *TMP*, *TMT*, *TTP*. The purpose of the weights is to be operated equally for each parameter to fitness function F_i.

Firstly, *TMP*(Total Message Processing time) is the summation of the processing times for request messages to be transferred. This parameter is defined by the following formula. The *ReMN* is the number of messages to be transferred. It means the number of bits set '1' in selected string. The objective of this parameter is to select a string with the fewest number of messages to be transferred.

$$TMP = \sum_{k \in x} \left(\text{Re } MN_k \times TimeUnit \right)$$

(where, x={i | v$_i$=1 for $0 \leq i \leq n$-1})

Secondly, *TMT*(Total Message Transfer time) means the summation of each message transfer times(*EMTT*) from the sender to processors corresponding to bits set '1' in selected string. The objective of this parameter is to select a string with the shortest distance eventually. So, we define the *TMT* as the following formula.

$$TMT = \sum_{k \in x} EMTT_k$$

(where x={i | v$_i$=1 for $0 \leq i \leq n$-1})

Last, *TTP*(Total Task Processing time) is the summation of the times needed to perform a task at each processor corresponding to bits set '1' in selected string. This parameter is defined by the following formula. The objective of this parameter is to select a string with the fewest loads. Load in parameter *TTP* is the volume of CPU queue length in the processor.

$$TTP = \sum_{k \in x} \left(Load_k \times TimeUnit \right)$$

(where x={i | v$_i$=1 for $0 \leq i \leq n$-1})

So, in order to have a largest fitness value, each parameter such as *TMP*, *TMT*, *TTP* must have small values as possible as. That is, *TMP* must have the fewer number of request messages, and *TMT* must have the shortest distance, and *TTP* should have the fewer number of tasks.

2.3.3 Algorithm

This algorithm consists of five modules such as Initialization, Check_load, String_evaluation, Genetic_operation and Message_evaluation. Genetic_operation module consists of three sub-modules such as Local_improvement_operation, Reproduction, Crossover. These modules are executed at each processor in distributed systems.

The algorithm of the proposed method for sender-initiated load redistribution is presented as fig 2.

An Initialization module is executed in each processor. A population of strings is randomly generated without duplication.

Algorithm : GA-based sender-initiated load redistribution algorithm

Procedure Genetic_algorithm Approach
  { Initialization();
    while (Check_load())
      if ($Load_i > T_{up}$) {
        Individual_evaluation();
        Genetic_operation();
        Message_evaluation(); }
      Process a task in local processor;
  }
Procedure Genetic_operation()
  { Local_improvement_operation();
    Reproduction();
    Crossover();
  }

Fig. 2. Proposed algorithm

A Check_load module is used to observe its own processor's load by checking the CPU queue length, whenever a task is arrived in a processor. If the observed load is heavy, the load redistribution algorithm performs the following modules.

A Individual_evaluation module calculates the fitness value of strings in the population.

A Genetic_operation module such as Local_improvement_operation, Reproduction, Crossover is executed on the population in such a way as follows. Distributed systems consist of groups with autonomous computers. When each group consists of many processors, we can suppose that there are *p* parts in a string corresponding to the groups. The following genetic operations are applied to each string, and new population of strings is generated:

(1) Local_Improvement_Operation

String 1 is chosen. A copy version of the string 1 is generated and part 1 of the newly generated string is mutated. This new string is evaluated by proposed fitness function. If the evaluated value of the new string is higher than that of the original string, replace the original string with the new string. After this, the local improvement of part 2 of string 1 is done repeatedly. This local improvement is applied to each part one by one. When the local improvement of all the parts is finished, new string 1 is generated. String 2 is then chosen, and the above-mentioned local improvement is done. This local_improvement_operation is applied to all the strings in population.

(2) Reproduction

The reproduction operation is applied to the newly generated strings. We use the "*wheel of fortune*" technique[4].

(3) Crossover

The crossover operation is applied to the newly generated strings. These newly generated strings are evaluated. We applied to the "one- point" crossover operator in this paper[4].

One-point crossover used in this paper differs from the pure one-point crossover operator. In pure one-point crossover, crossover activity generates based on randomly selected crossover point in the string. But boundaries between parts(p) are used as an alternative of crossover points in this paper. So we select a boundary among many boundaries at random. And a selected boundary is used as a crossover point. This purpose is to preserve an effect of the Local_improvement_operation of the previous phase. Therefore, the crossover activity in this paper is represented as fig 3.

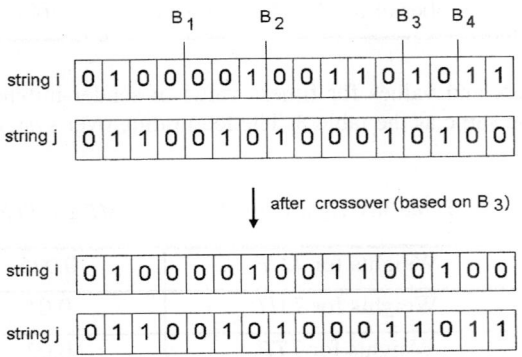

Fig. 3. Crossover Activity

Suppose that there are 5 parts in distributed systems. A boundary among the many boundaries(B_1, B_2, B_3, B_4) is determined at random as a crossover point. If a boundary B_3 is selected as a crossover point, crossover activity generate based on the B_3. So, the effect of the local_improvement_operation in the previous phase is preserved through crossover activity.

The Genetic_operation selects a string from the population at the probability proportional to its fitness, and then sends off the request messages according to the contents of the selected string.

A Message_evaluation module is used whenever a processor receives a message from other processors. When a processor P_i receives a request message, it sends back an accept or reject message depending on its CPU queue length.

3 Experiments

We executed several experiments on the proposed genetic algorithm approach to compare with a conventional sender-initiated algorithm

Our experiments have the following assumptions. Firstly, each task size and task type are the same. Secondly, the number of parts(p) in a string is four. In genetic algorithm, crossover probability(P_c) is 0.7, mutation probability(P_m) is 0.1. The values of these parameters P_c, P_m were known as the most suitable values in various applications[3]. Table 2 shows the detailed contents of parameters used in our experiments.

Table 2. Contents of parameter

number of processor	24
P_c	0.7
P_m	0.1
number of strings	50
number of tasks to be performed	5000

The parameters and values for fitness value of sender-initiated load redistribution algorithm are the same as the table 3. The load rating over systems supposed about 60 percent.

Table 3. Weight values for *TMP*, *TMT* and *TTP*

Weights for *TMP*	0.025
Weights for *TMT*	0.01
Weights for *TTP*	0.02

[Experiment 1] We compared the performance of proposed method with a conventional method in this experiment by using the parameters on the table 2 and table 3. The experiment is to observe change of response time when the number of tasks to be performed is 5000.

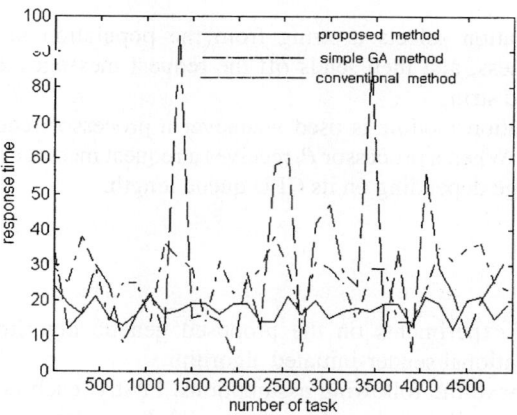

Fig. 4. Result of response time

Fig 4 shows result of the experiment 1. In conventional methods, when the sender determines a suitable receiver, it select a processor in distributed systems randomly, and receive the load state information from the selected processor. The algorithm determines the selected processor as receiver if the load of randomly selected processor is T_{low}(light-load). These processes are repeated until a suitable receiver is searched. So, the result of response time shows the severe fluctuation. In the proposed algorithm, the algorithm shows the low response time because the load redistribution activity performs the proposed genetic_operation considering load states when it determines a receiver.

[Experiment 2] This experiment is to observe the convergence of the fitness function for the best string in the population corresponding to a specific processor in distributed systems.

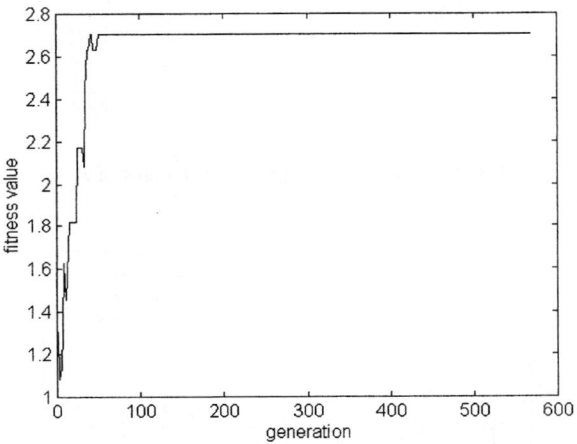

Fig. 5. Fitness value of the processor P_6

In this experiments, we observed the fact that the processor P_6 performs about 550tasks(550generations) among 5000 tasks, and the proposed algorithm generally converges through 50 generations. A small scale of the fluctuations displayed in this experiment result from the change of the fitness value for the best string selected through each generation.

[Experiment 3] This experiment is to observe the performance when the probability of crossover is changed.

Fig 6 shows the result of response time depending on the changes of P_c when P_m is 0.1. In accordance with value of P_c, It shows a different performance. But the proposed algorithm shows better performance than that of conventional algorithm and simple genetic algorithm approach.

[Experiment 4] This experiment is to observe the performance when the probability of mutation is changed.

Fig 7 shows the result of the response time depending on the changes of P_m when P_c is 0.7. In accordance with value of P_m, It shows a different performance. But the proposed algorithm shows better performance than that of conventional algorithm and simple genetic algorithm approach.

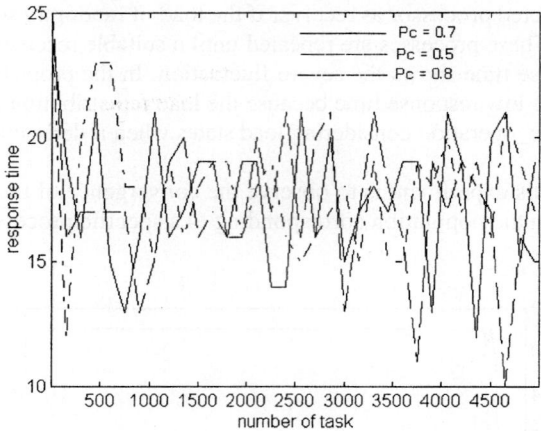

Fig. 6. Result depending on the changes of P_c

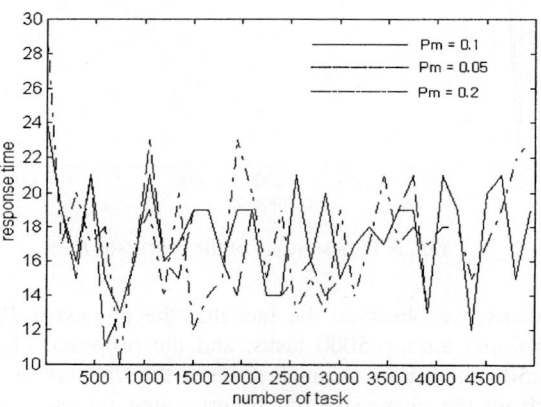

Fig. 7. Result depending on the changes of P_m

4 Conclusions

We propose a new dynamic load redistribution scheme in a distributed system, which is based on the new genetic algorithm with a local improvement operation. The proposed genetic algorithm is used to decide to suitable candidate receivers which task transfer request messages should be sent off. Through the various experiments, the performance of the proposed scheme is better than that of the conventional scheme and the simple genetic algorithm approach on the response time and the mean response time. The performance of the proposed algorithm depending on the changes of the probability of mutation(P_m) and probability of crossover(P_c) is also better than that of the conventional scheme and the simple genetic algorithm approach. But the proposed algorithm is sensitive to the weight values of *TMP*, *TMT* and *TTP*. In future, we will study on method for releasing sensitivity of the weight values

References

[1] A. Y. Zomaya and Y.H. the, "Observations on Using Genetic algorithms for Dynamic Load Balancing", IEEE Tr. On Parallel and Distributed Systems, vol.12 no. 9, pp.899-911 Sep. 2001.

[2] A. Hac and X. Jin, "Dynamic Load-Balancing in a Distributed System Using a Sender-Initiated Algorithm", Proc. 13th Conf. Local Computer Networks, pp. 172-180, 1988.

[3] J.Grefenstette, "Optimization of Control Parameters for Genetic Algorithms," *IEEE Trans on SMC*, vol.SMC-16, no.1, pp.122-128, January 1986.

[4] J.R. Filho and P. C. Treleaven, "Genetic-Algorithm Programming Environments," *IEEE COMPUTER*, pp.28-43, June 1994.

[5] T. Kunz, "The Influence of Different Workload Descriptions on a Heuristic Load Balancing Scheme," *IEEE Trans on Software Engineering*, vol.17, No.7, pp.725-730, July 1991.

[6] T.Furuhashi, K.Nakaoka, Y.Uchikawa, "A New Approach to Genetic Based Machine Learning and an Efficient Finding of Fuzzy Rules," *Proc. WWW'94*, pp.114-122, 1994.

[7] J A. Miller, W D. Potter, R V. Gondham, C N. Lapena, "An Evaluation of Local Improvement Operators for Genetic Algorithms," *IEEE Trans on SMC*, vol.23, No 5, pp.1340-1351, Sept 1993.

[8] N.G.Shivaratri and P.Krueger, "Two Adaptive Location Policies for Global Scheduling Algorithms," *Proc. 10th International Conference on Distributed Computing Systems*, pp.502-509, May 1990.

[9] [Terence C. Fogarty, Frank Vavak, and Phillip Cheng, "Use of the Genetic Algorithm for Load Balancing of Sugar Beet Presses," *Proc. Sixth International Conference on Genetic Algorithms*, pp.617-624, 1995.

[10] Garrism W. Greenwood, Christian Lang and steve Hurley, "Scheduling Tasks in Real-Time Systems using Evolutionary Strategies," *Proc. Third Workshop on Parallel and Distributed Real-Time Systems*, pp.195-196, 1995.

[11] Gilbert Syswerda, Jeff Palmucci, "The application of Genetic Algorithms to Resource Scheduling," *Proc. Fourth International Conference on Genetic Algorithms*, pp.502-508, 1991.

[12] Melanie Mitchell, *An Introduction to Genetic Algorithms*, MIT Press, 1996.

A Guided Search Method for Real Time Transcoding a MPEG2 P Frame into H.263 P Frame in a Compressed Domain

Euisun Kang[1], Maria Hong[2], Younghwan Lim[1,] Youngsong Mun[3], and Seongjin Ahn[4]

[1] Department of Media, Soongsil University, Seoul, Korea
{kanges, yhlim}@media.ssu.ac.kr
[2] Digital media Engineering, Anyang University, Kyonggi-do, Korea
maria@anyang.ac.kr
[3] Department of computer Science, Soongsil University, Seoul, Korea
mnu@computing.ssu.ac.kr
[4] Department of Computer Education, Sungkyunkwan University, Seoul, Korea
sjahn@comedu.skku.ac.kr

Abstract. Our objective is to enable a format transcoding between a heterogeneous compression format in real time mode and to enhance the compression ratio using characteristics of the compression frame. In this paper, we tried to transcode MPEG 2 digital contents having a low compression ratio into H.263 contents with a high compression ratio. After analyzing MPEG2 bit stream and H.263 bit stream of the same original video, we found that the number of intra coded macro blocks in MPEG2 data is much higher that the number of the intra coded blocks in H.263 data. In the process of P frame generation, an intra coded block is generated when a motion estimation value representing the similarity between the previous frame and the current frame does not meet a threshold. Especially the intra coded macro block has a great impact on the compression ratio. Hence we tried to minimize the number of intra coded macro blocks in transcoding the INTRA coded block into INTER coded block using the information about motion vectors surrounding the intra macro block in order to minimize the complexity of the motion estimation process. The experimental results show that the transcoding of MPEG2 into H.263 can be done in real time successfully.

1 Introduction

In ubiquitous communication environments, the devices for accessing a digital item have many different (even unpredictable) characteristics. Therefore the contents should be adopted according to the system device characteristics, network bandwidth and user preferences.

Digital item adaptation (DIA) is one of main parts in MPEG21. The goal of the DIA is to achieve interoperable transparent access to multimedia contents by shielding users from network and terminal installation, management and implementation issues. As shown in Fig.1, the combination of resource adaptation and descriptor adaptation produces newly adapted Digital Item.

DIA tools store all the necessary data of descriptor information [1]. There could be a variety of Digital Item that need to be managed in the DIA framework. Depending on the characteristics of digital item, the architecture or implementation method of Resource Adaptation Engine (RAE) may be different. Current research works for RAE are focused on the transcoding of single medium transcoder such as format transcoder, size transcoder, frame rate transcoders, MPEG2 into H.263 transcoder, or 3D stereoscope video into 2D video transcoder[2] etc. Also they provide a fixed single transcoder for each case of transcoding need.

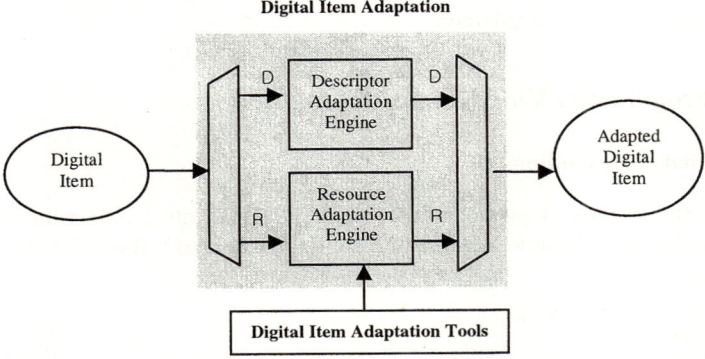

Fig. 1. Illustration of DIA

As mentioned above, researches for RAE are focused on the transcoding of single medium transcoder. And there are many research works about the adaptation with multimedia contents, especially MPEG series. Their concerns are on metadata for mobility characteristics[3], description tools[4] and metadata driven adaptation[5]. They provide a fixed single transcoder for each case of transcoding need. With these methods, if the QoS of source and the QoS of destination get changed, the same transcoder that was used for the adaptation at the last time cannot be used again. And in the previous studies, they suggested the frame rate, color depth and resolution in linking order of transcoders[6]. They considered network bandwidth and requiring bandwidth from a mobile host about multimedia content in mobile environment. In another study, they suggested the bit rate based transcoding with frame rate and resolution transcoding concurrently[7]. They considered network environment and capacity of server and client for multimedia service in real-time. But both of them did not consider the user preference about content occurring in real-time. And there was not clear explanation about a criterion to decide linking sequence between transcoders

One of critical problems of these works is that the transcoding method suggested can not be used to an application requiring a real time adaptation. In order to meet the real time, the MPEG2 contents should be transcoded in a compressed domain.

Comparing the MPEG 2 data and H.263 data for the same video, we found H.263 data has fewer I frames and more P frames than MPEG 2 data. More important difference we pay our attention is that the P frames of H.263 have much more INTER coded macro blocks than MPEG 2 P frame. The our main efforts for transcoding MPEG 2 data into H.263 data are spent to find a method of transcoding the INTRA coded blocks in P frame of MPEG 2 data into the corresponding INTER coded macro blocks in P frame of H.263. For generating corresponding INTER macro blocks in P frame of H.263, we must find motion vectors by any means of search method. Our key idea is that we can have the guiding information from the corresponding INTRA macro blocks in P frame of MPEG2. The guide information may accelerate search procedure.

In the following chapter, we describe the concept of transcoding MPEG2 into H.263 in compressed domain with problems. Then a guided search algorithm and experimental results are explained.

2 Heterogeneous Video Transcoding

2.1 Format Transcoding

There are three kinds of methods that convert MPEG2 into H.263 with characteristic of frame, I frame to I frame, I frame to P frame, P frame to P frame [10][11].

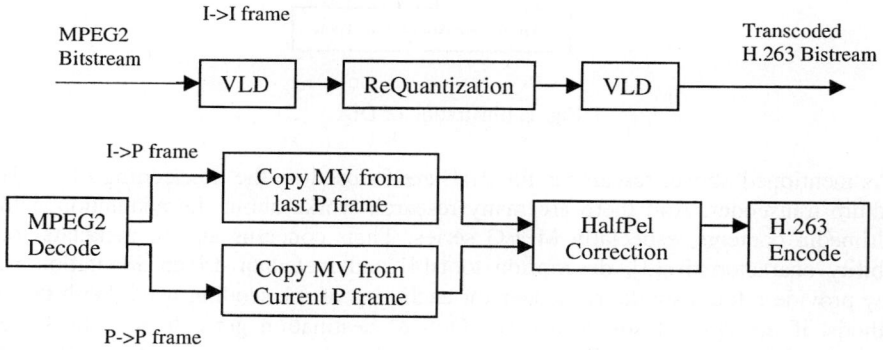

Fig. 2. The methods using the characteristics of frames

As depicted in Fig 2, it is fast to convert I frame to I frame due to avoid IDCT and DCT computations. The rest of conversion between frames requests a little complexity though it can look forward to being higher compressibility for re-sampling new motion vectors with Half-pel Search.

2.2 The Difference Between MPEG2 P Frame and H.263 P Frame

Each macroblock estimated by temporal compression for encoding a series video stream, is encoded in intra or inter mode. Inter mode makes compress ratio higher because of a motion vector that provides an offset from the coordinate position in the current picture to the coordinates in a reference picture. Otherwise, intra mode encoded similarly like spatial compress takes lots of bits than inter mode. Accordingly, we attempt to apply to real-time and enhance compression ratio using property of macroblock by temporal predictions. The following Figure shows the number of intra macroblock within MPEG2 and H.263 at temporal compression.

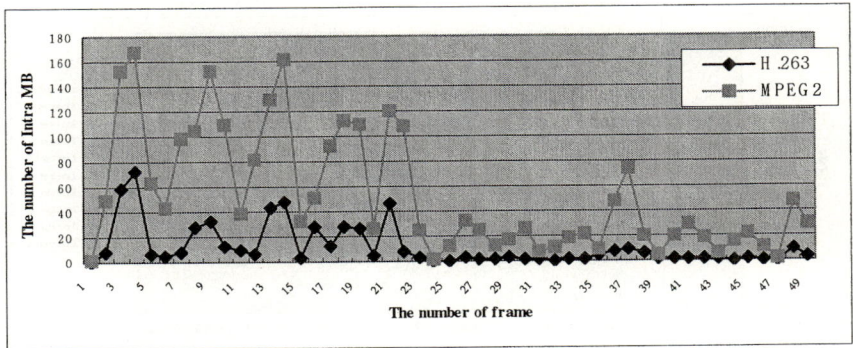

Fig. 3. The number of intra macroblocks in each frame of H.263 and MPEG2

In Fig 3, we can observe that MPEG2 has more intra coded blocks than H.263 has. The reason is that the value of threshold, used to decide whether the macroblock should be intra or inter coded, is different in temporal compression of MPEG2 and H.263. In this paper, in order to solve this problem we tried to convert intra macroblocks within P frame of MPEG2 into inter macroblocks of H.263 appropriately in real time, to enhance the compressibility by decreasing the number of intra macroblocks of MPEG2 and thus to prevent the damage of screen quality as much as possible after conversion.

2.3 Conversion of Macroblock in the Compressed Domain and the Problem

The simplest method of converting the intra macroblock within P frame of MPEG2 into the inter macroblock of H.263 is to carry out the process of motion estimation with the threshold of H.263 in the pixel domain. In this method, it is, however, computationally intensive for real time applications because all the processes of motion estimation are carried out. To alleviate this problem, we tried to apply information of motion vectors and modes of macroblocks within MPEG2 in compressed domain to analogize a new vector. This can reduce the computation all over due to perform without fully decoding and then re-encoding the video.

Figure 4 shows the result of PSNR in case of converting the intra macroblock into the inter macroblock. Intra 0 to Intra 8 indicate the number of adjacent macroblocks surrounding an intra macroblock. For example, "Intra 4" means that there are fewer than 4 Intra macroblocks.

If the motion vector points out wrong position, the video quality may be damaged because the motion vector refers to the most similar block from the previous frames. It can be seen through PSNR that the video quality decreases, as the number of adjacent intra macroblocks is larger in the above figure 3. The reason is that a newly calculated motion vector has a value of another specific point differently from the direction or the movement level of adjacent motion vectors. Additionally, when the block of the initially restored frame is broken in restoration, the current frames that refer to the previous frames for their restoration become broken more clearly, so that we cannot perceive to the naked eye.

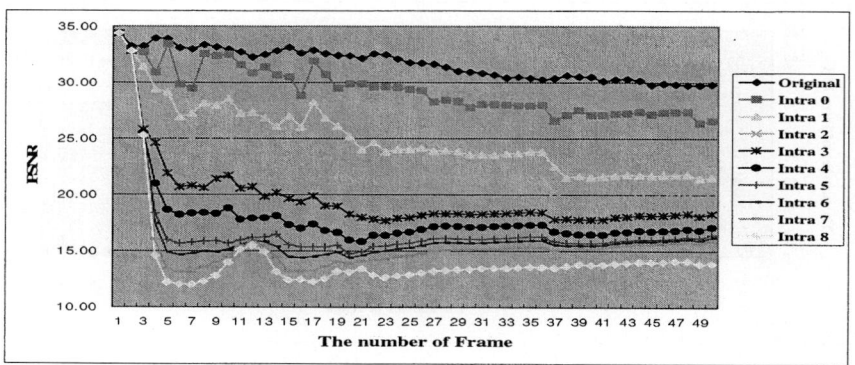

Fig. 4. PSNR in case of employing the adjacent Motion vectors

Therefore, we propose a guided search method for efficiently transcoding the intra coded block within MPEG2 into inter coded block within H.263. The guided search method reuse the information about motion vectors surrounding intra blocks and estimate motion about only intra macroblock in order to consider the problem of quality and minimize the cost of computation.

3 The Guided Search Algorithm Using the Adjacent Motion Vectors

As mentioned above, we can confirm that the problem of quality is generated by motion vectors pointing out wrong positions. Hence, we propose two efficient motion estimation algorithms, making use of neighboring motion vector for image quality, though it is a little more complicated temporally in converting macroblocks in the compressed domain. It can reduce the entire computation of conversion in comparison with the motion estimation process because it performs the motion estimation process not for all the macroblocks in frames but only for the intra macroblock within MPEG2 P frame. So far, many search algorithms [4] have been introduced to minimize a large amount of computation. This paper attempts to reduce the complexity of converting MPEG2 P frame into H.263 P frame by omitting step 1 in three-step search algorithm that has center-biased characteristics among many search algorithms of [4].

3.1 Two-Step Search Algorithm(TSSA)

Three-step search algorithm determines the direction of motion in step 1. Therefore, if the best direction has been decided, the step 1 can be avoided. TSSA starts with 8 motion vectors around an intra macroblock in order to determine the direction of motion and then the other points for comparison are selected based on the following algorithm.

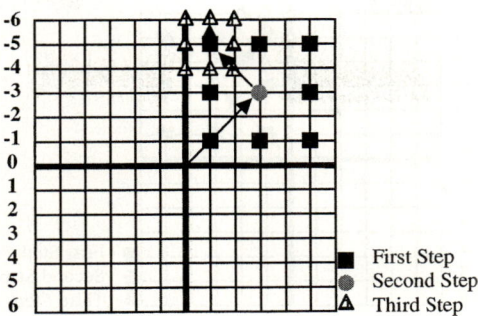

Fig. 5. Route of Two Step Search

[Algorithm 1] Two Step Search Algorithms

Step 1. Estimate the direction of motion using the adjacent 8 motion vectors. In order world, after setting the center of the whole search range, decide one of adjacent 8 locations included in 1/2 of the whole search area from the center. Among 9 points around search window center, find a point which the cost function is the smallest and proceed with step 2. If the direction of motion is 0 (in order words, the direction indicate the center), starts with step 1 of three-step search algorithm.

Step 2. Set adjacent 8 locations included in $1/2^2$ of the whole search range around the estimated direction. Among the 9 points including the estimated direction, the cost function is applied to new 8 points and decides an optimal point that has the minimum value.

Step 3. Find a point whose value of cost function is the fewest among the adjacent 8 macroblocks around the position found in step 2. In this step, if the smallest value of cost function is larger than the threshold of H.263, the current macroblock is decided by intra macroblock.

As seen in this process, because the direction is decided in first step of three-step search algorithm using the adjacent motion vectors, the algorithm can avoid 8 comparisons in the first step for finding the most similar position.

3.2 Axis Division Search Algorithm (ADSA)

TSSA can reduce all the frequencies of comparison considerably than three-step search algorithm because it reduces 9 times of comparisons that are performed in step 1 to 1 time. However, it cannot improve the compression ratio greatly than three-step search algorithm because its accuracy of finding the most similar macroblock is lower. ADSA can make the TSSA more accurate and increases the number of comparisons performed in step 1.

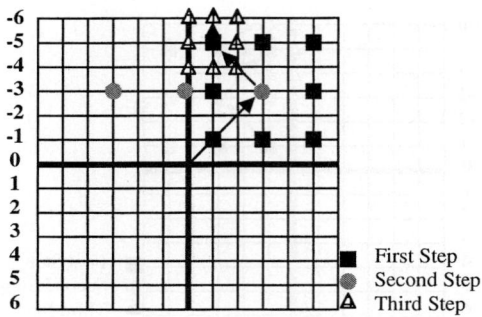

Fig. 6. Route of Axis Division Search Algorithm

[Algorithm 2] Axis Division Search Algorithm

Step 1. Estimates the direction of motion using the adjacent motion vector and set 3 positions horizontally or vertically in 1/2 of the whole search range. Then, decide a point whose cost function is the minimum and proceed with the next step. If the estimated direction is 0, starts with step 1 of three-step search algorithm.

Step 2. Set adjacent 8 positions included in $1/2^2$ of the search range around the point decided in step 1. Then, calculate cost function about 8 positions and select the best position whose weight of the cost function is the fewest.

Step 3. Find a point whose value of cost function is the fewest among the adjacent 8 macroblocks around the position found in step 2. In this step, if the smallest value of cost function is larger than the threshold of H.263, the current macroblock is decided by intra macroblock.

ADSA depends on 3 positions vertically or horizontally for solving the about deciding the direction of motion in step 1. Though the frequencies of comparison performed in step 1 are difficulty much more than TSSA, it can enhance accuracy and compression ratio.

4 Results of Experiment

In order to evaluate the performance, we captured 50 frames with CIF(352*288) resolution, the sequence were pre-coded as MPEG2 bit stream, and IPPP.... structure (i.e., the first frame was intra coded and the rest of the frames were inter coded.). These convert into H.263 with a lower bitrate by supported algorithms. We compare the performance of TSSA and ADSA with three-step search algorithm.

4.1 Frequency of Comparison

The frequency of comparison indicates the frequency of comparison for finding the most appropriate cost function for each step. The following figure shows the frequency of comparison for all frames using three-step, TSSA, and ADSA in order to convert the intra macroblocks within P frames of MPEG2 into the inter macroblocks of H.263. In this figure, the number of adjacent intra macroblocks of the x-axis indicates the number of cases that the type of macroblock adjacent to the converted macroblock is intra. For example, "4" means that there are 4 intra macroblocks among adjacent macroblocks.

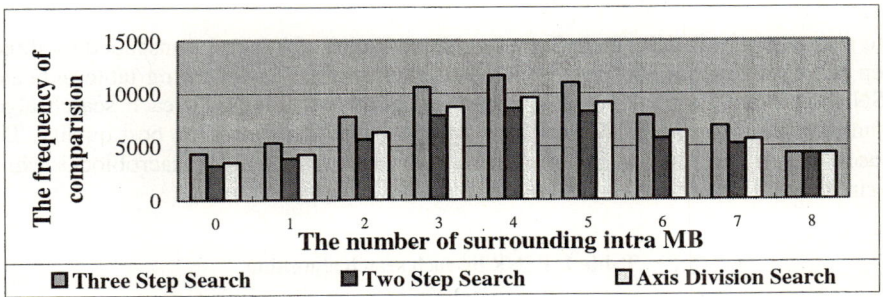

Fig. 7. The frequency of comparison by the number of adjacent Macroblocks according to the search algorithm

As a whole, the frequency of comparison is the most in three-step search algorithm, 9 times of comparison is performed in step 1, the second most in the ADSA, 3 times of comparison are performed in step 1, and the least in TSSA, step 1 is omitted. In case that neighboring intra macroblocks are less 8, the proposed algorithm can decrease the frequency of comparison up to 27% (TSSA) and 25% (ADSA).

4.2 Compressibility

The following figure shows compressibility for all frames in using the suggested guided search algorithm and three-step search algorithm. The search algorithm suggested in the above figure did not increase the compressibility than three-step search algorithm. The reason is that the suggested guided search algorithm was applied only to intra macroblocks. The number of intra macroblocks of TSSA is more than that of three-step search algorithm in comparing the threshold of H.263 because the direction of search is decided in step 1. And, the ADSA compares 3 times more than TSSA, so its number of intra macroblocks is less than that of TSSA, but it is more than the intra macroblocks of three-step search algorithm.

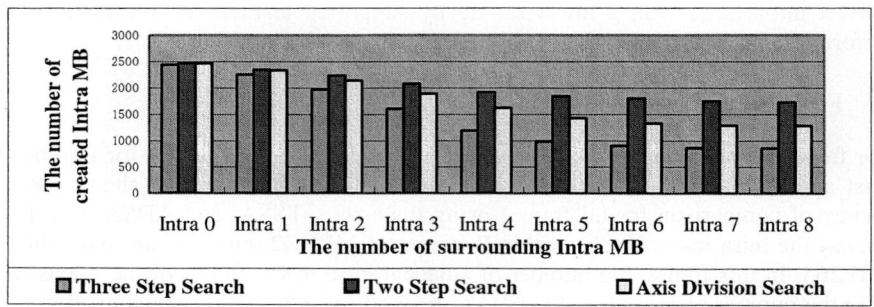

Fig. 8. The number of intra macroblocks generated after applying the guided search algorithm

4.3 Valuation of Image Quality

We also compare PSNR (Peak Signal to Noise Ratio) of images, transcoded by three-step search algorithm and proposed search algorithms. The following table indicates PSNR according to the number of adjacent intra macroblocks in each search algorithm. We confirm that TSSA of three search algorithms shows the best quality. The reason is that if all the neighboring macroblocks is intra coded, macroblocks which desire converting are not used to estimate the new motion vectors.

Table 1. PSNR by each search algorithm

	Three Step Search	TSSA	ADSA
Intra 0	31.60	31.61	31.61
Intra 1	31.53	31.55	31.56
Intra 2	31.43	31.53	31.51
Intra 3	31.30	31.48	31.44
Intra 4	31.19	31.44	31.37
Intra 5	31.10	31.42	31.31
Intra 6	31.07	31.41	31.29
Intra 7	31.06	31.40	31.27
Intra 8	31.06	31.40	31.27

5 Conclusion

Today, it is possible to acquire information anywhere, anytime by development of the wireless technology. However, the information cannot be found easily if not considering the environments of the sender (e.g Server) and the receiver (e.g device with small display, memory, power or etc.). To consider the environment of the receiver in this paper, we employed the type of video compression from the information of QoS(Quality Of Service) of the receiver and converted it appropriately, and to consider the environment of the network, we minimized the number of intra macroblocks within P frames of MPEG2 for increasing the compressibility. We proposed the guided search algorithm (TSSA and ADSA) using the motion vectors of 8 macroblock information surrounding an intra macroblock in order to reduce the complication of motion estimation. Besides we obtained the experiment result that the guided search algorithm has the less frequency of comparison than the existing algorithms and that the whole processing time can be reduced in converting P frames of MPEG2 into P frames of H.263.

Acknowledgments. This work was supported by grant No. R01-2004-000-10618-0 from the Basic Research Program of the Korea Science & Engineering Foundation.

References

1. Youn, J., Xun, J., and Ming, S., "Fast Video Transcoding Architectures for Networked Multimedia Applications," IEEE International Symposium on Circuits and Systems, pp.25-26, 2000.
2. Kang, E., S., "A Study on the Guided Search for Transcoding of MPEG2 P frame to H.263 P frame in a Compressed Domain," Department of Computing Graduate School Soongsil University., pp. 1-15, 2001.
3. Zhijun, L., and Georganas, N., D., "H.263 Video Transcoding for Spatial Resolution Downscaling," Proceedings of the International Conference on Information Technology: Coding and Computing, pp. 425-430, 2002.
4. Fung, K., T., and Siu, W., C., "Low Complexity and High Quality Frame-Skipping Transcoder," IEEE International Symposium on Circuits and Systems, Vol. 5, pp. 29-32, 2001.
5. Yang, J., and Edward J.. "A MPEG4 simple profile transcoder for low data rate wireless applications, "Proceedings of SPIE Visual Communications and Image Processing Part 1, pp. 112-123, 2002.
6. Kim, K., H., Park, D., S., and Whang, J., S., "Compiler," Korean Broadcasting University Press, pp. 96-112, 1998.
7. Huang, C., Liu, P., and Chang R., L., "QoS Streaming Based on a Media Filtering System," IEEE International Conference on Parallel and Distributed Systems, pp. 661-666, 2001.
8. Soam Acharya, Brain C. Smith. : Compressed Domain Transcoding of MPEG. ICMCS (1998) 295-304
9. G. Keesman, R. Hellinghuizen, F. Hoeksema, and G. Heideman. : Transcoding MPEG bitstream," Signal Processing: Image Communication, vol. 8 September (1996)
10. T. Shanableh and M. Ghanbari. : Heterogeneous video transcoding MPEG: 1,2 to H.263. in International Packet Video Workshop, New York City, NY, April (1999)
11. N. Feamster and S.J. Wee. : An MPEG-2 to H.263 Transcoder. SPIE Voice, Video, and Data Communications Conference, Boston, MA, September (1999)
12. Oliver Werner. : Requantiazation for Transcoding of MPEG-2 Intraframes. IEEE transaction on image processing, Vol, 8, No. 2, February (1999)
13. N. Bjork and C. Christopoulos. : Transcoder architecture for video coding. IEEE Trans. Consumer Electron., vol. 44, Feb. (1998) 88-98
14. B. Shen, I. K. Sethi and V. Bhaskaran. : Adaptive motion vector resampling for compressed video down-scaling. Proc. Of IEEE Int. Conf. On Image Processing, Oct. (1997)

Cooperative Security Management Enhancing Survivability Against DDoS Attacks

Sung Ki Kim, Byoung Joon Min, Jin Chul Jung, and Seung Hwan Yoo

Dept. of Computer Science and Engineering, University of Incheon,
177 Dohwa-dong Nam-gu, Incheon, Republic of Korea 402-749
{proteras, bjmin, smjiny, blueysh}@incheon.ac.kr

Abstract. In this paper, we propose a cooperative management method to increase the service survivability in a large-scale networked information system. We assume that the system is composed of multiple domains and there exists a domain manager in each domain, which is responsible to monitor network traffics and control resource usage in the domain. Inter-domain cooperation against distributed denial of service (DDoS) attacks is achieved through the exchange of pushback and feedback messages. The management method is designed not only to prevent network resources from being exhausted by the attacks but also to increase the possibility that legitimate users can fairly access the target services. Though the experiment on a test-bed, the proposed method was verified to be able to maintain high survivability in a cost-effect manner even when DDoS attacks exist.

1 Introduction

As the Internet becomes increasingly important as a business infrastructure, the number of attacks, especially distributed denial-of-service (DDoS) attacks continuously grows [2]. Most of networked information systems adopt intrusion prevention mechanisms such as firewalls, cryptography and authentication. Nevertheless, many successful attacks exploiting various vulnerabilities are found. Intrusion detection systems (IDSs) can effectively detect pre-defined attacks but have limitations in responding to continuously created novel attacks.

The size and complexity of a large-scale networked information system such as Internet makes it impossible to centrally manage the entire management process. Moreover, it is difficult for the systems configured with different management policies to control the system without imposing any limitations. We therefore adopt a distributed management approach.

We assume that the large-scale networked information system can be divided into multiple domains. Each domain can be defined as a group of networks that contain one or more autonomous management entities called domain managers. The term 'autonomous' means that a representative manager of a domain can make a decision on management policies and uniformly apply them to the network components of the domain.

Recently, there have been a lot of research efforts to defend DDoS attacks, which include rate-limiting, blackhole routing, and IP tracing-back [3,5,9,10,11,12]. These techniques are mainly to prevent network bandwidth from being exhausted by the DDoS attacks. Some of them have been adopted by Internet service providers (ISPs) and by network facility providers. However, not much of works have been studied to consider service survivability. Getting rid of DDoS attacks does not necessarily mean high survivability of services. Even though current measures can isolate a DDoS attack successfully, legitimate users may still suffer from being blocked to the access to target services.

In order to maintain high survivability of essential services, an inter-domain cooperation method against distributed denial of service (DDoS) attacks is proposed in this paper. The cooperation is based on the exchange of pushback and feedback messages among domain managers. This idea is not only to prevent network resources from being exhausted by the attacks but also to increase the possibility that legitimate users can fairly access the target services.

The rest of this paper is organized as follows. Section 2 summarizes related research results and explains the contribution of the research presented in the paper. In Section 3, in order to evaluate the performance of the management method, we define a survivability metric. Section 4 presents our distributed system architecture. Proposed mechanisms for inter-domain cooperative management are explained in Section 5. In order to verify the performance of the proposed mechanisms, a test-bed was implemented and several experiments were conducted. Section 6 presents the implementation and experimental results. Finally, Section 7 concludes the paper.

2 Related Works

This section is to provide background on what methods are currently available for protection against DDoS attacks and what their limitations are. Defense techniques against DDoS attacks include Access Control List (ACL), unicast Reverse Path Forwarding (uRPF), access rate limiting, traffic flow analysis, and remote triggered blackhole routing [5,9,10,11,12,13].

ACL is to cut the access off from the resources to be protected based on IP address, service ports, and contents. However, this method can be practical only when specialized hardware modules are equipped, otherwise it could be a big burden to the network facilities. It also requires access control policy to be updated in an efficient manner.

uRPF is to isolate IP spoofing attacks. As a packet arrives at a router, the router verifies whether there exists a reverse path to the source IP address of the packet. For most of DoS or DDoS attacks using IP spoofing, this technique is efficient. However, it has limitation when there are multiple routing paths. Besides, it only can prevent the IP spoofing.

When the amount of packets with a specific pattern increases up to a threshold, access rate limit technique limits the packets. This technique is also called rate filtering. The limitation of this technique is that it limits not only attacking packets but also normal packets.

Traffic flow analysis method is to monitor the source and destination addresses, the number of packets in each flow, and the upstream peer information. It can identify the interface from which spoofed traffics come. But, it requires access to other network facilities between the attacker and the victim.

Blackhole routing is to drop attacking packets toward a specific destination, by forwarding the packets to a virtual interface called Null0. Since this technique uses the forwarding function of the network facilities, it does not incur overload as ACL. However, it is confined only to layer 3 filtering.

In remote triggered blackhole routing, we need to install this function into edge routers. These routers are driven by blackhole routing servers in the same networks. The servers advertise it using Border Gateway Protocol (BGP) to multiple edge routers in order to forward packets with specific patterns to the blackhole IP block. This server can be designed to announce new routing information to other edge routers. It can be managed in Network operations centers (NOCs) or Security Operations Center (SOC) in order to manage novel attacks. This technique seems efficient in blocking DDoS attacks. But once an IP address is isolated, the service through the IP address is not accessible even by the legitimate users.

When we detect DDoS attacks, the most important step is how to react to the attacks. The common reaction to DDoS attacks is to put a filter in the router or the firewall where DDoS attacks are found. By filtering the malicious traffic, the particular website or local network could survive the attack. However, there are two aims for DDoS attacks. The first one is to flood a particular server and another one is to congest the network links. Although we can protect the server by blocking the malicious traffic locally, the attacker can still achieve his goal by flooding the network links. Thus, the best way is to push the filter back to the attack source. The closer the filter is to the source, the more effective is to protect the network link from being flooded. In this scheme, the downstream router needs to contact all its upstream neighbors and all the upstream neighbors need to estimate the aggregate arriving rate. This additional processing makes the router implementation much more complicated [4].

The contribution of this paper is demonstrating a cost-effective approach to support high survivability of essential services against DDoS attacks. We propose a cooperative management method based on the exchange of pushback and feedback messages among domain managers. The management method is designed not only to prevent network resources from being exhausted by the attacks but also to increase the possibility that legitimate users can fairly access the target services. Though the experiment on a test-bed, we have verified the performance of the method.

3 Survivability Metric

In this paper, we define the survivability metric. This is to measure the estimated survivability of a service.

If a legitimate user obtains results on service requests in a timely manner whenever he or she wants to access the service, we should say that the service survives in good shape. On the other hand, if the user cannot attain any result in spite of repeated requests in a long time span, the service should be considered as dead. Therefore, we define the metric as the average of aggregated ratios of the number of replies returned in a time limit to the number of requests sent to the server.

$$\text{Survivability} = \sum_{i=1}^{n} \left(\frac{\text{Reply}_i}{\text{Request}_i} \times P_i \right) \bigg/ \sum_{i=1}^{n} P_i$$

where, n is the number of users considered in a time slot, $Reply_i$ and $Request_i$ are the number of replies returned to the user i within the time limit and the number of requests sent to the server by the user i, respectively, and P_i is weight of user i. If all users are treated with the importance, P_i may be integer 1. Otherwise, it can be set differently for each user.

The metric should be between one and zero. When all the requests sent by every user are replied, it is one. If no reply is return from the server, it is zero, which can be interpreted as server crash.

4 Architecture for Cooperative Management

This section presents distributed system architecture. We need to redefine networked information system in order to fully support cooperative security management. The following requirement should be satisfied in such system architectures.

(1) Practically, the architecture should be applicable to the current information infrastructure. Heterogeneous resources including routers, switches, and network servers cannot be replaced at once. Apparently, drastic changes in the network would incur tremendous costs.

(2) High speed network performance should not be harmed too much. Degradation of network server performance should be acceptable at the cost of security management.

(3) The architecture needs to be suitable for automatic management process. We need to reduce the involvement of manual operations as much as possible.

We assume that the large-scale networked information system can be divided into multiple domains. Each domain can be defined as a group of networks that contain one or more autonomous management entities called domain managers as depicted in Figure 1 and Figure 2. Intra-domain architecture is illustrated in Figure 1. As shown in the figure, each domain can be further divided into sub-domains. The boundary of a domain defines autonomous management, which means that a representative manager of a domain can make a decision on management policies and uniformly apply them to the netw ork components within the domain. Figure 2 shows inter-domain relationship.

We define domain at a network system which can be managed autonomously. In a domain, there should be a representative manager which can assign management policies. A domain can be subdivided into multiple sub-domains.

Domains are connected each other through edge routers. An edge router is connected to a computing node which is able to monitor inbound and outbound traffics. This node is called a domain manager.

Within a domain, each node contains an agent, which is to monitor usages of resources such as CPU, memory, and network bandwidth. The agent is also responsible to trigger resource reallocation in the node and to report its situation to the Domain Manager.

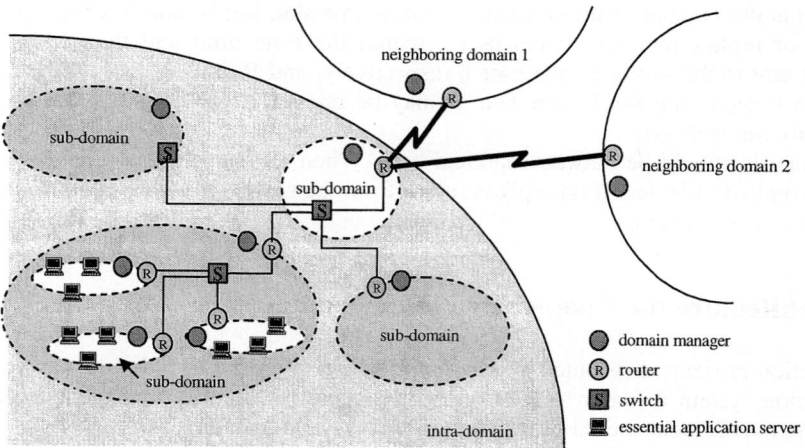

Fig. 1. Intra-Domain Architecture

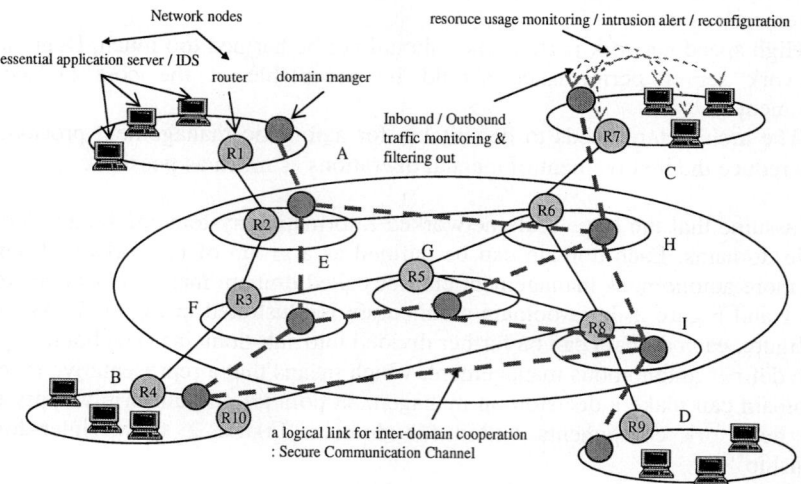

Fig. 2. Inter-Domain Architecture

5 Proposed Mechanisms to Enhance Survivability

This section is to explain management mechanisms proposed in this paper. They are in two categories. One for intra-domain, and the other is for inter-domain.

5.1 Management within a Domain

Since the number of network nodes is confined in a domain, it is relatively easy to treat DoS attacks. Therefore, it is necessary to monitor outbound traffics generated in a domain. The objectives of the monitoring are to detect abnormal outbound traffic flows and to provide essential services in the domain with enough bandwidth.

A domain manager collects packet headers periodically. From this information, it can detect IP spoofing and service port access violation. Statistics based on traffic flows also can be obtain in the process.

5.2 Inter-domain Cooperative Management

Inter-domain cooperation should be based on trust. Messages exchanged among domain managers are authenticated. In order not to be revealed to any attacker, the messages are encrypted and handled by the domain managers.

For this purpose, domain managers conduct inbound traffic monitoring. It is to detect abnormal traffics and to control bandwidth for essential services.

There are two types of messages exchanged among domain managers. One is the pushback message to cut off the traffic toward a certain victim node. The other is the feedback message. The feedback message is to increase the survivability as much as possible. Once an attack is controlled successfully by the virtue of the pushback message, the domain manager issues the feedback message back to the origin of the pushback message. Other domain managers receiving the feedback message cease the rate limit and return to the status before the corresponding pushback message was generated.

6 Implementation and Experimental Results

In this section, we explain implementation of a test-bed, experimental environment and the results.

6.1 Implementation of Test-Bed

TFN2K is a typical tool that is used to create a DDoS attack. It contains most of all kinds of DDoS attack methods. Master programs sending attack command messages communicate with agent programs by exchanging encrypted messages. The attacker can distribute attacking agents to computer systems with weak security measures while the attacker itself is hidden.

In Figure 3, the domain manager of S_A in which the victim V is contained forwards a pushback message to upstream domain manager of A. The pushback message requests rate-limit of packets which is directed to a certain service port of V. The domain manager A checks whether spoofed attacking packets exist. If the domain manager A cannot find them, it forwards the pushback message to the next hop domain manager C. This continues until the source of the attacking traffics. And then the corresponding domain manager isolates the attacker and generates a feedback message back to the origin of the pushback message. For example, once domain manager E detects and isolates A1, it forwards a feedback message through the pass of R9-R4-R3-R2-R1. This is to increase the survivability of the service to legitimate users.

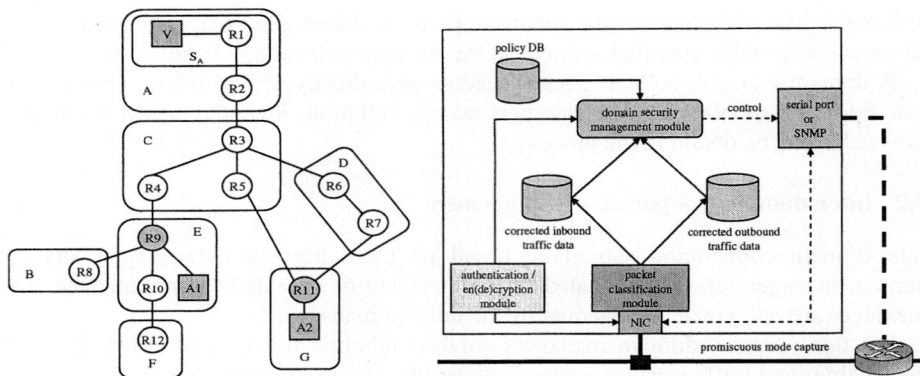

Fig. 3. DDoS Attack Situation **Fig. 4.** Structure of Domain Manager

A domain manager is closely coupled with a router to monitor inbound and outbound traffics. It logs IP source addresses, monitors available network bandwidth, and detect abnormal flows. Besides, it exchanges control and policy information with neighboring domain managers through secure communication channels. The messages exchanged among domain managers include pushback messages to filter attacking traffics toward a victim and feedback messages to recover traffic flow after the filtered situation made by the pushback messages. Figure 4 shows the structure of a domain manager.

The message structure includes an array storing 16 IP addresses, a source address table containing up to 5,000 collected addresses, authorization information, flag notating either pushback or feedback, and message identification. As the message passes by domain managers, each of them records its address into the array of the message. When the trace is over, the message is coming back to the origin of the message as a feedback. The message identification number is attached when it is created in a domain manager.

6.2 Experimental Results

Figure 5 depicts the experimental environment. It consists of three domains. In each domain, there is a domain manager. The domains are connected each other through Linux Routers.

We select the service provided by victim server as a file transfer. The average size is 130 M Byte. Domain managers take samples of packets in every 1 m sec. We use 6 attackers to simulate DDoS attacks. Spoofed ICMP packet flooding is generated with periods of 1 m sec, 10 m sec, 50 m sec, and 100 m sec. Figure 6 shows the raw data obtained from the experiments.

We measured the survivability metric defined in Section 3. By using the cooperation mechanism, the survivability can be increased from 0.2 to 1.0 in the best case when the service deadline is set to 140 seconds in the experiment.

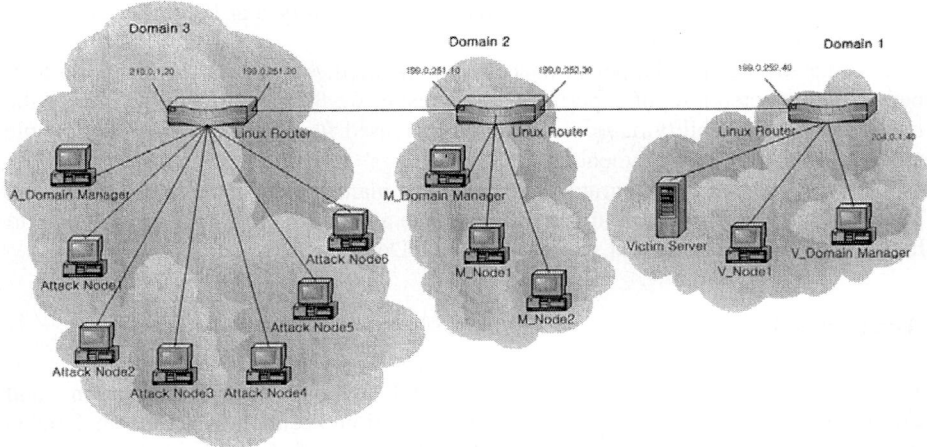

Fig. 5. Test-bed System

Packet Flooding interval	resource usage in victim's domain			spoofed packet detection rate (based on 1 hop)		file transfer completion times (130 M Byte file size, CBR stream)			
	observed the average # of flows in every a 5 second period	pps	bps	# spoofed packets	# detected Spoofed packets	condition without attacks	condition with attacks		
							condition without CM	condition with CM	elapsed time until PM triggered
1 ms	837	1330	6 Mbps	1000	916	135 sec	187 sec	139 sec	10 sec
10 ms	219	459	4.6 Mbps	1000	925		151 sec	137 sec	15 sec
50 ms	77	482	2.8 Mbps	1000	963		141 sec	137 sec	30 sec
100 ms	32	129	1 Mbps	1000	992		138 sec	138 sec	not measured

Fig. 6. Data Obtained from Experiments

7 Conclusion

In this paper, we have proposed a cooperative management method to increase the service survivability in a large-scale networked information system. The system is composed of multiple domains and there exists a domain manager in each domain, which is responsible to monitor network traffics and control resource usage in the domain. Inter-domain cooperation against distributed denial of service (DDoS) attacks is achieved through the exchange of pushback and feedback messages. The management method is designed not only to prevent network resources from being exhausted by the attacks but also to increase the possibility that legitimate users can fairly access the target services.

In order to evaluate the performance of the method, we have implemented a testbed, and conducted a set of experiment. As a result, we found that with the help of the method, the survivability of services can be increased fundamentally with reasonable amount of inherent management cost.

This method can be integrated with network management systems in the future. Through this work, we were able to demonstrate a cost-effective approach to support high survivability of essential services against DDoS attacks.

Acknowledgement

This research was supported by the MIC(Ministry of Information and Communication), Korea, under the ITRC(Information Technology Research Center) support program supervised by the IITA(Institute of Information Technology Assessment).

References

1. William Aiello, John Ioannidis, and Patrick McDaniel : Origin Authentication in interdomain routing, Proceedings of the 10th ACM conference on Computer and communications security, Oct.(2003)
2. Tatsuya Baba and Shigeyuki Matsuda : Tracing Network Attacks to Their Sources, IEEE Internet Computing, March-April(2002), 20-26
3. Andrey Belenky and Nirwan Ansari : On IP Traceback, IEEE Communications Magazine, July(2003)
4. John Ioannidis and Steven Bellovin : Implementing Pushback: Router-Based Defense Against DDoS Attacks, Proceedings of the Network and Distributed System Security Symposium, Feb(2002)
5. KICS of Korea Information Security Agency : Intercept and Analysis Technologies Against DDoS Attacks, Sep(2004)
6. Anukool Lakhina, Mark Crovella, and Christophe Diot : Characterization of Network-Wide Anomalies in Traffic Flows, IMC'04, Oct(2004)
7. Ratul Mahajan, Steven M.Bellovin, Sally Floyd, John Ioannidis, Vern Paxson, Scott Shenker : Controlling High Bandwidth Aggregates in the Network, ACM SIGCOMM Computer Communications Review, Vol. 32, No. 3, Jul(2002)
8. Byoung Joon Min, Sung Ki Kim, and Joong Sup Choi : Secure System Architecture Based on Dynamic Resource Reallocation, WISA 2003, LNCS2908, Aug(2003)
9. Tao Peng, Christopher Leckie, and Kotagiri Ramamohanarao : Defending Against Distributed Denial of Services Attacks Using Selective Pushback, Proceedings of the 9th IEEE Int'l Conference on Telecommunications, Jun(2002)
10. BGPExpert.com : How to Get Rid of Denial of Service Attacks, http://www.bgpexpert.com /antidos.php".
11. Cisco : Unicast Reverse Path Forw -ding(uRPF) Enhancements for the ISP-ISP Edge, ftp://ft-eng.cisco.com/cons /isp/security/URPF-ISP.pdf.
12. waterspring.org : Configuring BGP to Block Denial-of-Service Attacks, http://www.water springs.org/pub/id/draft-turk-bgp-dos-01.txt

Marking Mechanism for Enhanced End-to-End QoS Guarantees in Multiple DiffServ Environment[1]

Woojin Park, Kyuho Han, Sinam Woo, and Sunshin An

Department of Electronic and Computer Engineering, Korea University, 1, 5 Ga,
Anamdong, Sungbukku, Seoul, 136-701, Korea
{progress, garget, niceguy, sunshin}@dsys.korea.ac.kr

Abstract. Recent demands for real time applications have given rise to a need for QoS in the Internet. DiffServ is one of such efforts currently being developed by IETF. In a DiffServ network, previous researchers have proposed several marking mechanisms that are adapted to operate in intra-domain rather than in multi-domain. In this paper, we propose an enhanced marking mechanism based on a color-aware mode using the conventional tswTCM for flows crossing one or more DiffServ domains in order to guarantee effectively enhanced end-to-end QoS. The key concept of the mechanism is that the marking is carried out adaptively based on the initial priority information stored at the ingress router of the first DiffServ domain and the average queue length for promotion of packets. By statistics based on simulations, we show that the proposed marking mechanism improves an end-to-end QoS guarantee for most cases.

1 Introduction

Demands on Quality of Service (QoS) due to explosive growth of real-time traffics such as videoconferencing and video-on-demand services are persistently increasing in current Internet. Differentiated Services (DiffServ) is one of such efforts currently pursued by IETF [1], [2] to add QoS to Internet without fundamental change to the current IP networks. In DiffServ networks, service differentiation is provided based on the DiffServ Code Point (DSCP) field in the IP header and packets with the same DSCP are handled under corresponding forwarding discipline called Per-Hop Behavior (PHB) [3], [4].

DiffServ provides statistical QoS to a few predefined service classes instead of providing guarantees to individual flows. It provides service differentiation among traffic aggregates over a long time scale. A DiffServ network achieves its service goals by distinguishing between the edge and core network. It pushes all complex tasks, such as administration, traffic control, traffic classification, traffic monitoring, traffic marking etc., to the edge network where per flow based schemes may be used. Traffic passing through the edge network will be classified into different service classes and marked with different drop priorities. Core routers implement active queue management schemes such as RED with In and Out (RIO) [5], and provide service differentiation to the traffic according to preassigned service classes and drop priorities carried in the

---

[1] This work is supported by Korea Research Foundation under contract 2002-041-D00446.

packet header. RIO-like schemes achieve this objective by dropping low priority packets earlier with a much higher probability than dropping high priority packets.

In a DiffServ network, previous researchers have proposed several marking mechanisms that are adapted to operate in intra-domain rather than in multi-domain environment. In this paper, we propose an enhanced marking mechanism as an extension of tswTCM [8] to guarantee effectively enhanced end-to-end QoS for flows crossing one or more DiffServ domains. The key concept of the mechanism is that the marking is carried out adaptively based on the initial priority information stored at the ingress router of the first DiffServ domain and the average queue length for promotion of packets.

Our work is different from previous researches in that our enhanced marking mechanism based on a color-aware mode supports an end-to-end QoS guarantee in a multiple DiffServ environment. The contributions of this paper are as follows:

- We propose an enhanced tswTCM marking mechanism that can provide effectively an end-to-end QoS guarantee for flows crossing one or more DiffServ domains. In addition, our proposed marking mechanism operates in color-aware mode. The previously proposed tswTCM marker doesn't assume a specific mode.
- The proposed marking mechanism is compatible with existing markers.
- By statistics based on simulations, we show that the proposed marking mechanism can improve an end-to-end QoS guarantee in multiple DiffServ environment, in especial for middle load (20%-70% provision level).
- We compare our marking mechanism with the conventional tswTCM. Simulation results show that our marking mechanism achieves improved end-to-end QoS guarantee for most cases.

The remainder of the paper is organized as follows. After the introductory part, Section 2 gives the fundamental concept of traffic markers in DiffServ networks. In Section 3, we consider the problem of previous marking mechanisms in terms of end-to-end QoS guarantee in a multiple domain environment. In Section 4, we present the proposed marking mechanism as an enhanced version of the conventional tswTCM. Section 5 describes simulation topology and configuration parameters. In Section 6, we present and discuss simulation scenarios and results. Finally, we describe conclusion and future work in Section 7.

2 Basic Concept of Traffic Markers

The marker is intended to mark packets that will be treated by the AF (Assured Forwarding) PHB in down-stream routers, which is a component that meters an IP packet stream and marks it either green, yellow, or red. The color of the packet indicates its level of priority: red has the highest priority of rejection, followed by yellow and then green. Concerning the mechanism used to check the traffic conformity to the service profile, packet marking can be further classified in two broad categories: token-bucket based marking and average rate estimator based marking. The best-known markers are a single rate Three Color Marker (srTCM) [6], a two rate Three Color Marker (trTCM) [7], and a Time Sliding Window Three Color Marker (tswTCM) [8]. The first two markers in order exploits token-bucket based marking scheme and function in two modes: the color-aware mode in which the previous color of the packets is taken into account at the entrance to the DiffServ network, and the color-blind mode in which this color is ignored. And, the third marker uses average rate estimator based marking scheme and doesn't assume a specific mode. In the average rate estimator based category, marking is performed according to the measurement of the average arrival rate of aggregated flows. In this marker, the arrival rate is calculated according to the weighted average of the arrival rate over a certain time window/interval [5].

3 Consideration of Markers in a Multiple Domain Environment

The classification and marking on each packet are carried out at the boundary routers based on conformance to the contracted throughputs for the traffic flow. There is trade-off between color-aware and color-blind mode of markers in a multiple domain environment. In color-aware mode, the pre-colored green or yellow non-conform packets can be downgraded into either yellow or red according to the level of conformity because the marking information in the previous DiffServ domain affects the current domain. In the case of a multiple domain crossing, these downgraded packets are very likely to be lost in the following domains. This is because the downgraded packets cannot restore the state of the initial marking even though the current domain has the excess bandwidth. In color-blind mode, the marking is accomplished regardless of the previous marking information. In this case, the pre-colored green or yellow packets can be downgraded by the preemption of the pre-colored red packets if red conform packets arrive faster than green or yellow packets at the current domain. This is a fatal defect in a viewpoint of an end-to-end QoS guarantee.

To guarantee effectively an enhanced end-to-end QoS in a multiple domain environment of DiffServ networks, a new marker is needed. And this new marker has to function in a color-aware mode to protect the packets with high priority from the less important packets. In this paper, we focus on the color-aware mode of the conventional tswTCM.

4 Proposed Marking Mechanism

The DSCP field in the IP header (IPv4 or IPv6) is defined to allow differentiated processing based on the value of this field. It should be noted that the value of the DSCP uses only six bits of this field. As stated in previous Section, a new traffic marker is required to operate in a color-aware mode to provide an enhanced end-to-end QoS guarantee for flows crossing one or more DiffServ domains. However, the conventional tswTCM doesn't assume a specific mode. In this paper, we propose the enhanced marking mechanism to operate in color-aware mode. For this objective, we define CU field as IM (Initial Marking) field to provide effectively an end-to-end QoS guarantee for flows crossing one or more DiffServ domains as shown in Fig. 1.

To function effectively in a color-aware mode, the initial marking information of a packet is only stored at the first boundary router of DiffServ domain and doesn't be changed at boundary routers of DiffServ domains crossing to the destination. This additional marking information is to have their initial priority level restituted to protect the Packets with high priority from the less important packets when the current domain

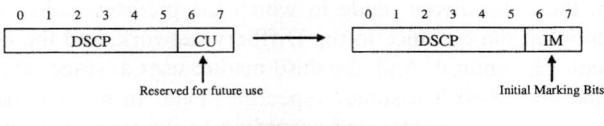

Fig. 1. Definition and usage of Initial Marking bits

has the excess bandwidth. The usage of IM bits as follows. If a packet has "00" as the value of IM field, it means that the packet doesn't be marked at other boundary routers. Also, this IM value enables our proposed marking mechanism to support the compatibility with existing markers. That is, because the previous markers don't use the IM field, the value of IM bits is always "00". If a packet has the other values except "00" as the value of IM field, it means that the packet has been marked at other boundary router. These values are utilized to restore their initial priority or to promote the higher priority level when the boundary router of other DiffServ domains has the excess bandwidth. The case of the higher priority promotion of packets is achieved based on the average queue length of the corresponding queue. Therefore, from the new definition and usage of this IM field, our proposed marking mechanism can support the enhanced end-to-end QoS guarantee than the previous marking mechanisms for flows crossing one or more DiffServ domains.

The pseudo-code in Fig.2 through 5 respectively describes the processing of the proposed marking mechanism according to the value of IM bits. The average queue length is calculated based on exponential weighted moving average (EWMA) [9].

```
avg-rate = Estimated Average Sending Rate of Traffic Stream
avg-qlen_n = Estimated Average Queue Length (%) of Queue N
IM = Initial Marking bits

01 : if (IM=="00")  // initial mode
02 :     if (avg-rate <= CIR)
03 :         the packet and IM are marked as green;
04 :     else if (avg-rate <= PIR) AND (avg-rate > CIR)
05 :         calculate P0 = (avg-rate - CIR) / avg-rate
06 :         with probability P0 the packet and IM are marked as yellow;
07 :         with probability (1-P0) the packet and IM are marked as green;
08 :     else
09 :         calculate P1 = (avg-rate - PIR) / avg-rate
10 :         calculate P2 = (PIR - CIR) / avg-rate
11 :         with probability P1 the packet and IM are marked as red;
12 :         with probability P2 the packet and IM are marked as yellow;
13 :         with probability (1-(P1+P2)) the packet and IM are marked as green;
```

Fig. 2. Pseudo-code of proposed marking mechanism when IM field is "00"

The processing of pseudo-code given in Fig. 2 is same as the marking strategy of the conventional tswTCM except that our proposed marking mechanism sets the value of the IM field based on conformance to the contracted throughputs for the traffic flow.

```
14 :  if (IM=="01")     // initial marking is green
15 :      if (avg-rate <= CIR)
16 :          the packet is marked as green;
17 :      else if (avg-rate <= PIR) AND (avg-rate > CIR)
18 :          if (q_len_0 <= α1)
19 :              the packet is marked as green;
20 :          else
21 :              calculate P0 = (avg-rate - CIR) / avg-rate
22 :              with probability P0 the packet is marked as yellow;
23 :              with probability (1-P0) the packet is marked as green;
24 :      else
25 :          if (q_len_0 <= α2)
26 :              the packet is marked as green;
27 :          else
28 :              calculate P1 = (avg-rate - PIR) / avg-rate
29 :              calculate P2 = (PIR - CIR) / avg-rate
30 :              with probability P1 the packet is marked as red;
31 :              with probability P2 the packet is marked as yellow;
32 :              with probability (1-(P1+P2)) the packet is marked as green;
```

Fig. 3. Pseudo-code of proposed marking mechanism when IM field is "01" (α1>α2)

The processing of pseudo-code shown in Fig. 3 is same as the marking strategy of the previous tswTCM except line 18-19 and 25-26. Also, this processing part is performed in boundary routers of other DiffServ domains crossing after a packet is marked at the first boundary router. In case that line 17 is true, the existing tswTCM downgrades the corresponding packet into Yellow. However, our proposed marking mechanism enables the packet to be marked to Green when the average queue length of the Green queue is smaller than α1. Line 25-26 operates similarly with line 17-18. In addition, pseudo-codes depicted in Fig. 4 and 5 are processed likewise. The specific values of average queue lengths (α1~α6, β1~β3) are handled in Section 5.

```
33 :  if (IM=="10")     // initial marking is yellow
34 :      if (avg-rate <= CIR)
35 :          if (q_len_0 <= α3)
36 :              the packet is marked as green;
37 :          else
38 :              the packet is marked as yellow;
39 :      else if (avg-rate <= PIR) AND (avg-rate > CIR)
40 :          if (q_len_0 <= α4)
41 :              the packet is marked as green;
42 :          else
43 :              calculate P0 = (avg-rate - CIR) / avg-rate
44 :              with probability P0 the packet is marked as red;
45 :              with probability (1-P0) the packet is marked as yellow;
46 :      else
47 :          if (q_len_1 <= β1)
48 :              the packet is marked as yellow;
49 :          else
50 :              calculate P1 = (avg-rate - PIR) / avg-rate
51 :              with probability P1 the packet is marked as red;
52 :              with probability (1-P1) the packet is marked as yellow;
```

Fig. 4. Pseudo-code of proposed marking mechanism when IM field is "10" (α3>α4)

In summary, the proposed mechanism is an extension of the conventional tswTCM. The key concept of our proposed mechanism is that we define unused CU field as IM field to store the initial marking information and use this information combined with the average queue length for promotion of packets. The value of IM bits is only marked at the first boundary router. This functionality is necessary for the restitution of the initial priority level of the packets to guarantee effectively an end-to-end QoS in color-aware mode and protects higher priority packets from being downgraded in other DiffServ domains. The proposed marking mechanism operates in the following manner: the boundary routers checks IM bits, and then if the bit is "00", the router marks DSCP and IM bits into its priority level according to the traffic conformity. Otherwise, it remarks DSCP bits based on the previous DSCP bits and IM bits. In addition, the proposed marking mechanism operates in two internal modes: the initial mode and remarking mode. The initial mode is performed whenever the IM bits are "00", its algorithm is equal to that of tswTCM. Otherwise, the remarking mode is operated, and the remarking of DSCP field about priority level is determined by the comparison of the previous priority level and the IM bits.

```
53 : if (IM=="11")          // initial marking is red
54 :     if (avg-rate <= CIR)
55 :         if (q_len_0 <= α5)
56 :             the packet is marked as green;
57 :         else if (q_len_1 <= β2)
58 :             the packet is marked as yellow;
59 :         else
60 :             the packet is marked as red;
61 :     else if (avg-rate <= PIR) AND (avg-rate > CIR)
62 :         if (q_len_0 <= α6)
63 :             the packet is marked as green;
64 :         else if (q_len_1 <= β3)
65 :             the packet is marked as yellow;
66 :         else
67 :             the packet is marked as red;
```

Fig. 5. Pseudo-code of proposed marking mechanism when IM field is "11" (α5>α6, β2>β3)

5 Simulation Topology and Configuration

Simulations have been performed using the ns-2 network simulator [10] and have been conducted using the topology depicted in Fig. 6 to study the performance of the proposed marking mechanism. With this topology we test the effectiveness and ad-

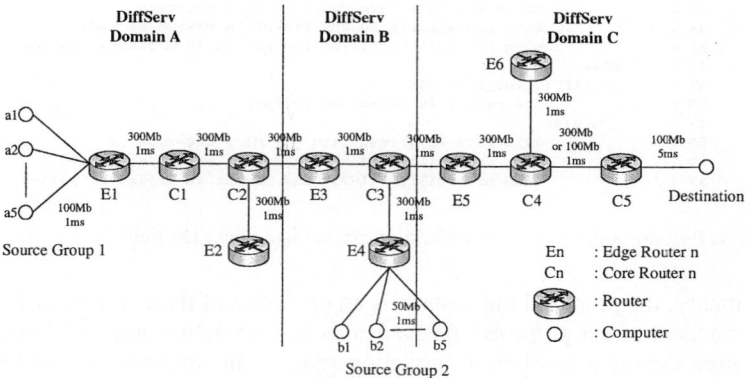

Fig. 6. Simulation network topology

vantage of using the proposed marking mechanism at the ingress router of DiffServ networks. The simulation configuration parameters are listed in Table 1 [11].

Table 1. Simulation configuration parameters

Parameter			Value
Test Time			10 sec
Packet Size			1000 bytes
Transmit Rate	a1~a5		30 or 50 Mbps
	b1~b5		15 Mbps
tswTCM	a1~a5	CIR	5 Mbps
		PIR	15 Mbps
	b1~b5	CIR	2.5 Mbps
		PIR	7.5 Mbps
MRED (RIO-D)	DP 0 min_th		100 packets
	DP 0 max_th		200 packets
	DP 0 min_Pb		0.02
	DP 1 min_th		50 packets
	DP 1 max_th		100 packets
	DP 1 min_Pb		0.1
	DP 2 min_th		25 packets
	DP 2 max_th		50 packets
	DP 2 min_Pb		0.2
Traffic (Exponential & Pareto)	Burst Time		300ms
	Idle Time		300ms
Edge & Core Router	Buffer Size		200 packets

Two UDP source groups (a1~a5, b1~b5), comprising two pareto distribution traffic sources, two exponential distribution traffic sources, and one CBR (Constant Bit Rate) source each, perform unidirectional data transmissions across links (from E1 or E4 to C5) to one corresponding destination. C1~C5 are core routers which implement RIO-D (RIO-Decoupled) [12]. And, simulations are executed using one or two UDP source groups according to test scenarios. For the conventional tswTCM and the proposed marking mechanism, PIR (Peak Information Rate) is set to CIR (Committed Information Rate) * 300%. The values that were used throughout the simulations do not necessarily correspond to an optimal configuration. However, from various simulation runs not described here, using a wide range of parameters, this configuration was found to be suitable for evaluating the proposed marking mechanism. Also, in our simulation, the specific values of average queue lengths ($\alpha 1$~$\alpha 6$, $\beta 1$~$\beta 3$) are shown in Table 2. In addition, these values can be adjusted from DiffServ domain administrators dynamically. The variation of parameters and their impact on the performance of the proposed marking mechanism under a number of different conditions, which will allows us to create more adaptive version of the proposed marking mechanism, is left as future work.

Table 2. Specific values of average queue lengths ($\alpha 1$~$\alpha 6$, $\beta 1$~$\beta 3$)

Constant	Value (%)	Constant	Value (%)	Constant	Value (%)
$\alpha 1$	50	$\alpha 4$	15	$\beta 1$	5
$\alpha 2$	30	$\alpha 5$	30	$\beta 2$	12.5
$\alpha 3$	20	$\alpha 6$	20	$\beta 3$	7.5

6 Simulation Results and Discussion

In this section, we will compare the performance of our proposed marking mechanism, which is a color-aware mode, with the conventional tswTCM using three scenarios. The results are presented by statistics based on simulations. We list the notations used in the figures as follows.

- TR = Number of Total Received packets
- TT = Number of Total Transmitted packets
- G, Y or R = Number of transmitted Green, Yellow, or Red packets

A. Scenario 1: Source group 1 has 5 flows (each 30Mbps) and source group 2 has no flows. No bottleneck (A-1) and bottleneck (A-2) in C4-C5 link

In this scenario, as all bandwidth of each link between routers is sufficiently large, a packet doesn't be dropped at the low or middle provision level (let's say under 70%). Fig. 7 shows time versus received or transmitted packet numbers evaluated in C4-C5 link for the proposed and conventional tswTCM. An edge router (E1) meters and marks equally packets in two mechanisms. However, edge routers, E3 and E5, in other domains, marks differently as described in Section 4. From Fig. 7, we can see that many yellow packets are promoted to the higher priority level in our proposed mechanism when the boundary routers of other DiffServ domains have the excess bandwidth. Results show that our proposed mechanism is more efficient than the conventional tswTCM under medium network provision level in terms of an end-to-end QoS guarantee.

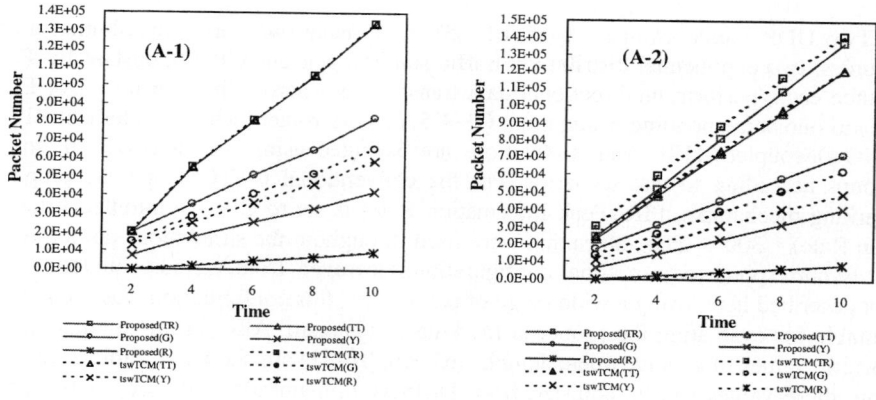

Fig. 7. Time versus received or transmitted packets. Source group 1 has 5 flows (each 30Mbps) and source group 2 has no flows. (A-1): no bottleneck in C4-C5 link (300Mbps), (A-2): bottleneck in C4-C5 link (100Mbps)

B. Scenario 2: Source group 1 has 5 flows (each 50Mbps) and source group 2 has no flows. No bottleneck (B-1) and bottleneck (B-2) in C4-C5 link

In this scenario, the throughput of two mechanisms is approximately 50% due to bottleneck link of C4-C5 (B-2). We can see that our proposed mechanism supports enhanced end-to-end QoS guarantee under the high provision level (let's say above 70%).

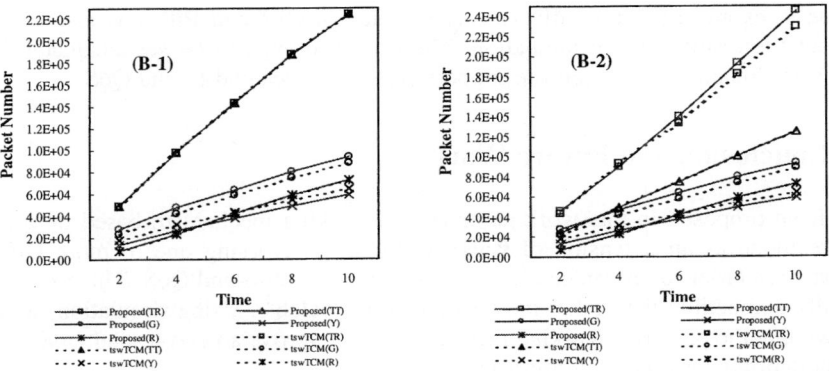

Fig. 8. Time versus received or transmitted packets. Source group 1 has 5 flows (each 50Mbps) and source group 2 has no flows. (B-1): no bottleneck in C4-C5 link (300Mbps), (B-2): bottleneck in C4-C5 link (100Mbps)

C. Scenario 3: Source group 1 has 5 flows (each 30Mbps) and source group 2 has 5 flows (each 15Mbps). No bottleneck (C-1) and bottleneck (C-2) in C4-C5 link

This scenario is same as scenario 1 except that the UDP traffics of source group 2 is generated between 3sec and 8sec. The results from this scenario show that many yellow packets is promoted to the higher priority level and prove that our proposed mechanism is operated more adaptively than the conventional tswTCM under a dynamic traffic environment.

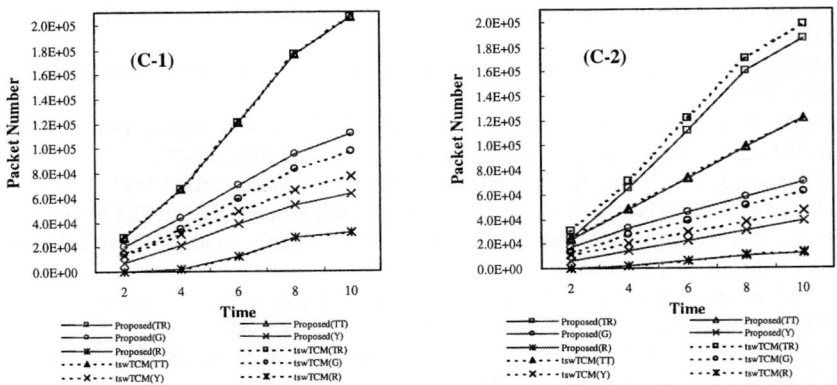

Fig. 9. Time versus received or transmitted packets. Source group 1 has 5 flows (each 30Mbps), and source group 2 has 5 flows (each 15Mbps). (C-1): no bottleneck in C4-C5 link (300Mbps), (C-2): bottleneck in C4-C5 link (100Mbps)

From extensive experiments not depicted graphically, it was found that our proposed marking mechanism performs better than the conventional tswTCM in most scenarios and the parameters for simulation are sensitive to the traffic source characteristics and link rates. In addition, the settings of RED-related threshold values and queue sizes are especially affected. The choice of CIR and PIR was also found to impact the results in our simulation. These parameters can be set adequately from DiffServ domain administrator to support more enhanced end-to-end QoS.

7 Conclusion and Future Work

We have proposed in this work an enhanced marking mechanism based on a color-aware mode as an extension of tswTCM for flows crossing one or more DiffServ domains in order to guarantee effectively enhanced end-to-end QoS. This mechanism is fully compatible with other markers and very scalable. Using simulation, we have shown that the proposed marking mechanism can improve an end-to-end QoS guarantee in multiple DiffServ environment.

As future work, we will develop more adaptive version of the proposed marking mechanism. Also, we will estimate and analyze this mechanism through the implementation on the IXP2400 network processor [13], [14] and devise improved marking mechanism to enforce fairness among different flows originated from the same subscriber network in DiffServ domain.

References

1. Blake, S., Black, D., Carlson, M., Davies, E., Wang, Z., and Weiss, W.: An architecture for differentiated services. RFC 2475 (1998)
2. Nichol, K., Blake, S., Baker, F., and Black, D.: Definition of the Differentiated Services Field (DS Field) in the IPv4 and IPv6 Headers. RFC 2474 (1998)
3. Jacobson, V., Nichols, K., and Poduri, K.: An expedited forwarding PHB. RFC 2598 (1999)
4. Heinanen, J., Baker, F., Weiss, W., and Wroclawski, J.: Assured forwarding PHB group. RFC 2597 (1999)
5. Clark, D. and Fang, W.: Explicit Allocation of Best Effort Packet Delivery Service, IEEE/ACM Transactions on Networking, Vol. 6. No. 4 (1998) 362-373
6. Heinanen, J. and Guerin, R.: A Single Rate Three Color Marker. RFC 2697 (1999)
7. Heinanen, J. and Guerin, R.: A Two Rate Three Color Marker. RFC 2698 (1999)
8. Fang, W., Seddigh, N., and Nandy, B.: A time sliding window three color marker (tswTCM). RFC 2859 (2000)
9. Floyd, S. and Jacobson, V.: Random early detection gateways for congestion avoidance, IEEE/ACM Transactions on Networking, Vol. 1 (1993) 397-413
10. Network simulator 2 (ns-2). University of California at Berkeley, CA, Available via http://www.isi.edu/nsnam/ns/ (2003)
11. Andrikopoulos, I., Wood, L., and Pavlou, G.: A fair traffic conditioner for the assured service in a differentiated services Internet. ICC 2000, Vol. 2, IEEE International Conference (2000) 18-22
12. Seddigh, N., Nandy, B., Pieda, P., Hadi Salim, J., and Chapman, A.: An experimental study of Assured services in a Diffserv IP QoS Network, Proceedings of SPIE symposium on QoS issues related to The internet, Boston (1998)
13. IXP2400 and IXP2800Network Processor Programmar's Reference Manual, Intel Corporation (2004)
14. IXP2400 Network Processor Hardware Reference Manual, Intel Corporation (2004)

An Efficient Handoff Mechanism with Web Proxy MAP in Hierarchical Mobile IPv6[1]

Jonghyoun Choi and Youngsong Mun

School of Computer Science, Soongsil University,
Sangdo 5 Dong, Dongjak Gu, Seoul, Korea
wide@sunny.ssu.ac.kr, mun@computing.ssu.ac.kr

Abstract. In Mobile IPv6, when a MN (Mobile Node) moves from home network to the foreign network, it configures a new Care-of-Address (CoA) and requests the Home Agent (HA) to update its binding. This binding process requires high signaling load. Thus, Hierarchical Mobile IPv6 (HMIPv6) has been proposed to accommodate frequent mobility of the MN and reduce the signaling load in the Internet. In this paper, we propose hierarchical management scheme for Mobile IPv6 where MAP (Mobility Anchor Point) has web proxy function for minimizing signaling load in HMIPv6. The performance analysis and the numerical results presented in this paper shows that our proposal has superior performance to the HMIPv6

1 Introduction

The Internet users have desire for high quality of service at anywhere. Mobile device users keep increasing by growth of mobile device and wireless techniques. Mobile IPv6 [1] proposed by IETF (Internet Engineering Task Force) provides a basic host mobility management scheme. Mobile IPv6 specifies routing support to permit IPv6 hosts to move between IP subnetworks while maintaining session continuity. Mobile IPv6 supports transparency above the IP layer, including maintenance of active TCP connections and UDP port bindings. To accomplish this, when a MN moves from home network to the foreign network, it configures a new Care-of-Address (CoA) and requests the HA to update it's binding. This binding allows a mobile node to maintain connectivity with the Internet as it moves between subnets. However, binding process requires high signaling load. Thus, HMIPv6 has been proposed to accommodate frequent mobility of the MN and reduce the signaling load in the Internet. In HMIPv6, when a MN moves into new AR domain, MN may perform one or two types of binding update procedures: both the global binding update and the local binding update (intra-MAP) or only the local binding update (Inter-MAP). However, HMIPv6 focused on the intra-MAP domain handoff, not on the inter-MAP domain handoff [4]. In this paper, we propose hierarchical management scheme for Mobile IPv6 where MAP has web proxy cache function for minimizing signaling load in HMIPv6.

---

[1] This work was done as a part of Information & Communication fundamental Technology Research Program supported by Ministry of Information & Communication in republic of Korea.

2 Overview of Hierarchical Mobile IPv6 System

The HMIPv6 protocol separates mobility management into intra-domain mobility and inter-domain mobility. A MAP in HMIPv6 treats the mobility management inside a domain. Thus, when a MH moves around the sub-networks within a single domain, the MN sends a BU message only to the current MAP. When the MH moves out of the domain or moves into another domain, Mobile IPv6 is invoked to handle the mobility.

The basic operation of the HMIPv6 can be summarized as follows.

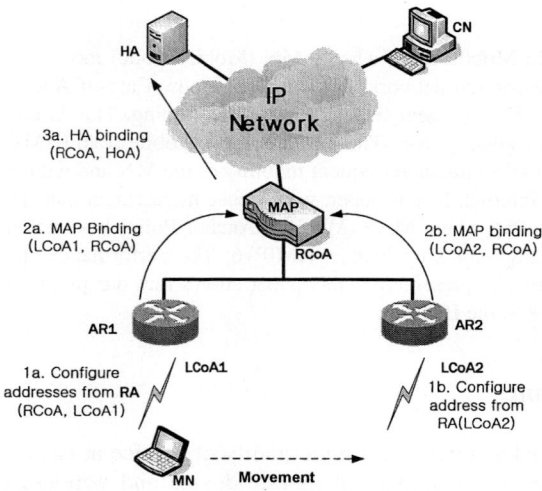

Fig. 1. The basic Operation of the HMIPv6

In HMIPv6, the MN has two addresses, an RCoA on the MAP's link and an on-link CoA (LCoA). When a MN moves into a new MAP domain, it needs to configure two CoAs: an RCoA on the MAP's link and an on-link CoA (LCoA). After forming the RCoA based on the prefix received in the MAP option, the MN sends a local BU to the MAP. This BU procedure will bind the MN's RCoA to its LCoA. The MAP then is acting as a HA. Following a successful registration with the MAP, a bi-directional tunnel between the MN and the MAP is established. After registering with the MAP, the MN registers its new RCoA with it's HA by sending a BU that specifies the binding (RCoA, Home Address) as in Mobile IPv6. When the MN moves within the same MAP domain, it should only register its new LCoA with its MAP. In this case, the RCoA remains unchanged.

3 MAP with Web Proxy in HMIPv6

In this paper, the MAP of proposed scheme has web proxy function to reduce signaling cost with CN. Thus, the MAP can transmit data to MN directly instead of CN.

This procedure reduces the number of connecting CN. The performance of proposed system depends on hit ratio of web proxy. Fig.2 shows proposed system model.

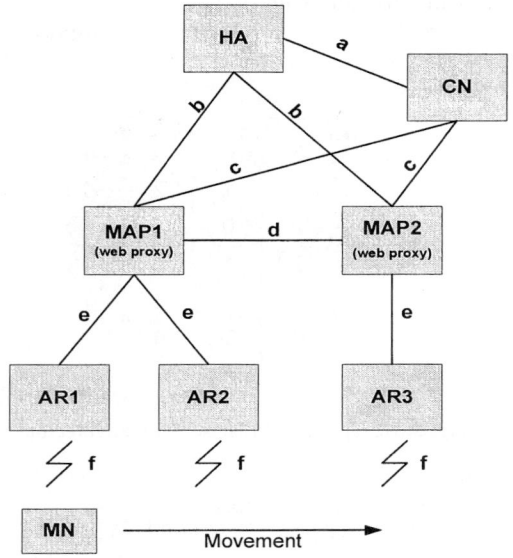

Fig. 2. System model of the proposed scheme

When MN moves from AR1 to AR2, the procedure of proposed system is same to HMIPv6's. While, MN moves from AR2 to AR3, proposed system operates as follows:

1. A MN detects movement by receiving router advertisement message from AR3.
2. A MN configures LCoA and RCoA.
3. A MN sends a local binding update to the MAP for binding new LCoA to new RCoA.
4. A MN sends a global binding update to the HA for binding new RCoA to Home Address (HoA).
5. In order to speed up the handoff between MAPs and reduce packet loss, a MN sends a local binding update to its previous MAP specifying its new LCoA.
6. The MAP1 can forward packets to a MN.
7. When a MN receives a tunneled from CN, a MN sends binding update to a CN. The probability of connecting to CN becomes smaller than normal HMIPv6's.

4 Performance Analysis

4.1 Mobility Model

In this paper, we assumed hexagonal cellular network architecture, as shown in Fig. 3. Each MAP domain is assumed to consist of the same number of range rings, R. Each range ring r (r ≥ 0) consists of 6r cells. The center cell is innermost cell 0. The cells labeled 1 formed the first range ring around cell "0," the cells labeled 2 formed the second range ring around cell 1 and so on. Therefore, the number of cells up to ring R, N(R) is calculated using the following Eq.(1).

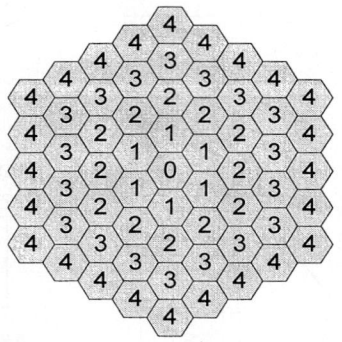

Fig. 3. Hexagonal cellular network architecture

$$N(R) = \sum_{r=1}^{R} 6r + 1 = \frac{6R(R+1)}{2} + 1 \qquad (1)$$

In terms of user mobility model, random-walk mobility model are taken into consideration as commonly used mobility model. The random-walk model is appropriate for pedestrian movements where mobility is generally confined to a limited geographical area such as residential and business buildings [4].

In terms of random-walk mobility model, we consider the two-dimensional Markov chain model used in [5]. In this model, the next position of an MN is equal to the previous position plus a random variable whose value is drawn independently from an arbitrary distribution [5]. In addition, an MN moves to another cell area with a probability of 1−q and remains in the current cell with probability, q. In the cellular architecture shown in Fig. 2, if an MN is located in a cell of range ring r (r > 0), the probability that a movement will result in an increase or decrease in the distance from the center cell is given by

$$p^+(r) = \frac{1}{3} + \frac{1}{6r} \text{ and } p^-(r) = \frac{1}{3} - \frac{1}{6r} \qquad (2)$$

We define the state r of a Markov chain as the distance between the current cell of the MN and the center cell. This state is equivalent to the index of a range ring where the MN is located. As a result, the MN is said to be in state r if it is currently residing in range ring r. The transition probabilities $\alpha_{r,r+1}$ and $\beta_{r,r-1}$ represent the probabilities of the distance of the MN from the center cell increasing or decreasing, respectively. They are given as follows:

$$\alpha_{r,r+1} = \begin{cases} (1-q) & \text{if } r = 0 \\ (1-q)p^+(r) & \text{if } 1 \leq r \leq R \end{cases} \qquad (3)$$

$$\beta_{r,r-1} = (1-q)p^-(r) \quad \text{if } 1 \leq r \leq R \tag{4}$$

where q is the probability that an MN remains in the current cell.

Let $P_{r,R}$ be the steady-state probability of state r within a MAP domain consisting of R range rings. As Eq.(3) and Eq.(4), $P_{r,R}$ can be expressed in terms of the steady state probability $P_{0,R}$ as follows:

$$P_{r,R} = P_{0,R} \prod_{i=0}^{r-1} \frac{\alpha_{i,i+1}}{\beta_{i+1,i}} \quad \text{for } 1 \leq r \leq R \tag{5}$$

With the requirement $\sum_{r=0}^{R} P_{r,R} = 1$, $P_{r,R}$ can be expressed by

$$P_{0,R} = \frac{1}{1 + \sum_{r=1}^{R} \prod_{i=0}^{r-1} \frac{\alpha_{i,i+1}}{\beta_{i+1,i}}} \tag{6}$$

where $\alpha_{r,r+1}$ and $\beta_{r,r-1}$ are obtained from Eq.(3) and Eq.(4)

4.2 Cost Functions

In order to analyze the performance of HMIPv6 [2] and proposed scheme, the total cost, consisting of location update cost and paging cost, should be considered. In normal HMIPv6, we divide the total cost into location update cost and packet delivery cost. Respectively, in proposed scheme, we divide total cost into new location update and packet delivery cost. The location update cost, new location update and the packet delivery cost are denoted by $C_{location}$, $C_{new-location}$, and C_{packet}, respectively. Then, the total cost of HMIPv6 (C_{total}) and proposed scheme ($C_{new-total}$) can be obtained as follows:

$$C_{total} = C_{location} + C_{packet} \tag{7}$$

$$C_{new-total} = C_{new-location} + C_{packet} \tag{8}$$

4.2.1 Location Update Cost

In HMIPv6, the MN has two addresses, an RCoA on the MAP's link and an on-link CoA (LCoA). When a MN moves into a new MAP domain, it needs to configure two CoAs: an RCoA on the MAP's link and an on-link CoA (LCoA). After forming the RCoA based on the prefix received in the MAP option, the MN sends a local BU to the MAP. This BU procedure will bind the MN's RCoA to its LCoA. The MAP then is acting as a HA. Following a successful registration with the MAP, a bi-directional tunnel between the MN and the MAP is established. After registering with the MAP, the MN registers its new RCoA with it's HA by sending a BU that specifies the binding (RCoA, Home Address) as in Mobile IPv6. When the MN moves within the same MAP domain, it should only register its new LCoA with its MAP. In this case, the RCoA remains unchanged.

When a MN moves into new AR domain, MN may perform one or two types of binding update procedures: both the global binding update and the local binding update or only the local binding update. C_g, C_{new-g} and C_l denote the signaling costs in the global binding update, the global binding update of proposed scheme and the local binding update, respectively. In the IP networks, the signaling cost is proportional to the distance of two network entities. C_g, C_{new-g} and C_l can be obtained from the below equations.

$$C_g = 2 \cdot (\kappa \cdot f + \tau \cdot (b+e)) + 2 \cdot N_{CN} \cdot (\kappa \cdot f + \tau \cdot (b+c)) \\ + PC_{HA} + N_{CN} \cdot PC_{CN} + PC_{MAP} \tag{9}$$

$$C_{new-g} = 2 \cdot (\kappa \cdot f + \tau \cdot (b+e)) + 2 \cdot N_{CN} \cdot (1 - HR_{proxy}) \cdot (\kappa \cdot f + \tau \cdot (b+c)) \\ + PC_{HA} + N_{CN} \cdot (1 - HR_{proxy}) \cdot PC_{CN} + PC_{MAP} \tag{10}$$

$$C_l = 2 \cdot (\kappa \cdot f + \tau \cdot e) + PC_{MAP} \tag{11}$$

where τ and κ are the unit transmission costs in a wired and a wireless link, respectively. As Fig. 2, b, c, e and f are the hop distance between nodes. PC_{HA}, PC_{CN} and PC_{MAP} are the processing costs for binding update procedures at the HA, the CN and the MAP, respectively. N_{CN} denotes the number of CNs which is communicating with the MN. HR_{proxy} denotes a hit ratio of web proxy. In proposed scheme, we reduce the probability of connecting to CN using web proxy. Thus, the number of N_{CN}, in proposed scheme, is smaller than normal HMIPv6's

In terms of the random walk mobility model, the probability that a MN performs a global binding update is as follows:

$$p_{R,R} \cdot \alpha_{r,r+1} \tag{12}$$

Specifically, if a MN is located in range ring R, the boundary ring of a MAP domain composed of R range rings, and performs a movement from range ring R to range ring R + 1. The MN then performs the global binding update procedure. In other cases, except this movement, the MN only performs a local binding update procedure. Hence, the location update cost of normal and proposed scheme per unit time can be expressed as follows:

$$C_{location} = \frac{p_{R,R} \cdot \alpha_{R,R+1} \cdot C_g + (1 - p_{R,R} \cdot \alpha_{R,R+1}) \cdot C_l}{T} \tag{13}$$

$$C_{new-location} = \frac{p_{R,R} \cdot \alpha_{R,R+1} \cdot C_{new-g} + (1 - p_{R,R} \cdot \alpha_{R,R+1}) \cdot C_l}{T} \tag{14}$$

where T is the average cell residence time.

4.2.2 Packet Delivery Cost

The packet delivery cost, C_{packet}, in HMIPv6 can then be calculated as follows:

$$C_{packet} = C_{MAP} + C_{HA} + C_{CN-MN} \tag{15}$$

In Eq.(15), C_{MAP} and C_{HA} denote the processing costs for packet delivery at the MAP and the HA, respectively. C_{CN-MN} denotes the packet transmission cost from the CN to the MN.

In HMIPv6, a MAP maintains a mapping table for translation between RCoA and LCoA. The mapping table is similar to that of the HA, and it is used to track the current locations (LCoA) of the MNs. All packets directed to the MN will be received by the MAP and tunneled to the MN's LCoA using the mapping table. Therefore, the lookup time required for the mapping table also needs to be considered. Specifically, when a packet arrives at the MAP, the MAP selects the current LCoA of the destination MN from the mapping table and the packet is then routed to the MN. Therefore, the processing cost at the MAP is divided into the lookup cost (C_{lookup}) and the routing cost ($C_{routing}$). The lookup cost is proportional to the size of the mapping table. The size of the mapping table is proportional to the number of MNs located in the coverage of a MAP domain [4]. On the other hand, the routing cost is proportional to the logarithm of the number of ARs belonging to a particular MAP domain [4]. Therefore, the processing cost at the MAP can be expressed as Eq.(17). In Eq.(17), λ_s denotes the session arrival rate and S denotes the average session size in the unit of packet. α and β are the weighting factors.

Let N_{MN} be the total number of users located in a MAP domain. In this paper, we assume that the average number of users located in the coverage of an AR is K. Therefore, the total number of users can be obtained as follows:

$$N_{MN} = N_{AR} \times K \tag{16}$$

$$\begin{aligned} C_{MAP} &= \lambda_s \cdot S \cdot (C_{lookup} + C_{routing}) \\ &= \lambda_s \cdot S \cdot (\alpha N_{MN} + \beta \log(N_{AR})) \end{aligned} \tag{17}$$

In MIPv6, using the route optimization, only the first packet of a session transmits the HA. Subsequently, all successive packets of the session are directly routed to the MN. The processing cost at the HA can be calculated as follows:

$$C_{HA} = \lambda_s \cdot \theta_{HA} \tag{18}$$

where θ_{HA} refers to a unit packet processing cost at the HA.

Since HMIPv6 supports the route optimization, the transmission cost in HMIPv6 can be obtained using Eq.(19). As mentioned before, τ and κ denote the unit transmission costs in a wired and a wireless link, respectively.

$$C_{CN-MN} = \tau \cdot \lambda_s \cdot ((S-1) \cdot (c+e) + (a+b+e)) + \kappa \cdot \lambda_s \cdot S \tag{19}$$

In proposed scheme, we reduce the probability of connecting to CN using web proxy. Thus, $C_{new-MAP}$ and $C_{new-CN-MN}$ can be calculated as follows:

$$\begin{aligned} C_{new-MAP} &= \lambda_s \cdot (1-HR_{proxy}) \cdot S \cdot (C_{lookup} + C_{routing}) + \lambda_s \cdot HR_{proxy} \cdot S \cdot (C_{proxy} + C_{routing}) \\ &= \lambda_s \cdot (1-HR_{proxy}) \cdot S \cdot (\alpha N_{MN} + \beta \log(N_{AR})) + \lambda_s \cdot (1-HR_{proxy}) \cdot S \cdot (\gamma N_{MN} + \beta \log(N_{AR})) \end{aligned} \tag{20}$$

$$\begin{aligned} C_{new-CN-MN} = &\tau \cdot \lambda_s \cdot (1-HR_{proxy}) \cdot ((S-1) \cdot (c+e) \\ &+ (a+b+e)) + \kappa \cdot \lambda_s \cdot (1-HR_{proxy}) \cdot S \end{aligned} \tag{21}$$

where Cporxy denotes processing cost of web proxy. A web proxy also is affected with the number of MN. γ is the weighting factors. Therefore, the packet delivery cost of proposed scheme can be calculated as follows:

$$C_{packet} = C_{new-MAP} + C_{HA} + C_{new-CN-MN} \tag{22}$$

5 Numerical Results

This section presents performance analysis of proposed scheme as compared with normal HMIPv6. The parameter values for the analysis were referenced from [4], [6] and [7]. They are shown in Table 1.

Table 1. Numerical simulation parameter for performance analysis

parameter	α	β	γ	θ_{HA}	τ	κ	A	b
value	0.1	0.2	0.05	20	1	2	6	6
parameter	c	d	e	f	N_{CN}	PC_{HA}	PC_{MAP}	PC_{CN}
Value	4	1	2	1	2	24	12	6

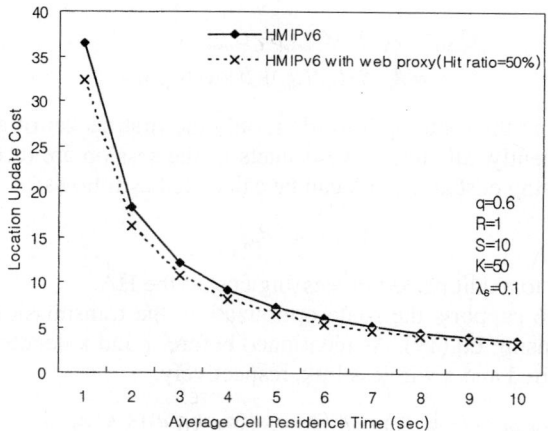

Fig. 4. Location update cost as a function of average cell residence time of MN

Fig 4 shows the variation in the location update cost as the average cell residence time is changed in the random-walk model. The location updates cost becomes less as the average cell residence time increases. This must be true because a MN becomes static by residing in a cell longer, the frequency of location update to HA become reduced. In a comparison of proposed scheme with HMIPv6, proposed scheme reduces the location update cost by 10% approximately.

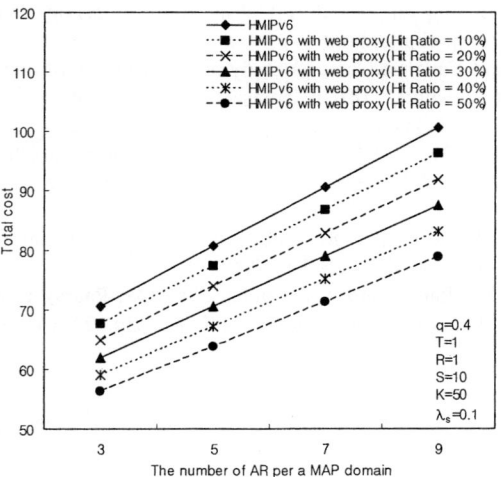

Fig. 5. Total cost as function of the number of AR per a MAP domain

In HMIPv6, the MAP needs to lookup the destination MN on mapping table (Binding Cache table). The cost for this lookup procedure depends on the number of MNs in a MAP domain. Therefore, the packet delivery cost increases as the number of MN in the MAP domain increases. In Eq.(16), the number of MN is $N_{AR} \times K$. Fig. 5 shows the impact of the number of AR per a MAP domain on the total cost in a random-walk model. As shown in Fig. 5, the total cost increases linearly as the number of AR increases. In a comparison of proposed scheme with HMIPv6, proposed scheme reduces the total cost by 4% (hit ratio = 10%) and 21% (hit ratio=50%) approximately.

6 Conclusions

HMIPv6 has been proposed to accommodate frequent mobility of the MNs and reduce the signaling load in the Internet. However, HMIPv6 focused on the intra-MAP domain handoff, not on the inter-MAP domain handoff [4]. In this paper, we propose MAP has web proxy cache function for minimizing signaling load in HMIPv6. The performance analysis and the numerical results presented in this paper shows that our proposal has superior performance to the HMIPv6 when hit ratio of web proxy on MAP is high. The proposed scheme reduces the location update cost by 10% and the total cost by 4% (hit ratio = 10%) and 21% (hit ratio=50) approximately.

References

1. D. B. Johnson and C. E. Perkins, "Mobility support in IPv6," IETF RFC 3775, June, 2004.
2. Hsham Soliman, Claude Castelluccia, Karim El-Malki and Ludovic Bellier, "Hierarchical MIPv6 mobility management," IETF Internet draft, draft-ietf-mipshop-hmipv6-03.txt (work in progress), October, 2004.

3. IETF MIPv6 Signaling and Handoff Optimization (mipshop) WG: http://www.ietf.org/html.charters/mipshop-charter.html
4. Sangheon Pack and Yanghee Choi, "A study on performance of hierarchical mobile IPv6 in IP-based cellular networks," IEICE Transactions on Communications, vol. E87-B no. 3 pp.462-469, Mar. 2004
5. I.F. Akyildiz and W. Wang, "A dynamic location management scheme for next-generation multitier PCS systems," IEEE Trans. Wireless Commun., vol.1, no.1, pp.178–189, Jan. 2002.
6. M. Woo, "Performance analysis of mobile IP regional registration," IEICE Trans. Commun., vol.E86-B, no.2, pp.472–478, Feb. 2003.
7. X. Zhang, J.G. Castellanos, and A.T. Capbell, "P-MIP: Paging extensions for mobile IP," ACM Mobile Networks and Applications, vol.7, no.2, pp.127–141, 2002.

A New Carried-Dependence Self-scheduling Algorithm

Hyun Cheol Kim

Dept. of Computer & Information, Jaineung College, 122, Songlim-Dong, Dong-Gu,
Incheon 401-714, Republic of Korea
hckim@mail.jnc.ac.kr

Abstract. In this paper we present an analysis on a shared memory system of five self-scheduling algorithms running on top of the threads programming model to schedule the loop with cross-iteration dependence. Four of them are well-known: self-scheduling (SS), chunked self-scheduling (CSS), guided self-scheduling (GSS) and factoring. Because these schemes are all for loops without cross-iteration dependence, we study the modification of these schemes to schedule the loop with cross-iteration dependence. The fifth is our proposal: carried-dependence self-scheduling (CDSS). The experiments conducted in varying parameters clearly show that CDSS outperforms other modified self-scheduling approaches in a number of simulations. CDSS, modified SS, factoring, GSS and CSS are executed efficiently in order of execution time.

1 Introduction

In many scientific applications, loops are the richest source of parallelism. Therefore, many loop scheduling schemes were proposed to exploit parallelism. In general, there are two major types of parallel constructs that are provided in all parallel languages: *Doall loops* (also called a parallel loop) and *Doacross loops* [1],[2][3]. However, most previous work for loop scheduling focused on *Doall loops* without cross-iteration dependence.

The dependence constraint among different iterations, called cross-iteration dependence, is our major concern. A cross-iteration dependence occurs if some data computed in one iteration is also used by another iteration. Data dependence analysis gives information about underlying data flow of a loop. To preserve those cross-iteration dependences, additional a code must be added to ensure proper synchronization between the processors executing different iterations [1],[2],[3].

This paper proposes a new self-scheduling method for parallel processing of a loop with cross-iteration dependence on shared memory systems. Also, we study the modification of several self-scheduling schemes using central queue in order to schedule the loop with cross-iteration dependence. Our scheme assigns loops efficiently in three-level considering the dependence distance of the loops. To adapt the proposed scheduling and modified self-scheduling schemes into various platforms, including a uni-processor system, we use Java thread for implementation and performance evaluation of five scheduling methods. A series of simulation results corresponding to various parameter changes are presented in this paper.

This paper is organized into the following sections. Section 2 revisits some well known loop scheduling schemes for shared memory multiprocessors. Then, carried-dependence self-scheduling (*CDSS*) scheme is introduced in Section 3 with an explanation of its working principles. Next, discussions on the their implementation and simulation results are presented in Section 4, and followed by a conclusion in Section 5.

2 Related Works

In this section we look at some of the dynamic loop scheduling algorithms, which have been proposed in the literature. These algorithms fall into two distinct classes: *central queue based algorithms* and *distributed queue based ones* according to the organization of take queue. According to the presence of a dedicated scheduler, algorithms can be classified the *central* and *self-scheduling method*.
A special case of scheduling through distributed control units is *self-scheduling* [4],[5],[6]. As implied by the term, there is no single control unit that makes global decisions for allocating processors, but rather the processors themselves are responsible for determining what task to execute next. In central queue based self-scheduling algorithms, such as *self-scheduling (SS)*, *guided self-scheduling (GSS)*, *factoring* and *chunked self-scheduling (CSS)*, iterations of a parallel loop are all stored in a shared central queue and each processor exclusively seizes some iterations from the central queue to execution. The major advantage of using a central queue is the possibility of optimally balancing the load. While keeping a good load balance, the central queue based algorithms differ in the way they reduce synchronization and loop allocation overheads [4],[5],[6]. In *SS* [4],[5] algorithms, each processor repeatedly executes iterations of the loop until all iterations are executed. *SS* achieves almost perfect load balancing. Unfortunately, this method incurs significant synchronization overhead. *CSS* [5],[6] reduces synchronization overhead by having each processor take k iterations instead of one, resulting in less synchronization overhead but load balancing is not as efficient as *SS*. *GSS* [7] changes the size of chunks at run-time. By allocating large chunks of iterations at the beginning of loops to processor, synchronization overhead can be reduced. In addition, allocating small chunks at the end of the loops gives rise to workload balance. Under *GSS* method, each processor is allocated to l/p iterations, where p is the number of processors and l is the number of remaining iterations. In *factoring* [8] algorithm, allocation of loop iterations to processors proceeds in phases. During each phase, only a subset of the remaining loop iterations is divided equally to processors. It balances workload better than *GSS* when the computation times of loop iterations are considerable. Other schemes in self scheduling models are *adaptive guided self-scheduling* [9], which includes a random back-off to avoid contention for the task queue, and assigns iterations in and interleaved fashion to avoid imbalance; *trapezoidal self-scheduling* [10], which linearly decreases the number of iterations allocated to each processor; *tapering* [11], which is suitable for irregular loops and uses execution profiles to select a chunk size that minimizes the load imbalance; *safe self-scheduling* [12], which uses a static phase where each processor is allocated a chunk of iterations and a dynamic phase during which the processors are self scheduled to even out the load imbalances; and *affinity scheduling* [13],[14][15], which takes processor affinity into account while scheduling. Under this scheme, all the processors are initially assigned an equal chunk taking into account data reuse and locality. Previous approaches to loop scheduling have been attempted to achieve the

minimum completion time by distributing the workload as even as possible and to minimize required synchronization overhead. However, all these previous works focused on loops without cross-iteration dependence. In Section 3, we explain our scheduling scheme for loops with cross-iteration dependence.

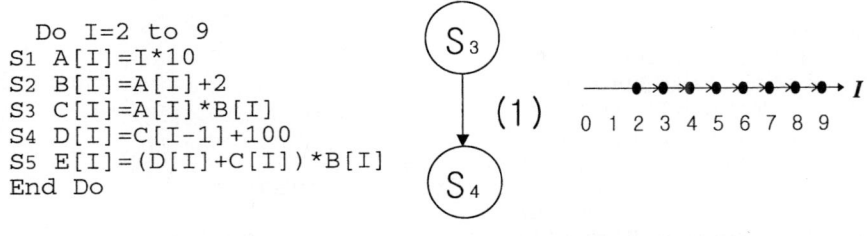

```
    Do I=2 to 9
S1  A[I]=I*10
S2  B[I]=A[I]+2
S3  C[I]=A[I]*B[I]
S4  D[I]=C[I-1]+100
S5  E[I]=(D[I]+C[I])*B[I]
    End Do
```

(a) An Example Program   (b) DDG   (c) Iteration Space Dependence Graph

Fig. 1. An Example Program and its Dependence Graph

3 The Proposed Algorithm

An example program of the loop with cross-iteration is Figure 1(a). Here, the value assigned to C in each iteration of S_3 is fetched by S_4 in the next iteration of the loop. Since some instances of S_4 depend on some instances of S_3, we write $S_3 \delta^f S_4$. There are many instances of each statement in the loop, but the *data dependence graph* (DDG) [1],[2],[3] contains only one node for each statement, as shown in Figure 1(b). This is a loop-carried, lexically forward flow dependence relation. The (1) annotation in Figure 1(b) represents the dependence distance of the loop. The dependence relations is represented by *iteration space dependence graph* [1][2][3], which contains one point for each iteration of the loop and the dependence is represented by an edge from the source iteration to the target iteration as shown in Figure 1(c). Also, the dependence relation between iterations is represented by loop dependence graph (LDG) [1],[2],[3].

Figure 2 illustrates the proposed carried-dependence self-scheduling (*CDSS*) algorithm where, *d* indicates dependence distance of the loops. The basic idea of *CDSS* is that it distributes the loop iterations into processors in three-level considering the dependence distance of the loops that have dependences between iterations while minimizing the loop allocation overhead (i.e. synchronization overhead to access exclusive shared variables). Also, *CDSS* uses a central task queue for scheduling of the loops. The data structures used in the algorithm are as follows. The central task queue, *TaskQueue*, has the role of identifier for the loops and *CrossDep* is an array that has *n* bits, each bit related dependency information corresponding to iterations of loops. It determines the execution of iteration j, which is dependent on iteration. If the value of a corresponding bit equals one, then the iteration can be executed. The initial values of the element of *CrossDep* array are set at 1 when the iterations have no incoming dependency arcs. Therefore, an iteration that has a value 1 of *CrossDep* would be executed immediately because it is an independent iteration. The remaining bits of *CrossDep* are set at 0. The shared variables, *CrossDep* and central task queue, which correspond to critical sections, are serialized by locking. The local variable Temp is used to reduce the access count to shared variables and the variable Rest represents the number of remaining iterations currently.

```
Entry Block :
/* get_routine */
1.   lock(TaskQueue)
2.      Temp =TaskQueue
3.   unlock(TaskQueue)
4.   if (Temp[front] == 1) then
/* get an iteration from front of queue */
5.      i = get_from_queue(1)
6.      Execute Block(i)
7.   else if (Rest > d-1)
/* assignment phase of middle of loop */
8.      start_chunk = get_from_queue(d)
9.   else
10.     start_chunk = get_from_queue(Rest)
/* assignment phase of end of loop */
11.  end if
12.  update Rest
/* exec_test_routine */
13.  for k=0, d-1
14.     i = start_chunk + k
15.     lock(CrossDep)
16.        Temp = CrossDep
17.     unlock(CrossDep)
Exec_test :
18.     if(Temp[i] == 1)
19.        Execute Block(i)
20.     else then
21.        waiting for system_defined interval
22.        reload Temp from CrossDep
23.        Exec_test
24.     end if
25.  end for
Execute Block(i) :
26.     execute the code of iteration i
Exit Block :
27.     j = detect(i)      /* j = i+d */
28.     lock(CrossDep)
29.        CrossDep =Fetch_Set(j, 1)
30.     unlock(CrossDep)
```

Fig. 2. Pseudo Code for Proposed CDSS Scheme

The proposed scheduling method is divided into three blocks according to functionality. *Entry Block* (lines 1-25) is inserted into processors to schedule the loops. In *get_routine*, the processor obtains the iterations from the front of the *TaskQueue* for execution. Here, the determination of the number of iterations is related to scheduling overhead. The scheduling police in *get_routine* is divided into roughly three phases. First, the processor obtains an iteration from the front of the queue only once (lines 4-6). Next, each idle processor obtains as much dependence distance as possible until the number of remaining iterations has less than d (lines 7-8). Finally, all remaining iterations are fetched by the processor (lines 9-10). In *exec_test_routine*, the processor tests whether the fetched iteration is ready for execution. The processor executes the iteration in *Execute Block*(line 26) when the iteration has satisfied dependence relations. If the dependent successor of the iteration is not finished, the processor will wait for execution of the iteration. When the iteration i is terminated, the value of the jth element of *CrossDep*, which has dependence information of the iteration depends on L_i, is set at 1 in *Exit Block* (lines 27-30). It synchronizes that iteration j is ready for execution. In order to schedule the loop with cross-iteration dependence using the existing self-scheduling method based on central queue, we have to modify them. First, the analyzing piece of the processor that tests the execution condition of the iterations must be inserted (lines 13-25). According to this operation, the iteration is running or waiting. Also, the dependence distance d in line 13 is changed according to different chunk sizes of each scheduling method. After the iteration executes, the routine to synchronize the dependence relations between iterations is also necessary (lines 27-30).

4 Implementation and Performance Evaluation

To adapt the proposed scheme and modified self-scheduling methods onto various platforms, including a uni-processor system, we use threads to perform processor activity for our implementation. Although only *SS*, *CSS*, *GSS* and *factoring* are multithreaded for our simulation study, *CDSS* is a general technique, which may be applied onto most self-scheduling schemes with reasonable chunk size. Other central queue based self-scheduling methods are all possible to be multithreaded to schedule the loop with iterations. We implanted five algorithms with the same form in thread level using a JDK1.2.2 programming environment and executed these methods with a Pentium–III 450MHz with 128Mb main memory (Redhat Linux 6.1).

In our simulated experiments, we compared the performance of the *CDSS* scheme with the modified four self-scheduling methods using central task queue in terms of overall execution time including scheduling overhead. For synchronization of critical sections, we use *synchronized* keywords and synchronization methods such as *wait* and *notifyAll* for implementation of locking [16]. The system parameter values in our experimentation are as follows: the number of threads (t) is 1 to 30 chosen randomly, the number of processors (p) is one in our simulation because our experimental tasks are fine grain and the scheduling operation time is relatively less than the execution time of a node. The application parameter including the number of iterations (n) is 60 to 180. The processing time of each iteration (e) was chosen to be 20, 70, 120, 170 and 220 miliseconds, respectively. And the dependence distance (d) of loops ranged from 2 to 4. Table 1 shows the overall execution times for varying parameter values in modified *SS* (*MSS*), modified *CSS* (*MCSS*), modified *GSS* (*MGSS*), modified *factoring* (*MFact*) and *CDSS*. As shown in the table, the proposed algorithm usually outperforms other modified self-scheduling algorithms in terms of total execution time including the scheduling costs and synchronization overheads. We analyzed the execution times effect on the system and application parameters.

Table 1. Execution Time for Scheduling Algorithm (msec)

t	d	n	e	Scheduling Algorithm				
				MGSS	MFact	MSS	MCSS	CDSS
2	2	60	20	1627	1537.2	888.8	1726.2	888.8
			120	7040.4	6656.8	3844.4	7540.6	3839.6
	3	120	170	19726.6	18284.4	10765.4	21158.4	10754.8
			220	25618	23390.6	13756.8	27052.4	13749
	4	180	20	4819.2	4508.2	2660.2	5202.6	2636.4
			120	20982.8	19624.6	11514.6	22606.6	11476.8
3	2	120	70	8792.2	8416.4	4775	9349.2	4768.4
			120	14199	13562.8	7742.4	15056.8	7714.8
	3	60	20	1386.6	1170.4	586.4	1646.4	588.8
			70	3727.8	3169	1588.6	4440.6	1584.8
	4	180	170	27668.8	25302.4	10759.8	31212.6	10764.8
			220	35322.8	32366.2	13766.2	39904.6	13754.8
4	2	60	70	4054.2	3734.4	2384	4522.8	2380.4
			170	9153.2	8434.6	5392	10225.6	5388.6
	3	120	20	2883.2	2665	1180.2	3371.4	1176.6
			120	12539	11543.4	5157.8	14628.2	5154
	4	180	20	4356.4	3976.2	1323.4	5052.8	1338.6
			220	33948.4	30974	10306.8	39220.8	10329.2
10	2	180	170	27252.2	25282.4	16174.6	30688.6	16172.2
			220	34845.8	32320.8	20680.2	39229.8	20669
	3	120	20	2281.8	1876.2	1179.8	2959.6	1170
			120	10044.6	8216.4	5159.4	13085.2	5144.4
	4	180	120	14763	12042.2	5834.8	19717	5800
			170	20463.8	16702.8	8080.4	27440	8050.6
20	2	120	20	2416.6	2110	1788.8	2972.8	1762.6
			70	6522.8	5653.6	4793.4	8025.2	4756.4
	3	60	20	698.6	765.2	592.8	652	592.6
			170	4301.2	4661.6	3588.2	3940.6	3576.8
	4	180	120	10900.2	8184.2	5845.4	15901.2	5835.8
			220	19287.2	14469.8	10341.2	28213.6	10340
30	2	180	20	3666.2	3173.8	2695.2	4495.8	2666.6
			70	9814.2	8471.6	7195.2	12034	7159.6
	3	120	170	9403.8	9000.4	7194.6	11117.6	7196.8
			220	11986	11487.8	9195.6	14223.8	9187.6
	4	60	120	2080.8	1951	1951.8	2076.4	1949.6
			170	2874	2699.8	2703	2880.8	2693.4

4.1 Effect of System Parameter

For parallel processing using threads, we created threads up to 30 on the same applications. Figure 3 shows the overall execution time of each scheduling scheme by varying the number of threads with various parameter values. Generally, it is clear

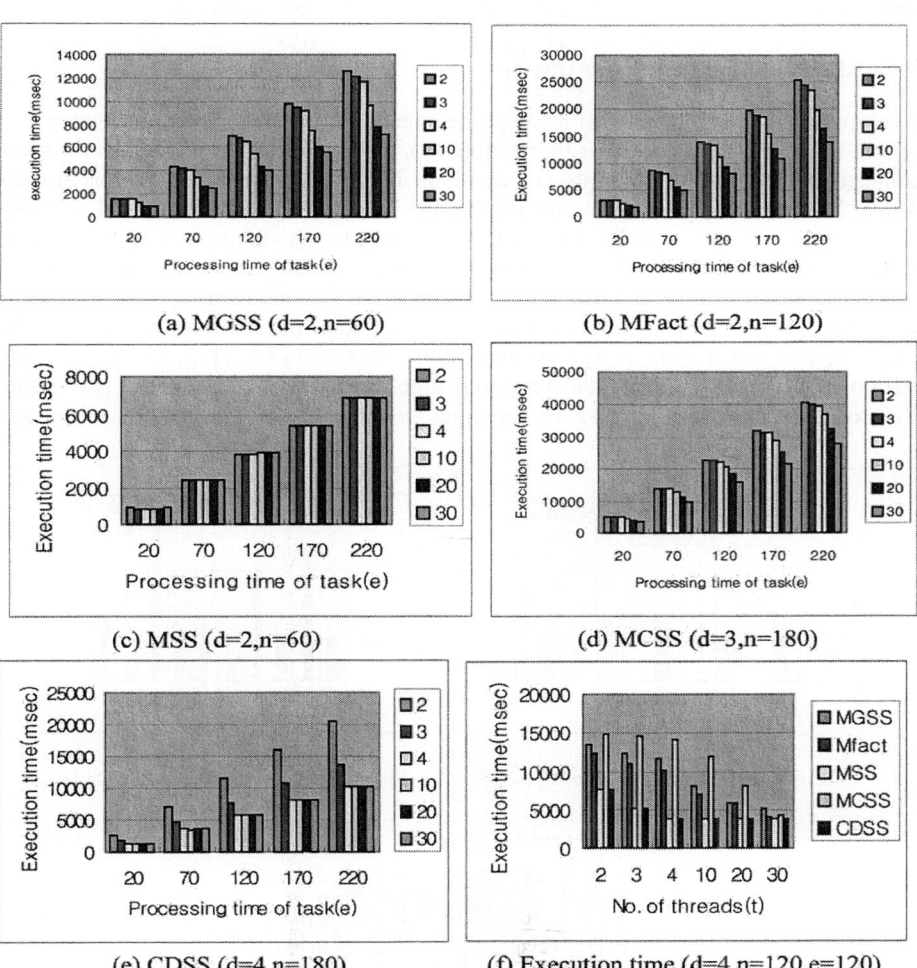

Fig. 3. Effect of the Number of Threads

that providing more threads can improve the performance substantially as shown in the Figure 3. Unfortunately, there are some cases in which some scheduling policies occasionally have poor performance when increasing the threads. In Figure 3(c), it shows performance degradation over a lot of threads when 30 threads were used. In *MGSS* (Fig.3(a)), *Mfact* (Fig.3(b)) and *MCSS* (Fig.3(d)), we have good performance by providing substantially more threads. However, *MSS* and *CDSS* show that a reasonable number of threads used to improve the performance depends on the dependence distance of the loops. Thus, we achieved the best performance using a few threads which equal the dependence distance d in most applications. We can observe that to get a reasonable number of threads for improved performance is very difficult but it could be achieved by repeating experiments on a dedicated scheduling scheme and application. Next, we analyze the effect of the application parameters.

4.2 Effect of Application Parameters

The application parameters also affected the execution times as shown in Figures 4, 5 and 6. By increasing the number of iterations (n), the execution time is long on the same number of threads. For example, as shown in Figure 4, the proposed algorithm has 1201, 2373 and 3577ms execution times on the variations of n with 60, 120 and 180, respectively, when $d=4$, $e=70$ and $t=30$. As a result, by doubling the number of iterations, we get about double the execution time. Next, we changed the processing time of task (e). Figure 5 shows the execution time on the variations of task sizes when $d=3$, $n=120$ and $t=10$. For example, when the execution costs per iteration varied from 20, 70, 120, 170 and 220ms, it had the execution times of 1170, 3173, 5144, 7173 and 9160ms, respectively, using the proposed algorithm. By increasing the processing time of the task, it has long execution times. Finally, Figure 6 shows the execution

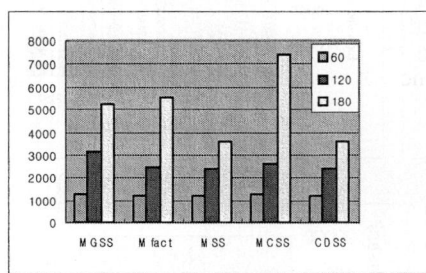
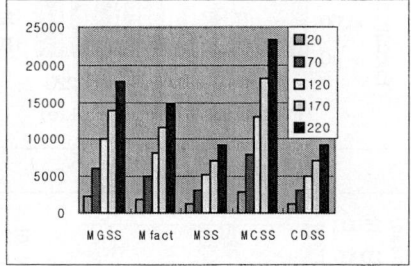

Fig. 4. Effect of n (d=4, e=70, t=30) **Fig. 5.** Effect of e (d=3, n=120, t=10)

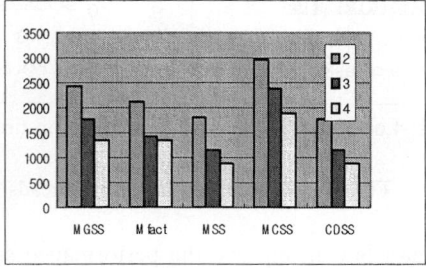

Fig. 6. Effect of d (n=120, e=20, t=20)

time on the effect of varying dependence distance (*d*) of loops with *n*=120, *e*=20 and *t*=20. In this experiment, the application with a large dependence distance can reduce the execution times by decreasing the overhead. For example, according to the variation of *d* with 2, 3 and 4, we can reduce the execution times to 1763, 1166 and 891ms, respectively, using *CDSS*. This is because the applications with large distances have few synchronization points for parallel execution and thus increase the thread parallelism. In various experimental environments, *CDSS* shows improved performance over *MSS*, *MFact*, *MGSS* and *MCSS* by about 0.02, 40.5, 46.1 and 53.6%, respectively, in our experimental parameter values.

5 Conclusions

We proposed a new scheduling method for efficiently execution of a loop with cross-iteration dependence on a shared memory multiprocessor. The proposed method is a self-scheduling algorithm and assigns the loops in three-level considering the synchronization point according to the dependence distance of the loops. Also, we studied the modification that converts the existing self-scheduling method based on the central task queue for parallel loops onto the same form applied to loops with cross-iteration dependence. To adapt the proposed and modified methods onto on various platforms, including a uni-processor system, we use thread for implementation. Compared to other assignment algorithms with various changes of application and system parameters, *CDSS* is found to be more efficient than other methods in overall execution time including scheduling overhead. With our new loop scheduling technique, the execution time of our experimental applications can be improved by 0.02%~53.6 compared to the modified methods.

References

1. Wolfe, M.: High Performance Compilers for Parallel Computing. Addison-Wesley (1996)
2. Quinn, M.J.: Parallel Computing -Theory and Practice. McGraw-Hill (1994)
3. Zima, H., Chapman, B.: Super Compiler for Parallel and Vector Computers. Addison-Wesley (1991)
4. Tang, P., Yew, P.C.: Processor Self-Scheduling for multiple nested parallel loops. Proc. 1986 Int. Conf. Parallel Processing (1986) 528-535
5. Fang, Z., Tang, P., Yew, P.C., Zhu, C.Q.: Dynamic Processor Self-Scheduling for General Parallel Nested Loops. IEEE Trans. on Computers, vol. 39, no. 7 (1990) 919-929
6. Kruskaland, C.P., Weiss, A.: Allocating independent subtasks on parallel processors. IEEE Trans. Software Eng., vol. 11, no.10 (1985) 1001 -1016
7. Polychronopoulos, C.D., Kuck, D.: Guided Self-Scheduling: A Practical Scheme for Parallel Supercomputers. IEEE Trans. on Computers, vol.36, no. 12 (1987) 1425-1439
8. Hummel, S.E., Schonberg, E., Flynn, L.E.: Factoring : A Method for Scheduling Parallel Loops. Comm. ACM, vol. 35, no. 8 (1992) 90-101
9. Eager, D.L., Zahorjan, J.: Adaptive guided self-scheduling. Tech. Rep. 92-01-01. Dept. of Comput. Sci. and Eng,, univ. of Wash (1992)
10. Tzen, T.H., Ni, L.M.: Trapezoid self-scheduling : A practical scheduling scheme for parallel computer. IEEE Trans. on Parallel and Distributed Syst., vol.4 (1993) 87-98

11. Lucco, S.: A Dynamic Scheduling Method for irregular parallel Programs. Proc. ACM SIGPLAN '92 Conf. Programming Language Design and Implementation (1992) 200-211
12. Liu, J., Saletore, V.A., Lewis, T.G.: Safe Self-Scheduling: A Parallel Loop Scheduling Scheme for Shared-Memory Multiprocessors. Int. Parallel Programming, vol.22, no. 6 (1994) 589-616
13. Markatos, E.P., LeBlanc, T.J.: Using Processor Affinity in Loop Scheduling on Shared-Memory Multiprocessors. IEEE Trans. on Parallel and Distributed Syst, vol. 5, no. 4. (1994) 379-400
14. Subramaniam, S., Eager, D.L.: Affinity Scheduling of Unbalanced Workloads. Proc. Supercomputing '94 (1994) 214-226
15. Yan, Y., Jin, C., Zhang, X.: Adaptively Scheduling Parallel Loops in Distributed Shared-Memory Systems. IEEE Trans. on Parallel and Distributed Syst., vol. 8, no.1 (1997) 70-81
16. Campione, M.: The Java Tutorial, Addison-Wesley (1999)

Improved Location Management Scheme Based on Autoconfigured Logical Topology in HMIPv6[*]

Jongpil Jeong, Hyunsang Youn, Hyunseung Choo, and Eunseok Lee

School of Information and Communication Engineering,
Sungkyunkwan University,
440-746, Suwon, KOREA +82-31-290-7145
{jpjeong, choo}@ece.skku.ac.kr
{wizehack, eslee}@selab.skku.ac.kr

Abstract. Though some studies involving general micro-mobility exist, micro-mobility research concerning a mobile node (MN) moving between Mobile Anchor Points (MAP) is lacking. In Hierarchical Mobile IPv6 (HMIPv6), a MN sends a binding update (BU) message to a MAP when the MN performs a handoff. This scheme reduces the handoffs and BUs performance costs. However, the total signaling cost of HMIPv6 rapidly increases with the number of corresponding nodes (CN) for a MN moving among MAPs. In this paper, we propose a scheme based on logical topology with autonomous configuration allowing a MN to move around between MAPs without performing a BU. We perform the performance evaluation for packet delivery and location update costs through a novel analytic approach. We show that the proposed scheme is approximately two times better than the existing one in terms of the total signaling costs through performance analysis. Furthermore, our scheme also reduces the location management cost effectively.

1 Introduction

Mobile IP (MIP) [1] is a de facto standard to address global mobility [2] for mobile users. It has been proposed and developed by the Internet Engineering Task Force (IETF). With the emergence of the global systems, such as IMT2000, MIP is becoming increasingly important for the macro-mobility support, though it contains some problems with location management. Specifically, when a mobile node (MN) moves its location, it delivers a changed Care of Address (CoA) to it's Home Agent (HA). Potentially leading to time delay as a result of the handoff requiring frequent registrations with the HA.

In the Hierarchical Mobile IPv6 (HMIPv6) [3], there is the MAP similar to the Gateway Foreign Agent (GFA) in MIPv4. A MN sends the message to a MAP only when the MN performs local handoffs. However the signaling cost of HMIPv6 rapidly increases as the number of corresponding nodes (CN) increases in a MN when the MN moves between MAPs globally. Thus, a new scheme capable of reducing the

---

[*] This work was supported in parts by Brain Korea 21 and the Ministry of Information and Communication in the Republic of Korea. Corresponding author: Prof. H. Choo.

signaling cost is required. In this paper, we specified in demonstrating a new location management scheme to solve the problem [7]. In this scheme, a MN does not register to CNs and HA, when the MN moves between adjacent MAPs through the wired connection. Though the scheme decreases location update costs it increases the packet delivery cost since the MAPs are connected to each other.

We propose a new location management scheme to support the macro-mobility allowing for smooth and fast handoffs. Each of the MAPs autonomously configures the logical topology with the neighboring MAPs. For a location update, it is not necessary to send signals from a MN to HA and CNs. A MN only performs the registration with a MAP to receive a Regional Care-of-Address (RCoA) and OnLink Care-of-Address (LCoA). In this scheme, the total signaling cost is lower than that of the HMIPv6 [6] and the forwarding scheme [7] up to 17 forwarding steps.

This paper is organized as follows. In section 2, the previous location management schemes are described. The motivation of our work and the new scheme are presented in section 3 based on HMIPv6. In section 4, the performance of the proposed scheme is evaluated. Finally, we conclude our paper, giving future direction.

2 Related Works

MIPv6 uses the same basic network entities as MIPv4 except that it is unnecessary for the Foreign Agent (FA). CoA is supplied to a MN using IPv6 without the support of AR, with a HA capable to process all of the data delivered to the destination in a tunnel. The data to be delivered to the MN is sent directly without passing through the HA. Hence, packet delivery routing is used efficiently.

HMIPv6 scheme has a new component called the MAP, along with some upgraded functionalities in MN and HA except CN [3]. If MN changes it's current address within a MAP, it performs a local BU. The MAP encapsulates all of the packets that are delivered to the MN. If the MN changes its local domain, the MN only registers the changed address to the MAP. During this time the RCoA is never changed, this becomes a merit in registrations. MAP's decreased signaling costs and can also be used for executing fast handoffs. If the MN moves into a new domain, the MN sends a new BU to all nodes on the list. Therefore the total signaling cost is proportionally increased depending upon the number of CNs. If we apply the scheme of [5], this problem does not occurr. Though in this scheme, Access Router (AR) undesirably has the same functionality as a MAP. In HMIPv6, a MAP offers specified functions such as an authentication, billing, and so on [3]. Therefore, AR cannot perform MAP's functions in the general system. Hence, we need a new scheme which is different from the method presented in [5].

Choi and Choo propose a scheme, which forwards the connection information to adjacent MAPs [7]. This scheme has some shortcomings on the process latency in tunneling, and it has demands that state information to be maintained by each MAP. This scheme is only operated within optimized hops since each MAP extracts location information depending on the delivered packets. Another approache is required to address the scheme.

In addition to the HMIPv6 specification [6], mobility-based MAP selection schemes have been proposed in the literature [10, 11]. In these schemes, a MN selects its serving MAP based upon its mobility. For example, the fastest MNs select the furthest MAP while the slowest MNs select the nearest MAP. In addition, an adaptive MAP selection scheme, where an MN selects a serving MAP by estimating session-to-mobility ratio (SMR), was proposed in [12]. The SMR is defined by the ratio of the session arrival rate to the mobility rate. The smaller a MN's SMR, the further the selected MAP will be from the MN. Note that a small SMR indicates that the MN is moving faster than the the rate of session arrivals. Additionally, the aim of [9] is to give a MN the information regarding the MAPs. The MN can select a favorite MAP out of these MAPs

3 The Proposed Scheme in HMIPv6

In this paper, we propose a scheme to solve the existing problem using the autoconfigured logical topology, while examining in detail the MN's operation. The scheme does not require a BU in HA and CNs to autonomously configure the logical topology between neighboring MAPs [8]. Subsequently applying the scheme to find neighboring MAPs for each MAP of an initial MN in HMIPv6.

3.1 Motivation

Each MAP configures by using the autoconfigured logical topology having route and cost information between MAPs. Each MAP also forecasts the neighboring MAP's movement location, and performs a BU. Through the autoconfigured logical topology, it is possible to support fast handoff based on anticipated information as well as location update and packet delivery costs. It is also possible to transfer the profile for authentication in advance since neighboring MAP's information on the security architecture, called AAA(Authentication, Authorization, Accounting) is known, and applying the QoS resource reservation.

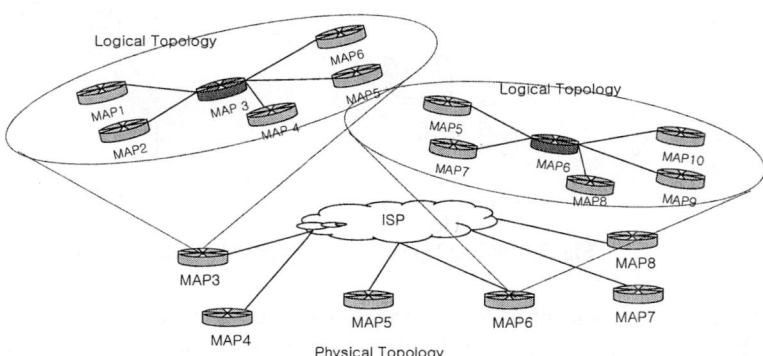

Fig. 1. Logical topology between MAPs

This paper is restricted to, on the practical side, the same administrator domain for the autoconfigured logical topology. We assume that the information between MAPs is known in advance. Fig. 1 shows the autoconfigured logical topology between MAPs [9]. Each MAP must be accessible to the information for neighboring MAPs to configure the autoconfigured logical topology with neighboring MAPs. If a MAP stores the information on a single static table, when a new MAP is added, the neighboring MAPs may have to update the table themselves in order to reconfigure the autoconfigured logical topology [9]. This is fairly burdensome. This paper highlights the fact that each MAPs uses the scheme, updating and configuring the logical topology autonomously.

3.2 The Proposed Scheme Based on Logical Topology

In this paper, the MAPs are named by sequence numbers such as MAP1, MAP2,..., MAPq for convenience. MAP1 already has a table to manage the neighboring MAPs and generates another table to manage the MN. In addition, MAP1 registers all CNs connected to HA and MN using its own RCoA. When HA and CN have the BU message, they store the MN's RCoA to use while connecting with a MN. A BU message does not transmit to HA and CNs when MN moves within MAP. Instead, MAP receives this message to record the information for the location update based on MN's LCoA.

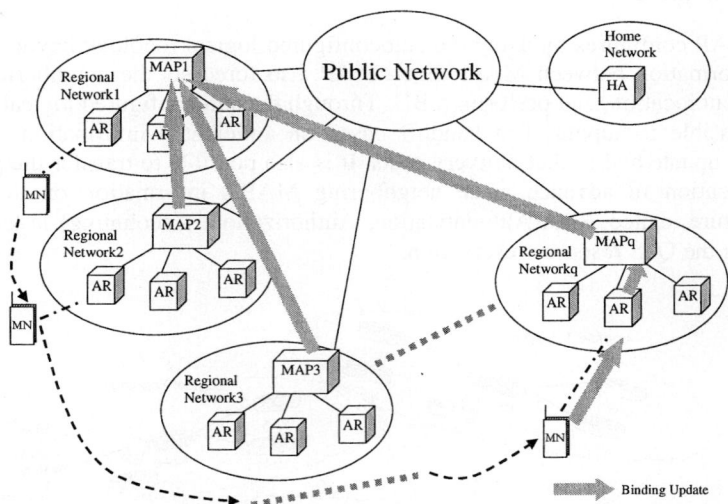

Fig. 2. Binding Update in the proposed scheme

A MN receives adjacent MAP information in the form of a table, which selects MAP2 because MAP2 has the smallest hop number in the table. For this step, we use the algorithm to select a MAP that has the shortest distance. MN sends MAP1's RCoA and initial MAP's RCoA to the MAP2 when MN registers to the MAP2. The MAP2 sends its RCoA and initial MAP's RCoA to the MAP1, and then the MAP1 compares the initial MAP's RCoA with itself. Therefore, when HA transmits data, the data is delivered by this route. This scheme saves signaling costs in comparison with the existing schemes, like HMIPv6. Similarly, when a MN moves from MAP2 to MAP3, the MN transmits the registration message (MAP2's RCoA, initial registration MAP's RCoA) to MAP3, and then MAP3 transmits two messages to MAP1. In this case, MAP1 deletes MAP2's RCoA in the MN's list since it contains the MAP's RCoA and records MAP3's RCoA. MAP1 and MAP3 are then linked through the registration. Therefore, the proposed scheme is not required to send the binding information to the HA and CNs.

As shown in Fig. 2, we consider a MN moving from MAP1 to MAPq as an example of global handoff. In the proposed scheme, the greatest achievment comes from a step reduction, while not continuously keeping links to several steps.In this paper, the maximum number of forwarding links allowed between MAPs is not fixed but optimized for each MN to minimize the total signaling cost. The optimal number is obtained, based on the operational difference between the existing scheme and the proposed one. MIPv6 allows MNs to move around the internet topology while maintaining the reachability and ongoing connections between the mobile and CNs.

To do this the MN sends the BU to its HA and all CNs communicating every time it moves. Hence, increasing the number of CNs influences the system performance in negative manner. However the proposed scheme is independent of the number of CNs while the MN moves in forwarding steps as we have discussed. In the next section, we evaluate the proposed scheme and calculate the optimal number of steps, comparing the number of forwarding links to the total signaling cost on the HMIPv6.

4 Performance Evaluation

In these sections, we compare and evaluate the performance of the proposed scheme based on HMIPv6, Forwarding scheme and HMIPv6. There is a signaling cost related to a location update and a packet delivery in the element, which influences the performance.

4.1 Modeling

4.1.1 Location Update Cost
Similar to [4-6], we define the following parameters for location updates.

- $C_{rm}/C_{pr}/C_{be}/C_{hp}/C_{cp}$: The transmission cost of the location update between the AR and the MN/ the MAP and the AR/ MAPs/ the HA and the MAP/ the CN and the MAP.
- $C_{Uh}/C_{Ur}/C_{Uc}/C_{Uq}$: The cost of the initial home registration/ the regional registration/ the CN registration/ signaling between two MAPs.
- C_{Up}: The home registration cost required to delete all previous forwarding links and create a new link.
- $a_r/a_p/a_h/a_c$: The processing cost for the location update at the AR/ MAP/ HA/ CN.

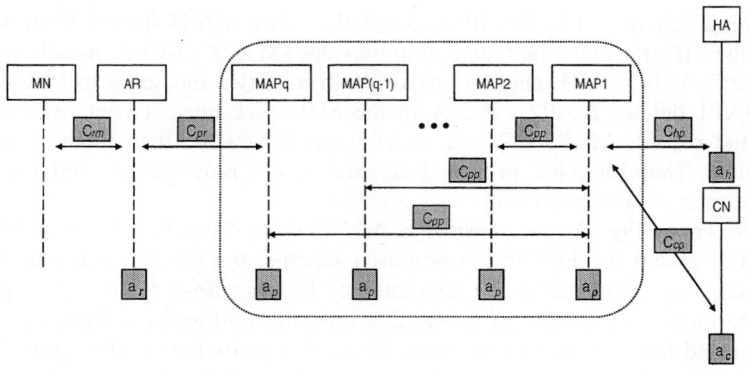

Fig. 3. Location update cost in the proposed scheme

In HMIPv6 scheme, if the MN moves out of the regional network, it must send the BU to CNs and the HA. On the other hand, Fig. 3 shows the location update procedure in the proposed scheme. It represents the signaling message flows for the location registration with the home network, regional location updates within a regional network, the connection between two adjacent MAPs, and the location update with the CN. According to the message flows, C_{Uh}, C_{Ur}, C_{Uc}, C_{Uq}, and C_{Up} are calculated as follows.

$$C_{Uh} = 2a_r + 2a_p + a_h + 2C_{hp} + 2C_{pr} + 2C_{rm}, \quad C_{Ur} = 2a_r + a_p + 2C_{pr} + 2C_{rm}$$
$$C_{Uc} = 2a_r + 2a_p + a_c + 2C_{cp} + 2C_{pr} + 2C_{rm}, \quad C_{Uq} = 2a_r + 2C_{rm} + 3a_p + 2C_{be} + 2C_{pr}$$
$$C_{Up} = 2a_r + 4a_p + a_h + 2C_{hp} + 2C_{pr} + 2C_{rm} + 2C_{be}$$

Let l_{hp}, l_{pr}, l_{cp}, and l_{be} be the average distances between the HA and the MAP in terms of the number of hops packets travel, the average distance between the MAP and the AR, the average distance between the CN and the MAP, and the average distance between the MAPs, respectively. We assume the transmission cost is proportional to the distance between the source and the destination routers, and the proportional constant is δ_U. Thus C_{hp}, C_{pr}, C_{cp}, and C_{be} can be expressed as $C_{hp} = l_{hp}\delta_U$, $C_{pr} = l_{pr}\delta_U$, $C_{cp} = l_{cp}\delta_U$, and $C_{be} = l_{be}\delta_U$, respectively. The transmission cost of the wireless link is generally higher than that of the wired link. We assume that the transmission cost per unit distance over the wireless link is ρ times higher than wired line one. The transmission cost between the AR and the MN can be written as $C_{rm} = \rho\delta_U$. The home registration, regional registration, the connection between the two MAPs, and location update cost with the CN can be expressed as follows.

$$C_{Uh} = 2a_r + 2a_p + a_h + 2(l_{hp} + l_{pr} + \rho)\delta_U, \quad C_{Ur} = 2a_r + a_p + 2(l_{pr} + \rho)\delta_U$$
$$C_{Uc} = 2a_r + 2a_p + a_c + 2(l_{cp} + l_{pr} + \rho)\delta_U, \quad C_{Uq} = 2a_r + 3a_p + 2(l_{be} + \rho + 2l_{pr})\delta_U$$
$$C_{Up} = 2a_r + 4a_p + a_h + 2(l_{hp} + l_{pr} + \rho + 2l_{be})\delta_U$$

Assume each MN may move randomly among N subnets and there are k subnets within a regional network. We use a discrete system to model the movements of each MN [4, 5]. Define a random variable M so that each MN moves out of a regional network at movement. At movement 1, MNs may reside in one of subnets 1 through N. At movement 2, MNs may move to any of the subnets. We assume MNs will move out to one of the other $N-1$ subnets with equal probability $1/{N-1}$. The probability that each MN moves out of the regional network, i.e. the probability of performing home registration and the registration of the CNs at movement m is $P_h^m = \frac{N-k}{N-1} \cdot \left(\frac{k-1}{N-1}\right)^{m-2}$, where $2 \leq m < \infty$. It can be shown that the expectation of M is

$$E[M] = \sum_{m=0}^{\infty} m P_h^m = 1 + \frac{N-1}{N-k}.$$

Assume within the regional network, the average time each MN stays in each subnet before making a movement is T_f. The signaling cost of the HMIPv6 is proportionally increased by the number of CNs(γ). Therefore, the location update cost per unit time is represented as below.

$$C_{LU}^{HMIPv6} = \frac{E[M] \cdot C_{Ur} + C_{Uh} + C_{Uc} \cdot \gamma}{E[M] \cdot T_f}$$

The average location update cost in the proposed scheme is represented as follows, when the MN moves from the MAP0 to the MAPq.

$$C_{LU}^{proposed} = \frac{E[M] \cdot C_{Ur} + q \cdot C_{Uq} + C_{Up} + C_{Uc} \cdot \gamma}{E[M] \cdot T_f}$$

In the worst case, Route Optimization problem occurs in schemes that forward continuously between MAPs, regardless of the MAP Selection algorithm and distance between MAPs, register initially from MN, when MAP is arranged on a straight line equally with forwarding scheme. Though, it could be reduced on the transmission delay cost for forwarding between MAPs and location update cost in MAPs.

4.1.2 Packet Delivery Cost(Tunneling Cost)

Due to tunneling, extra costs for the packet delivery exist under the proposed scheme. The packet delivery cost includes the transmission and processing costs required to route a tunneled packet from the MAP0 to the MAPq, and further forwarding it to the serving AR of the MN. Assume;

- $T_{pr}/T_{be}/T_{cp}$: The transmission cost of packet delivery between the MAP and the AR/ MAPs/ the CN and the MAP;
- $v_p^{proposed}$: The processing cost of packet delivery at the MAP in our scheme;
- v_p^{HMIPv6}: The processing cost of packet delivery at the MAP in the HMIPv6.

We assume the transmission cost from delivering data packets is proportional to the distance between the sending and the receiving routers with the proportional constant δ_D. Therefore, the cost of each packet delivery procedure can be expressed as shown below.

$$C^{HMIPv6}{}_{PD} = v^{HMIPv6}{}_{p} + T_{pr} = v^{HMIPv6}{}_{p} + l_{pr}\delta_{D}$$
$$C^{proposed}{}_{PD} = T_{pr} + T_{be} + v^{proposed}{}_{p} = \delta_{D} \cdot (l_{be} + l_{pr}) + v^{proposed}{}_{p}$$

The worst case occurs when a MAP is arranged on a straight line equally with the forwarding scheme. Though it could reduce the transmission delay cost coming from the forwarding between MAPs and packet delivery cost in MAPs, same as location update cost. The load($v^{proposed}{}_{p}$) on the MAP for processing and routing packets to each AR depends on the number of ARs(k) under the MAP, and the number of links (q) between the first and the last MAPs. When the number of forwarding links is 0, the packet processing cost function is as follows. We assume on average there are ω MNs in a subnet. Then the total number of MNs in an MAP serves in a regional network is ωk on the average. Therefore, we define the total packet processing cost function at the MAP as follows.

$$\zeta k \cdot \lambda_a (\alpha \omega k + \beta log(k))$$

where λ_a is the average packet arrival rate for each MN, α and β are weighting factors of visitor list and routing table lookups, respectively, and ζ is a constant which captures the bandwidth allocation cost at the MAP. On the other hand, when the number of forwarding links (q) among MAPs has more than 0, the processing cost function is proportional to q. We assume u is the average number of neighboring MAPs in a MAP. The cost at the proposed scheme is represented as shown below.

$$u \cdot \zeta \cdot \lambda_a (\alpha \omega k u + \beta log(u)), (q > 0)$$

The $v^{proposed}{}_{p}$ value is the sum of the two proposed scheme.

$$v^{proposed}{}_{p} = \{u \cdot \zeta \cdot \lambda_a (\alpha \omega k u + \beta log(u))\} + \zeta k \cdot \lambda_a (\alpha \omega k + \beta log(k))$$

4.1.3 Total Signaling Cost

From analysis we obtain the overall average signaling cost function in the proposed scheme and in the HMIPv6. First of all, the HMIPv6's total signaling cost is represented as.

$$C^{HMIPv6}(k, \lambda_a, T_f) = C^{HMIPv6}{}_{LU} + C^{HMIPv6}{}_{PD}$$

We assume that the MN moves from the MAP0 to the MAPq, and thus the total signaling cost is calculated as.

$$C^{HMIPv6}{}_{TOT}(k, \lambda_a, T_f, q) = q \cdot (C^{HMIPv6}{}_{LU} + C^{HMIPv6}{}_{PD})$$

The total signaling cost in the proposed scheme is acquired as follows.

$$C^{proposed}(k, \lambda_a, T_f) = C^{proposed}{}_{LU} + C^{proposed}{}_{PD}$$

Then, the total singling cost is calculated as shown below.

$$C^{proposed}{}_{TOT}(k, \lambda_a, T_f, q) = C^{proposed}{}_{LU} + q \cdot C^{proposed}{}_{PD}$$

4.2 Analytical Results

We analyze the performance of our scheme based on the above expression. Table 1 lists some of the parameters used in our performance analysis. The total number of subnets that MNs may access through the wireless channels are limited, and we assume $N=30$. For the evaluation, we assume $l_{hp}=20$, $l_{pr}=3$, $l_{cp}=10$, and $l_{be}=10$.

Table 1. System parameters

Pkt process cost				Proportional cost		Wireless multiple	# of MNs/subnet
a_h	a_p	a_r	a_c	δ_U	δ_D	ρ	ω
30	20	15	10	0.2	0.05	5	15
Weight		Pkt delivery const.				# of neighbor MAPs	
α	β	ζ	η			u	
0.3	0.7	0.01	10			5	

In Fig.4, we assume that $T_f=4$, $\lambda_a=0.3$, and $\gamma=5$. In the proposed scheme, the MN calculates the total signaling cost and compares with those of standard HMIPv6 and packet forwarding scheme [7]. The total signaling cost of the proposed scheme is smaller than those of the HMIPv6 and packet forwarding scheme. The total signaling cost of the packet forwarding scheme is smaller than the HMIPv6 up until 8 forwarding steps ($q \leq 8$). However when $q \geq 9$, the cost of the packet forwarding scheme becomes greater than the HMIPv6 one, and the MN sends the registration message to the HA and the CNs, removing all of the previous links among MAPs. Although at $q \geq 9$, the total signaling cost changes little. In the worst case, the total signaling cost of the proposed scheme is smaller than those of the HMIPv6 and the Forwarding up until 17 forwarding steps ($q \leq 17$). This scheme underlines the fact that it is more cost effective to manage the location of MN than both the HMIPv6 and the packet forwarding schemes.

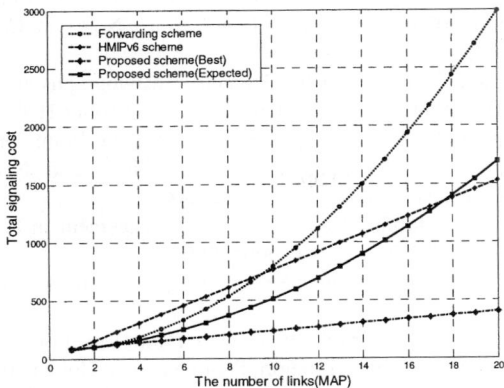

Fig. 4. The comparison of the total signaling costs

5 Conclusion

MIPv6 proposes optimal routing as a solution to the triangle routing problem in MIPv4. However, the signaling cost of HMIPv6 rapidly increases as the the number of CNs increase in a MN. Throughout this paper, we propose a new scheme through a logical topology, with a MN not required to register to CNs and the HA when the MN moves between adjacent MAPs. We use an algorithm to configure the logical topology. As a result of the logical topology, we are able to decrease the location update cost. In conclusion, we perform the simulations to depict the total signaling cost. We compared the proposed scheme to both the standard HMIPv6 and the forwarding scheme, subsequently illustrating the effectiveness of the proposed.

This paper assumes that the worst case is the average of the forwarding scheme when MAP is arranged on a straight line and the best cases of the proposed scheme. In the near future, we plan to conduct performance evaluation on our approach in detail.

References

1. C. E. Perkins, "IP mobility support," Request for Comments(RFC) 2002-2006.
2. P. Reinbold and O. Bonaventure, "A Comparison of IP Mobility Protocol," Tech. Rep. Infonet-TR-2001-07, University of Namur, Infonet Group, June 2001.
3. H. Soliman, C. Castelluccia, K. El-Malki, and L. Bellier, "Hierarchical MIPv6 mobility management (HMIPv6)," Internet Draft, Internet Engineering Task Force, draft-ietf-mobileip-hmipv6-06.txt, July 2002.
4. J. Xie, I. F. Akyildiz, "An optimal location management scheme for minimizing signaling cost in Mobile IP," Communications, 2002. ICC 2002. IEEE International Conference on, vol. 5, pp. 3313 –3317.
5. J. Xie, I. F. Akyildiz, "A distributed dynamic regional location management scheme for mobile IP," INFOCOM 2002. Twenty-First Annual Joint Conference of the IEEE Computer and Communications Societies. Proceedings. IEEE , vol. 2 pp. 1069 -1078
6. H. Soliman et al., "Hierarchical Mobile IPv6 mobility management (HMIPv6)," Internet draft, June 2003.
7. Daekyu Choi, Hyungseung Choo, "Cost Effective Location Management Scheme in Hierarchical Mobile IPv6," Springer-Verlag Lecture Notes in Computer Science, vol. 2668, pp. 144 - 154, May 2003.
8. F. Preparata, et al., "Computational Geometry (Monographs in Computer Science)," Springer-Verlag Berlin and Heidelberg GmbH & Co., October 1990.
9. Omae, K., Okajima, I., Umeda, N., "Mobility Anchor Point Discovery Protocol for Hierarchical Mobile IPv6," Wireless Communications and Networking Conference (WCNC) 2004, IEEE , vol. 4 pp. 21-25, March 2004.
10. K. Kawano et al., "A Mobility-Based Terminal Management in IPv6 Networks," IEICE Trans. Commun., vol. E85-B, no. 10, October 2002.
11. Y. Xu et al., "A Local Mobility Agent Selection Algorithm for Mobile Networks," in Proc. IEEE ICC, May 2003.
12. S. Pack et al., "An Adaptive Mobility Anchor Point Selection Scheme in Hierarchical Mobile IPv6 Networks," Technical Report, Seoul National University, 2004.
13. Y. Wang, W. Chen, and J. Ho, "Performance Analysis of Mobile IP Extended with Routing Agents," Technical Report 97-CSE-13, SMU, 1997.
14. C. Perkins, "Mobile IP," IEEE Communications Magazine, vol.40, pp.66-82, 2002.

Ontological Model of Event for Integration of Inter-organization Applications

Wang Wenjun, Luo Yingwei[*], Liu Xinpeng, Wang Xiaolin, and Xu Zhuoqun

Dept. of Computer Science and Technology, Peking University, Beijing, P.R.China, 100871
lyw@pku.edu.cn

Abstract. Inter-organization event handling process includes two levels' business processes: global business processes need multiple businesses of different organizations to work coordinately; local business processes are multiple activities contained in each business process. All these business applications are running in heterogeneous environments of different locates. How to describe business processes of different levels in event handling entirely, and have transactions and heterogeneous information from different organizations united semantically to satisfy requirements of different users, are the keys to successful event handling. This paper brings forward iEM based on the features of inter-organization integration as well as the multi-level and multi-facet characters in its event disposition. iEM provides stratified modeling of inter-organization integration events, accomplishes common expression of events among multiple domains, and gives common terminology for inter-domain information exchange. It can depict basic processes of events in detail by expression of events in multiple granularities and relationship specialization in the same granularity to meet different user groups' need. At last, Application of JERS (Joint Emergency Response System) using iEM is presented.

1 Introduction

Network information applications, like E-Government, E-Business, are often relevant to event disposition of multiple organizations. This kind of event disposition is characterized by multiple levels and granularities, which calls for network and communication technologies to have distributed, heterogonous application systems cooperate and integrate together. This is named as inter-organizational event disposition. Obviously, the corresponding inter-organizational event disposition is a processing procedure of multiple levels. An event disposition can be divided into one or more business processes (as shown in Figure 1).

Inter-organizational event disposition mainly associates with 2 types of business processes: general business process and part business process. General business process need business services from multiple organizations to act coordinately; each business service is also made up of multiple activities of self-organization arranged by certain business processes. While "business services" in a general business process

---

[*] Corresponding author: LUO Yingwei, lyw@pku.edu.cn. This work is supported by the 973 Program (2002CB312000); the NSFC (60203002).

are called "business processes" in the lower level part business processes. Applications that accomplish these business service functions are distributed at different locations, and run in heterogeneous environments.

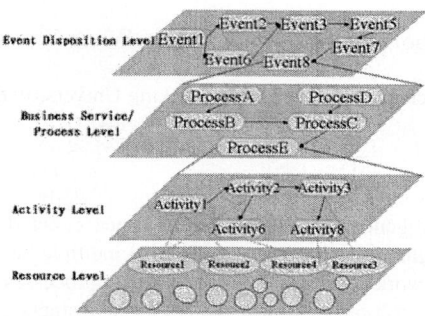

Fig. 1. Demonstration of Inter-organizational Event Disposition

Inter-organizational events have granularities of 3 levels as integral events, business processes and single activities. Different users may comprehend the same event information from different levels and aspects: (a) Granularity at event level: This caters to users who only care for schematic event information. For example, in the event that "Tom has bought a car by applying a loan", usually people only care for the time, the place and the type of car that Tom has bought; (b) Granularity at process level: This caters to users who are in need of information of processes included in an event. As for the example mentioned above, Tom himself should ascertain information of business processes as car selection, loan application, insurance application, car testing, license application, etc; (c) Granularity at activity level: This caters to users who want to have a detailed understanding of event dispositions. As for the example mentioned above, from the aspect of the bank, buying a car by applying a loan is a complex business process which associates with multiple sections, multiple activities and multiple missionaries; (d) Granularity at the comprehensive level: This caters to users who require information from the above 3 levels. As for the example mentioned above, many organizations need to know every activity of multiple processes.

The key to the event disposition lies in that the disposition of inter-organizational events should be described completely to have organization separated, distributed and heterogeneous transactions as well as information united semantically, which will thus meet requirements of different users.

Ontology [1] is an important strategy in describing semantic models. Ontology provides common comprehension of domain knowledge, determines commonly recognized terminologies within certain domains, and implements properties, restrictions and axioms in a formulated way at different levels. Ontology thus gains the characteristic features to provide explicit terminology definitions to express inter-organizational event.

In this paper, relevant studies of event and event models are introduced firstly, and the basis of iEM – ABC Ontology Model [2] is analyzed in emphasis. Finally, iEM is proposed with a typical example described by it.

2 Relevant Studies

2.1 Event and Event Model

Linguistics divides event statement into "Accomplishment Statement" and "Achievement Statement". Terence Parsons [3] and Qianduan Chen [4] think that an event in the aspect of linguistics has 4 forms: Accomplishment Event, Achievement Event, State Event and Activity Event.

Process, action and state [5] are other several concepts that are relevant to events. In the aspect of philosophy, there are sorts of viewpoints regarding event and process:

Tversky points out that, people identify events through their spatial structures and temporal structures, and action and event are two different concepts. Bestougeff [6] describes events according to class hierarchies. An event consists of a group of basic or composite processes, and each process is an action sequence to change entities.

In summary, there are 2 viewpoints in regard to comprehension of events: One is that events are composed of sub-events, but there may be no sequence relationship between these sub-events; another is that events (or event dispositions) are composed of a series of processes, which bear sequence orders.

During recent years, studies of ontology based event models are proceeded throughout the world. Typically, ABC Ontology Model presents an event-related common concept model.

2.2 Definition and Analysis for ABC Ontology Model

The basic part of ABC Ontology Model, which is a result of NSF sponsored Harmony Digital Library project. Relations among terminologies in ABC event model are expressed in Figure 2. Entity is the basic class of all other classes in this model. It belongs to no class, and represents any entities in the world. It has 3 subclasses: Temporality, Actuality and Abstraction to express time-related, noontime-related and abstract entities in the world.

ABC Event Model describes event-related concepts as event, situation, action, agent, and their relationships. An event marks the transformation between situations. A Situation is a context, which is a predicate to the existential aspect of an Actuality. McCarthy [7] describes Situation from the aspect of AI: "A snapshot of the world at some time points". An Action can be accomplished by one or some Agents in the context of an Event. Agent can be people, instruments or organizations.

ABC Ontology is a basic Ontology. It provides a basic model for domain-related or community-related development. But to describe inter-organizational events solely by ABC Event Model is not adequate: (a) Event in ABC can not reflect interaction and coordination needed by inter-organizational requirement conveniently; (b) ABC Event Model is not potent enough to express complex events which have level hierarchies and granularity divisions; (c) ABC Event Model uses discrete Situations as the trigger mechanism, and reflects the change of Situations through Actions participating in Events. This may be redundant for the condition where continuous changes of states are caused by continuous actions of a same agent; (d) ABC Event Model identifies consequence relationship between Events through atTime, precedes or follows properties. This way is not explicit enough when a complex event composed of subevents with complex temporal relationship.

Based on above analysis, we propose Inter-organizational Event Model (iEM) to express the multi-level and multi-granularity features.

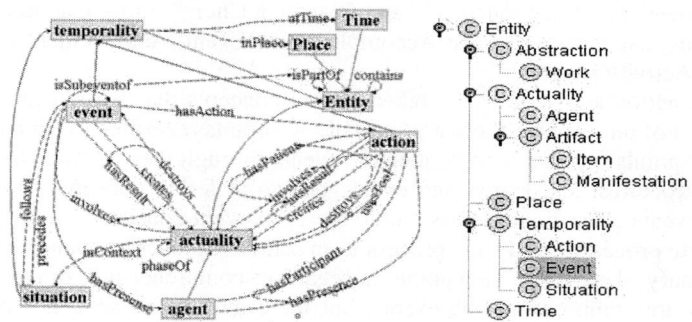

Fig. 2. Demonstration of ABC Ontology Model

3 Ontological Event Model - iEM

Events in ABC Ontology Model mark the transformations between Situations. These Events are actions with concrete parameters under certain temporal and spatial conditions. Agents are participants of these actions. Inter-organizational events all conform to this definition from iEvent, Process, Activity 3 levels' event-oriented semantic description: (a) iEvent has its time, place, action, participants of actions, starting state and ending state. Taking the above buying of a car by Tom as an example (in Section 1), it happens in X year X month X day. Before this event happens, Tom owns no car, while after it, Tom becomes a member of car-owners; (b) Process is also the subclass of Event. As for the example shown above, providing a loan for Tom - the process included in the event happens at some web site of the bank some time. Action in this level is providing a loan. Before the process, Tom does not owe the bank money, while after it he owes; (c) Activity is also subclass of Event. During above-mentioned process, credit investigation is a typical activity, which is proceeded by some functionary of bank.

iEvent, Process and Activity are all subclasses of Event in ABC Model, but they reflect actual event itself from different granularity. They have respective characters, and are inter-related such that they should be combined to express integral events. Multi-granularity oriented integral event description is the basic feature of iEM.

The prime functions of iEM are: (a) iEM is used to describe exchange information between business services in inter-organizational, and acts as common terminology of distributed application systems to dispose inter-organizational events. IEM can fully describe disposition procedures from iEvent level down to Activity level, and can have transactions and information from different organizations united semantically; (b) iEM orients multiple types of users, and expands the description of event granularity; (c) iEM is multi-organization related, which provides inter-organizational event model to implement semantic inter- organization sharing and interoperation.

3.1 Analysis Between ABC Ontology Model and iEM

ABC Event has 2 important concepts that can also easily be mixed up – Subevent and Action. ABC model uses "isSubEvent" to express integral-part relationship between Events. For example, this kind of relationship can exist between D-Day and the Second World War. But this relationship does not imply property "atTime" of a subevent is included in the range denoted by "atTime" property of its parent event. This is the same for the situations between an event and its subevent. There is obvious difference between subevents and actions of an event: An event and its subevents all mark the transformation points between their starting and ending states; actions only represent verbs of the relevant events which perform them. That is to say, Events and subevents have temporal and spatial restrictions, while actions have not.

As shown in Figure 3, in iEM, subevent of iEvent is Process, while action of iEvent is iEventAction, with isMemberProcessOf acting as the subproperty of ABC:isSubEventOf; subevent of Process is Activity, while action of Process is ProcessAction, with isMemberActivityOf acting as the subproperty of ABC:isSubEventOf; Activity has no subevent, and its action is iAction.

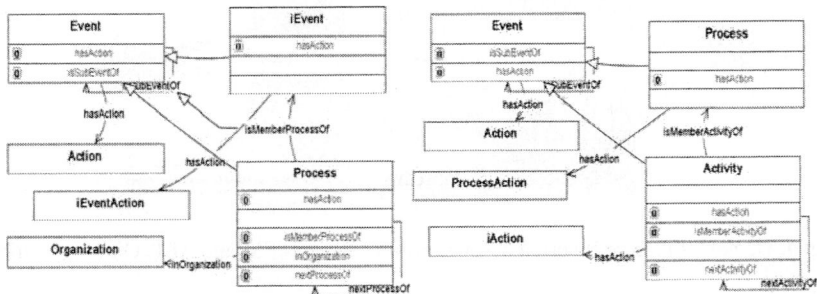

Fig. 3. Subevents and Actions of iEvent and Process

There are 2 viewpoints of differences between subevents of iEvent and ABC:Event: (a) Property atTime of former's subevent is included in the range of atTime of itself; (b) There are explicit relationships between subevents of the former: (i) NextProcessOf, subPropertyOf: InterProcessRelation; Domain: Process Range: Process, which describes consequence relationship between Processes of one iEvent; (ii) NextActivityOf, subPropertyOf: interActivityRelation; Domain: Activity Range: Activity, which describes consequence relationship between Activities of one Process.

Subevents and Process both reflect the first viewpoint that an event is composed of subevents and the second viewpoint that an event is made up of processes.

As shown in Figure 4, UML is used to describe inheritable relationship between classes in iEM and ABC Model.

(1) iEvent Level: (a) iEvent, subClassOf: Event. Similar to Event, iEvent has 2 Properties (beginWith and endWith) that associate to iSituation – subclass of Situation. The existence of iEvent should have its Processes included, which is the same of restriction on Property contains in ABC Ontology Model; (b) iSituation, subClassOf: Situation. It expresses the starting/ending state of iEvent, and describes states of rele-

vant objects associated in iEvent. Because iEM concentrates on the description of details in one integral event, iSituation is associated only twice through beginWith and endWith properties for the same iEvent, which is unlike to a series of Situation transformations represented by ABC:precedes and ABC:follows for one ABC:Event. (c) iEventAction, subClassOf: Action. It expresses an action participating in an iEvent. iEventAction associates with agents which perform actions through hasParticipant and hasPatient in ABC Ontology Model.

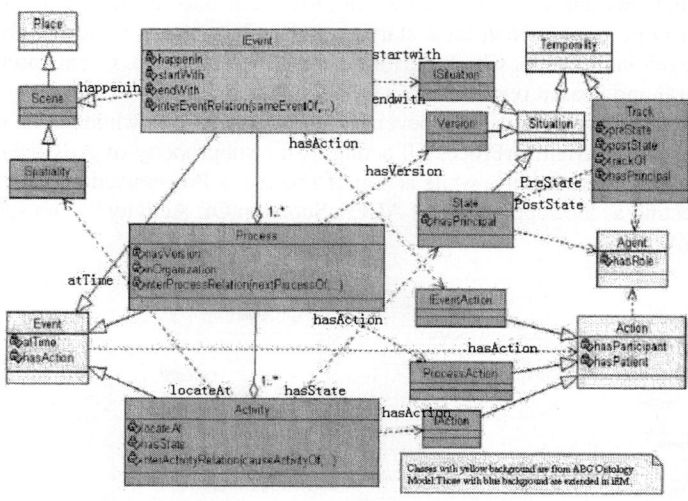

Fig. 4. Demonstration of Inheritage Relationship between iEM and ABC Ontology Model

(2) Process Level: (a) Process, subClassOf: Event. It expresses independent streamlines included in each organizations of iEvent. Similar to iEvent, the existence of Process needs to associate its Actions. Situations of a Process only care those objective environments that are changed by Process, which are called Versions in iEM. Version is associated to its Process through hasVersion property; (b) Version, subClassOf: Situation. It provides set of object states that are changed by corresponding Process in the organization. Different from the base class Situation, Version does not include states of organization-related objects that do not experience certain changes during Process. Thus each Process has its own Version that is linked by hasVersion, while Situations in ABC Ontology Model keep the changes of states through precedes and follows. The existence of Version must have corresponding set of States included; (c) ProcessAction, subClassOf: Action. It expresses Action of Process. Properties hasParticipant and hasPatient of ProcessAction can associate to Agents at that level.

(3) Activity Level: (a) Activity, subClassOf: Event. It expresses activities included in Process. Activity also has all Actions participating in it included to express an entire concept. (b) State, subClassOf: Situation. It expresses Situations that are changed by iAction. Each iAction is associated to its State through hasState property. Set of

States of all iActions in a Process form the main source of Process' Version; (c) IAction, subClassOf: Action. It expresses the single action contains in its Activity, or actually one action participating in above Process's behavior.

3.2 Inter-organizational Feature of iEM

Organization Model in iEM is one of the bases for iEM's inter-organizational feature (as shown in Figure 5). Event dispositions need many participants, which belong to many different departments of different organizations. Generally, participants act as certain roles. For example, credit investigator of bank is only a role. The actual person for the role may be Mike today, Jack tomorrow. The introduction of role improves the flexibility of event disposition. An agent can act as multiple roles.

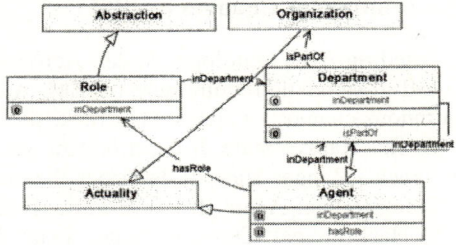

Fig. 5. Demonstration of iEM's Organization Model

Usually a participant belongs to certain department of certain organization. An organization may include multiple departments. Similarly, each department may include multiple sub-department. A department may have multiple roles. (1) Organization, subClassOf: Actuality, which is introduced into iEM. It describes organizations participating in iEvent. These organizations are corresponding to relevant processes, so they mark the organization feature of Processes. An Organization contains multiple Departments. (2) Department, subClassOf: Agent, which is introduced into iEM. It describes departments participating in event disposition. A Department may contain sub-Departments. Especially, principals of ProcessAction, also the domain of hasParticipant is a Department in an Organization included in a Process. That is, Agent at this level is represented by Department. (3) Agent is cited directly from ABC Ontology Model. In iEM, an Agent belongs to certain Department, and has one or more roles. (4) Role, subClassOf: Abstraction, which is introduced into iEM. It expresses the current role of an Agent when it participates into an event. It is associated to its Agent by Property hasRole. (5) InOrganization, Domain: Process Range: Organization. It associates Process to Organization it belongs to. (6) HasRole, Domain: Agent Range: Role. It describes Agent's role when it participates in iAction.

Each Process is related to an Organization, which shows that Activities contained in a Process are all performed in one Organization. An iEvent thus can have multiple processes, which naturally represented the inter-organizational feature of iEM (see Figure 6). In an iEvent, an Organization may have many Processes ordered by time.

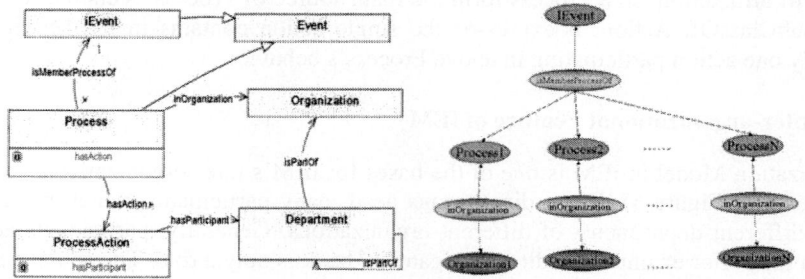

Fig. 6. Demonstration of iEM 's Inter-Organizational Feature

3.3 Ability to Describe Continuous Changes by iEM

As shown in Figure 7, based on ABC Ontology Model, Activity does some class and property extension in need of semantic requirement of iEM. A class Track is supplemented to describe continuous changes of states: (1) Track, subClassOf: Temporality; (2) PreState, Domain: Track Range: State. It describes the starting state of a Track; (3) PostState, Domain: Track Range: State. It describes the ending state of a Track; (4) TrackOf, Domain: Track Range: Agent. It associates a Track to its Agent which the Track describes; (5) HasPrincipal, Domain: State, Track Range: Agent. It describes the principals of State and Track. As for Track, this property acts as the inverse property of TrackOf.

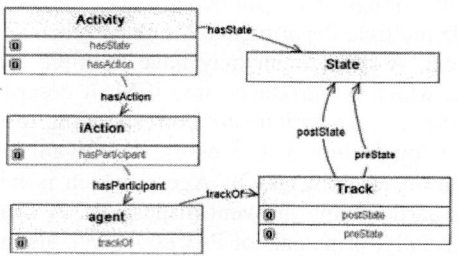

Fig. 7. Demonstration of Class Track in iEM

Track is the extension to Situation in ABC Ontology model. Because Situation in ABC Model can only describe discrete changes of an event, i.e., states before and after the event, while as for an inter-organizational event, continuous details between 2 states are required, a special class Track should be introduced. We cite this concept in Activity level. For example, in case of a fire control schedule, set-off of fire engines from their stations and reaching at the fire spot are 2 separate states. The continuous trails of these engines should be recorded through a Track. Track associates to its starting state and ending state by Property preState and postState respectively.

An Example – iEM-Based Emergency Event Model

Joint emergency response system (JERS) is a large-scale spatio-temporal system which integrates all sorts of emergency service resources and majors its features as common codes used for public emergency events reporting, integrated alarm reception and instruction and citizen oriented emergency service. JERS is a typical inter-organizational event disposition instance. A JERS case is "At 09:45, Sep 15th, 1998, joint emergency center of city A received the alarm from a citizen who reported that a heavy-duty truck hit to residential building located on No.1, Street B, No.40, Road C at 09:40. Due to the occurrence of traffic accident, the Center requires the sending of traffic police. While later it was also reported that due to the same accident, the involved building suffered a serious fire with great casualty. The committed truck was one stolen while the conductor – the thief had run away. The integration scheduler then commands the fire-control scheduler, the medical first-aid scheduler and the police scheduler to perform their respective responsibilities. What made the worse, some reported that electrical wires at No. 30 Road D were disjoined."

In the following, joint response is taken as the example to illustrate event disposition, as shown in Figure 8.

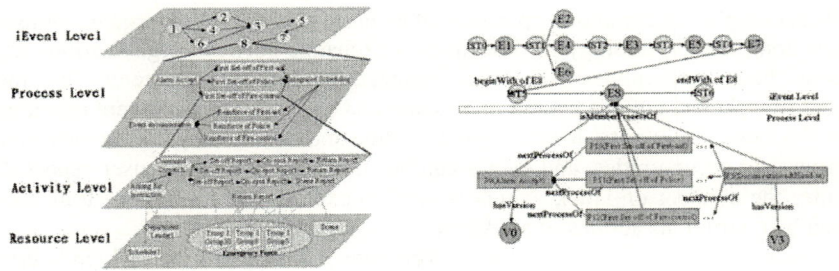

Fig. 8. Demonstration of Example Business   **Fig. 9.** Demonstration of iEM-based Instance(1)

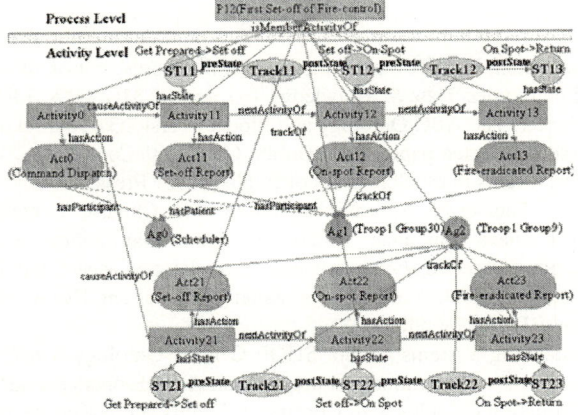

Fig. 10. Demonstration of iEM-based Instance (2)

(1) IEvent level: Joint disposition of emergency events have business processes from police stations, fire-control stations and governments involved. In Figure 9, up-to-date events in the emergency center include E1, E2, E3, E4, E5, E6, E7 and E8. E1 happens before E2, E4, E6; E3 happens after E1; E5 happens after E3; E7 happens after E5; the above hypothetical event E8 happens after E7. (2) Process Level: Organizations and related business processes in disposition of E8 include processes: alarm incept, documentation and hand-in of emergency center; first set-off, reinforcement of police scheduling center; first set-off, reinforcement of first-aid scheduling center; first set-off, reinforcement of fire-control scheduling center; (3) Activity Level: An Activity proceeded by an Agent is the minimal unit consisting a Process. As shown in Figure 10, Process "First Set-off" includes Activities as Act0, Act11, Act21, Act13, Act21, Act22, Act23.

5 Conclusion

iEM are composed of iEvent, Process and Activity. An event can be expressed from different granularity. These 3 levels contain events all in ABC model, except from different levels. They are interrelated, and should be used as a whole to describe integral events. iEM provides inter-organizational oriented event model built on different levels, accomplishes common expression of events among multiple organizations, and defines common terminology for information exchange between organizations. It can describe basic streamlines for inter-organizational events concretely by multi-granularity oriented complex event expression and specialization of event relationship under the same granularity. Multi-leveled, outline and detail compatible model fits for different user communities, and events can be shown from multiple user-aspect. IEM provides the concept of Track, which is used to express the continuous trail accomplished by continuous actions of an Agent.

We have designed iEM-based JERS information exchange format, which is used in the implementation of integration and coalition for City of Tianjin, China.

References

1. Jones, D.M. and Paton, R.C: Toward principles for the representation of hierarchical knowledge in formal ontologies, Data & Knowledge Troopering, 31(2): 99-113(1999).
2. C. Lagoze, J.Hunter: The ABC Ontology and Model (Version3), Journal of Digital Information, Special Issue - selected papers from Dublin Core 2001 Conference.
3. Terence Parsons: Events in the Semantics of English, MIT Press, 1990.
4. Qianduan Chen: Four-level Appearance System for Chinese Language, International Proseminar of Chinese Language System (Shanghai International Studies University), http://ling.ccnu.edu.cn/message/yyxlwx/chenqianrui_4storey_aspect.doc, 2003.
5. Kate Beard-Tisdale: A Review of Perspectives on Processes and Events, http://www.spatial.maine.edu/~bdei/BDEIevent.pdf, 2002.
6. Parallel Understanding Systems Group: SHOE General Ontology, University of Maryland at College Park, http://www.cs.umd.edu/projects/plus/SHOE/onts/general1.0.html, 2002.
7. McCarthy, J.: Situation Calculus with Concurrent Events and Narrative, http://www.formal.stanford.edu/jmc/narrative/narrative.html, 2000.

Secure XML Aware Network Design and Performance Analysis

Eui-Nam Huh[1], Jong-Youl Jeong[2], Young-Shin Kim[1], and Ki-Young Mun[3]

[1] Seoul Women's University, Division of Information and Communication, Seoul, Korea
[2] National Computerization Agency, BcN Team IT Infrastructure Division, Seoul, Korea
[3] Electronics and Telecommunication Research Institute,
Information Security Research Division, Daejun, Korea
{huh, amary46}@swu.ac.kr

Abstract. Currently, XML as a traffic type on the Internet is widely appeared one-commerce applications rather than HTML. XML based Denial of Service (XDOS) attacks are growing up tremendously. This paper presents a novel approach to manage XML attacks at the network layer efficiently and improves service performance on server side, while XML data is visible at the application layer. Thus it is clear that the server overhead becomes significant if a number of encrypted, signed, and malformed XML data are requested to the server. The proposed approach handles these issues efficiently and securely. The experiments show that the proposed XML Aware Network (XAN) platform is a necessary component for efficient Web Services.

1 Introduction

The use of e-commerce using Internet is increased unexpectedly and so the transaction is now digitalized. The e-commerce security must be kept from illegal transactions, issued privacy, unknown user's resource access and denial of service. Thus encryption for privacy, signed message disposition notification (MDN) for non-repudiation, and digital signature for authentication and integrity are the most important procedures in XML based e-commerce.

Recently, new standardization technology for the extended e-commerce has been developed such as ebXML, RosettaNet, and Web Services. The ebXML enhances basic XML security and authentication technology for e-commerce, and uses a basis of the standardized XML encryption and digital signature. In Web Services (WS), WS-Security is standardized to encrypt and sign the SOAP (Simple Object Access Protocol) message. In addition, WS-Security Policy, WS-Security Conversation, WS-Trust, and WS-Federation are standardized to support secure and scalable Web Services oriented e-business. The XML digital signature, XML encryption, SAML(Security Assertion Markup Language), XKMS(XML Key Management Specification) and XACML(XML Access Control Markup Language) are foundation class of those specifications. The distributed computing technology now like Grid enhances Web services technology to deploy of various field applications.

Hence, information technology will make a convergence of standardized services listed above for secure e-business and uses high performance network infrastructure for a convenient access to e-business services, which uses conventional XML based technologies.

However, XML oriented data should be handled on nodes efficiently in terms of performance, security, and standardized manner. Thus, this paper presents a novel approach to manage XML data including attacks and wrong format at the network layer efficiently and improves service performance on server side, while XML data is visible at the application layer. So this paper focuses mainly on the performance and security of the service provider domain in addition to consideration of the standardized manners.

2 Analysis of XML Based e-Business

2.1 XML Characteristics

Currently, many acceleration technologies for Web traffic such as HTTP load balancer, content cache equipments, SSL accelerators are developed but there are no capability of handling XML traffic. The XSLT(eXtensible Stylesheet Language Transformation) is designed to convert XML to HTML format, but the processing speed is too slow to serve efficiently. As shown in Fig. 1, transactions per second (denoted to TPS) and processing time (denoted to Latency) with data intensive XML document as shown in Fig. 1 becomes very small when XSLT is applied to server side.

Benchmark Run	TPS	Latency (mSec)
Direct XML to browser	45	1102
Server Convert, XSLTC	2	23300

Fig. 1. XML Processing Benchmark

As shown in Fig. 1, it is clear that secure XML traffic enhanced with encryption and digital signature technology will consume more CPU power. With above results, the secure e-business faces to performance problem. In addition to the above problem, there are more important issues as follows:

- SOAP (Simple Object Access Protocol) message spoofing occurs when the 'actor' element in SOAP message used to get authentication from URI value with ID and password is spoofed by malicious users.
- XML Denial of Service (XDOS) attacks may exist when malformed DTD or XSD, for instance, infinite loop, is delivered to the server side frequently.
- Application threat may exist in the case of wrong URI in XML or SOAP message indicating resource reference.
- Many resources are consumed to process secured Web Services messages as shown in Fig. 2. Given 1 unit to the "parsing" step, overally 25 units for one way XML processing are needed. (see [2])

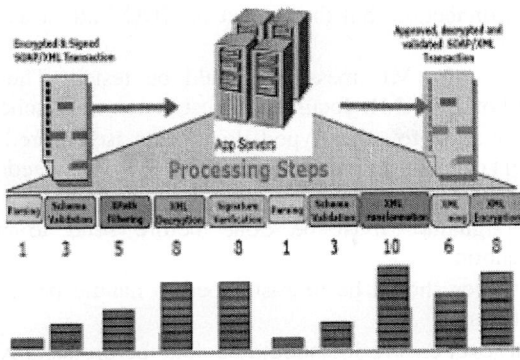

Fig. 2. The XML Processing Overhead

2.2 Related Work

Some XML products enabling digital signature and encryption are developed by IBM (Alpha Works) and Baltimore (X/Secure) for the secure e-business. The Alpha Works is a license free software, providing XML digital signature and its examples followed by the standardized specification [5]. Recently, Apache Group developed the digital signature module for web server called Apache-XML-Security. The .NET framework by Microsoft corporation integrates the XML digital signature for the secure Web Services platform [6]. The X/Secure of Baltimore company (see [6][7]) developed in part according to the standardized specification published by W3C and IETF [8]. The XML encryption as a recommendation standard is published by W3C [9]. Thus basic functions of XML based e-business for privacy, integrity, authentication, and non-repudiation are implemented using

3 XAN S/W Platform Design

In this chapter, firstly, ten design requirements of XML based e-business are discussed as the security in XML Web Services becomes more critical issues. Following lists are carefully considered for secure e-business s/w platform at the design time.

① Transport layer should be secure. Mostly SSL/TLS and VPN technology is used for the requirement and additionally users' certificates are also handled in the manner of hardware processing.
② XML data should be filtered. The content of XML document may contain DOS attack, so deep data inspection is required.
③ Internal resource should be masked. Like NAT(Network Address Translation) and Proxy server in IP service , a technology for XML Proxy is also required to secure internal domain resources.
④ Gateway against XDOS attacks is required. The number of attacks in XML is expected smaller than the number of the "sync flooding" attack

in TCP connection, but the impact of XDOS attack is much bigger than of it.

⑤ Validation of XML message should be tested. The element and the structure of any XML documents must match to its schema or DTD.

⑥ Transformation to other typed documents is required as user interface (browser) capacity might be different, so XSLT is needed as a component in the server side.

⑦ Digital signature must be done before send to inscribe the user identification.

⑧ All messages should be time-stamped to handle precisely the enterprise transactions.

⑨ Element level encryption should be done before send XML data as XML contains structure or meaning of value, for example, "<SSN>123-343-5678 </SSN>" exposure the social security number.

⑩ Finally, auditing system is a required component to audit a user and messages by analysing system logs, which may different from conventional logging approaches as XML signature includes the time-stamp.

Thus, considering the listed requirements, we design a following XAN platform consisting two main components, XAN-Sec (Security) and XAN-Acc (Accelerator) as shown in Fig. 3. XAN-Sec plays an important role in decryption and digital signature, and XAN-Acc does parsing, schema validation, transformation, and grammar validation.

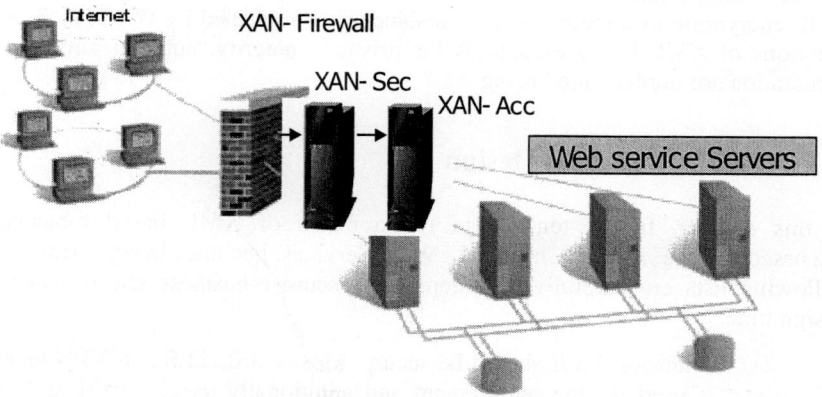

Fig. 3. XAN Platform Architecture

To design XAN-Sec and XAN-Acc platform, we used PCAP (Packet Capture) driver and modified TCP/IP stack modules. As shown in Fig. 4, XML document is now processed between the data link layer and the network layer.

Our detail process steps to handle the XML document are illustrated in Fig. 5. The XAN-Sec decrypts the encrypted XML delivered from clients and passes to the signature validation step. In each step, if errors are detected, then stores the client information, error information, the step name, and the document information to DB, which is used to compare, filter and reject the wrong documents that appeared repeatedly.

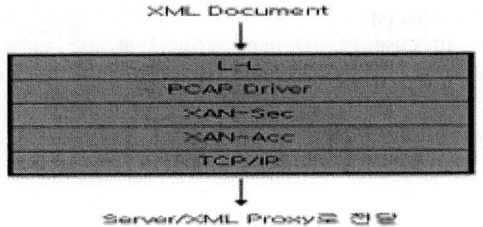

Fig. 4. XAN Flow Structure

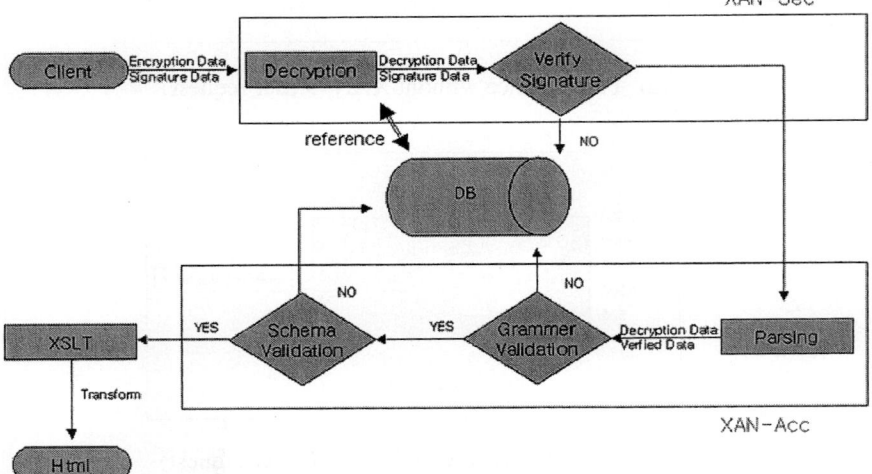

Fig. 5. XAN Processing Steps

In the XAN-Acc component, parse XML data part, test grammar validation using XPath and validate it with schema. The XSLT step converts XML to HTML. When all steps are completed without errors, the decrypted, validated document is routed to the Web Service Provider. If any error occurs in the step, the client information and the document are stored to DB.

4 Performance Analysis

We used two Pentium 4 personal computers to experiment XAN-Sec and XAN-Acc. The four types of XML documents with few elements by 4 to 10 are used. We test 15 times for each experiment to get the average performance.

The first experiment tested with single XML document to know whole processing time of XML. It takes 240msec in encryption and sign validation, and 120msec in parsing, grammar validation and schema validation. We observed performance in case of 5 user request at the same time. It takes 1800msec totally.

In the second experiment, we used separated XAN-Sec and XAN-Acc as a pipelined distributed processor. In case of 10 user request at the same time, 1750msec is measured. The Fig. 6 (a) and (b) shows the two experiments' results mentioned above.

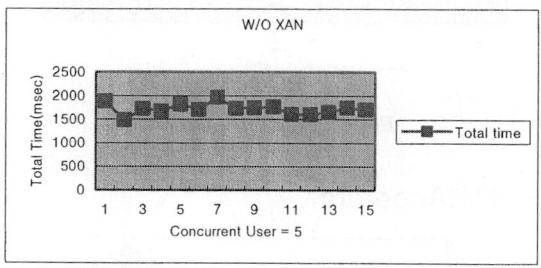

(a) Performance without XAN (5 user request)

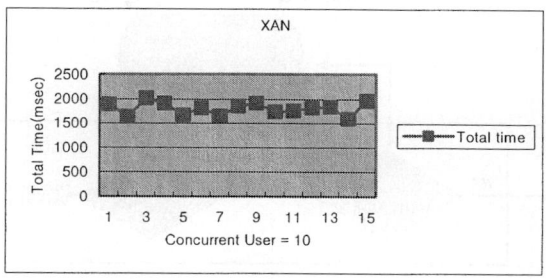

(b) Performance with XAN (10 user request)

Fig. 6. Effective Performance Comparison of XAN

Between the XAN-Sec and the XAN-Acc from the previous two experiments, performance in each component is different, which means load is not balanced. Thus we model performance of XAN to the time chart as shown in Fig. 7. We analyze in detail performance of XAN-Sec and XAN-Acc and found better model in order to fully utilize computing power to handle more documents. The first row with numbered in a rectangle indicates time(msec), and the second row illustrates the processing time of XAN-Sec. The third row indicates the processing time of XAN-Acc. There is many idle time (depicted to white space) in XAN-Acc as shown in Fig. 7.(a). Therefore the Fig. 7 (b) and (c) are proposed to utilize XAN-Acc fully.

The next experiment shows that the performance of the proposed models in Fig. 7. This experiments allows 10 concurrent users to request XML document. When not any optimized or load balance module applied, 1750 msec is required to handle the 10 XML documents. As shown in Fig. 8 (a), the Best-Fit model performs better than any other models. It takes 1600 msec for 10 XML documents, while 1900msec in the round-robbin model is required. From this result, the more user requests in the round-robbin model, the more waiting time occurs as the XML document need to process in sequence.

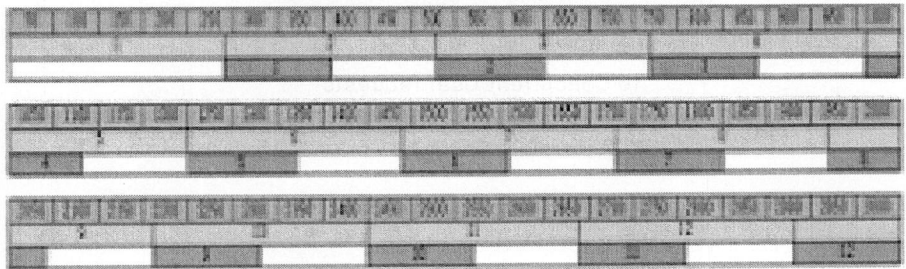

(a) Performance of XAN-Sec and XAN-Acc

(b) Performance of Round-Robbin applied to XAN

(c) Performance of Best-Fit applied to XAN

Fig. 7. Performance Time Chart of XAN-Sec and XAN-Acc

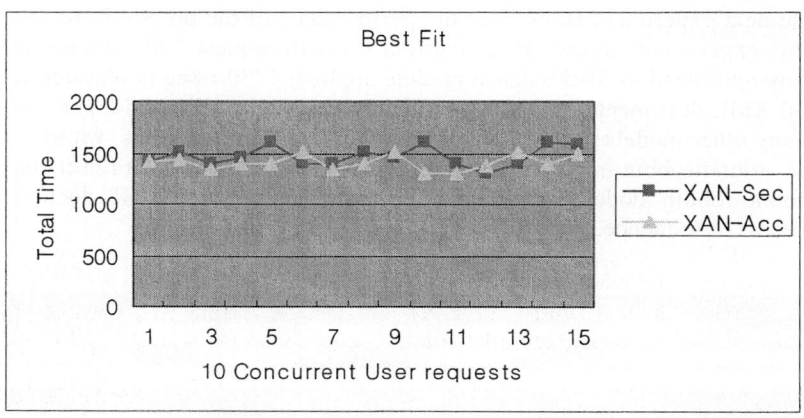

(a) Performance of Best-Fit XAN Model

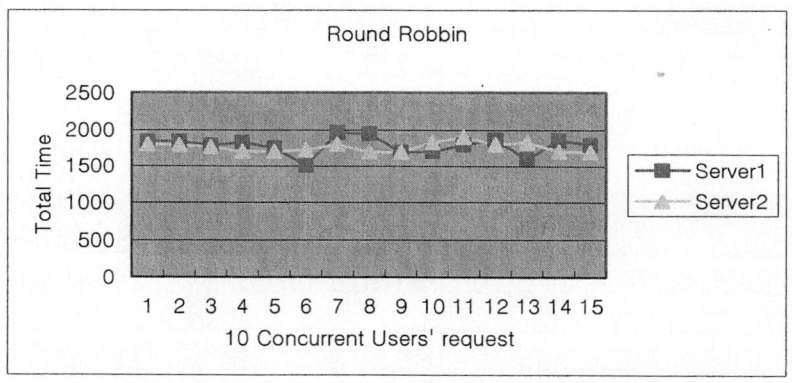

(b) Performance of Round-Robbin XAN Model

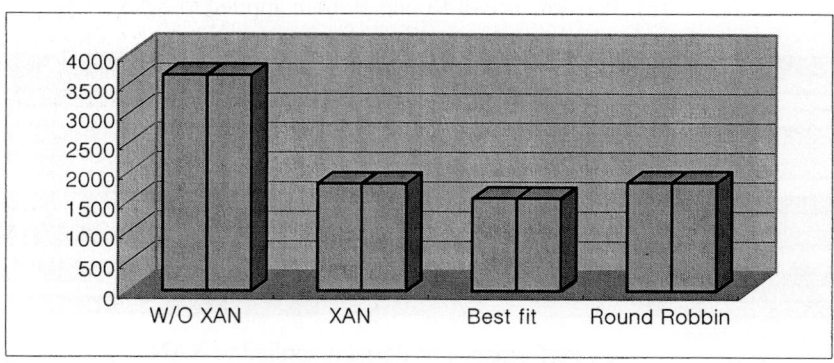

(c) Overall Performance Comparison

Fig. 8. Performance Improvement of XAN

5 Conclusion

This paper discusses many important issues in XML to enable the Web Services securely and efficiently. Especially, this paper focus on performance issues in XML based e-business by developing XML Aware Networking (XAN) platform S/W, which can diagnose and improves server side performance. Soon, in generation of ubiquitous, XML is widely used as a standard content format on many embedded platforms, However, still performance is a big question to handle XDOS attack and malformed XML request like SOAP message requiring deep content inspection. Therefore, this developed S/W will give experimental testing platform for Web Services. Also, in near future, XML aware processor might be on the market. The study to enhance speed of network equipment and processor will be continued.

References

[1] The World Wide Web Consortium(W3C)'s XML web page; http://www.w3.org/XML/
[2] Rick McGuir,. XML Acceleration Appliances. Emerging Internet Technologies, IBM Software Group. November 5, 2003
[3] DataPower web page; http://www.datapower.com
[4] XML Security Gateway web page:http://www.datapower.com/products/xs40.html
[5] IBM AlphaWorks Homepage, http://www.alphaworks.ibm.com/tech/xmlsecuritysuite
[6] Baltimore, "X/Secure White Paper," http://www.baltimoreinc.com/library/whitepapers/xsecure.html
[7] Baltimore, "X/Secure Developer's Guide," 1999.
[8] IETF/W3C, XML-SignatureRequirements (WorkingDraft)," Oct.1999, http://www.w3.org/ TR/1999/WD-xmldsig-requirements-19991014.html
[9] W3C XML Encryption WG, "XML Encryption Charter," http://www.w3.org, 2001.
[10] IETF/W3C, XML-Signature Syntax and Processing(Working Draft), Oct. 2000, http://www.w3.org/TR/2000/WD-xmldsig-core-20001012/
[11] xml-encryption@w3.org Mail Archives, http://lists.w3.org/Archives/Public/ xmlencryption/1

A Probe Detection Model Using the Analysis of the Fuzzy Cognitive Maps

Se-Yul Lee[1], Yong-Soo Kim[2], Bong-Hwan Lee[3], Suk-Hoon Kang[2],
and Chan-Hyun Youn[4]

[1] Department of Computer Science, Chungwoon University,
San 29 Namjang-Ri, Hongseong-Eup, Hongseong-Gun, Chungnam, 350-701 Korea
Pirate@cwunet.ac.kr
[2] Division of Computer Engineering, Daejeon University,
96-3 Yongun-Dong, Dong-Gu, Daejeon, 300-716 Korea
{kystj,shkang}@dju.ac.kr
[3] Department of Information & Communications Engineering, Daejeon University,
96-3 Yongun-Dong, Dong-Gu, Daejeon, 300-716 Korea
blee@dju.ac.kr
[4] School of Engineering, ICU,
119 Munjiro, Yuseung-Gu, Daejeon, 305-732 Korea
chyoun@icu.ac.kr

Abstract. The rapid growth of network-based information systems has resulted in continuous research of security issues. Intrusion Detection Systems (IDS) is an area of increasing concerns in the Internet community. Recently, a number of IDS schemes have been proposed based on various technologies. However, the techniques, which have been applied in many systems, are useful only for the existing patterns of intrusion. They can not detect new patterns of intrusion. Therefore, it is necessary to develop a new IDS technology that can find new patterns of intrusion. Most of IDS sensors provide less than 10% rate of false positives. In this paper, we proposed a new network-based probe detection model using the fuzzy cognitive maps that can detect intrusion by the Denial of Service (DoS) attack detection method utilizing the packet analyses. The probe detection systems using fuzzy cognitive maps (PDSuF) capture and analyze the packet information to detect SYN flooding attack. Using the results of the analysis of decision module, which adopts the fuzzy cognitive maps, the decision module measures the degree of risk of the DoS and trains the response module to deal with attacks. For the performance evaluation, the "IDS Evaluation Data Set" created by MIT was used. From the simulation we obtained the average true positive rate of 97.094% and the average false negative rate of 2.936%.

1 Introduction

The rapid growth of network in information systems has resulted in the continuous research of security issues. One of the research areas is IDS that many companies have adopted to protect their information assets for several years. In order to address the security problems, many automated IDS have been developed. However, between 2002 and 2004, more than 100 new attack techniques were created and published which exploited Microsoft's Internet Information Server (IIS), one of the most widely used web servers. Recently, several IDS have been proposed based on various technologies. A "false positive error" is an error that IDS sensor misinterprets one or more normal packets or activities as an attack. IDS operators spend too much time on distinguishing events. On the other hand, a "false negative error" is an error resulting from attacker is misclassified as a normal user. It is quite difficult to distinguish intruders from normal users. It is also hard to predict all possible false negative errors and false positive errors due to the enormous varieties and complexities of today's networks. IDS operators rely on their experience to identify and resolve unexpected false error issues.

Recently, according to the CERT-CC (Computer Emergency Response Team Coordination Center), hacking is increasing about 300% each year. A variety of hacking techniques are known: DoS, Buffer Overflow Attack, Probe Attack, Vulnerability Scan Attack and others. Among them, Vulnerability Scan Attack and Probe Attack are the two most frequently used methods. Port scan or vulnerability of network as abnormality intrusion of network is based on anomaly probe detection algorithms such as *scanlogd* [14], RTSD (Real Time Scan Detector) [15], and Snort [16]. Such open source programs have some problems in invasion probe detection. That is, *scanlogd* and RTSD can not detect slow scan, while Snort does not provide open port scan. Therefore, a new algorithm that can provide slow scan and open port scan is required.

The main objective of this paper is to improve the accuracy of intrusion detection by reducing false alarm rate and minimize the rate of false negative by detecting unexpected attacks. In an open network environment, intrusion detection rate is rapidly improved by reducing false negative errors rather than false positive errors. We propose a network based probe detection model using the fuzzy cognitive maps that can detect intrusion by the DoS attack detection method. A DoS attack appears in the form of the probe and SYN flooding attack, which is a typical example. The SYN flooding attack takes advantage of the vulnerable three-way handshake between the end-points of TCP [3-5, 7]. The proposed PDSuF [13] captures and analyzes the packet information to detect SYN flooding attack. Using the results of detection module, which utilizes the fuzzy cognitive maps, the detection module measures the degree of risk of the DoS and trains the response module to deal with attacks [6, 7].

The rest of this paper is organized as follows. The background and related work is summarized in Section 2. Section 3 describes the proposed new PDSuF model. Section 4 illustrates the performance evaluation of the proposed probe detection model. Conclusions and future work are presented in Section 5.

2 Related Work

Previous studies of DoS attack detection can be divided into three categories: attack prevention, attack source trace-back and attack identification, and attack detection and filtering. Attack prevention obviously provides avoidance of DoS attacks. With this method, server system may be securely protected from malicious packet flooding attack. There are indeed known scanning procedures to detect them based on real experience [1-2]. Attack source trace-back and identification is to identify the actual source of packet sent across network without replying to the source in the packets [8]. Attack detection and filtering are responsible for identifying DoS attacks and filtering by classifying packets and dropping them [10]. The performance of most of DoS detection is evaluated based on false positive error and false negative error. The detection procedure utilizes the victim's identities such as IP address and port number. Packet filtering usually drops attack packets as well as normal packets since both packets have the same features. Effectiveness of this scheme can be measured by the rate of the normal packet which is survived in the packet filtering. Among these schemes, attack prevention has to recognize how DoS attack is performed and detect attack pattern using predefined features [12]. Therefore, when a new attack detection tools are developed, new features that detect the pattern of attack needs to be defined. Current IP trace-back solutions are not always able to trace the source of the packets. Moreover, even though the attack sources are successfully traced, stopping them from sending attack packets is another very difficult task.

DoS attacked traffic is quite difficult to distinguish from legitimate traffic since packet rates from individual flood source are usually too low to catch warning by local administrator. Thus, it is efficient to use inductive learning scheme utilizing the Quinlan's C4.5 algorithm approach to detect DoS attack [11]. Inductive learning systems have been successfully applied to the intrusion detection. Induction is formalized by inductive learning using decision tree algorithm which provides a mechanism for detecting intrusion. The key idea of this approach is to reduce the rate of false errors. The false error rates of the known intrusion detection schemes are summarized in Table 1.

As shown in Table 1, FSTC (False Scan Tool and Clustering) provides the largest false negative error, while the Fuzzy ART scheme provides the smallest false negative errors and the largest false positive error. In the meantime, Inductive Learning System provides moderate false negative and false positive error on the average. From the above results, it is highly recommended to develop a new DoS detection scheme based on fuzzy cognition.

Table 1. False errors of IDS [2]

Methodology	False Negative Error	False Positive Error
FSTC	22.65%	20.48%
Inductive Learning System	9.79%	9.10%
K-Means (Average Value)	9.37%	20.45%
Fuzzy ART ($\rho = 0.9$)	6.03%	38.73%

3 PDSuF Model

3.1 PDSuF Algorithm

The PDSuF model is a network-based detection scheme that utilizes network data to analyze packet information. Based on the analysis of each packet, probe detection is performed. In order to determine intrusion detection, various features of packet is utilized including source IP address, source port number, destination IP address, destination port number, flags, data size, timestamp, and session pattern as given by (1).

Packet X = (src_ip, src_port, dst_ip, dst_port, flag, data, timestamp, pattern,…) (1)

Now it is needed to quantize each feature parameter based on comparison criterion to determine attack detection. The procedure to assign effect values can be summarized as follows.

[state 1] *Feature Equality*

$$FE(x) = \begin{Bmatrix} 0(x \neq a) \\ 1(x = a) \end{Bmatrix}$$

a: standard, x: comparison

[state 2] *Feature Proximity*

$$FP(x) = \frac{k}{|x-a|}$$

a: standard, x: comparison, k: constant

[state 3] *Feature Separation*

$$FS(x) = k|x-a|$$

a: standard, x: comparison, k: constant

[state 4] *Feature Covariance*

$$FC(x, y) = |\text{cov}(x(t), y(t))|$$

x, y: comparison, t: time, cov(): degree of dispersion

[state 5] *Feature Frequency*

$$FF(x) = \log_2 \frac{1}{\Pr(x)}$$

$\Pr(x)$: x's probability

Using the above state variables, the total degree of abnormality for a packet can be calculated as in (2).

$$A_{total}(x) = \omega_1 A_1 + \omega_2 A_2 + ... + \omega_n A_n$$

$$= \sum_{i=1}^{n} \omega_i A_i \qquad (2)$$

$A_{total}(x)$: Abnormality per packet
ω_i : Weight value of packet
A_i : Abnormality of packet
n : Total feature number of abnormality

If the total degree of abnormality for a packet is greater than the threshold of attack attempt, the associated packet is classified as abnormal.

3.2 PDSuF Architecture

The PDSuF architecture consists of network-based intrusion detection system and monitoring tool as shown in Fig. 1 [5, 9]. As monitoring tool, a protocol analyzer is used, whereas the detection system is directly connected to the router, which interconnects LANs. The PDSuF algorithm is obviously implemented on the detection system.

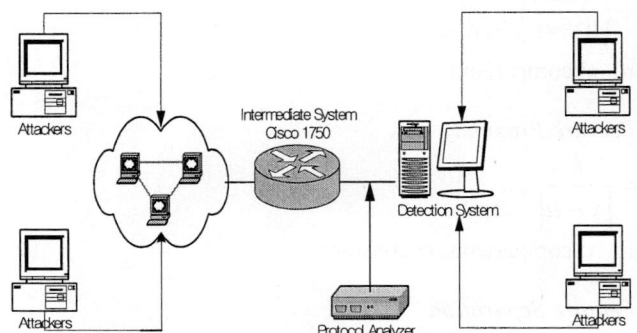

Fig. 1. PDSuF architecture

The detection module of the PDSuF is intelligent and uses causal knowledge reason in fuzzy cognitive maps. Fig. 2 shows the detection module using variable events that are mutually dependent. In detection module of Fig. 2, an optimal detection is

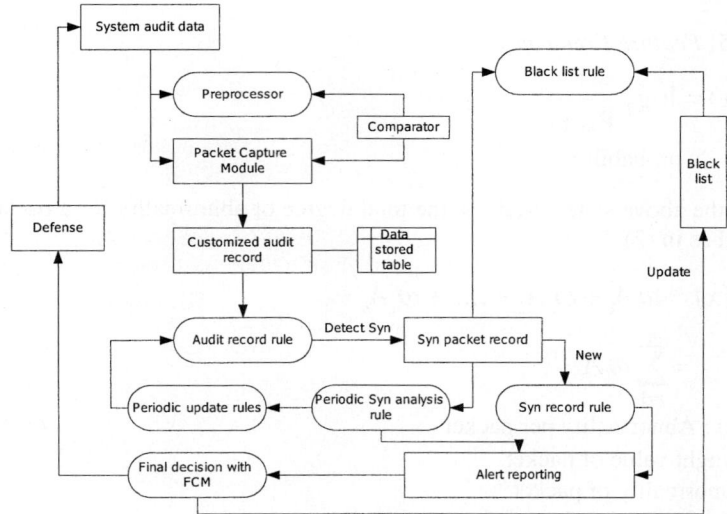

Fig. 2. Flowchart of detection module

provided by giving dependency to some events among several variable events. In addition, regarding the detected IP address as a probe, the detection module decides whether to save the IP address to the black list or not. The weight is the effect value of path analysis calculated using quantitative Micro Software's Eview Ver. 3.1. Fig. 3 shows the details of fuzzy cognitive maps (FCM) in Fig. 2. As the variable events dependent on the detection module, we can set the identity of IP address, the time interval of half-open state, the rate of CPU usability, the rate of memory, and SYN packet. For example, the weight between the two nodes is bigger than 0 since the rate of CPU usability increases in proportion to the size of SYN packet. In Fig. 3, each rectangular box represents feature event, while each number denotes effect value in FCM.

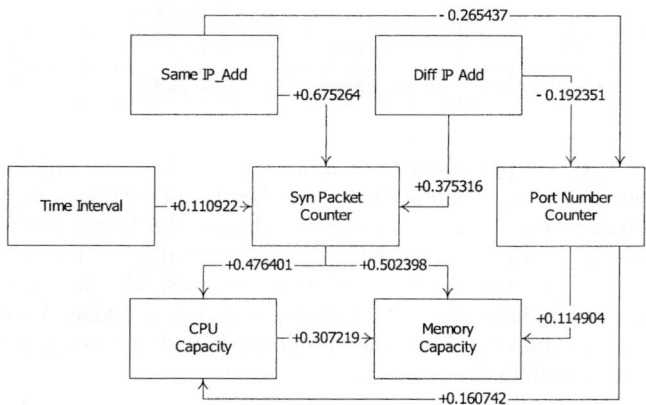

Fig. 3. Path model of the FCM

4 Performance Evaluation

For the performance evaluation of the proposed PDSuF model, we have used the KDD data set (Knowledge Discovery Contest Data) by MIT Lincoln Lab, which consists of labeled data (training data having SYN flooding and normal data) and non-labeled data (test data). Since the TCP SYN flooding attacks come from abnormal packets, detection of abnormal packets is similar to detection of SYN flooding attacks in TCP networks.

The best detection and false error rates are summarized in Table 2. The simulation results for the connection records of DoS attacks are collected for 10 days. The average rate of true positive is measured of 97.064%. According to the KDD'99 competition results, the best rate of the Bernhard's true positive is known as 97.1% [11]. Comparing Bernhard's true positive rate with that of PDSuF, we realized that the result of PDSuF is as good as Bernard's. In addition, the false negative rate of the proposed scheme, 2.936%, is considerably smaller than that of the Bernhard's, 3.91%.

Table 2. Best detection and error rates

Day	True Positive	False Positive	True Negative	False Negative
Day 1	95.623%	0.000%	100.000%	4.377%
Day 2	87.861%	0.000%	100.000%	12.139%
Day 3	96.098%	0.001%	99.999%	3.902%
Day 4	99.569%	0.000%	100.000%	0.431%
Day 5	100.000%	0.000%	100.000%	0.000%
Day 6	98.930%	0.000%	100.000%	1.070%
Day 7	100.000%	0.001%	99.999%	0.000%
Day 8	87.701%	0.000%	100.000%	12.299%
Day 9	100.000%	0.000%	100.000%	0.000%
Day 10	97.917%	0.000%	100.000%	2.083%
Average	97.064%	0.000%	99.999%	2.936%

Fig. 4. illustrates the performance of four different detection algorithms for both DoS and probing. The key difference between PDSuF and the others is that the former is resource based probe detection algorithm, whereas the latters are basically rule-based detection algorithms. Thus, the proposed algorithm is able to detect probe regardless of input patterns and the number of features. The key advantage of the PDSuF over the other alogoritnms is the ability of real-time update of effect values in FCM. Therefore, as shown in Fig. 4, the proposed PDSuF algorithm outperforms the other algorithms in both DoS and probe.

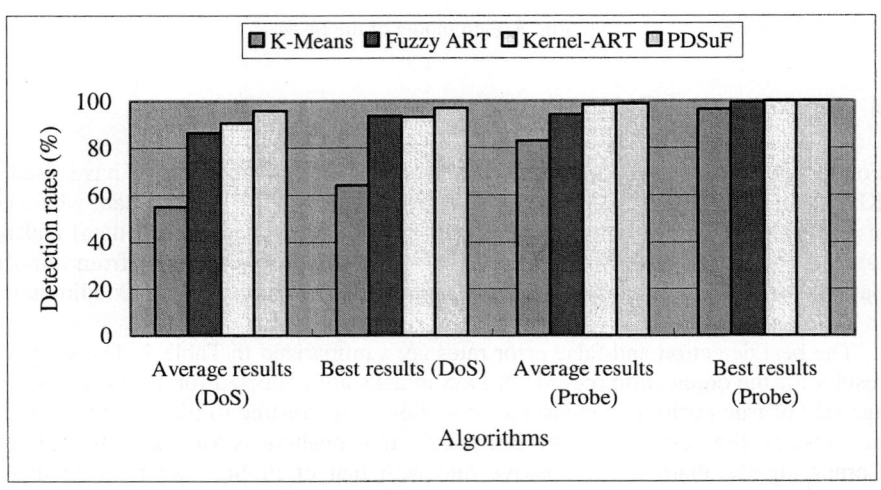

Fig. 4. Detection rates of DoS vs Probe

In order to evaluate the performance from the viewpoint of resource usage, system resource usage of the PDSuF is compared to that of Synkill, which is a well-known SYN flood attack detection tool developed by Purdue University[17]. Fig. 5 shows the system resource usage of both Synkill and PDSuF when DoS attack is applied at 100 seconds and the two detection tools are activated at 200 seconds. Both PDSuF and Synkill take care of the attack from 200 seconds to 350 seconds. In Fig. 5, we can see that resource usage of PDSuF drops drastically at about 250 seconds, while resource usage of Synkill drops rapidly at around 300 seconds. This results from the fact that the attack detection tools detect the attack and discard abnormal packets. Also, Fig. 5 illustrates that the proposed PDSuF outperforms Synkill using less system resources. The main reason that the PDSuF performs better than Synkill is that PDSuF is basically a probe detection scheme which is activated in advance for false errors, whereas Synkill is in operation after the attack, which results in longer time delay.

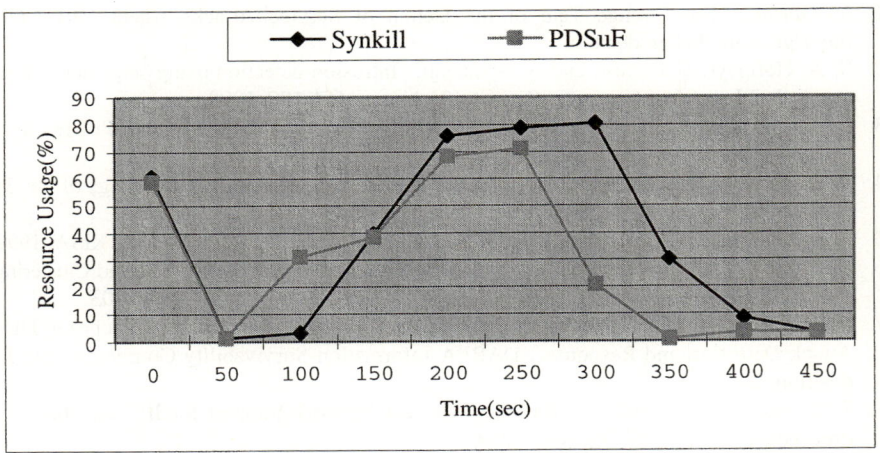

Fig. 5. Comparison of system resource usage

5 Conclusions

In this paper, we proposed a network based intrusion detection model using fuzzy cognitive maps which can detect intrusion by DoS attack. A DoS attack appears in the form of the intrusion attempt. The SYN flooding attack takes advantage of the weak point of three way handshake between the end points of TCP connections. The PDSuF model captures and analyzes the packet information to detect SYN flooding attack. Using the results of the FCM detection module, the detection module measures the degree of risk of the DoS and trains the response module to deal with attacks.

For the performance evaluation of the proposed model, the average rates of the true positive and false negative errors are measured. The true positive error rate of the PDSuF is similar to that of Bernhard's true positive error rate. However, the false negative rate of the proposed scheme is considerably smaller than that of the Bernhard's.

In addition, system resource usage of the PDSuF is compared to that of Synkill, which is a wellc-known SYN flood attack detection. The proposed PDSuF outperforms Synkill in system resource usage and time delay. The better performance results from the fact that the PDSuF is basically a probe detection scheme which is activated in advance for false errors. For further research, the PDSuF detection method needs to be extended to general purpose intrusion detection system.

Acknowledgements

This work was supported by University IT Research Center Project. It is also supported in part by KOSEF under grant No.(R05-2002-000-01008-0).

References

1. S. Gibson, "The Strange Tale of the Denial of Service Attacks Agent GRC.COM" http://grc.com/dos/grcdos.htm
2. S. A. Hofmeyr, S. Forrest, and A. Somayaji, "Intrusion detection using sequences of system calls," Journal of Computer Security, Vol. 6, pp.151-180, 1998.
3. R. Axelrod, "Structure of Decision: The Cognitive Maps of Political Elites," Princeton, NJ: Princeton University Press, 1976.
4. J. Cannady, "Applying Neural Networks to Misuse Detection," In Proceedings of the 21st National Information System Security Conference, 1998.
5. Hongik Univ., STRC, Intrusion Detection System and Detection Rates Report, KISA, 1999.
6. H. S. Lee, Y. H. Im, "Adaptive Intrusion Detection System Based on SVM and Clustering", Journal of Fuzzy Logic and Intelligent Systems, Vol. 13, No. 2, pp.237-242, 2003.
7. L. Feinstein, D. Schnackenberg, R. Balupari, D. Kindred, "Statistical Approaches to DDoS Attack Detection and Response", DARPA Information Survivability Conference and Exposition, 2003.
8. S. Savage., D. Wetherall, A. Karlin., "Practical Network Support for IP Trace-back," In Proceedings of ACM SIGCOMM, 2000.
9. A. Garg, A. L. Narasimha, "Policy Based end Server Resource Regulation," IEEE/ACM Transactions on Networking , Vol. 8, No.2, pp. 146-157, 2000.
10. P. Ferguson, D. Sene, "Network Igress Filtering: Defeating Denial of Service Attacks Which Employ IP Source Address Spoofing," RFC 2827, 2000.
11. W. Lee, S. J. Stolfo., "A Framework for Constructing Features and Models for Intrusion Detection Systems," In Proceedings of the 5th ACM SIGKDD International Conference on Knowledge Discovery and Data Mining, 2000.
12. K.C Chang, "Defending against Flooding-Based Distributed Denial of Service A Tutorial," IEEE Communications Magazine, 2002.
13. S. Y. Lee, "An Adaptive Probe Detection Model using Fuzzy Cognitive Maps", Ph. D. Dissertation, Daejeon University, 2003.
14. Solar, "Designing and Attacking Port Scan Detection Tools", Phrack Magazine, Vol. 8, Issue 53, pp. 13 – 15, 1998.
15. "Real-Time Scan Detector in real time networks," http://www.krcert.or.kr
16. S. Staniford, J. A. Hoagland, and J. M. Mcalerney, "Practical Automated Detection of Stealthy Portscans", http://silicondefense.com/software/spice/index.htm
17. C. L. Schuba, I. V. krsul, M. G. Kuhn., "Analysis of a Denial of Service Attack on TCP, " Proceedings of IEEE Symposium on Security and Privacy, pp. 208 – 223, 1997.

QoS Provisioning in an Enhanced FMIPv6 Architecture

Zheng Wan, Xuezeng Pan, and Lingdi Ping

College of Computer Science, Zhejiang University, Hangzhou, P.R.C, 310027
{chen-hl, xzpan, ldping}@zju.edu.cn

Abstract. Mobility management and QoS provisioning are both key techniques in the future wireless mobile networks. In this paper we propose a framework for supporting QoS under an enhanced "Fast Handovers for Mobile IPv6" (FMIPv6) architecture. By introducing the key entity called "Crossover Router" (CR), we shorten the length of packet forwarding path before the MN completes binding update. For QoS guarantee, we extend the FBU and HI messages to inform the NAR of the MN's QoS requirement and make advance resource reservation along the possible future-forwarding path before the MN attaches to the NAR's link. We keep RSVP states in the intermediate routers along overlapped path unchanged to reduce reservation hops and signaling delays. The Performance analysis shows that the proposed scheme for QoS guarantee has lower signaling cost and latency of reservation re-establishment, as well as less bandwidth requirements in comparison with MRSVP.

1 Introduction

Wireless devices are expected to increase in number and capabilities in the following years. Mobile and wireless access will become more and more popular. Thus Mobile IPv6 (MIPv6) protocol [1] is proposed to manage mobility and maintain network connectivity in the next generation Internet.

However, there are still two problems to be resolved in MIPv6 environment. Firstly, the handover latency and packet loss in basic MIPv6 protocol are not ideal, which raises the need for a fast and smooth handover mechanism. A number of ways of introducing hierarchy into IPv4 as well as IPv6 networks, and realizing the advanced configuration have been proposed in the last few years [2-4]. Secondly, as real-time services grow, the desire for high quality guarantee of these services becomes eager in MIPv6 networks. As we know, two different models are proposed to guarantee QoS in the Internet by IETF: the integrated services (IntServ) [5] and differentiated services (DiffServ) [6] models. However, only IntServ model which uses RSVP protocol [7] to reserve resources can provide end-to-end QoS.

In this paper we propose a scheme for QoS provisioning in an enhanced *"Fast Handovers for Mobile IPv6"* (FMIPv6) architecture [2]. Two enhancements are introduced to improve the performance of basic FMIPv6. To reduce tunnel distance between the *Previous Access Router* (PAR) and the *New Access Router* (NAR), we propose that the *Crossover Router* (CR) intercept packets destined to MN and forward them to the NAR. CR is the first common router of the old path and the new forwarding path. We also use an efficient mechanism to eliminate the long *Duplicate*

Address Detection (DAD) latency. As for QoS guarantee, we use extended FBU and HI messages to inform the NAR of the MN's QoS requirement. Upon receiving the information, the NAR initiates an advance reservation process along the possible future-forwarding path before the MN arrives the NAR's link. Again the CR is used to reduce the length of reservation path.

The rest of the paper is organized as follows. Section 2 presents some related work. Section 3 presents the overview of proposed scheme. Section 4 describes the detailed handover and resource reservation process. Section 5 gives the performance measurement, and Section 6 concludes the paper and presents some areas for future work.

2 Related Work

2.1 Fast Handover for MIPv6

FMIPv6 aims to decrease packet loss by reducing IP connectivity latency and binding update latency. The MN uses L2 triggers to discover available *access points* (APs) and obtain further information of corresponding *access routers* (ARs) when it is still connected to its current subnet. After that, the MN may pre-configure the *New CoA* (NCoA) and register it to the PAR to bind *previous CoA* (PCoA) and NCoA. Through these operations the movement detection latency and the new CoA configuration latency are reduced. To reduce the binding update latency, a bi-directional tunnel between the PAR and the NAR is used to forward packets until the MN completes binding update. When the MN moves to the new subnet link, it will announce its attachment to launch forwarding of buffered packets from the NAR.

However, there are two disadvantages in basic FMIPv6 protocol. One is that the tunnel between the PAR and the NAR is fairly long. The other is that the DAD procedure for NCoA validation causes large handover latency. We'll discuss the solutions later.

2.2 Techniques of QoS Provisioning

Due to host mobility and characteristics of wireless networks, there are several problems in applying RSVP to mobile wireless networks. In the past several years many RSVP extensions were proposed to solve the problems. Talukdar et al. [9] proposed the MRSVP protocol in which resource reservations are pre-established in the neighboring ARs to reduce the timing delay for QoS re-establishment. However, too many advance reservations may use up network resources.

Chaskar et al. [10] proposed a solution to perform QoS signaling during the binding registration process. This mechanism defines the structure of *"QoS OBJECT"* which contains the QoS requirement of MN's packet stream. One or more QoS OBJECTs are carried in a new IPv6 option called *"QoS OBJECT OPTION"* (QoS-OP), which may be included in the hop-by-hop extension header of binding update and acknowledgement messages. Fu et al. [11] applied QoS-OP in the *Hierarchical Mobile IPv6* (HMIPv6) [3] architecture. Both schemes make use of intrinsic mobility signaling and achieve faster response time for effecting QoS along the new path.

Moon et al. [12] explained the concept of CR, which is the beginning router of the common path. And the common path is the overlapped part of the new path and previous path. Fig. 1 presents an example of the common path and the CR. Shen et al. [13] presented an interoperation framework for RSVP and MIPv6 based on the "Flow Transparency" concept, which made use of common path by determining the "Nearest Common Router" (just like CR) too. In both schemes the CR ensures that reservation will not be re-established in the routers along common path. Thus the QoS signaling overheads and delays as well as data packet delays and losses during handover can be greatly reduced.

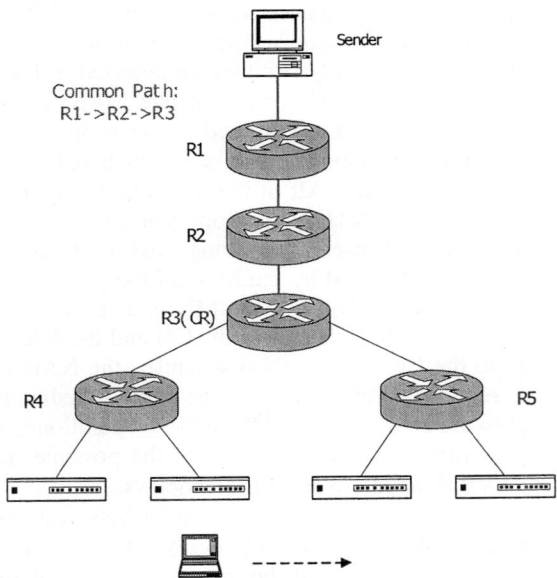

Fig. 1. Common Path and Crossover Router (CR)

3 Overview of Proposed Solution

The proposed solution includes two parts: some improvements to basic FMIPv6 and an efficient framework for end-to-end QoS guarantee in the enhanced FMIPv6 architecture.

Assuming that we have determined the location of CR, data forwarding path using the bi-directional tunnel of FMIPv6 would be CN-CR-PAR-CR-NAR. Obviously we can shorten the path to CN-CR-NAR. The method for CR determination will be introduced later. Though the bi-directional tunnel is eliminated in our scheme, a unidirectional tunnel from PAR to NAR is still included because the CR does not know when to intercept packets that destined to the MN's PCoA. When tunneling process begins, the PAR sends a TUN_BEGIN message which enables the CR to intercept the packets destined to the MN's PCoA and forward them to the NAR. In the opposite direction, the NAR directly sends packets with the CN's address filled in the destination address field. The CR intercepts these packets, sets the source address field to the MN's PCoA and forwards them to the CN.

A further modification to the basic FMIPv6 is the elimination of DAD procedure. We adopt the method of "Address Pool based Stateful NCoA Configuration" [8]. The NCoA pools are established at NAR or PAR. Each NCoA pool maintains a list of NCoAs already confirmed by the corresponding NAR. Thus the NCoA assigned to the MN at each handover event is already confirmed so that the DAD procedure can be ignored.

Now come to the part of QoS guarantee. As we know in FMIPv6 architecture, the NCoA is pre-established. Thus we can set up reservation along several possible future-forwarding paths (one or more NARs may be detected in FMIPv6) in advance when the MN still locates in the PAR's link. Just like MRSVP, *active* and *passive* Path/Resv messages and reservations are defined in our proposal. The NAR, which makes advance reservation and maintains soft state on behalf of the MN, acts as *remote mobile proxy*. To inform the NAR of the MN's QoS requirements, we extend the FBU and HI messages with QoS-OP in the hop-by-hop extension header.

Then we can initiate advance reservation along possible future path. Since there may be more than one NARs detected by the MN, all the possible future-forwarding paths must perform advance reservation. If the MN is a receiver, the CR issues the *passive Path* message to the NAR on behalf of the CN and the NAR in turn sends the *passive Resv* message to the CR. If the MN is a sender, the NAR issues the passive Path message. Upon receiving Path message, the CR immediately replies with a passive Resv message to the NAR. By performing these operations, the passive RSVP messages are restricted within the truly new part of the possible future path, which results in decreased RSVP signaling overheads and delays.

When the MN attaches to certain NAR's link, the packets sent from or destined to it can acquire QoS guarantee without any delay. At the same time advance reservations in other NARs' link must be released immediately. The modified FMIPv6 handover and resource reservation procedures when the MN acts as a receiver are depicted in Fig. 2.

In conclusion, our proposed QoS provisioning scheme has the following advantages:

1. The transmission of QoS requirement makes use of intrinsic mobility signaling of FMIPv6, which results in faster response time for effecting QoS along the new path.
2. The advance reservation along the possible future path decreases the delay of reservation re-establishment and provides QoS guarantee for the MN until it completes binding update.
3. The CR keeps reservation along common path unchanged. Thus the reservation delay and signaling cost can be minimized, which in turn minimizes the handover service degradation.
4. The duration of advance reservations in our proposal is much shorter than MRSVP.

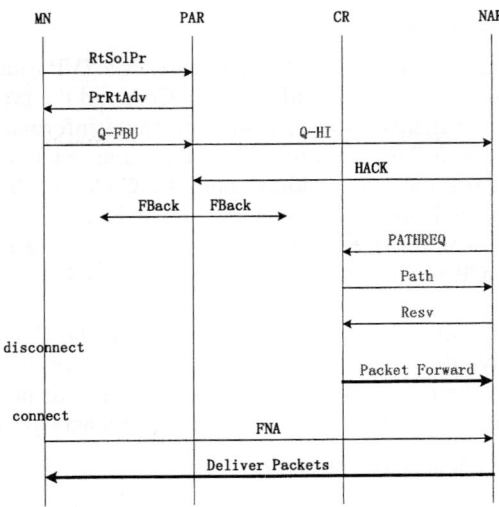

Fig. 2. Handover and Reservation Procedures of a Mobile Receiver

4 Detailed Operations

First of all, we assume that the MN moves into the boundary of the PAR so that the fast handover procedure launches. The procedures of proposed fast handover and resource reservation are as follows:

1) The MN discovers available APs using link-layer specific mechanisms and then sends a *Router Solicitation for Proxy* (RtSolPr) message including the identifiers of the APs to the PAR.
2) After the reception of the RtSolPr message, the PAR resolves the access point identifiers to subnet router(s) (i.e. the [AP-ID, AR-Info] tuples). Though several NARs may be discovered, the following description will just focus on the operations of certain NAR. Using the *"PAR-based stateful NCoA configuration"* proposed in [8], the PAR obtains a confirmed NCoA and responds the NCoA as well as the [AP-ID, AR-Info] tuple (via PrRtAdv) to MN.
3) In response to the PrRtAdv message, the MN sends a *Q-FBU* message to the PAR before its disconnection from the PAR's link. The Q-FBU message includes a QoS-OP (contains one or more QoS OBJECTs) in the hop-by-hop extension header. The QoS OBJECT may contain RSVP objects such as FLOW_SPEC, SENDER_TSPEC and FILTER_SPEC.
4) On reception of the Q-FBU message, the PAR again includes the MN's QoS requirement in the *Q-HI* message and sends it to the NAR. The Q-HI message should also contain the CN address corresponding to each QoS OBJECT, which will be used as the destination address of the *PATHREQ* message when the MN acts as a receiver.

Case 1. When the MN acts as a sender,

5a) The NAR directly issues the passive Path message. A RSVP router decides if it is the CR just by comparing the home address, the CoA and the previous RSVP hop carried in the passive Path message against the same information stored in the Path State. If there is a Path state related to the home address of passive Path message, and for the same home address both the CoA and the previous RSVP hop have been changed, then the router decides it is the CR. The binding of PCoA and NCoA is also included in a hop-by-hop extension header of the passive Path message. The CR will use the binding to prevent packet forwarding between the PAR and NAR.

6a) The CR does not forward the Path message further to the CN, but immediately replies with a passive Resv message to the NAR. By performing these operations, the RSVP states in the routers along the common path will not change. Fig. 3a describes the advance reservation process when the MN acts as the sender.

Case 2. Otherwise, the MN acts as a receiver,

5b) The NAR sends a PATHREQ message which has the CN's address as destination address (thus the CR can intercept this message) to request passive Path message. A RSVP router decides if it is the CR by searching the home address in PATHREQ against the same field in PATH state on the downlink direction. If there is a match of the home address in the PATH state in the downlink direction, then the router decides it is the CR. The PATHREQ message, which contains MN's home address and new CoA as introduced in [13], is extended to include the binding of PCoA and NCoA.

6b) The CR then issues the passive Path message to the NAR on behalf of the CN because the path between the CR and the CN is the common path and needn't any change. Finally the NAR will issue the passive Resv message towards the CR. Fig. 3b depicts the advance reservation process when the MN acts as the receiver.

7) At the same time as advance reservation process initiates, the NAR replies with a HACK message to the PAR, which may in turn issue the FBack message. The PAR may ignore sending this message because the NCoA is already confirmed.

8) When packet tunneling launches, the PAR will send a *TUN_BEGIN* message which has the CN's address as destination address. Upon receiving this message the CR begins to intercept packets destined to the PCoA and forward them to the NAR. Reversely, the NAR directly sends packets with the CN's address filled in the destination address field. The CR intercepts these packets, sets the source address field to the MN's PCoA and forwards them to the CN.

9) As soon as the MN attaches to the NAR, it sends the FNA message to the NAR. As a response, the NAR forwards buffered packets to the MN.

Finally, the MN can send a binding update to the HA and the CN. After it completes binding update, the CR stops intercepting packets sent from or destined to the MN. The packets will be forwarded with QoS guarantee along the new RSVP path.

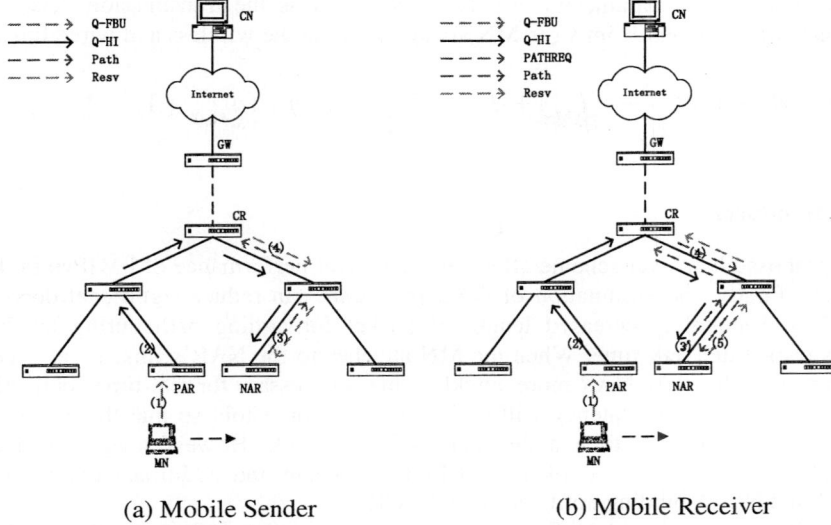

Fig. 3. Procedures of Advance Resource Reservation

5 Performance Analysis

In this section we study the performance of handover and resource reservation. We consider a network environment with a single domain made up of 16x16 square-shaped subnets and model the MN's mobility as a two-dimensional (2-D) random walk, which is similar to reference [14]. In a 2-D random walk, an MN may move to one of four neighboring subnets with equal probability. Under FMIPv6 architecture, only when the MN moves into the overlapped area of two or more APs, it may achieve information of the possible future NARs. Thus the number of the NARs is less than two. Under other simulated or real environments, the number of possible NARs is always less than the number of neighboring ARs in MRSVP.

Parameters:

N_p	average number of possible NARs in FMIPv6;
N_n	average number of neighboring ARs of current AR in MRSVP;
d_{x_y}	average number of hops between x and y;
B_w	bandwidth of the wired link;
B_{wl}	bandwidth of the wireless link;
L_w	latency of the wired link (propagation delay and link layer delay);
L_{wl}	latency of the wireless link (propagation delay and link layer delay);
P_t	routing table lookup and processing delay;
s_a	average size of a signaling message for resource reservation;
B_r	amount of the actual resource requirement of the handover MN;
t_r	average time the MN will resident in certain AR's link;
t_{pl}	time from completion of reservation to the beginning of L2 switch;
t_{l2}	time to complete L2 switch.

With the above parameters, we define $t(s, d_{x\_y})$ as the transmission delay of a message of size s sent from x (an MN always) to y via the wireless and wired links.

$$t(s, d_{x\_y}) = (\frac{s}{B_{wl}} + L_{wl}) + d_{x\_y} \times (\frac{s}{B_w} + L_w) + (d_{x\_y} + 1) \times P_t \qquad (1)$$

5.1 Handover

Our proposed handover scheme affects the handover performance of FMIPv6 in three aspects. Firstly, the elimination of DAD procedure can reduce significant delays in FMIPv6. Secondly, decreased length of packet forwarding path during handover saves packet delivery time. When the MN attaches to the NAR's link, it can receive these packets from the NAR more quickly. This is necessary for real-time applications for that more packets' latency will be less than the threshold so that the application can use them for real-time audio and video playback. However, we should also consider the signaling cost of TUN_BEGIN message and additional overheads of PCoA and NCoA binding notification to the CR.

Finally, our proposed QoS guarantee mechanism also influences the handover performance. The Q-FBU and Q-HI messages size is enlarged to hold QoS requirement. So the signaling cost is larger than the basic FMIPv6 protocol. Since the size of QoS requirement is small in proportion to the total signaling cost, the additional latency introduced by the Q-FBU and Q-HI messages can be ignored.

Further analysis is not presented and we focus the discussion on the performance analysis of resource reservation.

5.2 Resource Reservation

1) *Total signaling cost of resource reservation*: In our proposed scheme, signaling messages for resource reservation include Q-FBU, Q-HI, PATHREQ, Path and Resv messages. s_a is the average size of these messages. Q-FBU message travels from the MN to the PAR; Q-HI message from the PAR to the NAR; PATHREQ, Path and Resv messages from NAR to CR. The total signaling cost of resource reservation is denoted by C and is computed as the following.

$$C_{FMIPv6-R} = s_a \times (d_{MN\_AR} + d_{AR\_AR} + 3 \times d_{AR\_CR}) \times N_p \qquad (2)$$

If MN acts as a sender, the PATHREQ message is not used.

$$C_{FMIPv6-S} = s_a \times (d_{MN\_AR} + d_{AR\_AR} + 2 \times d_{AR\_CR}) \times N_p \qquad (3)$$

In MRSVP, Spec, MSpec, Path, active Resv and passive Resv message are the signaling messages for resource reservation. We consider the scenario that the sender acts as the *receiver_anchor* node [9].

$$C_{MRSVP} = s_a \times (d_{AR\_AR} \times N_n + d_{AR\_CN}) + 2s_a \times d_{AR\_CN} \times (N_n + 1) \qquad (4)$$

2) *Reservation establishment delay*: We compute the total delay since the MN issues Q-FBU to PAR. As the signaling cost, total delay of QoS establishment is affected by the same messages. The total delay of reservation establishment is denoted by D.

$$D_{FMIPv6-R} = t(s_a, d_{MN\_AR}) + t(s_a, d_{AR\_AR}) + 3 \times t(s_a, d_{AR\_CR}) \quad (5)$$

$$D_{FMIPv6-S} = t(s_a, d_{MN\_AR}) + t(s_a, d_{AR\_AR}) + 2 \times t(s_a, d_{AR\_CR}) \quad (6)$$

$$D_{MRSVP} = t(s_a, d_{AR\_AR}) + 3 \times t(s_a, d_{AR\_CN}) \quad (7)$$

Note that the delays we compute here are reservation establishment delays. Actually, except for switch operation between active and passive reservation, the resource reservation can be used immediately when the MN attaches to the new subnet both in our FMIPv6 based advance reservation mechanism and in MRSVP.

3) *Bandwidth requirements*: The duration of advance reservation along the paths between CR and possible NARs is t_{pl} plus t_{l2}. When the MN arrives certain NAR's link, reservation status on this link changes to active while other passive reservations are released. We use B to denote the total bandwidth requirements including active and passive reservation during the period a MN residents in certain AR's link.

$$B_{FMIPv6} = B_r \times (d_{AR\_CR} + d_{CR\_CN}) \times t_r$$
$$+ B_r \times d_{AR\_CR} \times (t_{pl} + t_{l2}) \times N_p \quad (8)$$

$$B_{MRSVP} = B_r \times d_{AR\_CN} \times t_r \times (N_n + 1) \quad (9)$$

Now we can compare the performance of our proposal and the MRSVP. Let's focus on three pairs of parameters: d_{AR_CR} against d_{AR_CN}, N_p against N_n, and $t_{pl}+t_{l2}$ against t_r. The comparison results are identical: the former is much less than the latter. Thus we can draw the conclusion that the total signaling cost of resource reservation and the reservation establishment delay, as well as bandwidth requirements in our scheme are much less than those in MRSVP.

6 Conclusion

This paper proposes a framework for QoS guarantee based on an enhanced FMIPv6 architecture. We introduce a key entity which called "Crossover Router" (CR) to reduce the length of packet forwarding path before the MN completes binding update. Furthermore we use "Address Pool based Stateful NCoA Configuration" mechanism to eliminate the long DAD latency. The proposed QoS guarantee scheme achieves low signaling cost and reservation re-establishment latency by making use of the FBU and HI signaling messages of FMIPv6 to transmit QoS requirements and adopting the idea of advance reservation and common path. Performance analysis shows that our proposal outperforms MRSVP in terms of signaling cost, reservation re-establishment delay, and bandwidth requirements.

The simulation based on NS2 [15] platform for our scheme will be done soon to achieve the further performance analysis under various environments. When and how to release passive reservations on other NARs' link after the MN attaches to certain NAR's link, should be considered. Furthermore, we are also making efforts to apply the idea of our QoS provisioning scheme to F-HMIPv6 architecture [4].

References

1. D. Johnson, C. Perkins, J. Arkko, "Mobility support in IPv6", IETF RFC 3775, June 2004.
2. R. Koodli (Ed.), "Fast Handovers for Mobile IPv6", Internet Draft, IETF, draft-ietf-mipshop-fast-mipv6-03.txt, October 2004.
3. H. Soliman, C. Castelluccia, K. El-Malki, L. Bellier, "Hierarchical Mobile IPv6 mobility management", Internet Draft, IETF, draft-ietf-mipshop-hmipv6-03.txt, October, 2004.
4. H.Y Jung, S.J. Koh, H. Soliman, K. El-Malki, "Fast Handover for Hierarchical MIPv6 (F-HMIPv6)", Internet Draft, draft-jungmobileip-fastho-hmipv6-04.txt, June 2004.
5. R. Braden, D. Clark and S. Shenker, "Integrated services in the Internet architecture: An overview", IETF RFC 1633, June 1994.
6. S. Blake et al., "An architecture for differentiated services," IETF RFC 2475, December 1998.
7. R. Braden, Ed., L. Zhang, S. Berson, S. Herzog, S. Jamin. "Resource reserVation protocol (RSVP) -- Version 1 Functional Specification", IETF RFC 2205, September 1997.
8. Hee Young Jung, Seok Joo Koh, Dae Young Kim, "Address Pool based Stateful NCoA Configuration for FMIPv6", Internet Draft, draft-jung-mipshop-stateful-fmipv6-00.txt, August 2003.
9. A.K. Talukdar, B.R. Badrinath and A. Acharya, "MRSVP: A resource reservation protocol for an integrated services network with mobile hosts", Journal of Wireless Networks, vol.7, iss.1, pp.5-19 (2001).
10. H. Chaskar, and R. Koodli, "QoS support in mobile IP version 6", IEEE Broadband Wireless Summit (Networld+Interop), May 2001.
11. X. Fu et al., "QoS-Conditionalized binding update in mobile IPv6", Internet Draft, IETF, draft-tkn-nsis-qosbinding-mipv6-00.txt, January 2002.
12. B. Moon and A.H. Aghvami, "Quality of service mechanisms in all-IP wireless access networks", IEEE Journal on Selected Areas in Communications, June 2004.
13. Q. Shen, W. Seah, A. Lo, H. Zheng, M. Greis, "An interoperation framework for using RSVP in mobile IPv6 networks", Internet Draft, draft-shen-rsvp-mobileipv6-interop-00.txt, July 2001.
14. Shou-Chih Lo, Guanling Lee, Wen-Tsuen Chen, Jen-Chi Liu, "Architecture for mobility and QoS support in all-IP wireless networks", IEEE Journal on Selected Areas in Communications, May 2004.
15. The Network Simulator - NS (version 2), http://www.isi.edu/nsnam/ns/

A Novel Hierarchical Routing Protocol for Wireless Sensor Networks[⋆]

Trong Thua Huynh[1] and Choong Seon Hong[2]

Department of Computer Science, Kyung Hee University,
1 Seocheon, Giheung, Yongin, Gyeonggi 449-701 Korea
Tel: +82 31 201-2532, Fax : +82 31 204-9082
htthua@networking.khu.ac.kr, cshong@.khu.ac.kr

Abstract. In this paper, we propose a novel hierarchical routing protocol for a large wireless sensor network (WSN) wherein sensors are arranged into a multi-layer architecture with the nodes at each layer interconnected as a de Bruijn graph and provide a novel hierarchical routing algorithm in the network. Using our approach, every sensor obtains a unique node identifier addressed by binary addressing fashion. We show that our algorithm has reasonable fault-tolerance, admits simple and decentralized routing, and offers easy extensibility. We also present simulation results showing the average delay, success data delivery radio in our approach. And we received acceptable results for some potential applications. Besides, to evaluate how well our protocol support for WSNs, we compare our protocol with other protocol using two these metrics (average delay, success data delivery radio).

1 Introduction

Wireless sensor networks are composed of a large number of sensors densely deployed in inhospitable physical environments. Dissemination information throughout such a network that requires fault-tolerance is a challenge. Although there are applications as described in [4], [9] which do not require addressing of sensors, we have argued many scenarios where addressing nodes is very essential. However, given that an addresses scheme needs to be build, IP-based addressing to this problem would not be a good solution. For one, IP-based addresses are global unique addresses but WSNs require local unique identifier. In our protocol, sensor nodes within a certain area interact with themselves in a distributed manner and come up with an addressing scheme in which each node obtains a unique local address. Motivated by above issues and the advantages of the de Bruijn graph admitting simple routing and possessing good fault tolerant capabilities in many interconnected networks, we introduce a multi-layer addressing architecture wherein sensor nodes are interconnected as de Bruijn graph at each

---

⋆ This work was supported by University ITRC Project of MIC, Dr. C.S.Hong is the corresponding author.

layer. To show how our algorithm supports simple routing, fault tolerance and scalability we introduce a hierarchical routing algorithm that is able to route packets along all possible paths between any pair of the source and destination nodes without performing explicit route discovery, repair computation or maintaining explicit state information about available paths at the nodes. The remainder of the paper is organized as follows. In section 2, we present the related work. The network architecture is shown in section 3. Section 4 provides routing algorithm for the multi-layer network. We present the performance evaluation in section 5. Finally, we conclude the paper in section 6.

2 Related Work

Many protocols have been proposed for WSNs in the last few years. In works addressed in [1], [3], [4], all sensors have been treated to be alike and are assumed to have similar functionality. In contrast, sensors in our protocol are heterogeneous. Data dissemination protocols are proposed for WSNs in [2], [3]. SPIN [3] attempts to reduce the cost of flooding data, assuming that the network is source-centric. Directed diffusion [2], on the other hand, selects the most efficient paths to forward requests and replies on, assuming that the network is data-centric. This approach, in company with works addressed in [1], [3], use a powerful concept of data-centric networking for sensor applications. Though this model is very interesting, it may not be applicable to many sensor applications. Certain applications like parking lot networks may require addressability for every sensor node and a method for routing packet to specific nodes. Clustering algorithms have also been proposed in many literatures, such as GAF [6]. SPAN [7] etc. The remarkable one among them is LEACH [8]. LEACH is an application-specific data dissemination protocol that uses clustering to prolong the network lifetime. However, LEACH assumes that all nodes have long-range transmission capability. This limits the capacity of protocol and application. Moreover, cluster head failure is also the problem in this approach. In contrast, our approach makes no assumptions like this and does not organize network as clustering architecture. Instead of this, our approach distributes sensors into several layers. Each layer is organized as a de Bruijn graph.

3 Network Architecture

In this section, we give a detailed description of the network architecture. In this architecture, we assume that the sensors are immobile. This assumption is reasonable especially for indoor applications such as smart home or smart building applications. Assume that we deploy a network with N sensors. The network architecture can be organized under a hierarchical model which consists of several layers with the node at each layer interconnected as a de Bruijn graph. For a detailed discussion on features of de Bruijn graph, see the paper by Sanmatham et al. [5]. Our architecture features is addressed as follows:

- Sensors are organized in a multi-layer architecture with the order of layer numbered increasingly starting from 1.
- k^{th} layer has 2^k sensor nodes.
- Sensors are addressed with binary address form, sensors in k^{th} layer use k bits for node addressing.
- Each node at k^{th} layer is connected to two children nodes at $(k+1)^{th}$ layer and is connected to its parent node at $(k-1)^{th}$ layer (k>1)
- Nodes in 1st layer and 2nd are connected completely. It means that, each node in the first layer has only one neighbor and 2 children. Each node in the second layer has 3 neighbors and 2 children nodes.
- Using graph theoretic notation, the de Bruijn graph BG(d,k) has $N = d^k$ nodes with diameter k and maximum degree 2d. We are interested in binary de Bruijn graph BG(2,k) which have $N = 2^k$. A node x addressed $x_{k-1}x_{k-2}\ldots x_1x_0$ in k^{th} layer (k>2) has 4 neighbors as follows:

$neig_1(\text{x}) = x_{k-2}\ldots x_1x_0x_{k-1}$; $neig_2(\text{x}) = x_{k-2}\ldots x_1x_0\bar{x}_{k-1}$
$neig_3(\text{x}) = x_0x_{k-1}\ldots x_2x_1$; $neig_4(\text{x}) = \bar{x}_0x_{k-1}\ldots x_2x_1$
where : $\bar{x}$ is complement of x.

The address of a node in k^{th} layer consists of two parts. One is k bit derived from $(k-1)^{th}$ layer. The other is 1 bit (0 or 1) added from right side. Node x addressed $x_{k-1}x_{k-2}\ldots x_1x_0$ has two children nodes and one parent node. These children are addressed as add($x_{k-1}x_{k-2}\ldots x_1x_0$,0) and add($x_{k-1}x_{k-2}\ldots x_1x_0$,1) while the parent is addressed as rmv($x_{k-1}x_{k-2}\ldots x_1x_0$). The following definitions describe two address transforming functions addition (add) and remove (rmv). Let K be a k-bit number and y be a binary number. Then

add(K,y) = Ky ;   rmv($x_{k-1}x_{k-2}\ldots x_1x_0$) = $x_{k-1}x_{k-2}\ldots x_1$
For example, add(001,1) = 0011; rmv(0110) = 011.

Figure 1 shows an example of the 3-layer network architecture followed by above mentioned features. Obviously, the network extension is not issue in this

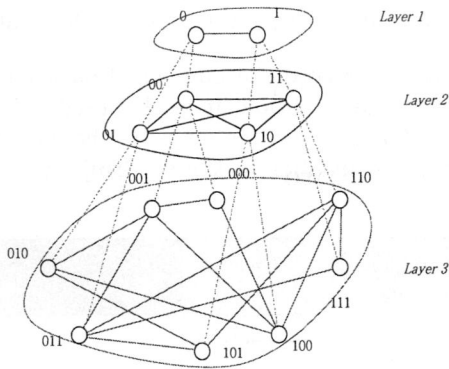

Fig. 1. A 3-layer hierarchical architecture. Node 01 (layer 2) is parent of two nodes 010 and 011 (layer 3) but is one of two children of node 0 (layer 1)

architecture. Whenever sensors need to be deployed into the network, they will be added at the highest layer of the hierarchical architecture. Thus extending network requires a fixed number of interconnections between the new nodes and the nodes at the last layer. If number of new nodes are more than that of the last layer can support, the remaining nodes will make up a new layer being last layer in new network architecture.

4 Routing Algorithm

In this section, we show that packets can be routed throughout the hierarchical architecture. We first consider routing within each layer and then consider routing across layers. To evaluate the routing complexity, we assume that a packet takes unit time to traverse a link.

4.1 Intra-layer Routing

Routing in the first layer and second layer takes unit time step since the nodes are completely connected. In this section, we describe a simple routing algorithm which is based on the construction of the Bruijn graph. Let a binary de Bruijn graph have $N = 2^k$ nodes and let $S = s_{k-1}s_{k-2} \ldots s_1 s_0$ be the source node that sends a packet to the destination node $D = d_{k-1}d_{k-2} \ldots d_1 d_0$. The packet consists of the data and packet header. The packet header contains the routing information. The packet format is depicted in figure 2.

The RF is a 2-bit binary number set to

- 00: the source and destination nodes are in same layer and the source node use Path 1 to route packet to the destination. (default)
- 01: it is the same as 00 but Path 2 is used instead of Path 1. Path 1 and Path 2 will be addressed later.
- 10: the source and destination nodes are in different layers. The source node is at a higher layer than the destination.
- 11: the source node is at a lower layer than the destination.

The counter c is used to record the number of packet hops from the source node to the current node. In addition, it is also used to generate the address of the next node in the path. Now, we describe the simple routing algorithm in each layer. From the construction of the de Bruijn graph, we know that the

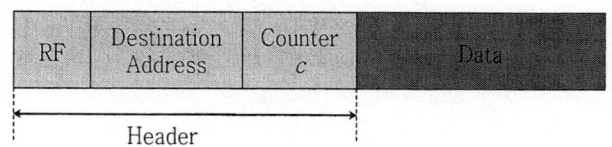

Fig. 2. Packet format contains the destination addresses, a counter (c), and a routing flag (RF)

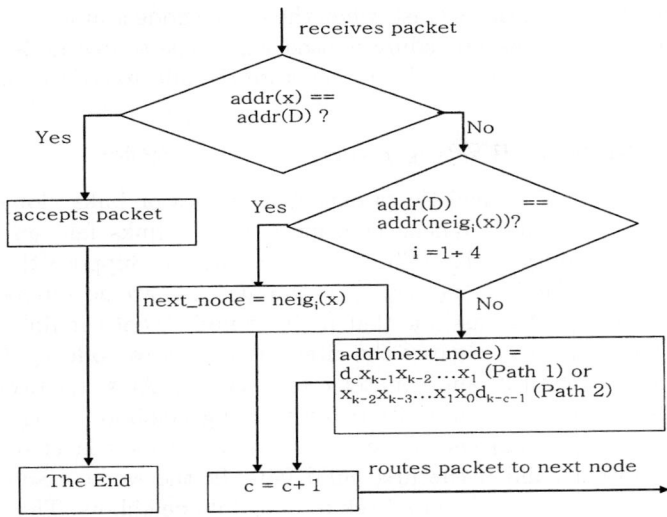

Fig. 3. Operations of node x. addr(x), $neig_i(x)$, and D present address, i^{th} neighbor of node x, and destination respectively

source node at k^{th} layer has the following neighbors - $d_0 s_{k-1} s_{k-2} \ldots s_1$ and $s_{k-2} \ldots s_1 s_0 d_{k-1}$. Using this property we can now generate two paths by appending successive bits of the destination node to the source address.

Path 1
(c=0) $s_{k-1} s_{k-2} \ldots s_1 s_0$(src address)
(c=1) $d_0 s_{k-1} s_{k-2} \ldots s_1$
(c=2) $d_1 d_0 s_{k-1} \ldots s_2$
.
.
(c=k) $d_{k-1} d_{k-2} \ldots d_1 d_0$(des address)

Path 2
(c=0) $s_{k-1} s_{k-2} \ldots s_1 s_0$(src address)
(c=1) $s_{k-2} s_{k-3} \ldots s_1 s_0 d_{k-1}$
(c=2) $s_{k-3} \ldots s_0 d_{k-1} d_{k-2}$
.
.
(c=k) $d_{k-1} d_{k-2} \ldots d_1 d_0$(des address)

Let $x_{k-1} x_{k-2} \ldots x_1 x_0$ be the address of the node x. The figure 3 describes steps executed by x.

4.2 Inter-layer Routing

In this session, we suppose that the source and destination nodes are in different layers. So, the RF is set to "10" or "11". In addition, the counter c is not incremented in order to maintain a proper value of the counter for intra-layer routing following the inter-layer routing. In case that the source node is at a higher layer than the destination, the source node first routes the packet to its parent rmv(S). This procedure is repeated recursively until the packet is received by a node at the same layer as the destination node. The RF is set to 00 or 01 now, and the source address is replaced by the address of the node that received the packet. The packet can then be routed to the destination using above intra-layer routing algorithm. In contrast, when the source node is at a lower layer than the destination, the same procedure is used where the source node first routes the packet to its child using add(S) to generate the address of the next node.

4.3 Fault Tolerant Routing Issue

In a large WSN, it is unrealistic to expect all nodes or links along a path to be free-fault at all times. Whenever some nodes or links fail, an alternative path that avoids faulty node or link must be derived. Suppose that nodes x_2 ($d_c x_{k-1} x_{k-2} \ldots x_1$) and x_3 ($x_{k-2} x_{k-3} \ldots x_1 x_0 d_{k-c-1}$) are neighbors of node x_1 ($x_{k-1} x_{k-2} \ldots x_1 x_0$). Also assume that if either node x_2 or the link between x_1 and x_2 has failed, then x_1 chooses the alternative path to node x_3. In addition, x_1 sets RF to 01 (Path 2) and counter c to 0 as well. At worst if both x_2 and x_3 fail, node x_1 routes to one of its two remaining neighbors $d_c x_{k-1} x_{k-2} \ldots x_1$ or $x_{k-1} x_{k-2} \ldots x_1 x_0 d_{k-c-1}$ and sets counter c to 0. For inter-layer routing, node x_1 chooses another child if the first child fails. In the worst case of both the children failing, node x_1 will route back to one of its neighbors. The approach is the same if parent of x_1 can not be reached. In all cases, counter c must be set to 0 and RF set to default.

5 Performance Evaluation

We developed a simulator based on SENSE simulation [10] to evaluate performance of our approach. The simulation use MAC IEEE 802.11 DCF that SENSE implements. Because network design choices: MAC scheme, network topology, node addressing, and packet routing vary among implementations, we do not compare our nodes to other solutions. In this simulation, we take account on two metrics: the end-to-end delay and the success data delivery radio with and without node or link failure. The end-to-end delay refers the time taken for a data packet to be generated until the time it arrives at the destination. The success data delivery radio is the rate of the number of successfully received data packets at the destination and the total number of packet generated by a source. Some of simulation parameters are listed in Table 1.

Table 1. Simulation Parameters

Parameter	Value
Network size	250x250
Number of sensors	254
Packet generating rate	1 packet/sec
Forward delay	0.01 sec
Data packet size	64 bytes
Simulation time	1000 sec
Radio of transmission range	20m

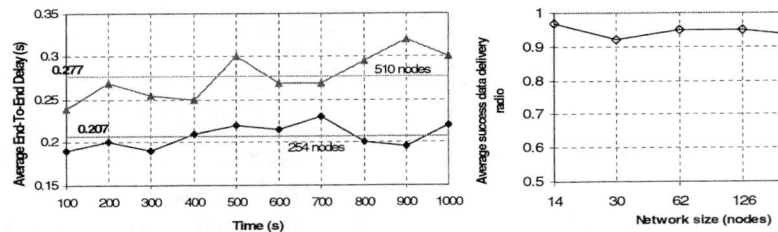

Fig. 4 a. Average end-to-end delay a function of simulation time

Fig. 4 b. Average success data delivery radio as a fucntion of network size

To illustrate the performance of our protocol, we choose randomly pairs of source and destination nodes in order to communicate each other. For 254 node network, the simulation results in figure 4a depict that the average end-to-end delay in this network is approximately 0.207 seconds. While for 510 node network, the average end-to-end delay is approximately 0.277 seconds. Obviously, this end-to-end delay is quite low and it is an acceptable value in many potential applications. In this simulation, no acknowledgement for data packets sending or forwarding are simulated. Therefore, no retransmission is incurred if a data packet lost because of collision or because a receiving node has returned to the sleep mode before the packet is received. Besides, the simplicity in routing algorithm such as no discovery phase, just passing packets to the next node in the path which is determined base on features of de Bruijn graph and logic of binary addresses reduces the time disseminate data from a source node to a destination node significantly. The simplicity and efficiency of our protocol are examined by simulation in figure 4b. We study metric average success data delivery radio as a function of sensor network size. To do this, we generate a variety of sensor fields of different sizes, ranging from 14 to 510 nodes in increments of 2^i (i = 4,5,6,7, and 8). When the number of node is changed, network size is also proportionally changed by scaling the square and keeping the radio range constant in order to approximately keep the average density of sensor nodes constant. Results show that the data is delivered successfully in the network is quite high (always more than 90%).

To indicate the fault-tolerance of our algorithm, we inject failures into the simulated network and evaluate the performance with 5%, 10% and 20% node/link failures one after the other. Results in figure 5a depict that although the average end-to-end delay increases when the network size increases, the increase of average delay is not significant (0.24 second delay for none node or link failure, 0.32 second delay for 20% node or link failures). Plus, figure 5b addresses the increase and decrease of the end-to-end delay as network size changes and there are node/link failures as well. These increase and decrease are insignificant. In figure 6a, we study metric average success data delivery radio as a function of node or link failures. And the results indicate that the success data delivery radio is decreased when the number of node or link failures increases. However, this decrease is not significant. And it still ensures this radio at quite high level accepted in a large amount of sensor network applications. When the next node

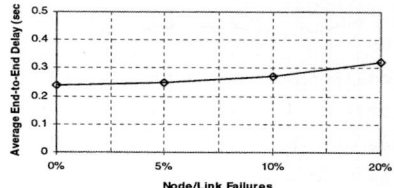

Fig. 5 a. Average end-to-end delay as a function of node/link failure

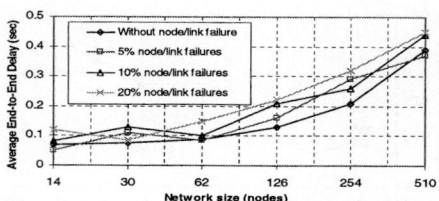

Fig. 5 b. Average end-to end dalay as a fucntion of node/link failure as network size changes

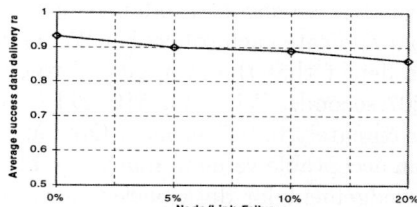

Fig. 6 a. Average success data delivery radio as a function of node/link failure

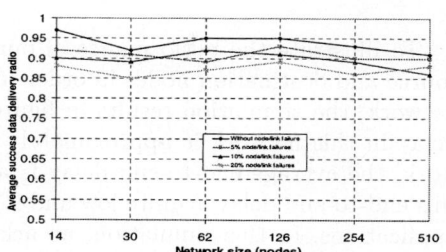

Fig. 6 b. Average success data delivery radio as a function of node/link failure as network size changes

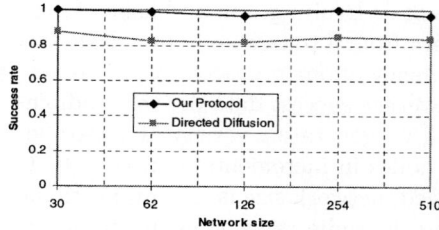

Fig. 7 a. Average success data delivery radio as a function of network size

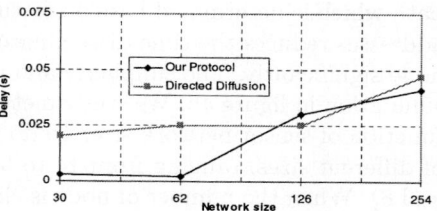

Fig. 7 b. Average end-to-end delay as a function of network size

or the link to next node of the node is fail, it is straightforward to the current node alters to another path (Path 1 or Path 2) without performing explicit route discovery, repair computation or maintaining explicit state information about available paths at that node. That is why the increase of average delay and the decrease of success data delivery radio are insignificant when the number of node/link failures increases. Figure 6b, one more again, proves that our protocol can support the fault-tolerance very well.

Here, we compare our protocol with another protocol (Directed Diffusion) using two these metrics average delay and success data delivery radio. In figure 7a, results show that the data is delivered successfully in the network is quite high and higher than that of Directed Diffusion. Besides, end-to-end delay is also factor showing the efficiency of our approach. As addressed in figure 7b, the end-to-end delay is quite low and it is a promising value in many potential applications. Because our protocol support the very simple routing algorithm to disseminate data from source to any destination while Directed Diffusion is the complexity protocol (discovery phase, reinforcement phase, routing phase) that create high end-to-end delay.

6 Conclusions

This paper describes a hierarchical architecture for WSNs wherein sensor nodes are distributed into layers. Nodes in each layer are interconnected as a de Bruijn graph. Our protocol easily solves fault tolerance and extensibility. We also provide a simple routing algorithm between any node pairs in the network. We created simulations based on SENSE [10] in order to illustrate the performance and bring out the main goal while proposing this approach. Because sensors have limited computation, in this paper, we do not investigate into the shortest path algorithm that consumes much energy than above mentioned simple routing algorithm. As a future work, since multi-year battery life is preferable for future sensors, we need to future study on optimal routing to improve the goodness of our protocol. We also need to study an optimal energy scheme in order to save energy for sensors.

References

1. Wendi R.Heinzelman et al., "Energy efficient communication protocol for wireless microsensor networks", in Hawaii International Conference on System Sciences, January, 2000.
2. C.Intanagonwiwat et.al., "Directed diffusion: A scalable and robust communication paradigm for sensor networks", in ACM Mobicom 2000.
3. J. Kulik, W. Heinzelman, and H. Balakrishnan, "Negotiation-based Protocols for Dissemination Information in Wireless Sensor Networks", in ACM/IEEE MobiCom '99, Seattle
4. Deborah Estrin et.al., "Next century challenges: Scalable coordination in sensor networks", in MOBICOM'99, Seattle.
5. M. R. Samantham, and D. K. Pradhan, "The de Bruijn Multi-processor Network: A Versatile Parallel Processing and Sorting Network for VLSI", IEEE Transaction on Computers, April 1989.
6. Y. Xu, J. Heidemann, and D. Estrin, "Geography-Informed Energy Conservation for Ad Hoc Routing," in ACM/IEEE on Mobile Computing and Networking (MOBICOM), Italy, July 2001.
7. B. Chen et al., "Span: An Energy-Efficient Coordination Algorithm for Topology Maintenance in Ad Hoc Wireless Networks," in ACM Wireless Networks, September 2002.
8. W. R. Heinzelman et al., "An Application-Specific Protocol Architecture for Wireless Microsensor Networks," IEEE Transactions on Wireless Communications, October 2002.
9. Jeremy Elson and Deborah Estrin, "An address free architecture for dynamic sensor networks", submitted for publication. http://www.isi.edu/ estrin/papers/.
10. Gang Chen, et al, "SENSE - Sensor Network Simulator and Emulator", http://www.cs.rpi.edu/ cheng3/sense/.

A Vertical Handoff Algorithm Based on Context Information in CDMA-WLAN Integrated Networks[†]

Jang-Sub Kim, Min-Young Chung, and Dong-Ryeol Shin

School of Information and Communication Engineering,
Sungkyunkwan University,
300 ChunChun-Dong, JangAn-Gu, Suwon, Korea
{jangsub, mychung, drshin}@ece.skku.ac.kr

Abstract. The integration of CDMA cellular network and wireless LAN (WLAN) has drawn much attention recently. In this integrated architecture, it is required to support a vertical handoff from the WLAN to the CDMA system when a user moves out of the WLAN coverage area and vice versa. We propose a context based handoff procedure and the corresponding mechanism from WLAN to CDMA system, and vice versa, based on wireless channel assignment. In other words, this paper focuses on the handoff decision which uses context information such as dropping probability, blocking probability, GOS, the number of handoff attempts and velocity. As a decision criterion, velocity threshold is determined to optimize the system performance. The optimal velocity threshold is adjusted to assign available channels to the mobile stations with various handoff strategies. The proposed scheme is validated using computer simulation. Also, the overflow traffic with *take-back* technique is evaluated and compared with non-overflow traffics in terms of GOS.

1 Introduction

The demand for better telecommunication services has led to the development of a number of wireless access technologies including Bluetooth, wireless LAN, wireless MAN, and Wireless WAN (2G, 3G, and 4G). They not only provide traditional voice services but also multimedia services with high bandwidth access. Each of these access technologies has a different data rate, network latency, interaction capability, mobility support, and cost per bit because they have been designed with specific services in mind. To meet the diverse and growing needs for telecommunication services, more specific and heterogeneous access technologies must be developed because there is no longer a multi-purpose access technology meeting all user requirements at a reasonable cost. New wireless access networks are required to laid over the existing ones. In such wireless overlay networks, integration of several access networks will effectively support a broad mix of services.

The combination of WLAN and CDMA technology uses the best features of both systems. They nevertheless tend to leverage the high-speed access of WLANs when-

---

[†] This work was supported by Korea Science and Engineering Foundation. (KOSEF-R01-2004-000-10755-0)

ever possible. However, CDMA and WLANs are based on different networking technologies. The integration of them, especially seamless roaming, thus becomes one of the critical issues in the ubiquitous environments some of which involve the interworking architecture, authentication, roaming services, seamless handoff, and implementation of a WLAN/CDMA2000 [1]. The motivation behind inter-technology for the hybrid mobile data networks arises from the fact that no one technology or service can provide ubiquitous coverage.

A handoff mechanism in an overlay CDMA and underlay WLAN should perform well so that the users attached to the CDMA just easily check the availability of the underlay WLAN. The decision criteria for handoff (or network selection) can be based on the maximum link speed, reliability, power utilization, billing, cost, user preference, mobile speed, and Quality of Service like bandwidth, delay, jitter, and loss rate, etc [2]. To simplicity, we do only consider the mobile speed in this paper. A good handoff algorithm is to be derived in order to minimize unnecessary handoff attempts. An appropriate handoff control is also an important issue in system management for the sake of the benefits above with the overlaid cell structures. This paper suggests four handoff strategies : a) no-overflow, b) Overflow − From WLAN to CDMA system, c) Overflow − From WLAN to CDMA system and vice versa, d) Take-back.

To efficiently support general applications, we present an optimization scheme which assumes no knowledge about specific channel characteristics. Furthermore this paper considers a mobile speed as a decision criterion and proposes a handoff procedure and the corresponding mechanism from WLAN to CDMA, and vice versa, based on wireless channel assignment. For the speed-sensitive handoff algorithm, different approaches have been proposed in [3]-[6]. In [6], new call and handoff attempts are overflowed from the speed-dependent preferred cell layer (WLAN) to an upper cell layer (CDMA). The overflowed call keeps its connection to the overflowed network in which it is traveling as soon as a channel becomes available in the cell. However, a flexible overflow mechanism with possible take-back of overflow traffic into the preferred cell layer has not been considered.

In short, we first propose a context based handoff procedure and the corresponding mechanism from WLAN to CDMA system, and vice versa, based on wireless channel assignment. Secondly, we present a handoff control scheme for a hierarchical structured network. As a decision criterion, velocity threshold is determined to optimize the system performance. The proposed scheme is validated using computer simulation. Also, the overflow traffic with take-back technique is evaluated and compared with non-overflow traffics in terms of GOS. The simulation results show that take-back strategy performs as good as other handoff strategy.

The rest of the paper is organized as follows. In Section 2, the system description and the required assumptions are presented. The role of the mobility model is viewed in the design phase of the networks. Section 3 explains the mobility model, performance parameters (i.e. new call blocking probability and handoff call dropping probability), and core part of algorithmic decision procedure for the optimal velocity threshold for the WLAN and CDMA selection schemes. Simulations are performed in Section 4 to validate the proposed approach. Finally, the summary of the result and the future related research topics are presented in the conclusion section.

2 System Description

We consider a large geographical area covered by contiguous WLANs.

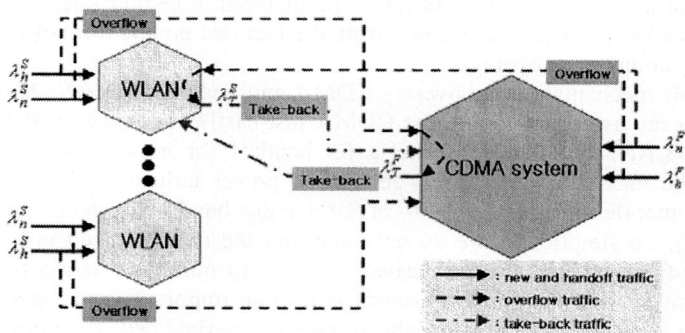

Fig. 1. Management of traffics in integrated system

Fig. 1 shows traffic flows between different wireless networks with related parameters. The WLAN constitutes the lower layer of the two-layer hierarchy. All the WLANs are overlaid by a large CDMA system. The overlaying CDMA system forms the upper cell layer. Each CDMA system is allocated c_0 traffic channels, and the number of channels allocated to the WLAN cell-i is c_i, $i = 1, 2, \hbar, N$. All channels are shared among new calls and handoff calls. In our system, mobile stations (MSs) are traversing randomly the coverage area of WLAN and CDMA system. We distinguish two classes of MSs, fast and slow MSs. We further assume that an MS does not change its speed during a call.

The operation of the system can be described as follows (see Fig. 1).

① A new call generated by a slow MS (or a fast MS) in WLAN (λ_n^S) (or CDMA system (λ_n^F)) : A new call is first directed to the camped-on WLAN (or CDMA system). If the number of traffic channels in use in the WLAN i (or CDMA system) is equal to c_i (or c_0), the new call may be overflowed to the overlaying CDMA system (or overlaid WLAN). The overflowed new call will be accepted by the CDMA system (or WLAN) if the number of traffic channels occupied in the CDMA system (or WLAN) is smaller than c_0 (or c_i); otherwise, the call will be lost.

② A handoff request of a slow MS (or a fast MS) in WLAN (λ_h^S) (or CDMA system (λ_h^F)) : A handoff request is first directed to the target WLAN (or CDMA system) independent of whether the current serving network is a neighboring WLAN or an overlaying CDMA system. If all traffic channels in the target WLAN (or CDMA system) are busy, the handoff request may be overflowed to the overlaying CDMA system (or neighboring WLAN). The overflowed handoff request will be served by the CDMA system (WLAN) only if there is any idle traffic channel; otherwise, the handoff request fails and the call is forced to terminate (dropped).

③ A take-back call of a slow MS (or a fast MS) in CDMA system (λ_T^S) (or WLAN (λ_T^F)) : Assume a slow MS (or a fast MS) roaming within a CDMA system (or WLAN) it is traversing. If this slow MS is engaged in a new or handoff call that has been successfully overflowed to the CDMA system (or WLAN), a take-back request is directed to the WLAN (or CDMA system) the MS enters at each border crossing time. This take-back request will be accommodated by the WLAN (or CDMA system) only if there is any idle traffic channel in the WLAN (or CDMA system). If all traffic channels in the target WLAN (or CDMA system) are busy, the slow MS (or a fast MS) will continue to be served in the CDMA system (or WLAN) (See Fig. 2).

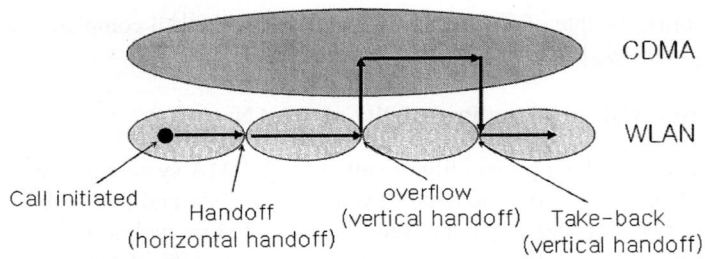

Fig. 2. Examples for take-back strategy

In this paper, all WLANs of the lower layer are treated equally to simplify the overflow and the take-back mechanisms.

3 Performance Measures and Analysis

The mobility models and perspective is given in [7]. We present analytical results for the proposed system. As stated, our objective is to focus on simple and tractable mechanism for which analytical results can give an insight into handoff between different networks. According to the velocity threshold, all the mobile users are divided into two groups; slower moving users (λ^S) and fast moving users (λ^F). In order to determine the value, which is one of the main goals of this study, a few assumptions related to mobility characteristics are made. The assumptions we employ in the mobility models are taken from [6] as cells are circular with radius R, mobiles are uniformly distributed in the system, mobiles making new calls in WLAN move in a straight line with a direction uniformly distributed between $[0, 2\pi)$, and mobiles crossing cell boundary enter a neighbor cell with the incident angle θ of distribution:

$f(\theta) = 1/2 \times \cos\theta \ \ for -\pi/2 < \theta < \pi/2$ and $f(\theta) = 0 \ \ for \ otherwise$.

WLAN cells compose of two types of new call traffics, represented by the call arrival rates λ_n^S and λ_h^S, respectively modeled by the Markov Modulated Poisson process (M/M/k/k, in voice traffic model) [8]. Let random variables X and Y denote the straight mobile path for new calls and handoff calls, respectively. With the assumption of the unique WLAN cell size and the same speed of the MS, WLAN cell boundary crossing rate per call (μ_B), provided that no handoff failure occurs [6]:

$$\mu_B = 1/E[T_Y] = 2E[V]/\pi R \tag{1}$$

Here, T_Y represents the time for a mobile to cross a WLAN cell with radius equivalent to R. $E[T_Y]$ is the cell sojourn time for a constant MS velocity. New calls assume to finish within the average call duration time, $1/\mu$, or the call handoffs to an adjacent cell. The proportion of the channel returned by the handoff is [7]

$$P_h = \mu_B/(\mu + \mu_B) \tag{2}$$

In other words, the rate of channel release and that of the call completion due to handoff are $\mu_B/(\mu+\mu_B)$ and $\mu/(\mu+\mu_B)$, respectively.

3.1 The New Call Blocking Probability of WLAN

We denote the blocking probability of calls from CDMA system and WLAN by P_{B0} and P_{B1}, respectively. And the handoff traffic from slow and fast mobiles is denoted as follows. The λ_{h0}^F and λ_{h0}^S is the rate of fast and slow mobile handoff traffic in a CDMA systems, respectively. The λ_{h1}^F and λ_{h1}^S is the rate of fast and slow mobile handoff traffic in a WLAN, respectively. And we denote the take-back traffic rate to CDMA system and WLAN by λ_{T0} and λ_{T1}, respectively. The P_{T0} and P_{T1} are the take-back probability from CDMA system and WLAN, respectively.

The aggregate traffic rate into the WLAN due to a slow MS is computed as follows:

$$\lambda_1^S = \lambda_{n1}^S + \lambda_{h1}^S + \lambda_{T1}^S \tag{3}$$

where the take-back traffic rate component is given as

$$\lambda_{T1}^S = (\lambda_{n1}^S + \lambda_{h1}^S + \lambda_{T1}^S)P_{B1}(1-P_{B0})P_T^S \tag{4}$$

The aggregate traffic rate into the WLAN due to fast MS is expressed as

$$\lambda_1^F = 1/N \times (\lambda_{n0}^F + \lambda_{h0}^F + \lambda_{T0}^F)P_{B0} + \lambda_{h1}^F \tag{5}$$

The generation rate of the handoff traffic of a slow mobile station in a WLAN is given as follows:

$$\lambda_{h1}^S = P_{h1}^S(\lambda_{n1}^S + \lambda_{h1}^S + \lambda_{T1}^S)(1-P_{B1}) \tag{6}$$

The generation rate of the handoff traffic of a fast moving MS in a WLAN is characterized as follows:

$$\lambda_{h1}^F = P_{h1}^F\{1/N \times (\lambda_{n0}^F + \lambda_{h0}^F + \lambda_{T0}^F)P_{B0}(1-P_{B1}) + \lambda_{h1}^F(1-P_{B1})\} \tag{7}$$

The parameter ρ is the actual offered load to a WLAN from the new call arrival and the handoff call arrival. Invoking this important property, we can use

$\rho_1 = \lambda_1^S / \mu_1^S + \lambda_1^F / \mu_1^F$ as the offered load to the WLAN, the Erlang-B formula calculates the blocking probability with the traffic ρ_1 and the number of channels c_1 [8]

$$P_{B1} = B(c_1, \rho_1) \tag{8}$$

where

$$P_B = B(c, \rho) = (\rho^c / c!) / (\sum_{i=0}^{c} \rho_i / i!)$$

3.2 The New Call Blocking Probability of CDMA System

The aggregate traffic rate into the CDMA system due to a slow MS is computed as follows:

$$\lambda_0^F = \lambda_{n0}^F + \lambda_{h0}^F + \lambda_{T0}^F \tag{9}$$

Here the take-back traffic rate component is given as

$$\lambda_{T0}^F = (\lambda_{n0}^F + \lambda_{h0}^F + \lambda_{T0}^F) P_{B0} (1 - P_{B1}) P_T^F \tag{10}$$

Thus, the aggregate traffic rate into the CDMA system due to a fast MS is given as

$$\lambda_0^F = N(\lambda_{n1}^S + \lambda_{h1}^S + \lambda_{T1}^S) P_{B1} + \lambda_{h0}^S \tag{11}$$

The generation rate of the handoff traffic of a slow MS in the CDMA system is calculated as

$$\lambda_{h0}^F = P_{h0}^F (\lambda_{n0}^F + \lambda_{h0}^F + \lambda_{T0}^F)(1 - P_{B0}) \tag{12}$$

The generation rate of the handoff traffic of a fast MS in the CDMA system is computed as

$$\lambda_{h0}^S = P_{h0}^F \{N(\lambda_{n1}^S + \lambda_{h1}^S + \lambda_{T1}^S) P_{B1}(1 - P_{B0}) + \lambda_{h0}^S (1 - P_{B0})\} \tag{13}$$

The probability of call blocking is given by the Erlang-B formula because it does not depend on the distribution of the session time. Invoking this important property, we can use $\rho_0 = \lambda_0^S / \mu_0^S + \lambda_0^F / \mu_0^F$ as the offered load to CDMA system, and blocking probability can be written as

$$P_{B0} = B(c_0, \rho_0) \tag{14}$$

3.3 The Handoff Call Dropping Probability of WLAN and CDMA System

Slow MSs are supposed to use WLAN channels. However, since handoff to CDMA system is also allowed, the probability of handoff call drop in WLAN can be calculated as follows. Let P_{10} denote the probability that a slow MS fails to be handoffed to a near WLAN. The probability of the calls, P_{B0}, in a WLAN denotes the probability of failed hand-up to the overlaying CDMA system due to the channel shortage. Then the handoff call dropping probability is

$$P_D^S \approx P_{10} P_{B0} + P_{10}(1-P_{B0})P_{F0}^S \tag{15}$$

Here P_{F0}^S is the probability that a slow MS handoff to CDMA system fail. The P_{10} is defined in such a way that the i th handoff request is successful but the $(i+1)$ th request is dropped:

$$P_{10} = f_1 + s_1 f_1 + s_1^2 f_1 + \hbar = f_1/(1-s_1) \tag{16}$$

where $f_1 = P_{h1} P_{B1}$, $f_0 = P_{h0} P_{B0}$, $s_1 = P_{h1}(1-P_{B1})$, and $s_0 = P_{h0}(1-P_{B0})$. f_i describe the probability that handoff fails due to channel shortage and the s_i is the probability of successful handoff. The overall probability of either dropping or handoff failure is

$$PD = R_S P_D^S + R_F P_D^F \tag{17}$$

where R_S and R_F is fraction of slow and fast MSs, respectively.

3.4 The Number of Handoffs

We will use the term handoff rate to refer to the mean number of handoffs per call. We use geometric models to predict handoff rates per call as cell shapes and sizes are varied. Approximating the cell as a circle with radius R and the speed of the mobile station with V, the expected mean sojourn time in the call initiated cell and in an arbitrary cell can be found [6], respectively

$$E[T_X] = 8R/3\pi E[V] \qquad E[T_Y] = \pi R/2E[V] \tag{18}$$

A user will experience a handoff if he moves out of the radio coverage of the base station with which he currently communicates. The faster he travels, probably the more handoffs he experiences. Using result from renewal theory, the expected number of handoffs given the speed of the user can be found.

$$E[N_h] = \frac{\pi E[V]}{4\mu R}(1 + \frac{4\mu R}{3\pi E[V] + 8\mu R}) \tag{19}$$

3.5 Grade of Service (GOS) and Network Selection

Among many system performance measures, *GOS* is most widely used. In fact users complain much more for call dropping than for call blocking. It is evaluated using the prespecified weights, *PB* and *PD*,

$$GOS = (1-\alpha)PB + \alpha PD \tag{20}$$

where *PB* and *PD* represent the blocking and dropping prob. of systems, respectively. The weight α emphasizes the dropping effect with the value of larger than one half.

A proposed perspective network selection procedure together with a few of preprocessing is shown in Fig. 3. For the estimation of the mobile speed, Global Positioning System (GPS) or Differential GPS can provide adequate location information. Using GPS and Time-of-Arrival (TOA) information from the user signal, we can estimate for user's velocity. We develop the selection algorithm based on an optimal velocity threshold. The problem here is to find V_T improving the GOS and decrease the number of handoff attempts (N_h) with the given traffic parameters and MS mobility; $f_\Lambda(\lambda)$ and $f_V(v)$. We have to find the velocity threshold satisfied the following equation.

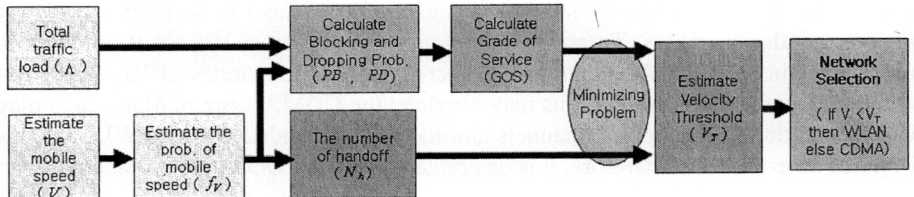

Fig. 3. Network selection Scheme

$$\min_{V_T}\{GOS(V), N_h(V)\} \quad (21)$$

The procedure is now concerned with the GOS in which the system wide new call blocking probability PB and the handoff call dropping probability PD are weighted to be averaged as in Equation (20). The GOS can be written as a function of V_T, and hence finding the optimum value of V_T minimizing the value of GOS and N_h is a typical minimization problem.

4 Numerical Examples

The proposed procedure is tested with a number of numerical examples for the overlaid structure. The test system consists of 10 WLANs in the CDMA system. The total traffic $\Lambda = \lambda_0 + n\lambda_1$, where λ_0 and λ_1 are the new call arrival rate for the CDMA system and the WLAN, respectively. The radius of the WLAN and the CDMA system are assumed 300m and 1000m, respectively. The average call duration is $1/\mu = 120$ sec. The number of channels in each CDMA system and WLAN is $c_0 = 30$, $c_1 = 10$ for the total $\Lambda = 60$ Erlang. Assume the traffic mobility distribution same as [6].

In operation phase use can draw a histogram to estimate the $\hat{f}_V(v)$, and the expected value of the mobile speed can be calculated by averaging the mobile speeds monitored by the system. Analytically we can obtain $E[V]$ for such a simple hypothetical velocity distribution [7]. And we consider four handoff strategies for comparison as follows.

    a) No overflow : A reference system where the two layers are kept completely independent.
    b) Overflow – From WLAN to CDMA system : A system where only overflow of new and handoff traffic for a slow MS to the CDMA system is allowed.

c) Overflow – From WLAN to CDMA system and vice versa : A system where overflow of new and handoff traffic for both slow and fast MS is allowed.
d) Take-back : Overflow of new or handoff traffic and take-back of both slow and fast MS to their appropriate layers.

Fig. 4 shows the plot of Equation (19) for the mobility distributions [6] of the MS in the system. As the velocity threshold increases, the number of handoff attempts in the system also increases. To achieve the goal of minimizing N_h, we want to place more users in the CDMA system, because crossing the boundaries of large cells becomes less frequent. However, this may overload the CDMA system. Many calls may be blocked due to the lack of channels and have to be handed down to WLAN. This imposes an extra cost. Therefore, it is desirable to keep the GOS in the system.

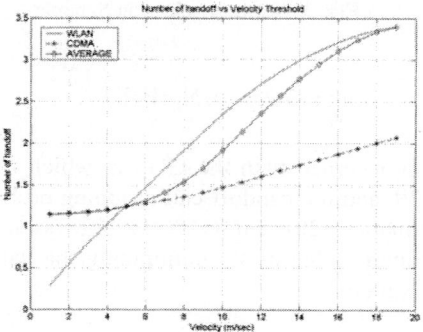

Fig. 4. The number of Handoff vs. velocity threshold

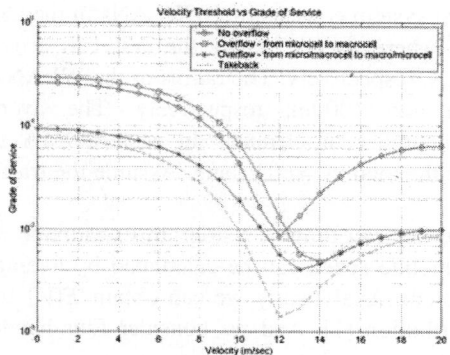

Fig. 5. Grade of Service vs. velocity threshold

We investigate the GOS, which is a function of both the traffic load and mobility distribution. Fig. 5 shows the plot of Equation (20) for the mobility distributions of the MS in the system. The vertical arrows in the figure show the range of the possible velocity thresholds at a certain load level. The lowest point in the range corresponds to the maximum allowable and optimal velocity threshold. Optimal V_T is 12m/sec, 14m/sec, 13m/sec, 12m/sec for case a), b), c) and d), respectively. Here the GOS of case c) and d) have minimums of nearly equal values, but V_T does different cases. Case d) is favorable (See Fig. 4) since V_T in the case d) is smaller than that of case c) and thus more users are serviced in the CDMA system while the WLAN serves the fewer users. As a results, the WLAN will give rise to a higher number of handoff requests for high-mobility users, and the corresponding number of handoff requests of the calls in progress may cause an excessive processing load in the network.

For the range exceeding the threshold, as V_T is smaller, more traffics can be accommodated for the increased P_{B1} for which more traffics are allocated to the WLAN. As the traffic increases, the V_T corresponding to the minimum P_{B0} becomes higher; more traffics should be assigned to the WLAN. For example, if the number of faster moving MS are more than those of the slower moving, the optimal V_T (in terms of GOS) lies in the relatively higher position of the region. From the mentioned figures above, when the overflow is likely to reduce the PD, we note that the traffic should be small enough as compared to the case without the overflow.

The take-back strategy provides the value of GOS nearly equal to case c) while it has the optimal velocity threshold smaller than that of case c). With all the observations in mind, the strategy we proposed has desirable characteristics, i.e., finding the optimal value of GOS and the number of handoff rate (See figure 4 and 5).

5 Conclusion

We have presented a handoff procedures with network selection deciding the optimal velocity threshold in order to improve the GOS and minimize the number of handoff attempts with the given traffic volume and four handoff strategies in WLAN and CDMA system. The simulation results show the dependency of the system performance upon the velocity threshold, V_T. The velocity threshold has shown to be an important system parameter that the system provider should determine to produce better GOS and lower handoff rate. From the simulation results we were able to validate the procedures determining the optimal V_T in which depends upon PD and PB as well as the number of handoff attempts. Furthermore, the take-back strategy is more favorable than other handoff strategies in this simulation environment.

References

1. Milind M. Buddhikot, Girish Chandranmenon, etal , Design and Implementation of a WLAN/CDMA2000 Interworking Architecture, IEEE Communications Magazine, November 2003.
2. Balasubramaniam, S., Indulska, J., Vertical Handover Supporting Pervasive Computing in Future Wireless Networks, Computer Communication Journal, Special Issue on 4G/Future Wireless networks. Vol 27/8, pp.708-719., 2003.

3. X. Lagrange and P. Godlewski, "Performance of a Hierarchical Cellular Network with Mobility-dependent handover strategies," in Proc. IEEE VTC '96, 1996.
4. C. W. Sung and W. S. Wong, "User Speed Estimation and Dynamic Channel Allocation in Hierarchical Cellular System," in Proc. IEEE VTC '94, 1994, pp. 91-95.
5. S. Rappaport and L. R. Hu, "Microcellular Communications Systems with Hierarchical Macrocell Overlays: Traffic Performance Models and Analysis," Proc. IEEE, Vol. 82, Sept. 1994.
6. Kwan L. Yeung and Sanjiv Nanda. "Channel Management in Microcell/Macrocell Cellular Radio Systems." IEEE Transactions on Vehicular Technologies, 45(4):601–612, November 1996.
7. JangSub Kim, WooGon Chung, HyungJin Choi and JongMin Cheong, Soon Park, "Determining Velocity Threshold for Handoff Control in Hierachically Structured Networks," PIMRC 98, 1998.
8. W. Fischer and K. S. Meier-Hellstern. "The Markov Modulated Poisson Process (MMPP) cookbook." Performance Evaluations, 18:149–171, 1992.

Scalable Hash Chain Traversal for Mobile Devices

Sung-Ryul Kim*

Division of Internet & Media and CAESIT
Konkuk University

Abstract. Yaron Sella recently proposed a scalable version of Jakobsson's algorithm to traverse a hash chain of size n. Given the hash chain and a computation limit m ($k = m + 1$ and $b = \sqrt[k]{n}$), Sella's algorithm traverses the hash chain using a total of kb memory. We improve the memory usage to $k(b-1)$. Because efficient hash chain traversal algorithms are aimed at devices with severely restricted computation and memory requirements, a reduction by a factor of $(b-1)/b$ is considered to be important. Further, our algorithm matches the memory requirements of Jakobsson's algorithm while still remaining scalable. Sella's algorithm, when scaled to the case of Jakobsson's algorithm, has a memory requirement of about twice that of Jakobsson's.

Keywords: efficient hash chain traversal, secure hash, pebbles.

1 Introduction

Many useful cryptographic protocols are designed based on the *hash chain* concept. Given a *hash function* $f()$ which is assumed to be difficult to invert, a hash chain is a sequence $< x_0, x_1, \ldots, x_n >$ of values where each value x_i is defined to be $f(x_{i-1})$. The hash chain is being used as an efficient authentication tool in applications such as the S/Key [2], in signing multicast streams [5], message authentication codes [5,6], among others.

Traditionally the hash chain has been used by one of two methods. One is to store the entire hash chain in memory. The other is to recompute the entire hash chain from x_0 as values are exposed from x_n to x_0. Both methods are not very efficient. The first one requires a memory of size $\Theta(n)$ while the second one requires $\Theta(n)$ hash function evaluations for each value that is exposed. The memory-times-storage complexity of the first method is $O(n)$ and it is $O(n^2)$ for the second method. As mobile computing becomes popular, small devices with restricted memory and computation powers are being used regularly. As these devices are being used for jobs requiring security and authentication, the memory and computation efficiency of hash chain traversal is becoming more important.

---

* Corresponding author: kimsr@konkuk.ac.kr

Influenced by amortization techniques proposed by Itkis and Reyzin [3], Jakobsson [4] proposed a novel technique that dramatically reduces the memory-times-storage complexity of hash chain traversal to $O(\log^2 n)$. The algorithm by Jakobsson uses $\lceil \log n \rceil + 1$ memory and makes $\lceil \log n \rceil$ hash function evaluations for each value that is exposed. The result has been further improved by Coppersmith and Jakobsson [1], to reduce the computation to about half of the algorithm in [4]. Both results are not intended for scalability. That is, the amount of computation and memory depends only on the length of the chain n and they cannot be controlled as such needs arise. For example, some device may have abundant memory but it may be operating in a situation where the delay between exposed hash values are critical. To address the issue Sella [7] introduced a scalable technique that makes possible a trade-off between memory and computation requirements. Using the algorithm in [7], one can set the amount m of computation per hash value from 1 to $\log n - 1$. The amount of memory required is $k\sqrt[k]{n}$ where $k = m + 1$. In the same paper, an algorithm designed specifically for the case $m = 1$ is also presented. The specific algorithm reduces the memory requirement by about half of that in the scalable algorithm.

We improve the memory requirement of Sella's scalable algorithm. While Sella's algorithm has the advantage of scalability from $m = 1$ to $\log n - 1$, the memory requirement is about twice that of Jakobsson's algorithm when $m = \log n - 1$. Our algorithm reduces the memory requirement of Sella's algorithm from $k\sqrt[k]{n}$ to $k(\sqrt[k]{n} - 1)$. This might not seem to be much at first. However, if k is close to $\log n$ the reduction translates into noticeable difference. For example, if $k = \frac{1}{2} \log n$, the memory requirement is reduced by a factor of $\frac{1}{4}$. If $k = \log n$, the memory requirement is reduced by half. Of particular theoretical interest is that our algorithm exactly matches the memory and computation requirements of Jakobsson's algorithm if we set $k = \log n$. Our technique has one drawback as the scalability is reduced a little bit. It is assumed that $m \geq \sqrt[k]{n}$, that is, our algorithm does not scale for the cases where m is particularly small.

We note here that, as it is with previous works, we count only the evaluation of hash function into the computational complexity of our algorithm. This is because the hash function evaluation is usually the most expensive operation in related applications. We also note that the computation and memory bounds we achieve (and in the previous works) are worst-case bounds (i.e., they hold every time a value in the hash chain is exposed).

2 Preliminaries

2.1 Definitions and Terminology

We mostly follow the same terminology and basic definitions that appear in [7]. A hash chain is a sequence of values $< x_0, x_1, \ldots, x_n >$. The value x_0 is chosen at random and all other values are derived from x_0 in the following way: $x_i = f(x_{i-1})$, $(1 \leq i \leq n)$, where $f()$ is a hash function. A single value x_i is referred to as a *link* in the chain. The chain is generated from x_0 to x_n but is exposed from x_n to x_0. We use the following directions terminology: x_0 is placed at the left end, and it is called the *leftmost* link. The chain is generated *rightward*, until we reach the *rightmost* link x_n. The links of the chain are exposed *leftward* from x_n to x_0. The time between consecutive links are exposed is called an iteration. The above hash chain has $n+1$ links. However we assume that at the outset, x_n is published as the public key. Hence, we refer to $X = <x_0, x_1, \ldots, x_{n-1}>$ as the full hash chain and ignore x_n.

The goal of the hash chain traveral is to expose the links of the hash chain one-by-one from right to left. The efficiency of a traversal is measured in both the amount of memory required and the number of hash function evaluations between each exposed link.

In the work by Sella the concept of a *b-partition* of the hash chain is used. It can be also be used to describe Jakobsson's algorithm. We repeat the definitions here.

Definition 1. *A b-partition of a hash chain section means dividing it into b sub-sections of equal size, and storing the leftmost link of each sub-section.*

Definition 2. *A recursive b-partition of a hash chain section means applying b-partition recursively, starting with the entire section, and continuing with the rightmost sub-section, until the recursion encounters a sub-section of size 1.*

The b-partition and recursive b-partition are easier to describe when we map the partition to a b-ary tree. We will use the basic terminology pertaining to trees without definition. The following is the definition of the mapping between the recursive b-partition and a b-ary tree. It is repeated also from [7].

Definition 3. *A recursive b-tree is a recursive mapping of a recursive b-partition onto a tree as follows.*

- *Each node in the tree is a section or a sub-section. In particular, the entire hash chain is represented by the root of the tree.*
- *For every section L that is partitioned from left to right by sub-sections $L_1, L_2, \ldots, L_b$, the corresponding node N is the parent of the corresponding nodes $N_1, N_2, \ldots, N_b$, and the children keeping the same left-to-right order.*

See Figure 1 for an example. It shows a recursive 4-tree with 3 levels. The root of the tree corresponds to the entire hash chain and the tree leaves correspond to single links in the hash chain. Even though each node is shown to be storing b links, only $b-1$ links will be the memory requirement for a node because the leftmost link is stored in the parent.

Sub-sections induced by recursive b-partition are *neighbors* if their corresponding nodes are at the same level and appear consecutively at the natural depiction of the recursive b-tree. Note that they may not share the same parent. Depending on its position in the recursive b-partition, a sub-section can have a *left neighbor*, a *right neighbor*, or both. A node in a recursive b-tree will sometimes be called a sub-section, meaning the corresponding sub-section in the recursive b-partition.

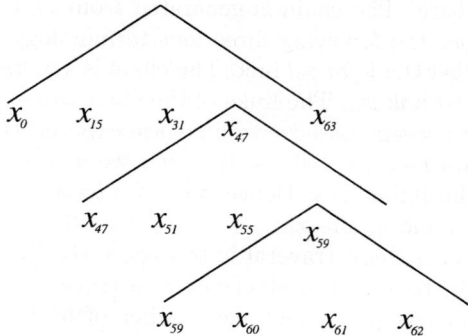

Fig. 1. Example 4-tree with 3 levels

In order to traverse the hash chain we adopt the notion of pebbles and stored links from previous works. A *pebble* is a dynamic link that moves from left-to-right in a hash chain. The mission of a pebble is to induce a b-partition of a subsection S. That is, it starts at a stored link and traverses the chain while storing the values corresponding to the b-partition of S. The initial stored link will be received from the (sub-)section that S is a part of. For a pebble that induces a b-partition $(x_{i_1}, x_{i_2}, \ldots, x_{i_b})$, the link x_{i_b} is called the *destination* because the pebble does not have to move any further after reaching x_{i_b}.

2.2 Jakobsson's Algorithm

Jakobsson's algorithm can be considered to be using the recursive 2-tree of the entire hash chain. There are $\lceil \log n \rceil + 1$ levels in the tree. One pebble is present at each level except at the root. When the algorithm starts, a pebble is placed at the mid-point of the sub-section S at each level.

When the mid-point is exposed the pebble is moved to the start of the left neighbor. From there it traverses the hash chain two links per exposed link. Because it (and all other pebbles) moves at twice the speed that the links are exposed, a pebble reaches the end of the left neighbor when all the links in S has been exposed. It can be shown that at most one of the pebbles in two adjacent levels are moving at any point. This leads to the above mentioned computation cost per iteration. We note that because the pebble starts to move only after its destination is reached and because there are only one link to store (the leftmost node is stored in the parent), the pebble never stores a link and moves to the right. Thus the memory requirement is the same as the number of pebbles (plus one at the root).

2.3 Sella's Algorithm

Sella's algorithm uses a recursive b-partition. As in the Jakobsson's algorithm, one pebble is placed at each level. However, the pebble is moved to the start of the left neighbor immediately when the rightmost link (not the rightmost stored link) in a section is exposed. The pebble moves at the same speed as the links are exposed.

The parameter b is set to be $\sqrt[k]{n}$ where $k = m + 1$. That is, the tree will have k levels. Because no pebble is needed at the root, the computational requirement per iteration is m. Initially, there are $b-1$ stored links at each level. However, because pebbles start to move before any stored link is exposed, one memory cell is required for each pebble. Thus, the total memory requirement amounts to $k\sqrt[k]{n}$. If we set $k = \log n$, the memory requirement becomes $2\log n$, which is about twice that of Jakobsson's algorithm.

2.4 Intuitions for Our Algorithm

In our proposed algorithm, each pebble starts to move after the rightmost stored value is used, thus removing the memory requirement for the pebbles. Because it started late (as in Jakobsson's algorithm) it makes accelerated moves. That is, it traverses b links (called *making a move*) while one link is exposed. It cannot be moving all the time because if so, the computational requirement per iteration will be $b \times m$. It makes a move about every $b/(b-1)$ times, enough to catch up to its late start but not at a regular interval as described later.

It is quite simple in Jakobsson's algorithm because there can be no overlap in computation for pebbles residing in two adjacent levels (in the b-tree). We have to consider b adjacent levels. There will be overlaps if we allocate the moves for different levels at regular intervals. So we have to carefully allocate the iterations where each pebble makes a move. The tricky part is to show that there always exists a way to allocate iterations such that a pebble is never too late to its destination.

3 Improved Algorithm

Assume that we have a hash chain $< x_0, x_1, \ldots, x_{n-1} >$ to traverse. Let k be $m+1$ and let b be $\sqrt[k]{n}$. We assume that $k-1$ is a multiple of m and we also assume that $m \geq b$.

3.1 Algorithm

We first form a recursive b-partition (also the corresponding recursive b-tree) off-line. We put a pebble at the rightmost stored link in each sub-section. Then, the following loop is repeated for n times.

- Exposure loop
  1. Expose and discard L, the current rightmost link in the hash chain.
  2. If $(L = x_0)$ stop.
  3. For each pebble p do
     - If L is a rightmost stored link in a sub-section S' and p is on L, then move p to the left end of S where S is the left neighbor of S' and call **Schedule Iterations** for p.

4. Advance all pebbles that are allocated the current iteration by b links.
5. Repeat from Step 1.

Now we describe the procedure **Schedule Iterations**. We define *level numbers* in the recursive b-tree. The lowest level is said to be at level 1 and the level number increases as we go up the tree. Thus, the root will be at level k. Assume that p is at level t. Let S be the sub-section where p is newly assigned. Let S' be the right neighbor of S. There are b^{t+1} links in S. We give a natural correspondence between a link in S and an iteration where a link in S' is exposed. The leftmost link in S corresponds to the iteration where the rightmost link in S' is exposed. And the correspondence proceeds towards the right in S and towards the left in S'. That is, the links in S and the links in S' are matched in reverse order. When we say that we allocate a link x_i in S to p it means that p will advance b times rightward at the time when the link in S' that corresponds to x_i is exposed. Note that the iterations corresponding to the b^t leftmost links in S have already passed. Let $(L_1, L_2, \ldots, L_b)$ be the b-partition of S.

Our strategy is to assign links to p as far right as possible. There are three conditions to consider.

1. A link should not be assigned to p if it is (or will be) assigned to a pebble in lower $b-1$ levels. This ensures that at any iteration, at most one pebble in consecutive b levels is making a move. There are many sub-sections in the lower $b-1$ levels but the schedule for each pebble will be the same regardless of the particular sub-section.
2. Pebble p should not move too fast. Because p has to store a link when it reaches a position of a stored link for S, one stored link in S' has to be exposed before we store one link in S.
3. The schedule assigned for the pebble q at level $t-1$. Pebble q cannot be moved until it receives the rightmost stored link in S'. Thus, the schedule for p must finish before the schedule for q starts. Although q is not in L_b yet, we know how it will be scheduled in L_b because the schedule is the same for all sub-sections that q is assigned to.

By the above conditions, the procedure first looks for the schedule of q and it starts at the left most link allocated for q. Then it moves leftward and allocates to p the links that are not allocated to any pebbles at the lower $b-1$ levels.

3.2 Correctness

We show by a series of lemmas that there always exists a schedule that satisfies the conditions mentioned in the algorithm. Let S be the sub-section at level t where p is newly assigned. Also let $(L_1, L_2, \ldots, L_b)$ be the b-partition of S. The following lemma shows that we have enough unallocated links to allocate for pebble p at level t and the $b-1$ pebbles in the lower $b-1$ levels (condition 1).

Lemma 1. *There are enough links in S for the b pebbles p and in lower $b-1$ levels.*

Proof. By the time p is moved to the start of S the iterations for the links in L_1 have already passed. So there are a total of $b^t(b-1)$ links to be allocated. p has to move across $b^t(b-1) = b^{t+1} - b^t$ links to reach its destination. Thus, it has to make $b^t - b^{t-1}$ moves (same number of links are to be allocated). At level $t-1$, there are $b-1$ sub-sections to schedule and in each sub-section there are $b^{t-1} - b^{t-1}$ moves to be made by the pebble in level $t-1$. So at level $t-1$ there are $b^t - 2b^{t-1} + b^{t-1}$ moves to be made in total. From then on to the level $t-b+1$, the same number of moves are to be made.

By adding them up, we can see that we need $b^{t+1} - 2b^t + 2b^{t-1} - b^{t-2}$ links to allocate to the pebbles. Subtracting this number from the number of available links $b^{t+1} - b^t$ we have the difference $b^t - 2b^{t-1} + b^{t-2}$, which is always positive if $b \geq 2$.

The following lemma and corollary shows that the memory requirement (condition 2) is satisfied.

Lemma 2. *There are enough links in L_b for the last b^{t-1} moves of p.*

Proof. There are b^t links that can be allocated in L_b. Pebble p at level t requires b^{t-1} moves by the condition of the lemma. The pebble at level $t-1$ has to make $b^{t-1} - b^{t-2}$ total moves. From then onto the next $b-2$ levels, the same number of moves are needed. Adding them up leads to $b^t - b^{t-1} + b^{t-2}$. Substracting this number from the available number of links b^t results in $b^{t-1} - b^{t-2}$, which is positive if $b \geq 2$.

Corollary 1. *Pebble p moves slow enough so that the memory requirement for level t is $b-1$.*

Proof. The above lemma shows that the last b^t links to be traversed by p can be actually traversed after all the stored links in the right neighbor of S. Because the first $b^t(b-2)$ moves are allocate as far to the right as possible and there are enough number of links to allocate for p, p will move at a regular pace at the worst. Thus, p moves slow enough to satisfy the memory requirement.

Even if we have shown that there are enough number of links to allocate for every pebble, it is not enough until we can show that condition 3 is satisfied. Consider a pebble r at level 1, because no pebble needs to be scheduled after r, r can be allocated the rightmost link in the sub-section. A pebble r' at level 2 has to finish its computation. So the leftmost link allocated to r' will be the b-th one from the rightmost link in a sub-section. We have to make sure that the leftmost link allocated to a pebble at every level is to the right enough so that the upper levels have enough room to finish its moves. The following lemma shows that to the left of the leftmost link allocated for pebble q at level $t-1$ (condition 3). Note that the final difference in the proof of the above lemma will be used in the proof of the next lemma.

Lemma 3. *There is enough links to allocate to p and the $b-1$ pebbles in the lower levels to satisfy condition 3.*

Proof. As mentioned above, the leftmost allocate link for level 1 is the first from the rightmost link in a sub-section. For level 2, it is the b-th link from the rightmost link. If $b = 2$, the leftmost allocated link for level 3 ls b^2-th one from the rightmost link. We can easily show that if $b = 2$, the leftmost allocate link for level m is b^{m-1}-th link from the rightmost link in a sub-section. However, if $b > 2$, the leftmost allocated link for level 3 is $(b^2 + 2b)$-th one from the rightmost link because $2b$ links that are already allocated to the pebble at level 1 cannot be allocated. We can prove by induction that the leftmost link for level m is at most the $(b^{m-1} + 2b^{m-2} + 2^2 b^{m-3} + \cdots)$-th one from the rightmost link in a sub-section. This number sums to at most $b^m/(b-2) < b^m$.

In the proof of lemma 2, we have $b^{t-1} - b^{t-2}$ links that need not be allocated to any pebble. Because we have allocate links to pebbles in levels t down to $t - b + 1$, we consider the leftmost link allocated to the pebble at level $t - b$. If we set m in the above calculation to $t - b$, then the leftmost link allocated to the pebble at level $t - b$ is at most the (b^{t-b})-th one from the rightmost link in a sub-section. Substracting from the available free links, we get $b^{t-1} - b^{t-2} - b^{t-b}$, which is nonnegative if $b \geq 2$.

Using the above lemmas, it is easy to prove the following theorem.

Theorem 1. *Given a hash chain of length n, the amount of hash function evaluation m, and $k = m + 1$, the algorithm traverses the hash chain using m hash function evaluations at each iteration and using $k(\sqrt[k]{n} - 1)$ memory cells to store the intermediate hash values.*

4 Conclusion

Given the hash chain and a computation limit m ($k = m + 1$ and $b = \sqrt[k]{n}$), we have proposed an algorithm that traverses the hash chain using a total of $k(b - 1)$ memory. This reduces the memory requirements of Sella's algorithm by a factor of $1/b$. Because efficient hash chain traversal algorithms are aimed at devices with severely restricted computation and memory requirements, this reduction is considered to be important. Further, our algorithm is matches the memory requirements of Jakobsson's algorithm while still remaining scalable. Sella's algorithm, when scaled to the case of Jakobsson's algorithm, has a memory requirement of about twice that of Jakobsson's.

References

1. D. Coppersmith and M. Jakobsson, Almost optimal hash sequence traversal, *Proc. of the Fifth Conference on Financial Cryptography (FC) '02*, Mar, 2002.
2. H. Haller, The S/Key one-time password system, *RFC 1760*, Internet Engineering Taskforce, Feb. 1995.
3. G. Itkis and L. Reyzin, Forward-secure signature with optimal signing and verifying, *Proc. of Crypto '01*, 332–354, 2001.

4. M. Jakobsson, Fractal hash sequence representation and traversal, *IEEE International Symposium on Information Theory (ISIT) 2002*, Lausanne, Switzerland, 2002.
5. A. Perrig, R. Canetti, D.Song, and D. Tygar, Efficient authentication and signing of multicast streams over lossy channels, *Proc. of IEEE Security and Provacy Symposium*, 56–73, May 2000.
6. A. Perrig, R. Canetti, D.Song, and D. Tygar, TESLA: Multicast source authentication transform, *Proposed IRFT Draft*, http://paris.cs.berkeley.edu/ perrig/
7. Yaron Sella, On the computation-storage trade-offs of hash chain traversal, *Proc. of the Sixth Conference on Financial Cryptography (FC) '03*, Mar, 2003.

A Rate Separation Mechanism for Performance Improvements of Multi-rate WLANs

Chae-Tae Im, Dong-Hee Kwon, and Young-Joo Suh

Department of Computer Science and Engineering,
Pohang University of Science and Technology (POSTECH),
San 31, Hyoja-Dong, Nam-Gu, Pohang, Korea
{chtim, ddal, yjsuh}@postech.ac.kr

Abstract. The fundamental access mode of the IEEE 802.11 MAC protocol is contention based. If the traffic load is heavy or the number of contending station is large, the number of collisions is increased and it leads to the performance degradation. In this paper, we propose a mechanism that tries to reduce the number of collisions by separating and grouping the contending stations and distributing those groups over time in multi-rate WLANs. For this, we issue the trade-off relationship between the throughput fairness and temporal fairness in multi-rate WLANs. Considering the trade-off relationship, we propose a Rate Separation (RS) mechanism in which the grouping is done based on the current transmission rates of contending stations. From our simulation study, we show that the proposed mechanism reduces the number of collisions and achieves improved performance over the IEEE 802.11b WLANs.

1 Introduction

The IEEE 802.11 standard defines the physical (PHY) and Medium Access Control (MAC) layers for both infra-structured and ad hoc networks [4]. The original standard supports the data rates of 1 and 2 Mbps. To provide higher bandwidth to users, the IEEE 802.11b standard [8] has been published. In this standard, a high-rate PHY extension for the direct sequence spread spectrum (DSSS) system is specified in the 2.4 GHz band and it provides additional 5.5 and 11 Mbps data rates. The IEEE 802.11a PHY extension [10] is a new standard that operates at the 5 GHz band with Orthogonal Frequency Division Multiplexing (OFDM) radio and provides the data rate ranging from 6 Mbps up to 54 Mbps.

All of the IEEE 802.11 extensions use the identical MAC protocol. The IEEE 802.11 MAC provides two channel access mechanisms, namely, Distributed Coordination Function (DCF) and Point Coordination Function (PCF). DCF is based on the Carrier Sense Multiple Access with Collision Avoidance (CSMA/CA) channel access mechanism while PCF is based on a simple polling mechanism. The contention based DCF is the mandatory access mode of IEEE 802.11 while PCF is an optional function for contention-free access mode. In this paper, only the basic access mode (DCF) is considered for discussion.

In the DCF mode, when there are many stations or massive load in a BSS (Basic Service Set), collisions occur frequently and much of the bandwidth is wasted. The probability of collisions is directly proportional to the number of contending stations and the collisions give significant negative impacts on the overall performance. [5,6,7] discussed this impact on the performance of IEEE 802.11 via numerical analysis and simulations. To tackle this problem, the authors in [6,7] proposed adaptive backoff algorithm which adaptively controls the contention window size based on the traffic condition of the network to reduce the collisions. But, these works limited by assuming the wireless network of a single transmission rate. In multi-rate WLANs, there is another problem affecting the network throughput due to the fact that the transmissions of stations at low rates take more time than those of stations at high rates [1,2,3]. In multi-rate WLANs, it seems to reasonable that the stations at high rates should transmit more data packets than those at low rates to enhance the network throughput. But, as the IEEE 802.11 MAC gives the equal number of channel accesses to each station regardless of its transmission rate, the network throughput is degraded when there are many stations at low rates. To enhance the network throughput in this case, a new scheme considering the transmission rate of stations is required.

In this paper, we propose a simple mechanism that tries to reduce the probability of collisions by grouping the contending stations and distributing them into several time periods. For the grouping of stations, we introduce a transmission rate based grouping strategy considering the characteristic of multi-rate capability of IEEE 802.11b. In this mechanism, only stations permitted by an AP (Access Point) can contend for the medium during a given time period. By doing this, the number of contending stations are dramatically decreased, and thus, the number of collisions is also reduced. This eventually leads to the improved performance. To provide fairness among the stations in terms of throughput and temporal share, the time periods are adaptively adjusted by the AP according to the network condition.

2 IEEE 802.11 DCF

The mandatory DCF, which is based on CSMA/CA, is the primary access protocol for the automatic sharing of the wireless medium between stations and APs having compatible PHYs. As described in [4], before a station starts transmission, it must sense whether the wireless medium is idle for a time period of Distributed InterFrame Spacing (DIFS). If the channel appears to be idle for a DIFS, the station generates a random backoff time, and waits until the backoff time reaches 0. The reason for this is to reduce the collisions occurred by the situation that many stations waiting for the medium to become idle can transmit frames at the same time. Thus, the distinct random backoff deferrals of stations can reduce the collision probability.

The DCF mode provides two different handshaking protocols for frame transmissions. In DCF, a sending station must wait for an ACK frame from the receiving station. This is due to the fact that the sending station cannot correctly detect a collision at the receiving station, and it cannot listen to the medium while it is transmitting due to the difference between the transmitted and the received signal power strengthes for the wireless medium. Thus, the basic handshaking procedure of DCF for data frame exchanges follows a DATA-ACK sequence. An optional handshaking procedure requires that the sending station and the receiving station exchange short RTS (Request-To-Send) and CTS (Clear-To-Send) control frames prior to the basic handshaking procedure. The RTS/CTS exchange provides a virtual carrier sensing mechanism in addition to physical carrier sensing to prevent the hidden terminal problem. Any stations hearing either a RTS or CTS frame update their Network Allocation Vector (NAV) from the duration field in the RTS or CTS frame. All stations that hear the RTS or CTS frame defer their transmissions by the amount of NAV time. The overhead of RTS/CTS frames exchange becomes considerable when data frame sizes are small. Thus, RTS/CTS frame exchange should be based on the size of a data frame. IEEE 802.11 defines a configurable system parameter, *dot11RTSThreshold*, for RTS/CTS exchange.

3 Proposed Mechanism

3.1 Traffic Separation by Transmission Rate

In this section, we present a mechanism termed Rate Separation (RS). We assume a network topology of a single cell (BSS: Basic Service Set) controlled by an AP (Access Point) where all traffics occur between the AP and stations. In the RS mechanism, the AP can group the stations based on their transmission rates and allocate proper resources to each group. Figure 1 illustrates the basic idea of the RS mechanism. As shown in the figure, a periodic super-frame consists of multiple sub-frames. Each sub-frame is started by the Separation Beacon Messages (SBMs). Figure 2 shows the structure of the SBM. In the figure, the 1-bit *SBM* flag is used to indicate that this beacon is SBM and the 16-bit *Duration* field is the duration of the current contention block (sub-frame period) initiated by the SBM. The 7-bit *Rate* field is used to indicate the transmittable rate(or multiple rates) in the sub-frame period. The transmittable rate means that only the stations at the same transmission rate specified in the SBM can join the contentions to get the medium. By the *Rate* field in the SBM, the contentions for the medium of each station is separated by its current transmission rate. For

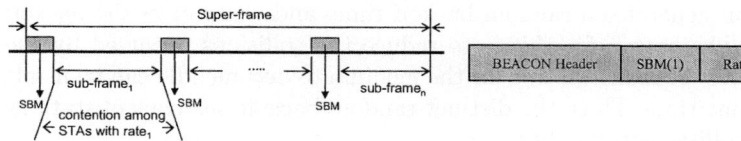

Fig. 1. Rate Separation Mechanism

Fig. 2. Separation Beacon Message Format

example, assuming there are 11Mpbs, 5.5Mbps and 2Mbps transmission rates, the first SBM can be used to announce that only the stations at 11Mbps have a chance for transmission, and the second and third SBMs are used for stations at 5.5Mbps and 2Mbps, respectively.

In the RS mechanism, to calculate the proper duration of each sub-frame the AP should maintain a network information table which includes station's ID, the current transmission rate, and the average packet size. The AP can know easily which stations are in its BSS through the association procedure between the AP and stations, and the current transmission rate of each station by measuring SNR (Signal-to-Noise Ratio). Since the transmission times for a data payload of the same length can be different according to transmission rates, the AP should know the average transmission time for each transmission rate group to calculate the value of the *Duration* field in the SBM. Based on the network information table, the AP calculates the duration of each sub-frame and super-frame, and allocates the resource for stations. The duration D_{Super_frame} and $D_{Sub\_frame_i}$ can be determined by the following equations.

Suppose that D_{Super_frame} and D_{Sub_frame} are the duration of the super-frame and the sub-frame, respectively, and let N_{Rate_i} be the number of stations at $Rate_i$ and T_{Rate_i} be the average transmission time of stations at $Rate_i$. Then, D_{Super_frame} and D_{Sub_frame} can be determined by the following equations.

$$D_{Super\_frame} = \sum_{Rate_i} D_{Sub\_frame_i} \qquad (1)$$

$$D_{Sub\_frame_i} = N_{Rate_i} \cdot T_{Rate_i} \cdot \alpha \qquad (2)$$

where α is the fraction factor which determines the duration of each sub-frame. If α value is larger than 1, statistically, all stations in a group have a chance to transmit a data frame at least once. Excessively or insufficiently allocated sub-frame duration may degrade the network performance. When the sub-frame duration is allocated excessively ($\alpha \gg 1$), the fairness problem can arise or the network utilization can be worsen due to the idle time wasted in the sub-frame if there are not enough stations. Contrarily, when the allocated sub-frame duration is insufficient ($\alpha \ll 1$), the number of collisions can increase and some stations can't transmit their frames and should wait until the next sub-frame. By Equations (1) and (2), stations in each group have different chances of getting the channel according to their transmission rates.

The key idea of the RS mechanism is to reduce the number of contending stations by separating them according to their transmission rates and distributing those groups over time. Suppose that the number of contending stations is n. In the IEEE 802.11 MAC, when the current transmission is finished, all n stations participate in the contention for the medium. As discussed in [6,7], the probability of collisions is directly proportional to the number of contending stations and the collisions give significant negative impacts on the overall performance. But, in the RS mechanism the contending stations are grouped and separated by their current transmission rates and the number of contending stations is limited for a given period. Thus, when the current transmission is finished, not all n stations but some among n stations participate in the contention for the medium. Since the other stations that do not currently participate in the contention have a chance to contend for the medium in the next period, the RS mechanism tries to reduce the probability of collisions by grouping the contending stations and distributing them into several time periods which can be dynamically adjusted by the AP according to the network condition. There can be different criteria for grouping the stations, but we choose the station's current transmission rate as a grouping criterion because it is one of the primary factors that directly affect the network performance [2,3].

3.2 Adaptive Sub-frame Allocation

In the RS mechanism, an improper channel allocation by the AP can arise due to several facts: (1) station's coming in and out of a BSS, (2) changes of stations' transmission rates due to mobility within a BSS, and (3) burst traffic generation at stations. (1) and (2) can be easily treated by the AP via the reassociation procedure and continually monitoring the transmission rate of each station. Thus, those changes can be easily updated to the network information table and the updated information can be announced in the next SBM frame. But, it is very difficult to solve (3). For this, we propose a delay sensing mechanism that the AP sends the next SBM frame immediately, after it detects that there is no traffic for a certain period of time. The following Equation (3) shows the timeout value for the delay sensing denoted by $T_{ds\_timeout}$.

$$T_{ds\_timeout} = T_{DIFS} + T_{delay\_sense} \quad (3)$$

where $T_{delay\_sense} = T_{slot} \cdot W_{delay}$ is the acceptable time to continue the subframe even if there is no traffic and W_{delay} is the window size for the delay sensing.

When there are few stations in the BSS, the RS mechanism may not be efficient since the probability of collisions is low and there are enough temporal resources for stations. Moreover, the frequent generation of SBM frames can be overhead, and may disturb the transmission of other stations. Thus, we adaptively use the RS mechanism. If the duration of a sub-frame is below a certain threshold value, the sub-frame is merged with the next sub-frame. As a result, if the number of sub-frames is more than two in a super-frame, the RS Mechanism is turned on. Figure 3 shows the algorithmic description of the adaptive RS mechanism.

3.3 Rate Fairness

In single transmission rate WLANs, the throughput fairness implies the fair channel occupancy time among stations. As the equal number of channel accesses is guaranteed by the random contention mechanism of CSMA/CA, each

```
for(all rates)
     if(duration of sub-frame < threshold)
          merge with the next sub-frame
     else
          allocate the sub-frame
if (number of sub-frames >= 2 )
          ON the rate separation mechanism
else
          OFF the rate separation mechanism
```

Fig. 3. Adaptive RS Algorithm

station can have the equal amount of time share and thus can achieve the equal throughput in single rate WLANs. But, in multiple rates IEEE 802.11b or IEEE 802.11a, they do not have the same meaning any more. If providing the throughput fairness is a primary concern, as the legacy 802.11 MAC provides, the wireless channel becomes under-utilized because a large portion of the channel (time) is occupied by the long low rate transmissions. Considering that the transmission time for 1 packet at 2Mbps gives the time for 5 packets transmissions at 11Mbps, we can easily expect that giving the same transmission opportunity to the stations regardless of their transmission rate will lead to overall network throughput reduction. On the other hand, if providing fair temporal share (temporal fairness) is the primary concern, stations at high rates should be given more chances to transmit to make up their relatively shorter transmission time compared to the long low rate transmission. In this case, achieving the temporal fairness (by some means) gives improved throughput performance, but the throughput fairness among the stations is broken.

To provide the fair share of the medium resource, we make use of the fact that the AP controls everything related to the medium resource in the RS mechanism. The AP can dynamically control the sharing of the channel by considering both the throughput fairness and the fair temporal share. This can be achieved by modifying the Equation (2) as follows.

$$D_{Sub\_frame_i} = \alpha \cdot \left(N_{Rate_i} \cdot T_{Rate_i} \cdot \beta + N_{Rate_i} \cdot T_{Rate_{highest}} \cdot (1-\beta) \right) \quad (4)$$

where $T_{Rate_{highest}}$ is the average transmission time of stations at the highest rate in the system, and β is the fairness preference factor and it has a value between 0 and 1. $N_{Rate_i} \cdot T_{Rate_i} \cdot \beta$ in Equation (4) is related to the throughput fairness because all stations in each transmission rate group have statistically have a chance to transmit a data frame. $N_{Rate_i} \cdot T_{Rate_{highest}} \cdot (1-\beta)$ is related to the temporal fairness because the transmission time is adjusted to that of the highest rate. For example, if $\beta = 0$, then the AP allocates the medium based on the throughput fairness while the resource is allocated by the temporal fairness basis when $\beta = 1$. In the next section, we investigate the impact of β on the performance.

4 Performance Evaluation

In this section, we present the results of our simulation study using the *NS2* simulator [9]. Figure 4 shows the network topology used in our simulation study.

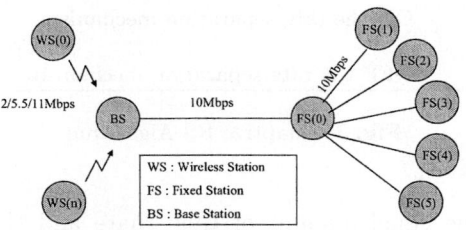

Fig. 4. Network Topology for Simulation

As shown in the figure, there are n wireless stations (*WS(0)* - *WS(n)*) in the wireless part and six fixed stations *(FS(0) - FS(5))* exist in the wired part. We assume that the wireless stations support data rates of 2, 5.5, and 11Mbps and use the DCF mode with the RTS/CTS handshake. Each *WS* is a CBR (Constant Bit Rate) traffic source to a *FS*. Each *FS* in the wired part can have maximally 6 traffic sinks at a time. Each *WS* generates the CBR traffic at the rate of 100Kbps or 150Kbps. For a given traffic load, we also vary the packet size to 512 and 1024 bytes. At the same traffic load, the 512 byte packet sending rate is twice faster than the 1024 byte packet sending rate. So, when the smaller packet size (512 byte)is used, there are more chances of collisions. Throughout the whole simulation runs, the number of stations at each transmission rate is equally divided. The RS mechanism starts activating when the number of stations is above 18 when the packet size is 512 byte and 12 when it is 1024 byte. The fraction factor α in Equation (4) is set to 1 and the rate fairness preference factor β values in Equation (4) is varied to 0, 0.5 and 1 to measure its impact on the performance.

Figure 5 shows the number of collisions observed during the whole simulation time as a function of the number of contending stations. Figure 5(a) shows the case when the IEEE 802.11 MAC is used. In the figure, we can see that the number of collisions rapidly increases as the number of contending stations are increased. By comparing the plot of 100Kbps-512bytes and 100Kbps-1024bytes, we can also verify that the number of collisions is more rapidly increased when the packet sending rate is higher. Figure 5(b) shows the number of collisions when the RS mechanism is used. Compared to Figure 5(a), the number of collisions is significantly reduced due to the fact that the RS mechanism tries to group the contending stations based on their transmission rates and temporally distribute them into several sub-frame durations. Note that we do not show the number of collisions in Figure 5(b) when the number of contending stations is less than the threshold at which the RS mechanism is activated (18 and 12 stations for

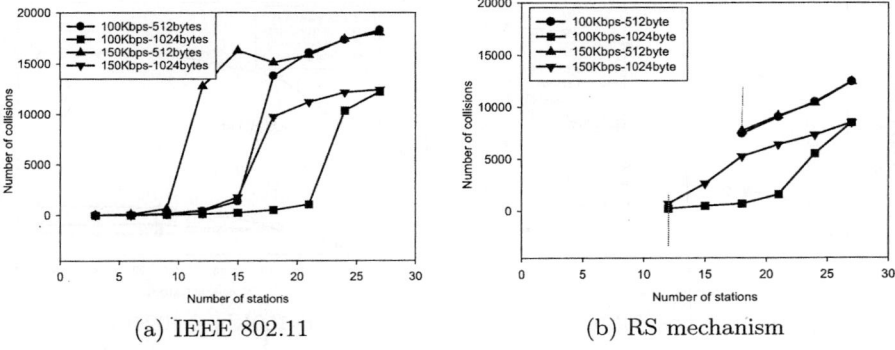

Fig. 5. Number of collisions

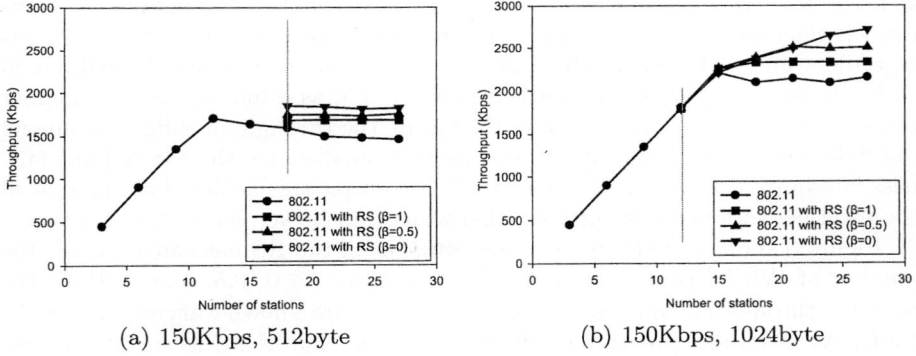

Fig. 6. Network throughput

512 and 1024 byte packets respectively), because the number of collisions is the same as the one in Figure 5(a) until the number contending stations reaches the thresholds. In the figure, we can see that the plots of the number of collisions are divided into two groups according to the packet size regardless of the packet sending rates: (100Kbps-512bytes, 150Kbps-512bytes) and (100Kbps-1024bytes, 150Kbps-1024bytes) groups.

Figure 6 compares the overall network throughput performance by varying β. Only the cases of 150Kbps-512bytes (Figure 6(a)) and 150Kbps-1024bytes (Figure 6(b)) are shown because other cases shows very similar behavior. As shown in the figure, the network throughput is improved regardless of the value of β when the RS mechanism is used. In the case of the IEEE 802.11 MAC, we can see that the network throughput is saturated, or even has a tendency to decrease, when the number of contending stations increases. When the RS mechanism is used, we can see that the highest network throughput is achieved when $\beta = 0$. As discussed, this is mainly due to fact that the temporal fairness is maximally achieved by giving more transmission opportunities to the stations

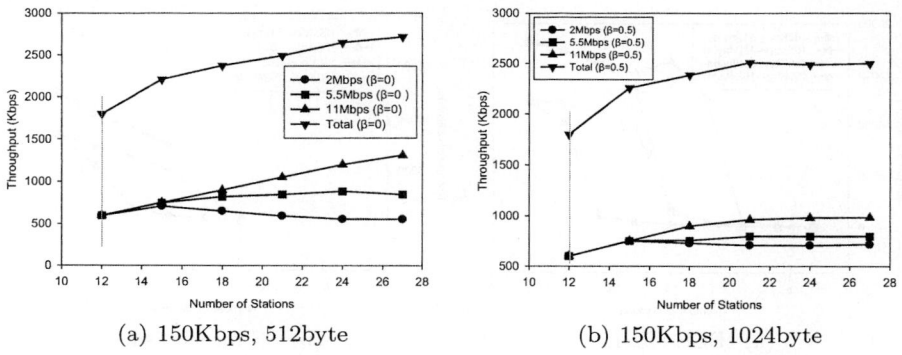

(a) 150Kbps, 512byte   (b) 150Kbps, 1024byte

Fig. 7. Group throughput

at high rates. When $\beta = 1$, we can see that the amount of throughput improvement is the smallest because equal transmission chances are given to all stations regardless of transmission rate. Figure 7 shows the aggregated throughput of each transmission rate group when β is 0 and 0.5, as a function of the number of contending stations. As shown in the figure, the throughput difference among the different transmission rate groups become smaller, i.e. the throughput fairness is improved as β is increased to 0.5. Note that this achieves at the cost of the total network throughput reduction which can be verified in the figure.

Figure 8 shows the throughput performance of the RS mechanism when the number of WS is fixed to 27 and β values varied to 0, 0.5, and 1. Both the network throughput and the group throughput are shown varying the traffic loads. As discussed, we can see from the figure that there is a trade-off between the network throughput and temporal fairness. One possible solution for resolving this trade-off is to compromise between the temporal fairness and the throughput fairness with the same preference factor ($\beta = 0.5$). In this case, we can achieve moderate network throughput improvement and temporal fairness. As the switching between emphasizing the network throughput and emphasiz-

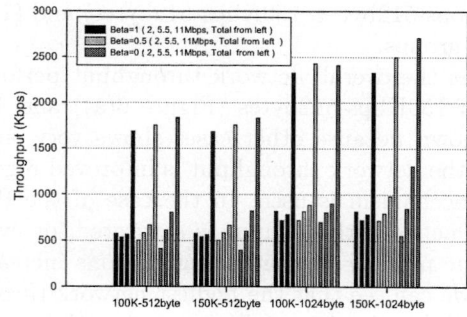

Fig. 8. Rate fairness, Number of Station = 27

ing the temporal fairness can be done easily by simply changing the sub-frame duration, the AP can dynamically control the network resource at any time it wants.

5 Conclusions

The performance of the IEEE 802.11 DCF is strongly dependent on the number of contending stations. In this paper, we issued the trade-off relationship between the throughput fairness and temporal fairness in multi-rate WLANs. We proposed a mechanism that tries to reduce the number of collisions by separating and grouping the contending stations, and distributing those groups over time. Considering the trade-off relationship above in multi-rate WLANs, as one of the grouping strategies, we proposed the Rate Separation (RS) mechanism in which the grouping is done based on the current transmission rates of contending stations. From the simulation study, we verify that the proposed RS mechanism reduces the number of collisions and achieves the improved performance.

References

1. M. Heusse, F. Rousseau, G. Berger-Sabbatel, and A. Duda, *Performance Anomaly of 802.11b*, INFOCOM'03, Apr. 2003.
2. S. Sheu, Y. Lee, and M. Chen, *Providing Multiple Data Rates in Infrastructure Wirelss Networks*. GLOBECOM'01. Global Telecommunications Conference, 2001.
3. K. Saitoh, Y. Inoue, M Iizuka, and M. Morikura, *An Effective Data Transfer Method by Integrating Priority Control into Multirate Mechanisms for IEEE 802.11 Wireless LANs*. VTC Spring 2002. Vehicular Technology Conference, 2002.
4. IEEE 802.11, *Part 11:Wireless LAN Medium Access Control (MAC) and Physical Layer (PHY) Specifications*, Standard, IEEE, Jun. 1997.
5. G. Bianchi, *Performance Analysis of the IEEE 802.11 Distributed Coordination Function*. IEEE Journal on Selected Areas in Communications, Mar. 2000.
6. G. Cali, M. Conti, and E. Gregori, *IEEE 802.11 Protocol: Design and Performance Evaluation of an Adaptive Backoff Mechanism*. IEEE Journal on Selected Areas in Communications, Sep. 2000.
7. F.Cali, M. Conti, and E. Gregori, *IEEE 802.11 Wireless LAN: Capacity Analysis and Protocol Enhancement*. INFOCOM 1998, Apr. 1998.
8. IEEE 802.11 WG, P802.11b, *Supplement to Standard IEEE 802.11. Higher Speed Physical Layer (PHY) Extension in the 2.4 GHz Band*. Sept. 1999.
9. K.Fall and K.Varadhan, *The ns notes and documentation*, available from http://www.isi.edu/ nsnam/ns, 2003
10. IEEE 802.11a, *Part 11:Wireless LAN Medium Access Control (MAC) and Physical Layer (PHY) Spedifications : High-Speed Physical Layer in the 5GHz Band*, Supplement to IEEE 802.11 Standard, Sep. 1999.

Improved Handoff Scheme for Supporting Network Mobility in Nested Mobile Networks*

Han-Kyu Ryu[1], Do-Hyeon Kim[2,**], You-Ze Cho[3], Kang-Won Lee[3], and Hee-Dong Park[4]

[1] Networks Lab. Samsung Electronics, Suwan, Korea
hankyu.ryu@samsung.com
[2] Dept. of Computer & Telecommunication Engineering, Cheju National University,
Jeju-do, 690-756, Korea
kimdh@cheju.ac.kr
[3] School of Electrical Engineering & Computer Science,
Kyungpook National University, Taegu. 702-701, Korea
yzcho@ee.knu.ac.kr, kw0314@palgong.knu.ac.kr
[4] Department of Computer Engineering, Pohang College, Pohang, 791-711, Korea
hdpark@pohang.ac.kr

Abstract. The IETF NEMO working group on network mobility is currently standardizing a basic support for moving networks. We need a handoff scheme for achieving the performance transparency during the movement of mobile routers in nested mobile networks. But the existed handoff schemes for a mobile host are not suitable for network mobility because packet transmission delay and traffic increase during mobile router's handoff in nested mobile networks. This paper proposes a new seamless handoff scheme using the bi-casting based on IPv6 in nested mobile networks. Compared with the handoff schemes using bi-directional edge tunneling for host mobility support on simulation, the results show that the proposed handoff scheme reduces the packet delay during a mobile router changes the point of attachment in nested mobile networks.

1 Introduction

There have been significant technological advancements in the areas of portable and mobile devices. And the rapid growth of wireless networks and Internet services drives the need for IP mobility. Traditional work in this topic is to provide continuous Internet access to mobile hosts only. Host mobility support is handled by Mobile IP and specified by the IETF Mobile IP working group [1].

Despite this, there is currently no means to provide continuous Internet access to nodes located in a mobile network. In this case, a mobile router changes its point of

---

* This work was in part supported by the University Fundamental Research Program of the Ministry of Information & Communication and the KOSEF (contract no. R01-2003-000-10155-0), Korea.
** Corresponding Author.

attachment, but there is a number of nodes behind the mobile router. The ultimate objective of a network mobility solution is to allow all nodes in the mobile network to be reachable via their permanent IP addresses, as well as maintain ongoing sessions when the mobile router changes its point of attachment within the Internet. The IETF NEMO working group is currently standardizing a basic support for moving networks. The basic NEMO protocol suggests a bi-directional tunnel between a mobile router and its home agent [2]. A unique characteristic of network mobility is nested mobility. Network mobility support should allow a mobile node or a mobile network to visit another mobile network (this is the example of a PAN in a train). Such scenarios may lead to multilevel aggregations of mobile networks. Although the NEMO basic solution with the mobile router-home agent tunnel supports nested mobility, it suffers from packet transmission delay and traffic increase in nested network in the topology, and transmission delays as the packets from the correspondent node are increased by all the home agents of the mobile routers in the nested mobility hierarchy. Therefore, in nested mobility, the support of handoff scheme is very important to allow packets between a correspondent node and a mobile network node.

A mobile network may comprise local mobile nodes or visiting mobile nodes and even other mobile networks. For instance, a bus is a mobile network whereas passengers are mobile nodes or even mobile networks themselves if they carry a PAN. The terminology document [3] describes nested mobility as a scenario where a mobile router allows another mobile router to attach to its mobile network. There could be arbitrary levels of nested mobility. The operation of each mobile router remains the same whether the mobile router attaches to another mobile router or to a fixed access router on the Internet. The solution described here does not place any restriction on the number of levels for nested mobility. But it should be noted that this might introduce significant overhead on the data packets as each level of nesting introduces another IPv6 header encapsulation.

Host mobility support is a mechanism which maintains session continuity between mobile nodes and their correspondents upon the mobile host's change of point of attachment. It can be achieved using Mobile IPv6 or other mobility support mechanisms. Network mobility support is a mechanism which maintains session continuity between mobile network nodes and their correspondent upon a mobile router's change of point of attachment. Solutions for this problem are classified into NEMO basic support, and NEMO extended support. NEMO basic support is a solution to preserve session continuity by means of bidirectional tunneling between mobile routers and their home agents much like what is done for mobile nodes when routing optimization is not used.

NEMO support is expected to be provided with limited signaling overhead and to minimize the impact of handover over applications, in terms of packet loss or delay. However, although variable delays of transmission and losses between mobile network nodes and their respective correspondent nodes could be perceived as the network is displaced, it would not be considered a lack of performance transparency [4].

To support handoff scheme for nested mobility, we apply the seamless handoff scheme using the bi-directional edge tunneling for host mobility support to the nested mobile network. And we propose the a new handoff scheme using the bi-casting based on an IPv6 address of new attachment mobile router in nested mobile networks. The proposed handoff scheme reduces the packet latency and packet loss during mobile router's handoff in nested mobile network. To support the proposed handoff scheme, mobile routers need fast to obtain an address of a new attachment mobile router in nested mobile networks. Compared with the seamless handoff schemes using the bi-directional edge tunneling in nested mobile network on simulation, the results show that the proposed handoff scheme reduces the packet delay during mobile router's handoff in mobile network.

The remainder of this paper is organized as follows. First, basic network mobility is introduced in section 2, then section 3 applies the seamless handoff scheme using the bi-directional edge tunneling to the nested mobile network. And section 4 presents the proposed handoff scheme and section 5 provides a performance analysis. Finally, the conclusion is given in section 6.

2 Basic Network Mobility

A mobile network is a network segment or subnet which can move and attach to arbitrary points in the routing infrastructure. A mobile network can only be accessed via specific gateways called mobile routers that manage its movement. Mobile networks have at least one mobile router serving them. A mobile router does not distribute the mobile network routes to the infrastructure at its point of attachment. Instead, it maintains a bidirectional tunnel to a home agent that advertises an aggregation of mobile networks to the infrastructure. The mobile router is also the default gateway for the mobile network. A mobile network can also consist of multiple and nested subnets. Fig. 1 shows a mobile network and mobile routers.

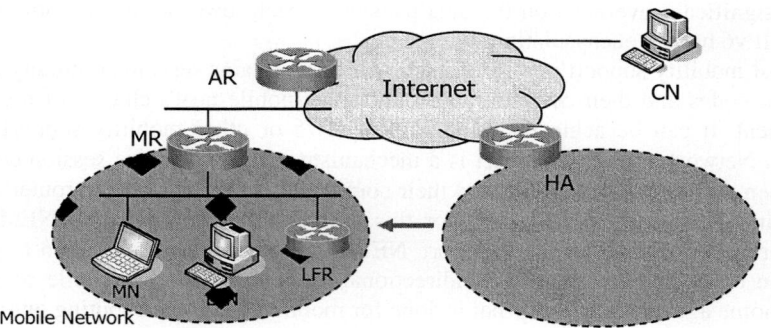

Fig. 1. Mobile network model

When the mobile router moves away from the home link and attaches to a new access router, it acquires a care-of address from the visited link. The mobile router can at any time act either as a mobile host or a mobile router. It acts as a mobile host as defined for sessions originated by itself, while providing connectivity to the mobile network. As soon as the mobile router acquires a care-of address, it immediately sends a binding update to its home agent as described. When the home agent receives this binding update it creates a binding cache entry binding the mobile router's home address to its care-of address at the current point of attachment [2].

The home agent acknowledges the binding update by sending a binding acknowledgement to the mobile router. A positive acknowledgement means that the home agent has set up forwarding for the mobile network. Once the binding process completes, a bi-directional tunnel is established between the home agent and the mobile router. The tunnel end points are mobile router's care-of address and the home agent's address. If a packet with a source address belonging to the mobile network Prefix is received from the mobile network, the mobile router reverse-tunnels the packet to the home agent through this tunnel. This reverse-tunneling is done by using IP-in-IP encapsulation. The home agent decapsulates this packet and forwards it to the Correspondent Node. For traffic originated by itself, the mobile router can use either reverse tunneling or route optimization.

When a data packet is sent by a correspondent node to a node in the mobile network, it gets routed to the home agent which currently has the binding for the mobile router. It is expected that the mobile router's network prefix would be aggregated at the home agent, which advertises the resulting aggregation. Alternatively, the home agent may receive the data packets destined to the mobile network by advertising routes to the mobile network prefix. The actual mechanism by which these routes are advertised is outside the scope of this document. When the home agent receives a data packet meant for a node in the mobile network, it tunnels the packet to mobile router's current care-of address. The mobile router decapsulates the packet and forwards it onto the interface where the mobile network is connected. The mobile router before decapsulating the tunneled packets, has to check if the source address on the outer IPv6 header is the home agent's address. However, this check is not necessary if the packet is protected by IPsec in tunnel mode. The mobile router also has to make sure the destination address on the inner IPv6 header belongs to a prefix used in the mobile network before forwarding the packet to the mobile network. Otherwise it should drop the packet. The mobile network could consist of nodes that do not support mobility and nodes that support mobility. A node in the mobile network can also be a fixed or a mobile router. The protocol described here ensures complete transparency of network mobility to the nodes in the mobile network. Mobile nodes that attach to the mobile network treat it as a normal IPv6 access network and run the Mobile IPv6 protocol [2].

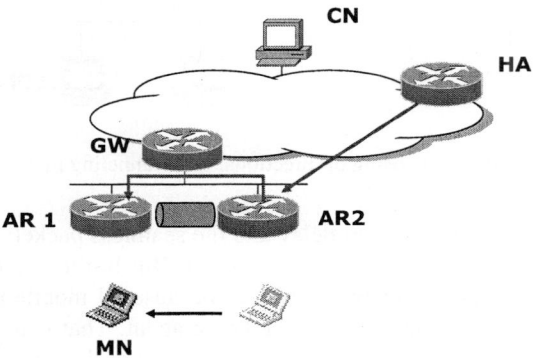

Fig. 2. Packet delivery using bi-directional edge tunneling

3 Handoff Scheme Using Bi-directional Edge Tunneling

Now IETF NEMO working group supports a basic network mobility as well as researching for efficient handoff in mobile networks. Handoff scheme for mobile networks considers the existed handoff schemes supporting mobile terminals presented on Mobile IP working group. Mobile IP working group is presented a fast handoff scheme using bi-directional edge tunneling based on IPv6 for supporting mobility of mobile hosts. We apply the seamless handoff scheme using the bi-directional edge tunneling for mobile hosts to the nested mobile network, And wd analyze the problem in terms of the transfer delay and packets loss. Bi-directional edge tunneling can delivers seamless packets between the old access router and the new access router in a handoff procedure [5, 6].

As shown Fig. 2, when a mobile node moves to the other attachment point, previously a mobile node notifies the information of the old access router to the new access router. And the old access router tunnels the new access router and forwards packets.

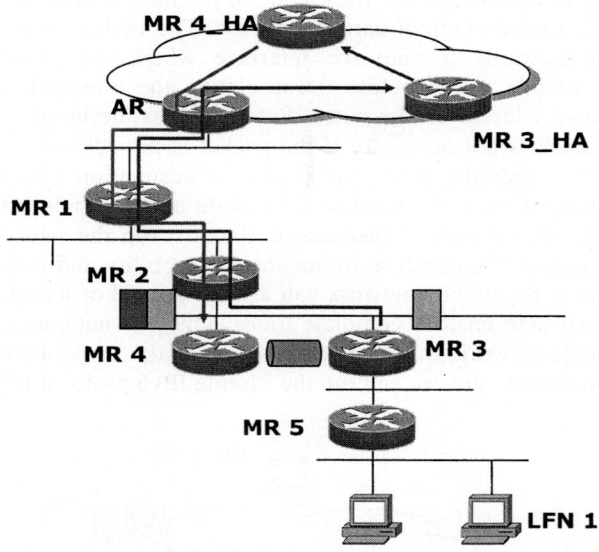

Fig. 3. Packet transfer procedure using bi-directional edge tunneling in nested mobile networks

This has the advantage of a short delay and the seamless packet delivery because a mobile route need does not notify its home agent. But bi-directional edge tunneling can not be efficient in nested mobile networks because all mobile routers deliver the packet to the corresponds node through the home agent. That is to say, An each mobile router sets up the tunnel via its home agent instead of the tunnel between the old access router and the new access router.

As shown Fig. 3, the MR5 moves in the MR3 to the MR4. And when a mobile router setups the tunnel between the MR3 and the MR4, packets deliver via theirs home agents. Therefore, MR4 has a long packet delay and a many packet loss.

4 The Proposed Handoff Scheme Using Bi-casting

To solve the problem of bi-directional edge tunneling, we propose the seamless handoff scheme using bi-casting based on IPv6 address of new attachment mobile router. In case that the mobile router transfers the binding update message to its home agent, the home agent delivers simultaneous packets both the old access router and the new access router. This method is called as the bi-casting [7].

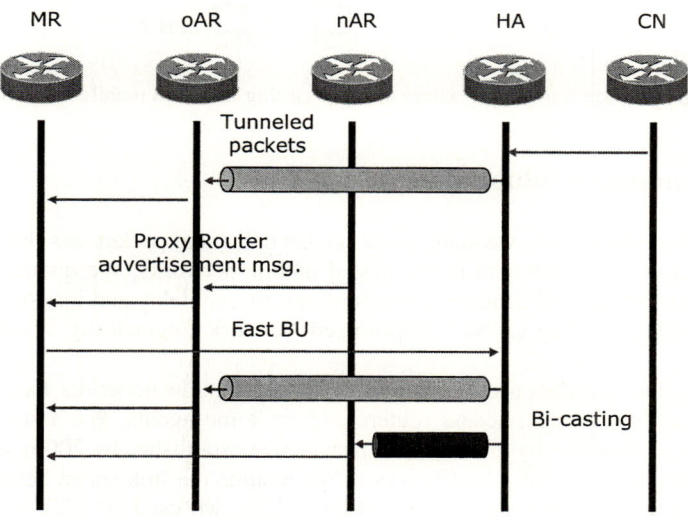

Fig. 4. Seamless handoff procedure using bi-casting in mobile networks

As shown Fig. 4, before the mobile router moves out of the service area, this router announces the information of the new access router to its home agent. And the home agent casts both the access routers at the handoff procedure.

Fig. 5 shows the example of bi-casting. When mobile router MR5 moves on the MR4, MR5 sends the router solicitation message to its home agent via MR3. The home agent sets the connection with the MR4. Bi-casting has the advantage of a short packet delay and a few packet losses. Compared with the bi-directional edge tunneling, bi-casting reduces the packet delay. However, the bi-casting cause to overload twice traffic between the home agent and two access router.

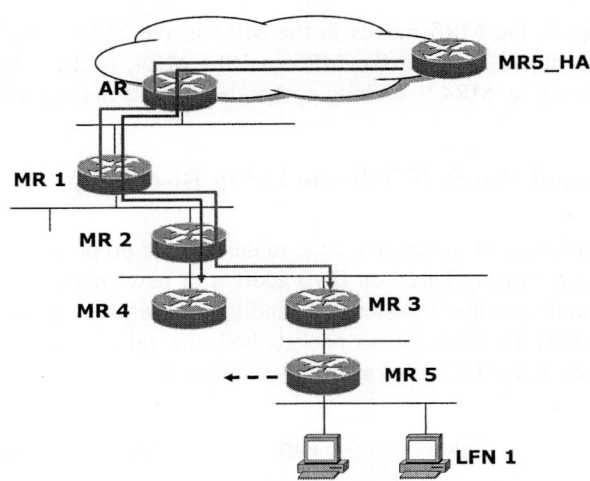

Fig. 5. Packet transfer procedure using bi-casting in nested mobile networks

5 Performance Evaluations

To verify the performance, we analyze the packet delay and packet loss that happen to change the point of attachment in the nested mobile networks. We quantatively analyze the seamless handoff scheme in terms of end-to-end delay and loss of packet and perform simulation using OPNET (Optimized Network Engineering Tool) by MIL3 company.

In Fig. 6, the simulation environment of nested mobile networks based on IPv6 consists of mobile routers, access routers and its home agents. We assume that the packet interval time sets by 20msec, and packet size establishes by 200bytes which is the real-time traffic size of a VoIP service. We assume the link speed between a mobile router and a access router is the 1.5Mbps. Also, we used OPNET for measurement of delay and loss.

Fig. 7 demonstrates the handoff latency according to depth using the basic Mobile IP in the nested mobile networks during a mobile router changes the point of attachment, where we don't use the seamless handoff scheme. As shown figure 7, the handoff latency is between 0.6sec and 1.75sec according to the change from 1 to 5 of the nested depth. The handoff latency factor is generated by the long packet transfer time between mobile routers and their home agents according to the nested depth during the mobile router's handoff in the nested mobile networks based on IPv6.

Fig. 8 describes the packet loss according to the nested depth during a mobile router's handoff in the nested mobile networks, where we don't use the seamless handoff scheme. In this figure, the packet loss is between 27 packets and 98 packets according to the change from 1 to 5 of the nested depth. Because the route can not be optimized, a packet loss happens as go via many home agents which suffered the overflow for the real-time services.

Fig. 9 shows the packet delay according to depth using the bi-directional edge tunneling and the proposed bi-casting based on IPv6 in the nested mobile networks during a mobile router changes the point of attachment. As shown in this figure, the packet delay is between 0.19sec and 1.65sec according to the change from 1 to 5 of the nested depth, where use the bi-directional edge tunneling. But the proposed bi-casting based on IPv6 has the packet delay from 0.03sec to 0.05sec during the mobile router's handoff in the nested mobile networks.

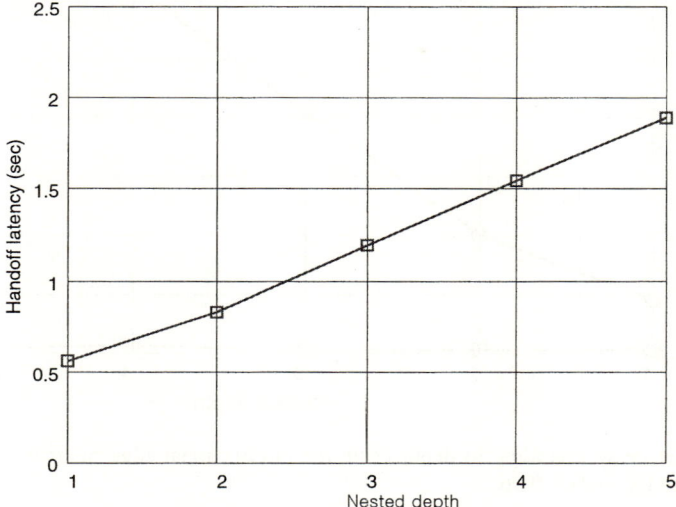

Fig. 6. Handoff latency according to the depth in the nested mobile networks

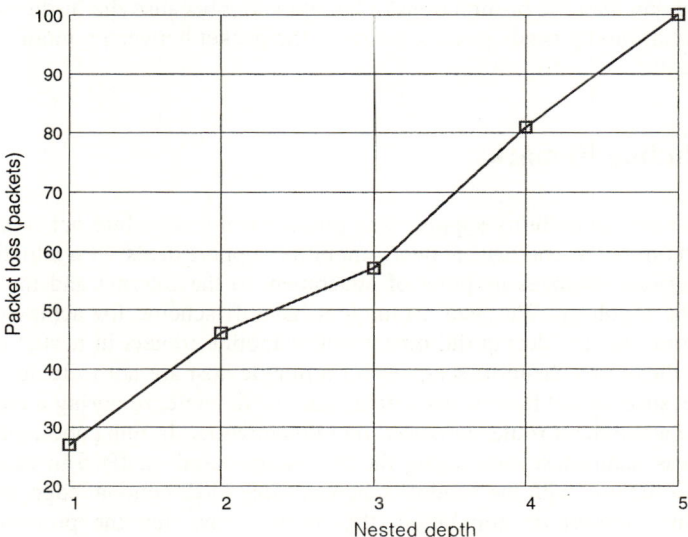

Fig. 7. Packet loss according to the depth during mobile router's handoff in the nested mobile networks

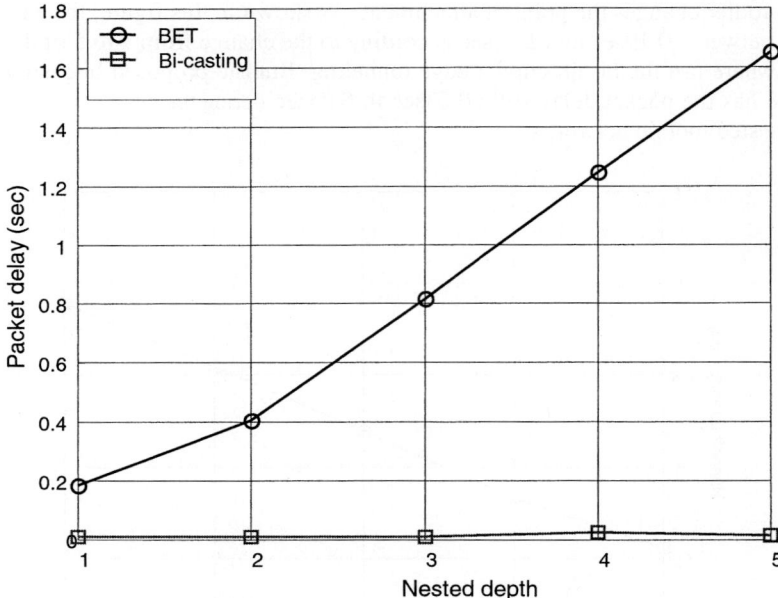

Fig. 8. Packet delay according to depth using the bi-directional edge tunneling and the proposed bi-casting based on IPv6

The simulation results demonstrate the delay of two handoff schemes in the nested mobile networks. Especially, the proposed bi-casting based on IPv6 can reduce more the packet delay than the bi-directional edge tunneling because the bi-directional edge tunneling additionally needs more to delivery the packet between a mobile router and it home agent.

6 Concluding Remarks

Traditional work on mobility support is to provide continuous Internet connectivity to mobile hosts only. In contrast, network mobility support deals with situations where an entire network changes its point of attachment to the Internet and thus its reachability in the topology. We need a seamless handoff scheme for achieving the performance transparency during the movement of mobile routers in nested mobile networks. But the existed handoff schemes for a mobile host are not suitable for network mobility because packet transmission delay and traffic increase during a mobile router changes the attachment router in nested mobile networks. In this paper, we propose a new seamless handoff scheme using the bi-casting based on IPv6 in nested mobile networks. Compared with the handoff schemes using bi-directional edge tunneling for host mobility support on simulation, the results show that the proposed handoff scheme reduces the packet delay during mobile router's handoff in nested mobile networks.

References

1. C. Perkins, "IP Mobility Support," RFC 2002.October 1996
2. Devarapalli, V., "Network Mobility (NEMO) Basic Support Protocol", draft-ietf-nemo-basic-support-03.txt, June 2004
3. T. Ernst and H.-Y. Lach. Network Mobility Support Terminology. Internet Draft, IETF. draft-ietf-nemo-terminology-02.txt. October 2004.
4. Ernst, T., "Network Mobility Support Goals and requirements", draft-ietf-nemo-requirements-03.txt, October 2004.
5. A. Yegin et al, "Fast Handovers for Mobile IPv6," Internet Draft, <draft-ietf-mobileip-fast-mipv6-02.txt>, July 2001.
6. J. Kempf et al, "Bidirectional Edge Tunnel Handover for IPv6," Internet Draft, <draft-kempf-beth-ipv6-02.txt>, September 2001.
7. K. E. Malki et al., "Low Latency Handoff Mobile IPv4," Internet Draft <draft-ietf-mobileip-lowlatency-handoffs-v4-01.txt>, May 2001.

A *Prompt Retransmit* Technique to Improve TCP Performance for Mobile Ad Hoc Networks[*]

Dongkyun Kim and Hanseok Bae

Department of Computer Engineering,
Kyungpook National University, Daegu 702-701, Korea
dongkyun@knu.ac.kr, bae@monet.knu.ac.kr

Abstract. In addition to research on MANET (Mobile Ad Hoc Networks) routing protocols, efforts to adopt TCP as a reliable transport protocol with some modifications have been made for the smooth integration with the fixed Internet. Basically, the standard TCP depends on the third duplicate ACK in order to retransmit a lost segment. However, when TCP operates on top of reactive routing protocols and is able to distinguish the segments transmitted over the old path from ones transmitted over a new path, TCP performance is improved by retransmitting the lost segment on the first duplicate ACK without waiting for the third duplicate ACK. As a result, it advances retransmission time thanks to our proposed *prompt retransmit* technique. Simulation work through GloMoSim shows that the *prompt retransmit* produces better TCP performance than the standard TCP.

1 Introduction

Recently, since research interest in the MANET (Mobile Ad Hoc Networks) has increased, the IETF MANET working group [1] has been trying to standardize its proactive and reactive routing protocols. In proactive protocols like OLSR (Optimized Link State Routing) [2] and TBRPF (Topology Broadcast based on Reverse-Path Forwarding) [3], all nodes should maintain their routing tables for all possible destinations, without regard to the actual desire for the route between source and destination nodes. However, in reactive routing protocols such as DSR (Dynamic Source Routing) [4] and AODV (Ad Hoc On-Demand Distance Vector) [5], only when a source node needs to send data packets to the destination node, it attempts to acquire the path in on-demand manner. Reactive approaches avoid the need of maintaining routing tables when there are no desires of routes between source-destination pairs.

In addition to the network layer protocols mentioned above, a transport layer protocol like TCP (Transmission Control Protocol) is also needed to provide reliable end-to-end message transmission. TCP is still needed for MANETs since

---

[*] This work was supported by grant No. R05-2004-000-10307-0 from Korea Science & Engineering Foundation.

it is widely used in the current Internet and we would like to achieve a smooth integration with the fixed Internet. However, earlier research work had confirmed that TCP cannot be directly used on MANETs due to the presence of the time-varying link characteristics and node mobility. Particularly, in MANET, since the TCP source cannot distinguish between segment[1] losses caused by congestion and that by node mobility, the degradation of TCP performance would reside.

Generally, the standard TCP performs the *fast retransmit* for a lost segment before it is retransmitted via a timeout event [11]. However, when TCP is served by on-demand reactive ad-hoc routing protocols like DSR and AODV and it can, furthermore, distinguish the segments transmitted over the old path from ones transmitted over a new path, it is enough to retransmit a lost segment as soon as the first duplicate ACK is received without waiting for the third duplicated ACK. Therefore, in this paper, we propose a technique to distinguish between the segments over the old path and ones over a new path using a notification from routing layer and an efficient retransmission scheme, called *prompt retransmit* technique, to improve TCP performance, accordingly.

The rest of this paper is organized as follows. We describe the issues on improving TCP performance for MANET in Section 2. We present our idea using the *prompt retransmit* technique in Section 3. We evaluate the technique in Section 4, which is followed by some concluding remark in Section 5.

2   Related Work on Improving TCP Performance

If a TCP source does not receive the acknowledgement segments from the destination in a timely fashion, timeout events for the transmitted segments will occur. Then, a TCP source assumes that congestion has occurred in the network and invokes the congestion control procedure, even though the actual segment loss is caused by node mobility. As a result, the TCP source performs a great reduction of transmission rate. This performance degradation is due to the TCP source's inability of knowing whether the actual segment losses is due to node mobility or network congestion.

To improve TCP performance for MANETs, several approaches have been proposed. In TCP-feedback [6], two special messages, RFN (Route Failure Notification) and RRN (Route Re-establishment Notification), are generated when route breakage occurs and a new path is acquired, respectively. On receiving the RFN message, TCP sender freezes all its variables such as timers and CWND (Congestion Window) size, and resumes the TCP process from the frozen state after receiving the RRN message. In the ELFN-based approach [7], ELFN (Explicit Link Failure Notification) message like the RFN message can be used to inform the TCP sender of the route breakage. However, instead of using RRN message as in TCP-Feedback, probe messages are sent regularly toward the destination in order to detect the route restoration. Additionally, in TCP-BuS (TCP

---

[1] For terminology, the terms, segment and packet, are used at TCP and routing layers, respectively.

with Buffering capability and Sequence Information) [8], the buffering mechanism at intermediate nodes helps to improve TCP performance, in addition to using explicit notification messages related to route breakage and restoration. Unlike the approaches mentioned above, ATCP [9] implements an intermediate layer between network and transport layers rather than imposing changes to the standard TCP. ATCP relies on ECN (Explicit Congestion Notification) mechanism for congestion control and ICMP protocol for detecting route failures. TCP-DOOR [10] utilizes only out-of-order packet information for detecting route failures.

3 Prompt Retransmit

3.1 Motivation

TCP's techniques [11] to reduce the transmission rate with the assumption that a network congestion occurred can be categorized into two types. One is that when a TCP source experiences timeout for the corresponding ACK, the TCP source considers that it occurred due to serious network congestion and reduces the next transmission rate radically. In the other technique, called the *fast retransmit* technique, the TCP source reduces the transmission rate by half when it receives the third duplicate ACK because the reception of the consecutive ACKs means that a network congestion can still exist, but it is not so serious. Therefore, the CWND (Congestion Window), which represents an amount of segments that the TCP source is allowed to transmit consecutively, is set to half of the previous value. The *fast retransmit* technique enables a TCP source to detect a lost segment and retransmit it earlier than the RTO (Retransmission Timeout) timer expires. In the current Internet, route changes according to network condition enable packets to be transferred over different routes from the route that the previous packets are being transmitted. Therefore, they can reach a TCP destination in out-of-order manner. That is why the TCP source waits for the third duplicate ACK instead of sending the lost segment immediately on receiving the first duplicate ACK.

However, note that when one of the on-demand reactive routing protocols such as DSR and AODV is utilized in MANET and TCP can, furthermore, distinguish the segments transmitted over the old path from ones transmitted over a new path, the TCP protocol becomes positioned on top of a kind of *connection-oriented* network at least until a route is broken. In other words, when the source desires to send packets to the destination, it acquires a connection-oriented path toward the destination before actual packet transmission and uses the path at least until a breakage of the acquired route. During the data transmission over the acquired connection, the source tries to achieve a route re-establishment when it is notified of route breakage. Hence, on detecting a hole of sequences of TCP segments transmitted over such a connection-oriented network, we need an efficient mechanism to retransmit the lost segment immediately without waiting for the arrival of additional ACKs.

3.2 Description of *Prompt Retransmit* Technique

To support *Prompt Retransmit*, a TCP source is required to utilize the sequence information of TCP segments on receiving explicit messages from routing layer. When the TCP source receives a route failure notification message, it stops further data transmission as in many techniques mentioned in Section 2 and keeps a record of the sequence information of the TCP segment which is sent latest, denoted by $LAST\_SEQ$. On the other hand, on receiving a route re-establishment notification message after route repair, the TCP source resumes its stopped transmission. With the $LAST\_SEQ$ sequence information, TCP can easily distinguish the on-the-fly TCP segments over the old path from ones traversed in in-order manner over the new path at least until a new route breakage occurs. The segments over the old path may arrive at the destination later than the TCP segments which are newly transmitted after a route repair.

If we consider a case without the technique based on $LAST\_SEQ$, the TCP source would be performing unnecessary early retransmissions even for the segments which will reach the destination a little bit later over the old path. Since the retransmission for the segments with out-of-order arrival is required on receiving the first duplicate ACK, it would result in generating duplicate segments in the network and it furthermore requires the TCP source to reduce its transmission rate.

For the TCP segments, therefore, whose sequence numbers are $\leq LAST\_SEQ$, we use the normal *fast retransmit* technique in which the third duplicate ACK forces the TCP source to retransmit a lost segment, because the TCP segments over the old path and a new path can reach the destination at different times under different traffic conditions.

On the other hand, for the TCP segments, whose sequence numbers are $> LAST\_SEQ$, it is enough to use the first duplicate ACK for the purpose of retransmitting the lost segment, because the TCP segments are definitely transmitted over a *connection-oriented path*. In this sense, at least for the packets traversed in in-order manner, the TCP source had better retransmit the lost segment as soon as possible when its lost segment is detected. Algorithm 1 describes the basic operation. Figure 1 shows a state transition diagram of our proposed *prompt retransmit* technique along with Algorithm 2. When TCP receives an RFN (Route Failure Notification) while it stays in $TRANSMITTING$ state (In $TRANSMITTING$ state, the TCP source is transmitting TCP segments), it determines $LASTE\_SEQ$ and goes to $HOLDING$ state in which the further transmission of TCP segments is stopped. During staying in $HOLDING$ state, when TCP receives an RRN (Route Re-establishment Notification), it resumes the stopped transmission of TCP segments. Note that in $TRANSMITTING$ state, Algorithm 1 is always applied.

For example, during the transmission of TCP segments whose sequence numbers are from 7 to 14, a route breakage has occurred (Figure 2 (a)). On being informed of the route breakage, the TCP source extracts a sequence information on the TCP segments that the TCP sender has sent recently, which is kept in $LAST\_SEQ$ variable (In this case, $LAST\_SEQ$ is 14). After a new route between

Algorithm 1 Our Retransmission Algorithm

(1) DUPACK stands for a duplicate ACK.
(2) SEQ_NUM represents a sequence number of a TCP segment.
if (SEQ_NUM of DUPACK > LAST_SEQ) **then**
  if (FIRST DUPACK received) **then**
    Retransmit a lost segment;
  end if
else
  if (THIRD DUPACK received) **then**
    Retransmit a lost segment;
  end if
end if

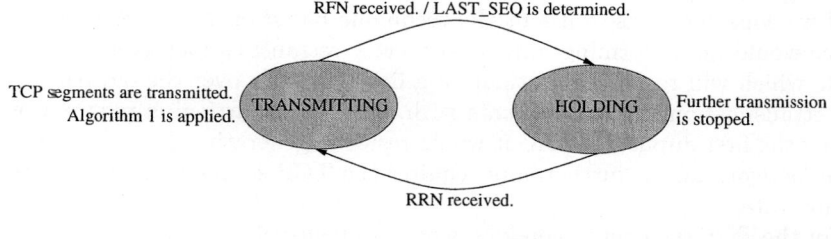

Fig. 1. Proposed *Prompt Retransmit* Mechanism Using *LAST_SEQ*

the source and destination nodes is acquired [2], the TCP segments whose sequence numbers are from 15 to 19 are transmitted over the new path (Figure 2 (b)). However, since the congestion condition of two paths is different, if the TCP segments of sequence numbers 15, 16 and others arrive at the destination before the TCP segments sequenced from 7 to 11 over the old path arrive actually there, the source will retransmit the TCP segments being traversed over the old path even if the TCP segments just arrive a little bit later over the old path due to different traffic condition. It obviously invoked an unnecessary retransmission with dropping of the transmission rate according to congestion control algorithm. However, according to the *LAST_SEQ*-based scheme above, the *LAST_SEQ* is set to 14 and we will apply the first duplicate ACK triggered retransmission technique only for the TCP segments greater than sequence number 14. For a lost segment whose sequence number is 17, the TCP source can early retransmit the lost segment using the first duplicate ACK technique. In addition, for the segments sequenced from 12 to 14, the timeout events will result

---

[2] In routing protocols, an intermediate node detecting the link breakage or a source can acquire a partial or new path, respectively. In this case, the source acquired a new path.

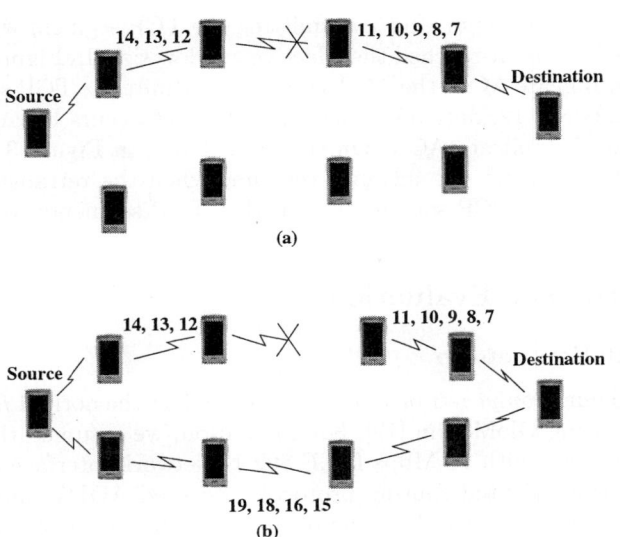

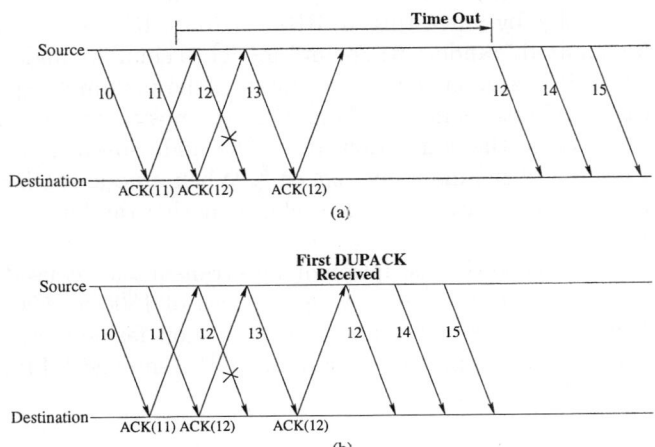

Fig. 2. (a) A route breakage occurs, and (b) after acquiring a new path

Fig. 3. (a) Fast Retransmit (b) Prompt Retransmit

in retransmitting them because they are discarded over the old path due to the route breakage.

In particular, we describe another advantage of using the first duplicate ACK. We assume a case where the TCP source does not transmit further TCP segments due to the reach of effective window size [3]. In Figure 3, the TCP source cannot

---

[3] The effective window size is determined by the minimum size of current CWND size and the receiver-side's available buffer size advertised by a receiver

send further TCP segments after transmitting the TCP segment whose sequence number is 13, because it reached the effective window size. In Figure 3 (a) relying on the third duplicate ACK, the TCP source retransmits the TCP segment whose sequence number is 12, only after the timeout event occurs because we cannot expect the third duplicate ACK. On the other hand, in Figure 3 (b) using the first duplicate ACK, we can advance the time when the retransmission of the lost segment and the TCP source sends further TCP segments continuously.

4 Performance Evaluation

4.1 Simulation Set-Up

We compared our *prompt retransmit* technique against the normal *fast retransmit* technique by using GloMoSim [12]. For simulation, we assumed that all mobile nodes are equipped with 11 Mbps IEEE 802.11 network interface cards.

As connection-oriented routing protocols, we used AODV and DSR as our underlying routing protocols [4]. In the reactive routing protocols, when a source needs to send data packets to the destination, it broadcasts an RREQ (Route REQuest) message. During this flooding of the RREQ message, a destination or an intermediate node which knows the path towards the destination responds to the RREQ message by unicasting an RREP (Route REPly) message back to the source. We adopted "random waypoint" model to simulate nodes movement, where the motion is characterized by two factors: the maximum speed and the pause time. Each node starts moving from its initial position to a random target position selected inside the simulation area. The node speed is uniformly distributed between 0 and the maximum speed [5]. When a node reaches the target position, it waits for the pause time, then selects another random target location and moves again.

In addition, we avoided the disruption of data transmission caused by network partitions by spreading 100 nodes in a square area of 1500 x 1500 meters and enabling each node to reach other nodes in multi-hop fashion. As a transport protocol and an application program, we used TCP-Reno and FTP application, respectively.

4.2 Simulation Results

First, we measured the performance improvement of *Prompt Retransmit* when applied to the AODV routing protocol. In this simulation, we evaluated the TCP throughput for static tandem network topologies where the route between the source and destination nodes is fixed and does not change. However, we allowed the neighboring nodes around the given path to move with much contention to channel access. Obviously, we could observe that the TCP throughput for

---

[4] Our approach can be applied to most reactive routing protocols.
[5] For the node mobility, we simulated the proposed approach by varying this maximum speed.

the case with 10 nodes is lower than that of the case with 5 nodes because the packet forwarded towards the destination experiences more contention over wireless links. We compared the case using the third duplicate ACK with the case using our first duplicate ACK in terms of TCP throughput, according to the degree of network congestion. In this simulation, the TCP segment loss rate represents the degree of network congestion.

As mentioned before, due to the early retransmission effect that our approach detects a segment loss when the first duplicate ACK is received and retransmits the lost segment immediately without additional waiting for the third duplicate ACK packet, we could obtained more TCP throughput for all congested networks as shown in Figure 4(a). Additionally, Figure 4(a) shows that the higher the network congestion is, the lower TCP throughput we obtained.

We also traced the progress of TCP segments transmitted by the source for each approach. As shown in Figure 4(b), our approach using the *prompt retransmit* shows faster progress of TCP segments' sequencing than the other. Using the third duplicate ACK-based retransmission technique, we could easily observe that timeout events prevented the source from increasing the sequence number of TCP segments transmitted. Our approach, however, enables the source to retransmit the lost segment immediately before timeout events occur.

Next, we investigated the performance using random network topologies. In this simulation, we measured the performance of TCP throughput with respect to node mobility. We varied the maximum speed of nodes according to the random waypoint model. In order to focus the effect of mobility on the performance, we fixed segment loss rate to 0.001. It was obvious that during TCP segments transmission, packet losses occurred randomly over wireless channel due to the channel contention and channel error.

As expected, we obtained lower TCP throughput when the node mobility was high. In particular, our approach showed better performance than the other regardless of node mobility, because the TCP source performed the early re-

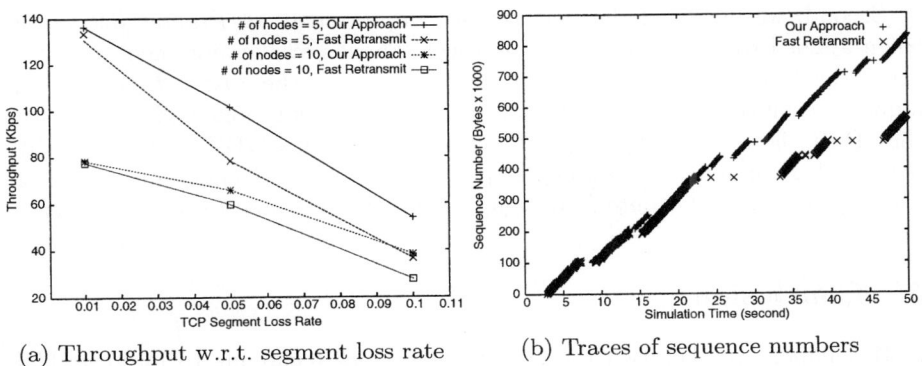

(a) Throughput w.r.t. segment loss rate (b) Traces of sequence numbers

Fig. 4. Static tandem network topologies using AODV

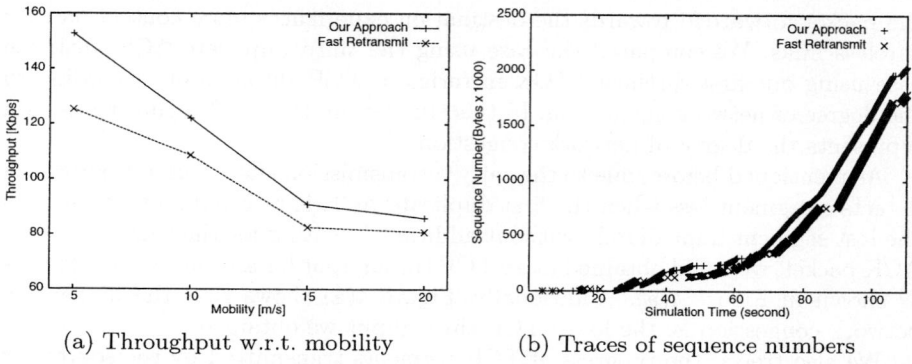

(a) Throughput w.r.t. mobility (b) Traces of sequence numbers

Fig. 5. Random network topologies with node mobility using AODV

Table 1. Throughput (bps) comparison using DSR. FR and PR are Fast Retransmit and Prompt Retransmit, respectively

Static tandem network (4 hops)			Random network		
Segment loss rate	FR	PR	Mobility (m/s)	FR	PR
0.01	133644	139370	5	103924	118444
0.05	91821	108319	10	99493	113010
0.1	33280	77322	15	19480	29388
			20	17051	22643

transmission as soon as it detects a hole of sequences of TCP segments, that is, the segment loss (Figure 5(a)). In addition, we observed the progress of TCP segments' sequencing at 10 m/s, for example. Our approach using the *prompt retransmit* showed faster progress of sequencing than the other using the *fast retransmit* (Figure 5(b)).

Second, we observed the performance improvement of *Prompt Retransmit* when applied to the DSR routing protocol. Thanks to the aforementioned features, the *Prompt Retransmit* technique out-performs the *Fast Retransmit* technique in terms of throughput without regard to network topology (Table 1). Note that DSR shows worse throughput performance than AODV at high mobility because high mobility increases the possibility that DSR chooses one of stale routes as a secondary route from route cache after a primary route is broken.

5 Conclusion

In this paper, we proposed the *prompt retransmit* technique to improve TCP performance in MANET. Basically, since the TCP in MANET is still put on top of the network layer which provides a connection-oriented type of routing service such as AODV and DSR, we don't need to delay the retransmission of a lost segment through the normal *fast retransmit* technique which utilizes the third duplicate ACK-based retransmission technique. Instead, using our first duplicate ACK-based retransmission approach, we could advance the retransmission time of lost segments. In particular, without regard to node mobility, our *prompt retransmit* scheme showed better performance than the normal *fast retransmit* technique. Furthermore, our approach is very simple to implement. With other techniques in the literature to improve TCP performance for MANETs, it seems to get more better performance, which is our future work.

References

1. Internet Engineering Task Force, "Manet working group charter," http://www.ietf.org/html.charters/manet-charter.html.
2. T. Clausen, P. Jacquet, A. Laouiti, P. Minet, P. Muhlethaler, A. Qayyum, and L. Viennot, "Optimized Link State Routing Protocol(OLSR)", IETF RFC 3626.
3. R. G. Ogier, F. L. Templin, B. Bellur, and M. G. Lewis, "Topology Broadcast Based on Reverse-Path Forwarding (TBRPF)", IETF RFC 3684.
4. D. B. Johnson, D. A. Maltz, and Y. Hu, "The Dynamic Source Routing Protocol for Mobile Ad-hoc Networks (DSR)", IETF Internet Draft, draft-ietf-manet-dsr-08.txt, February 2003.
5. C. E. Perkins, E. M. Belding-Royer, and S. R. Das, "Ad-hoc On-Demand Distance Vector (AODV) Routing", IETF RFC 3561.
6. K.Chandran, S. Raghunathan, S. Venkatesan, and R. Prakash, "A Feedback Based Scheme For Improving TCP Performance In Ad-Hoc Wireless Networks", IEEE ICDCS 1998.
7. G. Holland and N. H. Vaidya, "Analysis of TCP Performance over Mobile Ad Hoc Networks. 5th Annual International Conference on Mobile Computing and Networking", ACM Mobicom, August 1999.
8. D. Kim, C.-K. Toh, and Y. Choi, "TCP-BuS : Improving TCP Performance in Wireless Ad Hoc Networks", Journal of Communications and Networks, Vol. 3, No. 2, 2001.
9. J. Liu, S. Singh, "ATCP: TCP for Mobile Ad Hoc Networks", IEEE Journal on selected areas in communications, vol. 19, No. 7, July 2001.
10. F. Wang and Y. Zhang, "Improving TCP Performance over Mobile Ad-Hoc Networks with Out-of- Order Detection and Response", ACM Mobihoc02, June 2002.
11. M.Allman, V.Paxson, W.Stevens, "TCP Congestion Control," RFC 2581, Apr. 1999. http://www.rfc-editor.org/rfc/rfc2581.txt
12. http://pcl.cs.ucla.edu/projects/glomosim/

Enhanced Fast Handover for Mobile IPv6 Based on IEEE 802.11 Network*

Seonggeun Ryu[1], Younghwan Lim[1], Seongjin Ahn[2], and Youngsong Mun[1]

[1] School of Computing, Soongsil University, Seoul, Korea
sgryu@sunny.ssu.ac.kr,
{mun, yhlim}@computing.ssu.ac.kr
[2] Department of Computer Education, Sungkyunkwan University,
Seoul, Korea
sjahn@comedu.skku.ac.kr

Abstract. In MIPv6, whenever a mobile node changes its attached point, handover process should be followed to inform its home agent and correspondent of the mobile node's current location. The handover process is decomposed into layer 2 and layer 3 handovers again, and these two handovers are accomplished sequentially, which causes long latency problem. This problem is a critical issue in MIPv6. To make up for this, we propose an enhanced fast handover scheme to reduce the overall latency on handover, revising the fast handover [1]. Especially, several messages in layer 3 are sent in one frame during layer 2 handover. We use cost analysis for the performance evaluation. Compared to MIPv6 handover scheme, the proposed scheme gains 79% improvement while it gains 31% improvement, compared to the fast handover.

1 Introduction

As the advancement of wireless communication, requirements on the wireless Internet and mobility support are increasing. To support mobility, Mobile IPv6 (MIPv6) [2] has been proposed by the Internet Engineering Task Force (IETF). In MIPv6, when a mobile node (MN) moves to other sub network, it needs certain process, a handover, which causes long latency problem. There have been many researches to reduce the handover latency in MIPv6, such as a fast handover [1] and a hierarchical MIPv6 [3]. Still, aforementioned mechanisms are not appropriate to satisfy real-time traffic. We propose a enhanced fast handover scheme (EFH) to reduce the handover latency, modified from the fast handover scheme based on IEEE 802.11 [4].

In the fast handover, handover latency is smaller than the one of MIPv6, and packets from a CN to the MN are not lost during the handover because a tunnel is formed between a previous access router (PAR) and a new access router (NAR) before the layer 2 handover, and the NAR will buffer packets destined to the MN through the tunnel. The fast handover scheme, however, needs more signal packets and buffering.

---

* This work was supported by grant No. (R01-2004-000-10618-0) from the Basic Research Program of the Korea Science & Engineering Foundation.

In the EFH, handover latency and signal packets are reduced than the fast handover due to multiple signal packets in one frame.

The rest of this paper is organized as follows: in section 2, the fast handover and IEEE 802.11 are presented as related works. Enhanced fast handover scheme for MIPv6 is proposed in section 3, and in section 4, the EFH is evaluated using cost analysis. Finally, in section 5, we conclude discussion with future study.

2 Related Works

In wireless environments, typical technologies to support mobility are IEEE 802.11 [7] which supports wireless connection in a layer 2 (data link layer) and IETF MIPv6 which supports mobility in a layer 3 (network layer).

IEEE 802.11 wireless LAN is the layer 2 protocol and supports mobility in a link layer level. IEEE 802.11 handover takes place when a MN changes its association from one access point (AP) to another and its process consists of a search phase and an execution phase. The search phase presents a scan and execution phase consists of authentication and association. First the MN performs the scan to see what APs are available, and chooses one of the APs and performs a join to synchronize its physical and MAC layer timing parameters with the selected AP. Second the MN performs the authentication phase with the new AP (NAP), and performs the reassociation phase with the NAP. At the reassociation phase, the MN's link layer is connected to the NAP. In some 802.11 implementations, the scan performs far in advance of the handover and perhaps in advance of any real-time traffic. [4]

The fast handover is a scheme which improves a MIPv6 handover. In the fast handover, several portions of layer 3 handover are performed prior to the layer 2 handover. In other words, a MN performs layer 3 handover while it retains a connection to a previous access router (AR), and in this case, the PAR must have information about destined AR. The PAR establishes a tunnel between itself and a new AR (NAR) and verifies MN's new CoA (NCoA) through handover initiate (HI) message and handover acknowledge (HACK) message to the NAR. Packets that arrive at previous CoA (PCoA) are sent to the NAR through that tunnel during the handover. The tunnel is retained until the MN registers NCoA to CN.

The fast handover based on IEEE 802.11 wireless LAN has been studied. [4] shows a order that layer 2 and layer 3 handover processes are performed when the layer 2 protocol is the IEEE 802.11 wireless LAN and the layer 3 protocol is the MIPv6 fast handover. [5] havs been proposed by Y. Mun and J. Kim to improve the fast handover. They thought a scheme that mingles the layer 3 handover with the layer 3 handover, and propose that a MN encapsulates a BU message to HA in a layer 2 frame during layer 2 handover to reduce the handover latency. In this scheme, APs and a MN must have ability which can encapsulate and decapsulate an information element (IE) [6].

3 The Proposed Scheme (EFH)

When a MN moves into another network, a handover process should be followed not to lose connectivity. The handover process consists of a layer 2 and a layer 3 handovers. The layer 2 handover processes a scan phase, authentication phase and association phase. The layer 3 handover is decomposed into creating, verifying and registering a new address. The layer 2 and layer 3 handovers are performed sequentially in MIPv6 while several messages related to layer 3 handover are exchanged prior the layer 2 handover and the rest of the layer 3 handover in the fast handover. To perform the fast handover, several signal messages are needed. These additional messages make network traffic and use node's resources. In the EFH, the MN sends one frame containing several signal messages related to layer 3 handover during the layer 2 handover. Therefore, the EFH can register a new address to HA and CN more quickly than the fast handover. Fig. 1 shows messages of the EFH.

We assume that the MN must be able to encapsulate a BU destined to HA, a HoTI and a CoTI messages to an 802.11 reassociation request frame [6]. A NAP is assumed to decapsulate the 802.11 reassociation request frame and to forward the BU message to HA, the HoTI and the CoTI messages to CN. The EFH only can be performed when the fast handover is the predictive mode. Otherwise, the MN performs a MIPv6 handover (or the reactive mode of the fast handover). IPsec mechanism is employed between the MN and the HA, using current security association (SA) when the MN sends the BU message to HA and the HoTI message to CN.

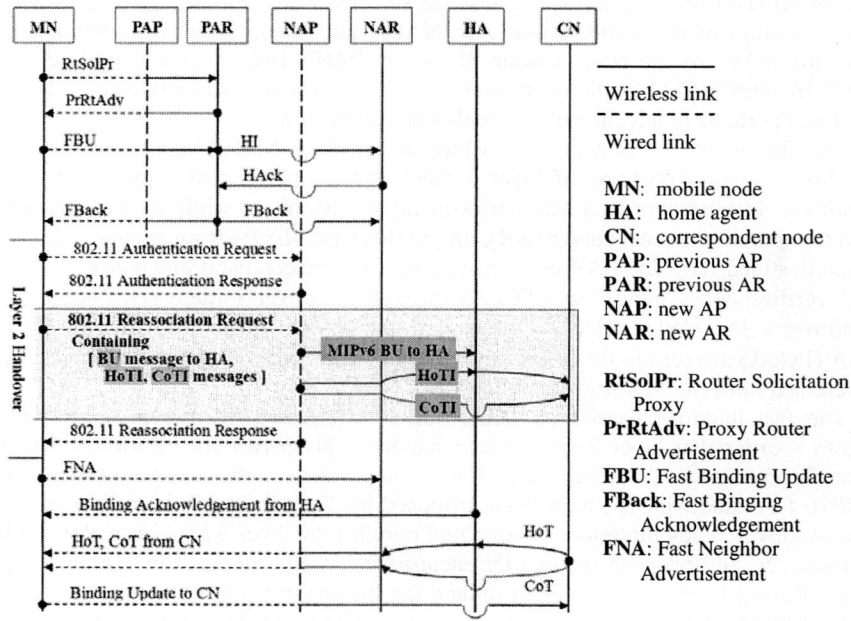

Fig. 1. The enhanced fast handover scheme (EFH)

The EFH detail is as follows:

Once authentication phase in the layer 2 handover is completed successfully, reassociation phase starts. A MN sends an 802.11 reassociation request frame, which contains three messages that are performed after the layer 2 handover; a BU, a home test init (HoTI) and a care-of test init (CoTI) messages. Source address of layer 3 handover messages is a NCoA. Once a NAP receives the reassociation request frame, it decapsulates that frame and forwards three messages. Finally the BU message gets to the HA via the NAR and HoTI and CoTI messages get to CN via the NAR. The rest of handover process is very similar to the fast handover.

4 Performance Evaluation

4.1 System Modeling

We assume that a CN generates data packets destined to a MN at a mean rate λ, and the MN moves from one subnet to another at a mean rate μ. We define packet to mobility ratio (PMR, p) as the mean number of packets received by the MN from the CN per movement. When the movement and packet generation processes are independent, stationary and ergodic, the PMR is given by $p = \lambda / \mu$. We assume that the cost for transmitting a control packet is given by the distance between the sender and receiver. The cost for transmitting a data packet will be l times greater, where $l = l_d / l_c$, where l_d is the average length of a data packet, and l_c is the average length of a control packet. The average processing cost for control packets at any node is assumed r. [8]

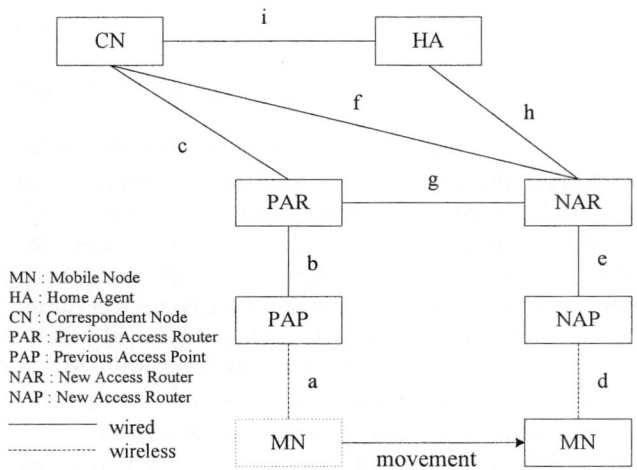

Fig. 2. System model

4.2 Cost Analysis

In this paper, we compare the EFH with a handover in MIPv6 and a fast handover, respectively. Overall handover latency is given by

$$C_{overall} = C_{signal} + C_{data},\qquad(1)$$

where C_{signal} is a sum of costs generating and transferring signal messages, and C_{data} is a sum of generating and transferring data messages during a handover.

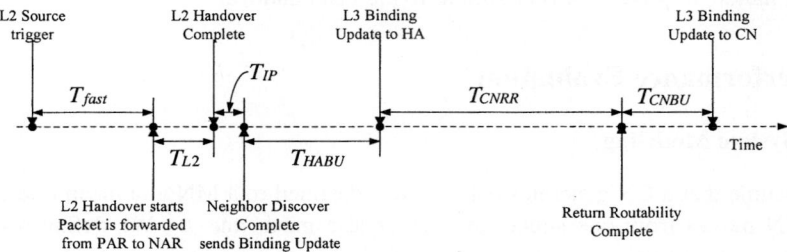

Fig. 3. Timing diagram in a fast handover

The fast handover consists of several phases (Fig. 3) [10], where T_{fast} is a delay that a MN gets and verifies a NCoA and a tunnel is established, T_{L2} is a delay during layer 2 handover, T_{IP} is a delay during informing a NAR of the NCoA, T_{HABU} is a delay during registering the NCoA to a HA, T_{CNRR} is a delay delay during RR procedure and T_{CNBU} is a delay during registering the NCoA to a CN.

The signal cost and the data cost are given by

$$C_{signal\_FH} = C_{sig.\_fast} + C_{sig.\_L2} + C_{IP} + C_{sig.\_HABU} + C_{sig.\_CNRR} + C_{sig.\_CNBU},\qquad(2)$$

$$C_{data\_FH} = P_{suc.} \times \lambda \times C_{tunnel\_FH} \times T_{tunnel\_FH} + \eta \times P_{fail} \times \lambda \times C_{loss} \times T_{loss}.\qquad(3)$$

A $P_{suc.}$ and a P_{fail} are denoted as the success probability and the failure probability of a predictive mode of the fast handover, respectively. The fast handover become a reactive mode if a MN's movement time within an overlapping cell area is longer than time establishing a tunnel (T_{fast}). In a case of the predictive mode, data packets generated by CN are tunneled by tunnel between a PAR and a NAR during a handover. In the other case, data packets generated by CN are lost during handover. To calculate the failure probability, we refer to section 4.3 in [5] which shows a failure probability of a predictive mode of the fast handover with MN's velocities and radiuses of a cell respectively, referring to [12]. The η is a weight factor, and the CN retransmits data packets when packets generated by CN are lost.

The signal cost of the fast handover, consist of 6 costs used for exchanged signal messages during respective periods in Fig. 3. The data cost is decomposed into a cost at success of predictive mode of the fast handover and a cost at failure. When the fast handover is predictive mode, C_{tunnel_FH} can be calculated by l times of a signal cost between a CN and the NAR, and $T_{tunnel} = T_{L2} + T_{IP} + T_{HABU} + T_{CNRR} + T_{CNBU}$. In the other case, C_{loss} can be calculated by l times of a signal cost between the CN and the MN, and $T_{loss} = T_{L2BASIC} + T_{IPBASIC} + T_{HABU} + T_{CNRR} + T_{CNBU}$, where $T_{L2BASIC}$ is a delay during a layer 2 handover consisting of a scan and a execution phases, and $T_{IPBASIC}$ is a delay during movement detection.

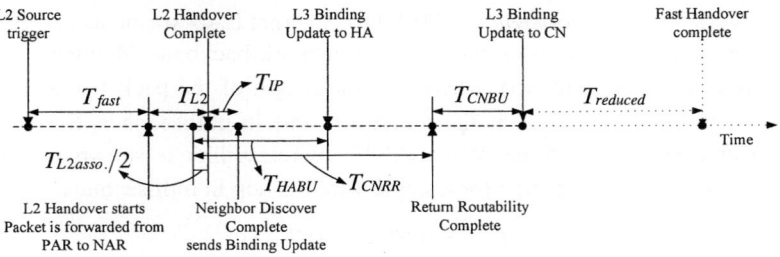

Fig. 4. Timing diagram in the EFH

Fig. 4 shows the time line of the EFH consists of several phases and is similar to the fast handover but $T_{L2asso.}$ which is a delay during association phase of a layer 2 handover. Overall handover latency of the fast handover is reduced by the EFH as shortly as $T_{reduced}$. Calculating the signal cost and the data cost of the EFH is similar to those of fast handover. In the EFH, the signal cost and the data cost are calculated like the fast handover, but the T_{tunnel_FH} is replaced by

$$T_{tunnel\_EFH} = T_{L2} + \max(T_{IP}, T_{CNRR} - \frac{T_{L2asso.}}{2}) + T_{CNBU}.$$

The rest of data cost is same to the fast handover.

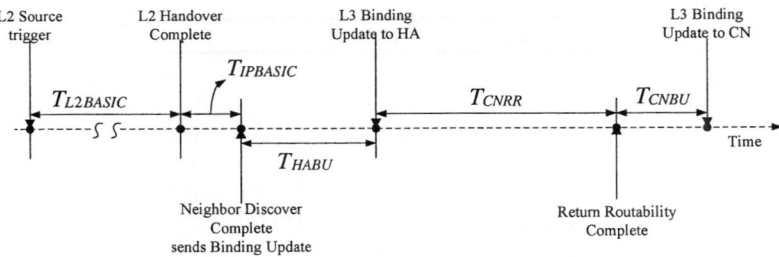

Fig. 5. Timing diagram in a MIPv6 handover

The MIPv6 handover consists of several phases (Fig. 5), and the signal cost is given by

$$C_{signal\_MIPv6} = C_{sig.\_L2BASIC} + C_{sig.\_IPBASIC} + C_{sig.\_HABU} + C_{sig.\_CNRR} + C_{sig.\_CNBU}, \quad (4)$$

where $C_{sig.\_L2BASIC}$ is a signal cost of frames of layer 2 handover consisting the scan and the execution phases, $C_{sig\_IPBASIC}$ is a signal cost of exchanging RS/RA between a MN and a NAR, and the rest of costs is same to (2). In MIPv6, data packets generated by a CN are lost during handover therefore, the data cost is same as (3) and P_{fail} is 1.

4.3 Numerical Results and Analysis

In this paper, we use formulas derived from empirical communication delay model suitable for the scenario where a CN, a HA and ARs are connected to a wired enterprise network consisting of switched 10 Mbps Ethernet LAN segments and IP routers, by carrying out experiments on the Bellcore network backbone. Regression analysis of the collected data yields (5), where k is the length of the packet in KB, h is the number of hops, and $T_{wired-RT}$ is the round-trip time in milliseconds. Similar experiments over a one-hop 2 Mbps WaveLAN^{TM} wireless link is shown in (6), where $T_{wireless-RT}$ is the round-trip time for a single wireless hop in milliseconds. [9]

$$T_{wired-RT}(h,k) = 3.63k + 3.21(h-1). \quad (5)$$

$$T_{wireless-RT}(k) = 17.1k. \quad (6)$$

T_{L2} and $T_{L2asso.}$ are derived where the AP is a Compaq EVOD 300v PC(Intel Celeron 1100MHz, 128MB RAM) using D-Link 520 PCI WLAN and the MN is a Dell Latitude CPx laptop(Intel Pentium III, 500MHz, 128MB RAM) using Lucent Orinoco Silver wireless LAN card. [11]

In Fig. 2 environment, we assume that a message processing cost is equivalent to the cost for communication over a single hop ($r=1$), a, b, c and d have a same value and c, f, g, h and i have a same value. System parameters specified in Table 1 is based on formula (5) and (6).

Table 1. System parameters

Parameter	Value
l_c	200 bytes
l_d	1024 bytes
T_{fast}	19.976 msec
T_{L2}	7.0 msec [11]
$T_{L2asso.}$	2.05 msec [11]
$T_{L2BASIC}$	84.35 msec
T_{IP}	3.524 msec
$T_{IPBASIC}$	6.549 msec
T_{HABU}	21.098 msec
T_{CNRR}	56.745 msec
T_{CNBU}	10.779 msec
T_{proc}	0.5 msec [12]
a, b, e, d	1
c, f, g, h, i	5
g	2

The variation of a signal cost and a data cost with variation of PMR. The signal cost shows fixed values with PMR because the signal cost is not affected by PMR. A MIPv6 handover has smallest signal cost because of additional signal messages in handover. The EFH uses smaller signal cost than a fast handover. **The data cost in three different schemes increases in proportion to PMR.**

Cost Ratio with Variation of PMR

Overall cost for a MIPv6 handover ($C_{overall\_MIPv6}$), overall cost for the fast handover ($C_{overall\_FH}$), and overall cost for the EFH ($C_{overall\_EFH}$) are calculated by (1)

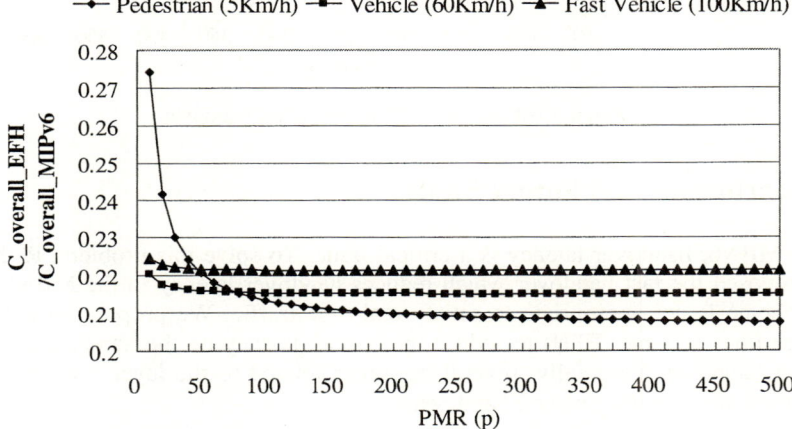

Fig. 5. Cost ratio of the EFH against a MIPv6 handover

$$\lim_{p \to \infty} \frac{C_{overall\_EFH}}{C_{overall\_MIPv6}} = \lim_{p \to \infty} \frac{C_{signal\_EFH} + C_{data\_EFH}}{C_{signal\_MIPv6} + C_{data\_MIPv6}} \approx 0.21, \quad (7)$$

where PMR > 500.

$$\lim_{p \to \infty} \frac{C_{overall\_EFH}}{C_{overall\_FH}} \lim_{p \to \infty} \frac{C_{signal\_EFH} + C_{data\_EFH}}{C_{signal\_FH} + C_{data\_FH}} \approx 0.69, \quad (8)$$

Where PMR > 500.

Fig. 5 and 6 shows the variation of cost ratio against the MIPv6 handover and the fast handover, respectively when radius of a cell is 100 meters. In a case of the pedestrian, the variation of cost ratio is high because it performs a handover infrequently at low PMR. At (7), the mean cost ratio is 0.21, and therefore, compared to the MIPv6 handover, the EFH gains 79% improvement. The mean cost ratio is 0.69 by (8), where the EFH saves the cost by 31% against the fast handover.

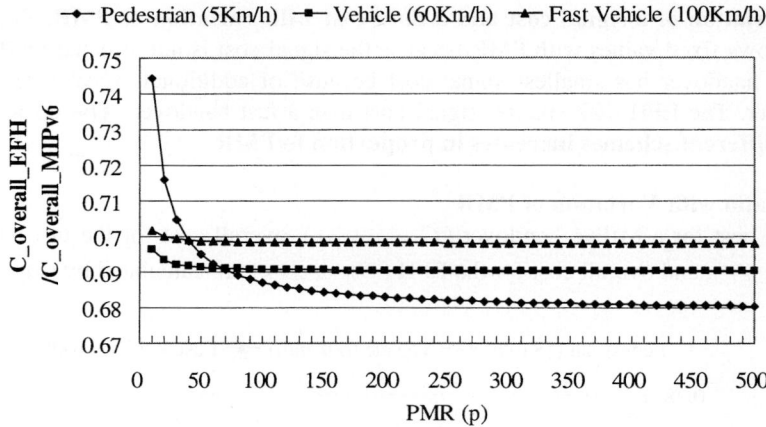

Fig. 6. Cost ratio of EFH against a fast handover

5 Conclusions and Future Study

In the MIPv6, handover latency is a critical issue. To solve this problem, R. Koodli has proposed the fast handover which reduces handover latency in the MIPv6. Still, this scheme is not appropriate to satisfy real-time traffic. We propose an enhanced fast handover scheme (EFH) to reduce the overall latency on the handover, revising the fast handover. Especially, several messages related to the layer 3 handover are sent in one frame during layer 2 handover.

We showed improved performance through comparison of costs, such as the cost ratio of the EFH against the MIPv6 handover and the fast handover. Compared to general MIPv6 handover scheme, the EFH gains 79% improvement while it gains 31% improvement, compared to the fast handover. For more accurate evaluation, we need to perform a simulation with real devices in the future because most parameters and formulas are derived by reference documents.

Reference

1. Koodli, R.: Fast Handovers for Mobile IPv6, work in progress (2004)
2. Johnson, D., Perkins, C., Arkko, J.: Mobility Support in IPv6, RFC 3775 (2004)
3. Soliman, H., Malki, K. El: Hierarchical Mobile IPv6 mobility management (HMIPv6), work in progress (2004)
4. McCann, P.: Mobile IPv6 Fast Handovers for 802.11 Networks, work in progress (2004)
5. Mun, Y., Park, J.: Layer 2 Handoff for Mobile-IPv4 with 802.11, work in progress (2003)
6. Kim, J., Mun, Y.: A Study on Handoff Performance Improvement Scheme for Mobile IPv6 over IEEE 802.11 Wireless LAN, work in progress (2003)
7. Ergen, M.: IEEE 802.11 Tutorial (2002)
8. Jain, R., Raleigh, T., Graff, C., Bereschinsky, M.: Mobile Internet Access and QoS Guarantees using Mobile IP and RSVP with Location Registers, in Proc. ICC'98 Conf. (1998) 1690-1695

9. Koodli, R., Perkins, C.: Fast Handovers and Context Transfers in Mobile Networks, ACM Computer Communication Review, Vol. 31, No. 5 (2001)
10. Pack, S., Choi, Y.: Performance Analysis of Fast Handover in Mobile IPv6 Networks, work in progress, IFIP PWC 2003, Venice, Italy (2003)
11. Vatn, J.: An experimental study of IEEE 802.11b handover performance and its effect on voice traffic, SE Telecommunication Systems Laboratory Department of Microelectronics and Information Technology (IMIT) (2003)
12. McNair, J., Akyildiz, I.F., Bender, M.D.: An Inter-System Handoff Technique for the IMT-2000 System, IEEE INFOCOM, vol. 1 (2000) 208-216

An Efficient Macro Mobility Scheme Supporting Fast Handover in Hierarchical Mobile IPv6*

Kyunghye Lee and Youngsong Mun

School of Computing,
Soongsil University, Seoul, Korea
lkhmymi@sunny.ssu.ac.kr, mun@computing.soongsil.ac.kr

Abstract. People who use wireless device for Internet are gradually increasing. For mobility service, MIPv6 has been proposed by IETF. With no restriction on node's geographical location, mobility is provided to a mobile node (MN) by MIPv6. However, MIPv6 has critical points, such as handover latency resulting from movement detections, IP address configurations and location updates which is unacceptable in real-time application. To make up for it, hierarchical MIPv6 (HMIPv6) has been proposed. HMIPv6 guarantees to reduce handover latency, because the MN only registers the new addresses at mobility anchor point (MAP) when the MN moves around access routers in the same MAP domain. HMIPv6 still has packet loss problem when the MN moves from one MAP to another. In this paper, we propose an efficient handover scheme which reduces packet loss when the MN moves between MAP domains. We adopt the fast handover method from FMIPv6 (fast MIPv6) for proposed scheme.

1 Introduction

People who use wireless devices for Internet are gradually increasing. For mobility service, MIPv6 has been proposed by IETF [1]. With no restriction on node's geographical location, mobility is provided to a mobile node (MN) by MIPv6. To keep connection with Internet from MN's movement, MIPv6 introduces a home agent (HA). Also care-of address (CoA) to indicate MN's current location is defined in MIPv6. The HA and correspondent node (CN) will keep MN's current location. Once HA and CN registrations end in a success, the MN will directly communicate with the CN. The weak point of MIPv6 is that the MN has handover latency when it moves between networks in MIPv6. Further, address configuration and location update on HA and CN to identify MN's current location considerably increase handover latency. To make up for it, hierarchical MIPv6 (HMIPv6) has been proposed [3].

HMIPv6 guarantees to reduce handover latency because a MN only registers its new addresses with a mobility anchor point (MAP) when the MN moves between

---
* This work was done as a part of Information & Communication fund amental Technology Research Program supported by Ministry of Information & Communication in republic of Korea.

access routers within the same MAP domain. In HMIPv6, two address types are defined; local care-of address (LCoA) and regional care-of address (RCoA). When the MN enters into one MAP domain, it creates new LCoA and RCoA. If the MN moves between access routers within the same MAP domain, it only configures new LCoA and registers the address at the MAP. With reduced registrations between the MN and HA and between MN and CN, the signaling costs is shortened.

To reduce handover latency between access routers within the same MAP domain, fast hierarchical MIPv6 (F-HMIPv6) has been proposed by IETF [4]. Since FMIPv6 [2] and F-HMIPv6 provide the fast micro mobility scheme between the access routers, they are not appropriate for the macro mobility between the MAP domains. Besides, the macro mobility handover using these schemes will not provide mobile devices with seamless service, which causes packet losses. Thus, we need to employ the fast handover scheme between the MAP domains to reduce packet loss and handover latency.

In this paper, we propose macro mobility scheme supporting fast handover on HMIPv6. Especially, we adopt the fast handover scheme from FMIPv6 for proposed scheme. The rest of this paper is composed as follows. In section 2, HMIPv6 and F-HMIPv6 operations are described. The proposed macro mobility scheme is introduced in section 3. The performance evaluation for proposed scheme is presented in section 4. Finally, we conclude the discussion in section 5.

2 Related Works

2.1 Hierarchical Mobile IPv6 (HMIPv6)

HMIPv6 can reduce signaling costs incurred when a MN moves around access routers within the same MAP domain. HMIPv6 has a MAP to manage the MN's location with no effect on the CN and HA. When the MN changes its default router within a MAP domain, it sends a local binding update (LBU) message to the MAP. Thus, the MAP can reduce signaling between the MN and HA and between MN and CN.

In HMIPv6, a MN has the RCoA and LCoA. When the MN moves into the new MAP domain, it creates the two addresses. Then, the MN registers the addresses at the MAP using the LBU message. The MAP received the LBU message performs a duplicate address detection (DAD). If the two addresses are available, then the MAP sends a local binding acknowledgement (LBACK) message to the MN. The MN received the LBACK message from MAP sends a binding update (BU) message to the HA and CN to register the new RCoA.

When a MN receives the proxy router advertisement (PrRtAdv) message from access router which belongs to the new MAP domain, it recognizes that enters into a new MAP domain. Then, the MN creates new LCoA and RCoA before entering into the new MAP domain using the subnet prefix and the MAP's IP address learned via the new access router.

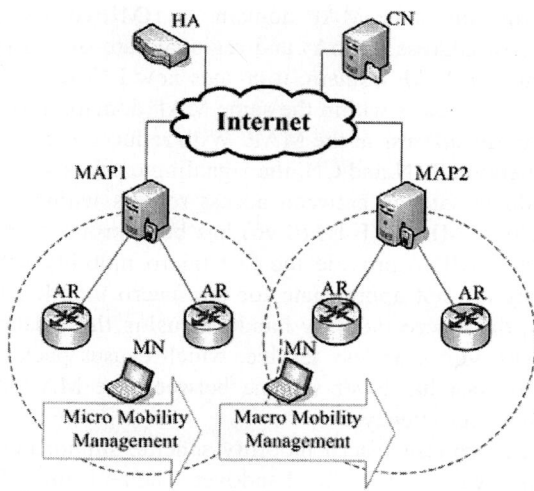

Fig. 1. This show the two types of handover when a MN moves around the networks in HMIPv6. The micro mobility handover is caused when the MN moves between access routers in the same MAP domain. In this case, the MN only creates a new LCoA. The macro mobility handover is caused when the MN moves from the one MAP domain to another. In this case, the MN creates the new LCoA and RCoA pair.

2.2 Fast Handover for HMIPv6 (F-HMIPv6)

In HMIPv6, when a MN moves between access routers within the same MAP domain, the handover between access routers is taken. To reduce handover latency between access routers within the same MAP domain, F-HMIPv6 has been proposed by IETF.

F-HMIPv6 uses the fast handover scheme of FMIPv6 to performing the handover between access routers within the same MAP domain. The MN only creates a new LCoA, and registers the new address at the MAP. The MAP binds the new LCoA with the MN's current RCoA. Thus, the MN do not need to register the new LCoA at the HA and CN. However, when a MN moves from one MAP domain to another, called the macro mobility handover, it creates the new LCoA and RCoA. The MN registers the new LCoA and RCoA at the MAP. Then, the MAP changes the MN's old RCoA to the new RCoA and binds the new LCoA with the new RCoA. The MN sends a BU message to the HA and CN to register the new RCoA. Thus, the macro mobility handover using HMIPv6 or F-HMIPv6 causes the packet loss and the long-term latency.

3 Proposed Scheme

In proposed scheme, a MN configures new two addresses before it moves into a new MAP domain. The address configuration is based on subnet prefix of the new MAP domain learned via the new access router which belongs to the new MAP domain. During the macro handover, the packets destined to MN's current address will be buffered at the new access router. Thus, the handover latency and packet loss problem can be settled. The MN informs new MAP's address of the current MAP, not its newly configured addresses. The proposed handover scheme is specified in Fig.3 in detail. As shown in Fig.2 a tunnel from the current MAP to the new access router through the new MAP is built, which allows the old MAP to transfer packets to MN's new location during handover. We assume that each access router has information of MAP within the same domain and shares the information with neighbored access routers. The procedure when a MN moves from MAP_1 domain to MAP_2 using the proposed method is shown in Fig.2.

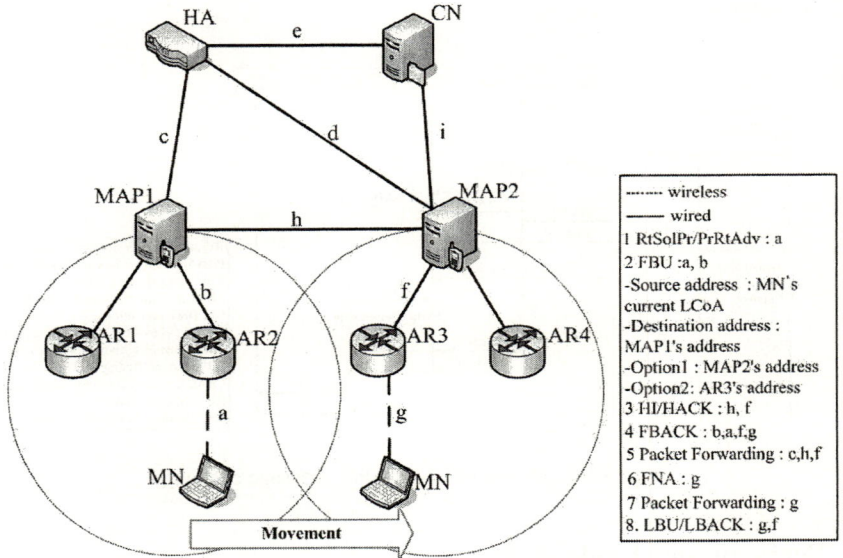

Fig. 2. The procedure when a MN moves from MAP1 domain to MAP2

1. The MN's new addresses configuration is based on subnet prefix of the new MAP domain learned via access routers which belong to the new MAP domain.
2. The MN sends the F-BU message to the MAP1 for fast macro mobility handover. The message format for FBU is shown in Fig. 2. This message should include MAP_2's IP address and the AR_3's IP address to establish the tunnel between the MAP_1 and the AR_3, as depicted in Fig. 2.
3. The MAP_1 sends the HI message to the MAP_2.
4. The MAP_2 performs DAD to verify the MN's new RCoA.
5. The MAP_2 sends the HI message to the AR_3.
6. The AR_3 performs DAD to verify the MN's new LCoA.
7. Then, the AR_3 sends the HACK message to the MAP_2.
8. The MAP_2 received the HACK message sends the HACK message to the MAP_1. Now, the MAP_1 and MAP_2 establish a tunnel between each other to transfer the packets during the macro mobility handover.

9. The MAP$_1$ sends the FBACK message to the MN to inform that MN's new addresses are available. Then, the layer-2 starts handover procedure and the layer-3 connection will be cut. During the handover, the packets destined to MN's current address are stored in buffer of the AR$_3$.
10. Before the MN enters into the MAP$_2$ domain, layer-3 handover is already done. Once the MN arrive new MAP domain, it sends the FNA message to the AR$_3$.
11. The AR$_3$ transfers the buffered packets to the MN.
12. Once the MN gets buffered packets from AR$_3$, it sends the LBU message to the new MAP.
13. When the MN receives the LBACK message from the new MAP, it sends the BU message to the HA.

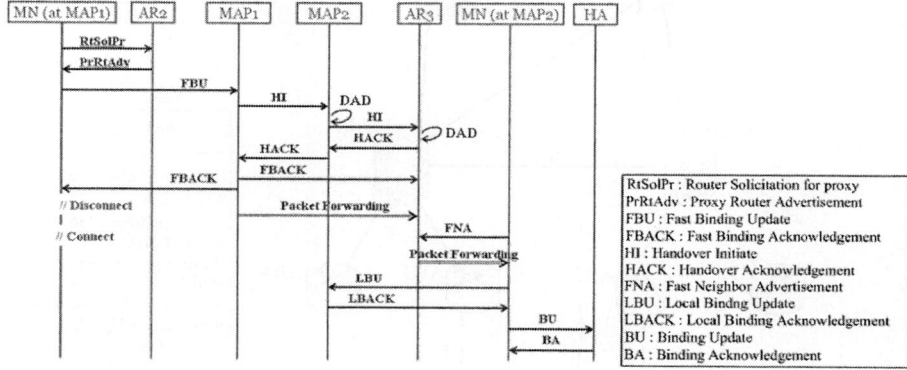

Fig. 3. The procedure of signaling message exchange

4 Performance Evaluation

In this paper, we assume that a CN creates the data packets at a mean rate, λ, and the MN moves from one subnet to another at a mean rate, μ [6,7]. We define packet to mobility ratio (PMR) as the mean number of the packets received by a MN from a CN per move and the PMR is given by $p = \lambda / \mu$. The average length of control packet and the average length of data packet are defined as l_c and l_d, respectively. The data packet to control packet rate is expressed as l. We assume that the cost of transferring a control packet is given by the distance between the sender and the receiver, and the cost of transferring a data packet is l times greater. The average cost of processing control packets at any node is assured r [6].

For performance evaluation, modeling the network architecture shown in Fig.2 is employed. The overall handover cost is decomposed as follows:

$$C_{total} = C_{signal} + C_{packet}, \tag{1}$$

where C_{signal} is a cost to handle signal packets during handover, and C_{packet} is a cost to handle data packets. Especially, C_{packet} is related to the packet loss rate during handover [7]. In the following subsections, signaling cost and packet delivery cost are discussed. The cost of proposed scheme and original HMIPv6 is considered together.

4.1 Signaling Cost

The proposed macro mobility scheme considers the additional signaling cost to reduce packet loss during the handover. When a MN enters into a new MAP domain, the signaling cost can be expresses as follows:

$$C_{signal-proposed} = C_{signal-fast} + C_{signal-IP} + C_{signal-Mreg} + C_{signal-Hreg}, \quad (2)$$

where $C_{signal-fast} = 2(2a+b)+3(h+f)+13r$, $C_{signal-IP} = g+2r$, $C_{signal-Mreg} = 2(g+f)+5r$, and $C_{signal-Hreg} = 2(d+g+f)+7r$.

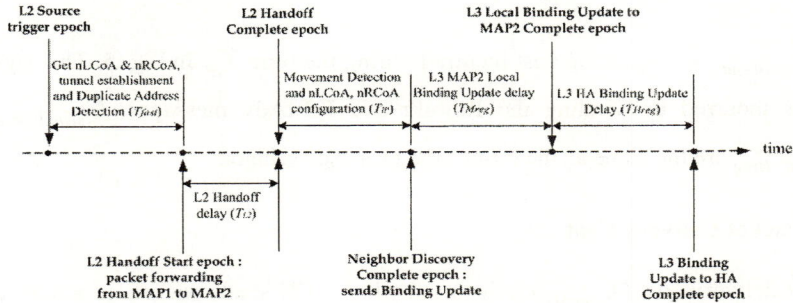

Fig. 4. Timing diagram in the proposed macro mobility scheme

In (2), $C_{signal-fast}$ is a signaling cost incurred during the time T_{fast}, $C_{signal-IP}$ is a signaling cost incurred during the time T_{IP}, $C_{signal-Mreg}$ is a signaling cost incurred during the time T_{Mreg} and $C_{signal-Hreg}$ is a signaling cost incurred during the time T_{Hreg} in Fig. 4.

In the original HMIPv6 when a MN moves from one MAP domain to another, the MN cuts connection with the current MAP for the handover between the MAP domains at first, and it establishes the connection with the new MAP at the new MAP domain in the original HMIPv6. Once connected new MAP, the MN performs the router discovery to obtain new two addresses. Thus, the total signaling cost can be expressed as follows:

$$C_{signal-original} = C_{signal-nd} + C_{signal-Mreg} + C_{signal-Hreg}, \quad (3)$$

where $C_{signal-nd} = 2a + 3r$.

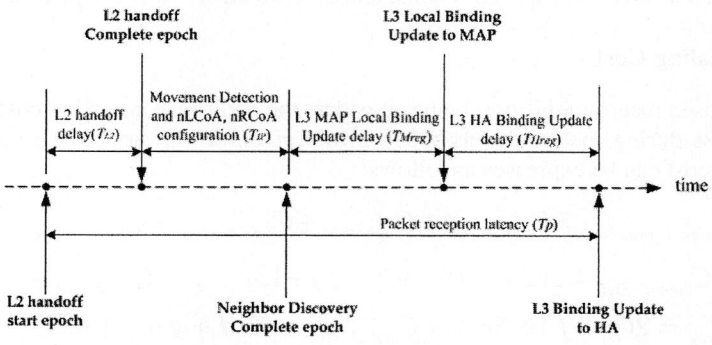

Fig. 5. Timing diagram in the original HMIPv6's handover

$C_{signal-nd}$ is a signaling cost incurred during the time T_{IP} in Fig. 5. This signaling cost is incurred by sending the RtSolPr and PrRtAdv messages. $C_{signal-Mreg}$ and $C_{signal-Hreg}$ are the same as the cost of the proposed scheme.

4.2 Packet Delivery Cost

Packet delivery cost (C_{packet}) is incurred when a CN sends packets to a MN. In this paper, we define the packet delivery cost as the sum of the packet tunneling cost (C_{tunnel}) and the packet loss cost (C_{loss}). Thus, the packet delivery cost can be expressed as follows:

$$C_{packet} = P_{suc.} C_{tunnel} + \eta P_{fail} C_{loss}, \quad (4)$$

where $P_{suc.}$ and P_{fail} are the probability of the handover success or the handover failure, η present the increase rate when a CN retransmits the packets to a MN.

$P_{suc.} C_{tunnel}$ is a packet delivery cost when the handover is success, $P_{fail} C_{loss}$ is a packet delivery cost when the handover is fail. The consideration of the probability presented in detail the section 4.3 of [10]. C_{tunnel} is signaling cost incurred during the time T_{L2}, T_{IP}, T_{Mreg} and T_{Hreg}. The packet tunneling cost (C_{tunnel}) can be expressed as follows:

$$C_{tunnel} = P_{suc} \times \lambda \times l \times C_{MAPtunnel} \times (T_{L2} + T_{IP} + T_{Mreg} + T_{Hreg}), \quad (5)$$

where $C_{MAPtunnel} = l(e+c+h+f+g) + 2r$.

The packet loss cost ($C_{MAPloss}$) presents a signaling cost incurred when the handover is fail. If the handover is fail, it performs the original HMIPv6 scheme. The packet loss time is the same that a MN performs the original HMIPv6. The packet loss can be expressed as follows:

$$C_{loss} = \eta \times P_{fail} \times \lambda \times l \times C_{MAPloss} \times (T_{L2} + T_{IP} + T_{Mreg} + T_{Hreg}), \quad (6)$$

where $C_{MAPloss} = l(e+c+b+a) + 5r$.

In Fig. 5, during the handover the packets destined to the MN's current addresses are lost until the MN registers the new two addresses with the HA. Thus, in the original HMIPv6 the packet delivery cost can be expresses as follows:

$$C_{packet} = C_{loss} = \eta \times P_{fail} \times \lambda \times C_{MAPloss} \times (T_{L2} + T_{IP} + T_{Mreg} + T_{Hreg}), \quad (7)$$

where P_{fail} value is 1.

4.3 Performance Analysis

In this paper, we use the uniform fluid model [7] to show the MN's mobility. In Fig. 1, the latency of the wired link (10Mbps LAN) can be expressed as follows [6]:

$$T_{RT-wired}(h,k) = 3.63k + 3.21(h-1), \quad (8)$$

where k is the size of packet (KB) and h is the number of hops. The results unit from (12) is millisecond. The latency time incurred during the time T_{L2} uses the value in [8]. In Fig. 2, we assume that the message processing cost is same the cost of communicating over a single hop ($r = 1$). We assume that distance cost in the same domain ($a = b = f = g$) is 1, and the distance cost between different domains ($c = d = e = i = h$) is 5. We define the system parameters as Table.1.

We assume that the retransmission of lost packets required one time by the MN. Thus, $\eta = 2$. The proposed scheme spends the signaling costs more than the original HMIPv6 handover because the signaling cost is gradually increasing regardless of the value of PMR. The proposed scheme spends the packet delivery cost less than the original HMIPv6 handover. The packet delivery cost is increasing in proportion of the PMR. Thus, the more PMR value is increasing, the more packet delivery cost is increasing. We define the total cost of the original HMIPv6 handover as the $C_{total-originalHMIPv6}$, and the total cost of the proposed scheme as the $C_{total-proposedHMIPv6}$. The Fig. 6 shows the variation of the cost ratio

$C = \dfrac{C_{total-proposed}}{C_{total-original}}$ with the PMR. The asymptotic value of the cost ratio is given by

$\lim\limits_{p->\infty} \dfrac{C_{total-proposedHMIPv6}}{C_{total-originalHMIPv6}}$. At high values of PMR ($p > 100$), the cost ratio approaches an asymptotic value

Table 1. System parameters

Parameters	Description	Value
l_c	The length of the control packet	200 bytes
l_d	The length of the data packet	1024 bytes
T_{fast}	The time during exchanging RtSolPr/FBU/FBACK	35.28 msec
T_{L2}	The performance time of the 2 layer handover [12]	7.00 msec
T_{IP}	The time from entering into the new MAP domain to receiving the FNA message	2.67 msec
T_{Mreg}	The time during registering new addresses at the new MAP	6.54 msec
T_{Hreg}	The time during registering the new addresses at the HA	21.10 msec

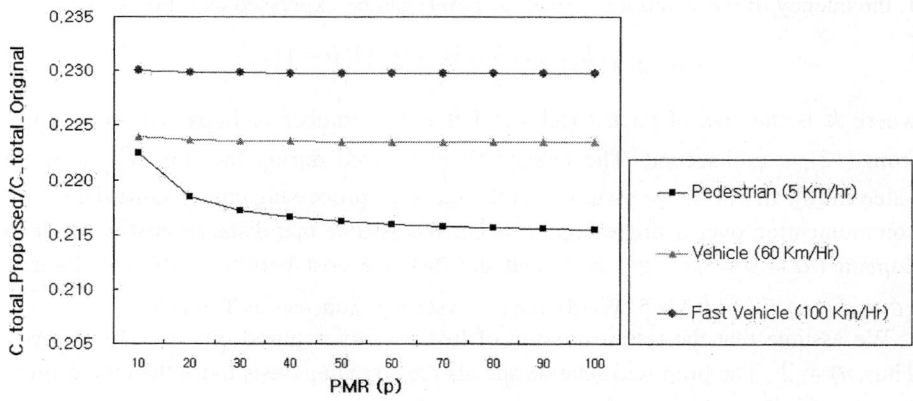

Fig. 6. Total cost ratio of the proposed scheme per the original HMIPv6 with the PMR

In the Fig. 6, the handover total cost ratio changes per the value of PMR. We observe that the total cost decreases when the PMR increases. The more speed of node slows, the more total cost decreases. The asymptotic value changes because the probability of the handover failure increases when the node's moving speed increases.

When the cost ratio approaches an asymptotic value, the cost ratio of the pedestrian is 0.215 and the cost ratio of the vehicle is 0.223 and the cost ratio of the fast vehicle is 0.229. Thus, the proposed scheme can be reduced the 78% of the total cost that the original HMIPv6 handover caused.

5 Conclusions

MIPv6 provides the mobility to mobile node with no restriction on node's geographical location. In MIPv6, to keep connection with internet from MN's movement, the care of address and home address introduced to manage the MN's location. The weak point of MIPv6 is that the MN has handover latency when it moves round networks. To reduce handover latency between access routers, HMIPv6 has been proposed. However, when a MN moves between access routers in the same MAP domain, the handover between access routers is taken. The macro mobility handover incurred between MAP domains using HMIPv6 or F-HMIPv6 causes the packet loss and the long-term latency.

In this paper, we propose the macro mobility scheme supporting fast handover between the MAP domains. We adopt the fast handover method from FMIPv6 (fast MIPv6) for proposed scheme to improve handover between MAP domains. Thus, the proposed macro mobility scheme can reduce the signaling cost and the handover latency incurred when the MN moves between the MAP domains. We observe that the proposed scheme reduces the 78% of the total cost that the original HMIPv6 handover caused.

References

1. Johnson, D., Perkins, C., Arkko, J.: Mobility Support in IPv6, RFC 3775, Internet Engineering Task Force (2004)
2. Koodli, R.: Fast Handovers for Mobile IPv6, work in progress (2004)
3. Soliman, Q.: Hierarchical MIPv6 mobility management, work in progress (2004)
4. Jung, H., Koh, S.: Fast Handover for Hierarchical MIPv6, work in progress (2004)
5. Vivaldi, I., Ali, B., Prakash,V., Sali, A.: Routing Scheme for Macro Mobility Handover in Hierarchical Mobile IPv6 Network, IEEE (2003)
6. Jain, R., Raleigh, T., Graff, C., Bereshinsky, M.: Mobile Internet Access and QoS Guarantees using Mobile IP and RSVP with Location Registers, in Proc. ICC'98 Conf., (1998) 1690-1695
7. Pack, S., Choi, Y.: Performance Analysis of Fast Handover in Mobile IPv6 Networks, in proc. IFIP PWC 2003, Venice, Italy (2003)
8. Vatn, J.: An experimental study of IEEE 802.11b handover performance and its effect on voice traffic, SE telecommunication Systems Laboratory Department of Microelectronics and Information Technology (IMIT) (2003)
9. McNair, J., Akyildiz, I., Bender, M.: An inter-system handoff technique for the IMT-2000 system, ACM Computer Communication Review, Vol.31, No.5. (2001)
10. Kim, J., Mun, Y.: A study on Handoff Performance Improvement Scheme for Mobile IPv6 over IEEE 802.11 Wireless LAN, Soongsil University (2003)
11. McCann, P.: Mobile IPv6 Fast Handovers for 802.11 Networks, work in progress, IETF (2004)

Study on the Advanced MAC Scheduling Algorithm for the Infrared Dedicated Short Range Communication

Sujin Kwag[1], SesangPark[1], and Sangsun Lee[2]

[1] Hanyang University, Division of Telecommunication Engineering,
Automotive Communication Research LAB,
133791 Seoul, Korea
sjkwag@ihanyang.ac.kr, pjesang@empal.com
http://oeic.hanyang.ac.kr
[2] Hanyang University,
Division of Electrical and Computer Engineering,
133791 Seoul, Korea
ssnlee@hanuang.ac.kr

Abstract. Dedicated Short Range Communication (IR-DSRC) is a communication system between Rode Side Equipment (RSE) and On Board Equipment (OBE) in a vehicle. Wireless resources shared among nodes should be efficiently allocated in the telecommunication system, and a MAC scheduling algorithm is needed. The purpose of the MAC scheduling algorithm is improving the efficiency after considering the QoS requirements of each node. IR-DSRC is limited to simple services and current MAC scheduling algorithm of it is similar to a Round Robin. Therefore, the efficiency of the system is inadequate for multimedia services. In this paper, we proposed a new MAC scheduling algorithm for the IR-DSRC system which improves the system efficiency and guarantees fairness among OBEs.

1 Introduction

As our society rapidly advances toward an information age, more and more people and their vehicles will depend on wireless technologies to keep them connected with others and to facilitate safe and efficient travel. Telematics, which is a new term for the convergence of Telecommunications and Informatics, provides some services to the passengers such as real time traffic information service, route guidance service, multimedia service, safety and emergency service and commercial vehicle operation service.

For the Telematics services, many kinds of technologies are needed such as telecommunication technologies, terminal platform technologies, traffic information technologies, database technologies, but the most important one of the technologies is telecommunication technologies.

IR-DSRC is a representative telecommunication system developed for the Telematics services and provides high data transmission connectivity in the fast moving vehicle (up to 160km/h). It currently provides some Telematics services such as Electronic Toll Collection System (ETCS), Bus Information System (BIS) and Commercial

Vehicle Operation (CVO). Because these services are simple and are provided in narrow area, a special MAC scheduling algorithm is not required. But as Telematics services becomes popular, variety of services will be provided using the IR-DSRC. Therefore, special MAC scheduling algorithms that use wireless resources more efficiently and guarantee fairness among OBEs will be needed.

In this paper, we propose an efficient MAC scheduling algorithm for IR-DSRC which considers both throughput and delay performance and thus attempts to improve over-all performance.

This paper is organized as follows: In section 2, we consider the IR-DSRC system. In section 3, we consider some pre-developed MAC scheduling algorithms. We propose a new MAC scheduling algorithm for IR-DSRC in section 4. In section 5, computer simulations are conducted and their results are discussed. Finally, concluding remarks are presented in section 6.

2   IR-DSRC

2.1   Overview of the IR-DSRC

The DSRC system is composed of a RSE and OBEs. The RSE is typically installed at a 5 m height of pole and the OBE is installed on the dashboard. Both pieces of Equipments operate in 800nm~900nm infrared and uses ASK-OOK (Amplitude Shift Keying-On Off Keying). This system supports full duplex transmission by the master-driven Time Division Duplex (TDD) MAC scheme, where communication from RSE-to-OBE and OBE-to-RSE strictly alternates, and also supports a data transmission speed up to 1 Mbps in a fast moving car. OBEs request an association to the RSE according to the slotted ALOHA scheme, and exchange data frames with RSE according to the asynchronous Time Division Multiple Access (TDMA) scheme.

2.2   Protocol of the IR-DSRC

IR-DSRC was designed based on the Open System Interconnection (OSI) reference model, and has 3-layer structure composed of an application layer, a datalink layer and a physical layer. The application layer is composed of a Transfer Kernel (T-KE), a Broadcast Kernel (B-KE) and an Initialization Kernel (I-KE). The datalink layer is divided into a Logical Link Control (LLC) sub-layer and a Medium Access Control (MAC) sub-layer. The LLC layer sets up the logical connection between the RSE and the OBE, and exchanges command PDUs (Protocol Data Unit) and response PDUs with another LLC layer. The MAC layer controls accesses to the shared medium, data flows and error.

In the IR-DSRC system, RSE and OBE use windows to transmit data frames. The window is the time period which is used by RSE or OBE to transmit a frame, and can be classified into a downlink window, a public uplink window or a private uplink window by their uses.

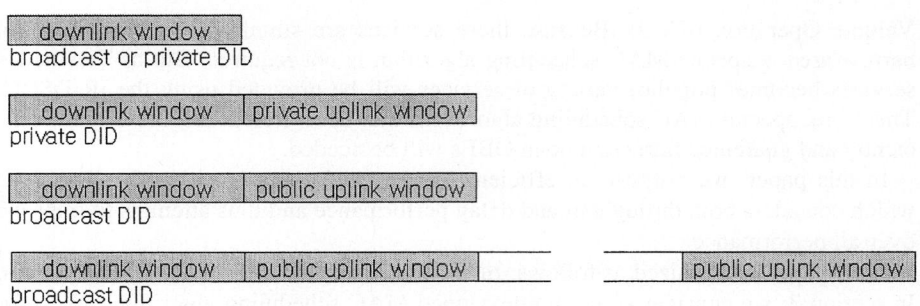

Fig. 1. Window management of the IR-DSRC

The RSE of the system is responsible for window scheduling for the communication with the OBEs. A downlink window is used by RSE to transmit data frame to more than one OBE. When OBEs request to connect to the RSE, it uses a public uplink window. Because the public uplink window can be used by any OBEs, there is some competitions that may cause collisions. After the OBE succeed in a connection request, it uses the private uplink window to transmit a data frame to the RSE. All uplink windows are allocated behind the downlink window

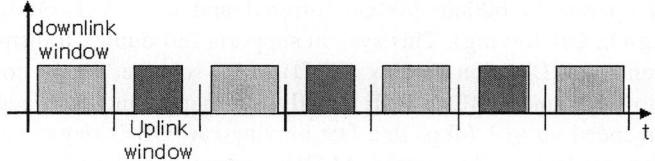

Fig. 2. Exchange of data frame between the RSE and OBEs

3 MAC Scheduling Algorithms

3.1 Definition and Necessity of the MAC Scheduling Algorithm

The nodes in the wireless communication system should share the restrictive wire-less resources with other nodes. Therefore, there is always competition to acquire more resources than other nodes and the drop in the efficiency happens. As a result, the MAC scheduling algorithm is essential to manage the wireless communication system efficiently. Because most problems happen when the nodes access the me-dium, most scheduling algorithms are implemented in the MAC layer. The best goal of the MAC scheduling algorithm is to guarantee to the nodes the best communication environment in consideration of the QoS requirements of each node such as band-width and delay bound.

An ideal MAC scheduling algorithm should be designed to satisfy following requirement. First, it should be easy to implement. The MAC scheduling algorithm should decide which node will get the right of transmission. Second, it should allocate the resources to nodes fairly. And misbehavior by one connection should not affect the performance received by other connections. Third, it should guarantee the per-form-ance bounds of each node. Therefore, it should provide better environment for the node which requires strict QoS requirement.

3.2 The Kinds of the MAC Scheduling Algorithm

3.2.1 Round Robin (RR)
Round Robin is the simplest MAC scheduling algorithm. The MAC scheduler in-vites each active queue in order and provide right of transmission. If there is no packet to transmit in the queue, the right passes to the next queue has packet to transmit. There-fore, the period of acquiring the right is decided by the number of active queues.

Round Robin scheduling algorithm is simple to implement. But it is not flexible be-cause the MAC scheduler invites each queue and gives the opportunity to send the same size packet. Moreover, it cannot guarantee the transmission of the packet which has variety of size

3.2.2 Weighted Round Robin (WRR)
Weighted Round Robin classifies the queues according to the QoS requirement and gives the weight to each queue.

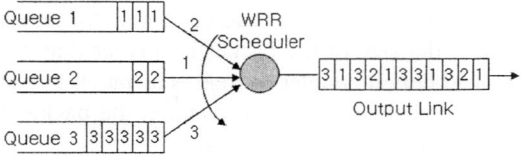

Fig. 3. Operation of WRR

The quantity of the traffic to allow can be controlled by changing the weight of each queue. MAC scheduler visits each queue in order and allows packets to send in propo-sition to the weight.

This algorithm can be seen a kind of extension of Round Robin which has a fixed weight. So it can protect the starvation problem and provide appropriate services. But this algorithm can't process packets which has various sizes, because weight means the number of packet to send. Because it processes packets based on the mean size of the packets, it can't guarantee the exact fairness among the nodes, and it is very diffi-cult to estimate the mean size of the packets.

3.2.3 Weighted Fair Queue (WFQ)
A Weighted Fair Queue uses the virtual clock and gives the index that means the ending time to the incoming packets and sorts the packets in the queue. And it ser-vices the packet which has the smallest ending time. WFQ can provide the differenti-ated services and can solve the starvation problem. WRR provides the services by packet unit and causes the problem when the packet size is various. On the contrary, WFQ provides the services by bit unit and it can use the weight more accurately. But

it is very complex to calculate the index and to sort the packets in the queue. Therefore, when there are many nodes much time delay occurs.

3.2.4 Deficit Round Robin (DRR)

Deficit Round Robin was made by modifying the WRR and can process the packets that have various sizes resulting in a reduction of a complexity of calculation.

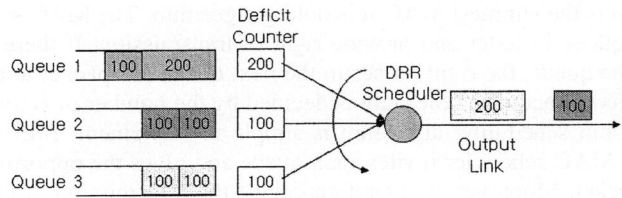

Fig. 4. Operation of DRR

DRR have 3 parameters such as a weight, a quanta, a deficit counter for each connection. The weight is the packet service rate for the output port bandwidth. The quanta in byte unit is proportioned to the weight of each queue and when the scheduler visits each queue, "deficit counter = deficit counter + quanta" is calculated. The quanta refers to the maximum bytes that can be sent at the scheduler's visit and is set '0' at the beginning. If the size of the packet is smaller than the deficit counter, the packet can be sent. And deficit counter decreases as much as packet's size. But if the size of the packet is larger than the deficit counter, the packet can't be sent and the surplus deficit counter is used at the next round.

DRR guarantees fairness because of the deficit counter. And because it has active list, the delay occurred when the scheduler visit each queue is reduced. But because the quanta are fixed, it can't process well when the large packet enter the queue. For instance, if a large packet generated by the multimedia services such as internet broadcast, VOD enters into the queue, the scheduler re-sets the deficit counter repeatedly until the it becomes lager than the packet's size. Therefore, the quality of services that require real-time processing drops by unnecessary round repeat. As a result, the scheme which sets the quanta dynamically is needed in the system environment that the mean size of packets is various for each service.

4 MAC Scheduling Algorithm of the IR-DSRC

4.1 Current MAC Scheduling Algorithm of the IR-DSRC

IR-DSRC is the master-driven TDD system that the RSE decides the transmission order of each node and resource allocation. Current MAC scheduling algorithm of the IR-DSRC is a type of the Round Robin and has similar weaknesses as the Round Robin. Therefore, modification is required to provide various multimedia services efficiently.

4.2 Advanced MAC Scheduling Algorithm for the IR-DSRC

Proposed MAC scheduling algorithm (MDRR) is a modified version of the DRR. This algorithm is designed to solve the delay problem in case a large packet enters the queue.

```
Initialize:
ActiveList = NULL;
Enqueue:
i = QueueInWhichPacketArrives
if(ExistsInActiveList(i) == FALSE) then
    Append queue I to ActiveList
    DCi = 0;
end if
Dequeue:
while(TRUE) do
    if(ActiveList ≠ NULL) then
        Remove head of ActiveList, say flow I;
        DCi = DCi + Qi
        while(QueueIsEmpty(i) == FALSE) do
            p = HeadOfLinePacketInQueue(i);
            if(size(p) > DCi) then
                if(size(p) < Dci + threshold) then
                    Transmit(p);
                    overdraft(i) = size(p) - DCi
                end if
                break
            end if
            Transmit(p);
            DCi = DCi - size(p);
            if(overdraft(i) > 0) then
                DCi = DCi - overdraft(i);
            end if
        end while
        if(DCi < 0) then
            DCi = 0;
        end if
        if(QueueIsEmpty(i) == FALSE) then
            Append queue i to ActiveList
        end if
    end if
end while;
```

Fig. 5. Operation of MDRR

That is, the number of round to re-set the deficit counter is smaller than the DRR. The variable such as over-transmission threshold (OTT), excess quantity (EQ) are supplemented Compared with the DRR. OTT is the maximum packet size that is allowed to send in case the packet is larger than the deficit counter.

EQ is a value which records the quantity in case a packet over the deficit counter is transmitted. The operation of the MDRR is followed. The MAC scheduler visits one queue. The packet in the queue is compared to the deficit counter. If the packet is smaller than the deficit counter it is allowed to send and the deficit counter is decreased as much as the packet's size after sending. But in case it is larger than the deficit counter, it is re-compared with the OTT. If the packet is smaller than the OTT, it is allowed to send the same as first case. And the deficit counter is set to 0 and the excess quantity (= packet size – deficit counter) is recorded at EQ. At the next round, the previous procedure is repeated. If deficit counter is lager than 0 and EQ is not 0, deficit counter is decreased as much as EQ and EQ is set 0.

5 Simulation of MDRR

5.1 Simulation Scenario for MDRR

IR-DSRC simulator has not been developed yet, so we simulated our algorithm via modified Bluetooth simulator, bluehoc2.0. This is because the form of the exchanging packets of the Bluetooth is very similar to the IR-DSRC's. We supposed that some OBEs are in the communication area and simulated 7 times from 1 OBE to 7 OBEs.

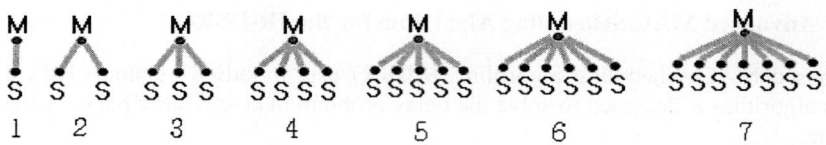

Fig. 6. Simulation scenario for MDRR

RSE and OBEs exchange the packets which have exponential distribution and mean size of 210 bytes for 20 seconds after each association. And we set the quanta 200 bytes in order to generate a delay by packets. Therefore, both throughput is much smaller than the data rate defined at the standard of IR-DSRC. We defined EQ as the relative size about Deficit Counter of each queue instead of allocating other value in each queue. We specified the dimension from 1.0 to 1.7 and executed simulation, and selected 1.1, 1.3, 1.5, 1.7 as a representative value and represented to the table. As a result, we drew the value that expresses maximum efficiency.

5.2 Analysis of Simulation Result of MDRR

In the case of 2 vehicles, when executed simulation using existing algorithm, the total throughput decreased much because of delay by repeated re-setting of the deficit counter. We applied MDRR algorithm in same simulation environment and changed OTT. As OTT increases, total throughput increased and when OTT is 1.7, total throughput increased 39.6%.

In case of 3 vehicles, it showed good result in all range of OTT at MDRR, and we can see that the total throughput is highest when OTT is 1.7.

In case of 4 vehicles the throughput trend according to the OTT was some different from 2 previous cases. In previous cases, as OTT increases the total throughput also increased. But in this case the total throughput was highest when OTT was 1.5 and decreased when it was 1.7. This maybe related to the number of car in the communication area. When a large packet enters queue, this packet can be sent quickly by MDRR but other packets are delayed by this packet. Therefore, OTT should not set high unconditionally.

In case of 5 vehicles, when OTT was 1.5 the total throughput was highest as previous case.

In case of 6 and 7 vehicles, when OTT was 1.3, the result of MDRR was highest than other cases. Each improvement compared to existing algorithm was 20.76 %, and 19.83 %. And as mentioned previous simulation, as the number of vehicle increases, the OTT that appears best total throughput was low. When OTT is large, some vehicles had starvation problem by another vehicle which exchanges large packets and this affected a negative effect on the total system efficiency.

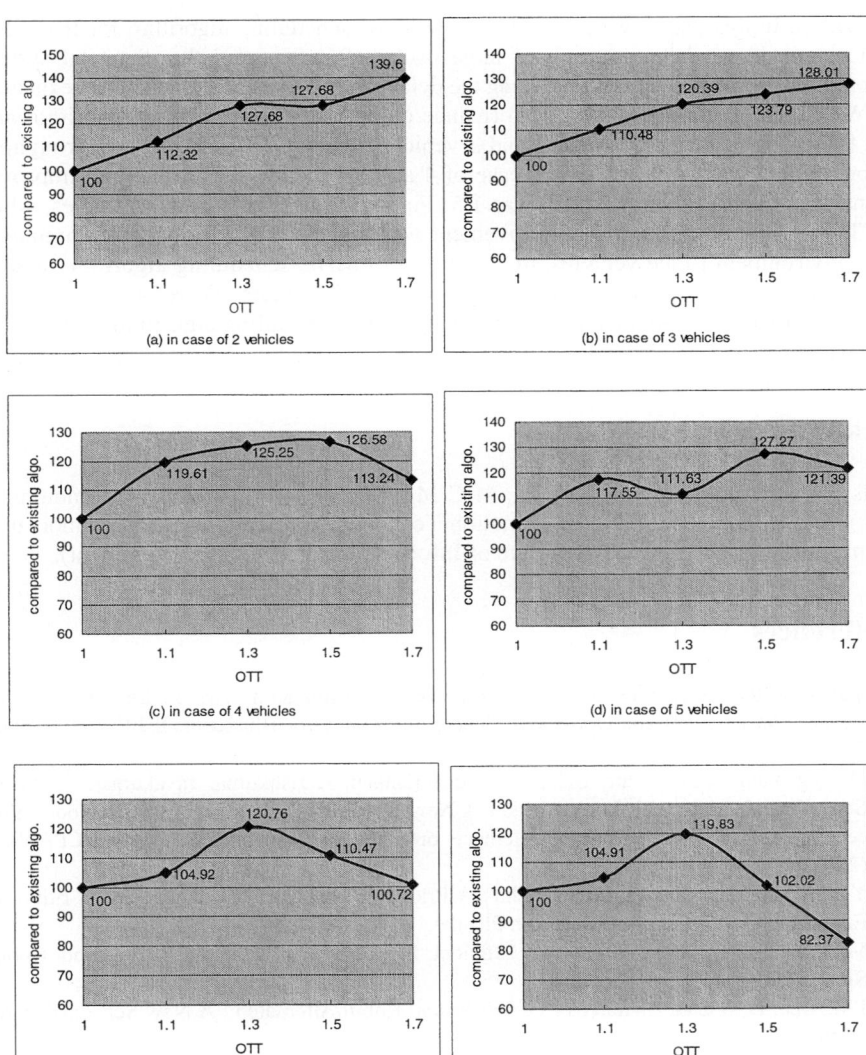

Fig. 7. Efficiency compared to existing algorithm

6 Conclusion

Current IR-DSRC is providing services such as ETC, BIS, CVO, EPC that use small quantity of packet in a narrow communication area. Therefore special MAC scheduling algorithm isn't needed in the RSE. But Telematics service will be popular and IR-DSRC should provide various multimedia services which have various packet sizes and it is difficult to operate the communication system efficiently via current MAC scheduling algorithm. In this paper, we proposed a new MAC scheduling algorithm for IR-DSRC that can improve the processing of large packet. We expected that the total system throughput would be better by reducing the delay caused by a large packet in the queue.

We executed simulation for performance comparison with existing algorithm and get following results. In case of 2 and 3 vehicles, when OTT was 1.7, the throughput improvement was highest. And in case of 4 and 5 vehicles, the throughput improvement was highest when the OTT was 1.5. Finally, in case of 6 and 7 vehicles, when OTT was 1.3, the throughput improvement was highest. To offer better communication environment to the vehicles, improvement of MAC scheduling algorithm to allocate resources is essential as well as improvement of data rate. Therefore, operation of the system will be more efficient by applying MAC scheduling algorithm proposed in this paper.

Acknowledgement

This research was supported by the MIC(Ministry of Information and Communication), Korea, under the ITRC(Information Technology Research Center) support program supervised by the IITA(Institute of Information Technology Assessment).

References

1. Yang-Ick Joo, Jong Soo Oh, Oh-Seok Kwon, Yongsuk Kim, Tae-Jin Lee, Kyun Hyon Tchah : An Efficient and QoS-aware Scheduling Policy for Bluetooth. IEEE (2002) 2445-2448
2. Daqing Yang, Gouri Nair, Balaji Sivaramakrishnan, Harishkumar Jayakumar, Arunabha Sen : Round Robin with Look Ahead : A New Scheduling Alogorithm for Bluetooth. Proceedings of the International Conference on Parallel Processing Workshop(ICPPW'02) (2002)
3. Yunxin Liu, Qian Zhang, Wenwu Zhu : A Priority-Based MAC Scheduling Algo-rithm for Enhancing QoS Support in Bluetooth Piconet.
4. M.Shreedhar, George Varghese : Efficient Fair Queuing using Deficit Round Robin. SIGCOMM'95 (1995) 231-242
5. J.M.Arco, D.Meziat, B.Alarcos : Deficit Round Robin Alternated : A New Schedul-ing Algorithm

Design and Evaluation of a New Micro-mobility Protocol in Large Mobile and Wireless Networks*

Young-Chul Shim, Hyun-Ah Kim, and Ju-Il Lee

Hongik University, Department of Computer Engineering,
72-1 Sangsudong, Mapogu, Seoul, Korea
{shim, hakim, eduard}@cs.hongik.ac.kr

Abstract. When a node becomes highly mobile, signaling overhead due to the registration becomes excessive in Mobile IP. To reduce this signaling overhead and provide smooth handoffs, many micro-mobility protocols have been proposed. But they still suffer from problems such as scalability and inefficiency in packet exchanges. To solve these problems we introduce a new micro-mobility protocol. In the proposed protocol, a domain is structured into two levels and both network-specific routing and host-specific routing are used to improve scalability. It also offers efficient mechanism for packet exchanges within a domain and clearly states the cooperation with Mobile IP during inter-domain packet exchange and handoffs. Through simulation, performance of the proposed protocol is compared with other protocols.

1 Introduction

In Mobile IP, whenever a mobile node enters into a new foreign network, it obtains a new care-of address, notifies it to the home network, and optionally notifies it to the correspondent node[1,2]. If a node becomes highly mobile, signaling overhead due to handoffs becomes excessive. To solve this problem, researchers have proposed micro-mobility protocols such as Cellular IP[3], HAWAII[4], and TeleMIP[5]. Figure 1 shows the Internet architecture with micro-mobility domains. A wireless access network called a domain consists of many cells and is connected to the Internet backbone through the router called a domain root router. There are many micro-mobility domains connected to the Internet backbone. Handoffs inside a domain are completely handled by a micro-mobility protocol while inter-domain handoffs are handled by the macro-mobility protocol that is Mobile IP. With this, intra-domain handoffs are hidden to the outside of a domain and much of signaling overhead is saved. Registration is required only during inter-domain handoffs.

Cellular IP and HAWAII route packets to a mobile node using host-specific routing methods. Therefore, a routing table size grows geometrically as the router gets closer to the domain root router. Moreover, for efficient operation, a tree-like topology is

---

* This research was supported by University IT Research Center Project, Korea.

preferred. So scalability can be a problem when the size of a domain grows. TeleMIP tries to solve this problem but still suffers from the problem that packet exchanges between two mobile nodes in the same domain are inefficient and the cooperation with MobileIP during inter-domain handoffs is not quite clear[6].

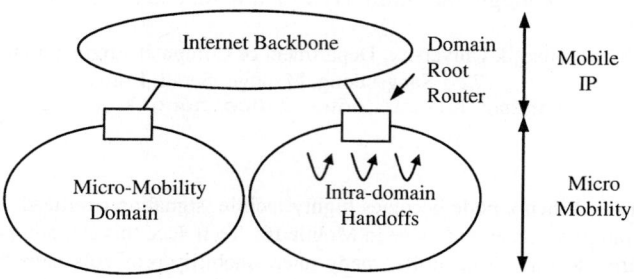

Fig. 1. Internet Architecture with Micro-Mobility Domains

In this paper we propose a new micro-mobility protocol that can be applied to large domains that may consist of several hundred thousand mobile nodes. A domain consists of many paging areas, each of which is comprised of many cells. In a network connecting paging areas to a domain root router, network-specific routing is used while host specific routing is used within a paging area. Paging mechanism is used to reduce both routing table size and signaling messages exchanged with mobile nodes. Packet forwarding mechanism is used between base stations to prevent packet losses during handoffs. Moreover, the necessary cooperation with Mobile IP is clearly stated for both inter-domain handoffs and packet exchanges with a node outside a domain.

The rest of the paper is organized as follows. Section 2 describes the network model adopted in our micro-mobility protocol. Section 3 explains the proposed protocol in detail. Section 4 compares the proposed protocol with other protocols through simulation and is followed by the conclusion in Section 5.

2 Network Model

Figure 2 depicts the network model on which our protocol is based. There are many domains and each domain is connected to the wired Internet backbone through a router called a domain root router (DRR). A domain consists of a collection of paging areas and they are connected to DRR through regular routers called connecting routers (CRs). A paging area is comprised of many cells, each of which is managed by a base station (BS). A collection of BSs belonging to a paging area forms a tree with the paging area router (PAR) at the root while DRR, CRs, and PARs can form an arbitrary topology.

When a mobile node, MN, moves from its home domain to a foreign domain, it gets two care-of-addresses (CoAs): a global CoA and a local CoA. The global CoA is the address of the foreign DRR and stored at MN's home DRR through the mobile IP registration protocol. If MN moves to another domain, this global CoA is changed to the IP address of the new DRR. MN gets its local CoA from PAR of the paging area in which MN resides. If MN moves to another paging area in the same domain, its local CoA is changed. Each paging area forms a subnet so IP addresses of its PAR

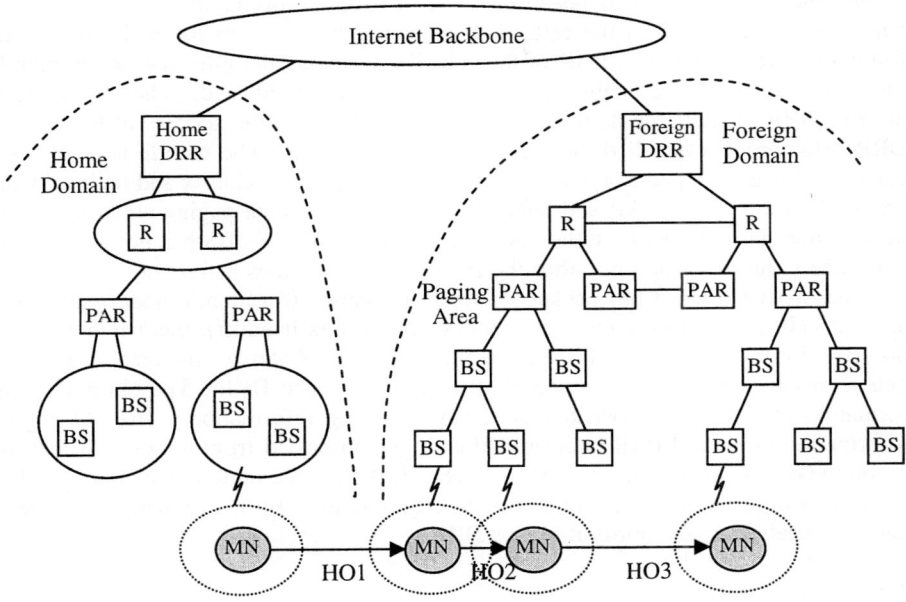

Fig. 2. Network Model

and all the BSs in this paging area have the same network prefix. The local CoA of a mobile node in this paging area is also chosen from the same subnet addresses. A local CoA is known to both PAR of the current paging area and DRR. PAR knows the local CoAs of all the mobile nodes in its paging area and stores the tuples (mobile node's home address, its local CoA) in its paging table. DRR knows the local CoAs of all the mobile nodes in its domain and stores the same kind of information in its paging table. DRR, CRs and PARs know all the subnetworks in the domain and have network specific entries for these subnetworks in their routing tables. Like HAWAII, the routing tables of PAR and all the BSs in a paging area have host specific entries for BSs and mobile nodes in this paging area. The host specific routing table entry for a mobile node is based on its local CoA. Note that a PAR's routing table has both network specific entries and host-specific entries. The default entry of any routing table stores the first node on its path toward DRR. So if a BS or router receives a packet but does not have the route, it sends the packet toward DRR.

There are three kinds of handoffs: inter-domain, inter-paging area, and intra-paging area. In Figure 2 they are depicted as HO1, HO3, and HO2, respectively.

3 The Proposed Micro-mobility Protocol

In this section we present the proposed micro-mobility protocol.

3.1 Movements of Idle Mobile Nodes

After a mobile node turns its power on in a cell within a foreign domain, it receives a beacon signal from BS of the cell. The beacon signal contains the addresses of the following three nodes: BS, PAR of the cell, DRR. From this information the mobile node knows in which domain, paging area, and cell it is located. The mobile node notifies DRR of its home domain of its global CoA that is the address of the foreign DRR. This process is called the registration in mobile IP. The mobile node also receives its local CoA and stores it. The mobile node's home address and its local CoA are notified to both its PAR and DRR. DRR knows in which paging area the mobile node is but the route between the PAR and the mobile node's BS has not been set up yet. The route is set up only after the mobile node becomes active. A mobile node becomes active when it has packets to send or receive. If a mobile node moves into another cell in the same paging area (intra-paging area handoff), there is nothing to do. But if it moves into a cell in a different paging area (inter-paging area handoff), it gets a new local CoA from its new PAR and notifies it to DRR. The old paging information in DRR is overwritten by the new paging information. The old paging information in the old PAR is cleared after it gets informed from the old BS that the mobile node has moved to other paging area. If the mobile node moves into a cell in a different domain (inter-domain handoff), it should perform the registration process, get new local CoA, and notify it to new DRR.

3.2 Packet Delivery

In this subsection we explain how packets are delivered from or to a mobile node using the following terminologies.

- MN1, MN2: mobile nodes 1 & 2
- CA1, CA2: local CoAs of mobile nodes 1 & 2
- PAR1, PAR2: PARs of mobile nodes 1 & 2
- DRR1, DRR2: DRRs of mobile nodes 1 & 2
- HA1, HA2: home agents of mobile nodes 1 & 2

Except CA1 and CA2, all notations can mean both a node and the IP address of this node. So MN1 can mean both the mobile node 1 and its home address.

We consider the case where a mobile node MN1 sends packets to another mobile node MN2. There can be three different cases regarding the locations of MN1 and MN2. They can be in the same paging area, in different paging areas of the same domain, and in different domains.

In the first case, MN1 and MN2 are in the same paging area and we assume that PAR1 is their common PAR. The packet is delivered as follows:

MN1 → PAR1 : | CA1, MN2 | MN1, MN2 | Data |
PAR1 → MN2 : | CA1, CA2 | MN1, MN2 | Data |

MN1 encapsulates the packet with the outside header whose source address is CA1. Because BSs in the paging area have routing entries based on only local CoAs and do not know the route to MN2, the packet is delivered to PAR1. PAR1 checks its paging table and finds that MN2 is in its paging area with the CoA CA2. If MN2 is idle, PAR1 finds the location of MN2 by the paging mechanism and a route between PAR1 and MN2 is set up. Then PAR1 changes the destination address of the outside header to CA2 and sends the packet to MN2. After receiving the packet, MN2 can send the following binding update to MN1.

MN2 → MN1 : | CA2, MN1 | MN2, MN1 | BU: DRR1, CA2 |

Note that the binding update message contains both the global CoA and the local CoA of MN2. After receiving this binding update packet, MN1 knows that MN2 resides in the same domain and, therefore, stores the local CoA CA2 for MN2, and can send packets directly to MN2 by using CA2.

In the second case, MN1 and MN2 are in the same domain and we assume that DRR1 is their common DRR. But they are in different paging areas. The packet is delivered as follows:

MN1 → PAR1 → DRR1 : | CA1, MN2 | MN1, MN2 | Data |
DRR1 → PAR2 → MN2 : | CA1, CA2 | MN1, MN2 | Data |

The encapsulated packet is delivered from MN1 to PAR1. Because PAR1 does not know about MN2, the packet is sent up to DRR1. DRR1 checks its paging table and finds the CoA of MN2 to be CA2. DRR1 changes the destination address of the outside header to CA2 and forwards the packet toward PAR2. Upon receiving the packet, PAR2 sends the packet to MN2. Upon receiving the packet, MN2 can send the same binding update message to MN1 as in the previous case. After receiving the binding update message, the packets from MN1 to MN2 now are delivered along, MN1, PAR1, PAR2, and MN2. The packet delivery from PAR1 to PAR2 is performed by PARs and connecting routers based upon the network prefix of CA2.

In the final case, MN1 and MN2 are in different domains. The packet is delivered as follows:

MN1 → PAR1 → DRR1 : | CA1, MN2 | MN1, MN2 | Data |
DRR1 → HA2 : | DRR1, MN2 | MN1, MN2 | Data |
HA2 → DRR2 : | HA2, DRR2 | MN1, MN2 | Data |
DRR2 → PAR2 → MN2 : | DRR2, CA2 | MN1, MN2 | Data |

The packet from MN1 travels through PAR1 to DRR1 because the domain 1 does not have the route to MN2. At DRR1, it changes the source address in the outside header to the global CoA of MN1, which is DRR1. The packet goes to MN2's home agent and then is tunneled to DRR2. DRR2 changes the source address and the destination address in the outside header to DRR2 and CA2. The packet is delivered to MN2 as explained in the previous case. Upon receiving this packet MN2 can send the binding update packet to MN1 as follows:

MN2 → DRR2 : | CA2, MN1 | MN2, MN1 | BU: DRR2, CA2 |
DRR2 → HA1 : | DRR2, MN1 | MN2, MN1 | BU: ... |
HA1 → DRR1 : | HA1, DRR1 | MN2, MN1 | BU: ... |
DRR1 → MN1 : | DRR1, CA1 | MN2, MN1 | BU: ... |

Because MN1 knows that MN2 is in a different domain by comparing MN2's global CoA in the binding update message and its global CoA. So MN1 stores the global CoA DRR2 for MN2 and uses this address when it sends packets directly to MN2.

Because the packet exchanges with a fixed node can be easily inferred from the previous explanations, we omit their discussion.

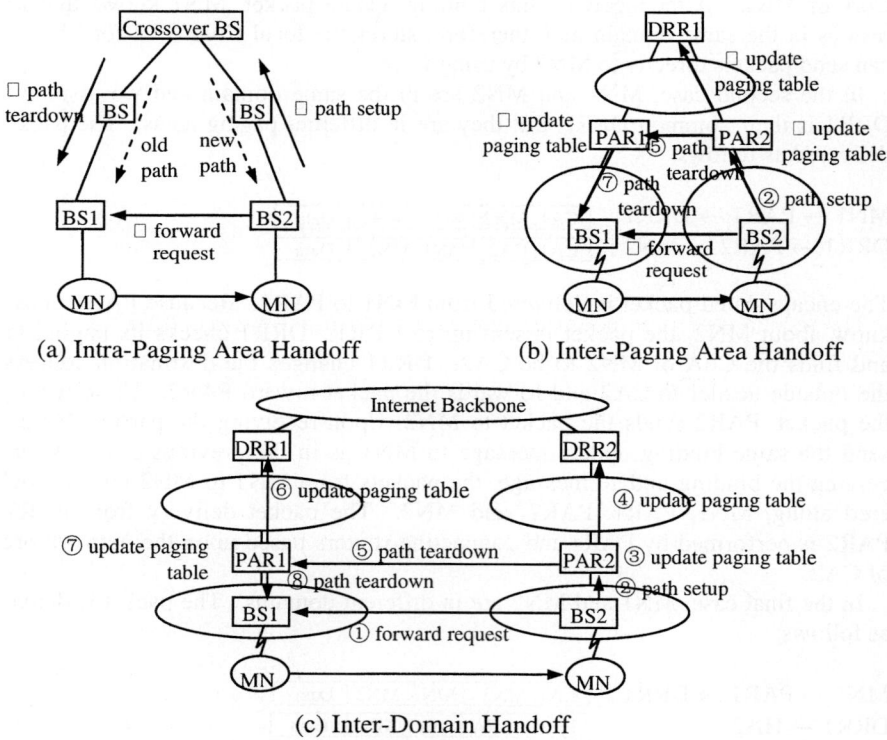

Fig. 3. Handling Handoffs

3.3 Handoffs of Active Mobile Nodes

In this subsection we explain how handoffs of an active mobile node is handled. We assume that a mobile node MN moves from an old cell managed by BS1 to a new cell managed by BS2. We use the following terminologies.

- MN: MN is also used as the home address of the mobile node
- CA1, CA2: CoAs of MN in the old and new cells
- PAR1, PAR2: PARs of the old and new cells
- DRR1, DRR2: DRRs of the old and new cells

We also assume that when a mobile node moves into a new cell, it receives from BS2 a beacon signal containing the addresses of (BS2, PAR2, DRR2). After checking with the addresses of (BS1, PAR1, DRR1) that the mobile node obtained from the old cell, it can find out what kind of handoff it is making.

Now we first consider the intra-paging area handoff. In this case PAR1 = PAR2 and DRR1 = DRR2. So MN does not need to get a new CoA. It just sends to BS2 a handoff signal containing the following information: (intra-paging area, CA1, BS1). Receiving this handoff signal, BS2 first sends a packet forward request to BS1 asking BS1 to forward packets destined for MN to BS2 so that BS2 can deliver those packets to MN in its cell. Then BS2 sends up a path setup signal until it reaches the crossover BS of the old path and the new path. The crossover BS can be PAR. The path setup signal sets up a new path for MN from the crossover BS to BS2 creating a routing table entry for MN in BSs on that path. From the crossover BS, a path teardown message is sent down to the old BS tearing down the old path for MN by deleting the old routing table entries for MN in BSs on that path. When the path teardown message reaches BS1, BS1 stops forwarding packets and the handoff process is completed. The intra-paging handoff process is shown in Figure 3 (a).

The second case is an inter-paging area handoff in the same domain. In this case DRR1 = DRR2 but PAR1 ≠ PAR2. MN in a new cell gets a new local CoA, CA2, and sends to BS2 a handoff signal containing the following information.

(inter-paging area, MN, CA1, BS1, PAR1, CA2)

Receiving this handoff signal, BS2 sends a packet forward request to BS1. Then it sends up a path setup signal toward PAR2 so that a new path for MN from PAR2 to BS2 can be set up. Upon receiving the path setup signal, PAR2 stores the tuple (MN, CA2) in its paging table, forwards the path setup signal to DRR1 so that DRR1 can also store the tuple (MN, CA2) in its paging table, and sends a path teardown signal to PAR1. Note that when DRR1 inserts a new tuple in the paging table, it also deletes the old tuple (MN, CA1). When it receives the path teardown signal, PAR1 first deletes the tuple (MN, CA1) from its paging table and sends down the path teardown signal toward BS1 so that the old path for MN from PAR1 to BS1 can be removed. When the path teardown message arrives at BS1, BS1 stops forwarding packets for MN. The inter-paging area handoff process is depicted in Figure 3 (b).

In case of inter-domain handoffs, PAR1 ≠ PAR2 and DRR1 ≠ DRR2. After sending a registration signal to its home DRR, MN gets a new local CoA, CA2, and sends to BS2 a handoff signal containing the following information.

(inter-domain, MN, CA1, BS1, PAR1, CA2)

Upon receiving this handoff signal, BS2 sends a packet forward request to BS1. After this, the steps 2, 3, 4 and 5 in Figure 3 (c) are the same as in the case of the inter-paging area handoff. Upon receiving the path teardown signal, PAR1 first sends this path teardown signal to DRR1 so that DRR1 can delete the tuple (MN, CA1) from its paging table. Then the steps 7 and 8 in Figure 3 (c) are the same as the steps 6 and 7 in Figure 3 (b). Finally BS1 stops forwarding packets for MN.

In the protocols handling handoffs, the packet forwarding scheme is used to avoid packet loss. New paths are set up bottom-up and old paths are torn down top-down to eliminate the possibility of making routing loops.

4 Performance Evaluation

In this section we compare the performance of the proposed protocol with other protocols through simulation. The simulators were implemented using NS2 2.1b6 on a Pentium IV PC running Redhat 9.0 and we performed two sets of experiments.

If MN exchanges packets with another node in the same domain, a simple analysis of the proposed protocol can show that our protocol is much more efficient than other protocols including HAWAII and TeleMIP. So we do not consider the intra-domain packet delivery and focus only on the inter-domain packet delivery where MN exchanges packets with another node that resides outside the domains over which MN roams. Because TeleMIP is similar to our protocol in this case, we compare with only HAWAII. But careful analysis reveals that the cooperation with the macro-mobility protocol required for the inter-domain packet delivery and handoffs is not clearly stated in TeleMIP but is lucidly defined in our protocol. Moreover TeleMIP does not provide mechanisms to avoid packet losses during handoffs. The comparison with mobile IP is also included because Mobile IP sets the basis of comparison.

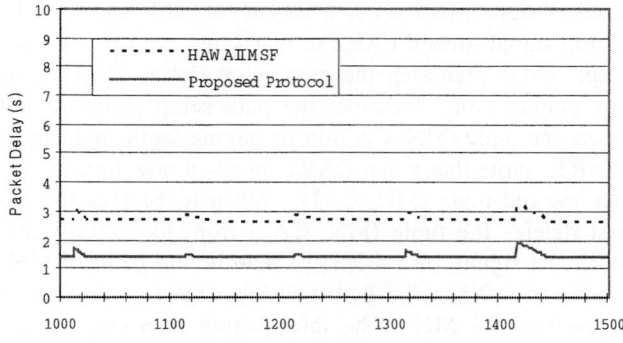

Fig. 4. Results for the 1st Set of Experiments with the Same Large Domain Size

In the first experiment we implemented our protocol and the HAWAII MSF schema on the network based on Figure 2. A paging area consists of 10 BSs and a large number of paging areas are connected to the DRR through 30 routers including PARs. Each link has the bandwidth of 10Mb and the transmission delay over a link ranges from 20ms to 130ms. There are 5000 mobile nodes in the foreign domain. A correspondent node (CN) residing outside the foreign domain sends a 160 byte UDP packet at every 20ms to a mobile node moving at the speed of 20m/s in the foreign domain. The packet delay from the correspondent node to the mobile node is measured and the simulation results are shown in Figure 4. The small peaks in the figure represent extra delays incurred during handoffs. The purpose of this experiment is to analyze how a micro-mobility protocol performs in a large domain. The figure shows that packet delay of HAWAII is much higher than our protocol. In HAWAII the size of a routing table grows geometrically as the router becomes closer to the domain root router because it uses host specific routing. As the domain size grows, the cost of updating routing tables and delivering packets for mobile nodes grows rapidly in HAWAII. So HAWAII performs poorly in a large domain.

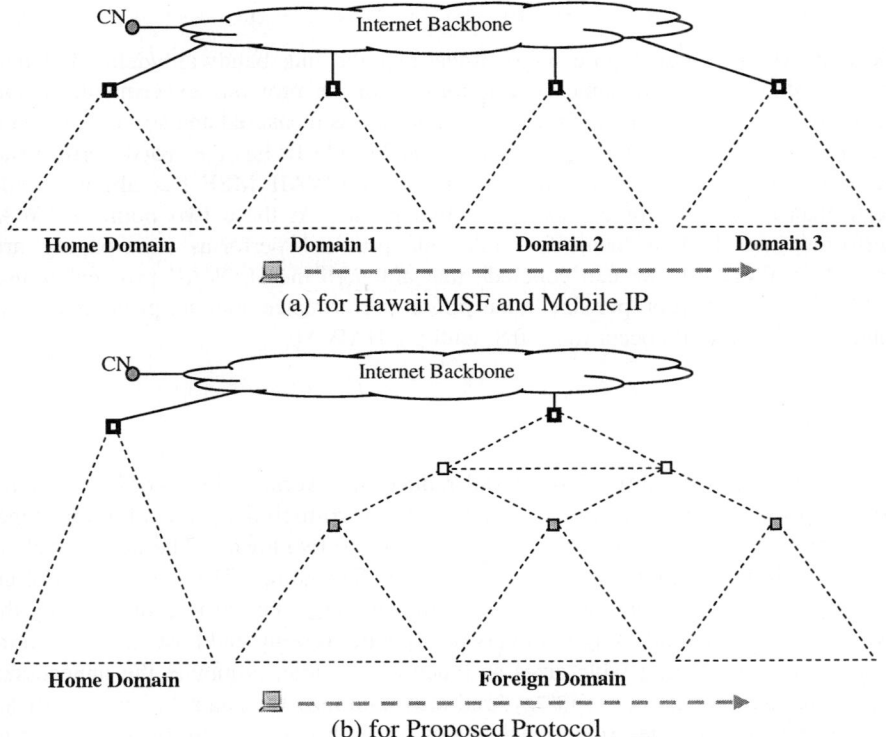

Fig. 5. Network Structures with Proper Domain Sizes

Since it is revealed that HAWAII is not suitable for large domains, in the second experiment we divided a large domain into three mid-size domains as in Figure 5 (a) and ran HAWAII MSF over them while our protocol was run on the large domain as in Figure 5 (b). To set the basis for comparison we also ran the Mobile IP protocol on

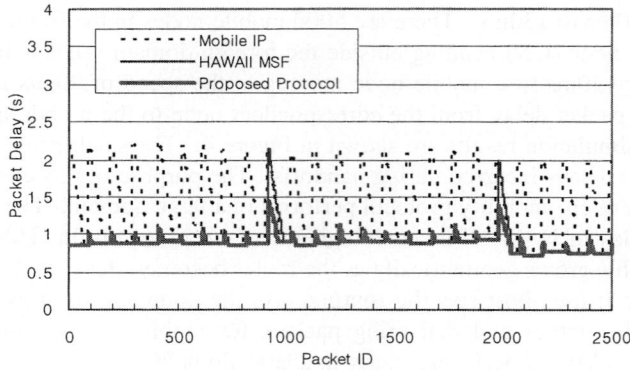

Fig. 6. Results for the 2nd Set of Experiments with Proper Domain Sizes

the network shown in Figure 5 (a). Regarding the link bandwidth/delay and node movement, we used the same assumptions as in the previous experiment. Packet delay from the correspondent node to the mobile was measured and simulation results are shown in Figure 6. The results show that Mobile IP has the worst performance because every handoff incurs a registration. HAWAII MSF has almost similar performance as our protocol except at two points. At these two points HAWAII performs inter-domain handoffs while our protocol performs inter-paging area handoffs. From this we can conclude that although the HAWAII protocol is used over a domain with a proper size, it has poorer performance than our protocol because inter-domain handoffs occur more frequently in HAWAII.

5 Conclusion

To solve the scalability problem and reduce the overhead in handoffs and intra-domain packet exchanges we introduced a new micro-mobility protocol in this paper. In the proposed protocol, a domain is structured into two levels. The upper level can take an arbitrary topology and uses network specific routing. The lower level is comprised of paging areas and uses host-specific routing. The purpose of adopting this two level hierarchy and using both network-specific routing and host-specific routing is to reduce the routing table sizes as much as possible, eliminate the unnecessary limitations on the network topology, and also minimize the control signal overheads in mobile nodes. We described mechanisms for both intra-domain packet exchanges and inter-domain packet exchanges. We showed that in the case of intra-domain packet exchanges our protocol exhibited better performance than other micro-mobility protocols. We then presented mechanisms for both intra-domain handoffs and inter-domain handoffs, showed how forwarding mechanism can be used to prevent packet losses during handoffs, and also explained the cooperation between the proposed micro-mobility protocol and the macro-mobility protocol for inter-domain handoffs. We showed the efficacy of the proposed protocol through simulation.

References

1. B.A. Forouzan, TCP/IP Protocol Suite: Second Edition, McGraw-Hill, 2003.
2. C. Perkins, IP Mobility Support, IETF RFC 2002, Oct. 1996.
3. A.G. Valko, Cellular IP: A New Approach to Internet Host Mobility, Comp. Commun. Rev., Jan. 1999, 50-65.
4. R. Ramjee et al, HAWAII: A Domain-Based Approach for Supporting Mobility in Wide-Area Wireless Network, IEEE/ACM Transactions on Networking, 10(3), June 2002, 396-410.
5. S. Das et al, TeleMIP: Telecommunications-Enhanced Mobile IP Architecture for Fast Intradomain Mobility, IEEE Personal Communications, August 2000, 50-58.
6. A.T. Campbell et al, Comparison of IP Micromobility Protocols, IEEE Wireless Communications, Feb. 2002, 72-82.

Performance Analysis of Transmission Probability Control Scheme in Slotted ALOHA CDMA Networks

In-Taek Lim

Pusan University of Foreign Studies,
Division of Computer Engineering, San 55-1,
Uam-Dong, Nam-Gu, Busan, 608-738, Korea
itlim@pufs.ac.kr

Abstract. This paper proposes a dynamic packet transmission probability control scheme in slotted CDMA system. In slotted CDMA system, multiple access interference is the major factor of unsuccessful packet transmissions. In order to obtain the optimal system throughput, the number of simultaneously transmitted packets should be kept at a proper level. In the proposed scheme, the base station calculates the packet transmission probability based on the offered load and then broadcasts it. Mobile stations, which have a packet to send, attempt to transmit with the received probability. Numerical analysis and simulation results show that the proposed scheme can offer better system throughput than the conventional one, and furthermore can guarantee a good fairness among all mobile stations regardless of the offered load.

1 Introduction

Spread spectrum code division multiple access (CDMA) technique has been widely used in military communication systems, and has been recently applied to third-generation mobile communication systems. This is shown in that most proposals for radio transmission technology (RTT) in IMT-2000 are based on CDMA [1][2][12]. Slotted ALOHA (S-ALOHA) random access protocol has been widely recognized for packet radio applications because of its simplicity in managing packet transmissions. Application of the conventional S-ALOHA protocol to CDMA technique, namely CDMA S-ALOHA system, offers relatively high system capacity [3][4].

In CDMA S-ALOHA system with a transmitter-based code method, a unique spreading code is assigned to each mobile station. If the number of simultaneously transmitting mobile stations increases above a certain threshold, almost all the packets received by the base station can be erroneous. Therefore, the number of simultaneously transmitting mobile stations causes unsuccessful packet transmissions. If the multiple access interferences can be remained close to the level that the system can support, it is expected to achieve the optimal system performance.

Many researches have recently appeared with the analysis of CDMA S-ALOHA system [4][5][7]. Most previous researches have been based on a fixed transmission probability without the transmission probability control scheme, and also have been done without distinguishing between new packet and retransmitted packet. In the case of a high transmission probability, packet errors will frequently occur due to the increased multiple access interferences as becoming the offered load high. On the other hand, if the transmission probability is low, the throughput will decrease because of the excessive restriction of packet transmissions. There are some researches to control the transmission probability aiming at improving the system throughput [8-11]. In these researches, mobile stations that fail to transmit a packet retransmit it with the decreased transmission probability. Continuously decreasing the transmission probability, the transmission probability of a specific mobile station becomes excessively decreased. As a result, the throughput can be decreased, the transmission delay can be increased, and moreover, fairness between mobile stations cannot be guaranteed.

This paper is intended to improve the performance and guarantee the fairness between mobile stations. For these purposes, this paper proposes a transmission probability control scheme for CDMA S-ALOHA system using transmitter-based code method.

This paper is organized as follows. The system model of CDMA S-ALOHA is presented in Section 2. The proposed backoff scheme is explained in Section 3, and an analytical model is presented in Section 4. The analytical and simulation results are presented in Section 5. Section 6 presents the conclusions.

2 System Model

2.1 Throughput of CDMA System

The number of simultaneously transmitted packets as well as the processing gain has a strong influence on the bit error probability of CDMA system. Accordingly, the system throughput, which is defined as the number of successful packets, can be affected by the bit error probability. When n packets are transmitted simultaneously, the probability $P_c(n)$ that a packet is successfully received and the number of successful packets $S(n)$ can be expressed as follows, respectively:

$$P_c(n) = \{1 - P_e(n)\}^L \tag{1}$$

$$S(n) = n \cdot \{1 - P_e(n)\}^L \tag{2}$$

where L is the packet length in bits and $P_e(n)$ is the bit error probability [6].

Fig. 1 shows the achievable throughput of CDMA system versus the number of simultaneously transmitted packets, where the packet length is 432 bits, the processing gain is 64, and E_b/N_0 is 15dB. As shown in Fig. 1, when the number of simultaneously transmitted packets is over 12, the throughput decreases as a result of packet errors due to the excessive multiple access interferences. Therefore, in order to achieve the maximum throughput in CDMA system, the

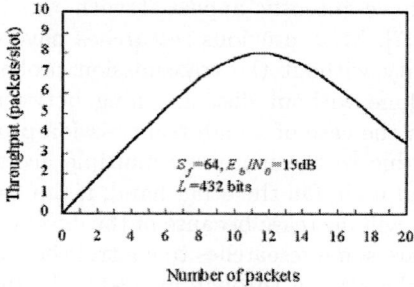

Fig. 1. The achievable throughput of CDMA system

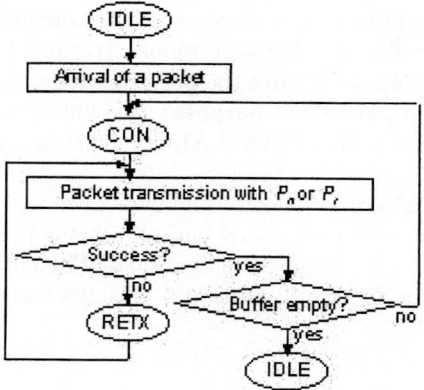

Fig. 2. Operation mode of mobile station

number of simultaneously transmitted packets should be controlled at a proper level.

2.2 CDMA S-ALOHA Protocol

The system consists of a single base station and N mobile stations, each with an infinite buffer capacity. Each packet has a fixed length of L bits. All the mobile stations synchronize their transmissions so that they transmit packets with a uniquely assigned transmitter-based code at the beginning of each slot. Every mobile station generates a packet in each slot with arrival rate λ. When the mobile station generates a packet, it is stored at the buffer. Stored packets are served on a first-in-first-out discipline. Fig. 2 shows the operation mode of each mobile station. As depicted in Fig. 2, all the mobile stations may be in one of three different operation states: idle state, contention (CON) state, and retransmission (RETX) state.

The mobile station, which does not have any packet in the buffer, is said to be in the idle state. When the mobile station in the idle state generates a

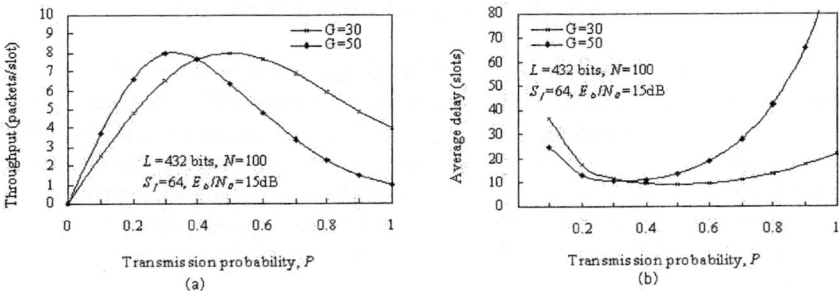

Fig. 3. (a) Throughput versus transmission probability. (b) Delay versus transmission probability

packet, it enters into the contention state and transmits a packet at the next slot with a certain transmission probability P_n. The mobile stations are informed as to whether or not the transmitted packets are successfully received by the base station in the form of acknowledgement through an error-free downlink channel. The mobile station that fails to transmit a packet or does not permitted to transmit changes to the retransmission state, and retransmits the packet at the next slot with a certain retransmission probability P_r. After successful transmission, the packet is removed from the buffer and the mobile station serves the next packet if exists. The retransmission process is repeated until the packet is successfully received.

In order to analyze the relationship between the transmission probability and the system performance, simulations were carried out in terms of the throughput and average delay according to the transmission probability for various values of offered load. Simulation results are depicted in Fig. 3, where G stands for the offered load, the packet length is 432 bits, the processing gain is 64, E_b/N_0 is 15dB, N is 100, and the transmission probabilities in the contention and retransmission state are set equal to P. From Fig. 3(a), the maximum throughput for the offered loads $G=30$ and $G=50$ can be achieved when $P=0.5$ and $P=0.3$, respectively. In these cases, the number of simultaneously transmitted packets is about 12. This is same as the number of simultaneously transmitted packets for achieving the maximum throughput as shown in Fig. 1. The delay for the offered loads $G=30$ and $G=50$ increases while decreasing the transmission probability when $P < 0.5$ and $P < 0.3$, respectively. The reason for this is that mobile stations excessively restrict their transmissions. Furthermore, at a high offered load ($G=50$), packet errors will frequently occur due to the increased level of multiple access interferences as the transmission probability increases. In this case, almost all of the mobile stations should retransmit their packets, and hence the average delay rapidly increases.

Therefore, in order to achieve the maximum performance, the number of simultaneously transmitted packets should be controlled at a proper level with the optimum value of transmission probability. For this purpose, the transmission probability control scheme is required in CDMA S-ALOHA system, which dynamically controls the transmission probability according to the offered loads.

3 Transmission Probability Control Scheme

The system model of transmission probability control scheme, which is named as the Proportional Backoff (PB) scheme, is presented in Fig. 4. In the proposed scheme, the base station controls the transmission probability of mobile stations in the centralized manner. Mobile stations in the contention state and retransmission state attempt to transmit packets with the transmission probability P_n and retransmission probability P_r, respectively. The base station calculates these probabilities based on the estimated traffic load and broadcasts over an error-free downlink control channel.

The mobile station that fails to transmit its packet at slot t retransmit with $P_r(t+1)$ at slot $(t+1)$, while the mobile station that enters into the contention state at slot t transmit with $P_n(t+1)$ at subsequent slot. The $P_n(t+1)$ and $P_r(t+1)$ are calculated as follows:

$$P_n(t+1) = \begin{cases} 1, & \text{if } N_r(t+1) \leq TH_m \\ 0, & \text{otherwise} \end{cases} \tag{3}$$

$$P_r(t+1) = \begin{cases} 1, & \text{if } N_r(t+1) \leq TH_m \\ \frac{TH_m}{N_r(t+1)}, & \text{otherwise} \end{cases} \tag{4}$$

where TH_m is the number of simultaneously transmitted packets at which the system throughput can be maximized, and $N_r(t+1)$ is the total number of mobile stations in the retransmission state at slot $(t+1)$. Let $N_f(t)$ be the number of failed mobile stations at slot t, and $N_b(t)$ the number of mobile stations not permitted to transmit at slot t. Then $N_r(t+1)$ can be derived as follows:

$$N_r(t+1) = N_f(t) + N_b(t) \tag{5}$$

The threshold value TH_m can be determined by (4), and $N_b(t)$ is given by

$$N_b(t) = N_n(t-1) \cdot \{1 - P_n(t)\} + N_r(t) \cdot \{1 - P_r(t)\} \tag{6}$$

$N_n(t)$ in (6) is the number of mobile stations that enter into the contention state at slot t, and can be derived by

$$N_n(t) = \{N - N_r(t) - N_n(t-1)\} \cdot \lambda \tag{7}$$

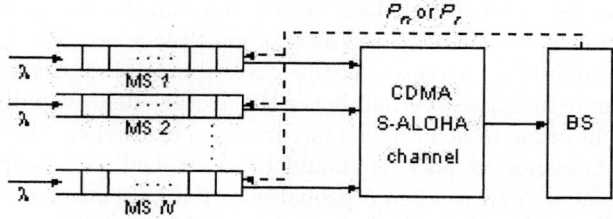

Fig. 4. System model of PB scheme

In (7), N is the total number of mobile stations in the system, and λ is the probability that each mobile station generates a packet in each slot. The base station cannot exactly know how many packets are generated in one slot. Therefore, λ is computed using a moving time average of the number of new packets that are successfully received.

In the proposed scheme, if the number of mobile stations in the retransmission state is less than TH_m, mobile stations in both the contention state and retransmission state should be allowed to transmit a packet. If the number of mobile stations in the retransmission state becomes more than TH_m, the base station sets P_n into 0 to suppress the transmission of new packets. Also, in this case, the base station sets P_r as the values at which the total number of simultaneously retransmitted packets becomes TH_m, in order to minimize the transmission delay.

4 Performance Analysis

Mobile stations in the idle state generate a new packet according to a Poisson process with the mean arrival rate λ, and changes to the contention state. Mobile stations in the contention state attempt to transmit a packet with probability P_n. Mobile stations change to the retransmission state when an attempt to transmit a packet fails or the transmission probability is not allowed, and try to retransmit at the subsequent slot with probability P_r.

With these considerations, all the mobile stations will remain in the idle or retransmission state at the beginning of slot. Let $N_B(t)$ denote the number of mobile stations in the retransmission state at the beginning of slot t. It is easy to show that the system state $\{N_B(t)\}$ is a finite-state discrete-time Markov chain, so the system performance can be determined by finding the state transition probability and steady state probability. The one-step transition probability P_{ij} from i to j can be defined as

$$P_{ij} = \{N_B(t+1) = j \mid N_B(t) = i\} \tag{8}$$

To evaluate P_{ij}, it will be useful to take into account that j results from the initial value i plus the number of mobile stations that enter into the contention state at the start of slot t, i.e., $N_A(t)$, minus the number of mobile stations transmitting successfully in the current slot, i.e., $N_S(t)$. If $N_S(t) = s$, $(j-i+s)$ out of $(N-i)$ mobile stations generate a packet at the beginning of slot t. If $N_T(t) = n$ and $N_B(t) = b$, b out of $(j-i+s)$ mobile stations in the contention state and $(n-b)$ out of i mobile stations in the retransmission state transmit their packet, and s out of n transmitted packets are correctly received. From these conditions, the state transition probability can be obtained as

$$P_{ij} = \sum_{x=0}^{N} \sum_{s=0}^{n} \sum_{b=0}^{j-i+s} B(N-i, j-i+s, \lambda) \cdot B(j-i+s, b, P_n) \cdot$$
$$B(i, n-b, P_r) \cdot B(n, s, P_c(n)) \tag{9}$$

where P_n and P_r are defined in (3) and (4), respectively, and $B(n, i, p)$ is defined as

$$B(n, i, p) = \binom{n}{i} p^i (1-p)^{n-i} \tag{10}$$

From the state transition probability, the steady state probability can be easily derived by

$$\pi_i = \sum_{j=0}^{N} \pi_j P_{ji}, 0 \leq i \leq N, \sum_{i=0}^{N} \pi_i = 1 \tag{11}$$

The above formula will be necessary to evaluate system performance analytically. In particular, if we define the throughput as the number of successfully transmitted packets in a slot, it can be obtained as

$$S = \sum_{n=0}^{N} \left[\sum_{s=0}^{n} s \cdot B(n, s, P_c(n)) \right] \cdot \sum_{i=0}^{N} \sum_{a=0}^{N-i} \sum_{m=max(0,n-i)}^{min(a,n)}$$
$$B(N-i, a, \lambda) \cdot B(a, m, P_n) \cdot B(i, n-m, P_r) \cdot \pi_i \tag{12}$$

5 Numerical and Simulation Results

All the parameter values for simulations are same as in Section 2. With theses assumptions, it can be seen that TH_m is equal to 12 from Fig. 1. Also, it is assumed that the total length of window used for computing the moving time average of packet arrival rate is set to 1,000 slots.

In order to validate the analysis model, this paper has performed computer simulations, and the result is depicted in Fig. 5. As the figure shows, the analytical result is consistent with the simulation result.

In this paper, we analyze the performance of the proposed PB scheme in comparison to the conventional Harmonic Backoff (HB) scheme [8][9]. The throughput and average delay versus the offered load are shown in Fig. 6. In HB scheme,

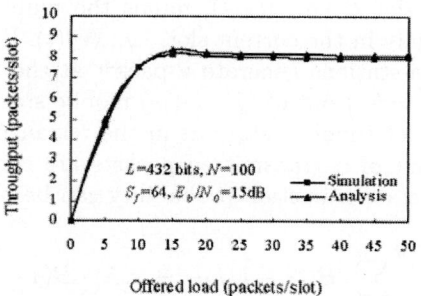

Fig. 5. Comparison between analysis and simulation result

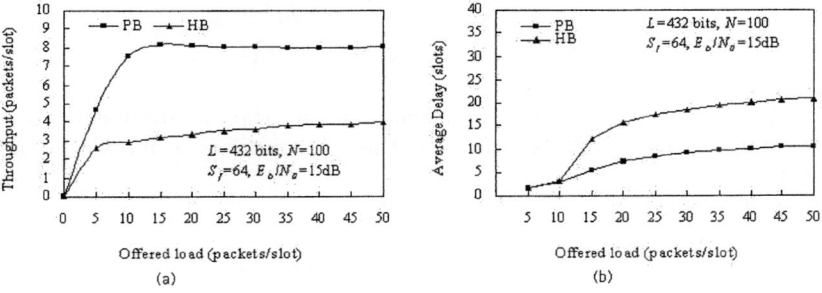

Fig. 6. (a) Throughput versus offered load. (b) Average delay versus offered load

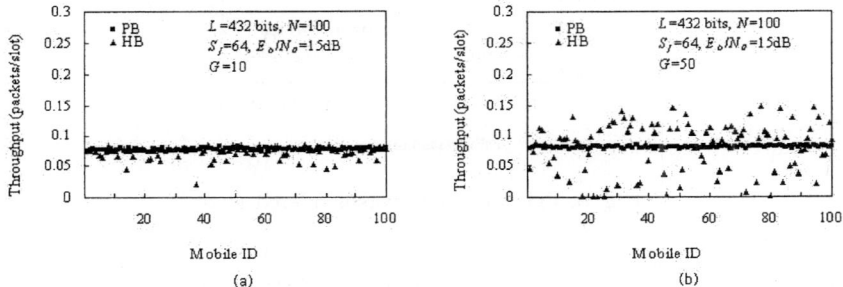

Fig. 7. Throughput of each mobile station (a) $G=10$. (b) ($G=50$)

the mobile station that fails to transmit a packet decreases its transmission probability independently with the number of mobile stations in the retransmission state. On the other hand, in the proposed PB scheme, the base station controls the transmission probability based on the traffic load and threshold TH_m. Therefore, PB scheme maintains the number of simultaneously transmitted packets at the given threshold TH_m in spite of the increased offered load. From Fig. 6, PB scheme gives better performance than HB scheme.

Fig. 7 shows the throughput of each mobile station when $G=10$ and $G=50$. With a light traffic load i.e. $G=10$, there are no significant differences in the throughput of each mobile station between PB and HB scheme. As shown in Fig. 7(a), both schemes provide a good fairness in transmitting packets when the offered load is low. In a heavy traffic load of HB scheme, mobile stations in the contention state are permitted to transmit a packet prior to mobile stations in the retransmission state. However, in PB scheme, when the number of mobile stations in the retransmission state is more than TH_m, mobile stations in the contention state are not permitted to transmit a packet. Therefore, as shown in Fig. 7(b), the proposed PB scheme provides fairer throughput for each mobile station than HB scheme when the offered load is high.

One of the interesting performance measures is the throughput fairness index, which is defined as follows:

$$Fairness = \frac{\left(\sum_{i=1}^{N} Y_i/Z_i\right)^2}{N \sum_{i=1}^{N} (Y_i/Z_i)^2} \qquad (13)$$

where Y_i and Z_i are the measured throughput and the fair share throughput of mobile station i, respectively, and N is the total number of mobile stations in the system.

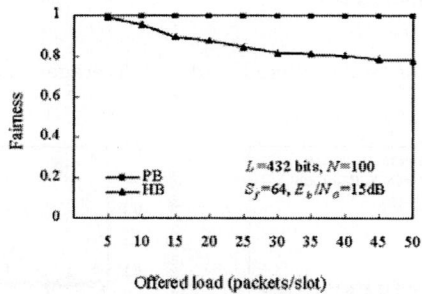

Fig. 8. Fairness index versus offered load

Fig. 8 compares the fairness index of PB and HB schemes according to the offered load. The fairness index of HB scheme decreases sharply as the offered load increases. However, the proposed PB scheme maintains the fairness index almost 1 regardless of the offered load. Therefore, PB scheme can guarantee fair packet transmissions among all the mobile stations even though the offered load increases.

6 Conclusions

This paper has proposed a MAC protocol for packet data services in the CDMA packet radio networks. The design objective is to improve the throughput and delay performance, and to guarantee the fairness among all the mobile stations. In the proposed scheme, the base station dynamically determines the packet transmission probability of mobile stations according to the offered load and then broadcasts this probability to all the mobile stations.

Simulation results show that the proposed scheme maintains the number of simultaneously transmitted packets at the optimum threshold, which can achieve the maximum throughput of CDMA S-ALOHA system. The throughput and delay performance of the proposed scheme outperforms by far those obtained with the conventional scheme. In particular, the proposed scheme can guarantee the fair packet transmission among all the mobile stations regardless of the offered load.

References

1. So, J. W., Han, I., Shin, B. C., Cho, D. H.: Performance analysis of DS/SSMA unslotted ALOHA system with variable length data traffic. IEEE J. Select. Areas Commun. **19(11)** (2001) 2215-2224
2. TIA TR45.5: The CDMA2000 RTT candidate submission to ITU-R. Draft Document (1998)
3. Makrakis, D., Sundaru Murthy, K. M.: Spread slotted ALOHA techniques for mobile and personal satellite communication systems. IEEE J. Select. Areas Commun. **10(6)** (1992) 985-1002
4. Raychaudhuri, D.: Performance analysis of random access packet switched code division multiple access systems. IEEE Trans. Commun. **29(6)** (1981) 859-901
5. Sallent, O., Agusti, R.: A proposal for an adaptive S-ALOHA access system for a mobile CDMA environment. IEEE Trans. Veh. Tech. **47(3)** (1998) 977-986
6. Lam, A. W., Ozluturk, F. M.: Performance bound for DS/SSMA communications with complex signature sequences. IEEE Trans. Commun. **40(10)** (1992) 1607-1614
7. Polydoros, A., Silvester, J.: Slotted random access spread-spectrum networks - an analytical framework. IEEE J. Select. Areas Commun. **5(6)** (1987) 989-1002
8. Choi, S., Shin, K. G.: An uplink CDMA system architecture with diverse QoS guarantees for heterogeneous traffic. IEEE/ACM Trans. Networking **17(5)** (1999) 616-628
9. Verikoukis, C. V., Olimos, J. J.: Up-link performance of the DQRUMA MAC protocol in a realistic indoor environment for W-ATM networks. Proc. VTC2000 (2000) 1650-1655
10. TIA/EIA/IS-95-A: Mobile station-base station compatibility standard for dual-mode wideband spread spectrum cellular systems. Telecommun. Indust. Assn. (1995)
11. Frigon, J. F., Leung, V. C. M.: A pseudo-Bayesian ALOHA algorithm with mixed priorities. ACM Wireless Networks **7(1)** (2001) 55-63
12. 3GPP TS 25.321, V5.4.0: MAC Protocol Specification. (2003)

RWA Based on Approximated Path Conflict Graphs in Optical Networks*

Zhanna Olmes, Kun Myon Choi, Min Young Chung,
Tae-Jin Lee, and Hyunseung Choo

School of Information and Communication Engineering,
Sungkyunkwan University, 440-746, Suwon, Korea
{zhanna, dilly97, mychung, tjlee, choo}@ece.skku.ac.kr

Abstract. Among many solutions to Routing and Wavelength Assignment (RWA) problems based on Edge Disjoint Paths (EDP), the Path Conflict Graph (PCG) algorithm shows outstanding performance in terms of wavelength. In this paper, we improve the PCG algorithm by imposing limitations on the EDPs length based on the fact that the EDPs are longer than the average length for the rarely selected demands. We conclude that the running time of the PCG algorithm can be reduced by half even in the worst case scenario while expending fewer wavelengths than or equal to that of the BGAforEDP and MAX_EDP algorithms by using the proposed PCG approximation technique.

1 Introduction

Telecommunications, broadcasting and the Internet are integrated to move into the era of digital convergence. These convergence services require fast transmission speed and as such the importance of high speed optical network technology is increasing. Change of equipments in order to keep up with the technological trends, however, encounters increasing expenditures that may not be feasible. Therefore, to efficiently manage the growing traffic demands, routing and wavelength assignment (RWA) in wave division multiplexing (WDM) optical networks becomes an essential area of research.

In general, the goal of routing and wavelength assignment is to minimize the number of wavelengths in static demands, or to minimize the blocking probability of dynamic demands with the fixed number of wavelengths and network cost. The RWA problem is NP-complete, so many heuristic methods are proposed. Depending on the objective of the algorithms, we can classify them into 1) minimizing the number of wavelengths [1], [2], 2) decreasing the failure ratio under the fixed number of wavelengths and dynamic demands environment [3], [4], 3) minimizing the network cost [5], [6], and 4) minimizing the execution time while keeping the number of wavelengths [7].

---

* This work was supported in parts by Brain Korea 21 and the Ministry of Information and Communication in Republic of Korea. Dr. H. Choo is the corresponding author.

Research on RWA has been usually performed under the specific network environment assumptions such as dynamic or static demands, existence of wavelength converters, and permitting multiple demands with the same source destination pairs. In this paper, we assume that there are static demands and no wavelength converters.

The RWA algorithm, which uses Path Conflict Graph (PCG) [9] to indicate the degree of path conflict, has improved performance remarkably in terms of the number of wavelengths compared to Bounded Greedy Approach for Edge Disjoint Paths (BGAforEDP) [1] and the Maximum Quantity of Edge Disjoint Paths (MAX_EDP) [7]. However, as either the network size or the number of demands increases, the execution time and the size of PCG grow exponentially. In this paper, we propose a new PCG-based RWA algorithm to improve execution time. The idea is to limit the length of the routing paths that have low selection possibilities in the process of building a PCG. We compared the performance of [1], [7], [9] with our proposed algorithm in the same environment. The results reveal that our algorithm, reduces execution time by half while the number of wavelengths is similar to that of [9] or less than that of [1], [7].

The remainder of this paper is organized as follows. Section 2 explains RWA algorithms based on Bounded Greedy Approach for EDPs [1], maximum EDPs [7], and Path Conflict Graph for EDPs [9] and related works. Section 3 provides a detailed description of the proposed algorithm by presenting a pseudo code and simple examples. Section 4 presents simulation results and analysis with respect to the number of wavelengths and execution time. The conclusion is found in Section 5.

2 Related Work

2.1 BGA for EDP Based Algorithm

A simple EDP scheme based on the shortest path algorithm is used in [1]. Let $G = (V, E)$ be the graph of the physical network, V and E be the set of vertices and the set of edges, respectively. Lastly, let D be the demand set, $D = \{(s_1, t_1), \ldots, (s_k, t_k)\}$. This scheme is executed with demand set to be connected, and a reasonable value d, where d is $Max(\sqrt{|E|}, diam(G))$. The sequence of the BGA for EDP algorithm is explained below.

First, the BGA for EDP algorithm randomly selects the lightpath D_i from D. It finds the shortest path P_i for this request (If we can not find the path in G, we proceed to the next demand). If the path length of P_i is less than d, we add D_i to the set of routed lightpaths $\alpha(G, D_{routed})$ and delete the edges used by P_i in G. If the path length of P_i is greater than d, the lightpath D_i is not routed. This is repeated for all requests in D. At the end of the algorithm, $\alpha(G, D_{routed})$ contains the lightpaths that are assigned the same wavelength. Next, remove $\alpha(G, D_{routed})$ from D. Then BGAforEDP is run on the original G and D to obtain the set of lightpaths that assign another wavelength. This is repeated until D becomes empty. The total number of assigned wavelengths is the result of this algorithm.

2.2 Find Maximum EDPs

A lookup-table and maximum EDP-based fast RWA algorithm is proposed in [7]. In a backbone network, the topology of the optical network does not change much and the demands of the same pattern appear repeatedly. In this case, an efficient solution is to use the lookup-table to store the entire available network EDPs. To get the maximum number of EDPs for a demand pair, the concept of Disjoint Path Selection Protocol (DPSP) [7] is used. In [7], a forward edge corresponds to an edge belonging to one of the paths and follows the direction from the source to destination in accordance with the path. A backward edge corresponds to an edge that belongs to one of the paths and follows the opposite direction to the corresponding forward edge. Here we use the shortest path, which passes only the unassigned forward edges along with the backward edges. Until we obtain the maximum number of EDPs, we run the shortest path algorithm and remove backward edges if they exist. We subsequently obtain some separate EDPs from the paths that share a backward edge. This is the basic idea of having the maximum number of EDPs and it is repeated until we find the minimum number of EDPs between the number of edges connected to the source and the number of edges connected to the destination. We can construct a lookup-table by using the EDPs and then choose the shortest path among the EDPs from the lookup-table and assign a wavelength to the selected path. If there are unassigned requests after considering all demand pairs, we try to connect the remainders with the increased wavelength number in G. This is repeated until D becomes empty. The total number of assigned wavelengths is the result of this algorithm.

2.3 PCG for EDP

In general network conditions, there exist one or more available paths for one demand. If one particular path is chosen, the others are then rendered incapable of being selected because of the fact that the demand for connection has already been met. In other words, the path that constitutes the least amount of conflict with other demands enables a higher a number of paths to be chosen for other demands and consequently, the number of wavelengths in use diminishes. We can use PCG as a data structure for efficiently determining the degree of conflict each path possesses. Process of the algorithm based on PCG is as follows; firstly, we find the maximum EDPs for all demands and calculate the path lengths. In order to make a PCG from a physical topology, we denote an EDP as a node and EDPs for the same demand as a partition. If there exists a conflict between the nodes, two EDPs share at least one link, we connect them. The number of edges connecting nodes in the PCG indicates the degree of overlapping between the EDPs. We start with a wavelength and select a node with the minimum degree and minimum path length. Then erase all the connected nodes with the node temporarily and its partition permanently. Delete the demand for the selected EDP from the demand set and append the EDP to the set of selected EDPs and the demand to the result set. If there are no more nodes to be chosen with one wavelength then consider another wavelength, and recover the temporarily erased nodes. Repeating the process until all the demands meet the requirements, we can acquire the appropriate number of wavelengths needed.

3 Proposed Algorithm

3.1 Basic Concepts

The algorithm based on PCG [9] is the best algorithm with respect to the wavelength among the solutions for the RWA problem [1], [7]. However, the time complexity and memory grow exponentially as the network size and demands between nodes are increased. That is because the algorithm needs extra memory for the node properties in addition to the connectivity information. Furthermore, more calculation is required on this information in the process of building a PCG. To remedy this shortcoming, this paper studies the techniques that can reduce the running by half while providing a reasonably small number of wavelengths.

When we are using the path conflict graph, the paths have a high probability of sharing the same edge, and in most cases, these paths are not assigned. This is because the PCG was built with the paths having a low probability of being used. In considering many EDPs, computation can be reduced significantly by removing only a small number of EDPs because the PCG size grows exponentially as the number of EDPs increases. In general, the probability of sharing the same edge with other paths is high when the length of the path is large. Based on this property, we can reduce running time by building PCG after removing long paths instead of considering all paths.

3.2 Algorithm Description

Figure 1 and 2 represents a pseudo code for the proposed algorithm. The input of the proposed algorithm is the physical network graph G and the demand set $D = \{(s_1, t_1), \ldots, (s_k, t_k)\}$ where s_k and t_k denote source and destination node of demand k, respectively. And N_λ denotes the total number of used wavelengths, τ denotes the set of all possible EDPs for the demands set, m_d denotes the minimum value of path conflict degree for EDPs, and m_l denotes the shortest path length among EDPs with degree m_d. Furthermore, $\tau_{same\_partition}$ denotes the set of EDPs in τ for the same demand and $\tau_{connected}$ denotes the

```
Build_PCG
01: For each pair (τ_i, τ_j)
02:         If (τ_i and τ_j share any edge)
03:             Connect (τ_i, τ_j) in G_p
04:         EndIf
05: EndFor
06: G_p' = G_p
```

Fig. 1. Pseudo code of generating Path Conflict Graph

```
INPUT: Network topology G and a demand set D
OUTPUT: Total number of wavelengths N_λ used for D
01: Find all possible EDPs and their lengths for each demand in D
02: Remove EDPs longer than the average length of EDPs + β (β ≥ 0)
03: Build_PCG
04: N_λ =1
05: While (D ≠ ∅) do
06:       If (EDP τ_i with m_d and m_l exists)
07:            D = D - (s_i, t_i)  /* (s_i, t_i) is the demand for τ_i*/
08:            τ = τ - ( τ_same_partition + τ_connected )
09:       else
10:            G_p = G_p'
11:            N_λ = N_λ + 1
12:       EndIf
13: EndWhile
14: Return N_λ
```

Fig. 2. Pseudo code of the proposed algorithm

Table 1. Notation

Notation	Meaning
G	Graph that models a network
N	Number of nodes
D	Demand set, $D = \{(s_1, t_1), ..., (s_k, t_k)\}$
τ	Set of EDPs for the demand set D, $\tau = \{\tau_1, \tau_2, ..., \tau_n\}$
P_l	Probability of a request between any node pair
P_e	Probability of edge existence between any node pair
N_c	Multiplicity of a single request
N_λ	Number of wavelengths
m_d	Smallest degree in τ
m_l	Smallest length with degree m_d in τ
$\tau_{same\_partition}$	Set of EDPs for the same demand
$\tau_{connected}$	Conflict path set
G_p	Path conflict graph
G_p'	Copy of path conflict graph
β	Offset value for getting upper bound of the EDPs length

set of EDPs sharing at least one link with any EDP. Table 1 summarizes these terminologies.

The most time consuming process, building PCG is shown in Figure 1. The initial path conflict graph can be constructed in this procedure.

Pseudo code of the proposed algorithm is shown in Figure 2 and the step by step description is provided. Proposed algorithm is divided into two categories according to the reference value to decide the proposed critical length value of the path in Step 2. Average Reduced Path Conflict Graph (ARPCG) is a method of measuring the average length of paths for the demand, while the Minimum

Reduced Path Conflict Graph (MRPCG) is used to determine the minimum length. To determine how execution time and the number of wavelengths change with respect to critical values, set parameter β regular values, and add average and the minimum length.

Step 1 (Line 01 in Fig. 2) Find all maximum EDPs for each demand in D using the method in [8]. For each EDP, physical path, path length, partition and conflict degree are obtained. The physical path represents a sequence of nodes and the path length denotes the number of hops on the physical path. The partition illustrates the demand pair corresponding to the EDP in path conflict graph.

Step 2 (Line 02 in Fig. 2) For each demand $\tau_i = (s_i, t_i)$ calculate the average length plus β, and remove paths longer than average length plus β from τ. Each demand must use only one EDP, so if we have only one EDP it is not removed. ($\beta \geq 0$)

Step 3 (Line 03 in Fig. 2) Build path conflict graph (G_p) and path conflict graph's copy ($G_p{'}$) for the remaining τ.

Step 4 (Line 04 in Fig. 2) Initiate the number of wavelengths N_λ by one.

Step 5 (Lines 06-08 in Fig. 2) Select τ_i whose conflict degree is m_d and length is m_l. If there are more than one EDPs, any EDP is randomly selected from EDPs with m_l and m_d. After one EDP τ_i is selected, $\tau_{same\_partition}$ and $\tau_{connected}$ are constructed. Then the demand (s_i, t_i) associated with the selected EDP (τ_i) is removed from D. For later restoration, remove τ connected with τ_i from G_p and $G_p{'}$, and remove τ connected with τ_i only from G_p.

Step 6 (Lines 09-12 in Fig. 2) If there is no EDP to be assigned, it restores G_p, and increases N_λ. Repeat steps 5 and 6, until D is empty.

Step 7 If D is empty, the value of N_λ is returned and the algorithm is finished.

3.3 Example

To investigate a process of proposed algorithm in detail, suppose that a network consists of 6 nodes and 9 edges as shown in Figure 3 and the demand set $D = $

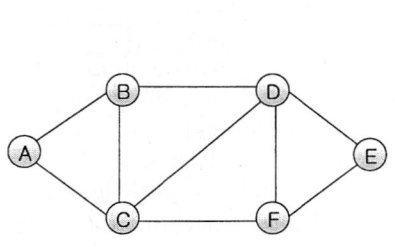
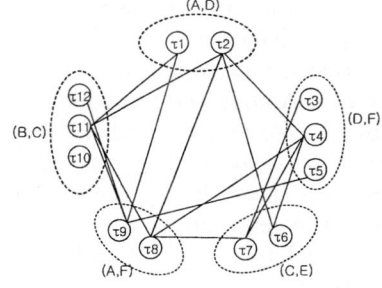

Fig. 3. A simple example of mesh network   **Fig. 4.** Path Conflict Graph

$\{(A,D), (D,F), (C,E), (A,F), (B,C)\}$. Firstly, all possible EDPs for each demand in D are found. There are 12 EDPs for demands (A,D), (D,F), (C,E), (A,F), and (B,C). Let $\tau_1, \ldots, \tau_{12}$ be the paths [A-B-D], [A-C-D], [D-E-F], [D-C-F], [D,F], [C-D-E], [C-F-E], [A-C-F], [A-B-D-F], [B-C], [B-A-C], and [B-D-C], respectively.

The path conflict graph for these 12 EDPs is shown in Figure 4. In making a path conflict graph by the proposed algorithm, we get an average length of EDPs in the same demand to find the demand with low selection probability. After adding a number (β) to the average length like in the pseudo code, we set a critical value. After the deletion of the long paths ($\tau_3, \tau_4, \tau_9, \tau_{11}, \tau_{12}$), we get a graph shown in Figure 5 (a).

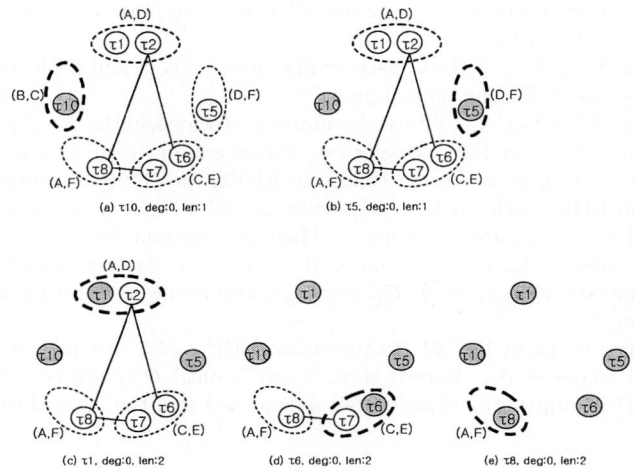

Fig. 5. Path Conflict Graph after the completetion of RWA for: (a) demand (B,C); (b) demand (D,F); (c) demand (A,D); (d) demand (C,E); (e) demand (A,F)

In Figure 5 it selects τ_{10} with the minimum overlapping degree $m_d(=0)$ and the path length $m_l(=1)$. The demand (B,C) for the selected τ_{10} is deleted from D. Because $\tau_{same\_partition}$ and $\tau_{connected}$ are an empty set, no EDPs are deleted from τ. Among the remaining EDPs, we choose τ_5 with the minimum overlapping degree $m_d(=0)$ and the path length $m_l(=1)$. We delete the demand (D,F) from D because $\tau_{same\_partition}$ and $\tau_{connected}$ are an empty set as well. There are no deleted EDPs from τ. Among the remaining EDPs, it also selects τ_1 with the minimum overlapping degree $m_d(=0)$ and the path length $m_l(=2)$. After deleting the demand (A,D) from D, delete $\tau_{same\_partition} = \{\tau_2\}$ from τ. After letting the demand (C,E) (A,F) be deleted, demand D becomes an empty set. The algorithm is then finished. As a result, the used number of wavelengths is 1.

4 Performance

4.1 Environment for the Simulation

In this section, we comparatively evaluate the performances of the suggested algorithm, BGAforEDP [1], MAX_EDP [7], PCG [9] from the view of running time and the number of total wavelengths used. Similar to the simulation environment of [7] and [9], we assume that the optical network rarely changes. The NSFNET shown in Figure 6 (one of the typical optical networks), was employed for the simulation. And randomly generated networks are also employed for the validation of the algorithm. To generate a random topology, we specify the number of nodes in the graph (N) and the probability of an edge existence (P_e) for any node pair. To guarantee the connected random graph we make a spanning tree with N nodes. We then recalculate P_e due to the connections for the N-1 edges in the spanning tree. Next, we randomly connect other edges based on the new P_e. The demand set is also generated randomly by populating the $N*N$ matrix with the number of requests. The set of the same demands can be generated under N_c, while each demand is generated with the probability of P_l.

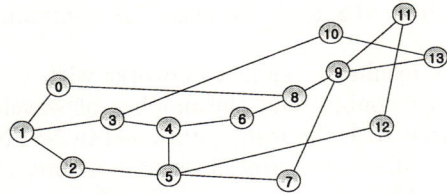

Fig. 6. NSFNET topology (node:14, edge:21)

4.2 Numerical Results

All simulations are done with the equal demand probability of 0.5 and 14 nodes. Figure 7 shows the total number of wavelengths and execution time in NSFNET

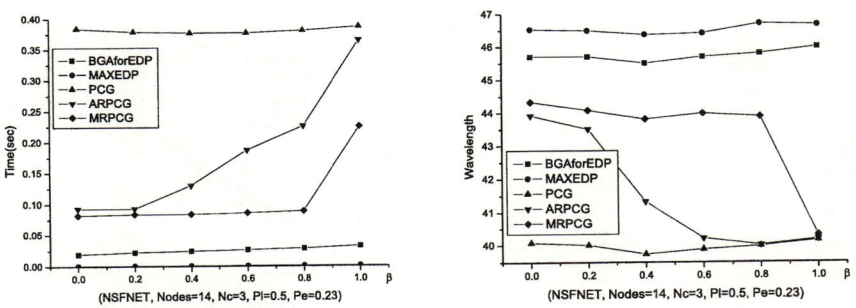

Fig. 7. Execution time and number of wavelengths under varying β in NSFNET (Node:14, edge:21)

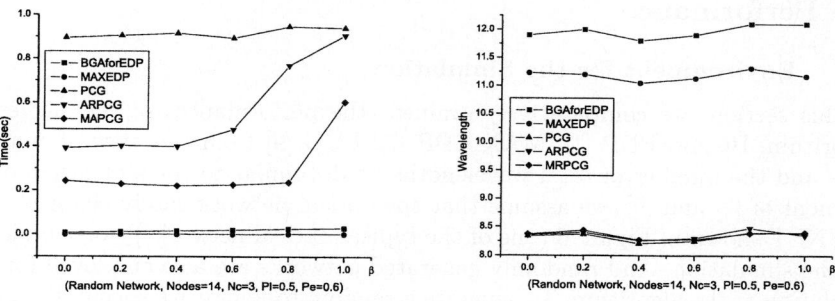

Fig. 8. Execution time and the number of wavelengths under varying β in random networks (Node:14, $\boldsymbol{P_e}$ =0.6, $\boldsymbol{P_l}$ =0.5)

under varying β. By varying the parameter β, we can see that there exists a tradeoff between the execution time and the number of wavelengths in ARPCG and MRPCG. The execution time is small with low β and the number of wavelengths is small with high β. The execution time of MRPCG is changed only when β is increased from 0.8 to 1, because the minimum length itself is an integer multiples.

Figure 8 shows the results in random networks with large number of edges. The execution time the doubles and the number of wavelengths decreases by about 1/4 in comparison with the results obtained in NSFNET. In general, the total number of EDPs affects the overall execution time of the algorithm and more EDPs can be found in a dense network. Therefore, in a dense network, more demands can be routed with smaller number of total wavelengths compared with a sparse network. In Figure 7 and 8, ARPCG and MRPCG shows outstanding performance and decreases execution time effectively compared with PCG. And even in the worst case, these algorithms show moderate performance. This is because the EDPs with low selection probability are excluded before making a PCG. As a result, the total number of EDPs is decreased. However, the excluded EDPs rarely influence the total number of wavelengths.

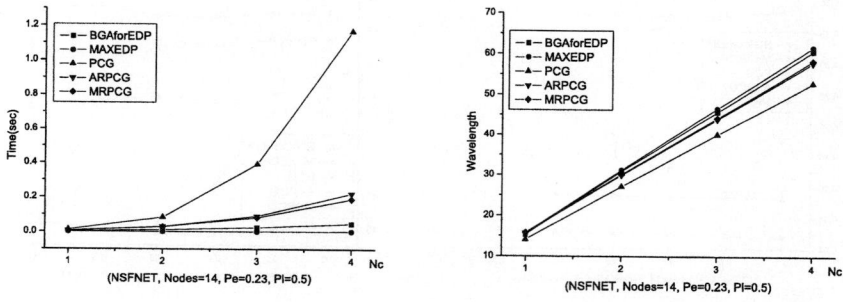

Fig. 9. Execution time and wavelength by $\boldsymbol{N_c}$ in NSFNET(node: 14, edge:21)

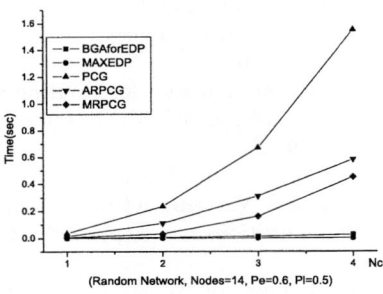 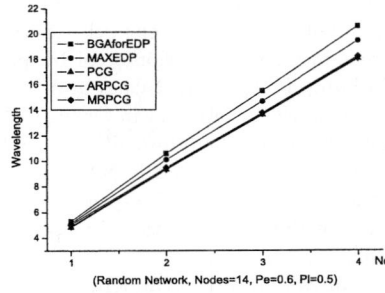

Fig. 10. Execution time and wavelength by N_c in Random Network(Node: 14, P_e =0.6, P_l =0.5)

Figure 9 and 10 shows the change of execution time and the number of wavelengths with increasing N_c in NSFNET and random networks, respectively. The number of PCG nodes is calculated by the sum of available EDPs for each demand. The parameter N_c affects the number of demands and the number of PCG nodes. As N_c linearly grows, the total number of wavelengths of ARPCG and MRPCG is close to that of PCG in comparison with NSFNET.

5 Conclusions

Among the many solutions to RWA problems based on EDP, the PCG algorithm shows excellent performance in terms of wavelength. However, PCG shows poor performance when the network size is large and the demand probability is high. This is due to the high time complexity and memory overhead. In this paper we propose a new algorithm based on an approximated path conflict graph from the fact that the shortest EDP has a high probability of being selected. Our algorithm removes paths according to their length before building the path conflict graph resulting in a considerably better performance when compared to the PCG algorithm. From the simulations we conclude that the proposed algorithm is more efficient in complicated or highly demanded networks. Moreover, the suggested algorithm resolves the large execution time problem, which was the original PCG algorithm's weakest point [9], without affecting the number of wavelengths. To reduce calculation time of PCG, we use the average and shortest length and propose ARPCG and MRPCG. We continue our research in order to find an optimized critical point, depending on various network conditions.

References

1. P.Manohar, D.Manjunath, and R.K.Shevgaonkar, "Routing and Wavelength Assignment in Optical Networks from Edge Disjoint Path Algorithms," IEEE Communications Letters, vol.5, pp. 211-213, May 2002.

2. D.Banerjee and B.Mukherjee, "Practical Approach for Routing and Wavelength Assignment in Large Wavelength-Routed Optical Networks," IEEE Journal on Selected Areas in Communications, vol.14, pp. 903-908, June 1996.
3. Y.Zhang, K.Taira, H.Takagi and S.K.Das, "An Efficient Heuristic for Routing and Wavelength Assignment in Optical WDM Networks," in Proc. of IEEE International Conference on Communications, vol.5, pp. 2734-2739, 2002.
4. M.D.Swaminathan and K.N.Sivarajan, "Practical Routing and Wavelength Assignment Algorithms for All Optical Networks with Limited Wavelength Conversion," in Proc. of IEEE International Conference on Communications, vol.5, pp. 2750-2755, 2002.
5. M.Alanyali and E.Ayanoglu, "Provisioning Algorithms for WDM Optical Networks," IEEE/ACM Trans. on Networking, vol.7, no.5, pp. 767-778, Oct. 1999.
6. I.Chlamtac, A.Farago, and T.Zhang, "Lightpath (Wavelength) Routing in Large WDM Networks," IEEE Journal on Selected Areas in Communications, vol.14, pp. 909-913, June 1996.
7. M. H. Kim and H. Choo, "A Practical RWA Algorithm Based on Lookup Table for Edge Disjoint Paths," Journal of KISS, Information Networking, vol.31, no.2, pp. 123-130, April 2004 (KOREAN).
8. G.Li and R.Simha, "The Partition Coloring Problem and its Application to Wavelength Routing and Assignment," in Proc. of First Workshop on Optical Networks, Dallas, TX, 2000.
9. D. H. Kim, M. Y. Chung, T.-J. Lee, and H. Choo, "A Practical RWA Algorithm Based on Maximum EDPs and Path Conflict Graph," in Proc. of TENCON 2004., vol.C, no.097, Nov. 2004.

Secure Routing in Sensor Networks: Security Problem Analysis and Countermeasures[1]

Youngsong Mun and Chungsoo Shin

School of Computing, Soongsil University, Seoul, Korea
mun@computing.ssu.ac.kr, endecha@sunny.ssu.ac.kr

Abstract. Recently, sensor networks support solutions for various applications. These sensor networks have limitations of system resource. Therefore, there are a number of proposals to solve these problems. Those, however, do not consider security problem. In the sensor networks, the security is a critical issue since an entire resource of sensor nodes can be consumed by attackers. In this paper, we will consider security problems about the MAC protocol which extends lifetime of node using active signal that prevents receiving duplicate messages, and the routing algorithm which considers node's lifetime. Then we propose countermeasures that establish the trust relationship between neighboring sensor nodes, checks the bi-directionality each nodes, and authenticates node's information messages.

1 Introduction

Sensor networks consisting of several thousands of computing devices, called sensors, are being widely used in situations where using traditional networking infrastructures are practically infeasible. Sensor networks often have one or more points of centralized control called base stations. A base station is typically a gateway to another network, a powerful data processing or storage center, or an access point for human interface. They might have workstation or laptop class processors, memory, and storage, AC power, and high bandwidth links for communication amongst themselves. However, sensors are constrained to use lower-power, lower bandwidth, shorter range radios, and so it is envisioned that the sensor nodes would form a multi-hop wireless network to allow sensors to communicate to the nearest base station. Therefore, there are a number of proposals to improve the sensor network protocols. [1]

Many sensor network routing protocols are quite simple, and for this reason are sometimes even more susceptible to attacks against general ad-hoc routing protocols. In the sensor networks, the security is a critical issue since an entire resource of sensor node can be consumed by attackers. To secure the sensor network, it is needed that trust relationship between neighboring nodes, and the authentication of messages.

---

[1] This research was supported by the MIC(Ministry of Information and Communication), Korea, under the ITRC(Information Technology Research Center) support program supervised by the IITA(Institute of Information Technology Assessment).

2 Protocols to Extend Lifetime and to Improve QoS

2.1 Efficient MAC Protocol Using Active Signal

In the sensor network, the nodes broadcast messages intended to be received by all neighboring nodes. In this environment, nodes may receive messages that they are already received by other node. This is the squandering of node's resources. Therefore it is need that prevents these receiving duplicate messages. One of the solutions is using an active signal.

The nodes are staying in sleep mode when it is not activity state. In this mode, the node does not act, so the node needs an only minimum power level to monitoring that some event. This active signal improves a node's power management. [2]

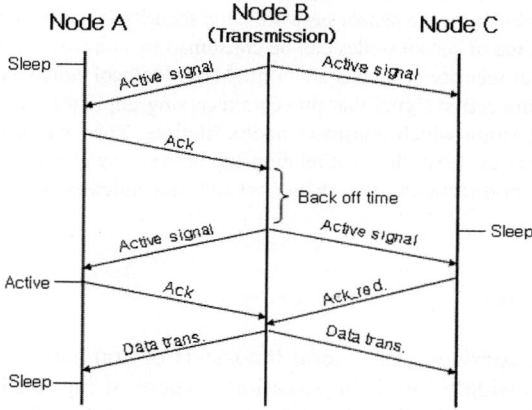

Fig. 1. Signal exchange

There is a Recognition label(R-label) field in an active signal. The R-label is randomly generated integer value. A sensor node can know a duplication of message through this field. If a value of active signal's R-label is in the buffer of sensor node, a sensor node sends an ack_red. signal message to the sender. And the sensor nodes keep the sleep mode. [3]

2.2 Node's Lifetime Considered Routing Algorithm

A sensor node decides a route considering neighboring sensor node's remained power and each path's cost. Each sensor nodes know power information of neighboring sensor nodes. A sensor node calculates the average of neighboring node's lifetime. The candidate nodes are the nodes which have a bigger lifetime value than the average value. A node selects a node as a next hop node that have a lowest link cost value of the candidate nodes. [3]

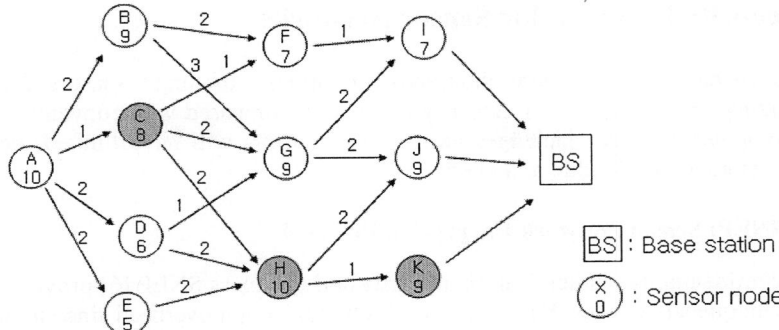

Fig. 2. Node's lifetime considered routing algorithm

3 Summary of Attacks

Many sensor network routing protocols are quite simple, and for this reason are sometimes even more susceptible to attacks against general ad-hoc routing protocols. [1]

Spoofed, altered, or replayed muting information: The most direct attack against a routing protocol is to target the routing information exchanged between nodes. By spoofing, altering, or replaying routing information, adversaries may be able to create routing loops, attract or repel network traffic, extend or shorten source routes, generate false error messages, partition the network, increase end-to-end latency, etc. [1]

Selective forwarding: Multi-hop networks are often based on the assumption that participating nodes will faithfully forward received messages .In a selective forwarding attack, malicious nodes may refuse to forward certain messages and simply drop them. However, such an attacker runs the risks that neighboring nodes will conclude that she has failed and decide to seek another route. A more subtle form of this attack is when an adversary selectively forwards packets. [1]

Sinkhole attacks: Sinkhole attacks typically work by making a compromised node look especially attractive to surrounding nodes with respect to the routing algorithm. For instance, an adversary could spoof or replay an advertisement for an extremely high quality route to a base station. [1]

The Sybil attack: In a Sybil attack, a single node presents multiple identities to other nodes in the network. The Sybil attack can significantly reduce the effectiveness of fault-tolerant schemes such as distributed storage, disparity and multi-path routing, and topology maintenance. [4]

Wormholes: In the wormhole attack, an adversary tunnels messages received in one part of the network over a low latency link and replays them in a different part'. The simplest instance of this attack is a single node situated between two other nodes forwarding messages between the two of them. However, wormhole attacks more commonly involve two distant malicious nodes colluding to understate their distance from each other by relaying packets along an out-of-bound channel available only to the attacker. [5]

4 Security Protocols for Sensor Networks

We envision a future where thousands to millions of small sensors form self-organizing wireless networks. Security is not easy compared with conventional desktop computers, severe challenges exist. These sensors will have limited processing power, storage, bandwidth, and energy.

4.1 SNEP: Sensor Network Encryption Protocol

Data confidentiality, authentication, integrity and freshness SNEP [6] provides a number of unique advantages. First, it has low communication overhead since it only adds 8 bytes per message. Second, like many cryptographic protocols it uses a counter, but we avoid transmitting the counter value by keeping state at both end points. Third, SNEP achieves even semantic security, a strong security property which prevents eavesdroppers from inferring the message content from the encrypted message. Finally, the same simple and efficient protocol also gives us data authentication, replay protection, and weak message freshness.

SNEP offers the following nice properties:

Semantic security: Since the counter value is incremented after each message, the same message is encrypted differently each time. The counter value is long enough that it never repeats within the lifetime of the node.
Data authentication: If the MAC verifies correctly, a receiver can be assured that the message originated from the claimed sender.
Replay protection: The counter value in the MAC prevents replaying old messages. Note that if the counter were not present in the MAC, an adversary could easily replay messages.
Weak freshness: If the message verified correctly, a receiver knows that the message must have been sent after the previous message it received correctly that had a lower counter value. This enforces a message ordering and yields weak freshness.
Low communication overhead: The counter state is kept at each end point and does not need to be sent in each message.

4.2 µTESLA: Authenticated Broadcast

Authenticated broadcast requires an asymmetric mechanism otherwise any compromised receiver could forge messages from the sender. Unfortunately, asymmetric cryptographic mechanisms have high computation, communication, and storage overhead, which make their usage on resource constrained devices impractical. µTESLA [6] overcomes this problem by introducing asymmetry through a delayed disclosure of symmetric keys, which results in an efficient broadcast authentication scheme.

The proposed TESLA protocol provides efficient authenticated broadcast [31, 30]. However, TESLA is not designed for such limited computing environments as we encounter in sensor networks.

SPINS design UTESLA to solve the following inadequacies of TESLA in sensor networks. TESLA authenticates the initial packet with a digital signature, which is too expensive for our sensor nodes. µTESLA uses only symmetric mechanisms. Disclosing a key in each packet requires too much energy for sending and receiving. TESLA discloses the key once per epoch. It is expensive to store a one-way key chain in a sensor node. µ TESLA restricts the number of authenticated senders.

4.3 Node-to-Node Key Agreement

The node A wants to establish a shared secret session key SK_{AB} with node B. Since A and B do not share any secrets, they need to use a trusted third party S, which is the base station in our case. In our trust setup, both A and B share a secret key with the base station, K_{AS} and K_{BS}, respectively. [6] The following protocol achieves secure key agreement as well as strong key freshness:

A → B : N_A, A
B → A : N_A, N_B, A, B, MAC(K_{BS}, N_A|NB|A|B)
S → A : {SK_{AB}}K_{AS}, MAC(K'_{AS}, N_A|B|{SK_{AB}}K_{AS})
S → B : {SK_{AB}}K_{AB}, MAC(K'_{BS}, N_B|A|{SK_{AB}}K_{BS})

The protocol uses our SNEP protocol with strong freshness. The nonce N_A and N_B ensure strong key freshness to both A and B. The SNEP protocol is responsible to ensure confidentiality through encryption with the keys K_{AS} and K_{BS} of the established session key SK_{AB}, as well as message authentication through the MAC using keys K'_{AS} and K'_{BS} to make sure that the key was really generated by the base station. [6]

5 Analysis Security Problems and Countermeasures

5.1 Efficient MAC Protocol

In the sensor network, the nodes broadcast messages intended to be received by all neighboring nodes. In this environment, nodes may receive message that it is already received by other node. This is the squandering of node's resources. Therefore it is need that prevents these receiving duplicate messages. One of the solutions is that it uses an active signal. [2]

The nodes are staying in sleep mode when it is not activity state. In this mode, the node does not act, so the node needs an only minimum power level to monitoring that some event. This active signal improves a node's power management. [3]

But there is a risk if an adversary uses this active signal. An adversary uses an active signal to cause that a node consumes its all power. Therefore the trust is needed between a sender and a receiver. In the sensor network, the nodes broadcast messages intended to be received by all neighboring nodes. A sender uses an active signal when it has messages to send to all neighboring nodes. In this case, it is needed that authenticated broadcast message. But to send an authenticated broadcast message is impractical to the constrained sensor nodes. [6] Therefore each sensor node must have a trust relationship with all neighboring sensor nodes.

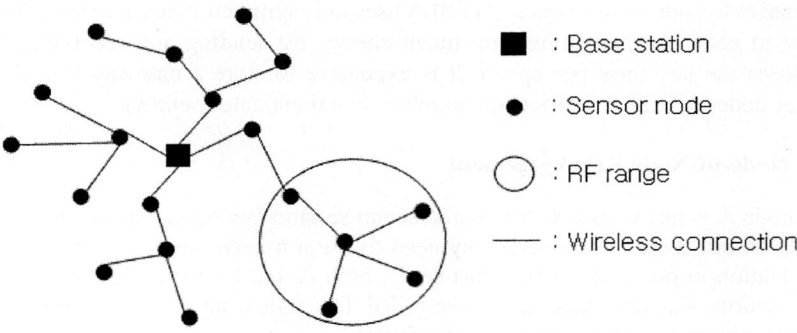

Fig. 3. Sensor network trust model

Above figure shows a sensor network topology. A sensor node communicates with other sensor nodes in the radio frequency range and sends packets to base station via neighboring nodes. A sensor node assumes that the neighboring nodes will forward its own packets to base station. The sensor nodes trust that the neighboring node will act as a legitimate sensor node. Therefore a trust relationship is needed between each sensor nodes.

It can be achieved by SNEP data authentication. [6] Node A generates nonce N_A randomly and sends it along with a request message R_A to node B.(1) Node B returns the nonce with the response message R_B in an authenticated protocol.(2) It can be optimized by using the nonce implicitly in the message authentication code(MAC) computation. Node A can trust node B by verifying the MAC.

$$A \rightarrow B : N_A, R_A \qquad (1)$$
$$B \rightarrow A : R_B, MAC(K_{mac}, N_A|C|R_B) \qquad (2)$$

A node which broadcasts an active signal to neighboring nodes and then waits until receives acknowledgment message. If a node does not receive the acknowledgment messages by all neighboring nodes, it resends an active signal after back off time. [3] In the dynamic sensor network environment, it is needed that the sensor node should check a bi-directionality with the neighboring sensor nodes. The sensor nodes will stop its activity when it consumed its entire power, or sensor nodes will move to other location. Also compromised sensor nodes will do not response to delay the packet. Therefore it is needed that limit the number of resending active signals.

If there is a sensor node that it does not respond to active signal, the sensor node must check the bi-directionality. There are sensor nodes which are node A and node B. If node B does not respond to node A's active signal, node A must check the bi-directionality with node B. It can be achievable by SNEP authentication. [6]

Generally, the sensor networks may be deployed in un-trusted locations. Since the base station is the gateway for the nodes to communicate with the outside world, compromising the base station can render the entire sensor network useless. Thus the base stations are a necessary part of trusted computing base. The sensor nodes can check the bi-directionality between each sensor nodes via base station.

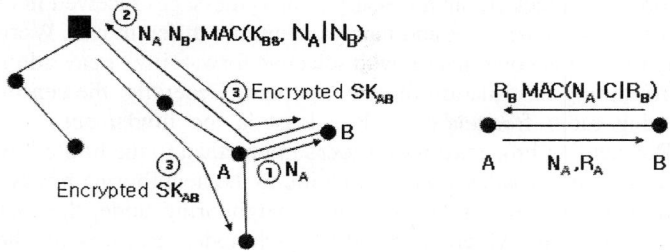

Fig. 4. Checking a Bi-directionality

5.2 Node's Lifetime Considered Routing Protocol

Many sensor network routing protocols are quite simple, and for this reason are sometimes even more susceptible to attacks against general ad-hoc routing protocols. Minimum cost forwarding [7] is an algorithm for efficiently forwarding packets from sensor nodes to a base station with the useful property that it does not require nodes to maintain explicit path information or even unique node identifiers.

Sensor networks may be deployed in hard to reach areas and be meant to run unattended on long periods of time. It may be difficult to replace the batteries on energy-depleted nodes or even add new ones.

In section 2.2, [3] a sensor node decides a route considering neighboring sensor node's remained power and each path's cost. Each sensor nodes know power information of neighboring sensor nodes.

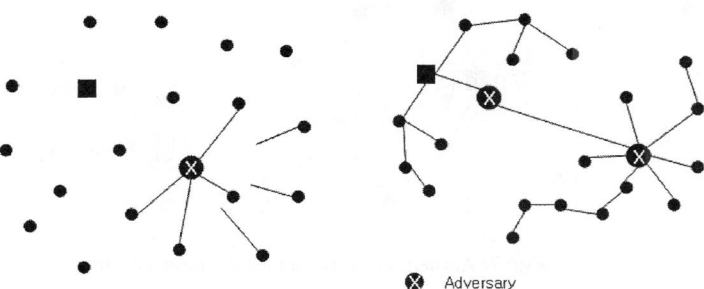

Fig. 5. Sinkhole attack   **Fig. 6.** Wormholes attack

Sinkhole attacks typically work by making a compromised node look especially attractive to surrounding nodes. [1] In the case, a laptop-class adversary with a powerfull transmitter can actually provide a high quality route by transmitting with enough power. One motivation for mounting a sinkhole attack is that it makes selective forwarding trivial. By ensuring that all traffic in the targeted area flows through a compromised node, an adversary can selectively suppress or modify packets originating from any node in the area.

In the wormhole attack [1], an adversary tunnels messages received in one part of the network over a low latency link and replays them in a different part. Wormhole attacks would likely be used in combination with selective forwarding or eavesdropping.

Therefore routing information must be secured. Generally, the sensor nodes communicate using radio frequency, so broadcast is the fundamental communication primitive. But sending broadcast messages is impractical to the limited sensor nodes.

Each sensor nodes broadcast its information to the neighboring nodes. When a sensor node send an information message to a neighboring node, the message can be protected by shared key which is shared by each nodes. But it is overhead to sensor nodes that send messages to each neighboring node using an each shared key.

$A \rightarrow B : \{D\}_{<K_{encr},C>}, MAC(K_{mac}, C|\{D\}_{<K_{encr},C>})$

(A, B: sensor node, D:A's information, K_{encr}, K_{mac}: Shared key, C : counter)

In the sensor network, the nodes can broadcast messages using μTESLA. But it is impractical to general sensor nodes. Since the node is severely memory limited, it cannot store the keys of a one-way key chain. Moreover, re-computing each key from the initial generating key K_n is computationally expensive. Another issue is that the node might not share a key with each receiver. Hence sending out the authenticated commitment to the key chain would involve an expensive node-to-node key agreement. [6]

In the SPINS, [6] a sensor node can broadcast authenticated messages via base station. Each sensor nods.

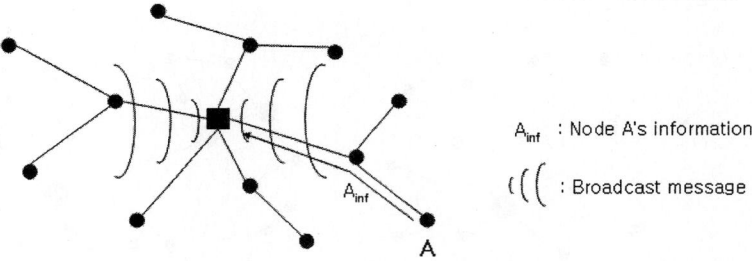

Fig. 7. Authenticated broadcast via base station

A sensor node A sends an information message to base station. Node A uses SNEP to send the information in an authenticated way to the base station. The base station receives a message from the node A. And then the base station broadcasts the message to the sensor network using μTESLA.

6 Conclusions

In the sensor networks which is constrained system, many sensor network routing protocols are quite simple, and for this reason are sometimes even more susceptible to attacks against general ad-hoc routing protocols. In the sensor networks, the security algorithms which are used by laptop-class devices are impractical. Recently, many solutions are proposed that improve the sensor node's performance, such as active signal, node's lifetime considered routing algorithm. Aforementioned solutions, however, are not considered by security problems. In this paper, we showed the security problems and proposed the countermeasures that establish the trust relationship between neighboring sensor nodes, check the bi-directionality each nodes, and authenticate node's information messages. Our proposals can protect the sensor nodes against the attackers that delay the packets, forward packets selectively and attract the neighboring sensor nodes. There are still some remained security problems which were not solved by us in this paper. Therefore we need the further studies which secure the sensor networks.

References

1. Chris, K., David, W. : Secure Routing in Wireless Sensor Network: Attacks and Countermeasures (2003) 113-127
2. Chunlong, G., Lizhi Charlie Z., Jan, M., " Low power distributed MAC for ad-hoc sensor radio networks, IEEE, (2001) 2944-2948
3. Jungahn, H., : Protocols to extend lifetime and to improve QoS in wireless sensor networks (2004)
4. Douceur, J. R., : The Sybil Attack, (2002)
5. Perring, A., Johnson, D. B., : Wormhole detection in wireless ad-hoc networks, (2002)
6. Adrian, P., Robert, S., Victor, W., David, C., J. D. Tygar : SPINS: Security protocols for sensor networks. (2001)
7. Ye, F., Chen, A., Lu, S., Zhang, L. : A scalable solution to minimum cost forwarding in large sensor networks. (2000) 243-254

Policy Based Handoff in MIPv6 Networks

Jong-Hyouk Lee, Byungchul Park, Hyunseung Choo, and Tai-Myung Chung

School of Information and Communication Engineering,
Sungkyunkwan University,
440-746 Suwon, Korea
jhlee@imtl.skku.ac.kr

Abstract. As the requirements on high-availability for multimedia intensified new applications increase the pressure for the higher bandwidth on wireless networks, just upgrading to the high capacity network is essential. However, that is not especially for solving the handoff-related problems. Policy-based network (PBN) technologies help MIPv6 networks ensure that network users get the QoS, security, and other capabilities they need. PBN has the intelligence, in the form of business policies, needed to manage network operations. In this paper, we propose the Policy Based Handoff (PBH) mechanism in mobile IPv6 (MIPv6) networks to have more flexible management capability on MNs. PBH employs dual routing policy so that it reduces the fault rate of handoff up to about 87% based on the same network parameters and solves the ping-pong problem. Experimental results presented in this paper shows that our proposal has superior performance comparing to other schemes.

1 Introduction

MIPv6 is a protocol that guarantees the mobility of mobile node (MN) within the IPv6 environment. In particular, its primary distinguishing features are that it provides route optimization and that it renders obsolete the Foreign Agents (FA), which formed a vital part of Mobile IPv4 (MIPv4). In the MIPv6, an MN is distinguished by its home address. When it moves to another network, it obtains a Care of Address (CoA), which provides information about the MN's current location. The MN registers its newly acquired CoA with its own Home Agent (HA). After that, it can directly communicate with its Correspondent Nodes (CN) [1,2].

In MIPv6, an MN sends a Binding Update (BU) message to both its CNs and HA when it moves to another network. It takes some time to send data to the changed CoA after CN receives it. If CN transmits data to MN during this time, the data are lost. To prevent this in MIPv6, MN sends a BU to the previous access router when MN moves to other networks. Then the previous access router knows the movement of MN through this message, and transmits to the new access router for MN. As a problem of this mechanism, the previous access router must save packets arrived for MN until receiving BU message from the new access router. Increase of the number of MNs leads to the required amount of buffers to be fairly increased. Therefore data lose occurs when buffers in access routers are not enough to hold data for MNs in handoff. This problem becomes more significant when MN density becomes high. While MN moves into other networks, it is possible to be in ping-pong state in a boundary area. Ping-pong effect occurs when there is not much difference in the strength of signal power between two networks and thus MN registration to someone is ambiguous at the moment. According to the increased handoff messages on ping-pong effect, two corresponding access routers should handle a certain amount of operations. The series of processing procedures burdens access routers and thus networks therefore the probability of handoff failure increases.

In this paper, we manage MN's handoff by employing policies based on Common Open Policy Service (COPS) protocol to solve the fault of transferring data from a CN to an MN. In the proposed PBH mechanism, edge Access Router (eAR) follows the policy, which is predefined by Policy Server (PS). When the ping-pong effect comes up, the corresponding eAR follows "Dual Routing Policy". The eAR passes data to both previous Access Router (pAR) and new Access Router (nAR). Both of them send the data to MN therefore the MN never loses its data. Consequently, the ping-pong effect is solved and the fault rate of handoff is also reduced up to about 87% based on the same network parameters.

The remainder of the paper is organized as follows. We begin in section 2 with a review of the problems associated with handoff in Policy Based Networks. In section 3, we define the PBH mechanism and describe how it works and how it is organized. In section 4, we evaluate the performance of the PBH mechanism. The final section gives our conclusions.

2 Related Works

2.1 Mobile IPv6 (MIPv6)

In MIPv6, MN typically acquires its CoA through stateless or stateful address auto-configuration based on the methods in IPv6 Neighbor Discovery. The CoA of the MN is changed when it moves to another network. When its CoA is changed, the MN sends a BU message to its HA and all of its CNs. If the MN demands an acknowledgement from the CN for this, the CN responds to the MN by using the binding acknowledgement option. The CN avoids the problem of the triangle routing by maintaining the binding update of MN whenever MN moves. However, MIPv6 still suffers from the buffering problem and problems induced by the ping-pong effect, and these problems have received a great deal of attention in the literature [3,4] as an important research goal.

Especially, [4] proposed a new MIPv6 handoff mechanism called Partial Dual Unicasting (PDU) based on the QoS supported RSVP signaling protocol. The PDU scheme reduces the handoff failure rate up to about 24% and guarantees the time for the handoff completion. However, the PDU scheme has the demerit that the CNs maintains an additional BU list of MN. In addition, the farther the merged Access Router (mAR) is from the CN, the more handoff failure occurs.

2.2 Policy-Based Network (PBN)

The Policy-based network architecture as defined in the IETF/DMTF consists of four basic elements which are the policy console, policy repository, Policy Decision Points (PDP), and Policy Enforcement Points (PEP). The policy console is used by an administrator to input the different policies that are active in the network. The policy console takes an input as the high-level policies that a user or administrator enters in the network and converts them to a much more detailed and precise low-level policy description that can be applied to the various devices in the network. The policy repository is used to store the policies generated by the policy console and retrieved by PDP or PEP. PDP translates the set of rules it retrieves from the policy repository to a format and syntax that is understood by the PEP function. PEP is responsible for executing the policies as defined by the PDP [5,6]. A policy console, PDPs, PEPs, and the repository can communicate with each other using a variety of protocols. Representative protocols are COPS and Light Weight Directory Access Protocol (LDAP). Based on such protocol it passes information to the system that manages policies [7,8].

The Policy Server (PS) that manages policies analyzes collected data and enforces the detailed command based on the policy regulation which administrator defines. PS may be located either in PEP or far away. In this paper, the PS serves as a PDP and the AR serves as a PEP. The AR enforces policies, which concern routings or access controls determined by the PS.

3 Policy Based Handoff Mechanism

Fig. 1 shows the case in which a MN moves from Network A to Network B. The MN receives a radio advertisement for the first time from nAR of Network B when it reaches P1. The advertisement message contains the nCoA value which is used by the MN when it moves to Network B.

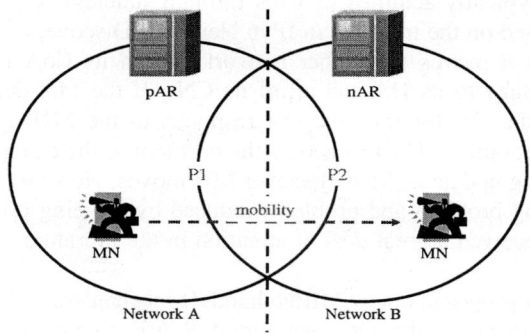

Fig. 1. Movement of MN

In the proposed PBH mechanism, a MN creates additional messages when it receives a Router Advertisement (RA) message from network B. These messages are called Additional Binding Update (ABU). The MN sends ABU message to both pAR and nAR at the same time. At that time, the pAR becomes aware of the transference of the MN and nAR is informed that the MN has come into its network. Both pAR and nAR notify the transference of the MN to PS which performs PDP function. The PS sends predefined policies to eAR, based on the COPS protocol. The policies enforce an eAR to send data to both pAR and nAR (The data are actually transferred from the CN to the MN). In other words, the data from the CN is duplicated by eAR and sent to both pAR and nAR.

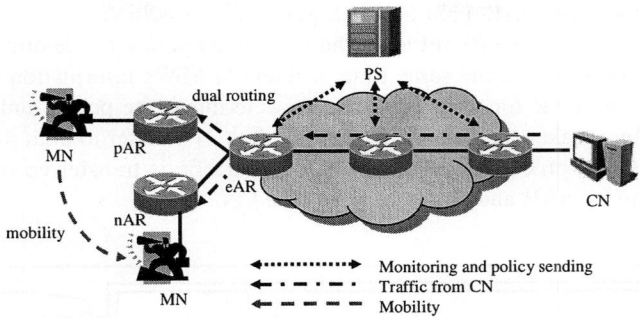

Fig. 2. Policy Server in Mobile Environment

The Fig. 2 describes the PBH mechanism that how the PS distributes the policies and how eAR follows the policies. The PS determines which policies to apply, based on the profile of the MN user. In addition, the PS causes eAR to execute these policies.

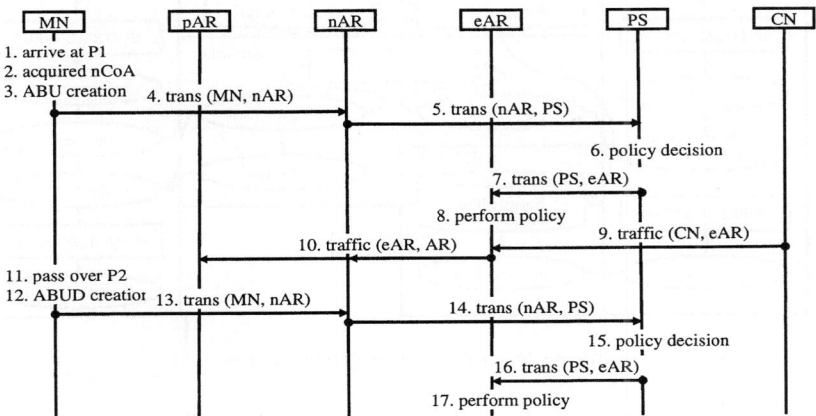

Fig. 3. Flow of entire messages

Each message in Fig. 3 is described as follows. (1) The MN arrives at P1. (2) The MN receives the message from nAR and generates nCoA. (3) An ABU message is created. (4) The ABU message is transmitted to both pAR and nAR at the same time. (5) Both pAR and nAR notify the PS that they have the ABU message from the MN, respectively. (6) The Policy decision for the dual routing. (7) The PS transmits the policy to eAR. (8) The eAR performs the policy. (9) Data traffic from CN is transmitted to eAR. (10) The traffic is transmitted to both pAR and nAR simultaneously based on the policy. (11) MN maintains the link to pAR until MN completely moves to nAR. (12) When MN passes over the point P2, it makes Additional Binding Update Deletion (ABUD) message. (13) ABUD message is sent to nAR. (14) eAR relays ABUD message to PS. (15) Policy decision to disable the dual routing. (16) The PS transmits the policy to eAR. (17) The eAR performs the policy.

As shown in Fig. 3, the PS retrieves the predefined policy for determining the specific action procedure. At the same time, it refers to MN's information which is sent by both pAR and nAR for retrieving. The PS determines the policy and then notifies eAR. The policy makes the data traffic sent by the CN transfer to both pAR and nAR. Following original process of MIPv6, CN's data traffic is transferred to nAR if MN finishes handoff to nAR and performs BU to the CN.

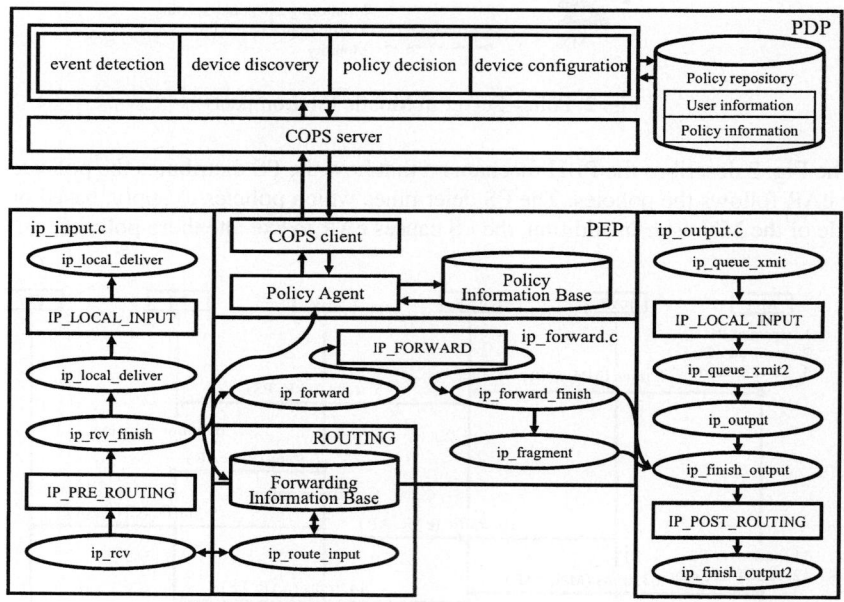

Fig. 4. Policy-based management architecture

Fig. 4 shows policy-based management architecture which is proposed here. Serving as a PS, The PDP decides policies. The PS is responsible for event detection, device discovery, policy decision, and device configuration. When the system state changes, an event is occurred and the information of the policy repository is changed. The PDP decides policies differently by events. For the flexible management, the PDP discovers devices and constructs the network topology automatically. The device configuration is responsible for gathering the information of devices with whom the PS communicates like what kind of device it is or control protocols.

4 Performance Evaluation

4.1 Basic Idea on Modeling

At each step, the time required to send a message is composed of the transmission time, the propagation time and the processing time, i.e. $M^i = \alpha^i + \beta^i + \gamma^i$, where i represents the step i. The transmission time α^i is computed by the size of the control message in bits over the bit rate of the link on which the message is sent, i.e. $\alpha^i = M_b/B$, where M_b is a control message assuming the fixed size. The B is the bandwidth of the link, B_l for the wired line, and B_w for the wireless case. The propagation time β^i varies depending on the transmission medium, i.e. β_l is the time for the wired line, and β_w is for the wireless one. The processing time γ^i has the same value at intermediate routers, MN, CN, pAR, and nAR. The wired medium is more stable than the wireless one, so the retransmission is not needed. Therefore, the physical transmission time T^i is represented by $M^i (= M_l)$. Later in each step the message processing time on the wired and wireless cases are represented as M_l and M_w, respectively. At the wireless link, the message retransmission is necessary because a message can be lost in any moment. MN retransmits the message when lost in the air transmission. By considering the number of link failures and the probability of link failure, we obtain the additional signal processing time at these steps in the wireless case, i.e. $T^w{}_i = 2M^w{}_i + T_{out}$ by [9]. And the additional message processing time else M_i may be required. It is assumed to be the message processing time T_{proc}.

Table 1. Performance Analysis Parameters

Variables	Definitions	Values
T_{proc}	extra processing time	0.5 msec
B_l	bits rate of wired link	155 Mbps
B_w	bits rate of wireless link	144 Kbps
M_b	control message size	50 bytes
β_l	wired link message propagation time	0.5 msec
β_w	wireless link message propagation time	2 msec
γ	message processing time	0.5 msec
T_{out}	message loss judgment time	2 msec
T_{pdt}	policy decision time	60 msec
q	probability of message loss	0.5
H_{a-b}	number of hops between a and b	-

4.2 Total Handoff Time

The system parameters used to analyze the system are listed in Table 1. Each value is defined based on the information provided in [9,11]. Fig. 3 represents the message flows. Based on the scheme shown in this figure, we compute the total handoff time. The handoff completion time is determined by summing I, II, III, and IV below.

I. Sum of the extra Processing Times (*PT*): Processing time is required in steps 2 and 3. So, $PT = T^2_{proc} + T^3_{proc}$. In the above case, T^i_{proc} has a fixed value (T_{proc}) at each step which is just as much as the processing time. Therefore,

$$PT = 2T_{proc} \qquad (1)$$

II. Sum of the Message transmission Times in the wired links (*MT$_l$*): The Message transmission in the wired line states takes place in step 5, assuming eAR and PS are in one system so that step 7 is not considered. In this case, the total time required for message transmission is $MT_l = min\{T^5_l \cdot H_{eAR\text{-}pAR}, T^5_l \cdot H_{eAR\text{-}pAR}\}$. We assume that each T^i_l ($i = 5$) has a fixed value (T_l), and thus

$$MT_l = T_l \cdot \min\{H_{eAR-pAR}, H_{eAR-nAR}\} \qquad (2)$$

III. Sum of the Message transmission Times in the wireless links (*MT$_w$*): The message transmissions in the wireless links take place in step 4. Here, the message transmission time is $MT_w = T^4_w$. From the equation T^i_w, MT_w becomes

$$MT_w = T_w = 2M_w + T_{out} \qquad (3)$$

IV. Policy Decision time (*PDT*): The Policy Decision time T_{pdt} is a system variable and is provided before the computation.

$$PDT = T^6_{pdt} = T_{pdt} \qquad (4)$$

Therefore, we obtain the total required time for the completion of handoff, by the summation of steps I to step IV.

$$T_{req} = PT + MT_l + MT_w + PDT \qquad (5)$$
$$= 2T_{proc} + T_l \cdot \min\{H_{eAR-pAR}, H_{eAR-nAR}\} + 2M_w + T_{out} + T_{pdt}$$

4.3 Handoff Failure Rate

The T is a random variable representing the time that the MN stays in the overlapped area and the T_{req} is the time required for the completion of handoff. Hence, the handoff failure rate is represented by $P = Prob(T < T_{req})$, where we assume that T is exponentially distributed. Thus,

$$P = prob(T < T_{req}) = 1 - \exp(-\lambda T_{req}) \qquad (6)$$

Where λ is the arrival rate of the MN into the boundary cell and its movement direction is uniformly distributed within the interval [0, 2π). Therefore, λ can be calculated from the equation $\lambda = VL/\pi S$ [10], where V is the expected velocity of the MN, which varies in the given environment, and L is the length of the boundary at the overlapped area assuming a circle with radius l, i.e. $L = (1/6) \cdot 2\pi l2 = (2/3)\pi l$. The area of the overlapped space S is $S = 2(1/6(\pi l^2 - (\sqrt{3}/4)l^2))$. Therefore, λ is calculated using equations involving l and V.

4.4 Experimental Results

We present graphs showing the probability of handoff failure, based on the analysis performed above. Firstly, we examine the relationship between the number of hops from eAR to CN and the handoff failure rate. In the basic handoff mechanism and the PDU mechanism, the handoff process must be performed within the overlapped area, so that the number of hops from eAR to CN affects the handoff failure rate. However, in the proposed mechanism, the number of hops from eAR to CN does not affect the handoff failure rate, because the handoff process is conducted between MN and eAR, where PS is located, as shown in the graph in Fig. 5. In the case where the number of hops is small, the handoff failure rate is rather high, because the time required to decide which policy to use is relatively large. However, in the case where the number of hops is larger than about 10, the proposed mechanism shows the best performance. We provide graphs representing different values of the velocity of the MN and the radius of the cell.

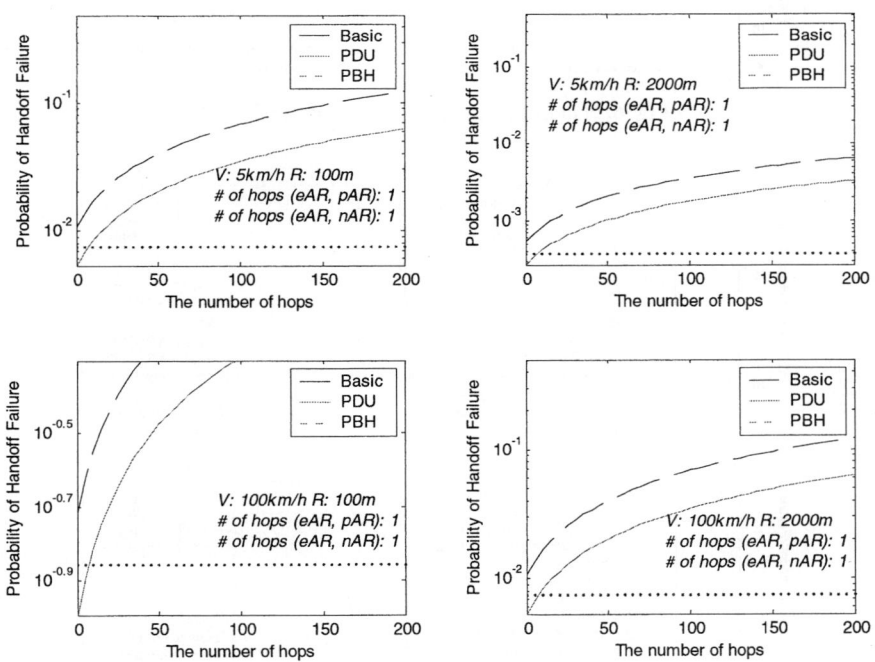

Fig. 5. Handoff failure rate by the hop count

The graph in Fig. 6 shows the handoff failure rate as a function of the radius of the cell. As the radius of the cell increases, the overlapped area becomes larger, so that it is easy to complete the handoff. As a result, the handoff failure rate decreases. Although the trend of the plots for all of the mechanisms is similar, the PBH mechanism shows the lowest handoff failure rate among all of the mechanisms, in the case where H_{eAR-CN} is larger than about 10.

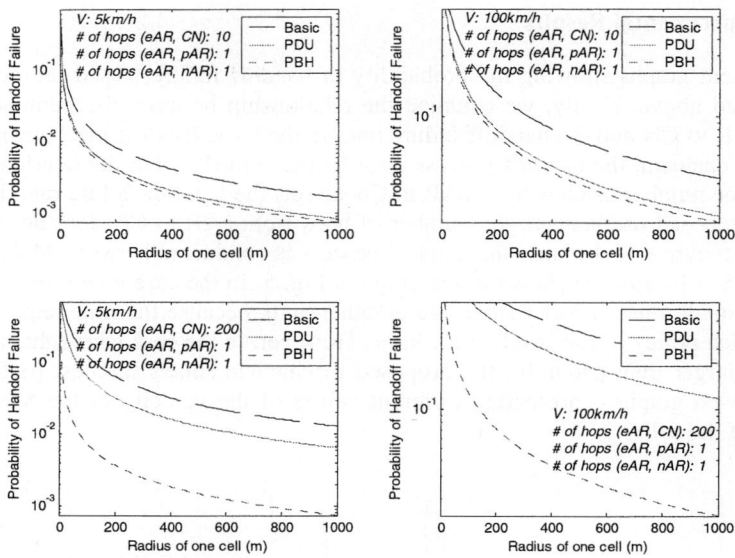

Fig. 6. Handoff failure rate by the radius of a cell

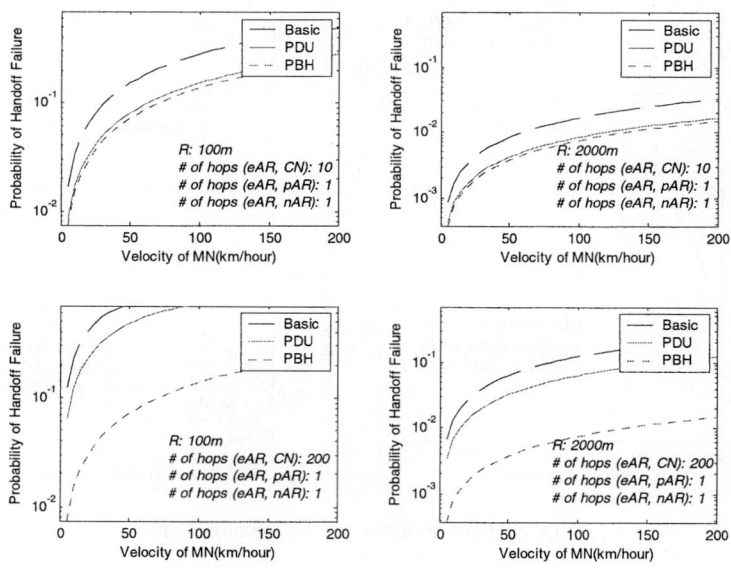

Fig. 7. Handoff failure rate by the velocity of MN

When handoff occurs with an MN which has a high velocity, it is difficult to complete the handoff, because the MN moves out of the overlapped area quickly. The graph in Fig.7 shows the relationship between the velocity of the MN and the handoff failure rate. The trend of the plots for all of the mechanisms is similar, and the PBH mechanism also shows the best performance in this case.

5 Conclusions

In this paper, we upgrade a PS to employ dual routing policy when an MN does handoff. An MN can receive traffic from CN promptly even if the MN is on the ping pong effect. In addition, by adopting PBN in mobile environments, it can be flexible to manage MNs. We have evaluated the proposed scheme that we can accomplish highly improved result. It reduces the fault rate of handoff up to about 87% based on the same network parameters and solves the ping-pong problem. In the future, we will describe the requirements for the efficient network management policy, and the access control and AAA for MN in the PBN.

References

1. D. Johnson, C. Perkins, J. Arkko, "Mobility Support in IPv6", RFC 3775, June 2004.
2. C. Perkins, "Mobility Support in IPv4", RFC 3344, August 2002.
3. Myung-Kyu Yi, Chong-Sun Hwang, "A Pointer Forwarding Strategy for Minimizing Signaling Cost in Hierarchical Mobile IPv6 Networks", EUC 2004, LNCS 3207, pp. 333-345, August 2004.
4. D. Choi and H. Choo, "Partial Dual Unicasting Based Handoff for Real-Time Traffic in MIPv6 Networks", ICCS 2003, LNCS 2660, pp. 443-452, August 2003.
5. L. Mark, "Policy-based Management for IP Networks", Bell Labs Technical Journal, pp. 75-94, October-December 1999.
6. B. Moore, E. Ellesson, J. Strassner, A. Westerinen, "Policy Core Information Model", RFC 3060, February 2001.
7. D. Durham, J. ED, "The COPS(Common Open Policy Service) Protocol", RFC 2748, January 2000.
8. M. Wahl, T. Howes, S. Kille, "Lightweight Directory Access Protocol", RFC 2251, December 1997.
9. J. McNair, I. F. Akyildiz, and M. D. Bender, "An inter-system handoff technique for the IMT-2000 system", IEEE INFOCOM, vol.1, pp. 208.216, 2000.
10. R. Thomas, H. Gilbert, and G. Mazziotto, "Influence of the moving of the mobile stations on the performance of a radio mobile cellular network", in Proceedings of the 3rd Nordic Seminar, 1988.
11. A. Ponnappan, Lingjia Yang, R. Pillai.R, "A policy based QoS management system for the IntServ/DiffServ based Internet", Proceedings of the Third International Workshop on Policies for Distributed Systems and Networks, Pages:159 - 168, June 2002.

An Effective Location Management Strategy for Cellular Mobile Networks*

In-Hye Shin[1], Gyung-Leen Park[1], and Kang Soo Tae[2]

[1] Department of Computer Science and Statistics, Cheju National University, Korea
{ihshin76, glpark}@cheju.ac.kr
[2] Department of Computer Engineering, Jeonju University, Korea
kstae@jeonju.ac.kr

Abstract. Reducing the location update cost has been a critical research issue since the location update process requires heavy signaling traffics. This paper proposes an effective location management strategy to reduce the location update cost in cellular mobile networks. The paper also develops analytical models to evaluate the performance of the proposed strategy as well as conventional schemes in various situations. The results of the evaluation show that the proposed strategy reduces the location update cost significantly.

1 Introduction

Many researchers have studied on various design problems of cellular mobile networks [1-3]. The typical architecture of the personal communication service (PCS) network is shown in Figure 1 [1]. The architecture has two types of databases: the home location register (HLR) and the visitor location register (VLR). The mobile switching center (MSC) associated with a specific VLR is in charge of several base station controllers (BSCs), lower control entities which in turn control several base stations (BSs). The MSCs are connected to the backbone wired network such as public switching telephone network (PSTN). The network coverage area is divided into smaller cell clusters called location areas (LAs). The VLR stores temporarily the service profile of the mobile station (MS) roaming in the corresponding LA. The VLR also plays an important role in handing call control information, authentication, and billing. The HLR stores permanently the user profile and points to the VLR associated with the LA where the user is currently located. Each user is assigned unambiguously to one HLR, although there could be several physical HLRs.

Finding the location of the roaming mobile station in order to setup the connection properly is called the location management [4]. The location management involves two major procedures – *location update* and *paging*. A base station periodically broadcasts its location area identifier (LAI) which is unique for each LA. When a mobile station enters a new LA, it receives a different LAI. Then, the mobile station sends a registration message to a new VLR. The new VLR sends a registration

---

* This research was supported by the MIC(Ministry of Information and Communication), Korea, under the ITRC(Information Technology Research Center) support program supervised by the IITA(Institute of Information Technology Assessment).

message to the HLR. Then the HLR sends a registration cancellation message to the old VLR and sends a registration acknowledge message to the new VLR. Now the HLR points the new VLR that has service profile and location information of the mobile station. This procedure is called the location update. When a call to a PCS user is detected, the corresponding HLR is queried. Once the HLR corresponding to the mobile station has been queried, the VLR/MSC currently serving the mobile station is known. Then paging is done through the LA where the MS is currently located.

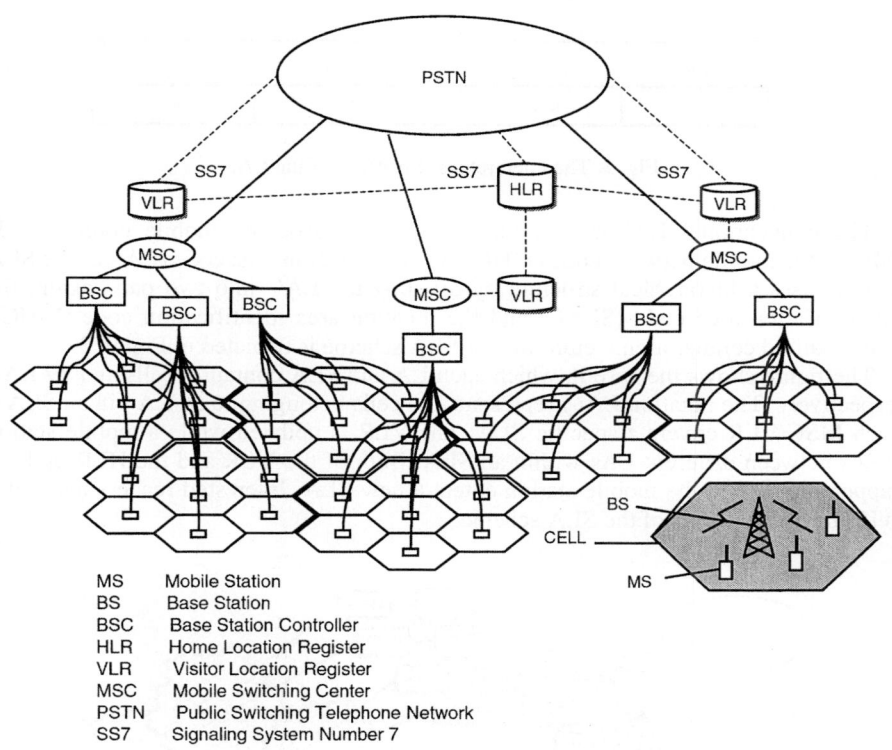

MS   Mobile Station
BS   Base Station
BSC   Base Station Controller
HLR   Home Location Register
VLR   Visitor Location Register
MSC   Mobile Switching Center
PSTN   Public Switching Telephone Network
SS7   Signaling System Number 7

Fig. 1. The Cellular Architecture in PCS

The total location update cost consists of the update cost of HLR and that of VLR. Since the HLR is connected to many VLRs as shown in Figure 1, it has heavy signaling loads. We already have proposed to employ the hierarchical structure for the system to reduce the location update cost of the HLR [5]. However, even in proposed scheme, frequent updates are required for the boundary LAs between different Super Location Areas (SLAs). The paper proposes an effective location management strategy in order to further reduce the location update cost. The paper also develops analytical models to evaluate the performance of the proposed strategy in various situations. The results from the evaluation show that the proposed strategy reduces the location update cost significantly. The proposed strategy is presented in Section 3 while Section 2 describes the related works. Section 4 develops analytical models to evaluate the performance of the proposed approach. The result of the performance evaluation is also shown in Section 4. Finally, Section 5 concludes the paper.

2 The Related Works

The SLA scheme [5] employs a *super location area*, a group of LAs. The size of an SLA could be various. An LAI structure also is modified as follows.

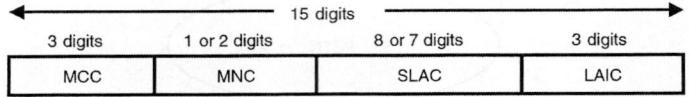

Fig. 2. The Hierarchical Structure of the LAI

The conventional LAI was divided into three parts: the mobile country code (MCC), the mobile network code (MNC), and the location area code (LAC). The SLA scheme uses a hierarchical structure by dividing the LAC into two parts again: the super location area code (SLAC) and the location area identification code (LAIC). The modified cellular architecture for the SLA scheme is depicted in Figure 3.

The figure shows the case in which each LA and SLA contain *7* cells and *19* LAs, respectively. The notation LAi-j represents the cells belonging to an LA j in an SLA i. Each MSC/VLR covers a specific SLA. The VLR is updated while a mobile station roams between different LAs within an SLA. Both of the VLR and the HLR updates happen only when the mobile station enters a new SLA. Interested readers may refer to [5] for more details of the SLA scheme.

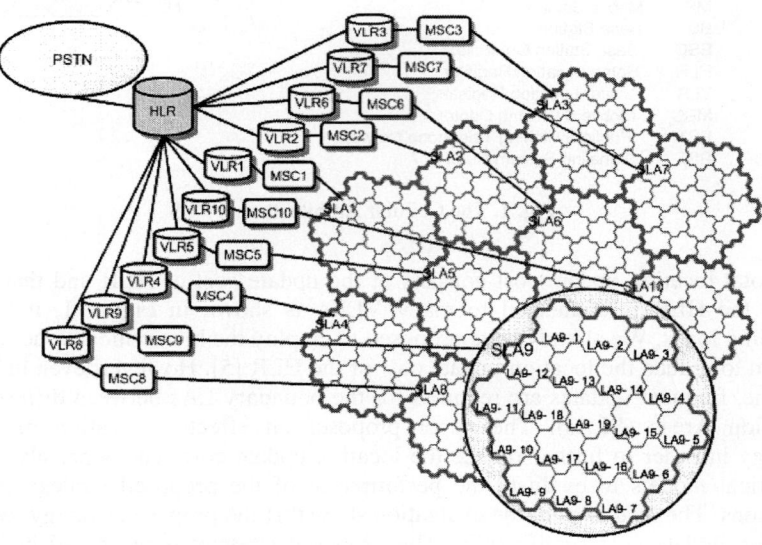

Fig. 3. The Cellular Architecture Employing the SLA

3 The Proposed Approach

The proposed approach employs *overlapping location areas* which are the common boundary LAs between different SLAs. That is, the proposed approach extends to the SLA scheme in the previous section by employing LA-level overlapping.

The modified cellular architecture for the proposed approach is depicted in Figure 4. The overlapping LAs are colored areas in the figure. The overlapping borders consist of twofold or threefold overlapping LAs. The overlapping LAs belong to all SLAs associated with them. The two-fold overlapping LA is included in the two SLAs and two different LAIs are broadcasted in the LA. The three-fold overlapping LA is included in three different SLAs and three LAIs are broadcasted in the LA. Note that the overlapping LAs are duplicately managed by corresponding all MSCs/VLRs. As an example, the *SLA9* out of SLAs shows an SLA with overlapping LAs in detail. When a mobile station enters to the overlapping LA, it must register only to one LA out of the overlapping LAs. The proposed approach requires both of the VLR and the HLR updates only when a mobile station enters the LA that belongs to the new SLA while does not belong to the old SLA. The VLR update occurs when it moves to a different LA within an SLA.

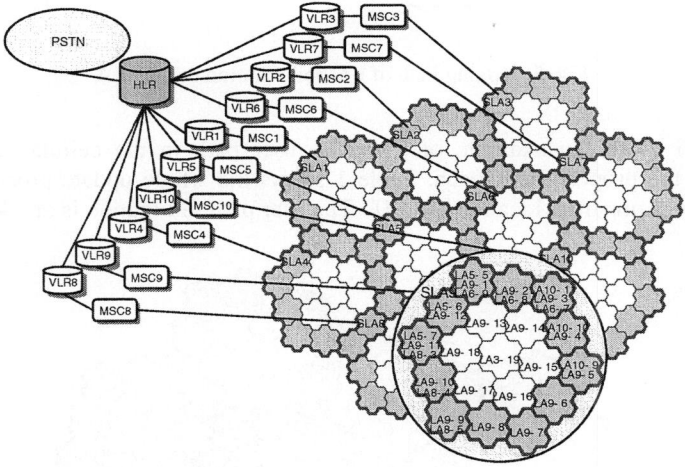

Fig. 4. The Cellular Architecture Employing The OSLA

Figure 5 shows an example of a mobile user path from location A to location U in the cellular architecture employing the SLA[5]. It is out of question that the LAI employed in SLA, divided into four parts, could be used like the traditional LAI, divided into three parts. Thus, the location update procedure of the conventional scheme could be described in the figure. In the conventional system, both of the VLR and the HLR are updated whenever a mobile station enters a new LA. So the location updates tend to occur frequently in the boundary cells of LAs [6-12]. In the SLA scheme, both of the VLR and the HLR updates occur when a mobile station moves to a different SLA like the path from location D to H in Figure 5. That is, the HLR updates also happen frequently in the boundary LAs of the SLAs.

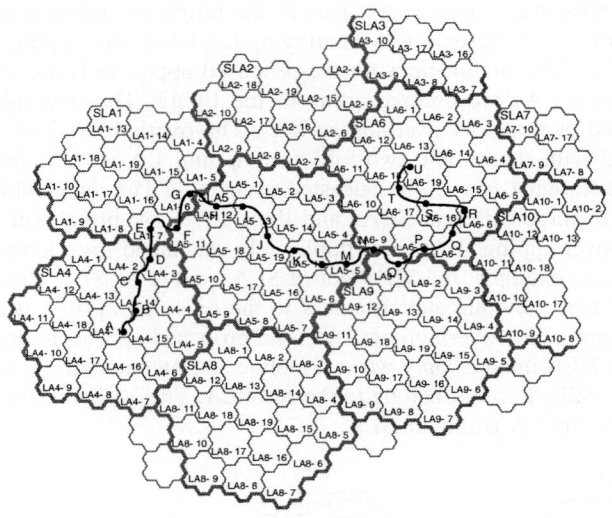

Fig. 5. Moving Path of a Mobile Station in the SLA

Figure 6 shows an example of a mobile user path in the cellular architecture employing the proposed approach. Table 1 shows the whole update process while a mobile user roams from location A to U. The user path in Figure 6 is an identical line

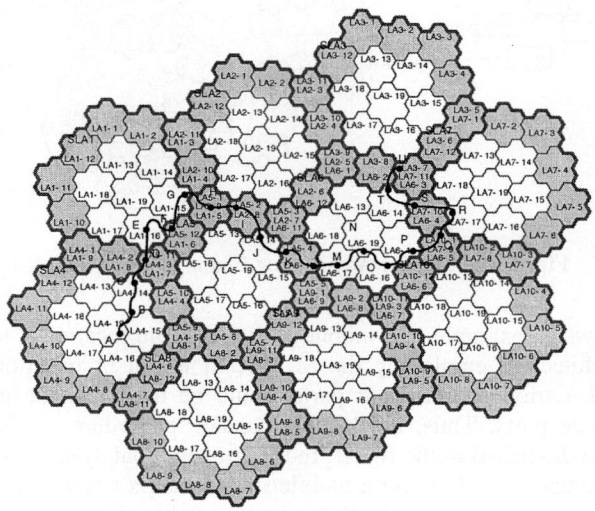

Fig. 6. Moving Path of a Mobile Station in the OSLA

of that in Figure 5. Also, the two service areas are the same. In overlapping areas, if one out of the currently broadcasted LAI belongs to the same SLA as that of the previously registered LA, a mobile user registers to it. For example, if a mobile user arrives at location D and H, the user registers to *LA4-3* out of two LAs and *LA1-5* out of three LAs, respectively. The *LA4-3* and *LA1-5* are LAIs belonging to the same SLA as LAI of the previously registered LA, respectively. In the proposed approach, both of the VLR and the HLR are updated only when a mobile user moves to the LA that belongs to the new SLA and does not belong to the old SLA at the same time. Particularly, in case of a path from location H to I, the mobile user has to register to either LA5-2 or LA2-8 as shown in the 8th row in Table 1. The decision of such registration is made by a random selection. According to the decision, the location update cost is different as shown in the 9th row in Table 1. Also, when the mobile user moves from location P to T, the HLR update is required only when the mobile user crosses the overlapping LAs like the path of P-Q-R or that of R-S-T. Thus, the HLR updates concentrated in boundary LAs are rather reduced and distributed.

Table 1. Comparison of The Location Update Cost

ID	The Path	The registered LA PCS,SLA/OSLA	The Register Update		
			PCS	SLA	OSLA
1	A→B	LA4-14/ LA4-14	VLR, **HLR**	VLR	VLR
2	B→C	LA4-14/ LA4-14	None	None	None
3	C→D	LA4-3/ LA4-3	VLR, **HLR**	VLR	VLR
4	D→E	LA1-7/ LA1-16	VLR, **HLR**	VLR, **HLR**	VLR, **HLR**
5	E→F	LA5-11/ LA1-6	VLR, **HLR**	VLR, **HLR**	VLR
6	F→G	LA1-6/ LA1-15	VLR, **HLR**	VLR, **HLR**	VLR
7	G→H	LA5-12/ LA1-5	VLR, **HLR**	VLR, **HLR**	VLR
8	H→I	LA5-13/(LA2-8 or LA5-2)	VLR, **HLR**	VLR	VLR, **HLR**
9	I→J	LA5-19/ LA5-14	VLR, **HLR**	VLR	VLR, **HLR** or VLR
10	J→K	LA5-15/ LA5-4	VLR, **HLR**	VLR	VLR
11	K→L	LA5-15/ LA5-4	None	None	None
12	L→M	LA5-5/ LA6-17	VLR, **HLR**	VLR	VLR, **HLR**
13	M→N	LA6-9/ LA6-19	VLR, **HLR**	VLR, **HLR**	VLR
14	N→O	LA9-1/ LA6-16	VLR, **HLR**	VLR, **HLR**	VLR
15	O→P	LA6-8/ LA6-15	VLR, **HLR**	VLR, **HLR**	VLR
16	P→Q	LA6-7/ LA6-5	VLR, **HLR**	VLR	VLR
17	Q→R	LA6-6/ LA7-17	VLR, **HLR**	VLR	VLR, **HLR**
18	R→S	LA6-16/ LA7-10	VLR, **HLR**	VLR	VLR
19	S→T	LA6-17/ LA6-14	VLR, **HLR**	VLR	VLR, **HLR**
20	T→U	LA6-19/ LA6-3	VLR, **HLR**	VLR	VLR
	The number of total updates		18 VLR **18 HLR**	18 VLR **7 HLR**	18 VLR **6 or 5 HLR**

4 Performance Evaluation

We have developed analytical models in order to compare the performance of three schemes: the conventional scheme, the SLA scheme, and OSLA (Overlapped SLA) which is presented in the paper. The notations used in the model are depicted in Table 2.

Table 2. The Notations Used in the Model

$\overline{K}$	The average number of mobile users in a cell
d	The size of an LA s The size of an SLA
$\overline{T_d}$	The average dwell time
N_c	The number of cells in an LA: $3d^2 - 3d + 1$
N_{la}	The number of boundary cells in an LA: $6(d-1)$
N_{Sla}	The number of LAs in an SLA: $3s^2 - 3s + 1$
N_{Sc}	The number of cells in an SLA: $(3s^2 - 3s + 1)(3d^2 - 3d + 1)$
N_{Sbc}	The number of boundary cells in an SLA: $6\{3d - 2 + (s-2)(2d-1)\}$
N_S	The total number of mobile users in an SLA
$\overline{R_{SLA}}$	The total location update rate for the given SLA
$\overline{R_{OSLA}}$	The total location update rate for the given OSLA
$\overline{R_{MS}}$	The average location update rate per mobile user

Figure 7 shows the details to develop the models. The size of an SLA is represented by the number of rings of LAs, s, as shown in Figure 7-(a). The size of an LA is represented by the number of rings of cells, d, as shown in Figure 7-(b). The models assume that a mobile user moves to one of neighboring cells with probability *1/6*.

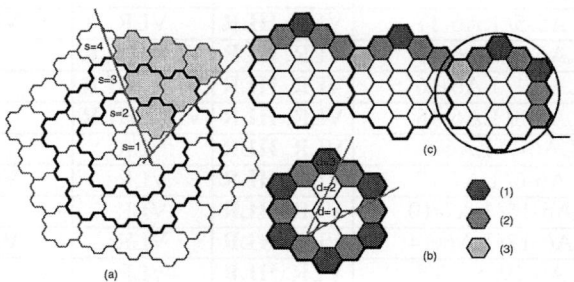

Fig. 7. The Boundary Cells in an LA or SLA

Figure 7-(1), 7-(2), and 7-(3) show the cells from which a mobile user moves to one of neighboring cells, causing the HLR update, with the probability of *3/6, 2/6,* and *1/6*, respectively. Even though we omit the details of the derivation due to the space limitation, interested readers may refer to [13].

$$N = N_{bc} \cdot \overline{K} = (3d^2 - 3d + 1) \cdot \overline{K} \tag{1}$$

$$\overline{R_{LA}} = 6 \left\{ \frac{2}{6} \cdot (d-2) + \frac{3}{6} \cdot 1 \right\} \cdot \overline{K} \cdot \frac{1}{\overline{T_d}} = (2d-1) \cdot \overline{K} \cdot \frac{1}{\overline{T_d}} \tag{2}$$

$$\overline{R_{MS}} = \frac{\overline{R_{LA}}}{N} = \frac{2d-1}{(3d^2 - 3d + 1) \cdot \overline{T_d}} \tag{3}$$

$$N_S = N_{Sc} \cdot \overline{K} = (3s^2 - 3s + 1)(3d^2 - 3d + 1) \cdot \overline{K} \tag{4}$$

$$N_{Sbc} = 6 \left[(s-2) \left\{ \frac{2}{6} 6(d-1) + 1 \right\} + \frac{3}{6} 6(d-1) + 1 \right] \tag{5}$$
$$= 6\{(2s-1)(d-1) + s - 1\} = 6\{2(s-2)(d-1) + s - 2 + 3(d-1) + 1\}$$

$$\overline{R_{SLA}} = 6 \left[\frac{2}{6} \cdot \{(2s-1)(d-2) + s - 1\} + \frac{3}{6} \cdot s + \frac{1}{6} \cdot (s-1) \right] \cdot \overline{K} \cdot \frac{1}{\overline{T_d}} \tag{6}$$
$$= \{(2s-1)(2d-1)\} \cdot \overline{K} \cdot \frac{1}{\overline{T_d}}$$

$$\overline{R_{MS}} = \frac{\overline{R_{SLA}}}{N_S} = \frac{(2s-1)(2d-1)}{(3s^2 - 3s + 1)(3d^2 - 3d + 1) \cdot \overline{T_d}} \tag{7}$$

$$\overline{R_{OSLA}} = 6 \left[(s-2) \frac{1}{2} \left\{ \frac{2}{6} \left(2(d-2) + 1 \right) + \frac{3}{6} \cdot 1 + \frac{1}{6} \cdot 1 \right\} + \frac{1}{3} \left\{ \frac{2}{6} \left(3(d-2) + 1 \right) + \frac{3}{6} \cdot 1 + \frac{1}{6} \cdot 1 \right\} \right] \cdot \overline{K} \cdot \frac{1}{\overline{T_d}} \tag{8}$$
$$= [(s-1)(2d-1)] \cdot \overline{K} \cdot \frac{1}{\overline{T_d}}$$

$$\overline{R_{MS}} = \frac{\overline{R_{OSLA}}}{N_S} = \frac{(s-1)(2d-1)}{(3s^2 - 3s + 1)(3d^2 - 3d + 1) \cdot \overline{T_d}} \tag{9}$$

Figure 8 shows the average HLR update rate according to the size of the SLA, when $d = 3$ and $T_d = 6$ minutes. It shows that the proposed scheme (OSLA) outperforms the SLA and the conventional scheme. Similarly, Figure 9 and Figure 10 show the average HLR update rate according to the size of the LA and the average dwell time, respectively. Both of them show that the proposed scheme outperforms the SLA and the conventional scheme.

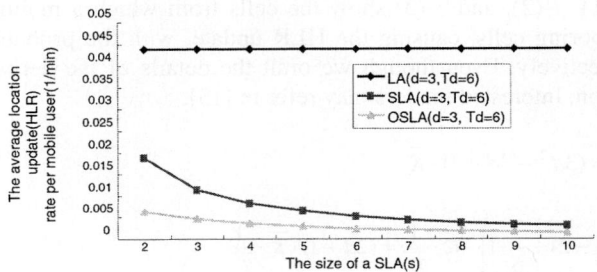

Fig. 8. Update Rate of HLR According to the Size of the SLA

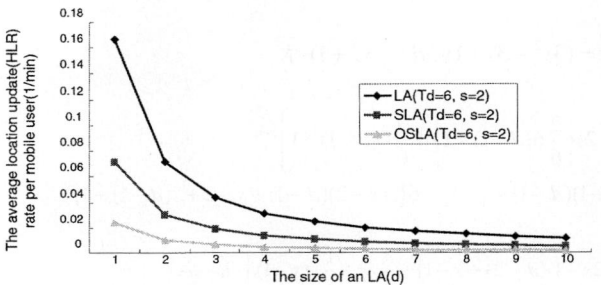

Fig. 9. Update Rate of HLR According to the Size of LA

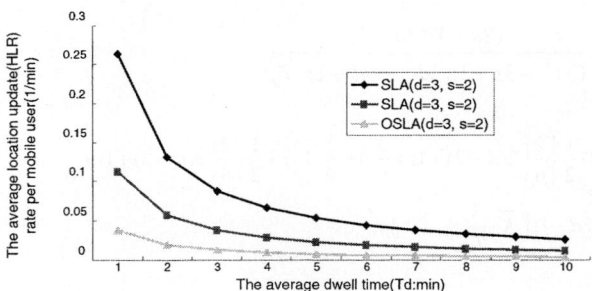

Fig. 10. Update Rate According to the Average Dwell Time

5 Conclusion

Since the location update process requires heavy signaling traffics, reducing the location update cost has been a critical research issue. We already have proposed to employ hierarchical structure in the cellular architecture in order to reduce the location update rate of the home location register. However, even in the proposed scheme, the boundary location areas still cause the frequent location updates. This paper proposed to apply overlapping location areas to further reduce the location update rates of the home location register. The paper also developed analytical models to evaluate the performance of the conventional scheme, the SLA scheme, and the proposed scheme in the paper.

The results from the performance evaluation show that the proposed scheme outperforms the SLA scheme which, in turn, outperforms the conventional scheme. The proposed approach will be viable considering the fact that the cost of communication channel is being increased while that of hardware is being decreased.

References

[1] N. E. Kruijt, D. Sparreboom, F. C. Schoute, and R. Prasad "Location Management Strategies for Cellular Mobile Networks," Electronics & Communication Engineering Journal, April 1998.
[2] Kumar, P.; Tassiulas, L. "Mobile multi-user ATM platforms: architectures, design issues, and challenges," *IEEE Network*, vol. 14, pp. 42-50. March-April 2000.
[3] Wang, K.; Liao, J.-M.; Chen, J.-M. "Intelligent location tracking strategy in PCS," Communications, *IEE Proceedings-* , vol. 147, pp. 63-68. Feb. 2000.
[4] S. Tabbane, "Location management methods for third-generation mobile systems," *IEEE Comm. Mag.*, pp. 72-84, August 1997.
[5] In-Hye Shin and Gyung-Leen Park, "On Employing Hierarchical Structure in PCS Networks ," *Lecture Notes in Computer Science*, vol 2, #2668, pp. 155-162, May 2003.
[6] D. Chung, H. Choo, and H. Y. Youn, "Reduction of Location Update Traffic Using Virtual Layer in [PCS," *Intl. Conf. On Parallel Processing*, pp. 331-338, Sept. 2001.
[7] D. Chung, H. Choo, H. Y. Youn, and J. K. Park, "Enhanced Virtual Layer Scheme for Mobility Management in PCS Networks," *IEEE Intl. Conf. on Parallel and Distributed Processing Techniques and Applications*, vol. 4, pp. 2218-2224, June 2001.
[8] T. P. Chu and S. S. Rappaport, "Overlapping Coverage with Reuse Partitioning in Cellular Communication Systems", *IEEE Trans. on Vehicular Technology*, vol.46, no.1, Feb. 1997.
[9] D. Gu and S.S. Rappaport, "Mobile user registration in cellular systems with overlapping location areas," Proc. VTC '99, pp.802-806, May 1999.
[10] C. -M. Weng and P. -W. Huang, "Modified group method for mobility management," Computer Communications 23, pp.115-122, 2000.
[11] L. –R. Hu and S. S. Rappaport, " An adaptive location management scheme for global personal communications," IEE Proc. Commun., vol. 144, no.1, pp.54-60, Feb. 1997.
[12] A. Bhattacharya and S.K. Das, "LeZi-Update: An Information-theoretic Approach to Track Mobile Users in PCS Networks," Proc. ACM/IEEE International Conference on Mobile Computing and Networking (MobiCom '99), pp.1-12, August 15- 19, 1999.
[13] In-Hye Shin and Gyung-Leen Park, "Performance Evaluation of SLA with overlapped LAs," Technical Report #4, Cheju National University.

On the Rila-Mitchell Security Protocols for Biometrics-Based Cardholder Authentication in Smartcards

Raphael C.-W. Phan[1] and Bok-Min Goi[2],*

[1] Information Security Research (iSECURES) Lab,
Swinburne Sarawak Institute of Technology,
93576 Kuching, Malaysia
rphan@swinburne.edu.my
[2] Multimedia University,
63100 Cyberjaya, Malaysia
bmgoi@mmu.edu.my

Abstract. We consider the security of the Rila-Mitchell security protocols recently proposed for biometrics-based smartcard systems. We first present a man-in-the-middle (MITM) attack on one of these protocols and hence show that it fails to achieve mutual authentication between the smartcard and smartcard reader. In particular, a hostile smartcard can trick the reader into believing that it is a legitimate card and vice versa. We also discuss security cautions that if not handled carefully would lead to attacks. We further suggest countermeasures to strengthen the protocols against our attacks, as well as to guard against the cautions highlighted. Our emphasis here is that seemingly secure protocols when implemented with poor choices of parameters would lead to attacks.

Keywords: Smartcards, biometrics, cardholder authentication, attacks.

1 Introduction

A protocol [3] is a set of rules that define how communication is to be done between two or more parties. In a common networked environment where the communication channel is open to eavesdropping and modifications, security is a critical issue. In this context, *security protocols* are cryptographic protocols that allow communicating parties to perform mutual authentication, key exchange or both. In [7], Rila and Mitchell proposed several security protocols intended for use with biometrics-based smartcard systems [6]. In this paper, we attack one of the protocols and show that it is insecure against man-in-the-middle (MITM) attacks, contrary to the designers' claims [7]. We also discuss

---

* The second author acknowledges the Malaysia IRPA grant (04-99-01-00003-EAR).

security cautions, namely how poor choices of security parameters would lead to attacks.

1.1 Standard Security Criteria

We describe standard security criteria expected of any security protocol:

Criterion 1: Mutual Authentication [7]. A smartcard reader must be assured that the smartcard inserted is a legitimate one, and vice versa.

Criterion 2: Resistance to Man-in-the-Middle (MITM) attacks [8]. An MITM attack is where an attacker places himself between two legitimate parties and can impersonate one or both of them. A security protocol should achieve this criterion else it entirely fails to achieve its standard objective of providing authentication between legitimate parties.

Criterion 3: Standard Collision Occurrence [8]. A collision in an n-bit value should only occur with the negligible probability of 2^{-n}.

In this paper, the above are the security criteria of interest to us since we will be showing in the ensuing sections situations where the Rila-Mitchell security protocols will fail to achieve them. The interested reader is further referred to [7] for details of other standard security criteria for protocols.

1.2 The Adversarial Model

The adversarial model used in our paper follows directly from the one considered by the designers themselves, Rila and Mitchell in [7]. They assumed that active attackers are allowed, namely those able to not only eavesdrop on communicated messages but also modify them to their liking. They also assumed that though an attacker can insert a hostile smartcard into a legitimate smartcard reader and also use a hostile smartcard reader to read legitimate smartcards, they claimed that such instances would be unsuccessful since their protocols are supposed to detect such violations. We later show in Section 3 that Protocol 3 does not fulfill this.

As is common with any security protocol, the following are assumed: An adversary could be an insider, i.e., a legitimate party in the network and who can initiate protocol sessions, introduce new messages, receive protocol messages from other parties intended for itself, etc. Note further that encryption only provides confidentiality but not integrity, meaning that though an attacker does not know the secret key used for encrypting any message parts, he could still replay previously valid encrypted parts.

We review in Section 2 the security protocols of [7]. In Section 3, we present our MITM attack. In Section 4, we discuss security cautions for the protocols and how these may cause attacks. We also suggest countermeasures to strengthen the protocols. We conclude in Section 5.

2 Rila-Mitchell Security Protocols

The notations used throughout this paper as follows:

C	The smartcard
R	The smartcard reader
N_A	The nonce generated by entity A (which may be C or R)
$BioData$	The captured fingerprint (biometric) image
EF	The extracted features from $BioData$
$\|$	Concatenation
$m_K(\cdot)$	A MAC function keyed by secret key, K
$LSB_i(x)$	The i least significant bits (rightmost) bits of x
$MSB_i(x)$	The i most significant bits (leftmost) bits of x
$x << y$	Cyclic shift (rotate) of x left by y bits
$x >> y$	Cyclic shift (rotate) of x right by y bits

Rila and Mitchell proposed in Section 3 of [7] two similar protocols that they admitted were secure only against passive attacks [3, 9], i.e. an attacker is unable to modify existing messages or create new messages. We remark that such an assumption is very impractical by today's standards because the communication link between the smartcard and the reader is commonly accessible by the public. This is true due since a smartcard could be used in various situations, and smartcard readers owned by diverse individuals. Ensuring that an attacker can only mount passive replay attacks is hence not feasible at all.

Therefore, henceforth we concentrate on Rila and Mitchell's suggestion in Section 4 of their paper [7], that their two protocols can be secured against active attacks by replacing the internal hash function, h with a message authentication code (MAC), m_K. We strongly feel that the MAC variants of the protocols are more practical than their hash function counterparts. In addition to these two, Rila and Mitchell also proposed a protocol to allow the reader to verify that the card inserted is a legitimate one. For lack of better names, we denote the three protocols in Section 4 of [7] as Protocols 1, 2 and 3.

Protocol 1 (Using Nonces and BioData)
Message 1: $R \to C$ N_R
Message 2: $C \to R$ $N_C \| BioData \| m_K(N_C \| N_R \| BioData)$
Message 3: $R \to C$ $EF \| m_K(N_R \| N_C \| EF)$

In the first step, the smartcard reader, R generates a random number, N_R and sends it to the smartcard, C. Then C using its built-in fingerprint sensor captures the fingerprint image, $BioData$, and generates a random number, N_C, and sends both these along with a MACed value of $N_C \| N_R \| BioData$ as message 2 to R. Next, R re-computes the MAC and verifies that it is correct. It then extracts the features, EF of $BioData$ and uses this to form message 3, along with a MAC of $N_R \| N_C \| EF$ to C. Then C re-computes the MAC and verifies that it is correct.

Protocol 2 (Using BioData as a Nonce)
  Message 1: $R \to C$ N_R
  Message 2: $C \to R$ $BioData \| m_K(BioData \| NR)$
  Message 3: $R \to C$ $EF \| m_K(EF \| m_K(BioData \| N_R))$

Protocol 2 is very similar to protocol 1 except instead of generating its own random number, C uses the captured fingerprint image, $BioData$ as a random number. Rila and Mitchell note that this relies on the assumption that two different measurements of the same biometric feature of the same person are very likely to be different [7]. Further, to assure that the smartcard has not been inserted into a hostile card reader and vice versa, Rila and Mitchell proposed a separate authentication protocol, as follows:

Protocol 3 (Using Nonces only, without BioData)
  Message 1: $R \to C$ N_R
  Message 2: $C \to R$ $N_C \| m_K(N_C \| N_R)$
  Message 3: $R \to C$ $m_K(N_R \| N_C)$

R generates a random number N_R, and sends it as message 1 to C. Then C generates a random number, N_C and sends this along with a MACed value of $N_C \| N_R$ as message 2 to R. Next, R re-computes this MAC and verifies its correctness. It generates a MAC of $N_R \| N_C$ which is sent as message 3 to C. Finally, C re-computes this MAC and verifies its correctness.

3 A Man-in-the-Middle (MITM) Attack on Protocol 3

We present a man-in-the-middle (MITM) attack on Rila and Mitchell's Protocol 3, showing that a smartcard reader can be bluffed by an inserted hostile smartcard into thinking it is legitimate, and vice versa. This disproves their claim in [7] that with this protocol the card reader can verify that the card inserted is a legitimate one.

An attacker places himself between a valid card, C and a valid reader, R. He puts C into a hostile cloned reader, R', and inserts a hostile smartcard, C' into R.

  $\alpha.1 : R \to C'$ N_R
  $\beta.1 : R' \to C$ N_R
  $\beta.2 : C \to R'$ $N_C \| m_K(N_C \| N_R)$
  $\alpha.2 : C' \to R$ $N_C \| m_K(N_C \| N_R)$
  $\alpha.3 : R \to C'$ $m_K(N_R \| N_C)$
  $\beta.3 : R' \to C$ $m_K(N_R \| N_C)$

Once C' is inserted into R, R generates a random number, N_R and issues it as message $\alpha.1$. This is captured by C' who immediately forwards it to R'. R' replays $\alpha.1$ as message $\beta.1$ to the valid card, C, which returns the message $\beta.2$. This is captured by R' and forwarded to C' which replays it as message $\alpha.2$ to R. R responds with message $\alpha.3$ to C', thereby the hostile card, C' is fully authenticated to the legitimate reader, R. C' forwards this message to R', which replays it as message $\beta.3$ to C, and the hostile reader, R' is authenticated to the legitimate card, C.

This MITM attack resembles the Grand Chessmaster problem [8] and Mafia fraud [2] that can be applied on identification schemes. One may argue that this is a passive attack and does not really interfere in any way since the protocol would appear to be the same whether the attacker is present or not. However, the essence of this attack is that both the legitimate card and reader need not even be present at the same place, but what suffices is that the MITM attack leads them to believe the other party is present. The hostile card and reader would suffice to be in the stead of their legitimate counterparts. This is a failure of mutual authentication between the legitimate card and reader, which should both be present in one place for successful mutual authentication. Thus, Protocol 3 fails to achieve criteria 1 and 2 outlined in Section 1.1.

4 Further Security Cautions and Countermeasures

We discuss further cautions on practically deploying the Rila-Mitchell security protocols. In particular, we show that when specifications are not made explicit, the resultant poor choices of such specifications during implementations may cause the protocols to fail criterion 3 of standard collision resistance, further leading to attacks that cause a failure of mutual authentication (criterion 1).

4.1 Collisions and Attacks on Protocol 1

We first present two attacks on Protocols 1 in this subsection, while attacks on Protocols 2 and 3 will be described in the next subsection.

Collision Attack 1. Let N_R be the random number generated by R in a previous protocol session, and n denotes its size in bits. Further, denote N_R' as the random number generated by R in the current session. Then for the case when the following two conditions are met:

$$N_R' = LSB_{n-r}(N_R), \qquad (1)$$
$$N_C' = N_C \| MSB_r(N_R) \qquad (2)$$

for $r \in \{0, 1, ..., n-1\}$, then the same MAC value and hence a collision would be obtained. This collision is formalized as:

$$m_K(N_C \| N_R \| BioData) = m_K(N_C' \| N_R' \| BioData). \qquad (3)$$

There are n possible cases for the above generalized collision phenomenon. Let m be size of $N_C \| N_R$ in bits. Then, the probability that the collision in (3) occurs is increased from $\frac{1}{2^m}$ to $\frac{n}{2^m}$. This is clearly an undesirable property since a securely used m-bit value should only have collisions with probability $\frac{1}{2^m}$. Under such cases, Protocol 1 would fail criterion 3.

We describe how this can be exploited in an attack. Let α be the previous run of the protocol, and β the current run. The attack proceeds:

$\alpha.1 : R \rightarrow C \quad N_R$
$\alpha.2 : C \rightarrow R \quad N_C \| BioData \| m_K(N_C \| N_R \| BioData)$
$\alpha.3 : R \rightarrow C \quad EF \| m_K(N_R \| N_C \| EF)$
$\beta.1 : R \rightarrow I_C \quad N'_R$
$\beta.2 : I_C \rightarrow R \quad N'_C \| BioData \| m_K(N_C \| N_R \| BioData)$
$\beta.3 : R \rightarrow I_C \quad EF \| m_K(N'_R \| N'_C \| EF)$

An attacker, I has listened in on a previous protocol run, α and hence has captured all the messages in that run. Now, he inserts a hostile smartcard into the reader, R and so initiates a new protocol run, β which starts with the reader, R generating and sending a random number, N'_R to the card, I_C. The attacker's card checks N'_R to see if it satisfies (4). If so, it chooses its own random number, N'_C to satisfy condition (5), and also replays the previously captured fingerprint image, $BioData$ as well as the previously captured MAC, $m_K(N_C \| N_R \| BioData)$ in order to form the message $\beta.2$. When the reader, R receives this message, it would re-compute the MAC, $m_K(N'_C \| N'_R \| BioData)$ and indeed this will be the same value as the received MAC in message $\beta.2$. It therefore accepts the hostile smartcard as valid and fully authenticated. Protocol 1 therefore fails in such circumstances to achieve criterion 1 of mutual authentication (see Section 1.1).

Collision Attack 2. Consider now the case when:

$$MSB_r(N_R) = MSB_r(BioData), \quad (4)$$
$$N'_R = N_R << r, \quad (5)$$
$$N'_C = N_C \| MSB_r(NR). \quad (6)$$

Then:

$$BioData' = LSB_{b-r}(BioData) \quad (7)$$

where b denotes the size of $BioData$ in bits. Let m be the size of $N_C \| N_R$, then since there are n cases of the above generalized collisions, the probability of such collisions occurring is $\frac{n}{2^{m+2r}}$ instead of the expected $\frac{1}{2^{m+2r}}$ as in the case of an $(m + 2r)$-bit value in equations (4) to (7).

We can do better than that and discard the restriction in (4). When:

$$N'_R = N_R >> r, \quad (8)$$
$$N_C = N'_C \| LSB_r(N_R) \quad (9)$$

are met, then:

$$BioData' = LSB_r(N_R) \| BioData. \quad (10)$$

The above generalization occurs with probability $\frac{n}{2^m}$ instead of $\frac{1}{2^m}$.

By exploiting either of the above generalizations which occur with a resultant probability of $\frac{n}{2^{m+2r}} + \frac{n}{2^m} \approx \frac{n}{2^{m-1}}$, our attack then proceeds similarly as

Attack 1. The steps in the initial four messages are the same. Then, prior to constructing the message $\beta.2$, the hostile smartcard checks N'_R to see if it satisfies (5) or (8). If so, it chooses its own random number, N'_C to satisfy condition (6) or (9) respectively, and also chooses the new fingerprint image, $BioData'$ as according to the condition (7) or (10) respectively. To complete message $\beta.2$, it replays the previously captured MAC, $m_K(N_C\|N_R\|BioData)$. When the reader, R receives this message, it would re-compute the MAC, $m_K(N'_C\|N'_R\|BioData')$ and indeed this will be the same value as the received MAC in message $\beta.2$. It therefore accepts the hostile smartcard as valid and fully authenticated. Protocol 1 therefore fails in such cases to provide authentication and allows a hostile smartcard to pass off with a fake $BioData$.

4.2 Collision and Attack on Protocol 2

Protocol 2 relies on the assumption that the fingerprint image, $BioData$ captured from the same person is random enough. However, the authors admitted that the difference between every two fingerprint captures would be small. We remark however that although $BioData$ on its own may be unique enough, the concatenation of $BioData$ and N_R may not be, i.e. if:

$$BioData' = MSB_r(BioData), \qquad (11)$$
$$N'_R = LSB_{b-r}(BioData)\|N_R, \qquad (12)$$

then the same MAC and hence a collision would result! This is given as:

$$m_K(BioData\|N_R) = m_K(BioData'\|N'_R). \qquad (13)$$

There are b possible such cases of collisions so the probability of this is $\frac{b}{2^m}$ instead of an expected $\frac{1}{2^m}$. Another scenario for this is when:

$$BioData' = BioData\|MSB_r(N_R), \qquad (14)$$
$$N'_R = LSB_{b-r}(N_R), \qquad (15)$$

then again a collision as in (13) results. There are similarly b possible such cases of collisions and hence the same probability of occurrence. The resultant probability for the two collision scenarios above is $\frac{b}{2^{m-1}}$. Protocol 2 therefore fails to achieve criterion 3. Our attack follows:

$\alpha.1 : R \to C \quad N_R$
$\alpha.2 : C \to R \quad BioData\|m_K(BioData\|N_R)$
$\alpha.3 : R \to C \quad EF\|m_K(EF\|m_K(BioData\|N_R))$
$\beta.1 : R \to I_C \quad N'_R$
$\beta.2 : I_C \to R \quad BioData'\|m_K(BioData\|N_R)$
$\beta.3 : R \to I_C \quad EF'|m_K(EF'\|m_K(BioData'\|N'_R))$

The steps in the first 4 messages are similar to the attacks in Section 3.1. Prior to constructing the message $\beta.2$, the hostile smartcard checks N'_R to see if it satisfies (12) or (15). If so, it chooses its fingerprint image, $BioData'$ to satisfy condition (11) or (14) respectively, and replays the previously captured MAC, $m_K(BioData\|N_R)$ in order to completely form the message $\beta.2$. When the reader, R receives this message, it would re-compute the MAC, $m_K(BioData'\|N_R')$ and indeed this will be the same value as the received MAC in message $\beta.2$. It therefore accepts the hostile smartcard as valid and fully authenticated. In this case, Protocol 2 therefore fails to achieve criterion 1 of mutual authentication, but instead allows a hostile smartcard to pass off with a fake $BioData'$.

4.3 Collision and Attack on Protocol 3

Protocol 3 is claimed in [7] to assure the smartcard that it has not been inserted into a hostile card reader, and vice versa. However, we have disproved this claim by mounting an MITM attack in Section 3. Here, we will further show how collisions occurring in Protocol 3 would allow for an additional attack to be mounted on this protocol.

Collision Attack. The attacker, who inserts a hostile smartcard into a valid reader, waits until the collision occurs:

$$N_R' = N_C. \tag{16}$$

He then chooses:

$$N_C' = N_R. \tag{17}$$

This allows him to replay previously captured MACs, as follows:

$\alpha.1 : R \to C \quad N_R$
$\alpha.2 : C \to R \quad N_C\|m_K(N_C\|N_R)$
$\alpha.3 : R \to C \quad m_K(N_R\|N_C)$
$\beta.1 : R \to I_C \quad N_R'$
$\beta.2 : I_C \to R \quad N_C'\|m_K(N_R\|N_C)$
$\beta.3 : R \to I_C \quad m_K(N_R'\|N_C')$

Here, a valid previous protocol run, α, whose messages are captured by the attacker. He then monitors every new message $\beta.1$ until the collision in (16) occurs, upon which he immediately chooses N_C' to satisfy condition (17). This he uses together with a replay of $m_K(N_R\|N_C)$ to form message $\beta.2$ to R. This MAC will be accepted as valid by R, who then returns with message $\beta.3$.

Further Cautions. We conclude by stating two other cautions regarding Protocol 3. Firstly, the case when $N_R = N_C$, then the MAC in message 2 can be replayed as message 3! Secondly, note that any attacker can use C as an oracle to generate $m_K(N_C\|x)$, where x can be any bit sequence sent to C as message 1! To do so, one would merely need to intercept the original message 1 from R to C, and replace it with x. Such an exploitation is desirable in some cases to mount knownz– or chosen−plaintext attacks [3] which are applicable to almost all cryptographic primitives such as block ciphers, stream ciphers, hash functions or MACs.

4.4 Countermeasures

The concerns we raised on Protocols 1 and 2 are due to their designers not fixing the length of the random numbers, N_R and N_C but left it as a flexibility of the protocol implementer. We stress that such inexplicitness can result in subtle attacks on security protocols [1, 4, 5, 10].

We also recommend to encrypt and hence keep confidential the sensitive information such as *BioData* and *EF* rather than transmitting them in the clear! This prevents them from being misused not only in the current system but elsewhere as such information would suffice to identify an individual in most situations. This improvement also makes the protocols more resistant to attacks of the sort that we have presented.

Figure 2 in the original Rila and Mitchell paper [7] also shows that the Yes/No decision signal to the Application component of the smartcard system is accessible externally and not confined within the card. This therefore implies that it will always be possible for an attacker to replay a Yes signal to the Application component regardless of whether an attacker is attacking the rest of the system. We would recommend that for better security, this Yes/No signal as well as the application component should be within the tamper-proof card, and not as otherwise indicated in Figure 2 of [7]. However, such a requirement poses additional implementation restrictions, especially when the application is access-control based, for instance to control access to some premises.

5 Concluding Remarks

Our attack on Rila and Mitchell's Protocol 3 shows that it fails to achieve the claim of allowing the reader to verify that an inserted card is legitimate and vice versa. Our cautions in Section 4 further serve as a general reminder to protocol designers and implementers that all underlying assumptions and potential security shortcomings should be made explicitly clear in the protocol specification, as has also been shown and reminded in [1]. We have also suggested some countermeasures to strengthen the protocols against such problems. However, our suggestions are not entirely exhaustive, hence further analysis needs to be conducted on them.

It would be interesting to consider how to secure the entire smartcard system to encompass the reader as well, since current systems only assure that the card is tamper-resistant while the reader is left vulnerable to tampering, and the communication line between the smartcard and the reader can be eavesdropped easily.

References

1. M. Abadi. Explicit Communication Revisited: Two New Attacks on Authentication Protocols. *IEEE Transactions on Software Engineering*, vol. 23, no. 3, pp. 185-186, 1997.
2. Y. Desmedt, C. Goutier, S. Bengio. Special Uses and Abuses of the Fiat-Shamir Passport Protocol. In *Proceedings of Crypto '87*, LNCS, vol. 293, Springer-Verlag, pp. 21-39, 1988.

3. N. Ferguson, B. Schneier. *Practical Cryptography*. Wiley Publishing, Indiana, 2003.
4. ISO/IEC. *Information Technology - Security Techniques (Entity Authentication Mechanisms Part 2: Entity authentication using symmetric techniques*, 1993.
5. G. Lowe. An attack on the Needham-Schroeder public-key protocol. *Information Processing Letters*, vol. 56, pp. 131-133, 1995.
6. L. Rila, C.J. Mitchell. Security Analysis of Smartcard to Card Reader Communications for Biometric Cardholder Authentication. *5th Smart Card Research and Advanced Application Conference (CARDIS '02)*, USENIX, pp. 19-28, 2002.
7. L. Rila, C.J. Mitchell. Security Protocols for Biometrics-Based Cardholder Authentication in Smartcards. *Applied Cryptography and Network Security (ACNS '03)*, LNCS, vol. 2846, Springer-Verlag, pp. 254-264, 2003.
8. B. Schneier. *Applied Cryptography: Protocols, Algorithms, and Source Code in C*, 2nd edn, John Wiley & Sons, New York, 1996.
9. D.R. Stinson. *Cryptography: Theory and Practice*, 2nd edn, Chapman & Hall/CRC, Florida, 2002.
10. P. Syverson. A Taxonomy of Replay Attacks. *7th IEEE Computer Security Foundations Workshop*, pp. 131-136, 1994.

An Efficient Dynamic Group Key Agreement for Low-Power Mobile Devices*

Seokhyang Cho, Junghyun Nam, Seungjoo Kim, and Dongho Won

School of Information and Communication Engineering,
Sungkyunkwan University, 300 Chunchun-dong, Jangan-gu, Suwon,
Gyeonggi-do 440-746, Korea
{shcho, jhnam}@dosan.skku.ac.kr, skim@ece.skku.ac.kr,
dhwon@simsan.skku.ac.kr

Abstract. Group key agreement protocols are designed to provide a group of parties securely communicating over a public network with a session key. The mobile computing architecture is asymmetric in the sense of computational capabilities of participants. That is, the protocol participants consist of the stationary server (application servers) with sufficient computational power and a cluster of mobile devices (clients) with limited computational resources. It is desirable to minimize the amount of computation performed by each group member in a group involving low-power mobile devices such as smart cards or personal digital assistants (PDAs). Furthermore, we are required to update the group key with low computational costs when the members need to be excluded from the group or multiple new members need to be brought into an existing group. In this paper, we propose a dynamic group key protocol that offers computational efficiency to the clients with low-power mobile devices. We compare the total communicative and computational costs of our protocol with others and prove its security against a passive adversary in the random oracle model.

Keywords: Group key agreement, mobile devices, multicast, CDH assumption.

1 Introduction

The basic requirement for secure group communications over insecure public channels is that all group members must agree on a common secret key. This shared secret key, called the *session key*, can later be used to facilitate standard security services, such as authentication, confidentiality, and data integrity. Group key agreement protocols are designed to meet this requirement, with the fundamental security goal being to establish the session key in such a way that no one except the group members can know the value of the session key.

---

* This work was supported by the University IT Research Center Project funded by the Korean Ministry of Information and Communication.

In key agreement protocols, more than one party contribute information to generate the common session key. In this paper we focus on *contributory* key agreement protocols in which the session key is derived as a function of contributions provided by all parties [1]. Therefore in our contributory key agreement protocols, a correctly behaving party is assured that as long as his contribution is chosen at random, even a coalition of all other parties will not be able to have any means of controlling the final value of the session key.

The mobile computing architecture we visualize is asymmetric in the sense of computational capabilities of participants. That is, the protocol participants consist of a stationary server (also called *application server* or *service provider*) with sufficient computational power and a cluster of mobile devices (also called *clients*) with limited computational resources. An asymmetric mobile environment is common in a number of applications such as Internet stock quotes, audio and music delivery, and so on [11].

In this paper, we propose a dynamic group key protocol that offers computational efficiency to the clients with low-power mobile devices. Our protocol combines the idea of E. Bresson *et al.* [6] with J. Nam *et al.*'s protocol [14]. The proposed protocol is suited for dynamic groups in which group members may join and leave the current group at any given time. Our protocol also achieves forward secrecy and is provably secure against a passive adversary under the computational Diffie-Hellman assumption. But the protocol in [6] does not achieve forward secrecy. Moreover, our protocol lowers computational costs by using a hash function and an exclusive OR (XOR) operation instead of multiplication in the protocol of [14].

Related Work. Ever since 2-party Diffie-Hellman key exchange was first proposed in 1976, a number of works [1, 4, 7, 8, 9, 10, 12, 13, 15, 16] have attempted to solve the fundamental problem of securely distributing a session key among a group of n parties. But unfortunately, all of them suffer from one or more of the drawbacks as $O(n)$ or $O(\log n)$ rounds of communication, $O(n)$ broadcasts per round, and lack of forward secrecy. In fact, most published protocols require $O(n)$ communication rounds to establish a session key, and hence become prohibitively expensive as the group size grows. Other protocols [4, 16], while they require only a constant number of rounds to complete key agreement, do not achieve forward secrecy.

In [10], Burmester and Desmedt presented a two-round protocol which provides forward secrecy but no proof of security in the original paper. Recently Katz and Yung [13] proposed a three-round protocol which provides a rigorous security proof against an active adversary in the standard model. However, an obvious drawback of this protocol is that communication overhead is significant with three rounds of n broadcasts. This means that each user in this protocol, in each of three rounds, must receive $n-1$ messages from the rest of the group before he/she can proceed to the next step. It is obvious that this kind of extreme connectivity inevitably delays the whole process of the protocol.

The initial work [9] proposed by Bresson *et al.* deals with the static case, and shows a protocol which is secure under the DDH assumption. Later works [7, 8] focus on the dynamic group key agreement to support membership changes that users join or leave and the session key must be updated whenever it occurs. More recently, Bresson and Catalano proposed a constant round key exchange protocol, based on secret sharing techniques that combines with ElGamal cryptosystem as underlying encryption primitive [5]. However, with increasing number of users, the complexity of the protocol goes beyond the capabilities of a low-power mobile device.

Our Contribution. Our group key agreement protocol is provably secure against a passive adversary. We provide a rigorous proof of security under the well-known Computational Diffie-Hellman (CDH) assumption in a formal security model [9]. In addition, in contrast to other asymmetric protocols [4,6], our protocol also provides perfect forward secrecy; i.e., disclosure of long-term secret keys does not compromise the security of previously established session keys.

Despite meeting all these strong notions of security, our construction is surprisingly simple and provides a practical solution for group key agreement in the asymmetric mobile environment. In a protocol execution involving mobile devices as participants, a bottleneck arises when the number of public-key cryptography operations to be performed by a mobile device increases substantially as the group size grows. It is therefore of prime importance for a group key agreement protocol to assign a low, fixed amount of computations to its mobile participants. To this end our protocol shifts much of the computational burden to the application server equipped with sufficient computational power. By allowing this computational asymmetry among protocol participants, our protocol reduces the computational cost of a mobile device to two modular exponentiations (plus one signature generation and verification) without respect to the number of participants.

2 The Protocol

Let p, q be two primes such that $p = 2q+1$. And let g be a generator of any cyclic group $\mathbb{G}$ of order q. By the notation U_n, we denote a special user called *server* whose role will become apparent in the description of the protocol. In the setup phase, any trusted party chooses $\mathbb{G}$ and g as defined above. The public parameters $\mathbb{G}$ and g are assumed to be known a priori to all parties. We also assume that each user knows the authentic public keys of all other users. Let $H : \{0,1\}^* \rightarrow \{0,1\}^\ell$ be a hash function modelled as a random oracle [3] in our security proof, where ℓ is the length of the session key to be distributed in the protocols.

We now present a dynamic key agreement scheme consisting of three protocols P_{ika}, P_{leave}, and P_{join} for initial group key establishment, user leave, and user join, respectively. First, the protocol P_{ika} proceeds as follows:

2.1 Initial Key Agreement: Protocol P_{ika}

Let $\mathcal{U} = \{U_1, U_2, \cdots, U_n\}$ be a set of n users who wish to generate a session key by participating in our group key agreement protocol P_{ika}. Then P_{ika} runs in two rounds, one with $n-1$ unicasts and the other with a single broadcast, as follows:

Step 1. Each U_i ($i \in [1,n]$) selects a random $r_i \in [1, q-1]$ and *precomputes* $z_i = g^{r_i} \bmod p$. Then each client U_i ($i \in [1, n-1]$) signs z_i to obtain signature σ_i such that $\sigma_i = Sign_{SK_i}(z_i)$. After that $U_i \neq U_n$ sends a message $m_i = (z_i, \sigma_i)$ to the server U_n.

Step 2. Upon receipt of each message, U_n verifies the correctness of the signature σ_i using the public key PK_i of U_i. The server U_n chooses a random $r \in [1, q-1]$ and computes $z(=g^r)$, $x_i = z_i^r \bmod p$. Then the server generates a nonce $\delta \in \{0,1\}^\ell$ where ℓ is a security parameter. After having received all the $n-1$ messages from the rest of the users, U_n computes X that

$$X = \bigoplus_{i \in [1,n]} H(\delta \parallel x_i)$$

where $x_i = z_i^r$. U_n also computes $Y = \{X_i \mid X_i = X \oplus H(\delta \parallel x_i), i \in [1, n-1]\}$ and then generates signature σ_n of message $\delta \parallel z \parallel Y \parallel \mathcal{U}$ using the secret key SK_n. Now U_n broadcasts the message $m_n = (\delta, z, Y, \mathcal{U}, \sigma_n)$ to the entire client group members. Lastly, U_n computes its session key K as $K = H(X, Y)$.

Step 3. Having received the broadcast message from U_n, each $U_i \neq U_n$ first verifies the correctness of the server's signature, and then computes

$$X = X_i \oplus H(\delta \parallel x_i)$$

where $x_i = z^{r_i}$ and its session key K as $K = H(X, Y)$.

2.2 User Leave: Protocol P_{leave}

Assume a scenario where a set of users $\mathcal{L}$ leaves the group $\mathcal{U}$ except for the server U_n. Then protocol P_{leave} is executed to provide each user of the new group $\mathcal{U} = \mathcal{U} \setminus \mathcal{L}$ with a new session key. Protocol P_{leave} requires only one communication round with a single broadcast and it proceeds as follows:

Step 1. The server U_n generates a new nonce $\delta_1 \in \{0,1\}^\ell$ and computes

$$X' = \bigoplus_{i \in \mathcal{U}} H(\delta_1 \parallel x_i).$$

And U_n also computes $Y' = \{X'_i \mid X'_i = X' \oplus H(\delta_1 \parallel x_i), i \in \mathcal{U} \setminus \{U_n\}\}$. Then U_n generates signature σ'_n of message $\delta_1 \parallel z \parallel Y' \parallel \mathcal{U}$ using SK_n. Now U_n broadcasts the message $m'_n = (\delta_1, z, Y', \mathcal{U}, \sigma'_n)$ to the entire client group members.

Step 2. Upon receiving the broadcast message m'_n, each U_i computes

$$X' = X'_i \oplus H(\delta_1 \parallel x_i)$$

Lastly, each user $U_i (i \in \mathcal{U})$ computes its session key K as $K = H(X', Y')$.

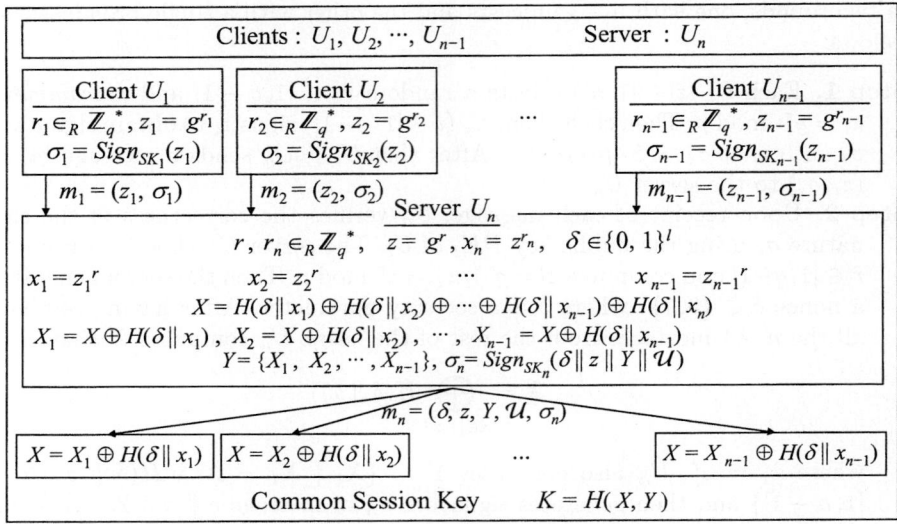

Fig. 1. Initial Key Agreement Protocol (P_{ika})

2.3 User Join: Protocol P_{join}

Assume a scenario in which a set of j new users, $\mathcal{J}$, joins the current group $\mathcal{U}$ to form a new group $\mathcal{U} = \mathcal{U} \cup \mathcal{J}$. Then the join protocol P_{join} is run to provide the users of $\mathcal{U}$ with a session key. P_{join} takes two communication rounds, one with j unicasts and the other with a single broadcast, and it proceeds as follows:

Step 1. Each $U_i \in \mathcal{J}$ selects a random $r_i \in [1, q-1]$ and computes $z_i = g^{r_i}$ mod p. $U_i \in \mathcal{J}$ then generates signature σ_i of $U_i \parallel z_i$, sends $m_i = (U_i, z_i, \sigma_i)$, and stores its random r_i.

Step 2. U_n proceeds in the usual way, generating a new random nonce $\delta_2 \in \{0,1\}^\ell$, computing X'', Y'' and $K = H(X'', Y'')$, updating the new z_i's. Then having received all the j messages from the new users, U_n computes X'' as

$$X'' = \bigoplus_{i \in \mathcal{U}} H(\delta_2 \parallel x_i)$$

where $x_i = z_i^r$. U_n also computes $Y'' = \{X_i'' \mid X_i'' = X'' \oplus H(\delta_2 \parallel x_i), i \in \mathcal{U}\setminus\{U_n\}\}$, and then generates signature σ_n'' of message $\delta_2 \parallel z \parallel Y'' \parallel \mathcal{U}$ using SK_n. Now U_n broadcasts the message $m_n'' = (\delta_2, z, Y'', \mathcal{U}, \sigma_n'')$ to the entire client group members. Lastly, U_n computes its session key K as $K = H(X'', Y'')$.

Step 3. Having received the broadcast message from U_n, each $U_i \neq U_n$ first verifies the correctness of the server's signature, and then computes

$$X'' = X_i'' \oplus H(\delta_2 \parallel x_i)$$

where $x_i = z^{r_i}$ and its session key K as $K = H(X'', Y'')$.

3 Efficiency

To analyze the communication complexity and computation cost, we now discuss the efficiency of the protocol introduced in the preceding section.

Communication Complexity. It is easy to see that our P_{ika} protocol runs only in two rounds of communication, requiring $n-1$ *unicasts* in the first step and a single broadcast in the second. Hence the total number of messages required by our P_{ika} protocol is n, which is optimal as shown in [2].

In contrast, the two-round protocol presented by Burmester and Desmedt [10] requires n broadcasts in each of two rounds, and therefore requires, in total, $2n$ *broadcast messages* to complete key agreement (as already mentioned, the protocol presented by Katz and Yung [13], in its basic form, is essentially the same as the BD protocol). More seriously, without the ability of broadcasting communication, this protocol requires $O(n^2)$ messages to be sent or received which turn out to be this protocol inefficient for many applications.

The protocol IKA.2 proposed by Steiner *et al.* [15] requires $2n-3$ unicasts and two rounds of broadcast in the initial key agreement. One notable drawback of IKA.2 is that $n-1$ unicasts messages are sent to the server U_n in the nth round. This might lead to congestion at U_n.

Computational Complexity. In addition to the cost of generating and verifying one signature, each U_i in our P_{ika} protocol computes two modular exponentiations in a group $\mathbb{G}$, except U_n who generates one signature, verifies $n-1$

Table 1. Complexity Comparison

Protocol	Complexity	Steiner *et al*'s IKA.2	Our protocol
P_{ika}	Rounds	$n^{1)}+1$	2
	Unicast	$2n-3$	$n-1$
	Broadcast	2	1
	Average computational cost (EXPs[2)])	3 ($O(n)$ for U_n)	2 ($O(n)$ for U_n)
P_{leave}	Rounds	1	1
	Unicast	0	0
	Broadcast	1	1
	Average computational cost (EXPs)	1 ($O(n)$ for U_n)	0
P_{join}	Rounds	$j^{\,3)}+1$	2
	Unicast	j	j
	Broadcast	1	1
	Average computational cost (EXPs)	1 or 2 ($O(n)$ for U_n)	0 or 2 ($O(j)$ for U_n)

1) The number of users in a newly updated group
2) Modular exponentiations in the group $\mathbb{G}$
3) The number of joining members in the group

signatures, and performs $n+1$ modular exponentiations. On the other hand, each client in the Steiner et al's initial key agreement protocol computes 3 modular exponentiations. Furthermore each client in the protocol IKA.2 newly has to compute one exponentiation except for the server to require $O(n)$ exponentiations when the members leave or join the current group. But, all users in our P_{leave} protocol do not need exponential computations by using the stored values and only the newly members in our P_{join} protocol compute two modular exponentiations.

In Table 1, we have compared the complexity of our protocol to that of the Steiner et al's protocol. As seen from the table, our protocol considerably outperforms the Steiner et al's IKA.2 protocol as an only two-round protocol, currently achieving forward secrecy.

4 Security Analysis

We now claim that the group key agreement protocol proposed in this paper is secure against passive adversaries provided that computational Diffie-Hellman (CDH) problem is hard.

Definition 1 (Computational Diffie-Hellman (CDH) Problem). Let $\mathbb{G}$ be a cyclic group $\langle g \rangle$ of prime order q and a, b are chosen at random in $\mathbb{Z}_q^*$. A (T, ϵ)-CDH-attacker in $\mathbb{G}$ is a probabilistic algorithm running in time T that given (g, g^a, g^b), outputs g^{ab} with probability at least ϵ. The advantage of any probabilistic, polynomial time algorithm $\mathcal{A}$ in solving CDH problem in $\mathbb{G}$ is defined to be :

$$\mathsf{Adv}^{\mathrm{CDH}}_{\mathcal{A},\mathbb{G}} = |\Pr[g^{ab} \leftarrow \mathcal{A}(\mathbb{G}, g, g^a, g^b) \mid g \in \mathbb{G}; a, b \in_R \mathbb{Z}_q^*]|$$

The CDH problem is (T, ϵ)-**intractable** if there is no (T, ϵ)-attacker in $\mathbb{G}$.

Definition 2 (Authenticated Group Key Agreement). The security of an authenticated group key agreement scheme P is defined in the following context. The adversary executes a protocol P_{ika}, P_{leave}, or P_{join} as many times as he/she wishes in an arbitrary order with P_{ika} being the first one executed. During executions of the protocols, the adversary $\mathcal{A}$, at any time, asks Test query to a fresh user, gets back an ℓ-bit string as the response to this query, and at some later point in time, outputs a bit b' as a guess for the secret bit b. Let CG (Correct Guess) be the event that the adversary $\mathcal{A}$ correctly guesses the bit b, i.e., the event that $b' = b$. Then we define the advantage of $\mathcal{A}$ in attacking P as

$$\mathsf{Adv}_{\mathcal{A},P}(k) = 2 \cdot \Pr[\mathsf{CG}] - 1$$

We say that a group key agreement scheme P is secure if $\mathsf{Adv}_{\mathcal{A},P}(k)$ is negligible for any probabilistic polynomial time adversary $\mathcal{A}$.

Theorem 1. Let $\mathsf{Adv}_P(t, q_h, q_{ex})$ be the maximum advantage in attacking P, where the maximum is over all adversaries that run in time t, and make q_h random oracle queries and q_{ex} execute queries. Then we have

$$\mathrm{Adv}_P(t, q_h, q_{ex}) \leq 2q_h q_{ex} \cdot \mathrm{Adv}_\mathbb{G}^{\mathrm{CDH}}(t'),$$

where $t' = t + O(nq_{ex}t_{exp})$ and t_{exp} is the time required to compute a modular exponentiation in $\mathbb{G}$.

Proof. Assume that an adversary $\mathcal{A}$ can guess the hidden bit b correctly with probability $1/2 + \epsilon$. Then we construct from $\mathcal{A}$ an algorithm that solves the CDH problem in $\mathbb{G}$ with probability ϵ/q_{ex}. Let us first define the following two distributions:

$$\mathrm{Real} = \left\{ (T, K) \;\middle|\; \begin{array}{l} r_1, r_2, \cdots, r_n, r \in_R \mathbb{Z}_q^*; \quad \delta \in \{0,1\}^l; \\ z_1 = g^{r_1}, z_2 = g^{r_2}, \cdots, z_n = g^{r_n}, z = g^r; \\ x_1 = g^{rr_1}, x_2 = g^{rr_2}, \cdots, x_n = g^{rr_n}; \\ h_1 = H(\delta \parallel x_1), h_2 = H(\delta \parallel x_2), \cdots, h_n = H(\delta \parallel x_n); \\ X = \oplus_{i=1}^n h_i; \quad y_i = X \oplus h_i, i \in [1, n-1] \end{array} \right\},$$

$$\mathrm{Rand} = \left\{ (T, K) \;\middle|\; \begin{array}{l} r_1, r_2, \cdots, r_n, r \in_R \mathbb{Z}_q^*; \quad \delta, w_1, w_2, \cdots, w_n \in \{0,1\}^l; \\ z_1 = g^{r_1}, z_2 = g^{r_2}, \cdots, z_n = g^{r_n}, z = g^r; \\ h_1 = w_1, x_2 = w_2, \cdots, x_n = w_n; \\ X = \oplus_{i=1}^n h_i; \quad y_i = X \oplus h_i, i \in [1, n-1] \end{array} \right\},$$

where $T = (z, z_1, z_2, \cdots, z_{n-1}, \delta, y_1, y_2, \cdots, y_{n-1})$ and $K = H(y_1, y_2, \cdots, y_n, X)$.

Lemma 1. *Let $\mathcal{A}'$ be an algorithm that, given (T, K) coming from one of the two distributions* Real *and* Rand*, runs in time t and outputs 0 or 1. Then we have :*

$$|\Pr[\mathcal{A}'(T, K) = 1 | (T, K) \longleftarrow \mathrm{Real}\,] - \Pr[\mathcal{A}'(T, K) = 1 | (T, K) \longleftarrow \mathrm{Rand}\,]|$$
$$\leq \frac{1}{q_h} \mathrm{Adv}_\mathbb{G}^{\mathrm{CDH}}(t + 2nt_{exp})$$

Proof. Assume that an algorithm $\mathcal{A}$ can distinguish between the two distributions with a non-negligible probability. Then, since H is a random oracle and a difference between the Real and the Rand is in a method of computing $h_i(i \in [1, n])$, we must find out at least one value of x_i to distinguish between them. Now, given an input $(g, A = g^r, B = g^\alpha \in \mathbb{G}^3)$ we construct an algorithm that outputs a value $C(= g^\beta)$ with $r\alpha = \beta \mod q$ as follows.

We first choose a random $\gamma_i \in_R \mathbb{Z}_q^*$ and define a random exponent r_i by $\alpha + \gamma_i \mod q$. We then can compute $z_i = Bg^{\gamma_i}$ and $X = \oplus_{i=1}^n h_i$ with a random $h_i \in \{0,1\}^l$, and finally construct $y_i = X \oplus h_i$. Consider the following distribution

$$\mathrm{Simul} = \left\{ (T, K) \;\middle|\; \begin{array}{l} \gamma_1, \gamma_2, \cdots, \gamma_n, x_i' \in_R \mathbb{Z}_q^*; \quad \delta, h_1, h_2, \cdots, h_n \in \{0,1\}^l; \\ r_1 = \alpha + \gamma_1, r_2 = \alpha + \gamma_2, \cdots, r_n = \alpha + \gamma_n; \\ z_1 = Bg^{\gamma_1}, z_2 = Bg^{\gamma_2}, \cdots, z_n = Bg^{\gamma_n}; \\ X = \oplus_{i=1}^n h_i; \quad y_i = X \oplus h_i, i \in [1, n-1] \end{array} \right\},$$

where T and K are as defined above. From this distribution, we have Rand $\equiv$ Simul since $z_i = g^{r_i}(= Bg^{\gamma_i})$ for all $i \in [1, n]$.

Given the transcript (T, K) from the distribution Simul as an input of $\mathcal{A}'$, we simulate a random oracle H at the same time. When $\mathcal{A}'$ finishes finally the execution, we selects a random $\delta \parallel x_i'$ that inputs in the random oracle simulation table. If $x_i' = x_i$ we can solve CDH problem $x_i = CA^{\gamma_i}$ to get $C = x_i'(A^{\gamma_i})^{-1}$. Therefore the algorithm $\mathcal{A}'$ having the transcript (T, K) provided by the simulation can not distinguish between two distributions. □

Lemma 2. *For any (computationally unbounded) adversary $\mathcal{A}$, we have :*

$$\Pr[\mathcal{A}'(T, K_b) = b | (T, K_1) \longleftarrow \mathsf{Rand} \; ; K_0 \longleftarrow \{0,1\}^l \; ; b \longleftarrow \{0,1\}] = 1/2$$

Proof. In experiment Rand, we represent from the transcript T the value y_i by the following $n-1$ equations and can rewrite that have the solution $(h_1, h_2, \cdots, h_n)$ as follows

$$y_1 = h_2 \oplus h_3 \oplus \cdots \oplus h_n = h_1 \oplus h_n \oplus y_n, \quad h_1 = y_1 \oplus y_n \oplus h_n$$
$$y_2 = h_1 \oplus h_3 \oplus \cdots \oplus h_n = h_2 \oplus h_n \oplus y_n, \quad h_2 = y_2 \oplus y_n \oplus h_n$$
$$\vdots$$
$$y_{n-1} = h_1 \oplus h_2 \oplus \cdots \oplus h_{n-2} \oplus h_n = h_{n-1} \oplus h_n \oplus y_n, \quad h_{n-1} = y_{n-1} \oplus y_n \oplus h_n$$

Therefore the adversary does not obtain any information about the value X from any one of transcripts since there are a lot of solutions as much as the independent variable h_n can take 2^l solutions. This implies that

$$\Pr[\mathcal{A}'(T, X_b) = b | (T, X_1) \longleftarrow \mathsf{Rand} \; ; X_0 \longleftarrow \{0,1\}^l \; ; b \longleftarrow \{0,1\}] = 1/2.$$

Since H is a random oracle, the statement of Lemma 2 immediately follows. □

Armed with the two lemmas above, we now give the details of the algorithm $\mathcal{B}$ from construction of the distribution Simul. Assume that an adversary $\mathcal{A}$ makes its Test query to an oracle activated by the δth Execute query. The algorithm $\mathcal{B}$ begins by choosing a random $d \in \{1, 2, \cdots, q_{ex}\}$ as a guess for the value of δ. $\mathcal{B}$ then invokes $\mathcal{A}$ and simulates the queries of $\mathcal{A}$. $\mathcal{B}$ answers all the queries from $\mathcal{A}$ in the obvious way, following the protocol exactly as specified, except for the case where a query is the dth Execute query. In this latter case, the algorithm $\mathcal{B}$ generates (T, K) depending on the distribution Simul and answers the dth Execute query of $\mathcal{A}$ with T.

The algorithm $\mathcal{B}$ outputs a random element in $\mathbb{G}$ if $d \neq \delta$. Otherwise, the algorithm answers the Test query of $\mathcal{A}$ with K. At some later point, when $\mathcal{A}$ terminates and outputs its guess b'. Applying the Lemma 1 and 2 together with the fact that $\Pr[b = b'] = 1/2$ and $\Pr[d = \delta] = 1/q_{ex}$, we obtain

$$\Pr[\mathcal{A}(T, K_b) = b | (T, K_1) \longleftarrow \mathsf{Real}; K_0 \longleftarrow \{0,1\}^l; b \longleftarrow \{0,1\}] = 1/2 + \epsilon,$$
$$\mathrm{Adv}_{\mathbb{G}}^{\mathrm{CDH}}(\mathcal{B}) = \epsilon/(q_h q_{ex}),$$

which immediately yields the statement of Theorem 1. □

5 Conclusion

In this paper we have proposed a dynamic group key agreement scheme with optimal message complexity; the protocol runs only in two rounds, one with $n-1$ unicasts and the other with a single broadcast. Therefore, due to its low communication cost, the protocol is well suited for low-power mobile devices. Furthermore, the protocol provides perfect forward secrecy and has been proven secure against a passive adversary under the computational Diffie-Hellman assumption. However, for practical purposes, more realistic and active attacks need to be captured into the security proof of the protocol, which we leave for further research.

References

1. G. Ateniese, M. Steiner, and G. Tsudik: New multiparty authentication services and key agreement protocols. IEEE Journal on Selected Areas in Communications, vol.18, no.4, pp.628–639, April 2000.
2. K. Becker, and U. Wille: Communication complexity of group key distribution. Proc. of CCS'98, pp.1–6, 1998.
3. M. Bellare and P. Rogaway: Random oracles are practical: A paradigm for designing efficient protocols. Proc. of CCS'93, pp.62–73, 1993.
4. C. Boyd and J.M.G. Nieto: Round-optimal contributory conference key agreement. Proc. of PKC'03, LNCS 2567, pp.161–174, 2003.
5. E. Bresson and D. Catalano: Constant round authenticated group key agreement via distributed computation. Proc. of PKC'04, LNCS 2947, pp.115–129, 2004.
6. E. Bresson, O. Chevassut, A. Essiari, and D. Pointcheval: Mutual authentication and group key agreement for low-power mobile devices. Proc. of MWCN'03, pp.59-62, 2003.
7. E. Bresson, O. Chevassut, and D. Pointcheval: Provably authenticated group Diffie-Hellman key exchange — the dynamic case. Asiacrypt'01, pp.290–309, 2001.
8. E. Bresson, O. Chevassut, and D. Pointcheval: Dynamic group Diffie-Hellman key exchange under standard assumptions. Eurocrypt'02, pp.321–336, 2002.
9. E. Bresson, O. Chevassut, D. Pointcheval, and J.-J. Quisquater: Provably authenticated group Diffie-Hellman key exchange. Proc. of CCS'01, pp.255–264, 2001.
10. M. Burmester and Y. Desmedt: A secure and efficient conference key distribution system. Eurocrypt'94, LNCS 950, pp.275–286, 1994.
11. Y. Huang and H. Garcia-Molina: Publish/subscribe in a mobile environment. Proc. of MobiDE'01, pp.27–34, 2001.
12. I. Ingemarsson, D. Tang, and C. Wong: A conference key distribution system. IEEE Trans. on Information Theory, vol.28, no.5, pp.714–720, September 1982.
13. J. Katz and M. Yung: Scalable protocols for authenticated group key exchange. Crypto'03, LNCS 2729, pp.110–125, August 2003.
14. J. Nam, S. Kim, S. Kim, and D. Won: Dynamic group key exchange over high delay networks. Proc. of ISPC COMM'04, pp.262-267, 2004.
15. M. Steiner, G. Tsudik, and M. Waidner: Key agreement in dynamic peer groups. IEEE Trans. on Parallel and Distributed Systems, vol.11, no.8, pp.769–780, 2000.
16. W.-G. Tzeng and Z.-J. Tzeng: Round-efficient conference key agreement protocols with provable security. Asiacrypt'00, LNCS 1976, pp.614–627, 2000.

Compact Linear Systolic Arrays for Multiplication Using a Trinomial Basis in $GF(2^m)$ for High Speed Cryptographic Processors

Soonhak Kwon[1], Chang Hoon Kim[2], and Chun Pyo Hong[2]

[1] Inst. of Basic Science and Dept. of Mathematics,
Sungkyunkwan University, Suwon 440-746, Korea
shkwon@skku.edu, shkwon@math.skku.ac.kr

[2] Dept. of Computer and Information Engineering,
Daegu University, Kyungsan 712-714, Korea
chkim@dsp.taegu.ac.kr, cphong@daegu.ac.kr

Abstract. Many of the cryptographic schemes over small characteristic finite fields are efficiently implemented by using a trinomial basis. In this paper, we present new linear systolic arrays for multiplication in $GF(2^m)$ for cryptographic applications using irreducible trinomials $x^m + x^k + 1$. It is shown that our multipliers with trinomial basis require approximately 20 percent reduced hardware resources compared to previously proposed linear systolic multipliers using general irreducible polynomials. The proposed linear systolic arrays have the features of regularity and modularity, therefore, they are well suited to VLSI implementations.

Keywords: finite field, systolic array, irreducible trinomial, VLSI.

1 Introduction

Finite field arithmetic in $GF(2^m)$ has received a considerable attention in recent years due to its applications in public-key cryptographic schemes and error correcting codes. In particular, two popular public key cryptosystems (elliptic curve cryptosystems and hyperelliptic curve cryptosystems) use arithmetic in $GF(2^m)$. One of the most important arithmetic operations in $GF(2^m)$ is multiplication since computing exponentiation and division can be performed by repeated multiplications. This paper focuses on the hardware implementation of fast and low complexity multipliers over $GF(2^m)$.

In cryptographic applications, since the field size m is very large, the global signals cause large fan-out, large wire delays, and complex routing. Therefore deterioration of the performance is inevitable. These problems are reduced in systolic architectures by using extra hardware resources. Consequently, a systolic architecture is a better choice than a non-systolic architecture for high speed VLSI implementations. Among the many systolic arrays for multiplication over

$GF(2^m)$, a polynomial basis is used in [1,2,6], a dual basis is used in [3,4,7], and a normal basis is used in [5]. If one does not want a basis conversion, the polynomial basis is preferred. The array type multiplication algorithms, when the polynomial basis is used, are classified as the LSB first (least significant bit first) and the MSB first (most significant bit first) scheme. The LSB first scheme processes the least significant bit of the multiplier operand first, while the MSB first scheme processes its most significant bit first. For $GF(2^m)$ multiplication, the multiplier implemented using the LSB first scheme has lower critical path delay compared to the multiplier based on the MSB first scheme due to increased parallelism among internal computations [8]. However the LSB first multiplier [2] uses two control signals while the MSB first multiplier [1] requires only one control signal. Therefore the hardware complexity of [2] is higher than that of [1] due to the extra latches (flip-flops).

In this paper, we modify the LSB first scheme used in [2] so that we eliminate the extra control signal. As a result, we obtain a bit serial systolic multiplier which has a comparable hardware complexity and shorter critical path delay than that of [1]. Also, by adjusting the multiplication algorithm, we find a systematic way of constructing a low complexity bit serial systolic multiplier when there is an irreducible trinomial $x^m + x^k + 1$. Based on the modified multiplication algorithm, we give explicit examples of new bit serial systolic multipliers for the following types of irreducible polynomial, $x^m + x^k + 1$, $k = 1, 2, 3, 4$. Analysis shows that the hardware complexity of the proposed multipliers are approximately 20 percent reduced from that of usual multipliers.

2 Bit-Level Multiplication Algorithm in $GF(2^m)$

Let $GF(2^m)$ be a finite field of 2^m elements, which is a vector space over $GF(2)$ of dimension m. We briefly explain basic finite field arithmetic and the LSB first scheme. Let $F(x) = f_0 + f_1 x + \cdots + f_{m-1} x^{m-1} + x^m \in GF(2)[x]$ be an irreducible polynomial over $GF(2)$ and let α be any root of $F(x)$. Then $\alpha \in GF(2^m)$ and $\{1, \alpha, \alpha^2, \cdots, \alpha^{m-1}\}$ is a standard polynomial basis over $GF(2)$. An element $A \in GF(2^m)$ is uniquely represented by $A = a_0 + a_1 \alpha + a_2 \alpha^2 + \cdots + a_{m-1} \alpha^{m-1}$ for some $a_0, a_1, \cdots, a_{m-1} \in GF(2)$. Now let $B = \sum_{i=0}^{m-1} b_i \alpha^i$ and $C = \sum_{i=0}^{m-1} c_i \alpha^i$ be other elements in $GF(2^m)$. We want to compute the product sum $AB + C$ by the LSB first scheme [2,8], $AB + C = C + A \sum_{i=0}^{m-1} b_i \alpha^i = C + \sum_{i=0}^{m-1} b_i A \alpha^i$. For a fixed i, let $A \alpha^i = \sum_{j=0}^{m-1} u_j \alpha^j$ and $A \alpha^{i+1} = \sum_{j=0}^{m-1} u'_j \alpha^j$. Then we have $\sum_{j=0}^{m-1} u'_j \alpha^j = A \alpha^i \alpha = \sum_{j=0}^{m-1} u_j \alpha^{j+1} = \sum_{j=1}^{m-1} u_{j-1} \alpha^j + u_{m-1} \alpha^m = \sum_{j=1}^{m-1} u_{j-1} \alpha^j + u_{m-1} \sum_{j=0}^{m-1} f_j \alpha^j = u_{m-1} f_0 + \sum_{j=1}^{m-1} (u_{j-1} + u_{m-1} f_j) \alpha^j$. Therefore we get

$$(u'_0, u'_1, \cdots, u'_{m-1}) = (u_{m-1} f_0, u_0 + u_{m-1} f_1, \cdots, u_{m-2} + u_{m-1} f_{m-1}),$$

which implies that $A\alpha^i$ can be recursively computed by above relation. Moreover for the same fixed i, let $\sum_{j=0}^{m-1} s_j \alpha^j = C + \sum_{j=0}^{i-1} b_j A \alpha^j$ and $\sum_{j=0}^{m-1} s'_j \alpha^j =$

Table 1. The LSB first algorithm

INPUT: $A = \sum_{k=0}^{m-1} a_k \alpha^k$, $B = \sum_{k=0}^{m-1} b_k \alpha^k$, $C = \sum_{k=0}^{m-1} c_k \alpha^k$
OUTPUT: $S = \sum_{k=0}^{m-1} s_k \alpha^k$ /* The result of $AB + C$. */
$(u_0, u_1, \cdots, u_{m-1}) \leftarrow (a_0, a_1, \cdots, a_{m-1})$
$(s_0, s_1, \cdots, s_{m-1}) \leftarrow (c_0, c_1, \cdots, c_{m-1})$ /* Initialize. */
for $i = 0$ to $m-1$ do
    for $j = m-1$ down to 0 do
        $s_j \leftarrow s'_j$, where $s'_j = s_j + b_i u_j$.
        $u_j \leftarrow u'_j$, where $u'_j = u_{j-1} + u_{m-1} f_j$ with $u_{-1} = 0$.
    end
end

$C + \sum_{j=0}^{i} b_j A \alpha^j$. Then we find $\sum_{j=0}^{m-1} s'_j \alpha^j = C + \sum_{j=0}^{i} b_j A \alpha^j = \sum_{j=0}^{m-1} s_j \alpha^j + b_i A \alpha^i = \sum_{j=0}^{m-1} s_j \alpha^j + b_i \sum_{j=0}^{m-1} u_j \alpha^j = \sum_{j=0}^{m-1} (s_j + b_i u_j) \alpha^j$. Therefore $C + \sum_{j=0}^{i} b_j A \alpha^j$ can also be recursively calculated by the relation,

$$s'_j = s_j + b_i u_j, \ 0 \leq j \leq m-1.$$

From above two observations regarding s_j and u_j for each i, we deduce that $AB + C$ can be computed by the following algorithm in Table 1.

3 Bit-Serial Systolic Arrays

Above algorithm is not a new one and one can find a similar algorithm in some other literature, for example, in [2,8]. But the novelty of our method is that we realize the algorithm very efficiently in a bit serial systolic arrangement consisting of m identical cells, where each of the cell has a new and simple design with only one control signal. In each ith cell, two operations $s_j \leftarrow s'_j$ and $u_j \leftarrow u'_j$ are done in parallel and both of them are serially computed from $j = m - 1$ to $j = 0$ as the data signals come in continuously. The following is a realization of $u_j \leftarrow u'_j$ in the ith cell where we choose $m = 4$ for simplicity.

The notation $\langle \ \rangle$ means that the next signals are repeated in this way. The signal θ is depending on τ (τ is f_0 except for the initial loading of the input 0.) and the bit value in the latch •. When the inputs are loaded serially in the first cell, the signal f_i is delayed by one clock than other inputs as shown in Fig. 1. This signal can be synchronized with other inputs by inserting one more latch on the signal flow of f_i. Also, since the output signal of u'_i is one clock delayed than other outputs, we may use latches appropriately to other output signals to synchronize all the outputs in the cell. The operation $s_j \leftarrow s'_j$ is easily combined in above circuit, where we still use only one control signal. The resulting circuit of ith basic cell is shown in Fig. 2(a). For convenience, we use the case $m = 4$.

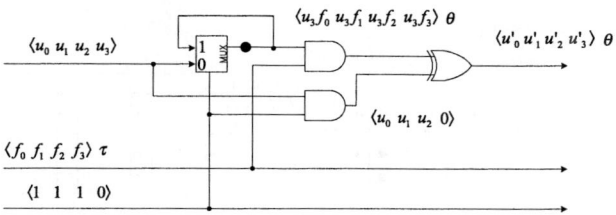

Fig. 1. The circuit explaining $u_j \leftarrow u'_j$

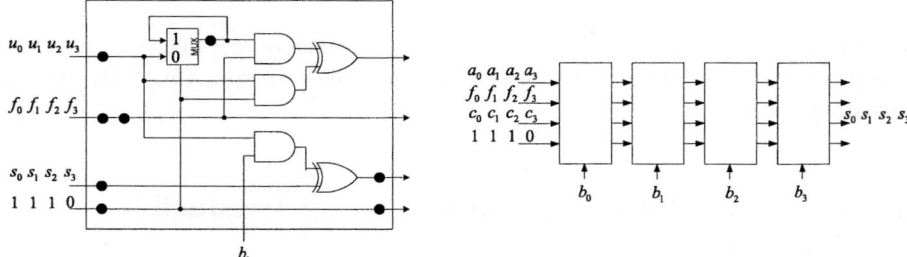

Fig. 2(a) and 2(b). The circuit of ith basic cell in $GF(2^4)$ and the corresponding systolic array

Fig. 2(b) is the description of the systolic array when the inputs b_i are ready in parallel.

Notice that our design needs only one control signal unlike the case of a similar design in [2] where they need two control signals, which necessarily increase the hardware complexity due to extra latches. If the input data come in continuously, it will yield output results at a rate of one per m clock cycles after an initial delay of $3m$ clock cycles. This is the same processing rate compared with previously proposed multipliers with same I/O format as described in Table 2. However, the multiplier in Fig. 2 has the shortest critical path delay among all multipliers. In addition, our systolic array can compute $AB + C$ whereas some other multipliers [1,3,4] can only compute AB. It should be mentioned that the dual basis multipliers in [3,4] need a basis conversion process which requires extra gates and wiring.

If the inputs b_i enter the first cell serially, then we need one more MUX (multiplexer) and two more latches to make a fully systolic array. The resulting array is shown in Fig. 3.

In Fig. 3(a), the loading operation of b_i occurs at ith ($0 \leq i \leq 3$) cell when the control signal is in logic 0 and it keeps the same value as long as the control signal is in logic 1. For example, since the inputs of b_i are one clock ahead of the inputs of the control signal at the second (1th) cell, the input signal b_1 and the control signal 0 is synchronized, and since the inputs of b_i are two clocks ahead of the inputs of the control signal at the third (2th) cell, b_2 and 0 are

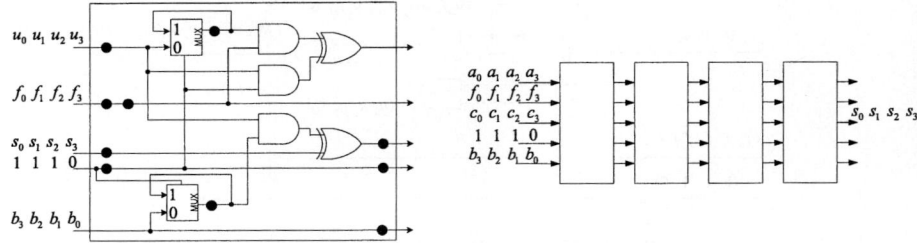

Fig. 3(a) and 3(b). A modified version of Fig. 2(a) and the fully systolic array

synchronized. Also note that Fig. 3(a) has a better or comparable hardware complexity and a critical path delay than those of corresponding fully systolic arrays of Table 2.

4 Low Complexity Multipliers Using Irreducible Trinomials

We have so far dealt with a multiplier with a standard polynomial basis with an irreducible polynomial $F(x) = f_0 + f_1 x + \cdots + f_{m-1} x^{m-1} + x^m \in GF(2)[x]$. However, if there is a specific type of irreducible polynomial, we may construct a multiplier which does not need the input signal $\langle f_0\ f_1\ f_2\ \cdots\ f_{m-1} \rangle$. For example, when there is an irreducible trinomial $x^m + x^k + 1$, we may generate the input signal of the polynomial by taking logical combinations of the control signal $\langle 1\ 1\ \cdots\ 1\ 0 \rangle$ and its shifted signals. First, note that it is easy to show that the polynomial $x^m + x^{m-k} + 1$ is irreducible over $GF(2)$ if and only if $x^m + x^k + 1$ is irreducible, since $1/\alpha$ is a root of $x^m + x^k + 1$ whenever α is a root of $x^m + x^{m-k} + 1$. Therefore the multiplication in the finite field $GF(2^m)$ is realized by using either one of the trinomials, $x^m + x^k + 1$ or $x^m + x^{m-k} + 1$. In this paper, we prefer to use $F(x) = x^m + x^{m-k} + 1$ because the input signal of the polynomial, $\langle f_0\ f_1\ f_2\ \cdots\ f_{m-1} \rangle = \langle 1\ 0\ \cdots\ 0\ 1\ 0\ \cdots\ 0 \rangle$, i.e. $f_0 = f_{m-k} = 1$ and $f_i = 0$ for $i \neq 0, m-k$, is easy to derive. To get this signal, note that the k-clock delayed (k-position left shifted) control signal is $\langle 1\ \cdots\ 1\ 0\ 1\ \cdots\ 1 \rangle$, where 0 appears only at the $m-k$th position. By taking a NAND gate of this signal and the control signal $\langle 1\ 1\ \cdots\ 1\ 0 \rangle$, we get the signal $\langle 0\ \cdots\ 0\ 1\ 0\ \cdots\ 0\ 1 \rangle$, where 1 appears only at the $m-k$th and the last (mth) position. Thus one clock right shifted signal is $\langle 1\ 0\ \cdots\ 0\ 1\ 0\ \cdots\ 0 \rangle$. This signal has the value 1 only at the first and $m - k + 1$th position, i.e. $f_0 = f_{m-k} = 1$ and $f_i = 0$ for other indices $i \neq 0, m-k$. Therefore we get the input signal of $x^m + x^{m-k} + 1$. Our idea is, of course, depending on the initial latch values of the circuit, so one should modify the design if necessary for each choice of trinomials. We mention the followings.

1. Our method is applicable for arbitrary irreducible trinomials $x^m + x^{m-k} + 1$, or equivalently $x^m + x^k + 1$. On the other hand, only the multipliers using the special trinomial $x^m + x + 1$ are reported in the current literature [4,7].
2. We omit the designs of fully systolic arrays from now on, since one easily get the corresponding designs by the same method of the explanation of Fig. 3.

4.1 $x^m + x^{m-1} + 1$ and $x^m + x^{m-2} + 1$

An irreducible polynomial $F(x) = x^m + x^{m-1} + 1 \in GF(2)[x]$ corresponds to input signal $\langle f_0\ f_1\ f_2\ \cdots\ f_{m-1}\rangle = \langle 1\ 0\ \cdots\ 0\ 1\rangle$. By using a suitable logic manipulation of the control signals as is explained before, we can make the signal $\langle f_0\ f_1\ f_2\ \cdots\ f_{m-1}\rangle$ occur without using the inputs f_i. That is, by taking a NAND gate of the control signal $\langle 1\ 1\ \cdots\ 1\ 0\rangle$ and one clock delayed (left shifted) control signal $\langle 1\ \cdots\ 1\ 0\ 1\rangle$, we have the signal $\langle 0\ \cdots\ 0\ 1\ 1\rangle$. We make this signals come in the cell one clock ahead of the other signals. The resulting signal is $\langle 1\ 0\ \cdots\ 0\ 1\rangle$ and the corresponding circuit is shown in Fig. 4 for the case $m = 4$ where $x^4 + x^3 + 1$ is irreducible over $GF(2)$. The signal $\langle 1\ 0\ \cdots\ 0\ 1\rangle$ and the signal $\langle u_0\ u_1\ \cdots\ u_{m-1}\rangle$ are synchronized in the sense that two inputs of the top AND gate in Fig. 4(a) are $\langle u_{m-1}\ u_{m-1}\ \cdots\ u_{m-1}\rangle$ and $\langle 1\ 0\ \cdots\ 0\ 1\rangle$. We omit a fully systolic array since the construction is similar to that of Fig. 3.

It should be mentioned that a multiplier using irreducible $x^m + x + 1$ is presented in [4] as a special case of the construction of a dual basis bit serial systolic multiplier. A similar construction (again using a dual basis) is also proposed in [7]. Since these multipliers do not need the inputs of f_i, one can save two latches (which amount to 10 AND gates) from a usual multiplier. It is clear from Table 2 that our multiplier with $F(x) = x^m + x^{m-1} + 1$ has a lower area complexity and a shorter or comparable critical path delay than those of dual basis multipliers with $F(x) = x^m + x + 1$ in [4,7], though the two polynomials are applicable to the same finite fields. Moreover since we are using a standard polynomial basis, we do not have to worry about the basis conversion process, which is usually required in dual basis multipliers.

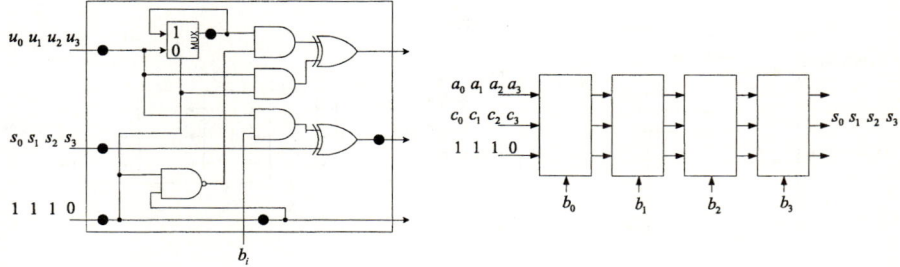

Fig. 4(a) and 4(b). The circuit of basic cell using irreducible $x^m + x^{m-1} + 1$ and the corresponding systolic array in $GF(2^4)$

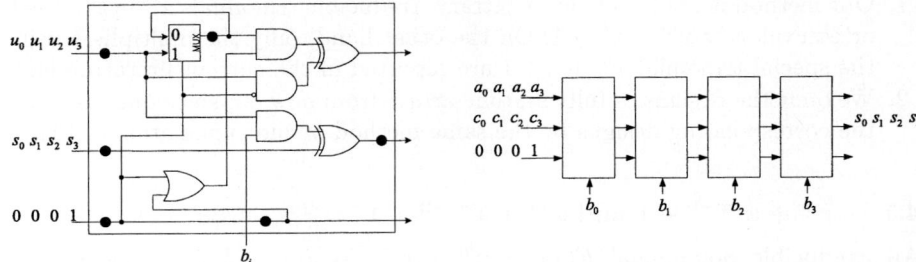

Fig. 5. An equivalent description of Fig. 4

A simple logic argument, NAND of (A, B) = OR of (NOR A, NOR B), shows that Fig. 4 and Fig. 5 are equivalent designs.

The reason why we are presenting above figure is that, in the forthcoming argument on the multiplier using $F(x) = x^m + x^{m-k} + 1$, $k = 2, 3, 4$, the delayed (left shifted) control signals are not what we want during the initial loading of the inputs if we use the control signal $\langle 1\ 1\ \cdots\ 1\ 0 \rangle$. This is because the initial latch values are all zero when the circuit starts to function. For example, when $k = 2$ and $m = 5$, the control signal $\langle 1\ 1\ 1\ 1\ 0 \rangle$ should have a two clock delayed (left shifted) signal $\langle 1\ 1\ 0\ 1\ 1 \rangle$. But this signal is in fact $\langle 1\ 1\ 0\ 0\ 0 \rangle$ when the loading process begins. Thus it gives a wrong answer $\langle 0\ 0\ 1\ 1\ 1 \rangle$ when one takes a NAND gate of the two control signals. The correct signal is $\langle 0\ 0\ 1\ 0\ 1 \rangle$. One may easily eliminate this problem by taking a control signal $\langle 0\ 0\ \cdots\ 0\ 1 \rangle$. Under such assumption, the two clock delayed control signal is $\langle 0\ \cdots\ 0\ 1\ 0\ 0 \rangle$ in any case. Therefore by taking an OR gate of the two control signals, we get the signal $\langle 0\ \cdots\ 0\ 1\ 0\ 1 \rangle$. Then one clock right shifted signal is $\langle 1\ 0\ \cdots\ 0\ 1\ 0 \rangle$, which is the input signal of the polynomial $F(x) = x^m + x^{m-2} + 1$. The algorithm is realized in the systolic arrangement shown in Fig. 6, where we choose $m = 5$ because $x^5 + x^3 + 1$ is irreducible over $GF(2)$.

We compare the cell complexity of our multipliers with other existing multipliers in Table 2. Note that, though there exists low complexity bit parallel multipliers [9,11] using an irreducible trinomial, and bit serial systolic multi-

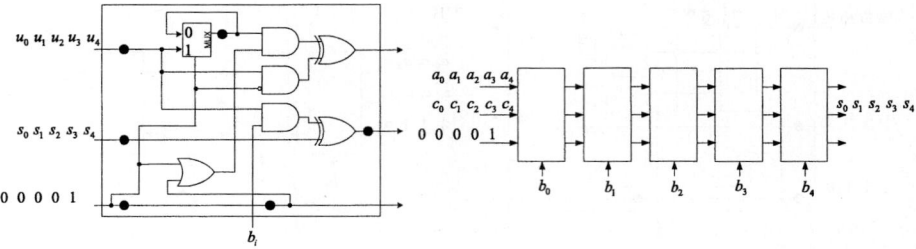

Fig. 6(a) and 6(b). The circuit of basic cell using irreducible $x^m + x^{m-2} + 1$ and the corresponding systolic array in $GF(2^5)$

Table 2. Comparison of our multipliers with other proposed designs

	$F(x)$	AND	XOR	3XOR	MUX	Latch	Critical Path Delay
[1, Wang]	polynomial	3	0	1	1	8	$D_A+D_{3X}+D_L$
[2, Yeh]	polynomial	3	2	0	1	10	$D_A+D_X+D_L$
[3, Fenn]	dual	2	2	0	2	8	$D_A+D_X+D_M+D_L$
[4, Wozniak]	dual	4	2	0	1	8	$D_A+D_X+D_L$
Fig. 2	polynomial	3	2	0	1	8	$D_A+D_X+D_L$
[4, Wozniak]	$x^m+x+1^\star$	4	3	0	1	6	$D_A+2D_X+D_L$
[7, Diab]	$x^m+x+1^\star$	3 + one NAND	2	0	2	6	$D_A+D_X+D_N+D_L$
Fig. 4	$x^m+x^{m-1}+1$	3 + one NAND	2	0	1	6	$D_A+D_X+D_N+D_L$
Fig. 5,6,7,8	trinomial	3 + one OR	2	0	1	6	$D_A+D_X+D_O+D_L$

All above multipliers support pipelined operation with latency m and throughput rate $1/m$. AND, XOR and MUX (multiplexer) mean 2-input gates and 3XOR means a 3-input XOR gate. $D_A, D_X, D_{3X}, D_M, D_N, D_O$ and D_L mean the delay time of an AND gate, a XOR gate, a 3XOR gate, a multiplexer, a NAND gate, an OR gate, and a latch, respectively. Also $\star$ denotes the use of a dual basis with the polynomial.

pliers [4,7] using the special polynomial $x^m + x + 1$, our construction of a bit serial multiplier is the first one which uses a systolic technique with arbitrary irreducible trinomials.

4.2 $x^m + x^{m-3} + 1$ and $x^m + x^{m-4} + 1$

By taking an OR gate of the control signal $\langle 0\ 0\ \cdots\ 0\ 1\rangle$ and 3-clock delayed (left shifted) control signal $\langle 0\ \cdots\ 0\ 1\ 0\ 0\ 0\rangle$, we get the signal $\langle 0\ \cdots\ 0\ 1\ 0\ 0\ 1\rangle$. Then one clock right shifted signal is $\langle 1\ 0\ \cdots\ 0\ 1\ 0\ 0\rangle$, which is the input signal of the polynomial $x^m + x^{m-3} + 1$. The corresponding systolic arrangement is shown in Fig. 7. For simplicity we use the case $m = 4$. The dotted line in Fig. 7(b) explains a global signal between the neighboring cells.

In a similar way, to get the multiplier with $F(x) = x^m + x^{m-4} + 1$, we take an OR gate of the control signal $\langle 0\ 0\ \cdots\ 0\ 1\rangle$ and 4-clock delayed control signal $\langle 0\ \cdots\ 0\ 1\ 0\ 0\ 0\ 0\rangle$, so that we have the signal $\langle 0\ \cdots\ 0\ 1\ 0\ 0\ 0\ 1\rangle$. Then one clock right shifted signal is $\langle 1\ 0\ \cdots\ 0\ 1\ 0\ 0\ 0\rangle$, which is the input signal of the polynomial $x^m + x^{m-4} + 1$. The resulting circuit is shown in Fig. 8. Here we choose $m = 7$ since $x^7 + x^3 + 1$ is irreducible over $GF(2)$.

4.3 Applicable Finite Fields

From Table 2, we see that the hardware complexity of our multipliers using trinomials is approximately 20 percent lower than that of the multipliers in [1,2,3,4] or Fig. 2. However, since a finite field $GF(2^m)$ may not always have a trinomial basis, there are some restrictions on the applicable finite fields. Thus we have to check the irreducibility of $x^m + x^k + 1$ over $GF(2)$. The irreducibility of $x^m + x^{m-1} + 1$ is well understood for moderately small values of m. It is known [10] that the values of $m \leq 1000$ for which $x^m + x^{m-1} + 1$ is irreducible over $GF(2)$

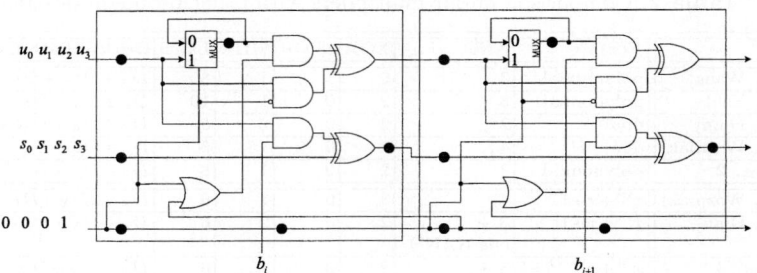

Fig. 7(a). The circuit of basic cell using irreducible $x^m + x^{m-3} + 1$

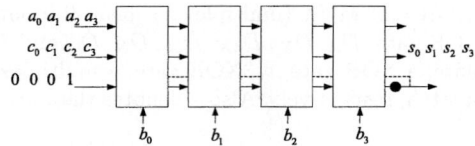

Fig. 7(b). Corresponding systolic array in $GF(2^4)$

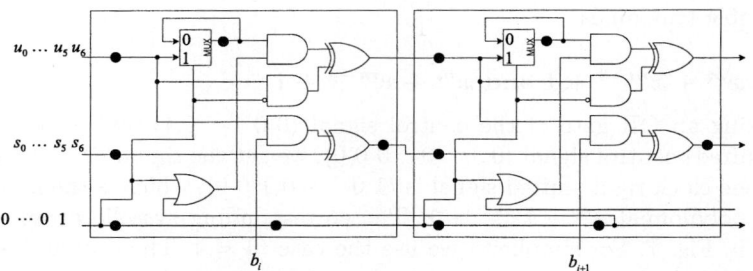

Fig. 8(a). The circuit of basic cell using irreducible $x^m + x^{m-4} + 1$

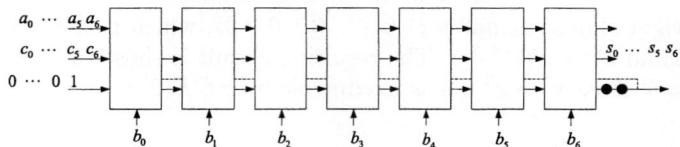

Fig. 8(b). Corresponding systolic array in $GF(2^7)$

are $m = 2, 3, 4, 6, 7, 9, 15, 22, 28, 30, 46, 60, 63, 127, 153, 172, 303, 471, 532, 865, 900$. There are 21 of them. We have no information, at this moment, about the existence of the tables of the irreducibility of $x^m + x^{m-k} + 1$ (or equivalently $x^m + x^k + 1$) for the values $k = 2, 3, 4$. However we tried those polynomials on the *Mathematica* for $m \leq 200$. The result is shown in Table 3.

Table 3. Irreducibility of $x^m + x^{m-k} + 1$ over $GF(2)$

polynomial	irreducible $m \leq 200$
$x^m + x^{m-1} + 1$	2, 3, 4, 6, 7, 9, 15, 22, 28, 30, 46, 60, 63, 127, 153, 172
$x^m + x^{m-2} + 1$	5, 11, 21, 29, 35, 93, 123
$x^m + x^{m-3} + 1$	4, 5, 6, 7, 10, 12, 17, 18, 20, 25, 28, 31, 41, 52, 66, 130, 151, 180, 196
$x^m + x^{m-4} + 1$	7, 9, 15, 39, 57, 81, 105

Though we do not have a plenty of experimental evidence, it is natural to guess that the trinomial $x^m + x^{m-k} + 1$ is much frequently irreducible when k is odd. Moreover, it is trivial to see that $x^m + x^{m-k} + 1$ is never irreducible when both k and m are even.

5 Conclusions

In this paper, we proposed low complexity linear systolic arrays for multiplication using arbitrary irreducible trinomials $x^m + x^{m-k} + 1$ with the LSB first scheme. Comparisons in Table 2 shows that our multipliers are significantly improved from other proposed designs. We presented explicit designs of the systolic arrays for $k = m-1, m-2, m-3, m-4$ or equivalently $k = 1, 2, 3, 4$. One can construct systolic arrays for other values of $1 \leq k \leq m-1$ by the same method of ours. In this case, there are global signals between ith and $i + \lfloor \frac{k-1}{2} \rfloor$th cells for the polynomial $x^m + x^{m-k} + 1$, $3 \leq k \leq \lfloor m/2 \rfloor$. Therefore if k is close to $m/2$, the delay time increases due to the global signals between the cells in long distance, which depends on the state of current VLSI technology. However, we do not have to worry about the global signal as long as there is an irreducible trinomial $x^m + x^k + 1$ (or equivalently $x^m + x^{m-k} + 1$) in $GF(2)[x]$ for small values of $k << m$. Finally, since our idea of the trinomial basis is independent of the LSB or MSB first scheme, we may construct similar multipliers of low complexity with a trinomial basis using the MSB first scheme also. But in the MSB case, the critical path delay is increased by the factor D_X.

Acknowledgements. This work was supported by Korea Research Foundation Grant (KRF-2004-015-C00004).

References

1. C.L. Wang and J.L. Lin, "Systolic array implementation of multipliers for finite fields $GF(2^m)$," *IEEE Trans. Circuits Syst.*, **38**, pp. 796–800, 1991.
2. C.S. Yeh, I.S. Reed and T.K. Troung, "Systolic multipliers for finite fields $GF(2^m)$," *IEEE Trans. Computers*, **C-33**, pp. 357–360, 1984.
3. S.T.J. Fenn, M. Benaissa and D. Taylor, "Dual basis systolic multipliers for $GF(2^m)$," *IEE Proc. Comput. Digit. Tech.*, **144**, pp. 43–46, 1997.
4. J.J. Wozniak, "Systolic dual basis serial multiplier", *IEE Proc. Comput. Digit. Tech.*, **145**, pp. 237–241, 1998.

5. C.Y. Lee, E.H. Lu and J.Y. Lee, "Bit parallel systolic multipliers for $GF(2^m)$ fields defined by all one and equally spaced polynomials," *IEEE Trans. Computers*, **50**, pp. 385–393, 2001.
6. B.B Zhou, "A new bit serial systolic multiplier over $GF(2^m)$," *IEEE Trans. Computers,* **C-37**, pp. 749–751, 1988.
7. M. Diab and A. Poli, "New bit serial systolic multiplier for $GF(2^m)$ using irreducible trinomials," *Electronics Letters*, **27**, pp. 1183–1184, 1991.
8. S.K. Jain, L. Song and K.K. Parhi, "Efficient semisystolic architectures for finite field arithmetic," *IEEE Trans. VLSI Syst.*, **6**, pp. 101–113, 1998.
9. H. Wu, M.A. Hasan and I.F. Blake, "New low complexity bit parallel finite field multipliers using weakly dual bases," *IEEE Trans. Computers*, **47**, pp. 1223–1234, 1998.
10. D.R. Stinson, "On bit serial multiplication and dual bases in $GF(2^m)$," *IEEE Trans. Inform. Theory*, **37**, pp. 1733–1736, 1991.
11. B. Sunar and C.K. Koc, "Mastrovito multipliers for all trinomials," *IEEE Trans. Computers*, **48**, pp. 522–527, 1999.
12. A.J. Menezes, *Applications of Finite Fields*, Kluwer Academic Publisher, 1993.

A Secure User Authentication Protocol Based on One-Time-Password for Home Network[*]

Hea Suk Jo and Hee Yong Youn

School of Information and Communications Engineering,
Sungkyunkwan University, 440-746, Suwon, Korea +82-31-290-7952
jojo@skku.edu, youn@ece.skku.ac.kr

Abstract. One-Time Password (OTP) authentication protocol can be used for authenticating a user by a server. It increases security by using a new password for each authentication while the previous password scheme iteratively uses a same password. In this paper we propose a secure user authentication protocol using a similar approach as S/Key, Lamport, Revised SAS and SAS-2 protocol but more secure than them. It employs a three-way challenge-response handshake technique to provide mutual authentication. Also, computation in the user device is reduced, resulting in less power consumption in the mobile devices. Compared with the S/KEY and Lamport protocol, the proposed protocol solves the problem of storing authentication data and limitation in the usage count. Moreover, the proposed scheme is practical to be used with Smart Card, and administration of authentication information is easy.

Keywords: Authentication, home network, one-time password, security, three-way handshake.

1 Introduction

Home Network (HN) has been considered as an acceptable solution for the digitally networked house, and security is one of the most important requirements in the HN in which many consumer devices are connected to the Internet [2]. The HN may suffer from attacks including eavesdropping and tampering of communication between the devices and remote server, masquerading the master of the device and controlling it, Denial of Service (DoS) attacks to the devices, etc. [1].

The computing resources inside a home or within an organization are usually protected by firewall or virtual private network (VPN)/IPsec in Home Gateway to prevent unauthorized access, which does not allow any remote access of home computers [6]. However, these security measures are not suitable to the HN since firewall allows data to enter the premise network after confirming the destination of the IP packets.

---

[*] This research was supported by the Ubiquitous Autonomic Computing and Network Project, 21st Century Frontier R&D Program in Korea and the Brain Korea 21 Project in 2004. Corresponding author: Hee Yong Youn.

Also, although the access technology using VPN is popular nowadays, it is more suitable to large network of high traffic than the HN with low traffic.

For this reason, HN needs a new secure authentication mechanism to protect the access to the data and correspondence. Traditional authentication requires the users to identify themselves by supplying a username and password [10,11]. This is because the frontline defense against attacker is usually an access control mechanism using password authentication. There exist many user/server applications that use passwords for authentication, ranging from basic telnet to remotely logging onto a server's database. In all these applications, it is possible for an attacker to intercept the password and then replay it to the server.

The replay problem can be overcome by using a system called One-Time Password (OTP) system [3] that was developed as an alternative to the traditional method of user authentication. In this paper we propose a secure user authentication protocol using a similar approach as S/Key [8], Lamport [7], Revised SAS [4] and SAS-2 [3], but more secure than them. It employs a three-way challenge-response handshake technique to provide mutual authentication. Also, computation in the user device is reduced, resulting in less power consumption in the mobile devices.

The remainder of the paper is organized as follows. Section 2 reviews the previous related work. Section 3 presents the proposed user authentication protocol, and Section 4 evaluates it. Section 5 concludes the paper.

2 Review of Previous Work

2.1 S/Key Protocol

The S/Key authentication is based on a secure hash function. A secure hash function is easy to compute in forward direction, but computationally infeasible to invert. If F is a secure hash function with input x and output y, then computing y given x, $y = F(x)$, is fast and easy, but finding an x_0 such that $y = F(x_0)$ for a given y is extremely difficult.

Here a user has to choose a secret password, which is concatenated with a random seed, and then hash function is applied to it. A sequence of one-time password is produced by applying the secure hash function multiple times. The $(N-i)th$ one-time password, p_{N-i}, is generated by running the user's secret password s ($s = K \oplus SEED$) through the hash function $(N-i)$ times (N: number of hashing, i: ith authentication).

$$p_{N-i} = F_{N-i}(s)$$

The password is stored in the host. The next time the user wants to login, the host asks for the password, p_{N-i+1}. The user computes it by applying the hash function $(i+1)$ times to the user's secret password. The host then verifies the password by applying the hash function one more time and comparing the result with the stored password (p_{N-i}). If the passwords match, the password p_{N-i+1} is stored in the host and the user can log in. This procedure is illustrated in Figure 1.

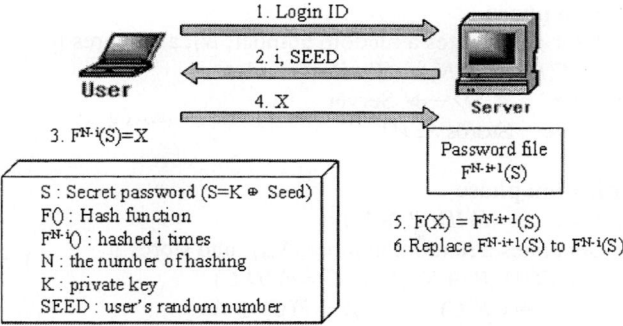

Fig. 1. The S/Key one time password system

2.2 Lamport's Protocol

In remotely accessed computer systems, a user identifies himself/herself by sending a secret password to the system. There exist three ways an intruder could learn the user's secret password and then impersonate the user when interacting with the system [7]:

- By gaining access to the information stored inside the system, e.g., reading the system's password file.
- By intercepting the user's communication with the system, e.g., eavesdropping on the line connecting the user's terminal with the system, or observing the execution of the password checking program.
- By the user's inadvertent disclosure of the password, e.g., choosing an easily guessed password.

In order to solve these problems, the Lamport's protocol consists of two phases; the registration phase and the authentication phase.

- **Registration phase**
  Step 1: User : Generates $F^N(x)$(means hashing N times) and sends it to the Server (F : hash function , x : password)
  Step 2: Server : Stores $F^N(x)$

- **Authentication phase (assume it is ith authentication)**
  Step 1: User : Calculates $F^{N-i}(x)$ and sends it to the Server
  Step 2: Server : Compares $F^{N-i+1}(x)$ with $F(F^{N-i}(x))$
  If they match, user is authenticated, and $F^{N-i}(x)$ is updated by $F^{N-i+1}(x)$
  Otherwise, user is rejected.

2.3 SAS-2 Protocol

The SAS-2 protocol [3] consists of two phases like the Lamport's protocol. The registration process is performed only once, while the authentication process is executed every time the user logs into the system.

- **Registration phase**
  Step 1: User : Generates a random number, N_1, and stores it
  $A=F(ID, P \oplus N_1)$ (P : Password)
  Step 2: User   : $A, ID \longrightarrow$ Server
  Step 3: Server : Stores A, ID

- **Authentication phase**
  Step 1: User : $A=F(ID, P \oplus N_i)$
  Generates a random number, N_{i+1}, and stores it
  $C = F(ID, P \oplus N_{i+1})$, $F(C)=F(ID,C)$
  $\alpha = C \oplus (F(C) + A)$, $\beta = F(C) \oplus A$
  $ID, \alpha, \beta \longrightarrow$ Server
  Step 2: Server : Calculates $F(C) = \beta \oplus A$, $C = \alpha \oplus (F(C) + A)$
  Compares $F(C)$ with $F(ID, C)$
  If they match, user is authenticated, and A is updated by C
  Otherwise, user is rejected.
  $r (= F(ID, F(C))) \longrightarrow$ User
  Step 3: User : Compares r with $H(ID, F(C))$
  If they match, server is authenticated.

3 The Proposed Protocol

3.1 The Overview

Home network consists of two or more devices interconnected to form a local area network (LAN) within the home. The number of users permanently connected to the Internet via DSL, TV cable, etc. continues to rise as permanent access with high bandwidth currently represents one of the fastest growing markets in the telecommunications business.

However, home network is exposed to various cyber attacks through Internet, and has security vulnerability such as hacking, malignancy code, worm and virus, DoS (Denial of Service) attack, communication network tapping as shown in Figure 2. As a result, technical development of home network with respect to security mostly focuses on putting the security functions on the home gateway to cope with cyber attack. Home gateway needs countermeasures against the attacks on main resources through illegal device connection or possibility of leakage of main data. Especially, in the premise of home network, vulnerability of component and data security exists in the wireless part needing authentication for accessing the component and data.

Security function is preferred to be loaded into home gateway that provides a primary defense against the external illegal attacks as a entrance guard that connects public network out of the house to the home network. The representative security functions loaded in home gateway are firewall, VPN (Virtual Private Network), etc. However, as mentioned earlier, they are not suitable to the HN because firewall allows data to enter the premise network if the destination is correct, and VPN is more suitable to a large network of high traffic.

Therefore, we propose a new scheme for this specific HN environment. The proposed scheme is for authentication of a user's identification who uses the device in the home network. Several techniques such as biometrics, password, certificate, or smart card can be used for user authentication in home network. However, such user authentication techniques must be examined before being employed in ubiquitous computing environment where the devices have low efficiency and performance. The user authentication technology for HN needs to allow accesses of the network resources remotely from outside of the network as well as from the premise network. Also, it needs to allow the users to get outside service such as internet banking from the premise network.

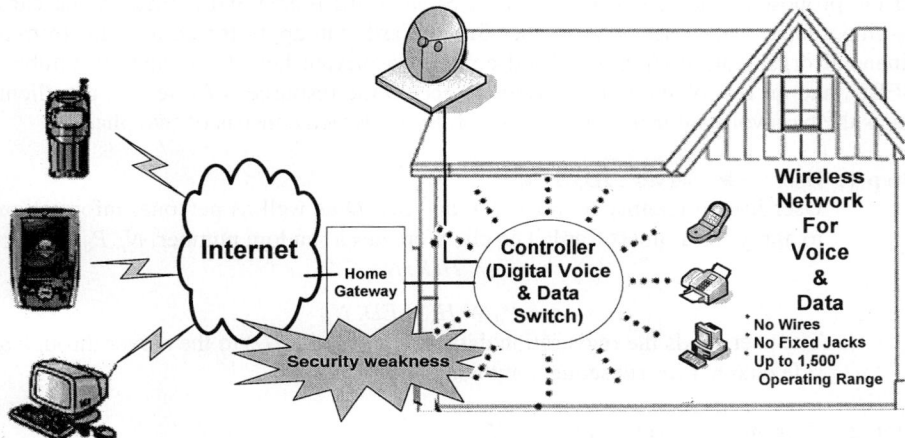

Fig. 2. Various cyber attacks through gateway in home network

In this paper a user authentication protocol is proposed using a similar approach as SAS-2 [3] but more secure than that. The main features needed to be considered in the design of an authentication protocol are for coping with the attacks like the followings [9,12]:

- **Eavesdropping attack:** This is the simplest type of attack. A host is configured to "listen" to and capture data not belonging to it. Carefully written eavesdropping programs can take usernames and passwords when users first login the network. Broadcast networks like Ethernet are especially vulnerable to this type of attack.
- **Replay attack:** An attack in which a valid data transmission is maliciously or fraudulently repeated either by the originator or by an adversary who intercepts the data and retransmits it, possibly as part of a masquerade attack.
- **Man-in-the-middle attack:** An attack in which an attacker is able to read and modify the messages between two parties at will without either party knowing that the link between them has been compromised. The attacker must be able to observe and intercept messages transferred between the two victims.

- **Stolen-verifier attack:** In most applications the server stores verifiers of users' passwords (e.g., hashed passwords) instead of the clear text of passwords. The stolen-verifier attack means that an adversary who steals the password-verifier from the server can use it directly to masquerade as a legitimate user during the user authentication phase.

The proposed scheme consists of three steps —registration, login, and authentication step.

3.2 Registration Step

In the proposed scheme every user has a Smart Card issued and signed by the card issuing center. Each user, with his/her Smart Card, can apply for an account from a financial organization. A Smart Card contains a private key, K, a random number, $SEED$, and an ID. When a client wants to access the resources of a server, the client takes the password authentication steps as follows, which consists of two phases.

Step 1: User ⟶ Server : ID, α, N_i
User has an identity number, ID, and $SEED$ as well as personal information to apply for a Smart Card. The client creates a random number, N_i, P_0, and α.
$$P_0 = H(K, N_i)$$
$$\alpha = P_0 \oplus H(SEED, N_i)$$
The user sends the registration data (α and N_i with ID) to the server through a safe channel for subsequent authentication.

Step 2: User ⟵ Server : $H(SEED, ID)$
Server searches for $SEED$ by the ID from the database, and calculates P_0 using $SEED$. Then, the server stores P_0 and N_i for subsequent authentication. Server sends $H(SEED, ID)$ to the user indicating safe delivery. The user authenticates the server using the received data and the data which was used for the user's authentication. The user calculates $H(SEED, ID)$ and compares it with the received data. If they do not match, the user tries registration process again. Otherwise, the server authentication is completed, Figure 3 shows the operation flow of registration step.

3.3 Login Step

In the login step, user informs and registers the user's own home network through the following steps. Refer to Figure 4.

Step 1: User ⟶ Server : ID
The user requests service (Login) to the server using the ID.

Step 2: User ⟵ Server : N_i, $SEED \oplus T_i$, $P_i \oplus T_i \oplus N_i$
After the server receives the user's ID, it calculates two data for the i-th authentication session. Firstly, the server calculates $SEED \oplus T_i$ using the time stamp T_i generated and $SEED$ stored in the user's Smart Card. Then it calcu-

lates $P_i \oplus T_i \oplus N_i$ where P_i is stored data. The time stamp T_i checks delay time, and is used for mutual authentication of the server. The server sends the authentication data (N_i, $SEED \oplus T_i$, $P_i \oplus T_i \oplus N_i$) to the user.

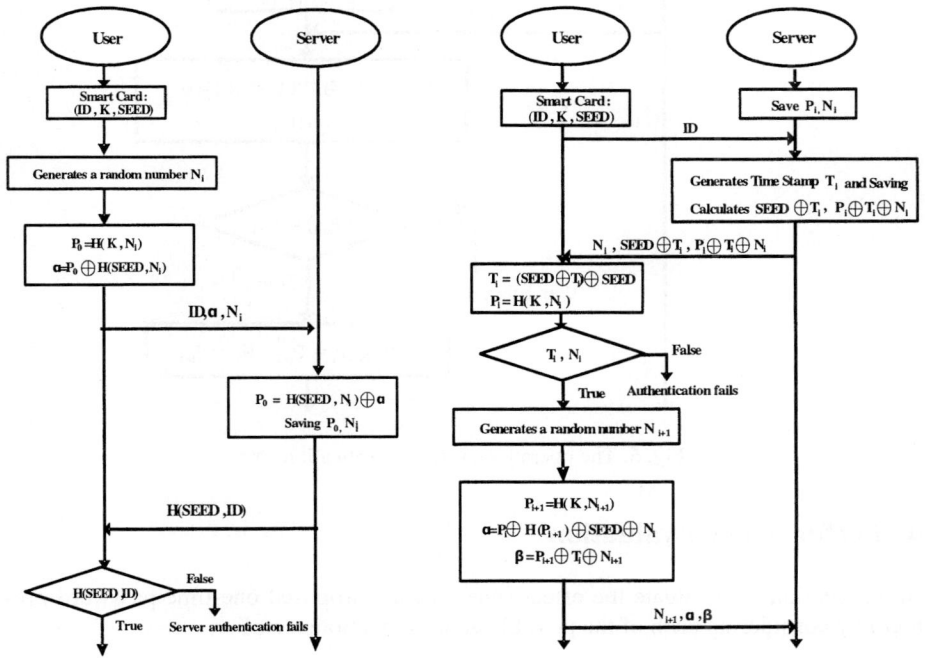

Fig. 3. Registration step   **Fig. 4.** Login step

Step 3: User ⟶ Server : N_{i+1}, α, β

After the user receives a message from the server, the user computes time stamp T_i by using $SEED$ stored in the user's Smart Card. The calculated T_i is checked if it is within some allowable range from the current time. If not, the user terminates the execution. Otherwise, the user assumes that the server is legal. Next, the user generates a random number, N_{i+1}, and computes $P_i = H(K, N_i)$ and $P_{i+1} = H(K, N_{i+1})$. Then computes $\alpha = P_i \oplus H(P_{i+1}) \oplus SEED \oplus N_i$ and $\beta = P_{i+1} \oplus T_i \oplus N_{i+1}$, and sends them to the server.

3.4 Authentication Step

After the server receives N_{i+1}, α, and β from the user, it obtains $H(P_{i+1})$ by $P_i \oplus SEED \oplus N_i \oplus \alpha$ where P_i is the stored verifier, N_i and P_{i+1} by $\beta \oplus T_i \oplus N_{i+1}$. Next, the server compares $H(P_{i+1})$ and P_{i+1}. If they do not match, the server terminates the execution. Otherwise, the server authenticates the user and updates P_i by P_{i+1} and N_i by N_{i+1} in order to compute the next session key as shown in Figure 5.

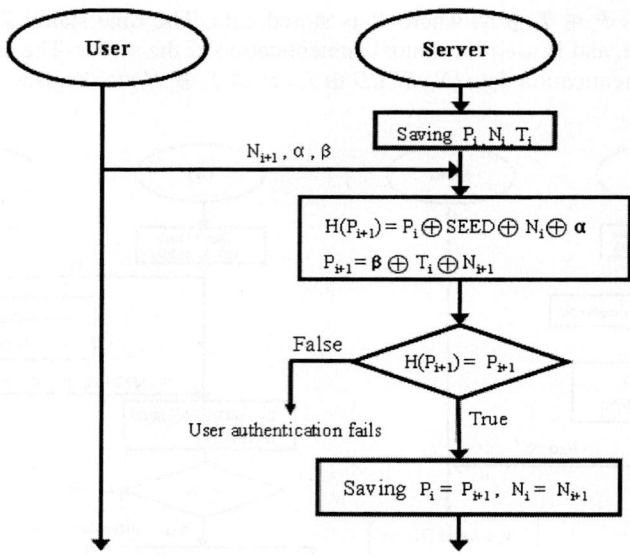

Fig. 5. The operation in the authentication step

4 Performance Evaluation

In this section we evaluate the effectiveness of the proposed one-time password protocol by considering each of the possible attacks mentioned earlier.

- **Replay attack:** In the proposed scheme, the private key, K, and one-time password sent to the server are hidden. As a result, an attacker cannot intercept the password and replay it to the server. Even an attacker could somehow reuse the authentication data, the attacker cannot pass the server authentication step since password is used only once and the attacker does not know the stored private key and SEED of the user or server.
- **Man-in-the-middle attack:** One of the key characteristics of the proposed one-time password protocol is that attacker cannot use password because it changes every time. Moreover, during transmission of authentication data, one can find out attack if authentication data (password) has been changed because data are sent along with the hash value such as K, N_i, and $H(K, N_i)$. Assume that an attacker has intercepted the following data from Internet and it changes N_i to $N_i{'}$.
N_i, $SEED \oplus T_i$, $P_i \oplus T_i \oplus N_i$, and $N_i{'} \leftarrow N_i$
When the user receives $N_i{'}$, the user computes time stamp T_i by using the user's SEED in Smart Card and compares user's $P_i \oplus T_i \oplus N_i$ by using the computed T_i, P_i, and received N_i. If they do not match, the user terminates the execution.
- **Denial of service attack:** Again the proposed protocol secures user authentication by changing the password, and the server updates the password to P_{i+1} only if the hashed result of the calculated $H(P_{i+1})$ is equal to once hashed P_{i+1}. Thus, the proposed scheme can resist the denial of service attack.

- **Stolen-verifier attack:** The user and server are safe since they share *SEED, K,* etc in Smart Card. Even though the seed value is stolen, the attacker cannot get the private key value because the value was hashed.
- **Mutual authentication:** The password scheme allowing user authentication may satisfy the security requirement. However, for those applications where confidential data are transmitted between the user and server, mutual authentication is necessary. The proposed scheme employs a three-way challenge-response handshake technique to provide mutual authentication. The server sends the authentication data (N_i, $SEED \oplus T_i$, $P_i \oplus T_i \oplus N_i$) to the user. Then the user checks time stamp T_i of delay time, and it can be used for mutual authentication.

We compare the proposed scheme with the previous protocols, S/Key, Lamport, Revised SAS, and SAS-2 in Table I. It shows that the proposed protocol allows more secure and efficient authentication although two message exchanges are needed between the user and server for mutual authentication. Notice that the Lamport and Revised SAS protocol require only one transmission since they do not allow mutual authentication. While the SAS-2 protocol allows mutual authentication like the proposed protocol, the user and server need one more round of hashing and the user needs to store random number N_i. Here V_i is the verifier in ith authentication session and M is maximum number of hash iterations.

Table 1. Comparison of the proposed protocol with other protocols

Performances Methods	User		Server		User->Server
	Hash Iterations	Data Storages	Hash Iterations	Data Storages	Transmission Iterations
S/Key [8]	$M - i$	i	1	i	2
Lamport [7]	$M - i$	i	1	i	1
Revice SAS [4]	5	N_i	2	ID, V_i	1
SAS – 2 [3]	4	N_i	2	ID, V_i	1
Proposed	3	None	1	ID, V_i, P_i	2

5 Conclusion

Home network is exposed to various cyber attacks such as hacking, malignancy code, worm and virus, DoS (Denial of Service) attack, communication network tapping, etc. Hence we have proposed a user authentication protocol for home network. It is based on one-time password scheme using Smart Card and hash function (no limitation of usage count) suitable to the home network environment. Compared with the S/KEY and Lamport protocol, the proposed protocol solves the problem of storing authentication data and limitation in the usage count of user authentication. Although the number of transmissions is increased by one compared with them, it allows safer authentication, simpler calculation, and mutual authentication. Moreover, the proposed scheme is practical to be used with Smart Card, and care and administration of the authentication information are easy.

References

[1] Hess, A., Schafer, G. ,"ISP-operated protection of home networks with FIDRAN", Consumer Communications and Networking Conference, 2004. CCNC 2004. First IEEE , 5-8 Jan. 2004
[2] Mahfuzur Rahman, P.Bhattacharya, "Remote Access And Networked Appliance Control Using Biometrics Features", IEEE Transactions on Consumer Electronics, Vol, 49, No.2, MAY. 2003
[3] Akihiro SHIMIZU, "A One-Time Password Authentication Method", Kochi University of Technology Master's thesis, January 2003.
[4] T. Tsuji, T. Kamioka, and A. Shimizu, "Simple and secure password authentication protocol, ver.2 (SAS-2)", IEICE Technical Report, OIS2002-30, vol.102, no.314, September 2002.
[5] Entrust, "Authentication: the cornerstone of secure identity management",Securing Digital Identities & Information, Apr.26.2004
[6] [Sang-hyun Kim, Chul-bum Kang, Hee-jin Jang, Sang-wook Kim,"A Secure Control of Home Network on a PDA", Korea Information Science Society, vol.29, Oct.2002
[7] L. Lamport, "Password authentication with insecure communication", Communications of the ACM, vol.24, no.11, 1981.
[8] N,Haller Bellcore,"The S/KEY One-Time Password System", Network Working Group, February 1995.
[9] http://www.faqs.org/docs/linux_network/x-082-2-firewall.attacks.html
[10] Pan-Lung Tsai, Chin-Laung Lei, Wen-Yang Wang, "A Remote Control Scheme for Ubiquitous Personal Computing", International Conference on Networking, Sensing & Control , March, 2004.
[11] Ching-Te Wang, Chin-Chen Chang, Chu-Hsing Lin, "Using IC Cards to Remotely Login Passwords without Verification Tables", Advanced Information Networking and Application March, 2004
[12] Faisal. RM, Ad Rahman Ahmad, "A Password Authentication Scheme With Secure Password Updating", Telematics System, Services and Applications 2004, May 2004

On AAA with Extended IDK in Mobile IP Networks*

Hoseong Jeon, Min Young Chung, and Hyunseung Choo

School of Information and Communication Engineering,
Sungkyunkwan University, 440-746, Suwon, Korea
+82-31-290-7145
{liard, mychung, choo}@ece.skku.ac.kr

Abstract. Mobile IP proposed by IETF supports continuous services for a mobile node (MN) based on its capability to roam around foreign domains [1]. Recently the rapid growth of wireless technology and its use in coordination with the Internet require a very careful look at issues regarding the security. As a large portion of 785-million world Internet users access such technologies in the context of providing security demanded services, it is essential to recognize the potential threats in wireless technologies. For this reason, IETF suggests that the existence of some servers capable of performing the authentication, authorization, accounting (AAA) services could help [4, 5, 6, 7]. In this paper, we propose an Extended IDentification Key (EIDK) mechanism based on IDK with Authentication Value (AV) that can reduce the number of signaling messages and thus signaling delay for services even in handoffs while maintaining the similar level of security to the previous works [10, 11]. The performance results obtained show that this method can provides a good solution to secure service procedures in mobile computing.

1 Introduction

Mobility in IP networks is a significant issue due to the increase of many portable devices such as notebooks, PCSs, and PDAs. Lots of popular applications including e-business require transmission of highly sensitive information often over wireless links. The mobility implies higher security risks than static operations in fixed networks, because the traffic may at times take unexpected network paths with unknown or unpredictable security characteristics. Hence, there is the need to develop technologies which will jointly enable IP security and the mobility over wireless links. Therefore, the integration of the Mobile IP and AAA protocol has been proposed [2].

By combining Mobile IP and AAA structure, the message on the Mobile IP network can be provided with additional security through AAA protocol.

---

* This work was supported in parts by Brain Korea 21 and the Ministry of Information and Communication in Republic of Korea. Dr. H. Choo is the corresponding author.

However, while an MN roams in foreign networks, a continuous exchange of control messages is required with the AAA server in the home network. The control message contains the confidential information to identify the privilege of the mobile user for the service. Standard AAA handoff mechanism has inefficient authenticating procedures that limit its quality of service (QoS). To resolve such problems, session key exchange mechanism [10] and ticket based mechanism [11] are proposed.

The session key exchange mechanism, basically, reuses the previously assigned session key. In this mechanism, the handoff delay can be decreased importantly. However, this mechanism requires that the trusted third party supports key exchanges between ARs. For this reason, it uses only the intra-handoff within the same domain. The ticket based mechanism using an encrypted ticket that can support authentication and authorization for the MN has been proposed. It reduces the delay and the risk on MN authentication in Mobile IPv6 (MIPv6). However, this mechanism generates additional signaling overheads and AAA server's overheads.

In the previous work, we have proposed IDK mechanism. This mechanism based on a pre-encrypted key reduces signaling delay at the authentication process. However, this mechanism just uses service requests due to the mobility of MNs [12]. For improving this shortage, we propose an EIDK based AAA mechansim. This mechanism with AV extends the effectiveness of IDK into the handoff process. According to the performance evaluation, the proposed mechanism compared to the previous mechanisms is up to about 20-40% better in terms of average latency that considers handoff latency and service latency.

The rest of the paper is organized as follows. In Section 2, an overview of the Mobile IP and AAA protocol is presented and the session key exchange mechanism and the ticket based AAA mechanism are given. Our proposed EIDK based AAA mechanism is discussed in Section 3. After that the performance is evaluated along with previous methods. Finally we conclude the paper in Section 5.

2 Related Works

Mobile IP has been designed by IETF to serve the increasing needs of mobile users who wish to connect to the Internet and to maintain communications as they move from place to place. In MIPv6, each MN is always identified by its home address regardless of its current point of attachment to the Internet. While away from its home IP subnet, the MN is also associated with a care of address (CoA), which indicates its current location. MIPv6 enables any IPv6 node to learn and cache the CoA associated with the home address of the MN, and then to send packets destined for the MN directly to it at the CoA based on the IPv6 routing header [3].

The IETF AAA Working Group has worked for several years to establish a general model for: Authentication, Authorization, and Accounting. AAA in mobile environment is based on a set of client and server (AAAF and AAAH)

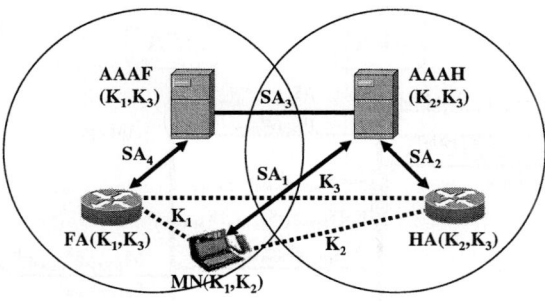

Fig. 1. AAA security association in Mobile IPv6

located in the different domains. AAA protocol operates based on the security associations (SAs) ($SA_s : SA_1, SA_2, SA_3$, and SA_4) as shown in Fig. 1. For the support regarding the secure communication, MN requires dynamic security associations. They are defined by sharing the session keys such as K_1, K_2, and K_3 between MN and Home Agent (HA), between HA and Foreign Agent (FA), and between FA and MN. Once the session keys have been established and propagated, the mobility devices can securely exchange data [4].

2.1 Session Key Exchange Mechanism

The session key exchange mechanism is based on a variant of Diffie-Hellman key agreement protocol instead of asymmetric key cryptography. The following premises are required in this scheme [10]:

- Regional registrations and reply messages are routed through old Foreign Agent(oFA) since that the MN is not directly connected to new Foregin Agent(nFA) prior to the Layer 2 handoff.
- For the network scalability, no pre-established security association should be required between oFA and nFA.
- To prevent session stealing attacks, session keys should be encrypted and exchanged in a secure fashion.
- It is assumed that FAs related to the handoff are trusted, that is, they are authenticated by Gateway Foreign Agent (GFA). Therefore, the impersonating attack is not considered in the method.

Fig. 2 shows the session key exchange procedures. This scheme is based on a variant of Diffie-Hellman key agreement protocol instead of public key cryptography. For the fast operations, this scheme reuses the previously assigned session keys, the session keys for FA(S_{MN-FA}, S_{FA-HA}). To ensure the confidentiality and integrity of the session keys, it uses the encryption and decryption under a short lived secret key, $K_{oFA-nFA}$, between oFA and nFA. The key is dynamically shared between them and can be created only two entities.

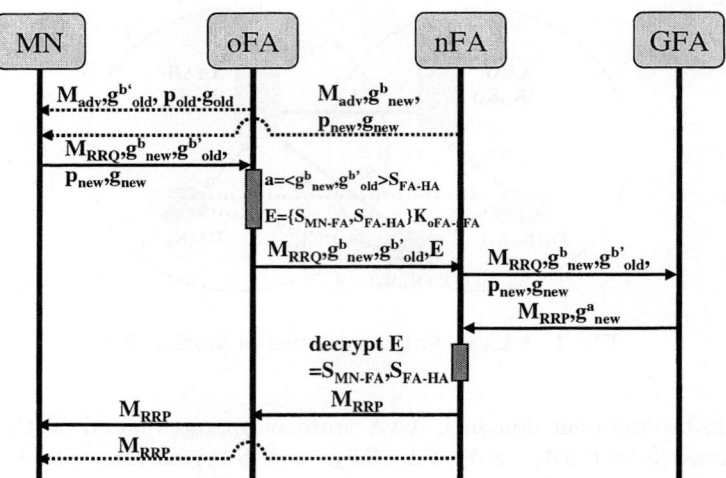

Fig. 2. Secure session key exchange procedure

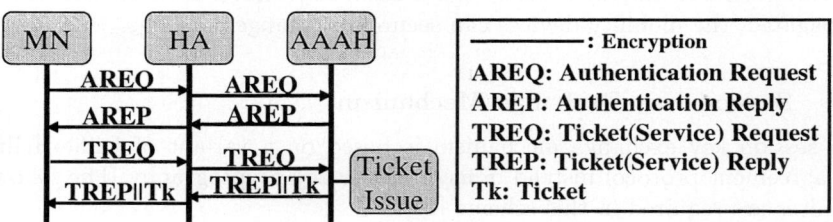

Fig. 3. Initial registration in ticket based AAA model

2.2 Ticket Based AAA Mechanism

A ticket based AAA mechanism reduces the overhead on the service request by utilizing the pre-encrypted ticket without intermediate encryptions and decryptions. If the MN wants to request a service, the MN sends a ticket to AAAH for the authentication of MN. The authentication of MN is performed by the Ticket Granting Service ASM (TGS ASM) in the AAA server. The result of authentication is returned to the MN, which allows the MN to request the service [11].

However, this mechanism has four additional signaling messages for the ticket issue. Fig. 3 describes exchanged additional signaling messages on initial registration. Four messages added are ticket(service) request message, AAA ticket (service) request message, AAA ticket(service) reply message, and ticket(service) reply message. The messages between MN and HA are based on the general Mobile IP protocol, and the messages between HA and Home AAA server (AAAH) are based on the extended AAA protocol in Mobile IP networks.

3 Extended IDentification Key Based AAA Mechanism

In the EIDK mechanism, we assume as follows: 1)An AAA Server authenticates and authorizes service subscribers, and verifies IDK. It also creates AV. 2)An AAA Client either HA or FA, which has the functionality to generate and to deliver AAA messages. 3)An AAA Broker (AAAB) authenticates MN instead of AAA Home (AAAH). 4)A Mobile Node generates IDK and delivers it.

In order to reduce the time for repeated encryptions and decryptions, an MN generates an encrypted information called IDK using authentication time (AT). This value represents the time at the initial registration of MN. The IDK consists of the following [12]:

- Network Access Identifier (NAI) of MN
- Address of the AAA server that provides services to the MN
- Service identifier allowed for the MN
- Home network address and IP address of the MN
- IDK lifetime
- A random number (128 bits)
- The session key shared by the MN and the AAA server
- CoA of next possible area expected to be moved (optional)
- Authentication time (AT).

The proposed mechanism reduces the authentication delay and signalings at the foreign domain by using AV. The AV contains an information for MN and session keys in FA. They are encrypted by SA between AAAH and AAAB [4]. It consists of following:

$$AV = SA_{AAAH-AAAB} \{ MN\ information\ ||\ FA's\ session\ keys \}$$

When MN moves to a foreign network AAAH creates an AV that is delivered to the AAAB. After that operation the AAAB authenticates MN instead of AAAH. As a result, the MN reduces authentication procedure and its delay in the foreign network since AAAB takes care of authentication job for MN on behalf of its AAAH. The detailed authentication procedure is described in the Section 3.3.

3.1 Initial Registration to AAAH

As indicated in Fig. 4, the sequence of message exchanges for each authentication mechanism is performed for the initial registration in the home network. We assume that there is no security associate between MN and HA. This is because we do not consider the pre-shared key distribution in AAA protocol in this work.

Fig. 4(a) shows the initial registration of the basic AAA model. And both the ticket based model and the proposed EIDK based one follow the basic AAA model in the initial registration. However, as you see in Fig. 3, additional signaling for issuing a ticket is required for faster services on requests in the ticket based model. Unlike the ticket based model scheme, MN receives AT along with the authentication reply message without further additional signaling in our scheme.

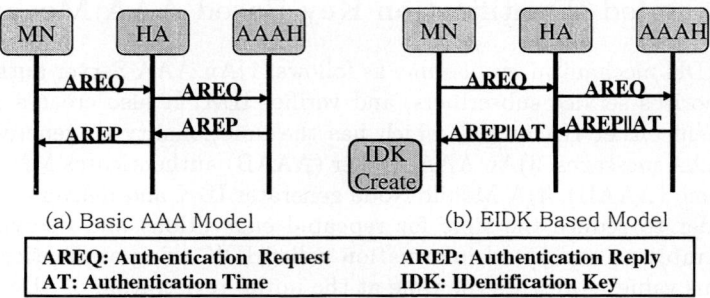

Fig. 4. Initial registration

Fig. 4(b) shows the initial registration procedure for the EIDK based mechanism. In the authentication reply procedure, AT is delivered to MN together with authentication reply message (AREP). Accordingly, both the MN and AAAH server share a secret value. This one is the arrival time of the request message for the MN at AAAH. The AT would be used as a part of the encryption key value on IDK by MN and later it is used as the decryption key in AAAH.

3.2 Service Requests

Fig. 5 shows the procedure routine of message exchanges for the service request in the home network. The service request message (SREP) is encrypted and decrypted by the key distributed from AAAH on the authentication process in the basic AAA model.

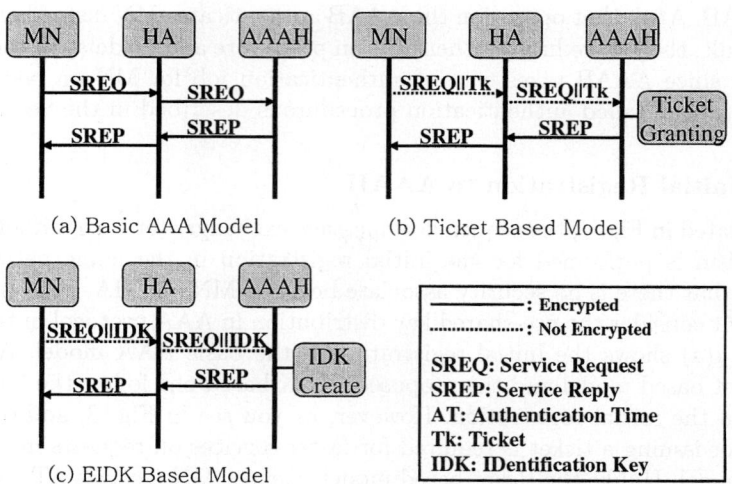

Fig. 5. Service request

As you see in Fig. 5(a), service request message (SREQ) and SREP are encrypted and decrypted at MN, HA, and AAAH when transmitting and receiving them, and these can be a significant overhead. Ticket based model in Fig. 5(b) reduces the overhead on the service request by utilizing the pre-encrypted ticket without intermediate encryptions and decryptions. This can be done by the extended AAA server structure. Also the model assumes that the time for ticket issuing and granting is not significant. However, this may not guarantee its superiority in the real world.

As shown in Fig. 5(c), the proposed EIDK based model does not need the extended AAA server structure, but just maintains the current one. Intermediate encryptions and decryptions are not necessary on the service request in our scheme. This is because we employ the pre-encrypted IDK which is created by MN beforehand. Unlike the basic AAA model, the EIDK based AAA model requires IDK creation and the time for it. But this scheme reduces the total delay since it eliminates the time for intermediate encryptions and decryptions.

3.3 Handoff Procedures

Fig. 6(a) represents the basic AAA handoff mechanism. In this mechanism, all authentication messages should be encrypted and decrypted for each security association. Fig. 6(b) shows session key reuse mechanism by using Mobility Anchor Point (MAP) instead of GFA. This is handled with detailed descriptions in Section 2.2. Basically, this scheme importantly reduces handoff latency, however there is a significant defect that is only applied to intra-handoff.

Lastly, Fig. 6(c) shows the proposed handoff mechanism. It eliminates encryption and decryption delay in the authentication procedure by using pre-encrypted IDK, and reduces the number of signalings due to the AV in the AAAB. If there is no AV to AAAB, the proposed scheme follows the basic AAA model procedures.

4 Performance Evaluation

In order to evaluate performance of our proposed algorithm, we make the following notations :

- E_{se}/E_{sd}: time required for symmetric key encryption/decryption of a message at MN/AR/AAAH/AAAF/AAAB
- AS: authentication time in AAAH
- Tk: ticket issuance and verification time in AAAH
- IDK: IDK creation and verification time in MN/AAAH/AAAB
- AV: authentication time using AV in AAAB

Authentication procedures can be classified into three case: initial registration, handoff and service request. And then handoff can be classified into three case by the position of the MN: intra handoff in home/foreign domain and inter handoff. Lastly, service request can be classified into two case: service request

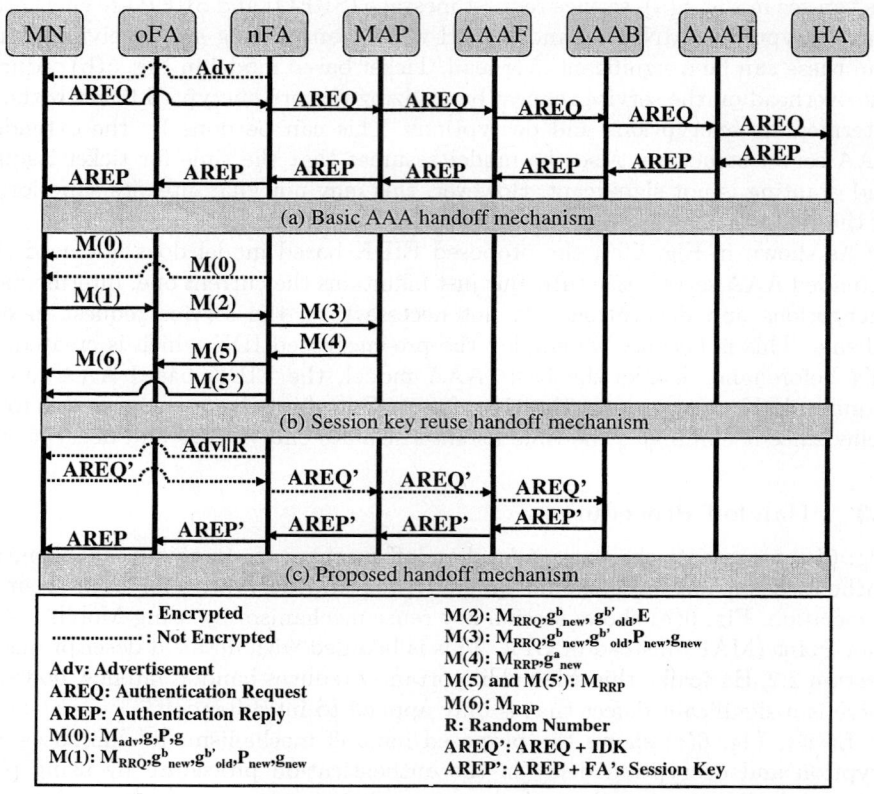

Fig. 6. Handoff procedures

Table 1. System parameters

Bit rates		Processing time	
Wire links	100 $Mbps$	Routers (HA,FA)	0.5 $msec$
Wireless links	2 $Mbps$	Nodes (MN)	0.5 $msec$
Propagation time		Tk	3.0 $msec$
Wire links	500 μsec	IDK	3.0 $msec$
Wireless links	2 $msec$	AS	1.0 $msec$
Data size		AV	1.0 $msec$
Message size	256 $bytes$	E_{se} and E_{sd}	1.0 $msec$

in home/foreign domain. We calculate times required in schemes we discuss (Fig 3,5,6,7, and 8) for performance evaluation based on above cases and the system parameter in Table 1 [9, 12], we compute the handoff probability and the average latency.

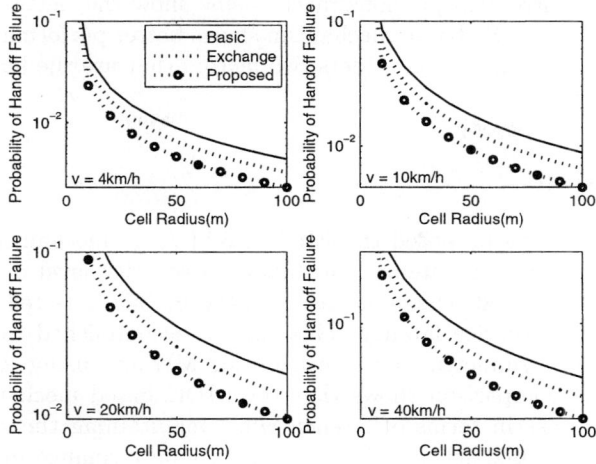

Fig. 7. The probability of handoff failure

We analyze the handoff procedure to obtain the handoff failure rate for each handoff mechanism. The handoff failure rate is influenced by few factors that are the velocity of MN and the radius of a cell. Fig. 7 shows the result of the handoff failure rate with intra handoff case. The proposed scheme consistently shows the better handoff failure rate comparison with previous mechanisms.

Fig. 8 shows the average handoff latency in terms of the service rate and the cell radius. In this result, the ticket based mechanism shows the better performance in the frequent service request because it consider the reduction of service

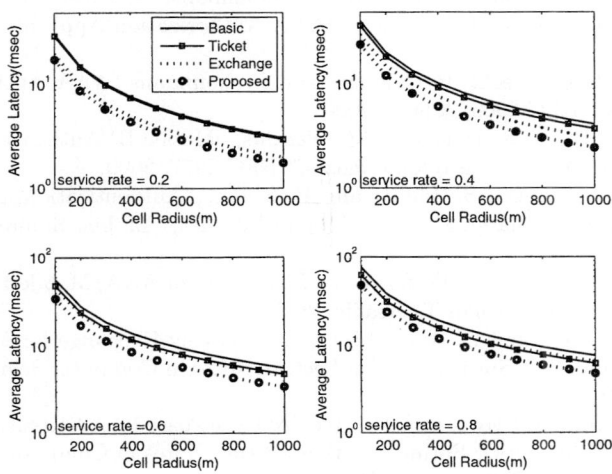

Fig. 8. Average handoff latency

request. Consequently, two previous mechanisms show the better performance in each factor. But, EIDK based mechanism shows better performance than previous mechanisms, because it considers two factors that are the handoff latency and service latency.

5 Conclusion

In this paper, we have proposed the EIDK based AAA mechanism using IDK and AV to provide reduce latency and maintain security level of the previous studies. The EIDK based scheme is based on the mobile initiated handoff, and reduce handoff and service latency. This can be accomplished by making the IDK based on the symmetric key algorithm for MN and using AV in AAAB. The performance comparison shows that the EIDK based mechanism is superior to previous ones in terms of latency while maintaining the security level. Especially this proposed scheme shows the better performance in the frequent handoff and service requests. We currently work on the analysis of security level.

References

1. C.E. Perkins, "IP Mobility Support," IETF RFC 2002.
2. C. Perkins, "Mobile IP Joins Forces with AAA," IEEE Personal Communications, vol. 7, no. 4, pp. 59–61, August 2000.
3. B. David, C. Perkins, and J. Arkko,"Mobility Support in IPv6," IETF draft, Internet Draft draft-ietf-mobileip-ipv6-17.txt, May 2002.
4. J. Vollbrecht, P. Cahoun, S. Farrell, and L. Gommans, "AAA Authorization Framework," RFC 2904, 2000.
5. J. Vollbrecht, P. Calhoun, S. Farrell, L. Gommans, G. Gross, B. debruijn, C.de Laat, M. Holdrege, and D. Spence, "AAA Authorization Application Examples", IETF RFC 2905.
6. S. Farrell, J. Vollbrecht, P. Calhoun, and L. Gommans,"AAA Authorization Requirements," RFC 2906, August 2000.
7. S. Glass, T. Hiller, S. Jacobs, and C. Perkins, "Mobile IP Authentication, Authorization, and Accounting Requirements," RFC 2977, 2000.
8. A. Hasan, J. Jahnert, S. Zander and B. Stiller, "Authentication, Authorization, Accounting and Charging for the Mobile Internet," Mobile Summit, September 2001.
9. A. Hess and G. Schafer, "Performance Evaluation of AAA/Mobile IP Authentication," 2nd Polish-German Teletraffic, 2002.
10. H. Kim, D. Choi, and D. Kim, "Secure Session Key Exchange for Mobile IP Low Latency Handoffs," Springer-Verlag Lecture Notes in Computer Science, vol. 2668, pp. 230–238, January 2003.
11. J. Park, E. Bae, H. Pyeon, and K. Chae "A Ticket-based AAA Security Mechanism in Mobile IP Network," Springer-Verlag Lecture Notes in Computer Science 2003, vol. 2668, pp. 210–219, May 2003.

12. H. Jeon, H. Choo, and J. Oh, "IDentification Key Based AAA Mechanism in Mobile IP Networks," ICCSA 2004 vol. 1, pp. 765–775, May 2004.
13. J. McNair, I.F. Akyildiz, and M.D Bender, "An inter-system handoff technique for the IMT-2000 system," INFOCOM 2000, vol. 1, pp. 203–216, March 2000.
14. J. McNair, I.F Akyildiz. and M.D Bender, "Handoffs for real-time traffic in mobile IP version 6 networks," GLOBECOM 2001.IEEE, vol. 6, pp. 3463–3467, November 2001.
15. C. Yang, M. Hwang, J. Li, and T. Chang, "A Solution to Mobile IP Registration for AAA," Springer-Verlag Lecture Notes in Computer Science, vol. 2524, pp. 329–337, November 2002.

Secure Forwarding Scheme Based on Session Key Reuse Mechanism in HMIPv6 with AAA*

Kwang Chul Jeong[1], Hyunseung Choo[1], and Sungchang Lee[2]

[1] School of Information and Communication Engineering,
Sungkyunkwan University, 440-746, Suwon, Korea
{drofcoms,choo}@ece.skku.ac.kr
[2] Department of Telecommunications,
Hankuk Aviation University, 200-1, Goyang, Korea
sclee@hau.ac.kr

Abstract. Due to an increasing number of portable devices over 785-million world Internet users, a support for quality of service (QoS) and security becomes a significant issue in Mobile IP networks. Therefore researchers have focused on the handoff with a great effect on QoS, and they proposed Low Latency Handoff and packet forwarding mechanism as representative fast handoff mechanisms. However when we consider a secure handoff using AAA, above two mechanisms implicit a falling-off of QoS because of a frequent session key distribution. Hence we propose a fast and secure handoff mechanism by authenticating MN at MAP instead of AAAH based on a hash function for reducing the frequency of location updates to Home Agent. We evaluate its performance by dividing packet delivery cost, location update cost, and encryption/decryption cost based on the novel analytical approach. The evaluation result shows the better performance with a lower Call-to-Mobility ratio.

1 Introduction

The Internet Engineering Task Force(IETF) developed the Mobile IP(MIP) de facto standard along with the wireless Internet based on the mobility which is an essential characteristic of the mobile network [10]. Also, due to the drastically increasing number of mobile users and speed of mobile nodes(MN), the area managed by one Access Router(AR) becomes smaller. And these trends naturally need the MN's handoff time to be shorter. Therefore Hierarchical Mobile IP(HMIP) scheme has been proposed to minimize the new Care of Address(CoA) registration cost to the Home Agent(HA) which is originated when the MN performs the handoff [7], and also the new packet forwarding scheme which registers the new CoA to the previous AR not to the HA has been proposed [2][3].

---

* This work was supported in parts by Brain Korea 21 and the Ministry of Information and Communication in Republic of Korea. Dr. H. Choo is the corresponding author.

As we all know, the operations based on the mobile networks are more risky than those in wired ones. So there is the need to develop the technology which satisfies the reliability and mobility concurrently, and finally Mobile IP adapting to the AAA has been proposed [12]. When the MN and agents transmit the data mutually, the AAAH server distributes the session keys to guarantee the reliability. But whenever the MNs move between the domains, they should be distributed new session keys for building a new Security Association(SA) with a corresponding AR. This fact results in severe problems that make the advantages of previously proposed schemes for reducing the handoff costs be removed. Therefore we propose the fast and reliable handoff scheme by integrating our new proposed authentication process based on the hash mechanism with the packet forwarding scheme through the registration process among the ARs.

The remainder of this paper is structured as follows. In Section 2, the comparison between the micromobility and the macromobility, and introduction of a forwarding scheme based on the AAA with LLH are presented. We discuss our proposed scheme in Section 3. After that performance is evaluated with the previous schemes in Section 4. Finally we conclude the paper in Section 5.

2 Related Works

2.1 Packet Forwarding Based on Mobile IP LLH with AAA

In basic Mobile IP, whenever the MN moves to another subnet even though it does not communicate with anyone, the MN should register a new CoA to HA. This problem becomes worse with the increasing population of mobile users, longer distances, and higher mobility patterns [4]. And even if the Call-to-Mobility Ratio(CMR) is quite low, a needless overhead has occurred because of the new CoA registration to the HA and CNs. Therefore, some schemes have been proposed to reduce this overhead.

At first, there is the Low Latency Handoff(LLH) scheme [7]. Basically the LLH focuses on the Regional Registration that performs only local registration within visited domain by putting Gateway Foreign Agent(GFA) or Mobility Anchor Point(MAP). That is, the GFA or MAP can reduce the CoA registration cost by substituting the registration procedure for the MN on behalf of the HA. The formal description of the LLH can be found in [7]. Secondly, there is a packet forwarding scheme [2][3]. In this scheme, the MN does not register the new CoA to the HA whenever it performs the handoff and the packets transmitted to the previous AR are tunnelled to the new AR by performing the proper registration between the ARs. Therefore when the MN performs the handoff, the packets can be transmitted to the new AR through the previous AR without the new CoA registration to the HA. And also, performance evaluation shows that the packet forwarding scheme is superior to the registration on the Mobile IPv6 in terms of costs [13].

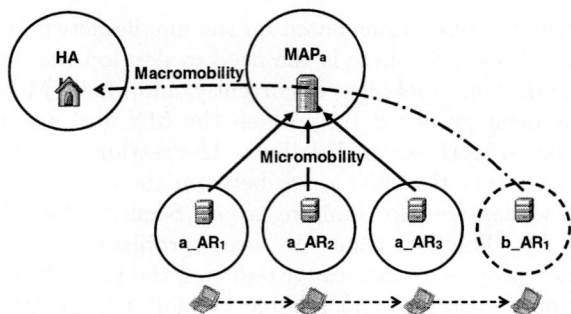

Fig. 1. Micromobility and Macromobility

2.2 Micromobility and Macromobility

User mobility in wireless networks that support IP mobility can be broadly classified with macromobility and micromobility. The macromobility protocols are for the case when an MN roams across different administrative domains of geographical regions. Generally the Macromobility occurs less frequent than the micromobility and involves longer time scales [5]. The micromobility protocols are designed for environments where mobile hosts change their point of attachment to the network quite frequently that the base Mobile IP mechanism introduces significant network overhead in terms of increased delay, packet loss, and signaling [1]. Therefore the MIP paradigm needs to be enhanced and many related works attempt to improve the performance of micromobility. The micromobility means the MN movement across multiple subnets within a single network domain. However due to the characteristic which occurs quite often, the MIP paradigm needs to be enhanced. Most of the related works attempt to improve the MIP micromobility handling capability [1].

3 Secure Forwarding Based on Session Key Reuse

We introduce a MAP-based fast authentication mechanism in case of the call forwarding between ARs under HMIPv6. In this paper, we assume that the MAP stores all hash results for the session keys which are shared among the every entity in the MAP administrative domain. This is possible because a delivery path for new session keys should pass by the Regional CoA(RCoA), the address of MAP, when the related entities are distributed from AAAH. However this also can make some problems. At first, data searching time becomes longer due to the large amount of data stored. So we propose the mechanism to handle this problem for the faster service.

Fig. 2 shows a storing type for the hash result of the session key and MN's ID, and MN's ID itself. And its hash result is k. That is, when the MAP receives the new session key information, it acquires the hash result based on the session key information and the MN's ID. The MAP stores the MN's ID by using that hash

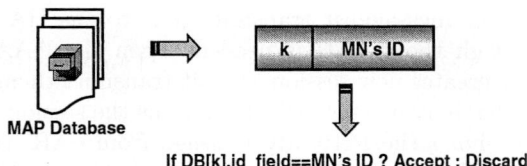

Fig. 2. Authentication using hash function

result as an index in the database. Although the MN's ID is opened to public, a pair of the MN's ID and k can be recognized uniquely due to a uniqueness of k. Therefore we can authenticate the MN easily under a simple procedure which is the verification for 'DB[k].id_field==MN's ID'. Only AR can generate k because the others do not recognize the hash mechanism shared in secret between the AR and the MAP.

Figure 3 through 5 show the handoff procedure employing the security in the HMIPv6. Fig. 3 shows a basic handoff procedure in HMIPv6, Fig. 4 shows the proposed IntraMAP handoff procedure, and Fig. 5 shows the proposed InterMAP handoff procedure.

In Fig. 3, on receiving the PrRtAdv message from the oAR, the MN performs the Address Autoconfiguration(AA) based on the MAC address and requests the registration to the nAR. The nAR requests the authentication to the AAAF for verifying the MN. After the AAAF checks a Network Access Identifier(NAI)

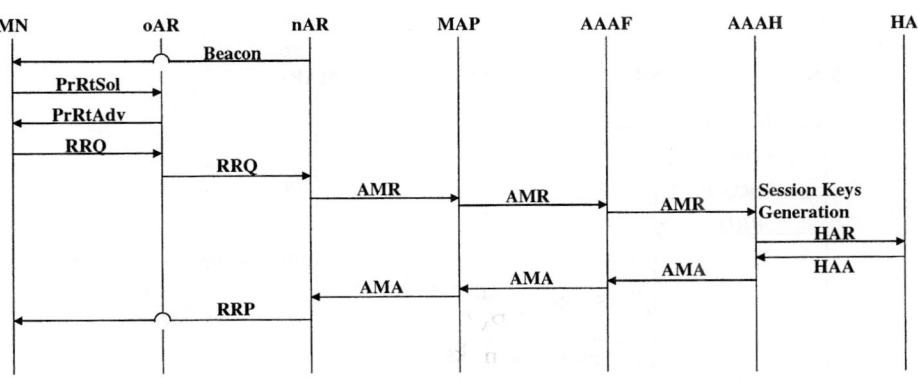

RRQ: Registration Request message
RRP: Registration Reply message
AMR: AA-Mobile-Node Request message
AMA: AA-Mobile-Node Answer message
HAR: Home-Agent-MIP-Request message
HAA: Home-Agent-MIP-Answer message
PrRtSol: Proxy Router Solicitation
PrRtAdv: Proxy Router Advertisement

Fig. 3. Basic Handoff procedure in HMIPv6

in the MN's Diameter message, it transmits NAI to the HA in case that the authentication through the AAAH is needed. Then the AAAH authenticates the MN and after it creates new session keys, it transmits them to the HA. And also the HA sends the reply message which contains the session keys to the nAR.

In Fig 4, on receiving the PrRtAdv message from oAR, the MN performs the AA using MAC address of oAR. The MN sends the new CoA to the oAR, and the oAR acquires the hash result k for the K_{oAR-MN} which is the session key between the oAR and the MN. Then the oAR encrypts k and the MN's ID based on the session key shared with the MAP, $K_{oAR-MAP}$, and the oAR uses an encrypted result as a temporary session key, $K_{oAR-nAR}$. On receiving the $K_{oAR-nAR}$, the MAP decrypts it and authenticates the MN based on k and the MN's ID. If the authentication is successful, the MAP encrypts $K_{oAR-nAR}$ based on the secret key shared with the nAR, $K_{nAR-MAP}$, and sends it to the nAR. Then the nAR detects that there are session keys to receive, and sends the session key request message to the oAR. Next, the oAR encrypts two session keys, K_{oAR-MN} and K_{oAR-HA} based on the temporary key $K_{oAR-nAR}$ and sends them to the nAR. At this time, the nAR can acquire two session keys successfully because the nAR had received the temporary key from the MAP readily. After that, the nAR performs the registration to the oAR for the data forwarding to the oAR.

In Fig. 5, on receiving the PrRtAdv message from oAR, the MN performs the AA and sends the new CoA to the oAR. Then the oAR acquires the hash results for the K_{oAR-MN} and the K_{oAR-HA} as described above. The oAR sends the session key, k, and MN's ID to the old MAP(oMAP). The k value and the

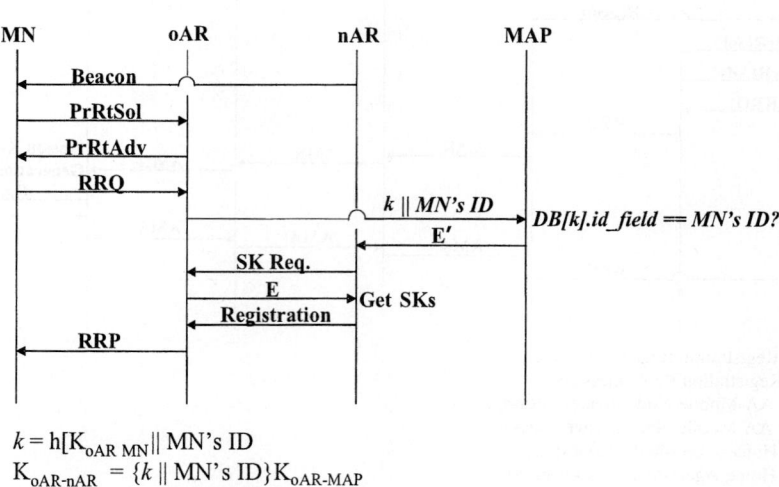

$k = h[K_{oAR\ MN} \| \text{MN's ID}$
$K_{oAR\text{-}nAR} = \{k \| \text{MN's ID}\}K_{oAR\text{-}MAP}$
$E = \{K_{oAR\ MN}, K_{oAR\ HA}\}K_{oAR\text{-}nAR}$
$E' = \{K_{oAR\text{-}nAR}\}K_{nAR\text{-}MAP}$

Fig. 4. Proposed IntraMAP handoff procedure in HMIPv6

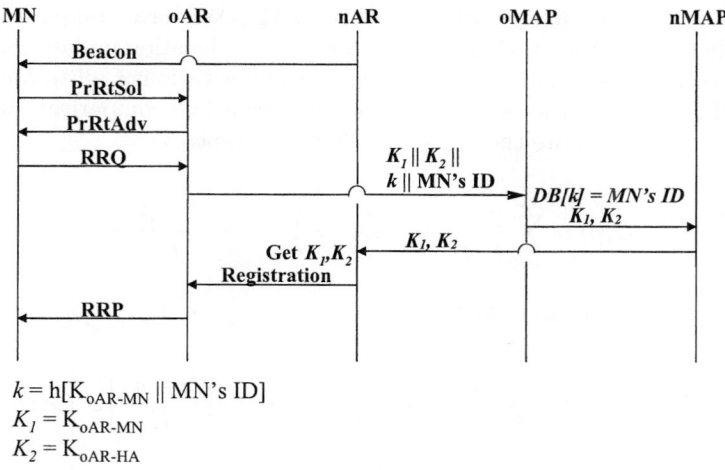

Fig. 5. Proposed InterMAP handoff procedure in HMIPv6

MN's ID are used to verify the MN, and session key is needed because the MAP domain where the MN stays now is different from the one where the MN would move from now on. If we assume that the ARs at the different MAP domain have a different cryptosystem, session keys should be transmitted through the new MAP(nMAP). After the oMAP authenticates the MN, oMAP transmits the session key to the nMAP and nMAP transmits that information to the nAR. Then the nAR acquires the needed two session keys by decrypting the messages based on the secret key shared with the MAP. Finally, the nAR also requests the registration to the oAR for the packet forwarding.

The fact that we can reduce unnecessary overhead that the MN should access to the AAAH whenever it performs the handoff in HMIPv6 is the main focus on the proposed scheme. That is, we can easily authenticate the MN using a hash function at the MAP not the AAAH, and prevent the MN from distributing new session keys whenever it performs the handoff.

4 Performance Evaluation

When we assume that $\alpha(i)$ denote the probability that an MN crosses i subnets between two consecutive packet arrivals, $\alpha(i)$ is defined in [6] and the total cost for location update, packet delivery, and the encryption/decryption under the basic MIP is described in Eq.(1). In the proposed scheme, the MN needs to update to the HA $\lfloor (i+\kappa)/K \rfloor$ times and update to the previous AR $i - \lfloor (i+\kappa)/K \rfloor$ times under given i, κ, and K. Especially the location update cost between ARs can be represented as S, and therefore we can compute the total location update cost like as Eq.(3). And when we compute the average packet delivery cost, we should consider additional packet tunneling cost between ARs. Hence we can represent the total packet delivery cost like as (4), the total encryption/decryption cost like as (6), and finally total cost like as (2). We have compared the proposed scheme with the previous one [11] for the total location update cost, packet delivery cost, and encryption/decryption cost under various Call-to-Mobility ratios(CMR) based on the system parameters. For a fair comparison, we fix the value $\kappa=3$, and compute the costs with $K=4$, $K=$opt.

$$C(\rho) = \sum_{i=0}^{\infty}\Big(iU\alpha(i) + iA\alpha(i)\Big) + F = \frac{U+A}{\rho} + F \qquad (1)$$

$$C'(\kappa, K, \rho) = M'(\kappa, K, \rho) + F'(\kappa, K, \rho) + A'(\kappa, K, \rho) \qquad (2)$$

$$M'(\kappa, K, \rho) = \sum_{i=0}^{\infty}\left(\left\lfloor\frac{(i+\kappa)}{K}\right\rfloor\right)U + \left(i - \left\lfloor\frac{(i+\kappa)}{K}\right\rfloor S\right)\alpha(i) \qquad (3)$$

$$F'(\kappa, K, \rho) = F + \sum_{i=0}^{\infty}\left(i + \kappa - \left\lfloor\frac{(i+\kappa)}{K}\right\rfloor K\right)T\alpha(i) \qquad (4)$$

$$A(\rho) = \sum_{i=0}^{\infty} iE\alpha(i) = \frac{E}{\rho} \qquad (5)$$

$$A'(\kappa, K, \rho) = \sum_{i=0}^{\infty}\left(\left\lfloor\frac{(i+\kappa)}{K}\right\rfloor\right)E + \left(i - \left\lfloor\frac{(i+\kappa)}{K}\right\rfloor E'\right)\alpha(i) \qquad (6)$$

And we define the following notation and parameters for our performance analysis in the rest of this paper. We refer to system parameters in [13] partially and add the cryptography parameters on them. According to the above definitions, the protocols of the MIP, and the proposed scheme described in Section 3, the MN registration cost(U) to the HA, the packet delivery cost(F) in the MIP protocol, the hierarchy setup cost(S), the packet delivery cost between ARs in the proposed scheme(T), the packet encryption/decryption and session key generation cost(E) in MIP, and the packet encryption/decryption cost(E') in the proposed scheme can be expressed as

$$U = 2m_{mr-u} + 2m_{rp-u} + 2m_{ph-u} + 2p_{mr-u} + 2p_{rp-u} + p_{ph-u}$$
$$S = 2m_{mr-u} + 2m_{rr-u} + 2p_{mr-u} + prr - u$$
$$T = m_{rr-d} + p_{rr-d}$$
$$F = m_{hp-d} + m_{rp-d} + m_{rm-d} + p_{hp-d} + p_{pr-d} + p_{rm-d}$$
$$E = 4c_{mr} + c_{rr} + 2c_{rp} + 2c_{ph} + G_{SK}$$
$$E' = 4c_{mr} + 2c_{rp} + 3c_{rr} + 2H_{hash} + 3E_{des}$$

Fig. 6(a) shows the total signaling costs for MIP and the proposed scheme under various CMR values. In this figure, we observe that when the CMR is small, the proposed scheme generates less traffic than the MIP scheme does

Table 1. Notation

Notation	Description
κ	Number of subnets crossed between the last packet arrival and the very last location update just before the last packet arrival
ρ	MN call-to-mobility ratio (CMR)
K	Threshold of the FA hierarchy level
U	Average MN location update cost to its HA
F	Packet delivery cost for a packet to an MN in a foreign network under the MIP scheme
F'	Average packet delivery cost for a packet to an MN in a foreign networks under the proposed scheme
M'	Total location update cost for an MN incurred between two consecutive packet arrivals under the proposed scheme
A	Encryption/Decryption cost for a packet in basic MIP
A'	Total encryption/decryption cost between two consecutive packet arrivals under the proposed scheme
T	Packet delivery cost between ARs in the proposed scheme
S	Hierarchy setup cost in the proposed scheme

Table 2. System Parameter

$p_{mr-d}=1$	$m_{mr-d}=5$	$p_{rr-u}=1$	$m_{rr-u}=50$	$c_{ph}=7.5$	
$p_{rr-d}=0.5$	$m_{rr-d}=25$	$p_{rp-u}=2$	$m_{rp-u}=100$	$G_{sk}=1$	
$p_{pr-d}=1$	$m_{rp-d}=50$	$p_{ph-u}=20$	$c_{mr}=3$	$H_{hash}=0.5$	
$p_{ph-d}=1$	$p_{mr-u}=50$	$m_{mr-u}=20$	$c_{rp}=c_{cr}=1.5$	$E_{des}=1.5$	

Table 3. Set of System Parameters

set	m_{ph-u}	m_{hp-d}	c_{ph}
1	250	125	7.5
2	500	250	15
3	1000	500	30

even without using the optimal values. This fact proves that the registration cost to the HA is quite bigger than the registration cost between ARs. However when we take a look at Fig. 6(b) and (c) which show the relative costs of the proposed scheme to that of the MIP scheme under various CMRs, it is obvious that the total cost for the proposed scheme with fixed threshold can exceed that for the MIP scheme when the CMR becomes larger. Because, the packets for the MN will be processed and tunneled through more ARs before it reaches the destination in the new scheme, and this trend becomes worse according to the increase of κ.

We can also see in Fig. 6(b) that schemes perform better when the relative cost to the HA is higher, and in Fig. 6(c) that schemes perform drastically worse when both K and ρ become larger. The reasons are quite intuitive.

Table 4. Meaning of System Parameters

A_{ab-c}	A	if $A = P \Rightarrow$ processing cost
		if $A = m \Rightarrow$ delivery cost
		if $A = c \Rightarrow$ encryption/decryption cost
	ab	Cost between a and b
	c	if $c = u \Rightarrow$ location update event
		if $c = d \Rightarrow$ packet delivery event
G_{SK}		Session key generation cost
H_{hash}		Hashing cost
E_{des}		Encryption/Decryption cost using DES. In case of wireless communication, we assign 2*E cost

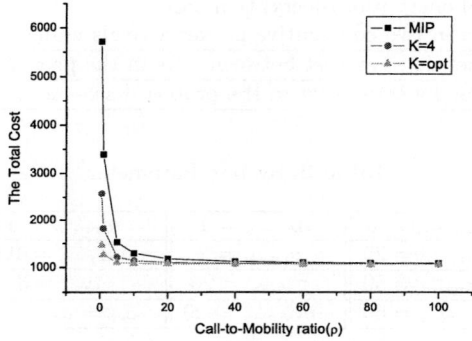

Fig. 6. Comparison of the total costs for different schemes

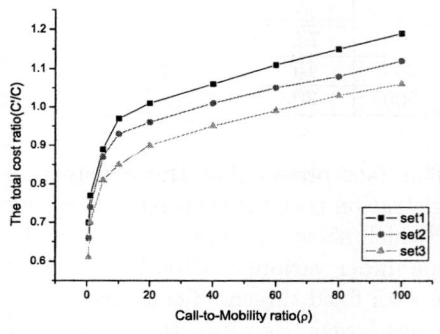

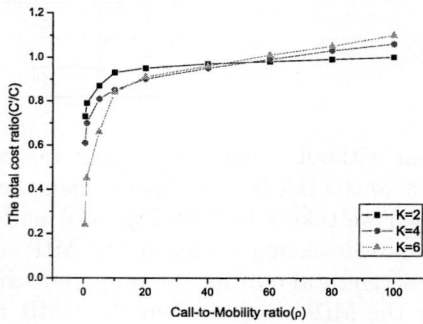

Fig. 7. Total costs for uniformly distributed κ under different parameters

Fig. 8. Comparison of the total costs under different K

5 Conclusion

As the importance of QoS-guarantee has increased, the registration cost to the HA become a main issue in the handoff process. To overcome a large overhead originated with the registration to the HA, the LLH [7] and packet forwarding scheme [2] have been proposed. However when we consider HMIPv6 adapted to the LLH for secure services, the MN should access to the AAAH server to be distributed new session keys whenever it performs handoff. Therefore this paper indicates this kind of a structural defect, and resolves the problem by our proposed fast authentication procedure using the hash mechanism at the MAP and the session key reuse mechanism. We evaluate the performance based on the analytical modeling, and we confirm that evaluation results show the better performance with a lower Call-to-Mobility Ratio.

References

1. A. T. Campbell, J. Gomez, S. Kim, C.-Y. Wan, Z. R. Turanyi, and A. G. Valko, "Comparison of IP micromobility protocols," IEEE WirelessCommun., vol. 9, pp. 72–82, Feb. 2002.
2. C.-H. Chu and C.-M. Weng, "Pointer forwarding MIPv6 mobility management," in Proc. GLOBECOM, Taipei, Taiwan, 2002, pp. 2133–2137.
3. DaeKyu Choi, Hyunseung Choo, and Jong-Koo Park, "Cost Effective Location Management Scheme Based on Hierarchical Mobile IPv6," ICCSA 2003, LNCS 2668, pp. 144–154, 2003.
4. R. Caceres and V. N. Padmanabhan, "Fast and scalable handoffs for wireless internetworks," in Proc. ACMMOBICOM96, 1996, pp. 56–66.
5. S. Das, A. Misra, P. Agrawal, and S. K. Das, "TeleMIP: telecommunications-enhanced mobile IP architecture for fast intradomain mobility," IEEE Pers. Commun., vol. 7, pp. 50–58, Aug. 2000.
6. Y. Fang, I. Chlamtac, and Y. B. Lin, "Portable movement modeling for PCS networks," IEEE Trans. Veh. Technol., vol. 87, pp. 1347–1384, Aug. 1999.
7. E. Gustafsson, A. Jonsson, and C. Perkins, "Mobile IPv4 Regional Registration, draft-ietf-mobileip-reg-tunnel-06.txt," work in progress, Mar. 2002.
8. S. Glass, T. Hiller, S. Jacobs, and C. Perkins, "Mobile IP Authentication, Authorization, and Accounting Requirements," RFC2977, 2000.
9. H. Kim, H. Afifi, "Improving Mobile Authentication with New AAA protocols" IEEE ICC 2003.
10. C. E. Perkins, "Mobile IP," IEEE Commun. Mag., vol. 35, pp. 84–99, May 1997.
11. C. E. Perkins, "Mobile IP Joins Forces with AAA," IEEE Personal Communications, vol. 7, no. 4, pp. 59–61, August 2000
12. Wenchao Ma and Yugang Fang, "Dynamic Hierarchical Mobility Management Strategy," Selected Areas in Communication IEEE Journal on, Volume: 22, Issue: 4, pp. 664-676, May 2004.
13. Yuguang Fang, Chlamtac, I., Yi-Bing Lin, "Portable movement modeling for PCS networks," Vehicular Technology, IEEE Transactions on, Volume: 49, Issue: 4, July 2000.
14. J. Xie and I. F. Akyildiz, "A distributed dynamic regional location management scheme for mobile IP," in Proc. IEEE INFOCOM, 2002, pp. 1069.1078.

A Hierarchical Authentication Scheme for MIPv6 Node with Local Movement Property[1]

Miyoung Kim, Misun Kim, and Youngsong Mun

School of Computer Science, Soongsil University,
Sangdo 5 Dong, Dongjak Gu, Seoul, Korea
mizero31@sunny.ssu.ac.kr, mskim@sunny.ssu.ac.kr,
mun@computing.ssu.ac.kr

Abstract. To reduce the amount of the signaling messages occurred in movement, Hierarchical Mobile IPv6 (HMIPv6) has been introduced as the hierarchical mobility management architecture for MIPv6 by regarding the micro movement. When approaching the visited link, the authentication procedure should be done successfully prior to any mobility support message exchanges. The Authentication, Authorization and Account (AAA) authentication service is applied gradually to the wireless LAN(Local Area Network) and Cellular networks. However, it may bring about the service latency for the sessions of needing the real-time processing due to not providing the optimized signaling in local and frequent movements. In this paper, we propose the authentication architecture with 'delegation' scheme to reduce the amount of signaling message and latency to resume for local movements by integrating it with HMIPv6 architecture. We provide the integrated authentication model and analyze materials comparing to the exiting authentication scheme. It cuts down the cost to 33.6% at average measurement.

1 Introduction

The popularization of the broadband hotspot service using wireless LAN technology and the increase of the needs for security infrastructure to authenticate an user with secure manner in cellular network result in the Diameter-based AAA service being applied to various kinds of related technologies. Moving into the visited network from the outside, visited network and mobile node should take a mutual authentication to prevent them from being vulnerable. It should be performed in prior to any mobility exchanges between the AAA servers in visited and home domain that is not adequate for HMIPv6 hierarchy[1] and its design rationale in inbound roaming which brings about the considerable service latency. In this paper, we propose the optimization of mobility signaling and authentication messages by adding the authentication scheme to the same hierarchy of HMIPv6 when inbound roaming occurs. To detail our idea, we introduce the 'Delegation' and the AAA

---

[1] This research was supported by the MIC(Ministry of Information and Communication), Korea, under the ITRC(Information Technology Research Center) support program supervised by the IITA(Institute of Information Technology Assessment)

O. Gervasi et al. (Eds.): ICCSA 2005, LNCS 3480, pp. 550–558, 2005.
© Springer-Verlag Berlin Heidelberg 2005

delegation entity is able to authenticate the node and distributes the keying-materials in behalf of MN(Mobile Node)'s Home Agent.

2 Proposed Authentication Scheme in HMIPv6

2.1 Related Works and Problems

Mobile IP WG defines the MAP(Mobility Anchor Point) acting a role of temporal HA which provides the MN with mobility service in behalf of HA in case of inbound roaming. The deployment report of it has been announced however the various kinds of vulnerabilities threats the HMIPv6 entities in location binding[2].

Moving the new MAP domain, MN receives the RA message containing the MAP information from the Access Router and determines which MAP is used for MN from the MAP list. If an intruder multicasts the forged RA(Router Advertisement), MN obtains the compromised MAP information that results in the packet redirection attack.

2.2 Propose Authentication Scheme in HMIPv6

The MN should be authenticated itself to get an access to the visited link by message exchanges with the Attendant in the current location which means that the AAA servers in home and visited link determine if it allows for the MN to use that link or not by DIAMETER messages exchanges between them. If the successful case of authentication, the MN carries out the home registration where it should generate the binding key to protect the registration messages first[3-6]. In this paper, by adding the 'delegation' function to the 'MAP+AAA' entity, the HAs role of management for MN's mobility of authenticating the MN and distributing the keying materials is delegated from MNs HA to the 'MAP+AAA' during the authentication progress and then the subsequent movements are handled by the 'MAP+AAA' in behalf of the MNs HA until the lifetime expires. Interacting with the 'MAP+AAA' in most case of the inbound roaming reduces the amount of signaling messages with outside the visited domain significantly.

For in-bound roaming, the service latency arising from the session handover should be minimized to resume the on-going session without service disruption. In this paper, we integrated the authentication with HMIPv6 by defining the extended entity of 'MAP+AAA' to process the authentication not to mention the MAP registration. Fig. 1. shows the authentication extension in HMIPv6 hierarchy.

The n denotes the number of CN communicating with the MN. R and n means the number of access router in current MAP domain and its domain boundary respectively. We get the total number of AR in the AR domain by following formula when adopting the 2-D Random-Walk model.

$$N(AR) = 1 + \sum_{1}^{n} n*6 = 3n(n+1) + 1 \qquad (1)$$

In Random-Walk mode, the next position is determined by two sum of the current position and the value selected in the discrete random set arbitrarily. If the state transition forms the discrete set, this could be a discrete time Random-Walk[7, 8].

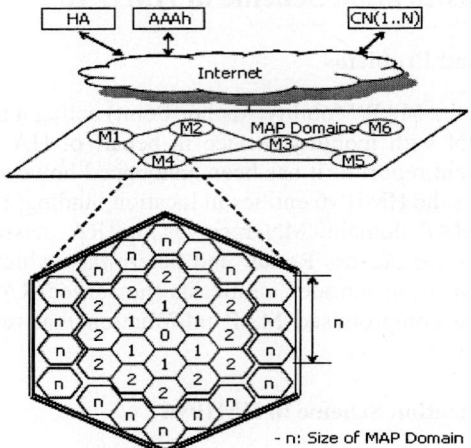

Fig. 1. Authentication architecture for MN in HMIPv6

3 Performance Evaluation

3.1 Analysis Model

Fig 2. depicts the set of states the MN may exist after movement. State 0 means the inner most access router and the number increase going outside. State n is the set of the outer mode access routers determining the size of boundary. The initial location of the MN is not relative to the specific state and the probability of the MN being in the state after random amount of time is defined as.

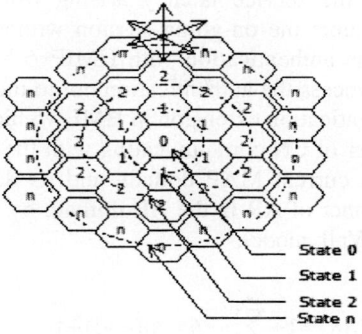

Fig. 2. MAP domain and states

$$\pi^{(n)} = \pi^{(n-1)} P = \pi^{(0)} P^n, n = 1, 2, \ldots \quad (2)$$

The Markov chain in Fig.3. shows that the each state with Random-walk model where $\pi_i(n)$ means the probability of the MN found in state-i after the time[9], n. So, the long-term Steady-State probability of the MN is $\sum_i \pi_i = 1$ [7].

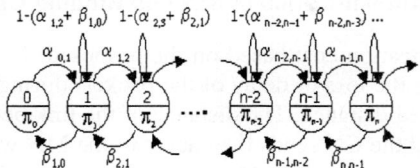

Fig. 3. State registration diagram using Random-Walk mobility model

$$P = \begin{bmatrix} 0 & 1 & 0 & 0 & 0 & . & 0 & 0 & 0 \\ 1/6 & 1/3 & 1/2 & 0 & 0 & . & 0 & 0 & 0 \\ 0 & 1/4 & 1/3 & 5/12 & 0 & . & 0 & 0 & 0 \\ 0 & 0 & 5/18 & 1/3 & 7/18 & . & 0 & 0 & 0 \\ 0 & 0 & 0 & 7/24 & 1/3 & . & 0 & 0 & 0 \\ . & . & . & . & . & . & & & . \\ 0 & 0 & 0 & 0 & 0 & . & 0 & \frac{(n-1)/3-1/6}{n-2} & 0 \\ 0 & 0 & 0 & 0 & 0 & . & \frac{(n-2)/3+1/6}{n-1} & 1/3 & \frac{n/3-1/6}{n-1} \\ 0 & 0 & 0 & 0 & 0 & . & 0 & \frac{(n-1)/3+1/6}{n} & 1/3 \end{bmatrix}$$

Fig. 4. Probability matrix for state transition

Steady-State probability can be calculated from the formula below. Given the transition matrix P, we define I as an identical matrix and let $Q = P - 1$. In this formula, 'e' is the n-vector of all 1's and b is the unit vector with 1's of the position (n+1) and 0's for the rest[10].

$$(Q \mid e)^T \pi = b \quad (3)$$

We define the $\alpha_{r, r+1}$ as the transition probability from a State near by the State 0 to outside and $\beta_{r, r-1}$ from a State in outside to inside. The probability matrix P from a State to another is in Fig.6. We can get the transition probability α and β from the formula 4 and 5 by generalizing transition matrix where the σ is assumed as probability of MN's movement.

$$\alpha_{i, i+1} = \begin{cases} \sigma & if\, i = 0 \\ \sigma\left(\dfrac{1/3(i+1) - 1/6}{i}\right) & if\, 1 \leq i \leq n \end{cases} \quad (4)$$

$$\beta_{i,i-1} = \left\{ \sigma\left(\frac{1/3(i-1) + 1/6}{i} \right) \right\} \quad if\ 1 \le i \le n \tag{5}$$

Also, the following derivation is from the formula 4 and 5.

$$\alpha_{i,i+1} + \beta_{i,i-1} = 2/3, 1 \le i \le n \tag{6}$$

3.2 Cost Analysis of Authentication of MN and Binding Update in HMIPv6

We analyze the authentication cost based on the proposed 'MAP+AAA' operation in HMIPv6 by considering the loss or delay of the packets during the time of processing at the node with weighted distance. The distance of the link between entities is shown as fig 2. We assume that the CN sends the packet to the MN with rate, λ and the MN moves to another subnet with rate μ. We distinguish the data packet from the control packet and let the mean length of it is as l_d and l_c respectively. Therefore the cost delivering the control packet is depending on the distance between sender and receiver and the cost of data packet is l times of control packet. We define the total cost as C_{All} which is the sum of the authentication cost(C_{Auth}), registration cost of LCoA and RCoA in HMIPv6 and packet processing cost sent from the CN(C_{HMIP}).

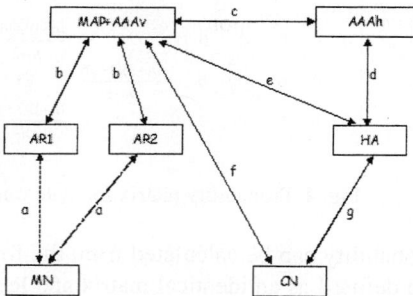

Fig. 5. Cost analysis model of binding registration and authentication

In Fig 5., when the MN moves to a new subnet, we can calculate the cost as the sum of searching the MAP, registering to MAP and authenticating the MN and visited link to register the RCoA to HA. In case of in-bound roaming, the cost is the sum of searching the MAP, registering to MAP and authentication of delegation mode. In this paper, we show the effectiveness of our proposal over the existing scheme considering the packet loss and delay.

$$C_{All} = C_{AUTH} + C_{HMIP} \tag{7}$$

We define the C_{AUTH-g} and C_{AUTH-l} as the authentication cost of DIAMETER exchange between visited and home domain to evaluate the cost when a movement occurs. In general scheme, the AAA servers in visited and home domain participate in the authentication process($C_{diameter(visit, \text{hom } e)}$) whereas only the AAA server in visited domain is for delegation scheme($C_{diameter(visit)}$).

$$C_{AUTH-g} = \frac{\pi_{\overline{u}} \cdot \sigma \cdot C_{diameter(visit, \text{hom } e)}}{E(T)}, 1 \leq i \leq n \qquad (8)$$

When applying the delegation scheme, most of the authentication process is performed by 'MAP+AAA' entity in visited domain and its cost is below.

$$C_{AUTH-l} = \frac{\pi_{\overline{u}} \cdot \sigma \cdot C_{diameter(visit)}}{E(T)}, 1 \leq i \leq n \qquad (9)$$

We can get the authentication cost C_{AUTH} in which the cost is same with general scheme at the first movement to the new MAP domain however the rest processing is more effective since the authentication is processed within a domain if delegation is used.

$$C_{AUTH} = \frac{\pi_n \cdot \alpha_{n,n+1} \cdot C_{diameter(visit, \text{hom } e)} + \pi_n \cdot (1-\alpha_{n,n+1}) \cdot C_{diameter(visit)}}{E(T)} \qquad (10)$$

$C_{diameter(visit, \text{hom } e)}$ is the authentication cost when the MN moves between the MAP domains where the visited and home AAA servers are participated in authentication process. By assumption of r and T_c as the packet processing cost and transmission delay respectively, the cost is below.

$$C_{AUTH} = \frac{\pi_n \cdot (1-\alpha_{n,n+1}) \cdot C_{diameter(visit)}}{E(T)}. \qquad (11)$$

$C_{diameter(visit)}$ is the authentication cost for in-bound roaming which is the sum of searching the MN in delegation list, delivering the Secure_Param_HA to MN and process the packet at each entity.

$$C_{diameter(visit)} = \omega(N(MN) + 2\varphi \log(N(AR)) + 5r, \text{ where } \omega \text{ and } \varphi \text{ are weight factors.} \qquad (12)$$

$N(MN)$ is the number of MNs in the domain and $N(MN) = N(AR) * X$ when we assume the mean number of the MN as X. This paper does not consider the transmission cost between the MN and access routers since it is small amount of constant value that can be ignored. C_{HMIP} consists of the registration cost to MAP(C_{BU-CoA}) and packet transmission cost (C_{packet}). When the MN moves to the outside of the domain, it should register the RCoA to it is HA whereas only LCoA to the MAP when in-bound roaming occurs. So, we can define the cost as below.

$$C_{BU-CoA} = \frac{\pi_n \cdot \alpha_{n,n+1} \cdot C_{RCoA} + \pi_n \cdot (1-\alpha_{n,n+1}) \cdot C_{LCoA}}{E(T)} \qquad (13)$$

If we consider the inbound roaming only, the formula 13 is reformed to formula 14.

$$C_{BU-CoA} = \frac{\pi_n \cdot (1-\alpha_{n,n+1}) \cdot C_{LCoA}}{E(T)} \qquad (14)$$

The packets sent from the CN are tunneled at the HA and delivered to the MN until the route optimization is completed after registering the RCoA to HA. After route optimization, CN takes the direct path to send the packets to MN. So, the packet transmission cost is calculated as below.

$$C_{packet} = C_{CN} + C_{MAP} \qquad (15)$$

The distance from the CN to MN is related on the hop count between them. When we define λ as the traffic ratio of the on-going session, we can get the C_{CN} as the cost of sending the packets by the CN.

$$C_{CN} = \lambda_d \cdot l_d((g+e)+3r) \qquad (16)$$

λ_d is the reception rate of date packet sent from the CN which depends on the traffic condition. t_{AUTH} and t_{MAP} are the time to complete the authentication and MAP registration and l_d is for mean length of the data packet. Receiving the data packet, the MAP searches the binding table(RCoA and LCoA) and routes it to MN with cost of C_{MAP}.

$$C_{MAP} = \lambda_d \cdot (\psi N(MN) + l_d \cdot \omega \log(N(AR)), \text{ where } \psi \text{ and } \omega \text{ are weight factors.} \qquad (17)$$

3.3 Performance Results

3.3.1 Cost Variation Depending on the Session Traffic

The first step of MN's movement is to authenticate itself with security entity in the visited link using the AAA procedure. During the authentication, the security entity takes blocking function in port control for outgoing packets from the MN and intercepts the incoming packets using the filtering function that results in the packet loss and delay until the authentication in finished. Therefore, the more sessions exist between the MN and CN, the more lost or delayed the packet becomes which takes account into the packet-loss cost as below.

$$C_{loss} = \lambda_d \cdot t_{AUTH} \cdot l_d(((g+e)+5r) + (\psi \log(N(AR)))) \qquad (18)$$

During the time (t_{AUTH}) the authentication procedure is underway, the MN receives the packet from the CN with a rate, λ_d where the mean length of the packet is l_d. The packets sent from the CN to MN are tunneled via HA and delivered to the MAP that in turn routes it to the MN. We can define a weighted-factor, ψ for packet processing and routing overhead at a node. When assuming the 'idle' state as the small value of ψ and 'busy' state as a double of 'idle', the fig.8. shows the cost variation of the packet loss or delay according to the traffic property and received packet ratio.

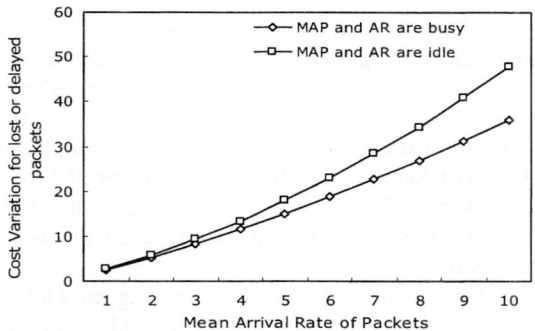

Fig. 6. Packet loss and delay cost in accordance with mean rate of packets received

The more the received ratio becomes high, the more increased rapidly the cost becomes. If the MAP is capable of processing the packets with high receiving ratio, the cost increases smoothly. On the other hand, if the receiving packet ratio is greater than the packet processing capability of MAP, the cost becomes increased more rapidly.

3.3.2 Cost Variation Depending on the Number of MN's

The mobility probability is in proportional to the number of visiting MN's in the area of AR. The movement is depending on the mobility property of the MN and the authentication cost increases in proportional to the number of MN.

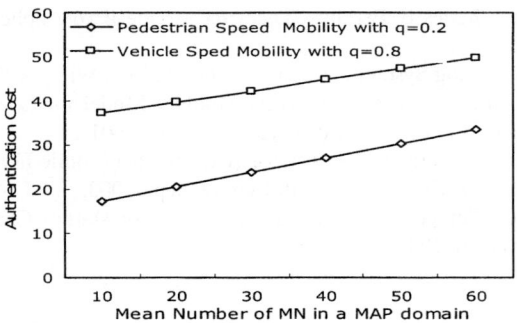

Fig. 7. Cost Variation depending on the MN's in MAP domain

With an assumption of the mean time the MN stays in a subnet, $E(T) = 2$, the fig.9. shows the cost variation depending on the average number of MN's in the MAP domain considering each of which moves with pedestrian ($q = 0.2$) and vehicle

($q = 0.8$) speed respectively. The domain size is assumed as 4(the number of ARs in this domain is 61). The more fast the MN moves with vehicle speed and more MN's exists, the more increased the cost becomes.

4 Conclusion

In this paper, we introduce the MAP+AAA entity with 'delegation' function to provide the authentication service for infra-domain movement with deploying it in the same hierarchy as HMIPv6. The MAP+AAA entity plays a role of the authentication and distribution of keying materials for the movements within the domain that it governs. The cost analysis is taken by considering the number of MN's, the size of MAP domain and traffic properties. The proposed scheme cuts the cost down for the high probability of movement and large amount of traffic on the sessions in comparing with the general approach.

As an adoptable to the real-time application requiring the QoS guarantee, the proposed scheme can be applied to the cellular network and Wireless Fidelity (WiFi) as well as the mobile ip to promote the security service.

References

1. H. Soliman, C. Castelluccia, "Hierarchical Mobile IPv6 mobility management(HMIPv6)", IETF Internet Draft, Mobile IP Working Group, June 2003.
2. A. Mankin, B. Patil, D. Harkins, E. Nordmark, P. Nikander, P. Roberts, T. Narten, "Threat Model introduced by Mobile IPv6", Internet Draft IETF, May 2001.
3. F. Dupont, J. Bournelle, "AAA for Mobile IPv6", Internet Draft, IETF, Nov. 2001.
4. M. Kim, Y. Mun, J. Nah, S. Sohn, "Localized Key Management for AAA in Mobile IPv6", IETF Internet Draft, Mobile IP Working Group, Oct. 2002.
5. Pat. R. Calhoun, Erik Guttman, Jari Arkko, "DIAMETER Base Protocol", Internet Draft, IETF, July 2002.
6. F, Le, B. Patile, Charles E. Perkins, "Diameter Mobile IPv6 Application", Internet Draft, IETF, Nov. 2001.
7. L. Kleinrock, Queueing Systems, Volume 1: Theory, John Wiley & Sons, 1975.
8. K. Chiang, N. Shenoy, "A Random Walk Mobility Model for Location Management in Wireless Network", in proc. IEEE PIMRC Conf. Sept. 2001.
9. S. Pack, Y. Choi, "Performance Analysis of Hierarchical Mobile IPv6 in IP-based Cellular Networks", in proc. IEEE PIMRC Conf. Beijing, Sept. 2003.
10. Waterloo Maple, "Computing the Steady-State Vector of Markov Chain", www.mapleapps.com, 2001.

An Effective Authentication Scheme for Mobile Node with Fast Roaming Property[1]

Miyoung Kim, Misun Kim, and Youngsong Mun

School of Computer Science,
Soongsil University,
Sangdo 5 Dong, Dongjak Gu, Seoul, Korea
{mizero31, mskim}@sunny.ssu.ac.kr,
mun@computing.ssu.ac.kr

Abstract. The long-deferred authentication brings about the latency of mobility service and is critical for the sessions requiring the real time processing. This paper proposes the scheme for providing the authentication service in a local domain with maintaining the design rationale of HMIPv6. Accepting the design logic of HMIPv6, the proposed scheme is able to reduce the overall service latency by reducing the authentication time. To materialize the idea, we analyze the authentication cost by considering the factors of the number of mobile node and its traffic properties.

1 Introduction

The fast moving MN needs to register its location frequently that results in security and performance degrading. To cope with this unexpected traffic, HMIPv6(Hierarchical Mobile IPv6) [1]reduces the amount of signaling messages and its latency for MIPv6 by considering the such a movement. When a MN moves to the visited link away from its home domain, the mutual authentication between the MN and visited link should be performed in prior to any kinds of mobility support messages to protect them being vulnerable to an attacker[2]. In general AAA scheme, the authentication is processed by messages exchange between the AAA servers in home and visited domain regardless of the micro mobility and causes the long-term latency depending on the authentication completion time.

In this paper, we make it possible for the MN to be authenticated with optimized signaling messages by integrating the AAA entity with the MAP(Mobility Anchor Point) of HMIPv6 and introducing the 'delegation' function to the AAA which enables the MN to be served by the AAA entity in local domain on behalf of the AAA server in its home domain until the delegation lifetime expires[3, 4].

---

[1] This research was supported by the MIC(Ministry of Information and Communication), Korea, under the ITRC(Information Technology Research Center) support program supervised by the IITA(Institute of Information Technology Assessment).

2 Authentication Scheme in HMIPv6

2.1 Authentication Approach

Mobile IP WG defines the MAP(Mobility Anchor Point) acting a role of temporal HA which provides the MN with mobility service in behalf of HA in case of inbound roaming. The deployment report of it has been announced however the various kinds of vulnerabilities threats the HMIPv6 entities in location binding[2].

Moving the new MAP domain, MN receives the RA message containing the MAP information from the Access Router and determines which MAP is used for MN from the MAP list. If an intruder multicasts the forged RA, MN obtains the compromised MAP information that results in the packet redirection attack.

2.2 Enhanced Authentication Scheme Proposed for HMIPv6

By adding the DIAMETER-based authentication entity to the basic hierarchy of the MIPv6 processing the micro mobility, it is possible that the MN is authenticated and generates the key within the visited domain when local movements has occurred. he AAAv server in visited domain is a DIAMETER-based authentication server located in the same level with the MAP(Mobility Anchor Point) where it may be in the same machine as MAP or not. The MN should be authenticated itself to get an access to the visited link by message exchanges with the Attendant in the current location which means that the AAA servers in home and visited link determine if it allows for the MN to use that link or not by DIAMETER messages exchanges between them. If the successful case of authentication, the MN carries out the home registration where it should generate the binding key to protect the registration messages first[5,6]. In this paper, by adding the 'delegation' function to the 'MAP+AAA' entity, the HAs role of management for MN's mobility of authenticating the MN and distributing the keying materials is delegated from MNs HA to the 'MAP+AAA' during the authentication progress and then the subsequent movements are handled by the 'MAP+AAA' in behalf of the MNs HA until the lifetime expires. Interacting with the 'MAP+AAA' in most case of the inbound roaming reduces the amount of signaling messages with outside the visited domain significantly.

For in-bound roaming, the service latency arising from the session handover should be minimized to resume the on-going session without service disruption. In this paper, we integrated the authentication with HMIPv6 by defining the extended entity of 'MAP+AAA' to process the authentication not to mention the MAP registration. Fig. 1 shows the authentication extension in HMIPv6 hierarchy.

The n denotes the number of CN communicating with the MN. R and n means the number of access router in current MAP domain and its domain boundary respectively.

$$N(AR) = 1 + \sum_{1}^{n} n*6 = 3n(n+1) + 1 \qquad (1)$$

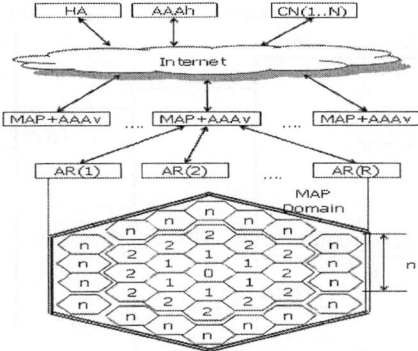

Fig. 1. An authentication mode in HMIPv6

The n denotes the number of CN communicating with the MN. R and n means the number of access router in current MAP domain and its domain boundary respectively.

$$N(AR) = 1 + \sum_{1}^{n} n*6 = 3n(n+1)+1 \qquad (1)$$

2.3 Authentication Procedure for Inter-/Intra-domain Movements

Since the security entity in visited link has no idea of how to authenticate the MN, the authentication message exchanges are sent out of the domain that results in the considerable amount of service latency although the in-bound roaming occurs. This section describes the authentication based on AAA infrastructure providing the strong security with its natural service. The messages are in [3,5,6].

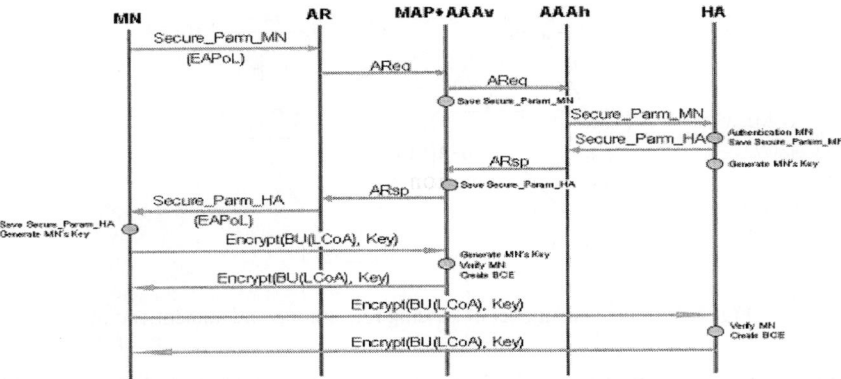

Fig. 2. The authentication and binding registration procedure for inter-domain movement

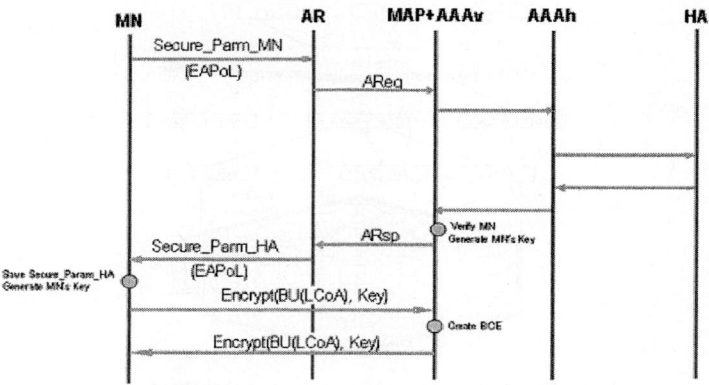

Fig. 3. Proposed authentication procedure using the delegation for intra-domain movement

3 Performance Evaluation

3.1 System Modeling and Cost Analysis

We analyze the authentication cost based on the proposed 'MAP+AAA' operation in HMIPv6 by considering the loss or delay of the packets during the time of processing at the node with weighted distance. The distance of the link between entities is shown as fig.4. We assume that the CN sends the packet to the MN with rate, λ and the MN moves to another subnet with rate μ. We distinguish the data packet from the control packet and let the mean length of it is as l_d and l_c respectively. Therefore the cost delivering the control packet is depending on the distance between sender and receiver and the cost of data packet is l times of control packet[8, 9].

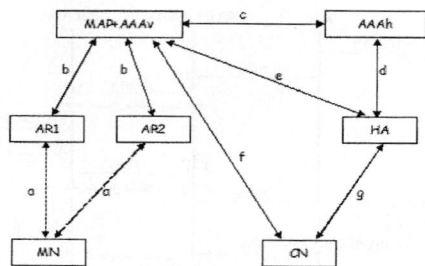

Fig. 4. Cost analysis model of binding registration and authentication

Fig.5. depicts the set of states the MN may exist after movement. State 0 means the inner most access router and the number increase going outside. State n is the set of the outer mode access routers determining the size of boundary. The initial location of

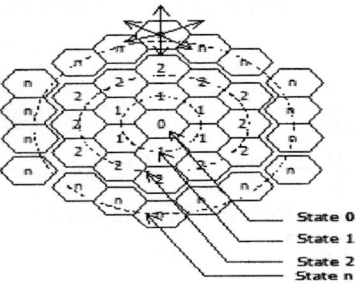

Fig. 5. MAP domain area and states

the MN is not relative to the specific state and the probability of the MN being in the state after random amount of time is defined as.

$$\pi^{(n)} = \pi^{(n-1)} P = \pi^{(0)} P^{n}, n = 1,2,... \tag{2}$$

The Markov chain in Fig.5. shows that the each state with Random-walk model where $\pi_{i(n)}$ means the probability of the MN found in state-i after the time[9], n. So, the long-term Steady-State probability of the MN is $\sum_i \pi_i = 1$ [7].

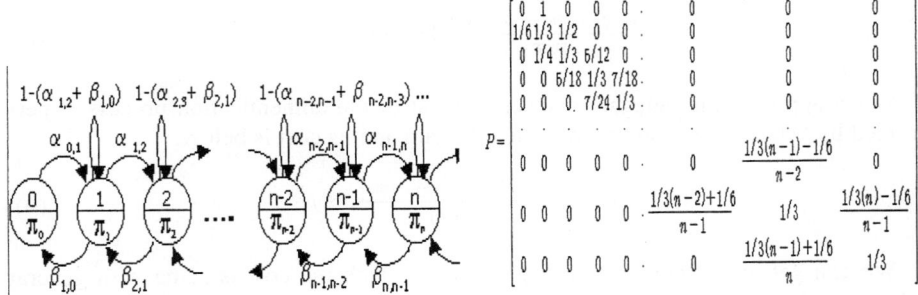

Fig. 6. State registration diagram using Random-Walk mobility model

$$P = \begin{bmatrix} 0 & 1 & 0 & 0 & 0 & \cdot & 0 & 0 & 0 \\ 1/6 & 1/3 & 1/2 & 0 & 0 & \cdot & 0 & 0 & 0 \\ 0 & 1/4 & 1/3 & 5/12 & 0 & \cdot & 0 & 0 & 0 \\ 0 & 0 & 5/18 & 1/3 & 7/18 & \cdot & 0 & 0 & 0 \\ 0 & 0 & 0 & 7/24 & 1/3 & \cdot & 0 & 0 & 0 \\ \cdot & \cdot & \cdot & \cdot & \cdot & \cdot & \cdot & \cdot & \cdot \\ 0 & 0 & 0 & 0 & 0 & \cdot & 0 & \frac{1/3(n-1)-1/6}{n-2} & 0 \\ 0 & 0 & 0 & 0 & 0 & \cdot & \frac{1/3(n-2)+1/6}{n-1} & 1/3 & \frac{1/3(n)-1/6}{n-1} \\ 0 & 0 & 0 & 0 & 0 & \cdot & 0 & \frac{1/3(n-1)+1/6}{n} & 1/3 \end{bmatrix}$$

Fig. 7. Probability matrix for state transition

Steady-State probability can be calculated from the formula below. Given the transition matrix P, we define I as an identical matrix and let $Q = P - 1$. In this formula, 'e' is the n-vector of all 1's and b is the unit vector with 1's of the position (n+1) and 0's for the rest[10].

$$(Q \mid e)^{T} \pi = b \tag{3}$$

We define the $\alpha_{r,r+1}$ as the transition probability from a State near by the State 0 to outside and $\beta_{r,r-1}$ from a State in outside to inside. The probability matrix P from a State to another is in Fig.6. We can get the transition probability α and β from the formula 4 and 5 by generalizing transition matrix where the q is assumed as probability of MN's movement.

$$\alpha_{i,i+1} = \left\{ q\left(\frac{1/3(i+1) - 1/6}{i}\right) \right\} \quad \begin{array}{l} if i = 0 \\ if 1 \leq i \leq n \end{array} \tag{4}$$

$$\beta_{i,i-1} = \left\{ q\left(\frac{1/3(i-1) + 1/6}{i}\right) \right\} \quad if 1 \leq i \leq n \tag{5}$$

Also, the following derivation is from the formula 4 and 5.

$$\alpha_{i,i+1} + \beta_{i,i-1} = 2/3, 1 \leq i \leq n \tag{6}$$

3.2 Authentication and Location Registration Cost of MN

In this Section, we calculate the cost to evaluate the effectiveness of our proposal in comparing with existing authentication scheme in HMIPv6. We define the total cost as C_{All} which is the sum of the authentication cost (C_{Auth}), registration cost of LCoA and RCoA in HMIPv6 and packet processing cost sent from the CN (C_{HMIP}).

$$C_{All} = C_{AUTH} + C_{HMIP} \tag{7}$$

We define the C_{AUTH-g} and C_{AUTH-l} as the authentication cost of DIAMETER exchange between visited and home domain to evaluate the cost when a movement occurs. In general scheme, the AAA servers in visited and home domain participate in the authentication process ($C_{diameter(visit, hom\,e)}$) whereas only the AAA server in visited domain is for delegation scheme ($C_{diameter(visit)}$).

$$C_{AUTH-g} = \frac{\pi_i \cdot q \cdot C_{diameter(visit,\,hom\,e)}}{E(T)}, 1 \leq i \leq n \tag{8}$$

When applying the delegation scheme, most of the authentication process is performed by 'MAP+AAA' entity in visited domain and its cost is below.

$$C_{AUTH-l} = \frac{\pi_i \cdot q \cdot C_{diameter(visit)}}{E(T)}, 1 \leq i \leq n \tag{9}$$

We can get the authentication cost C_{AUTH} in which the cost is same with general scheme at the first movement to the new MAP domain however the rest processing is more effective since the authentication is processed within a domain if delegation is used.

$$C_{AUTH-l} = \frac{\pi_n \cdot \alpha_{n,n+1} \cdot C_{diameter(visit,\,hom\,e)} + \pi_n \cdot (1 - \alpha_{n,n+1}) \cdot C_{diameter(visit)}}{E(T)} \tag{10}$$

$C_{diameter(visit,\,hom\,e)}$ is the authentication cost when the MN moves between the MAP domains where the visited and home AAA servers are participated in authentication process. By assumption of r and T_c as the packet processing cost and transmission delay respectively, the cost is below.

$$C_{diameter(visit,\,hom\,e)} = 2\varphi(\log(N(AR)) + T_c) + 9r, \text{ where } \varphi \text{ is the weight factor.} \tag{11}$$

$C_{diameter(visit)}$ is the authentication cost for in-bound roaming which is the sum of searching the MN in delegation list, delivering the Secure_Param_HA to MN and process the packet at each entity.

$$C_{diameter(visit)} = \omega(N(MN) + 2\varphi \log(N(AR)) + 5r, \text{ where } \omega \text{ and } \varphi \text{ are weight factors.} \quad (12)$$

$N(MN)$ is the number of MNs in the domain and $N(MN) = N(AR) * X$ when we assume the mean number of the MN as X. This paper does not consider the transmission cost between the MN and access routers since it is small amount of constant value that can be ignored. C_{HMIP} consists of the registration cost to MAP(C_{BU-CoA}) and packet transmission cost (C_{packet}). When the MN moves to the outside of the domain, it should register the RCoA to it is HA whereas only LCoA to the MAP when in-bound roaming occurs. So, we can define the cost as below.

$$C_{BU-CoA} = \frac{\pi_n \cdot \alpha_{n,n+1} \cdot C_{RCoA} + \pi_n \cdot (1-\alpha_{n,n+1}) \cdot C_{LCoA}}{E(T)} \quad (13)$$

If we consider the inbound roaming only, the formula 13 is reformed to formula 14.

$$C_{BU-CoA} = \frac{\pi_n \cdot (1-\alpha_{n,n+1}) \cdot C_{LCoA}}{E(T)} \quad (14)$$

The packets sent from the CN are tunneled at the HA and delivered to the MN until the route optimization is completed after registering the RCoA to HA. After route optimization, CN takes the direct path to send the packets to MN. So, the packet transmission cost is calculated as below.

$$C_{packet} = C_{CN} + C_{MAP} \quad (15)$$

The distance from the CN to MN is related on the hop count between them. When we define λ as the traffic ratio of the on-going session, we can get the C_{CN} as the cost of sending the packets by the CN.

$$C_{CN} = \lambda_d \cdot l_d((g+e)+3r) \quad (16)$$

λ_d is the reception rate of date packet sent from the CN which depends on the traffic condition. t_{AUTH} and t_{MAP} are the time to complete the authentication and MAP registration and l_d is for mean length of the data packet. Upon receiving the data packet, the MAP searches the binding table(RCoA and LCoA) and routes it to MN with cost of C_{MAP}.

$$C_{MAP} = \lambda_d \cdot (\psi N(MN) + l_d \cdot \omega \log(N(AR)), \text{ where } \psi \text{ and } \omega \text{ are weight factors.} \quad (17)$$

3.3 Result of Analysis

This section shows the cost variation calculated by the formulas with considering the mobile property, traffic condition and the number of MNs in the domain. To calculate the cost we define the cost and system parameters as shown in the Table 1[10].

Table 1. System Parameters

Param.	Param.	Param.
X =12	L_{RCoA} =20	c =6
ψ =0.1	t_{MAP} =2	d =2
ω =0.001	t_{AUTH} =2	e =6
T_c =80	l_d =0.1	f =6
r =0.00083	a =2	g =6
C_{RCoA} =100	b =2	

$$\pi = \begin{bmatrix} 0.0521571900 \\ 0.2123392793 \\ 0.2937039845 \\ 0.2729282637 \\ 0.1680969723 \end{bmatrix} \quad P = \begin{bmatrix} 0 & 1 & 0 & 0 & 0 \\ 1/6 & 1/3 & 1/2 & 0 & 0 \\ 0 & 1/4 & 1/3 & 5/12 & 0 \\ 0 & 0 & 5/18 & 1/3 & 7/18 \\ 0 & 0 & 0 & 7/24 & 1/3 \end{bmatrix},$$

Fig. 8. State transition and long-term Steady-State probability, where n=4

In Fig. 8 we define the size of the domain as 4(n=4). Then state transition matrix P is obtained by the formula 4 and 5 long-term Steady-State probability π is by the formula 3. By formula 6, the transition probability out of the domain ($\alpha_{4,5}$) is 3/8.

3.3.1 Cost Variation Depending on the Session Traffic

The first step of MN's movement is to authenticate itself with security entity in the visited link using the AAA procedure. During the authentication, the security entity takes blocking function in port control for outgoing packets from the MN and intercepts the incoming packets using the filtering function that results in the packet loss and delay until the authentication in finished. Therefore, the more sessions exist between the MN and CN, the more lost or delayed the packet becomes which takes account into the packet-loss cost as below.

$$C_{loss} = \lambda_d \cdot t_{AUTH} \cdot l_d(((g+e)+5r) + (\psi \log(N(AR)))) \tag{18}$$

During the time (t_{AUTH}) the authentication procedure is underway, the MN receives the packet from the CN with a rate, λ_d where the mean length of the packet

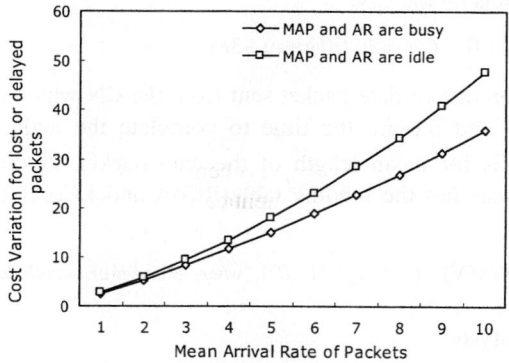

Fig. 9. Packet loss and delay cost in accordance with mean rate of packets received

is l_d. The packets sent from the CN to MN are tunneled via HA and delivered to the MAP that in turn routes it to the MN. We can define a weighted-factor, ψ for packet processing and routing overhead at a node. When assuming the 'idle' state as the small value of ψ and 'busy' state as a double of 'idle', the fig.8. shows the cost variation of the packet loss or delay according to the traffic property and received packet ratio.

The more the received ratio becomes high, the more increased rapidly the cost becomes. If the MAP is capable of processing the packets with high receiving ratio, the cost increases smoothly. On the other hand, if the receiving packet ratio is greater than the packet processing capability of MAP, the cost becomes increased more rapidly.

3.3.2 Cost Variation Depending on the MAP Size

Under taking a pre-assumption of the same mobile and traffic properties the MN has, the fig.10. shows the cost variation depending on the MAP domain size.

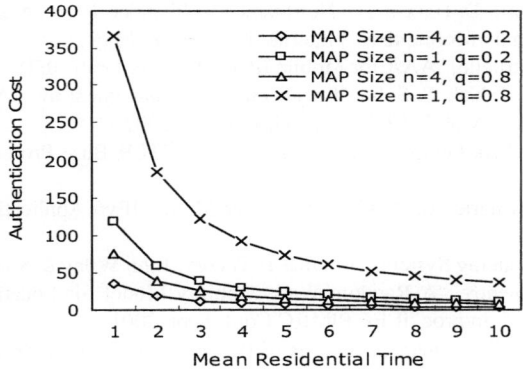

Fig. 10. Authentication Cost depending on the MAP size

The fig.10. shows the cost variation of authentication depending on the MAP domain size (n) for MN's movement with pedestrian ($q = 0.2$) and vehicle ($q = 0.8$) speed respectively. In assumption of the same mobility probability for each case, the message exchanges between the AAA servers in visited and home domain are decreased where the proposed authentication scheme is applied more often which causes the total cost off since the outgoing authentication packets is not necessary often in intra-domain authentication. If the MAP size is constant, the more the mobility probability becomes high, the more increased the total cost becomes since the authentication procedure happens more often for MN's movements.

4 Conclusion Remarks

As hierarchical management architecture to provide the MN in a visited MAP domain with seamless mobility service by the MAP entity, HMIPv6 is suitable for real-time application needed to be optimized and minimized in signaling overhead occurred in mobility service procedure. In the most deployment scenarios, the authentication between the MN and visited link should be performed first to provide the mobility service which is mandated as described in [2] to safeguard the sessions between them from the attacks.

The proposed scheme cuts down the authentication cost up to 57 percent in comparing with general approach by considering the performance factors such as the number of MNs and its traffic properties. As an adoptable to the real-time application requiring the QoS guarantee, the proposed scheme can be applied to the cellular network and WiFi as well as the mobile ip to improve the cost effective security service.

References

1. H. Soliman, C. Castelluccia, "Hierarchical Mobile IPv6 mobility management(HMIPv6)", IETF Internet Draft, Mobile IP Working Group, June 2003.
2. A. Mankin, B. Patil, D. Harkins, E. Nordmark, P. Nikander, P. Roberts, T. Narten, "Threat Model introduced by Mobile IPv6", Internet Draft IETF, May 2001.
3. F. Dupont, J. Bournelle, "AAA for Mobile IPv6", Internet Draft, IETF, Nov. 2001.
4. M. Kim, Y. Mun, J. Nah, S. Sohn, "Localized Key Management for AAA in Mobile IPv6", IETF Internet Draft, Mobile IP Working Group, Oct. 2002.
5. Pat. R. Calhoun, Erik Guttman, Jari Arkko, "DIAMETER Base Protocol", Internet Draft, IETF, July 2002.
6. F, Le, B. Patile, Charles E. Perkins, "Diameter Mobile IPv6 Application", Internet Draft, IETF, Nov. 2001.
7. L. Kleinrock, Queueing Systems, Volume 1: Theory, John Wiley & Sons, 1975.
8. K. Chiang, N. Shenoy, "A Random Walk Mobility Model for Location Management in Wireless Network", in proc. IEEE PIMRC Conf. Sept. 2001.
9. S. Pack, Y. Choi, "Performance Analysis of Hierarchical Mobile IPv6 in IP-based Cellular Networks", in proc. IEEE PIMRC Conf. Beijing, Sept. 2003.
10. Waterloo Maple, "Computing the Steady-State Vector of Markov Chain", www.mapleapps.com, 2001.

A Study on the Performance Improvement to AAA Authentication in Mobile IPv4 Using Low Latency Handoff[1]

Youngsong Mun and Sehoon Jang

School of Computer Science, Soongsil University,
1-1, Sangdo 5 Dong, Dongjak Gu, Seoul, Korea
mun@computing.ssu.ac.kr , rivside@sunny.ssu.ac.kr

Abstract. According to getting expand the importance of the security in mobile telecommunication environment. The security is under the research progress in mobileip working group. From now on, if the Mobile IPv4 technology will be applicable in the various fields and it is expanded, then the security must be core in trust side. In the meantime, the IPsec(Internet Protocol Security) was decided as the method for providing security in Mobile IPv4. And the related study has been progressed. But IPsec has drawbacks that the mobile devices such as cellular phone and PDA are weak in the performance of hardware. so the security processing is a big burden in the mobile devices and security level is not satisfied with mobile users. For working these problems, the integrated models with AAA infrastructure are being proposed. The purpose of AAA technology is to authenticate a mobile node in the foreign network. In this paper, our research has a focus on minimizing the authentication latency in AAA processing. Especially, we propose the model with Low-latency handoff scheme to reduce AAA authentication delay. And we carried out performance evaluation to prove it.

1 Introduction

Because the number of mobile device(such as cellular phone, PDA and notebook etc.) is increased and the technologies are rapidly developed, so providing the seamless services to the users on roaming with secure manner are requested. Currently, the mobileip WG of the IETF (Internet Engineering Task Force) was proposed a technology called Mobile IPv4. Besides, they treat Mobile IP Security with important. The standard document of Mobile IPv4 is proposed IPsec for providing security. If only with IPsec, it may possible to verify and authenticate the messages between mobile node and it's home agent by establishing the SA. However a mobile node is on roaming, it continues the ongoing sessions in the visited domain, it should be authenticated to access visited subnet. But IPsec can't conclude whether the mobile node visiting out of its home network is authenticated and registered in its home network or not.

---

[1] This research was supported by the MIC(Ministry of Information and Communication), Korea, under the ITRC(Information Technology Research Center) support program supervised by the IITA(Institute of Information Technology Assessment).

Also, when processing the packets between a sender and receiver, the IPsec may be loaded an additional burden such as SA lookup, SA creation, payload encryption and ICV calculation to battery-powered mobile node. i.e. currently, the hardware-performance-weaked mobile device(such as cellular phone, PDA) is not enough to process IPsec. To overcome the vulnerability of existing security scheme, the working group of the IETF is proposed the AAA technology. The AAA is infrastructure technology which is satisfying the secure authentication of mobile node. In this paper, we proposed a model based on Francis Dupont's approach and introduced the model combined with Low-latency Handoff to minimize the authentication latency and packet lost. Finally, we'll show the performance of the proposed method.

2 AAA for Mobile IPv6

In this section, we illustrate the representative draft submitted by Francis Dupont, "AAA for Mobile IPv6". This draft shows the integrated method with AAA infrastructure as the IKE message exchanges prior to Mobile IPv6 signaling hits the low performance in mobile environment. The AAA authentication and binding procedure is shown as bellow.

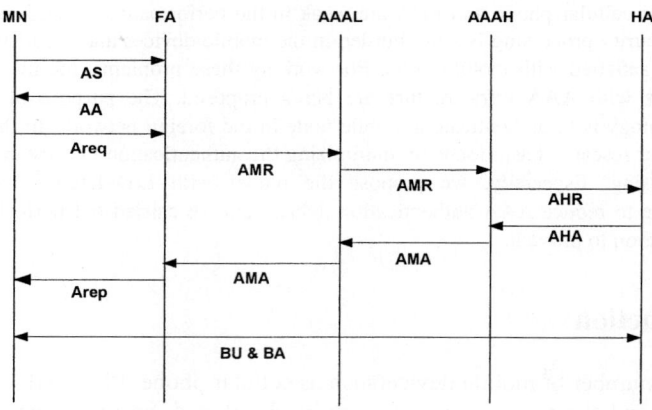

Fig. 1. AAA Authentication and Binding Update Processing

3 Low Latency Handoff for Mobile IPv4

Mobile IPv4 introduced low latency handoff to minimize the service delay occurred in handoff. Low latency handoff is classified according to registration order. The first category of the low latency handoff is pre-registration. It is assumed that bi-directional tunnel between oFA(old Foreign Agent) and nFA(new Foreign Agent). oFA can be obtained nFA's information from the tunnel. Before the MN performs layer2 handoff, the oFA transfers nFA's information to the MN through PrRtAdv message.And MN performs registration to HA.

The second category is post-registration. It is assumed that bi-directional tunnel is created between oFA and nFA previously. And the traffic destined to oFA is forwarding to nFA. The nFA is buffering the traffic packets. When the layer2 handoff of the MN is completed, the packet buffered in the nFA is forwarded to the MN. The Post-registration is divided into 2-party handoff and 3-party handoff. 2-party handoff is general case of the post registration. In this case, When MN is moving to nFA, the packet destined to oFA is forwarded to nFA directly. If the MN's velocity is slow, so it is easy to predict the MN's direction then a registration with nFA is certain. 3-party handoff adds the third FA, i.e. aFA to 2-party handoff architecture. If the MN's velocity is fast, then MN will be registered the nFA or will be moved another nFA. In this case, aFA handles the tunneling point to MN quickly. The message procedure is shown as follow.

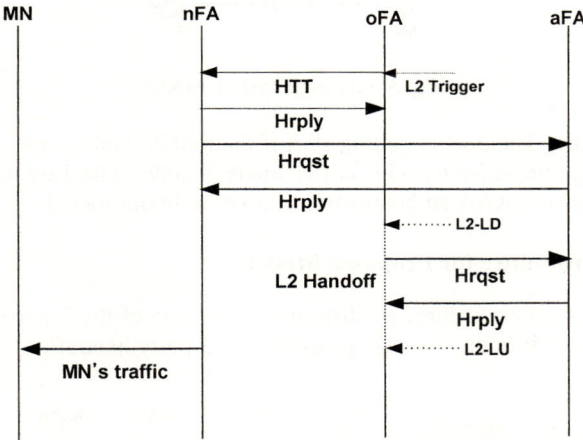

Fig. 2. Post-registration (3-party handoff)

4 Proposed Model

In this chapter, we propose a model based on analysis of related study. Our approach introduces the AAA authentication procedure to authenticate the mobile node in Mobile IPv4 and Low-latency Handoff procedure to minimize the authentication delay occurring from the AAA procedure.

4.1 Proposed System Model

To make proposed system model, we propose the method combined with Low-latency Handoff to reduce the processing time of AAA authentication procedure. In this model, the Foreign Agent plays as the AAA Attendant and Low-latency Handoff in standard documents. The system model applying Low-latency Handoff is as follow.

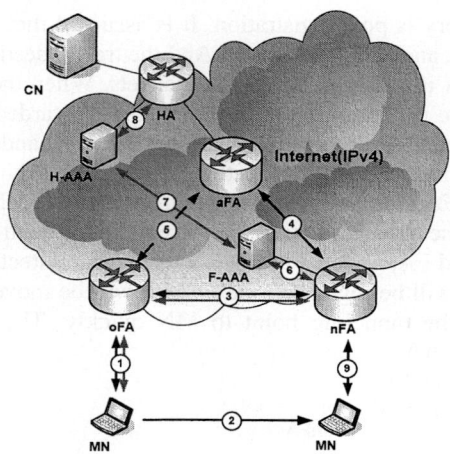

Fig. 3. Proposed System Model

The Low latency Handoff is getting start if the mobile node moves to another subnet with holding the sessions with correspondent node. The Low-latency Handoff procedure triggers the AAA authentication procedure in our model.

4.2 Message Procedure for Proposed Model

In proposed message procedure, we distinguish the case of the 3-party handoff in the post-registration. This is the message procedure in 3-party handoff.

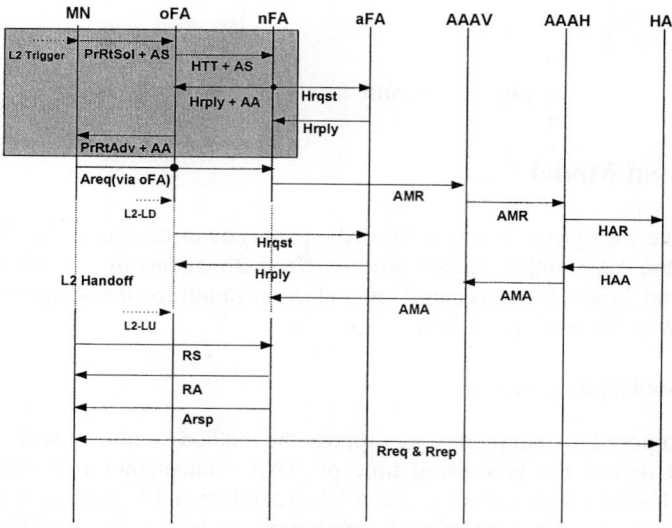

Fig. 4. Proposed Message (3-Party handoff)

Low-latency Handoff operation starts when the MN receives L2-trigger that recognizes a movement of MN. it sends PrRtSol message embedding AS(Attendant Solicit) message to oFA to set up the tunnel and obtained New CoA. At this time, by embedding the AAA message, AS into the PrRtSol, the AAA authentication procedure starts before MN occures Handoff.

The oFA should have the capability to send HTT messages including the AS message. Upon receiving the HTT message, the nFA processes the HTT and AS message and it responds to oFA with Hrply message embedding AA message as response. at this time, it starts tunneling procedure when received Hrply messages and oFA sends the PrRtAdv message including the AA message back to the MN.

MN starts AAA authentication procedure by sending AAA request message, the AReq. Starting the AAA authentication procedure during the L2 Handoff and RS/RA exchange, MN reduces the authentication delay since the rest of AAA procedure is performed after MN moves.

5 Performance Evaluation

5.1 Cost Analysis

The distances among the various entities involved in the proposed scheme are shown in Fig. 5. This system model is proposed for the cost analysis of our proposed scheme. We assume that a CN transmits data packets to the MN at a mean rate λ, and MN

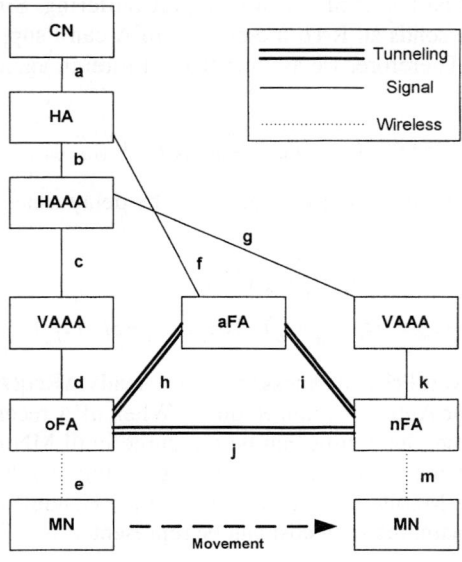

Fig. 5. System Model for Cost Analysis

moves from one subnet to another subnet at a mean rate μ. And we introduce Packet to Mobility Ratio (PMR) as the mean number of packets received by a MN from CN per move. The PMR is given by p = λ/ μ. Let the average length of a control packet be l_c and a data packet be l_d and we define their ratio as $l = l_c / l_d$. we assume that the cost of transmitting a control packet is given by the distance between the sender and receiver. and the cost of transmitting a data packet is l times greater. the average cost of processing control packets at any host is r.

5.2 Cost Analysis of Proposed Model

For the Mobile IPv4 Using Low latency Handoff and AAA, during the time interval when the MN moves to new domain, the total cost is defined as follow. It is given by (1).

$$C_{total} = C_{signal} + C_{delivery} \tag{1}$$

The signaling cost of MN authenticating at the new domain and the signaling cost of MN registering HA at the new domain is given by (2).

$$C_{signal} = 5m + 4i + 3(e+j) + 2(h+f+k+g+b) + 30r \tag{2}$$

Packet delivery cost can be represented by (3) in proposed model. Packet delivery cost is formed packet forwarding cost and packet loss cost. But in this paper, packet's loss isn't considered. The packets may be arrived at nFA before the MN attaches to nFA. These packets will be lost if nFA can't support buffering. Similarly, if the MN attaches to nFA and then sends an Rreq message, if oFA can't support buffering then the packets will be lost. Therefore, we assume that all foreign agents provide its own support for buffering.

$$C_{delivery-low} = (\lambda \times t_{delay-low} \times C_{dt-low}) + C_{loss-low} \tag{3}$$

Until packet is forwarding to MN at new domain, the delay time can be represented by (4).

$$t_{delay-low} = 2(t_m + t_f) + 4t_i + t_j + t_e + 12t_r$$
$$+ \max[t_e + t_j + t_m + 2(t_k + t_g + t_b) + 10t_r, 2t_m + 3t_r + t_{L2}] \tag{4}$$

At (4), The MN received PrRtAdv message and it sends AReq(Authentication Request) message to nFA for Authentication Request. When nFA receives HTT message from oFA, from these time, delay time can be measured until MN registers with HA. We calculated that larger one of the two delay time by using max function.

Therefore, When MN at Mobile IPv4 using Low latency Handoff and AAA Authetication moves to new domain, the total cost can be represented by (5)

$$C_{total-low} = 5m + 4i + 3(e+j) + 2(h+f+k+g+b) + 30r$$
$$+ (\lambda \times t_{delay-low} \times C_{dt-low}) + C_{loss-low} \tag{5}$$

5.3 Performance Evaluation of Proposed Model

Fig. 6 shows the packet delivery cost depending on the PMR value when the MN received PrRtAdv message. At a high values of PMR (p>100), the packet delivery cost is increasing. But proposed model is lower than general model in the cost. Fig. 7 shows the variation of the total cost ratio ($C_{total-low} / C_{total-g}$) depending on the PMR value. At a high values of PMR (p>100), the total cost ratio approximates 0.81. Therefore, proposed scheme is more cost effective than Mobile IPv4 handoff with AAA up to 19%.

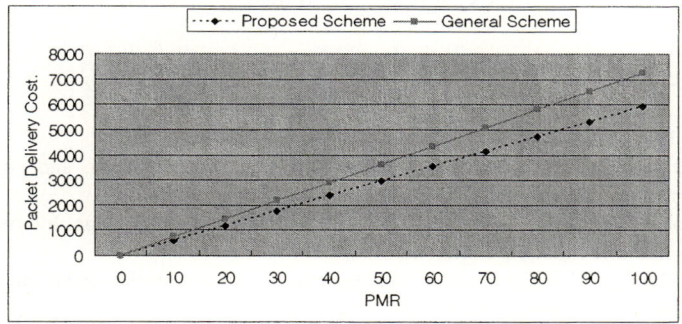

Fig. 6. The Packet Delivery Cost depending on the PMR

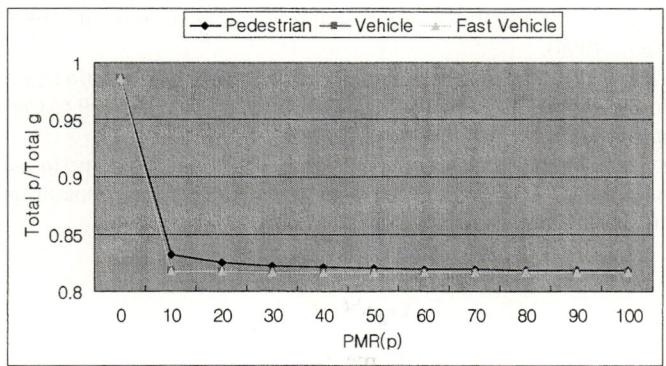

Fig. 7. The Cost ratio of $C_{total-low} / C_{total-g}$

6 Conclusions

Through the analysis of the vulnerability of Mobile IPv4 security in Mobile IPv4 working group, the authentication models integrated with AAA infrastructure are being proposed as the solution actively. Especially, the advanced AAA technology, the Diameter is being researched very much.

It is the peer to peer AAA protocol which has light weight protocol to provide the existing service such as PPP and extensible to the new coming service such as Mobile IP. However the MN is moving repeatedly within a short distance, the AAA authentication procedure may be a big burden to the MN.

In this paper, we introduce the model applying the AAA infrastructure as the solution of a weakness of security in the Mobile IPv4 and we propose Low latency Handoff to reduce the authentication delay for AAA message exchange. Our proposed model costs the lower than general model at the signaling and packet transmission. And it has a benefit from reducing the authentication time and packet loss by adopting the Low latency Handoff. After, if it can be predicted L2 trigger event then we can expect accuracy effect of the cost reduction in our proposed model.

References

1. Charles E. Perkins: IP Mobility Support for IPv4, RFC 3344 (2002).
2. S. Glass and C. Perkins: Mobile IP Authentication, Authorization, and Accounting Requirements, RFC 2977(2000).
3. .P.Calhoun, T. Johansson, and et al.:Diameter Mobile IPv4 Application, Internet draft, draft-ietf-aaa-diameter-mobileip-20(2004).
4. Francis Dupont, Maryline Laurent-Maknavicius, Julien Bournelle: AAA for Mobile IPv6, Internet Draft, draft-dupont-mipv6-aaa-01(2002).
5. C.Perkins and T. Eklund: AAA for IPv6 Network Access, Internet Draft, draft-ietf-perkins-aaav6-06(2003).
6. K. El Malki: Low Latency Handoffs in Mobile IPv4, Internet draft (2004).
7. R. Jain, T.Raleigh, C.Graff and M.Bereschinsky: Mobile Internet Access and QoS Guarantees using Mobile IP and RSVP with Location Registers, in Proc. ICC'98 Conf., pp.1690-1695, Atlanta.(1998).
8. Thomas, R., H. Gilbert and G. Mazzioto: Influence of the mobile station on the performance of a radio mobile cellular network, Proc. 3rd Nordic Sem., paper 9.4, Copenhagen, Denmark, Sep.(1988).
9. Jon-Olov Vatn: An experimental study of IEEE 802.11b handover performance and its effect on voice traffic, SE Telecommunication Systems Laboratory Department of Microelectronics and Information Technology (IMIT) (2003).

Authenticated Key Agreement Without Subgroup Element Verification*

Taekyoung Kwon

Sejong University, Seoul 143-747, Korea
tkwon@sejong.ac.kr

Abstract. In this paper, we rethink the security of authenticated key agreement and describe a simple protocol from the practical perspective. Our main focus is on reconsidering the need for real-time checking of random exchange to be in a proper subgroup, and on embedding identity assurance into implicitly authenticated keys. In spite that the result of this paper is not novel at present, it might be worth discussing the implication of authenticated key agreement not requiring extra computations or message blocks in run time. Trivial range checking is to be sufficient.

Keywords: Cryptographic Protocols, Authentication, Authenticated Key Agreement, Small Subgroup Attacks.

1 Introduction

Key agreement (or key exchange) protocols are necessary when two (or more) communicating parties wish to contribute information for establishing a new secret key. Neither party is allowed to predetermine the key before exchanging the information. Diffie-Hellman key agreement is the best flavor of public-key cryptography that allows basically two communicating parties *Alice* and *Bob*, without sharing any secret a priori, to derive a new secret key over a public channel [6].

During the past decades, Diffie-Hellman has been the most influential building block for such various cryptographic protocols from both theoretical and practical perspectives [2, 4, 10, 12, 15]. Let G be a cyclic group with generator g, for example, a multiplicative group $\mathbb{Z}_p^*$ where p is a large prime integer. *Alice* and *Bob* choose random integers x and y, respectively, where $1 \leq x, y \leq p-2$. They exchange $X = g^x \mod p$ and $Y = g^y \mod p$ for computing $Y^x \mod p$ and $X^y \mod p$, respectively. Hereafter, let us omit ' $\mod p$' from the expressions that are obvious in $\mathbb{Z}_p^*$. The derived secret is $Z = g^{xy} (= (g^y)^x = (g^x)^y)$. We often refer to the Computational Diffie-Hellman (CDH) problem with regard to security of cryptographic protocols from theoretical aspects. The CDH problem is to compute $Z = g^{xy}$ for given $X = g^x$ and $Y = g^y$ (say without x and y). It is widely

---

* This research was supported in part by University IT Research Center Project.

recognized that solving CDH is computationally difficult and is at most as hard as computing discrete logarithms in G [12, 15].

Authenticated key agreement protocols may add implicit or explicit key authentication to Diffie-Hellman [1, 3]. For implicit key authentication, each party should be assured that no other party aside from a specifically identified counterpart can possibly learn the secret key. For explicit key authentication, the party should additionally be assured that the counterpart may actually possess the secret key through key confirmation. The implicit key authentication can be achieved in two message passes, while the explicit key authentication may need three message passes. Thus, the implicit key authentication may be valuable when we consider channel efficiency in the protocol run. We mean by two-pass protocols that implicit key authentication is only provided while we do by three-pass protocols that explicit key authentication is provided for both parties, in light of Diffie-Hellman.

In this paper, we rethink the security of authenticated key agreement and describe a simple protocol from the practical perspective. Our concern is about two-pass authenticated key agreement protocols for implicit key authentication. The resulting protocol is actually similar to the famous MQV protocol in its structure but further considers security of the protocol. Our main focus is on reconsidering the need for real-time checking of random exchange to be in a proper subgroup, and on embedding identity assurance into implicitly authenticated keys. The former is for resisting small subgroup exploiting attacks, while the latter is against unknown key-share attacks in two-pass authenticated key agreement protocol. We discuss the implication of authenticated key agreement not requiring extra computations or message blocks in run time. Trivial range checking (e.g., $1 < X, Y < p$) is to be sufficient.

The rest of this paper is organized as follows: Section 2 describes the background of our study. Section 3 presents the simple protocol while Section 4 briefly analyzes it. This paper is concluded in Section 5.

2 Background

2.1 Desirable Properties of Authenticated Key Agreement

It is essential for secure authenticated key agreement protocols to withstand both *passive attacks* and *active attacks*. Passive attacks mean that an adversary attempts to prevent a protocol from achieving its goal by merely observing the communications between honest entities carrying out the protocol. Active attacks imply that the adversary additionally subverts the communications in any way possible, for example, by injecting, intercepting, replaying, or altering messages. In addition to key authentication and confirmation, there are a number of desirable *attributes* of authenticated key agreement protocols [3].

1. *Known session key*: A protocol should achieve its intrinsic goals even if an adversary learned previous session keys. The Denning-Sacco attack [5] is

the most famous attack disrupting this attribute, while such an attack is categorized as a known session key attack.
2. *(Perfect) forward secrecy*: The secrecy of previous session keys is not affected even if long-term secrets such as private-keys of one or more entities are compromised. Most protocols based on Diffie-Hellman have this property while some of them cannot guarantee it.
3. *Unknown key-share*: Communicating parties are desired to share a new session key with their specified counterparts only in their protocol run. However, an unknown key-share attack means that an adversary could make an honest party believe in sharing the key with herself, while the key is in fact shared with a different principal. For example, *Alice* believes the key is shared with some other party $Carol \neq Bob$ when the key is actually shared with *Bob*.
4. *Key-compromise impersonation*: Suppose that *Alice*'s long-term secret such as a private-key is compromised. An adversary is then able to impersonate *Alice* through the disclosed secret. However, this compromise should not enable the adversary to impersonate other parties to *Alice*.
5. *Loss of information*: Compromise of other information that would not ordinarily be available to an adversary may not affect the security of the protocol, for example, by loss of g^{ab} where a and b represent respective parties' long-term secrets.
6. *Key control*: Neither party is able to force the session key to a pre-selected value. For example, a small subgroup confinement [15] and key recovery attack [10] may break this attribute.

Desirable *performance attributes* are low overheads of computation and communication [3].

2.2 Small Subgroup Attacks

Aside from the theoretical security aspects of Diffie-Hellman, there are many practical concerns about protocol attacks. For example, an active attacker may replace X and Y respectively with X^q and Y^q where $p = Rq + 1$ for small R, so that Z is forced to lie in a small order subgroup. The attacker can also replace X and Y with arbitrary small subgroup elements, even if g is set as large prime order q. Most of such attacks can be prevented by *authenticating the random exchange* [2, 4, 15]. However, it is not enough in many cases.

In the key recovery attack, a small subgroup is exploited for finding partial bits of one's secret [10]. Suppose that $A = g^a$ and $B = g^b$ are respectively the certified public keys of *Alice* and *Bob*. *Alice* may set $X = \alpha g^x$ for a small subgroup element α, for example, in the MTI/A0 authenticated key agreement protocol that *Alice* and *Bob* should have agreed on $K = Y^a B^x$ and $K' = X^b A^y$ respectively. Thus *Alice* can find partial bits $\beta = b \mod ord(\alpha)$ from checking $K' \stackrel{?}{=} \alpha^\beta K$ in $O(2^{|ord(\alpha)|})$ steps, where $ord(\alpha)$ means the order of α [10]. In many protocols, this attack can be prevented by having g of large prime order and *checking the random exchange* to lie in the large prime order subgroup properly ($X^q \stackrel{?}{=} 1$), while such verification requires additional costs [4, 10].

2.3 Motivation

There are a large number of protocols designed for authenticated key agreement in two passes and three passes. Among them the MTI protocol [11] and the STS protocol [7] initiated two different formulations of Diffie-Hellman for authenticated key agreement, and influenced various successors such as MQV and SIGMA protocols [13, 9], respectively. One category is obtained by manipulating the agreed key in more complicated ways, while the other is constructed by adding extra message blocks. Our main concern is the former case; say, implicitly authenticated key agreement is obtained by manipulating the agreed key in more complicated ways. This formulation is valuable especially when we consider channel efficiency in the protocol run, since it can be run by exchanging Diffie-Hellman public keys in two passes without needing extra message blocks. If both communicating parties send the ephemeral public keys at the same time or commit the keys in the previous time period, respectively, the number of rounds can flexibly be reduced in this formulation. However, the protocols formulated in this category are more difficult to resist various kinds of protocol attacks, at least if the explicit key confirmation is omitted for efficiency. The representative attacks are the above-mentioned small subgroup attacks [10, 15] as well as the unknown key-share attacks [8], which may require extra computations or message blocks for resistance.

3 Secure Two-Pass Protocol

We design an authenticated key agreement protocol that is secure and efficient without verifying the random exchange to have large prime order, and satisfies all attributes for implicitly authenticated key agreement. Trivial range checking (e.g., $1 < X, Y < p$) is to be sufficient. Assume the public keys $A = g^a$ and $B = g^b$ are certified by an authority, and manipulated with respective identities I_A and I_B. On certifying the public keys, the authority may check and abort unless those keys are lying in the proper subgroup. *Alice* and *Bob* then can run the following protocol.

1. *Alice* and *Bob* exchange $X = g^x$ and $Y = g^y$, respectively, where $1 < X, Y < p$. It is trivial that both parties may check the range of received values.
2. They compute $K = (YB^{e_B})^{(x+ae_A)e}$ and $K' = (XA^{e_A})^{(y+be_B)e}$, respectively, where $e = h(0, I_A, I_B, X, Y)$, $e_A = h(1, I_A, I_B, X, Y)$ and $e_B = h(2, I_A, I_B, X, Y)$ for a strong one-way hash function $h(\cdot)$. Note that the bit-length of e_A and e_B can be adjusted to be shorter.

It is obvious that $K = K' = g^{(xy+xbe_B+aye_A+abe_Ae_B)e}$ and they are implicitly authenticated due to A and B. For explicit authentication and key confirmation, we can augment it to three passes in the way to exchange $h(3, I_B, I_A, Y, X, K')$ and $h(4, I_A, I_B, X, Y, K)$ [2]. Note also that the simultaneous exponentiation method [14] can be applied in the way that $K = Y^{(x+ae_A)e} \cdot B^{(x+ae_A)ee_B}$ and $K' = X^{(y+be_B)e} \cdot A^{(y+be_B)ee_A}$. Let us use · specifically when we imply simultaneous exponentiation.

4 Analysis

4.1 Security Analysis

We may observe security of our protocol informally. It might be easy to observe that our authenticated key agreement protocol satisfies all of the desirable security attributes listed in [2] and [4]. Among them, we first show that our scheme satisfies *known-key security* and *forward secrecy*, by making a simple reduction from the standard assumption on CDH.

Suppose P is a probabilistic polynomial time algorithm that breaks our protocol by answering K correctly to a query $\langle X, Y, A, B \rangle$ for A and B, with probability ϵ. Given random Diffie-Hellman instance X and Y, we show that we can compute $Z = g^{xy}$ by using P, with probability $\varepsilon \approx \epsilon$ in polynomial time.

Let r_i be chosen at random from $[1, p-2]$. We then input $\langle A^{r_0}, B^{r_1}, A, B \rangle$ to P so that P will produce K_1 by computing $K_1 = (g^{abr_0r_1} g^{abr_0e_{B1}} g^{abr_1e_{A1}} g^{abe_{A1}e_{B1}})^{e_1} = g^{abe_1(r_0r_1 + r_0e_{B1} + r_1e_{A1} + e_{A1}e_{B1})}$ for corresponding exponents $e_1 = h(0, I_A, I_B, A^{r_0}, B^{r_1})$, $e_{A1} = h(1, I_A, I_B, A^{r_0}, B^{r_1})$ and $e_{B1} = h(2, I_A, I_B, A^{r_0}, B^{r_1})$. We then compute $C_1 = K_1^{\{e_1(r_0+e_{A1})(r_1+e_{B1})\}^{-1}} = g^{ab}$.

Similarly, we input $\langle X^{r_2}, B^{r_3}, A, B \rangle$ and $\langle A^{r_4}, Y^{r_5}, A, B \rangle$ to P so that P outputs respectively $K_2 = (g^{xbr_2r_3} g^{xbr_2e_{B2}} g^{abr_3e_{A2}} g^{abe_{A2}e_{B2}})^{e_2} = g^{(xbr_2 + abe_{A2})(r_3 + e_{B2})e_2}$ and $K_3 = g^{(ayr_5 + abe_{A3})(r_4 + e_{B3})e_3}$ for exponents $e_2 = h(0, I_A, I_B, X^{r_2}, B^{r_3})$, $e_3 = h(0, I_A, I_B, A^{r_4}, Y^{r_5})$, and similarly for e_{A2}, e_{B2}, e_{A3}, and e_{B3}. We then compute the value $C_2 = K_2^{\{e_2r_2(r_3+e_{B2})\}^{-1}} C_1^{-r_2^{-1}e_{A2}} = g^{xb}$ and the value $C_3 = K_3^{\{e_3r_5(r_4+e_{B3})\}^{-1}} C_1^{-r_5^{-1}e_{A3}} = g^{ay}$.

Finally we input $\langle X, Y, A, B \rangle$ to P so that P outputs $K = (g^{xy} g^{xbe_B} g^{aye_A} g^{abe_Ae_B})^e$. We then compute $Z = K^{e^{-1}} C_1^{-e_Ae_B} C_2^{-e_B} C_3^{-e_A} = g^{xy}$, so as to solve the CDH problem for given X and Y with probability $\varepsilon \approx \epsilon$ in polynomial time. Thus we can say our scheme enjoys the benefit of Diffie-Hellman in a way that a compromised session key does not expose other session keys, while a compromised long-term key does not disclose previous session keys as well. This simple reduction might be used for full security argument in the further study.

Our protocol is secure against the *key-compromise impersonation* and *loss of information* attacks due to X^b and Y^a embedded in $K = g^{(xy + xbe_B + aye_A + abe_Ae_B)e}$. For example, an adversary compromising Alice's private key a cannot impersonate Bob to Alice due to the required computation of X^b, while loss of either information in $\langle K_{old}, A, B, g^{ab} \rangle$ does not affect the secrecy of K_{new}.

We examine the *unknown key-share* attack that allows *Malice* to make one party believe K to be shared with *Malice* while it is in fact shared with a different party [2, 4]. A common scenario is that *Malice* has $M = g^a$ certified without knowing the private key a of Alice, and uses it to talk with Bob as *Malice* while she poses as Bob to Alice simultaneously. Our protocol is secure against this attack because, for e, we have $h(I_A, I_B, X, Y) \neq h(I_M, I_B, X, Y)$ in computing each K. Note that the closest relative, MQV, is vulnerable to this attack [2, 4, 8].

Finally we show that our scheme is secure against *small order subgroup* attacks specifically without verifying the random exchange to have large prime order in real time. (1) A middle-person attacker may replace X and Y with X^w and Y^w, respectively, where $w = \frac{p-1}{r}$ for small factor r of $p-1$. However, Alice and Bob should respectively have to agree on $K = (Y^w)^{(x+ae_A)e}B^{(x+ae_A)e}$ and $K' = (X^w)^{(y+be_B)e}A^{(y+be_B)e}$. The attacker cannot guess K without obtaining g^{ab} and so on, while the keys are not eventually agreed, say $K \neq K'$. (2) An inside attacker, Alice without loss of generality, may set $X = \alpha g^x$ for a small subgroup element α. She then utilizes a message encrypted under K' or key confirmation $h(3, I_B, I_A, Y, X, K')$ of Bob for deriving partial bits in $O(2^{|ord(\alpha)|})$ steps. However, the partial bits might be set as $\beta = (y + be_B)e$ mod $ord(\alpha)$, not for the long-term private key b only, since Bob has obtained $K' = \alpha^{(y+be_B)e}g^{(xy+xbe_B+aye_A+abe_Ae_B)e}$ while $K = g^{(xy+xbe_B+aye_A+abe_Ae_B)e}$. It is negligible to derive partial bits of b without knowing y. It is also negligible to set $A = \alpha g^x$ intentionally since a certificate authority might have declined it. Note that the insider attack is actually defeated by the postulation of the certification authority.

4.2 Efficiency Analysis

It might be important to remove the obligation for the protocol parties to check the random exchange to have large prime order in real time, since such verification is expensive by requiring one modular exponentiation in G, for example, $X^q \stackrel{?}{=} 1$. This operation is enormous specifically when we set p as a safe prime such that $p = 2q+1$ for large prime q. Thus, our protocol might be very efficient as well as secure in this sense. The required computations are certificate verification of A and B, and modular exponentiations in G for computing $\langle X, K \rangle$ and $\langle Y, K' \rangle$ in respective sides, while X and Y can be pre-computed. Suppose that A and B are already verified, and X and Y are pre-computed by respective parties. This assumption can be considerable for various practical applications. Both Alice and Bob are then able to conduct authenticated key agreement in real time with only one simultaneous exponentiation in G, when we ignore a singular modular multiplication. Note that the simultaneous exponentiation is only about 25% more costly than a single exponentiation.

Though we have chosen $\mathbb{Z}_p^*$ or its large prime order subgroup for wide acceptance, it is also considerable to use a different cyclic group G such as an elliptic curve group for more spatial efficiency and easier manipulation with shorter private keys. Note that elliptic curve groups can be chosen to have prime order, say, without needing intrinsically subgroup element verification.

5 Conclusion

The main goal of this study is to rethink the security of authenticated key agreement and describe a simple protocol from the practical perspective. The two-pass minimization is considered. Our main focus is on reconsidering the need for real-time checking of random exchange to be in a proper subgroup, and on embedding identity assurance into implicitly authenticated keys. The resulting protocol is eventually close to MQV [2, 4, 8]. However, our protocol is secure against the unknown-key share attack as well as is released from real-time checking of random exchange to have large prime order. We believe these properties must be beneficial to practical use.

References

1. S. Blake-Wilson, D. Johnson and A. Menezes, "Key agreement protocols and their security analysis," In Proc. of IMA International Conference on Cryptography and Coding, December 1997.
2. S. Blake-Wilson and A. Menezes, "Authenticated Diffie-Hellman key agreement protocols," SAC '98, Lecture Notes in Computer Science, vol. 1556, pp.339-361, 1999.
3. S. Blake-Wilson and A. Menezes, "Unknown key-share attacks on the station-to-station (STS) protocol," PKC '99, Lecture Notes in Computer Science, vol. 1560, Springer-Verlag, pp. 154-170, 1999.
4. C. Boyd and A. Mathuria, "Protocols for authentication and key establishment," Springer-Verlag, pp.137-199, 2003.
5. D. Denning and G. Sacco, "Timestamps in key distribution protocols," Communications of the ACM, vol. 24, no. 8, pp. 533-536, August 1981.
6. W. Diffie and M. Hellman, "New directions in cryptography," IEEE Transactions on Information Theory, vol.22, no.6, pp.644-654, November 1976.
7. W. Diffie, P. Van Oorschot, and M. Wiener, "Authentication and authenticated key exchanges," Designs, Codes and Cryptography, pp.107-125, 1992.
8. B. Kaliski, "An unknown key-share attack on the MQV key agreement protocol," ACM Transactions on Information and System Security, vol.4, no.3, pp.275-288, August 2001.
9. H. Krawczyk, " SIGMA: The SIGn-and-MAc Approach to Authenticated Diffie-Hellman and Its Use in the IKE Protocols," Advances in Cryptology-CRYPTO'03, LNCS 2729, pp.400-425, 2003.
10. C. Lim, and P. Lee, "A key recovery attack on discrete log-based schemes using a prime order subgroup," CRYPTO '97, LNCS 1294, pp.249-263, 1997.
11. T. Matsumoto, Y. Takashima, and H. Imai, "On seeking smart public-key distribution systems," Trans. of IEICE, E69, pp.99-106, 1986.
12. U. Maurer, and S. Wolf, "Diffie-Hellman oracles," CRYPTO '96, LNCS 1109, pp.268-282, 1996.
13. A. Menezes, M. Qu, and S. Vanstone, "Some new key agreement protocols providing implicit authentication," In Proceedings of the 2nd Workshop on Selected Areas in Cryptography (SAC'95), pp.22-32, 1995.
14. A. Menezes, P. van Oorschot and S. Vanstone, *Handbook of applied cryptography*, CRC Press,Inc., pp.517-518, 1997.
15. P. Van Oorschot and M. Wiener, "On the Diffie-Hellman key agreement with short exponents," EUROCRYPT '96, LNCS 1070, pp.332-343, 1996.

Multi-modal Biometrics with PKIs for Border Control Applications*

Taekyoung Kwon and Hyeonjoon Moon

Sejong University, Seoul 143-747, Korea
{tkwon, hmoon}@sejong.ac.kr

Abstract. In this paper, we present a simple model that enables a low-cost but highly-scalable method for national border control applications. The proposed model could allow combining multi-modal biometrics with public key infrastructures (PKIs), in order to reduce the possibility of undesirable factors significantly. Instead of requiring tamper-resistant smart-card-level devices in a passport, we could print a barcode on it. The proposed model will be instantiated in our separate work.

1 Introduction

The significance of developing border control applications is rapidly growing and the deployment of biometrics is now observed by many countries [15]. Biometrics is actually the science of using digital technologies to identify or verify a human being based on the individual's unique biological (say physiological or behavioral) characteristic such as fingerprint, voice, iris, face, retina, handwriting, thermal image, or hand geometry [4, 8]. Among those various biometric features, fingerprint, iris pattern, facial image, and hand print are regarded as most suitable for border control applications for their relatively-accurate measuring, while it may cost high to deploy biometrics with regard to border control applications.

In this paper, we aim to present a simple model that enables a low-cost but highly-scalable method for combining multi-modal biometrics with public key infrastructures (PKIs), in order to reduce the possibility of undesirable factors significantly at nation's borders. Instead of requiring tamper-resistant smart-card-level devices in a passport, we could print a barcode on it. The proposed model will be instantiated in our separate work. We are currently developing a novel method to exploit the existing useful tools and will instantiate our model in the separate work [7]. Thus, this paper stresses the intention of our study in the proposed model rather than describing the specific schemes. The rest of this paper is organized as follows. Section 2 describes problems and requirements for a low-cost but highly-scalable method for national border control applications, while Section 3 and 4 will describe more details of the proposed model. Section 5 concludes this paper.

---

* This study was supported in part by a grant of the Korea Health 21 R&D Project, Ministry of Health & Welfare, Republic of Korea. (0412-MI01-0416-0002).

2 Preliminaries

2.1 Problems

Depending on biometrics one could face different challenges in border control applications.

- It may take high costs to process a huge amount of biometric information (on-line) for $1:n$ identification and to issue smart-card-enabled passports over the world for $1:1$ verification.
- Biometrics is still remaining as a useful technology in small scale applications, excluding world-wide border control applications.
- A passport holder may feel reluctant to provide his or her biometrics to an inspecting officer because of inconvenience and privacy infringement.

In order to deploy biometrics in border control applications, the firs and second problems must be resolved.

2.2 Requirements

In this subsection, we examine and discuss the requirements of our model to enable a low-cost but highly-scalable method for national border control applications. When we consider the deployment of biometrics in border control applications, we must let a passport holder present to an inspecting officer his or her biometrics along with the passport for $1:1$ verification.

The first requirement is that the existing passport issued by each national body over the world must be refined in low-cost, say without mandatory requirement of embedding any hardware device on it, for accommodating biometrics. In other words, we should be able to print a barcode on the passport along with human readable text, instead of requiring an expensive smart-card-level device for biometric information. The passport should still remain passive in that sense.

The second requirement is that a cryptographic scheme should be applied to the biometric information printed on the passport. A digital signature scheme is appropriate for providing integrity in a stringent way. We explore the most suitable schemes by presenting a formal model.

2.3 Further Problem with Barcodes

The barcode is the dominant automatic identification technology that fits our purpose [10]. Especially 2D codes provide much higher information capacity than conventional ones. The 2D barcode symbol (such as PDF 417 and QR code) can hold up to 4,300 alphanumeric characters or 3,000 bytes of binary data in a small area [2]. One drawback of storing biometric information in publicly readable form is its vulnerability to a potential biometric attack known as a *hill-climbing attack*. This attack could occur when an attacker has access to the biometric system and the user's template upon which (s)he wishes to mount a masquerade attack [12]. The attacker could exploit the compromised biometric template to produce a new image that exceeds the threshold of the biometric system and use that image again as input to the system to which the original template belongs. Therefore, care must be taken when deploying publicly readable barcodes along with biometrics.

3 Proposed Model

3.1 Assumption

Let us assume there exist a number of stable biometrics schemes for considering multi-modal biometrics in border control applications. We also postulate there exist secure digital signature schemes and PKIs for world-wide applications. Due to the rapid growing of the Internet and many business transactions, the cryptographic supports are widely recognized in today's computer applications. Under these assumptions, we devise a formal model for enabling a low-cost but highly-scalable method for national border control applications.

3.2 Formal Model

In order to authenticate a passport holder using biometrics without smart-card devices, we postulate that the passport holders can be screened with regard to their biometrics and demographic information in the passport that is not protected directly by hardware. So the passport holder is defined formally as $\mathcal{U} = \{\mathcal{B}, \mathcal{P}\}$ where $\mathcal{B}$ and $\mathcal{P}$ are defined as user's biometrics and possession (passport) respectively. $\mathcal{B}$ is regarded as a probabilistic algorithm returning user's biometrics while $\mathcal{P}$ is deterministic [6]. As for multi-modal biometrics, $\mathcal{B}$ can be regarded as a set of biometrics, for example, $\mathcal{B} = \{\mathcal{B}_0, \mathcal{B}_1\}$.

Based on our biometrics scheme, we manipulate the user's biometrics with regard to feature representation. We define the following transformation:

- $\mathcal{T}_0 : \mathcal{B}_0 \to \langle \mathcal{B}_{\mathcal{T}_0}, \mathcal{P}_{\mathcal{T}_0} \rangle$

where $\mathcal{B}_{\mathcal{T}_0}$ and $\mathcal{P}_{\mathcal{T}_0}$ are feature representation and eigenvector respectively.

Given a digital signature scheme Σ, we have to manipulate the key returned by $\mathcal{G}_\Sigma$ to be linked with both the user's biometrics and possession. Therefore, we define the following transformation:

- $\mathcal{T}_1 : \langle \mathcal{G}_\Sigma(1^\ell), \mathcal{G}_R(1^\kappa), \mathcal{B}_1 \rangle \to \langle \mathcal{B}_\mathcal{T}, \mathcal{P}_\mathcal{T} \rangle$ and
- $\mathcal{T}_2 : \langle \mathcal{B}_1, \mathcal{B}_\mathcal{T}, \mathcal{P}_\mathcal{T} \rangle \to \mathcal{G}_\Sigma$,

where $\mathcal{G}_R$ is a probabilistic algorithm returning a random integer from input 1^κ, and $\mathcal{B}_\mathcal{T}$ and $\mathcal{P}_\mathcal{T}$ are respective transformed values. We define $\mathcal{P} = \{\mathcal{B}_{\mathcal{T}_0}, \mathcal{B}_\mathcal{T}, \mathcal{P}_\mathcal{T}\}$ while $\mathcal{P}_{\mathcal{T}_0}$ is manipulated as an eigenvector that might be known to an inspector. Note that the inverse transformation is possible for $\mathcal{T}_0$, while it is computationally infeasible for $\mathcal{T}_1$ and $\mathcal{T}_2$. It is impractical for the latter transformations to measure $\mathcal{B}$ by feature extraction which cannot guarantee enough entropy.

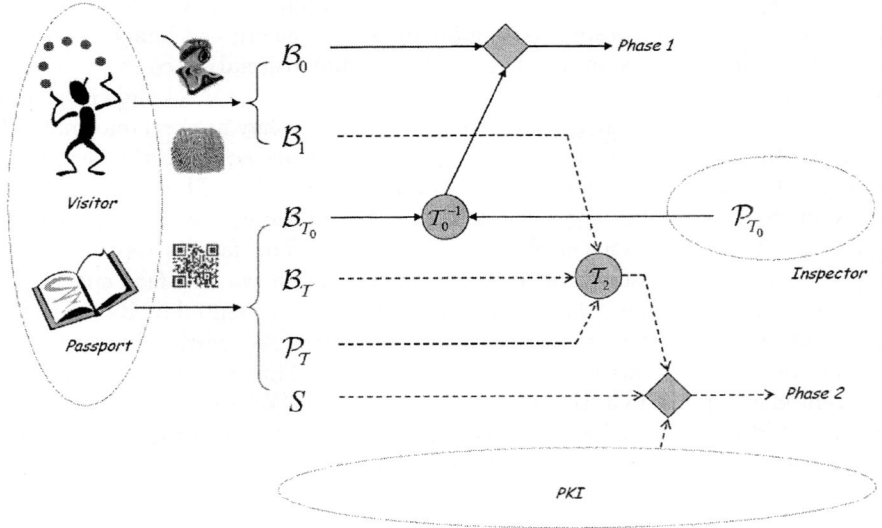

Fig. 1. Authentication Procedure in the Proposed Model

3.3 Authentication Procedure

Figure 1 describes the proposed authentication procedure. A passport holder (or visitor) $\mathcal{U}$ presents $\mathcal{B}$ and $\mathcal{P}$ to an inspecting officer who has $\mathcal{P}_{\mathcal{T}_0}$, a set of eigenvectors. In *Phase 1* (solid line), the inspecting officer retrieves an inverse transformation of $\mathcal{T}_0$ from $\mathcal{B}_{\mathcal{T}_0}$ and $\mathcal{P}_{\mathcal{T}_0}$ [14], and verifies its validity by retrieved information. In *Phase 2* (dashed line), the inspecting officer computes transformation $\mathcal{T}_2$ for deriving a corresponding cryptographic key, and verifies a digital signature S on the information including $\mathcal{P}$. A PKI supports the guaranteed verification of the keying values. In this paper, we technically define $\mathcal{B}_0$ as a face and $\mathcal{B}_1$ as a fingerprint. However, various biometrics can be applied to our scheme, for example, iris codes for $\mathcal{B}_1$ [1]. More details of Phase 1 and Phase 2 need specific schemes for concrete instantiation, while we aim to describe a formal model in abstract in this paper. Thus, more technical details will be studied in our separate work [7]. We discuss possible basic tools for our model in the following section.

4 Basic Tools

4.1 Principal Component Analysis

Principal component analysis (PCA) is a statistical dimensionality reduction method, which produces the optimal linear least squared decomposition of a training set [3, 5]. In a PCA-based face recognition algorithm, the input is a training set $t_1, ..., t_W$ of N images such that the ensemble mean is zero ($\sum_i t_i = 0$). Each image is interpreted as a point in $\Re^{n \times m}$, where the image is n by m pixels. PCA finds a representation in a $(W-1)$ dimensional space that preserves variance. PCA generates a set of $N-1$ eigenvectors $(e_1, ..., e_{N-1})$ and eigenvalues $(\lambda_1, ..., \lambda_{N-1})$. We normalize the eigenvectors so that they are orthonormal. The eigenvectors are ordered so that $\lambda_i > \lambda_{i+1}$. The λ_i's are equal to the variance of the projection of the training set onto the ith eigenvector. Thus, the low order eigenvectors encode the larger variations in the training set (low order refers to the index of the eigenvectors and eigenvalues). The face is represented by its projection onto a subset of M eigenvectors, which we call face space. It is actually a representation of face as a point. A face is represented by its projection onto a subset of eigenvectors in face space. In the face recognition literature, the eigenvectors can be referred to as *eigenfaces* [14, 9]. Thus the normalized face is represented as a point in a M dimensional face space. We conclude that the PCA scheme could be utilized for constructing very efficient face recognition method and providing the Phase 1 instantiation in our model.

4.2   2D Barcode: QRcode

Two-dimensional codes provide much higher information density than conventional barcodes. Due to the low density, conventional barcodes usually function as keys to databases. However, the increased information density of 2D barcodes enables the applications that require encoding of explicit information rather than a database key. QRcode is a 2D matrix symbol which consists of square cells arranged in a square pattern. It allows three models - Model 1, Model 2, and MicroQR. Model 1 and Model 2 each have a position detection pattern in three corners while the MicroQR has it in only one corner. The position detection pattern allows code readers to quickly obtain the symbol size, position and tilt. Model 2 is developed for enhanced specification with improved position correction and large volume of data capacity. MicroQR model is suitable for small amounts of data. A QRcode symbol can encode up to 7,089 characters (numeric data), 4,296 alphanumeric characters, and 2,953 8-bit bytes [2]. The symbol size is determined by the number of characters to encode. It can grow by 4 cells/side from 21x21 cells to 177x177 cells. The physical size of a symbol is determined by the cell pitch. The minimum cell pitch is the width of the smallest printed element that can also be resolved by the reader. With current printing and reader technology, the minimum cell pitch can be as low as 0.1mm and the signature data of 1024-bit(128 bytes) can be printed in an area less than 10 mm sq. The QRcode employs a Reed-Solomon algorithm to detect and correct data errors due to a dirtied or damaged area. There are four levels of error-correction capability that users can select. The error-correction level determines the maximum recoverable rate that is from 7% to 30% of the total code words. We conclude that the QRcode or similar methods are appropriate in our model with regard to their capacity.

4.3 Biometric Encryption and Digital Signature

Since it is not easy to derive a cryptographic key from biometric information which has variations, much work have been done to use an independent, two-stage process to authenticate the user through biometrics and release the key from hardware storage. Recently, an innovative technique has been developed by C. Soutar et al [13]. It links the key with the biometric at a more fundamental level during enrollment and then retrieve it using the biometric during verification. Subsequently, a novel technique that generates a digital signature from biometrics has been developed in our previous work [6]. Thus, we conclude that the digital signature scheme based on biometrics can be used for constructing very effective digital signature method and providing the Phase 2 instantiation in our model.

5 Conclusion

In this paper, we explore a simple model for enabling a low-cost but highly-scalable method in border control applications. In the proposed model, it is allowed to combine multi-modal biometrics with PKIs, in order to reduce the possibility of undesirable factors significantly at nation's border. Instead of requiring tamper-resistant smart-card-level devices in a passport, we could print a barcode on it for practical deployment in the existing infrastructure. The proposed model will be instantiated in our separate work [7].

References

1. J. Daugman, "High confidence personal identifications by rapid video analysis of iris texture," IEEE Conference on Security Technologies, pp.50-60, 1992.
2. Denso Inc., "QRmaker: User's Manual," *Denso Corporation*, Aichi, Japan, 1998.
3. K. Fukunaga, "Introduction to statistical pattern recongition," Academic Press, Orlando, FL, 1972.
4. A. Jain, L. Hong, and S. Pankanti, "Biometric identification," *Commun. of the ACM*, February 2000.
5. I. Jolliffe, "Principal Component Analysis," Springer-Verlag, 1986.
6. T. Kwon, "Practical digital signature generation using biometrics," Proceedings of ICCSA 2004, Lecture Notes in Computer Science, Springer-Verlag, 2004.
7. T. Kwon and H. Moon, Future Work, in submission, 2005.
8. V. Matyáš and Z. Říha, "Biometric authentication - security and usability", http://www.fi.muni.cz/usr/matyas/cms_matyas_riha_biometrics.pdf
9. H. Moon, "Performance Evaluation Methodology for Face Recognition Algorithms," Ph.D. Thesis, Dept. of Computer Science and Eng., SUNY Buffalo, 1999.
10. R. Palmer, "The Bar Code Book," *Helmers Publishing*, 3rd Ed., 1995.
11. P. Phillips and H. Moon and S. Rizvi and P. Rauss, "The FERET Evaluation Methodology for Face-Recognition Algorithms," *IEEE Pattern Analysis and Machine Intelligence*, (22), pp.1090-1104, 2000.
12. C. Soutar, "Biometric system performance and security," Manuscrypt available at http://www.bioscrypt.com/assets/bio_paper.pdf, 2002.

13. C. Soutar, D. Roberge, A. Stoianov, R. Golroy, and B. Vijaya Kumar, "Biometric Encryption," ICSA Guide to Cryptography, McGraw-Hill, 1999, also available at http://www.bioscrypt.com/assets/Biometric_Encryption.pdf
14. M. Turk and A. Pentland, "Eigenfaces for recognition," Journal of Cognitive Neuroscience, vol. 3, No. 1, pp. 71-86, 1991.
15. U.S. DoS, USVISIT, http://fpc.state.gov/20738.htm.

A Scalable Mutual Authentication and Key Distribution Mechanism in a NEMO Environment[*]

Mihui Kim, Eunah Kim, and Kijoon Chae

Department of Computer Science and Engineering, Ewha Womans University, Korea
{mihui,dmsk999}@ewhain.net, kjchae@ewha.ac.kr

Abstract. Long an issue of interest, network mobility technology is now being realized with the foundation of the NEMO(Network Mobility) Working Group (WG) in the IETF. Security problems for NEMO become more important because one or several MR(Mobile Router)s manage the mobility of an entire network as well as the nested mobile networks, but current NEMO lacks the defense mechanism against these problems. Thus, in this paper, we propose a scalable and ubiquitous mutual authentication and key distribution mechanism without TTP(Trusted Third Party) for a NEMO environment, that uses the threshold secret sharing technique for enhancing scalability and availability, and provides the low-processing for requirement of frequent mutual authentication by mobility. We simulated and analyzed our mechanism together with another general authentication mechanism in the view of scalability and processing delay. Our experimental results show that our mechanism provides a scalable security support for a NEMO environment.

1 Introduction

As the mobility of network has lately aroused interest, NEMO WG in IETF has standardized the basic support and the management schemes since 2001. This WG is concerned with managing the mobility of an entire network, which changes, as a unit, its point of attachment to the Internet and thus its reachability in the topology.

Cases of mobile networks include the following examples; networks attached to people(Personal Area Networks or PANs), networks of sensors and computers deployed in vehicles, access networks deployed in public transportation, and ad-hoc networks connected to the Internet via a MR[1]. These practical instances awake interest in the network mobility. Supporting of the mobility for network itself allows for session continuity for every node in the mobile network as the network moves. That is, nodes in a mobile network except MRs need not particularly features of mobile IP, and NEMO technology can solve a binding storm problem of mobile IP.

However, security problems in a NEMO become more important because one or several MRs manage the mobility of an entire network as well as the nested mobile networks. Also, compromise of a MR that performs basic NEMO operations can also compromise all the nodes under its charge in the NEMO. Thus, a compromised MR in

---

[*] This research was partially supported by University IT Research Center(ITRC) Project and Brain Korea(BK) 21 Project.

a NEMO can cause far more widespread damage than a compromised node in other wireless networks. To date, several drafts regarding NEMO security have been submitted; most relate to support of the secure basic NEMO protocol[2], or to the protection of control messages such as binding update (BU) messages. However, security mechanisms specific to NEMO are not yet considered in detail. Especially, mutual authentication is indispensable between parent-MR and MR, between Access Router(AR) and MR, or between MR and Visiting Mobile Node(VMN), because it is necessary that the MR/AR authenticate the served sub-MR/VMN, as well as that the sub-MR/VMN authenticate the parent-MR/AR forwarding their packet. However, AAA draft[3] for NEMO missed the parent-MR/AR authentication, although the attacker masquerading the parent-MR/AR could throw away the packets of served sub-MR/VMN, modify their packets, or act as the selfish node.

Most of certification services and some authentication usually use the trusted CA(Certification Authority) or TTP in order to sign the certificate and secret information. However, a CA/TTP can expose to any single point of compromise, single point of denial of service attack, or single point of failure, does not scale to large network size, and is not robust against channel errors in the wireless environment[4,5]. Our security architecture has four goals like following:

- **Scalability** and **availability** for many VMNs, and MRs.
- **Low processing** for requirement of frequent mutual authentication by mobility.
- **Long-term security** by periodically updating the signed information.
- **Ubiquitous** mutual authentication service.

Our architecture provides the following service for achieving these goals: localized and distributed TTP service using the threshold secret sharing, mutual authentication service without TTP, and key distribution service in the symmetric cryptosystem.

This paper is divided into five sections. In Section 2, we explain the basic operation of NEMO and threshold secret sharing as background. We introduce in Section 3 the proposed security mechanism. Next, we simulate our mechanism and explain our analysis of the simulation results, and a brief conclusion is finally presented.

2 Background

2.1 NEMO

A mobile network includes one or more MRs that connect it to the global Internet, in order to provide session continuity and access for all nodes in the NEMO, even as the network moves. The NEMO WG is extending Mobile IPv6 (MIPv6) for network mobility, providing backward compatibility with MIPv6.

There are three types of node in a NEMO: local fixed nodes (LFNs), local mobile nodes (LMNs), and visiting mobile nodes (VMNs). A node of any of these types may be either a host or a router. A LFN belongs to the mobile network and doesn't move topologically with respect to the MR. A LMN can move topologically with respect to the MR, and its home link belongs to the mobile network. A VMN can move topologically with the MR, and its home link doesn't belong to the mobile network. The MRs access the Internet from access routers (ARs) on visited links. The NEMO

WG is considering two special structures, nested NEMOs and multihoming structures. In a nested NEMO, a mobile network becomes attached to a larger mobile network. In a multihoming structure, more than one active interface are connected to the Internet. Figure 4 depicts the general composition of a NEMO.

To date, several drafts regarding NEMO security have been submitted; most relate to support of the secure basic NEMO protocol, or to the protection of control messages such as binding update (BU) messages. Chan-Wah Ng et al. [3] introduced the usage scenario and requirements of AAA service for supporting network mobility. These are rough AAA operation procedure between AR and MR, or between MR and MR/VMN. And this draft took the examples like IEEE 802.1x, PANA, or other EAP-variant, as "link-local" AAA protocol, and the examples like Diameter and RADIUS as the "global" AAA protocol. However, the suitable authentication service for NEMO was not described in detail.

Souhwan Jung et al. [2] described potential security threats to NEMO basic operations. The threats are mostly related to the integral use of IPsec and IP-in-IP tunnel between MR and HA. Threats related to the operations of nested MR and multi-homing are investigated. And security requirements for NEMO basic protocols and specific implementation notes were described. However, based on this threat analysis, the detailed security mechanisms appropriate to NEMO should be provided.

2.2 Threshold Secret Sharing

Informally, a secret sharing scheme is a system for distributing secret shares amongst a group of participants in such a way that only those coalitions belonging to a specified access structure can combine their shares to reconstruct a predetermined secret[6]. The threshold secret share assumes the limited number of coalition as specific K. This scheme has been studied in the cryptography context, and recently the proposals[4,5] adjusting this scheme to ad-hoc network were proposed in order to provide robust and ubiquitous security support.

3 Authentication and Key Distribution Protocol

3.1 Basic Concept

Each node in our protocol uses the signed secret information for mutual authentication and key distribution. However, the signing service of this secret information is distributed and localized through the threshold secret sharing mechanism, in order to enhance robustness, availability and ubiquity. That is, K-neighbor nodes perform the signing service, and a centralized TTP or Key Information Center need not at all for signing, which is different from [7]. In the only bootstrap, after a node deals out the secret share to K neighbor nodes, all nodes can get the secret share by the self-initialization algorithm like [4]. And, for the secure communication of each node, we assume the environment based on symmetric cryptosystem. Thus, after mutual authentication of each node, they can get the symmetric key with each other's signed secret information.

The structure of NEMO is a tree-like graph because of nested and multihoming structure, like figure 1. In sub-section 3.2, we'll explain our mutual authentication, signing service of secret information, and key distribution mechanism in detail.

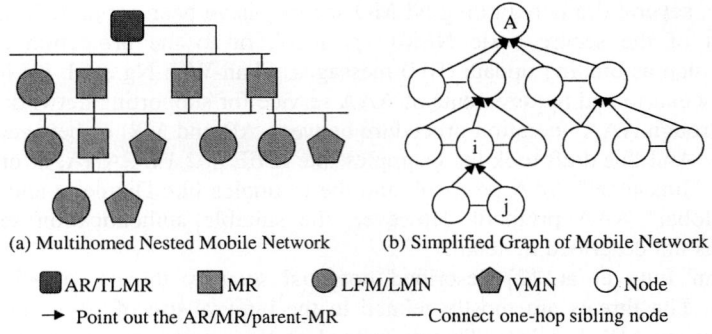

(a) Multihomed Nested Mobile Network   (b) Simplified Graph of Mobile Network

■ AR/TLMR   ■ MR   ● LFM/LMN   ⬟ VMN   ○ Node
→ Point out the AR/MR/parent-MR   — Connect one-hop sibling node

Fig. 1. General NEMO Structure and Simplified Graph

3.2 Protocol

Although our security architecture is basically the symmetric cryptosystem for the integrity and privacy of message communication, the secret information gains through the RSA-like structure and threshold secret sharing mechanism like [4,5]. However, K nearby neighbors of new node collaborate for signing the secret information of the new node, instead of K one-hop neighbors in the ad-hoc network[4,5], because the number of one-hop neighbors could be less than K by the structural characteristic of NEMO like figure 1. It is more available and robust than a remote CA serves, because wireless channel between MR and VMN/MR is error-prone, and a server could be one failure, compromise, and attack point. Thus, a pair of key {private key SK, public key PK} for TTP service need, and the signed secret information with SK for mutual authentication and key distribution is provided. The private key SK is $<d, n>$, and public key PK is $<3, n>$. This secret key d is shared as the partial secret of each node.

3.2.1 Authentication and Key Distribution

We will explain the signing service through the secret sharing mechanism with the figure 1(b). We assume that local nodes(LFM/LMN) need not to authenticate, and a node is authenticated to upper NEMOs if the node is authenticated to parent-MR. And we assume that VMN Vj moves in the NEMO of MR Vi, and Vj and Vi perform the following steps. These steps are similar with [7], but it needs not central TTP at all.

1. [RA Reception] Vj receives the RA(Route Advertise) message with some information for authentication from Vi whether it is periodical message or response message for the RS(Route Solicitation) message.

Vi→Vj : $RA \mid AUTHi$
$AUTHi = Xi \mid Yi \mid IDi \mid timei$
$Xi \equiv g^{3^{ri}} \pmod{n}$
$Yi \equiv Si \cdot timei \cdot g^{2^{ri}} \pmod{n}$

2. [Authentication for Parent-MR] Vj authenticates Vi with the received $AUTHi$. A value n is $p \cdot q$ such that p and q are large prime number, and integer g is a primitive element $GF(p)$ and $GF(q)$. A value ri is a random number generated by Vi at $timei$. A value g, a value n, ID extension function $f(ID)$ are public information, but Si is a secret information of Vi that was signed by neighbors. We will explain the signing scheme at the next sub-section. At first, Vj checks whether $timei$ is valid, that is the difference between the received time and $timei$ is reasonable, if the nodes can synchronize the clock. If they are not synchronized, it just skips. Next, Vj computes the $EIDi = f(IDi)$ for expanding the ID of Vi, and checks $EIDi \cdot timei^3 = Yi^3 / Xi^2$. If yes, MR authentication is success, but otherwise, Vj should find another MR.

3. [Authentication for New MR/VMN & Secret Information Acquisition] If Vj does not have its own secret information, it should obtain the information for later authentication and key distribution, at first. Thus, Vj requests it to parent-MR Vi. Vi receiving the secret information request checks the records for the Vi or authenticate it with the home network of Vj, if needed. This process with the home network for child VMN/MR authentication is beyond this paper. If granted, Vi broadcasts this request to all VMNs/MRs below the Top-Level MR(TLMR) V_A. If some nodes receiving this request are available for this service, they send Vi a commitment message that consists of their node ID and hop count from Vi. If Vi receives more (K-1) commitment messages, it decides the coalition of (K-1) members according to hop-count, and broadcasts the supposing coalition. Some nodes Vk included in the specified coalition generate the partial secret information Sv_k, and send it to Vi. After receiving all (K-1) partial secrets, Vi combines them with its own partial secret, and then derives the original secret information through the K-bounded Coalition Offsetting. This algorithm can be used as [4] is, except using 3 instead of exponent e. In order to encrypt the secret information before Vi sends it to Vj, Vi calculates a common key Kij with Xj that received together with the secret information request.

$$Xj \equiv g^{3 \cdot rj} \pmod{n}$$
$$Kij = Xj^{ri} = g^{3 \cdot ri \cdot rj}$$

Lastly, when Vj receives the encrypted Si, it calculates firstly a common key $Kji (= Xi^{rj} = g^{3 \cdot ri \cdot rj})$, and decrypts Si with Kji. Vj will use this Si for the later authentication and key distribution with other nodes, and skips following step 4 and 5.

4. [Authentication for New MR/VMN] If Vj has already a secrete information Si, it only sends Xj, Yj, IDj and $timej$ to Vi in order to authenticate itself.

$$Xj \equiv g^{3 \cdot rj} \pmod{n}$$
$$Yj \equiv Sj \cdot timej \cdot g^{2 \cdot rj} \pmod{n}$$

Upon receipt of the message, Vi calculates $EIDj = f(IDj)$, and checks the following equation; $EIDj \cdot timej^3 = Yj^3 / Xj^2$.

5. [Symmetric Key Generation] If the equation is true, Vi and Vj calculate the symmetric key Kij, and Kji respectively; $Kij = Xj^{ri} = g^{3 \cdot ri \cdot rj}$, $Kji = Xi^{rj} = g^{3 \cdot ri \cdot rj}$.

Figure 2 and 3 depict respectively authentication and key distribution procedure in the case of acquiring newly the secret information and of already having it.

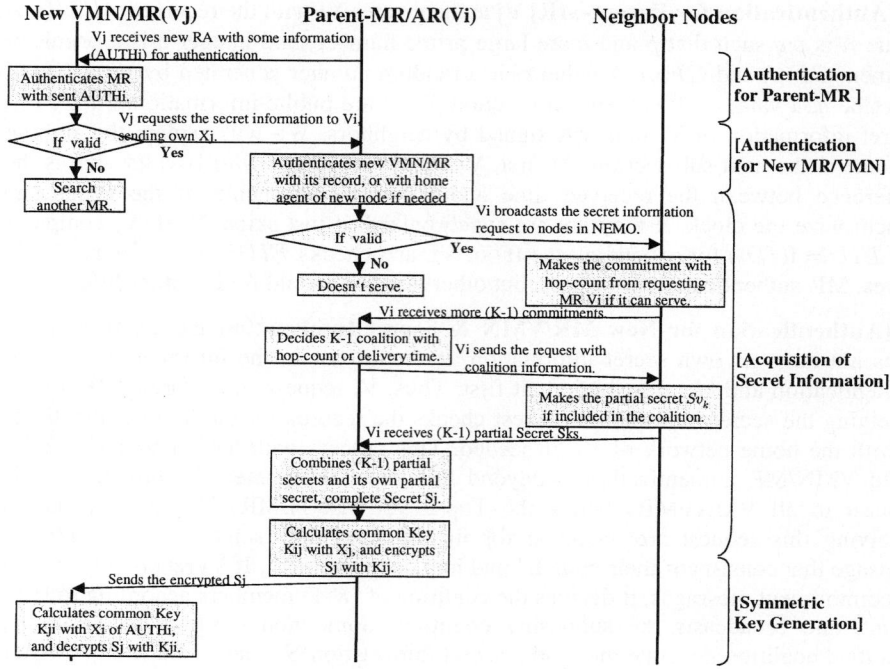

Fig. 2. Procedure in the case of acquiring newly the secret information

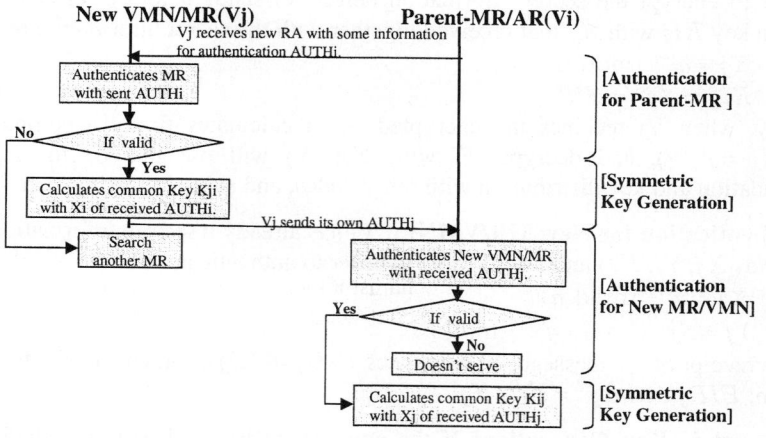

Fig. 3. Procedure in the case of already having the secret information

3.2.2 Singing Mechanism of Secret Information

As explained above, our signing mechanism is performed by the nearest neighbors under leading of parent-MR of requesting node in order to enhance availability, scalability, robust, and ubiquity. Because this process is performed after parent-MR authentication, requesting node can believe its parent. Also because parent-MR can have higher processing power than visiting node, parent-MR prefers to performing the combine process to make the complete secret information with partial signing information. And then this complete secret information is encrypted with common symmetric key between parent-MR and requesting node for secure delivery.

The basic operation is same as [4,5], but the parent-MR of requesting node should decide the coalition member according to hop count or delivery time.

The basic operation is like following: Each node Vi holds a polynomial share $Pvi = g(vi)$ (mod n) of the signing exponent SK according to random polynomial $g(x) = SK + \sum_{(j=1, k-1)} g_j \cdot x^j$ where $g_1, g_2, \ldots, g_{k-1}$ are uniformly distributed over a finite field. SK is d decided by $3 \cdot d$ (mod $(p-1) \cdot (q-1)) = 1$. p and q are large prime number, and $n = p \cdot q$.

After parent-MR receives the signing service request and decides to do this service, it computes $EIDi \equiv f(IDi)$ (mod 2^N) $\equiv (EIDi_1, EIDi_2, \ldots, EIDi_N)$ where N denotes the bit length of EID. And this parent-MR broadcasts the signing service request, it decides the coalition B consisted of (K-1) members according to commitment messages. Then, it broadcasts $EIDi$ and coalition B. If a node Vk is received the request that its own ID is included in coalition B, it computes partial secret information $Sv_k = (EIDk)^{(Pvk \cdot lvk(0) \bmod n)}$. The value $lv_k(0)$ is Lagrange coefficients[4]. Upon receiving such (K-1) partial secret information, the parent-MR multiples them with its own partial information to generate the candidate certificate as $S'i$. Finally, in order to recover complete Si from $S'i$, the parent-MR performs the k-coalition offsetting algorithm[4], applying 3 instead of public key e. Through the K-threshold secret sharing, parent-MR can finally get a secret information Si of Vi that is same with the value $(EIDi)^d$ (mod n). Lastly, parent-MR encrypts it with the common key generated by process explained in 3.3.1 sub-section, and sends it to Vi.

4 Evaluation

We have suggested a security architecture through mutual authentication and key distribution for the secure communication in a NEMO environment. In this section, we will evaluate the performance in the view of each design goal; scalability& availability, low processing, long-term security, and ubiquity. To analyze the performance, we implemented our mechanism(T-AUTH) like figure 2 and 3 with QualNet, and simulated it with a general configuration of NEMO like figure 1.

4.1 Scalability and Availability

In a NEMO environment, because cell phones and small routers also can become MRs that provide features of a NEMO, all mechanisms for NEMO should consider a scalability or availability problem. Especially, the big public transportation like a train or the nested structure of NEMO enlarges the size of a NEMO. For this problem, our security mechanism was designed to perform only between the served node and parent MR, or through nearby neighbors. This process structure would not depend on the size of a NEMO, in comparison with a process structure through a TTP or a CA.

To compare with the general authentication and key exchange mechanism in wireless IP network, we chose the W-SKE(Wireless Shared Key Exchange) protocol[8] and also implemented it with QualNet. W-SKE is a simple but general shared-key-based authentication and key exchange protocol. In the W-SKE, when the user roams to a portion of the network different from its home area, the authentication process involves a foreign AAA(F-AAA) server that eventually communicates to the user's home AAA (H-AAA) server. And, it uses a Message Authentication Code(MAC) for authentication, and a Pseudo-Random Function for key exchange, with a pre-shared key between H-AAA and user.

We measure two metrics, *Success Ratio* and *Average Delay*, as we increase the number of nodes in NEMO and vary the background traffic. *Success Ratio* measures the ratio of the number of successful authentication/key distribution services over the number of requests. And we use a delay limit(e.g. 5 sec) for judging the service result(success/failure). *Average Delay* measures the average latency of nodes that are successfully served.

Figure 4 shows the *Success Ratio*, as the simultaneous requesting node glows from 8 to 64. Mostly, the success ratio of T-AUTH is higher than W-SKE, even though there is no background traffic between NEMO and wired network. And we observe that W-SKE is far more susceptible to the number of simultaneous requesting node than T-AUTH, especially in the case (b) of generating background traffic, similar with real network. The *Average Delay* is shown in Figure 5. In the case (a) of no

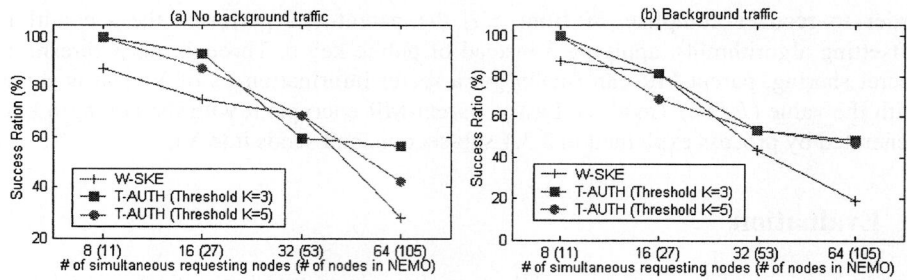

Fig. 4. Success ratio vs. the simultaneous requesting nodes #

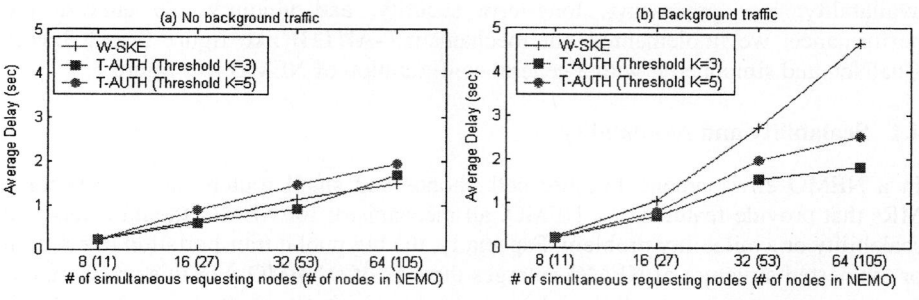

Fig. 5. Average delay vs. the simultaneous requesting nodes #

background traffic, T-AUTH is similar or a little higher than W-SKE. The main reason is that the flooded control messages of T-AUTH increase as the threshold K glows. However, in the case (b) of far realistic environment, T-AUTH is not affected by other normal traffic, and also is far less susceptible to the simultaneous requests.

4.2 Low Processing

At this sub-section, we measure the *Processing Delay(PD)* of our T-AUTH and W-SKE, when a mobile node roams to foreign networks in order to analyze the effect of the frequent handoff. For this experimental, we vary the NEMO size from 11 to 105, consisting of 2 NEMOs and 8 NEMOs respectively, and a mobile node moves to maximum 8 NEMOs sequentially. We measure the *Processing Delay* at each NEMO that include authentication and key exchange time. Figure 6 shows that the case (a) of T-AUTH requires a large time at only first served case, but the requirement incredibly decreases under 20% of the first latency as the node moves to other NEMOs. But the *PD* of W-SKE is similar even though the node moves to other NEMOs, and the average delay of W-SKE is bigger in all network configurations than T-AUTH.

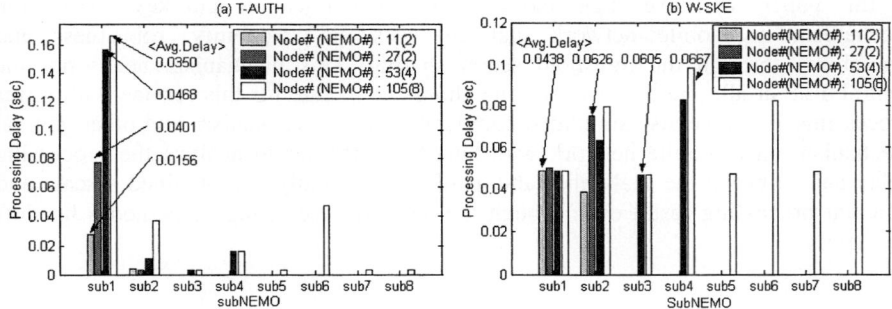

Fig. 6. Processing delay: authentication & key exchange time

This result is because our mechanism needs just two exponential computations(one for g^{rj}, and one for symmetric key Xi^{rj}) for each communication session like [7], once a node gets its signed secret information.

4.3 Long Term Security

To resist gradual break-ins over a long term period, the signed secret information Si of each node could be updated through the proactive update of [4,5]. Also, a common symmetric key between two nodes could renew periodically or every session once two nodes authenticate.

4.4 Ubiquitous Mutual Authentication

In a NEMO environment, mutual authentication is the most basic and important security scheme. In order to prevent that unauthorized node uses the resource of parent-MR/AR or compromises the served nodes in a NEMO through the MR/AR, the authentication for served nodes is indispensable. Contrary to this authentication, in order to prevent that bogus MR/AR monitors, modifies, or filters the sending data of served node, the authentication for parent-MR/AR is also essential. For this mutual authentication, our scheme used the secret information just like being signed by a central TTP.

However, from the viewpoint of NEMO characteristics, the frequent handoff and multi-hop wireless environment, a central TTP is difficult of providing the scalable, robust and ubiquitous signing service. Therefore, we localized and distributed the TTP service using the threshold secret sharing through the nearest neighbors. The proposal of the certificate service through the threshold secret sharing in the ad-hoc network[4,5] proved that the performance did not that much decrease by the secret sharing process in comparison with RSA-signature. We adapted the basic mechanism of [4,5] to the NEMO environment for signing the secret information, and we proved its robustness with Figure 4,5 and 6.

5 Conclusion

In this paper, we have suggested a mutual authentication and key distribution mechanism in mobile network that enhances the scalability, robustness and availability through the threshold secret sharing. This mechanism needs not the central TTP at all, and the nearest K-neighbors cooperate for this service with lower processing. We have also simulated comparatively our mechanism and other general mechanism on a mobile network with QualNet, in order to analyze the processing delay and to prove the scalability and availability. Finally, we obtained a scale and efficient processing result even though the network size is bigger or nodes handoff frequently.

References

1. 1 Thierry Ernst, "Network Mobility Support Goals and Requirements," IETF Internet Draft: draft-ietf-nemo-requirements-01.txt, May 2003.
2. Souhwan Jung, Fan Zhao, Felix Wu, HyunGon Kim, SungWon Sohn, "Threat Analysis for NEMO," IETF Internet Draft: draft-jung-nemo-threat-analysis-02.txt, February 2004.
3. Chan-Wah Ng, Takeshi Tanaka, "Usage Scenario and Requirements for AAA in Network Mobility Support," IETF Internet Draft: draft-ng-nemo-aaa-use-00.txt, October 2002.
4. Jiejun Kong, Petros Zerfos, Haiyun Luo, Songwu Lu, Lixia Zhang, "Providing Robust and Ubiquitous Security Support for Mobile Ad-Hoc Networks," IEEE ICNP 2001.
5. Haiyun Luo, et al., "Self-securing Ad Hoc Wireless Networks," 7th IEEE Symposium on Computers and Communications, 2002.
6. Mida Guillermo, et al., "Providing Anonymity in Unconditionally Secure Secret Sharing Schemes," Designs, Codes and Cryptography, Kluwer Academic Publishers, 2003.
7. Shiuh-Pyng Shieh, Wen-Her Yang, Hun-Min Sun, "An Authentication Protocol Without Trusted Third Party," IEEE Communications Letters, Vol. 1, No. 3, May 1997.
8. Luca Salgarelli, et al., "Efficient Authentication and Key Distribution in Wireless IP Networks," IEEE Wireless Communications, December 2003.

Service-Oriented Home Network Middleware Based on OGSA*

Tae-Dong Lee and Chang-Sung Jeong**

School of Electrical Engneering in Korea University,
1-5ka, Anam-Dong, Sungbuk-Ku, Seoul 136-701, Korea
lyadlove@snoopy.korea.ac.kr, csjeong@korea.ac.kr

Abstract. In this paper, we present a service-oriented home network middleware based on open grid service architecture. It supports two services: automatic distribution and dynamic migration schemes which shall be exploited efficiently in ubiquitous home networking environment. The deployment of Open Grid Service Architecture (OGSA) allows our system to have several advantages: adaptability, dynamic configuration, transparency, fault tolerance and performance. Automatic distribution service supports adaptability, dynamic configuration and transparency by providing automatic allocation and execution of home network platforms. Dynamic migration service supports fault tolerance and performance by migrating the processes in the failed server or with poor performance.

1 Introduction

Home network(HN) is now widely issued in the home consumer appliance industry. Since it is typically distributed, there is a growing demand for distributed computing technology to provide various services such as home automation services, home theater services, home gateway services, and so on. To provide these services, several middlewares, such as HAVi [1], Jini [2], and UPnP [3] have been introduced. In home networking environment, users should be able to access ubiquitously present appliances, and get services needed for devices and appliances anywhere and anytime. However, current middlewares for home network are localized inside home network, and typically assume a static and well administrated computing environment. Also, it is not easy to install home network services, middlewares and frameworks on a dynamically changing environment to deploy ubiquitous services for home networks. Moreover, there arises various problems in adaptability, heterogeneity, and security under heterogeneous environment which comprises a wide variety of organizations.

---

* This work has been supported by KIPA-Information Technology Research Center, University research program by Ministry of Information & Communication, and Brain Korea 21 projects in 2005.
** Corresponding author.

In this paper, we present SON-G(Service-oriented hOme Network Middleware based on open Grid service architecture) which addresses the problems arising in the home network environment by supporting two key services based on Open Grid Service Architecture(OGSA): automatic distribution and dynamic migration schemes. The former service supports adaptability, dynamic configuration and transparency by providing automatic allocation and execution of home network platforms, while the latter service supports fault tolerance and performance by migrating the processes in the failed server or with poor performance.

The paper is organized as follows: In section 2, we describe the related works, and in section 3, illustrate an architecture of SON-G. In section 4, we depict two main services offered by SON-G: automatic distribution service and dynamic migration service respectively. In section 5, we describe the experiment of SON-G, and in section 6, give a conclusion.

2   Related Works

The widely used HN middlewares include HAVi, Jini and UPnP[1, 2, 3]. HAVi provides a home networking standard for seamless interoperability between digital audio and video consumer devices. In other words, all audio and video appliances within the network will interact with each other and allow functions on one or more appliances to be controlled from another appliance, regardless of the network configuration and appliance manufacturer. Jini is a service discovery technology based on Java, developed by Sun Microsystems. Because of the platform-independent nature of Java, Jini can rely on mobile code for interaction between clients and services. Lookup services provide catalogs of available services to clients in a Jini network. On initialization, Jini services register their availability by uploading proxy objects to one or more of these lookup services. UPnP is a set of protocols for service discovery under development by the Universal Plug and Play Forum, an industry consortium led by Microsoft. UPnP standardizes the protocols spoken between clients and services rather than relying on mobile code.

Today, the existing home applications are based on a number of different APIs(such like HAVi, Jini and UPnP) that are often proprietary, incompatible, and normally just address subsets of the devices that exist in a home. This leads to a demand for a general API giving standardized access to all home devices independent of the network type used. There are several initiatives to define the necessary API. Among them we have OSGi (Open Service Gateway Initiative)[4]. OSGi service platform is a general-purpose, secure, managed Java software framework that supports the deployment of extensible and downloadable service applications known as bundles. The OSGi-compliant gateway can download and install bundles, when they are needed, and uninstall them when they are no longer needed.

While OSGi provides only standard API for services, Grid technology defines standard approaches to, and mechanisms for, basic problems that are common to a wide variety of Grid systems, such as communicating with other services, establishing identity, negotiating authorization, service discovery, error notification, security of virtual organizations (VOs) and managing service collections[6]. Grid technologies, Globus Toolkit [6] in particular, are evolving toward an OGSA which provides an extensible set of services which allows virtual organizations to be formed in various ways. Building on concepts and technologies from both the Grid and Web services communities, OGSA defines a uniform exposed service semantics and standard mechanisms for creating, naming, and discovering transient Grid service instances; provides location transparency and multiple protocol bindings for service instances; and supports integration with underlying native platform facilities. So far, no home network environment based on OGSA have been developed.

3 Architecture

SON-G provides HNE(Home Network Environment) with two key services on GCE(Grid Computing Environment). GCE consists of four modules: GRAM which allocates and manages the job in the remote hosts, MDS which provides information services [7], GridFTP which is used to access and transfer files, and GSI which enables authentication via single sign-on using a proxy. HNE comprises home network applications and middlewares. In HNE, each application joins to a corresponding middleware which enable any device to dynamically join a network, obtain an IP address, convey its capabilities, and learn about

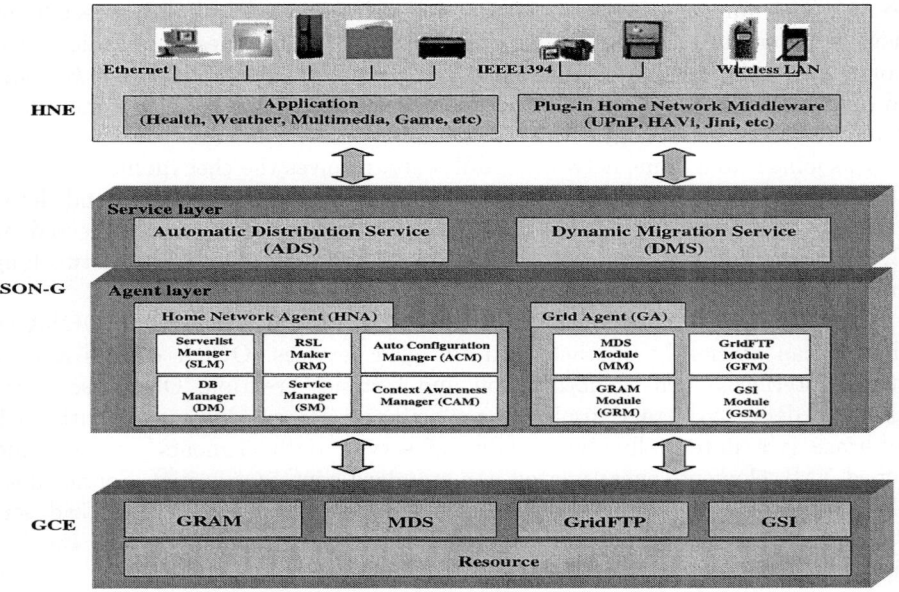

Fig. 1. SON-G architecture

the presence and capabilities of other devices automatically, and hence communicate with other devices directly with peer to peer networking. SON-G is an intermediate layer which is in the charge of a bridge between HNE and GCE. It is composed of service and agent layers. Agent layer consists of Grid Agent(GA) and Home Network Agent(HNA). GA makes use of the modules in GCE for resources allocation, file transfer, remote execution and security. HNA uses GA modules for deployment and execution of applications and other ubiquitous middlewares on local or remote hosts. Service layer consists of automatic distribution service(ADS) and dynamic migration service(DMS), which shall be explained in the subsequent section in detail.

GA consists of four managers: information manager(IM), file manager(FM), allocation manager(AM) and security manager(SEM). IM gathers the resource information, and FM transfers files to remote computers. AM allocates resources and executes the job in the remote hosts, and SEM provides the security to distributed computing environment. HNA is composed of six managers: serverlist manager(SLM), RSLMaker(RM), auto-configuration manager(ACM), context awareness manager(CAM),service manager(SM), and DB manager(DM). SLM makes the list of resources available in the corresponding virtual organization(VO). The number and performance of available hosts have great effect on the configuration of HNE. SLM periodically update and reference the serverlist of available resources using IM in GA. RM dynamically creates a RSL code to meet the needs of users and the requirements of HN. ACM automatically makes configuration files to provide information needed to initiate the applications and home network middleware on distributed resources in the serverlist. CAM provides information on the physical context, and reports relevant events to SM. The number of communication technologies used to interconnect home devices is on the rise, with an expanding list of technologies applicable in the home networking, including Bluetooth, IEEE 802.1x, IEEE 1394, Home RF, and so on. CAM has the mechanism to detect the new appliance using one of the communication technology.

SM joins into a home network middleware, receives the checkpoint data from the applications for dynamic migration in case of fault occurrence, and delivers them to DM. Also, it periodically receives and monitors data collected by CAM to check the registration and status of the devices in home networking environment.

Each module in GA and HNA supports an interface defined in OGSA so that it can efficiently interact with the grid services. OGSA defines a variety of interfaces such as notification and instance creation. Of course, users also can define arbitrary application-specific interfaces. Associated with each interface is a potentially dynamic set of service data elements - named and typed XML elements encapsulated in a standard container format. Service data elements provide a standard representation for information about Grid service instances. The important aspect of the OGSA model provides the basis for discovery and management of potentially dynamic Grid service properties.

4 SON-G Services

In this section, we describe two services in service layer of SON-G in more detail respectively.

4.1 Automatic Distribution Service

Automatic distribution service(ADS) enables the automatic execution of HN platforms and applications by allocating computing resources, transferring the executable files and carrying out them on the allocated computing resources. The automatic distribution allows the transparent use of computing resources, and the dynamic configuration used in the resource allocation for load balancing enables the better utilization of computing resources with the enhancement of scalability. The service is composed of three steps as shown in Figure 2(a):

Request/ Detection: User may directly request the installation of home network middleware and the execution of applications to ADS, or CAM detects the information of a new appliance, and requests the installation of the proper middleware and application on the new appliance to ADS.

Preparation: ADS distributes the home network middlewares and applications onto the distributed resources at preparation step. It consists of four stages: serverlist production, configuration, storage, and transmission. In serverlist production stage, ADS request SLM in HNA to create and maintain a host list about the available resources by making use of metadata on hosts which are registered using GIIS(Grid Index Information Service) of IM in GA. In configuration stage, ACM automatically creates a configuration file including all

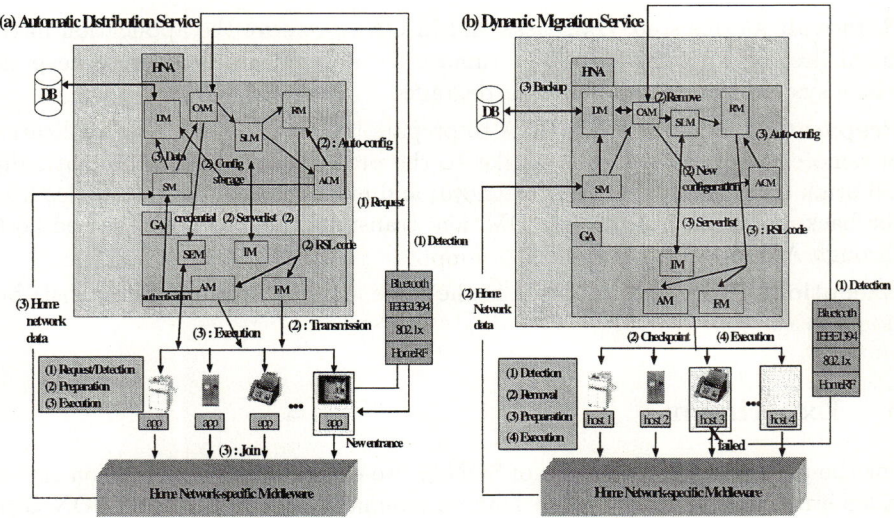

Fig. 2. SON-G services: Automatic Distribution and Dynamic Migration Services

the information required for initialization in remote hosts. RM automatically generates a RSL (Resource Specification Language) for resource allocation. In storage stage, the configuration file for each remote host is saved in DB by DM for later use in dynamic migration, and in transmission stage, the program files and configuration file are sent to the corresponding remote hosts through AM in GA.

Execution: ADS simultaneously executes application and home network middleware on the allocated remote hosts as indicated in RSL code through AM in GA. AM also provides the barrier which guarantees that all remote hosts start successfully. The application periodically delivers its own data to SM, which store thems in DB by DM in HNA to prepare for the dynamic migration which might be necessary in case of fault occurrence.

4.2 Dynamic Migration Service

Dynamic migration service(DMS) achieves two design goals: fault tolerance and performance by transferring the application in the failed host or the host with poor performance to the new host with the better one. DMS has four steps as shown in Figure 2(b):

Detection: CAM in HNA detects the fault of remote hosts by checkpoint, or finds out the remote servers with the degrading performance based on the information obtained by regularly retrieving the current status of remote hosts using IM in GA. Figure 2(b) shows three home appliances allocated in host 1, 2, and 3 at the preparation step of ADS. Suppose that CAM perceives the fault of host 3. Backup data related to application in host 3 are periodically stored into database by SM, and used when the application in host 3 is migrated into host 4.

Removal: At this step, DMS asks AM in GA to remove the application in the failed host or with the poor performance to be removed in order to keep the whole system from being halted or degraded.

Preparation: DMS requests HNA to prepare the creation of a new application in remote server. This step is similar to the preparation step in automatic distribution service, but that ACM retrieves and makes the configuration files from the backup data in DB through DM, and transmits them to the allocated hosts through AM in GA.

Execution: This step is executed in the same manner as in automatic distribution service.

5 Experiments

For the performance evaluation of SON-G, we evaluate the transmission rate of video streaming on home networking environment using UPnP and SON-G on Globus v3.0. UPnP is a HN-specific middleware which provides stable communication and interoperability for pervasive peer-to-peer network connectivity of

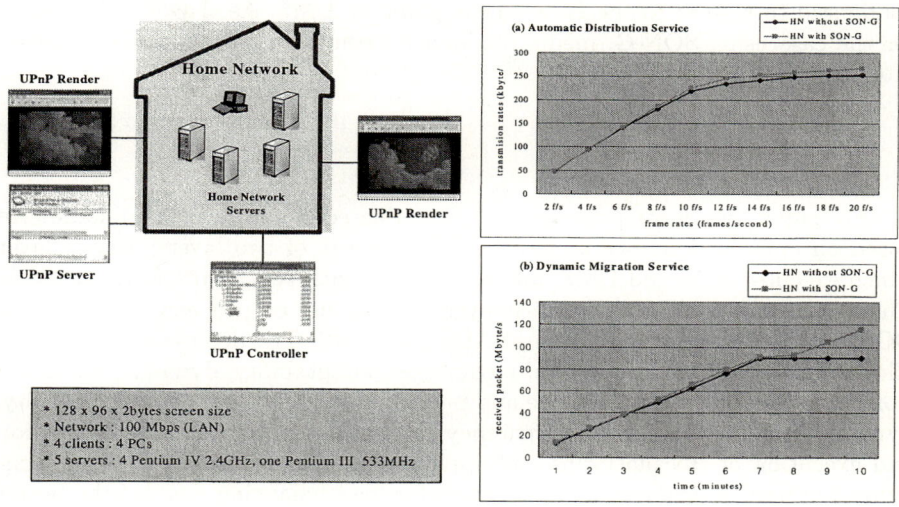

Fig. 3. Video streaming experiment about SON-G

intelligent appliances, wireless devices, and PCs. The experiment of SON-G is carried out using Intel UPnP tool which consists of three components: media server, media renderer and media controller. Media server is configured to share media files, reads metadata from audio tags and image formats, and makes it available on the network. Media Renderer adds a rich set of AV(Audio/Video) features to the Windows Media Player, ActiveX control, and supports multiple connections, media types, and playlists. Media controller sets up AV connections between UPnP media servers and media renderers. Our home network environment consists of 4 PCs as clients, and 5 servers with 1 Pentium IV 2.4GHz and 4 Pentium III 533MHZ, which can be regarded as one virtual organization.

We evaluated the performance of ADS and DMS by measuring the transmission rate with respect to the frame rate and the accumulated number of received packets with respect to time respectively. For each one, we considered two cases according to whether we incorporate SON-G onto HN using UPnP or not.

In the first experiment of ADS, the new devices like PDAs and mobile phones may be added into HNE, and the necessary UPnP components can be installed into the devices using ADS. For media server, Pentium III is selected for the case without SON-G, and Pentium IV for the case with SON-G with high probability, since ADS in SON-G allocates the resource with higher performance according to the information collected by SLM in HNA. Therefore, as shown in Fig. 3(a), the latter case shows higher transmission rate than the former case.

In the second experiment of DMS, we measured the accumulated number of packets received by media renderer after sent by media server at 10 frames/sec, each frame with 245.76 kbytes(128 x 96 x 2 bytes). Suppose a fault occurs in the media renderer after 7 minutes. Then, the new media renderer is selected, and configured to restore to the failed check point by DMS. As shown in Fig. 3(b), for the case using SON-G, media renderer continues to receive packets shortly after the fault occurrence.

6 Conclusions

In this paper, we have presented service-oriented home network middleware based on open grid service architecture. It consists of two layers: service layer which executes ADS and DMS, and agent layer with two agents HNA and GA. The agent layer enables the service layer to make use of grid services offered in OGSA for the efficient execution of ADS and DMS. The deployment of OGSA allows our system to have several advantages: adaptability, dynamic configuration, transparency, fault tolerance and performance. ADS supports adaptability, dynamic configuration and transparency by providing the automatic allocation and execution of HN platforms and applications. Especially, the dynamic configuration used in the resource allocation for load balancing enables the better utilization of computing resources with the enhancement of scalability. DMS supports fault tolerance and performance by migrating the processes in the failed server or with poor performance.

References

1. R.G Wendorft,.R.T Udink and M.P Bodlaender, "Remote execution of HAVi applications on Internet-enabled devices," Consumer Electronics, IEEE Transactions, Volume 47, Issue 3, Aug. 2001 pages 485-495
2. R. Gupta, S. Talwar, D.P. Agrawal, "Jini home networking: a step toward pervasive computing," Computer, Volume 35, Issue 8, Aug. 2002 pages:34- 40
3. D.S. Kim; J.M.Lee, W.H. Kwon and I.K. Yuh, "Design and implementation of home network systems using UPnP middleware for networked appliances," Consumer Electronics, IEEE Transactions, Volume 48, Issue 4, Nov 2002 pages:963-972
4. R.S. Hall, H. Cervantes, "An OSGi implementation and experience report," Consumer Communications and Networking Conference, 2004. CCNC 2004. First IEEE , 5-8 Jan. 2004 pages:394 - 399
5. I. Foster, C. Kesselman and S. Tuecke, "The Anatomy of the Grid: Enabling Scalable Virtual Organizations," International J. Supercomputer Applications, 15(3), 2001.
6. I. Foster, C. Kesselman, "Globus: A Metacomputing Infrastructure Toolkit," Intl J. Supercomputer Applications, 11(2):115-128, 1997
7. K. Czajkowski and I. Foster, "Grid Information Services for Distributed Resource Sharing," Proceedings of the Tenth IEEE International Symposium on High-Performance Distributed Computing (HPDC-10), IEEE Press, August 2001.

Implementation of Streamlining PKI System for Web Services

Namje Park[1], Kiyoung Moon[1], Jongsu Jang[1],
Sungwon Sohn[1], and Dongho Won[2]

[1] Information Security Research Division, ETRI,
161 Gajeong-dong, Yuseong-gu, Daejeon, 305-350, Korea
{namjepark,kymoon,jsjang,swsohn}@etri.re.kr
[2] School of Information and Communication Engineering, Sungkyunkwan University,
300 Chunchun-dong, Jangan-gu, Suwon-si, Gyeonggi-do, 440-746, Korea
dhwon@dosan.skku.ac.kr

Abstract. XKMS (XML Key Management Specification), one of web services security specification, defines the protocol for distributing and registering public keys for verifying digital signatures and enciphering XML documents of web service applications with various and complicate functions. In this paper, we propose XKMS-based streamlining PKI service model and design protocol component based on standard specification. Also describes the analysis and security method of PKI service for secure web services, paying attention to the features of XML based security service. This service model offers the security construction guideline for future global web services frameworks.

1 Introduction

The XML (eXtensible Markup Language) is a promising standard for describing semi-structured information and contents on the Internet. Some of the well-recognized bene-fits of using XML as data container are its simplicity, rich-ness of the data structure, and excellent handling of inter-national characters. The practical use of XML is increasing in proportion to spread speed of web services as global standard for Internet and Web Service. In this environment, a security mechanism for XML documents must be provided in the first place for secure web services. The security mechanism also has to support se-curity function for the existing non-XML documents, too.

The XML security standards define XML vocabularies and processing rules in order to meet security requirements. These standards use legacy cryptographic and security tech-nologies, as well as emerging XML technologies, to provide a flexible, extensible and practical solution toward meeting security requirements.

The Industry is therefore eager for XML and PKI (Public Key Infrastructure) to work together in fulfilling the widely held expectations for cryptographically secure, XML-coupled business applications. The best-known simplicity of XML is to provide portability of data between disparate business systems contrasts with the complexity of traditional PKI implementation. Therefore, a key architectural goal in the XML key management specification (XKMS) is to shield XML application developers from the complexity of traditional PKI implementation. It permits delegation of trust processing decisions to one or more specialized trust processors. It enables XML-based systems to rely on com-plex trust relationships without the need for complex or spe-cialized end-entity PKI application logic on the client plat-forms where XML processing is taking place.

The world recently, by way to offer certification about important transaction of this XML environment, is research-ing about XML key management to integration of PKI and public key certificate and XML application. At the same time setting a reference systems that embody this are developed. But, R&D for actually system that domestic can construct XKMS offer of Trust Service based on XML are insufficient. Therefore, further research activity is needed for the pro-gress and standardization of the XML key management technology, and it is necessary to develop XML key management system for the activation of the secure web services.

E-XKM (ETRI XML Key Management) system which will be introduced in this paper, is an XKMS-based system that has been implemented to support the processing, by a relying party, of key management associated with a XML digital signature, XML encrypted data, or other public key usage in an XML web application.

In this paper, we describe a design for XKMS-based streamlining PKI service model and we explain our analysis, service protocol component based on standard specification. First we investigate related work on XKMS in web services and then we explain overview of the service system structure. Then we propose a design for service model and explain analysis service protocol component. Finally, we explain function of protocol component and then we conclude this paper.

2 XKMS-Based Streamlining PKI Service Model

2.1 ETRI XKM Service Model

XKMS defines protocols for the registration and distribution of public keys[1,2,3]. The keys may be used with XML signatures, a future XML encryption specification, or other public key applications for secure messaging.

E-XKM system is comprised of the X-KISS protocol and the X-KRSS protocol. X-KISS allows a client application to delegate part or all of the tasks required to process an XML signature to a trust service. This is useful for developers who don't want to implement the signature checking them, or who want to delegate this functionality to an application service provider that may be optimized for signature checking. X-KRSS is an XML-based replacement for existing PKI file formats that are used when a user applies for a digital certificate[10,11]. XML brings the same advantages to PKI as it brings to other industries - open standards, platform independence, and human readability. Both protocols utilize SOAP, and WSDL is used to define message relationships. The X-KRSS and X-KISS protocols are expressed using the W3C's XML schema language[1]. Figure 1 shows E-XKM service model include X-KRSS service of W3C.

As shown in the figure, a key owner registers his key with an XKMS service provider who makes use of an underlying PKI to store and bind the keys with identification information. A commercial PKI typically contains a key registration authority, a certification authority, a key validation authority, and a secure keys directory in which all information is stored[12,14]. Any web services that wants to validate a <ds:KeyInfo> element it has received can invoke an XKISS service that once again makes use of the underlying PKI to complete the process.

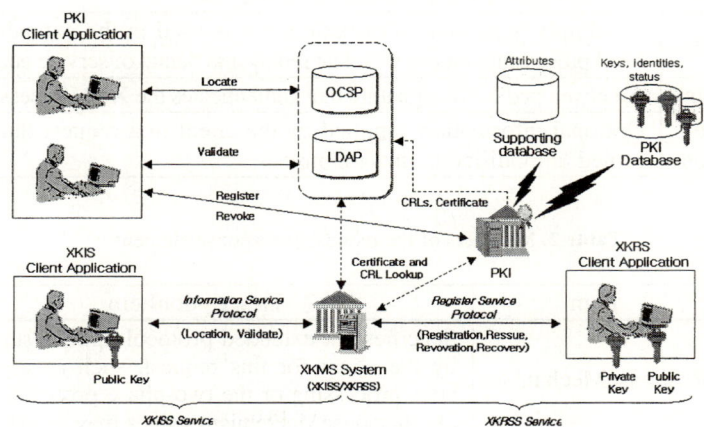

Fig. 1. XKMS-based Streamlining PKI Service Model

2.2 E-XKM Services Protocol

The XKMS protocol is essentially a request response pro-tocol layers on SOAP, with optional embellishments de-scribed at the end of the chapter [1,3].

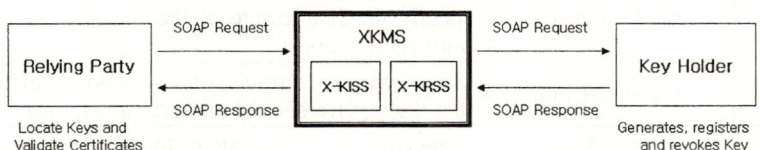

Fig. 2. XKMS Standard Protocol

The request and result messages used in the individual XKMS operations share a common format. These common members are defines in table 1 [1].

Additional members are defined for request messages, allows the client to specify the protocol options it supports, the types of and maximum quantity of information to be provided in the response, and additional information used in the extended protocol options. These additional members are described in table 2 [2]. Additional members are defined for request messages, allowing the service to specify the result of the operation (success, failure, etc) and binding the request to the response by means of the request Id. These additional members are described in table 2.

Table 1. Members Common to Request and Result Elements

Item	Description
Id@	A unique identifier for the message
Service@	The service URI of the XKMS service
Nonce@	Randomly generated information that is used in the extended protocol processing options to defeat replay and denial of service attacks
ds:Signature	An enveloped XML signature that authenticates the XKMS messages
Opaque Client Data	Optional information supplied by the client in a request that is returned unmodified in the response

Table 2. Members of the request & response element

	Item	DescriptionItem
Request Element	ResponseMechanism	Specifies ant extended protocol options supported by the client for this request, such as asynchronous processing or the two-phase protocol. Multiple ResponseMechanism values may be specified
	ResponseWith	Specifies a data type that the client requests be present in the response, such as a key value, an X.509 certificate, or a certificate chain. Multiple ResponseWith values may be specified
	PendingNotification	Optionally specifies a means of notifying completion of the operation when asynchronous processing is used
	OriginalRequestID@	This attribute is used in the extended protocol to specify the IDattribute of the initial request in a multistage request
	ResponseLimit	The maximum number of key binding elements that the service should return in a response
Response Element	ResultMajor	The principal result code of the XKMS operation
	ResultMinor	The secondary result code of the XKMS operation, giving additional information such as reason for the result
	RequestID	The IDattribute of the corresponding request

2.3 Analysis of E-XKIS Protocol

One of the major service of XKMS is XKISS defines protocols to support the processing by a relying party of key information associated with a XML digital signature, XML encrypted data, or other public key usage in an XML aware application [2].

Functions supported include locating required public keys given identifier information, and binding of such keys to identifier information.

XKISS defines three levels of key information service that is retrieval method, locate service, and validate service. It mentions the possibility of higher-level services, such as those dealing with long term trust relationships or the status of trust assertions.

The following fig.3 shows the locate service protocol. A client receives a signed XML document. The <KeyInfo> element in the signature specifies a retrieval method for an X.509 certificate. The client lacking the means to either resolve the URL or parse the X.509 certificate to obtain the public key parameters delegates these tasks to the trust service. The following fig.3 shows the validate service protocol. The client sends to the XKMS service a prototype containing some or all of the elements for which the status of the key binding is required. If the information in the prototype is incomplete, the XKMS service may obtain additional data required from an underlying PKI service. Once the validity of the key binding has been determined the XKMS service returns the status result to the client.

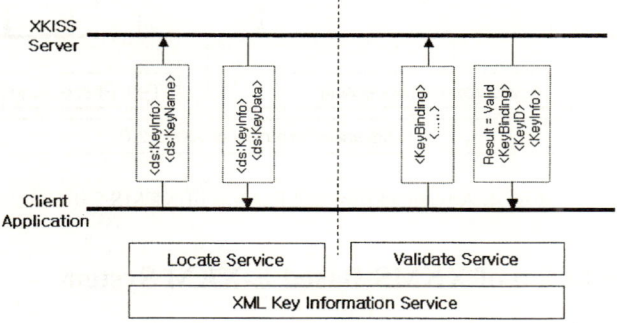

Fig. 3. E-XKISS Protocol

In XKMS 1.1, all operations consisted of a single request message followed by a single response. XKMS 2.0 specifies additional protocol options that allow a client to make mul-tiple XKMS requests simultaneously, allow an XKMS service to queue XKMS requests for later processing, and make it possible to defend against denial of service attacks [8].

First, asynchronous processing may be required because some form of operator intervention is required to complete an operation. Asynchronous processing is also desirable in cases where the request may take a long time to complete. Asynchronous processing involves two separate request/response pairs. Second, two phase request protocol providers protection against denial of service attacks by checking that the requestor can read IP packets sent to the purported source of the request.

The following fig.4 shows the Asynchronous processing. The client makes the first request specifying the response mechanism type <xkms:Asynchronous>. The service may return the actual response immediately or signal that the response will be returned asynchronously using the ResultMajor code <xkms:Pending>. Once the service has completed processing the request, the client obtains the result by issuing a pending request message [7]. The following fig.4 shows two-phase request protocol. The client sends an initial request to the service. Unless the ser-vice has reason to believe that the request is part of a denial of service attack, the service may respond with an immediate result. If the service has determined that it is under a denial of service attack and the request may be a part of that attack, it returns response with the ResultMajor code <xkms:Represent> that contains a nonce value. In order for the service to act on the request, the client must represent the request together with the previously issued none value.

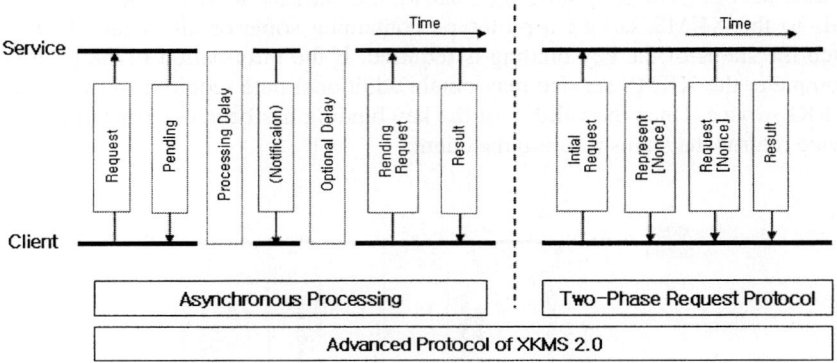

Fig. 4. Advanced Protocol features of XKMS 2.0

3   Implementation of XKMS-Based E-XKM System

3.1   E-XKM System Platform

E-XKM system has been implemented based on the design described in previous section. Package library architecture of XKMS based on CAPI(Cryptographic Application Programming Interface)[15] is illustrated in figure 5. Components of the E-XKM are XML security platform library, service components API, application program. Although XKMS service protocol component is intended to support XML applications, it can also be used in order environments where the same management and deployment benefits are achievable. E-XKM has been implemented in java and it runs on JDK ver 1.3 or more.

In case tools that is based on Java these advantage that can bring much gain economically, without porting process between kinds of machine. When develop program of multiplex platform environment. Specially, When develop client/server program. These can use same module, just as, it is in spite of that development environment can be different.

E-XKM system platform is a framework for the approaches about function of XKMS-based key management system and work for development based on java platform. XML security API is expressed by structure of java crypto library and XML paser, XSLT processor. And It includes service provide mechanism. SOAP security API supplies XML web services security. And XML security API and SOAP security

API supports key exchange and encryption. It supports XML signature and XML encryption function. Based on this, E-XKM service platform is composed. So, Service application program are achieved by component of service platform that is

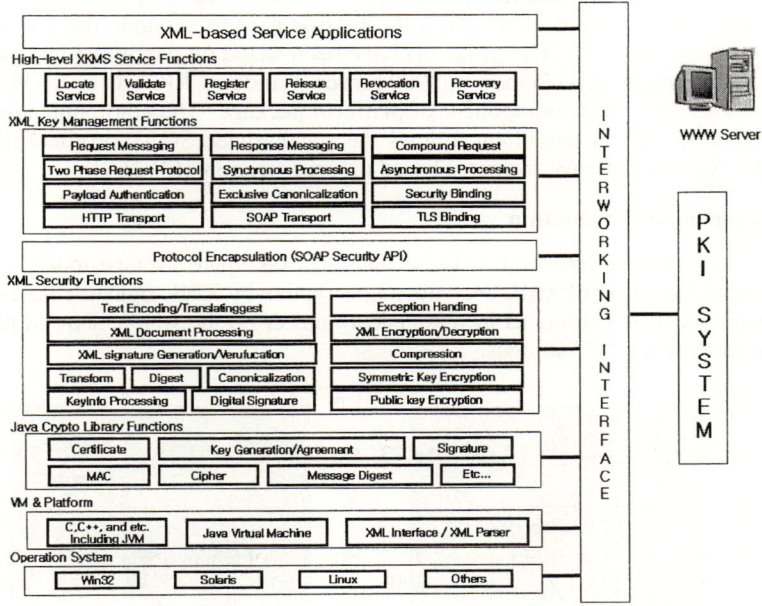

Fig. 5. Architecture of E-XKM System

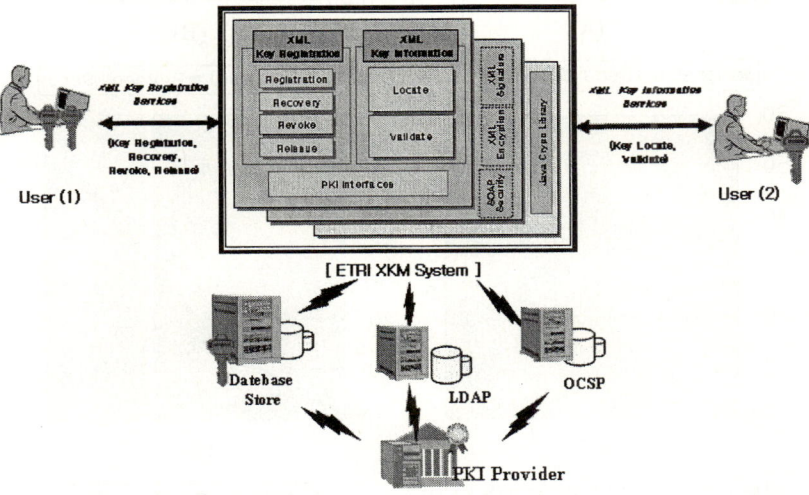

Fig. 6. Testbed of E-XKM system

constructed by each function. Other than system application, Many XML web application security can be provided using the XML security API and library that is provided from the E-XKM service platform. Figure 5 illustrates the architecture of service platform. Major components of platform are java crypto library, XML security API, SOAP security API, XML signature API, XML encryption API.

We use testbed system of windows PC environment to simulate the processing of various service protocols. The protocols have been tested on pentium 3 and pentium 4 PCs. It has been tested on Windows 2000 server, Windows XP. The E-XKM server is composed server service component of platform package. The communication protocol between the server and client follows the standardized SOAP protocol illustrated in figure. And the message format is based on specification of W3C.

3.2 Performance Evaluation

Figure 7-a showed difference for 0.2 seconds that com-pare average transfer time between client and server of XML encryption&decryption by XML signature base on XML security platform. According as increase client number on the whole, showed phenomenon that increase until 0.3 seconds.

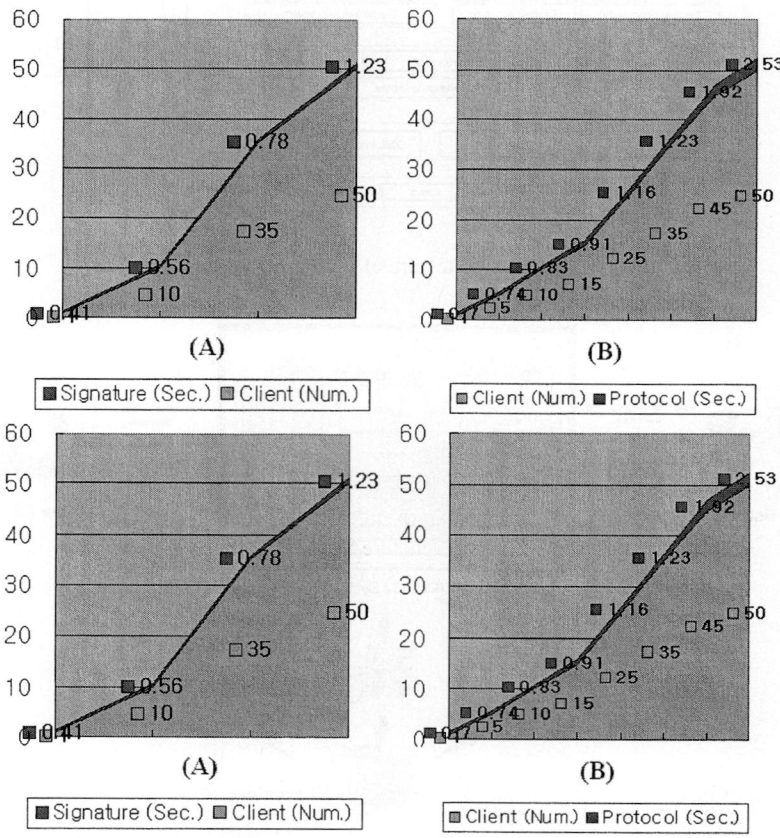

Fig. 7. Performance Evaluation

Figure 7-b is change of average transmission time according as increase client number in whole protocol en-vironment. If client number increases, we can see that average transfer time increases on the whole. And average transfer time increases rapidly in case of client number is more than 45. Therefore, client number that can process stably in computer on Testbed environment grasped about 40. When compare difference of figure 7-a and figure 7-b. Time of XML signature module is occupying and shows the importance of signature module about 60% of whole protocol time.

4 Conclusion

In this paper, we have proposed the XKMS-based streamlining PKI service model for secure global web services. And we designed a security platform based on XML that provides security services such as authentication, integrity and confidentiality for web services. It provides XML signature function, XML encryption function, java crypto library for securing XML document that are exchanged in the web services. And then we designed service component of E-XKM system based on XKMS standard specification of W3C. It provides function of XKISS and XKRSS service based on service protocol.

E-XKM system platform of this paper can be applied to various services that require secure exchange of e-document such as B2C, B2B and XML/EDI. Since it is developed in java program language, it can be ported easily to various platforms. And since XML signature, XML encryption, java crypto library is conforming to international standards, E-XKMS platform is compatible with many security platforms that conform to the standards.

Further research and development are needed on the integration between two system that E-XKM system and PKI system. And need continuous research for integration of XML signature & encryption technical development in mobile platform and XKMS based on wire/wireless system for web services of next generation web business environment.

References

1. W3C Note: XML Key Management(XKMS 2.0) Requirements. (2003)
2. W3C Working Draft: XML Key Management Specification Version 2.0. (2003)
3. W3C/IETF Recommendation: XML-Signature Syntax and Processing. (2002)
4. W3C Recommendation: XML Encryption Syntax and Processing. (2003)
5. RFC2459: X.509 Certificate and CRL Profile. (1999)
6. RFC2510: Certificate Management Protocol. (1999)
7. A Delphi Group: Web Services 2002:Market Milestone Report. A Delphi Group White Paper. IBM. (2002)
8. Steve Holbrook: Web Services Architecture-Technical Overview of the Pieces. IBM. (2002)
9. Jonghyuk Roh: Seunghun Jin and Kyoonha Lee, Certificate Path Construction and Validation in CVS. KICS-Korea IT Forum. (2002)
10. OASIS: Web Service Security, http://www-106.ibm.com/. (2002)
11. Jose L. Munoz et. Al.: Using OCSP to Secure Certificate-Using transactions in M-Commerce. Lecture Notes in Computer Science, Vol. 2846. Springer-Verlag (2003) 280-292

12. Namje Park et.al.: Development of XKMS-Based Service Component for Using PKI in XML Web Services Environment, Vol. 3043. (2004) 784-791
13. Sung-Min Lee et.al.: TY*SecureWS:An Integrated Web Service Security Solution Based on Java. Lecture Notes in Computer Science, Vol. 2738. (2003) 186-195
14. Yeonjeong Jeong, et. Al.: A Trusted Key Management Scheme for Digital Rights Management, ETRI Journal, V.27, No.1 (2005) 114-117
15. Stephen J. Elliott: fferentiation of Signature Traits vis-à-vis Mobile- and Table- Based Digitizers, ETRI Journal, V.26, No.6 (2004) 641-646

Efficient Authentication for Low-Cost RFID Systems

Su Mi Lee*, Young Ju Hwang, Dong Hoon Lee, and Jong In Lim

Center for Information Security Technologies(CIST),
Korea University, 1, 5-Ka, Anam-dong,
Sungbuk-ku, Seoul, 136-701, Korea
{smlee, pinpi}@cist.korea.ac.kr
{donghlee,jilim}@korea.ac.kr

Abstract. RFID (Radio Frequency Identification) technology is expected to play a critical role in identifying articles and serving the growing need to combat counterfeiting and fraud. However, the use of RFID tags may cause privacy violation of people holding an RFID tag. The main privacy concerns are information leakage of a tag, traceabiltiy of the person and impersonation of a tag. In this paper, we study authentication as a method to protect privacy, especially for low-cost RFID systems, which have much restrictions in limited computing power, low die-size, and low power requirements. Therefore, cost effective means of authentication is needed to deal with these problems effectively. We propose an authentication protocol, LCAP, which needs only two one-way hash function operations and hence is quite efficient. Leakage of information is prevented in the scheme since a tag emits its identifier only after authentication. By refreshing a identifier of a tag in each session, the scheme also provides a location privacy and can recover lost massages from many attacks such as spoofing attacks.

1 Introduction

RFID will surely be part of our everyday life in near future. Current researches in RFID technology have concentrated on an identification scheme of an RFID tag which makes the automated identification of products possible. Recently, secure and efficient identification protocols have received much attention with increasing applicability in various management systems of stocks, the classifications of goods in shops, animal identification, etc. For example, in a library, the use of RFID technology increases the efficiency of a job because a librarian can automatically manage inventories. Consequently, RFID technology for automatic object identification has a wide range of applications.

An RFID system consists of a radio frequency tag (transponder), a reader (transceiver), and a back-end database. A reader and a tag communicate by

---

* This work was supported by grant No. R01 − 2004 − 000 − 10704 − 0 from the Korea Science & Engineering Foundation.

RF signals, which make an RFID system vulnerable to various attacks such as eavesdropping, traffic analysis, spoofing and denial of service. These attacks may disclose sensitive information of tags and hence infringe on a person's privacy. Another type of privacy violation is traceability which establishes a relation between a person and a tag. If a link can be established between a person and the tag he/she holds, the tracing of the tag makes the tracing of the person possible [1]. To protect a person's privacy, a tag needs to authenticate a reader.

As noted in [3], spoofing is another possible attack to an RFID system. For example, an adversary may replace a tag of an expensive item with a bogus tag with data in a cheaper item. Thus this kind of impersonation allows an adversary to fool a security system into perceiving that the item is still present and may fool automated checkout counters into charging for a cheaper item. Thus a reader also needs to authenticate a tag in RFID systems to prevent a fake tag from impersonating a legitimate article. The scheme is said to provide mutual authentication if both a tag and a reader are assured that no adversaries can possibly make valid massages.

A tag is extremely limited in computation power, storage space, and communication capacities since an RFID chip with approximately 4,000 gates is considered to be of low-cost. This implies that classical authentication schemes in the literature is not suitable to low-cost RFID systems. Therefore it is of utmost importance to construct an efficient authentication scheme in consideration of these constraints in RFID systems. Designing a mutual authentication protocol for low-cost RFID systems is the main theme of the paper.

1.1 RFID System and Assumptions

An RFID system consists of three components and the characteristics of each component are as follows [4];

– *RFID tag (transponder)* is a small and low-priced device which consists of only a microchip with limited functionality and data storage, and an antenna to wireless communication with reading devices. RFID tags can be active or passive depending on powering techniques. While an active tag can generate power itself, a passive tag does not contain power supply on communication with reading devices and should only receive power from the reading devices when it is within range of some reading devices. For this reason the passive tag is cheap and suitable to low-cost RFID systems.
– *RFID reader (transceiver)* has an antenna and microchip for wireless communication. It can read and write tag data. The reader queries a tag to obtain tag contents though RF interface.
– *Back-end database* has a lot of information relevant to tags and is powerful in computational abilities.

The communication distance between a tag and a reader has a great difference between two communication directions (the reader-to-tag and the tag-to-reader communications). In the reader-to-tag communication, the information sent by a reader can be transmitted up to nearly a hundred meters, but in opposite direction the information sent by a tag reaches a few meters (e.g. 3 meters). Generally we assume that an adversary can monitor all messages transmitted between a reader and a tag. However in the reader-to-back-end database communication, we assume that the reader can establish a secure connection with the back-end database.

1.2 Related Works and Our Contributions

RELATED WORKS. The first step toward protecting user privacy in RFID systems was physical approaches such as *Kill the tag* [15], *Faraday cage* [10], and *Blocker tag* [9] techniques. *Kill the tag* technique is to restrict the use of a tag by removing its ID. This solution is simple and effective but has the weak point that a tag can not be reused in the RFID system. Also *Faraday cage* technique prevents a tag from hearing requests from readers by enclosing the tag. However this approach restricts the range of applications of RFID systems. The third solution, *Blocker tag*, aims at preventing a reader from determining which tags are present in its environment [1]. This is possible since a blocker tag is designed to answer any query from a reader and deceive a reader into believing that all possible tags exist. However a blocker tag may be a double-edged sword since it can be used to mount Dos attack on a reader. (We note that there exists 2^{96} possible "Electronic Product Code"(EPC) codes envisioned by the Auto-ID Center [2].)

Another approaches are to design an authentication protocol using cryptographic solutions. The scheme, called *Hash-Lock* in [15, 13, 14], prevents an exposure of tag *ID* by using cryptographic hash functions. Upon receiving a query from a reader, a tag first sends the hashed value of its key as a challenge to authenticate the reader. The tag reveals its *ID* only when the reader sends a pre-image (key) of the hashed value as a response. *Hash-Lock* scheme supports data privacy but can not protect location privacy of the tag since the fixed hash value is used in every authentication. The extended approach, *Randomized Hash Lock (RHLK)* [15], randomizes tag responses to a reader instead of a fixed tag response in order to protect location privacy. However this scheme still does not guarantee location privacy since a reader always responses with static tag *ID* obtained from its back-end database. Furthermore, this scheme is not scalable since the reader's computational workload is linear in the number of possible tags stored at the back-end database and a tag should be equipped with a random number generator as well as a one-way hash function. The scheme proposed recently, called hash-based *ID* variation scheme (*HIDV*) of [6], provides location privacy. However an adversary can query a tag and learn a valid tag response, which then allows the adversary to do a spoofing attack later to impersonate the tag. Therefore *HIDV* scheme does not provide mutual authentication.

Other approaches are based on re-encryption, where a ciphertext is encrypted again using asymmetric key cryptography [8] or symmetric key cryptography [5]. These approaches are more secure than the above presented approaches because of protecting a tag *ID* using asymmetric or symmetric key cryptography. An encryption-operation requires high computation cost, and is performed in a reader. However, as noted in [12], since an encrypted *ID* is constant [12], the data of each tag must be rewritten often. This makes re-encryption approach unsuitable to low-cost RFID systems.

OUR CONTRIBUTIONS. We propose a low-cost RFID authentication protocol LCAP which improves $HIDV$ scheme in both efficiency and security. To design more efficient authenticated protocol, we only uses a one-way hash function in a low-cost tag. LCAP ensures detection of variable vulnerabilities, greatly enhances the location privacy and can recover lost massages from many attacks. Furthermore, since a tag response is randomized in every session, our scheme prevents an adversary from performing a spoofing attack, which is not provided in $HIDV$.

In Table 1, we show efficiency analysis with respect to storage cost, computation cost, and security against variable vulnerabilities in $HIDV$, $RHLK$, and LCAP. As shown in the table, $RHLK$ is vulnerable to a spoofing attack by impersonating a tag to a legitimate reader and can only perceive loss of messages and replay attacks, but can not protect them. $HIDV$ can perceive spoofing attacks, but LCAP can even protect it before such attacks occur. Furthermore, LCAP is more efficient than $HIDV$ in all aspects.

— *Storage* : the storage cost of each entity.
— *Comp.* : the maximum computation cost of each entity during the execution of an authentication protocol.
— *Comm.* : the length of bits that a tag and a reader send during the execution of an authentication protocol. ORGANIZATION OF THE PAPER. The remainder of this paper is as follows: Section 2 shows possible security and privacy risks in RFID systems. Section 3 describes the proposed LCAP scheme and Section 4 analyzes the scheme in security and efficiency. Finally Section 5 concludes the paper.

Table 1. The analysis of efficiency and security

	Protocol	$RHLK$	$HIDV$	LCAP
Storage.	Tag	$2l$	$3l$	$1l$
	Reader	–	–	–
	Database	$2\frac{1}{2}l$	$10l$	$6l$
Comp.	Tag	$1h$	$3h$	$2h$
	Reader	–	–	–
	Database	–	$3h$	$2h$
Comm.	Tag-to-Reader	$2l$	$3l$	$1\frac{1}{2}l$
	Reader-to-Tag	$1l$	$2l$	$\frac{1}{2}l$
	Spoofing	×	△	○
	Loss of message	△	▽	▽
	Replay attack	△	○	○
	Location privacy	⊗	○	○

Notations of Table: l - the output size of a hash function or the length of ID, h - the cost of a hash function operation, Database - Back-end database; × - Attack, △ - Perception, ○ - Prevention, ▽ - Restoration, ⊗ - Traceability.

2 Security and Privacy Risk

We describe possible attacks which can infringe on a person's privacy in a RFID wireless communication; RFID tags have variable vulnerabilities to eavesdropping, traffic analysis, spoofing and denial of service. These attacks make it possibly to disclose sensitive information by an unauthorized reader. Another privacy violation is the linkability establishing a link between a person and a tag, which can be used to trace the tag and hence the person by an adversary. The infringement of location privacy is one of main security issues which should be considered for secure RFID systems. We describe these threats in detail below.

– *Information Leakage.* Some of tags a person holds are quite personal, provide sensitive information that the tag holder does not want to publish. Examples are a title of a book, health information of a person, expensive products, etc. However the disclosure of information arising during a transmission of data possibly reveals various personal details without awareness of the holder.
– *Impersonation.* Although an adversary can not easily access a tag (or reader) physically, she is able to interfere in wireless communication between a tag and a reader. When a reader queries a tag, the adversary can collect the information sent by the tag and the adversary can try a spoofing or reply attack to impersonate a target tag. Thus the adversary may substitute a fake tag for a legitimate article.
– *Traceability.* When a tag transmits a response to a reader, an adversary may try to distinguish whether the response is transmitted by the target tag or not. Once a link is established between the response and the target tag, the adversary who wants to know the person's location may achieve her goal.

To prevent a tag identification, an adversary may try to block or intercept the response message. Errors in message transfer such as loss, interception or blocking of messages should be detected to synchronize an identification protocol.

3 Our Low-Cost RFID Authentication Protocol (LCAP)

In this section we propose an authentication protocol using a challenge-and-response method, which is suitable to a low-cost RFID tag.

3.1 Initial Setup

Let $H : \{0,1\}^* \to \{0,1\}^l$ be a one-way hash function where a hash value space belongs to $\{0,1\}^l$. ID denotes identity of a tag and is a random value in $\{0,1\}^l$. Data fields of a tag and a reader are initialized to the following values:

– *Tag.* The data field of a tag is initialized to its own *ID*. The tag only stores *ID* value with l-bit string and needs a one-way hash function to execute a one-way hash function operation.
– *Reader.* A reader picks uniformly a random value r with $\{0,1\}^l$. A reader does not need to execute any operation.

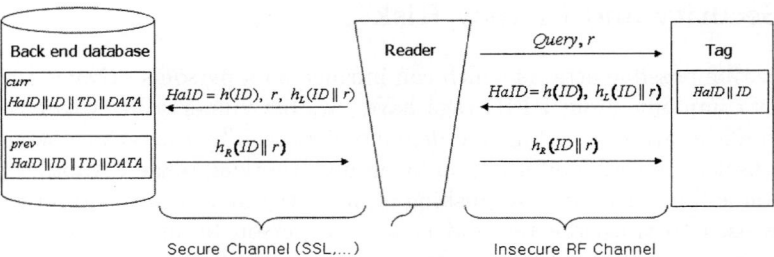

Fig. 1. LCAP protocol

- *Back-end database.* The data fields of a back-end database are initialized to HaID, ID, TD and DATA. HaID value is the hash value of ID used for identifying or addressing the tag. TD-entry is used to trace previous data information of a tag when loss of message occurs in the current session. DATA stores the information about an accessible tag. The back-end database maintains two rows; Prev for the previous session and Curr for the current session. Each row contains HaID, ID, TD, and DATA fields. In Prev, the back-end database records HaID and ID in the previous session. In Curr, it updates HaID and ID of Prev. TD-field of Curr has HaID value of Prev and TD-field of Prev contains HaID-value of Curr. TD is used to link between a previous and current sessions in order to synchronize the tag and the database in case of incompletion of the current session. The back-end database needs a one-way hash function to execute a hash operation.

3.2 LCAP Operation

When a tag enters an operating range of a reader, the reader starts a protocol for mutual authentication. LCAP scheme is operated as follows, as shown in Fig.1.

1. A reader picks a random value r and sends Query and r to an accessing tag.
2. When queried, the tag computes HaID=$h(ID)$ and $h(ID||r)$ using r and its own ID. The tag sends $h_L(ID||r)$ and HaID to the reader, where $h_L(ID||r)$ is a left half of $h(ID||r)$, so $h_L(ID||r)$ has the length of $\frac{1}{2}l$ bits.
3. The reader sends $h_L(ID||r)$, r, and HaID to the back-end database.
4. The back-end database checks if the value of HaID in Prev is matching to the value of HaID received from the reader. If successful, then the back-end database computes $h_R(ID||r)$ using r received from the reader and ID in Prev, where $h_R(ID||r)$ is a right half of $h(ID||r)$ of length $\frac{1}{2}l$ bits. Next, to update a ID value for the next session, the back-end database computes $h(ID \oplus r)$ and stores HaID=$h(ID \oplus r)$ and ID=ID$\oplus r$ in Curr. Here $h(ID \oplus r)$ is an ID of the tag in the next session. TD-field of Prev is filled with current HaID=$h(ID \oplus r)$. Finally the back-end database sends $h_R(ID||r)$ to the reader.
5. The reader forwards $h_R(ID||r)$ to the tag.
6. The tag checks a validity of $h_R(ID||r)$. If the message is valid, then the tag updates its own ID to ID$\oplus r$.

4 Security and Efficiency Analysis

4.1 Security Analysis

We analyze the security of our protocol against the threats introduced in Section 2.; *information leakage, impersonation,* and *traceability.*

Information Leakage. In LCAP, in order to get any information in a tag, an adversary must be authenticated. Without knowing ID and r, only way to be authenticated is to guess $h_R(ID||r)$ after collecting messages $HaID$, $h_L(ID||r)$, and r. However, because of one-wayness of h, the adversary can not get any information of $h_R(ID||r)$ from $HaID$, $h_L(ID||r)$, and r, and has to randomly pick a string from $\{0,1\}^{\frac{1}{2}l}$. The advantage of the adversary is at most $\frac{1}{2^l}$, which is negligible.

Impersonation. In our protocol, impersonation can be prevented by mutual authentication between a reader and a tag. In LCAP, whenever a tag reaches operating ranges of a reader (whenever the tag tries an identification), the reader queries a random value to the tag. An adversary collects a tag response and then she may try a spoofing attack to impersonate a valid tag. However without knowing ID of the target tag, the adversary can not compute $HaID = h_L(ID||r)$ that a valid tag can only generate. Therefore, it is not possible to impersonate the target tag by *spoofing attacks* in LCAP. We note that LCAP prevents *spoofing attacks* before it occurs, while *HIDV* scheme can only perceive a spoofing attack, but can not prevent it. *Replay attacks* can not compromise LCAP since the valid massage is freshed in each session by a random value r of $HaID=h(ID||r)$. LCAP can also perceive *errors in message transfer* and be restored using TD.

Traceability. Our scheme, as in *HIDV*, guarantees location privacy by using "dynamic" identifies, where a identifier is refreshed simultaneously by a tag and a back-end database in each session only after an identification successes. In [1], *Gildas Avoine* and *Philippe Oechslim* have described an attack based on refreshment avoidance. In the attack, an adversary always makes a tag unable to refresh its identifer and hence can traces the tag by tracing the static identifer of the tag. To perform this attack, the adversary should be able to priori predict every location of the tag to intervene each indentification process of the tag. This paradoxically implies that the adversary should trace the tag in order to obtain **traceability**. We note again that in LCAP, a identifier is only used once since the tag changes it by itself as soon as an identification is completed. Therefore refreshment avoidance attack does not need to be considered in LCAP.

4.2 Efficiency

We consider a storage cost, a communication cost, and a computation cost of entities. As compared with the previous in Table.1, LCAP is remarkably improved in a computation cost and a storage cost. A tag only stores its own ID of length l-bit, which is suitable to a low-cost RFID tag with extremely limited storage space. Since tags should complete the protocol in few seconds, an RF identification scheme should guarantee a low communication cost and a quick computation time without degrading security. A communication cost of our **LCAP** minimal; messages of tag-to-reader communication are $HaID$ and $h_L(ID||r)$ with a total $1\frac{1}{2}l$ bits and a message of reader-to-tag communication is $h_R(ID||r)$ with $\frac{1}{2}l$ bits. Since tags have an extremely limited computing power, a low computation cost is the most important contribution in RFID schemes. In **LCAP**, a tag and a back-end database perform only two one-way hash function operations, while $HIDV$ scheme needs three one-way hash function operations. **LCAP** maintains low-cost storage, communication, and computation and hence is well suitable to low-cost RFID systems.

5 Conclusions

In this paper, we have proposed an efficient and secure authentication method **LCAP** to protect privacy, especially for low-cost RFID systems. The proposed scheme needs only two one-way hash function evaluations and hence is quite efficient. Leakage of information is prevented in the scheme since a tag emits its identifier only after authentication. By refreshing a identifier of a tag in each session, the scheme also provides a location privacy and can recover lost massages from many attacks such as spoofing attacks.

References

1. G. Avoine and P. Oechslin. *RFID Traceability: A Multilayer Problem*, Financial Cryptography 2005.
2. Auto-ID Center, *860Mhz-960MHz Class I Radio Frequency Identification Tag Radio Frequency and Logical communication Interface Specification Proposed Recommendation Version* 1.0.0. Technical Report MIT-AUTOID-TR-007, AutoID Center, MIT, 2002.
3. R. Damith and E. Daniel and C. Peter. *Low-Cost RFID Systems: Confronting Security and Privacy*. Auto-ID Labs Research Workshop, 2004.
4. M. Feldhofer. *An Authentication Protocol in a Security Layer for RFID Smart Tags*. The 12th IEEE Mediterranean Electrotechnical Conference, MELECON04, IEEE, 2004.
5. P. Golle, M. Jakobsson, A. Juels, and P. Syversion, *Universal re-encryption for mixnets*. In Tatsuaki Okamoto, editor, RSA Conference Cryptographers' Track, LNCS 2964, pp.163-178, Springer-Verlag, 2004.
6. D.Henrici and Paul Muller. *Hash-based Enhancement of Location Privacy for Radio-Frequency Identification Devices using Varying Identifiers*. PerSec04 at IEEE PerCom, 2004.
7. A. Juels. *yoking-proofs" for RFID tags*. Workshop on Pervasive Computing and Communications Security, PerSec04, pp138-143,IEEE Computer Society, IEEE.
8. A. Juels and R. Pappu, *Squealing euros : Privacy protection in RFID-enabled banknotes*. In proceedings of Financial Cryptography -FC'03, 2003.

9. A. Juels, R. L. Rivest and M. Szudlo. *The Blocker Tag: Selective Blocking of RFID tags for Consumer Privacy*. In the 8th ACM Conference on Computer and Communications Security, pp. 103-111. ACM Press. 2003.
10. mCloak : Personal/corporate management of wireless devices and technology, 2003. http://www.mogilecloak.com.
11. M. Ohkubo, K. Suxuki and S. Kinoshita. *Efficient Hash-Chain Based RFID Privacy Protection Scheme*. Ubcomp04 workshop.
12. M. Ohkubo, K. Suzxuki and S. Kinoshita. *Cryptographic Approach to "Privacy-Friendly" Tags*. RFID Privacy Workshop, MIT MA USA, 2003.
13. S. E. Sarma, S. A. Weis and D. W. Engels. *Radio-frequency identification systems*. CHES02, vol.2523 of LNCS, pp.454-469, Springer-Verlag, 2002.
14. S. E. Sarma, S. A. Weis and D. W. Engels. *RFID systems, security and privacy implications*. Technical Report MIT-AUTOID-WH-014, AutoID Center, MIT, 2002.
15. S. A. Weis, S. E. Sarma, S. A. Weis and D. W. Engels. *Security and privacy Aspects of Low-Cost Radio Frequency Identification Systems*. First International Conference on Security in Pervasive Computing, 2003. http://theory.lcs.mit.edu/sweis/spc-rfid.pdf

An Efficient Performance Enhancement Scheme for Fast Mobility Service in MIPv6

Seung-Yeon Lee[1], Eui-Nam Huh[1], Sang-Bok Kim[1], and Y. Mun[2]

[1] Division of Information and Communication,
Seoul Women's University, Seoul, Korea
[2] School of Computer Science, Soongsil University, Seoul, Korea
{seungyeon, huh}@swu.ac.kr

Abstract. As secure mobility service is becoming a critical issue in the ubiquitous environment, the Mobile IP Working Group in IETF is preceding the research about it. If it provides weak security features to the mobile service, then the Mobile IPv6 will not be trusted. Although the IPSec (Internet Protocol Security) and RR (Return Routability) was selected as the methods for providing security supports and related works have been obligated, these approaches have drawbacks that the hand-held devices such as cellular phones and PDAs are battery-powered so the security processing is a big burden and security feature is not relatively abundant. To cope with these drawbacks, we propose the integrated models with AAA infrastructure as an alternative way to authenticate a mobile node by using the AAA authentication processing. In this paper, our research has a focus on minimizing the authentication latency in AAA processing and proposes the model with Fast Handoff scheme to make the better performance to AAA authentication delay.

1 Introduction

Mobile IPv6 proposes both IPSec (Internet Protocol Security) and RR (Return Routability) for providing security support. In using the IPSec alone, security association could be established between Home Agents (HAs), and messages are authorized. However, when the mobile node is on roaming at the time, Mobile Node (MN) should be authenticated by HA to access the visiting network.

To secure the mobile IPv6, there were AAA models based upon Diameter, the following studies typically use Diameter method-"Diameter Mobile IPv6 Application" by Chales E.Perkins and "AAA for Mobile IPv6" by Francis Dupont. In this paper, we propose a model based on Francis Dupont's approach and introduce the model combined with Fast Handoff to minimize the authentication latency and packet loss.

This paper is comprised of six chapters. In Chapter 2, we discuss introduction of authentication and binding procedure. As following Chapter 3, general process of AAA authentication procedures performed by Francis Dupont's model, which followed with fast handoff, is illustrated. Chapter 4 explains our proposed model to minimize typical authentication procedures by establishing Delegation Model combined with Fast Handoff. In chapter 5, the cost analysis between the proposed model and the comparison is simulated using mathematical model. In last chapter, we discuss further study and conclusion as well.

2 Security of Mobile IPv6

2.1 Authentication Header of IPv6

For authentication of security in IPv6, the authentication header in "Extended Header" of IPv6 (see Fig. 1) is used, and then 51, additionally transmitted with following IP datagram, accomplish the "Next Head Value".

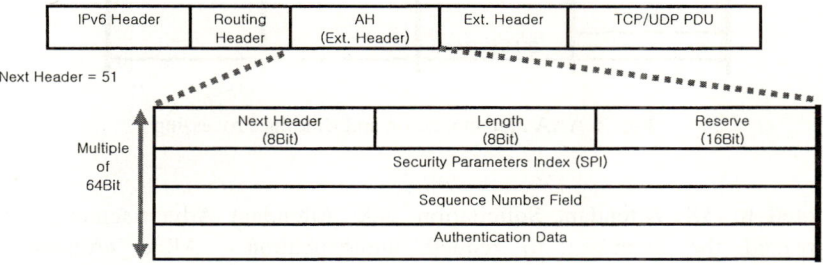

Fig. 1. Structure of the Authentication Header

2.2 Comparison of AAA Protocols

In order to provide authentication and authorization (AA) in AAA model, Radius method contains two features, authentication and authorization, while Diameter method provides AA separately along with PKI method. The Table 1 shows the comparison of two protocols.

Table 1. Comparison of AAA Protocol

	RADIUS	Diameter
Authentication, Authorization Verifiable	Merged	Separated
Packet inscription	Only user password	Whole packet payload
Transport layer Protocol	UDP	TLS/TLS over SCTP
PKI	Not Available	Provide
Safely Global Roaming	Limited provide	Basic Provide

2.3 AAA for Mobile IPv6

In this section, we present the whole Mobile IPv6 AAA as shown in Fig. 2, so MN configures IPv6 address by itself after receiving network prefix from visiting router,

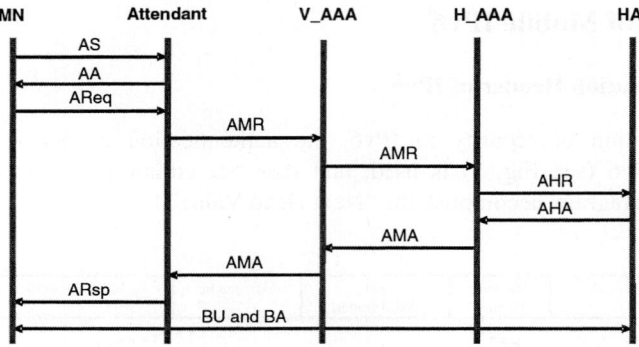

Fig. 2. AAA Authentication and Binding Processing

(denoted to AS: Attendant Solicitation, AA: Attendant Advertisement) when it recognized the attendant to request authentication AReq (Authentication Request).

Next phase, the Attendant received authenticated information by MN and send AMR(Authentication MN-Request) to Local AAA server. Local AAA Server transforms the requested messages to AAA protocol form, and transmits them to AAA server (H_AAA) of home agent domain. We assumed that between V_AAA and H_AAA, the roaming contract has already established.

H_AAA Server transmits AHR(Authentication HA-Request) messages to HA, and HA verifies authenticated message. Based on combined key between client and HA, HA produces the session key which will be used between MN and attendant and then key production materials will be returned to H_AAA. H_AAA Server sends AMA(Authentication MN-ACK) to V_AAA. Moreover, the attendant stores ARsp(Authentication Response) composed of session key and key production materials, which will later be sent to MN. After completing previous processes, MN and HA transmit BU (Binding Update) information and BA (Binding Acknowledge) to HA and MN, respectively.

3 Compared Model

This section describes the message procedure of AAA authentication process (Fast Handoff applied) in Mobile IPv6.

The typical message procedure starts if the MN moves to triggers Fast Handoff activity, when Fast Handoff has finished; following the AAA authentication is performed with message process of 12 steps. After completed the authentication of MN, the binding process to HA by sending the BU on MN is performed and finally new binding information is updated as BA received from HA.

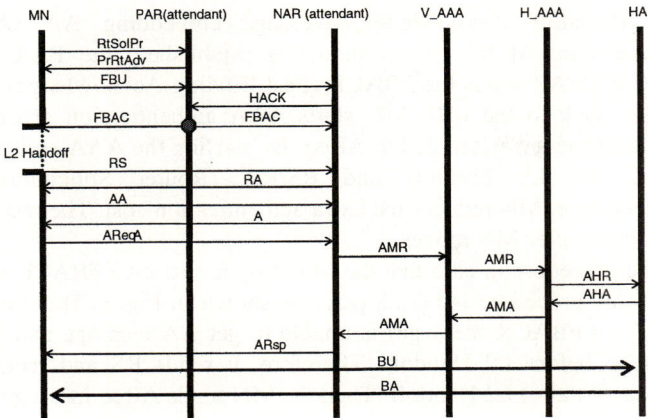

Fig. 3. Message Procedure of Compared Model

4 Proposed Model

4.1 AAA During Fast Handoff Using Authentication Header

In proposed message procedure, we propose the MN receiving FBACK message. Fig. 4 shows the message procedure in case that MN receives the FBACK message.

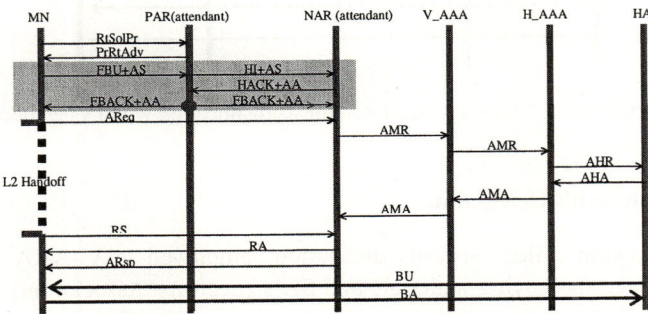

Fig. 4. Proposed Message Sequence for AAA with Authentication Header during Handoff

Fast Handoff starts operation when the MN moves to another link where it sends FBU message to PAR (previous agent router) to confirm the tunnel and obtained new CoA (Care of Address). At this time, by embedding the AAA message, AS (Authentication Solicitation) into the FBU (Fast Binding update), the AAA authentication procedure starts before the Handoff occurs.

The PAR should have the capability to send HI message containing the AS message. Upon receiving the HI message, the new AR (Authentication Request) processes the HI (Handoff Initiate) and AS message and responds to PAR with sending HACK (Handoff Acknowledge) message embedding AA (Attendant Advertisement) message. At this stage, tunnel is established, the PAR received HACK messages, and PAR sends the FBACK (Fast Binding Acknowledge) message containing the AA back to the MN. MN starts AAA authentication procedure by sending AAA request message, denoted to AReq. By starting the AAA authentication procedure during the L2 Handoff and RS/RA (Router Solicitation/Router Advertisement) exchange, MN reduces the extra authentication cost. The rest of AAA procedure is performed after MN moves.

Fig. 5 shows the procedure in case that the MN dose not receive FBACK message. In this case, the same procedure is taking place as shown in Fig. 4. However, if the MN (ie., MN with no FBACK message) is unable to get AA message and it cannot send AReq message before L2 Handoff. Therefore, it sends RS and receives RA message containing AA after L2 Handoff. Then the MN sends AReq message to NAR (New Agent Router).

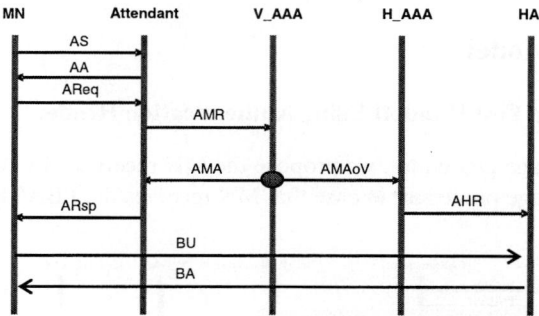

Fig. 5. Proposed Message Sequence for Authentication with Delegation

4.2 AAA Model with Delegation

An important option called "security delegation" employed to V_AAA in order to efficiently manage the keying materials and SA's (Security Association) context for MN. If MN sends the AReq message with delegation option, this message is relayed to the entity such as H_AAA or HA which plays a role of generating and distributing session key and keying materials.

If H_AAA has the right to message the keying materials, the security context used to authenticate and generate session_key for MN is transferred when H_AAA responds to the AMR with AMA (or AMAoV). The V_AAA will receive the AMA message with security context. V_AAA compares its capabilities with security context (SAs, algorithms, hash functions, etc.). If it has capabilities specified in the security context then V_AAA create an entry into the delegation entry list to accept and process delegation request. If it has insufficient capabilities, the delegation request is ignored and the message is processed as specified in [1] and [7].

If MN moves to another visited link in the same domain with MN's previous location and the delegation request option is set, V_AAA determines whether the MN is registered in delegation entry list.

If the entry exists then V_AAA authenticates the MN and generates the session_key according to the security context. After the delegation procedure is completed, V_AAA responds to AMR with AMA (or AMAoV), which contains session_key, keying materials and some security parameters.

The V_AAA maintains delegation entry until the lifetime is expired (the lifetime is specified by V_AAA, default values is 300sec). If the lifetime expires, the delegation entry is removed from the delegation list. The lifetime may be refreshed by a request from H_AAA (optional).

When the home agent has the right to manage the keying materials, the security context is transferred when the home agent responds to AHR with AHA. The rest of processing is identical to the procedure described above.

The procedure for the authentication request from MN in case of delegation is like as the Fig. 2.

4.3 Delegation Model Which Applying with Fast Handoff

The proposed message process assumes only the case that the mobile node receives the FBACK message. Thus, in our model, V_AAA performs the delegation mechanism as shown in Fig. 6.

There are advantages on our authentication model that delegation option is applied as described in section 4.2. It minimizes more process steps for authentication than our proposed authentication model introduced in section 4.1. The compared model consists of 12 steps in authentication processes from the Fast Handoff step to the Binding Update step. The two steps are decreased in authentication header model introduced in section 4.1.

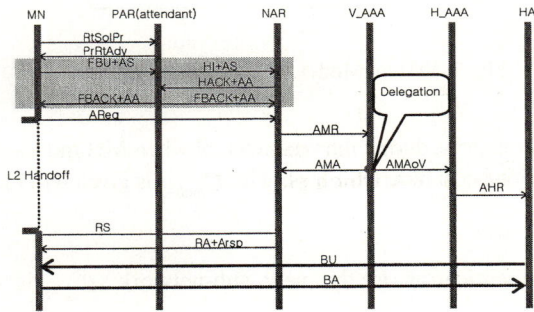

Fig. 6. Message process sequence of the proposed approach

As discussed in section 4.1, adapting the delegation mechanism discussed in section 4.2 can optimize the first model. Furthermore, the proposed model can reduce additionally two steps, AReq (Authentication Request) message to NAR, and ARsp (Authentication Response) message. Finally, we minimized four steps by discarding the authentication message that MN needs to send it to HA in previous approaches.

5 Costs and Performance Analysis

Each of the distances between the various entities involved in the proposed scheme is shown in Fig. 7. This system model is proposed for the cost analysis of Mobile IPv6 using Fast Handoff and AAA technology in the movement between sub-networks.

Assume that a CN transmits data packets to the MN at a mean rate, λ and MN moves from a sub-network to another at a mean rate μ. We introduce Packet to Mobility Ratio (PMR) as the mean number of packets received by a MN from CN per move. The PMR is given by $p = \lambda/\mu$. Let the average length of a control packet be l_c and a data packet be l_d and we define their ratio as $l = l_d/l_c$. We assume that the cost of transmitting a control packet is given by the distance between the sender and receiver and the cost of transmitting a data packet is l times greater. And we assume that the average cost of processing control packets at any host is r.

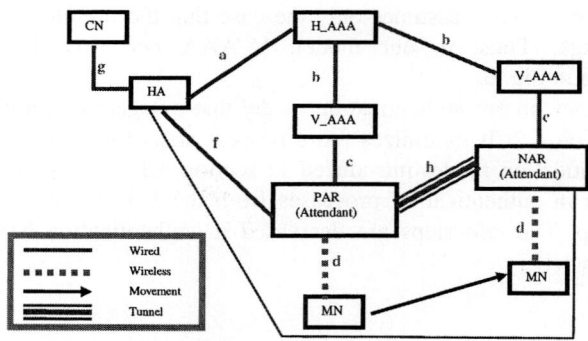

Fig. 7. System Model for Cost Analysis

For general authentication flow, during the time interval when MH moves within the sub-network in a domain, the total cost incurred a loss. The C_{auth-g} is given by (1).

$$C_{auth-g} = C_{rg-g} + C_{oldFA-g} \qquad (1)$$

The cost of MN's authentication at the new sub-network, C_{rg-g} is computed as follows.

$$C_{rg-g} = 2(a+b+c+e) + 3h + 12d + 20r \qquad (2)$$

The cost of data packets lost by delivery to the old domain during the authentication delay is given by (3).

$$C_{oldFA-g} = \lambda \times t_{auth-g} \times C_{dt} \qquad (3)$$

C_{dt} in the AAA of Mobile IP is $l(g+f)+r$ which means the cost of single data packet delivered from CN to MN via tunneling at HA. The time for authenticating delay can be represented as follows.

$$t_{auth-g} = 2(t_a + t_b + t_c + t_e) + 3t_h + 12t_d + 20t_r \tag{4}$$

Therefore, the total cost of general authentication case can be represented like (5).

$$C_{auth-g} = 2(a+b+c+e) + 3h + 12d + 20r + (\lambda \times t_{auth-g} \times C_{dt}) \tag{5}$$

First, (6) shows the total cost, C_{auth-a}, for the proposed model which applying Authentication Header during the time interval, when the MN moves between the sub-network in a domain.

$$C_{auth-a} = C_{rg-a} + C_{oldFA-a} \tag{6}$$

The cost of MN's authentication at the new sub-network in the proposed scheme C_{rg-a} is calculated like (7).

$$C_{rg-a} = 2(a+b+c+e) + 3h + 10d + 18r \tag{7}$$

The cost of data packets lost by delivery to the old domain during the authenticating delay is given by (8).

$$C_{oldFA-a} = \lambda \times t_{auth-a} \times C_{dt} \tag{8}$$

For the authenticating delay in proposed scheme, the time can be represented as follows.

$$t_{auth-a} = 2(t_a + t_b + t_c + t_e) + 3t_h + 10t_d + 18t_r \tag{9}$$

Therefore, the total cost of proposed scheme can be represented as follows.

$$C_{auth-a} = 2(a+b+c+e) + 3h + 10d + 18r + (\lambda \times t_{auth-a} \times C_{dt}) \tag{10}$$

The improved cost of the proposed scheme for general case can be calculated as follows.

$$I_{auth-a} = C_{auth-g} - C_{auth-a} = 2d + 2r + ((\lambda \times t_{auth-g} \times C_{dt}) - (\lambda \times t_{auth-g} \times C_{dt})) \tag{11}$$

Second, (12) shows C_{auth-d} during the time interval when the MN moves between the sub-networks in a domain, for the proposed delegated authentication flow, the total cost.

$$C_{auth-d} = C_{rg-d} + C_{oldFA-d} \tag{12}$$

The cost of MN's authentication at new sub-network in proposed scheme C_{rg-d} is computed as follows.

$$C_{rg-d} = a + b + 2(c+e) + 3h + 9d + 15r \tag{13}$$

The cost of data packets lost by being delivered to the old domain during the authenticating delay is computed as follows.

$$C_{oldFA-d} = \lambda \times t_{auth-d} \times C_{dt} \qquad (14)$$

For the authenticating delay in the proposed scheme, time is computed as follows.

$$t_{auth-d} = t_a + t_b + 2(t_c + t_e) + 3t_h + 9t_d + 15t_r \qquad (15)$$

Therefore, the total cost of the proposed scheme becomes as follows.

$$C_{auth-d} = a + b + 2(c+e) + 3h + 9d + 15r + (\lambda \times t_{auth-d} \times C_{dt}) \qquad (16)$$

The improved cost of proposed scheme for the general case computed as follows.

$$I_{auth-d} = C_{auth-g} - C_{auth-d} = a + b + 3d + 5r + ((\lambda \times t_{auth-g} \times C_{dt}) - (\lambda \times t_{auth-d} \times C_{dt})) \qquad (17)$$

Additionally, the enhanced cost between delegation model (section 4.2) and authentication header model (section 4.1) is obtained as follows.

$$I_{total} = C_{auth-d} - C_{auth-a} = a + b + d + 3r - ((\lambda \times t_{auth-d} \times C_{dt}) + (\lambda \times t_{auth-a} \times C_{dt})) \qquad (18)$$

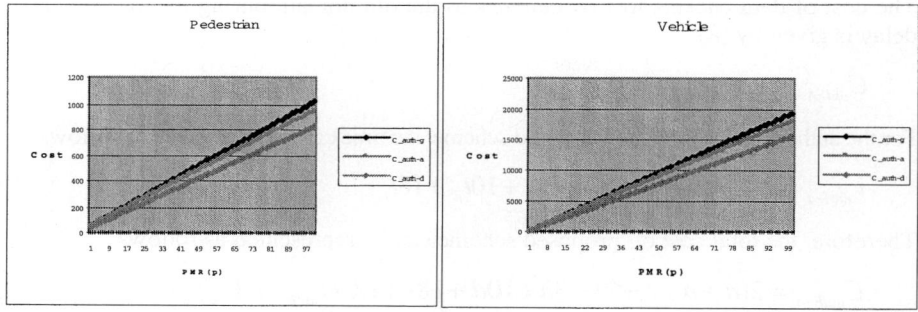

Fig. 8. PMR Analysis for Pedestrian and Vehicle

The cost is simulated for a slow mobile device (Pedestrian) and a fast mobile device (Vehicle), as shown in Fig.8. Our proposed model, *auth_d*, outperforms 20% rather than the previous general approach depicted as *auth_g* in the figure. The cost improvements of the proposed approach are 19% for slow mobile devices and 21% for fast one, respectively.

6 Conclusions

This study shows an effective cost by calculation of signaling cost and packet transmitting cost. Thus, this proposed model adapting Fast Handoff minimizes the time that MN is waiting for authentication and handoff. In addition, it reduces the packet loss occurred during Handoff in Mobile IPv6 and it gets advantage in minimizing the roaming time by reducing the authentication processes.

Acknowledgement

This work is supported in part by Ministry of Information and Communication in Korea.

References

[1] Charles E. Perkins and David B. Johnson, "Mobility Support in IPv6", Internet Draft, draft-ietf-mobileip-ipv6-24, December 2003.
[2] Charles E. Perkins and Thomas Eklund, "AAA for IPv6 Network Access", Internet Draft, draft-perkins-aaav6-06, May 2003.
[3] Charles E. Perkins, "Diameter Mobile IPv6 Application", Internet Draft, draft-le-aaa-diameter-mobileipv6-03, October 2003.
[4] S. Glass and C. Perkins, "Mobile IP Authentication, authorization, and Accounting Requirements", RFC 2977, October 2000.
[5] Stefano M. Faccin and Charles E. Perkins, "Mobile IPv6 Authentication, Authorization, and Accounting Requirements", Internet Draft, draft-le-aaa-mipv6-requirements-02, October 2003.
[6] R. Koodli et al, "Fast Handovers for Mobile IPv6", Internet Draft, draft-ietf-mobileip-fast-mipv6-06, March 2003.
[7] Sangheon Pack and Yanghee choi, "Performance Analysis of Fast Handover in Mobile IPv6 Networks", in proc. IFIP PWC 2003, Venice, Italy, September 2003.
[8] R. Jain, T. Raleigh, C. Graff and M. Bereschinsky, "Mobile Internet Access and QoS Guarantees using Mobile IP and RSVP with Location Registers," in Proc. ICC'98 Conf., pp. 1690-1695, Atlanta.
[9] Thomas, R., H. Gilbert and G. Mazzioto, "Influence of the mobile station on the performance of a radio mobile cellular network," Proc. 3rd Nordic Sem., paper 9.4, Copenhagen, Denmark, Sep

Face Recognition by the LDA-Based Algorithm for a Video Surveillance System on DSP

Jin Ok Kim[1], Jin Soo Kim[2], and Chin Hyun Chung[2]

[1] Faculty of Multimedia,
Daegu Haany University, 290, Yugok-dong,Gyeongsan-si,
Gyeongsangbuk-do, 712-715, Korea
bit@dhu.ac.kr

[2] Department of Information and Control Engineering,
Kwangwoon University, 447-1, Wolgye-dong,
Nowon-gu, Seoul, 139-701, Korea
chung@kw.ac.kr

Abstract. Face recognition is an important part of today's emerging biometrics and video surveillance markets. As face recognition algorithms move from research labs to real world product, power consumption and cost become critical issues, and DSP-based implementations become more attractive. Also, "real-time" automatic personal identification system should meet the conflicting dual requirements of accuracy and response time. In addition, it also should be user-friendly. This paper proposes a method of face recognition by the LDA (Linear Discriminant Analysis) Algorithm with the facial feature extracted by chrominance component in color images. We designed a face recognition system based on a DSP (Digital Signal Processor). At first, we apply a lighting compensation algorithm with contrast-limited adaptive histogram equalization to the input image according to the variation of light condition. While we project the face image from the original vector space to a face subspace *via* PCA (Principal Component Analysis), we use the LDA to obtain the best linear classifier. And then, we estimate the Euclidian distances between the input image's feature vector and trained image's feature vector. The experimental results with real-time input video show that the algorithm has a pretty good performance on DSP-based face recognition system.

1 Introduction

Human activity is a major concern in a wide variety of applications such as video surveillance, human computer interface, face recognition, and face image database management. And machine face recognition is a research field of fast increasing interest. The strong need for user-friendly systems that can secure our assets and protect our privacy without losing our identity in a sea of numbers is obvious. At present, one needs a Personal Identification Number (PIN) to get cash from an ATM, a password for a computer, a dozen others to access the internet, and so on. Although extremely reliable methods of biometric

Table 1. Typical applications of face recognition

Areas	Representative Works
Biometrics	Drivers Licenses, Entitlement Programs
	Immigration, National ID, Passports, Voter Registration
	Welfare Fraud
Information Security	Desktop Log-on (Windows, Linux)
	Application Security, Database Security, File Encryption
	Intra-net Security, Internet Access, Medical Records
	Secure Trading Terminals
Law Enforcement and Surveillance	Advanced Video Surveillance, CCTV Control
	Portal Control, Post-Event analysis
	Shoplifting and Suspect Tracking and Investigation
Smart Cards	Stored Value Security, User Authentication
Access Control	Facility Access, Vehicular Access

personal identification exist, e.g., fingerprint analysis and retinal or iris scans, these methods rely on the cooperation of the participants, whereas a personal identification system based on analysis of frontal or profile images of the face is often effective without the participant's cooperation or knowledge. Phillips [1] described in the advantages/disadvantages of different biometrics. Table 1 lists some of the applications of face recognition.

We implemented a fully automatic DSP-based face recognition system that works on video. Our implementation follows the block diagram shown in Fig. 1(b). Our system consists of a face detection and tracking block, an skin tone extract block, a face normalization block, and face classification blocks.

2 Face Detection System

Face detection is the most important process in applications such as video surveillance, human computer interface, face recognition, and image database management. Face detection algorithms have primary factors that decrease a detection ratio: variation by lighting effect, location and rotation, distance of object, complex background. Due to variations in illumination, background, visual angle and facial expressions, the problem of machine face detection is complex.

An overview of our face detection algorithm contains two major modules: 1) face segmentation for finding face candidates and 2) facial feature extraction for verifying detected face candidates. Our approach for face localization is based on the observation that human faces are characterized by their oval shape and skin color, also in the case of varying light conditions. Therefore, we locate face-like regions on the base of shape and color information. We employ the YC_bC_r color space by using the RGB to YC_bC_r transformation. We extract facial features based on the observation that eyes and mouth differ from the rest of the face in chrominance because of their conflictive response to C_b, C_r.

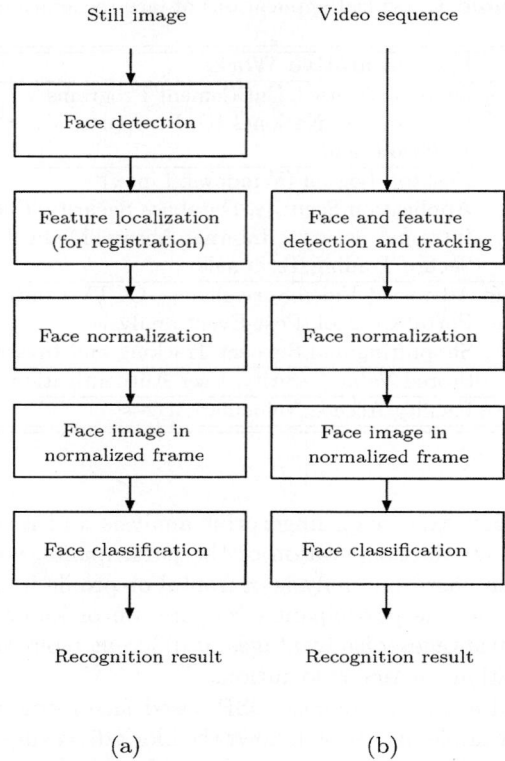

Fig. 1. Two approaches to face recognition: (a) Face recognition from still images; (b) Face recognition from video

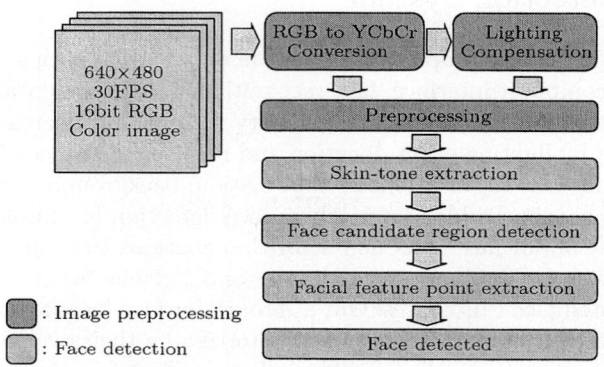

Fig. 2. Face detection algorithm

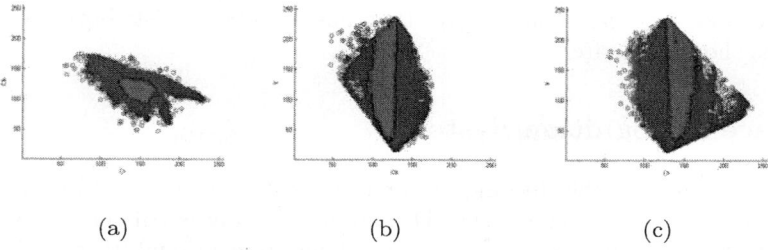

Fig. 3. The YC_bC_r color space and the skin tone color model (red dots represent skin color samples): (a) A 2D projection in the $CbCr$ subspace; (b) A 2D projection in the YCb subspace; (c) A 2D projection in the YCr subspace

2.1 Skin Color Modeling

We adopt the YC_bC_r space since it is perceptually uniform, is widely used in video compression standards (e.g., MPEG and JPEG), and it is similar to the TSL(Tint-saturation-luminance) space in terms of the separation of luminance and chrominance as well as the compactness of the skin cluster. Many research studies assume that the chrominance components of the skin-tone color are independent of the luminance component [2, 3, 4].

However, in practice, the skin-tone color is nonlinearly dependent on luminance. We demonstrate the luminance dependency of skin-tone color in different color spaces in Fig. 3, based on skin patches collected from IMDB [5] in the Intelligent Multimedia Laboratory image database. These pixels from an elongated cluster that shrinks at high and low luminance in the YC_bC_r space are shown in Fig. 3(b) and Fig. 3(c). Detecting skin tone based on the cluster of training samples in the C_bC_r subspace is shown in Fig. 3(a)

3 Face Region Segmentation

Region labeling uses two different process. One is region growing algorithm using seed point extraction and another one is flood fill operation [6]. The other method is shown Fig. 4(b).

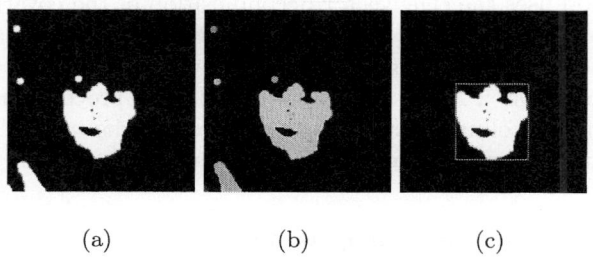

Fig. 4. Connected component labeling: (a) Segmented image; (b) Labeled image; (c) Facial candidate region

We choose face candidate region that has a suitable distribution of pixel. Result is shown Fig. 4(c).

4 Face Recognition System

There are many possible techniques for classification of data. Principal Component Analysis (PCA) and Linear Discriminant Analysis (LDA) are two commonly used techniques for data classification and dimensionality reduction.

Most face detection and recognition schemes can be divided into two different strategies. The first method is based on the detection of facial features [7], whereas the second approach tries to detect a face pattern as a whole unit [8, 9]. Following the second approach, each image pattern of dimension I by J can be considered as a vector $\mathbf{x}$ in a $N = IJ$ dimensional space. Obviously, images of faces will not be randomly distributed in this high dimensional image space. A suitable mean to reduce the dimensionality of the data set is the principal components analysis (PCA).

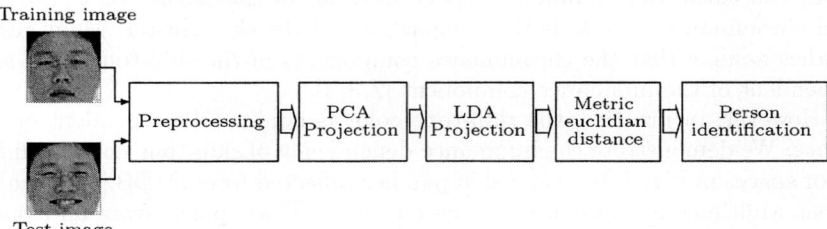

Fig. 5. The generalized LDA face recognition system

The central idea of principal components analysis is to find a low dimensional subspace (the feature space) that captures most of the variation within the data set and therefore allows the vest least square approximation [10]. When used for face detection and recognition, this principal subspace is often called the "face space" which is spanned by the "eigenfaces" [11].

4.1 Principal Component Analysis: PCA

The PCA method uses the eigenvector decomposition of the covariance matrix. The data then projected onto the eigenvector corresponding to several eigenvalues to reduce the data to a smaller dimension. If the images are of size $M \times N$, there will be a total of the $n = M \times N$ dimensional vector. In this process, the assumption is that all vectors are column vectors (i.e., matrices of order $n \times 1$). We can write them on a line of text simply by expressing them as $\mathbf{x} = (x_1, x_2, ..., x_n)^T$.

4.2 Linear Discriminant Analysis: LDA

Linear discriminant analysis easily handles the case where the within-class frequencies are unequal and their performances has been examined on test data

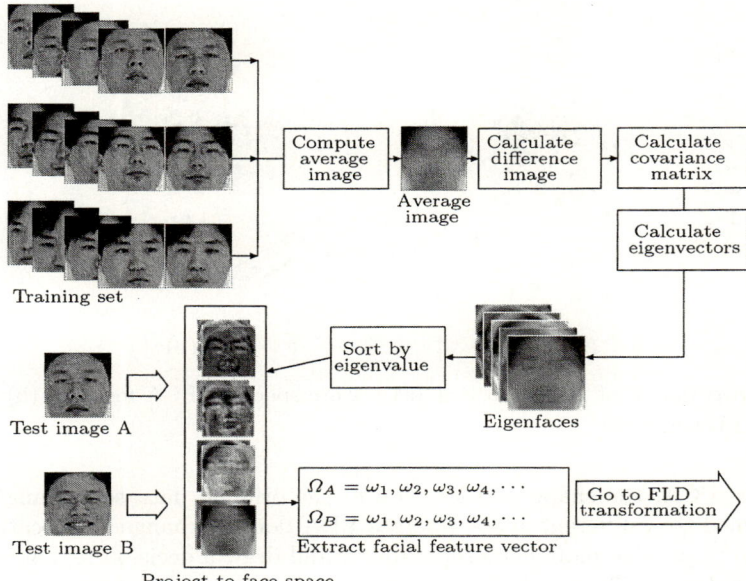

Fig. 6. Block diagram of the feature extract procedure by PCA method

generated randomly. This method maximizes the ratio of between-class variance to the within-class variance in any particular data set thereby guaranteeing maximal separability [12, 13].

The LDA for data classification is applied to classification problem in speech recognition and face recognition. We implement the algorithm for LDA in hopes of providing better classification. The prime difference between LDA and PCA is that PCA does more of feature representation and LDA does data classifi-

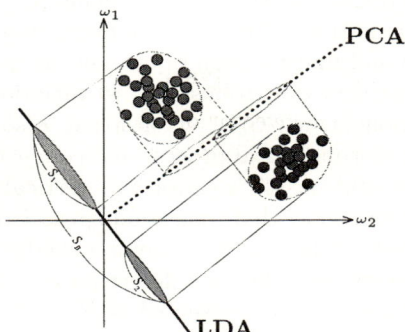

Fig. 7. A comparison of PCA and Fisher's linear discriminant (FLD) for a two class problem

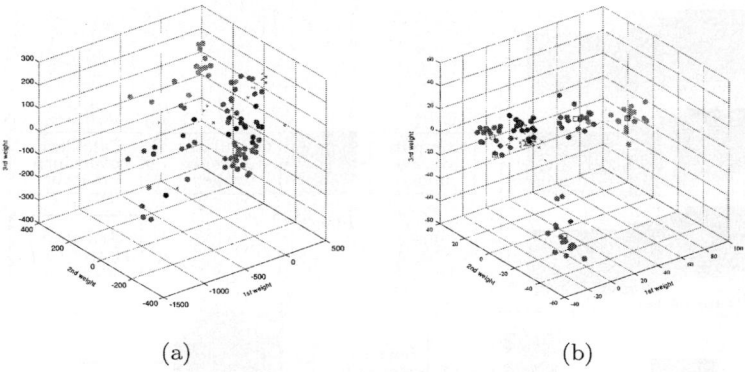

Fig. 8. Distribution of facial features: (a) feature space by PCA method; (b) feature space by LDA method

cation. In PCA, the shape and location of the original data sets change when transformed to a different space whereas LDA doesn't change the location but only tries to provide more class separability and draw a decision region between the given classes. This method also helps to better understand the distribution of the feature data. Figure 7 will be used as an example to explain and illustrate the theory of LDA.

5 Experimental Result

In our experiments, we have used the POSTECH face database [5], which consists of 1802 frontal images of 106 person (17 images each). Images are 256×256 pixels in full-color images. The test platform is a P4/2.4GHz computer with 512MB RAM under Windows 2K.

We realized the face region detection process using YC_bC_r color space. We use the response to a chrominance component in order to find eye and mouth. In a general, images like digital photos have problems (e.g., complex background, variation of lighting condition). Thus it is difficult process to determine skintone's special features and find location of eye and mouth. Nevertheless we can make to efficiency algorithm that robustness to variation lighting condition to use chrominance component in YC_bC_r color space. Also we can remove a fragment regions by using morphological process and connected component labeling operation. We find eye and mouth location use vertical, horizontal projection. This method is useful and show that operation speed is fast.

Figure 9 demonstrates that our face detection algorithm can successfully detect facial candidate region. The face detection and face recognition on the POSTECH image database [5] are presented in Fig. 10. The POSTECH image database contains 951 images, each of size 255 × 255 pixels. Lighting conditions (including overhead light and side lights) change from one image to another.

This system can detect a face and recognize a person at 10 fps. This speed was measured from user-input to final stage with the result being dependent on

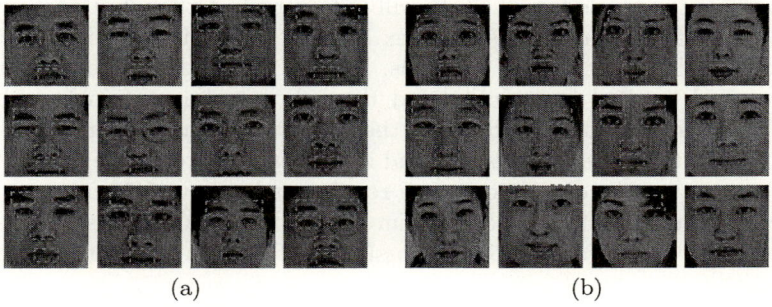

Fig. 9. Some face and facial component extraction: (a) Male (b) Female

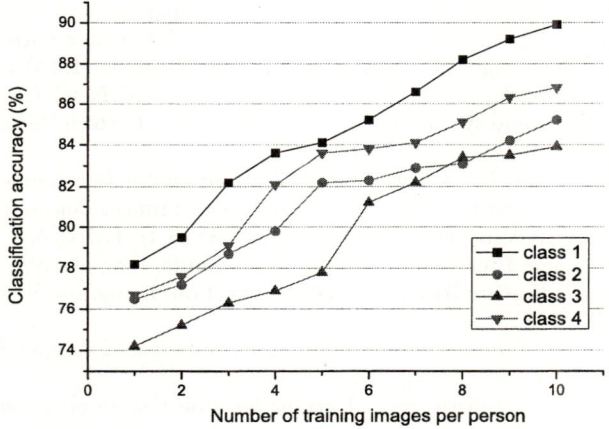

Fig. 10. Result of recognition experiments

the number of objects in an image. The system captures a frame through IDK, preprocessed it, detects a face, extracts feature vectors and identifies a person.

6 Conclusion

In this paper, a new feature extraction method for a real-time face recognition system is proposed. We represent a face detection system by using a chrominance component of skin tone and face recognition system which combines PCA and LDA. Our face detection method detects skin regions over the entire image, and then generates face candidates based on the spatial spatial arrangement of these skin patches. Our algorithm constructs maps of eye and mouth to detect the eyes, mouth, and face region. Detection results on several photo collections are shown in Fig. 9. The PCA and LDA method presented here is a linear pattern recognition method. Compared with nonlinear models, a linear model is rather robust against noise and most likely will not over-fit. Although it is shown that distribution of face patterns is complex in most cases, linear methods are still able to provide cost effective solutions. Fig. 10 shows that he effectiveness of the proposed method is demonstrated through experimentation by using the POSTECH face database. Especially, the results of several experiments in real life show that the system works well and is applicable to real-time tasks. In spite of the worst case, the complete system requires only about 100 ms per a frame. The experimental performance is achieved through a careful system design of both software and hardware for the possibility of various applications.

References

1. Phillips, P.J., McCabe, R.M., Chellappa, R.: Biometric image processing and recognition. Proceedings, European Signal Processing Conference (1998)
2. Menser, B., Brunig, M.: Locating human faces in color images with complex background. Intelligent Signal Processing and Comm. Systems (1999) 533–536
3. Saber, E., Tekalp, A.: Frontal-view face detection and facial feature extraction using color, shape and symmetry based cost functions. Pattern Recognition Letters **19** (1998) 669–680
4. Sobottka, K., Pitas, I.: A novel method for automatic face segmentation, facial feature extraction and tracking. Signal Processing : Image Comm. (1998) 263–281
5. IMDB. Intelligent Multimedia Laboratory, POSTECH, KOREA. (2001)
6. Mathworks: Image Processing Toolbox User's Guide. The MathWorks (2002)
7. C., Y.K., R., C.: Feature-based human face detection. Image and Vision Computing **15** (1997) 713–735
8. K., S.K., T., P.: Example-based learning for view-based human face detection. IEEE Trans. PAMI **20** (1998) 39–51
9. B., M., A., P.: Probabilistic visual learning for object representation. IEEE Trans. PAMI **19** (1997) 696–710
10. T., J.I.: Principal Component Analysis. Springer, New York, Berlin, Heidelberg, Tokyo (1986)
11. M., T., A., P.: Eigenspaces for recognition. Journal of Cognitive Neuroscience **3** (1991) 71–86
12. Belnumeur, P.N., Hespanha, J.P., Kriegman, D.J.: Eigenfaces vs. fisherfaces : Recognition using class specific linear projection. IEEE Trans. on PAMI (1997)
13. Duda, R.O., Hart, P.E., Stork, D.G.: Pattern Classification 2nd Ed. John Wiley & sons, New York (2001)

Weakly Cooperative Guards in Grids

Michał Małafiejski[1,*] and Paweł Żyliński[2,**]

[1] Department of Algorithms and System Modeling,
Gdańsk University of Technology, Poland
mima@eti.pg.gda.pl
[2] Institute of Mathematics,
Gdańsk University, Poland
impz@univ.gda.pl

Abstract. We show that a minimum coverage of a grid of n segments has $n-p_3$ weakly cooperative guards, where p_3 is the size of the maximum P_3-matching in the intersection graph of the grid. This makes the minimum weakly cooperative guards problem in grids NP-hard, as we prove that the maximum P_3-matching problem in subcubic bipartite planar graphs is NP-hard. At last, we propose a 7/6-approximation algorithm for the minimum weakly cooperative guards problem.

1 Introduction

The *guards problem in grids* is one of the art gallery problems (see [15, 16]) and it was formulated by Ntafos in 1986 [14]. A *grid* P is a connected union of vertical and horizontal line segments, an example of which is shown in Fig. 1(a). A point $x \in P$ can see point $y \in P$ if the line segment $\overline{xy} \subseteq P$. A *guard* g is a point of grid P. A set of points $S \subseteq P$ is a *guard set* for grid P if any point of P is seen by at least one guard in S. The *visibility graph* $VG(S)$ of a set of guards S in a grid is the graph whose vertex set is S and two vertices are adjacent if their points see each other. A set of guards S is said to be *cooperative* or respectively, *weakly cooperative*, if its graph $VG(S)$ is connected, and respectively, has no isolated vertices. The idea behind this concept is that if something goes wrong with one guard, all the others (resp. some other guard) can be informed.

Ntafos established that a minimum cover for a 2D-grid of n segments has $(n - m)$ guards, where m is the size of the maximum matching in the intersection graph of the grid, and it may be found in $O(n^{2.5})$ time [14]. The *intersection graph* $IG(P)$ of a grid P is defined as follows: each vertex of IG corresponds to a line segment of P and two vertices are connected by an edge if their corresponding segments cross. We say that a grid is *planar* if its intersection graph is planar. Let us recall that the intersection graph of a grid can be constructed in $O(n + m)$ time [6], where n is the number of segments and m is the number of

---
* Supported by FNP and TASK.
** Supported in part by the KBN grant 4 T11C 047 25.

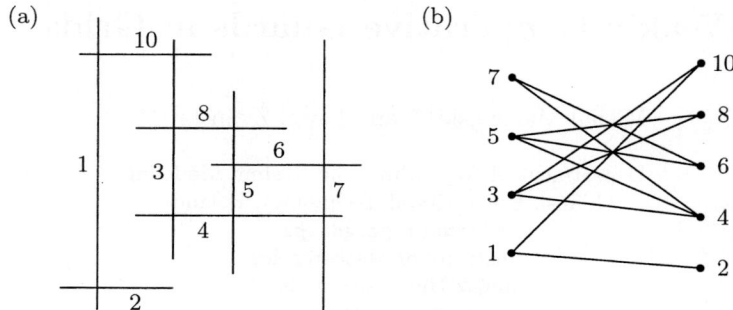

Fig. 1. (a) A grid of 9 segments and (b) its intersection graph

crossings. The intersection graph of the grid in Fig. 1(a) is shown in Fig. 1(b). In the case of 3D-grids, the problem of finding the minimum guard set is NP-complete [14].

The minimum cooperative guards (MinCG for short) problem in grids was solved by Nierhoff and Żyliński [13] who proved that the MinCG problem can be solved in a polynomial time for both two- and three-dimensional grids. In the first case, the MinCG problem corresponds to the problem of finding a minimum spanning tree in the intersection graph of a grid, thus an $O(n + m)$ time algorithm is obtained, where n is the number of segments and m is the number of intersections in the grid. In the latter case, the solution is obtained from a spanning set of a 2-polymatroid constructed from the intersection graph of the grid, and the constructed algorithm is of complexity $O(mn^{2.5})$. The minimum weakly cooperative guards (MinWCG for short) problem in grids has remained unsolved.

In this paper, our main results are the following theorems.

Theorem 1. *A minimum coverage of a grid of n segments has $(n - p_3)$ weakly cooperative guards, where p_3 the size of the maximum P_3-matching in the intersection graph of the grid.*

Theorem 2. *The maximum P_3-matching problem is NP-hard in subcubic bipartite planar graphs.*

Theorem 3. *The MinWCG problem in grids is NP-hard, even for planar grids in which any segment crosses at most three other segments.*

The organization of this paper is as follows. Section 2 is devoted to the study of the relation between a minimum cooperative guard coverage of a grid and a maximum P_3-matching in the intersection graph of the grid. In Section 3 we investigate the maximum P_3-matching problem in subcubic bipartite planar graphs. At last, we propose some approximation algorithms for the minimum weakly cooperative guards problem in grids based on Theorem 1.

2 Weakly Cooperative Guards, Spider Cover and P_3-Matching

Both the minimum guards problem in grids and the minimum cooperative guards problem in grids are solved by reductions to the intersection graphs of grids. As one can expect, the weakly cooperative guards problem also has its counterpart in intersection graphs.

From now on we will use the following notation. Let $G(V, E)$ be a (simple) graph. For vertex $v \in V$, let $E(v) \subseteq E$ denote the set of edges incident to v in G. Next, for any non-empty subset $E' \subset E$, by $G[E']$ we refer to the graph whose the vertex set contains all endpoints of edges in E' and whose the edge set is E'.

Let us recall that without loss of generality we can restrict guards to be located at crossings of segments of a grid [14]. Therefore, the placement of a guard at the crossing of two segments s_1 and s_2 in a grid corresponds to the edge $\{s_1, s_2\}$ in the intersection graph G. Hence there is a one-to-one correspondence between a minimum weakly cooperative guard set in the grid and a subset of edges E_S in graph G. The weakly cooperation of a guard set implies that its corresponding subset E_S of edges satisfies the following conditions:

(i) E_S covers all vertices in V,
(ii) there are no connected components isomorphic to K_2 in subgraph $G[E_S]$.

It follows that finding a minimum weakly cooperative guard set for a grid is equivalent to finding a minimum edge subset E_S of G satisfying (i) and (ii). Note that such a subset always exists as for any spanning tree of G conditions (i) and (ii) hold.

As we are looking for a minimum subset E_S of edges satisfying (i) and (ii), it is natural to ask about the structure of connected components in graph $G[E_S]$. The full characterization is given by the following lemma.

Lemma 1. *Let $G(V, E)$ be a graph and let $E_{min} \subset E$ be a minimum subset of edges satisfying (i), (ii). Let G_S be any connected component of graph $G[E_{min}]$. Then $G_S(V_S, E_S)$ has the following properties:*

(a) G_S is acyclic,
(b) $2 \leq diam(G_S) \leq 4$,
(c) there is at most one vertex of degree at least 3 in G_S and it is the center.

Proof. (a) By removing any edge from a cycle of $G[E_{min}]$, we get a smaller subset of edges satisfying (i)-(ii).

(b) The inequality $2 \leq diam(G_S)$ follows from (ii), so we only have to show that $diam(G_S) \leq 4$. On the contrary, suppose that $diam(G_S) > 4$. This implies that there is a path P_6 of order 6 in G_S. Let $P_6 = e_1 e_2 e_3 e_4 e_5$, where $e_i = \{v_i, v_{i+1}\}$, for $i = 1, \ldots, 5$. Then we get that edge e_3 is unecessary, thus contradicting the minimality of E_{min}.

(c) We have to consider two cases. First, let us suppose that there are at least two vertices of degree at least 3 in G_S, say v_1 and v_2. Then by removing any edge from a path joining v_1 and v_2 in G_S, we get a smaller subset of edges satisfying (i)-(ii).

In the case when there is only one vertex v of degree at least 3 and v is not a center of G_S, then by (b) there is a path $P_5 = v_3 v_2 v_1 v v_4$ with v_3 and v_4 as leaves in G_S. Clearly, v_1 is the center of G_S. But then edge $\{v_1, v\}$ is unecessary. □

From now on, we will call a graph G_S satisfying conditions (a)-(c) as a *spider*. Note that in the case $diam(G_S) = 2$, a spider is just a star-graph. Let us define the *maximum spider cover* (MaxSC for short) in a graph $G(V, E)$ as the problem of finding the maximum number of vertex-disjoint spiders that cover $V(G)$.

Lemma 2. *A family of spiders $S_1, \ldots, S_p$ is the solution to the MaxSC problem in the intersection graph of a grid P iff $\bigcup_{i=1,\ldots,p} E(S_i)$ is the solution to the MinWCG in grid P.*

Proof. Because the number of all edges of spiders $S_1, \ldots, S_p$ is $|V| - p$, then by Lemma 2.1, it is easy to observe that maximizing the number of spiders is equivalent to minimizing the number of guards. □

Before we proceed further, let us recall that the *maximum P_3-matching problem* in a given graph G is the problem of finding the maximum number of vertex-disjoint paths of order 3 in G.

Lemma 3. *Let p be the number of spiders in a maximum spider cover of a graph G and let p_3 be the cardinality of a maximum P_3-matching in G. Then $p = p_3$ and solutions of these problems are equivalent in the sense that one can be constructed from the other.*

Proof. ($p \leq p_3$) This enquality is obvious because all spiders are vertex-disjoint and each spider has a 3-vertex path as its subgraph.

($p \geq p_3$) Let V' be a set of vertices covered by maximum P_3-matching M in $G(V, E)$, and let $E' \subseteq E$ be a set of edges of all paths in M. By definition, E' satisfies the second condition (ii).

Claim. There are no vertices in G whose distance from M is 3 or more, otherwise M is not maximum.

Now, let $V_1, V_2 \subseteq V$ be subsets of vertices whose distance from V' is equal to 1 and 2, respectively. Let us connect vertices of V_1 with those of V' by adding edges to E', one edge per each vertex in V_1. The resulting graph $G[E']$ is acyclic, and the diameter of any its connected components is at most 4; the new set E' still satisfies (ii). Next, let us connect vertices of V_2 with those of the new V' by adding edges to the new E', one edge per each vertex in V_2. It is easy to see that the following observations hold:

- The number of connected components in the new $G[E']$ remains equal to p_3.
- The resulting graph $G[E']$ remains acyclic.
- For the resulting graph $G[E']$, the diameter of any of its connected components is now at most 4, otherwise matching M is not maximum.

- For the resulting graph $G[E']$, there is at most one vertex of degree three in any of its connected components, and this vertex is the center, otherwise matching M is not maximum.
- The new set E' still satisfies (ii).

By all these observations, the final E' is a set of spiders. And moreover, $G[E']$ covers all vertices of G by the above claim. □

The above reductions can be done in $O(|V|+|E|)$ time by checking neighbours of all vertices from the maximum P_3-matching,

Let us recall that a minimum (arbitrary) guard cover and a minimum cooperative guard cover of an n-segment grid has $n-m$ and $n-1$ guards, respectively, where m is the cardinality of the maximum matching in the intersection graph of the grid. We have analogous formula for weakly cooperative guards.

Theorem 4. *A minimum weakly cooperative guard cover for a grid P of n segments has $(n - p_3)$ guards, where p_3 is the cardinality of the maximum P_3-matching in the intersection graph of P.*

Proof. By Lemma 3, the number of edges in a maximum spider cover is equal to $2p_3 + (n - 3p_3) = n - p_3$. And, by Lemma 2, the thesis follows. □

We have just proved that the minimum weakly cooperative guards problem in a grid is equivalent to the maximum P_3-matching problem in the intersection graph of the grid. As we know, any bipartite planar graph is the intersection graph of a grid [2], and in the next section we will show that the maximum P_3-matching problem in subcubic bipartite planar graphs is NP-hard. Hence by Theorem 4, we get the following theorem.

Theorem 5. *The minimum weakly cooperative guards problem is NP-hard, even for planar grids in which any segment crosses at most three other segments.*

3 P_3-Matching in Subcubic Bipartite Planar Graphs

Generalized matching problems have been studied in a wide variety of contexts. One of the possible generalizations is to find the maximum number of vertex-disjoint copies of some fixed graph H in a graph G (*maximum H-matching*). In [8] Kirkpatrick and Hell show that any perfect matching is NP-complete for any connected H with at least three vertices (the same is for the maximum H-matching problem). In [1] it is shows that the maximum H-matching remains NP-hard even if we restrict ourselves to planar graphs.

In this section we show that the maximum P_3-matching problem in subcubic (maximum degree $\Delta \leq 3$) bipartite planar graphs is NP-hard. The idea of the proof is based upon reduction from the 3DM problem [7]. The three-dimensional matching problem can also be formulated in the way described below. Dyer and Frieze [5] proved that the 3DM problem remains NP-complete even for bipartite planar graphs. Hence from [5, 7] the following restricted 3DM problem ($\overline{\text{3DM}}$ for short) is NP-complete.

$\overline{\text{3DM}}$ PROBLEM:

Instance: A subcubic bipartite planar graph $G(V \cup M, E)$, where $V = X \cup Y \cup Z$, $|X| = |Y| = |Z| = q$. And, for every vertex $m \in M$ we have $deg(m) = 3$ and m is adjacent to exactly one vertex from each of the sets X, Y and Z.

Question: Is there a subset $M' \subseteq M$ of cardinality q covering all vertices in V?

Theorem 6. [5, 7] *The $\overline{\text{3DM}}$ problem is NP-complete.*

Let $G(V \cup M, E)$ be a subcubic bipartite planar graph, where $V = X \cup Y \cup Z$, $|X| = |Y| = |Z| = q$, every vertex $m \in M$ has degree 3, m is adjacent to exactly one vertex from each of the sets X, Y and Z. Let $G^*(V^*, E^*)$ be a graph obtained from G by replacing each vertex $v^i \in M$, $i = 1, \ldots, |M|$, (and all edges incident to it) with the graph $G_i(V_i, E_i)$ presented in Fig. 2, formally:

- $V_i = \{p_j^i\}_{j=1,\ldots,9} \cup \{x_j^i, y_j^i, z_j^i\}_{j=1,2,3}$
- $V^* = V \cup \bigcup_{i=1,\ldots,|M|} V_i$;
- $E^* = E \setminus E^- \cup E^+$, where:
  $E^- = \bigcup_{i=1,\ldots,|M|} \{\{x^i, v^i\}, \{y^i, v^i\}, \{z^i, v^i\}\}$,
  $E^+ = \bigcup_{i=1,\ldots,|M|} (E_i \cup \{\{x^i, x_2^i\}, \{y^i, y_2^i\}, \{z^i, z_2^i\}\})$,
  and x^i, y^i and z^i are neighbours of vertex v^i in graph G.

Note that $x^i = x^j$ iff $dist(v^i, v^j) = 2$. Clearly, graph G^* has $|V| + 18|M|$ vertices and $|E| + 17|M|$ edges, and $\Delta(G^*) = 3$.

Lemma 4. *There exists a solution of the $\overline{\text{3DM}}$ problem in graph $G(V \cup M, E)$ iff there exists a maximum P_3-matching of cardinality $q + 6|M|$ in graph $G^*(V^*, E^*)$.*

Proof. ($\Rightarrow$) Let M' be a solution of the $\overline{\text{3DM}}$ problem in graph $G(V \cup M, E)$, $|M'| = q$. A perfect P_3-matching in graph G^* consists of the following 3-vertex paths:

- if vertex v^i corresponding to graph G_i is in M', then we choose the following 3-vertex paths in graph G_i with attached vertices x^i, y^i and z^i (see Fig. 2): $x_1^i x_2^i x^i$, $y_1^i y_2^i y^i$, $z_1^i z_2^i z^i$, $x_3^i p_4^i p_3^i$, $y_3^i p_8^i p_9^i$, $z_3^i p_2^i p_1^i$, $p_5^i p_6^i p_7^i$;
- otherwise, if vertex v_j corresponding to graph G_j is not in M', then we choose the following 3-vertex paths in graph G_j with attached vertices x^j, y^j and z^j: $x_1^j x_2^j x_3^j$, $y_1^j y_2^j y_3^j$, $z_1^j z_2^j z_3^j$, $p_1^j p_2^j p_3^j$, $p_4^j p_5^j p_6^j$, $p_7^j p_8^j p_9^j$.

($\Leftarrow$) Let P be a maximum P_3-matching of cardinality $q + 6|M|$ in $G^*(V^*, E^*)$ (P is perfect, of course). Let us consider any subgraph G_i, together with vertices x^i, y^i and z^i. First, let us note that x^i (or resp. y^i or z^i) cannot be a center of a 3-vertex path in a perfect matching. Next, the following claim holds.

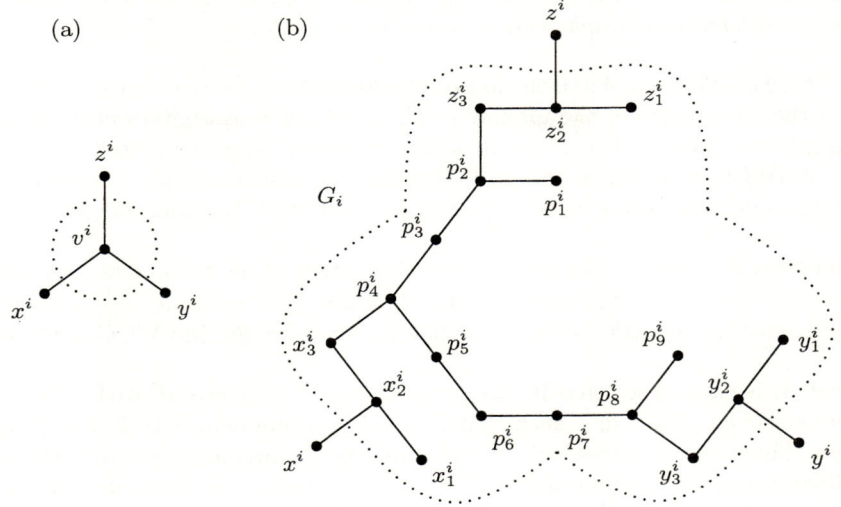

Fig. 2. (a) Vertex v^i is replaced by graph G_i (b)

Remark 1. If one of the paths $x_1^i x_2^i x^i$, $y_1^i y_2^i y^i$ or $z_1^i z_2^i z^i$ is in matching P, then all of them are in matching P.

Consider for example, $x_1^i x_2^i x^i \in P$ and $y_1^i y_2^i y^i \notin P$. Then $y_1^i y_2^i y_3^i$, $p_7^i p_8^i p_9^i$ and $p_4^i p_5^i p_6^i$ are in P, hence x_3^i is not covered by any path, a contradiction. The other cases can be proved analogously.

Therefore, our perfect P_3-matching P of graph G^* either (1) consists of paths $x_1^i x_2^i x^i$, $y_1^i y_2^i y^i$ and $z_1^i z_2^i z^i$ or (2) none of them is in this matching. So the following set $M' = \{v^i \in M : G_i$ is covered in way (1)$\}$ is a solution to the $\overline{3DM}$ in graph $G(V \cup M, E)$. □

By the above lemma and Theorem 6, we get the following theorem.

Theorem 7. *The perfect P_3-matching problem in subcubic bipartite planar graphs is NP-complete.*

Corollary 1. *The maximum P_3-matching problem in subcubic bipartite planar graphs is NP-hard.*

4 Final Remarks

Masuyama and Ibaraki [12] showed that the maximum P_i-matching problem in trees can be solved in linear time, for any $i \geq 3$. The idea of their algorithm is to treat a tree T as a rooted tree (T, r) (with an arbitrary vertex r the root) and to pack i-vertex paths while traversing (T, r) in the bottom-up manner. Hence by Theorem 4, we get the following corollary.

Corollary 2. *The minimum weakly cooperative guards problem for grids with trees as intersection graphs can be solved in linear time.*

Based on Theorem 4 and the following observation: $\frac{2}{3}n \leq n - p_3 \leq n-1$, where p_3 is the cardinality of the maximum P_3-matching in the intersection graph G of a grid, we have that any spanning tree of G is a $\frac{3}{2}$-approximated solution for the MinWCG problem in the grid. However, by disconnecting a spanning tree into spiders, this ratio can be improved for grids with bounded degree.

Theorem 8. *Suppose that the intersection graph of an n-segment grid with m crossings has a spanning tree of degree d. Then there exists a $(\frac{3(d-1)}{2d-1} + O(\frac{1}{n}))$-approximation algorithm of complexity $O(n+m)$ for the MinWCG problem.*

Proof. It is easy to see that in any n-vertex tree T of degree d, with $n \geq 2d+2$, there exists an edge e in T such that $T \setminus e$ has a component with $2 \leq m \leq 2d-2$ edges. Thus we can successively cut off connected components provided that each of them is a spider and has at most $2d-2$ edges, except for the last spider which has $2 \leq k \leq 2d$ edges. Hence all the spiders consist of at most $(2d-2)\frac{n-k-1}{2d-1} + k$ edges. As the number of edges in a maximum spider cover S is at most $\frac{2}{3}n$, we have the following inequalities

$$\frac{2}{3}n \leq n - |S| \leq (2d-2)\frac{n-k-1}{2d-1} + k \leq \frac{2d-2}{2d-1}n + \frac{2}{2d-1}.$$

By Lemma 2, this gives us a $(\frac{3(d-1)}{2d-1} + \Theta(\frac{1}{dn}))$-approximated solution for the MinWCG problem in the grid. □

But, again by Theorem 4, it follows that any approximation algorithm for the maximum P_3-matching problem can be applied to the MinWCG problem.

Lemma 5. *Let A be a r-approximation algorithm for the maximum P_3-matching problem. Then there exists a $\frac{3r-1}{2r}$-approximation for the MinWCG problem.*

Proof. Let p_3 and a be any optimal solution and any r-approximation for the maximum P_3 matching problem in the intersection graph of a grid, respectively. As $p_3/a \leq r$ and $\frac{2}{3}n \leq n - p_3 \leq n - a$, then

$$\frac{n-a}{n-p_3} \leq \frac{n - \frac{p_3}{r}}{n-p_3} \leq 1 + \frac{p_3\frac{r-1}{r}}{n-p_3} \leq \frac{3r-1}{2r}.$$

□

Up to now, the best known approximation algorithm for the maximum P_3-matching achieves $\frac{3}{2}$ ratio, and it is based upon the local improvements [3]. Hence we get the following theorem.

Theorem 9. *There exists a 7/6-approximation algorithm for the MinWCG problem in grids.*

Table 1. Guard problems in grids

Guard Type	Dimension	Complexity	Approximation Ratio
arbitrary	2	$O(n+m)$ [14]	
	3	NP-hard [14]	
cooperative	2	$O(n+m)$ [13]	
	3	$O(mn^{2.5})$ [13]	
weakly cooperative	2	NP-hard (Thm. 5) even for subcubic planar	$\frac{7}{6}$ (Thm. 9)

Finally, let us recall that the guards problem was also stated for three-dimensional grids. In the case of arbitrary guards, the minimum guard problem is NP-hard (reduction from the vertex cover problem), whereas for cooperative guards, the minimum cooperative guards problem is polynomial and a solution is obtained from a spanning set of a 2-polymatroid constructed from the intersection graph of a grid. Of course, we can ask about the minimum weakly cooperative guards problem in three-dimensional grids, however, by Theorem 5, the MinWCG problem in this class of grids is NP-hard.

We conclude with a table that summarizes results on guard problems in grids.

References

1. F. Berman, D. Johnson, T. Leighton, P.W. Shor, L. Snyder, Generalized planar matching, *J. Algorithms* **11** (1990), 153-184.
2. J. Czyżowicz, E. Kranakis, J. Urrutia, A simple proof of the representation of bipartite planar graphs as the contact graphs of orthogonal straight line segments, *Info. Proc. Lett.* **66** (1998), 125-126.
3. K.M.J. De Bontridder, B.V. Haldórsson, M.M. Haldórsson, C.A.J. Hurkens, J.K. Lenstra, R. Ravi, L. Stougie, Approximation algorithms for the test cover problem, *Mathematical Programming* **98** (2003), 477-491.
4. J.S. Deogun, S.T. Sarasamma, On the minimum cooperative guards problem, *J. Combin. Math. Combin. Comput.* **22** (1996), 161-182.
5. M.E. Dyer, A.M. Frieze, Planar 3DM is NP-complete, *J. Algorithms* **7** (1986), 174-184.
6. U. Finke, K. Hinchirs, Overlaying simply connected planar subdivisions in linear time. In *Proc. 11th Annu. ACM Sympos. Comput. Geom.* (1995), 119-126.
7. M.R. Garey, D.S. Johnson, *Computers and Intractability: A Guide to the Theorey of NP-completeness*, Freeman, New York (1979).
8. P. Hell, D.G. Kirkpatrick, On the complexity of general graph factor problems, *SIAM Journal of Comput.* **12** (1983), 601-609.
9. G. Hernández-Peñalver, Controlling guards, *Proc. of Sixth Canadian Conference on Computational Geometry* (1994), 387-392.
10. B. C. Liaw, R.C.T. Lee, An optimal algorithm to solve the minimum weakly cooperative guards problem for 1-spiral polygons. *Info. Proc. Lett.* **57** (1994), 69-75.

11. B.C. Liaw, N.F. Huang, R.C.T. Lee, The minimum cooperative guards problem on k-spiral polygons, *Proc. of Fifth Canadian Conference on Computational Geometry* (1993), 97-101.
12. S. Masuyama, T. Ibaraki, Chain packing in graphs, *Algorithmica* **6** (1991), 826-839.
13. T. Nierhoff, P. Żyliński, Cooperative guards in grids, *Third Annual CGC Workshop* (2003).
14. S. Ntafos, On gallery watchman in grids, *Info. Proc. Lett.* **23** (1986), 99-102.
15. J. O'Rourke, *Art Gallery Theorems and Algorithms*, Oxford University Press (1987).
16. J. Urrutia, *Art Gallery and Illumination Problems*, Handbook on Computational Geometry, Elsevier Science, Amsterdam (2000).

Mesh Generation for Symmetrical Geometries

Krister Åhlander*

Department of Information Technology,
Uppsala University, Uppsala, Sweden
krister.ahlander@it.uu.se

Abstract. Symmetries are not only fascinating, but they can also be exploited when designing numerical algorithms and data structures for scientific engineering problems in symmetrical domains.

Geometrical symmetries are studied, particularly in the context of so called equivariant operators, which are relevant to a wide range of numerical applications.

In these cases, it is possible to exploit the symmetry via the generalized Fourier transform, thereby considerably reducing not only storage requirement but also computational cost.

The aim of this paper is to introduce group theoretical aspects of symmetry, to point out the benefits of using these concepts when designing numerical algorithms, and to state the implications for the generation of the mesh.

Keywords: Noncommutative Fourier analysis, equivariant operators, block diagonalization.

1 Introduction

Many computational problems exhibit various kinds of symmetries, which may be exploited in order to improve the algorithms. In this paper, we consider two and three dimensional geometries which are invariant under finite groups of distance preserving transformations, i.e. rotations and reflections. For example, the so called Platonic solids (see Fig. 1) provide nice examples of this kind of symmetry.

The key to exploiting the symmetry of a structured way is to use information about the underlying *group*, and how the group *acts* on the physical as well as on the discretized geometry. There is, of course, a strong connection between symmetry and group theory, and a main purpose of this paper is to highlight this connection particularly when considering discretizations of symmetrical geometries.

The motivation of our interest in this topic comes from numerically solving partial differential equations (PDEs) in symmetrical geometries. The most time consuming part of such applications is usually the solution of a linear system of equations, $Ax = b$. The system matrix A will exhibit a certain structure, if the

---

* Partially supported by a grant from the Göran Gustafsson foundation.

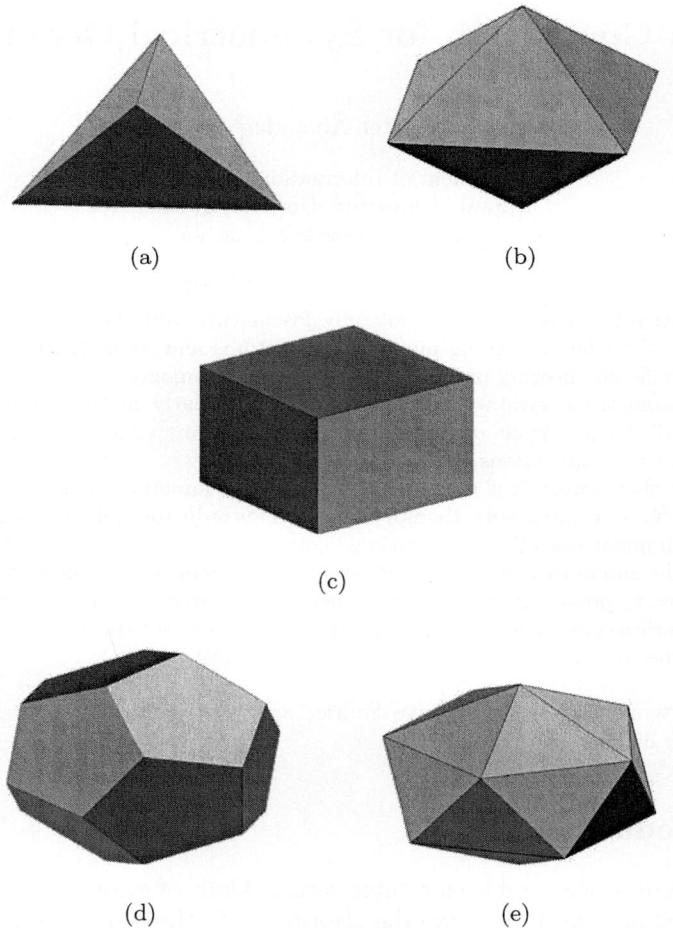

Fig. 1. The Platonic solids are (a) the tetrahedron, (b) the octahedron, (c) the cube, (d) the dodecahedron, and (e) the icosahedron. They are all invariant under certain groups of rotations and reflections, and were known already in ancient Greece

spatial PDE operator commutes with the symmetry group of the geometry. This operator can then be block diagonalized with the Generalized Fourier Transform, GFT.

Utilization of the GFT for numerical applications was pioneered about two decades ago [1, 2], but the technique remains relatively unknown in the scientific community. We believe that one reason for this is the lack of proper tools which support appropriate mesh generation for symmetrical geometries.

In this paper, we discuss various aspects of how group theory can be used for generating discretizations for symmetric geometries. One observation is that the structure of the underlying group should be utilized. For instance, if the underlying symmetry group is a *direct product* (to be explained below), the numerical algorithms – including the mesh generation – should exploit this information. We also want to stress the role of *fix points*, i.e. points which are fixed under some nonidentity symmetry group transformation. It is important to consider such issues both when designing the numerical algorithms and when generating computational meshes for symmetrical geometries.

This paper is organized as follows. In Section 2, we recapitulate some basic group theory and we briefly explain how the GFT is used in symmetry exploiting algorithms. In Section 3, we discuss some specific symmetry respecting discretizations. Finally, we summarize our observations on how group theory could be utilized when generating meshes for symmetrical geometries.

2 Background

We will first recapitulate the basic group theory and representation theory required for our discussion. We continue with a discussion on how the symmetry in the geometry corresponds to equivariance, a "symmetry" of the operator. We also introduce the GFT. Throughout, we stress the implications for the generation of the computational mesh.

Our presentation and notation is based on [3, 4]. For a more thourough discussion on representations, [5] is recommended. For an overview of GFTs, particularly fast GFTs, [6, 7] are good surveys.

2.1 Some Concepts from Group Theory

We present and exemplify the following basic concepts: group, subgroup, direct product, group action, isotropy subgroup, free action, and orbit.

A *group* $\mathcal{G}$ is a set of elements together with a binary operation such that the binary operation is associative, such that there is a unique identity e in the group and such that every group element g has a unique inverse denoted g^{-1}. Thus, for all $f, g, h \in \mathcal{G}$ we have:

$$f(gh) = (fg)h, \ ge = eg = e, \ gg^{-1} = g^{-1}g = e \ .$$

The number of elements in $\mathcal{G}$ is $|\mathcal{G}|$. If all elements in the group commute, it is called abelian.

Examples: All permutations on n elements form a group under composition, denoted $\mathcal{S}_n$, with size $|\mathcal{S}_n| = n!$. All reflections and rotations which map a regular polygon with n sides onto itself form a dihedral group with $2n$ elements, denoted $\mathcal{D}_n$. It contains n rotations and n reflections.

A *subgroup* $\mathcal{H}$ of $\mathcal{G}$ is a subset which is closed under the group operation.

Examples: An even permutation is a permutation which can be rewritten as an even number of transpositions, i.e. permutations which only permute two elements. All even permutations on n elements forms a group $\mathcal{A}_n$ which is a subgroup of $\mathcal{S}_n$. All rotations which map a regular polygon with n sides onto itself form a cyclic group, denoted $\mathcal{C}_n$. It is a subgroup of $\mathcal{D}_n$. $\mathcal{C}_n$ is also an example of an abelian group.

A direct product $\mathcal{G} \times \mathcal{H}$ of two groups $\mathcal{G}$ and $\mathcal{H}$ is the Cartesian product of the corresponding sets, equipped with the group operation

$$(g_1, h_1)(g_2, h_2) = (g_1 g_2, h_1 h_2), \ \forall g_1, g_2 \in \mathcal{G}, \ \forall h_1 h_2 \in \mathcal{H} \ .$$

Example: Consider a Cartesian 2D mesh with $M \times N$ nodes. Cyclic shifts in each direction form the $\mathcal{C}_M$ and $\mathcal{C}_N$ cyclic groups, respectively. $\mathcal{C}_M \times \mathcal{C}_N$ represents all cyclic shifts on the mesh.

In this paper, we are mostly concerned with finite groups of linear isometric transforms (i.e., rotations and reflections) which map symmetrical geometries onto themselves. These groups are subgroups of $\mathcal{O}_n$, the group of all reflections and rotations in n dimensions. In two and three dimensions, *all* finite subgroups are actually known, see e.g. [8]. We have

- Cyclic groups: $\mathcal{C}_n$ with n elements.
  $\mathcal{C}_n$ is isomorphic to the group of rotations which map a regular n-sided polygon onto itself.
- Dihedral groups: $\mathcal{D}_n$ with $2n$ elements.
  $\mathcal{D}_n$ is isomorphic to the group of rotations and reflections which map a regular n-sided polygon onto itself.
- $\mathcal{S}_4$ with 24 elements.
  $\mathcal{S}_4$ is isomorphic to the symmetry group (rotations and reflections) of the tetrahedron.
- $\mathcal{A}_5$ with 60 elements.
  $\mathcal{A}_5$ is isomorphic to the the group of rotations which map the icosahedron onto itself.

In addition, there exist direct groups of the above. The symmetry group of the cube and of the octahedron, for instance, is $\mathcal{C}_2 \times \mathcal{S}_4$, see [5], and the symmetry group of the dodecahedron and of the icosahedron is $\mathcal{C}_2 \times \mathcal{A}_5$, cf. [9].

In applications, group theory is particularly important in the context of a group acting on a set. A *group action* (from the right) is an operation $\mathcal{X} \times \mathcal{G} \to \mathcal{X}$ such that $xe = x$ and $x(gh) = (xg)h$ for all $x \in \mathcal{X}$ and all $g \in \mathcal{G}$. For example, permuting a number using a specific permutation is a group action, and so is the rotation of a polygon, acting on its corners.

Given a group action, the *isotropy subgroup* of $x \in \mathcal{X}$ is the subgroup $\mathcal{G}_x = \{g \in \mathcal{G} \mid xg = x\}$, i.e. all transformations whose action on the specified element x leaves x fixed. For example, if we consider the center of a square and the transformation group $\mathcal{D}_4$, we see that the the isotropy subgroup is the whole of $\mathcal{D}_4$ because every transformation leaves the center fixed. If we consider the isotropy subgroup of a point on a diagonal, the isotropy subgroup contains two elements, the identity transform and the reflection across that diagonal. If the isotropy subgroup only contains the identity transformation, it is called *trivial*.

The group action is *free* if every isotropy subgroup is trivial. For example, the action of $\mathcal{C}_5$ (which consists of 5 rotations) is free on the circumference of a pentagon, but the action of the full symmetry group $\mathcal{D}_5$ (which also includes 5 reflections) is not. Similarly, the action of $\mathcal{S}_3$ on the numbers $1, 2, 3$ is free, but the action of $\mathcal{S}_2$ is not.

The *orbit* $x\mathcal{G}$ of an element x contains all elements xg which are obtained when the group acts on the element. We note that the size of the orbit times the size of the isotropy subgroup of any element in the orbit equals the size of the group, $|x\mathcal{G}||\mathcal{G}_x| = |\mathcal{G}|$.

2.2 Equivariance—Symmetry of Operators

In many applications, the symmetry of a domain carries over to a symmetric structure in the operators. Consider a spatial operator, for instance the Laplacian, $\triangle$, and let $\mathcal{G}$ be a group of isometries. Many PDEs, for instance in structural mechanics, are such that the spatial operator commutes with isometric transformations. For instance, it does not matter if we first rotate a field and then apply the Laplacian, or if we do it the other way around. This property is called *equivariance*.

In discretized form, the spatial operator forms a matrix $\boldsymbol{A}$. We assume that the discretization is *symmetry respecting*, which means that the group action of g corresponds to a permutation of the indices $\mathcal{I}$ according to a multiplication with a permutation matrix $\boldsymbol{P} = \boldsymbol{P}(g)$. In the discretized domain, equivariance means that the discretized operator commutes with every such permutation matrix, $\boldsymbol{AP} = \boldsymbol{PA}$. Equivalently, this can be expressed as actions on the indices. A matrix $\boldsymbol{A}$ is equivariant w.r.t. $\mathcal{G}$ if

$$\boldsymbol{A}_{ig,jg} = \boldsymbol{A}_{i,j}, \ \forall i, j \in \mathcal{I}, \ \forall g \in \mathcal{G} \ .$$

Obviously, this property can be exploited to reduce storage requirements. If the group action is free on m orbits, the number of indices is $n = m|\mathcal{G}|$. Thus, the matrix has $m^2|\mathcal{G}|^2$ elements in total. However, thanks to equivariance, we need only to store (and compute) $m^2|\mathcal{G}|$ elements of $\boldsymbol{A}$. If we have equivariance under the symmetry of the cube, for instance, this means storage savings by a fraction of $|\mathcal{C}_2 \times \mathcal{S}_4| = 48$. We note the imortant prerequisite this puts on the mesh generation:

– The mesh should be symmetry respecting.

This property can also be used to reduce the storage of the mesh, see below. In addition, symmetry can be used to improve the efficiency of many algorithms, through the use of block diagonalizing equivariant operators. This is achieved using the GFT, which is the topic of next subsection.

2.3 The Generalized Fourier Transform

A well-known example of the GFT for a specific group is the Discrete Fourier Transform, DFT (implemented as the Fast Fourier Transform, FFT). The DFT is a GFT w.r.t. a cyclic group. In order to explain the GFT, we need to introduce the concept of a *representation*. A representation ρ of dimension d is a mapping from a group $\mathcal{G}$ to a matrix space $\mathbb{C}^{d\times d}$, such that $\rho(gh) = \rho(g)\rho(h)$ for all group elements g, h.

Two representations ρ and σ are *isomorphic* if there exists a coordinate transformation T such that $\rho(g) = T\sigma(g)T^{-1}$ for all $g \in \mathcal{G}$. A representation ρ is *reducible* if there exists a coordinate transformation such that $T\rho(g)T^{-1}$ is block-diagonal for all $g \in \mathcal{G}$.

To every finite group, there exists a complete list of irreducible non-isomorphic representations, $\mathcal{R}$. For the geometrical groups we consider, these can be found in the literature.

The GFT is formally an algebra isomorphism between a group algebra and a direct sum of matrix algebras, its Fourier space. In [3], we describe block versions of this, and how to transform a linear system of equations $\boldsymbol{Ax} = \boldsymbol{b}$ to a block diagonal matrix matrix equation $\hat{A}\hat{x} = \hat{b}$ in Fourier space. The block diagonalization implies that we obtain several independent systems of equations, which are easier to solve. For free actions, the GFT of the matrix is

$$\hat{A}_{i,j}(\rho) = \sum_{g\in\mathcal{G}} \boldsymbol{A}_{ig,j}\rho(g) ,$$

and for vectors we have

$$\hat{x}_i(\rho) = \sum_{g\in\mathcal{G}} \boldsymbol{x}_{ig}\rho(g) .$$

This is computed for each $\rho \in \mathcal{R}$ and for each i and j in a selection of indices $\mathcal{S}$, representing the orbits. In order to compute these formulas efficiently, we stress that the generation of the mesh should adhere to the following:

- The mesh nodes should be ordered such that there is a simple mapping from a specific $i \in \mathcal{S}$ to all indices in the same orbit.
- The ordering of indices in each orbit should match the group action.

For general actions, there is the issue of fix points. In this case, the mappings to the Fourier space need to be modified. This is discussed more in the next section.

3 Symmetry Respecting Discretizations

We illustrate the theoretical aspects by examples. Our first example discusses appropriate data structures in the presence of fix points. Our second example illustrates the benefits of using information about the group structure, particularly when the group is a direct product.

3.1 Discretizing an Equilateral Triangle

The symmetry group of an equilateral triangle, $\mathcal{D}_3$, is the smallest example of a non abelian group. It is therefore a good example with which to illustrate relevant mesh generation issues. The elements of the group are generated by a rotation α and a reflection β, see Fig. 2(a). In total, the group has six elements $\{e, \alpha, \alpha^2, \beta, \alpha\beta, \alpha^2\beta\}$.

Even if the group is small, we would like to stress the similarities between this example and more complex geometries. The surface of an icosahedron, for example, consists of twenty equilateral triangles. The discussion below applies to this case as well, but the potential gain is much bigger in the icosahedron case.

In order to be concrete, we assume that we want to solve Poisson's equation $\triangle u = f$, plus appropriate boundary conditions, where u and f are scalar fields. Again, this simple PDE illustrates relevant points which hold also for more complex PDEs, provided the operator commutes with the symmetry group. Note that we do not pose any symmetry restrictions on the right hand side, f.

The general approach for solving $\triangle u = f$ on the triangle is as follows:

1. Discretize a *fundamental region* of the domain. A fundamental region (also known as a computational cell) is any region from which it is possible to generate the whole domain by the group action. Consequently, the area of a fundamental region equals the area of the whole domain divided by the size of the group. For the triangle, a fundamental region is one sixth of its area, see Fig. 2(b). For the surface of the icosahedron, a fundamental region is only 1/120 of its total area.
2. The transformations in the symmetry group can be used to generate a symmetry respecting discretization of the whole domain.
3. Discretize the Laplacian $\triangle$ for the fundamental region.
4. Use the symmetries to obtain a discretized operator in the whole domain, thus obtaining the linear system of equations $\boldsymbol{Ax} = \boldsymbol{b}$, where $\boldsymbol{x}$ and $\boldsymbol{b}$ are discretizations of the fields u and f, respectively.

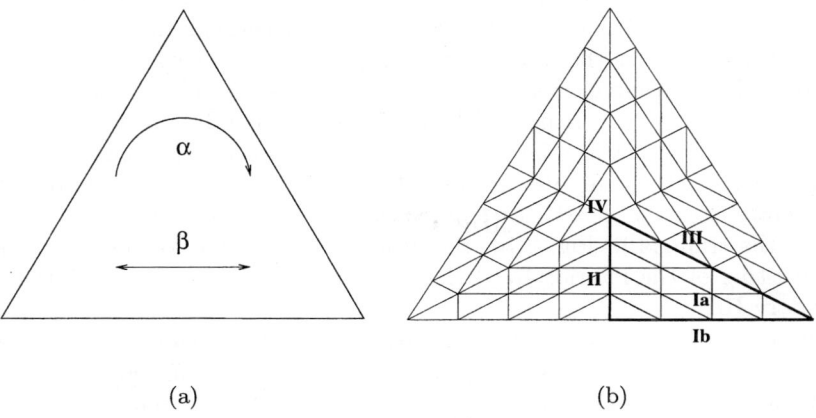

(a)   (b)

Fig. 2. (a) The symmetry group of the triangle is generated by the rotation α and the reflection β. (b) A symmetry respecting discretization of the triangle. A fundamental region is indicated, as well as regions of different isotropy subgroups

5. Use the GFT to obtain a corresponding system $\hat{A}\hat{x} = \hat{b}$ in Fourier space, where $\hat{A}$ is block diagonal. If the geometry has the symmetry of the triangle, we obtain three diagonal blocks, whereas an icosahedral symmetry would give ten diagonal blocks.
6. Solve the independent systems in Fourier space, and use the inverse GFT to obtain x.

The gain of this approach has been reported several times, see e.g. [10], particularly for N^3 algorithms such as equation solving with dense systems. The gain increases with the amount of symmetry. For the triangle, the gain is about 22, whereas the potential gain is over 3000 for the icosahedron [3].

Regarding the mesh generation, we point out that we only need to store the topology in the fundamental region, which reduces the memory consumption by the size of the group. For the nodes, we have a choice, though. We could store node coordinates for the fundamental region only, and compute other node coordinates "on demand", or we could store node coordinates for the whole domain. For some symmetries, such as the cube symmetry, the choice may seem obvious as motivated below. In general, however, the choice depends both on the application and on the computer platform. Thus, we argue that mesh generating tools should consider the following issues:

– Topological mesh information is needed for a fundamental region only.
– Physical mesh information (node coordinates) is needed for the fundamental region only—but in some applications it might be more efficient to construct the complete mesh.

Regarding the enumeration of the nodes, it was argued in Section 2 that the enumeration should take into account group action and the orbits. If the action is free, each orbit has $|\mathcal{G}|$ nodes. In this case, the set of node indices $\mathcal{I}$ equals the Cartesian product $\mathcal{S} \times \mathcal{G}$ where each index in the selection $\mathcal{S}$ represents one orbit. Thus, we identify $i \in \mathcal{I}$ with $s \in \mathcal{S}$ and $g \in \mathcal{G}$ through the group action sg. It is natural to order the indices lexicographically, either with the orbit index first or with the group element first. The ordering of the indices has implications for the efficiency of algorithms. When computing the GFT, it is clear that the memory layout is better adapted to the computations if elements in the same orbit are adjacent in computer memory. Thus, we confirm the observations above: *Enumerate free indices orbit-wise, according to the group action.*

For the triangle in Fig. 2(b), the group action is free for indices in region **I**, consisting of the interior **Ia** of the fundamental region as well as the outer boundary **Ib**.

However, orbits which contain fix points need special care, because they do not contain as many elements. In our example, there are fix points for indices belonging to regions **II**, **III**, and also for region **IV** which only has one node. There are at least two approaches to deal with this issue. The first approach is simply to extend the index domain $\mathcal{I}$ so that each fix point is counted several times, once for each element in the isotropy subgroup. Thus, indices in region **II** and **III** would be counted twice, since their isometry groups contain the identity plus one reflection. The index in region **IV** would be counted six times, once for every element in $\mathcal{D}_3$. A discretized vector $\boldsymbol{x}$ would similarly be extended to a vector in a larger vector space. An advantage of this approach is simplicity. When we extend the vector space, the group action becomes free, and the GFT is simpler to apply. A disadvantage, however, is that an equivariant matrix $\boldsymbol{A}$ would be transformed into a singular matrix $\hat{\boldsymbol{A}}$. In some applications this might not matter. For instance, if we are interested in the eigenvalues of $\boldsymbol{A}$, it does not matter much if we introduce a few spurios zero eigenvalues when transforming to $\hat{\boldsymbol{A}}$, see [4]. For other applications, such as solving dense linear systems of equations, the singularity is not acceptable.

In [3], we derive GFT formulas for orbits with fix points. The implication for the mesh is that we need to deduce, for each fix point index i, a subset $\mathcal{C}$ of group elements such that the set of indices $i\mathcal{C}$ equals the orbit $i\mathcal{G}$. For the triangle, it is clear that for indices in region **II** and **III**, such a subset is the rotations $\{e, \alpha, \alpha^2\}$, whereas the subset for the index in region **IV** could be chosen simply as $\{e\}$. With this information, we are able to transform a nonsingular $\boldsymbol{A}$ to a unique, nonsingular $\hat{\boldsymbol{A}}$.

Summarizing our observations on mesh generation for geometries with fix points, we find the following:

- Indices which share the same isotropy subgroup should be grouped together.
- It should be possible to extend the index domain in order to make the group action free.
- For each fix point, a subset $\mathcal{C}$ of $\mathcal{G}$ which fills its orbit should be identified, to facilitate the application of the GFT.

3.2 Discretizing a Cube

In this section, we discuss the discretization of a cube, see Fig. 3. As explained in Section 2, its symmetry group $\mathcal{K}$ is $\mathcal{C}_2 \times \mathcal{S}_4$, and our aim is to illustrate that it is relevant to take the direct product structure of the group into account.

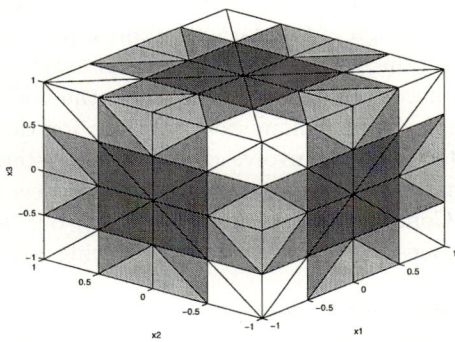

Fig. 3. A symmetry respecting discretization of the cube boundary. Elements of the same shade of gray belong to the same orbit

First, however, we discuss the issue of memory requirements for the nodes, cf. the discussion in the previous section.

As explained in [5], the elements of the symmetry group of the cube may be generated by the group of reflections $(x, y, z) \mapsto (\pm x, \pm y, \pm z)$, which is isomorphic to $\mathcal{C}_2 \times \mathcal{C}_2 \times \mathcal{C}_2$, and by $\mathcal{S}_3$, the permutations of the coordinates x, y and z. This implies that the application of any group element $g \in \mathcal{K}$ on any node with coordinates (x, y, z) in the fundamental region can be computed simply by a permutation and by a change of sign in up to three coordinates. Since these computations are very cheap, we argue that we in most cases need to store nodes only in the fundamental domain. This reduces the memory storage for the topology as well as for the physical coordinates with a fraction of 48, the size of the symmetry group.

Regarding the numbering of the indices in each orbit, we have already established that it should be related to the group action. By ordering the elements of $\mathcal{K}$ according to the direct product structure $\mathcal{C}_2 \times \mathcal{S}_4$ of the group, this is reflected in the numbering of the indices. We see, in fact, that each orbit $i\mathcal{K}$ is a Cartesian product $i\mathcal{C}_2 \times i\mathcal{S}_4$.

The advantage is clearly seen when considering the application of the GFT. The computation of the GFT for $\mathcal{K}$ can be carried out in two stages, first w.r.t. $\mathcal{S}_4$ for all orbits of $\mathcal{S}_4$, and then w.r.t. $\mathcal{C}_2$ for all orbits of $\mathcal{C}_2$. This is the same principle as when a 2D FFT is carried out, first in the x direction, and then in the y direction. The underlying explanation is the same: we are able to exploit the direct product structure of the underlying group. While this may seem obvious for the 2D FFT applied on Cartesian meshes, we find it interesting to see this benefit also in this context. As shown by the results in [3], the computational gain is almost a factor of two for the case of GFT for $\mathcal{K}$.

4 Conclusions and Future Work

We advocate the usage of group theory when discretizing geometries which exhibit symmetries. The very concept of symmetry is, in fact, best described by using group theoretical concepts. We argue that these underlying concepts should also be reflected in the design of the data structures, particularly the mesh. The support of symmetry in the data structures makes it easier to develop and utilize symmetry exploiting numerical algorithms such as the GFT.

Presently, we are developing numerical routines for the GFT, intended for large scale computations which involve dense linear systems of equations. We assume the system matrix $\boldsymbol{A}$ to be equivariant with respect to one of the groups listed in Section 2. Compared to similar efforts [1], our software is more devoted to efficiency issues, for example by considering parallelization and fast transforms.

There are many interesting related papers which provide directions for future work. For instance, if the equivariant system is sparse, the GFT of the matrix becomes particularly efficient, cf. [2, 11]. Allgower et al report on various ways of dealing with fix points [1]. Bonnet [10] describes an approach for "partially symmetrical" geometries. Tausch describes preconditioning of "almost equivariant" problems [12].

It is however, to all of these research directions, vital that the mesh generation takes the symmetries into account. Even though this is implicit in all of the above references, our aim with this paper is to make these issues explicit. Particularly, we want to make the following observations:

- The mesh should be symmetry respecting.
- The indices should be enumerated orbit-wise.
- The ordering of indices should match the group action.
- Topological mesh information is needed for a fundamental region only.
- Physical mesh information is needed for a fundamental region only.
- Indices which share the same isotropy subgroup should be grouped together.
- It should be possible to extend the index domain in order to make the group action free.
- For each fix point, subsets $\mathcal{C}$ of $\mathcal{G}$ which fills its orbit should be identified.
- Group structure, for instance when the group is a direct product, should be exploited.

Acknowledgements

I would like to thank André Yamba Yamba, Auden Trønnes, and Malin Ljungberg for their cooperation in this project. Specifically, I would like to thank Auden and Malin for their assistance in generating pictures for the icosahedron and for the discretized cube.

References

1. Allgower, E.L., Georg, K., Miranda, R., Tausch, J.: Numerical exploitation of equivariance. Zeitschrift für Angewandte Mathematik und Mechanik **78** (1998) 185–201
2. Bossavit, A.: Symmetry, groups, and boundary value poblems. a progressive introduction to noncommutative harmonic analysis of partial differential equations in domains with geometrical symmetry. Comput. Methods Appl. Mech. and Engrg. **56** (1986) 167–215
3. Åhlander, K. and Munthe-Kaas, H.: On Applications of the Generalized Fourier Transform in Numerical Linear Algebra. Technical Report 2004-029, Department of Information Technology, Uppsala University (2004)
4. Åhlander, K., Munthe-Kaas, H.: Eigenvalues for Equivariant Matrices. Journal of Computational and Applied Mathematics (2005) 14 pp. To appear.
5. Serre, J.P.: Linear Representations of Finite Groups. Springer (1977) ISBN 0387901906.
6. Maslen, D., Rockmore, D.: Generalized FFTs - a survey of some recent results. Technical Report PCS-TR96-281, Dartmouth College, Department of Computer Science, Hanover, NH (1996)

7. Rockmore, D.: Some applications of generalized FFTs. In Finkelstein, L., Kantor, W., eds.: Proceedings of the 1995 DIMACS Workshop on Groups and Computation. (1997) 329–369
8. Allgower, E.L., Böhmer, K., Georg, K., Miranda, R.: Exploiting symmetry in boundary element methods. SIAM J. Numer. Anal. **29** (1992) 534–552
9. Lomont, J.S.: Applications of Finite Groups. Academic Press, New York (1959)
10. Bonnet, M.: Exploiting partial or complete geometrical symmetry in 3D symmetric Galerkin indirect BEM formulations. Intl. J. Numer. Meth. Engng **57** (2003) 1053–1083
11. Bossavit, A.: Boundary value problems with symmetry and their approximation by finite elements. SIAM J. Appl. Math. **53** (1993) 1352–1380
12. Tausch, J.: Equivariant preconditioners for boundary element methods. SIAM Sci. Comp. **17** (1996) 90–99

A Certified Delaunay Graph Conflict Locator for Semi-algebraic Sets

François Anton*

Department of Computer Science,
University of Calgary, 2500 University Drive N.W.,
Calgary, Alberta, T2N 1N4, Canada
antonf@cpsc.ucalgary.ca

Abstract. Most of the curves and surfaces encountered in geometric modelling are defined as the set of common zeroes of a set of polynomials (algebraic varieties) or subsets of algebraic varieties defined by one or more algebraic inequalities (semi-algebraic sets). Many problems from different fields involve proximity queries like finding the nearest neighbour, finding all the neighbours, or quantifying the neighbourliness of two objects. The Voronoi diagram of a set of sites is a decomposition of the space into proximal regions: each site's Voronoi region is the set of points closer to that site than to any other site. The Delaunay graph of a set of sites is the dual graph of the Voronoi diagram of that set of sites, which stores the spatial adjacency relationships among sites induced by the Voronoi diagram. The Voronoi diagram has been used for solving the earlier mentioned proximity queries. The ordinary Voronoi diagram of point sites has been extended or generalised in several directions (underlying space, metrics, sites), and the resulting generalised Voronoi diagrams have found many practical applications. The Voronoi diagrams have not yet been generalised to algebraic curves or semi-algebraic sets. In this paper, we present a conflict locator for the certified incremental maintenance of the Delaunay graph of semi-algebraic sets.

1 Introduction

The geometric objects we encounter almost everywhere in the real world are curved objects in a three dimensional Euclidean space. Most of the curves and surfaces encountered in geometric modelling are defined as the set of common zeroes of a set of polynomials (algebraic varieties) or subsets of algebraic varieties defined by one or more algebraic inequalities (semi-algebraic sets). Examples of semi-algebraic sets include Bézier, Spline curves [Baj94] in

---

* The author wishes to thank Dr. David Kirkpatrick for many useful suggestions about this work and Dr. Jean-Pierre Merlet for introducing him to ALIAS as well as many useful suggestions with ALIAS.

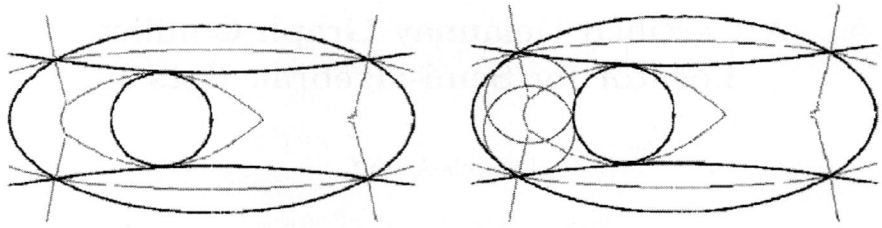

Fig. 1. Left: the Voronoi diagram (light lines) of a circle, an ellipse and a hyperbola (dark lines); right: the empty circles

geometric modelling, Computer Graphics and Aeronautics; the coupler curve [Mer96] in mechanism theory, workspace and singular configurations in robotics, etc. Many problems from different fields involve proximity queries like finding the nearest neighbour, finding all the neighbours, or quantifying the neighbourliness of two objects. One example is the retraction planning [ÓY85] problem (that addresses the optimal trajectory of a robot around obstacles) in robotics.

The Voronoi diagram [Vor08, Vor10] (see Figure 1) of a set of sites is a decomposition of the space into proximal regions (one for each site). Sites were points for the first historical Voronoi diagrams [Vor08, Vor10], but in this paper we will explore sites being conics and more generally semi-algebraic sets. The proximal region corresponding to one site (i.e. its Voronoi region) is the set of points of the space that are closer to that site than to any other site of the set of sites [OBS92].

Once the Voronoi region a query point belongs to has been identified, it is easy to answer proximity queries. The closest site from the query point is the site whose Voronoi region is the preceding found Voronoi region. The Voronoi diagram defines a neighbourhood relationship among sites: two sites are neighbours if, and only if, their Voronoi regions are adjacent. The graph of this relationship is called the Delaunay graph.

There have been attempts [OBS92] to compute Voronoi diagrams of curves by approximating curves by line segments or circular arcs, but the exactness of the Delaunay graph is not guaranteed [RF99]. Only approximations by conics can guarantee a proper continuity of the first order derivative at contact points, which is necessary for guaranteeing the exactness of the Delaunay graph [RF99]. Farouki et al. [RF99] have treated the case of parametric curves admitting a rational parameterisation, which excludes conics. Finally, the Voronoi diagram for curved objects does not necessarily satisfy the property of the abstract Voronoi diagrams (i.e., that the bisector of any pair of sites is an unbounded simple curve [Kle89]). Thus the Voronoi diagram of semi-algebraic sets cannot be computed using the randomised $O(n \log n)$ algorithm for the construction of abstract Voronoi diagrams.

A Certified Delaunay Graph Conflict Locator for Semi-algebraic Sets 671

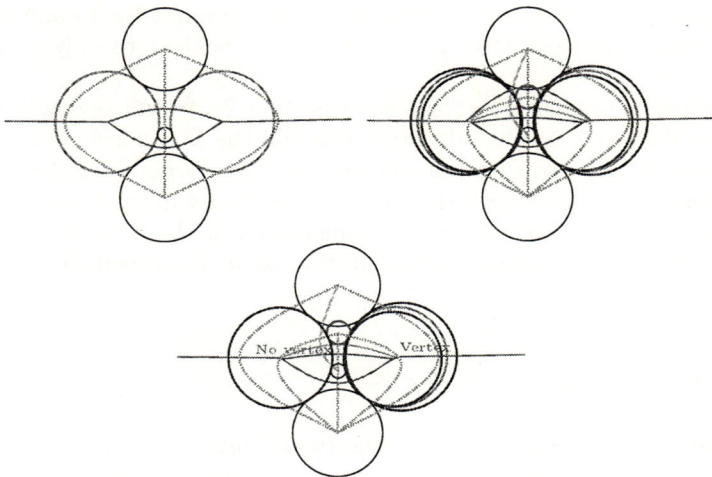

Fig. 2. The starting configuration (left), the real configuration after inserting the middle circle (centre) and the configuration computed by an approximate algorithm (right)

The certified computation of the Delaunay graph of semi-algebraic sets is important for two reasons. By *certified computation*, we mean a computation whose output is correct. First, the Delaunay graph is a discrete structure, and thus it does not lend itself to approximations. Second, the inaccurate computation of this Delaunay graph can induce inconsistencies within this graph, which may cause a program that updates this graph to crash. This is particularly true for the randomised incremental algorithm for the construction of the Voronoi diagram of semi-algebraic sets. We have illustrated such inaccuracies in an example. The starting configuration is shown on the left of Figure 2. There are three circles (drawn in plain lines). The Delaunay graph is drawn in dashed lines. The empty circles tangent to three sites (circles) have been drawn in dotted lines. The real configuration after addition of a fourth circle is shown on the centre of Figure 2. The configuration that might have been computed by an approximate algorithm is shown on the right of Figure 2: the difference between real and perceived situations has been exaggerated to show the difference. The old empty circles have been adequately perceived to be invalid with respect to the newly inserted circle. While on the right of the axis of symmetry two new Voronoi vertices have been identified as valid with respect to their potential neighbours, on the left of the axis of symmetry, only one Voronoi vertex has been identified as being valid with respect to its potential neighbours. While the new Voronoi edge between the middle and bottom circles can be drawn between the two new Voronoi vertices of the new, middle and bottom circles; the Voronoi edge between the top and new circles cannot be drawn, because there is no valid Voronoi vertex on the left. There is an inconsistency within the topology: there is one new Voronoi vertex (the Voronoi vertex of the top new and middle circles) that cannot be linked by a new Voronoi edge to any other new Voronoi vertex and thus, that Voronoi vertex is incident to only two Voronoi edges.

This paper is organised as follows. In Section 2, we will review the preliminaries: the definition of semi-algebraic sets and the Voronoi diagram for those sets. In Section 3, we will present our approach for formalising the conflict locator for the incremental maintenance of the Delaunay graph for semi-algebraic sets. In Section 4, we present the implementation of the evaluation of the Delaunay graph conflict locator. Finally, in Section 5, we present our conclusions and future work.

2 Preliminaries

2.1 Voronoi Diagram of Semi-algebraic Sets

Let us introduce the definition of semi-algebraic sets.

Definition 1. *(Semi-algebraic set, adapted from [BCR98–Definition 2.1.4]) A semi-algebraic set of $\mathbb{R}^N$ is a subset of the form*

$$\bigcup_{i=1}^{s} \bigcap_{j=1}^{r_i} \left\{ x \in \mathbb{R}^N | f_{i,j} \star_{i,j} 0 \right\},$$

where the $f_{i,j}$ are polynomials in the variables $x_1, ..., x_N$ with coefficients in $\mathbb{R}$ and $\star_{i,j}$ is either the "$<$" or the "$=$" symbol, for $i = 1, ..., s$ and $j = 1, ..., r_i$.

See [BR90] for an introduction on semi-algebraic sets.

Let $M = \mathbb{R}^N$, and δ denote the Euclidean distance between points. Let $\mathcal{S} = \{s_1, ..., s_m\} \subset M, m \geq 2$ be a set of m different subsets of M, which we call *sites*. The distance between a point x and a site $s_i \subset M$ is defined as $d(x, s_i) = \inf_{y \in s_i} \{\delta(x, y)\}$.

The one-dimensional elements of the Voronoi diagram are called Voronoi edges. The points of intersection of the Voronoi edges are called Voronoi vertices. The Voronoi vertices are points that have at least $N+1$ nearest neighbours among the sites of $\mathcal{S}$. In the plane, the Voronoi diagram forms a network of vertices and edges. In the plane, when sites are points, the Delaunay graph is known as the Delaunay triangulation. The Delaunay graph satisfies the following empty circle criterion: no site intersects the interior of the hyperspheres touching (tangent to without intersecting the interior of) the sites that are the vertices of any N−dimensional facet of the Delaunay graph (see Figure 1). The *Delaunay graph conflict locator* checks whether a newly inserted semi-algebraic set conflicts with (i.e. has points in the interior of) each one of the empty circles corresponding to any N−dimensional facet of the Delaunay graph formed by a given N−tuple of semi-algebraic sets and reports those conflicts.

2.2 ALIAS

ALIAS [Mer00] is an interval analysis based library for solving zero-dimensional systems of equations and inequalities. The analysis of the zero-dimensional system of equations and inequalities encountered in the Delaunay graph conflict locator computation (see Section 4) relies on results about the uniqueness of a root in an interval (Kantorovitch and Moore-Krawczyk [Mer00]).

All the general purpose solving procedures for zero-dimensional systems have been tested in the computation of the Delaunay graph conflict locator for semi-algebraic sets. These general purpose solving procedures are based on a bisection process on one (*single bisection*), several (*mixed bisection*) or all the variables (*full bisection*) using either:

- only the equations and inequalities of the system,
- the equations and inequalities of the system and the Jacobian of the system (Moore-Krawczyk test for finding "exactly" the solutions),
- the equations and inequalities of the system and the Jacobian and Hessian of the system (with Kantorovitch and Moore-Krawczyk tests).

The bisection process (single, mixed or full) can be set by setting the "ALIAS/single_bisection" parameter (to 2, 1 or 0) and in the case of mixed bisection, the number of bisected variables can be set by setting the "ALIAS/mixed_bisection" parameter. The variables that will be bisected will be the ones having the largest interval width.

The full bisection process on all the variables simultaneously may induce a combinatorial explosion (at each iteration, 2^N new interval vectors or *boxes* are produced). Instead of bisecting on all the variables simultaneously, it is possible to bisect only one variable at each iteration. This may reduce the computation time as the number of function evaluations may be reduced [Mer00]. Aside from these bisection processes, it is possible to use another bisection method called the $3B$ approach (by setting the "ALIAS/3B" parameter to 1 and the maximal range "ALIAS/Max3B" and minimal range "ALIAS/Delta3B" parameters). Each variable x_i and its range $\left[\underline{x_i}, \overline{x_i}\right]$ are considered in turn. Let x_i^m be the middle point of this range. First, the interval evaluations for the equations and inequalities in the system with the full ranges of the variables except for the variable i where the range is $\left[\underline{x_i}, x_i^m\right]$ are computed. Clearly, if one of the equations or inequalities is not satisfied, it is possible to reduce the range of the variable i to $[x_i^m, \overline{x_i}]$. If this is not the case, let's define a new x_i^m as the middle point of the interval $\left[\underline{x_i}, x_i^m\right]$ and repeat the process until either we have found an equation or an inequality that is not satisfied (inducing a new reduction of a variable interval) or the width of the interval $\left[\underline{x_i}, x_i^m\right]$ is lower than a given threshold δ. A similar procedure can be used to reduce the input interval on the right. We may additionally select a subset of equations and/or inequalities whose intervals will be evaluated. This can be done by setting the "ALIAS/SubEq3B" variable [Mer00].

3 The Formalisation of the Delaunay Graph Conflict Locator

We will first recall the concept of offset that is central to the approach we used to design the Delaunay graph conflict locator. In this section, we will focus on definitions in the plane. General definitions will be introduced in Section 4.

We denote by $V(f_1, ..., f_m)$ the set of points that are common zeroes of the polynomials $f_1, ..., f_m$. We will now introduce the true offset curve.

Definition 2. *(True offset) Let $C = V(f) \subset \mathbb{R}^2$ where f is a polynomial with real coefficients in the variables x and y, be an algebraic curve and $R \in \mathbb{R}^+$ be the offset parameter. The true R-offset curve to C is the locus of points being at the distance R from C (see left of Figure 3).*

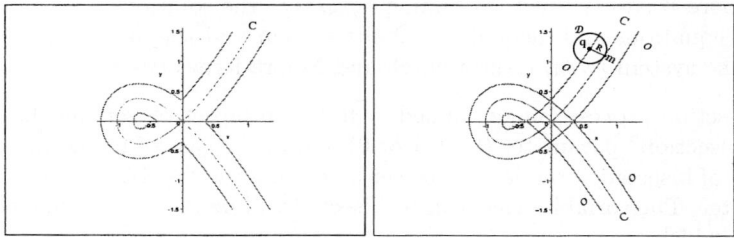

Fig. 3. The strophoid (C), its true offset (thick lines on the left), and its generalised offset (thick lines on the right

We will suppose that f has positive degree (i.e., f is not constant), which implies that C is not empty and not equal to $\mathbb{R}^2$. We will also suppose that $R \neq 0$ unless stated otherwise. Definition 2 is equivalent to saying that each point $q = (u, v)$ of the true offset curve is the centre of a circle $\mathcal{D}$ of radius R that is tangent to C, and does not contain any point of C in its interior.

Let us now introduce the generalised offset and emphasize its differences with the true offset.

Definition 3. *(Generalised offset, adapted from [ASS99]) The generalised offset to a hypersurface ν at distance R is the algebraic set containing all the intersection points of the spheres with centre on a non-singular (multiplicity of one) point of ν and radius R, and the normal lines to ν at the centre of the spheres.*

This is equivalent to saying that each point $q = (u, v)$ of the generalised offset curve $\mathcal{O}$ is the centre of a circle $\mathcal{D}$ of radius R that is tangent to C, but *may contain* points of C in its interior (see right of Figure 3). In [Ant04], we have studied the degree of the offset and in particular, the degree of the offset to the conics. These results are sumarised in Table 1. We have also studied the implicit equation of the generalised offset to a conic in [Ant04].

We are now ready to introduce the formalisation of the Delaunay graph conflict locator. We focus here on semi-algebraic sets in the plane in order to present

Table 1. The degree of the offset to conics

Conic	ellipse/hyperbola	parabola	circle	two lines
Offset degree	8	6	4	4

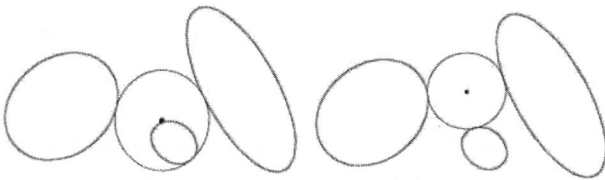

Fig. 4. A generalised Voronoi vertex (dot on the left) and a true Voronoi vertex (dot on the right) of three semi-algebraic sets (thick lines)

the Delaunay graph conflict locator in an easier way. Four semi-algebraic sets S_1, S_2, S_3 and S_4 are given: the first three are supposed to be the vertices of one or more triangles in the Delaunay graph, and the last one is the newly inserted semi-algebraic set. We consider now the notion of generalised offset to a semi-algebraic set (recalled earlier in this section), which can be seen as an expansion/shrinking of semi-algebraic sets. We introduce the notion of generalised Voronoi vertex:

Definition 4. *(generalised Voronoi vertex) A generalised Voronoi vertex of three semi-algebraic sets S_1, S_2, and S_3 is a point of intersection of the $r-$generalised offsets to S_1, S_2, and S_3 with the same offset parameter r (see Example on the left of Figure 4).*

The generalised Voronoi vertex shown on the left of Figure 4 is not a true Voronoi vertex because the circle centred on that vertex and touching the three semi-algebraic sets has points of one of the semi-algebraic sets in its interior. By opposition, the generalised Voronoi vertex shown on the right of Figure 4 is a true Voronoi vertex: the circle centred on that vertex and touching the three semi-algebraic sets has no points of any of the semi-algebraic sets in its interior.

The Delaunay graph conflict locator determines whether or not a newly inserted semi-algebraic set S_4 has conflicts with respect to (i.e. points in the interior of) any of the triangles of the Delaunay graph of $\mathcal{S}$ corresponding to the triple S_1, S_2, S_3 and reports all the conflicts. There are two possible outcomes to the conflict locator: either there are no conflicts and the triangles remain in the new Delaunay graph, or there are conflicts and the triangles corresponding to non-empty circles will not be present in the Delaunay graph any longer. We can see an example of the later case in Figure 5. The semi-algebraic set S'_4 has points in the interior of the "empty" circle corresponding to one of the two triangles $S_1 S_2 S_3$, thus that triangle must be removed (see Figure 5). The Delaunay graph conflict locator consists of determining if there are true Voronoi vertices of the semi-algebraic sets S_1, S_2, and S_3 at a distance (denoted as R) with respect

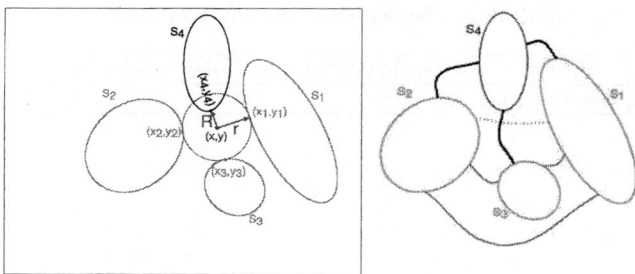

Fig. 5. The insertion of S_4 induces a conflict with one of the triangles $S_1 S_2 S_3$ (left). The new topology after insertion of S_4: spatial adjacency relationships in plain thin lines remain, those in dashed lines disappear, and those in plain thick lines appear (right)

to S_4 lower than the distance (denoted as r) with respect to S_1, S_2, and S_3 (see Figure 5) and to report them. The new triangles in the Delaunay graph are also considered and are the input to the Delaunay graph conflict locator in order to detect the conflicts of the new site with the edges of the old Voronoi diagram.

4 The Delaunay Graph Conflict Locator for Semi-algebraic Sets

4.1 The Algebraic Equations and Inequalities of the Delaunay Graph Conflict Locator

We will present here the system of algebraic equations and inequalities that defines the outcome of the Delaunay graph conflict locator. The definition of a semi-algebraic set has already been presented in Definition 1 in Section 2. Let $X_1, ..., X_{N+2}$, be semi-algebraic sets. Let us suppose that each semi-algebraic set X_i is defined as $\bigcup_{j=1}^{s_i} \bigcap_{k=1}^{r_{i,j}} \{x \in \mathbb{R}^N | f_{i,j,k} \star_{i,j,k} 0\}$, where $f_{i,j,k} \in \mathbb{R}[x_{i_1}, ..., x_{i_N}]$ and $\star_{i,j,k}$ is either the "$<$" or the "$=$" symbol, for $i = 1, 2, 3, 4$, $j = 1, ..., s_i$ and $k = 1, ..., r_{i,j}$.

Let us assume without loose of generality that each $\bigcap_{k=1}^{r_{i,j}} \{x \in \mathbb{R}^N | f_{i,j,k} \star_{i,j,k} 0\}$ for each X_i is defined by at least one nontrivial algebraic equation (i.e. different from the zero polynomial). We can make this assumption valid by adding the equations corresponding to $f_{i,j,k} = 0$ for each (i, j, k) such that j is the index of a component that is not defined as in the assumption and i is the index of the semi-algebraic set to which the component belongs. Let us denote V_i as the intersection of all the $V(f_{i,j,k})$ such that $\star_{i,j,k}$ is $=$ for each $i = 1, 2, 3, 4$. Let $\mathcal{N}_i$ be the normal space to V_i at the point $x_i = (x_{i_1}, ..., x_{i_N})$. Each f_{ijk} defining V_i induces $N - 1$ polynomials n_{ijkl} with $l = 1, ..., N-1$ that are the equations defining the normal to $V(f_{ijk})$ at

x_i. A point $q = (y_1, ..., y_N)$ belongs to $\mathcal{N}_i$ if its coordinates satisfy each one of the equations of the normal spaces to $V(f_{i,j,k})$ at x_i such that $\star_{i,j,k}$ is $=$.

For a given $q = (y_1, ..., y_N)$, let $\mathcal{M}_i$ be the the set of points $m_i = (z_{i_1}, ..., z_{i_N}) \in X_i$ such that q belongs to the normal space to V_i at the point m_i. In the general case, each set $\mathcal{M}_i$ is a finite set of points. However, if V_i contains a portion of hypersphere $PHS(q, \rho)$ centered on q, then $\mathcal{M}_i$ contains that portion of hypersphere. To get in all cases a finite set of points m_i of V_i, we use $\mathfrak{S}_i = \mathcal{M}_i$ when $\mathcal{M}_i$ is finite, and $\mathfrak{S}_i \cap PHS(q, \rho) = \{w_i\}$ for an arbitrary point w_i of $PHS(q, \rho)$ when V_i contains a portion of hypersphere $PHS(q, \rho)$ centered on q.

Let us consider the map $\pi : \mathbb{R}^{3N} \to \mathbb{R}^N$ defined by $\pi(x_i, q, m_i) = q$. The point q is at the distance r from the point x_i if, and only if, the distance between q and x_i is r. This is expressed algebraically by the equation $d_i(q, x_i) = (y_1 - x_{i_1})^2 + ... + (y_N - x_{i_N})^2 - r^2 = 0$.

The generalised r-offset $\mathcal{O}_i$ to X_i is the image by π of the points of $\mathbb{R}^{3N}$ defined by the following system of equations and inequalities:

$$\begin{cases} \exists j \in [1, s_i], \forall k \in [1, r_{i,j}], \\ \begin{cases} f_{i,j,k}(x_i) \star_{i,j,k} 0 \\ \text{if } \star_{i,j,k} \text{ is } "=", \end{cases} \\ \begin{cases} d_i(x_i, q) = 0 \\ \forall l = 1, .., N-1, n_{i,j,k,l}(x_i, q) = 0 \\ f_{x_{i_1}}(x_i) \neq 0 \text{ or } ... \text{ or } f_{x_{i_N}}(x_i) \neq 0 \end{cases} \end{cases}$$

The true r-offset to X_i is obtained as the difference of the generalised r-offset $\mathcal{O}_i$ to X_i and the union of each one of the images by π of the semi-algebraic sets defined by the following system of equations and inequalities for each point m_i of $\mathfrak{S}_i$:

$$\begin{cases} \exists j \in [1, s_i], \forall k \in [1, r_{i,j}], \\ \begin{cases} f_{i,j,k}(m_i) \star_{i,j,k} 0 \\ \text{if } \star_{i,j,k} \text{ is } "=", \end{cases} \\ \begin{cases} f(m_i) = 0 \\ \forall l = 1, .., N-1, n_{i,j,k,l}(m_i, q) = 0 \\ d(m_i, q) < 0 \end{cases} \end{cases}$$

It is obvious that a true Voronoi vertex of $X_1, ..., X_{N+1}$ is the intersection of the true r-offsets to $X_1, ..., X_{N+1}$ respectively. Now, what is left to write is first, that the true Voronoi vertex of $X_1, ..., X_{N+1}$ is at the distance R from X_{N+2}, or alternatively, that the true Voronoi vertex of $X_1, ..., X_{N+1}$ belongs to the true R-offset to X_{N+2}, and finally to evaluate the signs of the real values of $R - r$. Consider the $N+2$-dimensional points whose first N coordinates are the coordinates of a true Voronoi vertex of $X_1, ..., X_{N+1}$, and the remaining two are the distances r between that true Voronoi vertex and $X_1, ..., X_{N+1}$, and R between that true Voronoi vertex and X_{N+2}. The Delaunay graph conflict locator should report all those $N+2$-dimensional points such that $R - r < 0$.

4.2 The Formulation of the Delaunay Graph Conflict Locator with ALIAS

We have tested different solving procedures as well as different ALIAS parameters for computing the Delaunay graph conflict locator for semi-algebraic sets on conics. We first observed better results with the 3B consistency method (ALIAS/3B=1) than without (ALIAS/3B=0), both in terms of running time and in terms of number of searched boxes. Consequently, we tested only 3B based searching techniques. We can either apply the 3B consistency method on all the equations and/or inequalities, or we can specify a subset of equations and/or inequalities on which the 3B consistency method will be applied (the ALIAS/subeq3B list variable). Typically, one chooses the algebraic equations and inequalities with lowest degree for the SubEq3B subset. We can also select a maximal interval range (ALIAS/Max3B), that is the maximal range that the variables intervals should have in order for the 3B method to start being applied. By setting this maximum interval range to the maximum variable interval range, we can force the 3B method to be applied from the starting variables intervals. We can also select the tolerance interval range for the 3B method, by setting the ALIAS/Delta3B variable. The results are summarised in Table 2. The executable was run always on the same machine at off-peak hours (in the evening). The significant results is that single bisection is much faster than mixed bisection, which in turn is faster than full bisection. These running times illustrate the impact of the exponential growth of the number of boxes with the number of variables on the running times. The 3B consistency method improves the results obtained by single bisection. There is a trade-off between the additional time required by the additional 3B interval evaluations and the reduction of the variable intervals by the 3B method. The optimum in the example of Table 2 is to start the 3B method after the number of variable intervals has been reduced by 4 by the normal single bisection process.

Table 2. Some running time results with different ALIAS parameters on the system with the generalised offsets (see Table 5)

Bisection process	Running time for optimised GradientSolve
Full bisection	2 h 34 min 42 s
Mixed bisection (half the variables)	26 min 26 s
Single bisection + 3B	2 min 26 s
+ 3B with half Max3B	2 min 8 s
+ 3B with one quarter Max3B	2 min 2 s
+ 3B with one sixth Max3B	4 min 49 s
+ 3B with one eight Max3B	4 min 49 s

We have tested the full system of equations and inequalities corresponding to the Delaunay graph conflict locator as well as the partial system for identifying the generalised Voronoi vertices that are true or computing the conflict locator

Table 3. Some running time results for ellipses with the equations of the original curves

Running time	without subEq3B	with subEq3B
optimised General Solve	6 min 38 s	6 min 26 s
optimised Gradient Solve	2 h 56 min 10 s	2 h 55 min 4 s
optimised Hessian Solve	20 h 17 min 42 s	20 h 19 min 33 s
non-optimised Hessian Solve		

Table 4. Some running time results for ellipses with the equations of the original curves for the partial system

Running time	without subEq3B	with subEq3B
optimised General Solve	2 min 51 s	2 min 38 s
optimised Gradient Solve	26 min 52 s	27 min 8 s
optimised Hessian Solve	1 h 43 min 42 s	1 h 45 min 38 s

assuming we know which generalised Voronoi vertices are true Voronoi vertices. Note that we need to execute the program $N+2$ times instead of one if we use the partial system ($N+1$ times for identifying the true Voronoi vertices and once for computing the Delaunay graph conflict locator assuming the true Voronoi vertices have been identified). The results are summarised in Tables 3 and 4.

While there is a time disadvantage in running the solver on the full system with respect to running the solver (four times) on the partial systems for the Gradient and Hessian based solvers, running the General solver on the full system is less time consuming. What is important to observe at this point is that if the full system is not used, in addition to running four times the solver, we need to check that the coordinates and local distance to the defining semi-algebraic sets of each generalised Voronoi vertex corresponding to the solutions correspond to a true Voronoi vertex. The full system has all the constraints (inequalities in it). This is likely why the full system can be solved faster than the four partial systems. Moreover, the General solver is the fastest one on the full and partial systems based on the equations of the original curves. Finally, specifying a subset of equations and/or inequalities (ALIAS/subeq3B) on which the 3B bisection process will be applied improves the running time of the general solver while increases those of the gradient and hessian based solvers.

The general solver seems to be more efficient on the full system based on the equations specifying the semi-algebraic sets because the computations of the gradient and of the hessian are more time consuming when the number of variables increases. A way to improve the running time is to improve the computations of the intervals. What we have done is to use the parser to convert the Maple expressions into C++ code that uses the ALIAS C++ library in order to do interval computations and identify intervals with possible solutions. The interval computations resulting from this automatised process may not be optimal. This is true because the computation of the interval taken by a function depends on the way this function is written. We have tried to optimise the way functions are written for the interval computations on the general solver code. However, this did not lead to any significant improvement on the running time of the optimised code. It looks like there is a trade-off between the running time improvement owed to the change of expression of the function and the optimisation done by the C++ compiler (g++ version 2.95.2).

4.3 The Hybrid Symbolic/Scientific Computation of the Delaunay Graph Conflict Locator for Conics

In this section, we will present how we can use the implicit equation of the generalised offset to a conic in the ALIAS computations, and show some results that illustrate the considerable reduction of running time realised by using the equations of the generalised offsets instead of the equations of the original curves (see sample Maple program in [Ant04–Appendix D]). We have already studied and obtained an implicit equation of the generalised offset to the different types of conics in [Ant04–Appendix B]. Table 5 shows some results on the ellipses that were used for the tests of the previous section. These results show a decrease of running time except for the general solver with ratios varying between 35 and 331 (see Table 6). Similar improvements were obtained for hyperbolas and parabolas.

Table 5. Some running time results for ellipses with the equations of the generalised offsets

Running time	without subEq3B	with subEq3B
optimised General Solve	12 min 37 s	12 min 56 s
optimised Gradient Solve	2 min 26 s	4 min 57 s
optimised Hessian Solve	3 min 41 s	3 min 42 s

Table 6. The ratios of the running time results without/with the equations of the generalised offsets

Ratios (without/with generalised offset)	without subEq3B	with subEq3B
optimised General Solve	0.526	0.497
optimised Gradient Solve	72.40	35.37
optimised Hessian Solve	330.6	329.6

Another possibility of expression of the equations of conics is the replacement of the implicit equations of conics by their parametric equations in polar coordinates[1]. This did not lead to any running time improvement with any of the solving techniques used earlier.

---

[1] The general polar equation of a non degenerate conic with respect to one of its foci is: $r = \frac{p}{1-e\cos\theta}$, where e is the eccentricity of the conic and p is that eccentricity times the distance from the foci to the directrix.

5 Conclusions

The certified computation of the Delaunay graph of conics using interval analysis gradient and Hessian based solvers can benefit from the use of the implicit equation of the generalised offset to a conic (from 35 to 331 times faster, 2 min 26 s for ellipses, 6 min 20 s for hyperbolas). This result confirms the idea that knowing the structure of the set of solutions may help finding the solutions. Even though there is no known theoretical lower nor upper bound for the interval analysis based solvers, in practice those solvers can compute the Delaunay graph conflict locator much faster than the computation of the eigenvalues of the Schur complement of a submatrix of the sparse resultant matrix.

This paper has presented what we believe is the first computation of the Delaunay graph conflict locator for semi-algebraic sets (and in particular conics), and its application to a semi-dynamic algorithm for the construction of the Voronoi diagram of semi-algebraic sets.

References

[Ant04] François Anton. *Voronoi diagrams of semi-algebraic sets*. PhD thesis, The University of British Columbia, Vancouver, British Columbia, Canada, January 2004.

[ASS99] Enrique Arrondo, Juana Sendra, and J. Rafael Sendra. Genus formula for generalized offset curves. *J. Pure Appl. Algebra*, 136(3):199–209, 1999.

[Baj94] Chandrajit L. Bajaj. Some applications of constructive real algebraic geometry. In *Algebraic geometry and its applications (West Lafayette, IN, 1990)*, pages 393–405. Springer, New York, 1994.

[BCR98] Jacek Bochnak, Michel Coste, and Marie-Françoise Roy. *Real algebraic geometry*. Springer-Verlag, Berlin, 1998. Translated from the 1987 French original, Revised by the authors.

[BR90] Riccardo Benedetti and Jean-Jacques Risler. *Real algebraic and semi-algebraic sets*. Hermann, Paris, 1990.

[Kle89] Rolf Klein. *Concrete and abstract Voronoï diagrams*. Springer-Verlag, Berlin, 1989.

[Mer96] J.-P. Merlet. Some algebraic geometry problems arising in the field of mechanism theory. In *Algorithms in algebraic geometry and applications (Santander, 1994)*, pages 271–283. Birkhäuser, Basel, 1996.

[Mer00] Jean-Pierre Merlet. Alias: an interval analysis based library for solving and analyzing system of equations. In *SEA*, Toulouse, France, 14–16 June 2000.

[OBS92] Atsuyuki Okabe, Barry Boots, and Kōkichi Sugihara. *Spatial tessellations: concepts and applications of Voronoï diagrams*. John Wiley & Sons Ltd., Chichester, 1992. With a foreword by D. G. Kendall.

[ÓY85] Colm Ó'Dúnlaing and Chee-K. Yap. A "retraction" method for planning the motion of a disc. *J. Algorithms*, 6(1):104–111, 1985.

[RF99] Rajesh Ramamurthy and Rida T. Farouki. Voronoi diagram and medial axis algorithm for planar domains with curved boundaries. I. Theoretical foundations. *J. Comput. Appl. Math.*, 102(1):119–141, 1999. Special issue: computational methods in computer graphics.

[Vor08] G. F. Voronoï. Nouvelles applications des paramètres continus à la théorie des formes quadratiques. deuxième mémoire. recherches sur les paralléloèdres primitifs. première partie. partition uniforme de l'espace analytique à n dimensions à l'aide des translations d'un même polyèdre convexe. *Journal für die reine und angewandte Mathematik*, 134:198–287, 1908.

[Vor10] G. F. Voronoï. Nouvelles applications des paramètres continus à la théorie des formes quadratiques. deuxième mémoire. recherches sur les paralléloèdres primitifs. seconde partie. domaines de formes quadratiques correspondant aux différents types de paralléloèdres primitifs. *Journal für die reine und angewandte Mathematik*, 136:67–181, 1910.

The Offset to an Algebraic Curve and an Application to Conics

François Anton[1], Ioannis Emiris[2],
Bernard Mourrain[3], and Monique Teillaud[3]

[1] Department of Computer Science,
University of Calgary, 2500 University Drive N.W.,
Calgary, Alberta, T2N 1N4, Canada
antonf@cpsc.ucalgary.ca
[2] Department of Informatics and Telecommunications,
National Kapodistrian University of Athens,
Panepistimiopolis 15784, Greece
[3] INRIA Sophia-Antipolis, B.P. 93, 2004,
route des Lucioles, 06902, Sophia-Antipolis, France

Abstract. Curve offsets are important objects in computer-aided design. We study the algebraic properties of the offset to an algebraic curve, thus obtaining a general formula for its degree. This is applied to computing the degree of the offset to conics. We also compute an implicit equation of the generalised offset to a conic by using sparse resultants and the knowledge of the degree of the implicit equation.

1 Introduction

An important object in nonlinear computational geometry, Computer-aided Geometric Design and geometric modelling is the offset of a given curve or surface, which is defined as the locus of points at a given distance. A related question concerns the bisector of two curves or surfaces. Both problems have been addressed for some classes of curves or surfaces, mostly for their applications (see [1] for the application to the Voronoi diagram of semi-algebraic sets).

What Hoffmann and Vermeer [8] define as offset curves, Arrondo, Sendra and Sendra [2] define as generalised offset curves. The extraneous solutions (corresponding to a point of the curve or surface which corresponds to infinitely many points on the generalised offset) have been addressed in [8]. Hoffmann and Vermeer [8] did not address the offset computations, but they gave some numerical examples computed using Gröbner bases. Arrondo, Sendra and Sendra [2] computed the genus of the generalised offset curve.

In this paper, we will address the degree of the offset to an algebraic plane curve in its most general setting. Our main contributions are a general formula for the degree of the offset curve and its application for the determination of an implicit equation of the generalised offset to a conic. The conic is defined implicitly by a formal polynomial (i.e., a polynomial whose coefficients are formal constants). We used the general formula of the degree of the offset curve to eliminate the extraneous factors from the sparse resultant [3, 4] in order to get an implicit equation defining the generalised offset to a conic.

This paper is organised as follows: in Section 2, we study the equation of the offset. In Section 3, we study the algebraic properties of the offset to an algebraic curve in order to determine its degree. In Section 4, we show how to compute the implicit equation of the generalised offset to a conic.

2 Equations Defining the Offset

This section uses notions of algebraic geometry; for further background one may consult [9]. We assume the most general setting for the problem at hand, i.e. complex numbers and we let $k = \mathbb{C}$. In this paper, we assume that $V(f_1, \ldots, f_s) \subset E$ denotes the set of all the points of E whose coordinates $(x_1, \ldots, x_n)$ satisfy the polynomial equations $f_1(x_1, \ldots, x_n) = \cdots = f_s(x_1, \ldots, x_n) = 0$. If $E = k^n$ is an affine space, then the set V is called an affine variety. If $E = \mathbb{P}^n$ is a projective space over k, then the set V is called a projective variety. Distances are always measured in the Euclidean metric. The distance of a point to a curve is the smallest distance of the point from any point on the curve.

Definition 1. *(Offset curve) Let $C = V(f) \subset k^2$, for $f \in k[x, y]$, be an algebraic curve and $R \in \mathbb{R}^+$ be the offset parameter. The R-offset curve to C is the locus of points lying at distance R from C (see Figure 1).*

Definition 1 is equivalent to saying that each point $q = (u, v)$ of the offset curve is the centre of a circle $\mathcal{D}$ of radius R, which is tangent to C, and does not contain any point of C in its interior. We will suppose that f has positive degree (thus C is not empty) and $R \neq 0$ unless stated otherwise.

Definition 2. *(Generalised offset) The generalised R-offset curve to C denoted $\mathcal{O}$ is the locus of the centres $q = (u, v)$ of circles $\mathcal{D}$ of radius R, that are tangent to C (see Figure 1).*

The R-generalised offset to C is a superset of the true R-offset curve to C. Indeed, the circle centred on a point q' of the true R-offset curve to C of radius

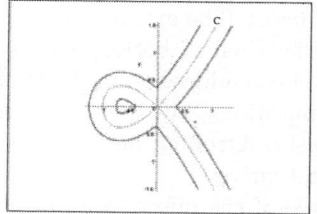 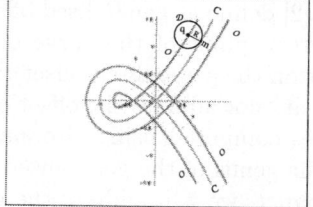

Fig. 1. The Maple plots of a strophoid (C), its true offset (left) and of its generalised offset O (right)

R is tangent to C. Its interior does not intersect C. The *Zariski closure* [9] of an affine (resp. projective) variety V is the smallest affine (resp. projective) variety containing X. The generalised offset to a hypersurface ν at distance R is the Zariski closure of the set of intersection points of the spheres with centre on ν and radius R, and the normal lines to ν at the centre of the spheres, cf. [2]. We will now establish the systems of equations and inequalities that define the generalised R-offset $\mathcal{O}$ and the true R-offset to an algebraic curve $C = V(f)$. The normal to C at a given point $p = (x, y) \in C$ is the set of points $q = (u, v)$ for which $n_{x,y}(u, v) = 0$, where $n_{x,y}(u, v) = -f_y(x, y) \cdot (u - x) + f_x(x, y) \cdot (v - y)$. Here f_x and f_y denote the partial derivatives of f. A point p of an algebraic curve $V(f) \subset k^2$ is called a *singular* point if, and only if, both partial derivatives of f vanish at p (i.e., $f_x(p) = f_y(p) = 0$). For a given $q = (u, v) \in k^2$, let $\mathcal{M} \subset \mathbb{R}^2$ be the the set of points $m = (\alpha, \beta) \in C \cap \mathbb{R}^2$ such that $n_{\alpha,\beta}(u, v) = 0$. This condition is achieved whenever the normal to $C \cap \mathbb{R}^2$ at m passes through q, or m is singular. In the general case, $\mathcal{M}$ is finite. However, if $C \cap \mathbb{R}^2$ is a circle centred on q, then $\mathcal{M} = C \cap \mathbb{R}^2$. More generally, if f is not square-free, there is an infinity of singular points. To get in all cases a finite set of points m of $C \cap \mathbb{R}^2$ such that $n_{\alpha,\beta}(u, v) = 0$, we use $\mathcal{S} = \mathcal{M}$ when $\mathcal{M}$ is finite, and $\mathcal{S} = \{w\}$ for an arbitrary point w of $C \cap \mathbb{R}^2$ when $\mathcal{M}$ is infinite.

Lemma 1. *The set of all the closest points on $C \cap \mathbb{R}^2$ from q is contained in $\mathcal{M}$.*

Proof. The polynomial n (considered here as a polynomial in α, β for fixed u, v) defining $\mathcal{M}$ expresses half of the differential of the scalar product $\overrightarrow{qm} \cdot \overrightarrow{qm}$ (which is equal to the square of the Euclidean distance $\delta(q, m)$) with respect to m. The closest points on $C \cap \mathbb{R}^2$ from q are global minima of the Euclidean distance $\delta(q, m)$, so, the differential (and thus, n) vanishes on them. □

The point $q = (u, v)$ on the generalised R-offset curve $\mathcal{O}$ can be constructed from a non-singular point $p = (x, y)$ on C as the intersection of the normal to C at p (given by $n_{x,y}$, as mentioned previously) and the circle $\mathcal{D}$ centred on p, and of radius R. The equation of this circle is $d_{x,y}(u, v) = 0$, where $d_{x,y}(u, v) = (u - x)^2 + (v - y)^2 - R^2$. Let us consider the map $\pi : \begin{array}{c} k^6 \to k^2 \\ (x, y, u, v, \alpha, \beta) \mapsto (u, v) \end{array}$. π can also be seen as the canonical projection of $k^4 \subset k^6$ onto k^2. If we consider the inclusion map from k^2 to k^4, $V(f)$ can be seen as a variety of k^4, which is the image of C by this inclusion map. Let $n(x, y, u, v) = n_{x,y}(u, v)$ and $d(x, y, u, v) = d_{x,y}(u, v)$ be considered now as polynomials in $k[x, y, u, v]$. $V(n)$ is a proper subset of k^4 provided that $p = (x, y)$ is not singular. If f is not square-free, there is an infinity of singular points. From now on, the affine and projective varieties will be considered in different underlying spaces. When necessary, we will specify the underlying space by an inclusion, e.g. $V(f) \subset k^4$ or $V(f) \subset k^2$. The generalised R-offset $\mathcal{O}$ is the image by π of the affine variety of k^4 defined by the following system of equations and inequalities:

$$\begin{cases} f(x,y) &= 0 \\ n(x,y,u,v) = 0 \\ d(x,y,u,v) = 0 \\ f_x(x,y) \neq 0 \text{ or } f_y(x,y) \neq 0 \end{cases}$$

$\mathcal{O}$ is the image of $(V(f) \cap V(n) \cap V(d)) \setminus V(f_x, f_y)$ by the canonical projection $\pi : k^4 \to k^2$ onto the (u,v)-plane. The *true R-offset* lies in $\mathbb{R}^2$ and is obtained as the difference of the generalised R-offset $\mathcal{O}$ and the union of the images by π of the sets defined by the following system of equations and inequalities, for each point $m = (\alpha, \beta)$ of $\mathcal{S}$:

$$\begin{cases} f(\alpha, \beta) &= 0 \\ n(\alpha, \beta, u, v) = 0 \\ (u-\alpha)^2 + (v-\beta)^2 - R^2 < 0 \end{cases}$$

3 The Degree of the Offset Curve

This section relies on several algebraic concepts, which can be found in more detail in [9]. Consider polynomials in the variables x, y, u and v with coefficients in k and the projective space $\mathbb{P}^4$ (of lines in k^5 passing through the origin). The homogenisation variable will be denoted by t. We consider the point $q = (u, v)$ on the generalised R-offset curve $\mathcal{O}$ constructed above from an arbitrary point $p = (x, y)$ on C. Let $\overline{g} = t^{\text{degree}(g)} g(\frac{x}{t}, \frac{y}{t}, \frac{u}{t}, \frac{v}{t})$ be the *homogenisation* of a polynomial g, $\overline{V}$ denote the *projective closure* of the variety V, i.e. the smallest projective variety containing V, and g^T denote the polynomial defining the *component at infinity* (i.e. the points with homogeneous coordinate equal to zero) of $\overline{V(g)}$. Note that $\overline{V(g)} = V(\overline{g})$ [9–p. 33]. Thus, $V(\overline{f}), V(\overline{n}), V(\overline{d}) \subset \mathbb{P}^4$.

We are now going to decompose the projective closure of $V(f, n, d) \subset k^4$ into the union of its component at infinity, its *singular component* (points induced by singular points p on C), and the generalised offset considered in k^4 (i.e. the affine variety $V(f, n, d) \setminus V(f_x, f_y) \subset k^4$). We will thus obtain the generalised offset considered in k^4 as a difference of projective varieties, and we will determine their dimensions and degrees in order to determine the degree of the generalised offset.

Let us define $W = \overline{V(f)} \cap \overline{V(n)} \cap \overline{V(d)}$, $W_a = W \setminus (\overline{V(f_x, f_y)} \cup V(t)) \subset \mathbb{P}^4$, $W_S = W \cap \overline{V(f_x, f_y)}$, and $W_\infty = W \cap V(t)$. We recall here the definition of a quasi-projective variety. In the Zariski topology, the only sets that are closed are algebraic (affine or projective) varieties. A *quasi-projective variety* is an open subset of a projective variety. W_a is a quasi-projective variety since $\overline{V(f)} \cap \overline{V(n)} \cap \overline{V(d)}$ and $\overline{V(f_x, f_y)} \cup V(t) \subset \mathbb{P}^4$ are projective varieties. W_S is a projective variety and $W_\infty = V(f^T, d^T, n^T, t)$ a projective subvariety. A variety X is called *reducible* if there exist varieties $X_1 \subset X$, $X_2 \subset X$, $X_1 \neq X$, $X_2 \neq X$, such that $X = X_1 \cup X_2$. A polynomial is *reducible* if it can be factorized into two polynomials of positive degree. Let $d_1 := -\iota(u-x) + (v-y)$, $d_2 := \iota(u-x) + (v-y)$, and let ι be a root of the equation $x^2 + 1 = 0$ in the algebraically closed field k. Then $V(d^T) = V(d_1) \cup V(d_2)$ is a *minimal* decomposition, i.e. into the smallest number of irreducible varieties. We consider the ring $k[x_1, \ldots, x_n]$ of polynomials in n variables with coefficients in k. Let $\langle f_1, \ldots, f_s \rangle = \{g_1 f_1 + \cdots + g_n f_n : \forall i, g_i \in k[x_1, \ldots, x_n]\}$ denote the ideal generated by $f_1, \ldots, f_s \in k[x_1, \ldots, x_n]$. Let $\sqrt{I} = \{f \in k[x_1, \cdots, x_n] : \exists n \in \mathbb{N}, f^n \in I\}$ denote the *radical* of an ideal I. Lastly, for any variety W, let $I(W)$ denote the set of all polynomials vanishing on W; this is easily shown to be an ideal. For any variety $W = V(p)$, $I(W) = \sqrt{(\langle p \rangle)}$ by Hilbert's Nullstellensatz [9]. Now, we will prove some lemmas that will be needed for the proof of Lemma 6. Let $V(f^T) = V(h_1) \cup V(h_2) \cup \cdots \cup V(h_s)$ be minimal: $\forall j \neq i, V(h_i) \not\subset V(h_j)$.

Lemma 2. *None of the irreducible components of $V(f^T) \subset \mathbb{P}^4$ is contained in an irreducible component of $V(d^T) \subset \mathbb{P}^4$. Equivalently, none of the h_i divides d^T.*

Proof. Let us first establish the equivalence of the two statements. For any i, we have $V(h_i) \subset V(d^T) \Leftrightarrow d^T \in I(V(h_i)) = \sqrt{\langle h_i \rangle}$. Since h_i is irreducible, the latter ideal equals $\langle h_i \rangle$. Hence, an equivalent expression is that h_i divides d^T. We will prove the second statement by contradiction. Assume $d^T \in I(V(h_i)) \subset k[x, y, u, v, t]$ for some $i \in \{1, 2, \ldots, s\}$. Thus, there exists $g \in k[x, y, u, v, t]$ such that $d^T = g \cdot h_i$. The sum of the degrees of g and of h_i equals 2. There are three cases, namely that the respective degrees equal 0,2 or 1,1 or 2,0. If the degree of h_i is 2, then h_i is reducible, therefore the first case is not possible. In the second case, $d^T = g \cdot (\alpha x + \beta y + \gamma)$, where α, β, γ are coefficients of k. This would imply that the terms in u^2 and the terms in v^2 of d^T must come from g. These terms induce terms in $u^2 x$, $u^2 y$, $v^2 x$, and $v^2 y$ which cannot be canceled. So the second case is also infeasible. In the last case, h_i cannot be a constant by definition. □

We consider the projective space $\mathbb{P}^N$ and we let $\xi \in \mathbb{P}^N$ be denoted by $(\xi_0 : \cdots : \xi_N)$, where $\xi_i \in k$, $i = 0, \ldots, N$ and not all the ξ_i are 0. Let $\mathbb{A}_i^N \subset \mathbb{P}^N$ consist of all the points for which $\xi_i \neq 0$ (thus, it is isomorphic to the affine space k^N). We now define regular functions, regular mappings of affine varieties, and regular mappings of quasi-projective varieties. Let X be an affine variety in k^N. A function g on X is *regular* [9] if there exists a polynomial G with coefficients in k such that $g(x) = G(x)$, $\forall x \in X$. Let Y be a variety of k^N. A mapping $g : X \to Y$ is *regular* if there exists N regular functions $g_1, \ldots, g_N$ on X such that $g(x) = (g_1(x), .., g_N(x))$, $\forall x \in X$. Let $f : X \to Y$ be a mapping of quasi-projective varieties and $Y \subset \mathbb{P}^N$. This *mapping* is called *regular* if for every point $x \in X$ and every open affine set $\mathbb{A}_i^N$ containing the point $f(x)$ there exists a neighbourhood U of x such that $f(U) \subset \mathbb{A}_i^N$, and the mapping $f : U \to \mathbb{A}_i^N$ is regular. The *dimension* of a variety X is $\sup \{n : X_0 \subsetneq \cdots \subsetneq X_n \subset X$ and $\forall i, X_i$ irreducible$\}$.

Theorem 1. *[9–theorem.8, Sect.1.6] Let $f : X \to Y$ be a regular mapping between projective varieties with $f(X) = Y$. Suppose that Y is irreducible and that all the fibres $f^{-1}(y)$ for $y \in Y$ are irreducible and of the same dimension, then X is irreducible.*

Lemma 3. *None of the irreducible components of $V(f^T, d^T) \subset \mathbb{P}^4$ is contained in $V(t) \subset \mathbb{P}^4$.*

Proof. Notice that $V(f^T)$ and $V(h_i)$ are not contained in the projective hyperplane at infinity $V(t) \subset \mathbb{P}^4$. By considering the intersection of two finite set unions, it is clear that the set $V(f^T, d^T)$ can be written as the union of all sets $V(h_i, d_j)$, for $i = 1, \ldots, s$, $j = 1, 2$. We shall now show that the $V(h_i, d_j)$ are irreducible. Consider the projection $\mathbb{P}^4 \to \mathbb{P}^2 : (t : x : y : u : v) \mapsto (t : x : y)$. It can be restricted to the following sequence of projections:
$$\pi_{ij}: \begin{array}{c} V(h_i, d_j) \to V(h_i) \\ (t:x:y:u:v) \mapsto (t:x:y) \end{array}.$$
Now, $\pi_{ij}(V(h_i, d_j))$ is trivially included in $V(h_i)$. The converse is also true: for any $q = (t_q : x_q : y_q) \in V(h_i)$, if we consider $q' = (t_q : x_q : y_q : u : v)$ such that $d_j(u, v) = 0$, then $q' \in V(h_i, d_j)$ and $\pi_{ij}(q') = q$. So, $\pi_{ij}(V(h_i, d_j)) = V(h_i)$. The π_{ij} are regular mappings of projective varieties. The fibres $\pi_{ij}^{-1}(\omega)$ for $\omega \in V(h_i)$ have dimension 1 since the points on these fibres have fixed x, y and t coordinates, and u and v are related by the equation of d_j. Thus, all the fibres have the same dimension. Also, each fibre is irreducible since $V(d_j)$ is irreducible. Then, by Theorem 1 we conclude that the $V(h_i, d_j)$ are irreducible. We will show that none of these $V(h_i, d_j)$ is contained in an irreducible component of $V(t)$. Let us suppose $t \in I(V(h_i, d_j))$ for some $i \in \{1, 2, \ldots, s\}$ and $j \in \{1, 2\}$. By Hilbert's Nullstellensatz, $I(V(h_i, d_j)) = \sqrt{\langle h_i, d_j \rangle}$. Since the $V(h_i, d_j)$ are irreducible, $\sqrt{\langle h_i, d_j \rangle} = \langle h_i, d_j \rangle$. Then there exists $a, b \in k[x, y, u, v, t]$ such that $t = ah_i + bd_j$. Since h_i and d_j don't have monomials with t nor constant terms, t in $ah_i + bd_j$ must come from a or b or both. Since h_i and d_j don't have constant terms, the monomial of least total degree containing t in $ah_i + bd_j$ must have a degree greater than or equal to 1 in the other variables. □

Lemma 4. *None of the irreducible components of $V(f^T, d^T, t) \subset \mathbb{P}^4$ is contained in an irreducible component of $V(n^T) \subset \mathbb{P}^4$ (or, equivalently, n^T does not vanish identically on any irreducible component of $V(f^T, d^T, t)$) if, and only if, $f_x^T + \iota f_y^T \notin \langle h_i \rangle$ and $f_x^T - \iota f_y^T \notin \langle h_i \rangle$ for $i = 1, 2, \ldots, s$.*

Proof. The fact that the statement in parenthesis is indeed equivalent can be proved in a manner analogous to that in the proof of Lemma 2. Now, consider the projection $V(h_i, d_j, t) \to V(h_i, t) : (t : x : y : u : v) \mapsto (t : x : y)$, which can be restricted to the following mappings, for $i = 1, \ldots, s, j = 1, 2$:
$$\pi_{ij}: \begin{array}{c} V(h_i, d_j, t) \to V(h_i, t) \\ (t:x:y:u:v) \mapsto (t:x:y) \end{array}$$
Notice that $V(f^T, d^T, t)$ is contained in the hyperplane at infinity $V(t)$, and so does $V(h_i, d_j, t)$. These mappings are regular mappings of projective varieties and $\pi_{ij}(V(h_i, d_j, t)) = V(h_i, t)$. Each fibre of these mappings is irreducible, and $V(h_i, t)$ is also irreducible. The fibres $\pi_{ij}^{-1}(\omega)$ have dimension 1 since the points on these fibres have fixed x, y, t coordinates and their v coordinate is related to u by d_j, and is therefore also fixed. Thus, all the fibres have the same dimension. Then, we can apply Theorem 1, and conclude that the $V(h_i, d_j, t)$ are irreducible, as in the proof of Lemma 3. We will prove

the first direction. Let us suppose $f_x^T + \iota f_y^T \in \langle h_i \rangle$ for some $i \in \{1, 2, \ldots, s\}$. We have to show that $n^T = 0$ on $V(h_i, d_1, t)$. Since $(f_x^T) + \iota(f_y^T) \in \langle h_i \rangle$, there exists a polynomial g such that $(f_x^T) + \iota(f_y^T) = g \cdot h_i$. But, $h_i = 0$ on $V(h_i, d_1, t)$. Thus, $(f_x^T) + \iota(f_y^T) = 0$, i.e. $f_y^T = \iota f_x^T$. Then, replacing in n^T, we get $n^T = -\iota f_x^T(u-x) + f_x^T(v-y) = (-\iota(u-x) + (v-y))f_x^T$ on $V(h_i, d_1, t)$. Since $d_1 = -\iota(u-x) + (v-y) = 0$ on $V(h_i, d_1, t)$, then $n^T = 0$ on $V(h_i, d_1, t)$. We can prove in the same way that if $(f_x^T) - \iota(f_y^T) \in \langle h_i \rangle$ then $n^T = 0$ on $V(h_i, d_2, t)$.
Reciprocally, let us suppose $n^T = 0$ on $V(h_i, d_1, t)$ for some $i \in \{1, 2, \ldots, s\}$. Since none of n^T, h_i, d_1 depend on t, $n^T = 0$ on $V(h_i, d_1)$. Since $d_1 = 0$ on $V(h_i, d_1)$, $\begin{cases} n^T = -f_y^T(u-x) + \iota f_x^T(u-x) = 0 \\ n^T = \iota f_y^T(v-y) + f_x^T(v-y) = 0 \end{cases}$ on $V(h_i, d_1)$. Since $(-f_y^T + \iota f_x^T) = \iota(\iota f_y^T + f_x^T)$, we can rewrite the last system of equations as $\begin{cases} (-f_y^T + \iota f_x^T)(u-x) = 0 \\ (-f_y^T + \iota f_x^T)(v-y) = 0 \end{cases}$ on $V(h_i, d_1)$. The subvariety $V(h_i, d_1, u-x, v-y) = V(h_i, u-x, v-y)$ is a one-dimensional variety in the two-dimensional variety $V(h_i, d_1)$. The set $V(h_i, d_1) \setminus V(h_i, d_1, u-x, v-y)$ is a dense open subset of $V(h_i, d_1)$. From the last system, we know that $(-f_y^T + \iota f_x^T)$ vanishes on this dense open subset. Now, we consider $V(h_i, d_1, -f_y^T + \iota f_x^T)$. This is a closed projective set. We know that it contains the open set $V(h_i, d_1) \setminus V(h_i, d_1, u-x, v-y)$. Thus, it contains also its Zariski closure i.e. $V(h_i, d_1)$. Thus, $(-f_y^T + \iota f_x^T)$ vanishes on $V(h_i, d_1)$. Therefore, there exists $a, b \in k[x, y, u, v]$ such that $-f_y^T + \iota f_x^T = ah_i + bd_1 = ah_i + b(-\iota(u-x) + v-y)$. Let a_{uv} be the sum of all the terms of a containing the variables u or v. Since $(-f_y^T + \iota f_x^T)$ does not depend on any of the variables u and v, so $a_{uv}h_i = b(\iota u - v)$. Since the left hand side of this equality lies in $\langle h_i \rangle$, so does the right-hand side, hence $b \in \langle h_i \rangle$. Thus, $-f_y^T + \iota f_x^T \in \langle h_i \rangle$. Thus, $f_x^T + \iota f_y^T \in \langle h_i \rangle$. In the same way, we can prove that if $n^T = 0$ on $V(h_i, d_2, t) \subset \mathbb{P}^4$ then $f_x^T - \iota f_y^T \in \langle h_i \rangle$. □

We are going to analyse the dimension of the one-dimensional component of W_∞. Let $l = \begin{cases} 1 & \text{if } f_x^T + \iota f_y^T \in \langle h_i \rangle \text{ or } f_x^T - \iota f_y^T \in \langle h_i \rangle \text{ for some } i \in \{1, 2, \ldots, s\} \\ 0 & \text{otherwise} \end{cases}$. We will now introduce hypersurfaces, necessary in the proofs of the next lemma and of Theorem 3. If X is a projective variety in $\mathbb{P}^N$ and $F \neq 0$ is a *form* on X (i.e. a real-valued homogeneous polynomial function on X), then we denote by X_F the sub-variety of X, known as a *hypersurface*, defined by $F = 0$.

Theorem 2. *[9–Theorem 4, p. 57] If a form F is not identically 0 on an irreducible projective variety X, then $\dim X_F = \dim X - 1$.*

In order to apply this theorem on the projective variety W_∞ whose irreducibility is not known, we have determined if none of its irreducible components is contained in an irreducible component of $V(F)$. Indeed, if an irreducible component $V(h_i)$ is contained in an irreducible component $V(F_i)$ of $V(F)$, then F_i vanishes on $V(h_i)$, and $\dim V(h_i) \bigcap V(F) = \dim V(h_i)$.

Lemma 5. *The dimension of $W_\infty = V(f^T, n^T, d^T, t) \subset \mathbb{P}^4$ is equal to l.*

Proof. We start in $\mathbb{P}^4$, which has dimension 4. By Lemmas 2, 3 and repeated application of Theorem 2 (from $V(f^T)$ to $V(f^T, d^T)$, and from $V(f^T, d^T)$ to $V(f^T, d^T, t)$) we get that $V(f^T, d^T, t)$ has dimension $4 - 3 = 1$. Thus, the dimension of W_∞ is 0 or 1. By Theorem 2, the dimension of W_∞ is 0 if, and only if, $l = 0$ by Lemma 4. □

The following two lemmas give the degrees of two one-dimensional components, when these exist. The one-dimensional component at infinity and the one-dimensional component of W_S. They will allow us to conclude with the degree of the offset in Theorem 3. For this theorem, but also more generally, we recall the notion of degree. The *degree* of a n-dimensional projective variety $X \subset \mathbb{P}^N$ is the maximum number of points of intersection of X with a projective linear subspace $\mathbb{P}^{N-n}$ in general position with respect to X (see page 234 in [9]). Thus, the degree of a projective variety is the degree of its maximal dimensional component. We recall the definition of localisation at a prime ideal $\mathfrak{P} \subset k[t_1, \ldots, t_m]$, where k is a field. We denote by $k[x_1, \ldots, x_m]_\mathfrak{P}$ the set of all rational functions f/g such that $f, g \in k[x_1, \ldots, x_m]$ where $g \notin \mathfrak{P}$.

Definition 3. *(adapted from [7–Def.A.8.16,p.480]) Let $F, G \in k[z, x, y]$ be homogeneous polynomials, let $p = (p_0 : p_1 : p_2) \in V(F) \cap V(G) \subset \mathbb{P}^2$, and let $\mathcal{M}_p = \langle p_0 x - p_1 z, p_0 y - p_2 z \rangle$ be the homogeneous ideal of p. Assume that $p_0 \neq 0$ then, the intersection multiplicity of $V(F)$ and of $V(G)$ at p is $\mu_p(F, G) := \dim_k(k[z, x, y]_{\mathcal{M}_p}/\langle F, G \rangle)$.*

Let C_∞ be the component at infinity of the projective closure of C, i.e. $C_\infty = V(f^T, t) \subset k^2$. Let C_S be the affine subvariety of C composed of all its singular points, i.e. $C_S = V(f_x, f_y, f) \subset k^2$.

Lemma 6. *The one-dimensional component of W_∞ has degree $2l \sum_{p \in C_\infty} \mu_{p'}(V(f^T), V(n^T))$, where p' is an arbitrary point of W_∞ whose projection on the projective (t, x, y)-plane is p.*

Proof. The variety's degree is given by that of the component with maximum dimension. By lemma 5, $l = 0$ iff $\dim W_\infty = 0$ and we say that the degree of the (non-existant) 1-dimensional component is 0. In the rest of the proof, we study the degree of the 1-dimensional component, assuming it exists, ie., for $l = 1$. Note that this component may not be connected. There exist $i \in \{1, \ldots, s\}, j \in \{1, 2\}$ in the above notation, such that $n^T = 0$ on $V(h_i, d_j, t)$, by Lemma 4. Hence, adding n^T to $\langle d^T, f^T, t \rangle$ does not change its radical. Thus, the points of $W_\infty = V(f^T, d^T, n^T, t)$ and $V(\sqrt{\langle d^T, f^T, t \rangle})$ are the same, but their multiplicities may differ. The one-dimensional component of $V(\sqrt{\langle d^T, f^T, t \rangle})$ is constituted of all the points p' whose projection on the projective (t, x, y)-plane is a point p of $C_\infty \subset \mathbb{P}^2$ and whose projection on the affine (u, v)-plane is the circle $V(d^T)$. The degree of the one-dimensional component of $V(\sqrt{\langle d^T, f^T, t \rangle})$ equals the product of the degree of d^T (which is 2) by the number of isolated points in C_∞, counted with multiplicities. The reason is that each point of C_∞ generates a one-dimensional component of W_∞ with the x, y coordinates of the point of C_∞, and where the t, u, v coordinates satisfy d^T. The degree of the one-dimensional component of W_∞ is twice the sum of the intersecting multiplicities of $V(f^T, n^T) \subset \mathbb{P}^4$ at the points p' for all the points p of C_∞. □

Lemma 7. *The degree of the one-dimensional component of W_S, defined at the beginning of this section, is:* $2\sum_{q \in C_S}(\mu_{q'}(V(\overline{f}), V(\overline{n})) - 1)$, *where q' is an arbitrary point of W_S whose projection on the affine (x, y)-plane is q. This component does not exist iff the degree vanishes.*

Proof. Each point q of C_S induces a trivial equation n. Thus, at the level of W_S, each point q of multiplicity m induces a one-dimensional variety that consists of all the points q' whose projection on the affine (x, y)-plane is q and whose projection on the projective (t, u, v)-plane is a projective circle centred at q and of radius R with multiplicity $m-1$ (the extraneous component may be simple). The only component at infinity of W_S are the points $(0 : x_0 : y_0 : 1 : y_0 \pm \iota(1 - x_0))$, where (x_0, y_0) is a common root of f^T, $f_x^T(x,y)$, and $f_y^T(x,y)$. It follows that W_S does not have a one-dimensional component at infinity. The points of W_S are the same as the points of $V(\sqrt{\langle \overline{f_x}, \overline{f_y}, \overline{f}, \overline{d} \rangle})$, but their multiplicities differ. The degree of the one-dimensional component of $V(\sqrt{\langle \overline{f_x}, \overline{f_y}, \overline{f}, \overline{d} \rangle})$ equals the product of the degree of $\overline{d}$ (which is 2) by the number of isolated points in C_S. The degree of the one-dimensional component of W_S is twice the sum of the multiplicities of $V(\overline{f}, \overline{n}) \subset \mathbb{P}^4$ at the points q' minus 1 for all the points q of C_S. □

Theorem 3. *Let p' and q' be defined as in Lemmas 6 and 7. The degree of the generalised R-offset $\mathcal{O}$ to an algebraic curve $V(f) \subset k^2$ of degree m, such as f is square free is:*

$$2m^2 - 2l \sum_{p \in C_\infty} \mu_{p'}(V(f^T), V(n^T)) - 2 \sum_{q \in C_S}(\mu_{q'}(V(\overline{f}), V(\overline{n})) - 1). \quad (1)$$

Proof. Bézout's Theorem [9] states that, for projective varieties, the degree of the intersection of two varieties is generically equal to the product of the varieties' degrees. Therefore, the degree of W is generically $2m^2$. The quasi-projective varieties W, W_a, W_∞, and W_S are related by $W_a = W \setminus (W_\infty \cup W_S)$. Since W is defined by three polynomials, its dimension is at least 1 according to Theorem 2. Since n has two more variables than f and the same degree, n can not identically vanish on $V(\overline{f})$. Hence, $\dim V(\overline{f}, \overline{n}) \leq 2$, by Theorem 2. Since d has coefficients that are polynomials in R, $R \neq 0$ and the coefficients of f and d do not depend on R, d can not identically vanish on $\overline{B} \cap \overline{N}$. Thus, the dimension of W is exactly one by the same theorem. Since W is one-dimensional, the degree of W is the sum of the degrees of its one-dimensional components. Thus, the degree of W_a equals the degree of W minus the degree of the one-dimensional component of $W_\infty \cup W_S$. Since W_S has no one-dimensional component at infinity (see proof of Lemma 7), the one-dimensional components of W_∞ and of W_S are disjoint, and the degree of the one-dimensional component of $W_\infty \cup W_S$ is the sum of the degrees of the one-dimensional components of W_∞ and of W_S. By Lemmas 6 and 7, the degree of W_a is given by Equation 1. By definition, t never vanishes on W_a. Finally, $\mathcal{O}$ is the image of W_a by the canonical projection $\pi : \mathbb{P}^4 \setminus \{(1:0:0:0:0)\} \to k^2 : (t:x:y:u:v) \mapsto (\frac{u}{t}, \frac{v}{t})$ that is a one-to-one mapping. Indeed, there is generically only one point of the curve corresponding to a point of the generalised offset. Thus, the result. □

By applying the results from this section, we get [1] the following degrees for conic offsets:

Conic	ellipse/hyperbola	parabola	circle	two lines
Offset degree	8	6	4	4

4 The Generalised Offset to a Conic

We will first review the main concepts and results about sparse resultants (see also [10, 3, 6]) that will be used hereafter. The resultant of $n+1$ polynomials $f_1, \ldots, f_{n+1}$ in $n+k$ affine variables is a polynomial in k variables that characterises the solvability of the system of polynomial equations $f_1(x_0, \ldots, x_{n+k}) = \cdots = f_{n+1}(x_0, \ldots, x_{n+k}) = 0$. Thus, it allows the elimination of n variables, and is therefore also called eliminant. The theory of sparse elimination has been widely covered in [10]. The degree of the classical projective resultant in the coefficients of each polynomial is the Bézout number of the other n polynomials (i.e., the product of their total degrees). Sparse resultants generalise the projective resultant and exploit the monomial structure of the polynomials expressed by the Newton polytope. The degree of the sparse resultant per polynomial, that is a sum of mixed volumes, is in general lower than that of the projective resultant.

The support A_i of a polynomial $f_i \in k[x_1^{\pm 1}, \ldots, x_n^{\pm 1}]$ is the set of exponent vectors in $\mathbb{Z}^n$ corresponding to non-zero coefficients, i.e. $f_i = \sum_{a \in A_i} c_a x^a, c_a \neq 0$. The Newton polytope Q_i of f_i in $\mathbb{R}^n$ is the convex hull of A_i, i.e. the smallest convex set containing A_i. For example, for the strophoid of equation $y^2 - x^2 - x^3 = 0$, the exponent vectors are: $(0,2)$, $(2,0)$, and $(3,0)$. A polytope $Q \subset \mathbb{R}^n$ has an n-dimensional volume, which will be denoted hereafter $Vol(Q)$. The Minkowski sum $A+B$ of point sets A and B in $\mathbb{R}^n$ is the point set $A+B = \{a+b | a \in A, b \in B\}$ (see example on Figure 2). Let n polytopes $Q_1, \ldots, Q_n \subset \mathbb{R}^n$ whose vertices belong to $\mathbb{Z}^n$. A polyhedral subdivision of a point set S is a collection of polyhedra whose union equals S, such that each intersection of two polyhedra of the same dimension is another polyhedron in the subdivision of lower dimension. Let $Q = Q_1 + \cdots + Q_m \subset \mathbb{R}^n$ be a Minkowski sum of polytopes, and assume that Q has dimension n. Then a subdivision $R_1, \ldots, R_s$ of Q is a mixed subdivision [4–Definition 6.5, page 344] if each cell R_i can be written as a Minkowski sum $R_i = F_1 + \cdots + F_m$ where each F_i is a face of Q_i and $n = dim(F_1) + \cdots + dim(F_m)$ (see example on Figure 2). The mixed volume is the sum of the volumes of the mixed cells [4–Theorem 6.7] (see example on

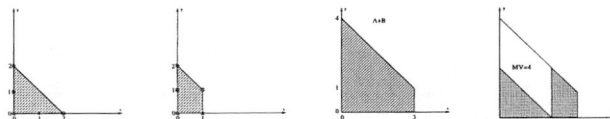

Fig. 2. The Minkowski sum (third diagram) of the point sets A (first) and B (second) and its mixed subdivision (filled-in cells are copies of Newton polytopes and mixed cells are unfilled)

Figure 2). If we replace a polytope Q_i by a polytope Q'_i included in it, the mixed volume of the Q_i does not increase.

A conic C can be defined implicitly as the variety defined by a second degree formal polynomial: $f(x,y) = \alpha x^2 + \beta xy + \gamma y^2 + \delta x + \epsilon y + \zeta = 0$. We shall compute an implicit equation of its generalised offset. The partial derivatives are $f_x = 2\alpha x + \beta y + \delta$ and $f_y = \beta x + 2\gamma y + \epsilon$. An equation of the normal $N = V(n) \subset k^4$ to the original conic at the point (x,y) is: $n(x,y,u,v) = -(\beta x + 2\gamma y + \epsilon)(u-x) + (2\alpha x + \beta y + \delta)(v-y) = 0$. If a conic is not degenerate (proper conic different from the union of two lines), then it has no singular points, and $S = \emptyset$. The generalised offset to a conic is the Zariski closure of the projection of the affine variety $V(f,n,d) \setminus V(f_x, f_y)$ onto the (u,v) plane. This is a subvariety of the Zariski closure of the projection of $V(f,n,d)$. Thus the equation of the generalised offset is a factor of the equation of the Zariski closure of the projection of $V(f,n,d)$, which is the resultant of f, n, d expressing the elimination of x, y. The objective in the computations is to simplify the polynomials, while producing a system of equations equivalent to the original one. In the case of the sparse resultant computation, this has been achieved by replacing one polynomial by a linear combination of polynomials, and having a Newton polytope inscribed in the Newton polytope of the original polynomial. The sparse resultants have been computed thanks to the sparse resultant software developed by Emiris [6, 5]. We computed the determinant of this matrix and its factorisation. The degree of the offset to conics allowed us to identify the factor that corresponds to an implicit equation of the generalised offset. In the case where α and β are different from 0, we call the conic "generic". Hereafter, all conics are generic.

The polynomial n can be rewritten in the following way: $n(x,y,u,v) = \beta x^2 - \beta y^2 + 2(\gamma - \alpha)xy + (-\beta u + \epsilon + 2\alpha v)x + (\beta v - \delta - 2\gamma u)y + (-\epsilon u + \delta v) = 0$. The monomial in x^2 can be eliminated if we replace $n(x,y,u,v)$ by $\alpha n(x,y,u,v) - \beta f(x,y)$ (see Figure 3). Similarly, if we replace $d(x,y,u,v)$ by $f(x,y) - \alpha d(x,y,u,v)$, the monomial in x^2 disappears (see Figure 3). The mixed volumes are $MV(f, \alpha n - \beta f) = MV(f, f - \alpha d) = 4$ (see Figure 4), and $MV(\alpha n - \beta f, f - \alpha d) = 3$. We get an equation for the generalised offset as the sparse resultant of $f, \alpha n - \beta f$ and $f - \alpha d$. Hereafter, all the implicit equations are publicly available at: http://pages.cpsc.ucalgary.ca/~antonf.

However, it is possible to further simplify the equation of a non-degenerate conic by using a coordinate system with origin at one of the foci (in the case of an ellipse or hyperbola) or the apex (in the case of a parabola), and one

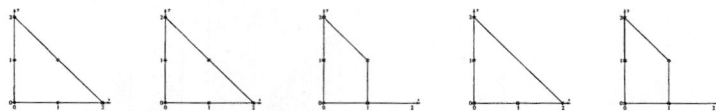

Fig. 3. The Newton polytopes of f, n, $\alpha n - \beta f$, d, and of $f - \alpha d$

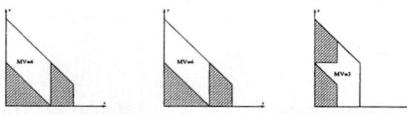

Fig. 4. The mixed volumes of f and $\alpha n - \beta f$, of $\alpha n - \beta f$ and $f - \alpha d$, and of f and $f - \alpha d$

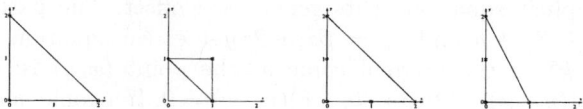

Fig. 5. The Newton polytopes of f, n, d and of $ad - f$ for ellipses and hyperbolas

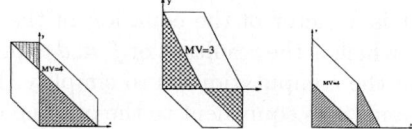

Fig. 6. A mixed subdivision and mixed volume computation of f, n, of $n, f - d$, and of $f, f - d$ for ellipses and hyperbolas

of the axes being the axis of the conic. By a simple change of coordinates, we can obtain easily the general equation of the generalised offset. Then, an ellipse or hyperbola simplifies to $\frac{x^2}{a^2} \pm \frac{y^2}{b^2} - 1 = 0$, assuming $ab \neq 0$. We can get an equivalent equation $x^2 + cy^2 + e = 0$, where $e \pm b^2 c = 0$. Let $f := x^2 + cy^2 + e$. It is easy to see from Figure 5 that no linear combination of f and d has terms in common with n. If we replace $d(x, y, u, v) = (u - x)^2 + (v - y)^2 - R^2 = 0$ by $f(x, y) - d(x, y, u, v)$, the monomial in x^2 disappears (see Figure 5). The mixed volumes are $MV(f, n) = MV(f, f - d) = 4$ and $MV(n, f - d) = 3$ (see Figure 6). We get an equation for the generalised offset as the sparse resultant of $f, n, f - d$ in the variables u and v. Examples are shown in Figure 7.

In the case of a parabola, the equation of the conic in a coordinate system with origin at the summit of the parabola, and one of the axes being the axis of the parabola, simplifies to $y^2 - 2px = 0$. The polynomial defining $N = V(n) \subset k^4$ can be rewritten in the following way: $n(x, y, u, v) = -2yu + 2xy - 2pv + 2py = 0$. It is easy to see from Figures 8 that no linear combination of f and d has terms

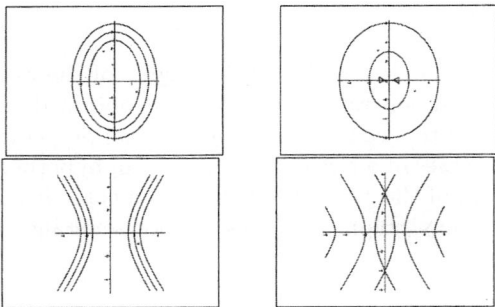

Fig. 7. Up: an ellipse with its 0.5−generalised offset (left) and its 3−generalised offset (right). Down: an hyperbola with its 0.5−generalised offset (left) and its 3−generalised offset (right)

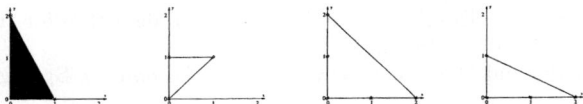

Fig. 8. The Newton Polytopes of f, n, d, and of $f - d$ for parabolas

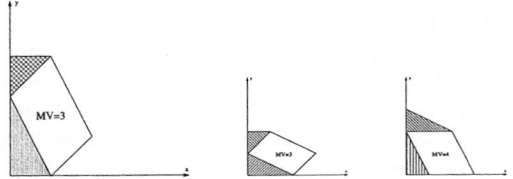

Fig. 9. The mixed volumes of f, n, of $n, f - d$ and of $f, f - d$ for parabolas

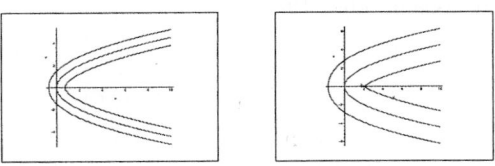

Fig. 10. A parabola and its 0.5−generalised offset. The same parabola and its 3−generalised offset

in common with n. If we replace $d(x, y, u, v) = (u - x)^2 + (v - y)^2 - R^2 = 0$ by $f(x, y) - d(x, y, u, v)$, the monomial in y^2 disappears (see Figure 8). The mixed volumes are $MV(f, n) = MV(n, f - d) = 3$ and $MV(f, f - d) = 4$ (see Figure 9). We get an equation for the generalised offset as the sparse resultant of $f, n, f - d$ in the variables u and v. Examples are shown in Figure 10.

5 Conclusions

We have obtained a general formula for the degree of the offset to an algebraic curve defined in its most general setting. We have obtained an implicit equation of the generalised offset to a conic defined by a formal polynomial, and simplified equations in the two cases of a circle, an ellipse or an hyperbola, and a parabola. This implicit equation of the generalised offset to a conic has been used in order to compute the Delaunay graph for conics and for semi-algebraic sets (see [1]).

References

1. F. Anton. *Voronoi diagrams of semi-algebraic sets*. PhD thesis, The University of British Columbia, Vancouver, British Columbia, Canada, January 2004.
2. E. Arrondo, J. Sendra, and J. R. Sendra. Genus formula for generalized offset curves. *J. Pure Appl. Algebra*, 136(3):199–209, 1999.
3. J. F. Canny and I. Z. Emiris. A subdivision-based algorithm for the sparse resultant. *J. ACM*, 47(3):417–451, 2000.
4. D. Cox, J. Little, and D. O'Shea. *Using algebraic geometry*. Springer-Verlag, New York, 1998.
5. I. Emiris. A general solver based on sparse resultants: Numerical issues and kinematic applications. Technical Report 3110, SAFIR, 1997.
6. I. Z. Emiris and J. F. Canny. Efficient incremental algorithms for the sparse resultant and the mixed volume. *J. Symbolic Comput.*, 20(2):117–149, 1995.
7. G.-M. Greuel and G. Pfister. *A **Singular** introduction to commutative algebra*. Springer-Verlag, Berlin, 2002.
8. C. M. Hoffmann and P. J. Vermeer. Eliminating extraneous solutions in curve and surface operations. *Internat. J. Comput. Geom. Appl.*, 1(1):47–66, 1991.
9. I. R. Shafarevich. *Basic algebraic geometry. 1*. Springer-Verlag, Berlin, second edition, 1994. Varieties in projective space, Translated from the 1988 Russian edition and with notes by Miles Reid.
10. B. Sturmfels. Sparse elimination theory. In *Computational algebraic geometry and commutative algebra (Cortona, 1991)*, Sympos. Math., XXXIV, pages 264–298. Cambr. Univ. Press, Cambridge, 1993.

Computing the Least Median of Squares Estimator in Time $O(n^d)$

Thorsten Bernholt[*]

Lehrstuhl Informatik 2, Universität Dortmund, Germany

Abstract. In modern statistics, the robust estimation of parameters of a regression hyperplane is a central problem, i.e., an estimation that is not or only slightly affected by outliers in the data. In this paper we will consider the least median of squares (LMS) estimator. For n points in d dimensions we describe a randomized algorithm for LMS running in $O(n^d)$ time and $O(n)$ space, for d fixed, and in time $O(d^3 \cdot (2n)^d)$ and $O(dn)$ space, for arbitrary d.

1 Introduction

A general problem in statistics is the characterization of a set of points $\mathcal{P}$ by a straight line. One well-known method is the ordinary least squares regression line, which is the line that minimizes the sum of the squared vertical point-line distances. The parameters of such a regression line are computed by calculating some sums, see e.g. [1]. Suppose a single point is moved towards infinity. As the sums are taken over all points, this point will have a massive impact on the regression line, so that, for example, one single measurement error could result in a totally wrong regression line. This leads to the definition of the breakdown point. Donoho and Huber [6] define: "The *breakdown point* is, roughly, the smallest amount of contamination that may cause an estimator to take on arbitrarily large aberrant values".

To cope with this problem the basic idea is to ignore a fraction of the points and base the regression line on the remaining "good" points. To decide which are the "good" points, each subset $S \subseteq \mathcal{P}$ is evaluated by a function $f : \mathbb{P}(\mathcal{P}) \longrightarrow \mathbb{R}^+$ and the subset with the best value is taken. $\mathbb{P}(\mathcal{P})$ denotes the set of all subsets of $\mathcal{P}$.

Let $\mathcal{P} = \{P_1, \ldots, P_n\}$ be a set of points with $P_i = (p_{i,1}, \ldots, p_{i,d})$ and $p_{i,j} \in \mathbb{R}$. For a given hyperplane L with the parameters $a_1, \ldots, a_d$ that is defined by

$$y = a_1 x_1 + \cdots + a_{d-1} x_{d-1} + a_d$$

let $r_i(L) = p_{i,d} - (a_1 p_{i,1} + \cdots + a_{d-1} p_{i,d-1} + a_d)$ be the residual of the point P_i with respect to the hyperplane L. A residual measures the vertical point-hyperplane distance. Let π be the permutation such that the sequence of the

---

[*] The financial support of the Deutsche Forschungsgemeinschaft (SFB 475, "Reduction of complexity in multivariate data structures") is gratefully acknowledged.

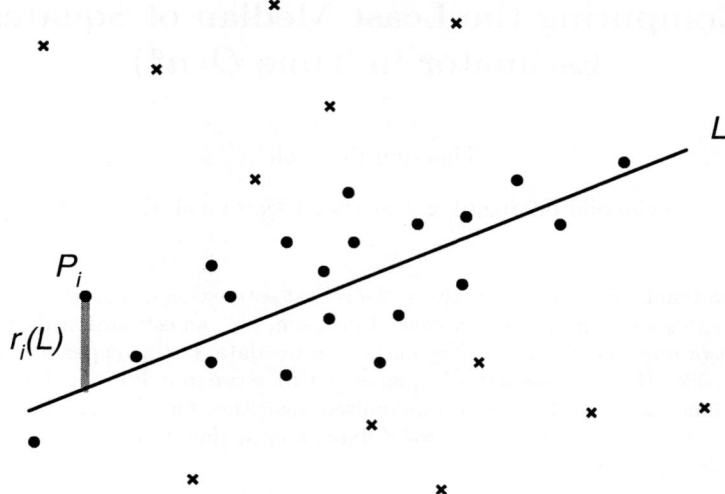

Fig. 1. The statistical point of view (Definition 1) is displayed. The residual $r_i(L)$ is the vertical distance from the hyperplane L to the point P_i. The points marked with ● have a residue less than r_i, the point marked with ✘ have a larger residual. The solution to LMS is defined as the hyperplane such that the squared value of the h-th residual is minimized

absolute values of all residues $|r_{\pi(1)}(L)|, \ldots, |r_{\pi(n)}(L)|$ is sorted. Hence, $r_{\pi(i)}(L)$ will denote the i-th residual in this order. Rousseeuw and Leroy [20] define the least median of squares estimators (LMS) (LMedS is also used as abbreviation in some papers) as follows:

Definition 1 (statistical point of view). *Given a set $\mathcal{P} = \{P_1, \ldots, P_n\}$ of n points, $P_i \in \mathbb{R}^d$ and a natural number h, with $\lceil n/2 \rceil \leq h \leq n$, find a hyperplane L, such that $r_{\pi(h)}(L)^2$ is minimized.*

The definition is illustrated in Figure 1. The highest breakdown point of 50% is achieved for $h = \lceil n/2 \rceil + \lceil (d+1)/2 \rceil$. One can choose $h = c \cdot n$ for a constant c with $0.5 \leq c \leq 1$, depending on the application and how many outliers are expected. This problem is also known as the least quantile of squares (LQS) estimator [21].

For a moment, choose $r = |r_{\pi(h)}(L)|$. Now, consider the two hyperplanes

$$L^+ : y = a_1 x_1 + \cdots + a_{d-1} x_{d-1} + a_d + r \tag{1}$$

$$L^- : y = a_1 x_1 + \cdots + a_{d-1} x_{d-1} + a_d - r \ . \tag{2}$$

The two parallel hyperplanes L^+ and L^- forms a *hyperstrip*. All points with a residual smaller than or equal to r are inside this hyperstrip. That means, replacing the "=" with a "≤" in Equation (1) and a "≥" in Equation (2), these

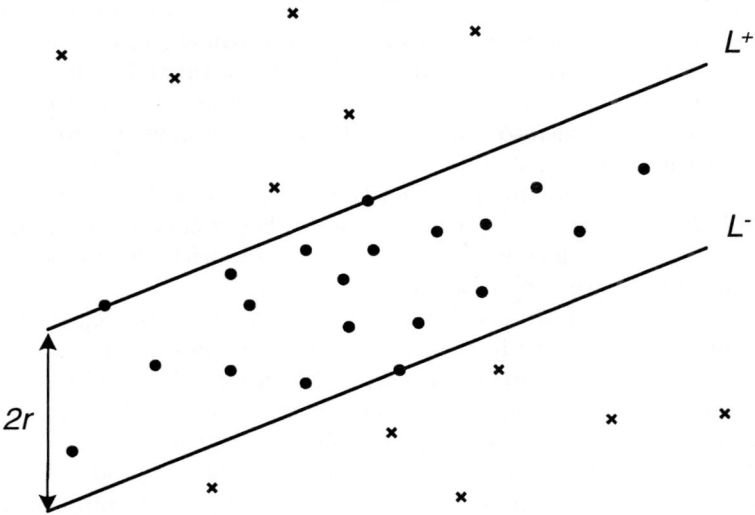

Fig. 2. The computational point of view (Definition 2) is displayed. The hyperstrip defined by L^+ and L^- divides the points in inlying points ● and outlying points ✗. The half width of the hyperstrip is r. The solution for LMS is the hyperstrip with the smallest r containing h points

points will fulfill these inequations. The *width* of this hyperstrip is $2r$. Denote by $subset(L^+, L^-) \subseteq \mathcal{P}$ the set of points that are inside the hyperstrip (L^+, L^-). We now rephrase the definition slightly, whereby both definitions describe the same problem:

Definition 2 (computational point of view). *Let a set $\mathcal{P} = \{P_1, \ldots, P_n\}$ of n points, $P_i \in \mathbb{R}^d$ and a natural number h, with $\lceil n/2 \rceil \leq h \leq n$, be given. Now find a hyperstrip (L^+, L^-) respectively its parameter $a_1, \ldots, a_d, r$, with $r \geq 0$, such that $|subset(L^+, L^-)| = h$ and the half width r of the hyperstrip is minimized.*

The definition is illustrated in Figure 2.

There is a wide interest in this topic. Stromberg [21] gives an exact algorithm for LMS running in time $O(n^{d+2} \log n)$. Erickson et al. [10] describe an algorithm for LMS with running time $O(n^d \log n)$. We describe a randomized algorithm with a running time $O(n^d)$ for LMS for d fixed, as commonly used in the literature. All mentioned algorithms deal with intersection of hyperplanes and therefore have to solve systems of linear equations. A linear system can be solved in time $O(d^3)$. For arbitrary dimension d, the algorithm, described here, has a runtime of $O(d^3 \cdot (2n)^d)$ and needs space $O(dn)$.

For two dimensions, Edelsbrunner and Souvaine [9] describe an algorithm running in time $O(n^2)$ using the topological sweep-line technique [19]. Mount et al. [15] use branch-and-bound to compute LMS in the plane and simulations show a running time of $O(n \log n)$. This algorithm in addition computes approximate solutions. There are two ways to approximate LMS: The first way is to find a solution with fewer than h points inside (quantile approximation) and the second way is to find a solution with a wider hyperstrip (width approximation). For fixed dimension d, Olson [16] describes a width approximation algorithm with a ratio of 2 running in time $O(n^{d-1} \log n)$. Mount et al. [14] present a quantile approximation algorithm with a ratio $1 - \epsilon$ and a running time of $O(n \log n + (1/\epsilon)^{O(d)})$, also for fixed d. For h close to n, an algorithm of Chan [3] is useful, which uses linear programming with violations and solves LMS in time $O(n \log(n - h) + (n - h)^2 \log^2(n - h))$.

The complexity of LMS is analyzed by Chien and Steiger [4]. They proved a lower bound of $\Omega(n \log n)$ in the model of algebraic decision trees. Gajentaan and Overmars [11] introduce the concept of 3-sum-hardness. Given n integer numbers, the 3-sum problem is to decide whether three distinct numbers sum up to zero. Besides reductions to other problems, they proved that, if one can solve LMS in $o(n^2)$, then the 3-sum problem, and others, can also be solved in $o(n^2)$. Erickson et al. [10] proof that if the affine degeneracy problem requires $\Theta(n^d)$ time, then the computation of width-LMS requires $\Omega(n^{d-1})$ time and of the exact LMS requires $\Omega(n^d)$ time, matching the running time of the algorithm presented in this paper, for d fixed.

The estimator is widley used, e. g., Plets and Vynckier [18] use the LMS-estimator to analyse 3-dimensional astronomical data. In [17] the estimator is used for mosaicing underwater images, moreover in [13] images of a mpeg stream are used to compose a panorama view.

In Section 2 we describe the main procedure of the algorithm. It uses procedures from Sections 3, 4 and 5.

2 Main Procedure

In this section, we give a short overview, how the algorithm works. For $d = 2$, we use the algorithm of Edelsbrunner and Souvaine [9] for LMS in the plane working in time $O(n^2)$ and space $O(n)$. For $d \geq 3$, at first, the d-dimensional points are mapped to $2n$ hyperplanes in a $d+1$-dimensional space. The details are given in Section 3. In this space, a solution is represented by a point. An optimal solution is a point L' in $\mathbb{R}^{d+1}$ with the following three properties (Lemma 2):

1. There are $n + h$ hyperplanes below or intersecting the point L'.
2. The last coordinate (w-axis) of the point L' is minimal over all points with Property 1.
3. The point L' is an intersection of at least $d + 1$ hyperplanes.

For two selected hyperplanes a *subproblem* consists of all points contained in the subspace described by these two hyperplanes. Therefore, we obtain $\binom{n}{2}$ subproblems $A_1, \ldots, A_{\binom{n}{2}}$. It is essential that the subproblems are permuted, i. e., they are in a random order. The optimal solution r^* of the first subproblem A_1

is computed using procedure SolveLMS, which has a runtime of $O(d^3 \cdot (2n)^{d-1})$. The details of the procedure SolveLMS are given in Section 4.

Consider that we have already computed the optimal solution of $A_1, \ldots, A_{i-1}$. For the succeeding subproblem A_i it is decided in time $O(d^3 \cdot (2n)^{d-2})$ whether in this subproblem a better solution than r^* exists. This is done by the application of the procedure DecideLMS, the details are given in Section 5. If a better solution exists, it is computed using SolveLMS, r^* is updated, and the algorithm continues with the next subproblem A_{i+1}, until all $\binom{n}{2}$ subproblems are processed. The time for all calls of DecideLMS are bounded by $O(d^3 \cdot (2n)^d)$.

To get the runtime for the calls of SolveLMS, we name the event, that r^* is improved, Ψ. As the subproblems are in a random order, the event Ψ only occurs $\log \binom{n}{2} = O(\log(n))$ times in the average case. This is a well-known result from [5] and also, e.g., discussed by Chan [2]. Therefore, in the average case the time for all calls of SolveLMS is bounded by $O(d^3 \cdot (2n)^{d-1} \cdot \log n)$.

For d=3, we have to reduce the number of calls to DecideLMS. Therefore, two levels of calls are needed. The first level processes subproblems defined by a single hyperplane and each call to DecideLMS can be computed in time $O(n^2)$. The second level remains as described above and it is called only $O(\log n)$ times in the average case. Hence, the number of calls of DecideLMS is reduced to $O(n \log n)$, and we get a runtime of $O(n^3)$. This proves the following theorem:

Theorem 1. *Given n points in $d \geq 2$ dimensional space, d not fixed, the LMS estimator can be computed in expected time $O(d^3 \cdot (2n)^d)$ and space $O(dn)$.*

A result from Karp [12] shows, that the probability, that the event Ψ occurs more than $2n$ times, is bounded by $\left(\frac{1}{2}\right)^n$. But, due to the lack of space, this is not explained in detail here.

3 Transformation of the Input

According to the point-hyperplane duality, we map a point $P_i = (p_{i,1}, \ldots, p_{i,d})$ from primal space to the hyperplane H_i defined by

$$v = p_{i,1} u_1 + \cdots + p_{i,d-1} u_{d-1} + p_{i,d}$$

In a second step we map from dual space to extended space with the axes $u_1, \ldots, u_{d-1}, v, w$ and map the hyperplane H_i to the two hyperplanes

$$H_i^+ : w = +p_{i,1} u_1 + \cdots + p_{i,d-1} u_{d-1} + v - p_{i,d} \qquad (3)$$
$$H_i^- : w = -p_{i,1} u_1 - \cdots - p_{i,d-1} u_{d-1} - v + p_{i,d} \qquad (4)$$

In the extended space, we say that the point $(q_1, \ldots, q_d, q_{d+1})$ is located *above* the hyperplane H if there exists a constant $c > 0$ such that the point $(q_1, \ldots, q_d, q_{d+1} - c)$ is located on the hyperplane H. The terms *below* is defined analogously.

What happens to a solution during the mapping? In the primal space, a solution consists of two parallel hyperplanes with h points between them. This is mapped to a vertical segment with h hyperplanes crossing it. Finally, the vertical segment is mapped to a point in $(d+1)$-dimensional space with $n+h$ hyperplanes below it. More precisely:

Lemma 1. *Let $L = (L^+, L^-)$ be a solution with value $r \geq 0$ in the primal space with*

$$L^+ : y = a_1 x_1 + \cdots + a_{d-1} x_{d-1} + a_d + r$$

and

$$L^- : y = a_1 x_1 + \cdots + a_{d-1} x_{d-1} + a_d - r \ .$$

Let L' be the solution in the extended space defined by the point $(a_1, \ldots, a_d, r)$. Now the point P_i is located between L^+ and L^- if and only if L' is located above or on H_i^+ and above or on H_i^-.

Proof. L' is above or on H_i^+

$\Leftrightarrow \exists c \geq 0 : (a_1, \ldots, a_d, r - c)$ is located on H_i^+
$\Leftrightarrow \exists c \geq 0 : r - c = p_{i,1} a_1 + \cdots + p_{i,d-1} a_{d-1} + a_d - p_{i,d}$
$\Leftrightarrow \exists c \geq 0 : p_{i,d} - c = a_1 p_{i,1} + \cdots + a_{d-1} p_{i,d-1} + a_d - r$
$\Leftrightarrow \exists c \geq 0 : (p_{i,1}, \ldots, p_{i,d-1}, p_{i,d} - c)$ is located on L^-
$\Leftrightarrow P_i$ is above or on L^-

An analogous calculation shows that L' is above or on H_i^- if and only if P_i is below or on L^+. □

An optimal solution is characterised by the following lemma:

Lemma 2. *An optimal solution for LMS in the extended space is a point L' with the following properties:*

1. *There are $n + h$ hyperplanes below or intersecting the point L'.*
2. *The last coordinate of the point (w-axis) is minimal over all points with Property 1.*
3. *The point is an intersection of at least $d+1$ hyperplanes.*

Proof.

1. Let the hyperstrip (L^+, L^-) be an optimal solution for LMS. Then there are h points that are located between L^+ and L^-. It follows from Lemma 1 that there are h pairs (H_i^+, H_i^-), such that both hyperplanes are below the point L'. As the w-coordinate of the point L' is $r \geq 0$, it follows from the Equations (3) and (4) that for all $i = 1, \ldots, n$ either a hyperplane H_i^+ or H_i^- is below the point. Therefore, there are $n + h$ hyperplanes below the optimal solution L'.
2. As stated in Definition 2 the solution of LMS is a hyperstrip of minimal width $2r$. As the half width of a hyperstrip and the w-coordinate of the dual point are equal to r, the solution is a point with a minimal w-coordinate with respect to Property 1.

3. If there is a point that is an intersection of less than $d+1$ hyperplanes, we can find a point with a smaller w-coordinate that is an intersection of at least $d+1$ hyperplanes. Thus, a point that is intersected by less than $d+1$ hyperplanes cannot be optimal. □

4 Procedure SolveLMS

The main algorithm has fixed two hyperplanes that define a lower-dimensional subspace and therefore define a subproblem A. To enumerate all solutions contained in the subspace of A, we loop through all subdivisions that are described by fixing $d-3$ hyperplanes in addition. This results in $\binom{2n-4}{d-3} \leq (2n)^{d-3}$ many subdivisions. In [7] an algorithm for enumerating all $(d-1)$-elementary subsets is given with a constant runtime per subset. If the hyperplanes, that define a subdivision, are linear independent, they describe a two-dimensional subspace. Otherwise, they describe a subspace with more than two dimensions, and such a subdivision can be ignored, as it is easy to see, that such a subdivision is covered by other linear independent subdivisions.

Each two-dimensional subspace is processed in the following way: The intersection of such a two-dimensional subspace with the remaining $n' = 2n - (d-1)$ hyperplanes results in a set of n' lines embedded in $\mathbb{R}^{d+1}$. Defining two axes in the two-dimensional subspace results in n' lines in the plane. We can now use the sweep-line algorithm from [8, 19] to enumerate all $\binom{n'}{2}$ intersection points. This algorithm works in time $O(n'^2)$ and space $O(n')$. For each intersection point we check if in the $d+1$-dimensional space exactly $n+h$ hyperplanes are below the point or intersecting it. To make this routine work in time $O(n'^2)$, we track a count for each line. The count for the first intersection point found on a line can be calculated in $O(n')$. The counts of the succeeding intersection points can be calculated by an update step in time $O(1)$, as the sweep-line algorithm reports the points in a topological order. The required space is $O(n')$ overall. The point with $n+h$ hyperplanes below and with the smallest w-coordinate is taken as the new optimum.

The procedure deals with $O((2n)^{d-1})$ intersection points. For each point we have to compute the intersection of up to $d+1$ hyperplanes. Each hyperplane is indeed an equation, therefore we have to solve a system of up to $d+1$ equations over $d+1$ variables. For d not fixed, this task can be performed in time $O(d^3)$, using, for example, Gaussian-elimination or other numerical algorithms. As a single point/hyperplane has $d+1$ entries, $O(dn)$ space is sufficient. Therefore, the procedure SolveLMS finds the optimal solution of a subproblem A in time $O(d^3 \cdot (2n)^{d-1})$ and space $O(dn)$ for $d \geq 3$.

5 Procedure DecideLMS

Given a subproblem A and a value r^*, we want to decide whether there exists a solution for LMS with a value better than r^* in A. Recall that the subproblem A is a $(d-1)$-dimensional subspace embedded in $\mathbb{R}^{d+1}$. This only holds if the two defining hyperplanes are linearly independent. Otherwise, the subproblem can be ignored, as discussed in Section 4.

We know from Lemma 2 that the w-axis measures the value of a solution. Therefore, we intersect this subspace with the hyperplane $H_{r*} : w = r^*$, resulting in a $(d-2)$-dimensional subspace. All points in this subspace represent solutions having the same value r^*, the question is whether there is a point with $n + h$ hyperplanes below it.

Lemma 3, which is presented below, shows that it is sufficient to enumerate all $\binom{2n-4}{d-2} \leq (2n)^{d-2}$ intersection points of the remaining hyperplanes with the subspace $A \cap H_{r*}$ and count the number of hyperplanes below. To enumerate all intersection points we use the same technique as in Section 4. Therefore, Procedure DecideLMS computes the decision, whether a better solution exists, in time $O(d^3 \cdot (2n)^{d-2})$ for $d \geq 4$. For calls to this procedure on the second level ($d = 3$) only a runtime of $O(n \log n)$ can be achieved since the considered subspace is one-dimensional and we have to sort the intersection points.

Lemma 3. *If the subproblem A contains a solution for LMS with a value smaller than r^*, then the intersection of A and H_{r*} contains an intersection point L' such that at least $n + h$ hyperplanes are below L'.*

Proof. In general, the set of $d + 1$ hyperplanes intersecting an optimal solution either contains two H^+- or contains two H^--hyperplanes. W.l.o.g. we focus on the case that the subproblem A is an intersection of the hyperplanes H_i^+ and H_j^+, as displayed in Figure 3. Let $L'' = (a_1, \ldots, a_{d-1}, v, \lambda)$ be the optimal solution in

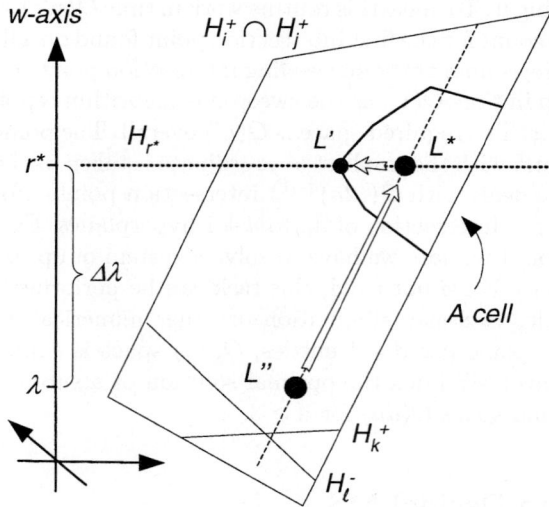

Fig. 3. The point L'' is moved towards L^*, the hyperplanes H_k^+ and H_ℓ^- remain below. The point is further moved towards L' in the corner of a cell, now intersecting $d + 1$ hyperplanes

A. As the optimal solution L'' is located in A, it intersects the two hyperplanes and we get from Equation (3)

$$\lambda = p_{i,1}a_1 + \cdots + p_{i,d-1}a_{d-1} + v - p_{i,d}$$
$$\lambda = p_{j,1}a_1 + \cdots + p_{j,d-1}a_{d-1} + v - p_{j,d} \ .$$

For clarity, we combine most of the terms into Q_i and Q_j and we get

$$\lambda = Q_i + v$$
$$\lambda = Q_j + v \ .$$

If we increase λ by $\Delta\lambda > 0$, we can reason that the point $L^* = (a_1, \ldots, a_{d-1}, v + \Delta\lambda, \lambda + \Delta\lambda)$ still intersects the two hyperplanes. We choose $\Delta\lambda$ such that $\lambda + \Delta\lambda = r^*$.

Let the hyperplanes H_k^+ and H_ℓ^- be located below L''. Then the point L'' satisfies the following two inequalities:

$$\lambda \geq Q_k + v$$
$$\lambda \geq -Q_\ell - v$$

A simple calculation shows, that L^* also satisfies the inequalities. Therefore, all hyperplanes located below L'' are also located below L^*. Since $\lambda > 0$, hyperplanes intersecting L'' are now below L^*. Well, there maybe some hyperplanes below L^* in addition. Hence, there are at least $n + h$ hyperplanes below L^*. If the point L^* is not an intersection of $d + 1$ hyperplanes, we move the point L^* towards a nearby intersection point L', such that it does not leave the subspace $A \cap H_{r*}$ and such that no hyperplane is crossed. □

6 Conclusions

In this paper, we have described a randomized algorithm for the least median of squares estimator for n points in d dimensions. It runs in time $O\big(d^3 \cdot (2n)^d\big)$ and space $O(dn)$. For d fixed, the runtime matches the lower bound of $\Omega(n^d)$.

Acknowledgement

I would like to thank Ingo Wegener for helpful suggestions concerning the presentation of the results and Claudia Becker, Thomas Fender and Ursula Gather for introducing me to the questions of robust statistics.

References

1. S. Boyd and L. Vandenberghe. *Convex Optimization*. Cambridge University Press, 2004.
2. T. M. Chan. Geometric applications of a randomized optimization technique. *Discrete & Computational Geometry*, 22:547–567, 1999.

3. T. M. Chan. Low-dimensional linear programming with violations. In *IEEE Symposium on Foundations of Computer Science*, pages 570–579, 2002.
4. H. Chien and W. Steiger. Some geometric lower bounds. In *ISAAC: 6th International Symposium on Algorithms and Computation*, pages 72–81, 1995.
5. T. H. Cormen, C. E. Leiserson, R. L. Rivest, and C. Stein. *Introduction to Algorithms (Second Edition)*. MIT Press & McGraw-Hill, 2001.
6. D. L. Donoho and J. Huber. The notion of breakdown point. In P. Bickel, K. Doksum, and J. L. Hodges, editors, *A Festschrift for Erich L. Lehmann*, pages 157–184, 1985.
7. P. Eades and B. McKay. An algorithm for generating subsets of fixed size with a strong minimal change property. *Information Processing Letters*, 19:131–133, 1984.
8. H. Edelsbrunner and L. J. Guibas. Topologically sweeping an arrangement. *Journal Computer and System Sciences*, 38:165–194, 1989, Corrigendum in 42:249–251,1991.
9. H. Edelsbrunner and D. L. Souvaine. Computing least median of squares regression line and guided topological sweep. *Journal of the American Statistical Association*, 85:115–119, 1990.
10. J. Erickson, S. Har-Peled, and D. M. Mount. On the least median square problem. In *ACM Symposium on Computational Geometry*, 2004.
11. A. Gajentaan and M. H. Overmars. On a class of $O(n^2)$ problems in computational geometry. *Computational Geometry*, 5:165–185, 1995.
12. R. M. Karp. Probabilistic recurrence relations. *Journal of the Association of Computer Machinery*, 41(6):1136–1150, 1994.
13. Y. Li, L-Q. Xu, G. Morrison, C. Nightingale, and J. Morphett. Robust panorama from mpeg video. In *IEEE International Conference on Multimedia and Expo 2003 (ICME '03)*, pages 81–84, 2003.
14. D. M. Mount, N. S. Netanyahu, C. D. Piatko, R. Silverman, and A. Y. Wu. Quantile approximation for robust statistical estimation and k-enclosing problems. *International Journal of Computational Geometry and Applications*, 10:593–608, 2000.
15. D. M. Mount, N. S. Netanyahu, K. Romanik, R. Silverman, and A. Y. Wu. A practical approximation algorithm for the LMS line estimator. In *Symposium on Discrete Algorithms*, pages 473–482, 1997.
16. C. F. Olson. An approximation algorithm for least median of squares regression. *Information Processing Letters*, 63(5):237–241, 1997.
17. O. Pizarro and H. Singh. Toward large-area mosaicing for underwater scientific applications. *IEEE Journal of Oceanic Engineering*, 28(4):651–672, 2003.
18. H. Plets and C. Vynckier. An analysis of the incidence of the vega phenomenon among main-sequence and post main-sequence stars. *Astronomy and Astrophysics*, 343:496–506, 1999.
19. E. Rafalin, D. Souvaine, and I. Streinu. Topological sweep in degenerate cases. In *International Workshop on Algorithm Engineering and Experiments*, pages 155–165, 2002.
20. P. J. Rousseeuw and A. M. Leroy. *Robust Regression and Outlier Detection*. Wiley, 1987.
21. A. J. Stromberg. Computing the exact least median of squares estimate and stability diagnostics in multiple linear regression. *SIAM Journal on Scientific Computing*, 14(6):1289–1299, 1993.

Pocket Recognition on a Protein Using Euclidean Voronoi Diagram of Atoms

Deok-Soo Kim[1], Cheol-Hyung Cho[1], Youngsong Cho[2], Chung In Won[2], and Dounguk Kim[2]

[1] Department of Industrial Engineering, Hanyang University,
17 Haengdang-Dong, Seongdong-Ku, Seoul, 133-791, South Korea
dskim@hanyang.ac.kr

[2] Voronoi Diagram Research Center, 17 Haengdang-Dong,
Seongdong-Ku, Seoul, 133-791, South Korea
{murick, ycho, wonci, donguk}@voronoi.hanyang.ac.kr

Abstract. Proteins consist of atoms. Given a protein, the automatic recognition of depressed regions, called pockets, on the surface of the protein is important for protein-ligand docking and facilitates fast development of new drugs. Recently, computational approaches for the recognition of pockets have emerged. Presented in this paper is a geometric method for the pocket recognition based on Voronoi diagram for atoms in Euclidean distance metric.

1 Introduction

Molecules such as protein, DNA, or RNA consist of atoms. Given the atomic structures of molecules, analyzing interactions between molecules is important for understanding their biological functions. An example is the interaction between a protein and a small molecule and this interaction is the basis of designing new drugs.

The study of molecular interactions, such as docking or folding, can be approached as a physicochemical and/or a geometrical point of view [13]. While the physicochemical approach is to evaluate and minimize the free energy between two molecules using, for example, the area of molecular surfaces, the geometric approach is to determine whether two molecules have geometrically meaningful features for interaction.

Interaction between a protein, called a *receptor*, and a small molecule, called a *ligand*, is usually done via some depressed regions on the surface of the receptor. Note that these depressed regions are usually called *pockets*. Since the recognition of pockets on proteins to find appropriate ligands from chemical data bases requires a huge amount of computation, a manual decision making does not make much sense. Hence, automatically recognizing pockets on proteins is important for protein-ligan docking for the development of new drugs [10]. While the efforts on the physicochemical approach have been done since the early days of science, the efforts to understand the geometry of biological systems have started very recently [3, 11, 12, 14].

In this paper, we provide the definition of pockets on the surface of a protein in the geometric point of view and present an algorithm to automatically recognize pockets. Given a protein, the proposed algorithm first computes a *Euclidean Voronoi diagram* of atoms. Then, we define a *pocket primitive* for each face of the convex hull of the protein. After pocket primitives are extracted, we evaluate the validity of boundaries between neighboring pocket primitives to test if two neighbors should be merged into a single pocket or not. Eventually, therefore, there will be a few pockets left on the surface of a receptor where each pocket corresponds to an appropriately depressed region.

2 Geometric Models of Protein and Related Terminologies

A protein, consisting of atoms, can be viewed as shown in Fig. 1. The spheres are van der Waals surfaces of atoms in the protein. In the model, there are two kinds of surfaces associated with the protein: *Solvent Accessible Surface(SAS)* and *Molecular Surface(MS)*. SAS consists of points on the space where the center of probe is located when the probe is in contact with the protein. The inner-most possible trajectories of points on the probe surface, then, define MS. MS, then, consists of two parts: *contact surfaces(CS)* and *reentrant surfaces(RS)* as shown in Fig. 1.

SAS usually defines a *free-space* that a small molecule can move around without interfering the protein and therefore plays the fundamental role for folding and/or docking [11]. On the other hand, MS, often called by another name *Connolly surface* after the name of the first researcher defined the surface, conveniently defines the boundary between interior and exterior volume of a protein so that the volume or the density of protein can be calculated [2].

Illustrated in Fig. 2 are two molecules interacting each other. The molecule labelled A is a receptor and the small molecule labelled B is a ligand. Two proteins interact each other as the protruded region in B has been geometrically inserted into the depressed region, which is called a *pocket* of A. Since pockets play important roles for the protein functions, the extraction of such pockets is inevitable.

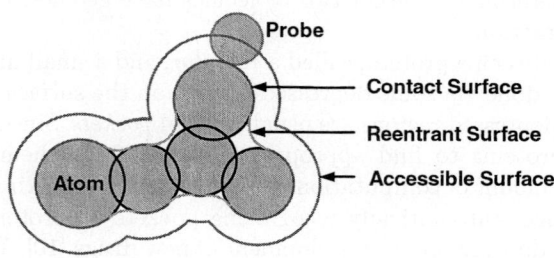

Fig. 1. Geometric model of a protein

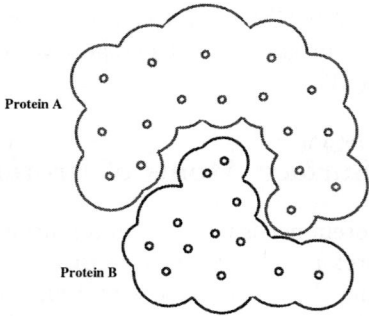

Fig. 2. A docking configuration between a receptor and a ligand

Let $A = \{a_1, a_2, \ldots, a_n\}$ be a protein consisting of a number of atoms $a_i = (c_i, r_i)$ where $c_i = (x_i, y_i, z_i)$ and r_i define the center coordinates and the radius of an atom, respectively. In addition, suppose that $S = \{s_1, s_2, \ldots, s_m\}$ be a small molecule which also consists of a number of atoms s_j, defined similarly to a_i, and S will be docking with A. Note that $m << n$ in general. Usually a small molecule is approximated by a spherical probe $R = (c_R, r_R)$ and most investigations on the geometric properties for the protein analysis are done using the probe. Note that R is defined as a minimum sized sphere enclosing all atoms of S to reflect an arbitrary orientation and locations of the small molecule without an interference against the protein A.

Let π_j be a pocket where $\pi_j = \{a_{j1}, a_{j2}, \ldots, a_{jk}\}$ and these atoms altogether define a depressed region on the boundary B of A. The boundary $B = \{b_1, b_2, \ldots, b_l\}$ is the subset of atoms of protein where some points on the van der Waals surface contribute to the molecular surface MS. In other words, B is a set of atoms $a_i \in A$ which is touched by the small molecule S without interfering the protein A. Hence, $\pi_j \subseteq B \subseteq A$ and the set $\Pi = \{\pi_1, \pi_2, \ldots, \pi_p\}$ is the set of all possible pockets on B.

3 Topology for Whole Protein

To effectively and efficiently answer to most geometric questions for a given protein, it is inevitable to have a convenient tool to represent the spatial structure of protein. In our research, we use Euclidean Voronoi diagram of atoms where the Euclidean distance metric is defined from the surfaces, instead of centers, of atoms.

While the ordinary Voronoi diagram of points and Voronoi diagram of spheres in a power metric have been studied quite extensively and efficient computational codes are available, its counterpart for Euclidean Voronoi diagram of spheres has not been studied as much as it has to be done. In many applications for proteins, the ordinary and power metric Voronoi diagram can be only approximations of what is actually needed. It is only very recently that the fast and robust construction of Voronoi diagram for circles and spheres with different radii became practical [6, 7, 9]. Once the Euclidean Voronoi diagram for spheres is available, many studies in geometrical perspective of a protein can be done quite efficiently [1, 4, 5, 8, 16].

4 Topology for Surface Atoms of Protein

Extracting pockets of protein needs to query information on the surface behavior of the protein, and therefore a convenient representation of the connectivity among atoms contributing the surface of the protein is essential. Even after the surface atoms are identified, recognizing pockets is still not an easy task at all in the computational point of view.

In the effort of transforming a computationally intractable problem of extracting true pockets into a tractable one, we define a simpler geometric structure called a *mesh* M on the surface of a protein A, where the topology and geometry of the mesh have the following implications. Let $M = \{V, E, F\}$ be a mesh defined on the surface of protein, where $V = \{v_1, v_2, \ldots\}$, $E = \{e_1, e_2, \ldots\}$ and $F = \{f_1, f_2, \ldots\}$ are sets of vertices, edges, and faces on M, respectively.

To define a mesh M on the surface of a protein A, we introduce a geometric operation called *blending* on among atoms. Blending surface over a protein A consists of two kinds of blends as shown in Fig. 3: *rolling blends* and *link blends*. A rolling blend γ is defined by rolling the probe R between two atoms, and a link blend λ is defined among three neighboring atoms by placing R on the top of the atoms.

Suppose that we apply a blending operation on A with a probe R and two atoms a_i and a_j are in the close neighborhood to produce a rolling blend γ_{ij}. If we define an edge e_{ij} between the centers c_i and c_j as the vertices for all such pairs of atoms, then we get the definitions of V and E for the *mesh* M on A. Note that $v_i \in V$ is identical to the center c_i of an atom $a_i \in A$. Even though V and E partially fill the required data for M, the missing data F can be easily computed as follows. Suppose that we define a face f_{ijk} among the centers c_i, c_j, and c_k of a triplet of atoms a_i, a_j and a_k which define a link blend. Then,

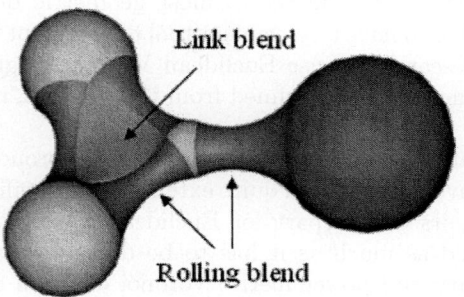

Fig. 3. Rolling blend and link blend

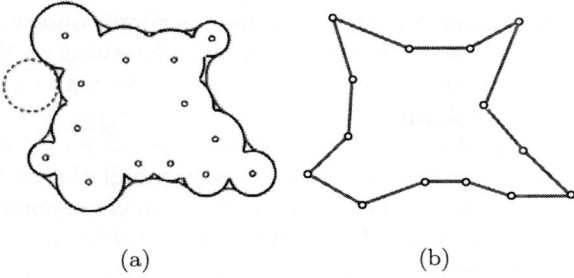

Fig. 4. An example of topology for surface atoms of a protein in 2D: (a) boundary of protein and (b) corresponding mesh

F of M is also obtained. Fig. 4 shows a 2D example illustrating a protein, the blending of the protein, and its mesh counterpart.

5 Extraction of Pocket Primitives

In this paper, let R_S be a probe for a small molecule S from which we want to define a pocket on the given protein A and let R_∞ be a hypothetical probe with an infinite radius. Let M_S and M_∞ be the mesh models defined by blending the protein A for probes with radii of R_S and R_∞, respectively. Then, M_∞ in fact corresponds to a mesh model bounded by faces defined by the centers of atoms with infinite Voronoi regions. Since $R_S << R_\infty$, $M_S \subseteq M_\infty$. The inclusion relationship in the above describes the volumetric relationship between two, not the inclusion between the data entities. In the data structure perspective, a reverse relationship holds.

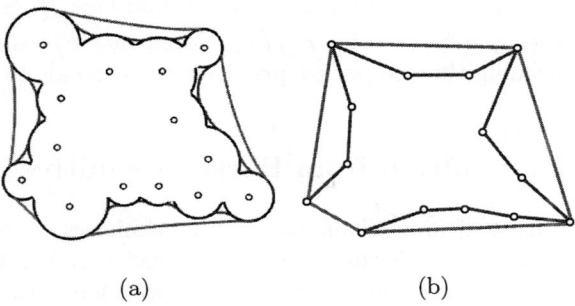

Fig. 5. Inner and outer mesh: (a) different protein boundaries for two probes and (b) corresponding meshes

Since $M_S \subseteq M_\infty$, let M_S and M_∞ be denoted by M_I and M_O meaning the inner and outer meshes, where $M_I = \{V_I, E_I, F_I\}$ and $M_O = \{V_O, E_O, F_O\}$.

Fig. 5 shows inner and outer meshes for a protein. From the figure, which is a 2D analogy for our problem in 3D, we can make a simple observation as follows: For each edge of convex hull, there is zero or one depression on the boundary of the protein. When an edge on the convex hull coincides with one of inner mesh, obviously no pocket is defined.

Even though the problem in 3D is not as simple as its 2D counterpart, we can make a similar observation. For a face of convex hull of protein, there can be a corresponding depressed region. A depressed region corresponding to a convex hull face, however, may or may not have a clearly defined boundary since a pocket may be related to a few convex hull faces. In such a case, the depressed region extracted from a convex hull face cannot be defined to form a complete pocket but a pocket has two or more convex hull faces which altogether define a single pocket. Hence, we first introduce *pocket primitive* as a depressed region on M_I corresponding to each face of convex hull.

A face $f_O^i \in F_O$ has three associated vertices $v_O^{i_1}$, $v_O^{i_2}$ and $v_O^{i_3}$. Since $M_O \subseteq M_I$ in the data structure perspective, there are always three vertices $v_I^{i_1}$, $v_I^{i_2}$ and $v_I^{i_3}$ which coincide $v_O^{i_1}$, $v_O^{i_2}$ and $v_O^{i_3}$, respectively. Let $\phi_{(i_1,\, i_2)}$ be the shortest path between $v_I^{i_1}$ and $v_I^{i_2}$ on the inner mesh M_I. The path from a vertex follows an incident edge and the distance between two neighboring vertices is defined as the sum of edge lengths connecting two vertices. Hence, the distance between two arbitrary vertices is the summation of edge lengths connecting two vertices.

The geometric meaning of the shortest path between two vertices is as follows: Since two vertices are on M_O as well, the other vertices on the shortest path define depressions on M_I from the corresponding face of M_O. Hence, the shortest path defines the most upward wall separating two relatively deep depressions. $\phi_{(i_2,\, i_3)}$ and $\phi_{(i_3,\, i_1)}$ can be similarly defined. Then, the face set $\tilde{F}_I^i$ consisting of $f_I^h \in F_I$, where f_I^h denotes a face of M_I and is interior to the three shortest paths $\phi_{(i_1,\, i_2)}$, $\phi_{(i_2,\, i_3)}$ and $\phi_{(i_3,\, i_1)}$. Note that $\tilde{F}_I^i$ forms a topologically triangular shaped depression on M_I from the boundary of M_∞. This depression is called a pocket primitive φ_i corresponding to a convex hull face $f_O^i \in F_O$ and is also represented another graph $\varphi_i = \{\tilde{V}_I^i, \tilde{E}_I^i, \tilde{F}_I^i\}$. Noted that $\tilde{F}_I^i$ can be a null set in the worst-case meaning that no pocket primitive corresponds to the face.

6   Pocket Recognition from Pocket Primitives

A pocket may consist of more than one pocket primitives. Hence, we check if two neighboring pocket primitives should be merged together to form a more meaningful depression based on an appropriate criterion. Let a *ridge* be the edge chain corresponding to the shortest path between two extreme vertices of a pocket primitive. Hence, a ridge plays the role of boundary between two incident pocket primitives. Let a *mountains* be the edge chain separating two pockets. If a ridge is sufficiently high, it can be regarded as a mountains. Note that a pocket primitive always has 3 ridges and a pocket is surrounded by 3 or more mountains. Therefore, the boundary of a pocket primitive may or may not be the boundary of pocket depending on conditions.

Suppose that a path, which is in deed a ridge, ϕ^k exists between φ^i and φ^j for an edge e_O^k of M_O. Note that there exists always a shortest path for an edge of M_O. Then, we can define a certain measure to determine the discrepancy between two chains e_O^k and ϕ^k. Depending on the measure and its prescribed threshold value, two pocket primitives sharing the chains may or may not be

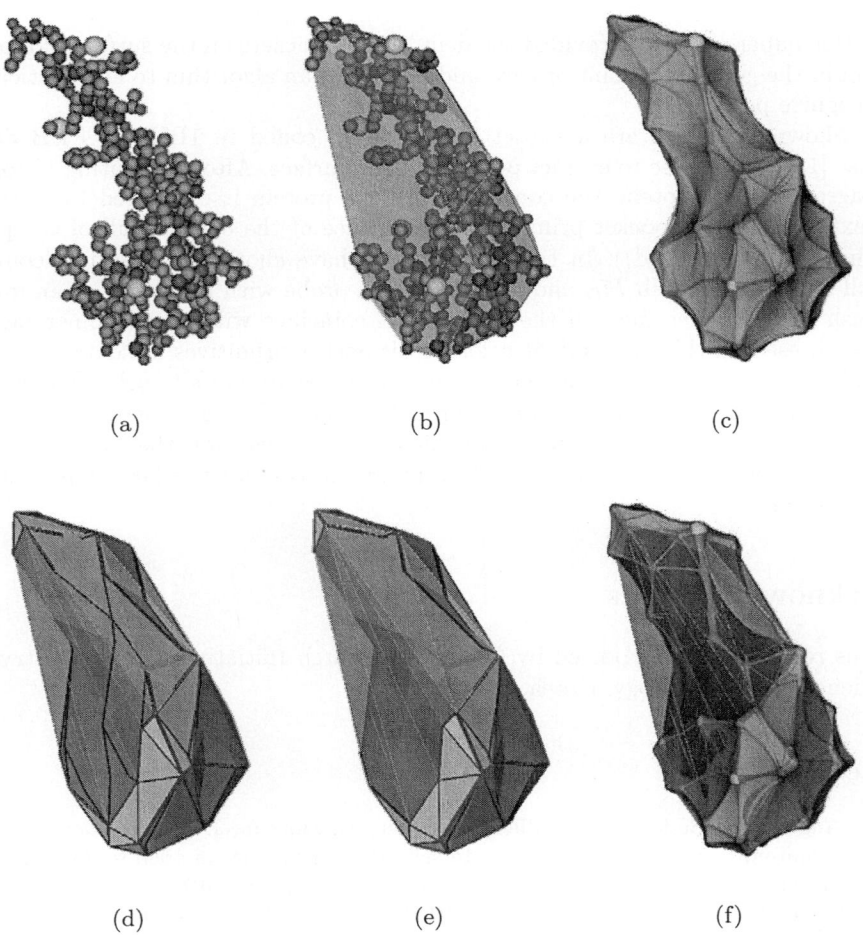

Fig. 6. Group A of the protein 1BH8 [15] downloaded from PDB data base and its use for the pocket extraction: (a) visualization of the 1BH8 protein, (b) the convex hull, (c) the molecular surface, (d) the mesh and pocket primitives, (e) the boundary of pocket after merges, and (f) the largest pocket on the molecular surface

merged. Even though there can be several ways to define such a measure, we use the concept of average distance between two chains.

Let δ_k be the average distance between e_O^k and ϕ^k. If δ_k is larger than a prescribed value, we merge two neighboring pocket primitives sharing the chains. Otherwise, we regard ϕ^k as a mountain chain. As such a threshold value, in this paper, we have chosen the average Euclidean distances between all edges in all ridges and their corresponding edges in outer mesh E_O. After all, atoms for vertices in merged pocket primitives define the pocket π_k.

7 Discussions and Conclusions

In this paper, we have provided the definition of pockets on the surface of a protein in the geometric point of view and presented an algorithm to automatically recognize pockets.

Shown in Fig. 6 are a subset of a protein, coded in 1BH8 in PDB data base [15], and its use to extract pockets on the surface. After computing Voronoi diagram of the protein, the convex hull of the protein is computed(Fig. 6(b)). Next, we define a pocket primitive for each face of the convex hull of the protein(Fig. 6(b) and (d)). In this example, we have chosen the mesh of convex hull as an outer mesh M_O and the mesh for a probe with radius 8Å as an inner mesh M_I. When a face on the outer mesh coincides with one of inner mesh, we represented the face as light color. After pocket primitives are extracted, we evaluate the qualities, the average depths, of ridges around all pocket primitives and merge the appropriate pocket primitive pairs. Fig. 6(e) shows the mesh structure on M_I of merged pockets while Fig. 6(f) illustrates the corresponding visualization using a molecular surface representation for the largest pocket in dark color.

Acknowledgments

This research was supported by Creative Research Initiatives from Ministry of Science and Technology, Korea.

References

1. Angelov, B., Sadoc, J.-F., Jullien, R., Soyer, A., Mornon, J.-P., and Chomilier, J.: Nonatomic solvent-driven Voronoi tessellation of proteins: an open tool to analyze protein folds. Proteins: Structure, Function, and Genetics. **49** (2002) 446–456.
2. Connolly, M.L.: Solvent-accessible surfaces of proteins and nucleic acids. Science. **221** (1983) 709–713.
3. Edelsbrunner, H., Facello, M., and Liang, J.: On the definition and the construction of pockets in macromolecules. Discrete Applied Mathematics. **88** (1998) 83–102.
4. Gerstein, M., Tsai, J., and Levitt, M.: The volume of atoms on the protein surface: calculated from simulation, using Voronoi polyhedra. Journal of Molecular Biology. **249** (1995) 955–966.

5. Goede, A., Preissner, R., and Frömmel, C.: Voronoi cell: new method for allocation of space among atoms: elimination of avoid- able errors in calculation of atomic volume and density. Journal of Computational Chemistry. **18** (1997) 1113–1123.
6. Kim, D.-S., Kim, D., and Sugihara, K.: Voronoi diagram of a circle set from Voronoi diagram of a point set: I. Topology. Computer Aided Geometric Design. **18** (2001) 541–562.
7. Kim, D.-S., Kim, D., and Sugihara, K.: Voronoi diagram of a circle set from Voronoi diagram of a point set: II. Geometry, Computer Aided Geometric Design. **18** (2001) 563–585.
8. Kim, D,-S., Cho, Y., Kim, D., Kim, S., Bhak, J.: Euclidean Voronoi Diagram of 3D Spheres and Applications to Protein Structure Analysis. International Symposium on Voronoi Diagrams in Science and Engineering, University of Tokyo, Tokyo, Japan. (2004) 13–15.
9. Kim, D.-S., Cho, Y., and Kim, D.: Edge-tracing algorithm for Euclidean Voronoi diagram of 3D spheres. Proc. 16th Canadian Conference on Computational Geometry. (2004) 176–179.
10. Kunts, I.D.: Structure-based strategies for drug design and discovery. Science. **257** (1992) 1078–1082.
11. Lee, B. and Richards, F.M.: The interpretation of protein structures: estimation of static accessibility. Journal of Molecular Biology. **55** (1971) 379–400.
12. Liang, J., Edelsbrunner, H., and Woodward, C.: Anatomy of protein pockets and cavities:Measurement of binding site geometry and implications for ligand design. Protein Science. **7** (1998) 1884–1897.
13. Parsons, D. and Canny, J.: Geometric problems in molecular biology and robotics. 2nd International Conference on Intelligent Systems for Molecular Biology, Palo Alto, CA. (1994) 322–330.
14. Peters, K. P., Fauck, J., and Frömmel, C.: The Automatic Search for Ligand Binding Sites in Protein of Know Three-dimensional Strucutre Using only Geometric Criteria. Journal of Molecular Biology. **256** (1996) 201–213.
15. RCSB Protein Data Bank. http://www.rcsb.org/pdb/, 2004.
16. Zimmer, R., Wöhler, M., and Thiele, R.: New scoring schemes for protein fold recognition based on Voronoi contacts. Bioinformatics. **14** (1998) 295–308.

Region Expansion by Flipping Edges for Euclidean Voronoi Diagrams of 3D Spheres Based on a Radial Data Structure

Donguk Kim[1], Youngsong Cho[1], and Deok-Soo Kim[2]

[1] Voronoi Diagram Research Center, Hanyang University,
17 Haengdang-dong, Seongdong-gu,
Seoul 133-791, Korea
{donguk, ycho}@voronoi.hanyang.ac.kr
[2] Department of Industrial Engineering, Hanyang University,
17 Haengdang-dong,Seongdong-gu,
Seoul 133-791, Korea
dskim@hanyang.ac.kr

Abstract. Voronoi diagrams have been known to have numerous applications in various fields in science and engineering. While the Voronoi diagram for points has been extensively studied in two and higher dimensions, the Voronoi diagram for spheres in three or higher dimensions has not been studied sufficiently. In this paper, we propose an algorithm to construct Euclidean Voronoi diagrams for spheres in 3D. Starting from the ordinary Voronoi diagram for the centers of spheres, the proposed *region expansion algorithm* constructs the desired diagram by expanding Voronoi regions for one sphere after another via a series of topology operations. Adopted data structure for the proposed algorithm is a variation of radial data structure. While the worst-case time complexity is $O(n^3 \log n)$ for the whole diagram, its expected time complexity can be much lower.

1 Introduction

Voronoi diagram has been known for its capabilities to handle various applications in science and engineering including computational geometry. The ordinary Voronoi diagram for point set and its construction have been studied extensively and the properties are well-known in 2 and higher dimensions [20]. However, the construction of the Voronoi diagram for spheres in the Euclidean distance metric, often called an *additively weighted Voronoi diagram* in the computational geometry community, has not been explored sufficiently even though it has significant potential impacts on diverse applications in both science and engineering [1, 8, 18, 21, 23].

Examples of the Euclidean Vornoi diagram of spheres are shown in Fig. 1. Fig. 1(a) shows the a Voronoi region, which corresponds to the large sphere in the middle, of the Euclidean Voronoi diagram of 15 spheres with three different sizes. In this example, unbounded Voronoi edges and faces are not rendered for the convenience of visualization. Fig. 1(b) shows the computed Voronoi diagram of 64 atoms, which is a subset of protein data downloaded from PDB(Protein Data Bank) [27], consisting of an α-helix.

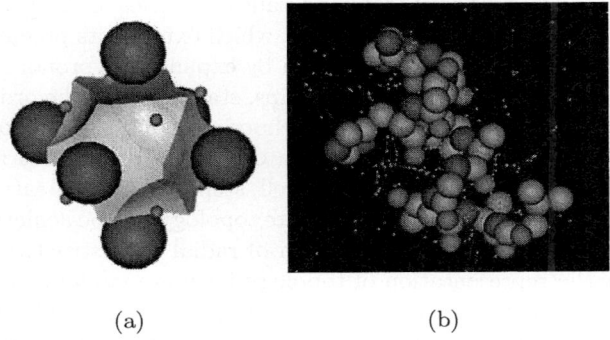

(a)　　　　　　　　　　(b)

Fig. 1. Examples of the Voronoi diagram for spheres: (a) 15 spheres, (b) 64 spheres

Note that the structural analysis of protein or RNA, which is the emerging field of research, requires an efficient computational tool to analyze spatial structures among atoms [8, 12, 21]. In the design of new material and the analysis of its properties, a similar analysis is fundamental as well [16, 19, 22]. However, due to the lack of appropriate algorithms and stable running codes for Euclidean Voronoi diagrams for spheres, most applications have instead adapted an ordinary Voronoi diagram for points, a power diagram, or an α-hull.

Unlike other diagrams, few reports are available on this problem. Aurenhammer discussed the transformation of the computation of the Euclidean Voronoi diagram for spheres in d-dimension to that of $(d+1)$-dimensional power diagram obtained from the convex hull in $(d+2)$-dimension [2]. Will wrote a Ph.D. thesis dedicated to the computation of Voronoi regions in the Euclidean Voronoi diagram for spheres [25] in 3D. In his decent work, Will showed that the Voronoi region of a sphere has a $\Theta(n^2)$ combinatorial complexity and proposed the *lower envelope algorithm* which takes an $O(n^2 \log n)$ expected time for a single Voronoi region, where n is the number of spheres. Will, by implementing the proposals by both himself and Aurenhammer, also provided experimental results on various data sets. Gavrilova, in her Ph.D. thesis, reported several important properties of Euclidean Voronoi diagram for spheres in arbitrary dimensions, including shapes of the Voronoi regions, nearest neighbors and empty-sphere properties [6, 7]. On the other hand, Luchnikov et al. proposed a practical idea of tracing edges which is simple yet powerful to obtain the desired diagram [17]. Recently, Kim et al. reported on the full implementation of the edge-tracing algorithm for constructing the whole Voronoi diagram, not a single region, with discussions on various applications including the analysis of protein structures [12, 13]. They showed that the whole Voronoi diagram can be constructed in $O(n^3)$ time in the worst-case. Boissonnat and Karavelas reported an algorithm to compute a Voronoi region using the convex hull of spheres transformed by inversion [3]. They showed that a Voronoi region can be constructed in $O(n^2)$ time in the worst-case which is dominated by the construction of the convex hull of n spheres in 3D.

In this paper, we present another algorithm to compute Euclidean Voronoi diagrams for 3D spheres with several important properties of the Voronoi diagram. The proposed *region expansion algorithm*, which extends its precursor in 2D [10, 11], constructs the whole desired diagram by expanding Voronoi regions for one sphere after another via a series of edge-flips, starting from the ordinary Voronoi diagram for the centers of spheres. After choosing a point generator and the corresponding Voronoi region, the algorithm continuously expands the point to a sphere, and the corresponding region as well, step by step. Repeating the process generator by generator, the correct ultimate topology can be achieved. The whole process is performed based on a variation of radial data structure [24] which is often used for the representation of topology for non-manifold solid models.

In this paper, we assume that the generators are in general position so that no five generators are cotangent to a vertex sphere and no four generators define an edge. For the convenience of notation, we will ignore the term *Voronoi* and call them simply a *vertex*, *edge*, or *face* unless it is necessary.

2   Overview of the Region Expansion

Let $S = \{s_1, s_2, \ldots, s_n\}$ be a set of spheres $s_i = (c_i, r_i)$, where $c_i = (x_i, y_i, z_i)$ and r_i denote the center and the radius of s_i, respectively. We assume that no sphere is completely contained inside another even though intersections are allowed between spheres. Associated with each s_i, there is a corresponding *Voronoi region* VR(s_i) for s_i, where VR(s_i) = $\{p \mid d(p, c_i) - r_i \leq d(p, c_j) - r_j, i \neq j\}$ and $d(\cdot)$ denotes L_2 Euclidean distance between two points. Then, EVD(S)= {VR(s_1), VR(s_2),..., VR(s_n)} is called a *Euclidean Voronoi diagram for S*. We call s_i a *generator* of VR(s_i).

Suppose that we choose a generator, which will expand to its full size starting from a point at its center, at a particular time. Then, the generator and the corresponding Voronoi region being expanded are called an *expanding generator* and an *expanding region*, respectively. Voronoi vertices on the boundary of an expanding region are called *on-vertices* and the others are called *off-vertices*. The Voronoi edges on the expanding region are called *on-edges* E_{on}, and the edges which has no on-vertex are called *off-edges* E_{off}. Then, the other edges are called *radiating-edges* and categorized into two groups: i) edges with an on-vertex and an off-vertex, and ii) edges with on-vertices at both ends. Similarly, faces are also grouped into three categories: *on-faces*, *off-faces*, and *radiating-faces*. A *vertex sphere* is the sphere simultaneously tangent to the generators defining a Voronoi vertex, and therefore its center is identical to the vertex.

Suppose that we are given with an ordinary Voronoi diagram for the centers of spheres. Then, each sphere is associated with a polyhedral Voronoi region which is an incorrect Voronoi region for the sphere itself. Given a set of spheres, we view each sphere grows, or *expands*, to its full size starting from a point initially shrunken to a point at its center. While the expanding sphere grows, the corresponding region expands simultaneously. If we can keep the topology, and the geometry as well if necessary, among vertices, edges, faces, and regions for the intermediate Voronoi diagram correctly and consistently, the complete Voronoi diagram can be computed by repeating the process generator by generator to the last. We call this process the *region expansion*.

It is obvious that growing the size of an expanding generator always increases the volume of the corresponding region. It can be easily shown that each on-vertex, during the region expansion, moves away from the initial expanding region by following the radiating-edge associated with the vertex. Similarly, each on-edge moves away from the initial expanding region by following the corresponding radiating-face. Note that a sufficiently small growth of expanding generator leaves the combinatorial structure of the diagram unchanged but causes changes only in the geometries of vertices, edges, and faces related to the region boundary.

However, certain changes may occur in topological structure at some point of time in the expanding process. For example, suppose that an on-vertex moves along a radiating-edge to meet a corresponding off-vertex on the edge. Then, the radiating-edge shrinks and degenerates to a point and disappears afterwards. We call an *event* for such a situation causing changes in the combinatorial structure.

Since Voronoi regions are always star-shaped, intersections between faces never occur interior to the face, but at the boundaries of faces. Hence, it is sufficient to consider only vertices and edges to detect new topological changes in the expansion process. Furthermore, it is only necessary to consider edges on radiating-faces since on-vertices and on-edges are always constrained to move along radiating-edges and radiating-faces, respectively. Since an event denotes a change in topology due to moving on-vertices or on-edges, the next event occurs at edges on the radiating-faces during the region expansion.

Note, however, that every edge except on-edges can be associated with an event if the size of expanding generator is sufficiently large. Since generator spheres have prescribed sizes, a subset of the events can be only realized while the others cannot. Note also that on-vertices and on-edges do not have any states associated.

3 Events in the Region Expansion

To devise the region expansion algorithm based on events, it is necessary to know the types of events, how to detect them, and how to handle necessary topological changes when an event is encountered.

Events can be classified in different points of view. Based on the conditions of a target edge, events can be classified as follows: i) *end-event*, ii) *mid-event*, and iii) *split-event*. An end-event denotes the case when an edge disappears at the end of the edge after the process of edge-shrinking. A mid event denotes the case that an edge disappears in the middle of the edge. This case occurs when both end vertices of an edge move toward interior of the edge to meet at a point. Similarly, a split-event denotes the case that a new vertex is created at a point in the middle of an edge so that the edge splits into two edges.

However, the classification can be further refined considering the topological structure among edges. Suppose that an end-event is defined for an edge e_i. Then, this end-event always occurs at the off-vertex v_{off} part of the edge e_i. Suppose that there is another edge e_j having an end-event at the same vertex v_{off}. In other words, e_i and e_j are incident to v_{off}. Then, the events of both e_i and e_j should take place at the same time at v_{off}. Therefore, it is necessary to consider the incident edges of an off-vertex together for a correct handling of an end-event. In 3D, it can be also easily shown that only one or two edges can be incident to a vertex v_{off} to realize simultaneous end-events at a given time of the region expansion process. Hence, we call the situations *one-end-event* or *two-end-event*, respectively. In the case of mid-event or split-event, however, no such a peculiarity occurs.

3.1 Detection of Events

The type of an event can be determined by observing configurations of associated generator spheres at each vertex of a Voronoi edge. At the moment that an expanding generator touches the vertex sphere, the topology of the vertex starts to change. Some edges incident to the vertex, before the expansion, may disappear after the expansion while the other incident edges remain in the topology structure with possibly changed connectivity. Note that these remaining edges start to shrink at this moment.

An edge, except on-edges, has a state either $++$, $+-$, or $--$, where each $+$ or $-$ symbol denotes the state of a vertex of the edge. When an edge disappears at a vertex (this vertex is always an off-vertex), we assign '$-$' as the state of the vertex for the edge. Otherwise, the vertex (it can be either off-vertex or on-vertex) is called in a '$+$' state. Note that a vertex can have multiple states depending on its contributions to edges.

Let *fixed tangent points* be tangent points between a vertex sphere, corresponding to a vertex v, and three generators defining an edge e. Let *reference tangent point* be the fourth tangent point on a vertex sphere against the generator, other than the edge defining three generators, corresponding to the vertex. Let $H(v, e)$ be a hyperplane which passes through three fixed tangent points. Let a *touch point* be the tangent point between a vertex sphere and an expanding generator. Let $H^+(v, e)$ be an open half-space which is limited by $H(v, e)$ and containing the reference tangent point. The opposite open half-space is denoted by $H^-(v, e)$.

It can be easily shown that if a touch point is in $H^-(v, e)$, an edge e disappears at the vertex v. On the other hand, if a touch point is in $H^+(v, e)$, the vertex v starts to move along the edge e toward the other vertex. By this observation, we can easily assign a state, $+$ or $-$, to a vertex of an edge. If the reference tangent point is located in $H^+(v, e)$, we assign a $+$ state. Otherwise, a $-$ state is assigned to a vertex.

If an edge has both + and − vertex states, the edge starts to shrink at the vertex with + and disappears at the vertex with −. Therefore, the edge should have an end-event. Note that the ordering of vertex states is insignificant. If an edge has ++, the edge starts to shrink from both end vertices during the region expansion, and therefore the edge has to disappear at a middle of the original edge. Hence, the edge has a mid-event. If an edge has −−, on the other hand, the edge should disappear at both end vertices. This means that the edge should go through a state of shrinking before it disappears. Since neither vertex can contribute to shrink, there has to be a point in the middle of the edge from which the shrinkage can start. Therefore, the edge should split at a point in the middle of the edge. Hence, if the edge has −−, it has a split-event.

3.2 Handling of Events

Depending on the events, different actions should be performed to keep the topology correctly and consistently. In the case of *one-end-event*, the event edge disappears and three new on-edges are born to constitute a new on-face. In addition, both end vertices of the event edge disappear, and three new vertices for the new on-face are born.

Two-end-event removes two edges having end-events simultaneously at the off-vertex. Then, the face which is bounded by both event edges also disappears. Therefore, three edges constituting the face disappear as well, but a new on-edge is born.

In the case of a *mid-event*, the following happens. The radiating-face bounded by the event edge and the corresponding on-edge disappears, and two on-faces are merged into one on-face. Therefore, this event decreases the number of on-faces. Since two on-faces are merged into one on-face, the corresponding two pairs of edges are merged into two edges.

Split-event differs from the others as follows. The event edge does not disappear but is divided into two edges, and each divided edge is regarded as a new edge to be tested for its event type. In this case, a new on-face bounded by two new on-edges appears. Since this face consists of two on-edges, two end vertices are born. In addition, the event edge and the corresponding on-edge are divided into two edges. Therefore, the number of edges are increased by four.

4 Topology Representation: A Radial Data Structure

EVD(S) is a set of mutually exclusive Voronoi regions. A Voronoi region in the Voronoi diagram is star-shaped with respect to the center of corresponding spherical generator. Hence, a region itself is considered as a manifold model. Each Voronoi region, unless it is an unbounded region, is bounded by a set of Voronoi faces. Two neighboring Voronoi regions share only one Voronoi face on their boundary. Hence, the whole EVD(S) defines a cell structure and it is one of the typical non-manifold models. It means that a non-manifold data structure should be used for the representation of the whole EVD(S) in a single structure.

The topology of EVD(S) can be distinguished from general non-manifold models as follow: (i) certain cases of non-manifold condition never occur in EVD(S). To be specific, isolated vertices, dangling edges and faces never occur in EVD(S); (ii) even though it depends on the implementation, the degrees of vertex and edge are assumed three and four, respectively. Therefore, it is natural to consider a compacter topology structure than one for a general non-manifold model such as a partial entity structure, a vertex-based representation, one based on coupling entities and so on [9, 14, 24, 26] In addition, there are also some topological differences between the Euclidean Voronoi diagram of spheres and that of points in 3D as follow:

- A Voronoi face of EVD(S) can have a number of islands.
- The dual of EVD(S), which we call a quasi-triangulation, is not a Delaunay triangulation in general.

Since the dual of Euclidean Voronoi diagram for points, EVD(P), is a Delaunay triangulation, the topology of EVD(P) can be represented in its dual representation which is very compact and memory efficient. Rigorous studies have been done for the topology representation of EVD(P) [4, 5, 15]. Since the dual of EVD(S) is not a Delaunay triangulation in general [13], we have to devise a different data structure. Note that EVD(S) consists of a number of edge graphs by the unconnected edges when its dual does not satisfy the condition for Delaunay triangulation.

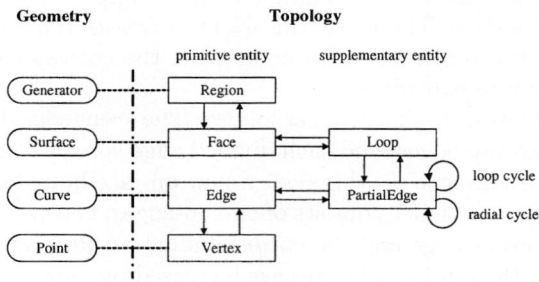

Fig. 2. Overview of data structure for EVD(S)

In this paper, we represent the topology for EVD(S) as a variation of the radial edge data structure, as shown in Fig. 2. The radial edge data structure is a non-manifold boundary topology representation based on edges [24]. It extends to represent manifold models to non-manifold ones with *use* of topological entities, i.e., vertex-use, edge-use, loop-use, and face-use. In the radial edge data structure, a face can be propagated over the edges to form correct shells with the cyclic ordering of edge-uses. To construct the adjacency relationships between Voronoi faces and Voronoi edges in our representation, we employ a partial edge entity corresponding to edge-uses from the partial entity structure [14]. The loop and radial cycles among three kinds of cyclic ordering in the topology for non-manifold models are explicitly represented in our data structure. Our representation does not represent some topological elements such as face use, loop use and so on due to differences between EVD(S) and a general non-manifold model. Hence, our representation is compacter than the original proposal of the radial edge data structure.

5 Algorithm

The region expansion algorithm for a single Voronoi region can be described as the following steps. The whole Voronoi diagram for spheres can be obtained after performing the following algorithm for all generators starting with the ordinary Voronoi diagram for the centers of the spheres. Note that an appropriate topology data structure can support the iteration in a coherent scheme.

Algorithm (Expand Region of s_i)

1. Find all the edges defining radiating-faces and insert into a set E.
2. For each edge $e \in E \setminus E_{\text{on}}$, determine its event time t_e and event type.
   If $t_e < r_i$, insert the edge e into the event queue Q which implements a priority queue.
3. Pop an edge from Q and check its event type.
   If the event type is end-event, test if it causes to one-end-event or two-end-events by checking the next edge in Q.
4. Perform an appropriate action for the detected event as explained in Section 3.2. After the treatment, find new edges bounding new radiating-faces and insert them into a set E^*.
   Perform Step 2 for all edges in the edge set E^*.
5. Repeat Step 3 and 4 until Q is empty.

In Step 2, it is necessary to compute the exact time when an event occurs. Due to space limit, we skipped to elaborate the details. However, the idea is rather simple as follows. While a generator is growing from the point at the center of the generator, it touches one of the empty tangent spheres on all Voronoi edges around the Voronoi region. The empty tangent spheres in fact denote a subset of the cyclide defined by three spherical generators. The time of contact is indeed the possible time of an event and the ordering of such possible event times around a Voronoi region determines the sequence of events. It can be further detailed to show that the actual computation can be rather simplified depending on the cases.

The computation necessary to expand a Voronoi region, starting from a center point to a complete sphere, takes $O(n^2 \log n)$ time in the worst-case since there can be $O(n^2)$ number of edges to be considered and sorting events is required according to the event time. Note that the combinatorial complexity of a single Voronoi region is known as $\Theta(n^2)$ [25] Therefore, the whole Voronoi diagram can be constructed by the region expansion algorithm in $O(n^3 \log n)$ time in the worst-case. We believe, however, that the expected time complexity to construct a whole diagram can be as lower as $O(n^2)$ in the region expansion algorithm. We want to mention that reducing the sizes of all generators by the smallest one results in the desired output much faster. However, the computation reduction is only a constant factor.

6 Conclusion

In this paper, a region expansion algorithm is given to construct Euclidean Voronoi diagrams for 3D spheres. The idea is simple as expanding Voronoi regions from the ordinary Voronoi diagram for the centers of the spheres via a sequence of topological operations. The topology representation, a variation of radial data structure, is also provided for the ease of implementation. While the algorithm computes the whole Voronoi diagram in $O(n^3 \log n)$ time in the worst-case, its expected running time can be much faster.

Acknowledgements

This research was supported by the Creative Research Initiatives from the Ministry of Science and Technology in Korea.

References

1. Angelov B, Sadoc J-F, Jullien R, Soyer A, Mornon J-P, Chomilier J. Nonatomic solvent-driven Voronoi tessellation of proteins: an open tool to analyze protein folds. Proteins: Structure, Function, and Genetics 2002;49(4):446–456.
2. Aurenhammer F. Power diagrams: properties, algorithms and applications. SIAM Journal of Computing 1987;16:78–96.
3. Boissonnat JD, Karavelas MI. On the combinatorial complexity of Euclidean Voronoi cells and convex hulls of d-dimensional spheres. in Proceedings of the 14th annual ACM-SIAM symposium on discrete algorithms 2003;305-312.
4. Brisson E. Representing geometric structures in d dimensions: topology and order. in Proceedings of 5th ACM Symposium On Computational Geometry 1989; 218–227.
5. Dobkin DP, Laszlo MJ. Primitives for the manipulation of three-dimensional subdivisions. in Proceedings of 3rd ACM Symposium on computational Geometry 1987;86–99.
6. Gavrilova M. Proximity and Applications in General Metrics. Ph.D. thesis: The University of Calgary, Dept. of Computer Science, Calgary, AB, Canada; 1998.
7. Gavrilova M, Rokne J. Updating the topology of the dynamic Voronoi diagram for spheres in Euclidean d-dimensional space. Computer Aided Geometric Design 2003;20(4):231–242.
8. Goede A, Preissner R, Frömmel C. Voronoi cell: new method for allocation of space among atoms: elimination of avoidable errors in calculation of atomic volume and density. Journal of Computational Chemistry 1997;18(9):1113–1123.
9. Gursoz EL, Choi Y, Prinz FB. Vertex-based representation of non-manifold boundaries, in: Wozny MJ, Turner JU, Preiss K eds., Geometric Modeling for Product Engineering, North Holland, Elsevier Science Publishers B.V. 1990;107–130.

10. Kim D-S, Kim D, Sugihara K. Voronoi diagram of a circle set from Voronoi diagram of a point set: I. Topology. Computer Aided Geometric Design 2001;18(6):541–562.
11. Kim D-S, Kim D, Sugihara K. Voronoi diagram of a circle set from Voronoi diagram of a point set: II. Geometry. Computer Aided Geometric Design 2001;18(6):563–585.
12. Kim D-S, Cho Y, Kim D, Cho C-H. Protein structure analysis using Euclidean Voronoi diagram of atoms, Proc. International Workshop on Biometric Technologies (BT 2004), Special Forum on Modeling and Simulation in Biometric Technology 2004;125–129.
13. Kim D-S, Cho Y, Kim D. Edge-tracing algorithm for Euclidean Voronoi diagram of 3D spheres, Proc. 16th Canadian Conference on Computational Geometry 2004;176–179.
14. Lee SH, Lee K. Partial entity structure: a compact boundary representation for non-manifold geometric modeling. ASME Journal of Computing & Information Science in Engineering 2001;1:356–365.
15. Lienhardt P. Subdivisions of n-dimensional spaces and n-dimensional generalized map, in Proceedings of 5th ACM Symposium on Computational Geometry 1989;228–236.
16. Luchnikov VA, Medvedev N, Naberukhin YI, Schober HR. Voronoi-Delaunay analysis of normal modes in a simple model glass. Physical Review B 2000;62:3181–3189.
17. Luchnikov VA, Medvedev NN, Oger L, Troadec J-P. Voronoi-Delaunay analyzis of voids in systems of nonspherical particles. Physical review E 1999;59(6):7205–7212.
18. Montoro JCG, Abascal JLF. The Voronoi polyhedra as tools for structure determination in simple disordered systems. The Journal of Physical Chemistry 1993;97(16):4211–4215.
19. Naberukhin YI, Voloshin VP, Medvedev NN. Geometrical analysis of the structure of simple liquids: percolation approach. Molecular Physics 1991;73:917–936.
20. Okabe A, Boots B, Sugihara K, Chiu SN. Spatial Tessellations: Concepts and Applications of Voronoi Diagrams. 2nd ed. Chichester: John Wiley & Sons; 1999.
21. Richards FM. The interpretation of protein structures: total volume, group volume distributions and packing density. Journal of Molecular Biology 1974;82:1–14.
22. Sastry S, Corti DS, Debenedetti PG, Stillinger FH. Statistical geometry of particle packings. I. Algorithm for exact determination of connectivity, volume, and surface areas of void space in monodisperse and polydisperse sphere packings. Physical Review E 1997;56:5524–5532.
23. Voloshin VP, Beaufils S, Medvedev NN. Void space analysis of the structure of liquids. Journal of Molecular Liquids 2002;96-97:101–112.
24. Weiler K. The radial edge structure: a topological representation for non-manifold geometric boundary modeling, in: Wozny MJ, McLaughlin HW, Encarnacao JL eds., Geometric Modeling for CAD Applications, North Holland, Elsevier Science Publishers B.V. 1988;3–36.
25. Will H-M. Computation of Additively Weighted Voronoi Cells for Applications in Molecular Biology. Ph.D. thesis, Swiss Federal Institute of Technology, Zurich; 1999.
26. Yamaguchi Y, Kimura F. Nonmanifold topology based on coupling entities. IEEE Computer Graphics and Applications 1995;15:42–50.
27. RCSB Protein Data Bank Homepage. http://www.rcsb.org/pdb/.

Analysis of the Nicholl-Lee-Nicholl Algorithm

Frank Dévai

London South Bank University, London, UK
fl.devai@lsbu.ac.uk

Abstract. The algorithm proposed by Nicholl, Lee and Nicholl (*Computer Graphics* **21**,4 pp 253–262) for clipping line segments against a rectangular window in the plane is proved to be optimal in terms of the minimum and maximum number of comparisons and the number of predicates used. It is also demonstrated that, due to its overhead, the algorithm in its compact form is slightly slower than simple algorithms. Though Nicholl et al proposed program-transformation techniques to expand the code to exploit the full potential of the algorithm, in some cases it takes more operations than simple algorithms, e.g., two intersections and three predicates instead of four intersections. While the algorithm is optimal on its own terms, it solves the clipping problem with the added restriction that only valid intersections are allowed to be calculated.

1 Introduction

The Nicholl-Lee-Nicholl (NLN) algorithm [12] is widely accepted as the theoretically best possible approach [5, 7, 8, 13] to determine the intersection of a line segment and a rectilinear rectangle, called the *window*, in the plane. This process is called *clipping*, a fundamental problem in computer graphics that has attracted much attention in the literature [1, 2, 3, 5, 6, 9, 16].

The NLN algorithm is discussed in textbooks [7, 8] used as a reference for new algorithms [5, 6, 10, 14] and recently has been extended to three dimensions [15]. The original algorithm is not published in the most efficient form; it has some overhead from geometric transformations and procedure calls. Automatic program-transformation techniques are proposed [11, 12] to expand the code to realize its full potential. Though its authors did not apply such a technique, Pitteway [13] used an algebraic package to implement an expanded variant.

Nicholl et al [12] observed that coding schemes used by earlier algorithms [7, 8, 17] are a source of inefficiency. Their objective was the development of an algorithm to manage all possible cases without going through a coding scheme. The second objective was to avoid computation of intersection points which are not endpoints of the output line segment. The NLN algorithm uses predicates to avoid unnecessary intersection calculations.

Based on operation counts and assumptions on the relative speeds of these operations (i.e., subtraction is slower than addition and division is slower than multiplication) Nicholl et al [12] demonstrated that, on average, their algorithm is faster than the Cohen-Sutherland [7, 8] and the Liang-Barsky [9] algorithms.

This model of computation, however, is no longer valid with instruction-level parallelism. Also all the cases averaged over symmetry tacitly assumes a uniform distribution of the input.

This paper takes higher level primitives, such as predicates and intersection calculations, and uses *equivalence partitioning* to analyse the algorithm independently of the probability distribution of the input and the size and the position of the window. The four straight lines along the sides of the window subdivide the plane into nine convex regions, and all the line segments with both endpoints falling in a particular region belong to an *equivalence class*. Each line segment within the same class can be processed by using the same amount of computational work independently of its size and position within the region. The computational work taken by a particular algorithm in the nine equivalence classes can be represented by a three-by-three matrix. In the case of the NLN algorithm we call this matrix the *acceptance-rejection matrix*.

In Sect. 2 the NLN algorithm is presented. In Sect. 3 lower bounds on acceptance-rejection tests, and in Sect. 4 lower bounds on predicates are established. In Sect. 5 first the acceptance-rejection matrix for the NLN algorithm is determined. The minimum and the maximum elements of this matrix are proved to be optimal. The algorithm is also proved to be optimal in terms of the number of predicates evaluated. In Sect. 6 conclusions are offered.

2 The Nicholl-Lee-Nicholl Algorithm

Cyrus and Beck [2] then Liang and Barsky [9] promoted the use of a parametric representation of the line segment, where the parameter value is 0.0 at the first endpoint (x_1, y_1) and 1.0 at the second endpoint (x_2, y_2). Let the bottom-left corner of the window represented by (x_L, y_B) and the top-right corner by (x_R, y_T), and let t_E and t_L be the parameter values at the points where the line segment enters and leaves the window respectively. Then $t_E = (x_L - x_1)/\Delta x$ and $t_L = (y_T - y_1)/\Delta y$, where $\Delta x = x_2 - x_1$ and $\Delta y = y_2 - y_1$.

Now imagine rotating the line segment in Fig. 1 counterclockwise around (x_1, y_1). Originally $t_E < t_L$ holds, then at the moment when the line segment

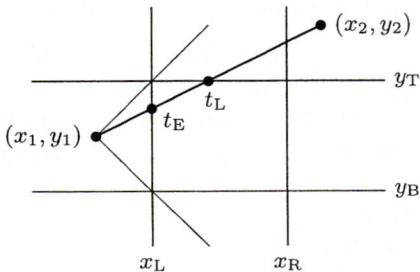

Fig. 1. Derivation of a predicate

hits the top-left corner, t_E will be equal to t_L, and when the line segment leaves the top-left corner, $t_E > t_L$ will hold. Therefore the condition $t_E < t_L$ is an appropriate *predicate* to decide if the line segment intersects the window.

Nicholl et al [12] use the predicate $(x_L - x_1)\Delta y < (y_T - y_1)\Delta x$, which is basically the same as above—only rearranged to use multiplications instead of divisions. This minor change, however, has the side effect of the dependence of the predicate on the assumptions of $x_2 > x_1$ and $y_2 > y_1$.

The original paper [12] presents the algorithm in the Pascal language, while it is presented here in a less cluttered and shorter pseudocode, with indentation alone indicating block structure. Parameters are passed to a procedure *by value*, unless a group of parameters is preceded by the reserved word **var**, which indicates parameter passing *by reference*. The main procedure is as follows.

procedure clip($x_L, y_T, x_R, y_B, x_1, y_1, x_2, y_2$)
    if $x_1 < x_L$ **then** *display* ← leftcolumn($x_L, y_T, x_R, y_B, x_1, y_1, x_2, y_2$)
    else if $x_1 > x_R$ **then**
        rotate180c(x_1, y_1); rotate180c(x_2, y_2)
        display ← leftcolumn($-x_R, -y_B, -x_L, -y_T, x_1, y_1, x_2, y_2$)
        rotate180c(x_1, y_1); rotate180c(x_2, y_2)
    else *display* ← centrecolumn($x_L, y_T, x_R, y_B, x_1, y_1, x_2, y_2$)
    if *display* **then** display the visible part of the line segment
end ▷ clip

The case when (x_1, y_1) is left to the line $x = x_L$ is processed as follows.

procedure leftcolumn(x_L, y_T, x_R, y_B, **var** x_1, y_1, x_2, y_2)
    if $x_2 < x_L$ **then** *display* ← FALSE
    else if $y_1 > y_T$ **then** *display* ← topleftcorner($x_L, y_T, x_R, y_B, x_1, y_1, x_2, y_2$)
    else if $y_1 < y_B$ **then**
        reflectaxis(x_1, y_1); reflectaxis(x_2, y_2)
        display ← topleftcorner($x_L, -y_B, x_R, -y_T, x_1, y_1, x_2, y_2$)
        reflectaxis(x_1, y_1); reflectaxis(x_2, y_2)
    else *display* ← leftedge($x_L, y_T, x_R, y_B, x_1, y_1, x_2, y_2$)
    return *display*
end ▷ leftcolumn

If (x_1, y_1) is in the top-left corner region, procedure topleftcorner is used.

procedure topleftcorner(x_L, y_T, x_R, y_B, **var** x_1, y_1, x_2, y_2)
    if $y_2 > y_T$ **then** *display* ← FALSE
    else if (x_2, y_2) is below the line through (x_1, y_1) and (x_L, y_T) **then**
        display ← leftbottomregion($x_L, y_T, x_R, y_B, x_1, y_1, x_2, y_2$)
    else
        reflectxminusy(x_1, y_1); reflectxminusy(x_2, y_2)
        display ← leftbottomregion($-y_T, -x_L, -y_B, -x_R, x_1, y_1, x_2, y_2$)
        reflectxminusy(x_1, y_1); reflectxminusy(x_2, y_2)
    return *display*
end ▷ topleftcorner

If (x_1, y_1) is in the top-left corner region, and (x_2, y_2) is below the line through (x_1, y_1) and (x_L, y_T), the line segment either enters the window at the left edge, or intersects the line $x = x_L$ below the bottom-left corner of the window. A second predicate is used to avoid this intersection calculation, and a third to decide if the line segment leaves the window at the bottom or the right.

procedure leftbottomregion($x_L, y_T, x_R, y_B,$ **var** x_1, y_1, x_2, y_2)
    if $y_2 \geq y_B$ **then**
        if $x_2 > x_R$ **then** calculate intersection with right edge
        calculate intersection with left edge
        display ← TRUE
    else if (x_2, y_2) is below the line through (x_1, y_1) and (x_L, y_B) **then**
        display ← FALSE
    else
        if $x_2 > x_R$ **then**
            if (x_2, y_2) is below the line through (x_1, y_1) and (x_R, y_B) **then**
                calculate intersection with bottom edge
            else calculate intersection with right edge
        else calculate intersection with bottom edge
        calculate intersection with left edge
        display ← TRUE
    return *display*
end ▷ leftbottomregion

If (x_1, y_1) is in the left edge region, the pseudocode is as follows.

procedure leftedge($x_L, y_T, x_R, y_B,$ **var** x_1, y_1, x_2, y_2)
    if $x_2 < x_L$ **then** *display* ← FALSE
    else if $y_2 < y_B$ **then** *display* ← p2bottom($x_L, y_T, x_R, y_B, x_1, y_1, x_2, y_2$)
    else if $y_2 > y_T$ **then**
        reflectaxis(x_1, y_1); reflectaxis(x_2, y_2)
        display ← p2bottom($x_L, -y_B, x_R, -y_T, x_1, y_1, x_2, y_2$)
        reflectaxis(x_1, y_1); reflectaxis(x_2, y_2)
    else
        if $x_2 > x_R$ **then** calculate intersection with right edge
        calculate intersection with left edge
        display ← TRUE
    return *display*
end ▷ leftedge

Procedure p2bottom[1] uses one predicate to decide if the line segment enters the window, then another to find the edge where the line segment leaves the window. If (x_1, y_1) is between the vertical lines $x = x_L$ and $x = x_R$, and if it is either in the top or bottom edge region, procedure leftedge can be reused. Otherwise (x_1, y_1) is inside the window. The pseudocode for procedures centrecolumn and inside are as follows.

---

[1] See myweb.lsbu.ac.uk/~devaifl/papers/nln for an expanded version of this paper.

procedure centrecolumn($x_L, y_T, x_R, y_B,$ **var** x_1, y_1, x_2, y_2)
 if $y_1 > y_T$ **then**
  rotate270c(x_1, y_1); rotate270c(x_2, y_2)
  $display \leftarrow$ leftedge($-y_T, x_R, -y_B, x_L, x_1, y_1, x_2, y_2$)
  rotate90c(x_1, y_1); rotate90c(x_2, y_2)
 else if $y_1 < y_B$ **then**
  rotate90c(x_1, y_1); rotate90c(x_2, y_2)
  $display \leftarrow$ leftedge($y_B, -x_L, y_T, -x_R, x_1, y_1, x_2, y_2$)
  rotate270c(x_1, y_1); rotate270c(x_2, y_2)
 else $display \leftarrow$ inside($x_L, y_T, x_R, y_B, x_1, y_1, x_2, y_2$)
 return $display$
end ▷ centrecolumn

procedure inside($x_L, y_T, x_R, y_B,$ **var** x_1, y_1, x_2, y_2)
 if $x_2 < x_L$ **then** p2left($x_L, y_T, x_R, y_B, x_1, y_1, x_2, y_2$)
 else if $x_2 > x_R$ **then**
  rot180c(x_1, y_1); rot180c(x_2, y_2)
  p2left($-x_R, -y_B, -x_L, -y_T, x_1, y_1, x_2, y_2$)
  rot180c(x_1, y_1); rot180c(x_2, y_2)
 else if $y_2 > y_T$ **then** calculate intersection with top edge
 else if $y_2 < y_B$ **then** calculate intersection with bottom edge
 return TRUE
end ▷ inside

If (x_1, y_1) is inside the window, and (x_2, y_2) is in an edge region, the edge intersecting the line segment can be found merely by comparisons. Procedure p2left decides if (x_2, y_2) is in the left (right) edge region or in a corner region. If it is in a corner region, p2left calls to p2lefttop, which uses one predicate to decide if the line segment leaves the window at a horizontal or at a vertical edge.

3 Lower Bounds on Acceptance-Rejection Tests

Consider the one-dimensional variant of the clipping problem: Given two intervals, $[x_1, x_2]$ and $[x_L, x_R]$, $x_L < x_R$, determine the part of $[x_1, x_2]$ inside $[x_L, x_R]$. Any algorithm for this problem should produce one of the following outputs.

1. $[x_1, x_2]$ ($[x_1, x_2]$ is totally inside $[x_L, x_R]$)
2. $[x_L, x_2]$ (first endpoint clipped to x_L)
3. $[x_R, x_2]$ (first endpoint clipped to x_R)
4. $[x_1, x_L]$ (second endpoint clipped to x_L)
5. $[x_1, x_R]$ (second endpoint clipped to x_R)
6. $[x_L, x_R]$ (both endpoints clipped, $x_1 < x_2$)
7. $[x_R, x_L]$ (both endpoints clipped, $x_1 > x_2$)
8. REJECT ($[x_1, x_2]$ is outside $[x_L, x_R]$)

We need at least three binary decisions to choose from the eight possible outputs, but no known algorithm can solve the problem in less than four comparisons. We demonstrate that four comparisons is actually a lower bound.

An *algebraic computation tree* on a set of variables $\{x_1, \ldots, x_n\}$ is a program consisting of two types of *statements*. One type of statements have the form of **return** OUTPUT, where OUTPUT is one of the possible results of the computation, e.g., in the case of the one-dimensional clipping problem, one of the outputs listen above. The other type of statements are similar to the if-then-else statements in high-level programming languages:

if $f(x_1, \ldots, x_n) \diamond c$ **then** statement 1 **else** statement 2

where $f(x_1, \ldots, x_n)$ is an algebraic function of the variables $x_1, \ldots, x_n$, $\diamond$ denotes any comparison relation, c is a constant, and 'statement 1' and 'statement 2' can be any one of the two types of statements. The execution of any **return** statement terminates the algorithm.

Each instance of the input variables $x_1, \ldots, x_n$ can be regarded as a point of the n-dimensional Euclidean space E^n. Each possible output identifies a subset of E^n, as shown in Fig. 2 for the clipping problem, where $n = 2$.

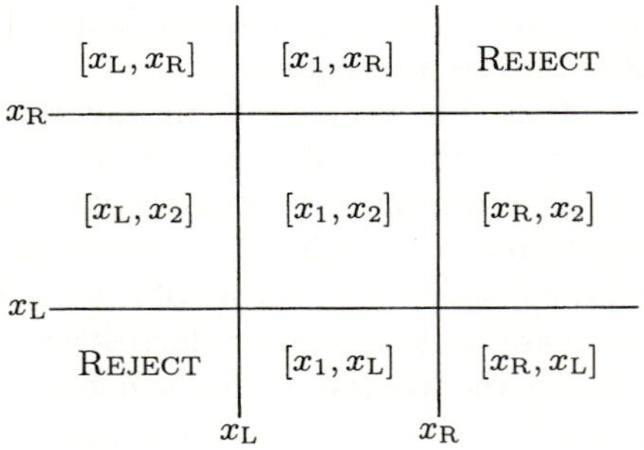

	$[x_L, x_R]$	$[x_1, x_R]$	REJECT
x_R	$[x_L, x_2]$	$[x_1, x_2]$	$[x_R, x_2]$
x_L	REJECT	$[x_1, x_L]$	$[x_R, x_L]$
	x_L	x_R	

Fig. 2. Partition of E^2 for the clipping problem

As three comparisons are anyway necessary, and a multiplication or a division is at least as expensive as a comparison, no one would reasonably use multiplications or divisions to save one comparison. Therefore we can assume that each $f(x_1, x_2)$ is a linear function. Then each subset of E^2 associated with any **return** statement must be a convex region, hence we need at least two **return** REJECT statements.

The tree must have at least nine leaf nodes, and must have depth at least four. As the x- and y-coordinates are independent from each other, the same argument applies for both. Hence any algorithm for two-dimensional acceptance-rejection tests must take at least eight comparisons.

4 Lower Bounds on the Number of Predicates

An input line segment may intersect the lines $y = y_B$, $x = x_R$, $y = y_T$ and $x = x_L$, called the *boundary lines*. An intersection is called *valid* if it is an endpoint of the output line segment; otherwise it is *invalid*. For simplifying the presentation we assume that the endpoints of the input line segment are disjoint, and that the input line segment never intersects a corner of the window (though the NLN algorithm can handle such cases at no extra cost). If only valid intersections are allowed to be calculated, we have the following possibilities.

If the first endpoint of the line segment is in any *edge region*, there are three possibilities. For example, if the first endpoint is in the left edge region, the line segment can intersect the line $x = x_L$ above the top-left corner, below the bottom-left corner or at the left edge of the window. We can check to see if $y_2 > y_T$ or $y_2 < y_B$, then one predicate is required decide if the intersection is above the top-left corner or below the bottom-left corner. Otherwise the line segment must intersect the left edge of the window.

Then the second intersection point is either inside the window, or the line segment leaves the window at one of the other three edges. Again we can check to see if $y_2 > y_T$ or $y_2 < y_B$, then another predicate is required decide if the intersection is on the top, bottom or right edge of the window. A similar argument can be used for the other three edge regions, hence if the first endpoint is in an edge region, two predicates are required.

If the first endpoint of the line segment is in any *corner region*, there are four possibilities. For example, if the first endpoint is in the bottom-left corner region, the line segment can either enter the window at the left edge, or intersect $x = x_L$ above the top-left corner, or enter the window at the bottom edge or intersect $y = y_B$ beyond the bottom-right corner of the window. To decide which one out of the four is the case, at least two predicates are required.

If the line segments enters the window, it can be decided by two comparisons, $x_2 \leq x_R$ and $y_2 \leq y_T$, whether the second endpoint remains inside the window. If $x_2 > x_R$ but $y_2 \leq y_T$, the line segment must intersect the right edge, and if $x_2 \leq x_R$ and $y_2 > y_T$, it must intersect the top edge of the window. In these cases no more predicates are required.

If, however, $x_2 > x_R$ and $y_2 > y_T$ hold, a third predicate is required to decide if the line segment intersects the right or the top edge of the window. A similar argument can be used for the other three corner regions, hence if the first endpoint is in a corner region, at least three predicates are required.

If the first endpoint of the line segment is in the *window*, no predicate is required if the second endpoint is beyond only one boundary line. If the second endpoint is beyond two boundary lines, one predicate is required.

5 Analysis of the Algorithm

First we establish the acceptance-rejection matrix for the NLN algorithm using the equivalence partitioning technique proposed in Sect. 1. If $x_1 < x_L$, procedure leftcolumn checks to see if $x_2 < x_L$ also holds, and rejects any line segment left to $x = x_L$ in two comparisons. From here it follows that each element in the first column of the matrix is 2. If $x_1 > x_R$, again procedure leftcolumn (with the window and the line segment rotated by 180 degrees) checks to see if $x_2 > x_R$, and rejects any line segment right to $x = x_R$ in three comparisons. Hence each element in the third column of the matrix is 3.

If both $x_1 < x_L$ and $x_1 > x_R$ are false, but $y_1 < y_B$ and $y_2 < y_B$ are true, procedure leftedge (in procedure centrecolumn) rejects any line segment in the bottom edge region in five comparisons. By similar reasoning any line segment in the top edge region can be rejected in four comparisons. If the line segment is inside the window, $x_L \leq x_1, x_2 \leq x_R$ and $y_B \leq y_1, y_2 \leq y_T$ must hold, hence trivial acceptance takes eight comparisons. Therefore the acceptance-rejection matrix for the NLN algorithm is $\begin{pmatrix} 2 & 4 & 3 \\ 2 & 8 & 3 \\ 2 & 5 & 3 \end{pmatrix}$.

We have already seen in Sect. 3 that any algorithm for two-dimensional acceptance-rejection tests must take at least eight comparisons, hence the matrix must have at least one element as large as 8. It takes at least one comparison to decide if an endpoint is on one side or another of a boundary line. The position of the second endpoint is independent of the first one, therefore each element of the matrix must be at least as large as 2.

The NLN algorithm is also optimal in the number of predicates used. If the first endpoint is in any edge region, the algorithm uses at most two predicates in procedure p2bottom. If the first endpoint is in any corner region, at most three predicates are used—one in procedure topleftcorner and another in procedure leftbottomregion. If the first endpoint is inside the window, at most one predicate is used in procedure p2lefttop. According to Sect. 4, that many predicates are required in each region.

Nicholl et al [12] argue that partial results of predicate calculations can be reused in intersection calculations. Though this can indeed be done in certain cases at the price of some overhead of passing on the partial results, three predicate calculations certainly cannot be reused in two intersection calculations.

For a machine-independent comparison we can establish acceptance-rejection matrices for other algorithms in a similar way. Equivalence partitioning, however, can also be used for experimental algorithm evaluation; it is sufficient to evaluate the running time for only one line segment in each of the nine equivalence classes. Though experimental algorithm evaluation in itself provides little insight into why one algorithm is faster than another, it can be used to measure the overhead associated with the elements of acceptance-rejection matrices.

The NLN algorithm was evaluated in timing experiments against the Cohen-Sutherland (CS), QuickClip (QC) and the Liang-Barsky (LB) algorithms. The C-code for the CS and QC algorithms were taken from Dévai [5]. All the code was written by the same person. The NLN algorithm was translated to C-code by this author from the Pascal code in Nicholl et al [12]. C functions were implemented

Table 1. Clock cycle counts

region/algorithm	CS	QC	LB	NLN
top-left corner	71	33	109	48
left edge region	93	33	109	48
bottom-left corner	93	33	109	48
top edge region	69	55	424	113
window	93	112	469	89
bottom edge region	91	66	319	125
top-right corner	49	22	214	115
right edge region	71	22	214	115
bottom-right corner	71	22	214	115
average	78	44	242	91

as macros wherever possible, e.g., the function calculating the four-bit codes for the CS algorithm, as well as transformation procedures, such as rotate90c, rotate180c etc., for the NLN algorithm. Clock cycle counts obtained on a 166 MHz Pentium MMX processor are given in Tab. 1.

Note that our analyses so far, except the bottom line of Tab. 1, made no assumptions on the input and the window. If we assume that the input line segments are given in a rectangular universe of width a and height b with sides parallel to the coordinate axes, and that all windows are equally likely, then the average window is a rectangle of width $a/3$ and height $b/3$ in the middle of the universe [4, 5]. Then all the regions in the equivalence partitioning have the same area. If all equivalence classes have the same number of line segments on the average, then it does make sense to average cycle counts over the equivalence classes. This, however, does not necessarily mean the assumption of evenly distributed input within the universe.

Though the acceptance-rejection matrix of the NLN algorithm indicates less comparisons than that of QuickClip [5], on average NLN takes more than twice as many clock cycles. The reason behind this is the overhead due to the complexity of the NLN algorithm. As we have already seen in Sect. 2, the predicates depend on the assumptions of $x_2 > x_1$ and $y_2 > y_1$. To handle this dependency the NLN algorithm uses 10 procedures, hence not only geometric transformations but also parameter passing contribute to the overhead. For example, while the NLN algorithm takes only two comparisons in the best case, the second comparison is performed in the procedure leftcolumn, with the overhead of passing eight floating-point parameters. Also the Boolean variable *display* is assigned and tested for each line segment. These 10 procedures are too complex to be implemented as macros.

6 Concluding Remarks

Nicholl et al [12] demonstrated that coding schemes, such as the one used by the Cohen-Sutherland algorithm, are a source of inefficiency. They also attempted to reduce the number of intersection calculations. The number of valid intersections is a trivial lower bound, and this paper demonstrated that the NLN algorithm is optimal in terms of the number of predicates evaluated.

We used equivalence partitioning to analyse the algorithm independently of the probability distribution of the input and the size and the position of the window. The minimum and the maximum elements of the resulting acceptance-rejection matrix were also found to be optimal.

The disadvantages of the NLN algorithm are its length and overhead compared with simple algorithms. Nicholl et al [11, 12] proposed a further expansion of the code in order to reduce the overhead. In certain cases, however, the NLN algorithm inherently needs more operations than simple algorithms: two intersection and three predicate calculations instead of four intersection calculations. The surprising conclusion is that, while the NLN algorithm is optimal on its own terms, it solves the clipping problem with the added restriction that only valid intersections can be calculated. Considering the above, the practical importance of the NLN algorithm is certainly less than its theoretical significance.

References

[1] J. F. Blinn. A trip down the graphics pipeline: Line clipping. *IEEE Computer Graphics & Applications*, 11(1):98–105, 1991.

[2] M. Cyrus and J. Beck. Generalised two- and three-dimensional clipping. *Computers & Graphics*, 3(1):2338, 1978.

[3] J. D. Day. A new two dimensional line clipping algorithm for small windows. *Computer Graphics Forum*, 11(4):241–245, 1992.

[4] F. Dévai. The average window is small. In *SIGGRAPH'96 Technical Sketches*, August 1996.

[5] F. Dévai. An analysis technique and an algorithm for line clipping. In *Proc. 1998 IEEE Conference on Information Visualization, IV'98*, pages 157–165, July 1998.

[6] V. J. Duvanenko, R. S. Gyurcsik, and Robbins W. E. Simple and efficient 2D and 3D span clipping algorithms. *Computers & Graphics*, 17(1):39–54, 1993.

[7] J. D. Foley, A. van Dam, S. K. Feiner, and J. F. Hughes. *Computer Graphics: Principles and Practice*. Addison-Wesley, 1996. (2nd ed in C).

[8] D. Hearn and M. P. Baker. *Computer Graphics with OpenGL*. Prentice Hall, 3rd edition, 2004.

[9] Y.-D. Liang and B. A. Barsky. A new concept and method for line clipping. *ACM Transactions on Graphics*, 3(1):1–22, 1984.

[10] G. Lu, X. Wu, and Q. Peng. An efficient line clipping algorithm based on adaptive line rejection. *Computers & Graphics*, 26(3):409–415, 2002.

[11] R. A. Nicholl and T. M. Nicholl. Performing geometric transformations by program transformation. *ACM Transactions on Graphics*, 9(1):28–40, 1990.

[12] T. M. Nicholl, D. T. Lee, and R. A. Nicholl. An efficient new algorithm for 2-D line clipping: its development and analysis. *Comput. Graph.*, 21(4):253–262, 1987.

[13] M. L. V. Pitteway. Personal communication, July 1998.

[14] N. C. Sharma and S. Manohar. Line clipping revisited: Two efficient algorithms based on simple geometric observations. *Comput. & Graph.*, 16(1):51–54, 1992.

[15] V. Skala and D. H. Bui. Extension of the Nicholl-Lee-Nicholl algorithm to three dimensions. *The Visual Computer*, 17(4):236–242, 2001.

[16] M. Slater and B. A. Barsky. 2D line and polygon clipping based on space subdivision. *The Visual Computer*, 10(1):407–422, 1994.

[17] M. S. Sobkow, P. Pospisil, and Y.-H. Yang. A fast two-dimensional line clipping algorithm via line encoding. *Computers & Graphics*, 11(4):459–467, 1987.

Flipping to Robustly Delete a Vertex in a Delaunay Tetrahedralization

Hugo Ledoux[1], Christopher M. Gold[1], and George Baciu[2]

[1] GIS Research Centre, School of Computing,
University of Glamorgan, Pontypridd,
CF37 1DL, Wales, UK
hledoux@glam.ac.uk, christophergold@voronoi.com
[2] Department of Computing,
Hong Kong Polytechnic University, Hong Kong
csgeorge@comp.polyu.edu.hk

Abstract. We discuss the deletion of a single vertex in a Delaunay tetrahedralization (DT). While some theoretical solutions exist for this problem, the many degeneracies in three dimensions make them impossible to be implemented without the use of extra mechanisms. In this paper, we present an algorithm that uses a sequence of bistellar flips to delete a vertex in a DT, and we present two different mechanisms to ensure its robustness.

1 Introduction

The construction of the Delaunay tetrahedralization (DT) of a set S of points in the Euclidean space $\mathbb{R}^3$ is a well-known problem and many efficient algorithms exist [1, 2, 3]. The 'inverse' problem — the deletion of a vertex v in a DT(S), thus obtaining DT$(S \setminus \{v\})$ — is however much less documented and is still a problem in practice. Most of the work on this topic has been done for the two-dimensional case and very little can be found for the three- and higher-dimensional cases. The problem has been tackled mostly by removing from the triangulation all the simplices incident to v and retriangulating the 'hole' thus formed (see Fig. 1 for the 2D case). Throughout this paper, we denote by star(v) the star-shaped polytope formed by the union of all the simplices incident to a vertex v in a d-dimensional Delaunay triangulation.

An optimal solution exists for the 2D problem [4], but sub-optimal algorithms are nevertheless usually preferred for an implementation because of their simplicity and because the average degree k of a vertex in a 2D Delaunay triangulation is only 6. The most elegant of these algorithms is due to Devillers [5], who transforms the problem into the construction of the convex hull of the points on the boundary of star(v) lifted onto the paraboloid in 3D. Mostafavi et al. [6] propose a simpler algorithm, where each triangle used for the retriangulation is tested against each vertex of star(v). These two algorithms have respectively a time complexity of O($k \log k$) and O(k^2).

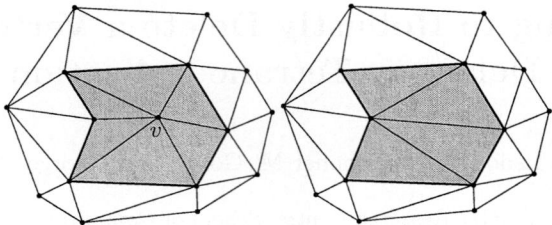

Fig. 1. Delaunay triangulations before and after vertex v has been deleted

Most algorithms in computational geometry assume that inputs are in 'general position', and the handling of degeneracies are usually left to the programmers. Modifying the original algorithm to make it robust for any input can be in some cases an intricate and error-prone task. Luckily, the deletion of a single vertex in a 2D DT does not have many special cases and its robust implementation is quite simple. However, the numerous degeneracies make the implementation of Devillers' and Mostafavi et al.'s algorithms impossible in 3D (impossible without the use of an extra mechanism that is), despite the facts that the former proved that his algorithm is valid in any dimensions, and that common sense suggests the latter algorithm generalises easily. The problems are caused by the fact that not every polyhedron can be tetrahedralized, as explained in Sect. 2. Devillers and Teillaud [7] recognised that and used perturbations [8] to solve the problem. Unfortunately, the major part of their paper is devoted to explaining the perturbation scheme and few details about the algorithm are given.

In this paper, we describe an algorithm to delete a vertex in a DT, and we show in Sect. 4 that, instead of creating a 'hole' in the tetrahedralization, it is possible to tackle the problem differently and use bistellar flips; the flips needed are described in Sect. 3. Flipping permits us to keep a complete tetrahedralization during the whole deletion process, and hence the algorithm is relatively simple to implement and numerically more robust. We also discuss in Sect. 5 two different methods to ensure the algorithm is robust against degenerate cases. The first one uses a symbolic perturbation scheme, and the second is an empirical method that requires the modification of some tetrahedra outside $star(v)$.

2   Delaunay Tetrahedralization

A Delaunay tetrahedralization (DT) of a set S of points in $\mathbb{R}^3$ is a set of non-overlapping tetrahedra, which have *empty* circumspheres, whose union completely fills the convex hull ($\mathcal{CH}$) of S. It can be constructed incrementally [3], with an algorithm based on the divide-and-conquer paradigm [2], or even by transforming the problem into the construction of a four-dimensional convex hull [9]. Another method consists of using bistellar flips (see Sect. 3) to modify the configuration of adjacent tetrahedra after the insertion of a new point. The major problem when designing flip-based algorithms in 3D is that even if some

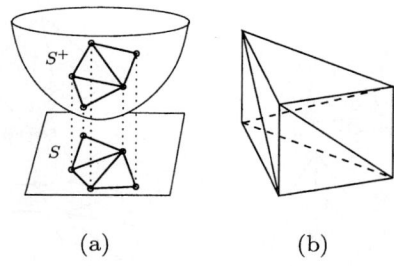

Fig. 2. (a) The parabolic lifting map of a set S of points in the plane. Each triangular face of the convex hull of the lifted set S^+ in 3D corresponds to a triangle of the 2D Delaunay triangulation of S. (b) The Schönhardt polyhedron is impossible to tetrahedralize

adjacent tetrahedra need to be modified, it is not always possible to flip them, as Joe [10] proves. He nevertheless later proved that if a single point v is added to a valid $DT(S)$, then there always exists at least one sequence of flips to construct $DT(S \cup \{v\})$ [1]. These results have also been generalised to $\mathbb{R}^d$ [11].

Constructing $DT(S)$ essentially requires two geometric predicates: *Orient*, which determines on what side of a plane a point lies; and *InSphere*, which determines if a point p is inside, outside or lies on a sphere. The *InSphere* test is derived from the well-known *parabolic lifting map* [9], which describes the relationship that exists between a d-dimensional Delaunay triangulation and a convex hull in $(d+1)$ dimensions (see Fig. 2(a)).

While any polygon in 2D can be triangulated, some arbitrary polyhedra, even if they are star-shaped, cannot be tetrahedralized without the addition of extra vertices, the so-called Steiner points. Fig. 2(b) shows an example, as it was first illustrated by Schönhardt [12]. In this paper, we are interested in a special case of the tetrahedralization problem: the polyhedron is star(v), a star-shaped polyhedron (not necessarily convex), formed by all the tetrahedra in $DT(S)$ incident to the vertex v. The tetrahedralization of star(v) is always possible, and moreover with locally Delaunay tetrahedra. Let $\mathcal{T}$ be a $DT(S)$. If v is added to $\mathcal{T}$, thus getting $\mathcal{T}^v = \mathcal{T} \cup \{v\}$, with an incremental insertion algorithm [1,3], all the tetrahedra in $\mathcal{T}$ whose circumspheres contain v will be deleted; the union of these tetrahedra forms a polyhedron P. Then, P will be retetrahedralized with many tetrahedra all incident to v. Now consider the deletion of v from $\mathcal{T}^v$. Notice that v could actually be any vertices in S, as a DT is unique and not affected by the order of insertion. The polyhedron star(v) is exactly the same as P, therefore $\mathcal{T}$ tetrahedralize P.

3 Three-Dimensional Bistellar Flips

A bistellar flip is a local topological operation that modifies the configuration of some adjacent tetrahedra. As shown in Lawson [13], there exist four different

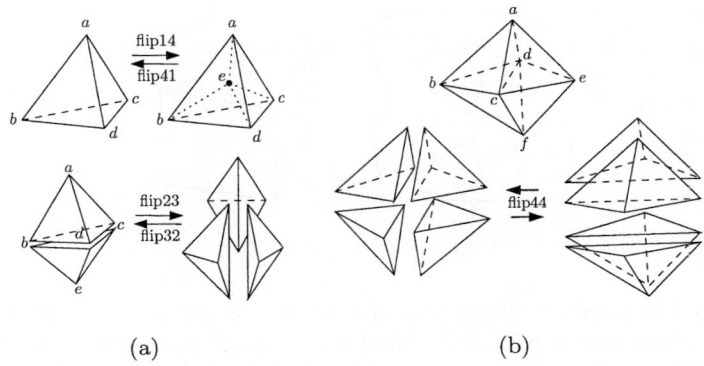

Fig. 3. (a) Three-dimensional bistellar flips. (b) Degenerate *flip44*

flips in $\mathbb{R}^3$, based on the different configurations of a set $S = \{a, b, c, d, e\}$ of points in general position: *flip14*, *flip41*, *flip23* and *flip32* (the numbers refer to the number of tetrahedra before and after the flip). These flips are illustrated in Fig. 3(a). When the five points of S lie on the boundary of $\mathcal{CH}(S)$, S can be tetrahedralized with two or three tetrahedra, and the *flip23* and *flip32* are the operations that substitute one tetrahedralization by another one. A *flip14* refers to the operation of inserting a vertex inside a tetrahedron, and splitting it into four tetrahedra; and a *flip41* is the inverse operation that deletes a vertex.

To deal with degenerate cases, other flips need to be defined. Shewchuk [14] defines, and uses for the construction of constrained Delaunay triangulations, degenerate flips. A flip is said to be degenerate if it is a non-degenerate flip in a lower dimension. It is used for special cases such as when a new point is inserted directly onto an edge or a face of a triangular face. Consider the set $S = \{a, b, c, d, e, f\}$ of points configured as shown in Fig. 3(b), with points b, c, d and e being coplanar. If S is tetrahedralized with four tetrahedra all incident to one edge — this configuration is called the *config44* — then a flip44 transforms one tetrahedralization into another one also having four tetrahedra. Note that the four tetrahedra are in *config44* before and after the flip44. A flip44 is actually a combination in one step of a flip23 (that creates a flat tetrahedron) followed immediately by a flip32 that deletes the flat tetrahedron; a flat tetrahedron is a tetrahedron spanned by four coplanar vertices (its volume is zero). We show in Sect. 5 why it is necessary when deleting a vertex.

4 Flipping to Delete a Vertex

This section describes an algorithm for deleting a vertex v in a Delaunay tetrahedralization $\mathrm{DT}(S)$ of a set S of points in general position. Unlike other known approaches, the polyhedron $\mathrm{star}(v)$ is not deleted from $\mathrm{DT}(S)$; $\mathrm{DT}(S \setminus \{v\})$ is obtained by restructuring $\mathrm{star}(v)$ with a sequence of bistellar flips.

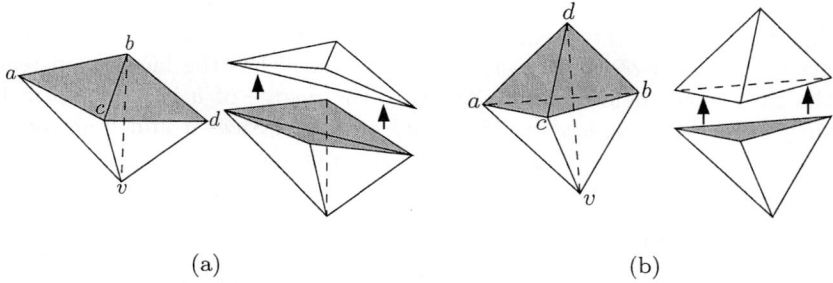

(a) (b)

Fig. 4. Flipping of an ear. In both cases, link(v), before and after the flip, is represented by the shaded triangular faces. **(a)** A 2-ear $abcd$ is flipped with a flip23. **(b)** A 3-ear $abcd$ is flipped by a flip32

4.1 Ears of a Polyhedron

Let P be a polyhedron that is made up of triangular faces. An ear of P is a potential, or 'imaginary' tetrahedron, that could be used to tetrahedralize P; this is the three-dimensional equivalent of the ear of a polygon as used in deletion algorithms (see e.g. [5]). Referring to the Fig. 4, we can affirm that there exist two kinds of ears for P: a *2-ear* is formed by two adjacent triangular faces abc and bcd sharing edge bc; and a *3-ear* is formed by three adjacent triangular faces abd, acd and bcd sharing vertex d. A 3-ear is actually formed by three 2-ears overlapping each other. On the surface of P, every 2-ear has four neighbouring ears and every 3-ear has three; in both cases, these neighbours can either be 2- or 3-ears. Not every pair of adjacent faces of P is considered as a *valid* ear. A 2-ear is valid if and only if the line segment ad is inside P; and a 3-ear is valid if and only if the triangular face abc is inside P. In the case of the deletion of a vertex v in a DT, P is a star-shaped polyhedron star(v). An ear of star(v) is valid if it is convex outwards from v; this can be tested with two *Orient* tests.

4.2 Flipping an Ear

Let link(v) be the union of the triangular faces on the boundary of star(v) which are not incident to vertex v. Consider an ear ε (formed by two or three adjacent triangular faces in link(v)) of the polyhedron star(v). Flipping ε means creating in the tetrahedralization the tetrahedron spanned by the four vertices of ε, by flipping the two or three tetrahedra that define ε. As shown in Fig. 4, different flips are applied to different types of ears:

1. if ε is a 2-ear $abcd$, defined by the tetrahedra $abcv$ and $bcdv$, then a flip23 creates the tetrahedra $abcd$, $abvd$ and $acvd$. The results of that flip23 are that, first, the tetrahedron $abcd$ is not part of star(v) anymore; and second, link(v) is modified as ε is replaced by an ear ε' formed by the triangular faces abd and acd. A 2-ear is said to be *flippable* if and only if the union of the two tetrahedra defining it is a convex polyhedron.

2. if ε is a 3-ear $abcd$, defined by the tetrahedra $abdv$, $bcdv$ and $acdv$, then a flip32 creates the tetrahedra $abcd$ and $abcv$. After the flip, the tetrahedron $abcd$ is not part of star(v) anymore, and link(v) has the face abc instead of the three faces before the flip32. Also, the degree of v is reduced by 1 by such a flip. A 3-ear is *flippable* if and only if vertices d and v are on each side of the face abc, i.e. that v should the 'outside' ε.

4.3 Deletion with *InSphere*

The algorithm we describe in this section, called DELETEINSPHERE, is a generalisation of Mostafavi et al.'s [6] in 3D, and proceeds as follows. First, all the ears of star(v) are built and stored in a simple dynamic list. The idea of the algorithm is to take an ear ε from the list (any ear) and process it if these three conditions are respected: ε is valid, flippable and locally Delaunay. An ear ε is locally Delaunay if its circumsphere does not contain any other vertices on the boundary of star(v); this is tested with *InSphere* tests. There is no particular order in which the ears are processed. If an ear respects the three conditions, it is flipped, and, as a result, star(v) is modified and a tetrahedron spanned by the four vertices of ε is added to DT($S \setminus \{v\}$). As flips are performed, star(v) 'shrinks' and the configuration of the tetrahedra inside it changes; thus non-valid ears become valid, and vice versa. If an ear does not respect one of the three conditions, the next ear in the list is tested. The time complexity of the algorithm is $O(t\,k)$: t tetrahedra are created to retetrahedralize star(v), and each of these tetrahedra must be tested against the k vertices on the boundary of star(v). A flip is assumed to be performed in constant time because only a finite number of adjacent tetrahedra are involved.

The correctness proof of the algorithm is omitted here, but here is the main idea. Let $\mathcal{T}$ be a DT, and let $\mathcal{T}^v = \mathcal{T} \cup \{v\}$ be the tetrahedralization after the insertion of v with a flip-based algorithm. Joe [1] proves that v can always be inserted in $\mathcal{T}$ by a sequence of bistellar flips; each flip will remove from $\mathcal{T}$ exactly one tetrahedron whose circumsphere contains v. Although some locally non-Delaunay tetrahedra will be impossible to flip, there will always be at least one possible flip at each step of the algorithm. Consider now the deletion of v in $\mathcal{T}^v$, thus getting $\mathcal{T}' = \mathcal{T}^v \setminus \{v\}$. Sect. 3 shows that each flip has its inverse, therefore reversing the flips used to construct $\mathcal{T}^v$ trivially constructs $\mathcal{T}'$. Also, notice that $\mathcal{T}'$ must be equal to $\mathcal{T}$ since we assume general position. This means that, to construct $\mathcal{T}'$, only the inverse flips of the flips used to construct $\mathcal{T}^v$ can be used. Thus, constructing $\mathcal{T}'$ can be seen as exactly the inverse of constructing $\mathcal{T}^v$, and, as a result, at each step of the process a locally Delaunay ear will be flippable.

5 Degeneracies

Degenerate cases occur when one of the following two conditions arises: the vertex v to be deleted lies on the boundary of $\mathcal{CH}(S)$; the set of points S is not in general position, that is four or more points are coplanar, and/or five or more points are cospherical.

The algorithm DELETEINSPHERE as described in Sect. 4 is not valid for deleting a vertex on the boundary of $\mathcal{CH}(S)$. This case can nevertheless be easily avoided by starting the construction algorithm of a DT with four non-coplanar points forming a tetrahedron big enough to contain all the points in S.

The coplanarity of points in S leads to the use of the degenerate bistellar *flip44*; we explain in Sect. 5.1 below how it is used.

When five or more points in S are cospherical, DT(S) is not unique. Let $\mathcal{T}$ be a DT(S) containing five or more cospherical points, and let v be a point located 'inside' the cospherical points. Consider the insertion of v in $\mathcal{T}$, thus getting $\mathcal{T}^v$, followed immediately by its deletion to get the tetrahedralization $\mathcal{T}'$. The tetrahedralization $\mathcal{T}'$, although being a valid DT, will not necessarily be the same as $\mathcal{T}$. Cospherical points will introduce an ambiguity as to which flips should be performed to delete v in $\mathcal{T}^v$. A flip used to delete v from $\mathcal{T}^v$, although possible and performed on a Delaunay ear, is not necessarily the inverse of a flip that was used to construct $\mathcal{T}^v$. Unfortunately, one or more of these 'wrong' flips can lead to a polyhedron star(v) that is impossible to tetrahedralize.

There exist two solutions to deal with this problem. The first one is to *prevent* an untetrahedralizable polyhedron by perturbing vertices to ensure that a DT is unique even for degenerate inputs. This is briefly described in Sect. 5.2. The main disadvantage of this method is that the same perturbation scheme must be used for all the operations performed on a DT. As a result, if one only has a DT and does not know what perturbation scheme was used to create it, then this method cannot be used. It is of course always possible to modify a DT so that it is consistent with a given perturbation scheme, but that could require a lot of work in some cases. Also, for some applications, using perturbations is not always possible. An example is a modelling system where points are moving while the topological relationships in the DT are maintained. It would be quite involved to ensure that the DT is consistent at all times with a perturbation scheme. The alternative solution consists of *recovering* from an untetrahedralizable star(v) by modifying the configuration of some triangular faces on its boundary. Such an operation requires the modification of some tetrahedra outside star(v). The implementation of this method is greatly simplified if a complete tetrahedralization is kept because the 'outside' tetrahedra can simply be flipped. The method is called 'unflipping' and is described in Sect. 5.3.

5.1 Handling Coplanar Points with the Degenerate *flip44*

Let ε be a 2-ear of a polyhedron star(v), and let τ_1 and τ_2 be the two tetrahedra defining ε. Consider τ_1 and τ_2 to have four coplanar vertices and ε to be Delaunay (e.g. in Fig. 4(a), *abdv* would be coplanar). A simple flip23 on ε is not always possible since it creates a flat tetrahedron; actually the flip23 is possible if and only if τ_1 and τ_2 are in *config44* with two tetrahedra τ_3 and τ_4 that are inside star(v) and define an ear ε_2 that is Delaunay. In that case, a flip44 will flip in one step both ε and ε_2. If τ_1 and τ_2 are not in *config44*, then ε cannot be flipped.

If four or more coplanar vertices are present in a DT, one must be aware of the presence of flat tetrahedra. They can be created during the process of updating a DT (after a flip), but no flat tetrahedron can exist in a DT (it violates the Delaunay criterion since the circumsphere of a flat tetrahedron is undefined). It is known that a DT without any flat tetrahedra always exists. In DELETE-INSPHERE, flat tetrahedra are permitted only if they are incident to v; an ear clearly cannot be flat because after being processed it becomes a tetrahedron of DT.

5.2 Symbolic Perturbations to Handle Cospherical Points

Perturbing a set S of points means moving the points by an infinitesimal amount to ensure that S is in general position. Unfortunately, moving points in $\mathbb{R}^d$ can have serious drawbacks: tetrahedra that are valid in the perturbed set of points can become degenerate (e.g. flat) when the points are put back to their original position. In the case of the deletion of a vertex v in a DT in $\mathbb{R}^3$, only cospherical points really cause problems — they can lead to an untetrahedralizable polyhedron — since coplanarity can be handled with the flip44. A method to perturb only cospherical points without actually moving them was proposed in Edelsbrunner and Mücke [15–Sect. 5.4]. It involves perturbing the points in $\mathbb{R}^{d+1}$ by using the parabolic lifting map. In $\mathbb{R}^3$, five points are cospherical if and only if the five lifted points are coplanar on the paraboloid in $\mathbb{R}^4$. Thus, each point of S is perturbed by a (very) small amount so that no five points in $\mathbb{R}^4$ lie on the same hyperplane. The method cannot be applied just for the deletion of a single vertex since the resulting tetrahedralization of star(v) would not necessarily be consistent with the tetrahedralization outside star(v). The main goal of using this method is having a unique DT even when five or more points in S are cospherical, so that there is a clear ordering of the flips to perform to delete v. The same perturbation scheme must therefore be used for every operation performed on the DT, including its construction.

For a very easy implementation of this perturbation scheme, see [16]. With this method, the amount by which each point is moved does not have to be calculated explicitly because the perturbations are implemented symbolically in the *InSphere* test. It should be noted that the `CGAL` library uses the same scheme [7], although the implementation is different.

5.3 Unflipping

We have tested the algorithm DELETEINSPHERE with many different datasets — points randomly distributed in a cube or a sphere, lying on the boundary of regular solids (spheres, cubes, paraboloid, etc.), and also geologic datasets with boreholes — and, in our experience, the case of an untetrahedralizable polyhedron occurs only when most of the points in S are both coplanar and cospherical, that is when the spacing between points in the $x - y - z$ directions is constant, as in a 3D grid. Unflipping means modifying some triangular faces on the boundary of star(v) so that a tetrahedralization of star(v) is possible. For example, the most 'common' untetrahedralizable star(v) during our tests was very similar to the Schönhardt polyhedron depicted in Fig. 2(b), except that the three quadrilateral faces are 'flat': they are formed by two coplanar triangular faces. Notice that if only one diagonal of a quadrilateral face is flipped (think of a *flip22* in a 2D triangulation) then the polyhedron can easily be tetrahedralized with three tetrahedra. Thus, to recover from an untetrahedralizable star(v), we propose flipping the diagonal of one flat 2-ear of star(v) and continue the deletion process as usual afterwards. This is an iterative solution: one flip might be sufficient in some cases, but if another untetrahedralizable star(v) is later obtained then another diagonal must be flipped. Flipping the diagonal of a flat ear obviously involves the modification of the tetrahedra, inside and outside star(v), incident to the diagonal edge. The only way to do this is with a flip44 involving the two tetrahedra τ_1 and τ_2 forming the flat ear and two tetrahedra τ_1' and τ_2' adjacent to them outside star(v). The assumption behind this method is the following. An untetrahedralizable star(v) occurs only when most of the points in S form a 3D grid (S can also be seen as being formed by many adjacent cubes formed by eight vertices), and therefore one of the flat ears of star(v) will be incident to one such cube. The modification of the tetrahedralization of a cube is allowed since its eight vertices are cospherical.

The configuration of tetrahedra outside star(v) and incident to a flat ear ε will not always be the same, as there are many ways to tetrahedralize a cube. The two most common configurations are as follows. In the first configuration, only two tetrahedra τ_1' and τ_2' are adjacent to ε (see Fig. 5(a)), thus forming a *config44*. A flip44 is then allowed if the five vertices of τ_1' and τ_2' are cospherical. In the second configuration, three tetrahedra are incident to ε, as in Fig. 5(b). Then if the six vertices of the three tetrahedra are cospherical, a flip23 on two of the three tetrahedra will modify the configuration such that only two tetrahedra are incident to ε. A flip44 is then possible.

Other configurations can occur and by flipping locally it is possible to obtain two tetrahedra incident to a flat ear. In our experience, when star(v) is an untetrahedralizable polyhedron, there is always a flat ear whose diagonal can be flipped such that DELETEINSPHERE makes progress towards the deletion of v.

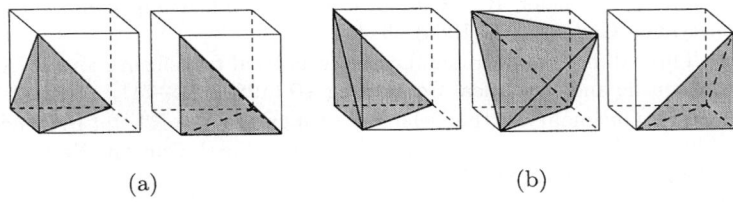

(a)   (b)

Fig. 5. Two possible configurations when unflipping. The flat ear ε is the bottom face of the cube. **(a)** Two tetrahedra are incident to ε. **(b)** Three tetrahedra are incident to ε

6 Discussion

In brief, to make DELETEINSPHERE robust for any configuration of data, several things must be done. First, the coplanarity of vertices must be handled with the degenerate *flip44*. Second, cospherical vertices, which may lead to an untetrahedralizable polyhedron, can be handled with either symbolic perturbations or with the unflipping method. It should be noticed that the implementation of a perturbation scheme is effective only if exact arithmetic is used for all the predicates involved.

Acknowledgements. Part of this work was done when the first author was at the Hong Kong Polytechnic University, and we thank the support of the Hong Kong's Research Grants Council (Project PolyU 5068/00E).

References

1. Joe, B.: Construction of three-dimensional Delaunay triangulations using local transformations. Computer Aided Geometric Design **8** (1991) 123–142
2. Cignoni, P., Montani, C., Scopigno, R.: DeWall: a fast divide & conquer Delaunay triangulation algorithm in E^d. Computer-Aided Design **30** (1998) 333–341
3. Watson, D.F.: Computing the n-dimensional Delaunay tessellation with application to Voronoi polytopes. Computer Journal **24** (1981) 167–172
4. Chew, L.P.: Building Voronoi diagrams for convex polygons in linear expected time. Technical Report PCS-TR90-147, Dartmouth College, Computer Science, Hanover, NH (1990)
5. Devillers, O.: On deletion in Delaunay triangulations. International Journal of Computational Geometry and Applications **12** (2002) 193–205
6. Mostafavi, M.A., Gold, C.M., Dakowicz, M.: Delete and insert operations in Voronoi/Delaunay methods and applications. Computers & Geosciences **29** (2003) 523–530
7. Devillers, O., Teillaud, M.: Perturbations and vertex removal in a 3D Delaunay triangulation. In: Proceedings 14th ACM-SIAM Symposium on Discrete Algorithms (SODA), Baltimore, MD, USA (2003) 313–319
8. Seidel, R.: The nature and meaning of perturbations in geometric computing. Discrete & Computational Geometry **19** (1998) 1–17
9. Edelsbrunner, H., Seidel, R.: Voronoi diagrams and arrangements. Discrete & Computational Geometry **1** (1986) 25–44
10. Joe, B.: Three-dimensional triangulations from local transformations. SIAM Journal on Scientific and Statistical Computing **10** (1989) 718–741
11. Rajan, V.T.: Optimality of the Delaunay triangulation in $\mathbb{R}^d$. In: Proceedings 7th Annual Symposium on Computational Geometry, North Conway, New Hampshire, USA, ACM Press (1991) 357–363
12. Schönhardt, E.: Über die zerlegung von dreieckspolyedern in tetraeder. Mathematische Annalen **98** (1928) 309–312
13. Lawson, C.L.: Properties of n-dimensional triangulations. Computer Aided Geometric Design **3** (1986) 231–246

14. Shewchuk, J.R.: Updating and constructing constrained Delaunay and constrained regular triangulations by flips. In: Proceedings 19th Annual Symposium on Computational Geometry, San Diego, USA, ACM Press (2003) 181–190
15. Edelsbrunner, H., Mücke, E.P.: Simulation of Simplicity: a technique to cope with degenerate cases in geometric algorithms. ACM Transactions on Graphics **9** (1990) 66–104
16. Shewchuk, J.R.: Sweep algorithms for constructing higher-dimensional constrained Delaunay triangulations. In: Proceedings 16th Annual Symposium on Computational Geometry, Hong Kong, ACM Press (2000) 350–359

A Novel Topology-Based Matching Algorithm for Fingerprint Recognition in the Presence of Elastic Distortions

Chenfeng Wang and Marina L. Gavrilova

Department of Computer Science, University of Calgary,
Calgary, AB, Canada
{cwang,marina}@cpsc.ucalgary.ca

Abstract. Fingerprint matching is a crucial step in fingerprint identification. In this paper, we propose a novel fingerprint alignment and matching scheme based on the geometric structure of minutiae set. The proposed scheme is transformation independent and is based on the Delaunay triangulation based technique. We also propose a novel application of RBF method, typically used for medical image matching, in order to deal with elastic deformations of the fingerprint. Finally, we develop an efficient global matching scheme based on the comparisons of minutiae sets and singular points to increase the validity of the matching results. Using the same amount of information on minutiae set as traditional minutiae based algorithms we achieve a faster performance and higher accuracy rates using our proposed algorithm.

1 Introduction to Fingerprint Matching

Fingerprint identification is one of the most reliable methods among biometric recognition technologies. It is widely applied for personal identification due to a high degree of security it provides. There are many fingerprint identification methods that appeared in literature in recent years.

In this paper, we will consider the minutiae-based method. Minutiae-based methods can be divided into two classes, which differ in feature information stored in the template. One stores the minutiae's characteristics, including their coordinates, directions and types, acquired during the off-line minutiae extraction [1, 2]. Then, the input fingerprint's minutiae are detected using the same minutiae extraction method on-line and then compared with the minutiae characteristics stored in the fingerprint template. Finally, the matching result is computed using some defined criteria. The other class of methods, in addition to minutiae, also records associated ridges between minutiae [3]. Minutiae and associated ridges are extracted from the input fingerprint and matched with the combined features stored in the template based on both minutiae and ridges information. However, the extra feature information may increase the complexity of computation as well as the storage space. In this paper, we concentrate on the general minutiae-based algorithms.

The minutiae-matching algorithm in this paper is inspired by algorithms by Jain et al [1] and Bebis [2], but it differs from them in three main aspects. First, for the choice of the matching index we use Delaunay edges rather than minutiae or whole minutiae triangles. Second, we use a deformation model, which helps to deal with the elastic finger deformations that sometimes cause big problems in fingerprint verification. Third, to get a better matching performance, we consider singular points and minutiae sets for matching and apply a new matching scheme to compute the matching score.

2 Voronoi Diagram and Delaunay Triangulation of Minutiae Set

The use of geometric information in biometric technology is a novel approach. The Voronoi diagram and Delaunay triangulation are two fundamental geometric structures in computational geometry. They can be used to describe the topology of the fingerprint, which is considered to be the most stable information for fingerprint matching purposes.

A Voronoi region associated with a feature is a set of points that are closer to that feature than to any other feature. Given a set S of points $p_1, p_2,...,p_n$, the Voronoi diagram decomposes the 2D space into regions around each point such that all points in the region around p_i are closer to p_i than to any other point.

Let $V(S)$ be a Voronoi diagram of a planar point set S. Consider the straight-line dual $D(S)$ of $V(S)$, that is, the graph obtained by adding an edge between each pair of points in S whose Voronoi regions share an edge. The dual of an edge in $V(S)$ is an edge in $D(S)$. $D(S)$ is a triangulation of the original point set, and is called the Delaunay triangulation. The Delaunay triangulation of a set of minutiae is shown in Fig. 3.

There are four reasons that we use Delaunay triangulation for the alignment of minutiae set. (1) The Delaunay triangulation is uniquely identified by the set of points. (2) Inserting a new point or dropping a point in a set of points only affects the triangulations locally, which means the algorithm can tolerate some error in minutiae extraction. (3) The total number of triangles in a set of n distinct points is $O(n^3)$, while the number triangles in Delaunay triangulation is only $O(n)$, which is very useful to speed our algorithm. (4) Delaunay edges are shorter on average than edges connecting two minutiae chosen randomly. Considering the elastic deformation of fingerprint, when the distance between two points in template image are further, the deformation is larger, so it is more difficult to match point pairs in input image even though the two images are corresponding images. Zsolt [5] have shown that local deformation of less than 10% can cause global deformation reaching 45% in edge length.

Bebis et. al. [2] used the complete Delaunay triangle as the comparing index. They assumed at least one corresponding triangle pair can be found between the input and template fingerprint images. The assumption may be violated due to the following factors: (1) low quality of fingerprint image (2) unsatisfactory robustness of the feature extraction algorithm (3) fingerprint deformation. We analyze three triangle edges corresponding to each triangle separately, so it is easier to find matching edges than complete triangles.

3 Fingerprint Verification

The purpose of fingerprint verification is to determine whether two fingerprints are from the same finger or not. In order to do this, we have to align the input fingerprint with the template fingerprint represented by its minutia pattern. The following rigid transformation can be performed:

$$F_{s,\Delta\theta,\Delta x,\Delta y}\begin{pmatrix}x_{templ}\\y_{templ}\end{pmatrix} = s\begin{pmatrix}\cos\Delta\theta & -\sin\Delta\theta\\\sin\Delta\theta & \cos\Delta\theta\end{pmatrix}\begin{pmatrix}x_{input}\\y_{input}\end{pmatrix}+\begin{pmatrix}\Delta x\\\Delta y\end{pmatrix} \qquad (1)$$

where $(s,\Delta\theta,\Delta x,\Delta y)$ represent a set of rigid transformation parameters: (scale, rotation, translation). Under a simple affine transformation, a point can be transformed to its corresponding point after rotating $\Delta\theta$ and translating $(\Delta x,\Delta y)$. In this paper, we assume that the scaling factor between input and template images is identical since both images are captured with the same device. We divide our algorithm into three stages as follows:

Step 1: Identification of feature patterns (based on triangle edges)
Step 2: Application of Radial Basis Function (RBF) to model the finger deformation and re-align images.
Step 3: Comparison of singular points and computation of the matching score.

3.1 Alignment of Minutiae Set — Local Matching

Let $Q = ((x_1^Q, y_1^Q, \theta_1^Q, t_1^Q)...(x_n^Q, y_n^Q, \theta_n^Q, t_n^Q))$ denote the set of n minutiae in the input image ((x,y): location of minutiae; θ: orientation field of minutiae; t: minutiae type, end or bifurcation;) and $P = ((x_1^P, y_1^P, \theta_1^P, t_1^P)...(x_m^P, y_m^P, \theta_m^P, t_m^P))$ denote the set of m minutiae in template image.

Table 1. Data structure used for comparing fingerprint images
(Y: Dependent on fingerprint transformation; N: Independent on it)

Feature	Fields					
Minutiae Point	x (Y)	y (Y)	θ (Y)	Type (N)		
Triangle Edge	Length (N)	θ_1 (N)	θ_2 (N)	$Type_1$ (N)	$Type_2$ (N)	Ridge count (N)

Table.1 shows features that we can use in local and global matching of fingerprints. (*Length*: length of edge; θ_1: angle between the edge and the orientation field at the first minutiae point. $Type_1$: minutiae type of the first minutiae; *Ridge count*: the number of ridges that these two minutiae points cross.)

As we discussed before, using triangle edge as comparing index has many advantages. For local matching, we first compute the Delaunay triangulation of minutiae sets Q and P. Second, we use triangle edge as our comparing index. To compare two edges, *Length*, θ_1, θ_2, *Type₁*, *Type₂*, *Ridgecount* (see Table. 1) values are used, all of which invariant of the translation and rotation. We assume two edges match if they satisfy the following set of conditions:

$$\frac{|Length_{input} - Length_{template}|}{\max(Length_{input}, Length_{template})} < threshold1$$

$$|\theta_{1input} - \theta_{1template}| < threshold2$$

$$|\theta_{1input} - \theta_{1template}| < threshold2 \qquad (2)$$

$$Type_{1input} = Type_{1template}$$

$$Type_{2input} = Type_{2template}$$

$$|Ridgecount_{input} - Ridgecount_{template}| < threshold3$$

Depending on the image size and quality, we can specify different *threshold*. When two edges match successfully, transformation $(\Delta\theta, \Delta x, \Delta y)$, which is used to align the input and template images, is obtained. If one edge from an input image matches two edges from the template image, we need to consider the triangulation to which this triangle edge belongs to and compare the triangle pair. For a certain range of translation and rotation dispersion, we pick for each transformation $(\Delta\theta, \Delta x, \Delta y)$, detect the peak in the transformation space, and record those transformations that are neighbors of the peak in transformation space.

We should note that these recorded transformations are close to each other, but not identical. If minutiae in the input image have the identical transform or we use the average transformation in global matching, that is the rigid transformation and we do not need to consider the deformation. Unfortunately, our fingers are elastic and deformation exists when images are taken, which can not ignored.

3.2 Modeling Deformation Using Radial Basis Functions

The deformation problem arises because of the inherent flexibility of the finger. Pressing or rolling a finger against a flat surface induces distortions which may vary from one impression to another. Such distortions lead to relative translations of features when comparing one fingerprint image with another. A good fingerprint identification system will always compensate for these deformations.

We want to develop a simple framework aimed at approximately quantifying and modeling the local, regional and global deformation of the fingerprint. We propose to use the Radial basic functions (RBF), which represent a practical solution to the problem of modeling of a deformable behavior. The application of RBF has been explored in medical image matching [6, 7] and also in image morphing. But to the best of our knowledge, it has not been applied for fingerprint verification.

For our fingerprint matching algorithm, deformation problem can be described as: knowing the consistent transformations of some points which we call control points in the minutiae set of input image, how to interpolate the transformation of other minutiae which are not control points. We do not consider all the transformations obtained by the local matching. We vote for them in the transformation space and pick those consistent transformations which form large clusters.

In the following, we briefly describe the basic idea of RBF method. Given the coordinates of a set of corresponding points (control points) in two images: $\{(x_i, y_i), (u_i, v_i) : i = 1,...,n\}$, determine function $f(x, y)$ with components $f_x(x, y)$ and $f_y(x, y)$ such that

$$u_i = f_x(x_i, y_i),$$
$$v_i = f_y(x_i, y_i), \quad i = 1,...,n. \quad (3)$$

Rigid transformation can be decomposed into a translation and a rotation (See Eq.(1)). In 2D, a simple affine transformation can represent rigid transformation as:

$$f_k(\vec{x}) = a_{1k} + a_{2k} x + a_{3k} y \qquad k = 1,2 \quad (4)$$

where $\vec{x} = (x, y)$. There are only three unknown coefficients in rigid transformation. The RBF is used in non-rigid transformation. In the two dimensional case, it is determined by $n+3$ coefficients in each dimension.

$$f_k(\vec{x}) = a_{1k} + a_{2k} x + a_{3k} y + \sum_{i=1}^{n} A_{ik} g(r_i) \qquad k = 1,2 \quad (5)$$

The first component is an affine transformation. The latter component is the sum of a weighted elastic or nonlinear basis function $g(r_i)$ which is related to the distance between $\vec{x}$ and the ith control point. The coefficients of the function $f_k(\vec{x})$ are determined by requiring that $f_k(\vec{x})$ satisfy the interpolation conditions:

$$f_1(\vec{x}_i) = u_j \quad \text{and} \quad f_2(\vec{x}_i) = v_i \qquad \text{for} \qquad j = 1,...,n \quad (6)$$

where n is the number of control points. Giving n linear equations together with the additional compatibility constrains:

$$\sum_{i=1}^{n} A_{ik} = \sum_{i=1}^{n} A_{ik} x_i = \sum_{i=1}^{n} A_{ik} y_i = 0 \qquad k = 1,2 \quad (7)$$

These conditions guarantee that the RBF is affine reducible, i.e. the RBF is purely affine whenever possible. There are $2(n+3)$ unknown coefficients and totally $2(n+3)$ functions in (6),(7). Solution of this linear system can be referred to [6,7].

Fig.1 (c) shows the alignment of two images Fig.1 (a) and (b) when we consider the transformation of input image as a rigid transformation. We tried every transformation in rigid transformation space and found that the maximum number of matching minutiae pairs between input image and template image is 6 (labeled by the big dashed circle in Fig. 1(c)). Circles denote minutiae of the input image after transformation, while squares denote minutiae of the template image. Knowing transformation of five minutiae (control points) in the input image, we apply the RBF to model the non rigid deformation which is visualized by the deformed grid in Fig. 1 (d). The number of matching minutiae pairs is 10, which greatly increased the matching scores of these two corresponding images.

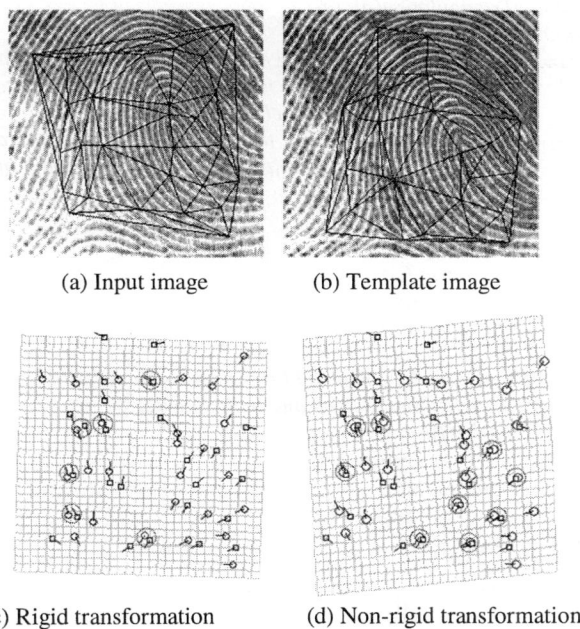

(a) Input image   (b) Template image

(c) Rigid transformation   (d) Non-rigid transformation

Fig. 1. Comparison of rigid transformation and non-rigid transformation in fingerprint matching

3.3 Global Matching

The purpose of global matching is to find the maximum number of matched features. After local matching procedure was performed and deformation model was applied to minutiae set, the numbers of paired and matched minutiae can be obtained. If two minutiae fall into the same tolerance box after identification, they are defined as paired in this paper. If paired minutiae have equal directions (within some tolerance), they become matched. So, now each minutia in both the template fingerprint and the input fingerprint is classified *as paired, matched, paired but unmatched or unpaired*. We propose our matching score:

$$M_{s1}(q,p) = \frac{100 \times m^r}{\sqrt{n'_q \times n'_p}} \qquad M_{s2}(q,p) = \frac{100 \times m^r}{N_{pair}} \qquad (8)$$

where m^r, N_{pair} represent the number of matched and paired minutiae respectively, n'_q and n'_p are the number minutiae in the overlap area of input and template image respectively.

```
Input:      Minutiae set q in input fingerprint image;
            Minutiae set p in template fingerprint image;
Ouput:      1 if the two images are corresponding images;
            0 if the two images are non corresponding images;
Begin
            DelaunayTriangulation(q);
            DelaunayTriangulation(p);
            For i=1 to M           (M is the number of triangle edge of input image)
                    For i=1 to N   (N is the number of triangle edge of template image)
                            If (edge(i) match edge(j))
                                    Save this match and record the transformation
                                    En(q,p)++;         (the number of matched edge)
                            End If
                    End For
            End For
            If       (En(q,p)<T_{e1})
                            return (0);
            else if  (En(q,p)>T_{e2})
                            return (1);
            Apply RBF to model the non rigid transform of fingerprint
            Compare the two minutiae sets and their singular points
            If       ( m^r <Tm1)
                            return (0);
            else if  ( m^r )>Tm2)
                            return (1);
            else if  (M_{s2}(q,p)<T_{s2})
                            return (0);
            else if  (M_{s1}(q,p)>T_{s1})
                            return (1);
            else
                            return (0);
            End If
End
```

Fig. 2. Sequential fingerprint matching algorithm

Even though the proposed formulas (8) is an improvement to the matching score in [3], how to utilize the information of matched minutiae is still not obvious and more research and experimentation should be devoted to this problem. To address the issue and get a more efficient matching algorithm, we [9] resort to the singular points to validate our matching. It is important to note that the minutiae matching and singular points validation method presented are statistically very robust. Even if some of the minutiae or singular points may not be extracted due to poor quality of image or not corrected matched, experiments show that the remaining minutiae are still sufficient in order to validate the identification. These minutiae or singular points may be matched and even validated, but they are statistically insignificant.

The sequential matching algorithm is shown in Fig.2. Note, $En(p,q)$ represents the number of matched triangle edges in local matching. T_{e1}, T_{e2} represents the two thresholds of matched triangle edges. Because we assume if input and template images come from same finger then at least one corresponding triangle edge pairs can be found between them, T_{e1} is often set to 1.

4 Experiments

We tested our matching algorithm on a fingerprint database from the University of Bologna. This fingerprint database contains 8 images per finger from 21 individuals for a total of 168 fingerprint images. The size of these images is 256 ×256. The captured fingerprint images vary in quality. The system is currently implemented using C++ programming environment and tested on a Celeron 1400MHZ CPU 512 RAM computer. Average time required for one matching is shown in Table 2.

Each fingerprint image in the test set was matched with other images except itself. There are totally 168 ×167 matching. The FRR (False Refused Rate) is the complement to one of the identification rates, i.e. the number of corresponding image pairs which have not been matched with sufficient accuracy divided by the total number of corresponding image pairs. The FAR (False Acceptance Rate) is the number of non corresponding image pairs which are matched by the system as a percentage of the total number of non corresponding image pairs.

Table 2. Some experiments reported by our matching algorithm
(Given $T_{e1} = 1, T_{e2} = 20, T_{m1} = 5, T_{m2} = 12, T_{s2} = 70$)

		FRR	FAR	Time cost for one matching
General minutiae based method		17.09%	0.841%	98 ms
Our method	T_{s1}=35	8.03%	0.32%	27 ms
	T_{s1}=40	8.48%	0.18%	
	T_{s1}=45	9.82%	0.09%	

We can estimate the performance of our matching algorithm by comparing it with the general minutiae based matching algorithm. As it can be seen from Table 2, the performance of our method is very good in terms of time cost, and it is also very efficient in terms of a space, since we use the triangle edge as our comparing index. The application of RBF to model finger deformation and our improvement in global matching can greatly increase the matching accuracy.

5 Conclusions

We have developed a novel minutiae matching algorithm, which is superior to traditional minutiae based method introduced in [1],[3]. The difference is that triangle edges of the Delaunay triangulation constructed for the minutiae set are used for matching. Triangle matching is fast and overcomes the relative non linear deformation

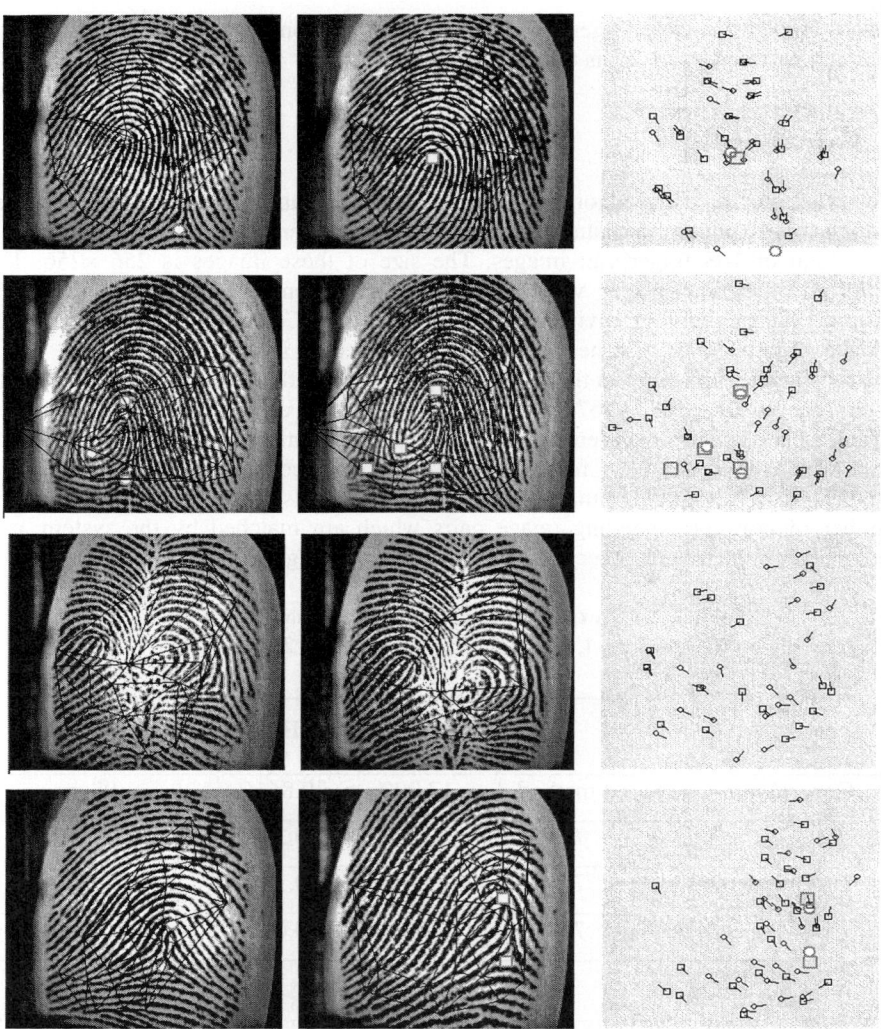

Fig. 3. Some experimental results for our matching algorithm. The first and second columns are input and template images respectively. The third column is the point pattern of template image plus point pattern of input image after transformation. Squares represent the minutiae or singular points from input image, while circles represent those from template image. Large circle or square denote singular points. The first two rows are correct matching. Rows three and four show unmatched images (FRR) and false matched images (FAR) respectively. (Fingerprint data base comes from University of Bologna.)

problems, because it is independent on the fingerprint rigid transformation. To overcome the relative deformation that is present in the fingerprint image pairs, we also proposed a novel RBF model that is able to deal with elastic distortions. Using RBF model, the algorithm handles all possible non linear distortions while using very tight bounding boxes. Compared with most methods that rely on rigid transformation assumptions, these modifications make our method more robust for non-linear deformation between two fingerprint images. Our experiments show that fingerprint images can be well matched using the minutiae based matching method, and that it is a fast, reliable and memory efficient technique.

Our future works will focus on two aspects. Experiments show that singular points can greatly help us to decrease FRR. Our objective for future research is to add some new features for our fingerprint verification technique. Second, we will concentrate on the fingerprint image enhancement algorithms that allow us to extract features more accurately.

Acknowledgements

Authors would like to acknowledge generous support from CFI, NSERC and GEOIDE Canadian granting agencies. We are also grateful to International Center for Voronoi Diagram Research (Korea) for helpful suggestions and comments on the paper.

References

[1] N. K. Ratha, K. Karu, S. Chen, and A. K. Jain, "A Real-Time Matching System for Large Fingerprint Databases", PAMI Vol.18, No.8, pp. 799-813, 1996.
[2] G. Bebis, T. Deaconu, M. Georiopoulous, Fingerprint identification using Delaunay triangulation, ICIIS99, Maryland, Nov, pp. 452-459, 1999.
[3] Jain, A., Lin, H., Bolle, R. On-line fingerprint verification, IEEE Trans Patt Anal 19 (4) pp. 302–313, 1997.
[4] Jiang, X., Yau, W.-Y, 2000. Fingerprint minutiae matching based on the local and global structures. In: Proc. 15th Internet. Conf. Pattern Recognition (ICPR, 2000) 2. pp.1042–1045.
[5] Zs. Miklos Kovacs-Vajna, "A Fingerprint Verification System Based on Triangular Matching and Dynamic Time Warping", IEEE Trans. on PAMI, Vol.22, No.11, 2000, pp.1266-1276
[6] M. Fornefett, K. Rohr and H.S. Stiehl, Radial basic functions with compact support for elastic registration of medical images, Image and Vision Computing, 19: pp. 87-96 (2001).
[7] Wirth, M.A., Choi, C., Jennings, A., A nonrigid-body approach to matching mammograms, in IEEE 7th International Conference on Image Processing and its Applications, Manchester UK, 1999, pp.484-487.
[8] A.W. Senior, R. Bolle. Improved Fingerprint Matching by Distortion Removal, In IEICE Transactions on Information and Systems, special issue on biometrics. Volume E84-D No. 7 July 2001 pp. 825-832.
[9] Chenfeng Wang and Marina L. Gavrilova, A Multi-Resolution Approach to Singular Point Detection in Fingerprint Images, International Conference of Artificial Intelligence, vol I, 2004, pp. 506-511.

Bilateral Estimation of Vertex Normal for Point-Sampled Models

Guofei Hu[1,*], Jie Xu[2], Lanfang Miao[1], and Qunsheng Peng[1]

[1] State Key Lab. of CAD&CG, Zhejiang University,
Hangzhou 310027, P.R.China
[2] School of Computer Science University of Waterloo,
Waterloo, Canada

Abstract. Vertex normal is an essential surface attribute for point-based rendering and modelling. Based on scale space theory, we propose a bilateral vertex normal estimation algorithm for point-sampled models. We adaptively construct multiple local polygonal rings for each vertex after a multi-layer neighbor decomposition, the vertex normals are then computed by applying a bilateral estimation scheme on the multi-layer neighbors. A brief survey of vertex normal estimation on point-sampled surfaces is also presented in this paper, and the detailed comparisons of the different methods on theoretical basis, time and space complexity, applied environments, numerical implementations and effects with different initial conditions are analyzed.

1 Introduction

Point-sampled models are normally generated by sampling the boundary surface of physical 3D objects with 3D scanning devices. Adhering to this kind of models, the objects are fully represented by the geometry of the sample points without additional topological connectivity and parametric information. As a straightforward representation form of highly complex sculptured object, the point-sampled model can be used in a variety of applications. In recent years, researchers developed various efficient techniques on representation[12], processing [16][17][27] and rendering[22][23][28] of point-sampled models taking points as effective primitive.

Besides the geometric position, the point-sampled model can also store a set of surface attributes at each sample point, such as surface normal, curvature, color, material properties or texture coordinates[18]. Vertex normal is of importance to point-based shading and rendering as well as point-based modelling and processing. In fact, Various surface processing techniques such as denoising[10], feature detection[18], shape modelling[19] and so on, involve vertex normal as part of input data. In this paper we will focus on the estimation of vertex normal on point-sampled surfaces.

---

* Supported by the National Natural Science Foundation of China (No.60103017) and the National Key Basic Research and Development Program (No.2002CB312101).

The classical way of estimating vertex normal for point-sampled models is to find a piece of surface approximating or interpolating a local set of unorganized points neighboring to the vertex, setting vertex normal as the normal of the counterpart vertex on the fitting surface. There are a variety of surfaces which have been adopted to serve this purpose, e.g., plane[8][17], triangulated polyhedron[13], NURBS patches[15], bivariate polynomial surface[1], or other higher order surface[26]. Since reconstruction of the local geometry with higher order surface are always computationally expensive, most previous works favor the planar fitting or polyhedral interpolating to the points when estimating the vertex normal.

Related work. Hoppe etal.[8] proposed a novel method to determine an oriented tangent plane at each sample point by introducing a signed distance function. They then converted the least squares best fitting problem to solving an eigensystem of a covariance matrix. Pauly et al.[17] extended the above method to estimate local surface properties such as normal and surface variation(an approximation to mean curvature) with a MLS kernel function. Gopi et al.[6] also proposed a different formulation to calculate the normals. Nevertheless our experimental data indicate that the vertex normals derived by these three methods are almost the same. Recently Mitra et al.[14] analyzed how the neighborhood size, curvature, sampling density, and noise affect the estimation of vertex normal by adopting a local least square fitting approach. Another approach of estimating vertex normal is to reconstruct unstructured cloud points locally or globally, then compute the vertex normal by averaging the normals of its neighboring triangles. Following this scheme, a variety of efficient triangulation approaches[2, 3, 6] accounting for some local properties(e.g.,Delaunay property) were developed in the past few years. Gouraud [7]presented the first vertex normal estimation algorithm adopting an equal mean weight in 1971. Thurmer et al.[25] and Max[13] proposed algorithms of mean weighted by angle and mean weighted by areas of adjacent triangles. Recently Jin et al. [9] developed a vertex normal recovery technique based on radial basis functions.

In this paper, we present a different approach, we adaptively reconstruct multiple polygonal rings at each vertex and then estimate vertex normal adopting a bilateral estimation scheme. The estimated normal can be applied to the reconstruction of polyhedral mesh[6], local parametric surface[18], global implicit surface[4] and so on.

2 Current Methods of Normal Estimation

In the remainder of this paper we use $P \subset R^3$ to denote the point set, $p_i = (x_i, y_i, z_i)$ a vertex in P and $N(p_i) = \{q_{i,1}, q_{i,2}, ..., q_{i,k}\}$ the k *nearest neighboring points* of p_i. Similarly for triangular mesh M, we adopt p_i to denote a vertex in $M \subset R^3$ and $N(p_i) = \{q_{i,1}, q_{i,2}, ..., q_{i,m}\}$ the m neighboring vertices connected with p_i by an edge.In particular, T_i will denote a triangle of the mesh, c_i the barycenter of T_i.

2.1 PCA Analysis

A natural idea of estimating normal at p_i is to find a plane approximating the unorganized points $N(p_i)$ by least-square fitting. Assume that the plane is $D: r_i \cdot n_i - d = 0$. n_i is estimated by solving the following least-square optimization:

$$min(\sum_{j=0}^{k}(q_{i,j} \cdot n_i - d)^2) \quad (1)$$

Based on the theory of Principal Component Analysis(PCA), Hoppe et al.[8] defined a 3×3 covariance matrix:

$$C_i = \begin{bmatrix} q_{i,1} - \overline{p_i} \\ q_{i,2} - \overline{p_i} \\ \cdots \\ q_{i,k} - \overline{p_i} \end{bmatrix}^T \begin{bmatrix} q_{i,1} - \overline{p_i} \\ q_{i,2} - \overline{p_i} \\ \cdots \\ q_{i,k} - \overline{p_i} \end{bmatrix} \quad (2)$$

where C_i is a symmetric positive semi-define matrix, $\overline{p_i}$ is the centroid of $N(p_i)$. The normal of the p_i is chosen to be the unit vector e_1 which corresponds to the minimal eigenvalue of C_i. It is a simple and efficient method for estimating vertex normal. Many researchers extended this method to various applications with improvements for simplicity and efficiency of implementation.

In 2000, Gopi et al. proposed another eigenvalue analysis method to estimate normal [6] by defining a new model of minimization:

$$min(\sum_{j=0}^{k}((q_{i,j} - \overline{p_i}) \cdot n_i)^2) \quad (3)$$

The above minimization problem can be converted to a singular value decomposition(SVD) problem:

$$min(\|A_i n_i\|_2) \quad (4)$$

where A_i is a $k \times 3$ matrix:

$$A_i = \begin{bmatrix} (q_{i,1} - p_i) - a_i \\ (q_{i,2} - p_i) - a_i \\ \cdots \\ (q_{i,k} - p_i) - a_i \end{bmatrix}, a_i = \frac{\sum_j (q_{i,j} - p_i)}{k} \quad (5)$$

2.2 Improved PCA Method

Pauly et al.[17][20]improved the PCA method by introducing weighted squared distances to the original minimization formula:

$$min(\sum_{j=0}^{k}(q_{i,j} \cdot n_i - d)^2 \phi_i), \phi_i = e^{-\frac{\|q_{i,j} - p_i\|}{\sigma^2 r^2}} \quad (6)$$

where ϕ_i is called MLS kernel function, it is typically set to a Gaussian kernel, r is the radius of the neighborhood sphere. Similarly, the method of Lagrange multipliers can be used to solve the constrained least-squares problem by reducing it to the eigenvector problem with the same form as Eq.(2), except that the centroid $\overline{p_i}$ is weighed average of $q_{i,j}$. Owing to the property of Gaussian kernel, the above process can remove some noise of scanned models when solving vertex normals with Eq.(6).

2.3 Mean Weighted Normal of Adjacent Triangles

Another class of methods are based on the local or global shape reconstruction of the scattered point set. It is a non-trivial task to reconstruct triangular mesh globally[2][4]. Local geometric reconstruction is relatively easy[12], but different polyhedral approximations result in different mesh connectivity, hence different vertex normal. Despite the slight difference in geometric reconstruction, the normal at a vertex p_i is calculated commonly as an average of the normals of the triangular faces adjacent to the vertex.

More complicated surfaces, such as NURBS or Bézier patches or implicit surface with RBF and so on, can also be used to fit or interpolate the unorganized points, but usually these methods are more complicated and time-consuming.

3 Bilateral Normal Estimation

Our bilateral normal estimation algorithm consists of the following steps:

- **Neighbors decomposition:** The neighbors of vertex are decomposed to multiple layers at different scale.
- **Normal initialization:** A simple scheme is applied to initialize the normals of all vertices.
- **Local polyhedron reconstruction:** A star-shape polygon is reconstructed for each layer of neighboring points.
- **Bilateral normal estimation:** Normals at each vertex is estimated in term of the contributions from triangles of different layers.

3.1 Neighbors Decomposition

Given a point-sampled model, we don't normally know the sampling density beforehand, moreover, its density may not be uniform on the different regions of surface. Therefore, the neighborhood size k is an essential parameter for good shape reconstruction results. In [18], Pauly et al. suggested to select a $k \in [6, 20]$. In general k should take different value for different applications and for different models. To reconstruct the local surface at each vertex properly based on the local sample point set, we decompose its neighboring points into multi-layers according to the Euclidean distance from $N(p_i)$ to p_i(see Fig.1), then compute vertex normal by accounting for weighted contributions from different neighbor layers. Comparatively, if k takes a large value, and all k neighboring points are

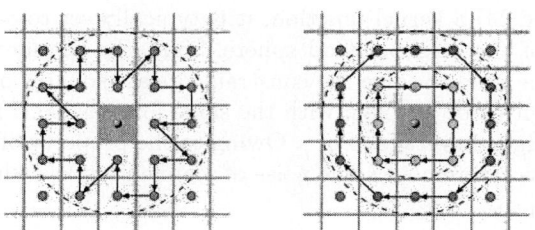

Fig. 1. multi-layer decomposition of local neighbors

organized into a single ring to compose a star-shaped polygon around the vertex, the boundary of the polygon must be dramatically zigzag and the resultant surface will no longer be smooth.

3.2 Initialization and Consistency Processing

We initialize the normal at vertex p_i as follows:

$$n_{i,0} = \frac{\sum_{j=1}^{k} w_j (p_i - q_{i,j})}{\sum_{j=1}^{k} w_j} \qquad (7)$$

where the weights satisfy $w_j = 1$. Obviously, according to Eq.(7), the direction of $n_{i,0}$ at vertices on the convex region is always opposite to that on the concave region, which would make the reconstructed point-sampled surface not orientable. To ensure that the point-sampled surface is an orientable surface, we must guarantee that, for arbitrary vertex p_i and its ε-neighborhood $N(q_{i,j})$, their normals satisfy $n_{p_i} \cdot n_{q_{i,j}} > 0$. Without loss of generality, we assume that the point-sampled surface is simple-connected. Thus, for each vertex p_i on P, if $n_{p_i} \cdot n_{q_{i,j}} < 0$, then turn over the direction of $n_{q_{i,j}}$ to be the same as n_{p_i}. The above process is conducted recursively with a seed. Like [6] and [8], we adopt a brute force scheme to orient the normals of all vertices to derive an orientable point-sampled surface.

3.3 Local Polyhedron Reconstruction

The initial normal $n_{i,0}$ determines a tangential plane $(n_{i,0})^{\perp}$ for (p_i). Let $N(q'_{i,j})$ denote the projection points of neighboring vertices onto the plane, we can construct a ring of $N(q'_{i,j})$ in clockwise order on the plane for each layer of neighboring points, we then reconstruct a ring of $N(q_{i,j})$ with the same topology as $N(q'_{i,j})$. Owing to the simplicity of the above reconstruction scheme, it is no longer necessary to consider Delaunay property of the embeded triangular mesh, which makes our approach less expensive. After the *local polyhedron reconstruction* at multi-levels for all vertices, we then estimate normal at p_i as described in the next subsection.

3.4 Scale Space Theory and Bilateral Estimation

In the computer vision field, various image filtering and edge detection techniques have been presented based on the scale-space theory[11]. In general, a new image can be obtained by convoluting Gaussian kernel with original image:

$$I(x,y,t) = I_0(x,y) * G(x,y,t) \tag{8}$$

where t is the scale parameter. Based on Eq (8), Perona et al.[21] presented an anisotropic edge detection method which was later extended to the applications of mesh smoothing and feature extraction. Recently Fleishman[5] and Jones[10] proposed two bilateral mesh smoothing algorithms. Here, we propose a vertex normal estimation technique by applying the scale space theory to point-sampled surface. We take areas of adjacent triangles and distances from the centroids of these triangles to the concerned vertex into account, and take the number of layers as scale parameter to estimate vertex normals as follows:

$$\boldsymbol{n}_i^j = \frac{\sum_k W_\sigma(\|\boldsymbol{p}_i - \boldsymbol{c}_{i,j}^k\|) W_\sigma(1/a_{i,j}^k) \boldsymbol{n}_{i,j}^k}{\sum_k W_\sigma(\|\boldsymbol{p}_i - \boldsymbol{c}_{i,j}^k\|) W_\sigma(1/a_{i,j}^k)} \tag{9}$$

where $W_\sigma(x) = e^{-x^2/2\sigma^2}$ is Gaussian error norm function, σ is the scale parameter. $\boldsymbol{n}_{i,j}^k$ is the normal of kth triangle adjacent to $\boldsymbol{p}_i$ in the jth neighbors layer, $\boldsymbol{c}_{i,j}^k$ is the centroid of the triangle, $a_{i,j}^k$ is its area, $\boldsymbol{n}_i^j$ is the bilateral estimation of normal from jth neighbors layer. After obtaining normals from various layers, we calculate vertex normal $\boldsymbol{n}_i$ at $\boldsymbol{p}_i$ by w_i^j-weighted average.

Three issues should be paied attention to while applying the above formulations : First, contributions from different layers of neighbors are not the same, so we must assign appropriate weights w_i^j for each layer. Second, noises are inevitably introduced in the process of scanning, our Gaussian kernel has the potentials to denoise out the model by controlling the scale parameter. Third, the initial normals may be less satisfactory, good results can be obtained after multiple iterations of the following three steps: consistency processing, local polyhedron reconstruction and bilateral estimation.

4 Experimental Results

We have implemented in Microsoft Windows Pro XP using the c++ language the six vertex normal estimation approaches as described in the previous section: PCA[Hoppe et al. 1992], Improved PCA[Pauly et al. 2002], Gopi's method[2000], mean weighted normals by areas of adjacent triangles after local polyhedron reconstruction, mean weighted normals based on the global reconstructed triangular mesh, and finally our bilateral normal estimation algorithm. The PCA and improved PCA solve the eigenvector problem by Jacobian while Gopis's method solves it by SVD(singular value decomposition). All functions run on a 1.6G Pentium IV with 512M of memory. To facilitate comparisons between different normal estimation methods, in the experiment, we sampled points directly from the original triangular meshes of the testing models, and calculated normal deviation error between the estimated normal $\bm{n}_i$ and the normal $\bm{n}_i'$ computed from original meshes as follows:

$$m_i = dot(\bm{n}_i, \bm{n}_i') \tag{10}$$

the mean and variation of $\{m_i\}$ are denoted by $\overline{m}$ and σ_m.

In Fig.2, we set $k = 30$ when estimating normals with different methods. The normal deviation introduced by local polyhedron reconstruction within single layer is the biggest and PCA and Improved PCA have almost the same result as the Gopi's method. For these methods, the vertices whose derived normals deviate greatly are most likely located at the regions with large curvature. The result derived by our bilateral normal estimation based on multi-layer reconstruction, on the other hand, is slightly different. If we assume the normal estimation error at all vertices as a 2D distribution function over the boundary surface of the model, the normal error distribution of our method is relatively dispersive.

Fig.3 shows that the normal estimation errors introduced by previous methods have a distribution similar to that of curvature of the point-sampled surfaces.

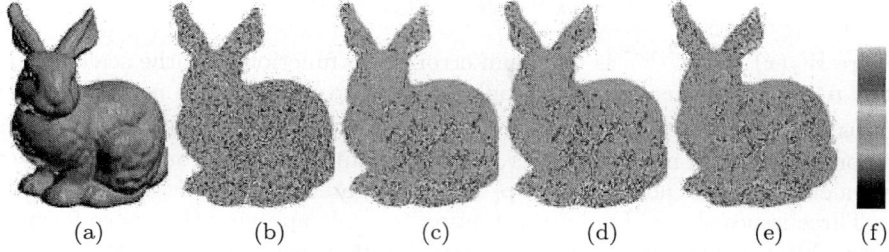

Fig. 2. (a) original bunny model; (b) color-coded normal error using a single layer reconstruction; (c) color-coded normal error using methods of Hoppe 1992 and Gopi 2000 respectively); (d) color-coded normal error using Improved PCA analysis(Pauly 2002); (e)color-coded normal error using our bilateral normal estimation; (f) color bar:white-red-green-blue-dark dithering in term of the normal error

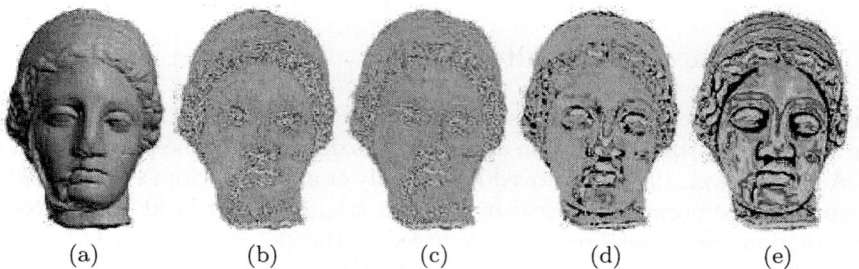

Fig. 3. color-coded visualizations of normal estimation error and curvature of vertex on point-sampled surfaces: (a) original Igea model; (b) PCA; (c) Our method; (d) Gaussian curvature; (e) mean curvature

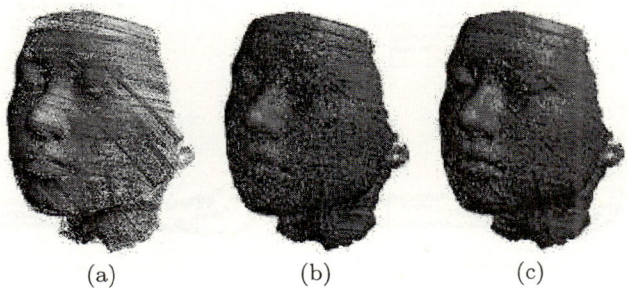

Fig. 4. (a) original holehole model; (b) mean weighted average of area of adjacent triangles after single layer reconstruction; (c) our bilateral estimation after multi-layer neighbors decomposition and reconstruction

$k = 3$ $k = 12$ $k = 30$ $k = 60$

Fig. 5. first row: normal estimation based on simple local reconstruction with single layer. second row: normal estimation based on PCA analysis(Hoppe 1992, Gopi 2000 and Pauly 2002). third row: our bilateral normal estimation based on multi-layer decomposition of neighbors

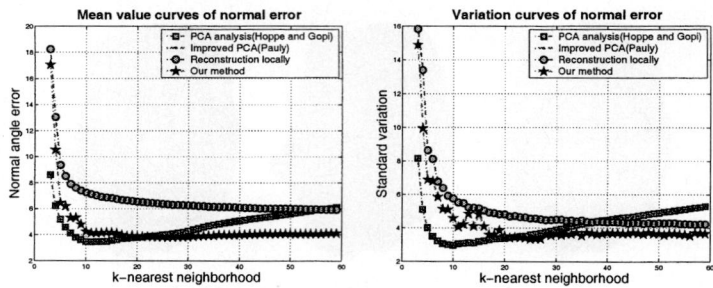

Fig. 6. comparisons of mean and variation of normal estimation error

Fig.4 shows a noisy model, while the appearance of reconstructed model with normal estimated by mean weighted average of area of adjacent triangles within single layer are still noisy, the result generated by our bilateral estimation method accounting for multi-layer neighbors is relatively smooth.

Fig.5 shows that PCA method gets an oversmoothing result when k increases, while our method is stable. PCA is the best, however, if k is small.

Fig.6 shows the mean and variation curves of estimated normal error by different methods in term of different k, PCA based methods(Hoppe,Gopi,Pauly et al.) have the lowest error near $k = 11$, and our method is more stable when $k > 11$. Although our approach is relatively computationally expensive because of local-polyhedron reconstruction at multi-levels, it results in the smallest error when k is big enough showed in Fig.5 and Fig.6.

5 Conclusion

We have presented a new normal estimation algorithm based on a multi-layer neighboring points decomposition and a bilateral normal estimation scheme. The approach is motivated by the image-based scale space theory as well as mesh denoising techniques. The aim of this work is to quickly estimate vertex normals based on the unorganized points and eliminate the affect of noises which embedded in the point set. This is achieved by approximating the local shape of the surface at multiple scales. The method induces little error when the neighboring area accounted for becomes large. The estimated vertex normals can be applied to other applications such as curvature estimation, point-based modelling,and point-based rendering.

References

1. M. Alexa, J. Behr, D. Cohen-Or, and S. Fleishman. Point set surfaces. *IEEE Visualization 01*, pages 21–28,2001.
2. N. Amenta, M. Bern, and M. Kamvysselis. A new voronoi-based surface reconstruction algorithm. *In Proc. of ACM SIGGRAPH 1998*, pages 415–422, 1998.

3. C. L. Bajaj, F. Bernardini, and G. Xu. Automatic reconstruction of surfaces and scalar fields from 3d scans. *In Proc. of ACM SIGGRAPH 1995*, pages 109–118, 1995.
4. J. C. Carr, R. K. Beatson, J. B. Cherrie, and T. J. Mitchell. Reconstruction and representation of 3d objects with radial basis functions. *In Proc. of ACM SIGGRAPH 2001*, pages 67–76, 2001.
5. S. Fleishman, I. Drori, and D. Cohen-Or. Bilateral mesh denoising. *In Proc. of SIGGRAPH03*, pages 950–953, 2003.
6. M. Gopi, S. Krishnan, and C. Silva. Surface reconstruction based on lower dimensional localized delaunay triangulation. *Computer Graphics Forum*, 19(3):467–478, 2000.
7. H. Gouraud. Continuous shading of curved surfaces. *IEEE Trans. on Computers*, 20(6):623–629, 1971.
8. H. Hoppe, T. DeRose, T. Duchamp, and J. McDonald. Reconstruction from unorganized points. *In Proc. of ACM SIGGRAPH 1992*, pages 71–78, 1992.
9. X. G. Jin, H. Q. Sun, J. Q. Feng, and Q. S. Peng. Vertex normal recovery using radial basis functions. *In Proc. of CAD&Computer Graphics 2003*, pages 29–31, 2003.
10. T. Jones, F. Durand, and M. Desbrun. Non-iterative, feature preserving mesh smoothing. *In Proc. of SIGGRAPH03*, pages 943–949,2003.
11. M. Kass, A. Witkin, and D. Terzopoulos. Snakes: Active contour models. *Proc. of IEEE Conference on Computer Vision,London, England*, pages 259–268, 1987.
12. L. Linsen. Point cloud representation. *Technical Report, Faculty of Computer Science, University of Karlsruhe*, 2001.
13. N. Max. Weights for computing vertex normals from facet normals. *Journal of Graphics Tools*, 4(2):1–6, 1999.
14. N. J. Mitra and A. Nguyen. Estimating surface normals in noisy point cloud data. *In Proc. of the nineteenth conference on Computational geometry,San Diego*, pages 08–10, 2003.
15. N. Park, I. D. Yun, and S. U. Lee. Constructing nurbs surface model from scattered and unorganized range data. *International Conference on 3-D Imaging and Modeling,Ottawa*, pages 312–320, 1999.
16. M. Pauly and M. Gross. Spectral processing of point-sampled geometry. *In Proc. of ACM SIGGRAPH 2001*, pages 379–386, 2001.
17. M. Pauly, M. Gross, and L. Kobbelt. Efficient simplification of point-sampled geometry. *IEEE Visualization 02* ,pages 163–170, 2002.
18. M. Pauly, M. Gross, and L. Kobbelt. Shape modeling with point-sampled geometry. *In Proc. of SIGGRAPH03, San Diego*,pages 641–650, 2003.
19. M. Pauly, R. Keiser, and M. Gross. Multi-scale feature extraction on point-sampled surfaces. *In Proc. of EUROGRAPHICS 03*, 2003.
20. M. Pauly, L. Kobbelt, and M. Gross. Multiresolution modeling of point-sampled geometry. *CS Technical Report #378, September 16, 2002*, 2002.
21. P. Perona and J. Malik. Scale-space and edge detection using anisotropic diffusion. *IEEE Transactions on Pattern Analysis and Machine Intelligence* ,pages 629–639,1990.
22. M. Pfister, J. Zwicker, V. Baar, and M. Gross. Surfels: Surface elements as rendering primitives. *In Proc. of ACM SIGGRAPH 2000*, pages 335–342, 2000.
23. S. Rusinkiewicz and M. Levoy. Qsplat: A multiresolution point rendering system for large meshes. *In Proc. of ACM SIGGRAPH 2000*, pages 343–352, 2000.
24. E. Shaffer and M. Garland. Efficient adaptive simplification of massive meshes. *IEEE Visualization 01*,pages 127–134,2001.

25. G. Thurmer and C. Wuthrich. Computing vertex normals from polygonal facets. *Journal of Graphics Tools*, pages 43–46, 1998.
26. W. Welch and A. Witkin. Free-form shape design using triangulated surfaces. *In Proc. of ACM SIGGRAPH 1994*, pages 247–256, 1994.
27. M. Zwicker, M. Pauly, O. Knoll, and M. Gross. Pointshop 3d: An interactive system for point-based surface editing. *In Proc. of ACM SIGGRAPH 2002*, pages 322–329, 2002.
28. M. Zwicker, H. Pfister, V. Baar, and M. Gross. Surface splatting. *In Proc. of ACM SIGGRAPH 2001*, pages 371–378, 2001.

A Point Inclusion Test Algorithm for Simple Polygons

Weishi Li, Eng Teo Ong, Shuhong Xu, and Terence Hung

Institute of High Performance Computing, 1 Science Park Road, #01-01 The Capricorn,
Science Park II, Singapore 117528
{liws, onget, xush, terence}@ihpc.a-star.edu.sg

Abstract. An algorithm for testing the relationship of a point and a simple polygon is presented. The algorithm employs a "visible edge" of the polygon from the point to determine whether the point is in the polygon or not according to the relationship of the orientation of the polygon and the triangle formed by the point and the visible edge. The algorithm is efficient and robust for floating point computation.

1 Introduction

Much of computational geometry, which is applied in geographical information systems, scientific visualisation, computer aided design/manufacturing, etc., performs its computations on geometrical objects known as simple polygons [11]. Point-in-polygon test is one of the most elementary tests in the applications. Several algorithms have been proposed and the most popular two are: counting ray crossing algorithm and computing winding number algorithm [7], [11-13].

The ray crossing algorithm is based on the number of the intersections between the polygon and a ray originating from the given point. An odd number of intersections indicate that the point is inside the polygon, and vice versa. The winding number algorithm is based on floating-point computations, particularly trigonometric computations. This makes it significantly slower than the ray crossing algorithm and also run into risk of computation errors [11].

Recent attempts to solve the point-in-polygon test problem were proposed by Feito et al.[5] and Wu et al.[14]. Feito et al. decomposed the polygon into a series of original triangles and carried out the test by testing the relationship between the given point and all the original triangles. They further improved the algorithm by considering the special case that the given point is located on the edges of the original triangles [4]. However, in terms of computation efficiency, it is still inferior to the ray crossing algorithm. The method proposed by Wu et al. is very similar to the ray crossing algorithm. But in the zero-crossing case, this method fails to decide the relationship between the given point and the polygon.

Above all, most of the publications are limited to polygons with integer coordinates. For polygons with floating point coordinates, even the ray crossing algorithm may also incur numerical errors due to rounding and cancellation errors [3], [6], [8].

In this paper, a new concept, namely visible edge, is put forward. An efficient visible edge searching algorithm, without solving any equation systems and/or using trigonometric functions, is developed. According to the orientation of the polygon and the triangle formed by the given point and the visible edge, the relationship between the given point and the polygon is determined.

The following section concerns the notation and basic facts used in this paper. The point-in-polygon test algorithm is presented in Section 3, followed by a conclusion that closes the paper.

2 Notations and Basic Facts

A simple polygon is defined as a finite collection of line segments forming a simple closed curve or the region of a plane bounded by the closed curve. Let $v_i = (x_i, y_i)$ for $i = 0, \ldots, n-1$ be n points in the plane and $e_0 = v_0 v_1$, $e_1 = v_1 v_2, \ldots$, $e_i = v_i v_{i+1}, \ldots$, $e_{n-1} = v_{n-1} v_0$ be n segments connecting the points. These segments bound a simple polygon if and only if the intersection between each pair of segments adjacent in the cyclic ordering is the single point shared between them, and nonadjacent segments do not intersect. In the following of this paper, we use "polygon" instead of "simple polygon" for convenience. Line segments, $e_0, e_1, \ldots, e_i, \ldots, e_{n-1}$, are called edges, and points, $v_0, v_1, v_2, \ldots, v_{n-1}$, are called vertices.

2.1 Orientation Test

A simple polygon is orientable, and the orientation can be theoretically decided by the winding number of an interior point [2]. Because the computation of winding numbers involves trigonometric functions, this method is error-prone and time consuming. Another method for deciding the orientation of a simple polygon makes use of the signed area of the polygon. If the area is positive, the polygon is oriented counter-clockwise. Otherwise, it is oriented clockwise. This approach is robust, but less efficient. A simpler method is to use a convex vertex of the polygon.

Consider a polygon P defined by $v_0, v_1, v_2, \ldots, v_{n-1}$, and v_i is a convex vertex. If $(v_j v_i \times v_i v_k) \cdot \mathbf{z} > 0$, then P is oriented counterclockwise. Otherwise, P is oriented clockwise. Here, $\mathbf{z} = (0, 0, 1)$, $j = (i+n-1) \% n$, and $k = (i+n+1) \% n$. It is obvious that the vertex with maximum/minimum coordinate(s) is a convex vertex if the vertex and its adjacent vertices are not collinear. Compared with the signed area method, the geometric meaning of this method is obvious and the computation is more efficient.

2.2 Signed Area of Triangles

For a triangle, $\Delta v_0 v_1 v_2$, its signed area can be computed as:

$$A(\Delta v_0 v_1 v_2) = ((x_1 - x_0)(y_2 - y_1) - (x_2 - x_1)(y_1 - y_0))/2. \tag{1}$$

If its area is positive, the triangle is counter-clockwise oriented and vice versa.

For an oriented edge $e_0 = v_0 v_1$ and a point p, if the area of triangle $\Delta v_0 v_1 p$ is positive, the point is on the left of the edge and vice versa. For two edges $e_0 = v_0 v_1$

and $e_i = v_i v_{i+1}$, if $A(v_0 v_1 v_i) > 0$ and $A(v_0 v_1 v_{i+1}) > 0$, e_i is on the left of e_0. Otherwise, if $A(v_0 v_1 v_i) < 0$ and $A(v_0 v_1 v_{i+1}) < 0$, e_i is on the right of e_0.

2.3 Ray/Edge Intersection

For an edge $e_i = v_i v_{i+1}$ and a ray l originating from a point $p = (x, y)$ and extending horizontally to the right, suppose $y_i < y_{i+1}$, if $y_i \leq y \leq y_{i+1}$ and p is on the left of e_i, then there must be an intersection between l and e_i on the right of p. To simplify the test, bounding box of the edge could be used. Obviously, if p is on the right of the bounding box, there is no intersection between l and e_i. If $y_i \leq y \leq y_{i+1}$ and p is on the left of the bounding box of e_i, there is an intersection between l and e_i on the right of p. Otherwise, if $y_i \leq y \leq y_{i+1}$ and p is in the bounding box, we use the method described in the Section 2.2, i.e. to test whether the area of the triangle $v_i v_{i+1} p$ is positive or negative, to determine the existence of the intersection.

3 Point-in-Polygon Test

Lemma 1. Consider a polygon P, and a point p that is not on the edges of the polygon, as shown in Fig. 1. Then there exists an edge, $e = v_l v_m$ $(m = (l + n + 1) \% n)$, which forms a triangle $\Delta v_l v_m p$ with point p, and the only intersections between the line segments $v_l p, v_m p$ and P are v_l and v_m, respectively. If the orientation of $\Delta v_l v_m p$ is the same as that of P, then p is inside P. Otherwise, p is outside P.

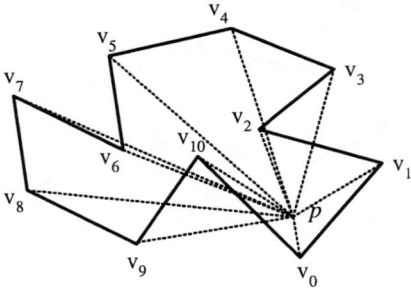

Fig. 1. Point-in-polygon test

Proof. The polygon P and the given point p can always be triangulated (Anglada, 1997). This means that there is at least one edge $e = v_l v_m$, such as $v_0 v_1$, $v_1 v_2$, $v_4 v_5$ and $v_{10} v_0$ shown in Fig. 1, which satisfies that the only intersections between the two line segments $v_l p$, $v_m p$ and P are v_l and v_m, respectively. Suppose that polygon P is oriented counterclockwise. If and only if point p is inside P, it is located in the left halfplane of e and the orientation of $\Delta v_l v_m p$ is counterclockwise. Otherwise, it is located in the right halfplane of e and $\Delta v_l v_m p$ is oriented clockwise. Similarly, suppose that P is oriented clockwise, if and only if point p is inside P, it is located in the right halfplane of e and $\Delta v_m v_l p$ is oriented clockwise. Otherwise, it is located in the left halfplane of e and $\Delta v_m v_l p$ is oriented counterclockwise. In summary, if the orientation of $\Delta v_l v_m p$ is the same as that of P, p is inside P. Otherwise, p is outside P.

Definition 1. (Visible edge). An edge e is defined as a visible edge of point p, if there exists a ray originating from p and intersecting with e at a non-vertex point v, and along the ray, v is the nearest intersection to p. As illustrated in Fig. 2, Edge $v_1 v_2$ is a visible edge of p.

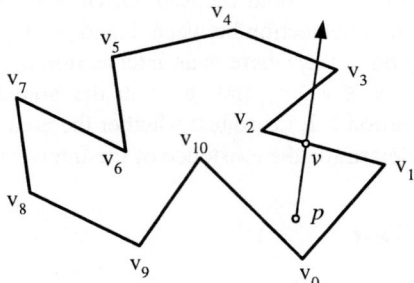

Fig. 2. Visible edge

Theorem 1. Edge $e = v_l v_m$ ($m = (l + n + 1) \% n$) is a visible edge of point p (such as $v_1 v_2$ shown in Fig. 3). If the orientation of triangle $\Delta v_l v_m p$ is the same as that of the polygon, then p is inside the polygon. Otherwise, p is outside the polygon.

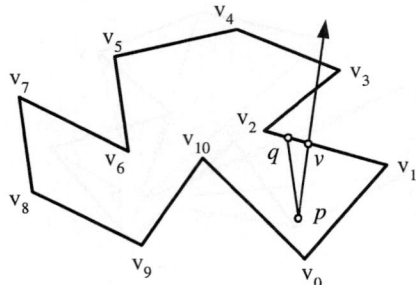

Fig. 3. Point-in-polygon test based on visible edge

Proof. As e is a visible edge, there must exist one ray originating from p and intersecting with e at a non-vertex point v, and line segment pv does not intersect with other edges of the polygon (see Fig.3). Choose a point q between vertex v and v_m on e. If $|vq| \to 0$, according to the definition of visible edges, there will be no intersection between pq and other edges of the polygon.

Obviously, inserting v and q into the polygon vertex set, between the endpoints of edge $v_l v_m$, will not change the intrinsic properties of the polygon (simplicity, orientation, etc.) and the relationship between p and the polygon. According to the **Lemma 1**, if the orientation of Δpvq is the same as that of the polygon, then p is inside the polygon. Otherwise, p is outside the polygon. Obviously, the direction of Δpvq is the same as that of $\Delta v_l v_m q$.

According to the theorem proven above, the key for point-in-polygon test is to find a visible edge of the given point. To speed up visible edge searching, a ray extending horizontally to the right is employed instead of an arbitrary ray originating from the given point. Only the edges with one vertex lying strictly above the ray while the other vertex lying on or below the ray are tested. We further identify the edges that intersect with the ray on the left of the given point. For two candidate edges, the relationship of their bounding boxes $A = (x_{min}, x_{max}, y_{min}, y_{max})$ and $B = (X_{min}, X_{max}, Y_{min}, Y_{max})$ can be:

(1) $X_{min} \geq x_{max}$, or

(2) $X_{max} \leq x_{min}$, or

(3) $x_{min} < X_{min} < x_{max}$, or

(4) $X_{min} < x_{min} < X_{max}$, or

(5) $x_{min} \leq X_{min}$ and $X_{max} \leq x_{max}$, or

(6) $X_{min} \leq x_{min}$ and $x_{max} \leq X_{max}$.

For case (5), one more condition, i.e. $y_{min} \leq Y_{min} < y_{max}$ or $y_{min} < Y_{max} \leq y_{max}$, should be satisfied due to the two edges are not intersecting with each other. Similarly for case (6), one more condition should be satisfied, i.e. $Y_{min} \leq y_{min} < Y_{max}$ or $Y_{min} < y_{max} \leq Y_{max}$.

As case (1) and (2), case (3) and (4), and case (5) and (6) are symmetric cases respectively, only case (1), (3) and (5) are illustrated in Fig. 4-6. In the following, the y-coordinate of the startpoints of the candidate edges is assumed to be less than that of the endpoints to facilitate discussion.

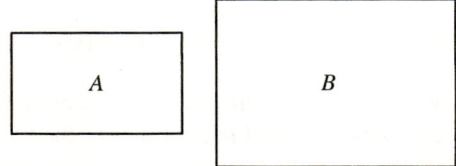

Fig. 4. Relationship of two bounding boxes: case (1)

For case (1), it is obvious that the intersection between the ray and the edge in bounding box A is nearer to point p, and similarly for case (2), the intersection between the ray and the edge in bounding box B is nearer to point p.

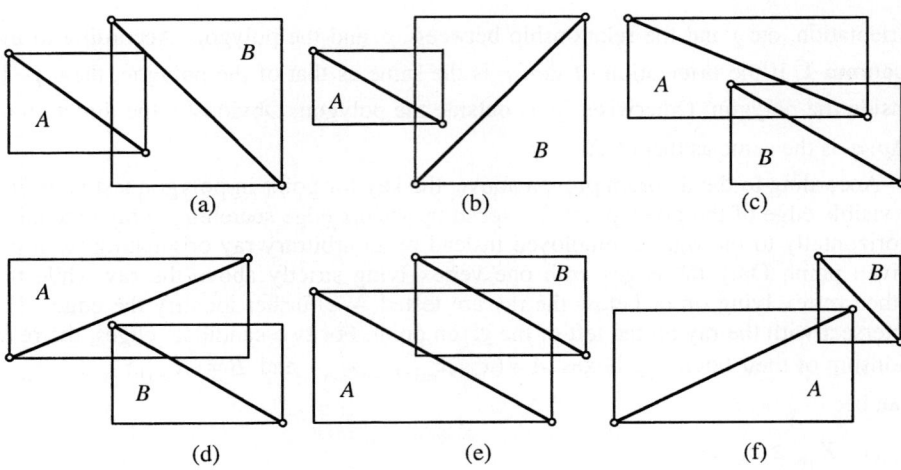

Fig. 5. Relationship of two bounding boxes: case (3)

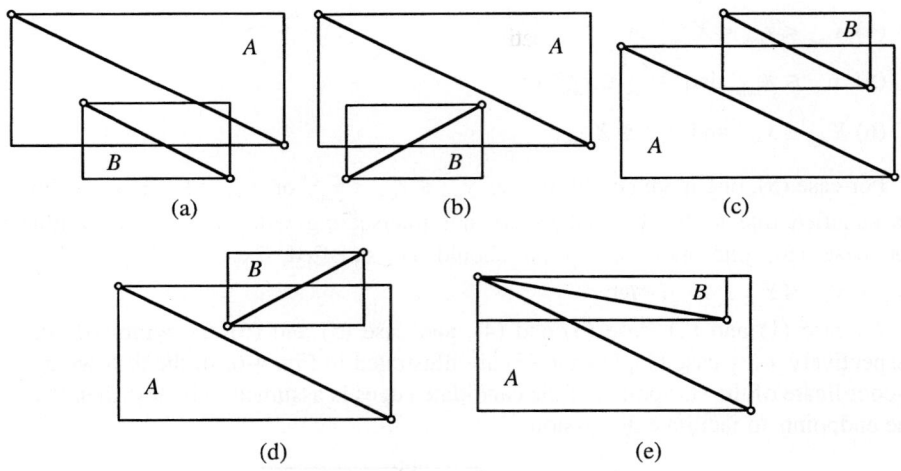

Fig. 6. Relationship of two bounding boxes: case (5)

For case (3), there are six possible subcases with respect to the relationship of y_{min}, y_{max} and Y_{min}, Y_{max} and the topological relationship of the two edges, as illustrated in Fig. 5(a-f). However, there exists one edge whose two vertices are on the same side of the other edge. If the two vertices are one the left of the other edge, then the intersection between the ray and the former edge is nearer to point p. Otherwise, the intersection between the ray and the latter edge is nearer to point p. Similarly, the edge whose intersection with the ray is nearer to point p can be found out for case (4).

For case (5), similar to case (3), there exists one edge whose two vertices are on the same side of the other edge if the two edges are not adjacent edges (see Fig. 6(a-d)).

Then, it can be process with the same method for case (3) to identify the edge whose intersection with the ray is nearer to point p. As for two adjacent edges (see Fig. 6(e)), the edge whose non-adjacent vertex on the left of the other edge is the visible edge as illustrated in Fig. 6(e). Similarly, the edge whose intersection with the ray is nearer to point p can be found out for case (6).

In practice, the bounding box of the polygon is usually used to pre-process the given data points when there are many points needed to be checked. Obviously, if a point is located outside the bounding box of the polygon, it cannot be inside the polygon. If none visible edges are found, p is located outside of the polygon.

4 Conclusion

A simple and robust method for point-in-polygon test has been presented in this paper. Inspired by constrained Delaunay triangulation, a new concept, called visible edge, is put forward and employed in the test. It is proved that the relationship between a point and a polygon can be determined by the relationship of the orientation of the polygon and the triangle formed by the given point and one of its visible edges. A simple method is presented to find the visible edge. The presented method is topology oriented and robust for floating point computation.

References

1. Anglada M. V.: Improved incremental algorithm for constructing restricted Delaunay triangulations. Computers & Graphics 21(1997) 215-223
2. Balbes R., Siegel J.: A robust method for calculating the simplicity and orientation of planar polygons. Computer Aided Geometric Design 8(1991) 327-335
3. Edelsbrunner H. and Mücke E. P.: Simulation of simplicity: a technique to cope with degenerate cases in geometric algorithms. ACM Transactions on Graphics 9(1990) 66-104
4. Feito F., Torres J. C.: Inclusion test for general polyhedra. Computers & Graphics 21(1997) 23-30
5. Feito F., Torres J. C., Urena A.: Orientation, simplicity, and inclusion test for planar polygons. Computers & Graphics 19(1995) 595-600
6. Forrest, A.R.: Computational geometry in practice. In: Earnshaw, E.A. (Ed.), Fundamental Algorithms for Computer Graphics. Springer-Verlag, Berlin, Germany (1985) 707-724
7. Haines, E.: Point in polygon straegies. In: Heckbert, P.(Ed.), Graphic Gems IV, Academic Press, Boston, MA, (1994) 24-46
8. Hoffmann, C.M., The problem of accuracy and robustness in geometric computation. IEEE Computer 22(1989) 31-41.
9. O'Rourke J.: Computational Geometry in C. 2nd edn. Cambridge University Press, Cambridge (1998)
10. Preparata, F.P., Shamos, M.I.: Computational Geometry - An Introduction. Springer, New York (1985)
11. Weiler, K.: An incremental angle point in polygon test. In: Heckbert, P.(Ed.), Graphic Gems IV, Academic Press, Boston, MA, (1994) 16-23
12. Wu H. Y., Gong J. Y., Li D. R., Shi W. Z.: An algebraic algorithm for point inclusion query. Computers & Graphics 24(2000) 517-522

A Modified Nielson's Side-Vertex Triangular Mesh Interpolation Scheme

Zhihong Mao, Lizhuang Ma, and Wuzheng Tan

Dept. of Computer Science and Engineering, Shanghai Jiao Tong University,
Shanghai 200030, PR China
{mzh_yu, ma-lz, tanwuzheng}@sjtu.edu.cn

Abstract. There are currently many methods for local triangular mesh interpolation that interpolates three end points and corresponding normal vectors of each input triangle to construct curved patches. With the curved patches we can eliminate the mismatch between the smoothness of the shading and the non-smoothness of the geometry that is particularly visible at silhouettes, showing as non-smooth edge junctions at the silhouettes vertices. In this paper we modify Nielson's side-vertex scheme to propose a hybrid interpolation scheme that integrates triangular mesh interpolation with subdivision technique. Using this scheme we can get curved patches with relatively good shape to improve the surface quality with little computation cost. Meanwhile we use adaptive normal vectors interpolation to achieve an efficient and qualitatively improved visible result for shading.

1 Introduction

The problem of fitting a surface through a set of data points arises in numerous areas of application such as medical imaging, scientific visualization, geometric modeling and rendering. There are many variations of the data-fitting problem. Tensor-product B-splines work well for modeling surfaces based on rectilinear control nets, but are not fit for more general topologies. Triangulated data can represent arbitrary topologies. In this paper, I will investigate using a local triangular mesh interpolation to fit a surface, which can be formulated as follows.

Given triangular meshes P in three dimension space, the given flat triangles are based only on the three vertices and three vertex-normals, construct curved patches that interpolates the vertices of P to substitute for the triangular flat geometry.

A lot of local parametric triangular surface schemes had been developed for many years. In general, while the methods satisfy the continuity and interpolation requirements of the problem, they often fail to produce pleasing shapes. Steve [1] gives a survey of parametric scattered data fitting by using triangular interpolants. He identified the causes of shape defects, and offered suggestions for improving the aesthetic quality of the interpolants. His investigations indicate that this poor shape is primarily an artifact of the construction of boundary curves. Because these schemes all have large number of free parameters that are set by using simple heuristics. By manually adjusting these parameters, one can improve the shape of the surface [2].

How to produce pleasing shape for the given surface data? Some global methods can improve the shape greatly. One way to improve the shape of the constructed surface automatically is to use variational methods [3]. The nonlinear optimization techniques minimize a fairness function while maintaining geometric continuity constrains. The function is the variation of curvature that can produce very fair free-form surfaces. It also allows the designer to specify technical and artistic shapes in a very natural way for a given design problem.

Another way to improve the shape of constructed surface is global subdivision [4,5,6,7,8]. Global subdivision, which utilizes a mesh of polygonal shapes, or a sequence of meshes, to describe a surface, is now becoming popular. The efficiency of subdivision algorithms and the flexibility with respect to the topology and connectivity of the control meshes makes this approach suitable for many applications such as surface reconstruction and interactive modeling [9]. The close connection to multi-resolution analysis of parametric surfaces provides access to the combination of classical modeling paradigms with hierarchical representations of geometric shape. But all these global methods rely on the availability of vertices of adjacent patches, so usually have a complex data structure and a high computational cost.

Recently Alex Vlachos[10] propose a local method that substitutes the geometry of a three-sided cubic Bézier patch that is re-triangulated into a programmable number of small(flat) sub-triangles for the triangle's flat geometry, and quadratically varying normal vectors for Gouraud shading. These curved patches require minimal or no change to existing authoring tools and hardware designs while providing a smoother, though not necessarily everywhere tangent continuous, silhouette and more organic shapes.

In this paper, I will consider a hybrid method that integrates local triangle mesh interpolation with subdivision technique. My method firstly modifies Nielson's side-vertex scheme [11] to construct a cubic Bézier patch net, then uses the degree elevation method of triangular Bézier patch [12,13] to approximate the triangular cubic Bézier limit patch that actually is a subdivision method within the triangular patch, finally uses adaptive normal vectors interpolation to achieve an efficient and qualitatively improved visible result for shading.

The paper is organized as follows. In section 2 we review the notation of the triangular Bézier patches. Section 3 discusses Nielson's side-vertex scheme. Section 4 proposes a modification of Nielson's method and discusses how to construct a triangular cubic Bezier patch. Section 5 applies an adaptive normal vector interpolation for shading. Section 6 summarizes our work.

2 Triangular Bézier patch

Bézier curves are expressed in terms of Bernstein polynomials, defined explicitly by

$$P(t) = \sum_{i=0}^{n} b_i B^n{}_i(t), \quad B^n{}_i(t) = \binom{n}{i} t^i (1-t)^{n-i}, \tag{2.1}$$

The two important properties of Bézier curves are (1) they interpolate their end points; (2) The difference of the first two (last two) control points gives the derivatives at the ends of the curve. The two main types of Bézier surface are rectangular and triangular Bézier patches. A triangular Bézier surface of degree n with control point $b_{i,j,k}$ defined by

$$P(u, v, w) = \sum_{i=0}^{n}\sum_{j=0}^{n-i} b_{i,j,k} B^{n}{}_{i,j,k}(u, v, w); \quad B^{n}{}_{i,j,k}(u, v, w) = \frac{n!}{i!j!k!} u^i v^j w^k$$
$$0 \leq u, v, w \leq 1; \quad u + v + w = 1; \quad i + j + k = n; \quad i, j, k \geq 0 \qquad (2.2)$$

The important properties of triangular Bézier patches(See Fig. 1) are (1) they interpolate their corner control points; (2) the boundary curves of the patch are cubic Bézier curves whose control points are the boundary points of the triangular control net, (3) the cross-boundary derivatives along one boundary are given by two layers of control points(e.g., for $t_0=0$, the cross-boundary derivatives are given by the control points P_i with $i = (0,i_1,i_2)$ and $i = (1,i_1,i_2)$).

3 Nielson's Side-Vertex Scheme

Nielson [11] developed a parametric side-vertex method to fit a piecewise smooth surface to a triangulated set of data. The method proceeds by first constructing three boundary curves, one corresponding to each edge of the input triangle. Three patches are created, one for each boundary/opposite vertex pair. The interior of each patch is constructed by passing curves from point along the boundary (or "side") to the opposite vertex, hence the name "side-vertex," as shown in Fig. 2. The three patches are then blended together to form the final patch (See Fig. 3).

All curves are constructed using a curve construction operator g_v that takes two vertices with normals and constructs a curve $g_v[V_0, V_1, N_0, N_1](t)$, such that $g_v(0) = V_0, g_v(1) = V_1, (g'_v(0), N_0) = 0, and (g'_v(1), N_1) = 0$, (,) denotes the dot product. Nielson's method also uses a normal field constructor g_n that constructs a continuous normal field along the curve g_v, where g_n is required to interpolate N_0 and N_1 at the end points. Nielson's scheme proceeds by building three patches G_i, $i \in \{p, q, r\}$ defined as:

$$G_i(b_p, b_q, b_r) = g_v[V_i, g_v[V_j, V_k, N_j, N_k](\frac{b_k}{1-b_i}), N_i, g_n[V_j, V_k, N_j, N_k](\frac{b_k}{1-b_i})](1-b_i) \qquad (3.1)$$

b_p, b_q and b_r are the barycentric coordinates of the triangle domain. Nielson notes the following two properties of G_i:

1. G_i interpolates all three of the boundaries.
2. G_i interpolates the tangent plane field of the boundary opposite vertex V_i.

The three surfaces are blended with rational functions to yield the final surface:

$$G[V_p, V_q, V_r, N_p, N_q, N_r] = \beta_p G_p + \beta_q G_q + \beta_r G_r, \qquad (3.2)$$

Where $\beta_i = \dfrac{b_j b_k}{b_p b_q + b_q b_r + b_r b_p}$.

Nielson shows that if the β_i are the blending coefficients of any three surfaces having the two above properties, then the resulting surface will interpolate all of the boundary curves and tangent fields. The selection of g_v and g_n has a large influence on the shape of the surface. Nielson's g_v operator constructs the cubic Hermite curve that interpolates the two points V_0 and V_1, and the two derivative vectors

$$g'_v(0) = aN_0 \times (V_1 - V_0) \times N_0, \quad g'_v(1) = bN_1 \times (V_1 - V_0) \times N_1, \qquad (3.3)$$

Where 'a' and 'b' are scalar degrees of freedom. Selection of 'a' and 'b' is a critical part of obtaining good shape. But there are no good criteria for setting these degrees of freedom [14].

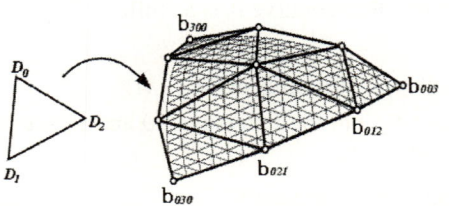

Fig. 1. Triangular cubic Bézier patch control nets

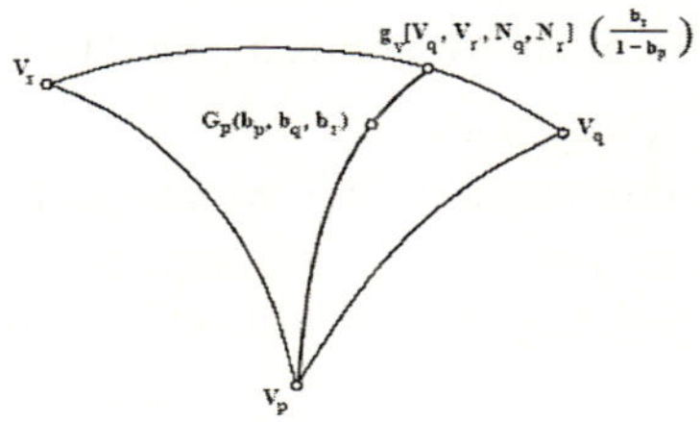

Fig. 2. Nielson's side-vertex method

Nielson used the following g_n:

$$g_n(V_0, N_0, V_1, N_1) = g'_v(t) \times [(1-t)N_0 \times g'_v(0) + tN_1 \times g'_v(1)] \qquad (3.4)$$

While we use these operators g_n to yield a C^1 continuity surface, the resulting surfaces maybe have poor shape. Because the normal vector computed by g_n is very coarse and different from the actual normal vector. Fig. 4 illustrates two surface shapes are decided by the configurations of side-vertex normals and the second can't satisfy the convex hull property of the surface. So while the method satisfies the continuity and interpolation requirements of the problem, it often fails to produce pleasing shapes.

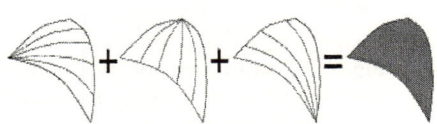
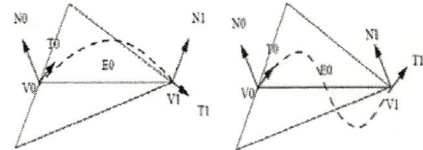

Fig. 3. Nielson's surface is a blend of three side-vertex surfaces

Fig. 4. Shape decided by the configuretions of surface normals

4 Modify Nielson's Method for Curved Patches

Here we introduce a hybrid triangle Mesh Interpolation method to construct triangular cubic Bézier patches. From section 2, we know a triangular cubic Bézier patch is decided by its ten control points $b_{i,j,k}$, We group the $b_{i,j,k}$ together as

End vertices: b_{300}, b_{030}, b_{003}.
Boundary vertices: b_{210}, b_{120}, b_{021}, b_{012}, b_{102}, b_{201}.
Center vertex: b_{111}.

4.1 Modify Nielson's Method to Decide the Curved Patch Net

Nielson gives an operator g_v to construct boundary curve, in his method he uses the cubic Hermite curve that interpolates the two end points and the two corresponding derivative vectors. This method introduces two free parameters. Selection of these parameters affects surface shape greatly. Here we use an alternative g_v operator for the representative Nielson scheme. We are given the points V_0, V_1 and normal vectors N_0, N_1, as shown in Fig. 5. Our choice of cubic Bézier curve b_i is as follow.

1. $b_0 = V_0$, $b_3 = V_1$
2. From equation 3.3, we can get tangent0 = $g'_v(0)$, tangent1 = $g'_v(1)$
3. P_1 and P_2 trisect the edge V_0V_1, project P_1 into the line tangent0 and P_2 into the line tangent1, we can get b_1 and b_2 respectively

$$b_1 = V_0 + \frac{[(V_1 - V_0) \bullet \text{tangent0}]}{3 \, | \, \text{tangent0} \, |} \frac{\text{tangent0}}{| \, \text{tangent0} \, |} \qquad (4.1)$$

$$b_2 = V_1 - \frac{[(V_1 - V_0) \bullet \text{tangent1}]}{3 \, | \, \text{tangent1} \, |} \frac{\text{tangent1}}{| \, \text{tangent1} \, |} \qquad (4.2)$$

This is roughly equivalent to setting Nielson's 'a' and 'b' shape parameters. Actually our method is a uniform chord length parameterization [15]. In general it can improve the surface shape. Thus we can decide three end vertices and six boundary vertices of the patch.

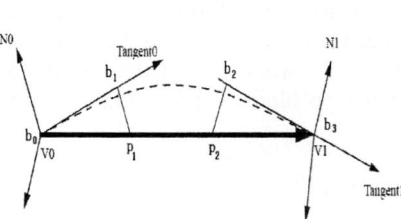

Fig. 5. Construct cubic Bézier curve

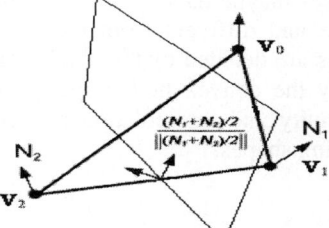

Fig. 6. Construction of the mid-edge control normal vector $n_{1,0,1}$ for quadratically varying normals

Another cause that affects the surface shape is that normal vectors constructed by operator g_n are coarse and different from the actual normal vectors. So while Nielson's method satisfies the continuity and interpolation requirements of the problem, it often fails to produce pleasing shapes. In order to improve the surface shape, we introduce Alex's method [10] to decide the center vertex of the cubic triangular Bézier patch b_{111}: Move the center vertex from its intermediate position V to the average of the six boundary vertices and continue its motion in the same direction for 0.5 the distance already traveled.

$$E = (b_{210}+b_{120}+b_{021}+b_{012}+b_{102}+b_{201})/6$$
$$V = (V_1+V_2+V_0)/3, \quad b_{111} = E + (E-V)/2. \tag{4.3}$$

Using this method, the curved Bézier patch doesn't deviate too much from the original triangle and preserves a good convex hull property. At each end point, every tangent vector is perpendicular to the corresponding normal, so at end points, the patches are C^1 continuity. But at other points, they are only C^0 continuity.

4.2 Subdivide the Curved Patch

The basic idea behind subdivision is to create a smooth limit function by infinite refinement of an initial piecewise linear function [6,7]. In order to eliminate the mismatch between the smoothness of the shading and the non-smoothness of the geometry that is particularly visible at silhouettes. In this paper, I will consider a hybrid method that integrates local triangular mesh interpolation with subdivision technique. My method is to employ the degree elevation equation to subdivide the cubic triangular Bezier patch into more programmable number of small sub-triangles, shown as Fig. 7.

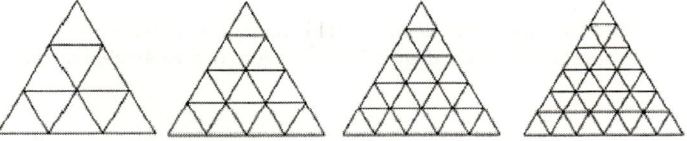

Fig. 7. Subdivision of a triangular patch domain

The process of degree elevation assigns a polygon εP to an original polygon P. We may repeat this process and obtain a sequence of polygons $P, \varepsilon P, \varepsilon^2 P, \varepsilon^3 P, \ldots$, etc. If we repeat this process of degree elevation again and again, we will see the polygons $\varepsilon^r P$ converge to the curve that all of them define:

$$\lim_{r \to \infty} \varepsilon^r P = \xi P$$

It is possible to write a triangular Bézier patch of degree n as degree n+1:

$$P(u,v,w) = \sum_{i=0}^{n}\sum_{j=0}^{n-i} b_{i,j,k} B^n_{i,j,k}(u,v,w) = \sum_{i=0}^{n+1}\sum_{j=0}^{n+1-i} b^*_{i,j,k} B^{n+1}_{i,j,k}(u,v,w) \tag{4.4}$$

For a triangular Bézier patch, we have u + v + w = 1 (see equation 2.2). We multiply the left hand of equation (4.4) by u + v + w:

$$\sum_{i=0}^{n}\sum_{j=0}^{n-i} b_{i,j,k} B^n_{i,j,k}(u,v,w)(u+v+w) = \sum_{i=0}^{n+1}\sum_{j=0}^{n+1-i} b^*_{i,j,k} B^{n+1}_{i,j,k}(u,v,w) \tag{4.5}$$

Use equation (2.2) to equation (4.5) and compare coefficients of like powers, we can get:

$$b^*_{i,j,k} = \frac{1}{n+1}(ib_{i-1,j,k} + jb_{i,j-1,k} + kb_{i,j,k-1}) \quad (4.6)$$

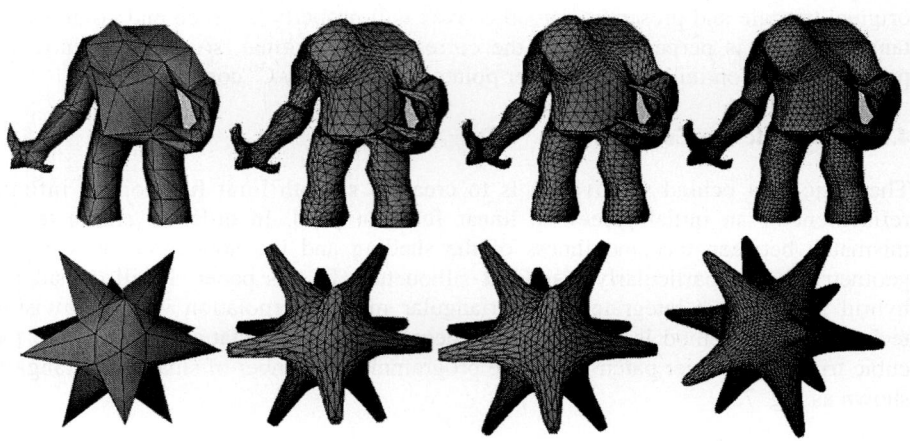

Fig. 8. A hybrid method that integrates local triangle mesh interpolation with subdivision technique, from left to right: initial mesh; degree 3 subdivision; degree 4 subdivision; degree 7 subdivision

The degree-elevated control net lies in the convex hull of the original one. The process generates a sequence of control nets that have the surface patch as their limit (see Fig. 8). From equation (4.6) we can find:

1. For three vertices of the input flat triangle, two of i, j, k equals 0. So we have $b_{n00} = V_1$, $b_{0n0} = V_2$, $b_{00n} = V_3$.
2. For boundary vertices, one of i, j, k equals 0, the boundary vertices are only decided by the two edge end points. So subdivision of the common edge of two adjacent triangles is similar. Using this method, no hole occurs between two adjacent triangles.
3. Vertices inside the triangle are decided by three end points.

5 Adaptive Normal Vector Interpolation

The normal to the geometry component of the triangles does not generally vary continuously from triangle to triangle. Therefore we define an independent linear normal variation by using degree elevation equation:

$$N^*_{i,j,k} = \frac{1}{n+1}(iN_{i-1,j,k} + jN_{i,j-1,k} + kN_{i,j,k-1}), N^* = \frac{N^*}{|N^*|}, \quad (5.1)$$

Where $N_{1,0,0} = N_0$, $N_{0,1,0} = N_1$, $N_{0,0,1} = N_2$. Linear variation of the normal approximates Phong shading. This is appropriate to the arch cases of cubic curves or quadratic curves, but for the serpentine cases of cubic curves, linear varying normals ignore inflections in the geometry as shown in Fig. 9, quadratic interpolation algorithm is appropriate for the serpentine cases [16,17]. To capture inflections a mid-edge control normal vector for the quadratic map is constructed following [10] e.g.: the average of the end-normals is reflected across the plane perpendicular to the edge as shown in Fig. 6. Recall that the reflection A' of a vector A across a plane with normal direction B is $A' = A - 2vB$, where $v = (B \bullet A)/(B \bullet B)$ and $\bullet$ denotes the dot product. Six control points of the normal patch are defined as follows:

$$n_{200} = N_0, \; n_{020} = N_1, \; n_{002} = N_2$$

$$v_{ij} = 2\frac{(V_j - V_i) \bullet (N_i + N_j)}{(V_j - V_i) \bullet (V_j - V_i)} \quad (5.2)$$

$$n_{110} = N_1 + N_2 - v_{12}(V_2 - V_1), n_{110} = n_{110} / \| n_{110} \|$$
$$n_{101} = N_1 + N_3 - v_{32}(V_1 - V_3), n_{110} = n_{101} / \| n_{101} \|$$
$$n_{011} = N_3 + N_2 - v_{23}(V_3 - V_2), n_{011} = n_{011} / \| n_{011} \|$$

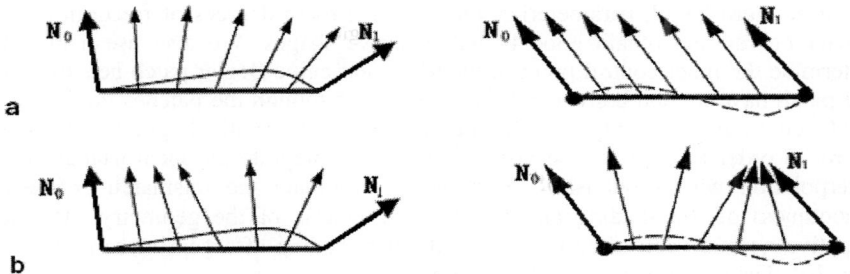

Fig. 9. Some examples of normal vector averaging over an edge. a. linear interpolation. b. quadratic interpolation. The dashed curve indicates the profile of the surefaces that should be simulated

After deciding the six control points, we also use the degree elevation equation (5.1) to compute the three end normals of each sub-triangle, here we have: $N_{2,0,0} = N_0$, $N_{0,2,0} = N_1$, $N_{0,0,2} = N_2$, $N_{1,1,0} = n_{1,1,0}$, $N_{1,0,1} = n_{1,0,1}$, $N_{0,1,1} = n_{0,1,1}$.

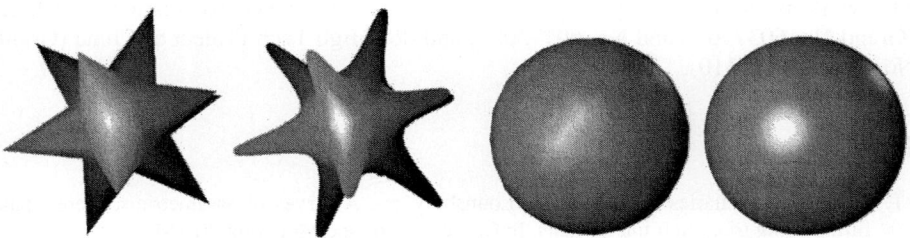

Fig. 10. Two examples shaded by adaptive normal vector interpolation. Left: Phong shading of initial mesh; Right: Shading by adaptive normal vector interpolation

We determine whether the curve is serpentine or not by

$$((N_0, \Delta) \geq 0) \oplus ((N_1, -\Delta) \geq 0)) \tag{5.3},$$

where Δ is the edge vector, $\oplus$ denotes logical exclusive OR, and (,) denotes the inner product. Thus the adaptive normal vector interpolation is expressed as [17]:

if $\{[((N_0, \Delta) \geq 0) \& \&(N_1, -\Delta) \geq 0)] \| [((N_0, \Delta) \leq 0) \& \&(N_1, -\Delta) \leq 0)]\}$

```
    apply linear interpolation for normal vectors
  else
    apply quadratic interpolation for normal vectors
```

6 Conclusion

Methods for local triangle mesh interpolation to triangulated, parametric data have existed for many years. The early schemes uniformly constructed surfaces that had some shape defects due to poor setting of degrees of freedom for the boundary curve and improper decision to the operator g_n for the normal vectors. In this paper we use a uniform chord length parameterization to set the extra degrees of freedom for cubic Bezier curves and obtain good boundary curve shapes. We also use a method to determine the inner control point of the triangular cubic Bézier patch nets that makes the patch have a good convex hull property. Even though the patches don't maintain the C^1 continuity, and only C^0 continuity, they can get a relatively good shape. Further more, in order to improve the visual quality, we integrate the local triangular mesh interpolation with subdivision technique to eliminate the mismatch between the smoothness of the shading and the non-smoothness of the geometry. At last the adaptive normal vector interpolation make the shading effect be more realistic (see Fig. 10). The properties that make our method attractive are:

➢ Do not require polygon adjacency information, all information is only limited in each PN triangle.
➢ Uses fairly simple data structures and the code executes very efficiently.
➢ Easy to integrate it into existing graphics applications and authoring tools.

Acknowledgements

The work is partially supported by national natural science foundation of China (Grand No. 60373070 and No. 60173035) and 863 High Tech Project of China (Grant No. 2003AA411310).

References

1. Steve Mann, Charles Loop, Michael Lounsbery, etc. A survey of parametric scattered data fitting using triangular interpolants. In Curve and surface Modeling. SIAM.
2. Stephen Mann, Surface Approximation Using Geometric Hermite Patches. PhD thesis, Uinversity of Washington, 1992.

3. Henry Moreton and Carlo Sequin. Functional optimization for fair surface design. In SIGGRAPH'92, 167-176, 1992.
4. J. Claes, K. Beets, and F. van Reeth. A corner-cutting scheme for hexagonal subdivision surfaces. In Proceedings of Shape Modelling International, pages 13-20, 2002.
5. Leif Kobbelt. sqrt(3) subdivision. In Siggraph 2000, Conference Proceedings, pages 103-112, 2000.
6. J. Warren and H. Weimer. Subdivision Methods for Geometric Design. Morgan Kaufmann, San Francisco, 2002.
7. C. Loop. Smooth subdivision surfaces based on triangles. Master's thesis, University of Utah, August 1987.
8. Denis Zorin. Smoothness of subdivision on irregular meshes. Constructive Approximation, 16(3), 2000.
9. D. Zorin, P. Schroder, W. Sweldens. Interactive Multiresolution Mesh Editing, SIGGRAPH 97 proceedings:259–269, 1997.
10. Vlachos Alex, Jörg Peters, Chas Boyd and Jason L. Mitchell. Curved PN Triangles. ACM Symposium on Interactive 3D Graphics :159-166, 2001.
11. Gregory Nielson. A transfinite, visually continuous, triangular interpolant. In G. Farin, editor, Geometric Modeling: Algorithms and New Trends: 235-245. SIAM, 1987.
12. G. Farin. Curves and Surfaces for Computer Aided Geometric Design. Academic Press, 1988.
13. Shi Fazhong. Computer Aided Geometric Design and NURBS. The higher education press, 2001.
14. Stephen Mann. An Improved Parametric Side-vertex Triangle Mesh Interpolant. In the graphics interface '98 proceedings:35-42, 1998.
15. J.M. Brun, S. Fouofu, A. Bouras, P. Arthaud. In search of an optimal parameterization of curves, COMPUGRAPHICS95: pp. 135-142, 1995.
16. Van Overveld CWAM, Wyvill B. Phong normal interpolation revisited. ACM Trans Graph 16(4): 397-419, 1997.
17. Yuan-Chung Lee, Chein-Wei Jen. Improved quadratic normal verctor interpolation for realistic shading. The Visual Computer(17): 337-352, 2001.

An Acceleration Technique for the Computation of Voronoi Diagrams Using Graphics Hardware

Osami Yamamoto

Department of Information Engineering,
Faculty of Science and Technology, Meijo University

Abstract. Drawing Voronoi diagrams with graphics hardware is a very easy and fast way of obtaining images of several types of Voronoi diagrams. Although graphics hardware is a good tool for making such images, its drawing speed is not so high as we expect when we draw it only using naive algorithms. This paper describes a technique for accelerating the drawing speed by reducing some polygons we do not need to draw. We focus on the algorithm for normal two-dimensional Euclidean Voronoi diagrams and segment Voronoi diagrams.

1 Introduction

The Voronoi diagram is an essential concept of computational geometry, and it is a strong tool in many areas in science and technology. It is a subdivision of a space into some regions for a set of given sites so that each point in the space belongs to the region of the nearest site to the point. Many types of Voronoi diagrams have been studied, and there are many effective algorithms for computing Voronoi diagrams in the Euclidean plane [Fo1,OI1,OB1,SH1].

Graphics hardware is a special kind of hardware for drawing images of objects in the three-dimensional Euclidean space on the computer display rapidly. Although it used to be attached to only expensive graphics workstations, it is getting common and is getting attached to ordinary PCs as the fever of realistic computer games rises recently. Using graphics hardware, we can easily obtain images of many types of Voronoi diagrams using the hidden-surface-removal function of graphics hardware [HC1,Ya1]. Moreover, two- and three-dimensional convex hulls can be computed using graphics hardware [Ya2]. Although graphics hardware draws objects in the three-dimensional Euclidean space rapidly, when we make an image of Voronoi diagrams of a lot of sites, the drawing speed is not so high as we expect. When we draw a Voronoi diagram of a lot of sites using graphics hardware, we have to draw a lot of polygons of wide area on the screen. It causes low speed of drawing.

There are many kinds of graphics hardware and the performance of the graphics hardware highly depends on specific graphics hardware. Therefore, it is difficult to discuss common ways of accelerating the drawing speed of Voronoi diagrams. However, it is valid that the bigger polygon we draw on the screen, the more time it takes. The algorithms for drawing Voronoi diagrams using graphics hardware draw a lot of polygons covering the screen. It takes a lot of time. The technique we describe in this paper is a reduction of the total area of the polygons we draw on the screen. We partition the screen into some number of buckets, and for each bucket, we choose the polygons we have to draw in the bucket using CPU. It drastically reduces the total area of the polygons we draw on the screen. However, at the same time the number of the polygons increases. If the number of the polygons we draw on the screen highly affects the running time of the graphics hardware we use, the total running time might not be shortened.

There are two types of algorithms for drawing Voronoi diagrams of sites. In this paper, we will focus on the algorithms. One of the algorithms uses planes to draw Voronoi diagrams; another algorithm uses cones. When we use the second algorithm, we have to approximate cones by some triangles. Therefore, it contains some error. On the other hand, the first type of algorithm are mathematically correct. However, it has a weakness for error when some sites are close to each other and it requires a more accurate depth buffer than the second algorithm [Ya1]. Both of the algorithms can be extended to algorithms for drawing Voronoi diagrams of sites and segments, which we will see in Section 2. However, the extended algorithm of the first algorithm is not suitable for drawing Voronoi diagrams of sites and segments because of the requirement of the accuracy of the depth buffer. In this paper we adopt the first algorithm for drawing Voronoi diagrams for sites and the second algorithm for drawing Voronoi diagrams for sites and segments. We experimented on both of the algorithms in Section 4. Our technique was effective in both of the algorithms. The result of the experiment is analyzed in the last section. Although there are many types of graphics hardware, we can explain the result by a simple model of drawing time for graphics hardware.

The technique described in this paper is very simple, and the idea may be applied to other types of Voronoi diagrams, convex hull problems and Delaunay triangulation.

2 Drawing Voronoi Diagrams with Graphics Hardware

We focus on two-dimensional Euclidean Voronoi diagrams of points and segments in this paper. Here is the definition of Voronoi diagrams. Suppose that some points (we call them *sites*) $p_1, \ldots, p_n$ and some segments $s_1, \ldots, s_m$ are given on the Euclidean plane. The Voronoi diagram of the points and the segments is a partitioning of the plane that consists of $n + m$ regions as follows:

$$R(b) = \left\{ x \in \mathbb{R}^2 \mid d(b,x) \leq d(p_j, x) \text{ and } d(b,x) \leq d(s_i, x) \right\}, \tag{1}$$

where b denotes one of the given points and the segments, and $d(b,x)$ denotes the Euclidean distance between b and the point x. Then, the equation $\mathbb{R}^2 = \bigcup_{b \in \{p_1,\ldots,p_n,s_1,\ldots s_m\}} R(b)$ holds, and the interiors of any two regions do not have any intersection.

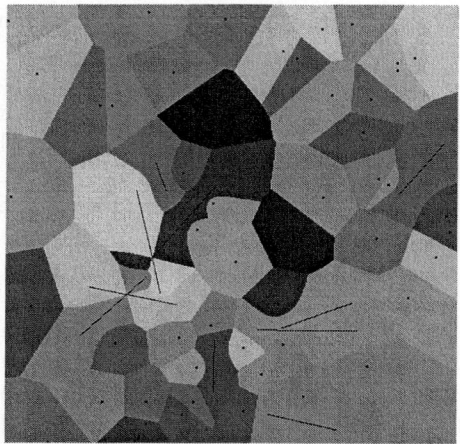

Fig. 1. A Voronoi diagram of 50 sites and 10 segments

Graphics hardware can make orthogonally projected images of polygons in the three-dimensional Euclidean space automatically. It is straightforward to make the image of Voronoi diagram of given points and segments. We define a surface in the three-dimensional Euclidean space for a given point or segment b as follows:

$$s(b) : p = (p_x, p_y) \in \mathbb{R}^2 \mapsto (p_x, p_y, d(b, p)) \in \mathbb{R}^3. \tag{2}$$

Then, it is quite obvious that we can obtain the Voronoi diagram of the given points and segments by drawing all the associated surfaces by different colors and looking at them from the point $(0, 0, -\infty)$ toward the origin [HC1]. If b is a site, the associated surface is a cone whose apex has the coordinates of $(p_x, p_y, 0)$. If b is a segment, the associated surfaces consists of two half cones and two pieces of planes. Because graphics hardware cannot draw curved surfaces directly, we have to approximate the surface by polygons when we actually implement the algorithm. Fig. 1 shows an image of a Voronoi diagram of 50 sites and 10 segments. The algorithm can draw Voronoi diagrams stably even if it contains some segments that intersect each other.

There is another way of obtaining the same images. It is obvious that the projected image of the surfaces

$$s(b) : p = (p_x, p_y) \in \mathbb{R}^2 \mapsto (p_x, p_y, d(b, p)^2) \in \mathbb{R}^3 \tag{3}$$

make the same image, though the surfaces are paraboloids, and they are difficult to draw using graphics hardware. However, if we apply the transformation

$$\varphi : (x, y, z) \in \mathbb{R}^3 \mapsto (x, y, z - x^2 - y^z) \in \mathbb{R}^3, \tag{4}$$

the paraboloids, which correspond to sites, are transformed to planes; surfaces corresponding to segments are transformed to surfaces that consist of two half planes and a ruled surface. The transformation does not change the drawn image, because x- and y-coordinates of the surface points are not changed and magnitude relations between the z-coordinates of the surface points are not changed by the transformation. Actually, the surfaces are denoted as follows:

$$s'(b) = \varphi \circ s(b) : p = (p_x, p_y) \in \mathbb{R}^2 \mapsto (p_x, p_y, d(b,p)^2 - \|p\|^2) \in \mathbb{R}^3. \quad (5)$$

If b is a site, $s'(b)$ is a simple plane because $d(b,p)^2 - \|p\|^2 = \|b\|^2 - 2(b,p)$, where (b,p) denotes the inner product of b and p. On the other hand, if b is a segment, computation of the z-coordinate is a little more complicated. Let p_1 and p_2 be the two end points of the segment b. Then,

$$d(b,p) = \begin{cases} \|p_1 - p\| & (t \leq 0), \\ \dfrac{\sqrt{\|p_1 - p\|^2 \|p_1 - p_2\|^2 - ((p_1 - p_2) \cdot (p_1 - p))^2}}{\|p_1 - p_2\|} & (0 < t < 1), \\ \|p_2 - p\| & (1 \leq t), \end{cases} \quad (6)$$

where $t = (p_1 - p) \cdot (p_1 - p_2)/\|p_1 - p_2\|^2$. Therefore, the z-coordinate $Z(p_x, p_y)$ of the surface point is denoted as follows:

$$Z(p_x, p_y) = \begin{cases} \|p_1\|^2 - 2(p_1, p) & (t \leq 0), \\ \dfrac{\|p_1 - p\|^2 \|p_1 - p_2\|^2 - ((p_1 - p_2) \cdot (p_1 - p))^2}{\|p_1 - p_2\|^2} - \|p\|^2 & (0 < t < 1), \\ \|p_2\|^2 - 2(p_2, p) & (1 \leq t). \end{cases}$$
$$(7)$$

The first and the third cases for the coordinate represent planes. The second one represents a surface that is not a plane. Because graphics hardware cannot draw arbitrary surface, we somehow have to partition the surface into some planes. When we use this method to draw a Voronoi diagram of sites, not including segments, the running time of the algorithm is a few times as short as the algorithm we have described previously because the associated surfaces are all planes. However, if we draw a Voronoi diagram of some segments using this algorithm, it is unstable, particularly when some segments intersect each other. This method requires a depth buffer of higher precision than the previous method for comparing the depths of surfaces properly [Ya1]. Therefore, this method is not suitable for the computation of Voronoi diagrams that involves segments. We only use this method for drawing Voronoi diagrams of sites in this paper.

As described in this section, we can obtain images of Voronoi diagrams of the given sites and the segments by drawing the planes and the surfaces in the three-dimensional space described above with different colors on the the screen. For each site and segment, we have to draw a plane or a surface that covers the screen. Although graphics hardware draws the planes rapidly, it takes a lot of time if we draw an image of a Voronoi diagram of a lot of sites and segments. Most parts of the planes and the surfaces are not drawn actually on the screen because they are hidden by other planes or surfaces. Therefore, we may be able to omit drawing some parts of the planes and the surfaces. In the next section, we describe an algorithm to find such parts of the planes and surfaces.

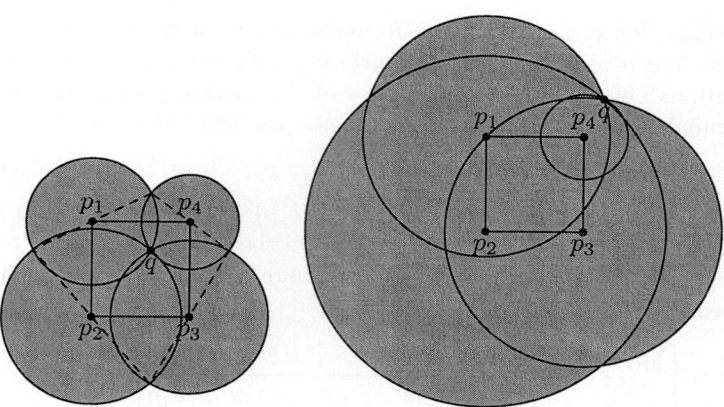

Fig. 2. The region for a site r whose Voronoi region may have an intersection with the bucket B. The figure shows two cases, a case the bucket includes the point q, and a case it does not

3 A Bucket Method for Accelerating the Algorithm for Drawing Voronoi Diagrams

In this section, we concentrate on a bucket method for accelerating the speed when we draw Voronoi diagrams with graphics hardware.

First, we consider a very simple situation. Let a point p be given on the plane. Let q and r be the two sites given on the plane. The two regions of the Voronoi diagram of the sites q and r are denoted by $R(q)$ and $R(r)$ respectively. It is obvious that $p \in R(q)$ if and only if $\|r - p\|^2 \geq \|q - p\|^2$. It means that $p \in R(q)$ if and only if r is outside the disk whose center is p and whose radius is $\|p - q\|$. Let us denote the disk $D(p, q)$.

Let p_i ($i = 1, \ldots, 4$) be the four vertices of a bucket B. Then, it is also obvious that B does not have any intersection with $R(r)$ if and only if $p_i \in R(q)$ for $i = 1, \ldots, 4$, which is equivalent to the condition $q \notin D(p_i, q)$ for $i = 1, \ldots, 4$. It means that $R(r)$ does not have any intersection with the bucket B if and only if r is not in the union of the disks $D(p_i, q)$, $i = 1, \ldots, 4$. (See Fig. 2). Here, we have the following lemma.

Lemma 1. *Let B be a rectangular bucket on the plane whose vertices are $p_1, \ldots, p_4$. Let q and r be the only given sites on the plane. Then, the region corresponding to r of the Voronoi diagram of these two sites has no intersection with B if and only if $r \notin \bigcup_{i=1}^{4} D(p_i, q)$.*

Let us reflect on the area of the region we discussed above.

Lemma 2. *The region $\bigcup_{i=1}^{4} D(p_i, q)$ contains the rectangular bucket B for any q on the plane. Let the bucket be a unit square whose vertices are $(0, 0)$, $(0, 1)$, $(1, 0)$, and $(1, 1)$. If q is in B, then the area A of the region is*

$$A = 2\pi \left[(q_x - 1/2)^2 + (q_y - 1/2)^2 + 2 \right] + 2, \tag{8}$$

where $q = (q_x, q_y)$. More generally the following inequality holds wherever q is located:

$$A < 4\pi \left[(q_x - 1/2)^2 + (q_y - 1/2)^2 \right]. \tag{9}$$

Proof. It is obvious that B is always contained in the region, because if r is in B, its Voronoi region obviously intersects with the bucket. If the site q is in B, the area A of the region is calculated as the sum of the area of the rectangle enclosed by dashed lines in Fig. 2 and the area of four half disks. The first area is always equal to two. Therefore, we obtain the result. The latter part of the lemma is obtained by the calculation of the sum of the area of the four disks that make the region. □

By Lemma 1, we can say that if we choose a point which is a site or a point in a segment, we can eliminate some sites and segments that cannot make a Voronoi region inside the bucket. Therefore, we do not have to draw the surfaces that correspond to such sites and segments when we make the image for the bucket. The smaller the area of the region is, the smaller the expected number of the sites and the segments we have to draw inside the bucket is. Therefore, by Lemma 2, if there are sites or segments inside the bucket, we should choose the closest point to the center of the bucket. Then, the region by the point has the smallest area. If there is no segments or sites inside the bucket, we should choose the point among the sites and segments that is closest to the center of the bucket. Although in this case the area of the region is not always the smallest one, it is only bounded by the inequality in Lemma 2.

The following lemma describes the bucket we have to check to find the sites and the segments the Voronoi regions of which may intersect with the bucket, when we can choose a point inside the bucket.

Lemma 3. *Let the screen consist of some buckets of square shapes of the same size (See Fig. 3). Let B be one of them. Let b be a site or a segment. If there is a site or a part of segment inside the bucket and b has no intersection between the region shown in Fig. 3, the Voronoi region for b does not appear in B.*

Proof. It is obvious from Lemma 1 and Fig. 3. □

Here is an algorithm for accelerating the drawing speed of the algorithm we described in the previous section.

Algorithm 1. Let the screen consist of some buckets of the same square shape. Input: sites and segments. Output: the image of the Voronoi diagram of the given sites and segments.

1. Make links from each buckets to the sites that are contained in the bucket. Similarly, make links from each buckets to the segments that pass the bucket.

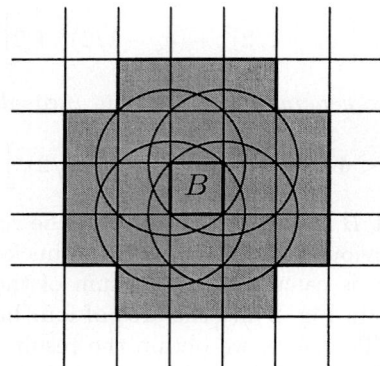

Fig. 3. The gray region shows the neighboring buckets that contain the sites or the segments that could make Voronoi regions in B, when B contains a site or a segment in it

2. For each bucket:
   (a) If the bucket contains a site or a part of a segment, find the one which is closest to the center of the bucket. Then, find sites and the segments whose Voronoi regions could appear in the bucket as follows: First, we search the sites and segments in the twenty neighboring buckets shown in Fig. 3. Then, we check if it could appear in the bucket using the result of Lemma 1.
   (b) If the bucket does not contain any sites or segments, find the closest one to the center of the bucket among the whole sites and segments.
   (c) Draw the surfaces corresponding to the sites and the segments which may appear inside the bucket using the clipping function of graphics hardware.

The wider region we draw on the screen with graphics hardware, the more time the graphics hardware needs. If we eliminate the surfaces by the method we described above, we can drastically reduce the total area of the surfaces we have to draw. Therefore, the algorithm speeds up the drawing process.

4 Experiments on the Technique

This section shows some experiments on the algorithm we have described in the previous section. We apply the acceleration method to two algorithms we have described in Section 2.

First, we show the running time for drawing Voronoi diagrams of sites by drawing planes using buckets and by drawing cones using buckets. In this paper, we used a computer with graphics hardware of the following specification:

CPU: IBM PowerPC 750, 800 MHz
RAM: 384 M bytes

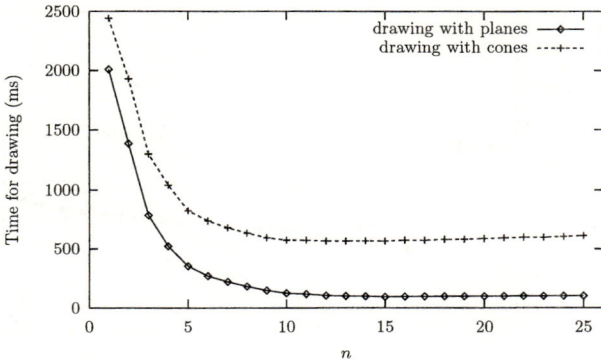

Fig. 4. The computation time for drawing a Voronoi diagram of 2000 random sites with $n \times n$ buckets on the screen of 600×600 pixels by planes and by cones

GPU (graphics hardware): ATI Mobility Radeon 7500
Video RAM: 32 M bytes
Number of color-buffer bitplanes: 24 (8 bitplanes for each RGB color)
Number of depth-buffer bitplanes: 24
Size of the screen: 600 pixels × 600 pixels

We drew a Voronoi diagram of 2000 random sites on the screen with $n \times n$ buckets of the same size, where $n = 1, \ldots, 25$. When we drew the Voronoi diagram with cones, the cones were approximated by pyramids that consist of 60 fragments of triangles. The time for drawing Voronoi diagrams is shown in Fig. 4. All the running times we measured were elapse time. We could not measure the time for GPU directly. When $n = 1$, just one bucket was used. That is, no acceleration was applied to the drawing process for that case. Therefore, compared with the case of $n = 15$, the technique accelerated the drawing speed by a factor of about 20 for the algorithm drawing planes. On the other hand, the technique accelerated the process by a factor of about 5, when we used the algorithm drawing cones. Generally, the effectiveness of this acceleration technique highly depends on the graphics hardware that is used. With some other graphics hardware, the technique accelerated the process by a factor of only a few times even for the algorithm using planes.

Next, we show a result on drawing Voronoi diagrams of sites and segments with cone shape surface and planes. In this experiment, we drew a Voronoi diagram of randomly distributed 2000 sites and 500 segments on the screen with $n \times n$ buckets, where $n = 1, \ldots, 25$. Fig. 5 shows the result. The technique accelerated the computation by a factor of about 3.5 times. Although the technique was still effective, it was not so effective as the case of drawing ordinary Voronoi diagrams of sites. Describing roughly, this result was caused by the lengths of segments. That is, because segments had lengths, the surfaces for the segments had to be drawn in many buckets unlike sites.

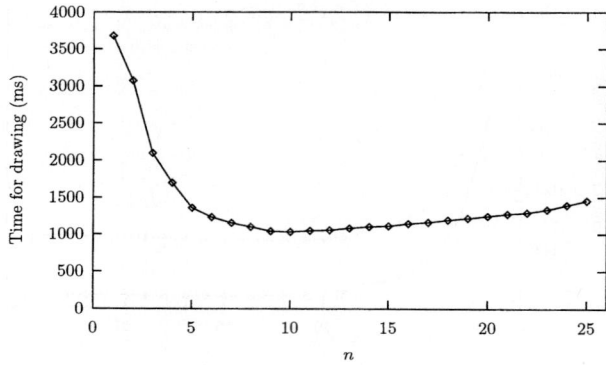

Fig. 5. The computation time for drawing a Voronoi diagram of 2000 random sites and 500 random segments with $n \times n$ buckets on the screen of 600×600 pixels and by cones

5 Discussion and Conclusions

First, we analyze the result of the previous section. The structure of graphics hardware is complicated. Typical graphics hardware has a pipeline structure in it [Ki1], and it is difficult to analyze the real runtime precisely. Here, we simplify the running time for graphics hardware when we draw simple polygons. Assume that it takes $t(s) = c_1 s + c_2$ of time to draw a simple polygon of area s on the screen using graphics hardware, where c_1 and c_2 are some positive constants. Assume that we draw N points on the screen with $n \times n$ buckets using planes. Let the size of the screen be 1×1. Then, the total expected computation time is roughly estimated as follows:

$$T_1 = t(1/n^2) \cdot (N/n^2) \cdot \alpha(N,n) \cdot n^2 = (c_1/n^2 + c_2) \cdot N \cdot \alpha(N,n), \qquad (10)$$

where $\alpha(N,n)$ denotes the expected ratio of the area of the region shown in Fig. 2 to the area of the bucket. Here, we do not count the effect of the boundary of the screen to the ratio. Although the ratio is not a constant, it does not change so rapidly as n changes when the sites are distributed densely.

On the other hand, if we draw the Voronoi diagram using the cones that are approximated by M triangles, the total expected computation time is

$$T_2 = \left(\sum_{i=1}^{M} t(a_i) \right) \cdot (N/n^2) \cdot \alpha(N,n) \cdot n^2 = (c_1/n^2 + Mc_2) \cdot N \cdot \alpha(N,n), \qquad (11)$$

where a_i denotes the area of clipped region of ith triangle. The sum of the area satisfies the relation $\sum_{i=1}^{M} a_i = 1/n^2$. The estimation explains the result shown in Fig. 4 very well. The estimation above implies that the difference $T_2 - T_1$ is proportional to $\alpha(N,n)$. Therefore, we can see that $\alpha(N,n)$ is almost constant in this case from Fig. 4.

The result in Fig. 5 is somewhat different from the previous result. Although the technique was still effective in this case, for $n > 10$, the running time increased gradually as n becomes bigger. Unlike sites, segments have lengths. Therefore, if we use a lot of small buckets, we have to draw the shapes for the segments in many buckets. Particularly when a segment is long, it crosses many buckets, and as a result, the shape for the segment has to be drawn in many buckets. It causes the growth of the number of polygons to draw on the screen and eventually the technique is not so effective as for the algorithm for drawing Voronoi diagrams of sites. Moreover, we just experimented on uniformly distributed sites and segments in this paper. It is not clear if it is still effective for arbitrarily distributed sites. According to further experiments for unevenly distributed sites, the technique accelerates on some level, though it is not so effective as for uniformly distributed sites. It is an area for future research.

In this paper, we chose the closest point to the center of each bucket and eliminated the points the corresponding Voronoi regions of which could not appear in the bucket. We may be able to use more than one points for eliminating such points, and they could eliminate more points. However, it does not imply that the method is totally faster than the technique we described in this paper. It is also an area for future research. Because of the simpleness of the technique, it may be able to be applied to computation of other types of Voronoi diagrams, Delaunay diagrams and convex hulls.

References

[Fo1] FORTUNE, S.: A sweepline algorithm for Voronoi diagrams. *Algorithmica* **2**, (1987) 153–174

[HC1] HOFF, K. E., CULVER, T., KEYSER, J., LIN, M., AND MANOCHA, D.: Fast computation of generalized Voronoi diagrams using graphics hardware. In *Proceedings of ACM SIGGRAPH Annual Conference on Computer Graphics*, ACM, (1999) 277–286

[Ki1] KILGARD, M. J.: *OpenGL Programming for the X Window System*. Addison-Wesley. (1996)

[OI1] OHYA, T., IRI, M., AND MUROTA, K.: Improvement of incremental method for Voronoi diagram with computational comparison of algorithms. *Journal of Operations Research Society of Japan* **27**, (1985) 306–336

[OB1] OKABE, A., BOOTS, B., AND SUGIHARA, K.: *Spatial Tessellations – Concepts and Applications of Voronoi Diagrams*. John-Willey. (1992)

[SH1] SHAMOS, M. I., AND HOEY, D.: Closest-point problems. In *Proceedings of Annual IEEE Symposium on Foundation of Computer Science*, IEEE, (1975) 151–162

[Ya1] YAMAMOTO, O.: Fast display of Voronoi diagrams over planes and spheres using graphics hardware. *Transactions of the Japan Society for Industrial and Applied Mathematics* **12**, 3, (2002) 209–234 (in Japanese)

[Ya2] YAMAMOTO, O.: Fast computation of 3-dimensional convex hulls using graphics hardware. *Proceedings of International Symposium on Voronoi Diagrams in Science and Engineering*, September 13-15, 2004, Tokyo, Japan, (2004) 179–190

On the Rectangular Subset Closure of Point Sets

Stefan Porschen

Institut für Informatik, Universität zu Köln, D-50969 Köln, Germany
porschen@informatik.uni-koeln.de

Abstract. Many applications like picture processing, data compression or pattern recognition require a covering of a set of points most often located in the (discrete) plane by rectangles due to some cost constraints. In this paper we introduce and study the concept of the *rectangular subset closure* of a point set M in the (discrete) plane which is aimed to provide some insight into the rectangular combinatorial structure underlying such a covering problem. We show that the rectangular subset closure of a set M is of size $O(|M|^2)$ and that it can be computed in time $O(|M|^2)$. The concepts and results are also generalized to the d-dimensional case.

Keywords: rectangular covering problem, closure operator.

1 Introduction

One often is interested in covering a given finite set of points for instance in the plane by some kind of (regular) objects like rectangles. For that it may be useful [5] to know the set of all rectangles tightly enclosing every subset of the input set. To that end, we introduce the notion of a *rectangular subset closure* of a given point set in the (discrete) plane. More precisely, given a finite set M of points distributed in the plane, we ask for the smallest set $\mathcal{R}(M)$ of all regular rectangles, such that for each subset S of M there exists a rectangle in $\mathcal{R}(M)$ that encloses S tight. $\mathcal{R}(M)$ is closely related to the rectangular subset closure of M, as it will be defined in the next section.

Besides the interest that this concept deserves from the structural (and computational) point of view, it also may be useful in some kind of preprocessing for geometric covering or clustering, and partition problems. Such problems in turn occur at the optimization core of front-end applications like image processing, data compression and pattern recognition [4, 6]. The points in the plane then represent pixels or some other binary data. From the computational point of view it has turned out that most of these covering problems are NP-hard [1–3, 5]. So exact bounds for such problems are expected to be exponential. However it may be possible by an appropriate preprocessing to decrease at least the polynomial pre-factors in such bounds, for which the rectangular subset closure could be helpful. For instance, for such a covering problem in [5] by dynamic programming a time bound has been obtained essentially of the form $O(kp(|M|)2^{|M|})$

(where $0 < k < 1$). Here p is a polynomial of degree 6. By the results in the present paper, the degree of the polynomial can be decreased in many situations.

However, the basic topic of this work is to exploit some combinatorial structure underlying rectangular covering problems. From a more general point of view, we develop the basic combinatorial theory around the subset closure. Specifically, we show that the size of the rectangular subset closure of a given point set M is not greater than $|M|^2$. We also show that it can be computed in time $O(|M|^2)$.

Finally, the presented concepts and results are also generalized to the d-dimensional case.

2 The Rectangular Subset Closure of a Point Set in the Plane

To fix the notation let $\mathbb{E}^2$ denote the euclidean plane viewed as the real vector space $\mathbb{R}^2$ equipped with the (orthogonal) standard basis $\boldsymbol{e}_x, \boldsymbol{e}_y \in \mathbb{R}^2$. Let an axis parallel integer lattice (simply called *grid*) $L_\lambda(2) = \mathbb{Z}\boldsymbol{e}_x\lambda + \mathbb{Z}\boldsymbol{e}_y\lambda$, for fixed real grid constant $\lambda > 0$, be embedded in $\mathbb{E}^2$. It is convenient to set $\lambda = 1$ (then $L_1(2) = \mathbb{Z}^2$) which from now on is assumed. Such a grid is *regular (or isothetic)* in the sense that it is in accordance with the orientation of the orthogonal basis in $\mathbb{E}^2$. A point $z \in L_1(2) =: L$ is determined by its coordinates $(x(z), y(z)) \in \mathbb{Z}^2$. For a fixed coordinate value x, we call $l(x) := L \cap \{(x, y) \in \mathbb{E}^2 : -\infty \leq y \leq \infty\}$ the (vertical) *grid line* through x. Similarly, we have (horizontal) grid lines $l(y)$ parallel to the $\boldsymbol{e}_x$-direction through a fixed coordinate value y. A finite *rectangular* grid region I is defined by $I := [A_x, A_y] \times [B_x, B_y] \cap L$ where $A_x, A_y, B_x, B_y \in \mathbb{Z}$ such that $A_x < A_y$, $B_x < B_y$. There is a natural (partial) order $\leq_L$ on the lattice given by $z_1 \leq_L z_2 \Leftrightarrow x(z_1) \leq x(z_2) \wedge y(z_1) \leq y(z_2)$ ($\forall z_1, z_2 \in L$). The (linear) lexicographic order on L is defined by $z_1 \leq_\ell z_2$ if either $x(z_1) < x(z_2)$ or $x(z_1) = x(z_2)$ and $y(z_1) \leq y(z_2)$. Recall that $\leq_L$ is not a linear order on the grid points as, for example, the points $z_1 = (2,5)$ and $z_2 = (3,4)$ are not comparable with respect to $\leq_L$, whereas $z_1 \leq_\ell z_2$ holds true. We call a $\leq_\ell$-sequence $\{z_1, \ldots, z_n\} \subset L$ a *chain* if $z_i \leq_L z_{i+1}, 1 \leq i \leq n-1$, and call it *strict* if $\leq_L$ is replaced by $<_L$ at any position. A $\leq_\ell$-sequence $\{z_1, \ldots, z_n\} \subset L$ is called an *antichain* if $x(z_i) < x(z_{i+1})$ and $y(z_{i+1}) < y(z_i), 1 \leq i \leq n-1$ (cf. Fig. 1).

Before we state precisely what the rectangular subset closure of a point set is, let us define the basic notion of a *rectangular base*. Roughly speaking, the rectangular base of a point set is the smallest rectangle enclosing it. More precisely, let $M \subset L$ be a finite set of points in the grid. To a subset $S \subseteq M$ we

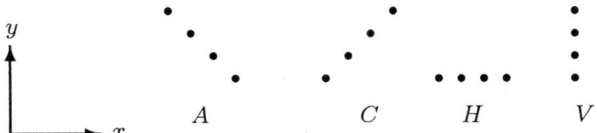

Fig. 1. An antichain A, a strict chain C, and two chains H, V (grid lines are omitted)

assign the grid points $z_d(S) := (x_d(S), y_d(S))$ and $z_u(S) := (x_u(S), y_u(S))$ defined by $x_d(S) := \min_{z \in S} x(z)$, $y_d(S) := \min_{z \in S} y(z)$ and $x_u(S) := \max_{z \in S} x(z)$, $y_u(S) := \max_{z \in S} y(z)$. Observe that the points $z_d(S)$ and $z_u(S)$ may not belong to S or even M but are always grid points.

Definition 1. *The unique set $r(S) := [x_d(S), x_u(S)] \times [y_d(S), y_u(S)]$ is called the* rectangular base *of S. The extremal points (which coincide for a single element set) $z_d(S), z_u(S)$ are called the* (rectangular) base points *of S. We write $b(S) := \{z_d(S), z_u(S)\}$ with the convention $b(\varnothing) := \varnothing$.*

Notice that $r(S)$ is the inclusion-wise smallest rectangular object containing S and that $b(S)$ just consists of its lower left and upper right diagonal vertices $z_d(S), z_u(S)$, respectively, so that $z_d(S) \leq_L z_u(S)$. For a chain S, we have a characteristic situation, namely $b(S) \subseteq S$. $r(S)$ either corresponds to a proper rectangle or it is degenerated to a point or a line segment, namely when S itself is a single point or a grid line segment.

To a set M of n grid points one can compute a set of 2^n rectangular bases, one for each subset of M. Observe that rectangular bases of different subsets may coincide which is exploited in the sequel. Let 2^M denote the power set of a finite set M. Notice that the above definition also makes sense without an underlying grid, i.e., when the points are distributed arbitrarily in the plane.

The rectangular subset closure of a point set M naturally appears as the smallest superset of M containing the base points of all subsets of M.

Definition 2. *For $M \subset L$ finite, the* rectangular subset closure $\mathrm{RS}(M) \subset L$ *is defined by* $\mathrm{RS}(M) := \bigcap \{L \supseteq M' \supseteq M : b(S) \subset M', \forall S \in 2^M\}$.

As an example consider Figure 2, where all *additional* base points, for a given set M, contained in the corresponding $\mathrm{RS}(M)$ are represented as white dots.

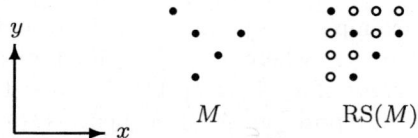

Fig. 2. A set M and its rectangular subset closure $\mathrm{RS}(M)$ (grid lines are omitted)

Definition 2 obviously is equivalent to:

Lemma 1. *For $M \subset L$ finite, we have $\mathrm{RS}(M) = M \cup \bigcup_{S \subseteq M} b(S)$.* □

The rectangular subset closure as introduced above gives rise to a closure operator defined for a fixed finite rectangular grid region $I \subset L$.

Proposition 1. $\mathrm{RS} : 2^I \to 2^I$ *is a closure operator.*

Remark 1. *As defined above, RS corresponds to a finite closure operator RS_I. Thus we obtain a family of finite closure operators $\{\mathrm{RS}_I : I \subset L\}$ where rectangular finite grid regions I are permitted only. Instead, one could also define $\mathrm{RS} : \mathcal{P}(L) \to \mathcal{P}(L)$ which corresponds to an infinite closure operator on the power set $\mathcal{P}(L)$ of L. For applications the finite version suffices, to which we restrict the subsequent considerations omitting subscript I.*

Proof of Proposition 1. Let $M \subseteq I$ be arbitrarily chosen. Because the finite grid region I is, by definition, rectangular, it is ensured that for every $S \subseteq M$ we have $b(S) \subset I$. Thus $\mathrm{RS}(M) \in 2^I$ so that RS as stated is a well defined function.

To verify that RS is, in addition, a closure operator, recall that a closure operator $\sigma : 2^I \to 2^I$ has the following defining properties: (i) $\forall S \subseteq I$ holds $S \subseteq \sigma(S)$, (ii) $\forall S_1, S_2 \subseteq I$ with $S_1 \subseteq S_2$ holds $\sigma(S_1) \subseteq \sigma(S_2)$, and (iii) $\forall S \subseteq I$ we have $\sigma(\sigma(S)) = \sigma(S)$.

It is quite obvious according to Lemma 1 that RS satisfies condition (i). Also it satisfies property (ii) since from $M_1 \subseteq M_2 \subseteq I$ follows $S \subset M_1 \Rightarrow S \subset M_2$ thus $b(S) \in \mathrm{RS}(M_1) \cap \mathrm{RS}(M_2)$. Hence, we obtain by Lemma 1 and property (i):

$$\mathrm{RS}(M_1) = M_1 \cup \bigcup_{S \subseteq M_1} b(S) \subseteq M_2 \cup \bigcup_{S \subseteq M_2} b(S) = \mathrm{RS}(M_2)$$

It remains to show (iii), i.e., that $\mathrm{RS}(\mathrm{RS}(S)) = \mathrm{RS}(S)$ holds true for every $S \subseteq I$. By definition we have $\mathrm{RS}(\emptyset) = \emptyset$ so we set $M := \mathrm{RS}(S)$ for arbitrary $S \subseteq I$ with $S \neq \emptyset$. It suffices to verify $(*) : \forall \emptyset \neq T \subseteq M : b(T) \subseteq M$ since then $\mathrm{RS}(M) = M$ holds according to Lemma 1. Let $T \in 2^M$ be non-empty with $b(T) = \{z_d(T), z_u(T)\}$. Observe that $z \in M$ if and only if there exists a subset $S_z \subseteq S$ such that $z \in b(S_z)$, since $\mathrm{RS}(M) = S$. Hence, we are done by identifying sets $S_{z_d}, S_{z_u} \subset S$ so that $z_d := z_d(T) \in b(S_{z_d}), z_u := z_u(T) \in b(S_{z_u})$.

To show that the lower base point $z_d \in b(T)$ is an element of M, notice that by definition there are grid points $z_x, z_y \in T$ such that $x(z_x) = x(z_d), y(z_x) \geq y(z_d)$ and $y(z_y) = y(z_d), x(z_y) \geq x(z_d)$. Moreover, there must exist non-empty sets $S_{z_x}, S_{z_y} \in 2^S$ with $z_x \in b(S_{z_x}), z_y \in b(S_{z_y})$. We even can assume that $z_x = z_d(S_{z_x})$ and $z_y = z_d(S_{z_y})$. But then for $S_{xy} := S_{z_x} \cup S_{z_y} \in 2^S$ holds $z_d(S_{xy}) = (x(z_x), y(z_y)) = (x(z_d), y(z_d)) = z_d \in M$.

The argumentation showing that also the upper base point $z_u \in b(T)$ is a member of M proceeds completely analogously, hence $b(T) \subseteq M$ implying $(*)$ and completing the proof. $\square$

By the way, the rectangular subset closure of a set M of grid points gives rise to a (plane-embedded) directed acyclic graph. Its vertex set is $\mathrm{RS}(M)$ and each chain $(z_d, z_u) \in \mathrm{RS}(M)^2$ forms an edge if and only if they appear as base points of some $S \subseteq M$. Such a graph has loops (z, z) corresponding to the single element subsets, i.e., to the points $z \in M$.

3 Computing the Rectangular Subset Closure of a Planar Point Set

Addressing the computational problem of constructing the rectangular subset closure of a set M of grid points, at first glance one might suggest that one has to run through all subsets $S \subseteq M$ and to add to M the corresponding base points $b(S)$. Obviously this would amount to an exponential time algorithm. But we can do much better.

For a fixed (finite) set M of lattice points, we have the following equivalence relation defined on its power set 2^M: $S_1 \sim S_2 \Leftrightarrow_{\text{def}} b(S_1) = b(S_2), \forall S_1, S_2 \in 2^M$ with classes $[S]$. We write $\mathcal{M} := 2^M / \sim$ for the corresponding quotient space. Defining

$$\sigma : 2^M \ni S \mapsto \sigma(S) := r(S) \cap M \in 2^M$$

($r(\emptyset) := \emptyset$) and $\mathcal{R}(M) := \{S \subseteq M : \sigma(S) = S\}$, we have:

Proposition 2. *For $M \subset L$ finite, $\sigma : 2^M \to 2^M$ is a closure operator; and there is a bijection $\mu : \mathcal{R}(M) \to \mathcal{M}$ defined by $S \mapsto \mu(S) := [S], S \in \mathcal{R}(M)$.*

Proof. By the definition of σ it is always ensured that $\sigma(S) \in 2^M$ for every $S \subseteq M$. To show that σ has the properties of a closure operator (cf. the proof of Proposition 1), we first observe that $S \subseteq r(S)$ implies $S \subseteq \sigma(S), \forall S \in 2^I$, i.e., the first condition is fulfilled. Also it is quite obvious that $r(S_1) \subseteq r(S_2)$ whenever $S_1 \subseteq S_2 \subseteq I$ thus we have also $\sigma(S_1) \subseteq \sigma(S_2)$ which is (ii). Addressing (iii) we have to show that if $T = \sigma(S)$ then $\sigma(T) = T$ for an arbitrary $S \subseteq M$. By definition we have $\sigma(S) = T = r(S) \cap M$ so that obviously $r(T) = r[r(S) \cap M] = r(S)$ implying $\sigma(T) = r(T) \cap M = r(S) \cap M = T$.

To prove that $\mu : \mathcal{R}(M) \ni S \mapsto [S] \in \mathcal{M}$ is a bijection we first have to show that $\forall S_1, S_2 \in \mathcal{R}(M) : S_1 \neq S_2$ implies $[S_1] \neq [S_2]$. But this is true, because $S_i = r(S_i) \cap M, i = 1, 2$, and $r(S_1) \cap M \neq r(S_2) \cap M \Leftrightarrow r(S_1) \neq r(S_2)$, thus $[S_1] \neq [S_2]$. Second, for $\emptyset \neq S \subset M$, let $[S] \in \mathcal{M}$, then $\sigma(S) \in \mathcal{R}(M)$ and $[\sigma(S)] = [S]$, because $r(S) = r[r(S) \cap M]$. □

Observe that, in general, for $S \subset M \subset L$, $S \cup b(S)$ does not equal $\sigma(S) = r(S) \cap M$, since $b(S)$ must not be contained in M. But even if $b(S) \subset M$ then the inclusion $S \cup b(S) \subseteq r(S) \cap M = \sigma(S)$ is proper in general. In this case, i.e. if $b(S) \subset M$, we have $S \cup b(S) = \sigma(S)$ if and only if in addition $S \in \mathcal{R}(M)$. In any case $b(T) = b(S)$ is true for every $T \in [S]$ regardless whether or not $b(S) \subset M$.

Now we come back to the problem of computing the rectangular subset closure of a point set:

Proposition 3. RS(M) *can be computed in time $O(|M|^4)$, for $M \subset L$ finite.*

Proof. For $\mathcal{B}_4(M) := \{\emptyset \neq T \subseteq M; |T| \leq 4\}$, we first prove that the map

$$\tau : \mathcal{B}_4(M) \ni T \mapsto \tau(T) := \mu^{-1}([T]) \in \mathcal{R}(M)$$

is well defined and is surjective where $\mu : \mathcal{R}(M) \to \mathcal{M}$ denotes the bijection according to Proposition 2. Assume $M \neq \emptyset$, since otherwise the situation is trivial. Assign to each $T \in \mathcal{B}_4(M) \subseteq 2^M$ its equivalence class $[T] \in \mathcal{M}$. According to Proposition 2 we have $\mu^{-1}([T]) =: \tau(T) \in \mathcal{R}(M)$. To complete the proof that τ is well defined, observe that each T lies in exactly one equivalence class $[T]$.

Now let $S \in \mathcal{R}(M)$ be arbitrarily chosen. Then either (i) $|S| \leq 4$ or (ii) $|S| \geq 5$ holds. In case (i) we have $S \in \mathcal{B}_4(M) \cap \mathcal{R}(M)$ and therefore $S = \mu^{-1}([S]) = \tau(S)$ thus $S \in \tau^{-1}(S)$. We observe that $\tau | \mathcal{B}_4(M) \cap \mathcal{R}(M) = \mathrm{id}_{\mathcal{B}_4(M) \cap \mathcal{R}(M)}$ where $\tau | \mathcal{B}_4(M) \cap \mathcal{R}(M)$ denotes the corresponding restriction of τ. Next, let S satisfy case (ii). We claim (∗): there exists $T \in \mathcal{B}_4(M)$ such that $r(T) = r(S)$. From (∗) the assertion follows since then $T \in [S]$ which is the same as $S \in [T]$ thus $S = \mu^{-1}([T]) = \tau(T)$. Hence, $T \in \tau^{-1}(S)$ which establishes that τ is surjective.

To justify (∗), let $z_d(S) = (x_d(S), y_d(S))$ and $z_u(S) = (x_u(S), y_u(S))$ be the base points of S, where $x_d(S) = \min_{z \in S} x(z), y_d(S) = \min_{z \in S} y(z)$ and $x_u(S) = \max_{z \in S} x(z), y_u(S) = \max_{z \in S} y(z)$. Hence, there must exist members $z_1, z_2, z_3, z_4 \in S$, at least one, for determining each of these extremal values: $x(z_1) = x_d(S), y(z_2) = y_d(S), x(z_3) = x_u(S), y(z_4) = y_u(S)$. In conclusion, for the set $T := \bigcup_{i=1}^{4} \{z_i\}$, holds $|T| \leq 4 \Rightarrow T \in \mathcal{B}_4(M)$ and by construction $r(T) = r(S)$ which has been claimed.

Finally, by surjectivity we obtain

$$\tau(\mathcal{B}_4(M)) = \mathcal{R}(M) \Rightarrow |\mathcal{R}(M)| \leq |\mathcal{B}_4(M)| = \bigcup_{i=1}^{4} \binom{|M|}{i} \in O\left(\sum_{i=1}^{4} |M|^i\right)$$

therefore $|\mathcal{R}(M)| \in O(|M|^4)$.

Thus, to compute $\mathrm{RS}(M)$, for given M, it is sufficient to add to M the base points of all sets in $\mathcal{B}_4(M)$ from which the theorem follows. □

Due to the last result, we immediately obtain a bound for computing $\mathcal{R}(M)$ also:

Corollary 1. *For $M \subset L$ finite, $|\mathcal{R}(M)| \in O(|M|^4)$ and $\mathcal{R}(M)$ can be computed in time $O(|M|^4)$.* □

It is a natural question whether the size of the rectangular subset closure of a point set M can be bounded by the cardinality of M itself. According to the proof of Proposition 3 the answer is yes, namely $|\mathrm{RS}(M)| \in O(|M|^4)$. But is this the best we can hope for? The next theorem tells us that $\mathrm{RS}(M)$ is of size $O(|M^2|)$.

Theorem 1. *For $M \subset L$ finite, we have $|\mathrm{RS}(M)| \leq |M|^2$, and $|\mathrm{RS}(M)| = |M|^2$ if and only if M is an antichain.*

Proof. It is not hard to see that if M is an antichain, then $|\mathrm{RS}(M)| = |M|^2$ (cf. Fig. 3). Next we show that antichains are extremal, i.e., $|\mathrm{RS}(M)| < |M|^2$, for any set M that is no antichain, from which the theorem follows. To that end, assume that M is sorted according to the lexicographic order $\leq_\ell$ in L.

We proceed by induction on $|M|$. For $|M| = 1$, obviously $\mathrm{RS}(M) = M$ and the assertion is true. Let $|M| = n \geq 2$ and assume that the assertion holds, for each set M' with $|M'| < n$. Let $z_0 = (x_0, y_0)$ be the largest element of M. That is, y_0 is the largest y coordinate value among all points of M which

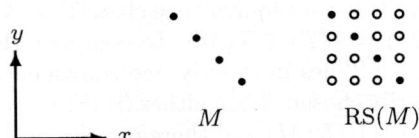

Fig. 3. An antichain M of $n = 4$ points yielding a rectangular subset closure $\mathrm{RS}(M)$ of $n^2 = 16$ points (grid lines are omitted)

are placed in the right most grid line $l(x_0)$ containing points of M, and let $M' := M \setminus \{z_0\}$. Let $\mathcal{Z}(M) := \{S \cup \{z_0\} : S \subseteq M'\}$ denote the set of all subsets of M containing z_0. Then obviously we have $2^M = 2^{M'} \cup \mathcal{Z}(M)$ as disjoint union and therefore $\mathrm{RS}(M) = \mathrm{RS}(M') \cup \{b(Z) : Z \in \mathcal{Z}(M)\}$. For the set $R := \bigcup_{Z \in \mathcal{Z}(M)} b(Z) \setminus \mathrm{RS}(M')$ of additional based points, we claim $(*)$: $|R| \leq 2|M'|+1$ and $|R| = 2|M'|$ iff M is an antichain. Observe that $(*)$ implies the theorem, because we have, $|\mathrm{RS}(M)| \leq |R| + |\mathrm{RS}(M')| \leq 2(n-1) + 1 + (n-1)^2 = |M|^2$, by the induction hypothesis $|\mathrm{RS}(M')| \leq |M'|^2$ (equality for M antichain).

To prove $(*)$, we show that (i): Each additional lower base point $z_d(Z) \in R$ contributed by a set $Z \in \mathcal{Z}(M)$ lies in the interval $[x_{\min}, x_0] \cap L$ of the grid line $l(y_0)$. (ii): Each additional upper base point $z_u(Z) \in R$ contributed by a set $Z \in \mathcal{Z}(M)$ lies in the interval $[y_0, y_{\max}] \cap L$ of the grid line $l(x_0)$. Here $x_{\min}$ denotes the smallest x-value occuring among the coordinates of the points in M' and $y_{\max}$ denotes the largest such y-value. Notice that claim $(*)$ follows from (i), (ii), since by (i) additional lower base points can only lie at those positions (x', y_0) in the grid part indicated in (i) for which exists $z' \in M'$ with $x(z') = x'$. Hence there can be at most $|M'|$ additional lower base points contributed by members from $\mathcal{Z}(M)$. The analogous argumentation holds for additional upper base points $z_u(Z)$ for $Z \in \mathcal{Z}(M)$. These can only lie on those positions (x_0, y') of the grid part indicated in (ii), for which exists $z' \in M'$ with $y(z') = y'$ yielding also at most $|M'|$ additional upper base points. Thus, adding point z itself, the number of additional base points is at most $2|M'|+1$ and exactly that number only if M is an antichain (in which case also M' is an antichain).

It remains to prove (i) and (ii). For $Z \in \mathcal{Z}(M)$, let $z_d(Z) = (x_d(Z), y_d(Z))$ and $z_u(Z) = (x_u(Z), y_u(Z))$, then obviously we have $x_d(Z) \leq x_0$, $y_d(Z) \leq y_0$ and $x_u(Z) \leq x_0$, $y_u(Z) \geq y_0$. To show (i) suppose $x_0 > x_{\min}$ (otherwise z_0 is the only additional lower base point) and further suppose that there is $Z \in \mathcal{Z}(M)$ such that $z_d(Z) \in R$ is not lying in the grid part indicated in (i). Hence, we have $y_d(Z) < y_0$ implying that there exists $z' \in Z' := Z \setminus \{z_0\} \subset M'$ such that $y_d(Z') = y_d(Z)$. But this means $z_d(Z) \in \mathrm{RS}(M')$, contradicting the assumption that it is an *additional* base point not yet occuring in $\mathrm{RS}(M')$, thus we have proven (i). For verifying (ii), assume that $y_0 < y_{\max}$ (otherwise z_0 is the only additional upper base point) and assume further that $Z \in \mathcal{Z}(M)$ such that $z_u(Z) \in R$ is not lying in the grid part indicated in (ii). Then we have $x_u(Z) < x_0$ and there is $z' \in Z' := Z \setminus \{z_0\} \subset M'$ such that $x_u(Z') = x_u(Z)$ yielding a contradiction as above. □

The proof of the last result can be mimicked to obtain a bound of quadratic time for computing the rectangular subset closure of a planar point set.

Corollary 2. *For $M \subset L$ finite, $\mathrm{RS}(M)$ can be computed in time $O(|M|^2)$.*

Proof. Let the elements of $M = \{z_1, \ldots, z_n\}$ be labeled by increasing lexicographic order in L and define $S_i = \{z_1, \ldots, z_i\} \subset M$. A corresponding sorting step takes time $O(n \log n)$. Initialize $\mathrm{RS}(\emptyset) = \emptyset$ and in step i of the algorithm compute $\mathrm{RS}(S_i) = \mathrm{RS}(S_{i-1}) \cup R_i$ as indicated in the proof of Theorem 1, where R_i is the set of additional base points contributed by subsets of S_i containing z_i. This step can be done in time $O(|S_i|)$. From which the corollary follows by summing over $i \in \{1, \ldots, |M|\}$. □

Finally, we show that the number of subsets of a given point set M yielding different rectangular bases, namely $|\mathcal{R}(M)|$, is of size $O(|\mathrm{RS}(M)|^2)$. According to Proposition 2, this also is the number of representatives of the classes determined by relation $\sim$, considering all subsets of points as equivalent that yield the same rectangular base.

Theorem 2. *For $M \subset L$ finite, we have $|\mathcal{R}(M)| \leq |\mathrm{RS}(M)|^2$.*

Proof. Consider the map

$$b : \mathcal{R}(M) \ni S \mapsto b(S) \in \{\{z\} : z \in M\} \cup \binom{\mathrm{RS}(M)}{2}$$

assigning to each $S \in \mathcal{R}(M)$ its base points $b(S) \subset \mathrm{RS}(M)$ with $|b(S)| \leq 2$. Because $S_1 = S_2 \Leftrightarrow b(S_1) = b(S_2)$ for all sets $S_1, S_2 \in \mathcal{R}(M)$ the mapping is injective and the assertion follows. □

Observe that this bound for $|\mathcal{R}(M)|$, in many cases, namely when $|\mathrm{RS}(M)| \in O(|M|)$ (which, e.g., is the case for chains) holds, is much better than that presented in Corollary 1. Moreover, observe that in case M is an antichain, i.e., the most extremely class regarding the size of $\mathrm{RS}(M)$, then for computing $\mathcal{R}(M)$ only all subsets of size two of M are needed. Hence, for an antichain M, holds $\mathcal{R}(M) \in O(|M|^2)$.

4 Generalization to the d-Dimensional Case

The setup described in the preceeding section will be generalized in the sequel to the d-dimensional case for $2 \leq d \in \mathbb{N}$. This generalization is not only interesting from an abstract point of view but it may be profitable also for modeling higher dimensional applications.

For fixed $1 < d \in \mathbb{N}$, let $\mathbb{E}^d$ denote the d-dimensional Euclidean space with fixed (orthogonal) standard basis $B^d = \{e_1, \ldots, e_d\}$. An (orthogonal) integer lattice in $\mathbb{E}^d$ then is given by $L_\lambda(d) = \mathbb{Z}e_1\lambda + \cdots + \mathbb{Z}e_d\lambda$ with lattice constant $0 < \lambda \in \mathbb{R}$. Again setting $\lambda = 1$ yields $L_1(d) =: L(d) = \mathbb{Z}^d$. A finite d-dimensional rectangular grid region is defined by $I(d) = ([A_1, B_1] \times \cdots \times [A_d, B_d]) \cap L(d)$, where $A_i, B_i \in \mathbb{Z}, A_i < B_i, 1 \leq i \leq d$. By $\leq_{\ell_d}$ we denote the restriction to

$L(d)$ of the (linear) lexicographic order in $\mathbb{E}^d$. And $\leq_{L(d)}$ is the natural partial order in $\mathbb{E}^d$ restricted to $L(d)$ which is the straightforward generalization of the corresponding order $\leq_L$ in the 2-dimensional case.

To generalize the basic term of a rectangular base to higher dimensions, suppose that we are given a set $M = \{\boldsymbol{m}_1, \ldots, \boldsymbol{m}_n\} \subset L(d)$ of grid points, where each $\boldsymbol{m}_i = (m_i^1, \ldots, m_i^d) \in L(d)$ is represented by its coordinates with respect to B^d. Then to each $S \subset M$ assign its d-base points $b_d(S) := \{\boldsymbol{m}_a(S), \boldsymbol{m}_b(S)\}$, where $m_a^i(S) := \min\{m^i | \boldsymbol{m} \in S\}$ and $m_b^i(S) := \max\{m^i | \boldsymbol{m} \in S\}, 1 \leq i \leq d$. Now the d-rectangular base $r_d(S)$ is determined by $r_d(S) = [m_a^1, m_b^1] \times \cdots \times [m_a^1, m_b^1]$. As in the 2-dimensional case, the d-base points $b_d(S)$ are the vertices incident to the main hyper-diagonal in $r_d(S)$ that is oriented along increasing coordinate values, i.e., $\boldsymbol{m}_a(S) \leq_{L(d)} \boldsymbol{m}_b(S)$. Thus, $r_d(S)$ is the smallest d-dimensional rectangular object enclosing S tightly. Now we are ready to define the d-dimensional version of the rectangular subset closure of a point set.

Definition 3. *The d-dimensional rectangular subset closure $\mathrm{RS}_d(M)$ of a finite point set $M \subset L(d)$ is defined as the smallest subset of $L(d)$ containing M that also contains the d-base points $b_d(S)$ of each subset $S \subseteq M$.*

Observe that, as in the 2-dimensional case, the d-dimensional rectangular subset closure has the following explicit form: $\mathrm{RS}_d(M) = M \cup \bigcup_{S \subseteq M} b_d(S)$. Analogously to the proof of Proposition 1, it can be shown that RS_d also gives rise to a closure operator (more precisely, to a family of finite closure operators, one for each finite rectangular grid region $I(d)$):

Proposition 4. *For each fixed finite rectangular grid region $I(d) \subset L(d)$,*

$$\mathrm{RS}_d : 2^{I(d)} \ni M \mapsto \mathrm{RS}_d(M) \in 2^{I(d)}$$

is a well defined (finite) closure operator. □

The equivalence relation $\sim$ on the power set 2^M for $M \subset L$ can also be generalized to the d-dimensional case where $M \subset L(d)$:

$$S_1 \sim_d S_2 \Leftrightarrow_{\mathrm{def}} b_d(S_1) = b_d(S_2), \forall S_1, S_2 \in 2^M$$

with classes $[S]_d$. Defining $\mathcal{M}_d := 2^M / \sim_d$ as well as

$$\sigma_d : 2^M \ni S \mapsto \sigma_d(S) := r_d(S) \cap M \in 2^M$$

$(r_d(\varnothing) := \varnothing)$ and $\mathcal{R}_d(M) := \{S \subseteq M : \sigma_d(S) = S\}$ we arrive at:

Proposition 5. *$\sigma_d : 2^M \to 2^M$ is a closure operator and there is a bijection $\mu_d : \mathcal{R}_d(M) \to \mathcal{M}_d$ defined by $S \mapsto \mu_d(S) := [S]_d, S \in \mathcal{R}_d(M)$.* □

Since each subset S contributes at most two base points to the d-dimensional rectangular subset closure we finally obtain by transferring the proofs of Theorem 1 and Corollary 2 (relying on the lexicographic order $\leq_{\ell_d}$ in $L(d)$):

Theorem 3. *For $M \subset L(d)$ finite, we have $|\mathrm{RS}_d(M)| \in O(|M|^2)$. Moreover, $\mathrm{RS}_d(M)$ can be computed in time $O(d|M|^2)$.*

Similarly, we can derive generalized results concerning the set $\mathcal{R}_d(M)$ of representatives according to relation $\sim_d$, for a finite set $M \subset L(d)$.

Theorem 4. *For $M \subset L(d)$ finite, we have $|\mathcal{R}_d(M)| \in O(|\mathrm{RS}_d(M)|^2)$.*

The proofs proceed analogously to those of Theorem 2. Observe that all computational time bounds provided in this section are polynomial since the dimension d is a fixed positive integer for every input set M.

5 Concluding Remarks and Open Problems

We introduced the concept of a rectangular subset closure for a given finite set of points in discrete Euclidean spaces, i.e., $\mathbb{Z}^d$, $d \geq 2$. Note that the concept and results can be straightforwardly generalized to finite lattices based on an arbitrary fixed lattice constant $\lambda > 0$. The results obtained also remain valid when the points are arbitrarily distributed in $\mathbb{E}^d$, $d \in \mathbb{N}, d \geq 2$ (for that instead of grid lines simply use the continuous lines containing them). An interesting question in the discrete case is, whether also for non-isothetical grids and/or (regular) objects that are not rectangular, a subset closure concept can be defined reasonably.

For an input set M of n points, the algorithm presented computes, based on a lexicographic sorting of M, the set $\mathrm{RS}(M)$ in time $O(n^2)$. Observe that there are many situations where $\mathrm{RS}(M)$ is much less than $|M|^2$. E.g. for an chain M, we even have $\mathrm{RS}(M) = M$. Thus, an open problem from the computational point of view is, whether there exists an algorithm computing $\mathrm{RS}(M)$ in time $O(\max\{n \log n, |\mathrm{RS}(M)|\})$ or even in optimal time $O(|\mathrm{RS}(M)|)$, for *every* input set M. The latter time bound especially would require an algorithm that does not need the input set to be ordered lexicographically which seems hard to achieve.

We proved that $|\mathcal{R}(M)| \in O(|\mathrm{RS}(M)|^2)$ and gave some hints that this could be improved, perhaps. In fact, the question remains whether a deeper analysis could yield $|\mathcal{R}(M)| \in O(|M|^2)$ (which is true at least for every antichain M).

References

1. E. Boros and P. L. Hammer, On Clustering Problems with Connected Optima in Euclidean Spaces, Discrete Mathematics 75 (1989) 81-88.
2. F. C. Calheiros, A. Lucena and C. C. de Souza, Optimal Rectangular Partitions, Networks 41 (2003) 51-67.
3. J. Hershberger and S. Suri, Finding Tailored Partitions, Journal of Algorithms 12 (1991) 431-463.
4. D. S. Hochbaum (ed.), Approximation Algorithms for NP-hard problems, PWS Publishing, Boston, Massachusetts, 1996.
5. S. Porschen, On Covering $\mathbb{Z}$-Grid Points by Rectangles, ENDM, Vol. 8, 2001.
6. S. L. Tanimoto and R. J. Fowler, Covering Image Subsets with Patches, Proceedings of the fifty-first International Conference on Pattern Recognition, 1980, pp. 835-839.

Computing Optimized Curves with NURBS Using Evolutionary Intelligence

Muhammad Sarfraz[1], Syed Arshad Raza[2], and M. Humayun Baig[1]

[1] Department of Information and Computer Science,
King Fahd University of Petroleum and Minerals,
Dhahran 31261, Saudi Arabia
{sarfraz, humayun}@ccse.kfupm.edu.sa
[2] The Accounting and Management Information Systems Department,
King Fahd University of Petroleum and Minerals,
Dhahran, 31261 Saudi Arabia.
saraza@kfupm.edu.sa

Abstract. In curve fitting problems, the selection of knots in order to get an optimized curve for a shape design is well-known. For large data, this problem needs to be dealt with optimization algorithms avoiding possible local optima and at the same time getting to the desired solution in an iterative fashion. Many evolutionary optimization techniques like genetic algorithm, simulated annealing have already been successfully applied to the problem. This paper presents an application of another evolutionary heuristic technique known as "Simulated Evolution" (SimE) to the curve fitting problem using NURBS. The paper describes the mapping scheme of the problem to SimE followed by the proposed algorithm's outline with the results obtained.

1 Introduction

In planar shape design problems, the main objective is to achieve an optimized curve with the least possible computation cost. For complicated shapes with large measurement data, the problem becomes dependent upon the selection of optimal knots. Algorithms based on heuristic techniques like genetic algorithms, simulated annealing, simulated evolution (SimE) etc., can provide us with an approach in finding optimal number of knots with reasonable cost. Since the data in such a problem cannot be approximated with a single polynomial, the application of splines, Bezier curves etc., are well known. Non-uniform Rational B-splines (NURBS) [4], providing more local control on the shape of the curve, gives a better approximation of the underlying data in shape design problems. In [9], knots corresponding to the control points have been optimized using a genetic algorithm. An approach based on Tabu search has been applied in [13]. An algorithm proposed in [11] discusses optimization of knots and weights using Simulated Annealing. The main contribution of this work is to propose a curve fitting algorithm based on SimE using NURBS.

The paper has been designed into various sections. The following section deals with the image contour extraction. Detection of significant points have been reported in Section 3. Section 4 gives a brief description of NURBS whereas the SimE algorithm has been discussed in Section 5. The proposed approached, with details of the evolutionary optimization curve technique has been explored and designed in Section 6. Demonstration of the executed results is given in Section 7 and Section 8 concludes the paper.

2 Image Contour Extraction

A digitized image is obtained from an electronic device or by scanning an image. The quality of digitized scanned image depends of various factors such as the image on paper, scanner type and the attributes set during scanning. The contour of the digitized image is extracted using the boundary detection algorithms. There are numerous algorithms for detecting boundary. We used the algorithm proposed by [8]. The input to this algorithm is a bitmap file. The algorithm returns a number of boundary points and their values.

3 Detection of Significant Points

Detection of significant points is the next step after finding out contour of the image. The significant points are those points, which partition the outline into various segments. Each segment is considered to be an element in our proposed approach. A number of approaches have been proposed by researchers [2]. In this paper, the detection of corner points has been implemented using the technique presented by Chetverikov and Szabo [2]. In [2] corner point is defined as a point where triangle of specified angle can be inscribed within specified distance from its neighbor points. It is a two pass algorithm. In the first pass the algorithm scans the sequence and selects candidate corner points. The second pass is post-processing to remove superfluous candidates.

4 NURBS

A unified mathematical formulation of NURBS provides free form curves and surfaces. NURBS contains a large number of control variable, because of those variables it is flexible and powerful. NURBS is a rational combination of a set piecewise rational polynomial of basis functions of the form

$$C(u) = \frac{\sum_{i=1}^{n} p_i w_i B_{i,k}(u)}{\sum_{i=1}^{n} w_i B_{i,k}(u)} \qquad (1)$$

where p_i are the control points and w_i represent the associated weights. u is the parametric variable and $B_{i,k}(u)$ is B-spline basis function. Assuming basis function of order k (degree $k-1$), a NURBS curve has $n+k$ knots and the number of control points equals to weights. t_i is in non-decreasing sequence: $t_1 \leq t_2 \leq ... \leq t_{n+k-1} \leq t_{n+k}$. The basis functions are defined recursively using non-uniform knots as

$$B_{i,1}(u) = \begin{cases} 1 & \text{for } t_i \leq u < t_{i+1} \\ 0 & \text{otherwise} \end{cases} \quad (2)$$

$$B_{i,1}(u) = \frac{u - t_i}{t_{i+k-1} - t_i} B_{i,k-1}(u) + \frac{t_{i+k} - u}{t_{i+k} - t_{i+1}} B_{i+1,k-1}(u) \quad (3)$$

The parametric domain is $t_k \leq u \leq t_{k+1}$. The NURBS knots are used to define B-spline basis functions implicitly. NURBS inherit many properties from B-spline [6], such as the strong convex hull property, variation diminishing property, local support, and invariance under affined geometric transformations. NURBS include weights as extra degrees of freedom, which are used for geometric design [5]–[7].

5 Outline of Simulated Evolution (SimE)

SimE is a powerful general iterative heuristic for solving combinatorial optimization problems [10], [12]. The algorithm consists of three basic steps: Evaluation, Selection and Allocation. These three steps are executed sequentially for a prefixed number of iterations or until a desired improvement in goodness is observed. The SimE algorithm starts with an initial assignment, and then seeks to reach better assignments from one generation to the next. SimE assumes that there exists a population P of a set M of n elements. A cost function is used to associate with each assignment of element m a cost C_m. The cost C_m is used to compute the goodness gm of element m for each $m \in M$.

The selection step partitions the elements into two disjoint sets P_s and P_r based on their goodness. The elements with bad goodness are selected in the set P_s and the rest of elements in the set P_r. The non-deterministic selection operator takes as input the goodness of each element and a parameter B, a selection Bias. Hence the element with high goodness still has a non-zero probability of being assigned to the selected set P_s. The value of the bias is application dependent. In our case the value of B has been taken as -0.5.

The Allocation step takes P_s and P_r and generates a new solution P' which contains all the members of the previous population P. The members of P_s are then worked upon so that their goodness could be enhanced in the subsequent iterations. The choice of a suitable allocation function is problem dependent [12].

6 Proposed Approach

The proposed approach to the problem is described here in detail.

6.1 Problem Mapping

In curve fitting problems, the solution space consists of the number of data points on the image boundary. Each and every data point is a candidate to be selected as a movable element. If M denotes the set of movable elements and D is the set of all data points on the image boundary then in our case:

$$M = \{d / d \in D\} \tag{4}$$

The initial solution (population) i is created using corner detection algorithm [2]. Mathematically we can say:

$$i = \{c / c \text{ is a corner point}, c \in M\} \tag{5}$$

In the initial solution, the corner points are the only end points of the segments (S) for a piece-wise polynomial fitting such that for n number of corner points we have n-1 segments. For each segment we need to calculate the parameters u, control points, knot vector and the weight of NURBS. One useful approximation for calculating parameter value for u uses the chord length between data points [6].

Weights of each segment are taken randomly between 0 and 1. A non-uniform knot vector is created using centripetal method [6]. After calculating all the parameters for each segment curve is fitted using NURBS. This fitted curve for each segment is considered as the initial solution for SimE.

The algorithm executes by taking the initial solution. The goodness g_i of each segment S_i is determined by

$$g_i = \frac{2(2l_i + k)}{AIC_i} \tag{6}$$

where l is the length of the knot vector, k is the order of the curve, AIC is the Akaike's Information Criteria [1]. The formula for the AIC (the Akaike's Information Criteria) is as follows:

$$AIC = N |\ln Q| + 2(2l + k) \tag{7}$$

where N is the total number of data points in each segment and Q is square error between the target and the fitted curves.

On the basis of the goodness g_i, the segments are then partitioned into two sets P_r and P_s. The P_s contains the segments with bad goodness that is those segments we need to operate on in the next generation. In our approach, in the Allocation step the segment having the lowest goodness in P_s undergoes a knot insertion process [3], thus increasing the number of segments by one. The two sets are then merged and then passed to the next generation (iteration). The segment undergone the knot insertion process in the previous generation is now having an improved goodness making it a better candidate to be selected in P_r. The total number of iterations is controlled by a parameter ξ in "(8)", where:

$$P_s \leq round(\xi \ P_r) \tag{8}$$

In this algorithm, the value of ξ has been taken as 0.2. This parameter gives us a better control on the algorithm's execution providing the results of desired percentage of fitness. The value of 0.2 thus gives us approximately 80% accuracy in the final (output) solution of the algorithm.

6.2 Algorithm Outline

The algorithm of the proposed scheme is contained on various steps. The outline of the algorithm together with details of all the steps is explained as follows:

Step 1: Input the digitized image
Step 2: Find the image contour [8]
Step 3: Find the corner points [2]
Step 4: For each segment i

    a) Find control points from data points using least square method [6]
    b) Find knot vector
    c) Find weights
    d) Fit the curve

    End for

Step 5: Initialize population P, Bias (B=-0.5)
Step 6: for j=1 to arbitrary number of iterations (say 100)

    a) Evaluation
        For each segment i in P
            Find goodness (g_i)
        End for
    b) Selection
        For each segment i in P
            If (Random [0, 1] <=1- g_i + B) then
                $P_s = P_s \ U \ \{i\}$
            Else
  $P_r = P_r \ U \ \{i\}$
        End for
    c) Allocation
        For a segment i of least g_i in P_s
            Insert a knot at midpoint of the segment i [3]
        End for
        $P = P_r \ U \ P_s$
    d) If $P_s \leq round(\xi \ P_r)$ then break
        End for

Step 7: Return the final solution
Step 8: End

7 Results

In this section, the results obtained by applying the proposed algorithm on various objects (jet plane, fork, slant U) have been presented. Though the proposed algorithm is independent of the degree, the segments of the objects have been approximated

Fig.1. Bitmap image of the object 'Jet plane'

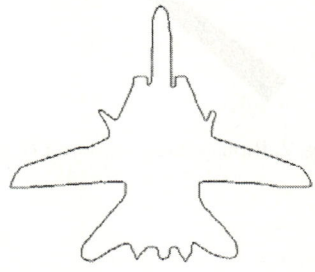

Fig. 2. Outline after boundary detection

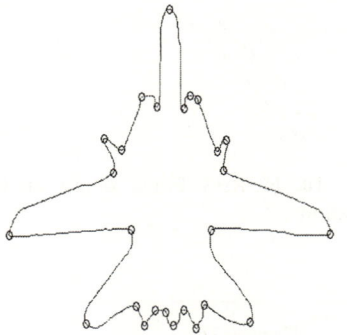

Fig. 3. Significant point detection

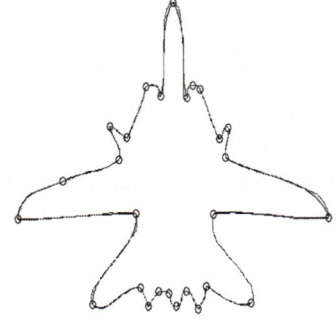

Fig. 4. NURBS Fitted object at First iteration

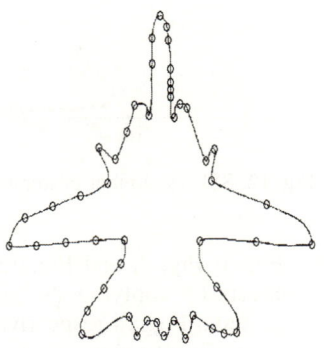

Fig. 5. Final NURBS Fitted object

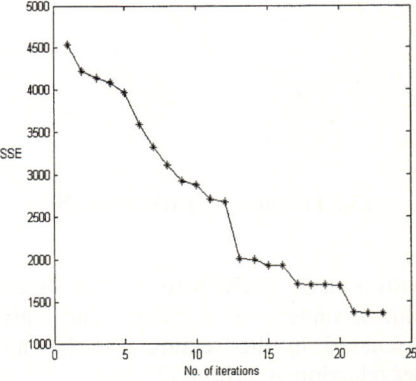

Fig. 6. SSE Vs Number of iterations

Fig. 7. Bitmap image of object (fork)

Fig. 8. The contour of the image obtained

Fig. 9. NURBS Fitted object at First iteration

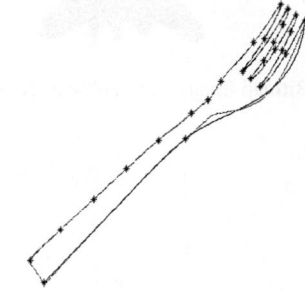

Fig. 10. NURBS Fitted object at 17th iteration

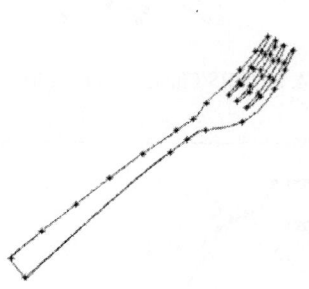

Fig. 11. Final NURBS Fitted object

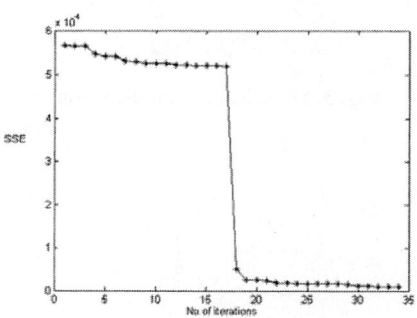

Fig. 12. SSE vs number of iterations

with a cubic NURBS to get the effective results. Fig. 1, Fig. 7, and Fig. 13 are the objects under consideration. The outline of the objects by applying the algorithm, discussed in the section 2, is shown in Fig. 2, Fig. 8 and Fig. 14 respectively. Cornerdetection algorithm [2] detected 26 segments (27 significant points) for jet plane, 9 segments (10 significant points) for fork.

Fig. 13. Bitmap image of object (slant U)

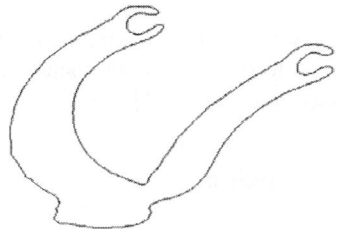

Fig. 14. The contour of the image obtained

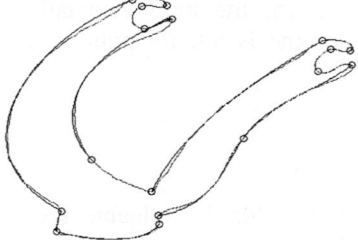

Fig. 15. NURBS Fitted object at First iteration

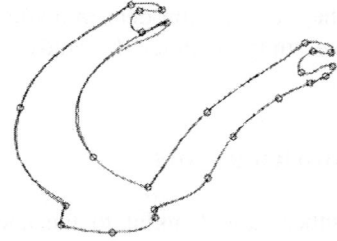

Fig. 16. NURBS Fitted object at eighth iteration

Fig. 17. Final NURBS Fitted object

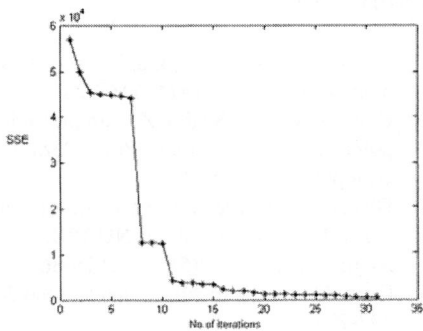

Fig. 18. SSE vs number of iterations

The proposed algorithm is run for 100 iterations. The bias value is taken as -0.5. The segment with the least goodness out of the total selected segments is added a knot using the middle point criteria [3]. The algorithm converged at 24th iteration for jet plane and at 34th iteration for fork is shown in Fig. 5 and Fig. 11 respectively. The Sum Square Error (SSE) between the boundary of the image and the NURBS fitted curve with respect to the number of iterations is as shown in Fig 12. Other results for slant U at different iterations (see Fig. 15, Fig. 16, Fig. 17, and Fig. 18) have also been shown. The results obtained using the proposed algorithm are found to be better compared to the results obtained using the genetic algorithm and simulated annealing discussed in [9] and [11].

8 Conclusion

The proposed approach is effective in the determination of appropriate number of knots using NURBS. The corner detection algorithm provides an initial solution to start with. The subsequent iterations evolutionarily approach towards the desired solution. Quite pleasing results have been obtained in a comparable amount of time with the existing non-deterministic approaches in the literature. A detailed comparative study is under the study of the authors and is left for publication elsewhere.

Acknowledgments

The authors are thankful to the anonymous referees for the valuable suggestions towards the improvement of this manuscript. This work has been supported by the King Fahd University of Petroleum and Minerals.

References

1. Akaike, H.: A new look at the statistical model identification," IEEE Transaction. Automatic Control, vol (1974) 716-723
2. Chetverikov, D., Szabo, Z.: simple and efficient algorithm for detection of high curvature points in planar curves. Proc. 23rd Workshop of the Australian Pattern Recognition Group. (1999) 175-184
3. Dierckx, P.: Curve and surface fitting with Splines. Clarendon Press (1993)
4. Farin, G.: From Conic to NURBS: A tutorial and survey. IEEE Computer Graphics and Applications. vol.12(5) (1992) 78-86
5. Farin, G.:Trends in curves and surface design. Computer-Aided Design, vol. 21(5) (1989) 293-296
6. Piegl, L., Tiller, W.: *The NURBS Book.* Springer-Verlag, New York, (1997)
7. Piegl, L.,Tiller, W.: Curve and surface reconstruction using rational B-splines. *Computer-Aided Design*, vol. 19(9) (1991) 485-498
8. Quddus, A.: Curvature Analysis Using Multi-resolution Techniques. PhD Thesis. Dept. Elect. Eng., King Fahd University of Petroleum & Minerals, Dhahran, Saudi Arabia(1998)
9. Sarfraz, M., Raza, S. A.: Capturing Outline of Fonts using Genetic Algorithm and Splines. The Proceedings of IEEE International Conference on Information Visualization-IV'2001-UK, IEEE Computer Society Press, USA, (2001) 738-743
10. Ralph Michael kling , Benerjee, P.: Empirical and Theoretical studies of Simulated Evolution Method Applied to standard cell Placement. IEEE Transactions on computer Aided Design vol. 10(10) (1991)

11. Riyazuddin, M.: Visualization with NURBS using Simulated Annealing optimization Technique. Master Thesis. Dept. ICS, King Fahd University of Petroleum & Minerals, Dhahran, Saudi Arabia (2004)
12. Sait, M.S., Youssef, H.: Iterative Computer Algorithms with Applications in Engineering: Solving Combinatorial Optimization Problems. IEEE Computer Society Press, California (1999)
13. Youssef, M.: Reverse Engineering of Geometric Surfaces using Tabu Search Optimization Technique. Master Thesis, Cairo University, Egypt (2001)

A Novel Delaunay Simplex Technique for Detection of Crystalline Nuclei in Dense Packings of Spheres

A.V. Anikeenko[1], M.L. Gavrilova[2], and N.N. Medvedev[1]

[1] Institute of Chemical Kinetics and Combustion SB RAS, Novosibirsk, Russia
[2] Department of Computer Science, University of Calgary, Calgary, AB, Canada
nikmed@kinetics.nsc.ru

Abstract. The paper presents a new approach for revealing regions (nuclei) of crystalline structures in computer models of dense packings of spherical atoms using the Voronoi-Delaunay method. A simplex Delaunay, comprised of four atoms, is a simplest element of the structure. All atomic aggregates in an atomic structure consist of them. A shape of the simplex and the shape of its neighbors are used to determine whether the Delaunay simplex belongs to a given crystalline structure. Characteristics of simplexes defining their belonging to FCC and HCP structures are studied. Possibility to use this approach for investigation of other structures is demonstrated. In particular, polytetrahedral aggregates of atoms untypical for crystals are discussed. Occurrence and growth of regions in FCC and HCP structures is studied on an example of homogeneous nucleation of the Lennard-Jones liquid. Volume fraction of these structures in the model during the process of crystallization is calculated.

1 Introduction

Investigation of structural transformations taking place during liquid, amorphous and crystalline phases is an important problem of modern material science. A characteristic feature of such processes is the structural heterogeneity, which means that the sample may contain regions of different structure, both crystalline and disordered. It is not an easy task to investigate these structural features. While the simulation of large computer models of atomic systems is rather routine problem, the analysis of regions of different structures requires development of special approaches. Recently, a considerable progress in this direction was achieved through the utilization of the Voronoi-Delaunay method [1,2]. An important aspect of implementation of the method is based on the Delaunay simplexes. The Simplex Delaunay is described by four atoms and represents a simplest three-dimension element (brick) of the structure. Any fragment of the structure can be presented as a cluster of Delaunay simplexes. Thus, one can determine regions of the required structure by obtaining simplexes of a given structural type [3-6]. The Delaunay simplexes can be used for more precise identification of regions of the given structure then method based on Voronoi polyhedra [7-9], spherical harmonics [10-11], or distribution of angle between geometrical neighbours [12-13]. Voronoi polyhedra and spherical harmonics characterize a nearest environment of the atom, i.e. the structural unit that consists of a rather large numbers of atoms (15 on average). It does not present a problem when large heterogeneities are studied. However it is not suitable for studying of small regions, such as nuclei (embryos) of a new phase consisted of a few atoms. For instance, a difference between face centered cubic (FCC) and hexagonal close packing (HCP) crystalline structures is visible on groups of 6 atoms, and main parts of icosahedron (fragments with 5-fold symmetry) is detected on groups of 7 atoms [6].

The method based on using Delaunay simplexes cannot be applied directly. First of all, we must point out that an individual simplex, as a rule, does not characterize the structure uniquely. For example, a good tetrahedron (closed to perfect shape) can belong to both FCC and HCP structures, as well as can be found in amorphous phase. On the other hand, crystalline structure is not always represented by one specific type of simplex. In particular, each densest crystalline structure consists of three types of Delaunay simplexes of a different shape: tetrahedron, quartoctahedron (a quarter of an octahedron), and, in a small proportion, simplexes close to flat square. All differences between the crystals are defined by the mutual arrangement of the above types of simplexes.

In our previous works related to identification of simplexes of a given shape, we have developed the methodology based on the measures of a simplex form T, Q and K. These measures are defined as special variances (dispersions) of lengths of edges of the simplexes, see [5,6,14] and below. To extract regions of a given structure, we studied arrangement of selected simplexes. Clusters of such simplexes allow revealing crystalline as well as specific non-crystalline aggregates of atoms. However, the type of the structure can be established only after a cluster of the simplexes is constructed. Thus, cycles (rings) of tetrahedra and quartoctahedra arranged in the form of rhombus are typical for FCC structure, and the trapezoid form if found in HCP structure [5,6]. Such analysis demonstrates presence of above structures in the model of a crystal but it is rather qualitative. An open problem is to determine the quantity of specific structures in a given sample. To address this problem, one needs a more specified, quantitative ascription of the simplexes to a given structure type.

In this paper, we suggest to characterize belonging of the simplexe to a given structure considering the shape of both a given simplex and its neighbors. We refer to this problem as a problem of identification of a structural type of a Delaunay simplex. As neighboring simplexes we propose to consider simplexes with adjacent faces. The structural unit, that identifies a type of a given structure, is an aggregate of eight atoms: four atoms of a given simplex and four atoms at its faces. However, only the central simplex is used for subsequent structure analysis.

2 The Voronoi-Delaunay Method

Using geometrical ideas of Voronoi and Delaunay for structural analysis of atomic systems is discussed in details in many articles (see, for example, [3-7,15]). A set of coordinates of all atoms {A} of the model is the basic data for structural analysis. At the first step, the Voronoi-Delaunay partitioning of the studied model is calculated. Actually, for our analysis we deal only with the Voronoi network, defined as a network of edges and vertices of a set of Voronoi polyhedra. The Voronoi network is represented by a set of coordinates of vertexes {D} and a table of their connectivity

{DD}. Every vertex of the Voronoi network is incident on four atoms of the system, which define of a Delaunay simplex. It means that every vertex of the Voronoi network determines position of one of the Delaunay simplexes of the system. Using coordinates of atoms, one can calculate any geometrical characteristics of the Delaunay simplex (in particular measures of shape). Next, using connectivity of the Voronoi network, it is convenient to study their mutual arrangement, and to define clusters of simplexes with a given structural characteristics [5, 16].

3 Simplex Shape Measures

3.1 T, Q and K Measures

Here we remind our definition of the simplex shape measures and make some minor unification of them.

A choice of simplex characteristic depends on the problem being studied. In our case we study dense packings of spherical atoms. The main configuration for this study is the tetrahedral configuration of four atoms. It is the densest local configuration and is preferable energy-wise for spherical atoms. In FCC and HCP crystals there are also octahedral configurations, which together with tetrahedral ones ensure translation symmetry of the crystal. Octahedral configuration is not a simplex, since it has six vertexes. A perfect octahedral configuration provides an example of degenerated configuration: all six vertices lie on a sphere. However, in computer simulation of physical systems, atoms are typically shifted from their ideal positions. Thus, every octahedral configuration is represented through Delaunay simplexes unambiguously. Usually, there appear four similar simplexes (quartoctahedra [14]). A perfect quartoctahedron has five equal edges, and the sixth edge is $\sqrt{2}$ times longer than five others. However, at some specific displacements of atoms, the perfect octahedron can be divided on five instead of four simplexes. The fifth simplex springs up from flat configuration of four atoms of octahedron. This simplex was found in the models of dense liquids and was called a simplex Kizhe [17]. It has two opposite edges (diagonals of a square) which are $\sqrt{2}$ times longer then other four edges. These simplexes are rare, however they also should be taken into account when studying crystalline structures.

To extract tetrahedral configurations closed to perfect we use measure T, called *tetrahedricity* [6,16]. It is the variance of the lengths of edges of the simplex

$$T = \sum_{i \neq j} (e_i - e_j)^2 / 15 <e>^2. \tag{1}$$

Here e_i and e_j are the lengths of the i-th and j-th edges, and $<e>$ is the mean edge length for a given simplex. The number 15 used as normalization factor is the number of possible pairs of six edges of the simplex. For a perfect tetrahedron, value T is equal to zero. A small value of T means unambiguously that the simplex is close to perfect tetrahedron.

For unambiguous extraction of a good quartoctahedron, we use special measure Q-quartoctahedrisity [6,16].

$$Q = \left(\sum_{\substack{i<j \\ i,j \neq m}} (e_i - e_j)^2 + \sum_{i \neq m} (e_i - e_m/\sqrt{2})^2 \right) / 15 <e>^2 \qquad (2)$$

This measure is similar to measure T, only now the computation of variance of edge lengths takes into an account that one edge is $\sqrt{2}$ times longer than the others. To compute Q, the longest edge m of a simplex needs to be found first, and then the calculation is carried out according to (2). It is obvious, that for almost perfect quartoctahedron, the value of measure Q approaches zero. The inverse is also true. Note, the value T defined according to (1) for perfect quartoctahedron is equal to 0.050. However, the same value can correspond many simplexes of other shape. Due to this fact, one needs to introduce a special measure for every different shape.

To extract simplexes Kizhe, a measure K was constructed following the same principle. Here we use the fact that two edges $\sqrt{2}$ times longer than the others [6,17].

$$K = \left(\sum_{\substack{i<j \\ i,j \neq m,n}} (e_i - e_j)^2 + \sum_{i \neq m,n} (e_i - e_m/\sqrt{2})^2 + \sum_{i \neq m,n} (e_i - e_n/\sqrt{2})^2 + (e_m - e_n)^2 \right) / 15 <e>^2 \qquad (3)$$

The meaning of this expression is when the value K approaches zero, a simplex degenerates into a square. To compute K, a pair of the longest opposite edges of the simplex needs to be found first as the edges m and n, and then the calculation is carried out according to (3).

The expressions (2) and (3) differ from analogous formulas derived in [3,6,17]. In those works, every component of the expression was normalized based on the number of pairs of edges. In our work we introduce a common normalization factor equal to 15, which is the total number of different pairs. This method of normalization is not crucial for selection of simplexes. However, for new characteristics of simplexes defined below, the unified expressions for measures of shape (1)-(3) are more convenient.

3.2 Calibration of the Measures

For ascription of Delaunay simplex to a given shape, one should indicate a boundary values of measures, T_b, Q_b and K_b, i.e. to make a calibration of the measures (1) – (3). Following [16,17], we calibrate our measures with the help of a known structure, namely FCC crystal at temperature below the melting point. Calibration models were generated by Monte Carlo method, and consisted of 10000 Lennard-Jones atoms in a cube with periodic boundary conditions. Initially the atoms were settled on sites of the perfect FCC lattice, and then the model relaxed. We generated models for two different temperatures: $T^*=0.48$ and $T^*=0.32$ (in reduced units). Melting temperature is equal approximately to 1.0. In both models atoms only fluctuate around their crystalline positions. Fig.1(a-c) demonstrate histograms for T, Q and K, calculated for all Delaunay simplexes in the crystals obtained. The peaks at small values of variables can be definitely related to the good tetrahedra (a), quartoctahedra (b) and simplexes Kizhe (c).

Boundary values which separate simplexes with appropriate shape from others can be chosen as location of minima on the histograms. We have assigned

$$T_{b'} = 0.018, \quad Q_b = 0.013 \quad \text{and} \quad K_b = 0.007 \tag{4}$$

So, the simplexes having one of the measures T, Q or K less then in (4) are of interest for our analysis. We will state that boundary values in (4) determine "a full set" of simplexes typical for crystal structure. Indeed, all of them can be found in the model which is, from a physical point of view, a good crystal. Obviously, decreasing of the boundary values gives us higher-grade shape of simplexes, but excess of them introduces simplexes which shape could not agree the crystal structure slightly distributed by thermal vibrations.

Note, that variations of boundary values (4) within 10-20% does not influence the obtained results, where we use not only the shape, but also the environment of the Delaunay simplexes (see below).

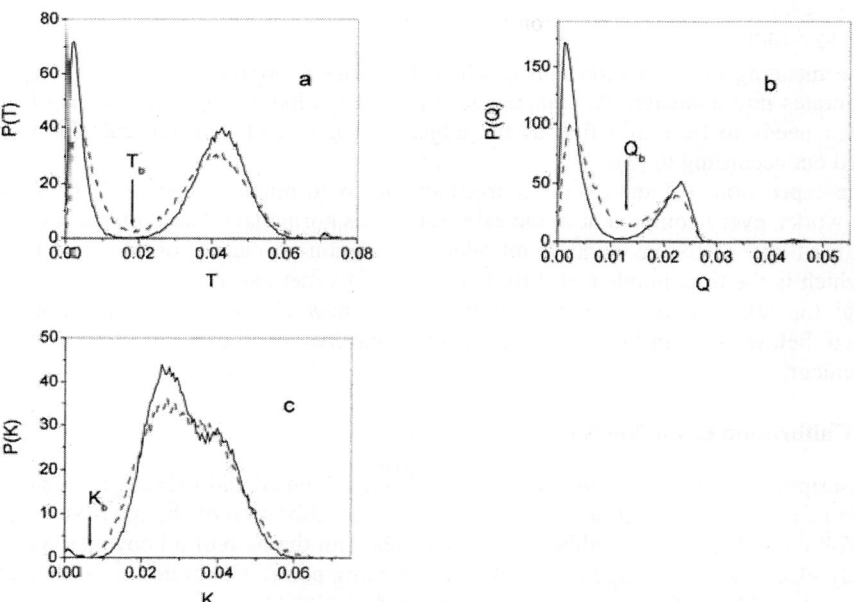

Fig. 1. Histograms of distribution of the Delaunay simplexes over different shapes for the models of FCC crystal: (a) tetrahedrisity T, (b) quartoctahedricity Q, (c) measure for simplex Kizhe K, see text, formulas (1)-(3). Solid lines for temperature $T^*=0.32$, dashed lines for $T^*=0.48$. Arrows show boundary values for extraction

4 Structural Types of the Delaunay Simplexes

4.1 Crystal Types

The essence of our approach for structural ascription of the Delaunay simplex is consideration of the environment of the simplexes together with their shape. In this work, the simplest step is realized in this direction: as an environment of the Delaunay simplex, only its neighbors adjacent by faces are taken into account.

In FCC crystal every tetrahedral configuration is adjacent over faces only to octahedra, and every octahedral configuration is adjacent only to the tetrahedra. Obviously, in terms of the Delaunay simplexes, it represents the following combinations of neighboring simplexes for a simplex of a given shape.

$$
\begin{aligned}
&\text{(I) } T: Q\,Q\,Q\,Q \\
&\text{(II) } Q: T\,T\,Q\,Q \\
&\text{(III) } Q: T\,T\,Q\,K \\
&\text{(IV) } K: Q\,Q\,Q\,Q
\end{aligned}
\qquad (5)
$$

Thus, thetrahedral simplex can be adjacent to four quartoctahedra (I). Quartoctahedron can be adjacent to two tetrahedra and two quartoctahedra (II) or to two tetrahedra, one quartoctahedron and one simplex Kizhe (III). Simplex Kizhe can be adjacent only to four quartoctahedra (IV). Situations (III) and (IV) arise in the case when octahedral configuration is divided onto simplexes Kizhe (see above).

The HCP has pairs of adjacent tetrahedra (trigonal bipyramids), and octahedra are organized in chains in which they are adjacent by faces. Thus, it is easily to formalize possible neighborhoods of the Delaunay simplexes:

$$
\begin{aligned}
&\text{(I) } T: T\,Q\,Q\,Q \\
&\text{(II) } Q: T\,Q\,Q\,Q \\
&\text{(III) } Q: T\,Q\,Q\,K \\
&\text{(IV) } Q: T\,T\,Q\,Q \\
&\text{(V) } Q: T\,T\,Q\,K \\
&\text{(VI) } K: Q\,Q\,Q\,Q
\end{aligned}
\qquad (6)
$$

Note, the combinations (I)-(III) are new ones, but (IV)-(VI) are the same as for FCC in (5). The similarity of some combinations is not surprising due to the inherent proximity of the densest crystalline structures. It also means that dissection of crystalline simplexes between FCC and HCP is not unambiguous in principle. Further classification of such "disputed" Delaunay simplexes requires additional considerations. The number of such questionable Delaunay simplexes can be decreased by the further analysis of the model. Indeed, if for instance disputed quartoctahedron ($Q: T\,T\,Q\,Q$) is adjacent to tetrahedra, which belong to FCC type, then it also can be classified as of FCC type. Therefore, after determination of the neighbors, we perform additional ascription of simplexes to crystalline types: if disputed simplex is adjacent to FCC type simplex (and does not have HCP type) then we assign it FCC type. Analogously, a disputed simplex neighboring HCP type and not of FCC type is classified as HCP type. If disputed simplex is adjacent both to FCC and HCP types, we keep it as disputed. Such cases take place at the bordering regions between FCC and HCP structures. We keep simplexes as disputed also is they do not have neighbors of either FCC or HCP types. This case happens in disordered phase for small aggregates of simplexes with crystalline shape. Note, the residuary disputed simplexes, nevertheless, represent regions of crystalline structure.

4.2 Non-crystalline Types

Proposed ideology to select Delaunay simplexes related to FCC and HCP may be extended to other structures, in particular, to non-crystalline ones. It is known that dense amorphous phase contains aggregates of good tetrahedra adjacent by faces (polytetrahedral aggregates). Such arrangement of more then two tetrahedra is extraneous for crystals, since they are incompatible with translational symmetry. For studying such aggregates, one should extract Delaunay simplexes with good tetrahedral shape having also at least two good tetrahedra in its neighborhoods:

$$T: T T * * \qquad (7)$$

The other pair of neighboring simplexes can have, in a general case, an arbitrary shape. Polytetrahedral clusters, and particularly, five-membered rings of tetrahedra (pentagonal bipyramids), are identified by simplexes (7).

Recently [6,9,18], a significant amount of pentagonal prisms was detected in models of dense packings of hard spheres and in frozen Lennard-Jones liquids. Existence of such configurations is not trivial. They are not crystalline but also are unnatural for amorphous phase. In this paper, we suggest to study them with the help of Delaunay simplexes of the following structure type:

$$T: T T Q * \qquad (8)$$

i.e. good tetrahedral, with two tetrahedra and at least one quartoctahedra in their neighborhoods.

5 The Model

We study a process of development of a crystalline phase on a model of rapidly cooled Lennard Jones liquid. The model was generated by the Monte Carlo method in NPT ensemble. It contains 10000 atoms in a cube with periodic boundary conditions. Initial configuration corresponded to the liquid phase at temperature $T^*=0.8$ and density $\rho^*=0.73$ (in reduced units). At every 500-th Monte Carlo step, temperature was decreased by $\Delta T = 0.00075$. It was found that 500 steps are enough for relaxation of the model at new temperature. This step by step cooling was continued to zero temperature. So we have got a set of successive configurations of the model on temperature interval from 0.8 to 0.0. Pressure was kept constant and equal to zero. During the process, density increased to $\rho^*=1.037$. The total number of Monte Carlo steps was 533000. In the result of such gradual cooling, a halfway crystallization happened. For full crystallization the slower cooling process would be required.

6 Results

Fig.2 shows a volume fraction for different structural components arising in our model in the process of crystallization. Here for determination of the structural types of the Delaunay simplexes according formulas (5)-(8), we used full set of crystalline shapes presented in (4). Volume of a given structural component (phase) was calculated as a sum of volumes of corresponding Delaunay simplexes. During the calculation we took into account all simplexes of a given type, i.e. as single as well as united in clusters, nuclei. We did not study individual nuclei in this paper. That is a topic for further investigation. Three upper curves in Fig.2 belong to crystalline phases. First of all, we see HCP is predominant in this case. It arises before FCC and exists up to complete freezing of the sample covering 30% of the volume. Total volume of crystalline phase (HCP, FCC and their disputed simplexes) occupy more then 60%, where 11% belong to disputed ones representing border regions between the crystal structures.

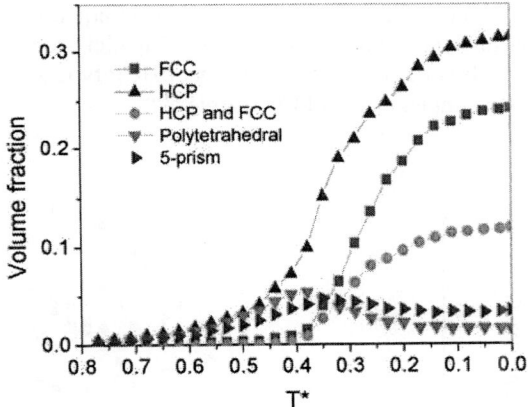

Fig. 2. Volume fraction of the Delaunay simplexes of different structural types as function of temperature in the process of crystallization of the model

Two lower curves belong to non-crystalline structural types, formulas (7), (8). The lowest curve shows fraction of volume occupied by polytetrahedral simplexes (7), except for volume of simplexes that correspond to pentagonal prisms (8). Simplexes of pentagonal prism type represent more then 3% of volume in completely frozen model, what is more then the other polytetrahedral simplexes. They consist, in particular, in central nuclei of five-fold twins in FCC phase.

The volume occupied by these non-crystalline simplexes demonstrates a maximum at the beginning of the intensive growth of crystal phase ($T^*=0.4$). This fact corresponds to the suggestion in the paper [6] that the polytetrahedral aggregates in liquids (together with embryos of pentagonal prisms) may initiate appearance of crystalline nuclei.

Approximately 30% of model volume is not related to mentioned structures types, and represent disordered structure, which is not recognized in this analysis.

Fig.3 demonstrates fraction of FCC and HCP structures in our model extracted with various criteria for quality of the Delaunay simplexe shapes. Pair of curves marked by 2 corresponds to boundary measures T, Q and K that are two times smaller then for the full set of crystalline simplexes (4). The pair of curves marked by 3 corresponds to boundary measures T, Q and K that are four times smaller. The corresponding curves from Fig. 2 for the full set of crystalline simplexes are also present (pair of curves marked by 1). For more perfect shapes, the volumes of extracted phases are obviously less. Non-trivial result obtained here is the fact that ratio between FCC and HCP phases is changed. If the curves 1 demonstrate predominance of the HCP simplexes, then the pair of curves 2 are practically coincide, and for pair of curves 3 the structure FCC becomes predominant. This result means that the simplexes of FCC structure are comprised of the more perfect simplexes then aggregates of HCP structure. It can shed a light on the fact, that HCP nuclei always present at the beginning stage of crystallization in spite of all systems of spherical atoms crystallize finally in FCC [19]. Interpretation of it is based usually on the suggestion that appearance of FCC and HCP is equally probable at the first stage. However, our analysis demonstrates that HCP nuclei are even more likely then FCC at the beginning. It is because the simplexes of HCP type are less affected by the shape of the Delaunay simplexes then FCC simplexes.

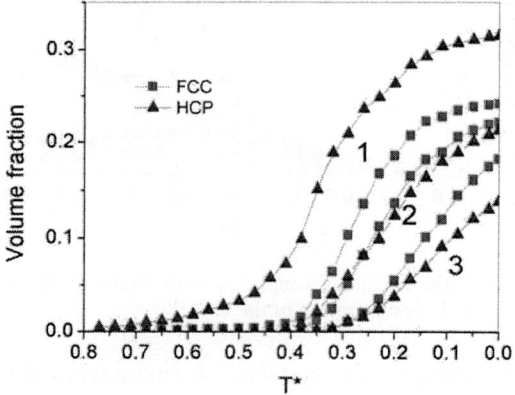

Fig. 3. Volume fraction of FCC and HCP structures for various criteria of shape quality of the used Delaunay simplexes. 1) *T<0.018, Q < 0.013, K < 0.007.* 2) *T<0.009, Q < 0.0065, K < 0.0035.* 3) *T<0.0045, Q < 0.00325, K < 0.00175*

Note that polytetrahedral nuclei (aggregates of adjacent tetrahedral), which are always present in dense disordered phase, also may correspond to HCP structure, because HCP, in contrary to FCC, have pairs of terahedra adjacent by face. The detailed analysis of spatial distribution of the Delaunay simplexes of HCP type demonstrates that at the first stage of crystallization they exist as small nuclei uniformly distributed over the model.

7 Conclusions

A novel method for extraction of crystalline nuclei for FCC and HCP structure using shape and mutual arrangement of the Delaunay simplexes is presented. The basic element is a Delaunay simplex of a shape similar to one of three characteristic forms of the Delaunay simplexes of the densest crystalline structures: a tetrahedton, a quarter of octahedron (quartoctahedron), and a flat square. A structure type of the simplex is determined by the shape of neighboring simplexes, adjacent by faces. Clusters of simplexes of a given structural type represent nuclei of a corresponding structure. The approach can be also applied for extraction of other specific structures, in particular, for polytertahedral aggregates typical for amorphous phase, and pentagonal prisms. The structure of Lennard-Jones liquid in the process of cooling is studied as part of experimentation. The non-trivial result stating that nuclei of HCP structure appear earlier than FCC nuclei is obtained using the proposed methodology.

Acknowledgements

The research is supported by the Grant CRDF No-008-X1, the RFFI 05-03-32647 Grant, the OTKA Grant, the CFI Grant and the NSERC Grant.

References

1. Voronoi G.F. Nouvelles applications des paremetres continus a la theorie des formes quadratiques. *J.Reine Andew.Math.*, Vol.134, 1908, pp.198-287; Vol.136, 1909, pp. 67-181.
2. Delaunay B.N., Sur la sphere vide. *Proc. of the Math. Congress in Toronto Aug.11-16,1924*, 1928, pp.695-700.
3. Naberukhin Y.I., Voloshin V.P., Medvedev N.N. Geometrical analysis of the structure of simple liquids: percolation approach. *Molecular Physics,* 1991, Vol. 73(4), pp. 917-936.
4. Brostow W., Chybicki M., Laskowski P., Rybicki J. Voronoi polyhedra and Delaunay simplexes in the structure analysis of molecular-dynamics-simulated materials. *Phys.Rev.B*, Vol. 57, No. 21, 1998, pp. 13448 -13452.
5. Medvedev N.N., *Voronoi-Delaunay method for non-crystalline structures*, SB Russian Academy of Science, Novosibirsk, 2000 (in Russian).
6. Anikeenko A.V., Gavrilova M.L., Medvedev N. N. Coloring of the Voronoi Network: Investigation of Structural Heterogeneity in the Packings of Spheres. *JJIAM* 2005
7. Finney J. Random packings and the structure of simple liquids. *Royal Society London* 1970, Vol. 319, pp. 479-494; pp. 495-507.
8. Hsu C.S., Rahman A.J. Crystal nucleation and growth in liquid rubidium. *Journal of Chemical Physics*, 1979, Vol. 70, pp. 5234-5240.
9. O'Malley B., Snook I., Crystal Nucleation in hard sphere system. *Phys.Rev.Lett.*, 2003, Vol. 90(8), 085702.
10. Steinhardt P.J., Nelson D.R., Ronchetti M. *Phys.Rev. B.* , 1983, Vol.28, pp.784-805.
11. Luchnikov V.A., Gevois A., Richard P., Oger L., Troadec J.P. Crystallization of dense hard sphere packings. *J.Mol.Liquids*, 2002, Vol.96-97, pp.185-194.

12. Kim Deok-Soo, Chung Yong-Chae, Seo Sangwon, Kim Sang-Pil, Kim Chong Min. Euclidean Crystal structure extraction in materials using Voronoi diagram and angular distributions among atoms. *Journal of Ceramic Processing Research.* 2005, in press.
13. Kim Deok-Soo, Chung Yong-Chae, Seo Sangwon, Kim Sang-Pil, Kim Chong Min. Distributions of BCC, FCC, and HCP structures in Al-Co composite materials. *Journal of Ceramic Processing Research .* 2005, in press.
14. Medvedev N.N., Naberukhin Yu.I., Shape of the Delaunay simplexes in dense random packings of hard and soft spheres, *J.Non-Chrystalline Solids*, 1987, Vol. 94, pp. 402-406.
15. Okabe, A., Boots, B., Sugihara, K. and Chin, S. *Spatial Tessellations: Concepts and applications of Voronoi diagrams*, Chichester, John Wiley, 2000
16. Medvedev N.N., Naberukhin Yu.I. Structure of simple liquids as a percolation problem on the Voronoi network. *J.Phys.A: Math.Gen.*, 1988, v.21, pp.L247-L252.
17. Voloshin, V.P, Naberukhin, Y.I, Medvedev, N.N. Can various classes of atomic configurations (Delaunay simplexes) be distinguished in random dense packing of spherical particles? *Molecular Simulation Journal*, 1989, Vol. 4, pp. 209-227.
18. Anikeenko A.V., Medvedev N.N., Bezrukov A., Stoyan D., Observation of fivefold symmetry structures in computer model of dense packing of hard spheres", *Journal of Non-Cryst. Solids*, 2005, in press.
19. Richard P., Gervois A., Oger L. and Troadec J.-P. Order and disorder in hard-sphere packings. *Europhys. Lett,* 1999, Vol.48 (4), pp. 415-420.

Recognition of Minimum Width Color-Spanning Corridor and Minimum Area Color-Spanning Rectangle

Sandip Das[1], Partha P. Goswami[2],
and Subhas C. Nandy[1]

[1] Indian Statistical Institute, Calcutta 700 108, India
[2] Calcutta University, Calcutta 700 009, India

Abstract. Given a set of n colored points with a total of m (≥ 3) colors in 2D, the problem of identifying the smallest color-spanning object is studied. We have considered two different shapes: (i) corridor, and (ii) rectangle of arbitrary orientation. Our proposed algorithms for the problems (i) and (ii) run in time $O(n^2 \log n)$ and $O(n^3 \log m)$ respectively.

1 Introduction

We are given a set S of n points in the Euclidean plane and m ($3 \leq m < n$) colors. Each point p_i is associated with a color c_i. A region is called *color-spanning* if it contains at least one point of each color. The motivation of studying different types of color-spanning region of smallest area/perimeter is described in [1]. Smallest color-spanning circle and axis-parallel square can be obtained in $O(mn \log n)$ time [2]. Two other important problems in this category are studied in [1] — (**P1**) finding the narrowest color-spanning corridor, and (**P2**) finding the smallest color-spanning axis-parallel rectangle. The time complexities for these two problems are $O(n^2 \log n + n^2 \alpha(m) \log m)$, and $O(n(n-m) \log^2 m)$ respectively.

We formulate problem **P1** using geometric duality, and propose an algorithm that runs in $O(n^2 \log n)$ time. Next, we generalize the problem **P2**, where the objective is to report the arbitrarily oriented color-spanning rectangle of minimum area. Our proposed algorithm runs in $O(n^3 \log m)$. Both the algorithms use $O(n)$ space. Our technique also improves the time complexity for solving problem **P2** (of [1]) to $O(n(n-m) \log m)$.

2 Narrowest Color-Spanning Corridor

A corridor is defined as an open region bounded by a pair of parallel lines. Its width is the perpendicular distance between the bounding lines. Given the set S, a corridor is said to be the *color-spanning corridor* (CSC) if it contains at least one point of each color from S in its interior. Our objective is to identify the narrowest CSC. We assume that the points in S are in general position, i.e., the line passing through each pair of points have distinct slope.

2.1 Formulation

The narrowest vertical CSC can be computed in $O(n)$ time after sorting the projections of the members in S on x-axis. From now onwards, by CSC we shall mean the non-vertical CSC. Theorem 1 characterizes the narrowest CSC.

Theorem 1. [1] *If a CSC bounded by a pair of parallel straight lines ℓ_1 and ℓ_2, is the narrowest then (i) one of ℓ_1 and ℓ_2 will contain two points of S and the other one will contain one point of S, (ii) color of these three points are different, and (iii) none of them appears inside the corridor.*

We formulate the problem using geometric duality. Let ℓ_1 and ℓ_2 be the two bounding lines of a non-vertical corridor C. In the dual plane, the corresponding points ℓ_1^* and ℓ_2^* have the same x-coordinate. Thus the dual of a corridor C can be represented by a vertical line segment $[\ell_1^*, \ell_2^*]$ in the dual plane, and is denoted by C^*. The width of C is $\frac{|y(\ell_1^*) - y(\ell_2^*)|}{\sqrt{1 + (x(\ell_1))^2}}$, and is referred to as the *dual length* of C^*.

Let $H = \{h_i = p_i^* \mid p_i \in S\}$, and $\mathcal{A}(H)$ denote the arrangement of H. Each line h_i is attached with the color c_i. A vertex of $\mathcal{A}(H)$ is said to be *bi-colored* if it is the intersection of two lines of different colors.

Observation 1. *Let C be a color-spanning corridor in the primal plane. In dual plane, C^* will intersect at least one line of each color.*

A vertical line segment in the dual plane satisfying Observation 1 is referred to as a *color spanning stick*, or *CS-stick*. Thus, recognizing the narrowest CSC is equivalent to finding a *CS-stick* of minimum *dual length*. We associate a $COLOR$ vector of length m with a *CS-stick*. Its i-th element indicates the number of lines of color i intersected by it. Theorem 1 suggests the following observation.

Observation 2. *Let C be a narrowest CSC bounded by ℓ_1 and ℓ_2. In the dual plane, (a) one of the points among ℓ_1^* and ℓ_2^* corresponds to a bi-colored vertex of colors say α and β, and the other one lies on a line of color say γ, where $\alpha \neq \beta \neq \gamma$, and (b) at least three elements (corresponding to the colors α, β and γ) of the $COLOR$ vector of the CSC will contain a value 1.*

We shall consider all the CSC satisfying the conditions stated in Observation 2, and report the one having minimum width.

2.2 Algorithm

We process the vertices of $\mathcal{A}(H)$ by sweeping a vertical line from left to right. At each bicolored vertex, we identify the narrowest CSC by computing the dual length of the *CS-stick above* and *below* that vertex. Our algorithm does not maintain the entire arrangement in memory. We now describe the method of computing *CS-stick* at a bicolored vertex v whose other end point is above v.

Data Structure: The sweep line status is maintained in an array B of size n. It contains the lines of H in the order in which they are intersected by the sweep line in its current position, from bottom to top. Each element $B[i]$ (representing a line ℓ) is attached with the following information:

- The *id* of line ℓ and its color.
- Two pointer fields, namely *prev* and *next*. The *prev* pointer points to a line ℓ^* of the same color appearing below ℓ in the current position of the sweep line. If ℓ^* is stored in $B[j]$, then $j < i$. Similarly, the *next* pointer points to a line ℓ^{**} of same color appearing above ℓ in the current position of the sweep line. If ℓ^{**} is stored in $B[k]$, then $k > i$. A NULL in any of these fields indicates the non-existance of such a line.
- A pointer, called *CS_ptr*. It points to a line ℓ' above ℓ in the array B such that the vertical line segment connecting ℓ and ℓ' at the present position of the sweep line is a *CS-stick*. If such a line does not exist then this field contains NULL.

The array B is initialized by the lines of H in increasing order of the ordinates of their intersections with the sweep line at $X = -\infty$. Next, a linear scan in the array B can be used to set all the pointers attached to the elements in array B.

Lemma 1. *Let a and b be two points of intersection of $\ell_1, \ell_2 \in \mathcal{A}(H)$ with the sweep line at its present position, and let the CS_ptrs of ℓ_1 and ℓ_2 point to ℓ'_1 and ℓ'_2 respectively. If a is below b then ℓ'_1 can not appear above ℓ'_2 along the sweep line at its present position.*

Lemma 2. *If the CS_ptr of two elements $B[i]$ and $B[j]$ point to the same element $B[k]$, then CS_ptrs of all elements $B[\alpha], \alpha = i, i+1, \ldots, j$ point to $B[k]$.*

The sweep process is guided by an event-queue Q realized as a min-heap. It stores at most n vertices of $\mathcal{A}(H)$ with respect to their x-coordinates. A vertex v_{ij} is in Q if the corresponding lines ℓ_i and ℓ_j (of colors say c_i and c_j) are consecutive entries in the array B and intersect (at v_{ij}) to the right of the sweep line. We attach (i) pointers with ℓ_i and ℓ_j that point to v_{ij} in Q, and (ii) a pair of pointers with v_{ij} in Q which indicates ℓ_i and ℓ_j in B.

Processing: During the sweep, the next event point v_{ij} is obtained from Q in $O(\log n)$ time. The updating of Q needs another $O(\log n)$ time using the method described in [3]. While processing of a bicolored vertex v_{ij}, we need to swap *line_id* fields along with the *prev* and *next* pointers attached to $B[\alpha]$ and $B[\alpha+1]$. In addition, the following steps are executed (see Figure 1).

- If the *prev* field of $B[\alpha]$ points to $B[\beta]$ prior to the processing of v_{ij}, then after processing of v_{ij} the *next* field of $B[\beta]$ will point to $B[\alpha+1]$.
- Similarly, if the *next* field of $B[\alpha+1]$ points to $B[\gamma]$ prior to the processing of v_{ij}, then the *prev* field of $B[\gamma]$ will point to $B[\alpha]$ after the processing of v_{ij}.

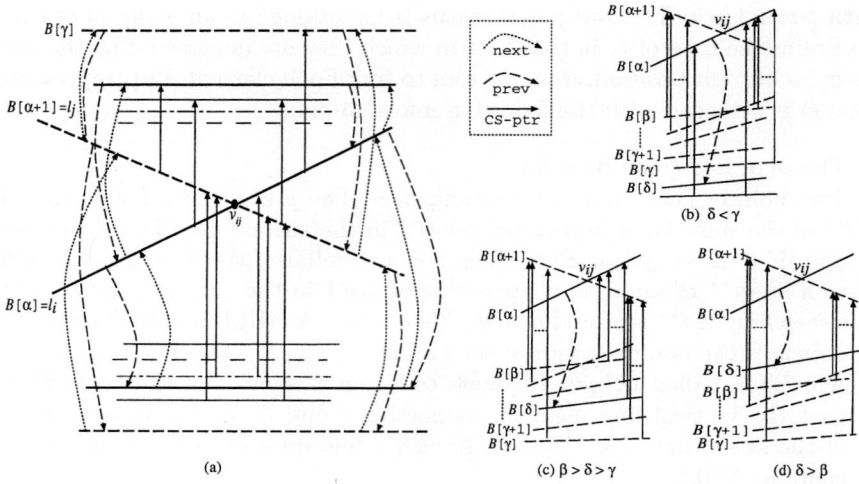

Fig. 1. Processing of vertex v_{ij}

- The *CS_ptr* of $B[\alpha]$ will remain same after the processing of v_{ij}. But the *CS_ptr* of $B[\alpha+1]$ may change since ℓ_j leaves from the *CS-stick* of ℓ_i after the processing of v_{ij}. In order to update this field, we need to observe the *next* field of $B[\alpha+1]$ prior to the processing of v_{ij}. Let it be pointing to $B[\gamma]$. If the *CS_ptr* of $B[\alpha+1]$ points to $B[\gamma']$ then the following three cases need to be considered:
  (i) if $\gamma' < \gamma$, then *CS_ptr* of $B[\alpha+1]$ will be changed to point $B[\gamma]$,
  (ii) if $\gamma' > \gamma$ and color_id of $B[\gamma']$ is c_i then *CS_ptr* of $B[\alpha+1]$ will be changed to point either $B[\gamma]$ or the *CS_ptr* of $B[\alpha]$ depending on which one is above.
  (iii) if $\gamma' > \gamma$ and color_id of $B[\gamma']$ is not c_i then *CS_ptr* of $B[\alpha+1]$ remains unchanged.
- Next, the *CS_ptrs* of the elements of B which point to $B[\alpha]$ and $B[\alpha+1]$ are to be updated as follows:

  The elements of B whose *CS_ptrs* point to $B[\alpha]$, will now point to $B[\alpha+1]$. Let $B[\gamma], B[\gamma+1], \ldots, B[\beta]$ be the elements of B whose *CS_ptrs* point to $B[\alpha+1]$ prior to the processing of v_{ij}. We use *prev* pointer of $B[\alpha]$ to locate a line $B[\delta]$ of color c_i below $B[\alpha]$. Here three cases need to be considered separately.

  $\delta < \gamma$: Here all the elements $B[\gamma], B[\gamma+1], \ldots, B[\beta]$ will point to $B[\alpha+1]$.

  $\gamma < \delta < \beta$: Here $B[\gamma], B[\gamma+1], \ldots, B[\delta]$ will point to $B[\alpha]$, and $B[\delta+1], B[\delta+2], \ldots, B[\beta]$ will point to $B[\alpha+1]$.

  $\delta > \beta$: Here all the elements $B[\beta], \ldots, B[\gamma]$ will point to $B[\alpha]$.

Thus, while processing each vertex in $\mathcal{A}(H)$, we may have to spend time linear in the number of elements in B whose *CS_ptrs* are getting changed. We can expedite this process by maintaining a height-balanced tree (namely *CS_tree*) instead of maintaining the *CS_ptr* field with each node in the array B.

A node of *CS_tree* corresponds to a tuple of indices $[i,j]$, such that the *CS_ptr* of all the edges stored in $B[i], B[i+1], \ldots, B[j]$ point to the same element in B (say $B[k]$). The index k is attached as integer field, called CS_ptr, with that node of the tree. By Lemmata 1 and 2, the intervals corresponding to the nodes in *CS_tree* are mutually exclusive and collectively exhaustive. Thus, the nodes of the *CS_tree* are ordered with respect to their *CS_ptrs*.

After initializing the array B, *CS_tree* can be constructed in linear time, and it supports the following queries: (i) given a $B[i]$, its *CS_ptr* can be obtained in $O(\log n)$ time from *CS_tree*, and (ii) while processing a vertex of $\mathcal{A}(H)$, the change in the value of *CS_ptr* of the elements in array B can be incorporated by splitting at most one node and/or merging at most one pair of nodes of *CS_tree*. This needs $O(\log n)$ time. Thus, we have the following theorem:

Theorem 2. *The worst case time complexity of the algorithm proposed for identifying the narrowest color-spanning corridor amidst a set of n points of m colors $(3 \leq m < n)$ is $O(n^2 \log n)$. The space complexity is $O(n)$.*

3 Minimum Area Color Spanning Rectangle

We now study the generalized version of Problem **P2**. Our objective is to identify the smallest (minimum area) color-spanning rectangle (*CSR*) of arbitrary orientation. A *CSR* is said to be *minimal* if there does not exist any other *CSR*, that is completely contained in it. Each side of a minimal *CSR* must be bounded by a point of distinct color and no point of those colors appear inside it.

For a given set of four points of S, it may either generate no *CSR* or infinite number of minimal *CSRs*. In the latter case, the one having minimum area is said to be a *prime minimal CSR*, or *PCSR* in short. We consider all possible *PCSRs*' and identify the one having minimum area.

Result 1. *One side of a PCSR must contain two points of distinct colors and it does not have these colors in its interior and on other boundaries.*

We consider each pair of points $p_a, p_b \in S$ of different colors, and identify all *PCSRs*' whose one side contains p_a and p_b. Let the colors of p_a and p_b be α and β ($\alpha \neq \beta$) respectively. Let ℓ_{ab} be the line joining (p_a, p_b), and p_c be a point of color γ ($\neq \alpha, \beta$) above ℓ_{ab}. We use $PCSR((p_a, p_b), p_c)$ to denote the set of *PCSRs* whose one side contains (p_a, p_b) and its parallel side contains p_c.

Result 2. *The number of $PCSR((p_a, p_b), p_c)$ is less than or equal to $m - 2$.*

We identify the *PCSRs* with bottom boundary aligned with ℓ_{ab} by sweeping a horizontal line upwards. At each encounter of a new point p_c, we inspect the existence of a $PCSR((p_a, p_b), p_c)$. We shall refer the line ℓ_{ab} as x-axis, and draw two vertical lines at p_a and p_b respectively. This split the region above x-axis into three strips, namely *LEFT*, *MID* and *RIGHT*.

Observation 3. *During the sweep, if a point of color α or β is recognized, then (a) if it is in MID then no other point encountered during further sweep can generate a PCSR, and (b) if it is in LEFT (resp. RIGHT), then no other point encountered to its left (resp. RIGHT) during further sweep can generate a PCSR.*

Observation 4. *If the color of newly encountered point p_c is γ ($\neq \alpha, \beta$), then in the following instances, no PCSR is generated:*
(a) there exist a point $p \in MID$ of color γ, and $y(p) < y(p_c)$,
(b) $p_c \in LEFT$, and there exists a point $p \in LEFT$ of color γ such that $x(p) > x(p_c)$ and $y(p) < y(p_c)$, and
(c) $p_c \in RIGHT$, and there exists a point $p \in RIGHT$ of color γ such that $x(p) < x(p_c)$ and $y(p) < y(p_c)$.

The implications of Observations 3 and 4 are as follows: During the sweep

- if the encountered point p_c is in *MID* and is of color α or β then sweep stops;
- if p_c is in *LEFT/RIGHT*, it prunes the search interval for *PCSR*. We use a pair of scalar variables $[\mathcal{L}, \mathcal{R}]$ as the sweep interval. If $p_c \in LEFT$ (resp. *RIGHT*) then by Observation 3(b), $\mathcal{L}$ (resp. $\mathcal{R}$) is set to $x(p_c)$. During further sweep, if a point encountered outside the interval $[\mathcal{L}, \mathcal{R}]$, it need not be considered.
- If a point of color γ is already encountered in *MID*, then a new point of color γ encountered anywhere during the sweep will not generate a *PCSR* (by Observation 4(a)), and is completely ignored.
- Let there be no point of color γ in MID and S_γ be the set of points of color γ in *LEFT* which are encountered by the sweep line up to the instant of time when p_c is encountered. Let $p_c \in$ LEFT, and $p^* \in S_\gamma$ be the point such that $x(p^*) = Max\{x(p) \mid p \in S_\gamma \, \& \, p \in LEFT\}$. Observation 4(b) says that, if $x(p_c) < x(p^*)$, the new point p_c will not generate a *PCSR*. Thus, projection of p^* on the x-axis defines the *left boundary of color γ* (denoted by $LB(\gamma)$). Similarly, Observation 4(c) defines $RB(\gamma)$. During the sweep, (i) $LB(i)$ and $RB(i)$ of each color i may change, and (ii) one needs to maintain the $LB(i)$ and $RB(i)$ of each color i, till it does not satisfy Observation 4(a). As soon as the color i appears in MID, $LB(i)$ and $RB(i)$ need not have to be maintained any more.

3.1 Data Structure

The colors ($\neq \alpha, \beta$), appeared in *LEFT*, *MID* and *RIGHT*, are stored in three data structures DS_{left}, DS_{mid} and DS_{right} as follows:

DS_{mid} is a bit-array of size m. $DS_{mid}[i] = 1$ (resp. 0) indicates a point of color i appeared (resp. not appeared) in *MID* strip during the sweep.

DS_{left} and DS_{right} are implemented using two AVL-trees of size at most m. For each color γ for which $DS_{mid}[\gamma] = 0$, $LB(\gamma)$ and $RB(\gamma)$ (if exists) are stored as the key value attached to γ in DS_{left} and DS_{right} respectively. The elements in DS_{left} (resp. DS_{right}) are ordered with respect to their key values.

We use an array $COLOR$ of size m whose each entry consists of three fields, namely bit, $left$ and $right$. If a point of color i appears in the search interval $[\mathcal{L}, \mathcal{R}]$, then $COLOR[i].bit$ is set to 1. $COLOR[i].left$ (resp. $COLOR[i].right$) points to the entry of color i in DS_{left} (resp. DS_{right}) (if present). Thus, if a color i appears in MID (i.e., $DS_{mid}[i] = 1$) then $COLOR[i].bit = 1$ but $COLOR[i].left = COLOR[i].right = NULL$. An integer variable $color\_count$ indicates the number of bit fields in the $COLOR$ array containing "1".

3.2 Theme of the Algorithm

The event points of this algorithm are the bi-colored vertices of $\mathcal{A}(H)$. These are obtained by sweeping a vertical line over $\mathcal{A}(H)$ from left to right (called $LEVEL\_1$ sweep). While processing a bi-colored vertex v_{ij}, we process the points in the primal plane above (resp. below) the line ℓ_{ij} (joining p_i and p_j) in order of their distances from ℓ_{ij}. These points can be obtained (in order) by sweeping of a line L (parallel to ℓ_{ij}) upwards (resp. downwards) in the primal plane. This will be referred as $LEVEL\_2$ sweep. During $LEVEL\_2$ sweep, if a point p_c of color α or β is encountered by the sweep line L and its x-coordinate lies in the interval $[\mathcal{L}, \mathcal{R}]$, then (i) if $p_c \in MID$, the $LEVEL\_2$ stops, and (ii) if $p_c \in LEFT$ (resp. $RIGHT$), then $\mathcal{L}$ (resp. $\mathcal{R}$) is updated as stated below.

The updating of $\mathcal{L}$ implies, deletion of all nodes in DS_{left} whose x-coordinates are less than $x(p_c)$. The corresponding $left$ pointers in $COLOR$ array are also set to NULL. If the $right$ pointer of the corresponding entry is also NULL, one needs to put "0" in the bit field of that color in the $COLOR$ array.

If the color of p_c is γ ($\neq \alpha$ or β), we take the following actions:

If $COLOR[\gamma] = 0$, then change it to 1, and add 1 to $color\_count$. If $p_c \in MID$, set $COLOR[\gamma].left = COLOR[\gamma].right = NULL$, $DS_{mid}[\gamma] = 1$, and remove $LB(\gamma)$ (resp. $RB(\gamma)$) from DS_{left} (resp. DS_{right}). If $p_c \in LEFT$, insert/update $LB(\gamma)$ ($= x(p_c)$) in DS_{left}, and accordingly set $COLOR[\gamma].left$. Similar actions work if $p_c \in RIGHT$. If $color\_count < m$, no $PCSR((p_i, p_i), p_c)$ is reported.

If $color\_count = m$, then we report new $PCSR((p_i, p_i), p_c)$. Without loss of generality, assume that $p_c \in LEFT$. Each element $\pi \in DS_{left}$ having x-coordinate $\leq x(p_c)$ is a candidate for the left boundary of such a $PCSR((p_i, p_i), p_c)$. Its right boundary is determined by the procedure **find_critical_color**.

find_critical_color: Identify a color θ such that (i) $\theta \in DS_{right}$, (ii) $\theta \notin DS_{left} \cup DS_{mid}$, and (iii) $RB(\theta)$ is the maximum among all colors satisfying (i) and (ii). If no such color θ exists then $\theta = \beta$.

The critical_color θ is obtained by scanning DS_{right} from its right-most element towards left. Next, the PCSRs' are reported by scanning DS_{left} from left to right until $x(p_c)$ is reached. We use two pointers ptr_1 and ptr_2. Initially, ptr_1 points to the leftmost element, say the color μ, in DS_{left}. The pointer ptr_2 points to $RB(\theta)$ in DS_{right}. The CSR determined by ptr_1 and ptr_2 will be a PCSR if $COLOR[\mu].right = NULL$ or $RB(\mu) > RB(\theta)$. We move ptr_1 to the next element in DS_{left} (keeping the critical_color unchanged) until a $PCSR$ is obtained. As soon as a $PCSR$ is reported with a color ν at its left boundary, the color ν, if present in DS_{right}, becomes the critical color. So ptr_2 is set to $COLOR[\nu].right$ (if $\neq$ NULL), and the same process continues. The process stops as soon as the move of ptr_2 fails, or ptr_1 reaches $LB(\gamma)$.

3.3 Complexity Analysis

While processing v_{ij}, the $LEVEL\_2$ sweep in both upward and downward direction is guided by the sweep-line status array of $LEVEL\_1$ sweep. When a point p_c is encountered by the $LEVEL\_2$ sweep line, three types of processing need to be considered - (i) updating $\mathcal{L}$ or $\mathcal{R}$ (if needed), (ii) updating DS data structures, and (iii) reporting of $PCSRs$. The time needed for (i) is $O(n)$ considering the entire $LEVEL\_2$ sweep. The time needed for (ii) is $O(\log m)$ if $p_c \in LEFT$ or $p_c \in RIGHT$; otherwise it is $O(1)$. Computing *critical_color* initially, and then reporting $PCSRs$ by scanning DS_{left} or DS_{right} also needs at most $O(m)$ time. Since the number of points to be encountered in $LEVEL\_2$ sweep is at most $O(n)$, the time complexity of processing a bicolored vertex is $O(mn)$.

Theorem 3. *The time and space complexities of identifying the PCSR of minimum area is $O(n^3 m)$ and $O(n)$ respectively.*

3.4 Further Improvement

In this section, we show that the generation of $CSRs$' which are not $PCSR$ can be avoided by using an integer field χ with each element in DS_{left} and DS_{right}, and using a new data structure *sequence-pair* as described below.

Let θ_L and θ_R be the critical colors in DS_{left} and DS_{right} respectively. Surely, the left boundary of a $PCSR$ can never be to the right of θ_L, and the right boundary of a $PCSR$ can never be to the left of θ_R. We now describe the procedure for setting the χ fields of the entries in DS_{left}. The same procedure works for setting the χ fields of DS_{right} entries.

Start from the left most entry of DS_{left} and scan towards right. Go on setting the χ fields to NULL till an entry μ is found such that $RB(\mu) \in DS_{right}$ and $RB(\mu) > RB(\theta_R)$. Set the χ field of μ to θ_R. Then again go on setting the χ fields of the entries in DS_{left} to NULL till an entry δ is found with $RB(\delta) \in DS_{right}$ and $RB(\delta) > RB(\mu)$. Set the χ field of δ to point μ in DS_{right}. This process will continue till $\theta_L \in DS_{left}$ is reached. If χ field of $\nu \in DS_{left}$ is last set prior to reaching $\theta_L \in DS_{left}$, then the χ field of θ_L is set to point $RB(\nu)$ in DS_{right}.

Observation 5. *If a color $\mu \in DS_{left}$ with non-NULL χ field, appears on the left-boundary of a CSR then it is a PCSR and its right boundary is indicated by the χ field attached to μ.*

The *PCSR*s can be generated using the χ fields as follows. Fix the left-boundary at θ_L and generate a *PCSR*. Its right-boundary corresponds to a color μ, indicated by the χ field of $LB(\theta_L)$. Fix the left boundary of next *PCSR* at $LB(\mu)$, and compute its right boundary using the χ field of $LB(\mu)$. Thus, the process of generating the *PCSR*s is guided by a sequence of tuples as defined below.

Definition 1. *The sequence pair for LEFT is a sequences of tuples $\{(u_1, w_1),$ $(u_2, w_2), \ldots, (u_k, w_k)\}$, where $\{u_1, u_2, \ldots, u_k\} \in DS_{left}$ and $\{w_1, w_2, \ldots w_k\} \in DS_{right}$, and these are obtained as follows:*
let $u_0 = LB(\theta_L)$; (* u_0 is not a member of sequence-pair *)
given u_{i-1} recursively compute $u_i = LB(\mu)$ and $w_i = RB(\mu)$, where μ is indicated by the χ field attached to u_{i-1}.
This recursive process continues till $w_i > RB(\theta_R)$.
If $w_i < RB(\theta_R)$ then set $k = i - 1$ (* length of the sequence-pair *).
Here $w_{k+1} = RB(\theta_R)$ (* w_{k+1} is not a member of sequence pair similar to u_0 *).

Note that, for two consecutive elements (u_{i-1}, w_{i-1}) and (u_i, w_i), for all the elements lying between w_{i-1} and w_i in DS_{right} (both exclusive) their corresponding colors appear to the right of $u_{i-1} \in DS_{left}$. Thus, w_i is the first element towards the left of w_{i-1} whose corresponding color in DS_{left} appears to the left of w_{i-1}.

Remark: The sequence-pair for the *RIGHT* slab is same as that of the *LEFT* slab as computed above.

We use an AVL-tree for representing the sequence-pairs for both *LEFT* and *RIGHT* slabs. Its each node stores a pair (u_i, w_i). During the *LEVEL_2* sweep of processing a vertex $v_{ij} \in \mathcal{A}(H)$, if $color\_count < m$, no *PCSR* exists. So, we shall study the role of *sequence-pair* for generating *PCSR* if $color\_count = m$, and a point p_c (of color γ say) is encountered by the sweep line.

Lemma 3. *If a PCSR exists with bottom and top boundaries defined by the line ℓ_{ab} and the point p_c respectively, then its left side will be defined by u_i, where (u_i, w_i) is a member of the sequence-pair. The right side of this PCSR will be w_{i+1} which corresponds to the next element (u_{i+1}, w_{i+1}) of the sequence pair.*

Updating Sequence-Pair

Let the existing sequence pair is $\{(u_1, w_1), (u_2, w_2), \ldots, (u_k, w_k)\}$ when the *LEVEL_2* sweep line encounters a new point p_c of color γ in *LEFT* slab. Now, the following cases need to be considered separately.

Case (i) *Points of color γ are present in LEFT but not in RIGHT:* After arrival of p_c, delete the existing $LB(\gamma)$ and insert new $LB(\gamma) = x(p_c)$ in DS_{left}. Moreover, if θ_L corresponds to color γ, we scan DS_{left} towards left to identify a color μ such that $COLOR[\mu].right = $ NULL and $DS_{mid}[\mu] = 0$, and then update $\theta_L = \mu$. Finally some new elements may be added in the *sequence-pair* if there exist some elements in DS_{right} beyond w_1 whose matching DS_{left} entries lie between u_1 and new θ_L.

Case (ii) *Points of color γ are present in both LEFT and RIGHT*: Let $u_i = LB_{old}(\gamma)$. Then observe that $\{u_1, u_2, \ldots, u_{i-1}, u_{i+1}, \ldots\}$, $\{w_1, w_2, \ldots, w_{i-1}, w_{i+1}, \ldots\}$ will remain as it is irrespective of the new value of $LB(\gamma) = x(p_c)$. Observe that $LB(\gamma) > u_i$. Here we first delete (u_i, w_i) from the sequence-pair. Note that, some (≥ 0) new pairs may be added in sequence-pair whose u-value will be within (u_{i-1}, u_{i+1}).

Case (iii) *Points of color γ are present in RIGHT but not in LEFT*: If θ_R does not correspond to color γ, then all points of color γ are to the left of θ_R in the *RIGHT* slab. So the sequence-pair remains unchanged. Otherwise, θ_R needs to be updated and some new elements needs to be inserted in the *sequence-pair*. This can be done as in Case (i).

Lemma 4. *If K new elements are added in the sequence-pair, then at most $K+1$ new PCSRs' will be generated with bottom and top boundaries passing through the line ℓ_{ab} and point p_c respectively.*

Complexity Analysis

Lemma 5. *The sequence-pair can be updated in $O(K \log m)$ time, where K is the number of new PCSRs reported.*

Proof. The updating of $LB(\gamma)$ in DS_{left} consists of setting the χ field of all the elements on the path from root up to the new position of $LB(\gamma)$ and needs $O(\log m)$ time. For each newly encountered point at most one element will be deleted needing $O(\log m)$ time, and at most $(K-1)$ elements will be added in the sequence-pair each requiring $O(\log m)$ time using χ field of the elements in DS_{left}. □

Lemma 6. *While processing a vertex $v_{ab} \in \mathcal{A}(H)$, the total time required for updating the sequence-pair is $O(n \log m)$ in the worst case.*

Proof. The sweep line faces at most $O(n)$ points in each half-plane of the line ℓ_{ab}. At the i-th point, we delete at most one element and add K_i elements in the sequence pair. Thus, total number of elements in the sequence pair at j-th step is $\sum_{i=1}^{j}(K_i - 1) \leq m$, for all $j = 1, \ldots, n$. So, $\sum_{i=1}^{n} K_i \leq m + n = O(n)$. Total number of deletions is also $O(n)$. Since deletion of an element needs $O(\log m)$ time and each insertion needs $O(\log m)$ time using χ field attached to the elements in DS_{left} (resp. DS_{right}), the time complexity follows. □

Theorem 4. *The time complexity of identifying a PCSR of minimum area is $O(n^3 \log m)$ in the worst case.*

Lemma 6 improves the time complexity of [1] for identifying the smallest color-spanning isothetic rectangle to $O(n(n-m)\log m)$ time using $O(n)$ space.

References

1. M. Abellanas, F. Hurtado, C. Icking, R. Klein, E. Langetepe, L. Ma, B. Palop and V. Sacristan, *Smallest color-spanning objects*, Proc. European Symp. Algorithms (ESA-2001), LNCS - 2161, pp. 278-289, 2001.
2. D. P. Huttenlochar, K. Kedem and M. Sharir, *The upper envelope of Voronoi surfaces and its applications*, Discrete Comput. Geom., vol. 9, pp. 267-291, 1993.
3. R. Janardan and F. P. Preparata, *Widest-corridor problems*, Nordic J. Comput., vol. 1, pp. 231-245, 1994.

Volumetric Reconstruction of Unorganized Set of Points with Implicit Surfaces

Vincent Bénédet, Loïc Lamarque,
and Dominique Faudot

Le2i Laboratory, University of Burgundy,
21078 Dijon, France
{Vincent.Benedet, Loic.Lamarque,
Faudot}@u-bourgogne.fr

Abstract. Many solutions exist to rebuild a three-dimensional object represented by a set of points. The purpose of our work is to provide an automatic reconstruction from an unorganized cloud, describing an unknown shape, in the aim to compute its volume. The approach employed in this paper consists in filling the object's interior with isosurfaces of potential fields and to use their fusion property in order to find the full volume and the continuous shape of the sampled object. Thus, the first step of our reconstruction is to search a correct interior for the object described by the set of points. Then, comes the positioning of implicit primitives into the cloud, deep inside of it and close to the boundary. A controlled fusion of the isosurfaces guarantees that no holes are present, such that we obtain a complete shape filling.

1 Introduction

The inherent motivation in our work is justified by the rebuilding of a scene containing one or more objects which the description is given to us by a set of points, $P = \{P_1, ..., P_n\}$. The cloud of dots thus defined represents a sampling in R^3 of the surface (if it is considered that there is only one, possibly including several connected components). It constitutes a raw data source for the reconstruction, for the only coordinates are known; we do not have further information about the vertices, they are neither sorted nor provided with a triangulation. We just assume that the connected components of the object are closed. Indeed, this assumption is necessary because our will is to rebuild its interior with the aim of being able to compute the volume and of deforming it with constant volume. In fact, the method presented here is an analogy with the filling of a container by a fluid which finally adopts its shape. The problem remains that we do not have a continuous contour. We thus should initially determine a correct interior from the sampled shape and then place implicit primitives inside the object until reaching its surface. The fusion property of the these implicit objects, that we will call blobs thereafter by abuse language, is used then to lead to a total filling of the object's volume. Obviously, this fusion must be controlled so that, on the one hand the filling does not comprise any holes, and on the other hand, the blobs do not leave the object.

2 Previous Works

Because of the significant number of practical applications requiring to rebuild a shape whose data source comes from a set of points, literature related with computer graphics provides a plethoric quantity of solutions about the reconstruction of an object described by a point cloud. Thus, Amenta et al. [1] propose an algorithm, based on a Voronoï diagram in three dimensions and bringing the concept of crust, allowing to find a triangulated and topologically correct surface. The method of alpha-shapes of Edelsbrunner [2] also makes it possible to find the shape of an object described by whole of points but requires one preliminary triangulation. Deviating from the methods based on the algorithmic geometry, implicit surfaces were also used for the rebuilding of object. So Hoppe et al. [3] provide a solution having the advantage of applying to a raw cloud without any additional information by considering a tangential plan at each point and an implicit function of distance. However, these methods don't allow to compute the volume of the object, so we cannot use them. A technique to rebuild a point cloud containing implicit surfaces generated by a point-based skeleton was for the first time suggested by Muraki [4] then improved by Tsingos et al. [5]. Nevertheless, this method presents the disadvantage of a manual initialization of the algorithm, which Bittar [6] cures while plunging the point cloud into a binary numerical volume to extract the median axis from it allowing, thereafter, the object reconstruction using implicit surfaces. In our case, the search for the median axis is not interesting because we wish to deform the object without recalculating this medial axis and with a constant volume. Although being connected with this form of rebuilding, our approach differs in the sense that the passage from the cloud to binary numerical volume, is not obtained from the median axis, but from the research of the form's interior in order to place our implicit primitives for the rebuilding. For that, it is necessary for us, as a preliminary, to define the discrete contour of the object.

3 Finding Contour from the Point Cloud

Talking about interior of an object defined by a set of dots in a continuous three-dimensional space obviously does not have any direction. A solution to alleviate impossibility of distinguishing the interior of the cloud in continuous space is to choose to discretize three-dimensional space by carrying out a voxelization of the cloud's bounding box. The set of points is therefore converted into binary numerical volume. To obtain an effective voxelization of a unstructured point cloud, in other words to lead to a binary volume correctly describing the topology of the object without creating nor removing overall locally related components or, it is advisable to determine an adequate length of voxels. Indeed, the choice of a significant voxel size will induce a connection of areas which should remain disjoined. Conversely, a too fine voxelisation will cause holes and will prevent from obtaining a closed contour.

For attempting to find a correct voxelization, Mari [7] thus presents a qualitative cloud study to adapt the voxels grid resolution to the characteristics of the set of points. Let $d_{related}$ defines the minimal distance between two points of two different local related components, d_{cloud+} the longest distance between two closest points and d_{cloud-} the smallest distance between two points.

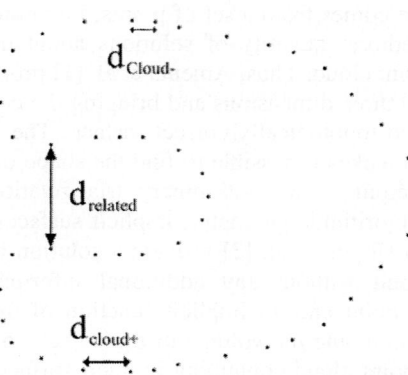

Fig. 1. Distances for a point cloud

The cloud we wish to rebuild must present characteristics of sufficient uniformity and density to obtain a respectful contour of the original form. In particular, d_{cloud+} distance must obviously be lower than $d_{related}$ to preserve related components, while remaining higher than d_{cloud-} in to obtain a closed surface.

A "good cloud" satisfies therefore two following inequalities for $\varepsilon > 0$:

$$d_{cloud+} - d_{cloud-} \leq \varepsilon \quad (1)$$
$$d_{cloud+} < d_{related} - \varepsilon$$

Nevertheless, only d_{cloud+} and d_{cloud-} quantities are computable, we cannot obviously determine the minimal distance between two points of two different local related components since we do not know a priori the object's nature.

In an empirical way, a choice that seems to give good results and can be retained to obtain an effective and righteous voxelization is to consider:

$$\varepsilon = d_{cloud-} \quad (2)$$

We can then consider $d_{related} > 2\, d_{cloud-}$ to raise any topological ambiguity. If we note l_v the voxel length, this one must respect:

$$d_{cloud+} < l_v < d_{related} \quad (3)$$

Consequently, the choice for l_v length must be as follows:

$$d_{cloud+} < l_v < 2\, d_{cloud-} \quad (4)$$

By adopting the selection criteria for the voxels length previously formulated, we can thus obtain a discrete closed contour of the object to rebuild. This outline is certainly only coarse, but allows us to lead to a sufficient result to determine a topologically correct interior.

4 Finding the Interior of the Object

To obtain the interior voxels of the object which we formed continuous contour, we employ a simple algorithm similar to that used by Bittar et al. [6]. Edge of voxels' box being initialized like outsides with the object (by grid construction), the technique consists in carrying out a sweeping of the voxels from a corner of the grid by marking the external voxels according to one 2-neighbourhood of voxels already treated. This method thus consists of a propagation of the external voxels around the surface of the object. Nevertheless, a second sweeping is done by starting from the opposite corner to take the concave zones of the object in account. This solution has the advantages of being simple and fast.

Now, we can separate the interior voxels into two layers. The *intermediate* voxels are those which remain close to the surface, the *deep* voxels are those located at the heart of the object to be rebuilt. To determine the intermediate voxels, we consider the interior voxels having a surface voxel in their 26-neighbourhood in three dimensions (8-neighbourhood in two dimensions).

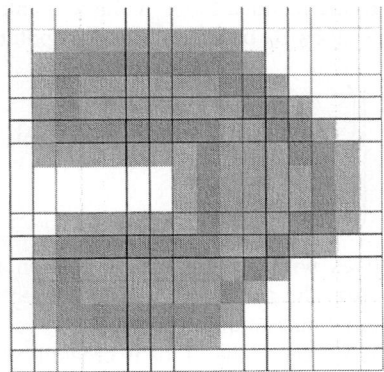

Fig. 2. *Deep* and *intermediate* voxels for the two-dimensional cloud shown above

Keeping in mind that the voxelisation is only one preliminary stage aiming at positioning implicit surfaces inside the object, one will seek to gather the deep voxels per packages in order to obtain voxels of higher size with the goal to minimize the number of blobs intended to replace them for the rebuilding. Our method's objective is to end at an adaptive implicit reconstruction. To obtain packages of voxels, one will carry out coding in octree of the intern voxels. The method consists starting from a voxel of the size of the grid of voxels and to subdivide it in eight voxels half the size of the grid. One judges then contents of each voxel. If it contains only deep voxels, or no, one stops there and one marks it consequently, if not one cuts out it in his turn in eight pennies voxels. The recurring process is thus applied until reaching the size of the initial voxels or if one obtains only full or empty voxels. Although we cannot reach a real minimum number of voxels by the means of this algorithm, one thus ensures oneself of a reduction in the quantity of intern blobs to place. Indeed, the regular and arbitrary cutting of the octree does not take account of the object's shape, therefore we cannot obtain an optimal regrouping. In addition, the fact of considering only cubic packages of voxels (square in 2D) is a choice based on the fact that they will be replaced by blobs with point-based skeleton. Consequently, we could gather the voxels differently and in a more effective way by using implicit objects with more complex skeletons like segments or plans with eventually anisotropic potential field functions.

5 Blobs Positioning

Since we seek to fully rebuild the object defined by the point cloud, namely the interior and the surface, the principle of our reconstruction by implicit surfaces breaks up into two parts. The first consists in replacing the interior voxels by implicit primitives of significant size, the second to positioning smaller implicit objects close to the surface. At the end, we blend these various elements in order to cover total volume. The goal of our step being to be in measurement in the long term to calculate the volume of the object represented by the group of dots, for we have solutions to carry out the computing of the blended blobs' volume thanks to Faudot et al. [8].

Therefore, we consider blobs B_i, based on Muraki's potential field function:

$$F_i = \begin{cases} \left(1 - \left(\frac{r}{R_i}\right)^2\right)^2 & \text{for } r \in [0, R_i] \\ 0 & \text{otherwise} \end{cases} \quad (5)$$

They are implicit primitives with a point-based skeleton whose potential field is a function of the distance from the center. They have moreover a limited influence, R_i, representing the ray of blob's maximum influence.

By considering that the fusion of the blobs is created by the sum of their potential field functions, implicit surface intended to rebuild the object is defined by:

$$F(P) = \sum_{i=1}^{n} \alpha_i F_i(P) \quad (6)$$

This surface must compulsorily pass by all the points of the cloud. Consequently, for a given threshold T, implicit surface must satisfy:

$$\forall P_j \in \{P_1, ..., P_N\}, \ F(P_j) = \sum_{i=1}^{n} \alpha_i F_i(P_j) = T \quad (7)$$

At first sight, it seems difficult to obtain a whole of blobs checking this condition. Nevertheless, the fact that each blob have a limited influence implies that for a given point P_j, the majority of the F_i will be null in this point. This makes it possible to place blobs close to a zone without calling into question the whole of the rebuilding. One can thus consider a local rebuilding of the cloud independently of the remainder of the blobs that we must place.

The first step of the blobs positioning relates to those in the heart of the cloud. We choose to replace each deep voxel by a blob sharing the same center and provided with an influence ray equal to half of the size of the considered voxel (computed starting from packaging voxel) added with the size of a basic voxel.

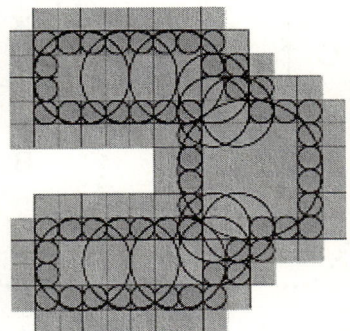

Fig. 3. Covering of the object's interior by blobs replacing deep and intermediate voxels

With regard to the intermediate layer of the blobs, one is satisfied to position them in the center of the corresponding voxels and to affect a ray R_i equals to $l_v / 2$. They have a plug role between the deep layer and the blobs placed at surface.

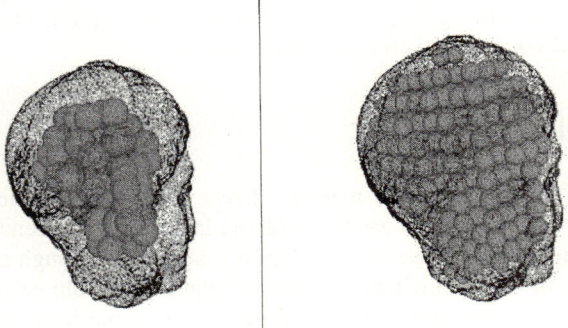

Fig. 4. Blobs (not blended) positioning for Stanford's Venus

To finalize the rebuilding we must blobs close to the surface of the object by ensuring that each point is touched by at least an implicit ball. For that, it is necessary to traverse the whole of the cloud and to create for each point a blob remaining inside the object whose surface touches the point. An intuitive approach of the problem consists, for a given point P, to determine the nearest interior voxel, with center B, then to place the ball in the center of the segment [PB] and to affect a ray with |PB|/2 length. Nevertheless, one realizes that for certain cases ("fold" in the surface corresponding to a local extremum) that the blob created will contain in its interior another point close to that which it is associated, it is thus necessary to re-examine its site. Let us suppose that the P' point is contained in the blob, the new center for the blob must check |C'P'|=|C'P| and be the segment [PB] to remain inside the object.

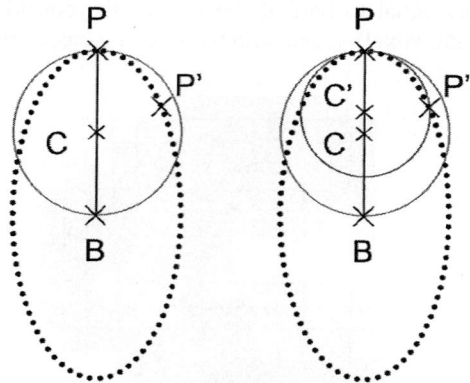

Fig. 5. Blob initially positioned contains another point. A new position is computed

The process of checking the inclusion of points in the blob must be continued until ensuring to obtain a ball not containing any point of the cloud. The disadvantage of this algorithm is its cost in computing times because we work on samples comprising a high number of points.

Thus created, these three different layers enable us to fill the object which we wish to rebuild. The following step is the fusion of these implicit primitives in order to guarantee a full volume as well as a regular surface.

6 Controlled Fusion

At this point, the course of our method of rebuilding is incomplete. Indeed, we are satisfied to place the blobs in each considered layer then we blend them without real theoretical justification. However, this fusion deserves a thorough study.

The fusion of the implicit primitives to obtain final volume is subjected to two constraints. First is to secure that the filling is carried out in an exhaustive way, i.e. fusion must be total and not leave voids between the blobs. The second condition is not to let the primitives leave the cloud.

To determine empty spaces after the fusion of implicit balls consists in fact to study the critical points of the density functions associated with the primitives. That returns to a research of the zeros of the final implicit function F. It is obviously impossible to determine formal solutions, also several analytical methods were developed to determine them. Thus Hart [9] proposes a technique based on the theory of Morse to locally find the critical points which he continues with Stander [10] by using a research by intervals of Newton.

A critical point of a real function F of one or more variables is a point X whose gradient vanishes. The function's value $F(X)$ at a critical point X is called a critical value. Given l_i i=1, ...,3 the eigenvalues of the function's Hessian, which is the matrix of second partial derivatives, each critical point X can be classified according to the signs of the three eigenvalues.

$$H(F) = J(\nabla(F)) = \begin{pmatrix} \dfrac{\partial^2 F}{\partial x^2} & \dfrac{\partial^2 F}{\partial x \partial y} & \dfrac{\partial^2 F}{\partial x \partial z} \\ \dfrac{\partial^2 F}{\partial y \partial x} & \dfrac{\partial^2 F}{\partial y^2} & \dfrac{\partial^2 F}{\partial y \partial z} \\ \dfrac{\partial^2 F}{\partial z \partial x} & \dfrac{\partial^2 F}{\partial z \partial y} & \dfrac{\partial^2 F}{\partial z^2} \end{pmatrix} \qquad (8)$$

If any of the eigenvalues is zero, then the critical point is called degenerate, otherwise it is non-degenerate and may be a maximum, a minimum or some kind of saddle point. In three dimensions, saddle points come in two varieties.

Table 1. Critical point classification according to the signs of the eigenvalues

l_1	l_2	l_3	Critical Point
-	-	-	Maximum Point
-	-	+	2-Saddle
-	+	+	1-Saddle
+	+	+	Minimum Point

We name index of the critical point the number of negative eigenvalues of $H(F)$. It is possible to classify the critical points according to the sign of the three eigenvalues, therefore according to the index.

A maximum point corresponds to a possible center of component, the three eigenvalues are negative. A "2-saddle" point corresponds to a point of possible connection between two components. A "1-saddle" point corresponds to a center of possible torus. A minimum point corresponds to a possible center of an air pocket, the three eigenvalues are positive. It results from this that one can punctually determine the topology of F.

So, in our case, we seek the "1-saddle" points and minima points witnesses to voids in our rebuilding. Thereafter, empty spaces must be filled with blobs.

By now, we are satisfied with visual checks and empirical corrections for the control of the fusion of the blobs. We did not implement yet the research of the critical points and the filling of the possible voids.

7 Conclusion

We have presented an original method for the reconstruction of an object defined by an unstructured point cloud using the implicit primitives with potential function and their possibility of blending. It constitutes an alternative to the methods based on preliminary research of a skeleton or a median axis which present the disadvantage of their instability. Nevertheless, it remains many improvements to be brought to it. First of all, it is advisable to optimize the placement of the blobs on the surface to reduce the costs of computing. The organization of the deep blobs must also be improved to minimize the number that the octree enables us to obtain; a non square regrouping per packages of voxels is possible to use implicit primitives with more advanced skeleton. Lastly, it remains essential to apply a rigorous technique of fusion control in order to ensure itself of a correct rebuilding. However, the method of rebuilding presented in this article has the originality to differ from those existing by its "interior" approach of the object. Moreover, it provides a solution to foresee the possibility of computing the volume of the rebuilt object and its deformation with constant volume, which the other techniques of rebuilding do not allow.

References

1. N. Amenta, M. Bern, M. Kamvysselisy. A new Voronoï-based surface reconstruction algorithm. Proceedings of ACM Siggraph '98, pages 415-421, 1998.
2. H. Edelsbrunner, E.P. Mücke. Three-dimensional Alpha Shapes. ACM Transactions on Graphics 13:43-72, 1994.
3. H. Hoppe, T. DeRose, T. Duchamp, J. McDonald, W. Stuetzle. Proceedings of ACM Siggraph '92, 71-78, 1992.
4. S. Muraki. Volumetric shape description of range data using "Blobby Model". Proceedings of ACM Siggraph '91, 227-235, 1991.
5. N. Tsingos and M.P. Gascuel. Implicit surfaces for semi-automatic medical organs reconstruction. Proceedings of Computer Graphics International '95, 3-15, 1995.
6. E. Bittar, N. Tsingos, M.P. Gascuel. Automatic reconstruction of unstructured 3D data: combining medial axis and implicit surfaces. Computer Graphics Forum, 14(3):457-468, 1995.
7. J.L. Mari. Modélisation de formes complexes intégrant leurs caractéristiques globales et leurs spécificités locales. Thèse. 2002.
8. D. Faudot, G. Gesquière, L. Garnier. An introduction to an analytical way to compute the volume of blobs. Int. Journal of Pure and Applied Mathematics, vol.11 n°1, 2004, pp 1-20
9. J.C. HART. Morse theory for computer graphics. Mathematical Visualization, H-C Hege and K. Polthier, Eds., Springer-Verlag, 257-268, 1998.
10. B. T. Stander, J. C. Hart. Guaranteeing the Topology of an Implicit Surface Polygonization for Interactive Modeling. Proceedings of ACM Siggraph '97, 279-286, 1997.

Guided Navigation Techniques for 3D Virtual Environment Based on Topic Map*

Hak-Keun Kim[1], Teuk-Seob Song[1,†], Yoon-Chul Choy[1], and Soon-Bum Lim[2]

[1] Dept. of Computer Science, Yonsei University
{air153,teukseob,ycchoy}@rainbow.yonsei.ac.kr
[2] Dept. of Multimedia Science, Sookmyung Women's University
sblim@sookmyung.ac.kr

Abstract. Navigating in a three dimensional (3D) virtual environment is difficult mainly because of limited navigational information that can only offer a screen-size visualization and low level imitation of other human senses. To address this problem, studies on various methods of navigation aid have been carried out. The navigation aid described in this paper was designed by applying the topic map technique, a semantic Web building technology, to a 3D virtual environment. A topic map builds a semantic link map, because it defines the link relations between topics. In the navigation aid's utilization experiment, in which the above mentioned linkage map was applied, looking for represented objects in detail, rather than looking for highly represented objects, was shown to be helpful in navigation. The provision of knowledge in and around the topic was confirmed to be effective for users' object selection during navigation under an unclear state of the object to be pursued.

1 Introduction

The most common task in 3D virtual environment is that of navigation around the space of the environment. The 3D virtual environment makes a user accept this environment as reality through stereoscopic images, and makes users feel as if they were onsite. The users can participate in the virtual environment seriously and actively by relying on such a "feeling." These strengths of the 3D virtual environment are utilized in various ways, such as to raise interest in entertainment programs, improve the impact of education and military training, and develop new technologies in the medical sector. The interactions between users and a 3D virtual environment should be conducted in a timely manner so that the user could feel a sense of reality and presence. These interactions can be classified into navigation, selection, and manipulation according to their applied stages and technologies [1]. Among these, navigation is the one that should be conducted in advance.

In this paper, the topic map [2], [3] technology was applied to a 3D virtual environment to solve such a problem. The topic map was approved by International Stan-

---

* This work was supported by KOSEF(Korea Science and Engineering Foundation) (R01-2004-000-10117-0(2004)).
† Corresponding Author.

dards Organization (ISO) [4], and it describes knowledge structure by connecting and organizing information resources according to semantic relevance between topics. It was suggested as a solution to efficiently navigate the large capacity of non-structured and non-organized information. The effects of the application of the topic-map-based navigation aid to a 3D virtual environment are following three important factors. First navigation using all topics that can be handled in the target environment of navigation is carried out. This is possible due to the topic map, which expresses the massive knowledge space with a standardized knowledge format called ontology [5]. Second, a user can continue to navigate by using a linked subject related to the current location, and does not need to go back to the original location of the first portal subject. In this way, the previous navigation experience can continue to be utilized in the navigation. The last one, each navigation stage provides a foundation on the purpose of navigation and on how it moves in the navigational process of going to the next stage. In this manner, users can conjecture where they are located in the virtual environment during the entire navigational process.

The composition of this paper is as follows. In Chapter 2, the navigation techniques that have been studied so far, and the topic map, are introduced, and in Chapter 3, a system design technique to apply the topic map to a 3D virtual environment is introduced. In Chapters 4 and 5, realization and evaluation are described. Finally, in Chapter 6, we describe the conclusion.

2 Related Works

In this chapter, the navigation aid that has been studied is described. A topic map is also introduced herein as a technology to be applied to the navigation aid in a virtual environment.

2.1 Navigation Aid

The first navigation aids to be proposed have been electronic analogues of the tools commonly used by people to navigate in unfamiliar real-world environments. From this perspective, the most common choice has been to come up with an overview of the environment available to the user. Besides this more traditional solution, novel navigation aids have recently been proposed by different authors [6], [5], [7].

Elvins [6] recognized that a landmark's legibility is an important factor in determining success or failure in a finding task. Landmarks not only express their own features, such as their 3D shapes, sizes, and textures, but they are also utilized by the user to understand the entire environment structure. With all these, it was proven that the utilization of adequate landmarks decisively promotes effective navigation.

Ramloll [5] proposed a new modeling method that uses the memories of a user's traveling experience instead of the space information of the coordinate system.

In his research, Ramloll divided 3D VE into small regions. He then collected noticeable images and assigned roles to each image, such as entrance to the area, exit from the area, and feature point of the area. By doing this, the user remembered the environment structure in an organized way, and reuse was also proven to be easy.

Active world [7] is a 3D virtual environment community that was commercialized in the Web environment. In the Active world, a menu table is provided as a navigation tool. The menu table binds the subjects with common things in groups, and is the navigation aid in a hierarchical type, along the level of the subjects. The hierarchically structured menu table has been designed to select the target, while narrowing down the target list from a broad scope to a narrow scope. To move the navigation scope to another group in the course of navigation, a detailed navigation, in which the scope is generalized up to the stage including the target/destination, and where the target list is then narrowed down, should be carried out.

The navigation aids mentioned above have a one-to-one linkage relation with a certain point in the environment. The process to retrieve this is designed to select the desired target/destination in the hierarchical selection list under a single subject. Accordingly, support of massive reference points is difficult and burdens a user cognitively due to a specific interface in each navigation aid.

2.2 Topic Map

The topic map is the knowledge-expressing technology of ISO/IEC standards, and presents a new theoretical background on the composition, drawing, and retrieval of information resources [2], [3]. It has a dual structure, composed of a knowledge layer and an information layer, for management of information space. Fig. 1 shows an example of a Web site structure based on a topic map. It shows the relationship between the knowledge layer and the information layer in an outlined structure.

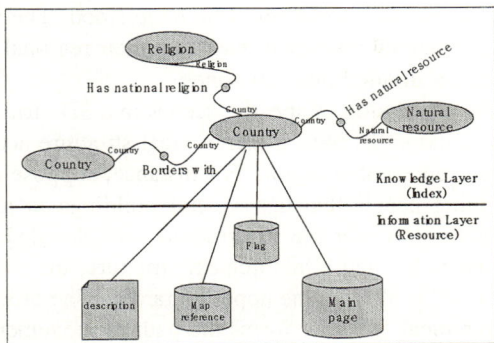

Fig. 1. Web Site Structure Based on a Topic Map [8]

The knowledge layer expresses the knowledge structure through topics and their association. The topic, which is the central factor of the knowledge layer, is the concept to be talked about in words. The topic may be any thing, regardless of whether it physically exists or it does not. The other element of the knowledge layer, Association, defines the relationship between these topics.

The information knowledge is the storage space of resources, which explains in detail and reifies the topics. Resources are various types of information, such as text, graphics, and audio. Occurrence links the topics of the knowledge layer and the resources of the information layer.

3 Guided Navigation Techniques Based on Topic Map

In this paper, the topic map technology, which is a semantic Web technology, was applied to the navigation of a 3D virtual environment. As the navigation aid suggested in this paper, it integrates subject knowledge, which can be handled in the world, and spatial knowledge, which can grasp the structure of the world. Using the topic map technology, the said navigation aid can help users navigate to their target/destination in the virtual environment as they follow the semantic linkage.

3.1 Knowledge Layer

The knowledge layer expresses the environment structure through linkage relations between topics in the topic map for a 3D virtual environment. Here, the structure of the environment includes not only the spatial structure that indicates the physical environment in the hierarchical structure, but also the knowledge structure regarding various subjects that can be handled in such an environment. Accordingly, two types of topics exist: one expresses the spatial structure and the other, the knowledge structure. These topics comprise the knowledge layer. These topics generate meanings through mutual linkage relations. The meanings here are used to call the target/destination points in the 3D virtual environment.

3.1.1 Spatial Knowledge
Spatial knowledge refers to the knowledge that organizes the virtual environment into a hierarchical structure so the world could be understood. The world is divided into several zones in the virtual environment, based on their regional features and classifications, and each zone is divided into sub-zones.

Landmarks are the expression of specific scenes in a 3D virtual environment in the form of 2D photos. Landmarks have a hierarchical structure according to the hierarchical position of the represented region. For instance, suppose a 3D virtual environment forms a shopping center that has several buildings in its substructure. Each building has several floors under the substructure, and detailed classification can be made in such a manner. In such a hierarchical structure, the subordinate structure is linked to the relation of "is a" from the upper hierarchy. The order linked according to the rank of the hierarchical structure forms the path of movement in the virtual environment. That is, when one moves from one point within a building to another point in another building, one comes out of the building through the current room and floor.

3.1.2 Subject Knowledge
Subject knowledge refers to categorized knowledge that organizes a variety of topics to be dealt with in the virtual environment. The subject may vary, depending on what the environment expressed. If the virtual environment is a shopping mall, goods, business types, and manufacturing companies can be the subjects, and if it is a museum, the relics, age classifications, excavation sites, and excavation materials can be the subjects.

The topic map navigation technique suggested in this paper linked a variety of subjects semantically through linkages between the topics. It linked the landmarks in the

3D virtual environment as well as the semantic linkages between the subjects. In this way, not only can users find the target/destination in the virtual environment through hierarchically built landmarks, but they can also navigate through the process of moving the subject words that are semantically linked.

3.1.3 Linkage of Spatial Knowledge and Subject Knowledge

The knowledge layer of the topic map is built by linkage, following the semantic relationships between topics. Fig. 2 shows an example of the knowledge layer that built a shopping mall in a virtual environment. In the figure, the landmarks consisting of spatial knowledge, and all the topics consisting of subject knowledge, compose the knowledge layer through association. Inside the spatial knowledge and subject knowledge, topics are linked through the "part of" relationship since they have a hierarchical structure. Beyond the spatial knowledge and subject knowledge, it is the relationship defines the semantic relationships, not the hierarchical relationships.

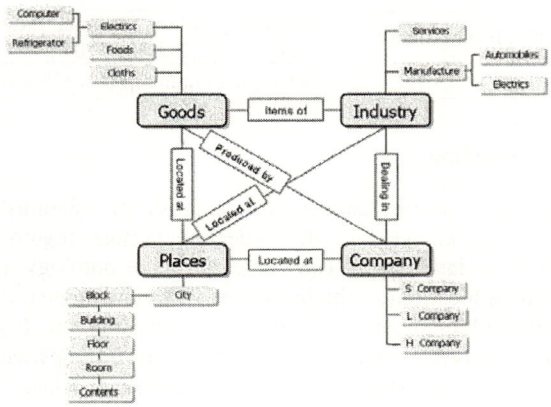

Fig. 2. Linkage of Spatial Knowledge and Concept Knowledge

The following shows the set association between topics that are beyond the categories of spatial knowledge and subject knowledge. S Company and A City are topics, and the relation of the association "located at" between these two topics exists. Likewise, the topics within the topic map are independent of one another and are simultaneously linked with a certain type of association relation. S Company is *located at* A City. A Computer is an *item of* Electrics. L Company *deals with* Electrics. B Computer is *produced by* S Company.

3.2 Information Layer

The information layer is the place where corporeal resource about the topics, which are selected according to the semantic linkage relations in the knowledge layer, is stored. The resources of the information layer include landmarks linked with the spatial knowledge and the information linked with the subject knowledge.

Resources related to the spatial knowledge are images that are worth giving meaning to from images in the environment. Among these images are the overview of a shopping mall, an intersection, a building, a lobby, a gallery, and an object. Resources related to the subject knowledge are topics that consist of concrete information of what one would deal with in the world. In the example of the shopping mall, examples of these topics would be products, companies, techniques, purposes, etc.

4 System Implementation

The development environment of the topic-map-based navigation aid system is shown in Table 1. To test the effect of this navigation aid, a 3D shopping mall was built. The shopping mall consisted of three buildings, and inside each building were 15 exhibition rooms in which various articles were displayed.

Table 1. Implementation Environment

3D Virtual Environment	Parallel graphics, Cortona VRML Client JDK 1.1.8, j2sdk1.4.1_04
Topic Map	Ontopia Omnigator [9], MSXML 4.0

4.1 Landmark Modeling

Ontology was prepared on the landmarks to express the relationship between spatial knowledge and subject knowledge, after selecting famous regional points in the 3D virtual environment as landmarks. The contents of the ontology are the name of the landmark, the camera location to obtain images and view angles, the landmarks in the upper and subordinate hierarchies and in the same hierarchy, etc. Fig. 3 shows an ontology that was built to express a relationship with the subject knowledge. The contents of the ontology are information about the producer, the delivery date, the exhibition place, the price, etc.

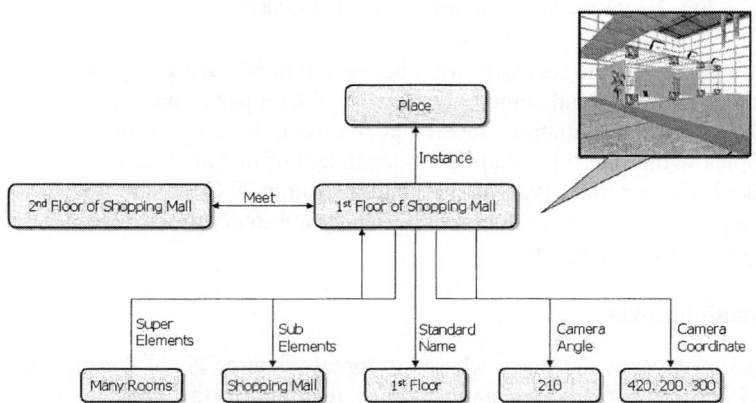

Fig. 3. Landmark Modeling to Express the Relationship between Subject Knowledge and Spatial Knowledge

4.2 XML Topic Map

The topic map consists of three basic factors, namely: topic, association, and occurrence. The topic may include any thing that can be handled in the 3D virtual environment. Examples of topics are a landmark's name, a good's name, a company's name, information on various subjects, etc.

```
<topic id = "ShoppingCenterA">                          <occurrence>
    <instanceOf>                                            <instanceOf>
        <topicRef xlink:href="#building"/>                      <topicRef xlink:href="#website"/>
    </instanceOf>                                           </instanceOf>
    <baseName>                                              <resourceRef xlink:href="http://hur77.cafe24.com/"/>
        <baseNameString>Yonsei department store</baseNameString>  </occurrence>
    </baseName>                                         </topic>
    <baseName>
        <scope><topicRef xlink:href="#childeren"/></scope>  <association id="opposite2">
        <baseNameString>Not availabe</baseNameString>           <instanceOf>
    </baseName>                                                 <topicRef xlink:href="#opposite"/>
    <baseName>                                                  </instanceOf>
        <scope><topicRef xlink:href="#adult"/></scope>      <member>
        <baseNameString>Availabe</baseNameString>               <roleSpec><topicRef xlink:href="#ShoppingCenterA"/></roleSpec>
    </baseName>                                                 <topicRef xlink:href="#building"/>
    <occurrence>                                            </member>
        <instanceOf>                                        <member>
            <topicRef xlink:href="navi.xtmp#Price"/></instanceOf>  <roleSpec><topicRef xlink:href="#ShoppingCenterB"/></roleSpec>
        <resourceData>                                          <topicRef xlink:href="#building"/>
            This is a cyber store in virtual world.         </member>
        </resourceData>                                 </association>
    </occurrence>
```

Fig. 4. A Sample Code of a Topic Map

The association sets up the relationships between topics. The hierarchical relations in the spatial structure and in the subject structure, and the linkage relationships between subjects, are expressed in the association. The occurrence designates the location of resources. The landmark name mentioned above refers to the location of the camera and the view angle that illuminates a specific point within the environment. Fig. 4 shows the part of the XML topic map devised on the basis of the ontology built according to the method explained in the previous chapter.

4.3 User Interface

A user can directly control movements in a 3D virtual environment using the VRML browser, and can also indirectly control movements in this environment via the topic map browser. In the topic map browser, the topics associated with the current topic are sorted by associated types. Moreover, users can move by selecting the desired target in the occurrence list that indicates the position where the current topic is located. Fig. 5 shows the overview of the user interface for the 3D virtual environment. On the left side of the screen is the start view of the environment. On the right side of the screen are electronic appliances retrieved as the current topic in the topic map browser. The association related to the topic suggests the topic of the goods for the super-element and two types of electronic goods for the sub-element. The names of the companies that produce electronic goods were provided. In the occurrence indicating the current topic's location, two locations were indicated. When a user selected one of these two locations, the system moved the scene to the selected location.

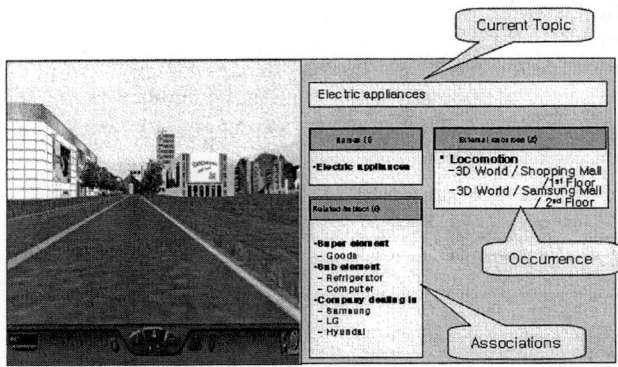

Fig. 5. The User Interface of Topic-Map-Based Navigation Aid

5 Evaluation and Discussion

An experiment was conducted to evaluate the navigation aid system suggested in this paper. Thirty high school students participated in this experiment. These students have experienced using computers but have not encountered a 3D virtual environment or a topic map. The participants were divided into three groups of ten persons each, and the experiment was conducted under different environments. Fifty computer models and fifteen automobile models were used as probable target objects.

The first group obtained assistance from the menu-table-type navigation aid, which is similar to Active World. Menu table is designed to present a sub-target list when the folder is selected, as in Windows Explorer. The final object, the target, can be reached on the third or fourth level in the search.

The second group obtained assistance from the topic map browser, and the third group was supposed to find the target through direct navigation, using the mouse and keyboard in the concerned environment. Each participant was given three targets. The following are the targets that were selected and the reasons for the selection of each of these targets.

The first each participant was search building. One of the three buildings located in the shopping mall. It is the most highly represented target. There was no visual obstacle from the portal location to the target.

Second each participant was search specific computer model. This was a target located in the computer exhibition room on the second floor of the electronic product's exhibition building. It is the lowest representation target. There were many visual obstacles from the starting location to the target. The experiment participants could reach target by entering the building and the global computer exhibition room. They had to make an adequate judgment on whether to enter the electronic product's exhibition building or the auto exhibition building in the task of finding a computer, by referring to visual hints during their navigation.

Last one each participant was search elegant car. In this experiment, there was no designated target that was presented to the participants, but they selected items

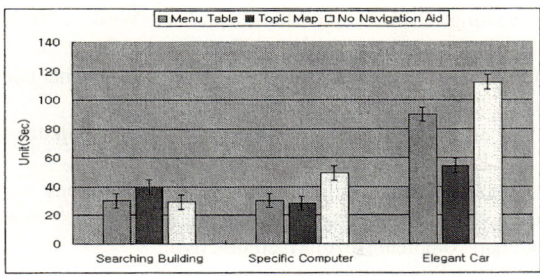

Fig. 6. Evaluation of Three User Interfaces

according to their own tastes. They tried to make a complete list of similar items, and selected from these items by comparing their strengths and weaknesses.

Fig. 6 shows the experiment results. The height of the bar graph shows the average and standard deviation of the length of time in which the ten participants completed the task. In the initial evaluation of the experiment, neither the menu table nor the topic map succeeded in helping the subjects find the auto exhibition building quickly. Rather, faster results were shown in the navigation that was conducted without using any navigation aid. In the task of finding a specific computer model, the subjects who were aided by the menu table and the topic map found the model faster.

There was a significant difference between the results of the topic-map-aided navigations and the navigations conducted without a navigation aid. Although the menu-table-aided navigation yielded better results than the no-aid navigation, the menu table could only present references about current interest items. Therefore, it could not present sufficient references about current interest items. The topic map, on the other hand, provided sufficient information on the current topic, along with movement information. The results show that the topic map helps users make prompt decisions.

6 Conclusion

This paper suggests a new navigation aid technique that would enable a user who is not familiar with the 3D virtual environment to easily find a target and move, using a topic map. This paper showed that the topic map can not only be utilized in the Web environment, but can also be applied as a navigation aid in a 3D virtual environment. The semantic linkage of the topic map expresses the hierarchical structure in the spatial structure, and semantically links the subjects related to the environment. The information in various directions regarding the topics selected in this manner was provided automatically.

References

1. J. Domingue, A. Stutt, M. Martins, H. Petursson and E. Motta, "Supporting Online Shopping through a Combination of Ontologies and Interface Metaphors," International Journal of Human Computer Studies, Vol. 59, Issue 5. pp. 699-723, 2003
2. A. Nordborg, "Topic Maps, " Master's thesis in Goteborg University. pp. 14-18, 2002.

3. J. Park, and S. Hunting, "XML Topic Maps," Addison-Wesley, 2003.
4. "Topic Map", available online at http://www.isotopicmaps.org/
5. R. Ramloll and D. Mowat, "Wayfinding in virtual environments using an interactive spatial cognitive map, " Information Visualization, London, 2001.
6. T. Elvins, D. Nadeau and D. Kirsh, "Worldlets - 3D Thumbnails for Wayfinding in Virtual Environments," In Proceedings of UIST'97, pp. 21-30, 1997.
7. "Active world", available online at http://www.activeworlds.com/
8. "A CIA World Factbook Topic Map", available online at http://www.ontopia.net/ omnigator/models/topicmap/complete.jsp?tm=factbook.hytm
9. "Ontopia", available online at http://www.ontopia.net
10. A. Car, "Hierarchical Spatial Reasoning: Theoretical Consideration and its Application to Modeling Wayfinding". Geoinformation, Vienna, 1996.
11. R. Darken and J. Silber, "Wayfinding Strategies and Behaviors in Large Virtual Worlds," In Proceedings of the ACM CHI 96, vancouver, pp. 142-149, 1996.
12. R. Downs and D. Stea, "Maps in Minds: Reflections on Cognitive Mapping," Harper & Row, New York. 1977.
13. C. Santos, P. Gros, P. Abel, D. Loisel, N. Trichaud and J. Paris, " Metaphor-Aware 3D Navigation, " INFOVIS Salt Lake, pp. 155-165, 2000.
14. P. Thorndyke and B. Hayes-Roth, "Differences in Spatial Knowledge from Maps and Navigation," Cognitive Psychology, pp. 560-589, 1982.

Image Sequence Augmentation Using Planar Structures

Juwan Kim[1] and Dongkeun Kim[2]

[1] Electronics and Telecommunications Research Institute,
161 Gajeong-dong, Yuseong-gu, Daejeon, 305-350, Korea
juwan@etri.re.kr
[2] Division of Computer Engineering, Kongju National University,
275 Dudae-dong, Cheonan, Chungnam, 330-717, Korea
dgkim@kongju.ac.kr

Abstract. In this paper, we present an image sequence augmentation that easily estimates camera parameters using a planar structure visible in an image sequence and superimposes 3D virtual objects on it. After constructing a virtual 3D model plane from the planar structure, we compute camera parameters recursively using the image of the absolute conic. Experimental results using simulated data and real data confirm that the proposed method is of clearly superior quality compared to that of traditional methods and performs better than existing ones. Our approach does not require any a priori information about the camera being used and metric coordinates in the scene. And the planar structure greatly simplifies the coordinate-system alignment problem, so the proposed method can easily be applied to cases where there is a planar surface visible somewhere in an image sequence and highest accuracy is not demanded.

1 Introduction

In augmented reality systems, one of the most basic challenges to overcome is the registration problem: the objects in the real and the virtual world must be properly aligned with respect to each other or the illusion that the two worlds coexist will be compromised. To effectively implement augmented reality in unstructured environments such as an image sequence, the primary requirement is accurate, fast and reliable camera tracking, which remains an enormous challenge till now [1].

Move-matching techniques can simultaneously estimate camera motion and 3D structure from imaged scenes [2],[3]. However, these techniques, which must generally batch-process the entire data sequence at once, are too time-consuming for interactive augmented reality applications. And also, it is not easy to align the system's arbitrarily chosen coordinate frame with that of the virtual object.

The possibility of calibrating cameras from views of planar objects is well known [4]-[8]. Zhang proposed the flexible technique for camera calibration that mixed the self-calibration method using camera motion and the photogrammetric calibration method using a precise calibration object [4]. However, metric coordinates of the planar pattern on the model plane have to be known in advance. Triggs presented the self-calibration method from views of planar scenes with unknown metric structure [5]. However, 9 or 10 views of the same plane are needed for reliable results. Simon *et al.* proposed the markerless camera tracking method in order to register virtual objects on uncalibrated image sequences [6]. In his research, only the focal length among the internal parameters was estimated from two sets of parallel lines and the estimated rotation matrix was not an orthogonal. Simon *et al.* proposed the pose estimation technique with several visible planes for registration with a calibrated camera [7].

Our approach is focused on a simple augmented reality system using a planar structure from an uncalibrated image sequence. The main idea of the proposed approach is to estimate the camera parameters using a rectangle on the planar structure for alignment of the real and virtual world. The proposed approach does not require any a priori information about the camera being used and metric coordinates in the scene. This paper is organized as follows: Section 2 describes the plane-to-image homography. Section 3 proposes more reliable and fast 3D virtual object registration method using planar structures. Section 4 provides the experimental results. Finally, conclusion is given in section 5.

2 The Plane-to-Image Homography

Let's assume that the plane π in 3D space is on Z=0 of the world coordinate system, as shown in figure 1. A 3D point **M** on the plane π can be denoted by $[X\ Y\ 1]^T$ and a 2D point **m** on the image plane π' which is projected by a camera C can be denoted by $[u\ v\ 1]^T$. The 3D point **M** and its projected point **m** are given by a planar homography **H**:

$$\mathbf{m} = \mathbf{HM} \text{ with } \mathbf{H} = [\mathbf{h}_1\ \mathbf{h}_2\ \mathbf{h}_3] = \begin{bmatrix} h_{11} & h_{12} & h_{13} \\ h_{21} & h_{22} & h_{23} \\ h_{31} & h_{32} & h_{33} \end{bmatrix} \quad (1)$$

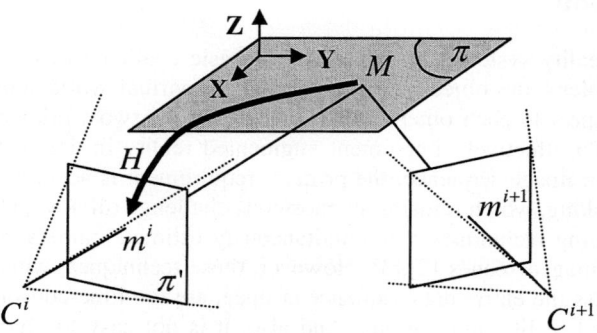

Fig. 1. A homography between a plane in 3D space and its projected planes

Using the planar homography, we consider the pinhole camera model, which relates the point **M** on the plane π to the point **m** on its projected plane π':

$$\mathbf{m} = \mathbf{P}\begin{bmatrix} X \\ Y \\ 0 \\ 1 \end{bmatrix} = \mathbf{K}[\mathbf{R}\mid\mathbf{t}]\begin{bmatrix} X \\ Y \\ 0 \\ 1 \end{bmatrix} = \mathbf{K}[\mathbf{r}_1\ \mathbf{r}_2\ \mathbf{r}_3\ \mathbf{t}]\begin{bmatrix} X \\ Y \\ 0 \\ 1 \end{bmatrix} = \mathbf{K}[\mathbf{r}_1\ \mathbf{r}_2\ \mathbf{t}]\mathbf{M} \cong \mathbf{H}\begin{bmatrix} X \\ Y \\ 1 \end{bmatrix} \quad (2)$$

The matrix **P** is the 3x4 projection matrix and the matrix **[R|t]** is the camera's rotation and translation parameters. The matrix **K**, called the camera's internal parameters, is given by

$$\mathbf{K} = \begin{bmatrix} \alpha & \gamma & u_0 \\ 0 & \beta & v_0 \\ 0 & 0 & 1 \end{bmatrix} \quad (3)$$

with (u_0, v_0) the coordinates of the principal point, α and β the scale factors in image u and v axes, and γ the parameters describing the skew of the two image axes.

The third column vector $\mathbf{r}_3$ of the rotation matrix **R** can be computed by the cross product $\mathbf{r}_1 \times \mathbf{r}_2$ using the knowledge that $\mathbf{r}_1$ and $\mathbf{r}_2$ are orthonormal vectors. As a result, if we know both the matrix **K** and the homography **H**, the rotation matrix **R** and the translation vector **t** are given by

$$\mathbf{r}_1 = \lambda \mathbf{K}^{-1}\mathbf{h}_1, \mathbf{r}_2 = \lambda \mathbf{K}^{-1}\mathbf{h}_2, \quad (4)$$
$$\mathbf{r}_3 = \mathbf{r}_1 \times \mathbf{r}_2, \mathbf{t} = \lambda \mathbf{K}^{-1}\mathbf{h}_3$$

with $\lambda = 1/\|\mathbf{K}^{-1}\mathbf{h}_1\| = 1/\|\mathbf{K}^{-1}\mathbf{h}_2\|$.

Figure 2 shows homography relationship between image frames in the image sequence. The homography between the ith image and the $(i+1)^{th}$ image is denoted by $\mathbf{H}_i^{i+1}$ and the homography between the virtual coordinate system for alignment 3D virtual objects and the ith image in the image sequence is denoted by $\mathbf{H}_w^i$.

3 Augmented Image Sequence Using Planar Homographies

In this paper, two procedures that precede estimation of the camera parameters and alignment of the real and virtual coordinate systems are as follows: One is to make sets of correspondences between adjacent frames that are used to determine H and a virtual 3D model plane used for computing the internal parameters of the camera. The other is to select at least 3 views of the same plane in the image sequence and select four points on a rectangle created by the reference axes for alignment of the real and virtual coordinate systems. Because we don't know metric coordinates in the scene, the four points of the rectangle are assigned the arbitrary world coordinates (0, 0), (1, 0), (1, δ), (0, δ), where δ is unknown aspect ratio of the rectangle.

3.1 Computing Internal Parameters

From the knowledge that $\mathbf{r}_1$ and $\mathbf{r}_2$ of the rotation matrix are orthonormal vectors, we have

$$\mathbf{h}_1^T \mathbf{K}^{-T} \mathbf{K}^{-1} \mathbf{h}_2 = 0$$
$$\mathbf{h}_1^T \mathbf{K}^{-T} \mathbf{K}^{-1} \mathbf{h}_1 = \mathbf{h}_2^T \mathbf{K}^{-T} \mathbf{K}^{-1} \mathbf{h}_2 \quad (5)$$

If we know metric coordinates of the model plane in 3D space and number of its images shown at a few different orientations is above 3, we can estimate the internal parameters by the closed-form solution using the image of the absolute conic $\mathbf{K}^{-T}\mathbf{K}^{-1}$ as follows.

$$\mathbf{B} = \lambda \mathbf{K}^{-T}\mathbf{K}^{-1} = \begin{bmatrix} \dfrac{1}{\alpha^2} & -\dfrac{\gamma}{\alpha^2 \beta} & \dfrac{v_0 \gamma - u_0 \beta}{\alpha^2 \beta} \\ -\dfrac{\gamma}{\alpha^2 \beta} & \dfrac{\gamma^2}{\alpha^2 \beta^2} + \dfrac{1}{\beta^2} & -\dfrac{\gamma(v_0 \gamma - u_0 \beta)}{\alpha^2 \beta^2} - \dfrac{v_0}{\beta^2} \\ \dfrac{v_0 \gamma - u_0 \beta}{\alpha^2 \beta} & -\dfrac{\gamma(v_0 \gamma - u_0 \beta)}{\alpha^2 \beta^2} - \dfrac{v_0}{\beta^2} & \dfrac{(v_0 \gamma - u_0 \beta)^2}{\alpha^2 \beta^2} + \dfrac{v_0^2}{\beta^2} - \dfrac{v_0}{\beta^2} \end{bmatrix} \quad (6)$$

where λ is an arbitrary scale factor and $\mathbf{B}$ is a symmetric matrix, defined by a 6D vector

$$\mathbf{x} = [\mathbf{B}_{11} \ \mathbf{B}_{12} \ \mathbf{B}_{22} \ \mathbf{B}_{13} \ \mathbf{B}_{23} \ \mathbf{B}_{33}]^T \quad (7)$$

From equation (5), we can be rewritten as 2 homogeneous equations in $\mathbf{x}$:

$$\begin{bmatrix} \mathbf{v}_{12}^T \\ (\mathbf{v}_{11} - \mathbf{v}_{22})^T \end{bmatrix} \mathbf{x} = 0, \quad (8)$$

$$\mathbf{v}_{ij} = [h_{i1}h_{j1} \quad h_{i1}h_{j2} + h_{i2}h_{j1} \quad h_{i2}h_{j2} \quad h_{i3}h_{j1} + h_{i1}h_{j3} \quad h_{i3}h_{j2} + h_{i2}h_{j3} \quad h_{i3}h_{j3}]^T$$

If we have at least 3 views of the same plane, we can determine $\mathbf{x}$ by stacking them such the above equation. Therefore, the internal parameters of the camera can be extracted from $\mathbf{B}$ [4].

$$\begin{aligned} v_0 &= (B_{12}B_{13} - B_{11}B_{23})/(B_{11}B_{22} - B_{12}^2) \\ \lambda &= B_{33} - [B_{13}^2 + v_0(B_{12}B_{13} - B_{11}B_{23})]/B_{11} \\ \alpha &= \sqrt{\lambda/B_{11}}, \quad \beta = \sqrt{\lambda B_{11}/(B_{11}B_{22} - B_{12}^2)} \\ \gamma &= -B_{12}\alpha^2\beta/\lambda, \quad u_0 = \lambda v_0/\beta - B_{13}\alpha^2/\lambda \end{aligned} \quad (9)$$

However, we don't know metric coordinates of the world rectangle for the reference axes in 3D space. Therefore, we estimate the aspect ratio of the rectangle using the property that the aspect ratio of the camera is affected by the aspect ratio of the world rectangle in the projected images by the camera.

At first, the four points of the rectangle are assigned the arbitrary world coordinates $(0, 0)$, $(1, 0)$, $(1, \delta)$, $(0, \delta)$. For an example, $\delta = 0.5$. Then, we can create a virtual 3D model plane in 3D space from n matching points using the homography between the user-selected image coordinates and the world coordinates of four points, where n is the number of common matching points shown in user selected views of the same plane in the image sequence. That is, the virtual 3D model plane is same as the model plane for camera calibration.

Using n matching points on the virtual 3D model plane and its image coordinates on the select images, we can determine the matrix $\mathbf{K}$ as describe above. If the aspect ratio of the computed camera is not 1, we will adjust Y-axis data of n matching points on the virtual 3D model plane multiplying by (β/α). And then repeat above procedures using the world coordinates of the adjusted virtual 3D model plane until the aspect ratio of the camera converges to 1.

3.2 Computing External Parameters

Once the internal parameters and the homography relating two images are known, we can easily compute the external parameters [**R**|**t**] as described in section 2. In general, however, the computed matrix **R** does not satisfy the properties of a rotation matrix because of various error sources. If the rotation matrix is orthogonal, it will be easy to render various graphic effects such as shadow without rendering distortion. So we obtain the best orthogonal rotation matrix **R'** from **R** using SVD (Singular Value Decomposition) method as follows.

$$\mathbf{UDV} = \text{SVD}(\mathbf{R})$$
$$\mathbf{R'} = \mathbf{U}\,\text{diag}(0,1,0)\mathbf{V}^T \qquad (10)$$

However, the matrix **R'** causes an alignment error of the image sequence and 3D graphic objects. In order to reduce the alignment error, we adjust the translation vector using the world coordinates and its projected coordinates of four points on the rectangle chosen for alignment the real and virtual coordinate systems by the least-square method.

4 Experimental Results

We developed the image sequence augmentation system using Microsoft Visual C++ and DirectX 9.0 and tested by two steps: camera calibration step and image augmentation step.

4.1 Camera Calibration Results

The proposed method has been tested on both computer-generated data and real data. In the calibration experiment with computer-generated data, a simulated camera has the following property: $f = 1000, skew = 1, u_0 = 258, v_0 = 254$.

We make three images shown at different orientations and random noises with average 1 pixel are added to the projected image points. In this experiment, the initial δ value for reconstructing the coordinates of the model plane is 0.5. Table 1 shows the computed calibration results that are considerably good even if there are small errors caused by random noise added in projected images.

In the calibration experiment with real data, we used Zhang's test data [4]. Table 2 shows the comparison of our method with other camera calibration methods. In this

Table 1. Camera calibration results with computer-generated data

iteration parameters	1	2	3	4
α	1,672.4	969.4	1,004.3	1,002.8
β	1,089.8	993.3	1,003.2	1,002.8
Skew	-6.4	1.3	1.3	1.3
u_0	96.7	263.7	258.1	258.3
v_0	-309.5	299.1	252.2	254.3

experiment, metric coordinates of the model plane were applied to only Zhang's method. The initial δ value used for reconstructing the world coordinates of the virtual model plane was 0.2. Our approach gives considerably good results even if it doesn't use any metric coordinates of the model plane.

Table 2. Comparison of calibration results with Zhang's test data

methods \ parameters	α	β	skew	u_0	v_0	metric coordinates of the model plane
our method	852.5	852.5	-0.7	305.1	216.6	No
Simon's method[6]	2,046.0	2,046.0	0.0	256.0	256.0	No
Zhang's method[4]	832.5	832.5	0.2	304.0	206.6	Yes

4.2 Image Sequence Augmentation

Figure 3 shows the results of applying the proposed method to the table sequence. To estimate the planar homographies, we tracked only the calendar region on the table and generated a set of correspondences between successive images as shown in figure 3 (a). Determining the camera parameter using edges of the calendar selected as the reference axes for alignment, we combined a 3D virtual cube, a virtual airplane and a latticed virtual XY plane with the image sequence. As shown in figure 3 (b), we could obtain good registration results without rendering distortion.

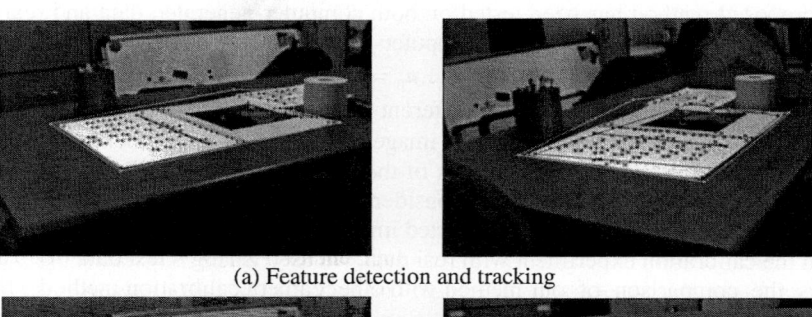

(a) Feature detection and tracking

(b) Augmented images

Fig. 3. Indoor sequences

The proposed method was applied to the outdoor sequence as shown in figure 4 (a). In this experiment, we tracked the playground as the reference plane and superimposed several 3D virtual objects on it as shown in figure 4 (b). Both the building and trees on the playground are fixed 3D virtual objects. On the other hand, tanks and airplanes are 3D virtual objects that roam about the playground and the sky. Registration results on this sequence were also good, with low jitter as shown in figure 4 (c).

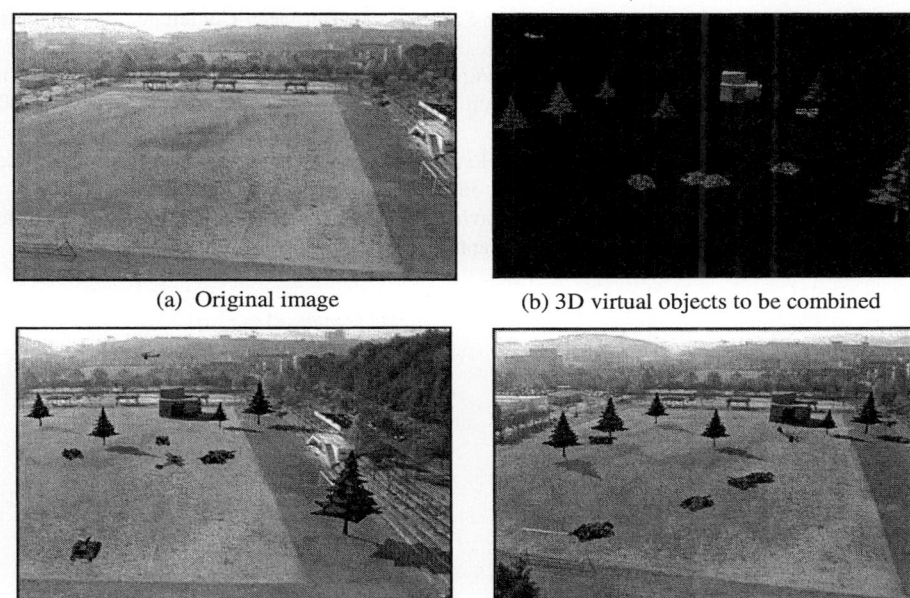

(a) Original image (b) 3D virtual objects to be combined

(c) Augmented images

Fig. 4. Outdoor sequences

5 Conclusion

In this paper, we have described an image augmentation method using a planar structure from an uncalibrated image sequence. Our approach does not require any camera parameters and metric coordinates in the scene. Compared with other techniques, however, our approach provides accurate and reliable results as shown in the experimental results. We think that the proposed method can easily be applied to cases where there is a planar surface visible somewhere in an image sequence and highest accuracy is not demanded.

References

1. Azuma, R., Baillot, Y., Behringer, R., Feiner, S., Julier, S., MacIntyre. B.: Recent Advances in Augmented Reality. IEEE Computer Graphics and Applications (2001) 34-47

2. Pollefeys, M., Koch, R., Van Gool, L.: Self-Calibration and Metric Reconstruction in spite of Varying and Unknown Internal Camera Parameters. International Journal of Computer Vision (1999), 32(1), 7-25
3. Fitzgibbon, A., W., Zisserman, A.: Automatic Camera Recovery for Closed or Open Image Sequences. Proc. European Conference on Computer Vision. Springer-Verlag (1998) 311-326
4. Zhang, Z.: A Flexible New Technique for Camera Calibration. IEEE Transactions on Pattern Analysis and Machine Intelligence (2000), 22(11), 1330-1334
5. Triggs, B.: Autocalibration from Planar Scenes. Proc. European Conference on Computer Vision, Springer-Verlag (1998) 89-105
6. Simon, G., Fitzgibbon, A., Zisserman, A.: Markerless Tracking using Planar Structures in the Scene. Proc. International Symposium on Augmented Reality, IEEE CS Press (2000) 137-146
7. Simon, G., Marie-Odile Berger: Pose Estimation for Planar Structures. IEEE Computer Graphics and Application (2002), No. 6, 46-53
8. Simon J. D. Prince, Ke Xu, Adrian David Cheok: Augmented Reality Camera Tracking with Homographies. IEEE Computer Graphics and Applications (2002), No. 6, 39-45

MultiPro: A Platform for PC Cluster Based Active Stereo Display System*

Qingshu Yuan[1], Dongming Lu[1,2], Weidong Chen[1], and Yunhe Pan[1,2]

[1] College of Computer Science and Technology, Zhejiang University,
[2] State Key Lab. of CAD and CG, Zhejiang University,
310027, Hangzhou, Zhejiang, P.R.China
{yuanqs, ldm, chenwd, panyh}@cs.zju.edu.cn

Abstract. Active stereo display system simulates the principle of human viewing, which generates two view images, one for each eye, switched by electronic glasses. Existing stereo development libraries, e.g. CAVELib, Avocano, require fully or partially understanding, at least caring of stereo display principles. Furthermore, most of existing libraries are used to develop a full application with all stereo display functions, e.g. multi-tasking, shared memory, stereo displaying, and so on. MultiPro is a PC cluster based system developed by C++ and Qt library. It's composed of following modules of controller, renderer, MultiPro Library, communication layer and application module. Benefiting from modular design, the application developer doesn't need to know anything about stereo display principles and the platform architecture. Besides, different applications can be switched at run-time of the platform, without the inconvenient switch of application processes. Three typical applications were developed and the results were excellent.

1 Introduction

Virtual Reality is an experience in which a person is surrounded by a three dimensional computer-generated representation, and is able to move around in the virtual world and see it from different angles [1].

To achieve this goal, many projection-based active stereo display systems were brought out, such as CAVE [2], Responsive Workbench [3, 4] and i-Cone [5] etc. Active stereo display system simulates the principle of human viewing, which generates two view images, one for each eye, switched by electronic glasses.

There are several aspects to be considered in active stereo display, including display synchronization, head tracking, multi-tasking (processing or threading), communication, stereo pair calculation, application development, and so on. Therefore, a good framework can reduce the burden of application developers. Benefiting from modular design and implementation, the platform provides developers with simple yet powerful development architecture. Furthermore, the platform can be configured for different active stereo systems by adjusting the screen configurations.

---

* The research was partially funded by National Basic Research Program of China (No. 2002CB312106), National Natural Science Foundation of China (No. 60273055) and China-US Million Book Digital Library Project.

2 Related Work

There are many active stereo display systems together with their software platforms brought forward during last decade.

CAVE has its motivation rooted in scientific visualization and was first demonstrated in SIGGRAPH'92. It's essentially a five (maybe four or six) sided cube, according to the head and eye positions of one participant, images are generated and projected onto the walls. The Responsive Workbench is a 3D interactive workspace in which computer-generated stereoscopic images are projected onto a horizontal tabletop display surface via a projector-and-mirrors system, and viewed through shutter glasses to generate the 3D effect. A 6DOF tracking system tracks the user's head, so that the user sees the virtual environment from the correct point of view. A pair of gloves and a stylus, also tracked by the system, can be used to interact with objects in the tabletop environment. i-Cone is a cylindrical 270-degree projection display system with high-resolution and reclining projection surfaces, developed as a Virtual Reality Display without corners and edges, which helps to avoid the geometric distortions and reflection effects on the walls.

With the overwhelming success of these devices, stereo display software platforms are also developed. Most of them are based on high-end SGI workstations and multi-pipeline graphic cards, but still some are based on PC clusters.

CAVE is driven by CAVELib [6], a widely used API for developing applications for immersive displays. It uses OpenGL or OpenGL Performer as the rendering engine. Items the CAVELib abstracts away for a developer are, window and viewport creation, viewer-centered perspective calculations, displaying to multiple graphics channels, multi-processing and multi-threading, cluster synchronization and data sharing, and stereoscopic viewing. i-Cone use Avocado [7] as the platform software. Avocado is an object-oriented framework for distributed, interactive VE applications. Data distribution is achieved by transparent replication of a shared scene graph among the participating processes of a distributed application. A sophisticated group communication system is used to guarantee state consistency even in the presence of late joining and leaving processes. Net Juggler and SoftGenLock [8] are two open source softwares, the association of which makes possible to run a system of active stereo and multi-displays. PCCAVE [9] is a PC Cluster with master-slave architecture. Its communication system is based on MPI to implement synchronization of parallel computing and intelligent information broker and both DirectX and OpenGL can be used as the rendering engine.

The development of existing systems require fully or partially understanding, at least caring of stereo display principles, e.g. display synchronization, head tracking, multi-tasking (processing or threading), network communication, stereo pairs calculation, application development, and so on. Furthermore, most of existing libraries are used to develop a full application with all stereo display functions.

3 MultiPro Platform

MultiPro, developed using C++ and Qt [10], is a system independent software platform for PC cluster based active stereo display systems. It hides stereo display, collaborative rendering, data sharing and head tracking details from users, in other words, developers using MultiPro just need to concern themselves of what they should do as in one PC.

3.1 MultiPro Overview

MultiPro platform is developed for PC cluster based active stereo display systems. Figure 1 shows a simple example system. PCs are connected by fast Ethernet. *Render nodes* are PCs with stereo display graphical cards and projectors connected. The *remote controller* is the interface of system administrators, in which the number of render nodes, network addresses and ports of the render nodes, network address and port of shared memory server, system resolution can be set. As the status of the scene should be shared by multiple render nodes, a *shared memory server* should be adopted. The projected images should be dynamically updated according to the head position of the participants which is tracked by a 6DOF head tracker. And the tracker can be connected to any of the render nodes.

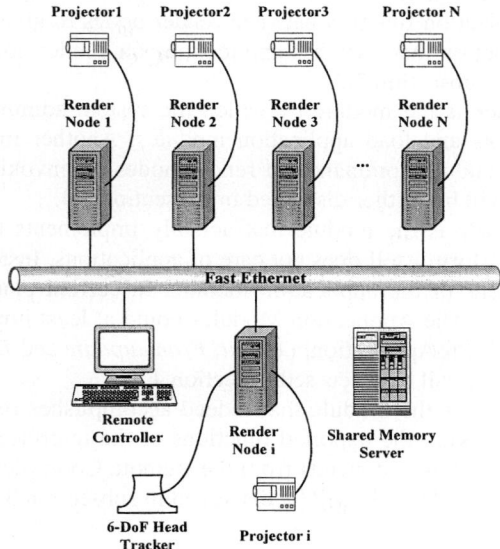

Fig. 1. System Overview

The whole system contains the following parts, Network Communication Layer, Remote Controller, Application Modules, Render Nodes, as illustrated in Figure 2.

MultiPro Library is compiled as a static library for developers (including Rendering module developers and MultiPro application module developers). It mainly offers the following functions, camera calculation, user navigation, interaction, shared memory and tracking. Details will be discussed in subsection 3.2.

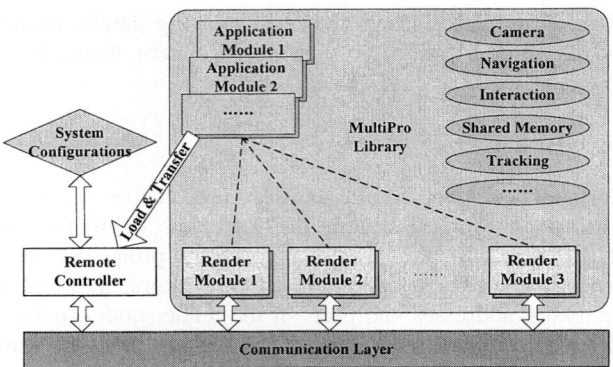

Fig. 2. Software Modules

Network Communication Layer is the module that deals with the issues of network communications between the Remote Controller and the Render Nodes, for example, load application function that transfer application module files to render nodes, the synchronization function between render nodes to give the user consistent illustrations, the function that switches applications for render nodes, and so on. Details will be given in subsection 3.3.

Remote Controller is the module by which the system administrator can modify system configurations and load application modules. Another important role of remote controller is sending commands to render nodes by invoking Communication Layer functions. It will be further discussed in subsection 3.4.

Application Module is the module that actually implements the rendering code, that is to say, the platform itself does not care of applications. Instead, it only invokes the exported functions of the application module. In current platform version, only OpenGL is supported. The Application Module should at least implement 3 functions of the interface IMultiProApplication, i.e. *Init*, *Frameupdate* and *Display*. Application (module) development will be discussed in section 4.

Rendering Module is the module that indeed accomplishes rendering task of the whole system by invoking the exported functions of the interface IMultiProApplication. This module receives commands from the Remote Controller and runs at all the render nodes. Rendering Module will be discussed in subsection 3.5.

3.2 MultiPro Library

MultiPro Library is the library that contains commonly used functions on stereo display and collaborative rendering, e.g. camera calculation, user navigation, interaction, shared memory and tracking and it is compiled as a static lib. Render modules and application modules are developed based on MultiPro Library.

Camera calculation provides the functions that calculate participant's viewing position, direction and frustum. Generally, this function is revoked by render modules, with project screen positions and eyes positions provided. The camera calculation results of each wall should be consistent since the walls are indeed for one scene.

User navigation provides users with capability to navigate freely in the virtual system since participants' movements are restricted in a very limited space in most VR systems. Three main functions are MultiProNavTranslate, MultiProNavRotate and MultiProNavScale, while other functions are used to convert between logical and physical coordinates. The navigation data is stored in the shared memory, so any render nodes can use it transparently and freely during rendering process.

Interaction provides users with all the functions related to the interaction, e.g. keyboard, mouse, wand, data glove. Keyboard and mouse are used to simulate user interactions (e.g. navigation simulation) and the data is stored in the shared memory. In this version, wand and data glove interactions are not implemented, but will be soon supported in the next version.

Tracking is used to track the head and eye positions of participants. Active stereo display requires the graphics cards to compute two different images, one for each eye, and display them alternatively. So the position and view direction of participant's eyes should be tracked. The Tracking subsystem of MultiPro Library provides the function of calculating and transforming coordinates of tracker to display system by inputting some sampling data. The tracking data is put in the shared memory, so, all render nodes can read the up to date data to render each frame.

Shared memory provides users with the data sharing functions. MultiPro platform is composed of multiple render nodes, and these processes are distributed in the network. Therefore, the platform needs to have a mechanism to pass the data to these processes, to share between these processes. For example, a moving car has its position and orientation which should be used to render frames during the rendering process. Under the shared memory mechanism, only one process needs to update the data, and others just read it transparently. In implementation, one shared memory server is needed, and the API for shared memory client seems transparent to users, that is to

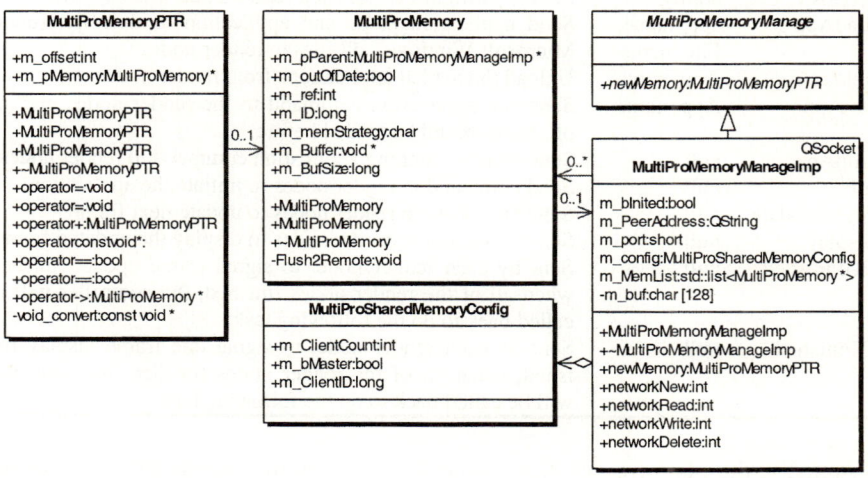

Fig. 3. Shared Memory Classes

say, the shared memory can be used like local memory. Figure 3 shows main classes for shared memory client. *MultiProMemory* is the class that stands for a shared memory instance, which records the ID of the memory, memory write strategy, memory reference count, and also contains a local memory with the same size as the shared memory. *MultiProMemoryManageImp* is the class that inherits QSocket (which is used to do network communication with memory server) and *MultiProMemoryManage* (which is the abstract interface that only contains a newMemory function). Each time when MultiProMemoryManageImp::newMemory is called, a MultiProMemory class will be newed with reference number 1, and a MultiProMemoryPTR class will be retured. MultiProMemoryPTR is a smart pointer class, which overloads such operators as void *, =, + and so on. Some operators will increase or decrease the reference number of the memory. The memory will automatically delete itself when the reference number is decreased to zero and the developer does not need to care about delete operations.

3.3 Network Communications, Rendering Barriers and Collaboration

Because rendering communications and its corresponding events, e.g. synchronization barrier, should be sent reliably over the Ethernet, TCP protocol is used in the platform. In our implementation, class QServerSocket is inherited as the server socket, and class QSocket is inherited as the client socket.

The main primitives and their descriptions are listed in Table 1.

Table 1. Main network communication primitives

Primitive	Parameters	Description
setHostCfg	config	Set the configuration of each render node when system starts.
loadApp	app_name, File_name	Send application name and application module file (under Microsoft Windows, .dll file) to render nodes.
unloadApp	App_name	Unload the loaded applications from render nodes.
runApp	App_name	Send run application command to the render nodes, the application should be loaded before.
stopApp	null	Send stop the running application command to render nodes.
init	null	Send command to render nodes to initiate the application.
frameUpdate	null	Send command to render nodes to update next frame.
display	null	Send command to render nodes to display the updated frame.
FUFinished	null	Sent by each render node, to signal frame update finished, when all of the render nodes finished, the controller will be called back to do the following task.
DFinished	null	Sent by each render node, to signal one frame display finished, when all of the render nodes finished, the controller will be called back to do the following task.

Actually, the platform will have separate processes in separate PCs, so *frameupdate* and *display* functions will be invoked by render nodes in parallel (The number of such render nodes depends on the configuration within the remote controller) which needs barriers to synchronize the rendering tasks.

Figure 4 illustrates the running flow chart of the system. For space limitation, only one render node is laid out in the figure, and other render nodes are alike. When the system starts, remote controller and all render nodes should be run. For render nodes, a TCP port should bind and listened. When all hosts are ready, the controller then connects to all the render nodes. After all the connections establish (Barrier 1, waits for all hosts connected), the controller invokes *loadApp* primitive and send app file to all render nodes. After that, the controller then invokes *runApp* and *init* primitive, all peers will invoke the *init* function exported by the application module. When all render nodes return (Barrier 2, waits for all return from *init*), the remote controller then alternatively invokes *frameUpdate* and *display* primitive. Because exactly one render node needs to update the frame data, other render nodes just ignore the frameUpdate command. During each loop, two barriers take effect, Barrier 3 waits for all render nodes returns from *Frameupdate* function, and Barrier 4 waits for all render nodes return from Display function.

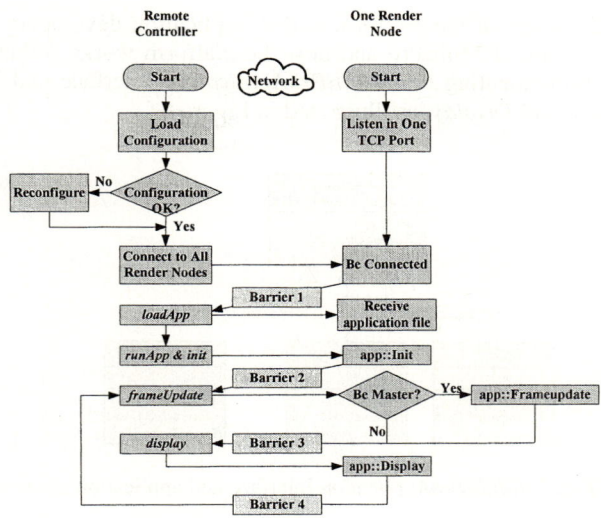

Fig. 4. Running Flow Chart

3.4 Remote Controller

The remote controller is the interface by which the system administrator can modify system configurations and load application modules. Another important role of it is sending commands to render nodes by invoking Communication Layer functions.

There exist 3 kinds of configurations. The first kind includes addresses and ports of each host, including render nodes' and the shared memory server's. The second kind includes those related to display, e.g. resolutions, color qualities, project screen positions for each host, and so on. The third kind includes those related to the platform itself, e.g. working directory, latest saved configuration file's path.

The Remote Controller is also responsible for the command sending task of the system, as explained in subsection 3.3.

3.5 Rendering Module

Rendering Module is run in each render node host which indeed accomplishes the rendering task. It gets the eyes' positions and receives the controller's commands and invokes the corresponding functions of IMultiProApplication interface.

What should be notable is that the rendering module should recalculate each eye's view frustum before each frame using functions provided by the *camera* part of MultiPro Library and invoke IMultiProApplication::Display two times per frame, one for each eye.

4 Application Development and Sample Applications

One obvious advantage of the platform is that application developers do not need to understand the details of MultiPro and how the platform works. What he/she should care of is only implementing the *IMultiProApplication* Interface and its 3 functions, *Init*, *Frameupdate* and *Display*, as illustrated in Figure 5.

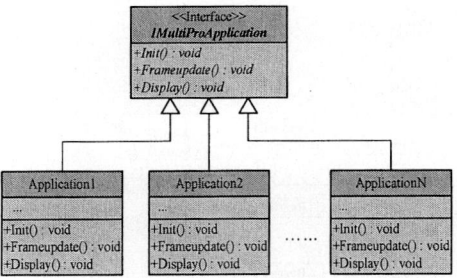

Fig. 5. *IMultiProApplication* Interface and application classes

For easy to use, example project files are made for Microsoft Visual C++ 6.0, Microsoft Visual C++ .Net and Linux QMake. Based on the platform and the library it provided, three applications were developed and will be discussed.

The first application is an empty application which does nothing in *Init*, *Frameupdate* and *Display* function. It is used to evaluate the platform's performance.

The second application is a rotary lined cube whose size is the same as CAVE device itself. It is used to exhibit the 3D effect of the active stereo display system and to verify the correct meets of screen borders. In the application's *Display* function, the cube is rotated r_angle degrees about X, Y and Z axis, and r_angle is increased by 5 degrees in the Frameupdate function.

The third application is developed for Dunhuang 285 cave by migrating from our existing system [11] developed for desktop PC. The only work to do is to copy the previous initiation and display codes to the *Init* and *Display* function.

5 Results

The platform is run under the following system environment, 4 PCs (one PC is initiated as the shared memory server, the controller, and the render node, details: Intel Pentium4 1700MHZ CPU, 1G Memory, Windows 2000 Professional with SP4, 3DLabs wildcat II 5110 display card, Intel 8255x-based PCI Ethernet Adapter), Marquee 6700 Ultra Projector, 4 sided CAVE with 10 feet by 10 feet screens,

On running the first application which does nothing except for the platform's overhead, the system gets consistent frame rate above 180 per second for each eye which seems excellent for human eyes.

On running the second application, a lined cube is correctly displayed in the screen, and it also gets the frame rate about 175 per second. When the cube's rotation angle is zero, the lines of the cube just lie in the screens (so called zero parallax), and at that frame, there will be no doubled images when taking off the glasses. Figure 6 (a) shows it.

On running the third application, the 285 cave is exhibited with the viewpoint in side it. The participant can freely tour though the cave and look at the frescos. Compared with the desktop versions, the CAVE version gives the user a full immersion, and it also gets rid of the difficulty of controlling the direction and tour path. Figure 6 (b) shows it.

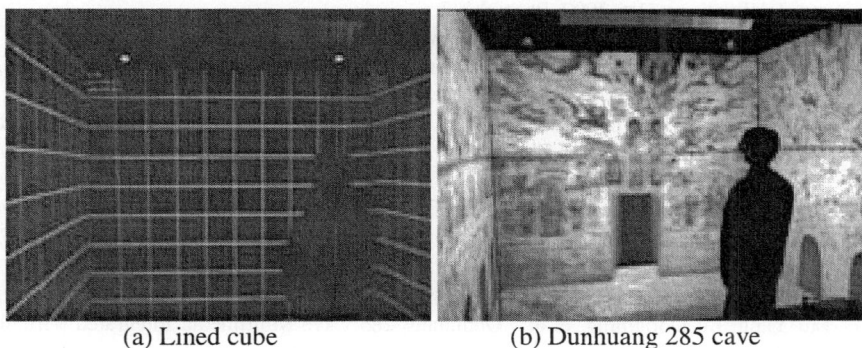

(a) Lined cube   (b) Dunhuang 285 cave

Fig. 6. Platform's Running Results

6 Conclusions and Future Work

The paper presents MultiPro, a platform for PC cluster based active stereo display system. Modular designed, the platform is composed by the following pars, controller, renderer, MultiPro Library, communication layer and application. The application developer does not need to understand the stereo display principles, that is, he or she just needs to write few codes on initiation, frame update, and display events. The design of the platform greatly reduces the burden of application developers. Based on the Qt library, the platform can be compiled and run at most of dominating operating systems. The network operations do bring out some system overhead, however, the result shows that it is acceptable.

For future work, a more transparent API architecture should be provided to developers, for example, distributed file access, wand and data glove interaction, scene management support, and so on. Second, though configurable, the platform should be tested in more stereo display devices and systems. Third, as stereo panorama is widely studied and, stereo panorama support should be added into the platform, which can be used to display large-scale outer scenes.

References

1. Howard Rheingold. *Virtual Reality*. Simon and Schuster Trade, New York (1991)
2. Carolina Cruz-Neira, Daniel J. Sandin and Thomas A. DeFanti. "Surround-Screen Projection-Based Virtual Reality: The Design and Implementation of the CAVE". *Proceedings of ACM SIGGRAPH*, August 1993, Page(s) 135-142.
3. Wolfgang Krueger and Bernd Fröhlich. "The responsive workbench". *IEEE Computer Graphics and Applciations*, 14(3), IEEE Computer Society Press, Los Alamitos, CA, USA, May 1994, Page(s) 12-15.
4. Maneesh Agrawala, Andrew C. Beers, et al. "The Two-User Responsive Workbench: Support for Collaboration Through Individual Views of a Shared Space". *Proceedings of ACM SIGGRAPH*, August 1997, Page(s) 327-332.
5. Andreas Simon and Martin Göbel. "The i-Cone™ A Panoramic Display System for Virtual Environments". *Proceedings of the 10th Pacific Conference on Computer Graphics and Applications*, Oct. 2002, Page(s) 3-7.
6. CAVE Library. VRCO Website. http://www.vrco.com/products/cavelib/cavelib.html
7. Henrik Tramberend. "Avocado: A Distributed Virtual Reality Framework". *Proceedings of IEEE Virtual Reality*, March 1999, Page(s) 14-17.
8. Jeremie Allard, Valérie Gouranton, et al. "Net Juggler: Running VR Juggler with Multiple Displays on a Commodity Component Cluster". *Proceedings of IEEE Virtual Reality*, March 2002, Page(s) 273-274.
9. YANG Jian, SHI Jiao-Ying, et al. "PCCAVE: Non-Expensive CAVE System Based on Networked PCS". *Journal of Computer Research & Development*. 38 (5), May 2001, Page(s) 513-518.
10. Trolltechm Web site. http://www.trolltech.com
11. Liu Yang, Lu Dongming, et al. "Dunhuang 285 Cave Multimedia Integrated Virtual Exhibit". *Journal of Computer-Aided Design & Computer Graphics*. 16 (11), Nov. 2004, Page(s) 1528-1534.

Two-Level 2D Projection Maps Based Horizontal Collision Detection Scheme for Avatar in Collaborative Virtual Environment[*]

Yu Chunyan[1, 2], Ye Dongyi[1], Wu Minghui[3], and Pan Yunhe[2]

[1] College of Mathematics and Computer Science,
Fuzhou University, Fuzhou, 350002, P.R.C
[2] Computer College, Zhejiang University, Hangzhou, 310027, P.R.C
[3] Department of Computer, Zhejiang University City College,
Hangzhou, 310015, P.R. C

Abstract. Collision-free locomotion of avatar is premise of its automatic constant navigation in CVE. This paper discussed how to design a horizontal collision detection scheme to prevent avatar from penetrating into all other virtual entities without stalling to achieve collision-free motion. First, it proposed several design considerations for a horizontal collision detection scheme. Second, it attached collision precaution and collision repulsive field to obstacles and divided process procedure of collision detection into three processes: collision determination, collision avoidance and path amendment. Furthermore, it presented a new collision detection scheme *PMBHCD* based on two level 2D projection maps, *Free Space Map* and *Detail Projection Map*. *PMBHCD* is discussed in three situations, applied different approaches, according to three types of obstacles: static obstacle, moving obstacle and other avatars. Finally, an implementation is described in details.

1 Introduction

An important feature of CVE is locomotion of virtual avatars, the facility for user to move through in a natural and easily controlled manner. Natural locomotion methods, involving using some input devices to control walk-through or fly-through motion [1,6], can contribute to a sense of self-mutual awareness. Locomotion related problems could be divided into three levels: path planning based on a map of virtual environment and an accessibility graph is at the highest level; and moving around obstacles when moving to an intermediated destination is at a higher level; the lowest but most essence level is collision detection [7], comparing avatar's current velocity with distance of obstacles in its path to avoid collision during next time step. And, constraints are extremely important component of locomotion to guarantee sense of reality [2,8]. A typical constraint is horizontal collision detection constraint keeping avatars from going through virtual objects such as walls directly.

---

[*] This project is supported by Fujian Technology Foundation No. K04005.

Darken [3,4] has researched about locomotion issues involving evaluating the use of maps, breadcrumbs and landmarks as tools in 3D VEs. The results suggest that users of large-scale virtual worlds require structure and augmentations such as maps and path restriction to improve collision detection greatly. Xiao [5] presented a new technique to control an avatar's locomotion in VE. It introduced artificial force fields, which acts upon user's virtual body such that it is guided around obstacles, rather than penetrating or colliding with them. Cohen [12] described efficient techniques for a case of collision detection between participant's avatar and the scene.

Focusing on the lowest level -- horizontal collision detection method in CVEs, one of the major problems for avatar locomotion, we proposes a new two-level Projection 2D Maps Based Horizontal Collision Detection (*PMBHCD*) approach, which prevents avatars from going through virtual objects and provides a simple and natural collision detection to generate collision-free motion. It is composed of three steps: Collision Determination, Collision Avoidance and Path Adjustment. The first step determinates whether a potential collision situation with virtual entity exists. The second step avoids an actual collision situation with a virtual object by systematically supporting avatar to by-pass obstacles without stalling. The last step supports avatar returning to its original locomotion direction after Collision Avoidance.

2   Overview of *PMBHCD*

2.1   Design Considerations

We mainly take following issues into consideration for a collision detection scheme.

- Real-time and Early Detection. CVE is a real-time system with collision occurring by accident. From the point of overall system, collision is a kind of non-determinate activities and cannot be eliminated in advance. From the point of an individual avatar, collision can be avoided at a certain time. Hence, a real-time collision detection scheme should be provided for each avatar to detect and recognize a collision as early as possible to avoid collision.
- User Transparency. The overall collision detection process should be transparent to each user. First, no users' intervention is required. Second, avatar should keep its movement instead of being stalled, and its path and velocity can be altered temporal to avoid a collision while its navigation destination shouldn't be altered.
- Naturalness. Collision detection should accord to real life to improve senses of reality and immersion; for instance, it should modify path as natural as possible.
- Efficient and complete. All collision should be absolutely avoided. Careful examining is time-consuming while real-time and early detection imply that no too much time is offered. Hence, detection scheme should be as simple as possible with small data space required and fast calculating to save time.
- Prevent user from spatial loss. If a user cannot point out own location, it is called spatial loss. The user will be lost after a series of collision detection with avatar's navigation velocity altered and cannot reach original destination at all. Therefore, collision detection should prevent user from spatial loss through amending avatar's path after collision avoidance process to guarantee spatial awareness.

2.2 Two-Level 2D Projection Map Based Collision Detection Scheme

Our *PMBHCD* approach provides a natural collision detection and avoidance method in CVE, closer to how we walk in the real world. There are three major processes: collision determination, collision avoidance and path amendment.

When an avatar locomotes with a *Free Space Map*, it is collision determination process's responsibility to monitor whether a situation with high collision probability happens. Generally, it calculates distance between avatar and obstacles to estimate collision probability and generates a collision avoidance event if a potential collision situation is detected. An avatar catching a collision avoidance event starts collision avoidance process, altering moving behavior to generate avoidance paths to avoid a potential collision. Original locomotion path are alerted for interposition of avoidance paths. The avatar will be disorientation and spatial loss without amendment. Hence, path amendment is called to guarantee it keeping its original locomotion through coming back to its original path with original orientation, speed and destination. Both collision avoidance and path amendment are based on *Detail Projection Map*.

Free Space Map and *Detail Projection Map* [9-11] is the basis of *PMBHCD*. The former, first level 2D projection map, is used to monitor and indicate moving track and current position of an avatar. It is the default assistant map of collision detection, ignoring all moving virtual objects. A representation of free regions in which an avatar can move is provided and a black dot is periodically displayed to indicate current position of avatar to improve spatial awareness; another black dot is used to designate locomotion destination. Free regions in it are just reachable but not available at a give time t to an avatar with a view to collision prohibition. The latter, the second-level map, loaded from a special projection map server real-time to display comprehensive surroundings of an avatar at time t when a potential collision is detected, is a composite map representing both free regions and obstacles involving static and moving virtual entities. It is a 2D cell projection map on the ground cell based on uniform cell grid. We define collision as a state that avatar contacts a bounding box of obstacle. Hence, all obstacles' projections are their bounding boxes' projections; in addition, moving obstacle's projection is snapshot at time t. Each cell grid records entity corresponding information involving types, velocity and so on.

3 Design of *PMBHCD*

3.1 Collision Detection with Static Obstacle

Collision detection with static obstacle is the simplest situation for obstacle's position is unaltered. The basic idea is changing direction while keeping speed value when necessary. Each obstacle O has two fields: collision precaution field P and collision repulsive field R which are two bounding boxes of O. And P contains R. We adopt sphere as the bounding box; thus, p_O and r_O are radiuses of two fields' projections and satisfy $r_O < p_O$ for each obstacle. A collision occurs once avatar breaks through repulsive field boundary; thus, avatar will be stalled to prevent from colliding with obstacle. Collision precaution field is to avoid above situation. We assume that there exists collision probability if an avatar enters precaution field. And a collision precaution event will be generated and passed to avatar to deal with. Hence, precaution field is to achieve early detection while repulsive field to realize complete.

Once the avatar catches a collision precaution event, its collision detection process starts. It loads a real-time *Detail Projection Map* in its visual field from a special server to replace original *Free Space Map*. For remarks, avatar has its own precaution field r_A; to simplify problem and reduce calculation complexity, obstacle's repulsive field is revised rather than original, that is, $r=r_A+r_O$. All following repulsive fields are revised repulsive fields. Calculate intersection points with repulsive field based on locomotion direction of avatar. If there are two or more intersection points, a collision avoidance event is generated to call back collision avoidance; and records intersection points P_3 and P_4. Otherwise, the avatar just attempts to cross through obstacle's precaution field without collision. Once collision avoidance process calls a collision avoidance event, *Distance (P_3, P_4)* is calculated. If no grid on *Line(P_3, P_4)* is higher than *Avatar.leglength* and *Avatar.step>Distance(P_3,P_4)*, the avatar step over obstacle and move from P_3 to P_4 directly. Otherwise, the avatar has to walk around obstacle. Then, calculate the intersection point P_2 of tangent line from avatar's current position P_0 with repulsive field boundary. Thus, modificative movement generated for avatar to avoid collision is to take P_2 as temporary destination and tangent as temporary direction θ_2. Grid P_1 records original velocity $\overrightarrow{Velocity}$ with direction θ_0, and path adjustment event is generated at P_2. (All direction in this paper is $\in \left[-\frac{\pi}{2}, \frac{3}{2}\pi\right)$.)

$$\theta_2 = \begin{cases} arctg(\frac{y_{P_2}-y_{P_0}}{x_{P_2}-x_{P_0}}) & x_{P_2}-x_{P_0} > 0 \\ arctg(\frac{y_{P_2}-y_{P_0}}{x_{P_2}-x_{P_0}})+\pi & x_{P_2}-x_{P_0} < 0 \\ \frac{\pi}{2} & x_{P_2}-x_{P_0} = 0 \,\&\, y_{P_2}-y_{P_0} > 0 \\ -\frac{\pi}{2} & x_{P_2}-x_{P_0} = 0 \,\&\, y_{P_2}-y_{P_0} < 0 \end{cases} \quad (1)$$

Once path amendment process catches a path adjustment event, it calculates path amend direction θ_3, and records amend velocity $\overrightarrow{Velocity_1}$ at P_2. Once avatar arrives at P_2, its velocity is revised to $\overrightarrow{Velocity_1}$ until it arrives at P_1; then, its velocity will revert to original $\overrightarrow{Velocity}$. The direction of $\overrightarrow{Velocity_1}$ is θ_3 and $|\overrightarrow{Velocity_1}|=|\overrightarrow{Velocity}|$ in which

$$\theta_3 = \begin{cases} arctg(\frac{y_{P_1}-y_{P_2}}{x_{P_1}-x_{P_2}}) & x_{P_1}-x_{P_2} > 0 \\ arctg(\frac{y_{P_1}-y_{P_2}}{x_{P_1}-x_{P_2}})+\pi & x_{P_1}-x_{P_2} < 0 \\ \frac{\pi}{2} & x_{P_1}-x_{P_2} = 0 \,\&\, y_{P_1}-y_{P_2} > 0 \\ -\frac{\pi}{2} & x_{P_1}-x_{P_2} = 0 \,\&\, y_{P_1}-y_{P_2} < 0 \end{cases} \quad (2)$$

Free Space Map is reloaded to replace *Detail Projection Map* after *PMBHCD* for collision detection with static obstacle finishes. During time interval from the moment that obstacle detects a potential collision to the moment modificative movement generated, avatar keeps its original moving direction and path to guarantee efficiency requirement. Figure 1 shows overall process of collision detection with static obstacle.

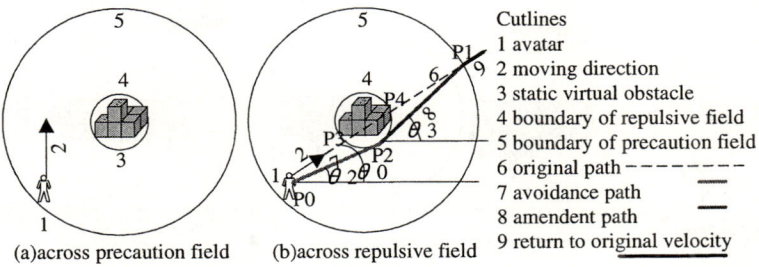

Fig. 1. Collision detection with static obstacle

3.2 Collision Detection with Moving Obstacle

Moving obstacle has velocity $\overrightarrow{D\_Velocity}$ making collision detection more complex.

Similar to static obstacle, moving obstacle has two fields: precaution and repulsive field. A collision precaution event is generated when avatar enters precaution field and obstacle's information passes to avatar includes moving direction and speed. The basic idea is same as collision detection with static obstacle that the speed value $|\overrightarrow{Velocity}|$ is holding and actual amend direction θ_2 and path amend direction θ_3 have to be calculated based on avatar's relative velocity to moving obstacle, relative amend direction R_θ_2 and path amend direction R_θ_3. However, a fatal shortage is complex calculations and frequent transform operations that are pretty time-consuming and violate real-time and efficient requirements. Hence, we adopt another solution: alter avatar's speed value to avoid collision rather than its moving direction.

In collision detecting process, *Detail Projection Map* loaded is a snapshot of avatar's visual field recording location of avatar at P_0 and location of moving obstacle at P_5 and its velocity vector $\overrightarrow{D\_Velocity}$ as well. At first, avatar's relative velocity to moving obstacle is calculated as $\overrightarrow{R\_Velocity} = \overrightarrow{Velocity} - \overrightarrow{D\_Velocity}$; then determination method in subsection 4.1 is used to determinate whether a potential collision exists. If it exists, a collision avoidance event occurs and calls collision avoidance process.

In collision avoidance, stepping over the obstacle is discarded. The process first compares avatars moving direction to that of moving obstacle and leads to three solutions: (a) $0 < \langle \overrightarrow{D\_Velocity}, \overrightarrow{Velocity} \rangle < \pi$ (b) $\langle \overrightarrow{D\_Velocity}, \overrightarrow{Velocity} \rangle = \pi$ (c) $\langle \overrightarrow{D\_Velocity}, \overrightarrow{Velocity} \rangle = 0$.

In situation (a), it is clear that avatar and moving obstacle move neither at the same direction nor at the opposite direction. Hence, collision must occur around the intersection point of their original moving track. For remark, collision is defined as contact between avatar and any part of repulsive field; so, avatar has to be outside of repulsive field around the intersection point when the obstacle is at the intersection point. In fact, we can find from figure 3 that repulsive field comes into being a zonal area called collision repulsive zone hereafter (it is shadowed in the figure 3). The best choose of avatar is to be far from this repulsive zone: if avatar is in this repulsive zone currently, it should leave this zone at shortest path as soon as possible; otherwise, we alter the speed value of avatar to postpone the enter time We discuss two situations separately. First, the intersection point P_6 is found out. There are two possibilities:

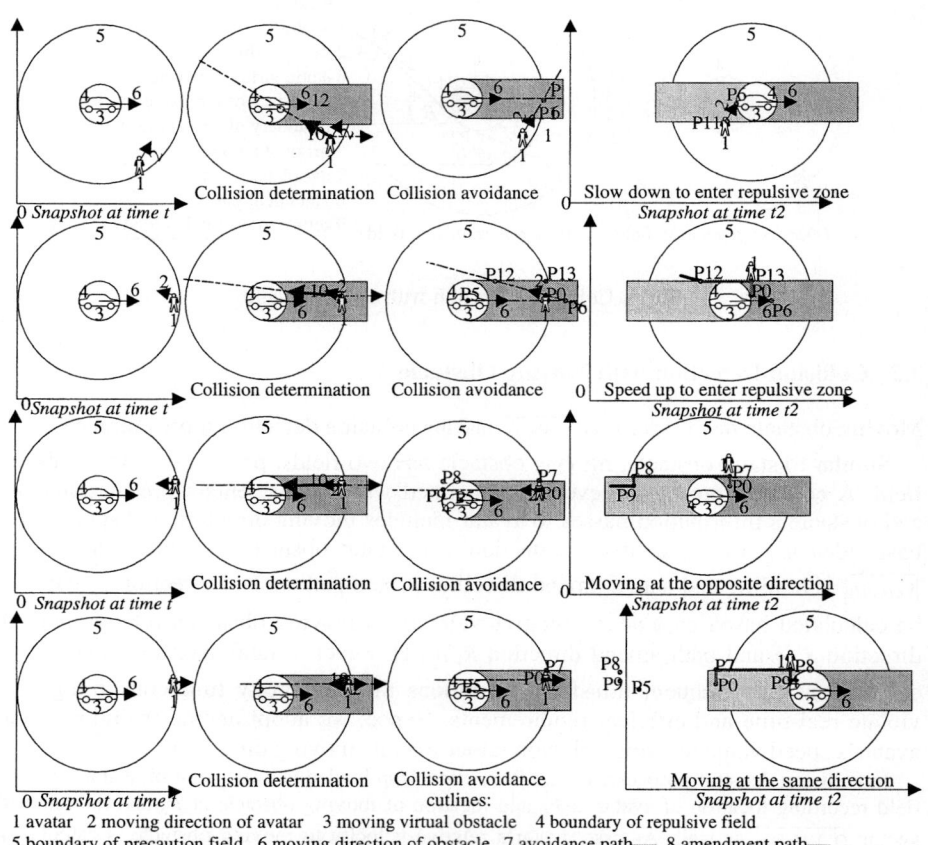

1 avatar 2 moving direction of avatar 3 moving virtual obstacle 4 boundary of repulsive field
5 boundary of precaution field 6 moving direction of obstacle 7 avoidance path━ 8 amendment path━
9 return to original velocity━ 10 relative moving direction 12 current repulsive zone▨ 13 original repulsive zone▨

Fig. 2. Collision detection with moving obstacle

(1) $Distance(P_0, P_5)\sin(\overrightarrow{P_5P_0}, \overrightarrow{D\_Velocity}) \geq r$. Avatar has to postpone or advance its entering time. Compare speed values of avatar and moving obstacle. If $|\overrightarrow{Velocity}| < |\overrightarrow{D\_Velocity}|$, avatar lows velocity to $\lambda_1 \overrightarrow{Velocity}$ to postpone arriving time when it arrives at P_6. Otherwise, avatar speeds up to $\lambda_2 \overrightarrow{Velocity}$ to advance arriving time when it arrives at P_6. At the same time, P_{11} is marked and records avatar's current velocity for the former and P_{10} for the latter. For the former, avatar returns to original velocity after it passed P_{11} while P_{10} for the latter.

$$\lambda_1 = \frac{Distance(P_6,P_5)(Distance(P_6,P_0)-r)}{Distance(P_6,P_0)(Distance(P_6,P_5)+r)}. \tag{3}$$

$$\lambda_2 = 1 + \frac{r}{Distance(P_6,P_0)}. \tag{4}$$

$$\begin{aligned} x_{P_{10}} &= x_{P_6} + r \times \cos\theta_0 \\ y_{P_{10}} &= y_{P_6} + r \times \sin\theta_0 \end{aligned}. \tag{5}$$

$$\begin{aligned} x_{P_{10}} &= x_{P_6} - r \times \cos\theta_0 \\ y_{P_{10}} &= y_{P_6} - r \times \sin\theta_0 \end{aligned}. \tag{6}$$

(2) $Distance(P_0,P_5)\sin(\overrightarrow{P_5P_0},\overrightarrow{D\_Velocity}) < r$. The avatar is in repulsive zone currently, and should leave before the obstacle arrives. Both moving direction and speed value should be altered. First, avatar moves at direction of perpendicular line of moving obstacle until distance between avatar and the line of moving direction of obstacle is equal to r; then, it moves at parallel direction to moving direction of moving obstacle. Detail indicants and calculating are as following: avoidance amend direction is θ_2, speed amend factor is λ_3, ending is P_{13}, path amend direction is θ_3, ending is P_{12}, speed value is the original one. Avatar returns to its original velocity vector after it passed cell grid P_{12}. The following are calculating formulas in details (moving direction of moving obstacle is D_θ_0):

$$\theta_7 = \begin{cases} D\_\theta_0 + \frac{\pi}{2} & D\_\theta_0 \in [-\frac{\pi}{2},\pi) \\ D\_\theta_0 - \frac{3}{2}\pi & D\_\theta_0 \in [\pi,\frac{3}{2}\pi) \end{cases}. \tag{7}$$

$$\theta_8 = \begin{cases} D\_\theta_0 + \frac{3}{2}\pi & D\_\theta_0 \in [-\frac{\pi}{2},0) \\ D\_\theta_0 + -\frac{\pi}{2} & D\_\theta_0 \in [0,\frac{3}{2}\pi) \end{cases}. \tag{8}$$

$$\theta_2 = \{\theta \mid \theta \in \{\theta_7,\theta_8\} \hbar\, (\langle\theta,\theta_0\rangle \leq \langle\theta_7,\theta_0\rangle) \hbar\, (\langle\theta,\theta_0\rangle \leq \langle\theta_8,\theta_0\rangle)\}. \tag{9}$$

$$\begin{aligned} x_{P_{12}} &= x_{P_6} + \frac{r}{\sin\langle\overrightarrow{Velocity},\overrightarrow{D\_Velocity}\rangle} \times \cos\theta_0 \\ y_{P_{10}} &= y_{P_6} + \frac{r}{\sin\langle\overrightarrow{Velocity},\overrightarrow{D\_Velocity}\rangle} \times \sin\theta_0 \end{aligned}. \tag{10}$$

$$\begin{aligned} x_{P_{13}} &= x_{P_0} + Distance(P_0,P_{12}) \times \sin(\langle\overrightarrow{Velocity},\overrightarrow{D\_Velocity}\rangle) \times \cos\theta_2 \\ y_{P_{13}} &= y_{P_0} + Distance(P_0,P_{12}) \times \sin(\langle\overrightarrow{Velocity},\overrightarrow{D\_Velocity}\rangle) \times \sin\theta_2 \end{aligned}. \tag{11}$$

$$\theta_3 = \begin{cases} arctg(\frac{y_{P_{12}}-y_{P_{13}}}{x_{P_{12}}-x_{P_{13}}}) & x_{P_{12}}-x_{P_{13}} > 0 \\ arctg(\frac{y_{P_{12}}-y_{P_{13}}}{x_{P_{12}}-x_{P_{13}}})+\pi & x_{P_{12}}-x_{P_{13}} < 0 \\ \frac{\pi}{2} & x_{P_{12}}-x_{P_{13}} = 0\ \&\ y_{P_{12}}-y_{P_{13}} > 0 \\ -\frac{\pi}{2} & x_{P_{12}}-x_{P_{13}} = 0\ \&\ y_{P_{12}}-y_{P_{13}} < 0 \end{cases}. \tag{12}$$

$$\lambda_3 = \max(\frac{Distance(P_0, P_{13})|\overrightarrow{D\_Velocity}|}{(Distance(P_6, P_5) + Distance(P_0, P_6))\cos(\langle\overrightarrow{P_0P_6}, \overrightarrow{Velocity}\rangle)|\overrightarrow{Velocity}|}, 1) \quad (13)$$

In fact, situations (b) and (c) belong to second type of (a). However, there is an intersection point P_6 in (a) while moving directions of avatar and obstacle are parallel in (b) and (c).

Situation (b) implies that avatar and obstacle are moving at opposite direction. The collision cannot be avoided through altering speed value merely. The avatar has to shift left or right to avoid collision. It is necessary to determine relative position relation F between obstacle and original path of avatar, that is, determine whether obstacle is at the right or left side of original path of avatar: calculate direction θ_4 of vector $\overrightarrow{P_0P_5}$, and compare it with direction θ_0 of avatar's velocity $\overrightarrow{Velocity}$ to determine value of F. If $F=LEFT$, avatar shifts right; otherwise, avatar shifts left and *Offset* gives out the offset, and avoidance amend direction is θ_2, ending is P_7. In order to guarantee validity, we use the approach of (a) to check the state after revision.

$$F = \begin{cases} LEFT & (\theta_4 \in [0, \frac{3}{2}\pi) \& \theta_4 - \theta_0 \geq 0) or (\theta_4 \in [-\frac{\pi}{2}, 0) \& \theta_4 - \theta_0 \leq 0) \\ RIGHT & others \end{cases}. \quad (14)$$

$$Offset = r - |\overrightarrow{P_0P_5}|\sin(|\theta_4 - \theta_0|). \quad (15)$$

$$\theta_2 = \begin{cases} \theta_0 - \frac{\pi}{2} & \theta_0 \in [0, \frac{3}{2}\pi) \& F = LEFT \\ \theta_0 + \frac{3}{2}\pi & \theta_0 \in [-\frac{\pi}{2}, 0) \& F = LEFT \\ \theta_0 + \frac{\pi}{2} & \theta_0 \in [-\frac{\pi}{2}, \pi) \& F = RIGHT \\ \theta_0 - \frac{3}{2}\pi & \theta_0 \in [\pi, \frac{3}{2}\pi) \& F = RIGHT \end{cases}. \quad (16)$$

$$\begin{array}{l} x_{P_7} = x_{P_0} + Offset \times \cos\theta_2 \\ y_{P_7} = y_{P_0} + Offset \times \sin\theta_2 \end{array}. \quad (17)$$

When avatar arrives at P_7, offset movement is over, and avatar returns to it original $\overrightarrow{Velocity}$ until it reaches P_8 and switches to path amendment process. Avatar needs to make another offset movement with the same offset while at the opposite direction until path amendment process finishes at cell grid P_9. The path adjustment direction θ_3 and P_8, P_9 are as follows:

$$\theta_3 = \begin{cases} \theta_2 - \frac{\pi}{2} & \theta_2 \in [0, \frac{3}{2}\pi) \& F = RIGHT \\ \theta_2 + \frac{3}{2}\pi & \theta_2 \in [-\frac{\pi}{2}, 0) \& F = RIGHT \\ \theta_2 + \frac{\pi}{2} & \theta_2 \in [-\frac{\pi}{2}, \pi) \& F = LEFT \\ \theta_2 - \frac{3}{2}\pi & \theta_2 \in [\pi, \frac{3}{2}\pi) \& F = LEFT \end{cases}. \quad (18)$$

$$\begin{array}{l} x_{P_8} = x_{P_5} + r \times \cos\theta_2 + r \times \cos\theta_0 \\ y_{P_8} = y_{P_5} + r \times \sin\theta_2 + r \times \sin\theta_0 \end{array}. \quad (19)$$

$$\begin{array}{l} x_{P_9} = x_{P_8} + Offset \times \cos\theta_3 \\ y_{P_9} = y_{P_8} + Offset \times \sin\theta_3 \end{array}. \quad (20)$$

Situation (c) implies that avatar and obstacle are moving at same direction. The basic idea is similar to that of situation (b) and only difference is calculating of P_8 (if $|Velocity| > |D\_Velocity|$, use formula (21); if $|Velocity| < |D\_Velocity|$, use formula (22)) (Condition $|Velocity| = |D\_Velocity|$ can not bring a collision at this situation):

$$x_{P_8} = x_{P_7} + \frac{(Distance(P_0,P_5) + r)|Velocity| + 2r|D\_Velocity|}{|Velocity| - |D\_Velocity|} \cos\theta_0 .$$
$$y_{P_8} = y_{P_7} + \frac{(Distance(P_0,P_5) + r)|Velocity| + 2r|D\_Velocity|}{|Velocity| - |D\_Velocity|} \sin\theta_0 \quad (21)$$

$$x_{P_8} = x_{P_7} + max(\frac{r|Velocity|}{|D\_Velocity| - |Velocity|}, \frac{(Distance(P_0,P_5) + r)|Velocity| - r|D\_Velocity|}{|D\_Velocity| - |Velocity|})\cos\theta_0 .$$
$$y_{P_8} = y_{P_7} + max(\frac{r|Velocity|}{|D\_Velocity| - |Velocity|}, \frac{(Distance(P_0,P_5) + r)|Velocity| - r|D\_Velocity|}{|D\_Velocity| - |Velocity|})\sin\theta_0 \quad (22)$$

4 Implementation of *PMBHCD*

As shown in figure 4, our proposed *PMBHCD* is implement in Java and interacts with the CVE scenes in VRML through EAI. Web server is responsible for updating data with client while projection map server is responsible for abstract and generate the two-level 2D projection maps for the overall CVE in real-time. Generally, a user loads a initial free space map from projection map server when he login the CVE and loads a initial detail projection map from projection map server when the client start its collision detection procedure.

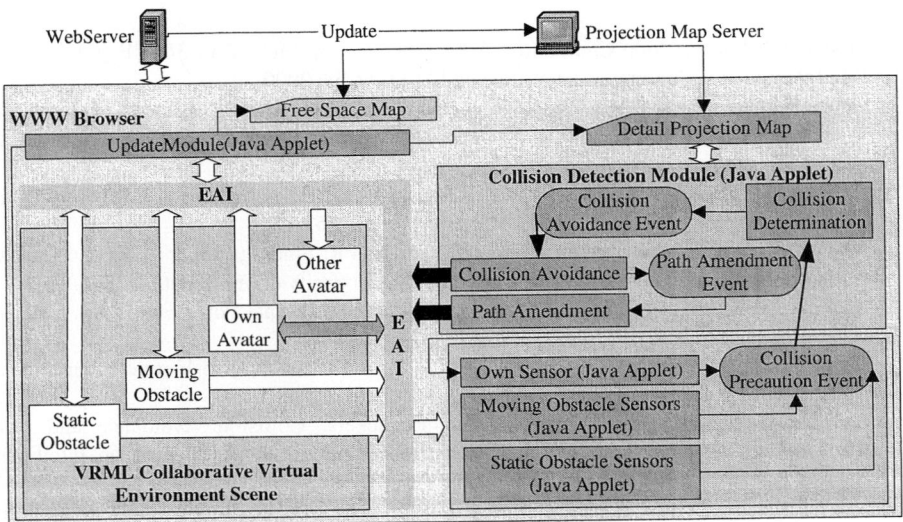

Fig. 3. Collision detection architecture

In our implementation, all collision sensors for all obstacle and avatar itself are implemented at client in Java Applet. We insist that this mode can improve distribution and real-time. More important, this mode permits set more reasonable collision precaution field and collision repulsive field according to the avatar's individual navigation speed.

5 Conclusion and Future Works

As the foundation of collision-free automatic navigation, collision detection is used to prevent avatar pass through other virtual entities directly to improve sense of reality, presence and immersion. Based on the three phases of collision detection and several design considerations, we propose a new two-level 2D projection map based collision detection scheme *PMBHCD*. This scheme introduced two field for each virtual entities: collision precaution field used to trigger a collision precaution event and guarantees prediction and collision repulsive field acts as the base of collision avoidance and path amendment. At current, the premise of our *PMBHCD* scheme is that both avatar and obstacle are moving with a constant velocity at a line, and this scheme pays no attention on collision cascade. Based on our *PMBHCD*, we will attempt to expand our research domain to collision detection with changing speed and curvilinear motion, collision cascade and cooperation between collision detection and path planning. Furthermore, we will study relations among collision precaution field and collision repulsive field and avatar's navigation speed to set more proper precaution field and collision field.

References

1. Tae-Wook Kwon, Yoon-Chul Choy. A New Navigation Method in 3D VE [C]. Proc. of 6th International Conference on Virtual and Multimedia, 2000.
2. Igor Pandzic, Tolga Capin, Nadia Magnenat-Thalmann et. al. A Versatile Navigation Interface for Virtual Humans in Collaborative Virtual Environments [C]. Proceedings VRST 97, Lausanne, Switzerland, ACM, 1997: 45-49.
3. Darken R., Sibert J. A Toolset for Navigation in Virtual Environment [C]. Proceedings of ACM User Interface Software and Technology, 1993: 157-165.
4. Darken R., Sibert J. Navigation and Wayfinding in VR: Finding Proper Tools and Cues to Enhance Navigation Awareness. Master's thesis, University of Washington, 1995.
5. Dongbo Xiao, Roger Hubbold. Navigation Guided by Artificial Force Fields [C]. CHI'98, 1998: 18-23.
6. Roger Young. Design a Group Navigation Tool for a Collaborative Virtual Environment. Master's thesis, University of Nottingham, 1996.
7. Patrick Monsieurs, Karin Coninx, Eddy Flerackers. Collision Avoidance and Map Construction Using Synthetic Vision [C]. Proceedings of the Workshop on Intelligent Virtual Agents VA'99 (Salford, September 1999), 33-46.
8. Al. Wilson, E. Larsen, D.Manacha et. al. Partitioning and Handling Massive Models for Interactive Collision Detection [C]. Proceedings of Eurographics 1999.

9. Bandi, S. and Thalmann, D. Space Discretization for Efficient Human Navigation [J]. Computer Graphics Forum, 1998, 17(3): 195-206.
10. Jed Lengyel et al., Real-Time Robot Motion Planning Using Rasterizing Computer Graphics Hardware [J]. Computer Graphics Proceedings, Annual Conference Series (August, 1990), ACM SIGGRAPH, 327-335.
11. Xu Wei-wei, Pan Zhi-geng, Zhang Ming-min. Decision Model in Intelligent Virtual Environment [J]. Journal of Image and Graphics, 2001, 6(A)(5): 496-501.
12. J. Cohen, M. Lin, D. Manacha et al. I-Collied: An Interactive and Exact Collision Detection System for Large-scale Environment [C]. Proceedings of 1995 Symposium on Interactive 3D graphics, Montrey, ACM press, 9-12.

A Molecular Modeling System Based on Dynamic Gestures

Sungjun Park[1], Jun Lee[1], and Jee-In Kim[2,*]

[1] Computer Science & Engineering, Konkuk University, Seoul, Korea
[2] Internet & Multimedia, Konkuk University, Seoul, Korea
jnkm@konkuk.ac.kr

Abstract. We propose a molecular modeling system which visualizes three dimensional models of molecules, presents them using a large stereoscopic display and allows scientists observe and manipulate the molecular models using their gestures using hands and arms. The system consists of a three dimensional stereoscopic display, data gloves, and motion tracking devices. Scientists can examine, magnify, translate, rotate, combine and split the molecular models in natural and convenient ways using gestures. The proposed system is based on "Dynamic Gestures". It means that the system utilizes direct manipulation and reflects real-time behaviors of users with two hands. The system also allows us to observe and manipulate not only the whole structure of molecular models but also specific regions of molecular models in detail. With the proposed system, we can rotate two chemically bonded molecular models in opposite directions simultaneously using two hands. On the other hand, typical input devices such as a mouse cannot allow us to perform and manipulate such operations simultaneously and effectively. In order to simulate bonded molecular models precisely and in real-time using the dynamic gestured-based system, we implement an energy minimization algorithm and suggest a new data structure for showing the three dimension molecular models.

1 Introduction

Molecular modeling includes analyses of three dimensional structures of molecules. One dimensional character strings of molecular structures are translated into three dimensional structures of molecules. In developing a new drug, the molecular modeling is used to design a practicable candidate molecule. The process conducts a docking simulation where a receptor is combined with a ligand at a specific position called as an active site. During the simulation, we may be required to rotate the molecular models simultaneously. It is important to make the structures of molecular models stable. The simulation is required to do so, because it can computationally prove or disprove if such a chemical operation is possible. The simulation is basically calculating energy minimization equations.

There have been many researches and tools for molecular modeling. Most of them focus on visualizing structures of molecules in three dimensions. Molecular modeling

---

* Corresponding author.

procedures require scientists to examine and manipulate three dimensional models of molecules. During a docking process, the three dimensional structures of a receptor and a ligand are visually examined by scientists. Also, many molecular models should be processed. For each case, the distances between the models must be measured. Since three dimensional structures of most molecular models look quite similar, it is very difficult for scientists to differentiate the structures using views projected on conventional two dimensional monitors. As for their input devices, most tools provide a mouse and a keyboard. However, it is not easy to complete molecular modeling procedures using such devices. Molecular modeling operations are essential when we exercise simulations of energy minimization in order to verify the stability of the result of the docking procedure. The operations are required to compute parameters such as the distance between molecular models, the angles of rotations of the models for docking. Therefore, more natural input methods are required. Just a mouse and a keyboard are not sufficient enough.

We propose a molecular modeling system in this paper. The system adopts a large, stereoscopic display device. The stereoscopic views are more realistic and helpful for scientists to understand and manipulate three dimensional structures of molecules. The system provides data gloves and motion tracking devices rather than a mouse and a keyboard. So, scientists can use their two hands simultaneously in examining and manipulating molecular models. The operations include translation, rotation, zoom-in and out, selection, separation, combination, etc. They are used in assembling and disassembling procedures, and docking procedures. It is expected that scientists would feel more natural and comfortable with the dynamic gestures than a mouse and a keyboard. Therefore, the molecular modeling procedure becomes easier and more productive.

2 Related Works

RASMOL[1], QMOL[2] and VMD[3] are molecular modeling tools and they visualize three dimensional structures of molecules. RASMOL and QMOL are widely used, because they provide fast and simple ways of examining three dimensional structures of molecules. VMD supports stereoscopic views so that scientists can utilize polarized glasses or HMD (Head-Mounted Display) in order to examine three dimensional structures of molecules. Accelrys developed Insight II[4], a tool for molecular modeling. It is the most popular tool and used in the fields of biology, new drug design, etc.

Those tools are based on conventional input and output devices such as a two dimensional monitor, a mouse, and a keyboard. For example, Insight II users perform molecular modeling by examining a front view and a side view, because the tool does not provide "real" three dimensional views. It is not easy to find active sites, where binding a receptor and ligand can occur, with the two dimensional views of the front and the side. Though scientists could examine the front and the side views, it is not intuitive to have the depth information for docking procedures from the visualized models. Also, it is impossible for mouse users to rotate two molecular models simultaneously. GROPE[5] is a molecular modeling system using a haptic device. Though GROPE can improve perception and find docking sites easily, it also has some drawbacks. The system consists of expensive equipments. It is inconvenient and complicated for scientists to manipulate GROPE.

3 System Overview

The proposed system is called as VRMMS(VR-based Molecular Modeling System). It consists of five components: File Manager, Operation Manager, Rendering Engine, Computing Engine and Sensor Manager. Information of molecules is stored in PDB(Protein Data Bank) files. *File Manager* reads data from PDB files and exercises parsing the data. *Operation Manager* arranges the parsed data in order to compute energy equations. The results are arranged to be properly displayed by *Rendering Engine*. *Sensor Manager* handles input signals from sensing devices such as a mouse, a keyboard, data gloves, etc. Rendering Engine visualizes[6] three dimensional models of molecules. *Computing Engine* computes energy equations which are essential in the simulation.

In addition to visualization, VRMMS supports various functions such as docking procedures for molecules using multiple loading, non-bonding atoms and amino acids from molecules, bonding molecules based on peptide bonds, etc. Figure 1-(a) shows a docking procedure by multiple loading molecules. An amino acid is extracted and assembled as shown in Figure 1-(b). VRMMS supports these operations by rendering molecular models fast and computing energy equations in real-time. Since the docking and the assembling/disassembling operations change structures and status of molecular models, VRMMS needs a new data structure[7] to support such changes in real-time. The design concept of the data structure is based on scene graph[8,9].

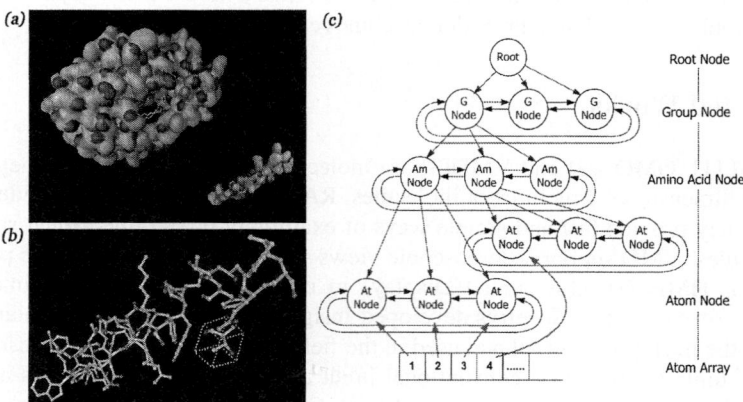

Fig. 1. Procedures of (a) Docking (b) Assembling and/or Disassembling using VRMMS (c) Data Structures for molecular modeling

VRMMS has a large (72 inch) display system[10] and generates stereoscopic views. The stereoscopic display helps scientists in examining three dimensional structures of molecular models[11]. A conventional two dimensional monitor cannot provide them with useful views which are realistic enough to perform three dimensional observations and manipulations. HMD(Head Mounted Display) could generate three dimensional views. We used 5DT's Data Glove as a tool of expressing gestures[12] and Polhemus' 3DSpace Tracker as a motion tracking device. However, the device is designed for a single user. On the other hand, the large display device could allow participating in molecular modeling procedures for multiple users at the same time.

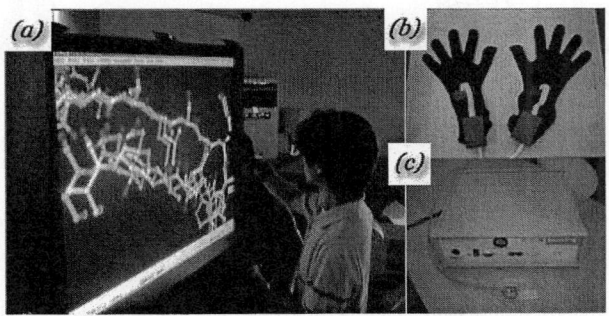

Fig. 2. (a) Stereoscopic Display (b) Data Gloves (c) Motion Tracking Device

4 Molecular Modeling

4.1 The Composition of a Candidate Molecule

When a new drug is developed, the first step is design candidate molecules through a docking simulation. We can build a candidate molecule by combining small molecules. It is important to ensure that the structures of the combined molecules are stable. It means that such a combination is chemically possible. To explain the stability of molecules, we exercise molecular torsion of molecular models. Such experimental operation is dominantly used in a real world, especially in laboratories, and it is consequent.

As shown in Fig.3-(a), revolution from α-Carbon to N is a counter clockwise movement and a clockwise to C. That is, placing α-Carbon at the center where it can make opposite direction to N and C, the left movement to N from α-Carbon is possible. Fig.3-(b) visualizes a chemical structure in three dimensions. With both hands, the rotation is conducted. For our convenience, we use 3D-cursors for each molecular model. As shown in Fig. 3-(b), a 3D-cursor looks like a box with three arrows which represent Cartesian coordinates and are labeled as 'x', 'y', and 'z'. A 3D-cursor is assigned to a molecular model and can be used to express the orientation of the molecular model during rotation. We can more easily understand the orientations of molecular models with 3D-cursors, because the arrows of 3D-cursors explicitly show the orientations.

How can we know if we select a molecular model with a 3D-cursor? We can do a collision detection procedure. Let $BminX$, $BminY$ and $BminZ$ denote minimum x, y and z values of a bounding box of a molecular model and $BmaxX$, $BmaxY$, $BmaxZ$ represent maximum values x, y and z values of the bounding box. If Hx, Hy, Hz represent the position of a 3D-cusor, the following expressions are used for the collision detection procedure:

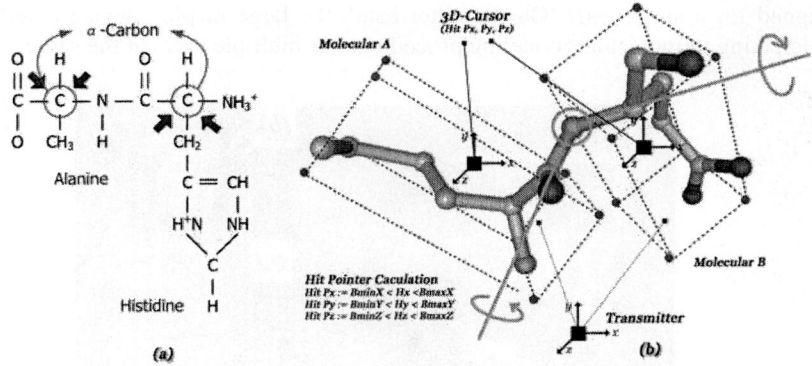

Fig. 3. (a)Specify points which can be rotated (b) Rotate the molecules with both hands.

$$BminX < Hx < BmaxX$$
$$BminY < Hy < BmaxY \quad (1)$$
$$BminZ < Hz < BmaxZ$$

When a user checks whether a combined molecular structure is stable, he/she would like to rotate their component molecular models in opposite directions (clockwise and anti-clockwise) simultaneously with user's both hands. In order to rotate molecular model at α-Carbon with both hands, we exercise two steps of experiment. First, we obtain an angle() value from transformation information of a tracker, which is $T[x, y, z, Azimuth(\Psi), Elevation(\rho), Roll(\phi)]$. Second, we rotate the molecular models from the angle value obtained from the first step.

M_A and M_B denote coordinate positions of α-Carbon.

$$M_A := (Ax, Ay, Az)$$
$$M_B := (Bx, By, Bz) \quad (2)$$

Then, we can get three components of the distance of two molecules as follows.

$$X_{AB} := |B_x - A_x|, Y_{AB} := |B_y - A_y|, Z_{AB} := |B_z - A_z| \quad (3)$$

Define $Max(D)$ and Θ as follows:

$$Max(D) := X_{AB}, \theta := T_{azimuth}$$
$$Max(D) := Y_{AB}, \theta := T_{elevation} \quad (4)$$
$$Max(D) := Z_{AB}, \theta := T_{roll}$$

Among the elements, the greatest value is a base axis, and the value of the rotation (*Azimuth, Elevation, Roll*) is gained from an angle eligible to the base axis and the rotation values are obtained from the 6 DOF input values from a tracker.

4.2 Docking Simulation

Once we have candidate molecules from the first step, we can exercise a docking simulation. The docking simulation aims to find a candidate molecule (a ligand) which can be combined with a receptor at a specific position called an active site. Energy between a receptor and a ligand should be calculated in real time, when the receptor and the ligand are combined during the docking simulation. As shown in Fig 4, the receptor and the ligand approach and the energy is calculated in real-time as the user's two hands approach.

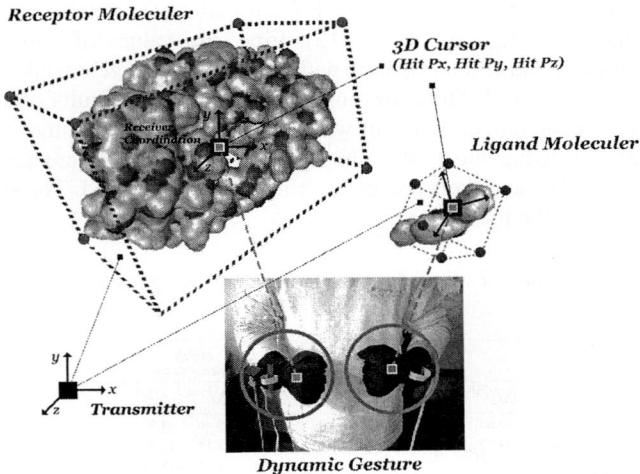

Fig. 4. Docking simulation with both hands

The input values from a tracker are very important. T_t represents the axis value of moving x, y, and z, and T_r is defined as rotation angles toward x axis, y axis, and z axis. Such axis information is transformed by VRMMS. It is because a coordinate system of the tracker and a coordinate system of OpenGL are different. C is a transformation matrix. M_R, and M_L are defined as each axis information of a receptor and a ligand. Consequently, M_R and M_L indicate that C is transformed by VRMMS system and T has axis information from a tracker.

$$M_R = CT_R$$
$$M_L = CT_L \tag{5}$$

4.3 Energy Minimization

When scientists perform the docking procedures by combining a receptor and a ligand, VRMMS executes a simulation and checks if such a bonding is chemically possible. In other words, VRMMS computes energy minimization equation for the docking. The energy stability computation for docking is defined as a sum of electrostatic energy ($E{elec}$) and Van der Waals energy ($Evdw$) as follow :

$$E_{total} = E_{elec} + E_{vdw} \tag{6}$$

5 Experiments

We compared VRMMS with Insight II which is the most popular tool in the field of molecular modeling. To apply the system to experiments, HIV-1 (Human Immunodeficiency Virus) was selected as a receptor. Fifteen materials related to reproduction of HIV-1 were chosen as ligands [11,12, 13]. First, the values of computing energy equations for binding the fifteen ligands with the receptor were calculated. The comparison would show how different or similar the simulation results of Insight II and VRMMS were. The second experiment was designed to compare times for exercising docking procedures using the tools. The docking procedures were performed with Insight II and VRMMS and their processing times were measured. These values could be used in evaluating the tools.

Table 1. Comparision of values from the simulations using InsightII and VRMMS

PDB code	Insight Energy Value				MMVR Energy Value ($T = Elec + Vdw$)			
	ΔE^{Vdw}	ΔE^{elec}	ΔE^{T}	RMSD (A)	ΔE^{Vdw}	ΔE^{elec}	ΔE^{T}	RMSD (A)
1gno	-7.63	-0.32	-7.95	1.02	-9.08	-0.46	-9.54	0.98
1hbv	-14.73	-1.24	-15.97	0.92	-14.21	-1.14	-15.35	0.86
1hps	-16.87	0.74	-16.13	2.41	-14.64	1.10	-13.54	3.15
1hpv	-10.15	-0.93	-11.08	0.36	-10.28	-0.74	-11.02	0.42
1hvj	-11.85	-0.11	-11.96	1.25	-10.85	-0.21	-11.06	1.28
1hvk	-16.25	0.55	-15.70	0.37	-14.21	0.65	-13.56	0.89
1hvl	-15.43	-1.20	-16.63	0.35	-15.35	-0.98	-16.33	0.39
1hvs	-12.31	-0.24	-12.55	1.66	-11.28	-0.34	-11.62	1.93
1hte	-1.24	-0.23	-1.47	0.39	-1.89	-0.65	-2.54	0.98
1htf	-22.61	-2.30	-24.91	0.32	-18.87	-2.15	-21.02	0.94
1htg	-17.46	-1.23	-18.69	0.49	-18.31	-1.24	-19.55	0.44
1pro	-9.95	0.67	-9.28	1.04	-9.70	0.62	-9.08	1.26
1sbg	-11.29	0.08	-11.21	2.01	-12.99	0.13	-12.86	1.36
2upj	-10.80	0.49	-10.31	1.59	-10.87	0.98	-9.89	1.89
4phv	-17.43	-0.98	-18.41	0.67	-15.64	-1.12	-16.76	0.92

The results from the first experiment are listed in Table-1 and graphically presented in Fig.5. Table-1 present energy values of fifteen ligands using Insight II and VRMMS. The results show that there are no significant differences between the values of calculating energy equations using Insight II and those using VRMMS. Therefore, the results of docking procedures using the two tools would not be significantly different. Table-1 and Fig.5 show the energy values of docking procedures which aims to combine the HIV-1 receptor with fifteen ligands (expressed in the PDB code).

In the second experiment, we compared times for completing docking procedures of Insight II and VRMMS. The subjects were never exposed to neither Insight II nor VRMMS. With ten subjects, we asked to perform docking procedures five times with

Insight II and VRMMS, respectively. The results are summarized in Table-2. The average times of docking procedures were computed and charted in Fig.6-(a).

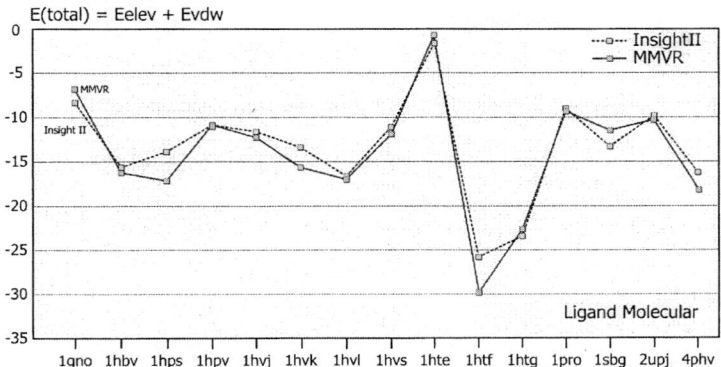

Fig. 5. Energy values of fifteen ligands

Table 2. Comparison of measured times of completing docking procedures with InsightII and VRMMS

Subject	Insight II					MMVR				
	1	2	3	4	5	1	2	3	4	5
A	5:36	2:30	2:01	2:32	1:58	15:12	11:20	8:20	3:01	1:38
B	10:33	5:20	3:21	3:12	2:58	12:30	7:20	6:22	3:53	1:28
C	6:58	4:30	2:36	2:20	2:01	5:50	3:16	2:18	2:23	1:35
D	5:33	4:13	3:20	3:13	2:37	12:24	7:30	4:48	2:20	2:36
E	7:20	6:37	5:10	3:24	2:53	10:12	3:30	1:02	1:05	1:50
F	7:48	7:13	4:22	4:01	3:20	7:55	6:30	6:02	1:45	1:38
G	5:48	4:20	3:15	3:20	2:44	10:35	3:22	2:48	3:30	2:56
H	8:12	5:12	4:23	4:40	3:48	10:42	6:18	2:20	2:18	1:54
I	5:32	4:30	3:22	3:07	2:23	10:23	3:48	1:30	2:33	1:49
J	7:32	4:48	3:20	2:28	2:32	12:28	3:20	1:42	1:03	1:12

They could exercise docking procedures faster, as they experienced and learned more about the tools. We could find a learning effect. As Fig.6-(a) shows, the subjects did it faster with Insight II in the beginning. However, they did it faster with VRMMS in the end. The average docking time of the last trial was reduced to 76% of the first trial with Insight II and 86% of the first one with VRMMS. The results can be interpreted as follows: The subjects were more familiar with a mouse than data gloves. Therefore, the docking procedure using Insight II showed better performance during the first trial. As the subjects try more docking procedures and learn more about the gesture-based interaction method, they performed better with data gloves. The learning curve of VRMMS was better than that of Insight II.

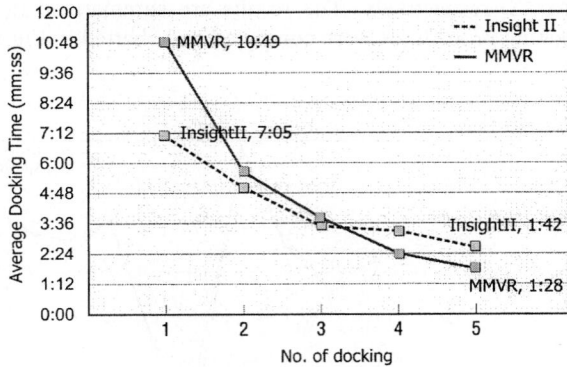

Fig. 6. Comparison of average time for docking procedures as number of docking increases

6 Concluding Remarks

We proposed a molecular modeling system based on dynamic gestures. It is expected that the system can solve constraints of existing molecular modeling tools. The system has the following features. (i) A conventional monitor was replaced by a stereoscopic display device. A user can have more useful depth values of molecular models. This gives not only more realistic views of molecular models but also more accessible views for multiple users. (ii) The dynamic gestures of users are recognized and used in assembling and disassembling procedures, and docking procedures. The input method of the gestures requires data gloves and motion tracking devices. (iii) A new data structure for fast rendering and efficient computations of energy minimization was developed. A couple of experiments showed that the proposed system could be used in docking simulations. The system was compared with the existing molecular modeling system. The computed values from the proposed system were similar to those of the existing system. The performance of the proposed system was good.

In the future, we would like to remove data gloves and motion tracking devices. We would use computer vision technologies to replace data gloves and trackers. It would be more economical and convenient way of viewing and manipulating three dimensional structures of molecular models.

Acknowledgements

This work was supported by the Ministry of Information & Communications, Korea, under the Information Technology Research Center(ITRC) Support Program.

References

1. RASMOL Http://www.umass.edu/rasmol/
2. Gans J, Shalloway D, : Qmol: A program for molecular visualization on Windows based PCs, Journal of Molecular Graphics and Modeling, Vol. 19, (2001) 557-559.

3. John E.Stone, Justin Gullingsrud, Kaus Schulten, : A System for Interactive Molecular Dynamics Simulation, ACM S. (2001) 191-194. Http://www.ks.uiuc.edu/Research/vmd/
4. Insight II Http://www.Accelrys.com
5. Frederick P. Brooks, Jr., Ming Ouh-Young, James J. Better, P. Jerome Kilpatrick, : Project GROPE-Haptic Display for Scientific Visualization, ACM Computer Graphics, Vol. 24, (1990) 177-185.
6. Szymin Rusinkiewicz, : QSplat: A Multiresolution Point Rendering System for Large Meshes, ACM, (2000) 343-352.
7. Noritaka OSAWA, Kikuo ASAI, Yuji Y.SUGIMOTO, : Immersive Graph Navigation Using Direct Manipulation and Gesture, ACM, (2000) 147-152.
8. Sense 8, WorldToolkit Technical Overview, (1998)
9. Martin Naef, Edouard Lamboray, Oliver Staadt, Markus Gross, : The blue-c Distributed Scene Graph, ACM, (2003) 125-133.
10. Seonhyng Shin, Dongsik Jo, Changseok Cho, Gun A. Lee, Namkyu Kim, Gerard Jounghyun Kim, Chan-Mo Park, Youngyeol Chu, : Interaction for Large Display based Data Visualization System, HCI Conference.
11. Bram Stolk, Faizal Abdoelrahman, Anton Koning, Paul Wielinga, Marc Neefs, Stubbs, An de Bondt, Peter Leemans, Peter van der Spek, : Andrew Mining the Human Genome using Virtual Reality, ACM, (2002).
12. C. Shahbi, : AIMS: An Immersidata Management System, presented at VLDB First Biennial Conference on Innovative Data Systems Research(CIDR2003), Asilomar, CA, (2003)
13. Earl Rutenber, Eric B.Fauman, Robert J.Keenan, Susan Fong, Paul S.Furth, Paul R.Ortiz de Montellano, Elaine Meng, Irwin D.Kuntz, Dianne L.DeCamp, Rafael Salto, Jason R.Rose, Charles S.Craik, Robert M.Stroud, : Structure of a Non-peptide Inhibitor Complexed with HIV-1 Protease, The Journal of Biological, Chemistry, Vol. 268, No21, (1993) 15343-15346
14. Norio Yasui-Furukori, Yoshimasa Inoue, Misturu Chiba, Tomonori Tateishi, : Simulataneous determination of haloperidol and bromperidol and their reduced metabolites by liquid-liquid extraction and automated column-switching high-performance liquid chromatography, Journal of Chromatography B, Vol. 805, (2004) 174-180
15. Junmei Wang, Paul Morin, Wei Wang, and Peter A. Kollman, : Use of MM-PBSA in Reproducing the Binding Free Energies to HIV-1 RT of TIBO Derivatives and Predicting the Binding Mode to HIV-1 RT of Efavirenz by Docking and MM-PBSA, Journal of American Chemical Society, Vol 123, (2001) 5221-5320

Face Modeling Using Grid Light and Feature Point Extraction

Lei Shi, Xin Yang, and Hailang Pan

Institute of Image Processing and Pattern Recognition,
Shanghai Jiaotong University, Shanghai, 200030, China
{sl0030322014, yangxin, panhailang}@sjtu.edu.cn

Abstract. In this paper, an algorithm for extracting three-dimensional shape of human face (Face Modeling) from 2D images using grid light is proposed. The grid light is illuminated by common white light instead of laser light in order to protect the human eyes or skin and reduce cost. A simple and uncoded grid pattern is projected on human face to solve the problem of correspondence between a pair of stereo images. The grid stripes are extracted and thinned by applying first smoothing and then a marker watershed segmentation algorithm. For the sake of providing more details for facial model, feature point extraction is introduced. The set of matched feature points will be added to the set of matched points. The final set of matched points is further used to calculate three-dimensional depth information of face. Experimental results have shown the feasibility of the proposed method.

1 Introduction

Generating a facial model of human is an important problem in many multimedia applications, such as teleconferencing, virtual reality, animation and face recognition, etc... Many attempts have been made in order to deal with the problem.

At present, face modeling has several major approaches .

Starts with manually-constructed B-spline surfaces and then applies surface fitting and constraint optimization to the surfaces, DeCarlo et al.[1] use the anthropometric measurements to generate a general face model.

In the second approach, facial model is directly acquired from 3D laser scanners or structured light range sensors. Water's face model [2] is a well-known model from the kinds of equipment. In many methods of face modeling, the facial model is all regarded as a generic model. Kawai et al.[3] presented a method of range data integration based on region segmentation and extraction of feature parameters.

The third approach, facial model is reconstructed by digital equipment, such as low-cost and passive input devices (video cameras or digital camera). For instance, Chen and Medioni [4] build facial models from a pair of stereo images. However, currently it is still difficult to extract sufficient information about the facial geometry only from 2D images. This is the reason why Guenter et al. [5] utilize a large number of fiducial points to acquire 3D facial geometry.

Even though we can acquire 3D information from expensive 3D laser scanners or structured light range sensors, it still takes too much time to scan and the system must remain stable during the whole process. For sake of overcoming these shortcomings, recently, some researchers try to incorporate some prior knowledge of facial geometry or make use of a generic facial model. For instance, Ansari, A.-N. et al. [6] deformed a 3D generic model from two orthogonal views (frontal and profile views) to acquire facial model. Zhang [7] deforms a generic mesh model to an individual's face based on two images. However, generic facial model that can be provided is mostly from occidental. For other races, these generic models must be modified. On the other hand, the number of those feature points which is used to deform the generic model is far less than that of the whole set of points of generic facial model.

In case a generic facial model is not appropriate or can't be provided, some methods integrating structure light and computer vision can be applied on human face modeling, such as Andrew Naftel et al.[8].

In 2004, Philippe Lavoie[9]proposes a new method for reconstructing 3D face model from the left and right two-dimensional (2-D) images of an object using a grid of pseudorandom encoded structured light. The proposed method can offer three distinctive advantages over a conventional stereo system:1) without new textures;2) less computational intensive;3) solve easily correspondence problem. Based on the system from Philippe Lavoie, we propose a new method.

Until recently, most methods of 3D capture have usually required the use of laser scanning equipment for 3D modeling. Such equipment, although highly accurate, is not suitable for some of the application scenarios in mind. Firstly, laser scanning may damage the eye, plus a scan can take several seconds to complete, requiring the subject to remain perfectly still during this time. For these reasons our system attempt to extract 3D structure information using shape approaches by common white light source[10,11].

Based on an uncoded grid light, our new method allows for the introduction of a new procedure for grid extraction and grid intersection location, the procedure can determine with precision a set of points on the object surfaces on the left and right images. The grid pattern projected onto a head model is a regular geometrical pattern. Some details will be lost if the process above is applied only. In order to overcome the shortcoming, facial features extraction is introduced. Firstly, a pair of image projected by grid light will be obtained; secondly, a pair of color image without new texture from grid light will be obtained at the same angle. The former will be used to construct rough 3D face model. The latter will be used to extract feature points to increase the details of rough 3D face model.

In what follows, we will describe the main algorithms of this procedure. In the Section 2, we will discuss the design and extraction of the projected grid pattern. Section 3 is a description of the extraction process of grid information. Section 4 describes the process of feature point extraction. While in Section 5, 3D reconstruction of corresponding points from the right and left image is described, and experimental results from different angles show the feasibility of our method. Eventually, in Section 6, a conclusion section reviews the main steps and the unique features of this system.

2 Uncoded Grid Light

A grid pattern projected onto an object can capture the whole view of the object, unlike a line or a dot pattern which requires scanning over the object to ensure that the object's surface is covered with the pattern.

In this paper, we have tried a simple scheme that fixes grid stripes along a pair of cross axes, the cross axes are distinguished very easily from others stripes in the grid pattern because of its gray intensity and stripe width. A description of how the uncoded and structured light grid projected onto a face is shown in Fig. 1.

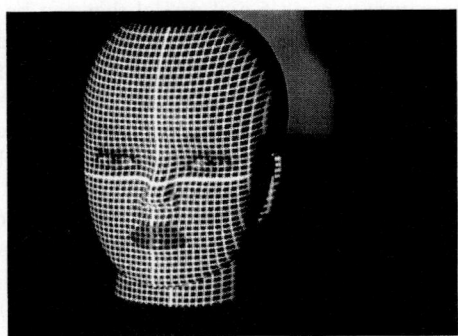

Fig. 1. Uncoded grid light

The extraction process of interest point is done in the two steps: extracting the grid and extracting those intersections.

The extraction of the grid from the image is done in the two steps: smoothing and an extraction process. The smoothing process consists of applying 2D Gaussian filter with standard deviation of $\sigma = 6.5$ (and kernel size 3×3). The filtering is used to eliminate noise signal in original image.

The quality of our grid stripe is worse than that of classical structure light system because grid light is illuminated by common white light instead of laser light in our system. The classical methods of extraction and thinning can't meet our requirements. Considering these conditions, our grid extraction process starts with the application of a marker watershed algorithm[12-14]. The watershed algorithm can find contiguous edges in an image accurately but suffers from the over-segmentation problem. Selected markers are usually employed to overcome over segmentation.

In this paper, we propose a maker watershed algorithm for grid stripe segmentation. The watershed algorithm is applied on the gradient magnitude image. A fast immersion-based algorithm developed by Vincent and Soille[12] is employed. All pixels in the gradient image are sorted by increasing gray-level values. Flooding is performed beginning from the marker image. Once the image is completely flooded, the watershed lines will be obtained.

$$M = I < 128 \qquad (1)$$

Where I is the original image, and M is the maker image.

The grid stripe from the watershed algorithm has exactly one pixel wide. The extracted grid stripe, as obtained by applying the watershed method, is shown in Fig.2 The original image is shown in Fig.1.

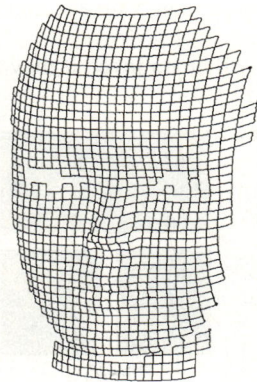

Fig. 2

3 Extracting Grid Information

In this section we describe the image processing methods and algorithms used to locate grid intersections in the grid image and to label these detected grid intersections.

More grid information, such as the precise location of these intersection points, their connectivity and their labels, are needed after the grid has been exacted by the marker watershed algorithm. To obtain all information, some steps are required.

A. Intersection Detection and Location

This section describes the algorithm used to detect and locate grid intersections after segmentation.

We have proposed a new algorithm based on graph connectivity. The candidates for intersection detection are those pixels lying on the extracted grid stripes. These intersections are detected based on a set of conditions on the set of nonzero pixels in a 3 by 3 square neighborhood centered about the candidate points. We call these nonzero pixels the border point.

Firstly, these conditions require that the set of nonzero border point in the square centered about the candidate points consists of three or four connected border points, shown in Fig.3(a) and (b). If there are intersections, they will be divided into different classes according to the distances between each two intersections. Then the average coordinates of these intersections from the same class will be calculated. The Euclidean distances between the average point and all the intersections of a class will be computed. The intersection with minimum Euclidean distance in the class will be regarded as the candidate point.

For smooth areas, there is one intersection for each node of grid stripes. But for not so much smooth areas, there are one or two intersections for the same node, shown in Fig.3(c). Secondly, we will compute the average of these intersections' coordinates if two or more intersections from the same node are so close to each other that the distance between them is no more than 5 pixels, and then replace these intersections by new candidate from the mean.

Thirdly, those intersections will be removed if they are from noise signals.

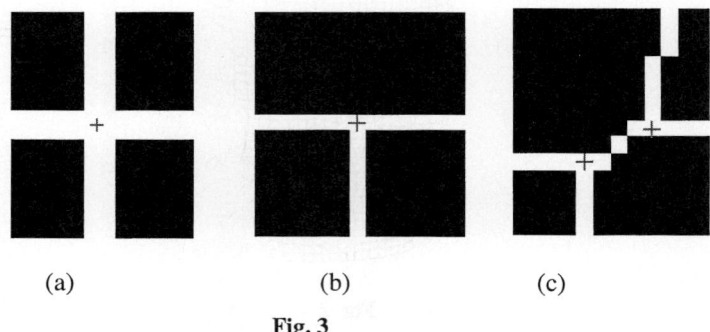

(a) (b) (c)

Fig. 3

B. Intersection Labeling

A key step in the 3-D reconstruction using structured light is to solve the intersections' labeling problem (the stereo correspondence problem). We solve the problem by labeling all grid intersections before matching the left and right image.

Labeling grid intersections is done in three steps:

1) locating the cross axes;
2) finding out all grid intersections on the cross axes, and then label these grid intersections;
3) labeling all grid intersections based on labeled grid intersections.

The first step is to locate the cross axes in the original image. We can locate very easily the cross axes because its gray intensity and stripe width is more than that of other stripes. The problem is solved by erosion and dilation operator in our method.

The second step is to find out intersections on the cross axes. By compared the sets of all grid intersections with the locations of cross axes, we can find out that the set of all grid intersections on the cross axes. The center intersection on the cross axes will be regarded as the principle point in the set; other intersections on the cross axes will be labeled based on the principle point.

The final step is to label all grid intersections based on labeled grid intersections in the extracted grid stripe. Since we have labeled the grid intersections on the cross axes, other grid points can be labeled.

The label method assumes an intersections of X row and Y column on the extracted grid stripes as (X,Y), then the grid intersection on X+1 row and Y column will be labeled as (X+1,Y). We will label all grid intersections based on the principle. A set of labeled intersections can be acquired by the method.

4 Facial Feature Extraction

The grid pattern projected onto a head model is a regular geometrical pattern. Thus, in many applications such as face recognition, accurate feature can't be obtained only by grid light. For the sake of overcoming the shortcoming, facial features extraction is introduced. Firstly, a pair of image projected by grid light will be obtained; secondly, a pair of color image without new texture from grid light will be obtained at the same angle. The latter will be used to extract feature points.

Our feature extraction method extracts 12 feature points from each view (eye and mouth). The *YCbCr* color space is applied in our method, where *Y* is the luminance component and *Cb*, *Cr* are the chrominance components.

After establishing the feature fields for two eyes and mouth, we extract these feature points by building two likelihood maps: the skin likelihood map and the mouth likelihood map. The skin likelihood map assigns to each pixel *v*, the probability *P(Cb(v),Cr(v)|skin)*, where the probability density function *P(Cb,Cr|skin)* is a Gaussian distribution.

The Gaussian distribution model is as follows:

$$P(Cb.Cr) = \exp\left[-\frac{1}{2}(v-\mu)^T \sum{}^{-}(v-\mu)\right] \quad (2)$$

Where $\mu = (\overline{Cb}, \overline{Cr})$.

In Eq.(2), $\overline{Cb}$ is mean vector of *Cb* and $\overline{Cr}$ is mean vector of *Cr*. Also, *v* is the input pixel value. Skin region is selected by thresholding, that is :

$$P(Cb.Cr \mid skin) = \begin{cases} 1, P > Thresthod \\ 0, otherwise \end{cases} \quad (3)$$

It is well-known that the eye regions have low probabilities because there is no skin in these regions. With the help of the information about the eye centers locations and thresholding using Otsu's method [15] , the regions of the eyes can be separated from the whole facial image. Then the boundaries are enhanced by performing erosion followed by dilation.

We then find out the smallest convex polygon points[16] on the perimeter that contains the region. From these points we can create a binary mask which resembles the shape of the eye. The shape of each eye is then approximated by an ellipse and the major and minor axes are computed. The two ends of the major and minor axes determine the four feature points for each eye.

By repeating the above process for the mouth except that *P(Cb,Cr|skin)* is replaced with *P(Cb,Cr|mouth)* , we can obtain the four feature points for the mouth.

After finding out all feature points from eye and mouth, we should match these points. The set of matched feature points then will be added to the set of matched feature points from grid light.

5 Calculating 3D Information

In most classical structure light system, 3-D information is acquired as result of a triangulation procedure. For the sake of avoiding the risk of occlusion, binocular vision is applied to acquire 3D information of the object in our system.

Original left image and right image are acquired from two digital cameras. The resolutions of camera are 1024×768 in our system. The two cameras' intrinsic parameters, extrinsic parameters and the relative position between the two cameras can be acquired by calibration. The results from binocular vision will be used to show the feasibility of our system. Original right and left image projected by grid light are shown in Fig. 4.

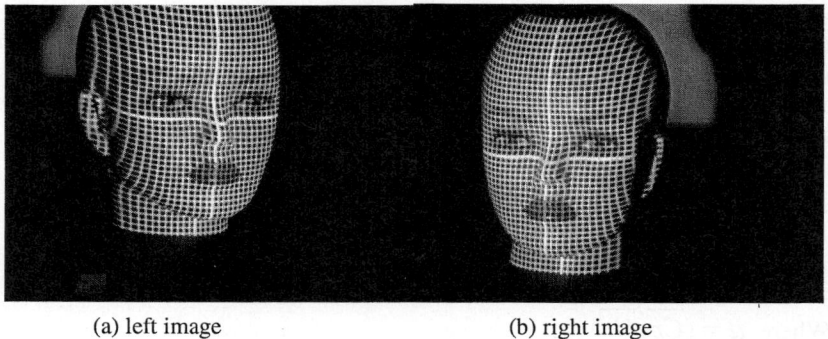

(a) left image (b) right image

Fig. 4

3D face information is calculated by binocular vision[17,18]. When two pairs of images of face are taken, a 3D information map can be easily obtained. The disparity of a point gives a scaled version of its 3D information. The result from authors' system is shown in Fig.5 (a) and (b). These green cross lines illustrate the result of feature point extraction in Fig. 5 (a).

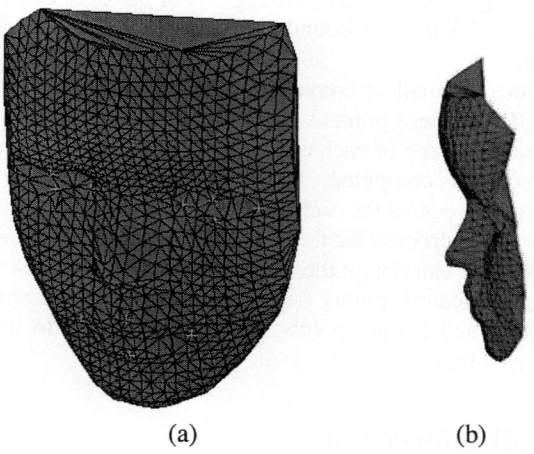

(a) (b)

Fig. 5. The final 3D reconstruction results

6 Conclusion

The problem with recovering the 3-D information of human face from two 2-D images of the same face is solved in this paper by using a common white light system consisting of a projector and two digital cameras. The grid pattern in our system facilitates the matching of the similar points situated on the two 2-D images. The sets of matched points help in determining the depth map and, eventually, a scaled value of the 3-D information of each point on the facial surface. On the other hand, feature point extraction is introduced to improve the quality of facial model. This precision can be considerably improved by further processing the two images.

The unique features of this system are:

1) white light source;
2) uncoded grid pattern;
3) grid extraction and labeling method;
4) integrating feature point extraction.

These preliminary results show that it is feasible to use grid light and new uncoded grid pattern to reconstruct 3D facial model. We can conclude stating that the system described above is also adapted to 3D reconstruction of the object surface.

References

1. D. DeCarlo, D. Metaxas, and M. Stone, An anthropometric face model using variational techniques. *In Proc. SICGRAPH,* pages 67-74, July 1998
2. F.I. Parke and K. Waters, Appendix 1: Three-dimensional muscle model facial animation. *Computer Facial Animation.* September 1996
3. Y. Kawai, T. Ueshiba, T. Yoshimi, and M. Oshima, Reconstruction of 3D Objects by Integration of Multiple Range Data. In *Proc.11th International Conference Pattern Recognition,* I: 154-157, 1992
4. Q. Chen and G. Medioni, Building human face models from two images. In *Proc.IEEE 2nd Workshop Multimedia Signal Proces*sing, pages 117-122, December 1998.
5. B. Guenter, C. Grimm, D. Wood, H. Malvar, and F. Pighin, Making faces. In *Proc. SIGGRAPH,* pages 55-66, July 1998.
6. A-Nasser Ansari , 3D face modeling using two orthogonal views and a generic model. In *Proc. International Conference on Multimedia and Exp*o, 3:289-92, July 6-9, 2003
7. Z. Zhang, Image-based modeling of objects and human faces. In *Proc. SPIE,* Volume 4309, January 2001
8. A. Naftel and Z. Mao, Acquiring Dense 3D Facial Models Using Structured-Light Assisted Stereo Correspondence. *Technique Report,* Department of Computation, UMIST. 2002
9. P. Lavoie, D. Ionescu, and E. M. Petriu, 3-D Object Model Recovery From 2-D Images Using Structured Light. *IEEE Transactions on Instrumentation and Measurement,* 53(2):437-443, April 2004
10. J.Y. Bouguet and P. Perona, 3D Photography on Your Desk. In *Proc. International Conference on Computer Vision,* Bombay, India, January 1998
11. J.Y. Bouguet, Visual methods for three-dimensional modeling. *Ph.D paper,* Computer Vision Research Group Dept of Electrical Engineering, California Institute of Technology, July 1997

12. V. Luc and P. Soille, Watersheds in Digital Spaces: An Efficient Algorithm Based on Immersion Simulations. *IEEE Transactions on Pattern Analysis and Machine Intelligence*, 13(6):583-598, 1991
13. Paul R. Hill, C. Nishan Canagarajah and David R. Bull, Image Segmentation Using a Texture Gradient Based Watershed Transform. *IEEE Transactions on Image Process*, 12(12):1618-1634, 2003
14. Roerdink and Meijster, The Watershed Transform: Definitions, Algorithms and Parallelization Strategies. *FUNDINF: Fundamental Information*, 41:187-228, 2001
15. N. Otsu, A Threshold Selection Method from Gray-Level Histograms, *IEEE Transactions on Systems, Man, and Cybernetics*, 9(1): 62-66, 1979
16. A.V.Melkman, On-line construction of the convex hull of a simple ployline. *Information Processing Letters*, 25(1):11-12,1987
17. J. Wang and Y.F. Li, 3D object modeling using a binocular vision system. In *Proc. the 16th IEEE Instrumentation and Measurement Technology Conference*, 2: 684 – 689, May 24-26, 1999
18. R. Srinivasan, M. Shridhar and M. Ahmadi, Computing surface information from monocular and binocular cues for vision applications. In *Proc. the 27th IEEE Conference on Decision and Control*, 2:1085 – 1089, December 7-9, 1988

Virtual Chemical Laboratories and Their Management on the Web

Antonio Riganelli[1], Osvaldo Gervasi[2], Antonio Laganà[1], and Johannes Froehlich[3]

[1] Department of Chemistry, University of Perugia, Italy
`{auto, lag}@impact.dyn.unipg.it`
[2] Department of Mathematics and Computer Science, University of Perugia, Italy
`osvaldo@unipg.it`
[3] Institute of Organic Chemistry,
Technical University of Vienna, Austria
`johannes.froehlich@tuwien.ac.at`

Abstract. The paper illustrates the efforts spent and the results obtained when introducing virtual reality in education and training both in academic and industrial environment. The project has been developed on the VMSLab-G portal (http://www.vmslab.org) and is part of an ongoing effort in exploiting Virtual Reality, at both meter (HVR, Human Virtual Reality) and nanometer (MVR, Molecular Virtual Reality) for training and education.

1 Introduction

The virtual dimension represents already an important part of our daily experience and turns out to be very useful for research activities. The use of virtual reality approaches in research, in fact, allows the construction of realistic simulation environments which lead to an enhancement of molecular intuition in complex phenomena such as those occurring in the domain of chemistry and physics [1]. An example of this activity developed in our laboratory is SIMBEX [2] (Simulation of Molecular Beam Experiments). SIMBEX is an application-simulator that, by making leverage on the possibilities offered by grid environments, allows a virtual molecular reality representation of processes relevant to atmospheric, combustion and plasma chemistry.

Recently, within the ELCHEM working group of COST in Chemistry [2bis], we have been working at extending virtual reality approaches to practice laboratories to be attended on the web [3] for education and training in Chemistry and molecular science related problems.

From the educational point of view, active and experiential learning (learning by doing) has been recognized as extremely important in natural sciences in general and in Chemistry in particular. The evolution of Information Technology (IT) and network bandwidth allows to associate real chemical practice laboratories with virtual laboratory sessions. This makes the process of acquiring chemical knowledge a personally scheduled, ubiquitous and continuously accessible process particularly suited also for lifelong education and training. Moreover, simulated laboratory environments, such as the VMSLab-G developed in our laboratory [4], allow reversibility of the user's action (i.e., the user can always "redo" or "undo" an action or a move), which is not always possible in a real environment. Furthermore, it makes it possible to run remote and/or unfeasible experiences. Finally, virtual experiential learning properly structured with the help of pedagogy experts increases the motivation and the pleasure of learning.

The management of virtual laboratories and of related knowledge and procedures has prompted the development of web based suitable Learning Management Systems (LMS)s [5]. A key technology for this development is semantic web [6].

This work addresses the problems arising both in developing virtual laboratories (of which we give a couple of examples) and in embedding them in a semantic web environment.

This paper is organized as follows. In section 2 we describe the virtual laboratory environment implemented using virtual reality technologies. In section 3 we illustrate how semantic web approaches are introduced. In section 4 the main conclusions are drawn.

2 The Virtual Laboratory Environment: Meter and Nano-meter Scale

The technical characteristics of the VMSLab-G portal has been given elsewhere [4]. Here we report only a brief description and the innovations introduced.

The application makes extensive use of VRML [7], X3D [8] and Java [9]. The first two languages are used for the modeling of the virtual objects whereas Java is used as the computational engine of the calculations needed for elaborating data and driving the interaction among virtual objects.

The entire portal was realized using standard web technologies. In particular we used open source technologies such as PHP (version 5.0.0, see ref. [10]) to make the programmers work transparent and MySQL (version 3.23.49, see ref. [11]) to manage the underlying database. The application runs on a machine with an Apache web server (version 1.3.27). Other technical details will be given in the context of the description of the experiment described later.

Two distinct features characterize the VMSLab-G application:

a. the possibility of accessing virtual molecular descriptions;
b. the possibility of accessing a virtual representations of experiments and procedures.

The first feature is the possibility of dealing with the molecular level that is implemented by making use of the already mentioned SIMBEX simulator. SIMBEX acts as the computational engine of the simulation and provides virtual reality tools with the information needed for the visualization and the walk-through the molecular aggregates under investigation.

The second approach feature is the possibility of dealing with the human level that provides an environment for pre-Lab (to familiarize with the apparatuses before physically accessing the laboratory) and post-Lab (to replay the experience and rationalize it by analyzing the results) operations.

In this section we discuss two virtual experiments. The first experiment shows the possibility of providing users with an environment able to assist them in running step by step a physical chemistry protocol. The second experiment shows the possibility of mastering safety and security issues when dealing with a potentially hazardous environment.

2.1 The Laser Refractometry Experiment

Laser refractometry is an analytical technique used for qualitative and quantitative chemical analysis. The virtual experiment presented here follows a real protocol [12] designed to be executed as a classroom demonstration. The purpose of the experiment is to estimate the refractive index (n) of different liquids using a 25-mL beaker and a pocket laser pointer.

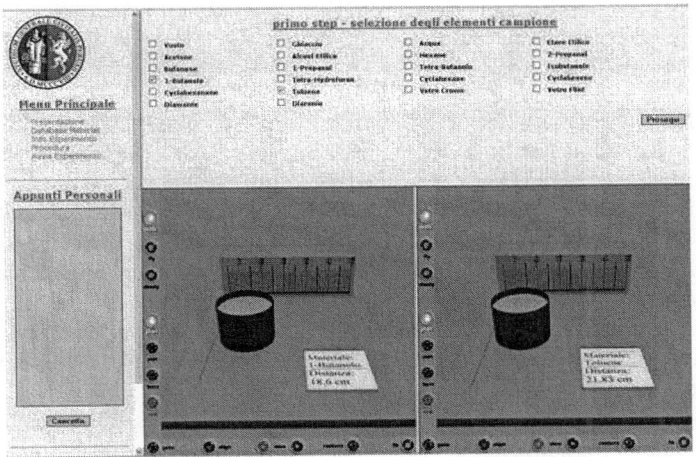

Fig. 1. The virtual laser refractometry experiment

The workflow of the experiment is articulated into 3 steps. In the first step, a couple of chemically related substances (whose refractive index (n) is given in handbooks) is selected. Then, the points of the screen paper hit by the laser beam after being refracted by the chosen substances are reported on a graph. The application starts from a minimum database of substances (ethanol, 1-propanol, cyclohexane, ethyl ether, etc.) for which the refractive index is known. The virtual experiment gives the possibility of adding, delivering, deleting or changing any item of the database.

The second step is devoted to the calculation of the linear correlation between n and the difference in position of the laser beam light point on the screen (the displacement). This is realized by a least squares fitting of the linear relationship to the measured displacements that the web application performs on background.

In the third step, the user can perform two distinct actions. In the first action, the user can choose a known sample from the database, perform the virtual experiment and, after reporting in the plot the value of the refractive index n versus the displacement, mark on the graph the measured displacement from the fitted line. In the second action, the user can pick up an unknown sample, perform the measurement and on the basis of the fitted line predict the type of substance using the value of the measured refractive index.

As shown in Figure 1, each step of the experiment is presented with a panel sheet on the left hand side panel. This allows both to record the values of the measurements (for example, in the first step, materials and corresponding displacements) and to draw the linear plot arising from the fitting line determined by the least squares procedure (with the corresponding statistical numerical details obtained in the second step), or to record all the values corresponding to the guesses of the unknown substance (in the final step). Moreover, the student can write at any moment notes on the virtual sheet (see the left hand side panel of the Figure). This allows to have, at the end of the experiment, a printable PDF document describing, step by step, the entire experiment. This document is dynamically produced by using the ClibPDF library [13].

2.2 Virtual Reality and Safety

The virtually unlimited (in terms of quantity, quality, space and time) domain of knowledge and relationships that can be dealt by virtual reality approaches embedded into semantic web technologies, has pushed us to link virtual environments to chemical safety, in particular in the context of the virtual assemblage of an apparatus and of the control of chemical processes. After all, safety in addition to being one of the main concerns and cost source for industry it is also an important educational issue in realizing practice hands-on laboratory at any level of education.

In this, virtual reality, with its intrinsic ability to simulate complex systems such as the interaction between human beings and machinery, enormously facilitates learning and training of personnel involved in risky operations.

Designing learning environments for intrinsically hazardous situations and operations with complex equipments is crucial for chemical and chemistry related industries since real-on-the-field training is often impossible in many cases (see Ref. [14] for a review) while virtual reality offers a risk free environment.

The work presented here is realized in agreement with the IChemEdu project [15].The main aim of the IChemEdu project is to provide students with a web tool, IChemLab, enabling the students to virtually practice on synthetic organic chemistry ("wet chemistry"). IChemLab has a database made of 275 detailed synthetic experimental protocols.

One of the risky situations in a real laboratory environment is the assemblage of the experimental setup before starting a practice session. Because of the potential danger of the related operations, this activity is conducted by a tutor who is also responsible for the final validation of the apparatus (the tutor certifies that the student is able to assemble the apparatus).

In our virtual reality approach the student has the possibility of activating a virtual session (in this respect the virtual session is used as a pre-laboratory) to practice in the building of the apparatus before attending the laboratory session. Here we discuss the assemblage of the setup needed for the protocol of the synthesis of 1-Naphtoic acid. To carry out this synthesis the student needs to build up a sufficiently simple apparatus that (though potentially dangerous) consists of the assemblage of several pieces of equipment (a sketch of a piece of equipment drawn using VRML is shown in Figure 2):

a reflux condenser, a three-necked flask, a dropping funnel, a washing tower, a drying tube, a stopper, a quickFit, a Glass tube and a carbon bon dioxide gas bomb.

Fig. 2. A three-necked flask realized in VRML. The Java engine classifies the connections (see also Fig. 3) as FL1 (left), FL2 (middle) and FL3 (right)

The main purpose of the application is to provide a tool for the assemblage of the apparatus. This means assembling the objects in the right order, choose the right orientation of the pieces, and so on. In practice, the student picks up the objects and moves them in the virtual world. If during these operations a not allowed operation is performed, the application alerts the student and stops him/her asking for a repetition of the procedure. The computational engine is able to discriminate between several types of forbidden operations. The alert function is for example activated only when the user performs a not allowed physically action that could prevent the prosecution of the operations. At the same time, when the student performs a physically possible but illogical action that could expose him/her to danger (e.g. two components which fit together but are not the right ones) the assemblage can be continued yet the error is notified when validation phase is reached.

An important point to emphasize here is that solutions may not be unique. In the cases in which more than one solution is, indeed, possible (more than one apparatus can be built out of the given components) though only one combination is marked as the "optimum" (that in our case is sketched in figure 3) but it could also be FL1 to RF2, FL2 to QT1, and FL3 to FL2.

Other protocols can use different apparatuses whose building blocks could be either the same as those described here be of different complexity. However, regardless the complexity of the setup, the implementation strategy of the apparatus is always the same: the equipment components are elements of a database and the Java algorithm may be customized to the particular needs of the protocol.

3 Semantic Web Approach

The possible use of virtual reality in conjunction with semantic web technologies is still in an early stage. As a matter of fact it is mainly being explored in the context of the possibility of reducing the complexity associated with indexing and searching information in a web environment. This is particularly true for chemistry, in particular, and life science, in general, since they are overwhelmed with information and a great deal of work is under progress to address this issue (see the recent W3C workshop for life science [17]).

A more advanced use of semantic web technologies that is very recently emerging [17bis] is the creation of standards for the management and delivery of learning units in a web environment incorporating semantic technologies.

In our work we tackle this domain from three different sides: i) the use of metadata to describe scientific information such as chemistry and molecular science knowledge; ii) the use of web technologies such as RDF (Resource Description Framework) to build intelligent assessment tool; iii) the use of metadata in the context of building virtual environments to deal with safety problems like those described in the previous section.

The first approach is used in the present project for MVR in which XML domain related subjects (such as MathML and CML [18]) are used to represent mathematical formalism and chemical structural form, respectively. The advantage of using such a kind of representation has been discussed in detail elsewhere [4].

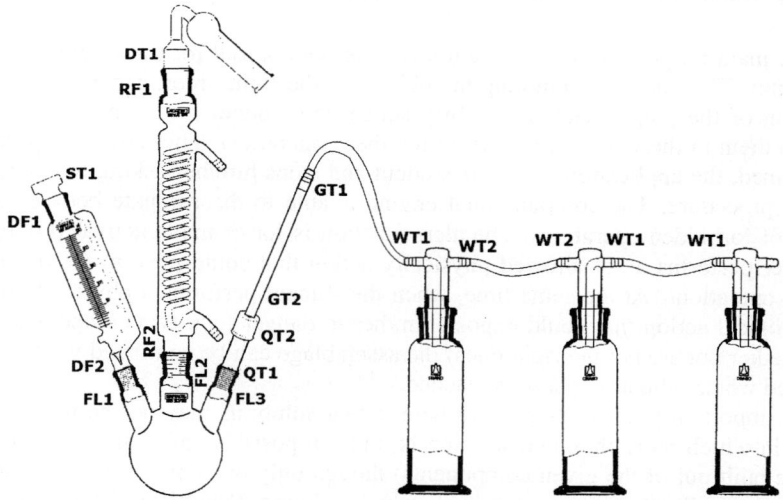

Fig. 3. Sketch of the apparatus assemblage representing the optimum solution. Each element is identified by a unique set of identifiers used by the Java engine to allow the assemblage

The second approach is the key feature of a project developed in our group called SELE (SEmantic LEarning) [19]. SELE is a Learning Management System (LMS) based on an application of semantic web technologies to e-learning.

In the context of the VMSLab-G project, SELE can be seen as an abstract container of Learning Units which make use of virtual reality approach.

The last approach, i.e. the use of metadata for managing virtual environments to deal with safety problems, represents a project under development and details will be given elsewhere. Here we illustrate only just the basic ideas. As we have seen in the previous section, the process of building a chemical apparatus (even a relatively simple one as that shown in figure 3) is complex from the algorithmic point of view. This complexity derives, to a large extent, from the difficulty in defining (in a fashion understandable to the machine) the components of the equipment and how a virtual user, or an automa when acting inside an immersive virtual reality environment, could interact with them.

4 Conclusions

Virtual reality technologies are today mature to be used as the building block for the development of advanced e-learning environment that is to design and implement tools for acquiring, experiencing and assessing knowledge. Moreover, the work showed how semantic web technologies can be profitably used both to catalogue knowledge and to realize ad hoc assessment tool instruments that can be adapted to the user profile and learning path. This exploits the main features of the Web infrastructures since it makes intensive use of related programming environments and facilities.

References

1. Riganelli, A., Gervasi, O., Laganà, A., Albertí, M.: Multi scale virtual reality approach to chemical experiments. Lecture Notes on Computer Science, 2658 (2003) 324-330.
2. Gervasi, O., Laganà, A.: SIMBEX: A portal for the a priori simulation of crossed beam experiments. Future Generation Computer Systems, 20(5) (2004) 703-715; http://coschemistry.epfl.ch
3. Laganà, A., Riganelli, A., Gervasi O., Yates, P., Wahala K., Salzer, Varella, E., R. Frohelich, J.: ELCHEM: A metalaboratory to develop grid e-learning technologies and services for Chemistry, this book
4. Gervasi, O., Riganelli, A., Pacifici, L., Laganà, A.: VMSLab-G: A Virtual Laboratory prototype for Molecular Science on the Grid. Future Genaration Computer Systems, 20(5) (2004) 717-726
5. Brase, J., Nejdl, W.: Ontologies and Metadata for eLearning in S. Staab & R. Studer (Eds.) Handbook on Ontologies. Springer-Verlag, (2004)
6. Lassila, O.: Web Metadata: A matter of semantics. IEEE Internet Computing 2(4), (1998) 30-37
7. Ames, A. L., Nadeau, D.R., Moreland, J.L.: VRML 2.0 Sourcebook. Wiley Computer Publishing, New York Tokio (1997)
8. http://web3d.org
9. http://java.sun.com/docs/books
10. http://www.php.net
11. http://www.mysql.com

12. De Vicente Finageiv Neder, A., García E., Viana L.N.: The Use of an Inexpensive Laser Pointer to Perform Qualitative and Semiquantitative Laser Refractometry. J. Chem. Edu. 78 (11) (2001) 1481-1482
13. http://www.phpcafe.net/index.php/pg/php_manual/get/ref.cpdf.php
14. Nasios, K.: Improving chemical plant safety training using virtual reality. PhD thesis, University of Nottingham (2003)
15. http://www.ichemlab.at - Participating institutions: Institute of Chemical Technologies and Analytics; Vienna University of Technology; Institute of Applied Synthetic Chemistry / Division Inorganic Chemistry, Vienna University of Technology; Institute of Applied Synthetic Chemistry / Division Macromolecular Chemistry, Vienna University of Technology; Institute of Organic Chemistry, Technical University Graz; Institute of Inorganic Chemistry, Technical University Graz.
16. Nilsson, M.: The semantic web: How RDF will change learning technology standards. Center for User-Oriented IT-design. Royal Institute of Technology, Stockholm (September 27, 2001). http://www.cetis.ac.uk/content/20010927172953
17. http://www.w3.org/2004/07/swls-cfp.html; Proceedings of the 2nd Europen Web-Based Learning Environment Conference (WBLE), Lund Sweden (Oct 2001)
18. http://www.xml-cml.org and http://www.w3.org/Math
19. Gervasi, O., Riganelli, A., Catanzani, R., Fastellini, F., Laganà, A.: A Learning Management System based on a Semantic Web approach, ExpoE-learning. (2004); Gervasi, O., Riganelli, A., Laganà, A., Integrating learning and assessment using the Semantic Web, this book
20. Gervasi, O., Riganelli A.,, Pennicchi D., Lagana ,A.: (in preparation)

Tangible Tele-meeting System with DV-ARPN (Augmented Reality Peripheral Network)

Yong-Moo Kwon and Jin-Woo Park

39-1 Hawalgog-dong, Sungbuk-ku,
Imaging Media Research Center,
Korea Institute of Science & Technology,
Seoul, 136-791, Korea
`ymk@kist.re.kr`, `jwp@imrc.kist.re.kr`

Abstract. In this paper, we introduce our researches on tele-meeting system that supports the network interaction with augmented reality. Our system supports networked haptic handshakes for the greeting, networked AR (Augmented Reality) for the augmenting the remote real object with the virtual object, and the network interaction share to the virtual object among participants using VRPN (Virtual Reality Peripheral Network). First, we describe a haptic-based network handshake that enables us to use the sense of sight, hearing, and touch all together with haptic device and video conferencing. Second, we present a networked AR system that supports augmentation of remote real object with local virtual object using ARToolKit and DV video streaming technique using DVTS. This makes AR system be possible with the situation that real object does not exist local area. Third, we propose a networked interaction with AR that supports the share of interactions to the virtual object that is augmented to the real object using AR technique. This concept provides interaction to virtual object at any participants and supports the share of the interaction result with all the participants in the network. The AR and network interaction concepts are integrated and implemented using ARToolKit, VRPN and DVTS. Here, we call it as DV-ARPN (Augmented Reality Peripheral Network).

Keywords: Multi-modal, Tangible, Tele-meeting, Augmented Reality, Haptic, Network Interaction

1 Introduction

For the remarkable advance of the internet, network environment is not only increasing in volume but creating working communication network. And a distant computer supported collaborative work environment is on its way to an international scale. Therefore effective oversea tele-meeting is needed for an important facility for business with geographically distributed operations. There are many examples for tele-meeting system such as the Microsoft NetMeeting, the Virtual Rooms Video Conferencing Service (VRVS), the Virtual Auditorium and so on [1]. Based on high-speed networks, these systems offer high-quality audio and video equipment. Ongoing study of tele-meeting for supplement reality results for instance, AR conferencing, that the remote collaborators can become part of any real world surroundings [2], cAR/PE!, which allows participants to discuss virtual 3D geometry on a physical table for remote collaboration [3]. However still, these are focusing on the video and sound for reality and immersive factor or interaction tools are lacking.

In this paper, we address the network interaction with augmented reality for the tangible tele-meeting. Our system supports audio/video, and haptic interaction for the greeting handshake, augmented reality with mark interaction and network interaction.

2 Tangible Tele-meeting System

Our tangible tele-meeting system is consisted of three stages. First stage is a greeting stage which people can have an eye contact with handshake using haptic device for greetings. Second is a main meeting stage that the main conference is held. The main 3D content is augmented with participants and they can have interaction with it using their joysticks, haptic devices, etc. The final stage is a farewell stage to say goodbye among participants and the tele-meeting is finished. The system overview is illustrated in Fig. 1.

Fig. 2 shows the concept of DV-ARPN. Fig. 3 shows AG based tangible tele-meeting system.

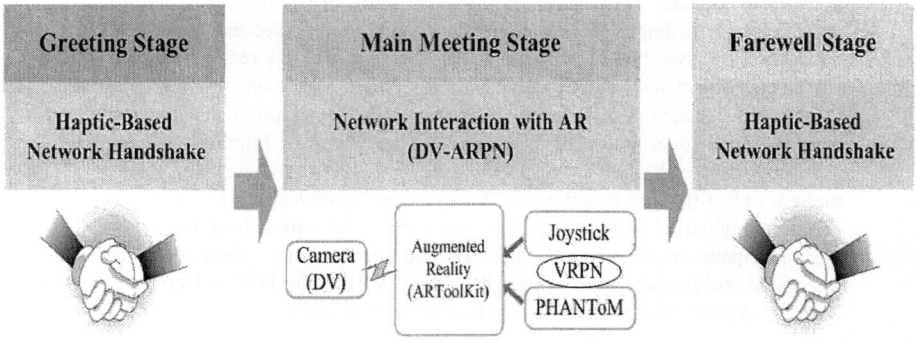

Fig. 1. Tangible tele-meeting system overview

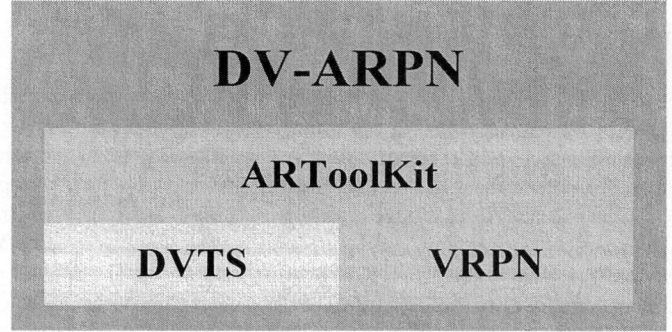

Fig. 2. DV-ARPN

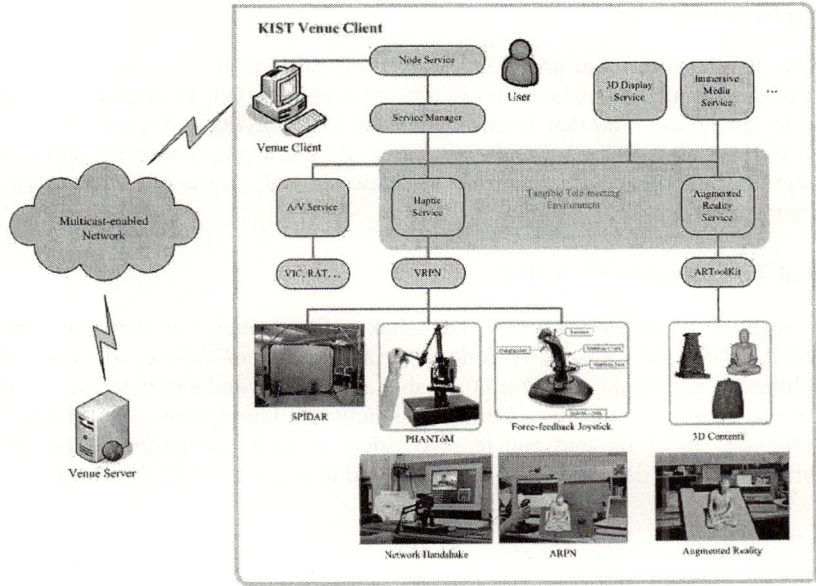

Fig. 3. AG based tangible tele-meeting system

The main features of our system are summarized as follows:

(1) Networked Haptic Media
    - Haptic Handshake for Tele-meeting through Network
    - Supports Phantom Premium 1.0, Force Feedback Joystick
(2) Multimodal tele-meeting
    - A/V for video conference
    - Haptic for handshake
(3) DV-ARPN (Digital Video - Augmented Reality Peripheral Network)
    - DV Streaming for Augmented Reality
    - Augmentation through Network
        (Virtual Object + Streaming Video of Real-Environment)
    - Network Interaction to Augmented Virtual Object
        * Provide interaction through network
        * Share of interaction result

3 Network Handshake

Network handshake is based on the network haptic device with digital video to integrate the sense of sight and touch. Through this technology, our tele-meeting system can immerse people with watching visual scene of other, hearing other's voice and feeling of other's hand from the beginning. We use the DVTS (Digital Video Transport System) for visual scene with sound through network and SenseAble PHANToM® for the haptic device for handshake.

3.1 Haptic Interaction

The word 'haptic' means 'of or relating to or proceeding from the sense of touch'. A haptic interface is a device which allows a user to interact with a computer by receiving tactile feed back. This feedback is achieved by applying a degree of opposing force to the user along the x, y, and z axes. We use the haptic device for participants to interact with the other participants on the network in the sense of touch, especially for the action, handshake.

3.2 Real-Time Remote Handshake

At the recent work, real-time remote handshake with haptics was demonstrated between Switzerland and Canada [4]. A demonstration has been developed, which illustrates a bilateral tele-haptics platform to enable a local user and a remote user to shake hands in real time over a telecommunications network. Our system is similar to this, except we use DVTS to send and receive video streaming and sound. The system configuration and the demonstration are illustrated in Fig. 4 and Fig. 5 in turn.

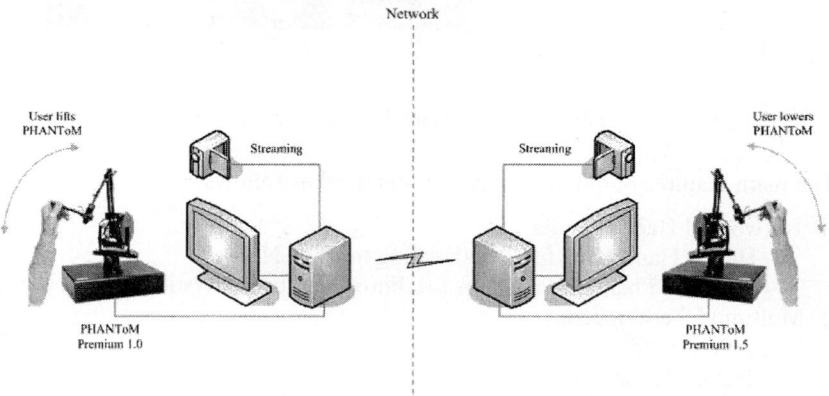

Fig. 4. Haptic-based network handshake system overview

4 Network Interaction with AR

The basic goal of an AR system is to enhance the user's perception and interaction with the real world through supplementing the real world with 3D virtual objects that appear to coexist in the same space as the real world [5]. But most of research is not considering the case that the real world from remote. But if the real world is too far to go, or too dangerous to be existed with users, or, like our tele-meeting system, the real world needs to be broadcasted to many others, the AR system cannot be working in the local alone. We can solve this problem easily by sending streaming data from real world through network. Also, if we can have interaction with the augmented 3D contents in the real world, tele-meeting will be more effective and attentive. There are many kind of the interacting device that we can consider, such as joystick, mouse, tracker, and haptic devices, etc.

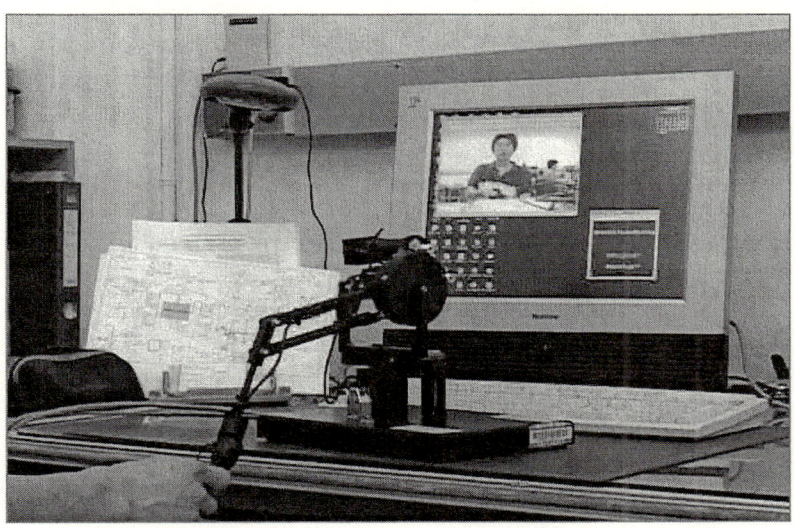

Fig. 5. Network handshake system demonstration

There are powerful tools to achieve our network interactive AR system, including ARToolKit [6], DVTS [7], and VRPN [8]. ARToolKit is a freely-available software toolkit for rapidly building AR applications and developed by Marc Billinghurst and Hirokazu Kato. It has many advantages such that, just a simple monocular camera is needed, the system is very simple and it works very fast. The DVTS (Digital Video Transport System) is developed by Akimichi Ogawa of WIDE project with the DVTS consortium. And the VRPN (Virtual Reality Peripheral Network) is developed by Russell M. Taylor II.

Basically, our tele-meeting system consists of a leader who is the main speaker and one or more participants who are distant from the leader and from each other as well. The conceptual figure of this is shown in Fig. 6. Under the circumstance of tele-meeting, the participants are able to share the main contents in the remote environment at each of own local environment by receiving the video and audio streaming including marker information. So whatever the main content is, we can augment and discuss about it in the local with the appropriate environment without visiting some places. Having an interaction with the same contents among the participants will make the tele-meeting more efficient. So, our system supports the interaction to the virtual object while each local system has 3D virtual object that is augmented to the received remote real environment video at the local system. In more detail, our system augments virtual object using ARToolKit while streaming with DVTS and have interaction with it using VRPN, so we call it DV-ARPN. There can be many scenarios of ARPN for instance the leader may have interaction with his virtual object, or the local participant may have interaction with the remote's virtual object. To avoid confusing, all of these scenarios have just one VRPN server in the network and the others are all VRPN clients which are showing the interaction result. In our experiments, we test with Sony DCR-PC115 digital camcorder with 17 frames/sec of frame rate which was enough to augment the 3D model in real-time.

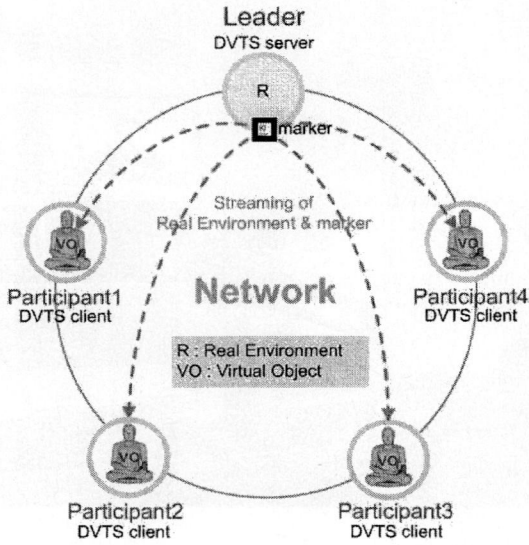

Fig. 6. Network Interaction overview

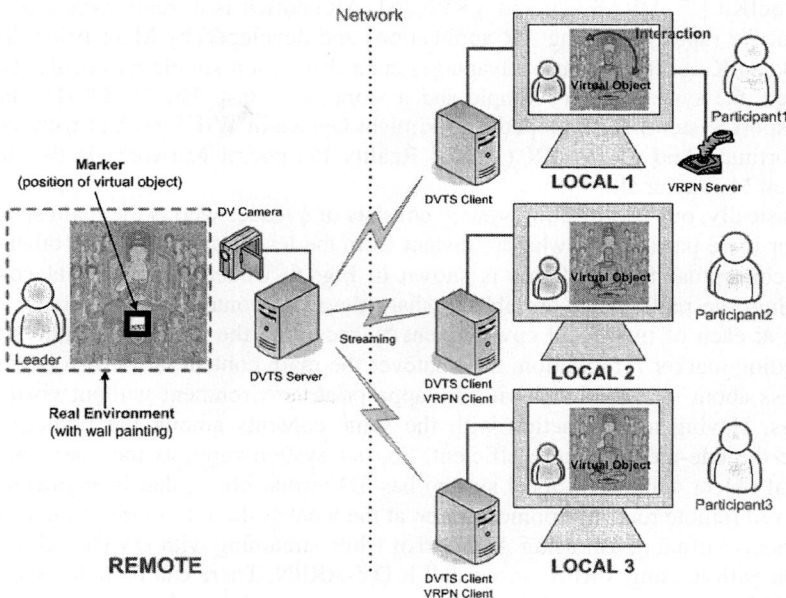

Fig. 7. DV-ARPN system overview

The concept of DV-ARPN is shown in Fig. 7. As shown in Fig. 7, the 3D Buddha model is augmented in the local and the joystick is used for the tool of interaction to the virtual 3D model. The marker position of the remote real environment is a place which the Buddha model should have been sitting. Any of participants can have interaction with the virtual object as VRPN server. In the illustrative example shown in Fig. 7, the remote real environment is a still image. However, our system also supports the remote environment including moving object because our system is based on the video AR processing. Fig. 8 shows a screen capture of DV-ARPN demonstration with 3D virtual Buddha model and joystick.

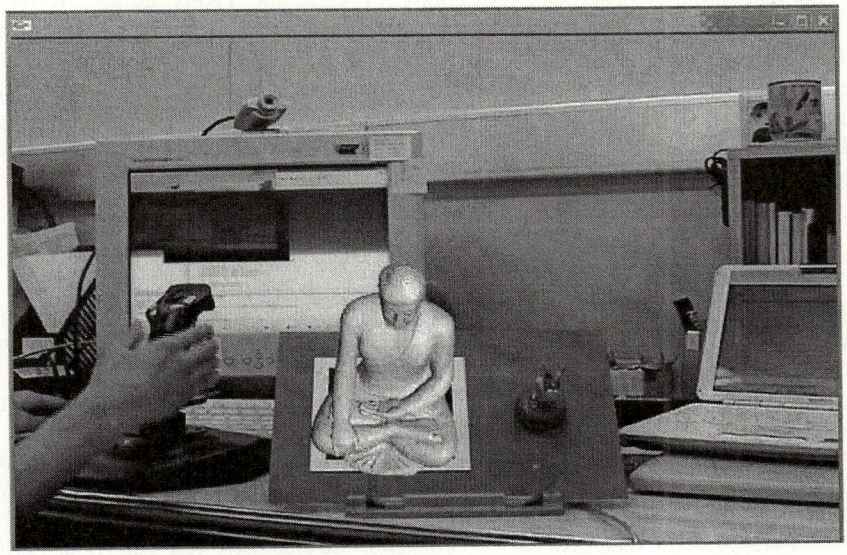

Fig. 8. Demonstration of DV-ARPN

For clarity, the main features of DV-ARPN are compared with the related tools in Table 1.

Table 1. Comparison with the related tools

	ARToolKit	VRPN	DV-ARPN
Augmented Reality (AR)	Support	No	Support
AR with remote video	No	No	Support
Interaction to virtual object through Network	No	Support	Support
Share of interaction result to virtual object	No	Support	Support

5 Conclusion

This paper addressed the lacking of tele-meeting system in the view of immersive and interaction and expands it as multi-modal system using many tools. Our system is composed of network handshake and interactive tele-meeting parts. The first part is about haptic-based network handshake which enables us to use the sense of sight, hearing, and touch all together with DVTS and haptic device. The second part is the network interaction with AR which makes it possible to interact with the main conference contents among participants. ARToolKit, DVTS, and VRPN are used to receive the remote scene of real world and augment 3D virtual object on it with interaction.

The original contribution of this paper is a tangible tele-meeting system architecture that supports network haptic interaction, multimodal tele-meeting concept, combining augmented reality with network interaction. We are currently developing multi-modal tele-meeting system with 3D model of hand for the realistic network handshake, and developing another interaction interface for the interactive AR.

References

1. Francesco Isgro, Emanuele Trucco, Peter Kauff, Oliver Schreer, "Three-Dimensional Image Processing in the Future of Immersive Media," IEEE Trans. Circuits and Systems for Video Tech., vol. 14, no. 3, 2004.
2. Mark Billinghurst, Adrian Cheok, Simon Prince, Hirokazu Kato, "Real World Tele-meeting," Proc. Of IEEE Computer Graphics and Applications, 2002.
3. H. Regenbrecht, C. Ott, M. Wagner, T. Lum, P.Kohler, W. Wilke, E. Mueller, "An Augmented Virtuality Approach to 3D Videoconferencing," Proc. Of the 2nd IEEE and ACM ISMAR, 2003.
4. David Wang, Kevin Tuer, Liya Ni, Pino Porciello, "Conducting a Real-Time Remote Handshake with Haptics," Proc. Of HAPTICS, 2004.
5. R. Azuma, Y. Baillot, R. Behringer, S. Feiner, S. Julier, B. MacIntyre, "Recent Advances in Augmented Reality," Proc. Of Computers & Graphics 2001, 2001.
6. ARToolKit, http://www.hitl.washington.edu/research/shared_space/download/
7. Digital Video Transport System, http://www.sfc.wide.ad.jp/DVTS/
8. Virtual Reality Peripheral Network, http://www.cs.unc.edu/Research/vrpn/

Integrating Learning and Assessment Using the Semantic Web

Osvaldo Gervasi[1], Riccardo Catanzani[1],
Antonio Riganelli[2], and Antonio Laganà[2]

[1] Department of Mathematics and Computer Science,
University of Perugia, via Vanvitelli, 1, I-06123 Perugia, Italy
ogervasi@computer.org, riccardo@woodie.unipg.it
[2] Department of Chemistry, University of Perugia,
via Elce di Sotto, 8, I-06123 Perugia, Italy
{auto, lag}@dyn.unipg.it

Abstract. An integrated Learning Management and Web assessment system, based on a Semantic Web approach is presented. The use of ontology allows not only the implementation of an innovative Learning Management system in which the learning objects may be produced through the local Content Management subsystem or acquired from URI pointing to external resources, but also of an "intelligent" assessment system that will take into account the learning path followed by the student. The proposed architecture is suited to implement a blended learning system in which the frontal lessons are reinforced by e-learning modules.

1 Introduction

The present work illustrates an e-learning environment based on an application of Semantic Web technologies to e-learning. The functionalities of the Learning Management (LM) system have been conceived so as to satisfy the needs of higher education systems especially at University level. Semantic Web approaches [1, 2] consist of a set of new technologies able to manage knowledge contents using formal ontology and to enable the assemblage of Web-based information and services that are understandable and reusable by machines.

In this work the Semantic Web technologies are used for the definition and the description of the various learning units introduced by the teachers, to make these units reusable and retrievable, and to follow the student during his "walk around" the various learning units. The adoption of Semantic Web technologies in the e-learning universe opens new opportunities for the interoperability of the LM system with the various learning facilities and the on-line assessment systems, like EOL[3].

The Learning Management system, called SELE, has been developed on a Linux environment making use of Freesoftware and Opensource components.

The paper is organized as follows. In section 2 the architecture of the system is described. In sections 3, 4 and 5 the SELE User Interface, the SELE Content

Management and the EOL assessment systems are presented. In section 6 how the self-assessment tests are produced is discussed.

2 The SELE LM System Architecture

In Figure 1 the architecture of the SELE LM system is represented. As illustrated in the figure, the SELE system is made of three main subsystems, that will discussed in the next sections: the SELE User Interface, the SELE Content Management and the on-line assessment system EOL that provides the teacher not only with the final test results but also with partial results produced by the students during the self-assessment phase of the learning process.

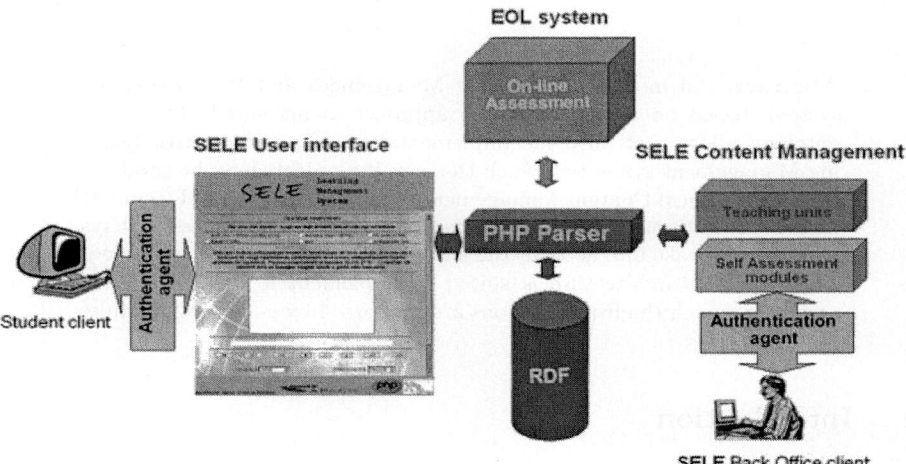

Fig. 1. The SELE Learning Management system architecture

The architecture of SELE system is suitable for implementing a *blended learning system*, in which the student may reinforce with the e-learning facilities and the interaction with the teacher the experience of the frontal lessons. SELE system is also designed to facilitate the teacher in the student's assessment process.

3 SELE User Interface

After the registration phase, the student has the possibility of selecting the courses he/she wants to follow by entering a personal code previously given by the teacher or by the system administration. The personal code guarantees the teacher that the student has granted the right to access the e-learning environment.

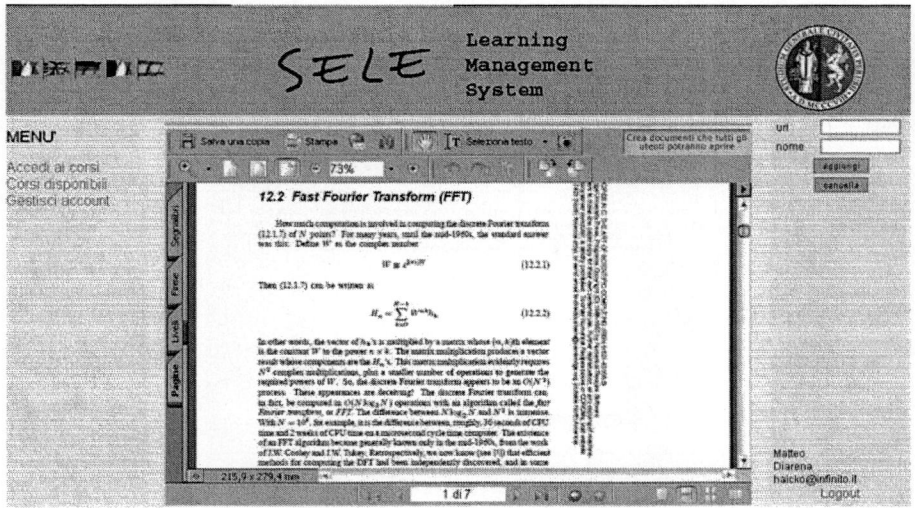

Fig. 2. SELE User Interface

After having accessed the course material the student will learn the lessons. A critical cross point of the process is the moment the student declares, after having gone through all the learning units, that he/she is ready to carry out a self- assessment session. It is up to the teacher to decide whether the test needs to take into account the learning path followed by the student. If this is the case, the system will privilege questions related to the specific optional arguments taken by the student during his/her preparation. On the contrary, only questions with no dependence on other learning units will be selected for all other students.

The Web page structure, that is homogeneous through the various SELE environments (see Figure 2), shows in the left top part of the page the list of languages available. By clicking on the flag corresponding to the desired language, the user will automatically change the language of all the contents and of the dynamic Web pages. On the right bottom panel the user details are provided together with the "logout" link to be used to leave the application and drop all session variables.

The page is divided in three columns. The left column gives the list of all the active functionalities for the user. The central column gives the principal working area devoted to content presentation, self-assessment test, etc. The right column allows the user to customize a personal environment in a way that he/she can add links to interesting sites, to useful documentation, etc.

4 SELE Content Management Subsystem

In SELE LM there is an ad hoc designed environment devoted to the production of learning units to be inserted in the SELE application: the SELE Content

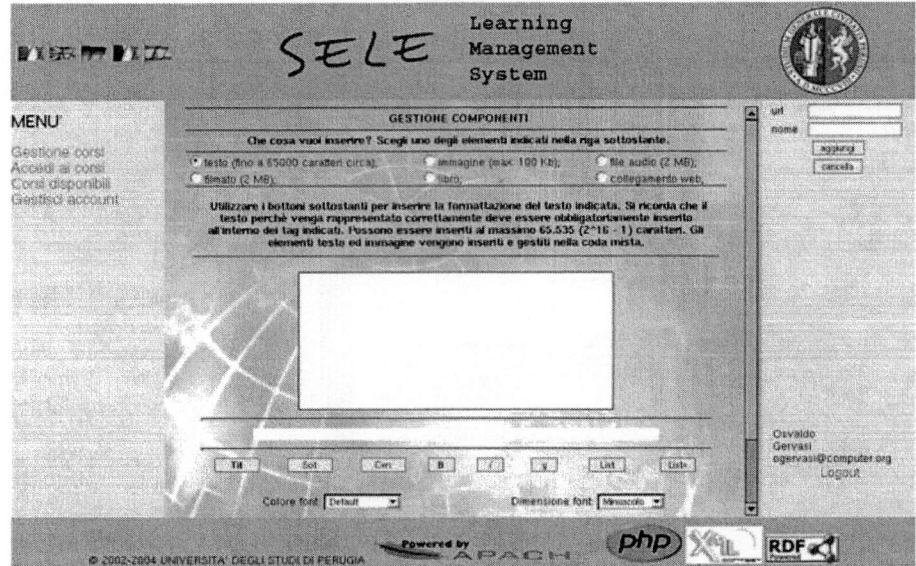

Fig. 3. Content Management subsystem of SELE

Management subsystem. That component allows the teachers to produce and maintain the learning units they are responsible of.

The insertion of learning material stored in different Web servers is carried out by specifying the corresponding URI and by defining the fields needed to issue RDF (Resource Description Framework) [6] statements. The RDF statements are used in this context to classify the learning unit, to indicate its possible use in other courses, to offer the user the possibility of introducing personal annotations about the learning material and to build an intelligent system of self-assessment. Moreover, this technology allows the development and the distribution of learning material in a real collaborative fashion by exploiting Grid tools and linking molecular virtual reality instruments[7].

Figure 3 shows the development environment allowing the assemblage of Web pages in HTML. Instruments for the upload of digital material (images, movies, audio file, Internet resources, etc.) and text are made available to the teacher. The text introduced can be formatted with the help of standard tools in a WYSIWYG fashion. This subsystem makes use of an RDBMS for the management of the objects of each page. By exploiting the object oriented properties of the PHP language, the programming has been carried out independently from the specific database utilized. In this way, the direct interaction with the data base (at the moment the Firebird RDBMS has been adopted) is managed via an interface subsystem.

5 The EOL System

The EOL system allows to carry out on-line evaluation tests in a tutor controlled environment (like a didactic laboratory). The evaluation test is made of a predefined number of questions (fixed by the teacher) of closed answer type. Questions associated with each course are divided in subsets of increasing complexity among which the asked questions in each individual test are fished in a random way so as to obtain the desired average difficulty.

Questions can be of multiple choice (MC) or multiple response (MR) type. In the MC case, 5 answers are associated with each question (only one right answer). In the MR case still five answers are associated with each question though there are more than one right answers.

At the end of the test, the list of questions, the related answers and the associated evaluation is made available to the teacher together with the statistical error. No negative scores are given for wrong answers though in the MR case the error bar widens if some of the answers are incorrect. In general MR questions belong to higher levels of difficulties.

The EOL system is also multilingual: the teacher can formulate the questions and the answers in the languages of preference. An option of EOL made possible by the use of Semantic Web technologies is the production, for each candidate, of a set of personalized questions pivoted by the data available in the other SELE subsystems (this option is not yet fully implemented).

6 The Self-Assessment Test

As already mentioned above, the teacher has the possibility of linking to each page of the learning unit a set of questions allowing the student to check the level of knowledge and competences he/she acquired going through the units. It is also possible to include questions focused to related learning units. In this case, these questions will be supplied to the student only if he/she has gone through those units. The related check is implemented in a totally transparent way by using Semantic Web technologies (see also [8]). In this respect, it is of great help for the teacher the possibility of setting sequences of dependent learning units using a specific environment that summarizes those learned by the student and the results of the self-assessment tests he/she has undertaken). The same kind of information is available to the EOL system in a way that allows the teacher, when planning the on-line exams, to enrich (both from the qualitative and quantitative point of view) the profile of the candidate.

7 Conclusions

The adoption of Semantic Web technologies allowed the design and the assemblage of a flexible Learning Management system able to extend its functionalities in a simple manner. This is in the spirit of the Tim Berners-Lee [9] work stressing the importance of introducing machine understandable metadata in a Web environment and in the e-learning system. The practical implementation of these research lines, assembled in our laboratory has been designed to satisfy the needs of higher education systems, the Universities in the first place. It is however extensible to deal with education systems of a different level.

The key features of the SELE system are the full reusability of the collected information and its links with Grid computing and virtual reality environments. This makes SELE a valid potential tool for ubiquitous experiential learning (learning by doing) that is extremely important in education.

SELE system is a blended learning system that helps the student during the learning percourse and facilitates the teacher performing the assessment of the students.

Acknowledgement

Financial support from the European Union through Leonardo 2 and the COST action D23 (ELCHEM: e-learning technologies for Chemistry) is acknowledged. Thanks are also due to the members of the Mutalc group of ECTN for interesting discussion.

References

1. Hendler, J., Berners-Lee, T., and Miller, E., Integrating Applications on the Semantic Web, Journal of the Institute of Electrical Engineers of Japan, Vol 122(10), October, 2002, p. 676-680; http://www.w3.org/2002/07/swint;
2. Berners-Lee, T., Hendler J., and Lassila, O., The Semantic Web, Scientific American, May 2001;http://www.scientificamerican.com/article.cfm?articleID=00048144-10D2-1C70-84A9809EC588EF21\&catID=2
3. Gervasi, O. and Laganà, A., EoL: A Web-Based Distance Assessment System, Lecture Notes on Computer Science, 3044, 2004, pp. 854-862
4. http://www.dasp.unipg.it;http://leonardo.cec.eu.int/pdb/Detail_fr.cfm?numero=831&annee=97;
5. Gervasi, O., Giorgetti, F., Laganà, A., Distance Assessment System for Accreditation of Competencies and Skills Acquired Through in-Company Placements(DASP), *INET99: The Internet Global Summit*, S.Jose, CA (USA), 1999; http://www.isoc.org/inet99/proceedings/posters/216/index.htm;
6. M. Nilsson, The Semantic Web: How RDF will change learning technology standards Center for User-Oriented IT-design, Royal Institute of Technology, Stockholm September 27, 2001; http://www.cetis.ac.uk/content/20010927172953;

7. Gervasi, O., Riganelli, A., Pacifici, L., and Laganà, A., VMSLab-G: A Virtual Laboratory prototype for Molecular Science on the Grid, Future Genaration Computer Systems, 20(5), 2004, pp. 717-726
8. Shen, R.M., Tang, Y.Y., Zhang, T.Z., The intelligent assessment system in Web based distance learning education, Frontiers in Education Conference, 2001. 31st Annual, Oct. 200, 17-11
9. Berners-Lee, T., and Miller, E., The Semantic Web lifts off, W3C. ERCIM News No. 51, October 2002;http://www.ercim.org/publication/Ercim_News/enw51/berners-lee.html;

The Implementation of Web-Based Score Processing System for WBI

Young-Jun Seo[1], Hwa-Young Jeong[2], and Young-Jae Song[1]

[1] Department of Computer Engineering, Kyung Hee University,
1 Seocheon, Giheung, Yongin, Gyeonggi 449-701 Korea
{yjseo, yjsong}@khu.ac.kr
[2] Faculty of General Education, Kyung Hee University,
130-701, 1, Hoegi-dong, Dongdaemun-gu, Seoul, Korea
jhymichael@empal.com

Abstract. The Web-Based instruction system has implemented according to traditional process until present. But, these methods have inefficiency in system development and give trouble of operation and administration after development. Therefore, it needs induction of component based development in Web-Based instruction system. In this paper, we constructed score processing system by component composition. We embodied component by Java Beans and used C2 architecture at the composition method. In this paper, we propose application possibility of component based development in Web-Based instruction system.

1 Introduction

Recently the World Wide Web is proposing new ways of education for many people. Educational programs and learning materials installed and supported in one place can be used by thousands of students from all over the world[1]. Web-based learning offers many advantages like flexibility, accessibility, self-paced format and so forth to their study[2]. Most systems of web-based learning areas including web course-ware have followed traditional development process because of application environment and characteristic of web server language. But these existing development processes have long development time and complicate operating process and, after the system was developed, give difficulty in operation and administration of system. Also, it could not cope with various requirements for additions and modifications of learning contents and functions. Therefore, also in web-based system development process, the changeover from static system development to dynamic system development is required.

As alternative technique for this purpose, component based development technique is being proposed. Component based development paradigm is the development tech-nique that creates another new component or constructs completed software through creation, selection, assembly/composition and evaluation of software module which is reusable component. In order to precisely composite and operate components, archi-tecture-based component generation and composition process must be achieved. Several architecture-based technologies exist: Pipe-and-filter architecture of UNIX, Blackboard architecture[4] that has been widely used over the years in general appli-cations, Style-based Unicon[5], Aesop[6], C2[7, 14], Symantec-model based Wright, Rapide, and Domain Spec based ROOM[8]. Among these, C2 style architecture is one of the architectures that support asynchronous interaction through messages instead of using direct method-call method among components. Kim[3] did not propose the concrete process about the application of component based development method to education field, but he mentioned the necessity of introduction in full. As the applica-tion of component in learning area, the research about generation of small unit component like Jeong[9] is proceeding, but the real system construction by component composition is not accomplished yet.

In this paper, we proposed the construction technique for whole score processing system. For this purpose, we implemented components which can be used in e-Learning area and composed each component with C2 architecture. Each component was implemented with Java beans as component model, and MS SQL was used as database.

2 Related Work

2.1 Web-Based Score Processing System

Web-based score processing system is widely used in educational administration sys-tem or learning result processing of web course-ware. [15] proposed the score processing in score input part among educational support system and [17] proposed the learning result processing in decision part for correct/wrong answers of web course-ware questions. But, as above proposed researches were implemented by traditional development process, modification, correction and addition of processing functions and reuse of developed system are difficult. As advanced research about score proc-essing, [18] suggested the method that enhanced the reliability of score processing about automatic grading by setting several patterns of correct answers for subjective questions and measuring similarity. But this research has a problem. For example, additional correction and modification is required and the reuse of grading functional module is difficult because this research implemented the functional module about grading using C and PHP.

2.2 Component-Based Development

CBD software systems are built by assembling components already developed and prepared for integration. CBD has many advantages. These include more effective management of complexity, reduced time to market, increased productivity, improved quality, and a greater degree of consistency[10].

These improvements result from reuse: building software from existing well tested building blocks is more effective than developing similar functionalities from scratch[11]. That is, CBD is a state-of-the-art software development paradigm and is the technique that makes it possible to develop high quality software rapidly and effi-ciently. Java beans, EJB, CORBA and .NET are good examples as representative component platform. Developed components have interface to connect each other and they are assembled and composed according to the architecture-based composition specification. Unicon, C2, Aesop, ACME and Wright are component architectures to assemble and composite components.

To ensure that a composed application is consistent with respect to the expectations and assumptions of its constituent components, interfaces can be used. The most basic type of interface only lists the services that a component provides or requires[12]. The interface of a component should be all we know about the component. Therefore it should provide all the information on what the component does, i.e. its operations, and how we can use the component, i.e. its context dependencies [13].

2.3 C2 Architecture in the CBD

C2[14] architecture supports message-based communication between components, multi thread, independency of each layer, component connecting structure through Message Routing Connector and GUI software requirement. C2 style architecture is also suitable for distributed or heterogeneous environment, ap-

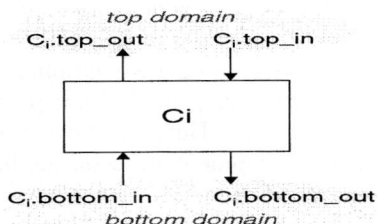

Fig. 1. C2 component domain

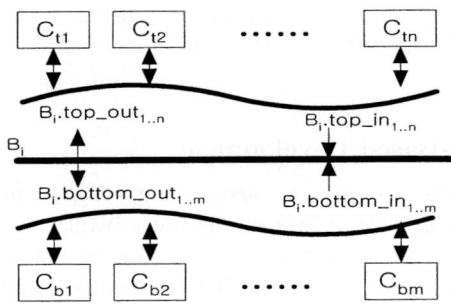

Fig. 2. C2 Architecture Domain

plying component in an environment of divided address space, multi-user and multiple tool kit, and dynamic structure[4]. Basic structure of C2 consists of component and 2 types of blocks having top port and bottom port. Exchange of message is carried out by Request message through top port and Notification message through bottom port. As shown in Figure 1, Request message is sent to upper level through top port and Notification message is received from upper level through bottom port of the upper level. Central to the architectural style is a principle of limited visibility or substrate independence: one compo-nent within the hierarchy can only be aware of components above it, and is completely unaware of the components beneath it. We can only receive the final result only through Notification messages sent from the lowest level.

Components of each layer are independent and the result can be verified with Notification messages of component in the lowest layer. Figure 2 shows component composition domain by C2 architecture [4].

3 Web-Based Score Processing Component by C2 Architecture

In this paper, we designed and implemented the score processing function module as component and used UML for design and analysis. The whole system was constructed by composition of implemented components on the basis of C2 architecture. Figure 3 shows the internal assembling structure of server side of this system.

Each functions classified into 4 parts were implemented as Java beans components according to corresponding process. Each component receives requests, handles the requests with database through JDBC and returns results. Figure 4 represents use-case diagram.

This system is divided into student mode and professor mode. After authentication, student can inquire the grade and professor can carry out score processing including grade input, grade calculation etc. Figure 5, 6 shows sequence diagram of student and professor mode representing message flow between each functional component.

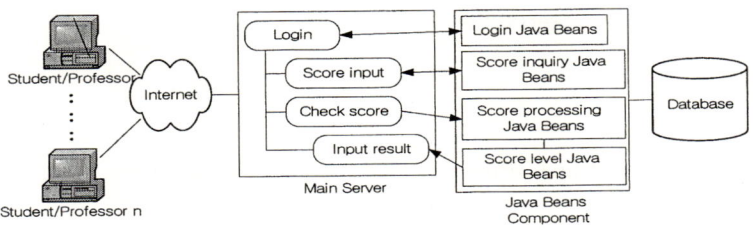

Fig. 3. Internal Assembling Structure of Server

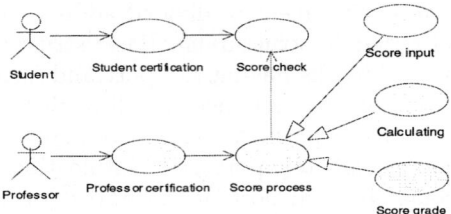

Fig. 4. Use-case Diagram

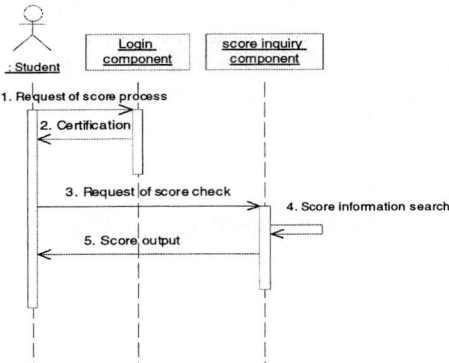

Fig. 5. Student Mode Sequence Diagram

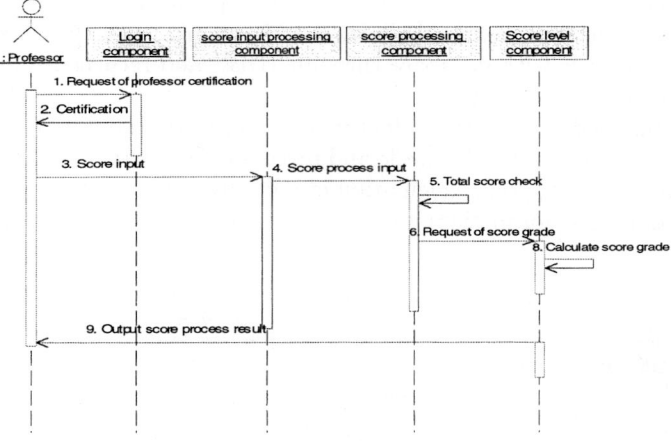

Fig. 6. Professor Mode Sequence Diagram

When authentication of log-in is verified, student can request inquiry of score. Score inquiry component searches grade information in database and returns the results to student.

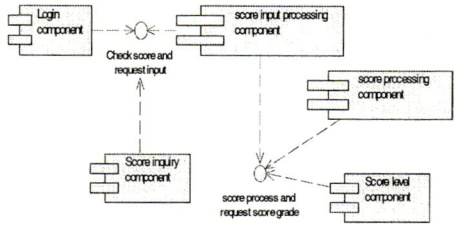

Fig. 7. Component Diagram

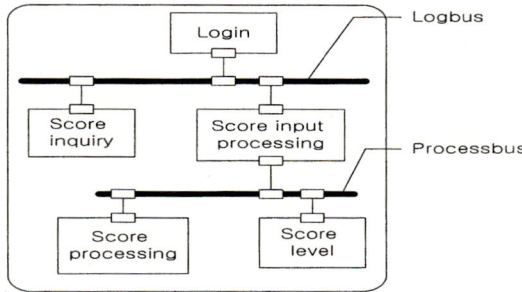

Fig. 8. Score Processing Component Composition Structure by C2 Architecture

Professor can input score and verify grade after authentication of log-in. For this purpose, score input processing component receives inputted grade information and passes on score processing component. Score processing component calculates total score on the basis of inputted score and checks whether each sum of score is over 100 points. Score level component calculates grade level according to grade level mark and returns the results to professor who can verify inputted score. Figure 7 shows relationship among components.

Interfaces representing linking structure between components are 'score inquiry/input request' and 'score process/level request'. That is, after log-in, they request score inquiry component or score input processing component to process the request according to student and professor mode. Score processing and score level component also request processing according to score processing and calculation of grade level after score input. Figure 8 represents the composition structure of implemented com-ponents on the basis of C2 architecture. As a connector playing a role as an interface between components in C2 architecture, we designed Logbus connector linking upper components and Processbus connector linking lower components.

According to these composition structure, each component connection shows as follows.

```
Log-in LogHnd = new Log-in("Login"); // log-in
Check-scr ChkscrHnd = new
    Check-scr("Checkscore");
Input-scr InscrHnd = new Input-scr("Inputscore");
    :                :
Scr-grd ScrgrdHnd = new Scr-grd("Scrgrade");
ConnectorThread Log-bus =
            new ConnectorThread("Logbus");
ConnectorThread Process-bus =
            new ConnectorThread("Processbus");
addComponent (LogHnd);
addComponent (ChkscrHnd);
addComponent (InscrHnd);
    :          :
addComponent (ScrgrdHnd);
addConnector (Log-bus);
addConnector (Process-bus);
weld (LogHnd, Log-bus);
weld (Log-bus, ChkscrHnd);
weld (Log-bus, InscrHnd);
    :          :
weld (Process-bus, ScrgrdHnd);
start()
```

Each component instance is created by new operator. Logbus and Processbus con-nector are created by ConnectorThread. Generated components and connectors are registered in C2 architecture by addComponent and addConnector and connected according to composition structure.

4   Score Processing System Constructions

This system was implemented with JSP and Java beans under Window XP environ-ment. User can select student or professor log-in mode and professor

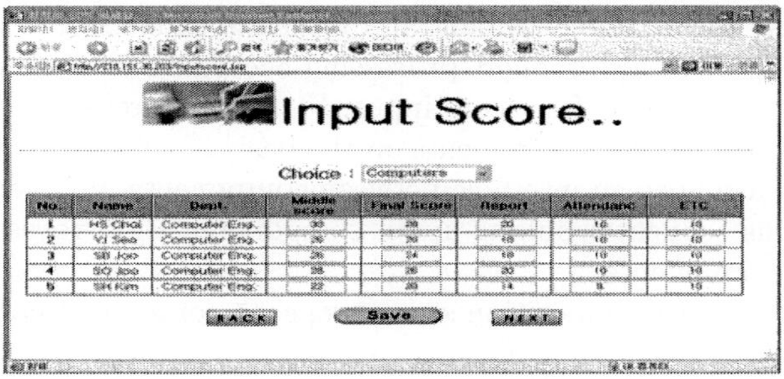

Fig. 9. Score Input Screen of professor Mode

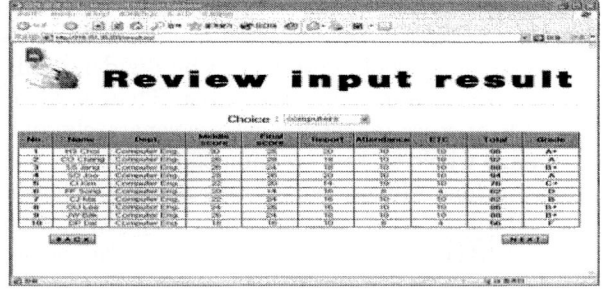

Fig. 10. Results of Inputted Score in professor Mode

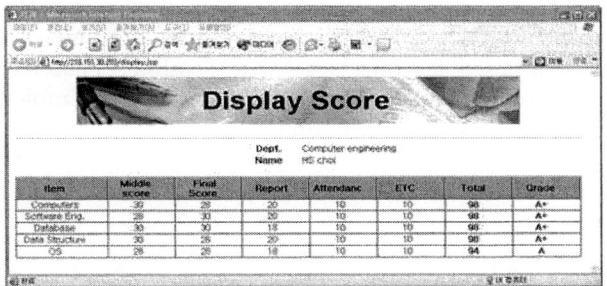

Fig. 11. Student's Grade Inquiry Screen

can perform grade input, grade verification and so on. Figure 9 shows grade input screen of profes-sor mode.

Each sum of midterm examination score, final examination score, report, attendance and so on can not be over 100 points and score processing component verifies this and calculates total score. Figure 10 shows processing results of inputted score in profes-sor mode.

This system allows student only to inquire the results of processed score. Figure 11 shows student's grade inquiry screen by processing of score inquiry component after log-in.

5 Conclusion

In this paper, we constructed score processing system by component composition. For this purpose, we implemented every functional logic of grade processing with Java beans which is Java based component model and completed whole system by composing embodied components on the basis of C2 architecture. Compared to existing web-based learning systems constructed by traditional development techniques, this technique has lots of advantages as follows.

First, the efficient development is possible. As every functional logic is implemented with component which can be processed independently and connected by composition specification, whole process can be divided easily and simultaneous embodiment of each component is possible. That is, we can implement each functional module of score processing system separately and at the system construction time, make whole system by composing these.

Second, the substitution of functional module is easy. If the changes of grade level calculation standards or methods are required in this score processing system, we can modify just the corresponding component and substitute it in that part of whole composition structure.

Last, the reuse is convenient. When new system requires same functional module, each score processing component can be adopted without modification to new composition structure.

In this paper, score processing system implemented just a few simple functions in order to show that the system construction through component composition in the web-based learning system is possible. Therefore, in order to perform practical score processing functions, the additional implementation and application of more functions are required.

References

1. P. Brusilovsky, E. Schwarz and G. Weber, "A Tool for Developing Adaptive Electronic Textbooks on WWW ", proc of WebNet-96, Association for the Advancement of Computing in Education (AACE), 1996.
2. Kay M. Perrin and D. Mayhew, "The Reality of Designing and Implementing an Internet-based Course", Online Journal of Distance Learning Administration, Vol 3, Number 4, 2000
3. Jae-Seng Kim, "Development processing modeling of web education component in distrib-ute computer environment", Journal of Korea Information Education Vol 6, Number 2, 2002.
4. R. N. Taylor, N. Medvidovic, K. M. Anderson, E. J. Whitehead, Jr., Robbins, J. E., Nies, K, A., Oreizy, P. and Dubrow, D. L., "A Component-and Message-Based Architectural Style for GUI Software", IEEE Transactions on Software Engineering, Vol.22. No.6, June, 1996.
5. M. Shaw, R. DeLine, D. V. Klein, T. L. Ross, D. M. Young, and G. Zelesnik, "Abstractions for Software Architecture and Tools to Support Them", IEEE Transactions on Software En-gineering, Vol. 21, No. 4, April 1995.
6. D. Garlan, R. Allen, and J. Ockerbloom, "Exploiting Style in Architectural Design Environments", Proceedings of SIGSOFT '94 Symposium on the Foundations of Software Engineer-ing, December 1994.
7. N. Medvidovic, P. Oreizy, and R. N. Taylor, "Using Object-Oriented Typing to Support Architectural Design in the C2 Style", Symposium on the Foundations of Software Engi-neering (FSE4), San Francisco, CA, Oct. 1996
8. B. Selic, G. Gullekson, and P. T. Ward, "Real-Time Object-Oriented Modeling", John Wiley and Sons, Inc., 1994.
9. In-Kee Jeong, "Development of S/W component for searching algorithm education", Journal of Korea Information Education Vol 6, Number 2, 2002.

10. Ivica Crnkovic, "Component-based Software Engineering - New Challenges in Software Development", Software Focus, John Wiley and Sons, December, 2001.
11. Miguel Goul?o, "CBSE: a Quantitative Approach", Proceeding of ECOOP 2003.
12. Chris L?er, Andr? van der Hoek, "Composition Environments for Deployable Software Components," Technical Report 02-18, Department of Information and Computer Science, University of California, Irvine, August, 2002.
13. Kung-Kiu Lau, "The Role of Logic Programming in Next-Generation Component-Based Software Development", Proceedings of Workshop on Logic Programming and Software Enginering, July, 2000.
14. The C2 Style, "http://www.isr.uci.edu /architecture/c2.html", Information and Computer Science, University of California, Irvine. 2003.
15. N. Medvidovic and R. N. Taylor, "A Classification and Comparison Framework for Soft-ware Architecture Description Languages", IEEE Transactions on Software Engineering, Vol. 26, No. 1, January 2000.
16. Hwa-Young Jeong, Young-Jae Song,"Development of ASP Based Score Processing System uing UML", Proceeding of Korea Information Science Society, Vol.28 No.02, 2001.
17. Jin-Kyung Lee, Uo-Chon Jeon, "Design and Implementation of Estimate System for Web based Instruction", Journal of Korea Information Education Vol 4, Number 1, 2000.
18. Heo-Jeong Bak, Won-Seok Kang, "Design and Implemention of Question Marking System Using Synonym Dictionary", Journal of Korea Computer Education Vol 6, Number 3, 2003.

ELCHEM: A Metalaboratory to Develop Grid e-Learning Technologies and Services for Chemistry

A. Laganà[1], A.Riganelli[1], O. Gervasi[2], P.Yates[3], K. Wahala[4], R. Salzer[5], E. Varella[6], and J. Froehlich[7]

[1] Department of Chemistry, University of Perugia, Via Elce di Sotto, 21,
06123 Perugia, Italy
{auto, lag}@impact.dyn.unipg.it
[2] Department of Mathematics and Computer Science, University of Perugia, via Vanvitelli, 1,
06123 Perugia, Italy
osvaldo@unipg.it
[3] Staff Development and Training Centre, Keele University, UK
p.c.yates@chem.keele.ac.uk
[4] Laboratory of Organic Chemistry, University of Helsinki, Finlad
kristiina.wahala@helsinki.fi
[5] Institute of Analytical Chemistry, Dresden University of Technology, Germany
reiner.salzer@chemie.tu-dresden.de
[6] Department of Chemstry, University of Tessaloniki, Greece
varella@chem.auth.gr
[7] Institute of Organic Chemistry, Technical University of Vienna, Austria
johannes.froehlich@tuwien.ac.at

Abstract. The integration of collaborative and networking activities with problem solving environments and web technologies on the grid allows the development of advanced environments for training and education. This paper describes some activities performed within ELCHEM (a Metalaboratory of the D23 (METACHEM: Metalaboratories for complex computational applications in chemistry) COST in Chemistry action) using the outcome of MUTALC (the multimedia working group of ECTN (European chemistry thematic network)) to develop grid technologies for research, projects and services for Chemistry and Molecular Science activities.

1 Introduction

ELCHEM [1] is a Metalaboratory (a cluster of distributed laboratories connected on the network sharing hardware, software and know-how) of the COST in chemistry [2] action D23. The goal of ELCHEM is to develop educational technologies and services using a Metalaboratory approach founded on the outcomes of the activities of MUTALC [3], the multimedia working group of the European Chemistry Thematic Network ECTN [4]. This is achieved by combining the know how of chemistry teachers with the skills of computer science experts.

The role of the computer experts is to build into the educational practice of the working group the use of the most advanced information and communication technologies (ICT). This use is strategic to the end of designing modern education and training activities. In fact, the impressive evolution of telematic networks is making the access to the World Wide Web regular for increasingly larger shares of the population. At the same time the fast growing implementation of grid infrastructures is making the exploitation of distributed know how and computer resources increasingly more efficient [5].

This has led to a focusing of the activities of ELCHEM on the design and the development of interoperable and ubiquitous educational tools as typical of Open and Distance Learning (ODL) approaches. These ODL tools are being designed by making leverage on the concepts of the virtual campus [6] and of the virtual practice laboratory [7,8].

At the same time, due to the characteristics of the above mentioned Metalaboratory organization, the activities of ELCHEM are also distributed among the geographically dispersed physical partner laboratories and articulated on three levels.

The first (basic) level of activity of ELCHEM is that of internally funded research and new teaching and learning material production in which accessibility to the work and products of all laboratories as well as the circulation of ideas and results are full and completely free. The second level of activity of ELCHEM is that of externally funded research and applied educational projects in which activities are coordinated by the laboratory heading the funding application while the other laboratories provide either part of the needed work or are engaged in similar or complementary tasks. The third level of activity of ELCHEM is that of grid and web services in which the Metalaboratory is engaged in assembling virtual campus and virtual laboratory grid services to be offered on the web. In this case different components of the service are provided by the partner laboratories and, whenever it is the case, by third parties.

The paper is organized as follows. In section 2, first level progress on designing applications to Science of markup languages and semantic web approaches are discussed. In section 3, second level ongoing collaborative projects aimed at designing, developing and implementing tools for teaching and learning chemistry are described. In section 4, the services being assembled to offer on the Web the outcomes of the first two levels of activities of ELCHEM are illustrated.

2 Level 1: Research Activities

In this section ongoing ELCHEM research activities on the use of mark up languages and semantic web tools in dealing with molecular science knowledge is discussed.

2.1 Markup Languages for Science

The management of scientific knowledge in a distributed environment is a key challenge for next generation web technologies. This is particularly true for domains such as chemistry, physics and mathematics in which the knowledge is represented by non standard textual information and symbolic representation plays an important role. This makes the added value of semantic web environments [9] crucial.

A first step along this direction is the use of extensible markup language (XML) [10] and its domain specific markup languages like the Chemistry Markup Language (CML) [11] and the Mathematical Markup languageMathML [12].

Further research efforts are being focused on the implementation of the semantic environment and on the definition of the ontologies for various classes of problems.

The use of the web for handling chemical knowledge is complicated by the large variety of the information required. It is an obvious statement that it is almost meaningless to represent chemical structures using the (text based) HTML environment. In this context, in fact, they are treated as a single graphic piece of information which means losing the structural details;

Further difficulties arise when expressing the chemical processes in terms of equations and formulae since also this information goes lost when using images in HTML. Not to mention the unsuitability of HTML representation for visually impaired users since they are not resizable and their content cannot be read by screen readers.

As already mentioned, MathML is a markup language specific to the construction of mathematical web pages. This implies also the adoption of authoring tools and of a MathML enabled browser. These aspects are being investigated by the Keele ELCHEM laboratory to assemble e-learning tools devoted to incoming students with weak mathematical ability and experience [13].

2.2 Semantic Web Approach to e-Learning

A more general approach is implemented in the SELE (SEmantic LEarning) application [14]. SELE is a Learning Management System (LMS) [15] based on an application of semantic web technologies to e-learning. The semantic web approach consists of a set of new technologies aimed at managing knowledge contents by using formal ontologies and assembling web-based information and services understandable and reusable by machines.

In our work the semantic web technologies are used to define and describe the various learning units produced by the teachers, to make these units reusable and retrievable to implement innovative self- assessment tools, and to follow the evolution of the students' acquisition of knowledge while accessing the various learning units.

This makes the LMS interoperable on different e-learning facilities. The SELE system is made of three main subsystems: the *SELE User Interface*, the *SELE Content Management* and the *EOL on-line assessment* system [16] which provide the teachers not only with an e-learning tool and the final test results but also with the partial results produced by the students during the self-assessment phases of the learning process.

The SELE Content Management subsystem allows the teachers to produce and maintain the learning units they are responsible of. The insertion of learning material stored in different Web servers is carried out by specifying the corresponding URI and by defining the fields needed to issue the RDF (Resource Description Framework) statements. The RDF statements are used in this context to classify the learning units, to indicate their possible use in other courses, to offer the user the possibility of introducing personal annotations about the learning material and to build an intelligent system of self-assessment.

As already mentioned, the teacher has the possibility of linking to each page of the learning unit a set of questions allowing the student to check the level of knowledge and competence he/she has acquired while going through the learning units.

3 Level 2: Project Activities

Level 2 project activities have concentrated on the design of practice laboratories. In particular a first project deals with the student approach to the analytical instrumentation and a second project deals with the design of organic synthetic chemistry protocols and the handling of related knowledge for a practice laboratory. Finally a third project deals with the assemblage of a virtual laboratory environment for a practice laboratory and the molecular virtual reality rationalization of related experiments

3.1 The "Network for Education – Chemistry" Project of the Dresden University of Technology

The Institute of Analytical Chemistry of Dresden University of Technology is developing within the framework of the German lead project "Network for education - Chemistry" [17] an internet-based system of multimedial learning material aimed at builiding a "digital *Pre*Lab". The project is intended integrate (not to replace!) the real lab exercises in chemistry. The material developed ranges from theoretical explanations to virtual instruments. The Dresden group is one of 16 project contributors to the project "Network for education - Chemistry".

The users find on the web a broad range of educational material about the most important chemical methods of analysis in a clearly structured form. The theoretical background as well as typical application areas for each analytical method (e.g., chromatography, spectroscopy) is illustrated with the support of expressive multimedia animations. The process of learning is easied by the insertion of interesting problems and attractive exercises. Both problems and exercises enable the learner to execute the newly acquired knowledge, to apply it to new problems, and to assess the learning progress.

One of the most effective aspects of the project is the training based on virtual instruments aimed at providing a distinct surplus for the educational process rather than replacing real lab exercises. The current virtual analytical laboratory offers virtual instruments of the most used analytical methods: gas chromatography, infrared spectroscopy, Raman spectroscopy, NMR spectroscopy and mass spectrometry. These virtual instruments are designed emphasize basic operations since they do not reproduce any feature of commercially available instruments designed by a particular manufacturer. Thus the virtual instruments may remain unchanged for a number of years, while the design of real instruments changes from time to time.

Virtual instruments have many advantages: they are quite safe, cheap and quick (though the digital *Pre*Lab based on virtual instruments cannot be taken as an alternative to real chemical experiment). Main advantages of the digital *Pre*Lab are: (i) the learners understand chemical processes easier, (ii) they learn the operate expensive high-tech lab instrumentation and (iii) they acquire detailed experience to exploit modern communication techniques.

3.2 The iChemEdu Project of the Technical University of Vienna

iChemEdu is a project of the Institute of Applied Synthetic Chemistry (Division Organic Chemistry) of the Vienna University of Technology in collaboration with other institutions (see Ref. [18])

The main scope of the project is to provide students with a basic web environment tool, IChemLab, to practice on synthetic organic chemistry ("wet chemistry") that plays an important role in chemistry study curricula. Experimental procedures and theoretical backgrounds are mostly taken from printed literature, but have to be adapted to site specific general conditions. Due to frequent changes of supervising staff, laboratory information, as well as knowledge about experienced practical problems, get regularly lost. This is a subsidiary scope of iChemLab that has developed an internet-based laboratory information system to preserve knowledge and maintain data.

For this reason, a database application containing 275 detailed synthetic experimental protocols has been developed for iChemLab by extracting and revising information from some 5000 work reports collected over the last 5 years.

To supply students with the possibility of time- and location-independent access to the system, iChemLab is designed as a web based application with JAVA applets e.g. for the presentation and input of structural and spectroscopic data. All acquired data is held in a MS SQL-Database-Server. Thus the student has the option to prepare for the experiments / submit obtained data / write reports outside of (cost intensive) course hours utilizing the PCs in the lab, the e-learning room of the Department or a private computer via internet connection.

The experiments and procedures stored in the web-accessible database are regularly improved by the feed-back entered by students and assistants during lab courses. Observed problems can immediately be announced and improvements are made accessible to the colleagues utilizing the system. Since all results and data obtained are also recorded (like chemical yield, physical properties, NMR-spectra and chromatographic data etc.) the achievement of a single student, which is the basis for marking, can now be objectively compared with the results of the colleagues (and not, as practised before, exclusively with literature data).

A second corner stone of the "e-information triangle" iChemEdu, which is connected to iChemLab, is iChemExam, a self assessment area to check knowledge required for performing the experiments. Besides the most common answering methods such as multiple choice selections and checks for numeric values, the Vienna group has developed new tools to answer and proof questions via input of chemical structures (e.g. to answer organic synthetic questions or structural problems). The supplied answer-structures are screened for correctness via a specially developed algorithm called SEICO (Spherical Environment Integer Code). SEICO is able to discriminate almost all the most popular organic compounds by combining in a single integer code the information about the molecular geometry, the type of bonds, the stereo-chemical properties and other molecular features.

iChemLecture, the third part of the system, is an "e-collection" of textbooks and contents for chemists converted from e.g. printed material by optical character recognition and full text indexing or generated by creating specific contents (e.g.: TeachMe products: http://teachme.tuwien.ac.at) with an appropriate author systems (e.g. Coimbra: http://www.coimbra.at).

iChemEdu is a hypertext system that interconnects iChemLab, iChemExam and iChemLecture via bi-directional software interfaces, which enable direct linking / input for queries from one part into the other as well as to further (external) databases. Experimental procedures contain JAVA script supported links to thesauri supplying extended information like theoretical background of the desired reaction, applicable labware, safety data, information about waste disposal etc.

iChemEdu has a modular architecture to enable also, in the future, smooth integration of new information categories and content.

3.3 The VMSLab Project of the University of Perugia

The VMSLab [8] is the result of a collaboration between the Departments of Chemistry and Computer Science of the University of Perugia.

The main object of the project is to introduce virtual reality (VR) techniques as the building blocks of the construction of realistic simulation environments for both research and education. Concentrating on the last aspect, we have shown that VR tools do not only facilitate the approach of the user to the experiment by offering an effective way of virtually exploring its components but it also offers him/her an environment in which everywhere and at any time the can accompany studying, learning, training and teaching with molecular virtual reality tools. This is realized with the VMSLab portal that focuses on the management of a virtual laboratory.

The scope of the portal is twofold. First, it represents the main entrance to the virtual world in its human VR (HVR) and molecular VR (MVR) components. Second, it provides an environment aimed at collecting and managing molecular science related knowledge for the specific use of a virtual laboratory for a virtual chemical laboratory.

The additional particularly valuable feature of VMSLab is the unique combination of HVR and MVR to support the simulation of the chemical laboratory with an immersion in the microscopic world of the molecules and their laws. To this end well established information technologies (such as JAVA and XML) offer a new generation of simulation and knowledge management facilities coupling a realistic representation of the events with the management of more traditional textual type information.

The HVR component has been devoted to the realization of chemistry practice laboratories taken from the organic and chemical physics domain. The virtual practice session is structured so as to provide the user with a work-flow template outlining the sequence of steps to be performed during the experiment. This kind of virtual experiments were built mainly with the purpose of offering pre and post-laboratory assistance to the student. The virtual experiments are based on program scripts which simulate quantitatively the motion of the objects and produce images emulating th real experiment according to a well established laboratory protocol.

The MVR machinery of VMSLab is that of SIMBEX [19]. It makes use of Molecular Dynamics approaches to produce a priori quantitative treatments of the atomic and molecular processes. The complexity and the duration of related calculations makes it necessary to distribute the computational efforts on the Grid and use parallel computing techniques.

At present MVR technologies have been particularly developed as Virtual Monitors of SIMBEX in which the outcome of the molecular dynamics calculations are preprocessed with Java or PHP tools. SIMBEX Virtual Monitors may also produce VR representations of complex molecules.

4 Level 3: Service Activities

The third level of activity of ELCHEM addresses the problem of providing services on the web since research and project activities of ELCHEM have led to the development of skills and products that can be utilized on the web as services. To this end links are being established with some academic spinoff companies.

4.1 The Assessment Activities

As already mentioned, one of the skills developed by the ELCHEM laboratories is the ability to support electronic assessment processes. In particular, IChemTest (developed within the IChemEdu project) and EOL (developed within the VMSLab project) have already implemented some of their tools as web services.

We have already mentioned the SEICO algorithm developed by the Vienna laboratory to screen the accuracy of chemical structures returned as answers to ICHemExam questions. Here we give some details on the demo site of EChemTest [20] an assessment web service being built by ECTN.

EChemTest is organized as follows. It is available as a web application running on a windows 2000 server of the unipg.it domain. Security is implemented at different levels. At data level use is made of a Raid-1 system to guarantee the back-up and retrieval of data. An additional level of security is provided by a local firewall (to protect the server) and by a perimetrical firewall (to protect the entire net of the unipg.it domain). The perimetrical firewall is configured in a way that allow a personalization on the basis of the IP source/destination, service and protocol parameters for each sub net.

4.2 The Virtual Campus Activities

The University of Helsinki has launched the Finnish Virtual University (FVU) service to offer selected e-learning products from the various universities (including courses and laboratories) for use by students and teachers. This service supports 20 Finnish universities to harmonize their information systems, to develop compatible practices, to allow students, teachers, researchers and administrators to share useful products and practices, provide national support and databases for on online courses and counselling activities. Among the services provided there are those offered by the center for ICT education that allows students and teachers to use free facilities for producing and using e-learning products. It includes the use of virtual learning environments, learning management systems, digital educational material, videoconference (support service and training), managerial support and strategic planning. There is also a coordination structure (4 central experts) for the campus experts (1 expert per campus) who carry out local training and technical support activities.

A particularly innovative aspect of these services is the use of the IQ (Intelligent Questionnaire) tools which are intelligent ICT tools for individual learners and groups. These tools are IQ learn (a tool for assessing and developing learners individual qualities and learning skills based on self regulating theory) and IQ team (a tool for assessing and developing group processes for collaborative learning and knowledge creation). They are made of an interactive test bank with three questionnaire sets for students' self-evaluation, a tutoring set with a hypertext structure for each subcomponent tutoring students towards self-regulation with related guidelines for teachers and a learning diary with a collection of learners' experience and test profiles. Another key service of the FVU is NetLab. NetLab provides services for the students to assemble authonomous and personal learning paths in the laboratory through an ad hoc web interface. The interface provides instructions to carry out syntheses (including animations for molecular processes), to safely deal with equipments and materials, to write reports, etc.. NetLab provides, for example, 3D models for synthesis problems together with an electronic feedback. Other services of NetLab consist in activating teaching supports for e-learning processes, personal study plans, motivation of the students; visualization of the microscopic processes; interaction between lectures and practical work; illustration of the laboratory procedures (including physical aaspcts of instrumentations and spaces), safety measures and economic implications; planning the personal schedule and resources; search for chemical information.

4.2 Web and Satellite Support

The Department of Chemistry, Aristotle University of Thessaloniki, is coordinating a project on the use of multimedia in the diagnosis and preservation of Mediterranean cultural heritage. The project is involving the following institutions: Malta Centre for Restoration, Rey Juan Carlos University (Madrid), Avignon University, Ca Foscari University (Venice), Sidi ben Abdellah University (Fez, Morocco), Yarmouk University (Irbid, Jordan). It was developed as a cooperation of the Working Group on Chemistry and Cultural Heritage, European Chemistry Thematic Network, and the EUROMED project IKONOS, offering cultural heritage preservation courses via satellite induced interactive communication.

The project combines web and satellite based distance learning for basic chemistry like instrumental analysis applied to cultural heritage issues and physicochemical properties of materials in artefacts and monuments.

References

1. COST working group D23/005/001 "ELCHEM: A Metalaboratory to develop e-learning technologies for Chemistry"
2. http://costchemistry.epfl.ch
3. Mutalc: "Multimedia in teaching and learning Chemistry" working group of ECTN
4. http://www.cpe.fr/ectn
5. Gervasi, O., Riganelli, A., Laganà, A.: Virtual Reality Applied to Molecular Sciences. Lecture Notes in Computer Science (ICCSA2004) 3044 (2004) 827-836.
6. http://www.chemistry.helsinki.fi/eChemicum
7. http://www.vmslab.org
8. Gervasi, O., Riganelli, A., Pacifici, L., and Laganà, A.: VMSLab-G: A Virtual Laboratory prototype for Molecular Science on the Grid. Future Generation Computer Systems 20(5) (2004) 717-726
9. Lassila, O.: Web Metadata: A Matter of Semantics. IEEE Internet Computing, **2**, No. 4, July (1998) 30-37
10. http://www.w3c.org/xml

11. http://www.xml-cml.org
12. http://www.w3.org/Math
13. Yates, P.C.: Teaching Mathematics to First Year Undergraduate Chemists in the Context of their Discipline. Maths for Engineering and Science, LTSN MathsTEAM, (2003) 40-41
14. Gervasi, O., Riganelli, A., Catanzani, R., Fastellini, F., and Laganà A.: A Learning Management System based on a Semantic Web approach, ExpoE-learning. (2004) Ferrara; Gervasi, O., Riganelli, and Laganà A.: Integrating learning and assessment using the Semantic Web (submitted, this book)
15. Avgeriou, P., Papasalouros, A., Retails, S., and Skordalakis, M.: Towards a pattern language for Learning Management System. Eucational Technology & Society 6(2) (2003) 11-24
16. Gervasi, O., Laganà, A.: EoL: A Web-Based Distance Assessment System. Lecture Notes on Computer Science, 3044, (2004) 854-862
17. http://www.vs-c.de
18. http://www.ichemlab.at
19. Gervasi, O., Laganà, A.: SIMBEX: a portal for the a priori simulation of crossed beam experiments. Future Generation Computer Systems 20(5) (2004) 703-715
20. http://www.echemtest.unipg.it

Client Allocation for Enhancing Interactivity in Distributed Virtual Environments

Duong Nguyen Binh Ta and Suiping Zhou

School of Computer Engineering,
Nanyang Technological University, Singapore 639798
{pa0236892b, asspzhou}@ntu.edu.sg

Abstract. Distributed Virtual Environments (DVEs) are distributed systems that allow multiple geographically distributed clients to interact concurrently in a shared virtual world. In practice, a multi-server architecture is usually employed as the communication architecture for DVEs, and the virtual world is partitioned into several zones to distribute the load among servers. A new problem, termed *client allocation*, arises when assigning the participating clients to servers. Current approaches usually assign clients to servers according to the locations of clients in the virtual world, i.e., clients interacting in a zone will be assigned to the same server. However, if the network delay from a client to its assigned server is large, the interactivity of the application may be greatly degraded. In this paper, we formulate the client allocation problem, and propose some allocation algorithms to assign clients to servers in DVEs in a more efficient way, taking into account both the virtual locations of clients and the network delays between clients and servers. Simulation results show that our algorithms are effective in enhancing the interactivity of DVEs.

1 Introduction

Recent advances in high-speed networking technologies, computer graphics and CPU processing power have enabled the development of Distributed Virtual Environments (DVEs). DVEs are distributed systems that allow multiple geographically distributed clients to explore and interact with each other in real-time within a shared, 3D virtual world [1], in which each client is represented by an avatar. A client controls the behavior of his/her avatar by various user inputs, and changes in an avatar's behavior need to be propagated to other clients to support the interactions among clients. Applications of DVEs can be seen in many areas, such as collaborative design [2], military simulation [3], e-learning [4] and multi-player games [5, 6].

Typically, various types of resources are needed in a DVE system, e.g., network bandwidth, CPU cycle, memory, etc. In addition, DVE systems are usually connected to the Internet. The large, variant Internet latency may damage the interactivity and consistency of DVE systems. In practice, to support DVEs, usually a *multi-server* communication architecture [7] is employed. In this architecture, multiple servers are connected to each other, and a client is connected to one of these servers.

If the entire virtual world is replicated at *all* servers in the system, a client may select to connect to its closest server to reduce the communication delay to the system. This approach is regarded as the mirrored server architecture [8]. However, the mirrored server architecture is not scalable, and only suitable for small-scale virtual worlds, with a few tens of clients. In order to deal with large-scale virtual environments with hundreds, or even thousands of clients interacting concurrently, which is the focus of this paper, usually the virtual world is spatially partitioned into several distinct *zones*, with each zone handled by only one server, as in [9]. Clients only interact with other clients in the same zone, and may move to other zones. A server only needs to handle one or more zones, not the entire world, thus the system becomes more scalable. In this paper, we refer to such a partitioning approach as the zone-based approach.

Traditionally, in the zone-based approach, all clients in a zone connect to the same server. However, due to the fact that clients in a DVE system is geographically distributed and the heterogeneous nature of the Internet, clients in a zone will have different network delays to the server of that zone. For a zone, we are interested in the *maximum round-trip* client-to-server delay of all clients in that zone [10], since if the round-trip delay of a client in a zone is high, the interactivity of the application for *all* clients in that zone may be damaged. In this paper, we propose some simple yet effective allocation algorithms to assign clients to servers to reduce the round-trip client-to-server delay, thus improving the interactivity of the application. In addition, we extend the algorithms to deal with dynamic cases, when clients may change their zones while exploring the virtual world.

The rest of the paper is organized as follows. Section 2 discusses the client allocation problem for zone-based virtual worlds and some related work. Section 3 describes the proposed algorithms. Experiments and results are described in section 4, and section 5 concludes the paper.

2 Client Allocation Problem

2.1 Definitions

In this paper, we focus on DVEs that adopt a multi-server architecture and the zone-based approach. The servers are interconnected via well-provisioned network links with low network delays. The connections between clients and servers may have high delays. We assume that network delays have known upper bounds.

As shown in Fig. 1(a), suppose that we have two servers s_1 and s_2 hosting two zones z_1 and z_2, respectively. The avatars of client c_1 and c_2 are in zone z_1, while the avatar of client c_3 is in zone z_2.

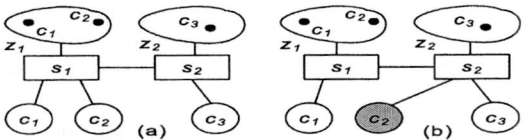

Fig. 1. Multi-server architecture with zone-based approach

We define the following important concepts:

Group: A group g_i consists of all clients that are interacting in the same zone z_i of the virtual world. For example, in Fig. 1(a), c_1 and c_2 are in the same group g_1.

Contact server: A *contact server* of a client is the server that this client directly connects to. Clients only send requests to their contact servers. The contact server may execute the request and response to the client if it is hosting the client's zone, or it may forward the request to another server which is hosting the client's zone. For example, in Fig. 1(a), s_1 is the contact server of c_1 and c_2.

Target server: A *target server* of a client is the server that is hosting the client's zone. Requests from a client will be forwarded to its target server. The target server may respond to the client directly if it is also the contact server of the client, or it may respond indirectly via the client's contact server. All clients in a group have the same target server, while they may have different contact servers.

For example, in Fig. 1(a), s_1 is both the contact and target server of c_1 and c_2. In Fig. 1(b), we switch c_2 to server s_2 (but the avatar of c_2 is still in zone z_1), the target server of c_1 and c_2 is still s_1, but the contact server of c_1 is s_1, while the contact server of c_2 is now s_2. Requests from c_2 are forwarded to s_1 by s_2.

Response time: The response time measures the interactivity of DVEs. To determine the response time for client $c_i \in g_j$, we need to consider all other clients in g_j in order to ensure consistency and fairness for all clients interacting concurrently in g_j [10]. The target server s_k for g_j needs to wait for requests from the farthest (in terms of network delay) client in g_j before executing the requests of other clients in g_j. Then, the resulted responses must not be presented to any client in g_j until the farthest client receives its response [10]. Therefore, the response time for all clients in g_j depends on the client with highest round-trip delay, and is determined by:

$$RT_{g_j} = \max_{c_i \in g_j} d_{c_i s_k}$$

where $d_{c_j s_k}$ is the round-trip delay from a client $c_i \in g_j$ to its target server s_k.

For example, in Fig. 1(a), assume that $d_{c_2 s_1} > d_{c_1 s_1}$, the response time for c_1 and c_2 can be calculated as $d_{c_2 s_1}$.

System response time: The system response time measures the maximum response time for all groups in the system, and is denoted as SRT:

$$SRT = \max_{g_i} RT_{g_i}$$

Delay bound: The delay bound D for a DVE system indicates the minimum requirement of SRT in order to maintain application interactivity. For different types of DVEs, there are different delay bound requirements. For example, multiplayer Real-Time Strategy games typically require a delay bound of $500ms$, while First-Person Shooter games require a delay bound of $100ms$ [5].

Server load: Load of a server in the DVE system can be divided into network-related load (receiving, sending packets) and application-related load (executing requests) [11]. The target server of a client needs to handle both network-related and application-related load generated by the client, while a contact server of the client only needs to handle network-related load. In this paper, the load of a server s_i is denoted as L_{s_i}, and can be expressed by the following:

$$L_{s_i} = N_{s_i}\alpha + M_{s_i}\beta$$

where N_{s_i} is the number of clients in all the zones that server s_i is hosting, M_{s_i} is the number of clients that have s_i as their contact server but not the target server, α is the load for handling a client (both network-related and application-related load), and β is the network-related load for handling a client.

For example, in Fig. 1(b), assume $\alpha = 1, \beta = 0.5$ we have $L_{s_1} = 2$, and $L_{s_2} = 1 + 0.5 = 1.5$.

In this paper, we assume that servers are identical, and has the same load threshold, denoted as T. If the load of a server exceeds T, it is saturated, and may affect the interactivity of DVE greatly, as shown in [12].

2.2 Problem Formulation

We have observed that a client which is far (in terms of network delay) from its target server will degrade the interactivity of the whole group. In order to meet the interactivity requirement of the application, we must seek some client allocation mechanisms that satisfies the delay bound D, and at the same time, avoids server saturation, i.e.,:

$$SRT \leq D \wedge L_{s_i} \leq T, \forall s_i$$

Note that in practice, sometimes we may not be able to find an allocation that satisfies the delay bound. In that case, we should try to find an allocation that minimizes the SRT, and avoids system saturation.

2.3 Related Work

To our knowledge, there is not much existing work that directly addresses the client allocation problem. Instead, most of the existing work is based on the approach of assigning clients to servers according to clients' virtual locations [13, 14, 15], which is regarded as virtual allocation (VA) approach in this paper. If a client in a group is far from its target server, the response time for that group may become very high, thus the interactivity of the application may be damaged.

In [16], the authors proposed a server selection algorithm that considers both network delay and virtual location of clients. However, this work uses the average delay between clients and servers as the main performance metric. To ensure consistency and fairness, the maximum round-trip delay is a more appropriate performance metric.

In a more recent work [10], the authors proposed a distributed algorithm for clients to selects the best server in a mirrored architecture for online games, taking into account the network delay between clients and servers. The mirrored architecture replicates the entire virtual world at every server, hence it is not scalable. Our work focuses on large-scale virtual worlds, thus the mirrored architecture is not appropriate.

3 Client Allocation Algorithms

In this section, we describe some simple yet effective allocation algorithms for assigning clients to servers. Our first algorithm is called Enhanced Allocation (EA), while the second algorithm is an improved version of EA using a round-robin refinement strategy, the Round-Robin EA (RREA). Then, we extend the proposed algorithms to deal with dynamic cases, in which clients may change their zones dynamically.

Our proposed algorithms are centralized, and can be executed in a master server that manages the whole server system. On joining the virtual world, clients may probe all servers to determine the network delays, and the probing results are sent to the master server to execute the allocation algorithms.

3.1 Enhanced Allocation Algorithm

The proposed EA algorithm exploits the fact that well-provisioned links between servers usually have much lower delay than connections between clients and servers. Therefore, we can achieve a better response time if a client do not send requests directly to its target server. Instead, the client selects the closest server as its contact server, and requests are forwarded to its target server via the low-delay link between its contact server and target server. However, this should be done with care, since allocating a client to a contact server which is different from its target server will increase the system load, due to the extra network-related load that the contact server has to handle. The EA algorithm has two parts: *initial assignment* and *refined assignment*.

Initial assignment: In the initial assignment, for each group g_i we find a target server s_j that minimizes the response time $RT_{g_i} = \max_{c_k \in g_i} d_{c_k s_j}$, and does not saturate server s_j, i.e., $L_{s_j} + N_{g_i}\alpha \leq T$, where N_{g_i} is the number of clients in g_i. Then, the contact server and target server of each client in g_i are set to s_j. The purpose of this initial assignment process is to select the best target server for each group.

Refined assignment: In the refined assignment, for each group g_i, if $RT_{g_i} > D$, we find a list of all clients $c_k \in g_i$ that exceed the pre-specified delay bound D.

The list is sorted in descending order according to the round-trip delay between each client and its target server. Then, we repeat the following until the list is empty: remove a client c_k from top of the list (the client with highest round-trip delay), and try to find a server s_l that satisfies the delay bound, i.e., $d_{c_k s_l} + d_{s_l s_j} \leq D$, where s_j is the target server of c_k, and $d_{s_l s_j}$ is the round-trip delay between s_l and s_j.

To find the server s_l for c_k, we sort the list that consists of all servers in the system in ascending order according to the round-trip delay between c_k and each server. Then the first server s_l in the list (the closest server to c_k) is picked. If s_l will be saturated due to assigning c_k to it ($L_{s_l} + \beta > T$), or if the delay bound cannot be satisfied by using s_l, we select the next server in the list and so on. If there's no server that satisfies the delay bound, a non-saturated server s_l that provides client c_k the minimum round-trip delay to c_k's target server will be selected. The new contact server of c_k is now s_l, while the target server of c_k is unchanged.

3.2 Round-Robin Enhanced Allocation Algorithm

The refined assignment process in the EA algorithm will sequentially reassign all clients that exceed the delay bound in each group. If the system load is high, we may not be able to reassign all clients exceeding the delay bound. In that case, only some groups in the system can be refined. Therefore, the response times for the groups may vary greatly: some refined groups have low response time, while the rest unrefined group have high response time. This will be demonstrated in Section 4.

The RREA algorithm is proposed as an improvement of the EA to the above problem. In general, RREA algorithm is basically the same as EA algorithm. However, RREA's refined assignment process follows a round-robin way: at each iteration we reassign only *one* client per group. For example, suppose that the number of clients exceeding the delay bound in group g_1 and g_2 is 3 and 2, respectively, and the system is near saturation so it only allows the reassignment of 3 clients. With EA, only 3 clients in g_1 are reassigned, hence RT_{g_2} is still high, which leads to high system response time SRT. With RREA, 2 clients in g_1 and 1 client in g_2 are reassigned, hence the SRT is lower, compared to EA. Thus, the RREA algorithm may improve the SRT better than the EA algorithm.

3.3 Dynamic Allocation

When interacting with the virtual world, clients may move from zone to zone (changing their target server). In this case, we need to dynamically allocate clients according to the allocation algorithm used. One of the most important problems in dynamic allocation is how to minimize the number of client migration from one server to another server, since client migration may affect the interactivity of the application [14]. Basically, a client needs to disconnect from its current contact server and establish a new connection to another contact server, which in general may take a long time compared to the delay bound of the application.

The VA algorithm will switch client to another server whenever the client changes the zone, since in VA the contact server and target server are the same. On the contrary, our proposed algorithms, EA and RREA, only need to switch a client to another server if the round-trip delay from the current contact server to the new target server exceeds the delay bound, otherwise the contact server of the client only needs to re-route this client's requests to the new target server. Hence, our algorithms is able to achieve a lower number of migrations than the VA algorithm.

4 Experiments and Results

4.1 Experiment Parameters

In the experiments, the round-trip delay between servers are randomly selected in the range $[40-70ms]$, and the round-trip delay between clients and server are randomly selected in the range $[120-300ms]$. These values are taken from [17]. The load threshold T of a server is set to 50 clients. The network-related load and application-related load is assumed to be equal, i.e., $\alpha = 2\beta = 1$, as shown in [11]. Each zone in the virtual world has 25 clients interacting concurrently. We measure the performance results by executing each algorithm 100 times.

4.2 Results and Discussions

In this section, we discuss the experimental results to compare our proposed algorithms, EA and RREA, with the VA algorithm. The response time of a group (RT_{g_i}), system response time (SRT) as well as the *system utilization* (measuring how full the system is, and is calculated as the server load divided by the load threshold) are of interest in the analysis.

The response times for *each* group (RT_{g_i}) in a 50-servers system with the delay bound $D = 200ms$ are shown in Fig. 2, in which Fig. 2(a) shows a 40-groups system (1000 clients), and Fig. 2(b) shows a 80-groups system (2000 clients). In Fig. 2(a), it is observed that EA and RREA are able to provide an allocation that guarantees the delay bound $200ms$ $(SRT \leq 200ms)$, while the VA algorithm is failed in doing so. In addition, Fig. 2(a) shows that EA and RREA have similar performance. This is because the system is not fully utilized in this case, as shown in Table 1, thus the EA is able to refine client assignment for all groups. With 1000 clients, the system utilization in EA and RREA is around 0.5, which is slightly higher than the utilization in VA (which is 0.4). This is due to the extra network-related load that contact servers in EA and RREA have to handle.

When the system load is high, RREA has better performance than EA, as shown in Fig. 2(b). In this figure, both EA and RREA are only able to provide an allocation that *approximates* the delay bound of $200ms$, since the system is very near to saturation (the system utilization is near to 1), as shown in Table 1, hence the EA and RREA can't continue to reassign clients' contact servers to met the delay bound. The system utilizations of EA and RREA in this case are 0.9887 and 0.9782, respectively.

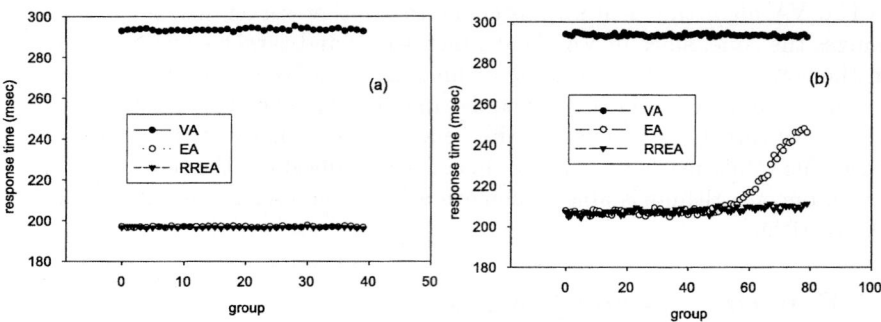

Fig. 2. RT for each group, 50 servers, $D = 200ms$ (a)1000 clients, (b) 2000 clients

Table 1. System utilization, 50 servers

Algorithm	$D = 200ms$		$D = 150ms$	
	1000 clients	2000 clients	1000 clients	2000 clients
VA	0.4	0.8	0.4	0.8
EA	0.4973	0.9887	0.5301	1
RREA	0.4972	0.9782	0.574	0.985

However, in this case, we noted that the SRT in RREA is better than in EA (the improvement is around $30ms$), due to the round-robin refined assignment process. The EA can only approximate the delay bound for a number of groups, but it failed to do so with remaining groups, as shown in Fig. 2(b). Thus, the SRT of EA in this case is higher than the SRT of RREA, as shown in Table 2. We may conclude that although EA and RREA have similar system utilization, RREA shows better performance when system has high load.

In addition, RREA can approximate the delay bound much better than EA, if tighter delay bound is required, i.e., the interactivity requirement of the application is higher, which means there may be more clients exceeding the delay bound that need to be reassigned. As shown in Table 2, when the delay bound is $150ms$, the RREA is able to achieve a much better SRT than EA in high system load (2000 clients, the improvement is around $65ms$). Note that in this case, the delay bound $150ms$ can't be guaranteed, due to the chosen delay ranges in the experiment, and RREA can only approximate the bound better than EA. Hence, we may conclude that the RREA algorithm is able to achieve an allocation with better SRT than the EA algorithm, especially when the system is in high load, and interactivity requirement is high.

We have also experimented our proposed algorithms with the dynamic case, in which clients move from zone to zone. In the experiment, for each group, we change the zone for a random number of clients in the range [0, 25]. Table 3 shows the effectiveness of the EA and RREA algorithm over the VA algorithm. Our algorithms are able to provide a better SRT than that with VA algorithm, and at the same time, the number of client migrations due to the reallocation process

Table 2. System response time, 50 servers

	$D = 200ms$		$D = 150ms$	
Algorithm	1000 clients	2000 clients	1000 clients	2000 clients
VA	294.81	294.02	293.87	294.23
EA	197.71	248.02	182.65	278.82
RREA	197.52	211.2	177.39	211.39

Table 3. SRT and client migration, 1000 clients, 50 servers, $D = 200ms$

Algorithm	SRT	Number of migrations
VA	294.62	95.82
EA	197.11	27.36
RREA	196.98	32.78

in our algorithms is much smaller than the VA algorithm. More specifically, the number of migrations in our algorithms is around 1/3 the number of migrations in the VA algorithm. Thus, our algorithm may help to reduce the impact of client migrations on application interactivity.

In summary, EA and RREA are much better than the existing VA algorithm in terms of interactivity. The RREA algorithm yields better performance than EA algorithm, especially when the system is in high load, and the interactivity requirement is high.

5 Conclusions and Future Works

In this paper, we have formulated the client allocation problem for DVEs. The interactivity of the DVE system can be achieved if the allocation mechanism could satisfy the delay bound requirement and at the same time, avoid system saturation. We have proposed two algorithms, EA and RREA, to address the client allocation problem. Experiment results show that our algorithms are much better than the traditional VA algorithm in terms of interactivity performance.

In our future work, we will investigate how to reduce the system utilization in our EA and RREA algorithm, while maintaining the interactivity of the DVE system.

References

1. S. Singhal and M. Zyda: Networked Virtual Environments. Addison-Wesley, New York (1999)
2. J. Dias, R. Galli, A. Almeida, C. Belo and J. Rebordao: mWorld: A Multiuser 3D Virtual Environment. IEEE Computer Graphics Vol 17(2) (1997)
3. D. Miller, J. Thorpe: SIMNET: The advent of simulator networking. Proc. of the IEEE Vol. 83(8) (1995)

4. T. Nitta, K. Fujita and S. Cono: An Application of Distributed Virtual Environment to Foreign Language. IEEE Education Society (2000)
5. J. Smed, T. Kaukoranta, and H.Hakonen: Aspects of Networking in Multiplayer Computer Games. Proc. of the International Conference on Application and Development of Computer Games in the 21st Century (2001)
6. Zona Inc. and Executive Summary Consulting Inc.: State of Massive Multiplayer Online Games 2002: A New World in Electronic Gaming. Available at http://www.zona.net (2002)
7. T. K. Das, G. Singh, A. Mitchell, P. S. Kumar, and K. McGee: NetEffect: A Network Architecture for Large-Scale Multi-User Virtual Worlds. Proc. of the ACM VRST (1997)
8. E. Cronin, B. Filstrup and A. Kurc: A Distributed Multiplayer Game Server System. Technical Report, University of Michigan (2001)
9. Sony Online Entertainments: Everquest. Available at http://eqlive.station.sony.com.
10. K. W. Lee, B. J. Ko and S. Calo: Adaptive Server Selection for Large Scale Interactive Online Games. Proc. of NOSSDAV (2004)
11. A. Abdelkhalek, A. Bilas and A. Moshovos: Behavior and Performance of Interactive Multi-player Game Servers. Special Issue of Cluster Computing: the Journal of Networks, Software Tools and Applications (2002)
12. P. Morillo, J.M. Orduna, M. Fernandez and J. Duato: On the Characterization of Distributed Virtual Environment Systems. Proc. of Euro-Par (2003)
13. J. Lui and M. Chan: An Efficient Partitioning Algorithm for Distributed Virtual Environment Systems. IEEE Transaction on Parallel and Distributed Systems Vol. 13(3) (2002)
14. Duong N. B. Ta and S. Zhou: A Dynamic Load Sharing Algorithm for Massively Multi-Player Online Games. Proc. of the 11th IEEE International Conference on Networks (2003)
15. W. Cai, P. Xavier, S. Turner and B. S. Lee: A Scalable Architecture for Supporting Interactive Games on the Internet. Proc. of the 16th Workshop on Parallel and Distributed Simulation (2002)
16. K. Fujikawa, M. Hori, S. Shimojo and H. Miyahara: A Server Selction Method based on Communication Delay and Communication Frequency among Users for Networked Virtual Environments. Proc. of ASIAN (2002)
17. V. Cardellini, M. Colajanni and P. S. Yu: Geographic Load Balancing for Scalable Distributed Web Systems. Proc. of MASCOTS (2000)

IMNET: An Experimental Testbed for Extensible Multi-user Virtual Environment Systems

Tsai-Yen Li, Mao-Yung Liao, and Pai-Cheng Tao

Computer Science Department, National Chengchi University, Taipei, Taiwan
{li, g9105, g9310}@cs.nccu.edu.tw

Abstract. Multi-user virtual environment (MUVE) systems enable virtual participation in many applications. A MUVE usually is a complex system requiring technologies from 3D graphics and network communication. However, most current systems are designed to realize specific application contents and usually lack system extensibility. In this paper, we propose an extensible architecture for a client-server based MUVE system called IMNET. This XML-based MUVE system allows function modules to be flexibly plugged into the system such that network or user interface experiments can be easily incorporated. We will use two examples to illustrate how to flexibly change the system configurations on the server and client sides to enhance system functions or to perform experiments. We believe that such an experimental test-bed will enable a wider range of researches to be carried out in a more efficient way.

1 Introduction

A multi-user virtual environment (MUVE) system is a system allowing many users to share the same 3D virtual world through the network and participate in the activities in the world as avatars. It allows a user to interact with other users or the environment via textual or visual communications. A snapshot of the user interface in a virtual environment is shown in Fig. 1. The feature of not being constrained by physical existence allows such a system to have a great potential value in applications that cannot be easily realized in the real world. For example, a 3D role-playing game allows its users to act as a fictional characters in an ancient world. A MUVE can also be adopted to simulate military activities. In addition, many examples have demonstrated that it can also be used to visualize or perform scientific experiments that cannot be easily explained by texts and figures [3].

Many MUVE systems have been proposed in the literature. In early years, most systems were developed for research purposes. However, in recent years, one can see more commercial systems being designed to host such a virtual environment for general purposes or for special purposes such as on-line games. Designing a MUVE is a complex task requiring multi-discipline trainings involving networking and 3D technologies. Most MUVE systems are packaged as a standalone application or a program module that can be embedded in a web page. Although some of them may have external application programming interface (API) for integration with other programs [12], most of them cannot be extended at design time or configured at run time. In this paper, we propose an experimental MUVE test-bed, called *IMNet (Intelligent Media*

Fig. 1. An example dialog scene in a virtual environment

Network), that is designed to be extensible for incorporating other function modules such as message filters or user interface components. IMNet adopts a client-server architecture and uses XML as the base language for server and client configurations as well as the message protocols for MUVE. We will demonstrate the extensibility of our system by two examples incorporated into the server and client programs, respectively.

The rest of the paper is organized as follows. We will review the work pertaining to MUVE systems. In the third section, we will describe the proposed extensible system architecture for the server and client programs. We will describe the message protocol and its encoding in the fourth section. Two examples will then be given to illustrate the functions of the experimental test-bed. Finally, we will conclude the paper with some future extensions.

2 Related Work

The MUVE related research proposed in the literature has various aspects. Some of them focus on system architecture and message protocols [1][5][6] while others focus on applications such as in military simulation and education [3][2]. In terms of system architecture, most systems fall into two types: *client-server* and *peer-to-peer*. The client-server architecture is the most widely used one. For example, RING [4] by UC Berkeley and AT&T Bell lab, Community Place [7] by the Computer Science laboratory and Architecture laboratory of SONY, Blaxxun Community Server [13] by Blaxxun, and ActiveWorlds [12] system by ActiveWorlds are all examples that adopt a client-server architecture. VNet is another MUVE system with a client-server architecture that opens its source for cooperative development [10]. In this type of systems, since messages must be routed through the server, the server can easily become the bottleneck. Therefore, many researches try to address the problem of reducing the amount of data transmission by data filtering or dead reckoning techniques. In addition, some research proposes the idea of using multiple servers to distribute the load on the server side [11]. This type of research aims to increase the scalability of a MUVE system such that more users can be served at the same time. The experiments in much of this research are done by modifying a specific MUVE system, and the implementation cannot be easily ported to other systems.

In addition to the communication issues, standards for 3D animation and display on the client side also attract many attentions. Many recent MUVE systems use open

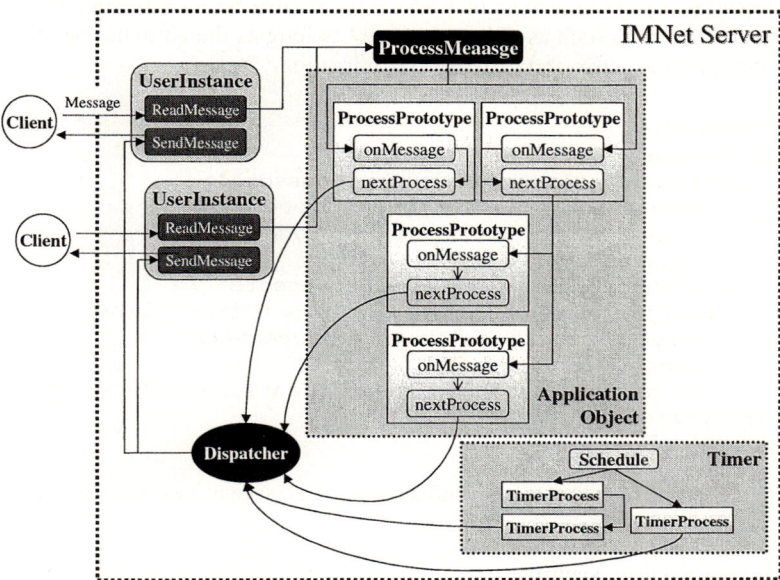

Fig. 2. System architecture of IMServer with the pluggable message processing mechanism

standards such as VRML [15] to model the geometry and animation of a virtual scene while the other use proprietary formats. The 3D display program is usually packaged as a 3D browser that can be embedded in a web page. The browser could be stand-alone or integrated with other external programs such as a Java Applet to provide application-specific functions [13]. Due to the extensibility of XML (eXensible Markup Language), the VRML standard is migrating into the XML-based X3D language [18]. However, most MUVE systems use a fixed proprietary format as the application protocol for message passing [10]. [9] is an example that attempts to change the message format of VNet to an XML-based protocol. Although the system designer can easily design new tags to enrich the functions of a MUVE, the programs on the client and server sides needs to be modified to accommodate the changes. The system is not designed to incorporate plug-in modules that are specified at design time or even at run time. In addition, although the protocol is more extensible, the size of a same message could be larger if raw XML strings are used for transmission.

3   System Description

In this section, we will describe the system architecture of the server and client programs in the IMNet virtual environment system. IMNet is a client-server based MUVE system adopting XAML (eXensible Animation Modeling Language)[8] as the language for 3D display and animation. The server program is called *IMServer* while the client is called *IMClient*. XAML is an animation scripting language that is designed to specify animations in a range of abstractions. For example, it can be used to specify low-level joint values as in VRML. It can also be used to specify high-level goal-oriented motions such as "Move to Café" as long as the animation engine knows how to interpret the script and generate the animation.

```
<serverConfig>
   <processors>
      <processor class="example.processorA">
         <processor class="example.processorC" />
         <processor class="example.processorD" />
      </processor>
      <processor class="example.processorB" />
      <timerprocessor class="exampleTestTimer" delay="5000">
         <processor class="example.processorE" />
      </timerprocessor>
      <timerprocessor class="exampleTestTimer" delay="2000" />
   </processors>
</serverConfig>
```

Fig. 3. Example of server configuration on message processing structure

3.1 Server System Architecture

According to [7], a MUVE system consists of four modules, Client, Server, Application Object, and Server Client Protocol. The Application Object (AO) module is responsible for interpreting the messages and managing the application contexts (for example, virtual shopping mall). Data filtering routines such as dead reckoning algorithms can also be implemented in the AO module to improve the performance of the server by filtering out unnecessary information for the clients.

The system architecture of IMServer including the AO module is shown in Fig. 2. When a client logs into the system, a UserInstance is created on the server to take care of the message input and output for the client. All messages are sent to the ProcessMessage routine for data processing in the AO module. Each data processing unit in the AO module is called ProcessPrototype. All ProcessPrototype's in the AO module are organized as a tree structure to process the message data in parallel or in sequence. The onMessage method of the first ProcessPrototype in each tree branch is called to process the message data and decide if it will pass the data to the next ProcessPrototype or simply filter them out. Each of the leave ProcessPrototype's in a tree may generate messages to the dispatcher for distribution to other clients. In addition to being driven by the incoming message events, the server can also produce messages voluntarily through the timer service. The processing units of the service are also organized in a tree structure such that they can work together in parallel or in sequence.

With best extensibility in mind, we have designed a mechanism to set up the above processing tree at run time on the server side. This mechanism is described as an XML configuration file, as the example shown in Fig. 3. The java class is specified in the "class" attribute of each process. In this example, process A and B are the start of

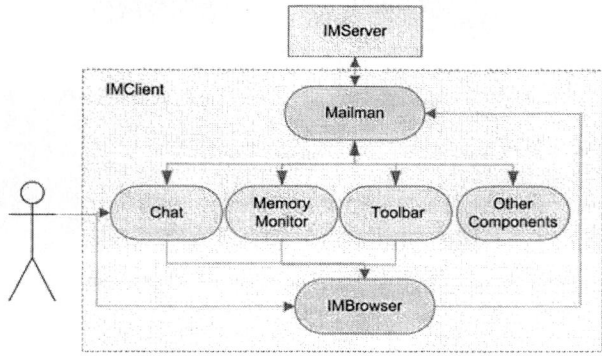

Fig. 4. System architecure of IMClient

```
<imclient>
 <components>
    <component class="Mailman" name="mailman"/>
    <component class="SpChat" name="chat"/>
    <component class="SpActionButton" name="actionButton"/>
    <component class="SpMemoryMonitor" name="memoryMonitor"/>
    <component class="SpIMBro" name="browser"/>
 </components>
 <eventDispatcher name="mailman">
    <connect ip="127.0.0.1" port="62266"/>
    <eventlistener name="chat"/> ...
 </eventDispatcher>
 <toolbar>
    <button name="memoryMonitor"/> ...
 </toolbar>
 <toolbar>
    <buttonContainer name="actionButton">
       <xamlButton name="bow" text="Bow"
file="Behavior/Bow.xml"/>
       ...
    </buttonContainer>
 </toolbar>
 <panels>
    <panel name="browser"/>
    <panel name="chat"/>
 </panels>
</imclient>
```

Fig. 5. An example of configuration file for IMClient

the two branch processes. Process A first filters the messages and pass them to process C and D s equentially. Since both processes B and D are the last process in their braches, they may produce messages that will be distributed to other clients. In addition, independent timer processors can be evoked periodically according to the specified delays. Since the processing routines' classes are organized and bound at run time, experiments can be done easily by specifying appropriate filtering or processing routines in the configuration file without recompiling the server's code.

Fig. 6. A snapshot of IMClient user interface for the configuration in Fig. 5

3.2 Client System Architecture

The client side program of IMNet is called IMClient. The system architecture of IMClient is depicted in Fig. 4. The program consists of two major components: Mailman (communication module) and IMBrowser (3D animation engine), and other GUI components such as textual chat and action buttons. The program is updated according to two types of events: messages from the server and actions from the user. A message from the server is first processed by the Mailman module and passed to all other interested modules. Corresponding components of the screen will be updated according to the type of the message such as a movement or a chat message. The user can also create events, such as entering chat messages or clicking on action buttons, to be sent to the server via the Mailman module.

A main feature of IMClient is that the components comprising the program can be configured at run time. The program is set up by loading a configuration file at initialization such as the one shown in Fig. 5. In this file, each class module is defined as a named component which may or may not contain a GUI widget. The relations among these components are set up according to the event dispatcher and listener model. The latter part of this file describes how the components are connected to the GUI widgets. For example, both IMBrowser and Chat implement a panel widget to be arranged in the client window as shown in Fig 6. Two types of toolbars are also used in the client window: static toolbar and dynamic toolbar. A static toolbar contains buttons that must be initialized at start-up time while dynamic toolbar allow buttons to be created and inserted at a later time. For example, an action button of an avatar for a canned motion in a MUVE can be downloaded from the server as long as the canned motion is described by a XAML script.

```
<IMNet from="userA" to "userB">
  <Chat> See you later. </Chat> <!-textual>
  <AnimItem> <!- XAML script>
     <AnimImport src="Bye"/>
  </AnimItem>
</IMNet>
```

Fig. 7. An example message in IMNet with textual and animation contexts

4 Message Protocol

4.1 IMNet Message Protocol

The message protocol used in IMNet adopts XAML as the base animation scripting language. The protocol needs to deliver messages containing information such as user login events, movements, and animation. Since complex and extensible animations can be embedded in an XAML script, the remaining message types for the virtual environment application can be kept minimal. In the current design, the additional tags include the following: `<IMNet>`, `<Chat>`, `<Login>`, `<Logout>`, and `<UserMove>`. In Fig. 7, we show an example message about a user A whispering to user B while performing an animation described in an XAML script, stored in a separated file. In addition, a login message uses the format of `<Login id="userC" url="wrl/avatar_03.wrl">` while a user movement message uses the format of `<UserMove x="10" y="20">`. The latter message can also be described in an XAML script, but we make it a standalone message to optimize this type of frequently used actions.

4.2 Message Encoding

Although messages in the XML format have the advantage of being extensible, they also have the drawback of being large in size. The problem gets worse for a MUVE system when the animation gets more low-level and complex. Similar problems also arise in the WAP (Wireless Application Protocol) [16] application. The WAP development community proposed an encoding method called WBXML (WAP Binary XML) [17] to convert XML string into a concise binary format. We have also adopted such an encoding method to deliver IMNet messages. However, instead of converting an XML string to WBXML, we generate WBXML directly from an internal DOM (Document Object Model) for efficiency.

We have done experiments to compare the encoding and decoding performance of different methods as well as with the original XML format. The experimental data, as shown in Fig. 5, were measured on a personal computer with an AMD XP2500 processor. Note that the WBXML encoding method outperforms the Java serialization method and the common ZIP compression method in encoding time as well as decoding time. However, the string size after the WBXML encoding is 2.5 times larger than the one with ZIP compression on average although the size has been reduced by 4.2 times on average compared to the Java serialization method.

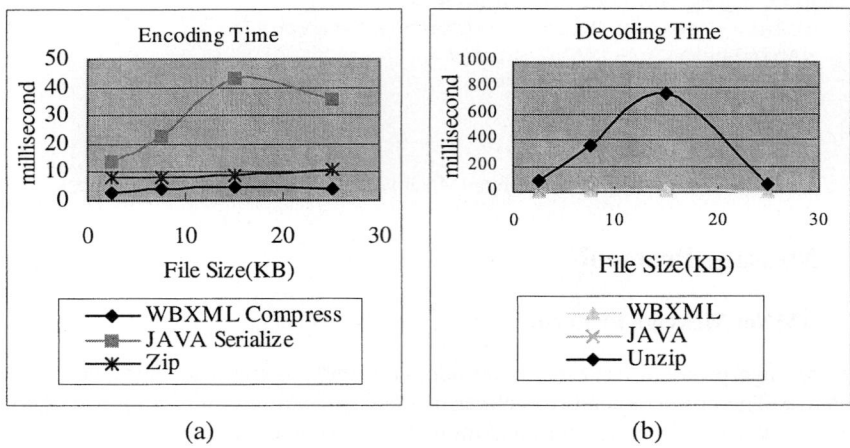

Fig. 8. Comparisons of encoding and decoding performances

5 Examples of System Extensibility

5.1 User-Centric Throughput Adjustment Experiments

For a client-server based MUVE system like IMNET, the server is commonly considered as a bottleneck for message exchanges. Therefore, much research has proposed to use the idea of data filtering to reduce the amount of traffic that needs to be transmitted across the network. The decision is usually made by some intelligent modules such as view culling and dead reckoning on the server side to filter out unnecessary information according to each client's configuration. In fact, each client's ability in display and network I/O may vary greatly, and a uniform policy is not going to fit every clients need and may waste the server's resources in sending out unnecessary messages that the clients cannot digest. Other factors such as message types and user activities may also imply the demands for customized filtering policies according to the user's model. The server should also adjust its message update frequency according to its own CPU and network I/O performance. We call a server with this type of capability a server that can perform user-centric throughput adjustment.

Fig. 9 shows the system configuration, similar to Fig. 2, for this experiment. The performance monitoring module subscribes the incoming messages and monitors the CPU and I/O performance of the server. These data are maintained as the system states for the server and clients. The other branch of the message flow starts from the dead reckoning module, which filters out messages for the clients keeping the expected moving direction. The filtered messages will be passed to the throughput control module which determines whether the messages should be sent to each specific client or not according to their system states and the server's current loading. The filtered messages are sent out to the clients via the dispatcher module. Note that setting up a system experiment like this does not require the designer to recompile the program since the message flows are set up at run time according to a system configuration file similar to the one shown in Fig. 3. This feature allows the designer to insert or remove different experimental modules and treat the system as a flexible

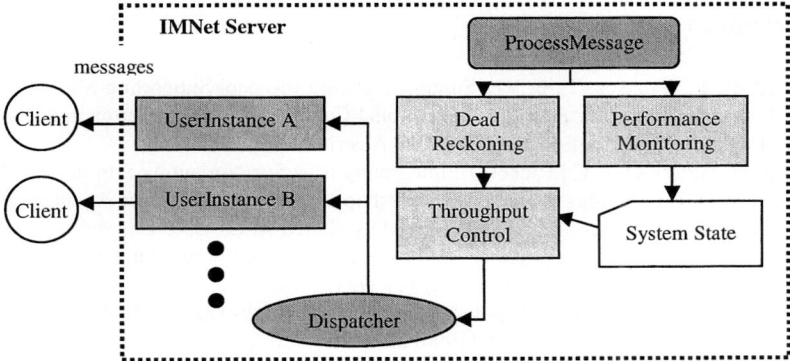

Fig. 9. An example of server configuration for the adjustable throughput control experiment

or remove different experimental modules and treat the system as a flexible experimental test-bed.

5.2 Extension to Voice-Enabled User Interface

One can also extend the functions of IMClient by specifying appropriate components in the configuration file. We will use a voice-enabled user interface as an example to illustrate how to add new system functions and user interface modules into the system. The voice-enabled user interface is based on a research aiming to incorporate voice dialogs in MUVE [8]. The system allows two avatars in the same scene to talk to each other via voice dialogs and embedded animations while allowing a third user to observe the dialog. The protocol that the system has used is called XAML-V since it acts as a plug-in extending the XAML animation scripting language. Assume that this new module is called the VUI module. It implements a text panel for displaying voice dialogs and subscribes to the incoming messages via the Mailman component. This new module can be hooked up to IMClient with ease by modifying the XML-based configuration file in Fig. 5.

6 Conclusions

In this paper we have described an experimental testbed for MUVE. This system adopts an XML-based protocol that allows an extensible animation scripting language to be efficiently embedded in the messages. The extensibility of the system is also shown in the system configurability of the server and clients by two examples. We believe that the extensibility of this system will enable more research to be conducted in a more efficient way.

Acknowledgement

This work was partially supported by a grant from National Science Council under a contract NSC 93-2213-E-004-001.

References

1. Bouras, C., Tsiatsos, T.: pLVE: Suitable Network Protocol Supporting Multi-User Virtual Environments in Education. In: International Conference on Information and Communication Technologies for Education, Vienna, Austria (2000) 73-81
2. Elliott, C, Lester, J. C., Rickel, J.: Integrating affective computing into animated tutoring agents. In: Proceedings of IJCAI '97 workshop on Intelligent Interface Agents (1997)
3. Fellner, D.W., Hopp, A.: VR-LAB - A Distributed Multi-User Environment for Educational Purposes and Presentations. In: Proceedings of the Fourth Symposium on the Virtual Reality Modeling Language, Germany (1999) 121-132
4. Funkhouser, T.: Network Topologies for Scalable Multi-User Virtual Environments. In: Proceedings of IEEE VRAIS'96 (1996) 222-228
5. Greenhalgh, C.: Implementing Multi-user Virtual Worlds: Ideologies and Issues. In: Proceedings of the Web3D-VRML 2000 fifth Symposium on Virtual Reality Modeling Language (2000) 149-154
6. Huang, J. Y., Fang-Tsou, C. T., and Chang, J. L: A Multi-user 3D Web Browsing System. In IEEE Internet Computing, Vol. 2, 5, (1998) 70-79
7. Honda, Y., Matsuda,, K., Rekimoto, J., Lea, R.: Virtual Society: extending the WWW to support a multi-user interactive shared 3D environment. In: Proceedings of the First Symposium on Virtual Reality Mmodeling Language (1995) 109-166
8. Li, T.Y., Liao, M.Y., Liao, J.F.: An Extensible Scripting Language for Interactive Animation in a Speech-Enabled Virtual Environment. In: Proceedings of the IEEE International Conference on Multimedia and Expo (ICME2004), Taipei, Taiwan (2004)
9. Liu, Y.L., Li, T.Y.: A Multi-User Virtual Environment System with Extensible Animations. In: Proceedings of the Web3D 2003 Symposium (2003)
10. Robinson, J., Stewart, J., and Labbe, I.: MVIP-audio enabled multicast VNet. In: Proceedings of the Web3D-VRML 2000 Fifth Symposium on Virtual Reality Modeling Language (2000) 103-109
11. Smed, J, Kaukoranta, T., Hakonen, H.: A Review on Networking and Multiplayer Computer Games. In :Technical Report 454, Turku Centre for Computer Science (2002)
12. ActiveWorlds, http://www.activeworlds.com
13. Blaxxun, http://www.blaxxun.com/
14. VoiceXML, http://www.w3.org/TR/voicexml20/
15. VRML, http://www.web3d.org/x3d/specifications/vrml/vrml97/
16. WAP, http://www.wapforum.org/
17. WBXML, http://www.w3c.org/TR/wbxml/
18. X3D, http:///www.web3d.org/x3d

Application of MPEG-4 in Distributed Virtual Environment

Qiong Zhang[1], Taiyi Chen[1], and Jianzhong Mo[2]

[1] College of Computer Science and Technology, Zhejiang University, China
[2] eMedia Center, University of Wisconsin-Madison, USA

Abstract. MPEG-4 is a multimedia standard which basically defines an innovative object-based encoding/decoding scheme for audio-visual scene. It provides a set of powerful tools for content authors, network service providers and end users. In this article, we discussed the possible applicability of MPEG-4 technologies in the Distributed Virtual Environment (DVE), based on the analysis of the essential requirements of DVE. An application example, i.e., Collaborative Virtual Disassembly is introduced afterward.

1 Introduction

Distributed Virtual Environment provides users with the illusion that they are in a shared virtual world where they could collaborate, communicate, and interact with other participants as well as the environment itself [1]. In order to achieve above function, several additional requirements not provided by stand-alone virtual environment must be met. Singhal and Zyda [2] pointed out several common features of networked virtual environments, i.e., a shared sense of space, a shared sense of presence, a shared sense of time, a way to share, and a way to communicate.

MPEG-4 (i.e. ISO/IEC 14496) became International Standard in the beginning of 1999. It builds on the proven success of three fields [3], i.e., digital television, interactive graphics applications and interactive multimedia. This paper focus on the application of MPEG-4 based technology in distributed virtual environment. Section 2 discusses the applicability of MPEG-4 technologies in DVE. Section 3 gives a simple DVE example, i.e. collaborative virtual disassembly, followed by a conclusion in Section 4.

2 MPEG-4 Tools and Its Application in DVE

MPEG-4 standard primarily includes four parts, i.e. Systems, Visual, Audio and DMIF (Delivery Multimedia Integration Framework). Roughly speaking, Systems part is responsible for the scene description (that is the spatio-temporal positioning of audio-visual objects as well as their behavior in response to interaction) and synchronization, identification, description and association of stream content. Visual and Audio part provide a set of technologies to represent natural or synthetic, 2D or 3D video and audio objects respectively. DMIF part defines a generic interface to the data stream delivery layer functionality.

MPEG-4 uses a client-server model. An MPEG-4 client (or player, or browser) as illustrated in Fig 1 contacts an MPEG-4 server, asks for content, receives the content, and renders the content. This "content" can consist of video data, audio data, still images, synthetic 2D or 3D data, or all of the above.

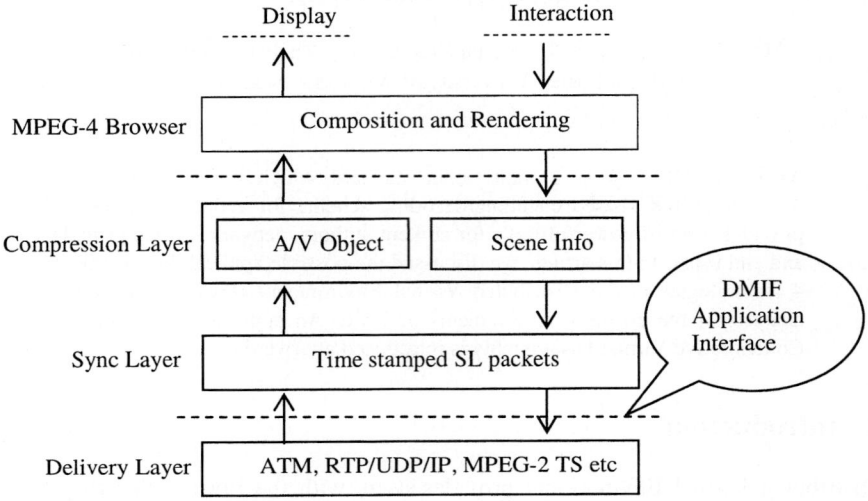

Fig. 1. Typical MPEG-4 Client Architecture

Although MPEG-4 is not attached to any particular application in mind, many of its features or tools can easily applied to the development of DVE. Based on the requirement analysis in the previous section, we show how MPEG-4 tools fit those requirements and help the development of DVE.

The shared sense of space among all the users is achieved by the scene structure of Binary Format for Scenes (BIFS), a key part of MPEG-4 Systems. It consists of an encoded hierarchical tree of nodes with attributes and other information. Essentially, the scene structure provides a common virtual space where user can join in and resign from.

The shared sense of presence can be represented by individual avatar in the common scene structure. Every avatar has status, position, geometry attributes, view direction and other properties. It can be realized with BIFS in a very intuitional way. All the users have their corresponding node in the scene structure. User join-in or resignation behavior is simply equivalent to the node add or deletion operation.

The shared sense of time requires the ability to change the scene and inform/update all users in time. MPEG-4 Systems have a set of technologies to deal with scene update and real time issues of multiple streams (i.e. stream synchronization) and single stream, which include BIFS Scene update, synchronization between streams, and so on.

The way to share can be grouped into two major categories: client-side interaction and server-side interaction. Client-side interaction involves content manipulation, which is handled locally at end-user's terminal. Server-side interaction involves content manipulation that occurs at the transmitting side, initiated by a user action. This requires a back channel support. Furthermore, the client-side interaction can take several forms. Generally, interactivity mechanisms are integrated with the scene description information, in the form of linked event sources and targets (i.e. routes) as well as sensors (special nodes that can trigger events based on specific conditions). However, most application-dependent interactions that usually triggered by specific keyboard key presses or mouse movements are beyond the MPEG-4 scope. Just like server-side interaction, those types of interaction need not be standardized.

There are possible three ways for users to communicate each other in MPEG-4, including BIFS command, back channel, and MPEG-J.

3 Example – Collaborative Virtual Disassembly

Collaborative virtual disassembly is a typical DVE application. In this environment, engineers must share data and work together to disassemble a product. MPEG-4 technology can help create collaborative environment for virtual disassembly. Fig 2 illustrates the basic architecture of this system. It is a client/server-based system,

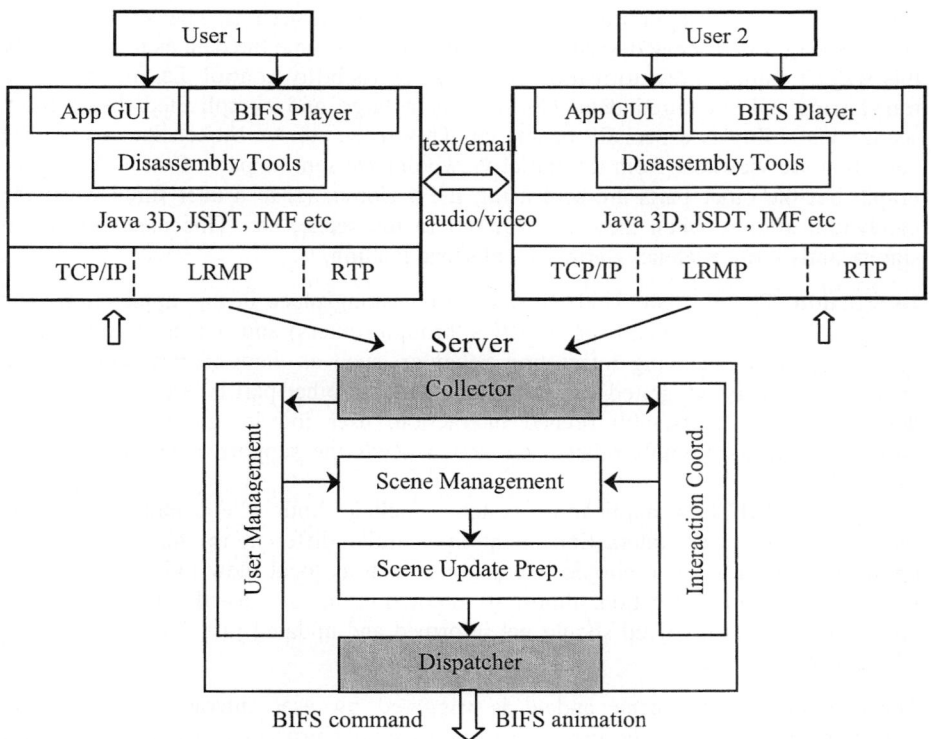

Fig. 2. Architecture of the Collaborative Virtual Disassembly System

including one server and variable number clients. Please note we only apply a portion of possible MPEG-4 tools discussed in section 2 into this prototype system. In particular, we use BIFS as the disassembly scene structure representation and large part of the interaction is based on embedded sensor nodes and route definition. BIFS command and animation is used as scene update method to dynamically update scene structure at client site. The main functionalities and implementation of this system can be described as follows.

Scene Description. BIFS is used for scene description in virtual disassembly environment. All the users involved have their own avatar node representation in the scene structure. User join-in or resignation is accomplished by node insertion or deletion.

BIFS builds largely on the concepts from Virtual Reality Modeling language (VRML) in terms of both its structure and the functionality of object composition nodes and extends it to fulfill MPEG-4 requirements.

Compared with VRML, the efficiency of BIFS representation is largely due to binary and context-dependent encoding scheme. Through context-dependent encoding and quantization, there's generally single or double digits compression rate gain dependent on the encoded objects.

Besides, since BIFS not only describes the spatial arrangements of the objects in the scene, but also temporal update relationship between streams. The server side can stream the encoded scene description and receivers consume the data as it arrives. By this way, it's quite direct to realize awareness or visibility control, i.e., the ability to transferring only required data. Considering a large and complicated disassembly scene, this ability is especially beneficial. This scene can be broken into parts. The parts that are viewable from the initial viewpoint are sent as part of the initial scene graph, but the other parts are sent using BIFS Command at a later time: when the bandwidth allows, when they are required in the scene, etc. This mechanism can significantly reduce latency during initial scene loading.

Interaction. There're two kinds of interaction taking place in this application. The first part is already standardized by BIFS through route(s) and sensor(s). Some most often frequently occurring interaction behavior, such as view change and position change, can easily integrated into this framework. Another part is largely application dependent, e.g. disassembly-related interaction, user join-in and resignation etc. Among them, disassembly-related interaction needs the support of virtual assembly tools [5].

Common MPEG-4 application systems include both client and server side interaction. However, interaction situation is quite different in this typical DVE application. Except some simple interactions, such as local view point change only need one way communication, almost all interactions are server-side interaction since all the other users involved should get informed and updated just like the local user who triggers this event.

Scene Update. The scene update is triggered by user interaction. After user interaction approved by the consistence management process at server side, server will then prepare corresponding BIFS command or BIFS animation frames and send them to all the other users for scene update.

Regarding the disassembly interaction, it usually can be represented with disassembly sequence/path. If server translates this kind interaction into BIFS commands, a sequence of BIFS commands are required to accomplish this process. During this process, some data redundancy will occur due to the fact that every BIFS command need to specify which node and which field to update. On the contrary, BIFS animation provide a more efficiently way to communicate continuous changes in a scene. For example, BIFS animation can achieve better compression than BIFS command for changing always the same field. The gain comes from the fact that the NodeID and fieldID needs not to be specified at each frame in the case of BIFS animation and from the specific quantization, prediction and entropy encoding used.

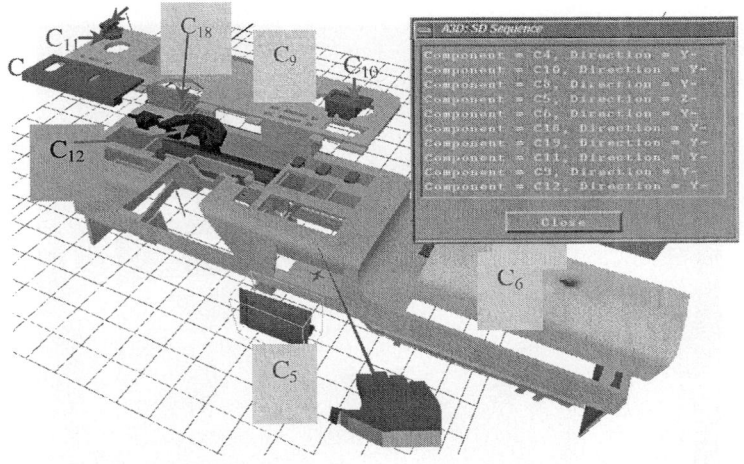

Fig. 3. An example of disassembly sequence

All the other interaction is prepared with BIFS command, which is very suited for one-time update.

Communication. The communication between users can use email/attachment, text-based chatting. Furthermore, with the integration with media streams like audio and video, audio chatting and video conference/communication can be seamless

Server. Basically, server plays a role as the central coordinator for all the user interactions and command issuer for scene updates. It maintains two primary data structures. One is the dynamic BIFS scene structure implemented as a directed acyclic graph. It describes the relationships between nodes and includes an active ROUTE table. Another one is the active user list that points to corresponding node in the scene structure and records the user status and other administration data.

As shown in Fig 2, the Collector module is implemented as a daemon process and responsible for collecting all the client interaction requests. Dependent on the interaction request type (user related interaction or any other interaction), the request will be distributed to User Management or Interaction Coordination module for processing. User Management module will determine whether to allow user join in, resign, change status or privilege and so on. Interaction Coordination module will perform all the consistence check between different clients if there's a interaction from one client has effect on the interaction(s) coming from other client(s) and determine which interactions to occur based on the user role/privilege and arriving time. After that, dynamic BIFS scene structure and/or active user list will be updated through Scene Management module and Scene Stream and Update Message Preparation module is driven to prepare BIFS command or BIFS animation frames for clients. At last, the Dispatcher module will send corresponding scene update messages to all the clients.

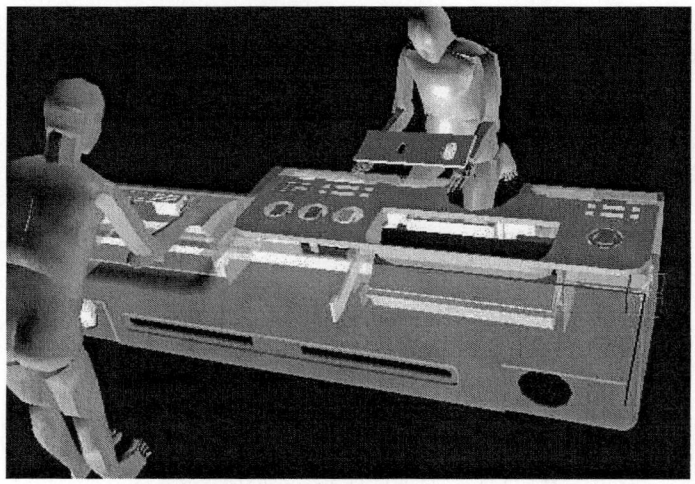

Fig. 4. Collaborative disassembling a dashboard in DVE

Fig 4 illustrates a sample scenario in this DVE application: two users are involved in the disassembling a dashboard. Either user has an avatar representing his current status (including position, view angle etc) in this environment.

3 Conclusion

This article gives an analysis of the applicability of MPEG-4 tools in DVE applications. For every aspect of DVE common features, suitable MPEG-4 candidate tools are introduced and discussed. The following DVE example Collaborative Virtual Disassembly further illustrates efficiency of MPEG tools in this typical application. We eventually can get to the conclusion that many of MPEG-4 technologies or tools are ready for use and can be easily applied to the development of DVE applications.

References

1. Churchill, E.F., Snowdon, D., Munro, A. (ed): Collaborative Virtual Environments, Springer-Verlag, Springer Verlag London (2001)
2. Singhal, S., Zyda, M.: Networked Virtual Environments – Design and Implementation, Addison-Wesley, New York (1999)
3. Koenen, R.: MPEG-4 Overview, ISO/IEC JTC1/SC29/WG11 N4668, March (2002)
4. ISO/IEC 14496-1, Information Technology-Coding of audio-visual objects, Part 1: Systems, January (1999)
5. Mo, J., Zhang, Q., Gadh, R.: Virtual Disassembly. International Journal of CAD/CAM, vol 2 (2002) 29-37

A New Approach to Area of Interest Management with Layered-Structures in 2D Grid[*]

Yu Chunyan[1,2], Ye Dongyi[1], Wu Minghui[3], and Pan Yunhe[2]

[1] College of Mathematics and Computer Science, Fuzhou University, Fuzhou, 350002, P.R.C
[2] Computer College, Zhejiang University, Hangzhou, 310027, P.R.C
[3] Department of Computer, Zhejiang University City College, Hangzhou, 310015, P.R. C

Abstract. Area of Interest Management, which essence is a communication filter, is an important component to improve scalability of CVE. With some special design considerations presented, this paper proposed a new approach called *GLAOIM* which is based on 2D grid with layered-structure. It is divided into two phases: message possible-influence area determination and message receiver determination. Whether a user could receive update message from a sender depends on the result of coactions of spatial relations and corresponding message transmission characteristics. Hence, 2D-Grid is applied to localize spatial position and solve spatial correlated problems quickly while layered structure is applied to combine all kinds of media's transmission characteristics, rules, etc. Moreover, *GLAOIM* introduces a new indicator *Interest Visibility* and provides an interior-outward traverse to calculate each grid's *Interest Visibility* value to determine receivers. Also, it provides a generic solution with better extensibility for messages with different kinds of media.

1 Introduction

As we know, entities in CVE produce a mass of update packets although in which only a minority of the users are interested. An obvious way to save bandwidth is to disseminate update packets only to the users who are interested in them [3]. For example, a user in a room with a single closed door does not need to know what is happening in the hallway. This technique to limit amount of messages a user receives when it takes part in a collaborative session is called area of interest management [5]; accordingly, an expression of data interest is called area of interest (AOI). AOI management (AOIM), of which essence is to transmit messages to users who are indeed interested in it instead of all users to reduce messages number exchanged in system, is an important approaches to enable CVE scalable [1,2]. Actually, it is a communication filter. AOI expression is the base and AOI filtering scheme is the core. Some papers used communication visibility to describe this problem [4].

The initial motivation of our work is that a user prefers to receive all kinds of information including video and audio packets, update messages, etc, from his adjacent users and virtual objects rather than those in distance. We propose an AOIM approach called *GLAOIM*, to handle message transmission among users and all virtual

---

[*] This project is supported by Fujian Technology Foundation No. K04005.

objects in a more scalable manner. It is based on a segmentation of the space into static and discrete two-dimension grids [6], each of which is associated with several layered structures including media-type-dependent transmission rules of update messages. Both users and message source are localized in certain 2D-grids and applied rules in layered structures relevant to their occupied and adjacent grids. Grid occupied by message source and its adjacent cooperates with their associated layers to select a subset of users in virtual world to create an AOI communication group. Each message source only sends messages to those in its AOI communication group.

This paper is organized as follows. Section 2 summarizes various existing approaches. Section 3 and 4 describe our 2D grid with layered structure based and message source position centric AOIM approach – *LGAOIM*. Section 5 describes an experiment and analyzes advantages. The conclusion is given in section 6.

2 Related Work

Existing AOIM approaches are broadly divided into two categories: region partition and aura management. The former still has two types: spatial and functional partition. In NPSNET [7,8], the world is subdivided into hexagonal regions. All participants in one cell are assumed to be interested in their own cell and its six neighbor cells. Papers on NPSNET expressed the importance of enhancing message culling using functional groups, such as tanks or foot soldiers, could have their messages included or culled as a group, in addition to standard cell-based spatial partition [8]. The COMIC model, used in the DIVE and MASSIVE systems, involves concepts of focus and nimbus [9,10]. Focus represents the area of space in which a user is interested for a certain medium while nimbus represents the area in which that entity can be detected. And an entity's aura is a bounding region containing both the focus and nimbus for purposed of determining interaction: if and only if two entity's auras overlap is there a possibility of interaction between these two entities. Hence, itis both message receiver and sender centric [12]. WAVE [11] supports both region partition and aura management. The world is partitioned into several regions for each of which there is an aura manager. Since hierarchically structured aura managers are, the movement of a participant from one region to another is performed via a parent aura manager. Although all the above approaches have achieved communication filtering, they have own deficiencies. Region partition is static and is not appropriate to dynamic CVE system. Aura management is based on a aura-nimbus information model: if a simple model which neglects surroundings in CVE is adopted, it cannot reflect the actual interest relationship and if a overall model takes all factors into consideration, it must be too complex to calculate and cannot guarantee real-time.

3 Overview of AOI management

Generally, AOIM is a process that determines which users are interested in current update message and whether receive it or not. Fig.1 illustrates its basic idea.

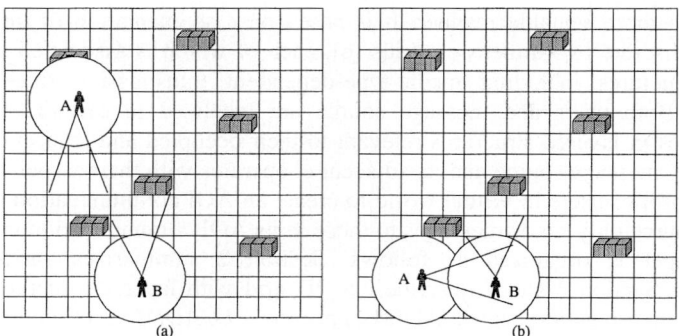

Fig. 1. This figure illustrates problem of AOIM. Circle represents area avatar can be detected and triangle represents area avatar can detect. In (a), neither A nor B is interest in each other for there is no intersection among circles and triangles. In (b), A's triangle interacts with B's circle. So, B is detected by A. That is, A is interested in B and should receive B's update messages

When designing an AOIM scheme, we mainly take following issues into consideration besides general requirements such as efficient, real-time and dynamic.

- Communication filtering. As mainly motivation and ultimate objective of AOIM, it should guarantee that a virtual entity's messages only be sent to those users who are interested in them to decrease communication traffic. This requirement has two aspects: complete and low redundancy. A user should receive all messages from the entities which he is actually interested in and all the users interested in a certain entity must receive all messages from this entity. To guarantee complete mentioned above, it is inevitable that there must be some redundant communication messages existing. Hence, another aspect is to achieve low redundancy.
- Message sender and receiver transparent and unsymmetrical. AOIM scheme should be transparent to both entity emitting message and user receiving message rather than user explicit requests or entity explicit assigns like a user registers an entity to request its update messages. Besides, message sender and message receiver are unsymmetrical, that is, an entity needn't send message to another one although maybe it receives messages from it simultaneously.
- Generic, Independent and Extensible. There are many kinds of message involved in AOIM; at least two kinds: one is visual dependent, and the other is aural dependent. Each kind of message has its own transmission characteristics. AOIM should provide a generic mechanism, independent of the message kind, to all kinds of message although they have different characteristics. AOIM should be extensible especially for messages of new media.

Different network architectures have different requirements for communication and an AOIM scheme should match the fundamental architecture to achieve best effect. Our experimental CVE system is web-based and adopts a kind of hierarchical client-server architecture as Figure 2 shows, which can greatly reduce communication traffic through dividing all users into several logical groups mastered by corresponding sub-servers. And now, *LGAOIM*, aiming at providing a natural, quick and user transparent scheme, is applied to manage AOI under a single sub-server, based on hierarchical client-server architecture and a set of 2D-Grids with layered structures.

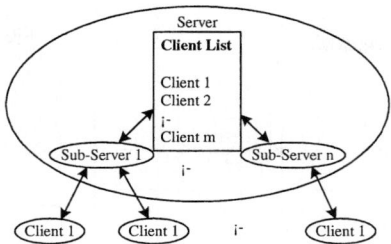

Fig. 2. Illustration of network architecture

4 AOI Management Based on 2D-Grids with Layered Structures

Whether a user should receive message from a sender depends on the result of coactions of spatial relations and corresponding message transmission characteristics. The spatial relations include distance, orientation, and shelter relation among sender, user and all virtual entities around sender, etc. Message transmission characteristics that are media-type-dependent include surroundings transmission characteristics and message per se characteristics. For example, it involves amplitude, spectrum, audio penetrability of certain surrounding entity, etc., for audio update message.

We divide AOIM into two steps: message possible influence area determination and message receiver determination. First, determining the maximum possible influence area of message based on message's media type and corresponding maximum transmission distance regardless of surroundings of itself. All users in this possible influence area have possibility to receive update message. Whether a user should receive messages is lied on whether these messages could be conveyed along surroundings of sender to user's position obeying corresponding media transmission rules. Hence, second step takes surroundings' transmission characteristics into consideration to calculates whether messages can arrived at those users' position: if and only if it can arrive a position, will it be transmitted to user at that position. We apply 2D-Grid with layered structure to implement this basic idea: 2D-Grid is applied to localize position and implement spatial correlated problems quickly while layered structure to combine all kinds of media transmission characteristics, rules, etc.

4.1 Some Definitions

Definition 1. Message Type *MediaType* determines the media type of message, such as audio, visual, etc.
Definition 2. Message Source *MS*, a particle as a substitute for the original message source -- the overall entity, Generally is the centroid of entity.
Definition 3. Grid, a two-dimensional unit square, is a tuple of <$i, j, Status, EntityID, Layers$> where i is its X-coordinate, j is its Y-coordinate, *Status* indicates its occupied

state, *EntityID* is associated with an entity *ID* if it is occupied, and *Layers* are a set of pointers to its associated multi AOIM-dependent layers, denoted as $G_{i,j}$.

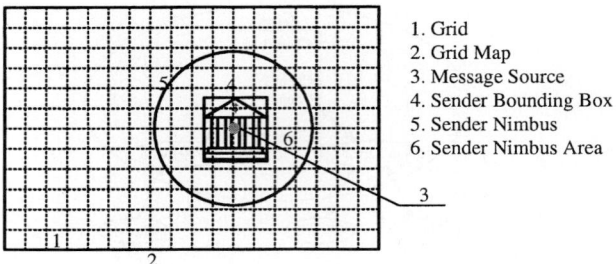

1. Grid
2. Grid Map
3. Message Source
4. Sender Bounding Box
5. Sender Nimbus
6. Sender Nimbus Area

Fig. 3. Illustration of some visual concepts

Definition 4. A Grid-based Map *GM*, proportionately segments the 2D projective map of a CVE scene into a set of orthogonal cells, is defined as a matrix on grids that

$$GM_{M \times N} = \begin{bmatrix} G_{1,1} & \cdots & G_{1,N} \\ \vdots & \ddots & \vdots \\ G_{M,1} & \cdots & G_{M,N} \end{bmatrix}.$$

Definition 5. Sender Bounding Box *SBB*, a most compact rectangle covering entity *E*, is defined as a tuple of <*length, width, center, angle*> where *length* and *width* are length and width of *SBB*, *center* is the intersection point of two diagonals, and *angle* is a radian value of inclination of length and horizontal line.

Definition 6. Sender Nimbus *SN*, boundary of possible influenced area of message sender regardless of other entities, is a closed curve *ClosedCurve(x,y)*=0. For grid-based definition, *SN* is a set of grids around message source which form a closed curve: $SN = \{G_{i,j} | \exists y((y \in [j, j+1)) \cap (ClosedCurve(i, y) = 0))\}$.

Definition 7. Sender Nimbus Area *SNA*, a closed region having a boundary of *SN*, $SNA = \{(x, y) | ClosedCurve(x, y) \leq 0\}$. For grid-based definition, it is a finite set of grids as follows: $SNA = \{G_{i,j} | (G_{i,j} \in SN) \cup (ClosedCurve(i, j) \leq 0\}$

Definition 8. Message Transmission Attributes Set *MSAS* is a set of message type correlated attributes influencing message transmission, such as amplitude for audio.

Definition 9. Message Transmission Attributes Set *MTAS* is a set of message type and grid material correlated attributes that determines message transmission characteristics, such as penetrability for visual message on a certain grid's material.

Definition 10. Message Transmission Rule Set *MTRS* is a set of message type correlated rules that determines message transmission path, distance and so on, i.e. attenuation rule and superpose rule for audio message.

Definition 11. Interest Visibility *IV* is an indicator of message's visibility to a grid. If message is visible to a grid and this grid is occupied by an agent, this message should be sent to this agent; otherwise, it is unnecessary.

Definition 12. AOI Traverse *AOIT*, an action executed to determine which agent should receive the current update message, traverses all surrounding grids to calculate their *IS* values from *MS* until it reaches *SN*.

Definition 13. AOI Traverse Box *AOITB*, a structure to advance *AOIT* from *MS* to *SN*, is a set of grids forming a closed curve around the *MS*.

Definition 14. AOI Traverse Rules *AOITR* defines how to educe *IS* values for grids in a ew external *AOITB* from those of an adjacent internal *AOITB* to advance *AOIT*.

Definition 15. AOI Layer, a kind of structure associated with each grid, includes attribute sub-layer and rule sub-layer layers for each kind of media type. The former includes *MSAS* and *MTAS* and the latter includes *MTRS* and *AOITR*.

4.2 Implementation of *GLAOIM*

Our experimental CVE system is hierarchical client-server architecture. That is, a message is sent to sub-server first. It is sub-server's responsibility to transmit this message to other users or other sub-servers and system server. At the same time, each sub-server maintains real-time status of the system. We consider a sub-server to be an affirmant for message receiver. On each sub-server, the overall CVE scene is segmented into a set of two-dimensional grid and calculated to form an overall grid-based map *GM* maintained by sub-server in real-time. Each grid is free or occupied by one of users, static entities, moving entities at time *t*. Generally, a user occupies one grid at time *t*. Although an entity could occupy server grids, a grid is selected to represent it as message source *MS*. Each grid is associated with several AOI Layers as defined above. This hierarchical multi-layers representation allows for an aggregation of different components of the expected AOIM, such as audio- or visual-dependent AOIM; also, it is feasible to extend for new media-dependent AOIM.

In our *GLAOIM* approach, the basic measurement unit is grid; so, it applies grid to locate and uses the number of grids to represent *length*, *width*, etc. When sub-server gets an update message, the message type of which is *MediaType*, its *LGAOIM* is started to localize *MS* and calculate *SBB* for the corresponding entity, that is $MS = G_{S_i, S_j}$ and $SBB = <L, W, G_{C_i, C_j}, \theta>$. At the same time, attribute sub-layer of *MS.Layers[MediaType]* is updated. Based on *MS* and *SBB*, applied with the local rules included in *MS.Layers[MediaType]*, *SN* and *SNA* are determined.

From definition 6 and 7, the major job of first step of *GLAOIM* is to determine *SN* and *SNA*. The first important problem is how to express *SN*. As defined, *SN* is a closed curve; Hence, it is appropriate to use formula to express nimbus, such as a circle around the source. This kind of expression is absolutely precise; but its implementation would be complex and the required computation would be difficult. In *LGAOIM*, the space is divided into static and discrete grids. Obviously, the nimbus curve, corresponding to a set of grids, is jagged as a result of discretization. To simplify this jagged nimbus, we approximately extend the actual boundary with a square which has a center of *MS*. Figure 4 shows source nimbus be expressed using formula, grids and extents. Applied with the extended expression, *SN* is defined as a set of grids forming a square around *MS* and *SNA* is defined as a matrix as follows:

$$SN = \begin{matrix} \{G_{N_1, N_j}, G_{N_1, N_j+1}, \cdots, G_{N_i, S_j}, \cdots, G_{N_1, N_j+K}, G_{N_1+1, N_j+K}, \cdots, G_{S_i, N_j+K}, \cdots, G_{N_1+L, N_j+K}, \\ G_{N_1+L, N_j+K-1}, \cdots, G_{N_1+L, S_j}, \cdots G_{N_1+L, N_j}, G_{N_1+L-1, N_j}, \cdots G_{S_i, N_j}, \cdots, G_{N_1+1, N_j}\} \end{matrix}$$

$$SNA_{L \times K} = \begin{bmatrix} G_{N_i,N_j} & \cdots & G_{N_i,S_j} & \cdots & G_{N_i,N_j+K} \\ \vdots & \ddots & \vdots & \ddots & \vdots \\ G_{S_i,N_j} & \cdots & G_{S_i,S_j} & \cdots & G_{S_i,N_j+K} \\ \vdots & \ddots & \vdots & \ddots & \vdots \\ G_{N_i+L,N_j} & \cdots & G_{N_i+L,S_j} & \cdots & G_{N_i+L,N_j+K} \end{bmatrix}$$

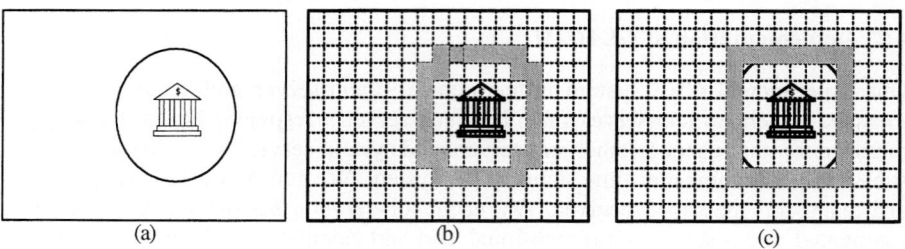

Fig. 4. Source Nimbus can be expressed using formulate, grids and extents. In (a), nimbus is expressed by a precise formulate, that is, a circle. In (b), nimbus is divided into static and discrete grids. In (c), nimbus is extended with a square

For message receiver determination, we present an interior-outward AOI traverse from *MS*: according to corresponding traverse rules, an AOI traverse process expands structure *AOITB* from *MS* to *SN* to calculate the value of *IV* for each grid in *SNA*. The user, occupying a grid with *IV* value greater than zero, is the message receiver. Thus, the problem of receiver determination is translated into interest visibility calculation.

For all types of update messages, *IV* value for any grid *G* in *SNA*, accurately calculated according to the corresponding penetration factors of all grids on the special transmission track from *MS* to *G*, can be deduced only based on *IV* values of the grids belong to that adjacent interior *AOITB* for our interior-outward traverse mechanism. To appropriate to our *SNA* which is a square taking *MS* as its center, our *LGAOIB* approach also applies squares to represent *AOITB*. We divide *AOITB* process into two steps including *Expand* and *IVTransmission* functions as follows:

$AOITB_{init} = \{G_{S_i-1,S_j-1}, G_{S_i,S_j-1}, G_{S_i+1,S_j-1}, G_{S_i+1,S_j}, G_{S_i+1,S_j+1}, G_{S_i,S_j+1}, G_{S_i-1,S_j+1}, G_{S_i-1,S_j}\}$

$Expand : AOITB_{Internal} \xrightarrow{Expand} AOITB_{AdjacentExternal}$

$IVTransmission : \{IV[G_{ij}] | G_{ij} \in AOITB_{Internal}\} \xrightarrow{IVTransmission} \{IV[G_{ij}] | G_{ij} \in AOITB_{AdjacentExternal}\}$

Being taken example for visual update message, *IV* value for any grid *G* in *SNA* is accurately calculated according to the corresponding penetration factors of all grids on a line from *MS* to *G*. Now it can be deduced only based on *IV* values of *G*'s adjacent grids on that adjacent interior *AOITB*. To simplify calculation, the *IVTransmission* function we adopt is given as follows: Let $PF(Direction)$ be a proper penetration factor of grid *G* for current traverse direction,

$$IV[G_{ij}] = \begin{cases} IV[G_{i-\text{sgn}(i-S_i),j}] \times G_{ij} \cdot MTAS \cdot PF(Direction) & i \neq S_i, \left|\dfrac{j-S_j}{i-S_i}\right| \leq \dfrac{1}{3} \\ IV[G_{i,j-\text{sgn}(j-S_j)}] \times G_{ij} \cdot MTAS \cdot PF(Direction) & j = S_j, \left|\dfrac{i-S_i}{j-S_j}\right| \leq \dfrac{1}{3} \\ IV[G_{i-\text{sgn}(i-S_i),j-\text{sgn}(j-S_j)}] \times G_{ij} \cdot MTAS \cdot PF(Direction) & |i-S_i| = |j-S_j| \\ \min(G_{i,j-\text{sgn}(j-S_j)}, G_{i-\text{sgn}(i-S_i),j-\text{sgn}(j-S_j)}) \times G_{ij} \cdot MTAS \cdot PF(Direction) & 1 < \left|\dfrac{j-S_j}{i-S_i}\right| < 3 \\ \min(G_{i-\text{sgn}(i-S_i),j}, G_{i-\text{sgn}(i-S_i),j-\text{sgn}(j-S_j)}) \times G_{ij} \cdot MTAS \cdot PF(Direction) & \dfrac{1}{3} < \left|\dfrac{j-S_j}{i-S_i}\right| < 1 \end{cases}$$

Furthermore, matrix *IVM*, introduced to accomplish AOI traverse cooperated with *SNA*, records *IV* value for all grids in *SNA* and is initialized based on *SNA* as follows:

$$IVM_{L \times K} = \begin{bmatrix} -1 & \cdots & -1 & \cdots & -1 \\ \vdots & \ddots & \vdots & \ddots & \vdots \\ -1 & \cdots & 1 & \cdots & -1 \\ \vdots & \ddots & \vdots & \ddots & \vdots \\ -1 & \cdots & -1 & \cdots & -1 \end{bmatrix}$$

for i = 1 to L
  for j = 1 to K
    if *SNA*[i][j].*EntityID* = *MS.EntityID*
      IVM[i][j] = 1
    else
      IVM[i][j] = −1

With the expansion of *AOITB*, *IV* value of each grid in the current *AOITB* is calculated; also, the corresponding values in the *IVM* are updated. When *AOITB* is equal to *SN*, *IV* values for all grids in *SNA* are obtained; that is, *IVM* is a complete matrix for interest visibility for current update message. Thus, for all users within *SNA*, check their located grids' *IV* values in *IVM*: if the value is greater than zero, this user is interested in the update message; otherwise, this user ignores the update message. So far, all problems of AOIM are solved.

5 Experiment and Analysis

GLAOIM is implemented with a Java application at server and a set of Java applets interacting with VRML scene through EAI at clients. The experiments are carried out on PCs running IE with VRML plug-in and aims at AOIM visual update messages.

Figure 5 illustrates an experiment to test *GLAOIM* in a CVE scene. (a) is a simple CVE scene in VRML file from a viewpoint independent of all users includes two avatars with a vitreous baffle-wall between them, which is particular for it has different penetration factors for lights emitted from its right side to left. The penetration factor that is represented by *Transparency field* in VRML file for lights from right side to left is 0.5 and the corresponding scene is shown in (b). (c) shows the other penetration factor from left side to right is 0; so users at right side can observe those users at left side while a contrary observation is invisible. To apply *LGAOIM*, an overall grid map *GM* is deduced from an aerial view of (a) shown in (d). In (e), two hatched squares represent *SNA* for each user respectively, which covers the other one. Using *Expand* and *IVTransimission* functions for visual update message, the penetration factor of the grids of vitreous baffle-wall is 0.5 when the left user is taken as a message source while that is 0 when the right user as message source. (f) and (g) give out part of *IV* values for these two users respectively. As (f) and (g) show, B has interest in A while A has no interest in B.

We analyze advantages of *LGAOIM* in four perspectives based on the experiment.

First, *GLAOIM* achieves communication filtering. As experiment shows, it ignores transmission of all update messages from B to A while reserving those from A to B.

Second, it ensures message sender and receiver transparent and unsymmetrical. *GLAOIM* is based on the overall *GM* maintained at server site and all users and virtual entities are localized at a certain grid on the *GM* according to real-time position update message from client sites. The whole implementation requires no explicit request. Furthermore, it can determine right receiving under unsymmetrical situation.

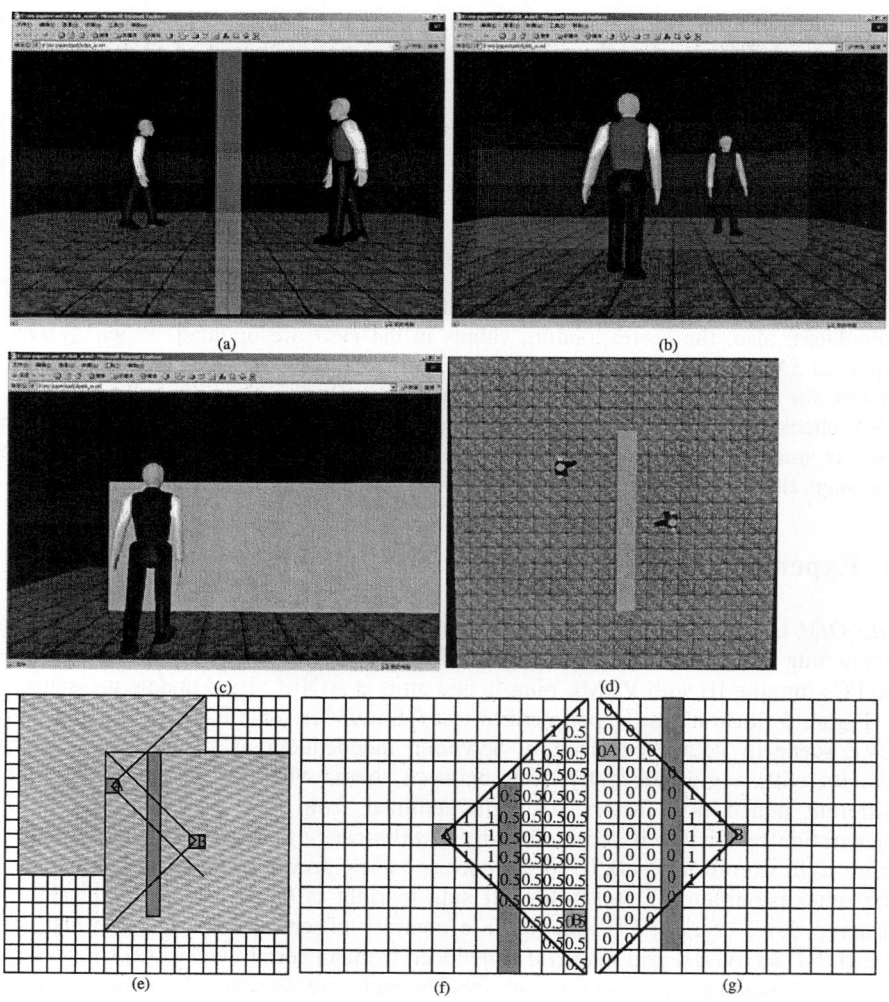

Fig. 5. Example of *GLAOIM* for visual message

Third, it provides good extensibility for new kind of media type. Although the above experiment is about visual update message, AOIM is carried out in similar manner as long as there are proper *MSAS*, *MTAS*, *MTRS* and *AOITR* layers for other kind of media types. Hence, appending new *MSAS*, *MTAS*, *MTRS* and *AOITR* layers to each grid can implement overall AOIM for a new kind of media.

Fourth, our scheme is pretty simple and easy to implement. All information including rules and characteristics are recorded in the layered structures of 2D grids at server site and updated in real time. Calculation of *Interest Visibility* for a grid can be quickly accomplished based on those values of its adjacent grids along with interior-outward traverse *AOITB*. It mainly requires simple arithmetic operations; so the process time is short and the data space required is also very small.

6 Conclusion Remarks

As an effective communication filter, AOIM maintains transmitting update messages of all kinds media types to those users who are indeed interested in these messages rather than all users to reduce number of messages exchanged in the system. In this paper, we have presented a new 2D grid with layered structure based and position centric AOIM approach. This new scheme divided AOIM into two steps: message sender possible influence area determination and message receiver determination; and it takes spatial position and transmission characteristics as two determination factors. Hence, it applies grid to represent position information and uses layered structure to combine transmission characteristics with position information. Furthermore, it adopts a quick interior-outwards AOI traverse to determine interest visibility for all grids in an influence area. This AOIM scheme, which is message sending source position centric rather than message receiver or message sender per se centric, is independent of all media type and can be easily extended for new media type.

References

1. S. Singhal, and M. Zyda. Networked Virtual Environments Design and Implementation. Addison Wesley, July 1999.
2. Mingyu Lim, Dongman Lee. Improving Scalability Using Sub-Regions in Distributed Virtual Environments. ICAT'99, 1999.
3. J. Smed, K.Timo, H. Hakonen. A Review on Networking and Multiplayer Computer Games. Turku Centre for Computer Science TUCS Technical Report No 454 April 2002.
4. Michael Capps, Seth Teller. Communication Visibility in Shared Virtual Worlds.
5. P. L. Hartling. Octopus: A Study in Collaborative Virtual Environment Implementation.
6. F. Tecchia, C. Loscos, R. Conroy et.al. Agent Behaviour Simulator (ABS): A Platform for Urban Behaviour Development.
7. M.R. Macedonia, M.J. Zyda, D. Pratt et. al. NPSNET: A Network Software Architecture for Large Scale Virtual Environments," Presence, Vol. 3, No. 4, Fall 1994, pp. 265-287.
8. M. Macedonia, M. Zyda, D. Pratt et.al. Exploiting Reality with Multicast Groups. IEEE Computer Graphics & Applications, September 1995, pp. 38-45.
9. O. Hagsnad. Interactive Multiuser VEs in the DIVE system, IEEE Multimedia, Springer 1996, 30-39.

10. S. Benford, C. Greenhalgh. Introducing Third Party Objects into the Spatial Model of Interaction. Fifth European Conference on Computer Supported Cooperative Work (ECSCW'97), 8-10, Lancaster, September 1997.
11. R. Lea, Y. Honda, K. Matsuda et al. Issues in the design of a scalable shared virtual environment for the internet. HICSS-30, Hawaii, Jan 1997.
12. B. Logan, M. Fraser, D. Fielding et al. Keeping in Touch: Agents Reporting from Collaborative Virtual Environments.

Awareness Scheduling and Algorithm Implementation for Collaborative Virtual Environment[1]

Yu Sheng[1], Dongming Lu[1, 2], Yifeng Hu[1], and Qingshu Yuan[1]

[1] College of Computer Science and Technology, Zhejiang University
[2] State Key Lab. of CAD and CG, Zhejiang University,
310027, Hangzhou, Zhejiang, P.R. China
mintbaggio@hotmail.com, ldm@cs.zju.edu.cn,
{yuanqs, hyf78}@hotmail.com

Abstract. The limitation of network resource reduces the awareness capability of CVE system, which becomes the main bottleneck for applications in Internet. In this paper, we study the relationship between the awareness capability of system and network parameters, such as latency, jitter, and data loss rate; also, according to the features and preferences of user awareness, a network-status-based solution is proposed for self-adaptable awareness scheduling. Moreover, the scheduling algorithm is implemented and verified by a prototype system, results from which indicate that awareness scheduling is useful for decreasing the loss of the awareness capability of system in a limited or unstable network.

1 Introduction

Awareness is an important concept in Computer Supported Cooperative Work (CSCW). It is always defined as understanding what happens in the environment. Awareness was defined in distributed work group that it provides a view of one another in the daily work environments [1]. In the real world, awareness is the first step to understand the surrounding environment, and the beginning of all activities.

Collaborative Virtual Environment (CVE) makes use of Virtual Reality technology, which offers reality and immersion to collaborative users, and improves the awareness capability of system. Internet broadens the area of CVE applications. However, the limitation of network resource reduces the awareness capability of Virtual Environment (VE), and becomes the primary bottleneck for Internet applications. There are many methods to solve this problem, all of which focus on improving the system performance by decreasing transmission cost, such as LOD, Dead Reckoning[2][3], scheduling based on visual scope and priority [4], etc. However, the common problem is that the decrease of transmission cost cannot completely equal the improvement of the awareness capability. So we introduce a solution named *Awareness Scheduling*, which schedules awareness objects in VE according to network status and user preference.

---

[1] The research was partially funded by National Basic Research Program of China (No. 2002CB312106), National Natural Science Foundation of China (No. 60273055), China-US Million Book Digital Library Project, and project of Science and Technology Department of Zhejiang Province (No. 2004C23035 and No. 2004C33083).

In the next section, we will introduce the concept of awareness scheduling, and then we will elaborate awareness scheduling in the following steps. First, the evidence of scheduling is analyzed. Then the awareness model is defined. Thirdly, the scheduling algorithm is implemented, which is experimented and verified in the fourth section. Finally, the work of this paper is summarized, and the future work is described.

2 Awareness Scheduling in CVE

2.1 The Features of User Awareness

There are mainly three awareness objects in CVE: scene environment, virtual object, and others' social activity. So, user's awareness effect, which is equal to the awareness capability of system, is the sum of all awareness objects of the three types. For example, in a virtual meeting system, the user's awareness effect includes the awareness of surrounding environment, meeting content and other participators, etc.

However, the contribution of each awareness object doesn't equal. To different users, CVE has the characteristics of individual configuration and Area of Interest (AOI). Users may be interested in different awareness objects with different expressing precision and priority definition. For example, in virtual meeting, some users only want the awareness information of avatar and sound, others may want the video. Furthermore, users may have different AOI in different period of time. Take the virtual meeting above for example; users may be only interested in their talking partners when communicating, but more interested in the subject and other speakers during the meeting.

2.2 Influences of Network Status to Awareness Capability

Network status can be described by several parameters, such as latency, data loss rate and jitter. Consequently, we can study these parameters to analyze the influences of network status to awareness capability. In CVE, the increase of latency will weaken the awareness capability, e.g., the lag of real-time action and voice. It is indicated by research that jitter influences much more to awareness capability than latency, e.g., there is no significant difference in overall performance between networks having latencies of 200 milliseconds without jitter and 10 milliseconds with jitter [5]. Data loss rate indicates the current congestion level of network, which will cause the data to be discrete and incomplete, and influence the awareness capability.

2.3 Requirements of Media Form to Network Parameters

Different media form has different requirement to the three network parameters. For example, stream media allows a specific level of data loss rate, but is sensitive to jitter. As objects in VE are concerned, the data must be transferred properly and the latency must be reduced as much as possible. While user interactive data, which calls for a strong real-time processing ability, is especially sensitive to latency. It is indicated in the DIVE system that the interactive data package must be received in 100 milliseconds; otherwise, there will be a serious lag of awareness [6].

Consequently, awareness objects in CVE contribute differently to users, and call for different network performance. Different media form also requires different network performance. In a limited or unstable network, we can adjust the awareness objects based on contributions, media forms, and user preference to adapt the network and make user's awareness effect the highest.

3 Awareness Scheduling and Algorithm Implementation

3.1 Evidences of Awareness Scheduling

According to the analysis above, we get three evidences of awareness scheduling: features and preferences of user awareness, network status, and media forms. They affect the awareness scheduling differently, and must be completely concerned in the scheduling algorithm.

3.2 Awareness Model

Based on the features of user awareness, awareness model is defined to be composed of awareness object, awareness priority, awareness level, awareness QoS and awareness correlation [7]. They are explained as follows:

Awareness Object: It is a complete object to be identified, operated and controlled by users in CVE.

Awareness Priority: It indicates user preference of an awareness object, and the rate that it contributes to user's awareness effect. It's determined by user preference.

Awareness Level: Define different levels for an awareness object using a quantitative approach. It's quantitated in three hierarchies: awareness, identification, and comprehension. Awareness is the lowest level, which means user can be conscious of the object's existence, but can't identify its concrete attribute. In the hierarchy of identification, user can identify the object's attribute, but can't operate it. On the highest level, comprehension, user can not only identify the object, but operate it.

Awareness QoS: It describes the network service quality required by awareness media on each awareness level. Different kind of media at the same level may require different network service quality. When the described QoS parameters can't be satisfied, the system will lower its corresponding awareness level.

Awareness Correlation: According to the feature of human being's cognition, there is correlation in contributions of different awareness object, which depends on its intrinsic attribute, such as similarity, and user awareness configuration. There are three kinds of correlation: positive correlated, non-correlative, and negative correlated. Positive correlated indicates that one object enhances the awareness capability of the other, while negative correlated means a negative effect. Non-correlative means no relevance in awareness capability between two objects.

3.3 Denotation of the Awareness Capability of System

Before the definition of awareness capability is given, we denote several elements in awareness model as follows:

Awareness Priority P_i: P_i is the awareness priority of awareness object i. For each P_i, it satisfies,

I. $0 < P_i < 1$, and

II. $\sum_{i=1}^{n} P_i = 1$

Awareness Capability W_i: W_i is the awareness capability of an object at awareness level L. Suppose that an object's W_i is 1 at the highest level, then the values on other levels will be in the range of 0 and 1. We can denote W_i as a function of L, which is $W_i = F_i(L)$. Function F_i depends on the attributes of awareness object, which may be different from each other.

Awareness Correlation C_{ij}: C_{ij} is awareness correlation coefficient of awareness objects i and j, which satisfies,

I. $C_{ii} \equiv 1$

II. $0 < C_{ij} < 1$, when i, j are positive correlated

III. $C_{ij} = 0$, when i, j is non-correlative

IV. $-1 < C_{ij} < 0$, when i, j are negative correlated.

According to the contents above, for an awareness object i in CVE, its singular awareness capability is $P_i * W_i$. Take awareness correlation into account, its awareness capability will be $A_i = \sum_{j=1}^{n} C_{ij} * P_i * W_i$. So, for a system containing n objects, its awareness capability A is expressed by the following formula,

$$A = \sum_{i=1}^{n} A_i = \sum_{i,j=1}^{n} C_{ij} * P_i * W_i \tag{1}$$

Consequently, scheduling for awareness objects is transformed to the procedure of searching the maximum value of A.

3.4 Awareness Scheduling Algorithm

Suppose a scheduling event occurs because of resource fluctuation or change of user preference, or because that the user's demand can't be satisfied, which means certain awareness capability must be lost. As mentioned above, P_i and C_{ij} depend on the user awareness preference and inherent attributes of awareness object, while W_i is the function of awareness level. So when the scheduling event occurs, there is a possible awareness capability loss AL_i for each awareness object, which is,

$$AL_i = ((\sum_{j=1}^{n} C_{ij}) * P_i * (F_i(L) - F_i(L-1)))\qquad(2)$$

The calculation of A is valid only under the condition that awareness QoS is guaranteed, so we have to find the awareness objects $K_1, K_2, ..., K_m$ which will be scheduled to make $\sum_{i=1}^{m} AL_{ki}$ minimum and the awareness QoS of each object is satisfied. However, for each awareness object i, its QoS satisfaction is interactive. So the key to the problem is to find the reason that the awareness capability decreases, according to which we can adjust the awareness level of corresponding object.

(1) The Reason That Awareness Capability Decreases
We analyze the relationship between the three network parameters that influence awareness QoS and the parameters that describe the network status, which are,

I. Latency has a compact relationship with bandwidth and utilization.
II. Jitter is independent of bandwidth when the network load is low. But the correlative factor can be 0.95 when network overloads [8].
III. Data package is lost for three reasons. First of all, active loss, e.g., data loss is used to guarantee the best effect in jitter compensation algorithm. Secondly, network overloads, e.g., there will be a 50% data loss when an application that calls for 10Mbps network is applied on a 5Mbps network. Thirdly, there will be a 0.5% data loss when network is fully loaded.

It's concluded from the analysis above that data traffic of each kind of awareness objects is an important factor of influence to VE. Therefore, decreasing appropriate data traffic will be an effective method to guarantee the awareness QoS to be satisfied.

(2) How to Decrease Data Traffic
Suppose that the data traffic to be decreased is V. We sort all the awareness objects from low to high according to loss of awareness capability, as shown in Figure 1.

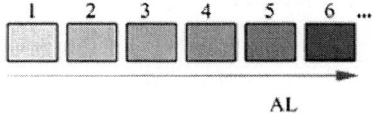

Fig. 1. Incremental series of AL

According to the QoS parameters of an awareness object, the data traffic V_i can be calculated while the awareness level of the object is lowered by 1. Then we will find a descendent series of awareness objects $K_1, K_2, ..., K_m$, which can satisfy:

I. The decreasing data traffic produced by lowering the awareness level of the objects $K_1, K_2, ..., K_m$ is larger than V, that is, $\sum_{i=1}^{m} V_{Ki} > V$.

II. In all the series $K_1, K_2, ..., K_m$, which satisfy $\sum_{i=1}^{m} V_{Ki} > V$, it makes the total awareness capability loss minimum, that is, $\sum_{i=1}^{m} AL_{ki}$ is minimum.

It is a typical knapsack problem, which can be solved by dynamic programming [9].

(3) Algorithm Description
According to the analysis above, the algorithm is described as follows.

Step1. Calculate the data traffic, network traffic, and the data traffic ΔV that needed to be decreased, according to the requirement of QoS.

Step2. Examine the status of each awareness object, and calculate the awareness capability loss AL of each awareness object when its awareness level is lowered by 1, and the network traffic V it saves.

Step3. Find the awareness object Obj, whose V_{Obj} is bigger than ΔV, and AL_{Obj} is the smallest.

Step4. Use dynamic programming to find the awareness objects list ObjList which satisfies the two conditions in Step3, and its V is smaller than ΔV.

Step5. Compare AL_{Obj} with $sum(AL_{ObjList(i)})$, the smaller one is the result object or objects list.

In Step 1, current network traffic can be obtained by network inspection, and data traffic can be calculated from awareness QoS [7][10][11] of each object in CVE.

The algorithm describes the scheduling when QoS of awareness objects can't be satisfied. When CVE detects that the network performance is sufficient for improving awareness level; correspondingly, the awareness object (list), which can mostly improve the awareness capability from the network resource available, will be found. The procedure is similar to the algorithm above. Therefore, the algorithm is self-adaptable.

4 Experiment and Verification

We've designed and developed an experimental system to support the technology of awareness scheduling above, using V-NET [12] as a prototype.

4.1 System Framework

The system consists of 3 layers: user layer, awareness management layer, and network layer.

(1) User Layer
User layer is made up of two parts. One is the user interface, which is the input and output of the system; the other is user awareness configuration, including user's AOI parameters, awareness priority of awareness objects, etc.

the instance '$height = 174cm, weight = 65kg, and\ Nationality = China$' is not identified by the signature. But if only any M(=2) features are utilized, the signature will identify the instance.

In summary, our main contributions in this paper are as follows. (1) A novel intrusion detection methodology is introduced briefly. (2) Model generalization led to by M of N features in a signature is discussed, and then an evaluation methodology on it is designed. In addition, an average detection cost function is defined to quantify the detection performance for intrusion detection.

The remaining parts of this paper are organized as follows. Section 2 talks about the related work. Section 3 describes the formal intrusion detection methodology in brief. In section 4, an evaluation methodology is also designed for the M of N features scheme. Experiments in section 5 reveal the useless of the scheme on intrusion detection. Lastly, we draw conclusions in section 6.

2 Related Work

Our research work in this paper is generally related to model generalization in the behavior model. First, the intrusion signatures in SID techniques can be generalized to cover more behavior space. In [1], using a fitness function which depends on false positive rate and detection rate, the generalized signatures (represented by a *finite state transducer*) is optimized by the evolution programming. In general, the model generalization on intrusion signatures can solve the intrusion variations detection problem partially.

Secondly, the normal behavior model of AID techniques can be generalized as well to detect novel instances, and it can be done in several ways. Based on a distance metric and a distance threshold [3][9], the instances in the existing audit trails are clustered unsupervisedly, and the new instances are labeled by the existing instances in their clusters. In statistical methods for intrusion detection [6] [7] [9], the (statistical) resource usage profiles are mined from the existing audit trails. The novel instances are detected according to whether they fall into these profiles. Among these two styles, the existing audit trails are modeled inexactly to accommodate more resource behaviors in the profiles, and thus to achieve the model generalization.

Most of past works only credited the overall efficiency of an intrusion detection technique to such model generalization, and there is hardly any evaluation of the effect of the model generalization. This is partially due to the difficulty in pinpointing the contribution of model generalization to the overall efficiency. Fortunately, our methodology not only overcomes the problem, it also allows one to adjust the extent of model generalization.

3 Brief Introduction: An Intrusion Detection Methodology

Any intrusion detection system builds the behavior models of the resources using a set of features, or a *feature vector* $FV = \{F_1, F_2, \ldots, F_n\}$, where F_i is a feature in the feature set. Every feature in the feature vector can be one of these types: A feature associated with an instant of time (e.g., *the fields in the current packet*), or with a time interval (e.g., *the number of SYN packets within 2 seconds*), or with the context of a current event (e.g., *the system-call events in stide [4], the state events in STAT [8]*). The context is defined over the timeline proceeding the point in time when the event in question happens. In general, an atomic feature F_i in the feature vector can be categorized into *nominal*, *discrete* or *continuous* one. A feature vector for intrusion detection can contain any number of nominal, discrete, and/or continuous features. In addition, a feature can also be as complex as a compound feature (see Section 3.3).

In this methodology, we assume that there is a training audit trail indexed by its timestamp, in which the normal and the intrusion audit trails are labelled correctly. We also assume that the training audit trails represent our known knowledge about the computing resource. Then, the instances of the feature vector are collected from the training audit trails as $\{I_{FV}^1, I_{FV}^2, I_{FV}^3, \ldots\}$. For each feature instance I_{FV}^i, there is a **status** that indicates the label of audit trails where it is collected. For example, if an instance I_{FV}^i is left by an intrusion 'Nimda', its status is 'Nimda'.

3.1 Basic Concepts and Notations

For a feature F, several of its basic concepts are defined formally as follows.

- Its feature space $Dom(F)$ is the defining domain of the computing resource.
- Any value in $Dom(F)$ is defined as a feature value v_F, and $v_F \in Dom(F)$. In general, there are many feature values in the feature space $Dom(F)$.
- A feature range $R_F(v_F^1, v_F^2)$ is the range between any two feature values v_F^1 and v_F^2 in its feature space, which includes all feature values falling between v_F^1 and v_F^2. For a discrete or continuous feature, $R_F(v_F^1, v_F^2) = [v_F^1, v_F^2]$. For a nominal feature, every feature value is independent. Thus, each nominal feature value is referred to as a feature range in this methodology so that for a nominal feature F, $[v_F^i] = [v_F^i, v_F^i]$. If a feature value v_F^j is within the bounds of a feature range, we say that it falls within the feature range, denoted as $v_F^j \in R_F(v_F^1, v_F^2)$. The concept of *feature range* is used to treat uniformly every (nominal, discrete, or continuous) feature in our methodology. For $R_F(v_F^1, v_F^2)$, we further define $upper(R_F) = v_F^1$ and $lower(R_F) = v_F^2$.

Notations. In reality, some feature values of a feature will never occur, and thus unreasonable. As mentioned above, there is a series of feature instances $\{I_{FV}^1, I_{FV}^2, \ldots\}$ collected from existing audit trails. The feature values in every feature instance I_{FV}^i are *reasonable* as it occurs. The following notations are given (Note that in order to avoid cluttering the expressions, we have dropped the subscript of F): if $F \in FV$,

- $v(I_{FV}^i, F)$ is the feature value of the feature F in the feature instance I_{FV}^i.
- $I(v_F, F)$ is the set of feature instances whose values of F are equal to v_F.

$$I(v_F, F) = \{I_{FV}^k | v_F = v(I_{FV}^k, F)\}$$

- $I(R_F, F)$ is the set of feature instances whose values of F fall in R_F.

$$I(R_F, F) = \{I_{FV}^k | v(I_{FV}^k, F) \in R_F\}$$

3.2 NSA Label

Definition 1 (NSA label of a feature value). *If a feature value v_F occurs only in the normal audit trails, it is normal. If it occurs only in the intrusive audit trails, for example, intrusion signatures, it is labeled as anomalous. Otherwise, i.e., if it occurs in both normal and intrusive audit trails, it is labeled as suspicious. For brevity, we will refer to the normal, suspicious, or anomalous label as the **NSA label** of the feature value v_F, denoted as $L(v_F) = \{`N', `S', `A'\}$.*

Note that a feature value is either normal or anomalous in a specific feature instance, but its NSA label is collected from all related feature instances. We will further extend the concept of **NSA label** to feature ranges of a feature.

NSA Labels of Feature Ranges. With respect to a user-defined splitting strategy, the feature space $Dom(F)$ can be split into a set of mutually exclusive feature ranges $\{R_F^1, R_F^2, \ldots, R_F^m\}$, such that **(1)** there is no common feature value v_F, which falls in R_F^j and R_F^k at the same time $(j \neq k)$, and **(2)** $I(R_F^i, F) \neq \Phi$ $(i \geq 1)$. Then, the concept of NSA labels can be extended to these feature ranges as follows. For the feature range R_F,

$$L(R_F) = `N' \Leftrightarrow \forall i(v(I_{FV}^i, F) \in R_F \to L(v(I_{FV}^i, F)) = `N')$$
$$L(R_F) = `A' \Leftrightarrow \forall i(v(I_{FV}^i, F) \in R_F \to L(v(I_{FV}^i, F)) = `A')$$
$$L(R_F) = `S' \Leftrightarrow \exists i \exists j(v(I_{FV}^i, F) \in R_F \land L(v(I_{FV}^i, F)) = `A')$$
$$\land (v(I_{FV}^j, F) \in R_F \land L(v(I_{FV}^j, F)) = `N')$$

Feature Subspaces. After grouping the feature ranges of a feature F based on NSA labels, we can partition the feature space $Dom(F)$ into *three* feature subspaces: *normal, suspicious and anomalous*, denoted as $N(F)$, $S(F)$ and $A(F)$, respectively. Thus we have,

$$N(F) = \{R_F^j | j \geq 1, L(R_F^j) = `N'\}$$
$$S(F) = \{R_F^j | j \geq 1, L(R_F^j) = `S'\}$$
$$A(F) = \{R_F^j | j \geq 1, L(R_F^j) = `A'\}$$

In the following description, we denote $\Omega(F) = N(F) \cup S(F) \cup A(F)$.

3.3 Combining: Compound Feature

Definition 2 (compound feature). *With respect to two features F_1 and F_2, a compound feature F_{12} is defined as a subset of the cartesian product of $\Omega(F_1)$ and $\Omega(F_2)$, such that each element in this set actually represents at least one feature instance in the audit trails. In other words, if an ordered pair $(R_{F_1}^a, R_{F_2}^b)$ is a compound feature range (i.e., $(R_{F_1}^a, R_{F_2}^b) \in \Omega(F_{12})$), then $I((R_{F_1}^a, R_{F_2}^b), F_{12}) \neq \Phi$.*

$$\Omega(F_{12}) = \{(R_{F_1}^a, R_{F_2}^b) | R_{F_1}^a \in \Omega(F_1), R_{F_2}^b \in \Omega(F_2), I((R_{F_1}^a, R_{F_2}^b), F_{12}) \neq \Phi\}$$

Based on the definition of *cartesian product*, for any feature instance $I_{FV}^i \in I((R_{F_1}^a, R_{F_2}^b), F_{12})$, it will be recognized by feature ranges $R_{F_1}^a$ and $R_{F_2}^b$ as well (i.e., $I_{FV}^i \in I(R_{F_1}^a, F_1)$ and $I_{FV}^i \in I(R_{F_2}^b, F_2)$), and vice versa. Therefore, $I((R_{F_1}^a, R_{F_2}^b), F_{12}) = I(R_{F_1}^a, F_1) \wedge I(R_{F_2}^b, F_2)$. For the sake of ambiguity, we will refer to a single feature as an *atomic* feature.

Theorem 1. *The feature ranges of a compound feature are mutually exclusive, i.e., for two different feature ranges of a compound feature $(R_{F_1}^a, R_{F_2}^b)$ and $(R_{F_1}^c, R_{F_2}^d)$, there is no such feature instance I_{FV}^i so that $I_{FV}^i \in I((R_{F_1}^a, R_{F_2}^b), F_{12})$ and $I_{FV}^i \in I((R_{F_1}^c, R_{F_2}^d), F_{12})$.*

Similar to atomic features, every feature range of F_{12} has an NSA label, and all of its feature ranges are mutually exclusive (**Theorem 1**, please see its proof in the appendix). A compound feature space can be partitioned into three feature subspaces like an atomic feature, i.e., $\Omega(F_{12}) = N(F_{12}) \cup S(F_{12}) \cup A(F_{12})$.

Compounding More Features. In summary, the compound feature built from two atomic features shows the same properties as any of its component atomic features. Therefore, we can treat the compound feature as an atomic one to build higher order compound features. Using this recursive procedure, the feature vector FV for intrusion detection can be converted into an equivalent n-order compound feature $F_{1...n}$ with normal $N(F_{1...n})$, suspicious $S(F_{1...n})$ and anomalous $A(F_{1...n})$ feature subspaces. We will rewrite the compound feature ranges according to the following rule: $(R_F^a, (R_F^b, R_F^c)) = (R_F^a, R_F^b, R_F^c)$.

3.4 Behavior Signature

In our methodology, the behavior models of the resource are constituted by (*normal, suspicious and intrusion*) behavior signatures, which are defined as:

Definition 3 (behavior signature). *Assuming that there exists a feature vector $FV = \{F_1, F_2, \ldots, F_n\}$, and that the feature ranges of every feature are determined beforehand. A behavior signature $\overline{Sig_{FV}^i}$ is a feature range of the compound feature $F_{1...n}$ with its NSA label. In other words, the behavior signature is the combination of feature ranges of all features in FV labelled by its statuses of corresponding feature instances in the existing audit trails.*

As indicated in the above definition, every behavior signature[1] represents a state of the resource at a specified time point. According to NSA labels of signatures, the behavior model can be split into three parts: normal, suspicious, and intrusion behavior models. In anomaly-based intrusion detection, only the normal behavior model is utilized, but signature-based intrusion detection identifies intrusions based on the intrusive behavior model. However, the best scenario is to do intrusion detection using the complete behavior model.

---

[1] For brevity, '*behavior signature*' will be simplified as '*signature*' within the context of this paper.

4 Intrusion Detection via Signatures

4.1 Building Behavior Models

To use our methodology for intrusion detection, an splitting strategy is designed as follows to build the feature ranges for every feature. For nominal features, the splitting strategy do nothing except building one feature range for every feature value. For every discrete/continuous feature, an initial feature range is built for every feature value. Two initial feature ranges are *neighboring* if there are no feature values between them in the audit trails.

Specific for every discrete feature, the unknown feature subrange between any two neighboring initial feature ranges is split and combined into these two initial feature ranges as follows. If the size of the unknown feature subrange is an odd number n, $(n-1)/2$ of it will combine into every side, and the left 1 is assigned to one side randomly. If the size is an even number n, $n/2$ of it will combine into every side. In contrast, specific for every continuous feature, the unknown feature subrange between any two neighboring initial feature range will be split equally and combined into both sides.

In the following step, if two neighboring feature ranges have the same NSA label, they will be combined into a single feature range by expanding its range size. This can economize the storage space for the ultimate behavior models.

4.2 Detecting Behaviors Using M of N Features in a Signature

In our evaluation methodology, an instance in the test audit trails will be detected as follows. Utilizing the feature ranges of every feature, a temporal signature will be formed for the instance. If it matches any signature in the behavior model with M among N features, the status list of the signature will be inserted into the status list of the temporal signature. Obviously, the status list of the temporal signature is empty initially. After comparing with all signatures in the behavior model, the detection results for the instance is aggregated into its status list.

Then, the average cost for every instance in the test audit trails is calculated to quantify the detection performance. For a normal behavior, it will be detected as an anomaly if the status list include other status(es) other than 'normal'. Otherwise, it is detected as 'normal'. For a intrusive behavior, it will be detected as the same intrusion if the status list is identical to the status of the behavior, and it will be detected as normal if the status list only include the 'normal' status. Otherwise, it will be detected an an 'anomaly'.

4.3 To Measure the Detection Performance

The two main objectives of intrusion detection are (1) to detect the intrusions correctly (as anomalies), and (2) to identify the behaviors correctly (i.e. normal behaviors or its original intrusions). With respect to the detection results, every instance in the test audit trails will be assigned a *cost* value as the detection performance of the behavior model to it [5]. Specifically, if the *normal* instance is detected as 'normal', the cost is 0, otherwise, the cost is 3. Simultaneously, if the intrusive instance is identified as its original intrusion label, the cost is 0. If the intrusive instance is detected as an anomaly, the cost is 1. If the intrusive instance is detected as 'normal', the cost is 3.

Suppose that there are T instances in the test audit trails. According to the detection results, several statistics are further defined as follows.

- $\#_{(N,N)}(M)$: the number of *normal* instances detected as 'normal';
- $\#_{(N,A)}(M)$: the number of *normal* instances, but detected as 'anomalies';
- $\#_{(N,*)}(M)$: the number of *normal* instances in the test audit trails;
- $\#_{(I,I)}(M)$: the number of *intrusive* instances detected as their original intrusions;
- $\#_{(I,A)}(M)$: the number of *intrusive* instances detected as 'anomaly';
- $\#_{(I,N)}(M)$: the number of *intrusive* instances detected as 'normal';
- $\#_{(I,*)}(M)$: the number of *intrusive* instances in the test audit trails.

Where, it is obvious,

$$\#_{(N,*)}(M) = \#_{(N,N)}(M) + \#_{(N,A)}(M) \quad (1)$$
$$\#_{(I,*)}(M) = \#_{(I,I)}(M) + \#_{(I,A)}(M) + \#_{(I,N)}(M) \quad (2)$$
$$T = \#_{(N,*)}(M) + \#_{(I,*)}(M) \quad (3)$$

With respect to specific M, the average cost of every instance in the test audit trails is defined as:

$$cost(M) = (\#_{(N,A)}(M) \times 3 + \#_{(I,N)}(M) \times 3 + \#_{(I,A)}(M) \times 1) \times \frac{1}{T} \quad (4)$$

From above equation, with the increase of $cost(M)$, the detection performance with the parameter M becomes worse. Obviously, the average cost at $M = N$ is the baseline for the detection performance. If $cost(M) > cost(N)$, the efficiency for intrusion detection has been degraded by such M of N scheme. Otherwise, it is useful for intrusion detection. An efficient intrusion detection technique will cause smaller average cost for every instance.

5 Experiments

We have chosen a typical dataset for network intrusion detection from KDD CUP 1999 contest, in which every record is an instance of a specific feature vector collected from the audit trails. This is because the dataset meets the requirements of our formal framework: labeled audit trails and intrusion-specific feature vector. The number of records in the datasets are: *training-4898431 records, test-311029 records*. For a detailed description of the datasets, please refer to '*http://www-cse.ucsd.edu/users/elkan/clresults.html*'.

5.1 Evaluating the M of N Scheme

In our experimental evaluations, $N = 41$ and the parameter M is variable from 41 to 30. The behavior model is first built from the training audit trails. Then, every instance in the test audit trails is detected with respect to specific M, and the detection performance is quantified by the average cost of every instance within the detection results.

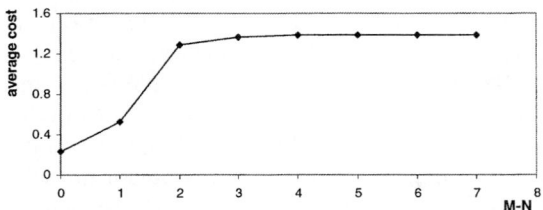

Fig. 1. The influence of M of N features scheme

Experimental Results The detection performance baseline with $M = 41$ is $cost(41) = 0.228$. For the sake of comparison, we use $N - M$ as the horizontal axis in Figure 1, in which the influence of M of N features scheme on intrusion detection is illustrated. It is obvious that the average cost of every instance is decreased with the increase of $N - M$, i.e. with the decrease of M. In other words, even though the M of N features scheme can generalize the behavior model, it will degrade the detection performance for intrusion detection. Therefore, the scheme should not be utilized to enhance the detection performance in intrusion detection.

The Statistics About the Detection Results In Table 1, the detection results are detailed with respect to varying M. In general, an efficient intrusion detection technique should identify most (normal and intrusive) behaviors, and the identification ability is indicated by the numbers in $\#_{(N,N)}(M)$ and $\#_{(I,I)}(M)$. However, in Table 1, these two numbers are decreased with the decrease of M. In other words, with the decrease of the parameter M, the identification capability is degraded, and most normal and intrusive behaviors will be detected as 'anomalies'. As an extreme case, all the behaviors will be detected as 'anomalies'. This case will also occur if the behavior model is empty. In other words, the behavior model with more generalization caused by the M of N features scheme is almost no use for intrusion detection. In summary, the M of N feature scheme will largely degrade the detection performance for intrusion detection.

Table 1. The detection results with respect to M

M	$\#_{(N,N)}(M)$	$\#_{(N,A)}(M)$	$\#_{(I,I)}(M)$	$\#_{(I,A)}(M)$	$\#_{(I,N)}(M)$	$cost(M)$
41	57102	3491	215835	21635	12966	0.228293825
40	53739	6854	134578	102487	13371	0.524587739
39	8067	52526	10225	238353	1858	1.290892489
38	2571	58022	39	249817	580	1.368435098
37	26	60567	3	250342	91	1.389953991
36	2	60591	0	250354	82	1.390137254
35	1	60592	0	250418	18	1.389735362
34	0	60593	0	250429	7	1.389674275
33	0	60593	0	250436	0	1.389629263
32	0	60593	0	250436	0	1.389629263
31	0	60593	0	250436	0	1.389629263
30	0	60593	0	250436	0	1.389629263

6 Conclusions and Future Work

In this paper, we first present a formal intrusion detection methodology based on a general feature vector. Using the framework, the M of N feature scheme is evaluated with respect to the detection performance for intrusion detection. To achieve it, we also propose a average cost function to quantify the detection performance for intrusion detection. The experimental results show that, even though the M of N scheme can generalize the behavior model to cover more unknown behaviors, it will degrade the detection performance for intrusion detection by triggering more false alarms. More specifically, the identification ability of every signature will be lost with the decrease of M, i.e., with more generalization in the behavior model. The conclusion is critical for intrusion detection since all the features in a signature should be used to identify a (normal/intrusive) behavior, which does not follow our intuition.

In the future work, we will further utilize the formal framework to analyze the problems in intrusion detection, and try to propose solutions or suggestions for these problems. At the same time, we will propose an efficient intrusion detection methodology based on the behavior signatures of the computing resource.

References

1. K.P. Anchor, J.B. Zydallis, G.H. Gunsch, and G.B. Lamont. Extending the computer defense immune system: Network intrusion detection with a multiobjective evolutionary programming approach. In *ICARIS 2002: 1st International Conference on Artificial Immune Systems Conference Proceedings*, University of Kent, 2002.
2. H. Debar, M. Dacier, and A. Wespi. A revised taxonomy for intrusion detection systems. *Annales des Telecommunications*, 55(7–8):361–378, 2000.
3. E. Eskin, A. Arnold, M. Prerau, L. Portnoy, and S. Stolfo. A geometric framework for unsupervised anomaly detection: Detecting intrusions in unlabeled data. In D. Barbara and S. Jajodia (editors), *Applications of Data Mining in Computer Security*, Kluwer, 2002.
4. S.A. Hofmeyr, S. Forrest, and A. Somayaji. Intrusion detection using sequences of system calls. *Journal of Computer Security*, 6(3):151–180, 1998.
5. W. Lee, M. Miller, and S. Stolfo. Toward cost-sensitive modeling for intrusion detection. Technical Report No. CUCS-002-00, Computer Science,Columbia University, 2000.
6. W. Lee and S.J. Stolfo. A framework for contructing features and models for intrusion detection systems. *ACM Transactions on Information and System Security*, 3(4):227–261, Nov. 2000.
7. M.V. Mahoney and P.K. Chan. Learning Nonstationary Models of Normal Network Traffic for Detecting Novel Attacks. In *SIGKDD 2002*, July 23-26 2002.
8. G. Vigna and R.A. Kemmerer. NetSTAT: A Network-based Intrusion Detection System. *Journal of Computer Security*, 7(1):37–71, 1999.
9. K. Wang and S.J. Stolfo. Anomalyous payload-based network intrusion detection. In *Proceedings of RAID*, 2004.

Theorem 1. *The feature ranges of a compound feature are mutually exclusive, i.e., for two different feature ranges of a compound feature $(R_{F_1}^a, R_{F_2}^b)$ and $(R_{F_1}^c, R_{F_2}^d)$, there is no such feature instance I_{FV}^i so that $I_{FV}^i \in I((R_{F_1}^a, R_{F_2}^b), F_{12})$ and $I_{FV}^i \in I((R_{F_1}^c, R_{F_2}^d), F_{12})$.*

Proof. We prove this by contradiction. If there exists a feature instance I_{FV}^i so that $I_{FV}^i \in I((R_{F_1}^a, R_{F_2}^b), F_{12})$ and $I_{FV}^i \in I((R_{F_1}^c, R_{F_2}^d), F_{12})$.

$$I_{FV}^i \in I((R_{F_1}^a, R_{F_2}^b), F_{12})$$
$$\Leftrightarrow I_{FV}^i \in I(R_{F_1}^a, F_1) \wedge I(R_{F_2}^b, F_2)$$
$$\Leftrightarrow v(I_{FV}^i, F_1) \in R_{F_1}^a, v(I_{FV}^i, F_2) \in R_{F_2}^b \qquad (5)$$

Similarly,

$$I_{FV}^i \in I((R_{F_1}^c, R_{F_2}^d), F_{12})$$
$$\Leftrightarrow v(I_{FV}^i, F_1) \in R_{F_1}^c, v(I_{FV}^i, F_2) \in R_{F_2}^d \qquad (6)$$

Recall that there is no common feature value v_F, which falls into R_F^j and R_F^k simultaneously ($j \neq k$). From (1) and (2), we can get $R_{F_1}^a = R_{F_1}^c$ and $R_{F_2}^b = R_{F_2}^d$. Thus, $(R_{F_1}^a, R_{F_2}^b) = (R_{F_1}^c, R_{F_2}^d)$. This contradicts the assumption that the two feature ranges are different.

High-Level Quantum Chemical Methods for the Study of Photochemical Processes

Hans Lischka[a], Adélia J.A. Aquino[a], Mario Barbatti[a], and Mohammad Solimannejad[b]

[a] Institute for Theoretical Chemistry, University of Vienna,
Währingerstrasse 17, A-1090 Vienna, Austria
{Hans.Lischka, Adelia.Aquino, Mario.Barbatti}@univie.ac.at
[b] Quantum Chemistry Group, Department of Chemistry, Arak University,
38156-879 Arak, Iran.
m-solimannejad@araku.ac.ir

Abstract. Multireference configuration interaction calculations have been performed on the excited state energy surfaces of the methyleneimmonium cation using recently developed methods for the computation of analytic gradients and nonadiabatic coupling terms. Excited-state structures and minima on the crossing seam have been determined. It was found that the topology of the methyleneimmonium surfaces is qualitatively different from that of the isoelectronic ethylene. In the former case a conical intersection between the S_1 and ground states is found for the twisting motion around the CN bond, whereas a more complicated motion including pyramidalization and hydrogen-transfer is needed in case of ethylene.

1 Introduction

The understanding of the photodynamics of molecular processes is of utmost importance in Chemistry and Molecular Biology. Theoretical methods based on quantum chemical approaches allow the detailed modeling of reaction mechanisms at the molecular basis. These investigations require the knowledge of accurate energy surfaces in excited states and appropriate methods for dynamics calculations. Even though a large variety of efficient quantum chemical methods is available for electronic ground-state calculations, the situation is much more difficult when excited-state energy surfaces and surface crossings should be treated. Analytic energy derivatives and nonadiabatic coupling vectors are crucial tools for the determination of energy minima, saddle points and conical intersections.

In the last years we have developed theoretical methods and computer programs for the analytic calculation of energy derivatives and nonadiabatic couplings at the multireference configuration interaction (MR-CI) level [1-3]. This method is conceptually at a much higher level than the complete active space self-consistent field (CASSCF) approach, which has been mostly used so far for the optimization of excited-state structures. The MR-CI method allows a simultaneous and balanced description of a multitude of excited states. Moreover, it is a very stable method, which allows the investigation of extended sections of energy surfaces including dissociation processes. An efficient program implementation of the aforementioned analytic gradients and nonadiabatic couplings has been developed within the COLUMBUS program system [4,5].

In a series of applications we have demonstrated the power of these new approaches e.g. for the calculation of energy surfaces for Woodward-Hoffmann processes in the electronic ground state (Diels-Alder reaction [6]) or for the bond-stretch isomerism in the benzo[1,2:4,5]dicyclobutadiene system [7]. However, the real power of our methods becomes apparent for excited-state calculations, where very difficult situations, such as the simultaneous calculation of valence and Rydberg states have been treated [8-10].

Even though the focus in this paper will be laid on the calculation of energy surfaces, we want to stress the importance of dynamics investigations based on these highly accurate surfaces. In a recent study we have investigated the excited states of ethylene [11] and have discussed in particular the topology of the intersection seam and its relevance for the photodynamics of this system. Surface-hopping dynamics [12] is a very efficient method for the description of the nonadiabatic dynamics. On-the-fly techniques are practically the only way to treat larger molecular systems since tedious global surface-fittings can be avoided in this way. However, inexpensive methods for the computation of the energy surfaces have to be used since very large numbers of energy points are calculated. In order to combine the quality of our MR-CI calculations with the efficiency requirements of the on-the-fly approach, the semiempirical AM1 method had been used for the surface-hopping dynamics as developed by Persico et al. [13]. It turned out that a careful fit of the AM1 parameters to our MR-CI surfaces was crucial. Extensive dynamics calculations were performed in this way and a detailed picture of the photodynamics, especially of the ethylene lifetime has been given elsewhere [14].

The purpose of this contribution is to present our recent results on the excited-state surfaces of polar double bonds where the methyleneimmonium cation $CH_2NH_2^+$ has been chosen as example. Silaethylene and fluorethylene have already been investigated by us previously [15,16]. The methyleneimmonium cation is of interest since it is the first member in the series of protonated Schiff bases leading to retinal, which is the key compound in the primary process of vision. The $CH_2NH_2^+$ molecule has also been studied previously by Bonačić-Koutecký et al. [17] by means of MRD-CI calculations. The present calculations are planned as first benchmark investigations, which will be used to guide subsequent AM1 surface-hopping dynamics calculations in the same way as they have been performed for ethylene.

2 Computational Details

State-averaged (SA)-CASSCF calculations were performed for the determination of the molecular orbitals (MOs). A CAS(2,2) in the π space was used for the description of the valence space of ethylene and a CAS(6,4) (six electrons in two σ and two π orbitals) was used for the methyleneimmonium cation. Rydberg orbitals are taken into account by an auxiliary (AUX) space, into which single excitations from the valence CAS are constructed. For ethylene, the ground state (N), the $V(\pi\pi^*)$, $Z(\pi^{*2})$ and (π-3s,3p) Rydberg states are included in the state-averaging.

For methyleneimmonium, the ground state, the V state, two $\sigma\pi^*$ states, the (π-3s,3p) and σ-3p$_y$ Rydberg states were included. The same active space was used as reference space in subsequent MR-CI calculations with singles and doubles (MR-CISD). The carbon and nitrogen 1s orbitals were frozen in the CI calculations. The interacting space restriction was used in all calculations. Size-extensivity corrections were computed via the generalized Davidson correction [18] and are denoted by the label +Q. As basis set the d-aug-cc-pVDZ [19-21] and aug-cc-pVDZ [19,20] basis sets were chosen.

Two sets of calculations were performed. In the first set valence and Rydberg states were included. These calculations are labeled as MR-CISD/SA-7-CAS(2,2)+aux/d-aug-cc-pVDZ for ethylene and MR-CISD/SA-9-CAS(6,4)+aux/d-aug-cc-pVDZ for methyleneimmonium, indicating the number of states included in the state-averaging, the active space and the basis set. The second set of calculations (denominated as MR-CISD/SA-3-CAS(2,2)/aug-cc-pVDZ and MR-CISD/SA-2-CAS(6,4)/aug-cc-pVDZ, respectively) included the valence states only.

3 Potential Curves and Conical Intersections

3.1 Ethylene

In Figure 1 the potential energy curves for the rigid rotation around the CC bond in ethylene are displayed. Valence and Rydberg states have been computed. For more

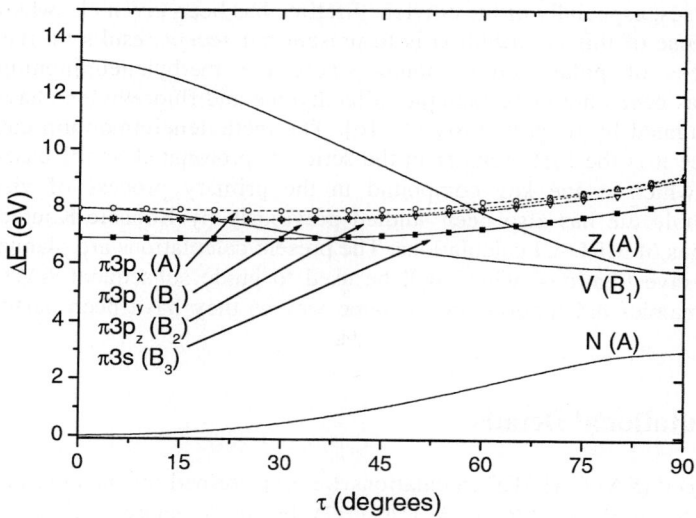

Fig. 1. Potential energy curves for the torsion coordinate calculated at the MR-CISD/SA-7-CAS(2,2)+aux/d-aug-cc-pVDZ level

details see Ref. 11 and references therein. Curves are plotted in a diabatic way following the character of the wavefunction. As one can see from the figure, the two valence-excited states are strongly stabilized by the torsion whereas the Rydberg states are destabilized. For the orthogonal structure ($\tau = 90°$) the V and Z states are almost degenerate. However, there exists still a gap of about 2.5 eV to the ground state N. The path to the conical intersection with the ground state requires additional pyramidalization and hydrogen-migration motions [22-24]. This fact leads to lifetimes of the excited states of about 100-150 fs (see [14] and reference therein).

3.2 The Methyleneimmonium Cation

In Figure 2, torsional potential energy curves are shown for $CH_2NH_2^+$ in analogy to the ethylene case depicted in Figure 1. Figure 2 shows a completely different behavior of the excited states on torsion as compared to ethylene. It is interesting to note that the $\sigma\pi^*$ state is the lowest singlet excited state. This state leads to a direct intersection with the ground state (see also Ref. 17).

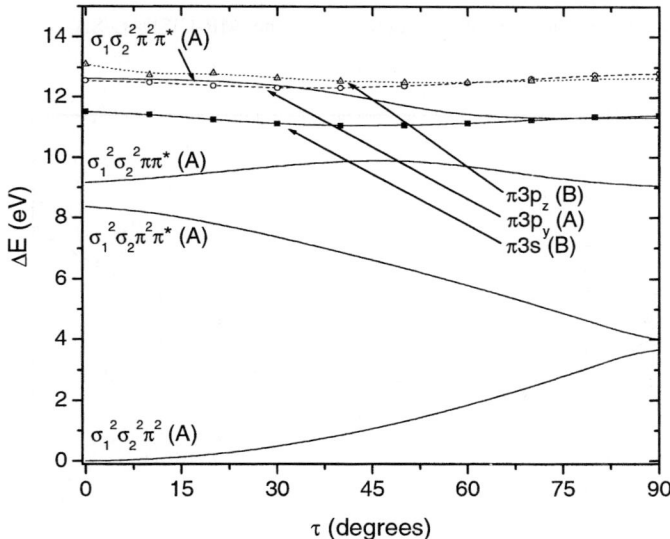

Fig. 2. Potential energy curves for the torsion coordinate calculated at the MR-CISD/SA-9-CAS(6,4)+aux/d-aug-cc-pVDZ level

In order to characterize the orthogonal structure in more detail, CH_2 and NH_2 pyramidalization curves have been computed starting from the twisted orthogonal structure. In case of the NH_2 pyramidalization (Figure 3) only a very shallow minimum for the ground state curve is observed. For the CH_2 pyramidalization (Figure 4) a destabilization of all curves is found. A crossing of the $1^1A''$ ground and the $1^1A'$ states occurs at a pyramidalization angle of around 30°.

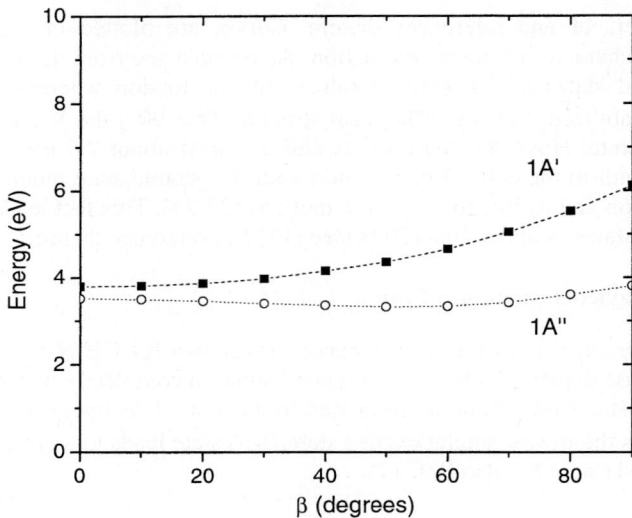

Fig. 3. Rigid NH_2 pyramidalization calculated at the MR-CISD+Q/SA-2-CAS(6,4)/aug-cc-pVDZ level

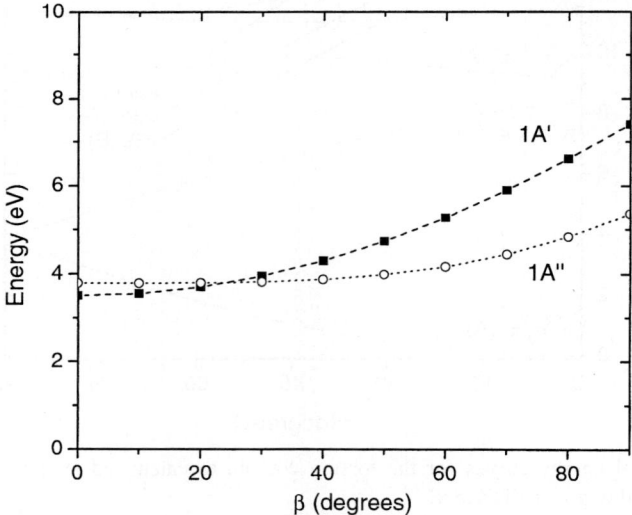

Fig. 4. Rigid CH_2 pyramidalization calculated at the MR-CISD+Q/SA-2-CAS(6,4)/aug-cc-pVDZ level

The structure of the minimum of the crossing seam (MXS) is displayed in Figure 5. It shows an orthogonal structure with a torsional angle of 90°. Starting from a distorted, nonorthogonal geometry in the MXS search led back to the orthogonal structure. Thus, the 90° form is a true MXS.

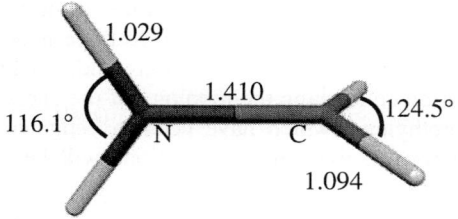

Fig. 5. The twisted-orthogonal MXS structure computed at the MR-CISD/SA-2-CAS(6,4)/aug-cc-pVDZ level. Bond distances given in Å, angles in degrees

Vertical excitation energies and relative stabilities of selected optimized structures are given in Table 1. The vertical excitation energy of 8.35 eV is somewhat higher than the one of ethylene (7.7 eV). The energies of the twisted-orthogonal S_0 and S_1 structures are close to each other. Practically identical structures were obtained as a result of the optimizations of the twisted-orthogonal structure of the S_1 state and the MXS. From this fact we conclude that only a MXS exists for this structure and the geometry optimization procedure led to the same structure.

Table 1. Energy differences (eV) for selected methyleneimmonium structures computed at the MR-CISD+Q/SA-2-CAS(6,4)/aug-cc-pVDZ level

Structure	State	Energy (eV)
Planar (S_0)	S_0	0.00
	S_1	8.35
Twist.-orthog. (S_0)	S_0	3.50
	S_1	3.77
Twist.-orthog. (S_1)	S_0	3.60
	S_1	3.66
MXS	S_0/S_1	3.68

4 Conclusions

The calculations show that the topology of the energy surfaces leading to a conical intersection with the electronic ground state and thus to rapid, radiationless transfer to the ground state is completely different in ethylene and the methyleneimmonium cation. The polarity of the CN bond destroys the quasi-degeneracy of the S_1 and S_2 states found for ethylene. In case of the methyleneimmonium cation, following the torsion around the CN bond on the S_1 surface leads directly to the twisted-orthogonal MXS. The energy gain between the vertically excited structure and the conical intersection is about 4.7 eV for the methyleneimmonium cation (twisted–orthogonal, see Table 1) and is about 3.2 eV (twisted-pyramidalized) in case of ethylene [11].

Thus, for the methyleneimmonium cation the lifetime of the S_1 state is expected to be significantly shorter as compared to ethylene since there is no need for complicated pyramidalization and hydrogen-migration motions and the kinetic energy is much larger. A similar situation concerning the topology of the energy surfaces is found for silaethylene and fluorethylene, which have been investigated in our group [15,16] also. Based on these results, dynamics calculations will be carried out in order to verify these predictions.

Acknowledgments

The authors acknowledge support by the Austrian Science Fund within the framework of the Special Research Program F16 and Project P14817-N03. Mario Barbatti thanks for the financial support from the Brazilian funding agency CNPq and Mohammad Solimannejad for travel funds from the University of Vienna.

References

1. Lischka, H., Dallos, M., Shepard, R.: Analytic MRCI Gradient for Excited States: Formalism and Application to the n-π* Valence- and n-(3s,3p) Rydberg States of Formaldehyde. Mol. Phys. 100, (2002) 1647-1658
2. Lischka, H., Dallos, M., Szalay, P. G., Yarkony, D. R. Shepard. R.: Analytic Evaluation of Nonadiabatic Coupling Terms at the MR-CI level. I: Formalism. J. Chem. Phys. 120 (2004) 7322-7329
3. Dallos, M., Lischka, H., Shepard, R., Yarkony, D. R., Szalay, P. G.: Analytic Evaluation of Nonadiabatic Coupling Terms at the MR-CI level. II. Minima on the Crossing Seam: Formaldehyde and the Photodimerization of Ethylene. J. Chem. Phys. 120 (2004) 7330-7339
4. Lischka, H., Shepard, R., Pitzer, R. M., Shavitt, I., Dallos, M., Müller, Th., Szalay, P. G., Seth, M., Kedziora, G. S., Yabushita, S., Zhang, Z.: New High-level Multireference Methods in the Quantum-Chemistry Program System COLUMBUS: Analytic MR-CISD and MRAQCC Gradients and MR-AQCC-LRT for Excited States, GUGA Spin-Orbit CI, and Parallel CI Density. Phys. Chem. Chem. Phys. 3 (2001) 664-673
5. Lischka, H., Shepard, R., Shavitt, I., Pitzer, R M., Dallos, M., Müller, Th., Szalay, P.G., Brown, F. B., Ahlrichs, R., Böhm, H. J., Chang, A., Comeau, D. C., Gdanitz, R., Dachsel, H., Erhard, C., Ernzerhof, M., Höchtl, P., Irle, S., Kedziora, G., Kovar, T., Parasuk, V., Pepper, M., Scharf, P., Schiffer, H., Schindler, M., Schüler, M., Zhao, J.-G.: COLUMBUS, An Ab Initio Electronic Structure Program, release 5.9 (2004)
6. Lischka, H., Ventura E., Dallos, M.: The Diels-Alder Reaction of Ethylene and 1,3-Butadiene: An Extended Multireference Ab Initio Investigation on Structures and Energies of the Concerted and Nonconcerted Pathways. ChemPhysChem 5 (2004) 1365-1371
7. Antol, I., Eckert-Maksić, M., Lischka, H., Maksić, Z. B.: On the Bond-Stretch Isomerism in the Benzo[1,2:4,5]dicyclobutadiene System - An Ab Initio MR-AQCC Study. ChemPhysChem 5 (2004) 975-981
8. Müller, Th., Dallos, M., Lischka, H.: The Ethylene 1^1B_{1u} V State Revisited. J. Chem. Phys. 110 (1999) 7176-7184

9. Dallos, M., Müller, Th., Lischka, H., Shepard, R.: Geometry Optimization of Excited Valence States of Formaldehyde Using Analytical MR-CISD and MR-AQCC Gradients, and the conical intersection formed by the $1^1B_1(\sigma\text{-}\pi^*)$ and $2^1A_1(\pi\text{-}\pi^*)$ states. J. Chem. Phys. 114 (2001) 746-757
10. Müller, Th., Lischka, H.: Simultaneous Calculation of Rydberg and Valence States: Excited States of Formaldehyde. Theor. Chem. Acc. 106 (2001) 369-378
11. Barbatti, M., Paier, J., Lischka, H.: Photochemistry of Ethylene: A Multireference Configuration Interaction Investigation of the Excited-state Energy Surface. J. Chem. Phys. 121 (2004) 11614-11624
12. Tully, J. C., Molecular Dynamics with Electronic Transitions. J. Chem. Phys. 93 (1990) 1061-1071
13. Granucci, G., Persico, M., Toniolo, A.: Direct Semiclassical Simulation of Photochemical Processes with Semiempirical Wave Functions. J. Chem. Phys. 114 (2001) 10608-10615
14. Barbatti, M., Granucci, G., Persico, M., Lischka, H.: Semiempirical Molecular Dynamics Investigation of the Excited State lifetime of Ethylene. Chem. Phys. Lett. 401 (2005) 276-281
15. Pitonak, M., Lischka, H.: Excited-State Potential Energy Surfaces of Silaethylene: a MRCI Investigation. Mol. Phys., in press
16. Barbatti, M., Aquino, A. J. A., Lischka, H.: A Multireference Configuration Interaction Investigation of the Excited-State Energy Surfaces of Fluoroethylene. In Preparation
17. Bonačić-Koutecký, V., Schöffel, K., Michls, J.: Critically Heterosymmetric Biradicaloid Geometries of Protonated Schiff Bases. Theor. Chim. Acta 72 (1987) 459-474.
18. Bruna, P. J., Peyerimhoff, S. D., Buenker, R. J.: The Ground State of the CN^+ ion: a Multi-reference CI Study. Chem. Phys. Lett. 72 (1981) 278-284
19. Dunning, T. H. Jr.: Gaussian Basis Sets for Use in Correlated Molecular Calculations. I. The Atoms Boron through Neon and Hydrogen. J. Chem. Phys. 90 (1989) 1007-1023
20. Kendall, R. A., Dunning, T. H. Jr.: Electron Affinities of the First-row Atoms Revisited. Systematic Basis Sets and Wave Functions. J. Chem. Phys. 96 (1992) 6796-6806
21. Mourik, T. v., Wilson, A. K., Dunning, T. H. Jr.: Benchmark Calculations with Correlated Molecular Wavefunctions. XIII. Potential Energy Curves for He_2, Ne_2 and Ar_2 using Correlation Consistent Basis Sets through Augmented Sextuple Zeta. Mol. Phys. 96 (1999) 529-547
22. Ohmine, I.: Mechanisms of Nonadiabatic Transitions in Photoisomerization Processes of Conjugated Molecules: Role of Hydrogen Migration. J. Chem. Phys. 83 (1985) 2348-2362
23. Freund., L., Klessinger, M.: Photochemical Reaction Pathways of Ethylene. Int. J. Quantum Chem. 70 (1998) 1023-1028
24. Ben-Nun, M., Martínez, T. J.: Photodynamics of Ethylene: Ab Initio Studies of Conical Intersections. Chem. Phys. 259 (2000) 237-248. Ben-Nun, M., Martínez, T. J.: Ab initio Molecular Dynamics Study of Cis-trans Photoisomerization in Ethylene. Chem. Phys. Lett. 298 (1998) 57-65.

Study of Predictive Abilities of the Kinetic Models of Multistep Chemical Reactions by the Method of Value Analysis

Levon A. Tavadyan, Avet A. Khachoyan, Gagik A. Martoyan, and Seyran H. Minasyan

Institute of Chemical Physics National Academy of Sciences of Republic of Armenia,
5/2 Sevak Str, Yerevan 375014, Republic of Armenia
Tel/Fax: (3741) 28-17-42
tavadyan@ichph.sci.am

Abstract. A numerical method of researching kinetic models' predictive abilities is suggested. As a criterion of predictive abilities of the kinetic model were chose the degree of coincidence of the kinetic models reduced by the value numerical analysis for experimentally studied and predicted conditions.

The present approach is illustrated by means of the kinetic model of ethylbenzene oxidation at butylated hydroxytoluene (BHT) inhibition.

1 Introduction

The important characteristics of the kinetic model is its predictive ability. Revealing the predictive ability of the kinetic model of chemical reaction is especially topical at chemical reaction control. It is obvious, that controlling influences on the reaction with the use of its kinetic model are more appropriate to be carried out within those limits of change of controlling parameters where the model preserves its predictive ability.

Before, in research of predictive abilities of kinetic models, the method of parametric sensitivity analysis was used [1]. It is obviously important, however, in addition to the used approaches to have development of new criteria characterizing the predictive ability of the kinetic models. It allows to update more successfully the model and to project the experiments directed to the improvement of the predictive ability of reaction mechanisms.

The objective of the present paper is the development of a new criterion of the kinetic models' predictive ability evaluation based on the value revelation of the significance of individual steps and species [2-5].

2 Results

2.1 Evaluation Strategy and the Improvement of the Kinetic Model's Predictive Abilities

As an evaluation criterion and that of the improvement predictive ability of the kinetic model was suggest the degree of coincidence of the sets of value contributions of chemical species and individual steps of the kinetic model. These would be determined under conditions of model adequacy compared to experimental results and predicted conditions. Such "chemically structured" index of the predictive abilities of the kinetic model significantly facilitates the actions directed to the improvement of this model index in the following way:

- kinetic model reconstruction
- new experiments projecting at the description of which the kinetic model of reaction with higher predictive ability is devised.

2.2 Characteristic of the Kinetic Significance of Individual Reaction Steps and Species via Value Magnitudes and Their Calculation

Value Parameters. The kinetic significance of reaction steps and species included in the reaction mechanism was determined using the approach of value magnitudes, which have introduced before. We made the comparison of the results of the value analysis with the classical sensitive analysis in reference [3].

The value of G_i step determined as the relation of response by a chosen characteristic magnitude of the reaction system $(F(t))$, at any time t, to perturbation of i reaction step rate (r_i) at the t_o initial time is:

$$G_i(t) = \frac{\partial F[r_1(t),...,r_n(t)]}{\partial r_i(t)} \bigg/_{r_i=r_i(t_0),\ i=1,2,...,n} \qquad (1)$$

Similarly, the value of reaction system species (ψ_j) is determined as the relation of output value F response, at a time t, to perturbation of species concentration change rate $(f_i(t))$ at the initial time:

$$\psi_j(t) = \frac{\partial F[f_1(t),...,f_n(t)]}{\partial f_j(t)} \bigg/_{f_j=f_j(t_0),\ j=1,2,...,m} \qquad (2)$$

Value contribution of reaction steps (h_i) and species accumulation is:

$$h_i(t) = r_i(t)G_i(t) \quad \text{and} \quad b_j(t) = f_j(t)\psi_j(t) \qquad (3)$$

Account of Value Parameters. Uniform chemical system at isothermal conditions can be presented by the system of ordinary differential equations:

$$\frac{dc_i}{dt} = f_i(\mathbf{k},\mathbf{c}), \qquad \mathbf{c}(t_0) = \mathbf{c}^0 \qquad (4)$$

Where $\mathbf{c}(t)$ is m-vector of species concentration, and $\mathbf{k}$ is the n-vector of rate constants.

The approach of Hamiltonian systematization for dynamic systems was used to calculate the value parameters.

The objective function of the reaction system is chosen, which is represented in the form of integral:

$$I(t) = \int_{t_0}^{t} F(t)dt \qquad (5)$$

Furthermore, according to (5) the corresponding Hamiltonian H is written:

$$H = \psi_0 F + \sum_{i=1}^{m} \psi_i f_i, \qquad H = \text{const}, \qquad \psi_0 = \pm 1 \qquad (6)$$

Kinetic trajectories of value magnitudes (ψ_i, G_j) simultaneously with the concentrations of species c_i are determined by solving the systems of differential equations:

$$\frac{dc_i}{dt} = \frac{\partial H}{\partial \psi_i} = f_i, \qquad i = 1, 2, \ldots, m$$

$$\frac{d\psi_i}{dt} = -\frac{\partial H}{\partial c_i} \qquad (7)$$

$$G_j = \sum_p a_{jp} \psi_{jp}(t) - \sum_e a_{je} \psi_{je}(t)$$

where a_{jp}, a_{je} are stoichiometric coefficients of j, the individual step, p and e, the indexices relating to the products and to the initial substances of that step, respectively.

3 Illustrative Examples
The Revelation of the Kinetic Models' Predictive Abilities of Ethylbenzene Oxidation Reaction Inhibited by BHT

3.1 Kinetic Model of Reaction

The choice of oxidation-inhibited reactions of organic compounds by molecular oxygen is caused by the topicality of the increase of their kinetic models' predictive abilities. The result of the action of oxidation-antioxidant chain reaction inhibitors is very often necessary for predicting the conditions outside the domain of reaction experimental study. Hence, the descriptive ability of the similar reaction kinetic models is tested by experiments. Whereas the practical use of antioxidants requires the predictions be made at high initial concentrations of the inhibitors to which corresponds significantly long-term chemical conversion.

The mechanism of BHT inhibition action in oxidation reactions of organic substances is thoroughly studied [6]. The expanded kinetic model of this ethylbenzene oxidation reaction, consisting of 34 steps is represented in Table 1. This model adequately (with an error of 7% and less) describes the experimental kinetic curves of oxygen uptake at various temperatures (60 ^{0}C, 120 ^{0}C) and initial BHT concentrations in the range of 6.5×10^{-5} M to 6.4×10^{-4} M.

Table 1. Kinetic model of ethylbenzene oxidation liquid-phase reaction inhibited by BHT

No	Reactions	Rate Constants $T=60\ ^{\circ}C$	$T=120\ ^{\circ}C$
1	$2RH + O_2 \rightarrow 2R^{\bullet} + H_2O_2$	9.26×10^{-13}	7.7×10^{-10}
2	$R^{\bullet} + O_2 \rightarrow ROO^{\bullet}$	8.75×10^{8}	1×10^{9}
3	$ROO^{\bullet} + RH \rightarrow ROOH + R^{\bullet}$	2.74	39
4	$RO^{\bullet} + RH \rightarrow ROH + R^{\bullet}$	2.32×10^{6}	5.85×10^{6}
5	$HO^{\bullet} + RH \rightarrow H_2O + R^{\bullet}$	1×10^{9}	1×10^{9}
6	$HOO^{\bullet} + RH \rightarrow H_2O_2 + R^{\bullet}$	7.62	60
7	$HOO^{\bullet} + ROOH \rightarrow H_2O_2 + ROO^{\bullet}$	1.05×10^{3}	3×10^{3}
8	$ROO^{\bullet} + H_2O_2 \rightarrow ROOH + HOO^{\bullet}$	1.68×10^{2}	4.4×10^{2}
9	$RO^{\bullet} + ROOH \rightarrow ROH + ROO^{\bullet}$	4.9×10^{8}	6.43×10^{8}
10	$ROOH + RH \rightarrow RO^{\bullet} + H_2O + R^{\bullet}$	1.28×10^{-10}	2.72×10^{-7}
11	$ROOH + RH \rightarrow R(-H)O + H_2O + RI$	3.83×10^{-10}	4.08×10^{-7}
12	$H_2O_2 + RH \rightarrow R^{\bullet} + H_2O + HO^{\bullet}$	1.06×10^{-10}	1.62×10^{-7}
13	$ROO^{\bullet} + ROO^{\bullet} \rightarrow 2RO^{\bullet} + O_2$	5.5×10^{6}	1×10^{7}
14	$ROO^{\bullet} + ROO^{\bullet} \rightarrow ROH + R(-H) + O$	1×10^{7}	3.5×10^{7}
15	$ROO^{\bullet} + HOO^{\bullet} \rightarrow ROOH + O_2$	3×10^{8}	3×10^{8}
16	$ROO^{\bullet} + HOO^{\bullet} \rightarrow R(-H)O + H_2O +$	1×10^{8}	1×10^{8}
17	$ROO^{\bullet} + InH \rightarrow ROOH + In^{\bullet}$	2.04×10^{4}	6.94×10^{4}
18	$HOO^{\bullet} + InH \rightarrow H_2O_2 + In^{\bullet}$	6.12×10^{4}	2.08×10^{5}
19	$RO^{\bullet} + InH \rightarrow ROH + In^{\bullet}$	4.54×10^{6}	2.1×10^{7}
20	$ROOH + InH \rightarrow RO^{\bullet} + H_2O + In^{\bullet}$	1.75×10^{-7}	6.7×10^{-5}
21	$InH + O_2 \rightarrow HOO^{\bullet} + In^{\bullet}$	5.8×10^{-12}	8.77×10^{-9}
22	$In^{\bullet} + In^{\bullet} \rightarrow InH + In(-H)$	1.1×10^{4}	3.3×10^{4}
23	$In^{\bullet} + In^{\bullet} + O_2 \rightarrow QP_1$	4.9×10^{3}	4.9×10^{3}
24	$In^{\bullet} + ROO^{\bullet} \rightarrow QP_2$	3×10^{8}	3×10^{8}
25	$In^{\bullet} + HOO^{\bullet} \rightarrow InH + O_2$	1.5×10^{8}	1.5×10^{8}
26	$In^{\bullet} + HOO^{\bullet} \rightarrow QP_2$	6.5×10^{8}	6.5×10^{8}
27	$In^{\bullet} + ROOH \rightarrow InH + ROO^{\bullet}$	1	15.23
28	$In^{\bullet} + RH \rightarrow InH + R^{\bullet}$	1.91×10^{-6}	6.96×10^{-5}
29	$QP_2 \rightarrow 2RO^{\bullet}$	2.09×10^{-8}	4.13×10^{-5}
30	$QP_1 \rightarrow 2RO^{\bullet}$	2×10^{-7}	4.5×10^{-4}
31	$HOO^{\bullet} + HOO^{\bullet} \rightarrow H_2O_2 + O_2$	3.5×10^{8}	3.5×10^{8}
32	$ROOH + InH \rightarrow P + H_2O$	4.73×10^{-7}	1.8×10^{-4}
33	$In^{\bullet} + H_2O_2 \rightarrow InH + HOO^{\bullet}$	1	15.23
34	$InH + H_2O_2 \rightarrow In^{\bullet} + HO^{\bullet} + H_2O$	1.75×10^{-7}	6.7×10^{-5}

Note: for reaction (1) the dimension of rate constant is [mol^{-2}l^{2}c^{-2}], RH is ethylbenzene and InH is BHT;

3.2 Calculation of the Kinetic Trajectories of Value Contributions of Individual Steps

Numerical calculations have been performed using the computer program VALKIN developed by us. The algorithm of the program is developed on the basis of the Hamiltonian systematization of kinetic models. For the integration of differential equations the ROW-4A algorithm developed by Gottwald [7] is used.

At calculation of value contributions of individual steps as objective functional of reaction (F), the rate of ethylbenzene (RH) is r_{RH} expenditure is chosen, in this case the objective functional (5) will look like:

$$I(t) = \Delta[\text{RH}]_t = \int_{t_0}^{t} r_{\text{RH}} dt \qquad (8)$$

Using corresponding equations (6) and (7), the kinetic trajectories of value contributions of the steps are designed in the reaction induction period τ_{ip}.

Value ranking of steps has allowed to reveal the significant and insignificant steps. As a result, the base - minimum mechanism of reaction has emerged, shown in Table 2. As a criterion of step insignificance is chosen the condition according to which, at simultaneous exception of the revealed "low-value" steps from the reaction kinetic model, the change of significances of the current concentrations RH, InH, ROOH, ROO$^{\bullet}$, In$^{\bullet}$ does not exceed 3.5 %. This control was realized for two different time points of the reaction that meet BHT conversions equal to 10-20% and 60-70%. Preliminary ranking of steps according to their value contributions was made in an interval of times $10^{-7} c$ to τ_{ip}. As "candidates" on exception from the kinetic model were chosen the steps with small significances of the resultant value contributions $\left(\overline{h_j} < 10^{-5}\right)$ in the specified interval of time. Here $\overline{h_j} = h_j \left/ \left(\sum_{1}^{34} h_j^2\right)^{1/2}\right.$.

3.3 Studies of the Predictive Ability of the Kinetic Model

The evaluation of the predictive abilities of the kinetic model was carried out according to the criterion represented above. For this purpose the base mechanism, which is a set of significant steps describing the experimental data was compared with the base mechanism determined in the predicted conditions. As mentioned above, the greater degree of coincidence these base kinetic models have, the more reliable is the prediction.

As it follows from the data shown in Table 2, satisfactory coincidence of the set of "dominant" steps of these base models is observed at 120 ^{0}C. It can be considered as an index of good ability of the kinetic model for predicting at this temperature.

On the contrary, essential differences between the base mechanisms determined for the experimental conditions of the initiated oxidation is observed at 60 ^{0}C, and between those base mechanisms determined under the predictive conditions, in case of non-initiated oxidation reaction at BHT inhibition, it testifies to much smaller reliability in case of predictions made at 60 °C.

It seems that, the value analysis allows estimating opportunities of the forecast "chemically". Moreover, it is very important, that on this basis it becomes possible to plan new experiments in which the kinetic model provides more reliable predictions. In this case by kinetic model constructing is meant both the refinement of rate constant significances of the individual steps and the addition of the model with new steps.

Table 2. Base mechanisms of ethylbenzene oxidation liquid-phase reaction inhibited by BHT under various initial conditions, including those for experimentally studied and predicted reaction conditions

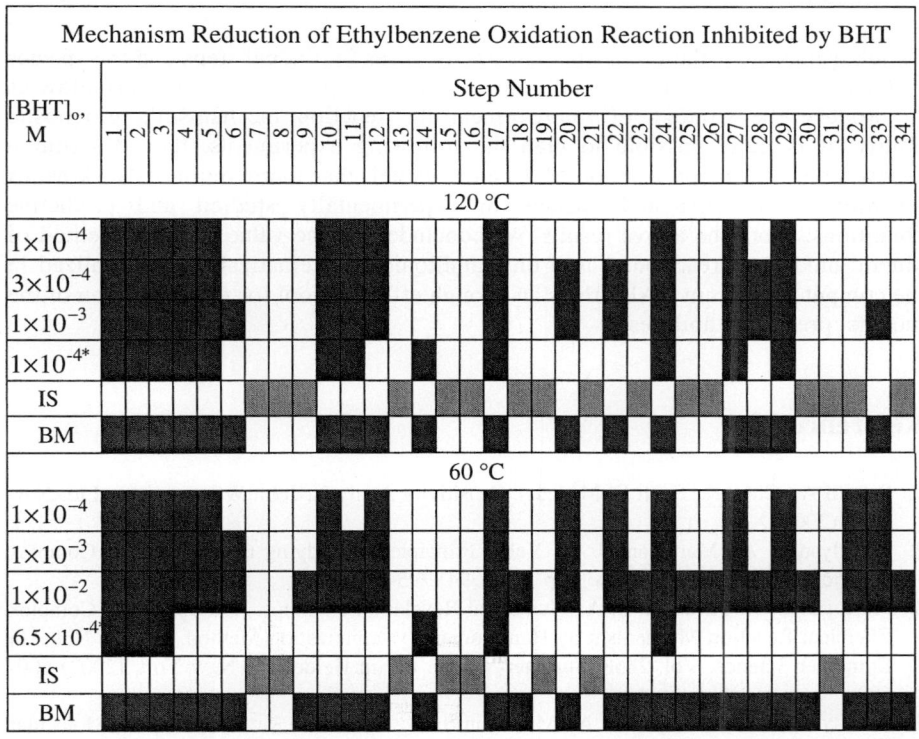

Notes: ☐ - insignificant steps of the kinetic model under the given particular conditions

▨ - insignificant steps in the model for the chosen range of the initial data (IS)

■ - steps in the base model (BM)

* - experimental initial conditions

Therefore, for example, it follows from Table 2, that at 120 ^{0}C the step (20) of autoinitiation with the participation of an inhibitor molecule is insignificant in experimental kinetic conditions, whereas its kinetics is significant in predicting conditions. To purpose to increase the predictive ability of the model based on these data new experiments are planned to reveal step (20). For this purpose experiments have been carried out at much high initial hydroperoxide concentration $([ROOH]_0 = 10^{-2} M)$. Under these new conditions, the value contribution of the step (20) is vital importance and as a whole the kinetic model has adequately described the kinetic curve of oxygen consumption. This result allows to come to the conclusion on the accuracy of the chosen magnitude of the rate constant of step (20) and on the greater confidence in predictive ability of the kinetic model of chemical reaction.

4 Conclusions

Simultaneously revelation of kinetics significances individual steps and the species of kinetics model in experimentally investigated and predictive conditions allow us to produce the strategy of improvement the reaction mechanism's prognostic ability. It is quite possible to plan the following experiments: they description demand on such reconstruction of kinetics model, that more similarity to kinetics magnitudes of steps and species in experimentally studied and predictive conditions. From the above results, we conclude that the value analysis method of kinetic models of reactions based on Hamiltonian Systematization and realized in the computer program VALKIN is an enough effective tool for the evaluation of the models' predictive abilities.

References

1. Saltelli A., Chan K., Scott E. M. (eds): Sensitivity Analysis. John Wiley & Sons Ltd. New York (2000)
2. Tavadyan L. A., Martoyan G. A.: Value Principle of Studying the Kinetics of Complex Kinetic Reactions. Chem. Phys. Rep. **13** (1994) 793-797
3. Martoyan G. A. Tavadyan L. A.: Numerical Revelation for Steps. And Species in Complex Chemical Reaction Mechanism by Hamiltonian Systematization Method. Lecture Notes in Computer Science. Vol. 2658. Springer-Verlag, Berlin Heidelberg New York (2003) 600-609
4. Tavadyan L. A., Martoyan G. A., Minasyan S. H.: Numerical Revelation of the Molecular Structure for Reaction Effective Stimulator or Inhibitor by Method of Hamiltonian Systematization of Chemical Reaction System Kinetic Models. Lecture Notes in Computer Science. Vol. 2658. Springer_Verlag, Berlin Heidelberg New York (2003) 593-599
5. Martoyan G. A., Tavadyan L. A.: Numerical Revelation and Analysis of Critical Ignition Conditions from Branch Chain Reactions by Hamiltonian Systematization Method of the Kinetic Models. Lecture Notes in Computer Science. Vol. 3044. Springer_Verlag, Berlin Heidelberg New York (2004) 313-320

6. Denisov E. T. Azatyan V. V.: Inhibition of the Chain Reactions. RAS, Chernogolovka, 1997 (*in Russian*)
7. Gottwald B. A.: A Digital Simulation System for Coupled Chemical Reactions. Simulation. **33** (1981) 169-173

Lateral Interactions in O/Pt(111): Density-Functional Theory and Kinetic Monte Carlo

A.P.J. Jansen and W.K. Offermans

Laboratory of Inorganic Chemistry and Catalysis,
ST/SKA, Eindhoven University of Technology,
P.O. Box 513, 5600 MB Eindhoven, The Netherlands
tgtatj@chem.tue.nl,
http://www.catalysis.nl/~theory/

Abstract. It has become clear over the last couple of years that interactions between adsorbates are very important for the kinetics of processes in heterogeneous catalysis. We show that we can calculate the nearest-neighbor, next-nearest-neighbor, and linear 3-particle interaction between oxygen atoms on a Pt(111) surface using density-functional theory. Trying to compute more interactions leads to overfitting. Kinetic Monte Carlo simulations show that the calculated interactions give a good description of the desorption kinetics.

1 Introduction

Interactions between adsorbates, so-called lateral interactions, on transition metal surfaces have been known for a long time, as they can lead to structured adsorbate layers (adlayers) at low temperatures. It is only recently that the importance of these interactions on the kinetics of catalytic processes at higher temperatures has been acknowledged. It has been shown that they may change reaction rate constants by an order of magnitude or more. This has led to an increasing interest in the quantitative determination of such interactions. It is very difficult to determine lateral interactions from experiments, because experimental results generally depend in a very complicated way on these interactions. Various groups have therefore looked at quantum chemical calculations, in particular density-functional theory (DFT), to get values for lateral interactions. The question here is if it is at all possible to get reliable results from such calculations. Interactions of the order of the thermal energy $k_\mathrm{B}T$ will affect the rates of reactions appreciably, because they change the activation energy E_act, which affects rate constants via a factor $\exp[-E_\mathrm{act}/k_\mathrm{B}T]$. Energies of this magnitude at temperatures between room temperature and 1000 K are smaller than the accuracy of DFT calculations. Larger lateral interactions are also found when adsorbates get close to each other. Such interactions might be calculated more accurately. However, they may not be so relevant. They are repulsive and the adsorbates will avoid them, unless there are many adsorbates and they are forced close together. In this paper we determine for oxygen atoms on a Pt(111) surface which lateral interactions we can calculate with confidence. We do this by fitting various models for the lateral interactions to adsorption energies obtained from DFT calculations, and determining which parameters improve the models and which lead to overfitting. We also look at how useful the reliable lateral interactions are for the kinetics using kinetic Monte Carlo (kMC) simulations of oxygen desorption.

2 Computational Details

2.1 Density-Functional Theory

DFT calculations have become very popular for doing quantum chemical calculations, as DFT combines efficiency with accuracy. We have done DFT calculations with the VASP code,[1] which solves the Kohn-Sham equations with a plane wave basis set and the (relativistic) ultrasoft pseudopotentials introduced by Vanderbilt and generated by Kresse and Hafner.[2,3] The Generalize Gradient Approximation of Perdew and Wang (PW-91) has been used, because it generally yields good bond energies.[4] All calculations on O/Pt(111) were done with a surface model consisting of a supercell with a slab of five metal layers separated by five metal layers replaced by vacuum. A $5 \times 5 \times 1$ grid for Brillouin zone sampling obtained via the Monckhorst-package was used, and a cut-off of 400 eV. This yielded adsorption energies per oxygen atom converged to within

Table 1. All adlayer structures for which the adsorption energy of the oxygen atoms have been determined. Also given are the coverage θ and the adsorption energy per oxygen atom E_{ads} (in kJ/mol)

1×1	2×2-3O	$\sqrt{3} \times \sqrt{3}$-2O	2×1	3×2-3O	$\sqrt{3} \times \sqrt{3}$	3×2-2O
$\theta = 1.00$	0.75	0.67	0.50	0.50	0.33	0.33
$E_{\text{ads}} = -301$	-333	-350	-367	-365	-381	-379
2×2	$2 \times \sqrt{3}$	$\sqrt{7} \times \sqrt{7}$-2O	3×3-2O	3×2	$\sqrt{7} \times \sqrt{7}$	3×3
$\theta = 0.25$	0.25	0.25	0.22	0.17	0.13	0.11
$E_{\text{ads}} = -402$	-387	-391	-387	-393	-395	-421

5 kJ/mol with respect to k-point sampling, energy cut-off, number of slab and vacuum layers, and cell size. The supercells are shown in table 1. These structures allow us to calculate various lateral interactions. Oxygen atoms adsorb on fcc sites. Other sites are energetically less favorable by at least 50 kJ/mol.

2.2 Kinetic Monte Carlo Simulations

Our kMC is based on a master equation that describes the evolution of the system, and is given by[5]

$$\frac{dP_\alpha}{dt} = \sum_\beta [W_{\alpha\beta} P_\beta - W_{\beta\alpha} P_\alpha], \qquad (1)$$

where α and β refer to the configuration of the adlayer, the P's are the probabilities of the configurations, t is real time, and the W's are rate constants. $W_{\alpha\beta}$ corresponds to the reaction that changes β into α. A configuration can loosely be regarded as the way the adsorbates are distributed over all sites in the system. The derivation of the master equations shows that the rate constants can be written as[5]

$$W_{\alpha\beta} = \nu \exp\left[-\frac{E_{\text{act}}}{k_B T}\right], \qquad (2)$$

with k_B the Boltzmann-constant, T temperature, ν the prefactor, and E_{act} the activation barrier of the reaction that transforms configuration β into configuration α.

The kMC simulations form a powerful numerical method to solve the master equation exactly. For systems with rate constants that depend on lateral interactions and on time, which is the case here, we have found that only the first-reaction method is efficient.[6,7,8] It generates an ordered list of times at which a reaction takes place, and for each time in that list the reaction that occurs at that time. A kMC simulation starts with a chosen initial configuration. The list is traversed and changes are made to the configuration corresponding to the occurring reactions.

2.3 The Reaction Models

We have simulated the temperature-programmed desorption (TPD) of oxygen from Pt(111). The only reaction that can occur in the TPD experiment in our case is

$$2\text{O(ads)} \rightarrow \text{O}_2(\text{gas}) + 2*, \qquad (3)$$

where $*$ stands for a vacant site. The two sites involved in the reaction are nearest or next-nearest neighbors. Oxygen atoms can also diffuse by hopping to neighboring sites. We have included lateral interactions for desorption and diffusion through the Brønsted-Polanyi relation for the activation barrier:[9]

$$E_{\text{act}} = E_{\text{act}}^{(0)} + \alpha(\Delta\Phi_{\text{final}} - \Delta\Phi_{\text{initial}}), \qquad (4)$$

where $E_{\text{act}}^{(0)}$ is the activation barrier without lateral interactions, and $\Delta\Phi_{\text{final}}$ ($\Delta\Phi_{\text{initial}}$) is the change in adsorption energy in the final (initial) state due to lateral interactions. The coefficient α is the Brønsted-coefficient. We optimized it for the desorption, but for diffusion we used $\alpha = 1/2$. The prefactor ν and activation energy $E_{\text{act}}^{(0)}$ for desorption was also optimized (see section 3). For diffusion a minimal value for $E_{\text{act}}^{(0)}$ was chosen so that E_{act} in equation (4) was always positive whatever the occupation of neighboring sites. The value was determined for each model of the lateral interactions separately. By taking a minimal value for $E_{\text{act}}^{(0)}$ the variation of the diffusion rate as a function of temperature was minimized. Changing $E_{\text{act}}^{(0)}$ does not change the equilibrium of the adlayer in any way, because this depends only on the lateral interactions. The prefactor for diffusion was also given a minimal value, but large enough so that the adlayer was equilibrated at all times.[10]

3 Results and Discussion

Table 1 shows all adlayer structures that have been used to obtain the lateral interactions. For adlayer structure m ($m \geq 1$) we have fitted the adsorption energy per oxygen atom $E_{\text{ads}}^{(m)}$ with

$$E_{\text{ads}}^{(m)} = E_{\text{ads}}^{(0)} + \sum_n c_n^{(m)} \varphi_n, \tag{5}$$

where $E_{\text{ads}}^{(0)}$ is the adsorption energy for an isolated oxygen atom, φ_n is a lateral interactions parameter, and $c_n^{(m)}$ is the number of such interactions per oxygen atom in adlayer structure m. We looked at models for the lateral interactions that included pair interactions φ_{NN}, φ_{NNN}, and φ_{NNNN}, and 3-particle interactions φ_{linear}, $\varphi_{\text{triangle}}$, φ_{bent}, and all possible subsets of these interactions (see figure 1).

Table 2 shows results of a number of models for the lateral interactions. They are obtained by a linear regression procedure, in which we minimized

$$\chi^2 = \sum_m \left[E_{\text{ads}}^{(m)} - E_{\text{ads,calc}}^{(m)} \right]^2. \tag{6}$$

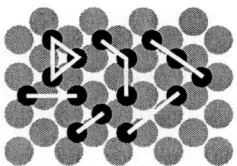

Fig. 1. Definition of the pairwise interactions φ_{NN} (bottom-middle), φ_{NNN} (bottom-left), φ_{NNNN} (bottom-right), and the 3-particle interactions φ_{linear} (top-right), $\varphi_{\text{triangle}}$ (top-left), φ_{bent} (top-middle).

as a function of $E_{\text{ads}}^{(0)}$ and the lateral interactions parameters. Here $E_{\text{ads}}^{(m)}$ is given by equation (5) and $E_{\text{ads,calc}}^{(m)}$ are the energies in table 1 obtained from the DFT calculations. The results agree well with those of Tang et al.[11] The main question we are interested in is how reliable the models are: i.e., if we use a model to compute the energy of another adlayer how much can we trust the result. To answer that question we use the leave-one-out method.[12] Instead of minimizing χ^2 we leave out one adlayer structure and then determine the lateral interactions that minimize

$$\chi_k^2 = \sum_{m \neq k} \left[E_{\text{ads}}^{(m)} - E_{\text{ads,calc}}^{(m)} \right]^2. \quad (7)$$

We have done this for each adlayer structure k, and then looked at the leave-one-out error

$$R_{\text{loo}}^2 = \frac{1}{N_{\text{str}}} \sum_k \left[E_{\text{ads,pred}}^{(k)} - E_{\text{ads,calc}}^{(k)} \right]^2, \quad (8)$$

where $E_{\text{ads,pred}}^{(k)}$ is the energy of the adlayer structure k obtained with the lateral interactions that have been determined without that structure (i.e., it minimizes χ_k^2), and N_{str} is the number of adlayers structures in the summation. This error indicates how well a model predicts the energy. Adding parameters will not necessarily decrease this error. An increase indicates overfitting.

Figure 2 shows how the leave-one-out error changes with the model of the lateral interactions. It is clear that the nearest-neighbor and next-nearest-neighbor interactions are the most important. These two interactions already provide a good description of all the adlayer structures. The next-next-nearest-neighbor interaction does not improve the model, but remarkably the linear 3-particle interaction does. Our results indicate that it is not justifiable to determine more lateral interactions as was done for this system before,[11] and also for the similar O/Ru(0001).[13] Table 2 gives the lateral interactions for some important models. It also gives the best model that includes the next-next-nearest-neighbor interaction. The reason to include this model is that this interaction has been used to explain island formation that has been observed for this system.[14] The interaction in the table is indeed attractive, but this is not the case for all models that include it. Moreover, even if it is attractive it is definitely too weak to explain the island formation at the experimental temperatures of up to 500 K, as has been pointed out before by Zhdanov and Kasemo.[15] The reason why

Table 2. Adsorption energy for an isolated oxygen atom, lateral interactions, and the leave-one-out error for different models of the lateral interactions. If a lateral interaction is not specified it is not included in the model. All quantities are in kJ/mol

$E_{\text{ads}}^{(0)}$	φ_{NN}	φ_{NNN}	φ_{NNNN}	φ_{linear}	R_{loo}
-396.3 ± 1.2	19.9 ± 2.3	5.5 ± 0.9		6.1 ± 2.0	3.1
-393.9 ± 1.5	20.8 ± 2.0	5.0 ± 0.8	-1.9 ± 0.9	7.2 ± 1.8	3.7
-396.8 ± 1.6	26.0 ± 1.5	4.8 ± 1.2			4.6
-393.3 ± 2.1	29.7 ± 1.8				6.1

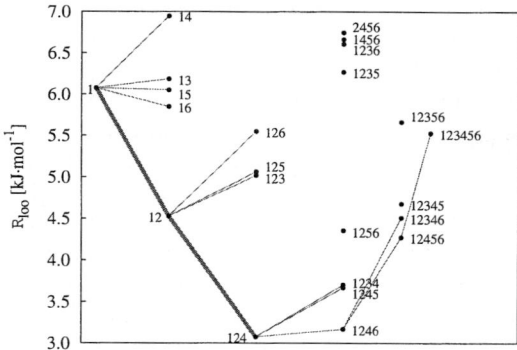

Fig. 2. The leave-one-out error (in kJ/mol) as a function of the model of the lateral interactions. The numbers indicate the different lateral interactions: $\varphi_{\text{NN}}(=1)$, $\varphi_{\text{NNN}}(=2)$, $\varphi_{\text{NNNN}}(=3)$, $\varphi_{\text{linear}}(=4)$, $\varphi_{\text{triangle}}(=5)$, and $\varphi_{\text{bent}}(=6)$. The lines are guides to the eye. They connect models that differ by one lateral interactions parameter. A fat line indicates that adding a parameter improves the model

the other lateral interactions cannot be determined is that they are too weak and consequently would have huge relative errors.

The root-mean-square error of the model with only nearest-neighbor and next-nearest-neighbor interactions is 3.9 kJ/mol. The one of the model that also include the linear 3-particle interaction is 2.9 kJ/mol. Using these errors as estimates for the accuracy of the DFT calculations gives the error estimates of the lateral interactions in table 2. These should be regarded as lower estimates. If we would use a larger estimate for the accuracy of DFT then the errors of the lateral interactions would increase proportionally. Note that, because the term $E_{\text{ads}}^{(0)}$ is always the same in the expressions for $E_{\text{ads}}^{(m)}$ independent of the adlayer structure, possible systematic errors in DFT only affect $E_{\text{ads}}^{(0)}$ and not the lateral interactions.

To further assess the quality of the models of the lateral interactions that we have developed, we have used kMC to simulate TPD experiments. In such an experiment an adsorbate, in our case atomic oxygen, is deposited on a catalyst at a low temperature when no reactions occur. Then the temperature is raised, generally linearly, and the rate of desorption is measured. A peak in the desorption rate is normally interpreted as desorption from a particular type of site: the temperature of the peak depends on the bonding energy. Lateral interactions can complicate the interpretation of TPD spectra enormously, although here they mainly cause only a shift of the peak maximum temperature to lower values with increasing initial coverage.

At high coverages it may be possible that adsorbates occupy less favorable adsorption sites to avoid strong repulsive interactions. For O/Pt(111) the fcc sites are the ones that have the highest adsorption energies. Other sites are less favorable by at least 50 kJ/mol. From the results in table 2 it seems obvious that such sites will only be occupied if at least one, possibly more, nearest-neighbor interactions need to be avoided. There have been reports of other sites than fcc becoming occupied starting at $\frac{1}{4}$ ML.[17] Indeed if we use our model with only fcc sites and simulate TPD spectra with initial coverages above $\frac{1}{4}$ ML, we find desorption at much lower temperatures than experimentally. We have therefore restrict ourselves to coverages below $\frac{1}{4}$ ML.

The main kinetic parameters for desorption are not the lateral interactions, but the activation energy and the prefactor for desorption. From a given set of lateral interactions we have determined these parameters by fitting the simulated TPD spectra to the experimental ones using Differential Evolution in a procedure described for CO desorption from Rh(100) elsewhere.[10, 18] Adsorption experiments indicate that there is a precursor for adsorption.[19] This also means that there is a transition state not far from the surface. Consequently, the adsorption energy and the activation energy for desorption will be different, which is why we fit the activation energy. Another consequence of this transition state is that lateral interactions may affect the energy of this transition state differently from the way they affect the adsorbates on the surface. In general we determine this by fitting the Brønsted-Polanyi parameter α (see equation (4)) just like the prefactor and the activation energy.

Figure 3 shows the experimental and simulated TPD spectra for four initial coverages. All kinetic parameters that have been used for the simulations shown in this figure can be found in table 3. The model without lateral interactions shows a shift of the peak maximum temperature to lower temperatures with increasing initial coverage that is much too small. Also the prefactor and activation energy are much too small for associative desorption. For the model with the nearest- and next-nearest-neighbor interactions the simulated spectra already agree reasonably well with the experimental ones. The shift of the peaks to lower temperature are still somewhat too small, and at low initial coverage the simulated peak seems too broad, whereas at higher coverages it is too narrow. There seems to be some hypercorrection in the prefactor and activation energy, which are now a bit too large for associative desorption. Including the linear 3-particle interaction improves the model even more with reasonable values for the prefactor and activation energy.

The reason why the model with the linear 3-particle interaction is better is not this interaction itself, but the fact that this model has a lower nearest-neighbor repulsion. Because the nearest-neighbor interaction is large compared to the thermal energy, there is only a small probability at low coverage that two oxygen atoms are nearest neighbors. It is even less likely that three oxygen atoms will line up to feel the linear 3-particle interaction. We have shown that in such a situation the TPD spectra will not depend on the corresponding interaction parameter.[10] If we fit all the kinetic parameters we find that the nearest-neighbor repulsion should be about 8 kJ/mol smaller than the values of the DFT models to give good TPD spectra if the oxygen atoms desorb from nearest-neighbor positions. This need not mean that the DFT values are incorrect. It can also mean that the oxygen atoms may desorb when farther apart.

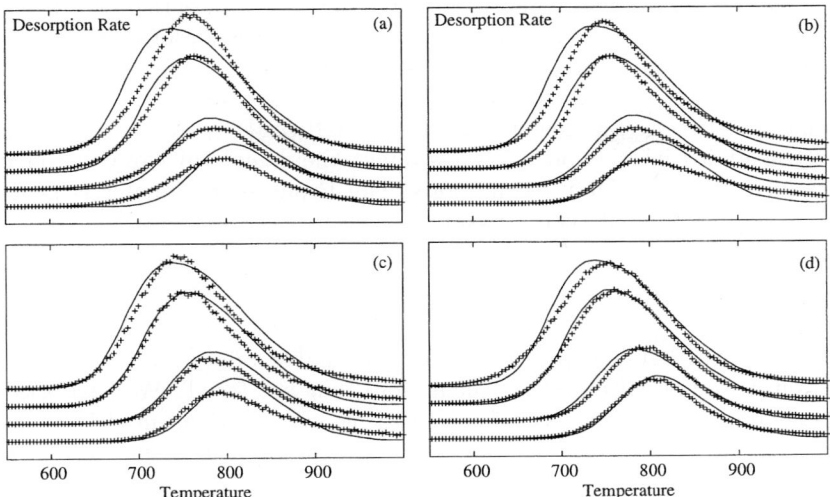

Fig. 3. Simulated and experimental temperature-programmed desorption spectra for O/Pt(111). The solid lines are experimental spectra.[16] The crosses indicate simulated spectra for different models of the lateral interactions: (a) a model without lateral interactions, (b) a model with only nearest and next-nearest pair interactions, (c) and (d) models with also a linear 3-particle interaction. In (c) the O_2 is formed from two atoms at nearest-neighbor positions. In (d) they are at next-nearest-neighbor positions. The kinetic parameters for the different models can be found in table 3. In each plot the curve from top to bottom are for initial oxygen coverage of 0.194, 0.164, 0.093, and 0.073 ML, respectively. The heating rate is 8 K/sec

Table 3. Kinetic parameters for the different models for which temperature-programmed desorption spectra are shown in figure 3. The lateral interactions for figure 3a have been set to 0.0. The lateral interactions for figures 3b, 3c, and 3d have been determined from the DFT calculations. All other parameters have been determined by fitting the simulated spectra to the experimental ones. Energies are given in kJ/mol and the prefactor ν in sec^{-1}

α	E_{act}	ν	φ_{NN}	φ_{NNN}	φ_{linear}	see figure
1.000	129.1	$1.3 \cdot 10^8$	0.0	0.0	0.0	3a
0.879	258.5	$7.2 \cdot 10^{18}$	26.0	4.8	0.0	3b
0.747	224.0	$7.6 \cdot 10^{15}$	19.9	5.5	6.1	3c
0.773	206.4	$2.5 \cdot 10^{13}$	19.9	5.5	6.1	3d

To investigate this idea we also did simulations in which O_2 can be formed from two oxygen atoms at next-nearest-neighbor positions. For these simulations the agreement between the experimental and simulated spectra becomes very good if we take the lateral interactions from DFT (see figure 3d). They are as good as the best simulated spectra with desorption from nearest-neighbor position but with fitted lateral interactions. In particular the high-temperature tails of the spectra become much better. Varying the lateral interactions for this new desorption mechanism does not improve the simulated spectra appreciably.

We note when we allow the lateral interactions to vary to obtain the best agreement between experimental and simulated TPD spectra that the next-nearest-neighbor interaction is always about the same and close to the value determined from the DFT calculations. The Brønsted parameter is generally large, which means that there is a late transition state.

4 Conclusions

We have shown that it is possible to obtain reliable values for some lateral interaction parameters from DFT calculations, but not for all. We have used the leave-one-out method to check for overfitting when various models of lateral interactions in O/Pt(111) are used. It seems that nearest- and next-nearest-neighbor pair interactions can be determined accurately, and also a 3-particle interaction with three oxygen atoms in a row at neighboring sites. If we determine more interaction parameters we see that the strongest interactions are the ones that we can determine reliably. The next-next-nearest-neighbor interaction, which has been used to explain island formation of the oxygen atoms, cannot be determined reliably. The same holds for other multiple-particle interactions.

To determine how useful the models of the reliable lateral interactions are for describing kinetics, we did kMC simulations of TPD experiments of oxygen desorption. The models seem to give a good agreement between simulated and experimental TPD spectra. The most important next-nearest-neighbor interaction from fitting the spectra agrees well with the one obtained from DFT calculations.

Acknowledgment

This work was supported by the Deutsche Forschungsgemeinschaft (DFG) in the framework of the DFG-priority program SPP 1091 "Bridging the gap between ideal and real systems in heterogeneous catalysis."

References

1. G. Kresse and J. Furthmüller. Efficient iterative schemes for ab initio total-energy calculations using a plane-wave basis set. *Phys. Rev. B*, **54** (1996) 11169–11186.
2. D. Vanderbilt. Soft self-consistent pseudopotentials in a generalized eigenvalue formalism. *Phys. Rev. B*, **41** (1990) 7892–7895.
3. G. Kresse and J. Hafner. Norm-conserving and ultrasoft pseudopotentials for first-row and transition elements. *J. Phys.: Condens. Matter*, **6** (1994) 8245–8257.
4. J. P. Perdew. Unified theory of exchange and correlation beyond the local density approximation. In P. Ziesche and H. Eschrig (eds.), *Electronic Structure of Solids '91*, p. 11 (Akademie Verlag, Berlin, 1991).

5. R. J. Gelten, R. A. van Santen, and A. P. J. Jansen. Dynamic Monte Carlo simulations of oscillatory heterogeneous catalytic reactions. In P. B. Balbuena and J. M. Seminario (eds.), *Molecular Dynamics: From Classical to Quantum Methods*, pp. 737–784 (Elsevier, Amsterdam, 1999).
6. D. T. Gillespie. A general method for numerically simulating the stochastic time evolution of coupled chemical reactions. *J. Comput. Phys.*, **22** (1976) 403–434.
7. A. P. J. Jansen. Monte Carlo simulations of chemical reactions on a surface with time-dependent reaction-rate constants. *Comput. Phys. Comm.*, **86** (1995) 1–12.
8. J. J. Lukkien, J. P. L. Segers, P. A. J. Hilbers, R. J. Gelten, and A. P. J. Jansen. Efficient Monte Carlo methods for the simulation of catalytic surface reactions. *Phys. Rev. E*, **58** (1998) 2598–2610.
9. R. A. van Santen and J. W. Niemantsverdriet. *Chemical Kinetics and Catalysis* (Plenum Press, New York, 1995).
10. A. P. J. Jansen. Monte Carlo simulations of temperature-programmed desorption with lateral interactions. *Phys. Rev. B*, **69** (2004) 035414(1–6).
11. H. Tang, A. V. der Ven, and B. L. Trout. Phase diagram of oxygen adsorbed on platinum (111) by first-principles investigation. *Phys. Rev. B*, **70** (2004) 045420(1–10).
12. D. M. Hawkins. The problem of overfitting. *J. Chem. Inf. Comput. Sci.*, **44** (2004) 1–12.
13. C. Stampfl, H. J. Kreuzer, S. H. Payne, H. Pfnür, and M. Scheffler. Towards a first-principles theory of surface thermodynamics and kinetics. *Phys. Rev. Lett.*, **83** (1999) 2993–2996.
14. J. L. Gland and E. B. Kollin. Carbon monoxide oxidation on the Pt(111) surface: Temperature programmed reaction of coadsorbed atomic oxygen and carbon monoxide. *J. Chem. Phys.*, **78** (1983) 963–974.
15. V. P. Zhdanov and B. Kasemo. Simulation of oxygen desorption from Pt(111). *Surf. Sci.*, **415** (1998) 403–410.
16. D. H. Parker, M. E. Bartram, and B. E. Koel. Study of high coverages of atomic oxygen on the Pt(111) surface. *Surf. Sci.*, **217** (1989) 489–510.
17. D. I. Jerdev, J. Kim, M. Batzill, and B. E. Koel. Evidence for slow oxygen exchange between multiple adsorption sites at high oxygen coverages on Pt(111). *Surf. Sci.*, **498** (2002) L91–L96.
18. D. Corne, M. Dorigo, and F. Glover. *New Ideas in Optimization* (McGraw-Hill, London, 1999).
19. C. T. Campbell, G. Ertl, H. Kuipers, and J. Segner. A molecular beam study of the adsorption and desorption of oxygen from a Pt(111) surface. *Surf. Sci.*, **107** (1981) 220–236.

Intelligent Predictive Control with Locally Linear Based Model Identification and Evolutionary Programming Optimization with Application to Fossil Power Plants

Mahdi Jalili-Kharaajoo

Young Researchers Club, Islamic Azad University,
Tehran, Iran
mahdijalili@ece.ut.ac.ir

Abstract. In this paper, an intelligent predictive control algorithm based on Locally Linear Model Tree (LOLIMOT) is implemented to control a fossil fuel power unit. The controller is a non-model based system that uses a LOLIMOT identifier to predict the response of the plant in a future time interval. An evolutionary programming (EP) approach optimizes the identifier-predicted outputs and determines input sequence in a time window. This intelligent system provides a predictive control of multi-input multi-output nonlinear systems with slow time variation.

1 Introduction

Power plant control has been a subject of many classic and modern control studies. Dealing with its multi-input multi-output (MIMO) characteristics, a power plant is a challenge in adaptive and optimal control approaches. In addition, power plant has a wide-range of nonlinear operations that makes the linear model based controller to be complicated. In recent years, intelligent control systems have been used extensively to solve control problem of nonlinear systems, which are able to cover a wide range of operations and plant variation by using learning algorithms. Model reference controller is a widely used intelligent control structure [1] that has a complex implementation in MIMO case. Predictive control has been applied in power plant and process control extensively [2,3]. Computational optimization methods, such as Genetic Algorithms (GA) and Evolutionary Programming (EP) have begun a new area in training and adapting control systems to the plant variations. In this paper, development of an intelligent predictive controller system is introduced to control a boiler/turbine unit. A LOLIMOT identifier [4,5] performs as the plant model to anticipate the output response in a prediction time interval. This control scheme uses EP in optimization of control inputs.

2 Predictive Control Structure

A predictive control anticipates the plant response for a sequence of control actions in future time interval, which is known as prediction horizon [6,7]. The control action in this prediction horizon should be determined by an optimization method to minimize the difference between set point and predicted response. Model-based predictive control (MPC) [7] has been extended to a limited class of nonlinear systems [7,8]. Development of an adequate nonlinear empirical model for a complex system such as power plant is a challenging problem. Intelligent systems may replace the empirical model of the plant in predictive control methodology. The structure of an intelligent predictive control system is shown in Fig. 1. A non-model based identifier predicts response of the actual plant in prediction horizon. This identifier is a LOLIMOT identifier. The optimization block fined the sequence of inputs to minimize a cost function for future time, but only the first value is applied to the plant.

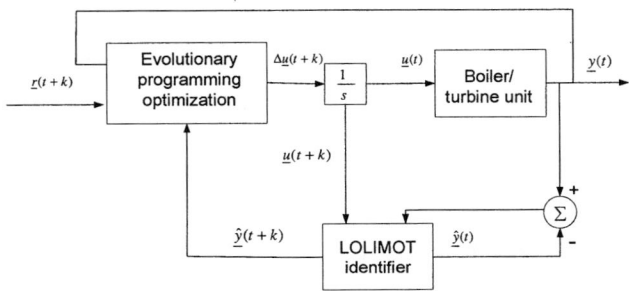

Fig. 1. Intelligent predictive control of a power unit

The power plant simulation is performed by a model that is developed by Bell and Astrom [9]. This is a 3rd order nonlinear model, derived by physical and empirical methods, as in the following

$$\frac{dP}{dt} = -0.0018 u_2 P^{\frac{9}{8}} + 0.9 u_1 - 0.15 u_3 \tag{1}$$

$$\frac{dE}{dt} = \frac{((0.73 u_2 - 0.16) P^{\frac{9}{8}} - E)}{10} \tag{2}$$

$$\frac{d\rho_f}{dt} = \frac{(141 u_3 - w_s)}{85} \tag{3}$$

$$w_s = (1.1 u_2 - 0.19) P \tag{4}$$

$$L = 0.05(0.13073 \rho_f + 60 \alpha_{cs} + 0.11 q_e - 67.995) \tag{5}$$

$$\alpha_{cs} = \frac{(1 - 0.001538 \rho_f)(0.8 P - 25.6)}{\rho_f (1.0394 - 0.0012304 P)} \tag{6}$$

$$q_e = (0.85 u_2 - 0.147) P + 45.6 u_1 - 2.5 u_3 - 2.1 \tag{7}$$

where P is drum steam pressure (Kg/cm^2), E is electrical power (MW), w_s is steam mass flow rate (Kg/s), L is water level deviation about mean (m), ρ_f is fluid density (Kg/m^3), u_1, u_2 and u_3 are normalized fuel, steam feed-water valve positions, α_s is steam quality (mass ratio), and q_e is evaporation rate (Kg/s). In addition, the actuator dynamics of control valves are also modeled by

$$\left|\frac{du_1}{dt}\right| \leq 0.007 \ (s^{-1}) \ \ 0.0 \leq u_1 \leq 1.0 \tag{8}$$

$$-2.0 \leq \left|\frac{du_2}{dt}\right| \leq 0.02 \ (s^{-1}) \ \ 0.0 \leq u_2 \leq 1.0 \tag{9}$$

$$\left|\frac{du_3}{dt}\right| \leq 0.05 \ (s^{-1}) \ \ 0.0 \leq u_3 \leq 1.0 \tag{10}$$

to limit the rate if change in valve positions.

3 Control Input Optimization

The intelligent predictive control system does not depend on the mathematical model of the plant. Therefore, the optimization cannot be implemented by conventional methods in MPC. The search engine-based on evolutionary programming (EP) [10] is used to determine the optimized control variables for a finite future time interval. The EP performs a competition search in a population and its mutation offspring. The members of each population are the input vector deviations that are initialized randomly. The mutation and competition continue making new generations to minimize value of a cost function. The output of the optimizer block is the control valve deviations that are integrated and applied to the identifier and power unit. The EP population consists of the individuals to present the deviation of control inputs. This population is represented by the following set

$$U_n = \{\Delta U_{1,n}, \Delta U_{2,n}, ..., \Delta U_{n_p,n}\} \tag{11}$$

such that U_n is the nth generation of population, and n_p is the population size. The ith individual is written by

$$\underline{\Delta U}_{i,n} = [\underline{\Delta u}_1^{i,n}, ..., \underline{\Delta u}_m^{i,n}], \ for \ i = 1, 2, ..., n_p \tag{12}$$

where m is number of inputs. The $\underline{\Delta u}_j^{i,n}$ is the jth vector of the ith generation as in the following:

$$\underline{\Delta u}_j^{i,n} = [\Delta u_j^{i,n}(1) ... \Delta u_j^{i,n}(n_u)]^T, \ for \ j = 1, ..., m \tag{13}$$

such that n_u is the number of steps in the discrete-time horizon for the power unit input estimation that is defined by

$$n_u = N_u - N_1 \qquad (14)$$

where N_1 is the start time of prediction horizon and N_u is the end time of the input prediction. The individuals of input deviation vector belongs to a limited range of real numbers

$$\Delta u_j^{i,n}(.) \in [\Delta u_{j,\min}, \Delta u_{j,\max}] \qquad (15)$$

In the beginning of EP algorithm, population is initialized randomly chosen individuals. Each initial individual is selected with uniform distribution from the above range of corresponding input.

The EP with adaptive mutation scale has shown a good performance in locating the global minima. Therefore, this method is used as it is formulated in [11]. The fitness value of each population is determined with a cost function to consider the error of predicted input and output in prediction time window. The cost function of the ith individual in the population is defined by

$$f_{i,n} = \sum_{k=1}^{n_y} \left\| r(t+k) - \hat{\underline{y}}_{i,n}(t+k) \right\|_R^2 + \sum_{k=1}^{n_u} \left\| \Delta U_{i,n}(k) \right\|_Q^2 + \left\| r(t) - \hat{\underline{y}}(t) \right\|_R^2 \qquad (16)$$

where $r(t+k)$ is the desired reference set-point at sample time of $t+k$, and $\hat{\underline{y}}_{i,n}(t+k)$ is the discrete predicted plant output vector which is determined by applying $\Delta U_{i,n}(k)$ into the locally-linear fuzzy identifier for time horizon of $n_y = N_2 - N_1$.

The $\Delta U_{i,n}(k)$ in (16) is the kth of the ith individual in the nth generation. The input deviation vectors is determined in a smaller time window of n_u as in (14) such that $n_u \leq n_y$. The inputs of the identifier stay constants after $t + n_u$.

The maximum, minimum, sum and average of the individual fitness in the nth generation should be calculated for further statistical process by

$$f_{\max}|_n = \{ f_{i,n} | f_{i,n} \geq f_{j,n} \; \forall f_{j,n}, j = 1,...,n_p \} \qquad (17)$$

$$f_{\min}|_n = \{ f_{i,n} | f_{i,n} \leq f_{j,n} \; \forall f_{j,n}, j = 1,...,n_p \} \qquad (18)$$

$$f_{sum}|_n = \sum_{i=1}^{n_p} f_{i,n} \qquad (19)$$

$$f_{avg}|_n = \frac{f_{sum}|_n}{n_p} \qquad (20)$$

After determining the fitness values of a population, the mutation operator performs on the individuals to make a new offspring population. In mutation, each element of the parent individual as in (13) provides a new element by adding a random number such as

$$\Delta u_j^{i+n_p,n}(k) = \Delta u_j^{i,n}(k) + N(\mu, \sigma_{i,j}^2(n)) \tag{21}$$

for $i = 1,2,...,n_p \quad j = 1,2,...,m \quad k = 1,2,...,n_u$

such that $N(\mu, \sigma_{i,j}^2(n))$ is Gaussian random variable with mean $\mu = 0$ and variance of $\sigma_{i,j}^2(n)$. The variance of the random variable in (21) is chosen to be

$$\sigma_{i,j}^2(n) = \beta(n)(\Delta u_{j,\max} - \Delta u_{j,\min}) \frac{f_{i,n}}{f_{\max}|_n} \tag{22}$$

where $\beta(n)$ is the mutation scale of the population such that $0 < \beta(n) \leq 1$. After mutation, the fitness of offspring individuals are evaluated and assigned to them.

The generated new individuals and old individuals produce a new combine population with size of $2n_p$. Each member of the combined population competes with some other members to determine which one is valuated to survive to the next generation. For this purpose, the ith individual $\Delta U_{i,n}$ competes with j^{th} individual $\Delta U_{j,n}$, such that $j = 1,2,...,p$. The number of individuals to compete with is a fixed number p. The p individuals are selected randomly with uniform distribution. The result of this competition is a binary number $v_{ij,n} \in \{0,1\}$ to represent *lose* or *win*, and is determined by

$$v_{ij,n} = \begin{cases} 1 & if \lambda_{j,n} < \frac{f_{j,n}}{f_{j,n} + f_{i,n}} \\ 0 & otherwise \end{cases} \tag{23}$$

such that $\lambda_{j,n} \in [0,1]$ is a randomly selected number with uniform distribution, and $f_{j,n}$ is the fitness of the jth selected individual. The value of $v_{ij,n}$ will be set to 1 if according to (23) the fitness of the ith individual is relatively smaller than the fitness of the jth individual. To select the survived individual, a weight value is assigned to each individual by

$$w_{i,n} = \sum_{k=1}^{p} v_{ij,n} \quad for \quad i = 1,2,...,2n_p \tag{24}$$

The n_p individuals with the highest competition weight $w_{i,n}$ are selected to form the $(n+1)^{th}$ generation. This newly formed generation participates in the next iteration. To determine the convergence of the process, the difference of maximum and minimum fitness of the population is checked against a desired small number $\varepsilon > 0$ as in

$$f_{\max}|_n - f_{\min}|_n \leq \varepsilon \tag{25}$$

If this convergence condition is met, the mutation scale with the lowest fitness is selected as sequence of n_u input vectors for the future time horizon. The first vector is applied to the plant and the time window shifts to the next prediction step.

Before starting the new iteration, the mutation scale changes according to the newly formed population. If the mutation scale is kept as a small fixed number, EP may have a premature result. In addition, a large fixed mutation scale will raise the possibility of having a non-convergence process. An adaptive mutation scale provides a change of mutation probability according to the minimum fitness value of n_p individuals in the $(n+1)^{th}$ generation. The mutation scale for the next generation is determined by

$$\beta(n+1) = \begin{cases} \beta(n) - \beta_{step} & if \ f_{min}|_n = f_{min}|_{n+1} \\ \beta(n) & if \ f_{min}|_n < f_{min}|_{n+1} \end{cases} \quad (26)$$

where n is generation number, β_{step} is the predefined possible step change of the mutation scale in each iteration.

4 Locally Linear Model Tree (LOLIMOT) Identification of Nonlinear systems

In the following, the modeling of nonlinear dynamic processes using LOLIMOT models is described. The network structure of a local linear neuro-fuzzy model [4,5] is depicted in Fig. 2. Each neuron realizes a local linear model (LLM) and an associated validity function that determines the region of validity of the LLM. The validity functions form a partition of unity, i.e., they are normalized such that

$$\sum_{i=1}^{M} \varphi_i(\underline{z}) = 1 \quad (27)$$

for any model input $\underline{z}$.

The output of the model is calculated as

$$\hat{y} = \sum_{i=1}^{M} (w_{i,o} + w_{i,1}x_1 + ... + w_{i,n_x}x_{n_x})\varphi_i(\underline{z}) \quad (28)$$

where the local linear models depend on $\underline{x} = [x_1,...,x_{n_x}]^T$ and the validity functions depend on $\underline{z} = [z_1,...,z_{n_z}]^T$. Thus, the network output is calculated as a weighted sum of the outputs of the local linear models where the $\varphi_i(.)$ are interpreted as the operating point dependent weighting factors. The network interpolates between different Locally Linear Models (LLMs) with the validity functions. The weights w_{ij} are linear network parameters.

The validity functions are typically chosen as normalized Gaussians. If these Gaussians are furthermore axis- orthogonal the validity functions are

$$\varphi_i(\underline{z}) = \frac{\mu_i(\underline{z})}{\sum_{j=1}^{M} \mu_j(\underline{z})} \quad (29)$$

with

$$\mu_i(\underline{z}) = \exp(-\frac{1}{2}(\frac{(z_1 - c_{i,1})^2}{\sigma_{i,1}^2} + ... + \frac{(z_1 - c_{i,n_z})^2}{\sigma_{i,n_z}^2})) \tag{30}$$

The centers and standard deviations are *nonlinear* network parameters.

In the fuzzy system interpretation each neuron represents one rule. The validity functions represent the rule premise and the LLMs represent the rule consequents. One-dimensional Gaussian membership functions

$$\mu_{i,j}(z_j) = \exp(-\frac{1}{2}(\frac{(z_j - c_{i,j})^2}{\sigma_{i,j}^2})) \tag{31}$$

can be combined by a t-norm (conjunction) realized with the product operator to form the multidimensional membership functions in (29). One of the major strengths of local linear neuro-fuzzy models is that premises and consequents do not have to depend on identical variables, i.e. $\underline{z}$ and $\underline{x}$ can be chosen independently.

The LOLIMOT algorithm consists of an outer loop in which the rule premise structure is determined and a nested inner loop in which the rule consequent parameters are optimized by local estimation.

1. *Start with an initial model:* Construct the validity functions for the initially given input space partitioning and estimate the LLM parameters by the local weighted least squares algorithm. Set M to the initial number of LLMs. If no input space partitioning is available a-priori then set $M = 1$ and start with a single LLM which in fact is a global linear model since its validity function covers the whole input space with $\varphi_i(\underline{z}) = 1$.
2. *Find worst LLM:* Calculate a local loss function for each of the $i=1,...,M$ LLMs. The local loss functions can be computed by weighting the squared model errors with the degree of validity of the corresponding local model. Find the worst performing LLM.
3. *Check all divisions:* The LLM l is considered for further refinement. The hyperrectangle of this LLM is split into two halves with an axis-orthogonal split. Divisions in each dimension are tried. For each division $\dim = 1,...,n_z$ the following steps are carried out:
   (a) Construction of the multi-dimensional MSFs for both hyperrectangles.
   (b) Construction of all validity functions.
   (c) Local estimation of the rule consequent parameters for both newly generated LLMs.
   (d) Calculation of the loss function for the current overall model.
4. *Find best division:* The best of the n_z alternatives checked in Step 3 is selected. The validity functions constructed in Step 3(a) and the LLMs optimized in Step 3(c) are adopted for the model. The number of LLMs is incremented $M \to M+1$.
5. *Test for convergence:* If the termination criterion is met then stop, else go to Step 2.

For the termination criterion various options exist, e.g., a maximal model complexity, that is a maximal number of LLMs, statistical validation tests, or information criteria. Note that the *effective* number of parameters must be inserted in these termination criteria.

Fig. 3 illustrates the operation of the LOLIMOT algorithm in the first four iterations for a two-dimensional input space and clarifies the reason for the term "tree" in the acronym LOLIMOT. Especially two features make LOLIMOT extremely fast. First, at each iteration not all possible LLMs are considered for division. Rather, Step 2 selects only the worst LLM whose division most likely yields the highest performance gain. For example, in iteration 3 in Fig. 3 only LLM 3-2 is considered for further refinement. All other LLMs are kept fixed. Second, in Step 3c the local estimation approach allows to estimate only the parameters of those two LLMs which are newly generated by the division. For example, when in iteration 3 in Fig. 3 the LLM 3-2 is divided into LLM 4-2 and 4-3 the LLMs 3-1 and 3-3 can be directly passed to the LLMs 4-1 and 4-3 in the next iteration without any estimation.

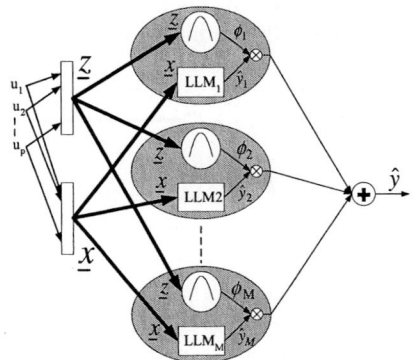

Fig. 2. Network structure of a local linear neurofuzzy model with M neurons for nx LLM inputs x and nz validity function inputs z

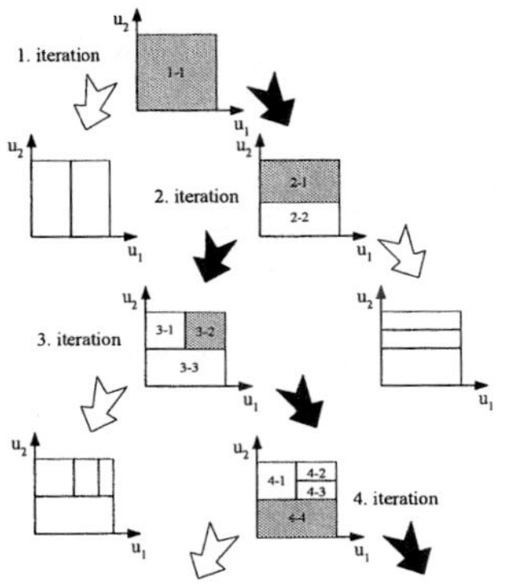

Fig. 3. Operation of the LOLIMOT structure search algorithm in the first four iterations for a two-dimensional input space ($p = 2$)

5 Controller Implementation

The identifier is trained and initialized before the control action starts. The training data is obtained from the mathematical model of the power unit. A set of valve position inputs is chosen to excite the power unit model in a staircase of electrical power operating points. The input vector of the identifier includes control valve positions, delayed drum pressure and level deviation.

After the initial training, identifier is engaged in the closed loop of the predictive control as in Fig. 1. The parameters of prediction horizon is selected to be $n_y = 100$ and $n_u = 70$, with time step of $\Delta t = 10 \sec$. Population size is chosen to be $n_p = 10$. In the first simulation test, the electrical power is increased from low to medium power (15.26 MW to 66.65 MW). Both steam pressure and electrical power set-points are ramped up on the test. The transient response of the drum pressure, P, and electrical power, E, are shown in Figs. 4 and 5 respectively. In addition the mathematical model of the plant replaces the identifier, and the response of this set up is also illustrated in these figures for comparison. As it can be seen in these figures, steam pressure starts following the ramp variation with a fast slop, and that is due the prediction horizon. The pressure response takes an overshoot and settles to constant reference input after 75 prediction steps. The electrical power performs faster and makes a ringing before settling down. The on-line training of the identifier explains the ripples in steady state. To improve the transient response, one may consider a larger prediction time, or a larger population size in EP.

Figs. 6 and 7 show transient of the drum pressure and electrical power in response to step change set-point. The electrical power and drum pressure are stepped up from medium power to the high power operating points (66.65 MW to 128.9 MW). The pressure response is slower than the electrical power transient, and settles after a ringing in 150 prediction steps. However, electrical power shows a transition that reflects the discrete nature of control actions. The envelope of this power transient is similar to an over-damped transient response, settling in 100 prediction steps.

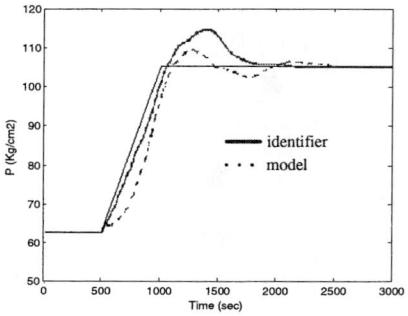

Fig. 4. Response of drum pressure in power and pressure ramp test

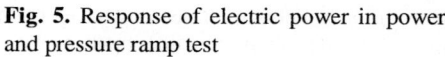

Fig. 5. Response of electric power in power and pressure ramp test

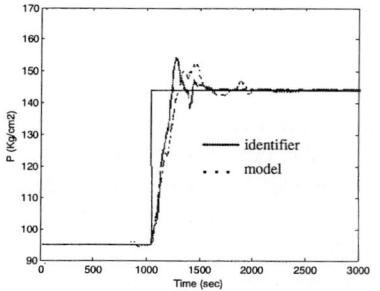

Fig. 6. Response of drums pressure in power and pressure step test

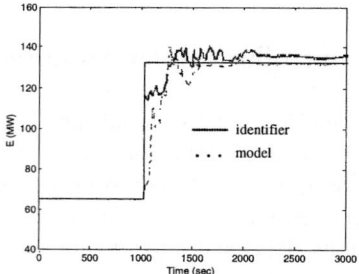

Fig. 7. Response of electric power in power and pressure step test

6 Conclusion

Implementation of an intelligent predictive control is introduced for a boiler/turbine power unit. This controller uses a locally linear model tree (LOLIMOT) as an identifier to predict the plant outputs in response to a sequence of valve positions in future time window. This sequence of plant inputs is optimized with EP. Simulation of this predictive control in power/pressure step and ramp tests shows a fast transient response to the input changes. The obtained identifier is trained for local and wide range training data. Overall, the intelligent predictive control is suitable for nonlinear MIMO systems with slow varying dynamics as complex as power plants.

References

1. Astrom, K.J. and B. Wittenmark, Adaptive Control. Reading, MA, Addison-Wesley, 1989.
2. Tolle, H. and E. Ersu, Neurocontrol; Learning Control Systems Inspired by Neural Architectures and Human Problem Solving. Lecture notes in control and identification sciences, Berlin, Springer-Verlag, 1992
3. Saint-Donat, J., N, Bhat and T.J. McAvoy, Neural-Net based model predictive control. Advanced in Intelligent Control , C.J. Harris (Ed), London, Tailor and Francis, 1994.
4. Nelles, O., Nonlinear system identification: From classical approaches to neural networks and fuzzy models. Springer, 2001
5. Nelles, O., Local linear model tree for on-line identification of time variant nonlinear dynamic systems. Proc. International Conference on Artificial Neural Networks (ICANN), pp.115-120, Bochum, Germany, 1996.
6. Clarke, D., Advanced in model-based predictive control. Oxford University Press, New York, 1994.
7. Camacho, E.F.,and C. Bordons, Model Predictive Control. Springer, Berlin, Germany, 1998.
8. Garduno-ramirez, R. and K.Y. Lee, Wide range operation of a power unit via feedforward fuzzy control. IEEE Trans. Energy Conversion, vol.15, no.4, pp.421-426, 2000.
9. Astrom, K.J., K.J. and R.D. Bell, Dynamic models for boiler-turbine-alternator units: Data logs and paramater estimation for a 160 MW unit. In: Report LUTFD2/(TFRT-3192)/1-137/(1987), Department of automatic control, Lung Institute of Technology, Lung, Sweden, 1987.
10. Fogel, L.J., The future of evolutionary programming. Proc. 24th Asilomar Conference on Signals, Systems and Computers. Pacific Grove, CA, 1991.
11. Lai, L.L., Intelligent system application in power engineering: Evolutionary programming and neural networks. John Wiley & Sons Inc., New York, USA, 1998.

Determination of Methanol and Ethanol Synchronously in Ternary Mixture by NIRS and PLS Regression*

Q.F. Meng[1], L.R. Teng[1], J.H. Lu[1], C.J. Jiang[1], C.H. Gao[1],
T.B. Du[2], C.G. Wu[2], X.C. Guo[2], and Y.C. Liang[2,**]

[1] College of Life Science, Jilin University,
Changchun 130012, China
[2] College of Computer Science, Jilin University,
Changchun 130012, China
liangyc@ihpc.a-star.edu.sg

Abstract. This paper reports the usefulness of partial least squares (PLS) in the analysis of NIR spectra. Based on this work, the ethanol and methanol could be measured synchronously in the chemical industry. It is demonstrated that the proposed technique is quite convenient and efficient. PLS regression has been applied to establish a satisfactory calibration model via adopting optimum wavelength. In this model, the correlation coefficient in methanol determination is 0.99991 and the root mean square error of calibrate (RMSEC) is 0.431; in ethanol determination, the correlation coefficient is up to 0.99998, and the RMSEC is 0.193. Compared with the measured value of GC in the sample determination, the range of relative error is from 0.721% to 3.505%.

1 Introduction

NIR spectroscopy is the measurement of the wavelength and intensity of the absorption of near infrared light by a sample. Near infrared light spans the 800 - 2500 nm range. It is typically used for quantitative measurements of compound containing organic functional groups, especially O-H, N-H, and C-H. Over the last few years, near infrared (NIR) spectroscopy has rapidly developed into an important and extremely effective method of analysis. In fact, for certain research areas and applications, ranging from material science and chemistry to life sciences, it has become an indispensable tool because this fast and cost-effective type of spectroscopy provides qualitative and quantitative information not available from any other technique.

Partial least squares (PLS) regression is an extension of the multiple linear regression model, and PLS is probably the one with least restrictions among the various multivariate extensions of the multiple linear regression model. This flexibility allows

---

* This work was supported by the science technology development project of Jilin Province of China (Grant No. 20020503-2).
** Corresponding author.

it to be used in any situations where the use of traditional multivariate methods is severely limited, for example, the situation where observations are less than predictor variables. PLS regression has been used in various disciplines such as chemistry, economics, medicine, psychology, and pharmaceutical science where predictive linear modeling, especially with a large number of predictors, is necessary.

Several methods have been used for the determination of the alcoholic content in beverages. Such as refratometric methods [1] and density methods [2], potassium dichromate, HPLC methods [3], electrochemical methods. These procedures are usually slow, consume large amount of sample and reagents or use expensive instruments of difficult implementation. Barboza [4] and Engelhard [5] had applied near infrared spectroscopy to determination of alcoholic content in beverages separately. Tipparat et al [6] measured ethanol content using near infrared spectroscopy with flow injection. Mendes et al [7] measured ethanol content in fuel ethanol and beverages by NIR. However, the methods measured ethanol content only. In this study, we try to develop a calibration method for the determination of alcoholic contents simultaneously in methanol, ethanol and water ternary mixtures using the combination of NIR spectroscopy and PLS regression, which is efficient, non-destructive and inexpensive.

2 Experimental

2.1 Sample and Instrumentation

Methanol and ethanol (both HPLC grade) were purchased from Tianjin Guangfu Fine Chemical Research Institute. Water was purified by an Ultrapure Water System model Elix 10 (Millipore, USA). The resistance of the finally prepared water was greater than 18.2 M . The ternary mixtures of water, methanol and ethanol were prepared in 24 standards with ethanol contents of 0-100 % (v/v) and methanol contents of 0-100 % (v/v). The NIR spectra in the 850-1870 nm were measured at ambient temperature with an UV-Vis-NIR spectrometer (UV-3150, SHIMADZU Corporation, Japan), and the scan may go in 1nm steps increment with a 2.0 nm slit width. The mixture was contained in a quartz cuvette cell having a pathlength of 1.0 mm. We took the spectrum of the empty cell as a background, and the NIR spectra of ternary mixtures were measured three times.

2.2 Data Analysis

The partial least squares calibration model was obtained by the TQ Analyst applicative software included in the Omnic (Thermo Nicolet). Eighteen standards were used for calibration and the other six were separated for validation.

Four unknown samples were prepared with ethanol, methanol and water in different ratios. Their NIR spectra were obtained, and as a reference value, their contents were analyzed by GC on a Hewlett-Packard 6890 gas chromatograph.

3 Results and Discussion

3.1 NIR Spectra of Water, Methanol and Ethanol

The NIR spectra of water, methanol and ethanol ternary mixture in the 850-1870nm are shown in Fig. 1 (A). Part of the spectra (1560-1750nm) is shown in Fig.1 (B). As is seen, there is a small band around 920nm referred to as the third overtone C-H. [4]. A broad band centered near 1450nm is composed of a number of overlapped bands due to the combinations of O-H antisymmetric and symmetric stretching modes of various water species and first overtones of the O-H stretching modes of free and hydrogen-bonded species of methanol and ethanol. Weak features in the region between 1667-1754 nm are due to the first overtones and combinations of the C-H stretching modes of alcohol [8]. The spectrum of water crosses the spectrum of methanol at 1568 nm, 1631 nm, 1665 nm and 1743 nm, respectively. As shown in Fig. 1, in the region of 1568 ~1743 nm, the spectra contain the most information of the alcohol and the least information of the water.

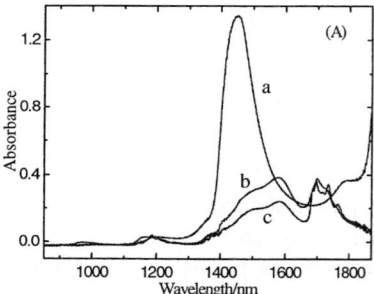 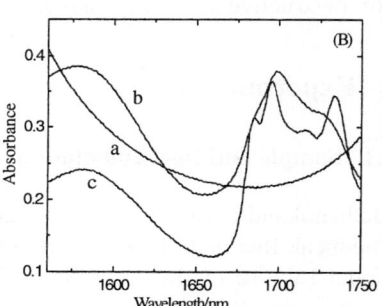

Fig. 1. NIR spectra of pure water (a), methanol (b) and ethanol (c) (A) in the 850~1870 nm region; (B) in the 1560~1750 nm region

3.2 Influence of Spectral Region

Basing on the feature of the spectra shown in Fig. 1, different part of the spectra was chosen to construct different models. To evaluate the error of each calibrate model, root mean squares error of calibrate (RMSEC) was used, calculated as eq. (1).

$$RMSEC = \sqrt{\frac{\sum (y_p - y_r)^2}{n}} \quad (1)$$

where y_p is the predicted value, y_r is the reference value measured by GC method and n is the number of samples, and the correlative coefficient is also used for evaluating.

The results are shown in Table 1. The data indicate that correlative coefficient of both ethanol and methanol is fairly high and their RMSEC values are fairly low in the 1667-1754nm region. Even so, the result is not considered satisfactory compared with the coefficients and RMSEC values produced in the region of 1568-1743nm. All the data shown in Table 1 support that the calibration model in 1568-1743 region should be the optimum one. Adopting this region to establish the model would get more realistic contents of methanol and ethanol.

Table 1. Analysis results of different wavelength regions

Wavelength regions(nm)	Ethanol		Methanol	
	Corr. Coeff.	RMSEC	Corr. Coeff.	RMSEC
850~1870	0.99967	0.737	0.99971	0.767
1667~1754	0.99994	0.314	0.99988	0.419
1568~1743	0.99998	0.193	0.99991	0.431

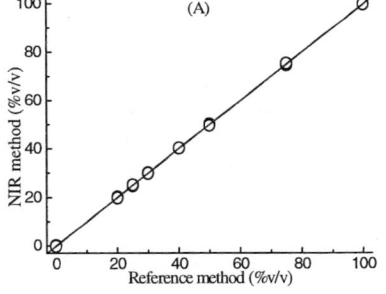
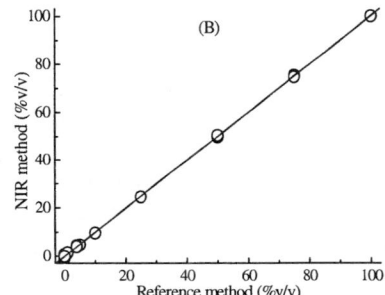

Fig. 2. Reference versus PLS regression plot for ethanol (A) and methanol (B) content prediction

Fig. 2 shows the calibrate results of PLS regression, the values obtained by the reference method are plotted versus NIR method for alcoholic contents. The number of partial least squares compounds (i.e. factors) is chosen according to software outputs and were 5 of ethanol and 4 of methanol.

3.3 Results of the Sample Determination

To investigate the actual applicability of the developed PLS model to practical applications, the best NIR PLS calibration model was used to determine the alcoholic content of 4 unknown mixtures. The results were evaluated by the GC method. The relative errors of them were calculated to estimate the effect. As shown in Table 2, when both alcoholic contents of methanol and ethanol are in 0~100%, the contents values predicted by NIR PLS calibration model correlate well with that of GC method. The relative errors are not more than 3.505%.

Table 2. Comparison of determination results between NIR and GC

Sample ID	Methanol			Ethanol		
	GC (%)	NIR (%)	RE[a] (%)	GC (%)	NIR (%)	RE (%)
1	12.31	12.10	-1.706	15.78	15.98	1.267
2	40.24	40.53	0.721	45.52	44.98	-1.186
3	4.85	4.68	-3.505	82.74	81.33	-1.704
4	69.92	70.87	1.359	7.75	7.93	2.322

[a] relative error.

4 Conclusions

In this study, it is verified that the near infrared spectroscopy and PLS regression can be applied to measure the alcoholic contents of the ternary mixture of methanol, ethanol and water in random ratios. Being a non-destructive and an efficient method, it allows the determination of both alcoholic contents synchronously. It proves to be a good substitute for traditional method and can be employed in online monitoring.

References

1. AOAC Official methods of analysis of AOAC international 950.04, 16th edn. AOAC International, Gaithersburg (1997).
2. AOAC Official methods of analysis of AOAC international 942.06, 16th edn. AOAC International, Gaithersburg (1997).
3. Yarita, T., Nakajima, R., Otsuka, S., Ihara, T.A., Takatsu, A. and Shibukawa, M.: Determination of ethanol in alcoholic beverages by high-performance liquid chromatography-flame ionization detection using pure water as mobile phase. *Journal of Chromatography A.*, Vol. 976, (2002) 387-391.
4. Barboza, F.D. and Poppi, R.J: Determination of alcohol content in beverages using short-wave near-infrared spectroscopy and temperature correction by transfer calibration procedures. *Analytical and Bioanalytical Chemistry*, Vol. 377, (2003) 695-701.
5. Engelhard, S., Lohmannsroben H.G. and Schael F: Quantifying ethanol content of beer using interpretive near-infrared spectroscopy. *Applied Spectroscopy*, Vol. 58, (2004) 1205-1213.

6. Tipparat, P., Lapanantnoppakhun, S., Jakmunee, J. and Grudpan, K., Determination of ethanol in liquor by near-infrared spectrophotometry with flow injection. *Talanta*, Vol. 53, (2001) 1199-1204.
7. Mendes, L. S., Oliveira, F.C.C., Suarez, P.A.Z. and Rubim J.C: Determination of ethanol in fuel ethanol and beverages by FT-near infrared and FT-Raman spectrometries. *Analytica Chimica Acta*, Vol. 493, (2003) 219-231.
8. Katsumoto, Y., Adachi D., Sato, H. and Ozaki, Y: Usefulness of a curve fitting method in the analysis of overlapping overtones and combinations of CH stretching modes. *Journal of Near Infrared Spectroscopy*, 10, (2002) 85-91.

Ab Initio and Empirical Atom Bond Formulation of the Interaction of the Dimethylether-Ar System

Alessandro Costantini[1], Antonio Laganà[1], Fernando Pirani[1], Assimo Maris[2], and Walther Caminati[2]

[1] Dipartimento di Chimica, Università di Perugia, Perugia
{alex, lag, pirani}@dyn.unipg.it
[2] Dipartimento di Chimica Fisica Ciamician, Universita di Bologna, Bologna
{maris, caminati}@ciam.unibo.it

Abstract. The main static features of the dimethylether-Ar cluster have been obtained by fitting spectroscopy measurements (with the support of ab initio calculations). The resulting estimates of the cluster structural features are compared with those obtained from a recently proposed atom-bond pairwise additive formulation of the interaction.

1 Introduction

The increasing availability of computing power on distributed platforms is making it more and more feasible to carry out extended dynamical calculations for complex molecular systems [1].

These calculations, usually, rely on the integration of classical equations of motion for the nuclei under the action of the potential energy surface (PES) associated with the electronic structure of the molecular system. Most often the PES of the system under investigation is generated empirically by making the assumption that the intermolecular interaction can be formulated as a sum of few-body terms [2].

Unfortunately, there is no alternative to this procedure since the study of complex systems cannot be undertaken using rigorous quantum techniques. This means that the properties of the molecular systems predicted by the calculations are to be taken with reservation since, in general no checks for the accuracy of the few-body formulation of the interaction can be made.

Accuracy checks can be performed only for those systems for which ab initio estimates of the electronic energy can be calculated and a cross comparison of the properties computed using the adopted PES can be compared with information derived by spectroscopy and/or scattering experiments. This is, indeed, the case of the study discussed in this paper that reports on the work carried out to investigate the properties of the gas phase system dimethylether-Ar (DME-Ar) adduct formed in a molecular beam experiment (see Fig.1) and analyses the formulation of the related PES.

In particular, in the paper a comparison is made of the properties of DME-Ar derived by fitting spectroscopy measurements to a one dimensional model interaction derived from ab initio calculations of the electronic energies with those obtained from a recently proposed [3] [4] atom-bond pairwise additive formulation of the interaction.

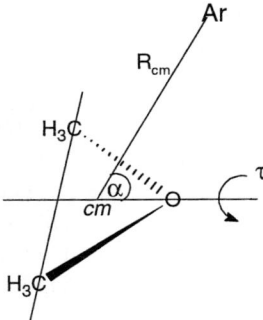

Fig. 1. Sketch of the DME-Ar adduct geometry

The paper is organised as follows: in section 2 the properties derived from the experiment are analysed; in section 3 the atom bond formulation of DME-Ar system is discussed; in section 4 some conclusions are drawn.

2 The Rationalization of the Experimental Findings

The experiment consists of the generation of the DME-Ar complex via a supersonic expansion at conditions optimized for the formulation of the 1:1 adduct and of the measurement of the related rotational transition frequencies to estimate the spectroscopic constants of the system.

Two different experimental setups were used: a millimetre-wave free jet absorption spectrometer and a molecular beam Fourier transform microwave spectrometer [5]. In the first setup the adduct is formed by mixing Ar and DME at 298K and expanding the gas from 800 and 10 mbar, respectively, to about 5×10^{-3} mbar through a pulsed nozzle (repetition rate 5 Hz) with a diameter of 0.35 mm. The spectrometer covers the range 60-78 GHz. In the second setup, the Ar gas at 2 bar was mixed with the DME vapor at 20 bar and the mixture was expanded into a nozzle having a diameter of 0.8 mm. The spectrometer covers the range 6-18.5 GHz.

The estimates of both the distance, R_{cm}, separating the Ar atom from the center of mass (cm) of the adduct, and the angle, α, that formed by R_{cm} with the O-cm line at equilibrium (see Fig.1), derived from the experiment, are given in the left hand side column of Table1.

Table 1. Experimental and ab initio geometrical properties DME-Ar at equilibrium

	Exper.	Theory
R_{cm} / Å	3.58	3.53
α / deg	77.5	73

Other information obtained from the experiment are the dissociation energy of the complex (210 cm^{-1}) and the barrier to inversion by tunneling of the Ar-atom from above to below (and viceversa) the C-O-C plane. The first data was obtained by assuming the Ar stretching motion to be isolated from the other motions, the complex to result from the assemblage of Ar and a rigid dimethylether and the related interaction to be described by a Lennard-Jones potential. The second data was obtained by scaling the minimum energy path (MEP) for the inversion process worked out from an analysis of the electronic energies calculated using an MP2/6-311++G** ab initio method. The calculations were performed using the GAUSSIAN 98 software package [6] and considering the dimethyl ether a rigid molecule at its equilibrium geometry.

To illustrate the main features of the resulting model ab initio PES we plot in Fig. 2 its isoenergetic contours (taken at the minimum along R_{cm}) as a function of α and τ.

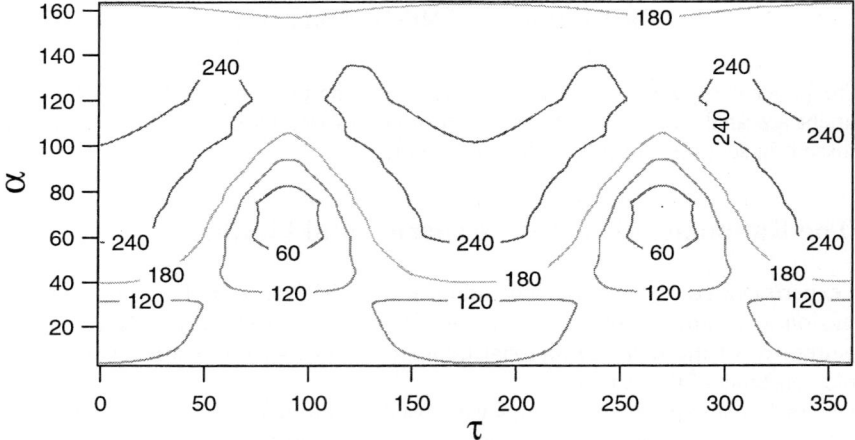

Fig. 2. Isoenergetic contours of the ab initio PES (taken at the minimum along R_{cm}) plotted are function of α and τ. The energy zero is set at the equilibrium energy of the complex and spacing between contours is of 60 cm^{-1}

The two dimensional (2D) contours of the model ab initio single out two deep wells associated with the most stable (with Ar sitting either above or below the C-O-C plane) adduct geometry. Relevant R_{cm} and α value are given in the right hand side column of Table1. A more complete set of quantitative information on the stationary points of the ab initio PES is given in Table 2.

Table 2. Characteristics of the main stationary points of the 2D cross section of the ab initio PES taken at the minimum along R_{cm}

E / cm^{-1}	R_{cm} / Å	α / deg	τ / deg
0	3.526	73	± 90
91	3.910	23	0/180
126	3.941	0	All
125	4.168	180	All
214	4.843	120	0/180

Information on the inversion barrier were derived from the ab initio data by further modeling the MEP of the ab initio PES as

$$V(\tau)/cm^{-1} = 45.58 + 4.99\cos(2\tau) - 29.63\cos(4\tau) + 9.84\cos(6\tau) \quad (1)$$

with the associated values of R_{cm} and α formulated as functions of τ by the equations

$$R_{cm}(\tau)/\overset{\circ}{A} = 3.7984 + 0.1421\cos(2\tau) \\ - 0.0766\cos(4\tau) + 0.0228\cos(6\tau) \quad (2)$$

$$\alpha(\tau)/\deg = 53.94 - 17.50\cos(2\tau) \\ + 4.75\cos(4\tau) + 1.21\cos(6\tau). \quad (3)$$

It is important to stress out here that to match theoretical information with the experiment the value of $V(\tau)$ had to be scaled down by a factor 0.47.

3 The Atom-Bond Formulation of the DME-Ar Interaction

To build a more realistic PES we used the recently proposed functional formulation of the interaction as a pairwise sum of atom-bond interaction terms [3] [4].

This formulation makes use of the bond polarizability additivity to represent the intermolecular repulsion and attraction components as a sum of atom-bond pair contributions

$$V = \sum_i^{atom} \sum_j^{bond} V_{ij}(r_{ij}, \vartheta_{ij}) \quad (4)$$

where r_{ij} being the atom (i) to the center of bond (j) distance (for the sake of simplicity both i (atom) and j (bond) labels will be dropped hereafter) and ϑ_{ij} the angle formed by the bond axis and r_{ij}. The interaction for each ij pair is given as

$$V(r,\vartheta) = \varepsilon(\vartheta)\left[\frac{m}{n(r,\vartheta)-m}\left(\frac{r_m(\vartheta)}{r}\right)^{n(r,\vartheta)} - \frac{n(r,\vartheta)}{n(r,\vartheta)-m}\left(\frac{r_m(\vartheta)}{r}\right)^m\right] \quad (5)$$

In eq.(5) r_m is the equilibrium value of r, ε is the energy associated with the atom-bond interaction at r_m, m is equal to 6 for neutral-neutral systems, and $n(r, \vartheta)$ is calculated using eq.(6)

$$n(r,\vartheta) = \beta + 4.0\left(\frac{r}{r_m(\vartheta)}\right)^2 \quad (6)$$

with β being a constant parameter within a broad class of systems and characterizing the nature and hardness of the interacting particles (in our case $\beta = 10$) [4].

The angular dependence of both $r_m(\vartheta)$ and $\varepsilon(\vartheta)$ is usually parameterized in terms of parallel (||) and a perpendicular ($\perp$) contributions as given in eq. (7) and (8). In fact in eq. (7) and (8) $\varepsilon_{||}$, $r_{m||}$, $\varepsilon_\perp$ and $r_{m\perp}$ represent the well depth and location for the parallel and the perpendicular approaches, respectively, of the atom to the considered diatom (bond) of the molecular frame.

$$\varepsilon(\vartheta) = \varepsilon_\perp \sin^2(\vartheta) + \varepsilon_{||} \cos^2(\vartheta) \tag{7}$$

$$r_m(\vartheta) = r_{m\perp} \sin^2(\vartheta) + r_{m||} \cos^2(\vartheta) \tag{8}$$

The values for the atom-bond pairs of DME-Ar have been worked out using the procedure to ref. [7] which exploits the combined use of bond polarizability tensor components (to represent the electronic charge distribution around the molecular frame) and its effects on the attractive and repulsive interaction components.

The values estimated in this way are shown in Table 3.

Table 3. Parameters for the atom bond pairs of DME-Ar

Atom-bond	$r_{m\perp}$ / Å	$r_{m||}$ / Å	$\varepsilon_\perp$ / cm^{-1}	$\varepsilon_{||}$ / cm^{-1}
C-O···Ar	3.400	3.790	25.02	38.82
C-H···Ar	3.641	3.851	38.85	32.14
LP···Ar	3.490	3.510	19.96	22.92

As can be seen from row 3 of table 3 a specific care was put to model the interaction associated with the oxygen lone pairs (LP). The length of the related polarization ellipsoid has been taken to be 0.45 Å and the value of the angle formed by the two lone pairs (γ) has been set to be 108° (see left hand side of Fig.3).

Also in this case the molecule of the dimethylether has been constrained to be rigid and have the minimum energy geometry. To mimic this constraint, two C-H dummy bonds have been interposed in between the two methyl groups, as shown in the right hand side of Fig. 3.

Fig. 3. Sketch of the two lone pairs of the oxygen in DME-Ar (left hand side) and of the two C-H dummy bonds (right hand side)

The C-H dummy bonds are characterized by a purely repulsive nature. This makes it possible to mimic the interaction of the Van der Waals spheres of the methyl groups when they come to closer contact because of the related rotovibrational motions. The properties of the resulting PES are illustrated in Fig. 4 by showing the isoenergetic contours in a fashion similar to that of Fig. 2. More quantitative information on the stationary points of the surface are given in Table 4 in a fashion similar to that of Table 2. The absolute value of the global minimum of the surface has been estimated to be 267 cm^{-1}.

The energy values incorporate that of the dipole moment (E_μ) of the DME using the follow expression (9)

$$E_\mu / cm^{-1} = \frac{7372.88}{R_{cm}^6} \cdot [3(\cos\alpha^2)+1] \qquad (9)$$

where R_{cm} is given, as elsewhere in this paper, in Å.

A comparison of the energetic contours of the ab initio potential with those of the atom-bond approach (compare Fig. 2 and 4) shows that the latter are lower than the former by a factor of about 0.5 that coincides with the factor of 0.47 by which the ab initio MEP had to be lowered to fit experimental data. This can be due to the trigonometric nature of the model adopted to rationalize the experiment.

On the contrary the geometric characteristics of the atom-bond PES at the stationary points (given in a more quantitative fashion in table 4 where the characteristics of the main stationary points of its 2D cross section are given) are in good agreement with the corresponding values of the ab initio PES given in Table 2.

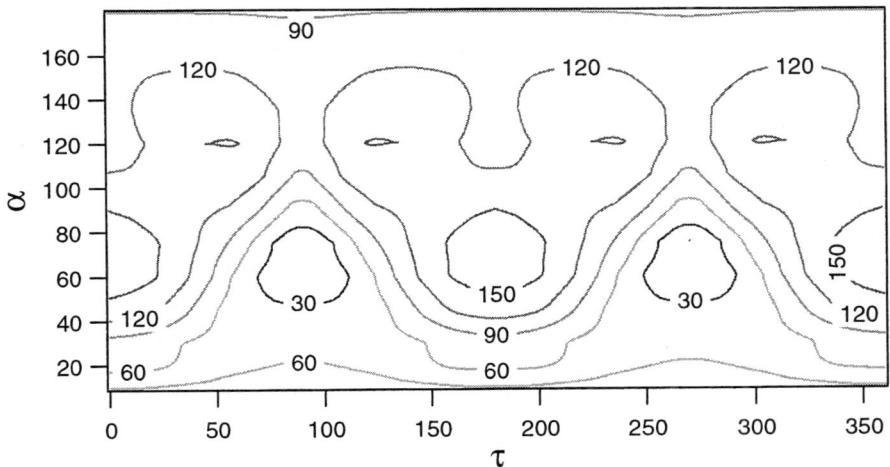

Fig. 4. Isoenergetic contours of the atom-bond PES (taken at the minimum along R_{cm}) plotted as a function of α and τ. The energy zero is set at the equilibrium energy of the complex and spacing between contours is of 30 cm^{-1}

As to energetic of the atom bond PES, from an analysis of Figures 2 and 4 and Tables 2 and 4 it is apparent its substantial agreement with the ab initio surface in the region around the minimum value on the oxygen side.

Table 4. Characteristics of the main stationary points on the 2D cross section of the atom-bond PES values at the minimum along R_{cm}

E / cm^{-1}	E$_{abs}$ / cm^{-1}	R_{cm} / Å	α / deg	τ / deg
0	-267	3.59	70	±90
61	-206	3.95	23	0 / 180
69	-198	3.92	0	All
91	-176	4.18	180	All
112	-155	4.86	120	0 / 180

The difference between the two surfaces increases while going away from the region of minimum. This suggests that the model used to rationalize the experiment with the help of the ab initio values is more sensitive to the oxygen region well than to the methyl one making the latter too repulsive.

This has prompted the necessity for performing extended dynamical calculations aimed at evaluating the dynamic properties of the system so as to have a more direct (model independent) evaluation of the surface. For this reason, we are carrying out distributed calculations of the properties of the system using DL_POLY [8] on the Grid prototype infrastructure[1] of ref. [9].

4 Conclusions

We have tackled the study of the DME-Ar cluster using an alternative formulation of the interaction in order to analyze the validity of the proposed PES and to check the suitability of atom-bond pairwise formulations of the potential energy for this type of systems.

The information derived from the experiment fairly well agree with the result for our atom-bond approach especially about the well that stabilizes the adduct.

On the contrary, going away from this region, the difference between the two surfaces increases with being atom-bond PES being less repulsive of about a factor 0.5. In other words, the PES of ref. [5] is found to be excessively rigid (probably due to the analytical formulation adopted for the model treatment of the measured data). As a matter of fact, scaling factor coincides with the factor of 0.47 applied to ab initio values to rationalize experimental findings.

This discrepancy has suggested an extension to carry out molecular calculations in order to obtain theoretical estimates of the observables of the system on the atom-bond PES.

Acknowledgments. Thanks are due for financial support to ASI, MIUR and COST in Chemistry.

---

[1] Developed within the FIRB project GRID.IT: Piattaforme abilitanti per griglie computazionali ad alte prestazioni orientate a organizzazioni virtuali scalabili finded by MIUR and coordinated by Prof. M. Vanneschi.

References

[1] Kesselman, C., Foster, I.: The Grid: Blueprint for a Future Computing Infrastructure. Morgan Kaufmann Publisher, USA (1998);
Baker, M., Buyya, R., Laforenza, D.: Grids and Grid Tecnologies for Wide Area Distrbuted Computing Software Practice and Experience, 32 (15) Wiley Press, USA (2000).
[2] Murrell, J.N., Carter, S., Farantos, S.C., Huxley, P., Varandas, A.J.C.: Molecular Potential Energy Surfaces, Wiley, New York (1984).
[3] Albertì, M., Castro, A., Laganà, A., Pirani, F., Porrini, M., Cappelletti, D.: Chem. Phys. Lett. 392, 514-520 (2004).
[4] Pirani, F., Albertì, M., Castro, A., Moix Teixidor, M., Cappelletti, D.: Chem. Phys. Lett. 394, 37-44 (2004).
[5] Ottavini, P., Maris, A., Caminati, W., Tatamitani, Y., Suzuki, Y., Ogata, T., Alonso, J.L.: Chem. Phys. Lett. 361, 341-348 (2002).
[6] Frisch, M.J., et al.: GAUSSIAN 98, Revision A.7, Gaussian, Inc., Pittsburgh, PA (1998).
[7] Pirani, F., Cappelletti, D., Liuti, G.: Chem. Phys. Lett. 350, 286-296 (2001).
[8] Smith, W., Forester, T.R.: DL_POLY2: A general-purpose parallel molecular dynamics simulation package. J. Mol. Graph., 14 (3), 136-141 (1996).
[9] Storchi, L., Manuali, C., Gervasi, O., Vitillaro, G., Laganà, A., Tarantelli, F.: Lecture Notes in Computer Science 2658, 297-306 (2003).

A Parallel Framework for the Simulation of Emission, Transport, Transformation and Deposition of Atmospheric Mercury on a Regional Scale

Giuseppe A. Trunfio, Ian M. Hedgecock, and Nicola Pirrone

CNR – Institute for Atmospheric Pollution, c/o UNICAL,
Rende, 87036, Italy
{a.trunfio, i.hedgecock, n.pirrone}@cs.iia.cnr.it

Abstract. The cycle involving emission, atmospheric transport, chemical transformation and deposition, is often a major source of mercury contamination in places distant from anthropogenic sources. Therefore atmospheric modelling and simulation are important to evaluate the association between emissions and potential effects on human health. Unfortunately, once released into the atmosphere, mercury is subjected to a variety of complex physical-chemical processes before its deposition. Thus performing detailed mercury-cycle simulations on the large spatial scale required implies the use of parallel computing. In this work, chemistry, photolysis, emissions and deposition models have been linked to open-source atmospheric meteorological / dispersion software. The result is a parallel modelling system, capable of long term simulations and mercury emission reduction scenario analyses.

1 Introduction

As it is well known mercury (Hg) is a chemical element that has severe effects on human health. The problem of mercury pollution became evident about two decades ago, when a significant increasing of the mercury content in some ecosystems was observed. In particular, strong pollution was often discovered in places relatively distant from mercury sources, and therefore long-range transport of atmospheric mercury and subsequent deposition was hypothesized to be an important cause of Hg pollution in water and soil. In this context, atmospheric simulation is important to advancing the mercury cycling knowledge state, especially in order to evaluate the association between anthropogenic emissions and local Hg pollution. On the other hand, once released into the atmosphere, from natural or anthropogenic sources, mercury is subjected to a number of physical-chemical processes before its deposition. This makes it difficult to develop a proper modelling tool of the Hg cycle, and it is computationally expensive to perform detailed simulations on the required large scale.

In recent years, a number of numerical simulation models of atmospheric mercury have been developed in order to improve the understanding of the mercury atmospheric pathway and to help the formulation of effective emission reduction strategies (e.g. [1, 2]). In this paper an integrated mercury model development has been performed within the atmospheric meteorological/dispersion model RAMS (Regional

Atmospheric Modeling System). The RAMS code is currently released under GPL (General Public Licence) by ATMET (ATmospheric, Meteorological and Environmental Technologies) [3]. The resulting framework runs in parallel and permits the incorporation of almost any type of source (point or diffuse source), gas and aqueous phase chemistry, altitude and cloud cover dependent photolysis rate calculation, wet and dry deposition and air-water exchange processes. The modelling system is potentially capable of long term simulations of regional deposition patterns, and emission reduction scenario analysis.

2 The Mercury Integrated Modelling System

RAMS has a number of features which make it extremely useful for air quality studies. First of all MPI (Message Passing Interface) based parallel processing is implemented in RAMS by the method of domain decomposition. Dynamic load balancing is available where computational load differs between subdomains of a grid or if computer nodes differ in computing capacity and, even with the extra computational load from the mercury processes modules, RAMS results in very good parallel efficiency.

Besides, RAMS has numerous options for the boundary conditions, two-way grid nesting capabilities, terrain following coordinate surfaces and non-hydrostatic time-split time differencing, a detailed cloud microphysics parameterization, various turbulence parameterization schemes, radiative transfer parameterizations through clear and cloudy atmospheres and a detailed surface-layer parameterization. In particular, the representation of cloud and precipitation microphysics in RAMS includes the treatment of each water species (cloud water, rain, pristine ice, snow, aggregates, graupel, hail) [4].

The surface heterogeneities connected to vegetation cover and land use are assimilated and described in detail in RAMS by means of the LEAF (Land Ecosystem Atmosphere Feedback) model [4]. This model represents the storage and vertical exchange of water and energy in multiple soil layers, including the effects of freezing and thawing soil, temporary surface water or snow-cover, vegetation, and canopy air. The surface domain meshes are further subdivided into patches, each with a different vegetation or land surface type, soil textural class, and wetness index to represent natural sub-grid variability in surface characteristics. Each patch contains separate prognosed values of energy and moisture in soil, surface water, vegetation, and canopy air, and calculates exchange with the overlying atmosphere weighted by the fractional area of each patch. The LEAF model assimilates standard land use datasets to define the prevailing land cover in each grid mesh and possibly the patches, then parameterizes the vegetation effects by means of biophysical quantities.

It is worth noting that two relevant reasons for using RAMS are its full microphysical parameterization for wet processes, which greatly influence the wet mercury chemistry and deposition, and the detailed parameterization of surface processes which aids proper descriptions of the mercury air-surface exchange processes.

2.1 Model Components, Input and Output

RAMS allows a desired number of prognostic scalar fields to be added to the model simulation, which are then automatically advected and diffused forward in time. In the present version of the model four additional scalars have been added: *elemental mercury* ($Hg^0_{(g)}$), *oxidised mercury* ($Hg^{II}_{(g)}$), *particulate mercury* (Hg^P) and the sum of oxidised Hg species in the aqueous phase ($Hg^{II}_{(aq)}$). In the absence of liquid water, for instance after the evaporation of non-precipitating cloud droplets, $Hg^{II}_{(aq)}$ is transported as if it were Hg^P. However, unlike Hg^P, it is considered soluble so that should it once again be in the presence of liquid water it is assumed to be scavenged and returns to the aqueous phase.

In order to model accurately the dynamics of mercury, a number of *ad-hoc* modules were linked to RAMS. These modules account for major processes (see Fig. 1) that affect the fate of mercury in the atmosphere: natural and anthropogenic mercury emissions, dry and wet deposition, chemical transformation, air-water exchange, air-soil exchange and canopy emissions.

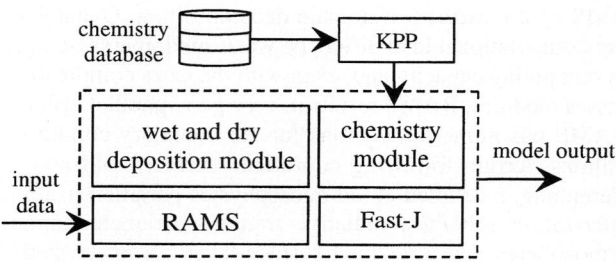

Fig. 1. Major components of the mercury modelling system. KPP [7] is the Kinetic Preprocessor which gives the chemistry integrator source code from a chemistry database (i.e. symbolic reactions and constant rate values); the module Fast-J [6] computes the photolysis rate constants for each model column during the simulation

Standard model inputs (see Fig. 2) are the USGS (United States Geological Survey) global land use data and topography dataset at 30'' (i.e. about 1-km resolution); the global monthly climatological sea surface temperature data at 1-degree resolution (i.e. about 100 km). The European Centre for Medium-Range Weather Forecasting (ECMWF) meteorological data have been used. The data is available at 6-hourly intervals, with a spatial resolution of $2.5° \times 2.5°$ latitude and longitude, and 10 pressure levels. At the moment the database for anthropogenic mercury emissions in Europe, compiled for 1995 [5], is employed in the model. The Hg emission inventory provides annually averaged emissions from point and area sources and Hg speciation (i.e. the proportions of Hg^0, Hg^{II} and Hg^P). Given the importance of Hg oxidation by OH, the production of OH needs to be well described as does the O_3 concentration. Therefore the initialisation of RAMS has been adapted to include 6-hourly O_3 concentration fields from ECMWF.

Model output includes the concentration fields of Hg^0, Hg^{II} and Hg^P, deposition velocities, emissions and deposition fluxes, and the mercury concentration in rain.

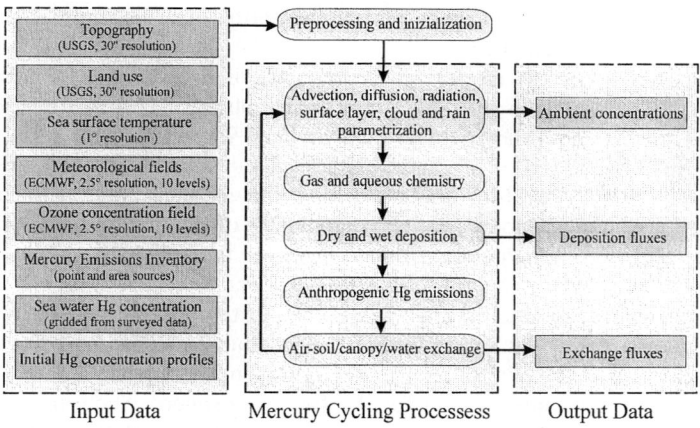

Fig. 2. Input-output data and flow diagram of our mercury cycle simulation system

2.2 Chemical Processes in the Atmosphere

The Hg atmospheric chemistry model used in the integrated modelling system has been designed specifically for the purpose and includes both gaseous and aqueous phase chemistry relevant to Hg. Oxidation is fundamental to Hg deposition as oxidised species are much less volatile and much more soluble than Hg^0. In order to well describe the production of OH, which greatly affects the Hg oxidation, an on line photolysis rate constant calculation procedure is used. The program Fast-J [6] has been linked to RAMS in order to provide photolysis rate constants as a function of latitude, longitude, date, time of day, altitude and optical depth.

In order to reduce the calculation time, operator splitting between the meteorological model and the chemistry model has been employed. RAMS typically uses 90s time steps for the grid resolution chosen for our simulations, but the chemistry model is called every 15 minutes. The chemistry model is run for each cell, whilst Fast-J is called for each column, the altitude dependent photolysis rate constants being stored in a temporary array. While $Hg^0_{(g)}$, $Hg^{II}_{(g)}$, $Hg^{II}_{(aq)}$ and Hg^P are all transported by the model, other chemical species are not transported but their concentrations stored 'locally' for each cell after each chemistry time step.

The non-Hg chemistry currently used in the preliminary simulations includes detailed HO_x and SO_x chemistry in both the gas and aqueous phases, in order to obtain reasonable values for the OH concentration and the cloud droplet pH. The inclusion of the chemistry and mass transfer model has been incorporated into the overall modelling framework in such a way that the chemistry model may be updated with new reactions and coefficients separately from the rest of the model. In particular the updated chemistry database (i.e. the reactions and constant rates) is recompiled using the open source program *Kinetic PreProcessor* (KPP) [7] which is released under the GPL and can produce a proper integrator source code (based on the second-order Rosenbrock method in this application). As illustrated in Fig. 1, the integrator is then linked to RAMS, thus avoiding any necessity to manually alter the RAMS code to include new chemical species or reactions. This was considered to be of great importance as the model was conceived with the intention of testing numerous and various chemical modelling schemes.

In the near future the model will be tested including detailed NO_x emissions and chemistry. The inclusion of a VOC emissions database is also foreseen; chemistry models based on both the CBM-IV and SAPRC 99 mechanisms have already been prepared. The coupling of a parameterised sea salt aerosol production model to the integrated model will allow the inclusion of a detailed MBL (Marine Boundary Layer) photochemical model [8]. The complexity of this model will necessarily have a major influence on calculation times, however the full MBL model will only be applicable over part of the modelling domain, and investigations are under way to refine the integrated model in order to use the most appropriate chemistry models in a given situation, using for example different chemical mechanisms for the MBL, urban/industrial boundary layer and remote continental boundary and possibly a separate chemistry mechanism again for the free troposphere. Such an approach would give the best possibility of ensuring accurate chemistry modelling whilst making the most efficient use of computing resources. This methodology has already been tested in the current version of the modelling system where two chemical mechanisms have been used, one for 'dry' situations and one for 'wet', the difference being the inclusion of aqueous phase chemistry if the model cell had a liquid water content above a threshold value. The method allows a significant time saving.

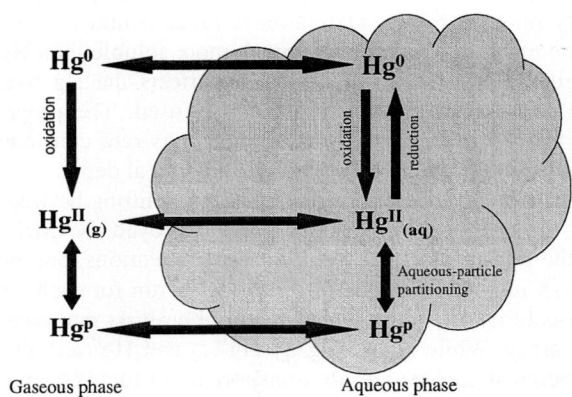

Fig. 3. The modelled chemical processes

The present version of the chemistry model contains over 100 reactions and mass transfer between the gas and aqueous phase of 10 species. The Hg chemistry included in the model is summarised below and in Fig. 3 (more details can be found in [8]):

1. **Gas phase oxidation:** The reactions of Hg with O_3, H_2O_2, OH, HCl and Cl_2 are included, the reaction with Br will be included in the MBL chemistry mechanism.
2. **Aqueous phase oxidation:** The oxidation reactions included are those with O_3, HOCl / OCl-, and OH.
3. **Aqueous phase reduction:** The reduction of Hg^{II} compounds to Hg^0 by reaction with HO_2 / O_2^-, are included as are the reduction of $HgSO_3$ and $(HgSO_3)_2^{2-}$.

4. **Gas – Aqueous Phase Equilibrium**: The mass transfer between the gas phase and cloud droplets of Hg^0, HgO, and $HgCl_2$ are considered.
5. **Aqueous Equilibria**: The equilibria between Hg^{2+}, and the OH^-, Cl^-, and SO_3^{2-}, are included. For Cl^- complexes up to $HgCl_4^{2-}$ are considered. Br^- complexes will be included for the MBL mechanism.
6. **Aerosol – Liquid Equilibria**: $soot_{(air)} - soot_{(aq)}$ is accounted for according to the method used in [9].
7. **Aqueous – Soot Equilibria**: The partitioning of Hg compounds between the aqueous phase and the solid phase within droplets ($Hg^{II}_{(soot)} - Hg^{II}_{(aq)}$) (see [9]).

2.3 Dry and Wet Deposition

The low solubility and relatively high volatility of $Hg^0_{(g)}$, mean that dry deposition of elemental mercury is unlikely to be a significant pathway for removal of atmospheric mercury, although approximately 95% or more of atmospheric mercury is elemental mercury. Therefore, as in a number of other modelling studies the $Hg^0_{(g)}$, dry deposition is assumed to be zero in the model.

A surface and meteorological variable dependent deposition velocity is used to calculate the $Hg^{II}_{(g)}$ deposition flux. In RAMS the surface grid cells are divided into sub-grid patches, each with a different vegetation or land surface type, in order to to represent natural sub-grid variability in surface characteristics. The overall deposition velocity is therefore computed as:

$$\overline{v}_d = \sum_{i=1}^{np} v_i f_i \qquad (1)$$

where np is the number of patches (which is specified during RAMS pre-processing phase) and f_i is the fractional index coverage of the i-th patch. The patch deposition velocity v_i is calculated using the classic Wesely's resistance model [10] as:

$$v_i = (r_a + r_d + r_c)^{-1} \qquad (2)$$

where r_a is the atmospheric resistance through the surface layer, r_d is the deposition layer resistance and r_c is the canopy or surface resistance. The resistance r_a, which represents bulk transport by turbulent diffusion through the lowest 10 meters of the atmosphere, is obtained, as usual, according to the formula reported in [10]. Also the deposition layer resistance, which represents molecular diffusion through the lowest thin layer of air, is parameterized in terms of the Schmidt number by the equation developed in [10]. Canopy or surface resistance over land is computed according to the Wesely's formulation:

$$r_c = \left[(r_{st} + r_m)^{-1} + r_{uc}^{-1} + (r_{dc} + r_{lc})^{-1} + (r_{ac} + r_{gs})^{-1} \right]^{-1} \qquad (3)$$

where $r_{st} + r_m$ is the leaf stomata and mesophyllic resistance, r_{uc} is the upper canopy resistance, r_{dc} is related to the gas phase transfer by buoyant convection in canopies, r_{lc} is the lower canopy resistance, r_{ac} is a resistance that depends on canopy height and density, and r_{gs} is the ground surface resistance. All the latter contributions are computed according to the Wesely's formulae. It is worth noting that many of these resistances are season and landuse dependent and some are adjusted using solar radiation, moisture stress and surface wetness provided by the specific RAMS submodels.

Surface dry deposition velocity of particulate mercury $Hg_{(p)}$ is largely dependent on particle size whose distribution is assumed to be log-normal with a geometric mass mean diameter of $0.3 \mu m$ and a geometric standard deviation of $1.5 \mu m$. The deposition velocity is determined by dividing the distribution into a fixed number of size intervals, calculating the velocity v_i for each interval and aggregating them in a weighted mean. The resistance approach has been adopted for v_i calculation. Particulate matter does not interact with vegetation as gases do, and in particular, particles are usually assumed to stick to the vegetation surface and therefore $r_c \approx 0$. Thus we assume [11]:

$$v_i = \frac{1}{r_a + r_b + r_a r_d v_g} + v_g \qquad (4)$$

where $v_g \approx 0$ is the gravitational settling velocity, proportional to the square of particle diameter and which is negligible for $Hg_{(p)}$; the aerodynamic resistance r_a expression is identical to that used for $Hg^{II}{}_{(g)}$ dry deposition, while the resistance to diffusion through the quasi-laminar sub-layer r_b is parameterised in terms of the Schmidt number (see [8] for details).

Wet deposition is an important removal process for both oxidised and particulate mercury while, because of its low solubility, direct wet removal of $Hg^0{}_{(g)}$ is negligible. Obviously the aqueous phase oxidation of $Hg^0{}_{(g)}$ is important here, and so is therefore the rate at which droplets take up $Hg^0{}_{(g)}$. This process is included as two reactions in the chemistry model describing uptake and out-gassing of $Hg^0{}_{(g)}$ and thereby representing the equilibrium between the gas and aqueous phases.

A relevant fraction of particulate mercury is removed by both in-cloud scavenging (rainout) and below-cloud scavenging (washout). The rate of particle wet removal depends upon ambient concentration, cloud type, rainfall rate and particle size distribution, but usually the Hg^P wet deposition is determined by a synthetic scavenging efficiency coefficient from the ambient concentration. Wet deposition of the insoluble fraction of particulate Hg is modelled assuming that the local depletion rate is proportional to the concentration C (mercury mass per unit volume of air), introducing a wet scavenging rate Λ, dependent on the precipitation intensity: $\partial C / \partial t = -\Lambda C$. The scavenging rate is used to reflect the propensity of mercury to be removed by the current precipitation, including all possible below-cloud and in-cloud processes.

2.5 Canopy Emissions $Hg^0{}_{(g)}$

As pointed out by many researchers (e.g. [12]) in vegetated areas the Hg^0 dissolved in soil water is transported to leaves and emitted to the atmosphere via the transpiration stream. As in [13] the elementary mercury plant emissions flux F_c is determined, assuming a constant Hg^0 concentration in the evapotranspiration stream [14], as $F_c = E_c C_s$, where E_c is the evapotranspiration rate and C_s is the $Hg^0{}_{(g)}$ concentration in the soil water.

The evapotranspiration rate E_c is directly derived from the detailed LEAF RAMS submodule [4]. The latter is able to take into account correctly the type of vegetation in each cell and its minimum stomatal resistance. It also takes into account stomatal closure caused by excessively warm temperatures or cold (freezing) temperatures, lack of solar radiative flux (which is separately parameterized in RAMS), lack of water in the soil layers comprising the plant root zone and the canopy water vapour mixing ratio at the leaf surface which effects evaporation from the stomata.

2.6 Air-Soil and Air-Water Exchange of $Hg^0_{(g)}$

$Hg^0_{(g)}$ natural soil emission is mainly dependent on soil temperature, soil moisture and solar radiation. Hereby the elementary mercury net flux from soil F_s (ng m^{-2} h^{-1}) is parameterized as a function of soil temperature T_s (°C) using the relation found in [15] as:

$$\log F_s = 0.057\ T_s - 1.7 \tag{5}$$

In modelling $Hg^0_{(g)}$ air-water exchange the film theory has been used for mass transfer calculations at the interface (e.g. see [9, 13]). The $Hg^0_{(g)}$ evasional flux F_w is driven by the fugacity difference between the overlying air and surface water:

$$F_w = K(C_w - C_g / \bar{H}) \tag{6}$$

where K is the overall mass transfer coefficient, C_w and C_g are respectively the Hg^0 concentration in water and air and $\bar{H}$ is the dimensionless inverse Henry's Law constant of Hg. All quantities are computed using parameterisations from the literature (see [8] for details).

3 First Results

The model has been run from the 15th January 1999 until the 31st of March, the first four weeks are a spin-up period during which the results are still influenced by the initial conditions. The modelling region covers the Mediterranean and Europe as can be seen in the Figs. 4-6. The modelling grid is made up of 130×125, 50 km × 50 km cells and 25 vertical levels (406250 cells in total). The ten week simulation took less than 24 hours on a Linux cluster based on 25 3-Ghz Xeon processors.

The Figs. 4-6 have been chosen to illustrate the interplay between meteorology, photochemistry, and the $Hg^0_{(g)}$ and $Hg^{II}_{(g)}$ concentration. Figs. 4 and 5 show the $Hg^0_{(g)}$ (ng/m^3) and $Hg^{II}_{(g)}$ (pg/m^3) concentrations in the bottom model layer respectively, whilst Fig. 6 shows the cloud cover. All the figures refer to midday on the 8th of March. The correlation between $Hg^{II}_{(g)}$ production and the factors that influence O_3 and OH production can be seen. O_3 production is favoured by high temperatures and O_3 photolysis leading to OH production is a function of cloud cover.

Also Figs. 4 and 5 reproduce a phenomena which became evident during the MAMCS and MOE projects [16, 17]; that in spite of the higher $Hg^0_{(g)}$ concentrations to be found in the north of Europe, $Hg^{II}_{(g)}$ concentrations were found to be consistently higher in the Mediterranean region.

In Fig. 7 the model output can be compared to the daily average $Hg^0_{(g)}$ data obtained at Fuscaldo (Italy, 39° 25'N, 16°,00'E) during the 2nd MAMCS measurement campaign (15th Feb - 1st March, 1999). The results look quite good, especially considering the high uncertainties in the model parameters and input data.

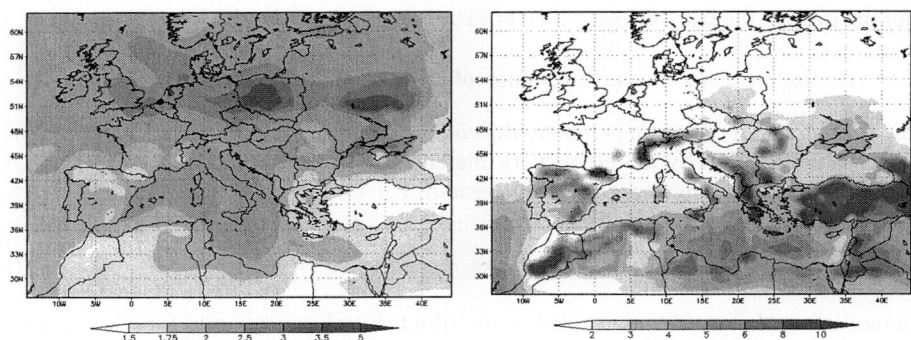

Fig. 4. The modelled $Hg^0_{(g)}$ concentration   **Fig. 5.** The modelled $Hg^{II}_{(g)}$ concentration

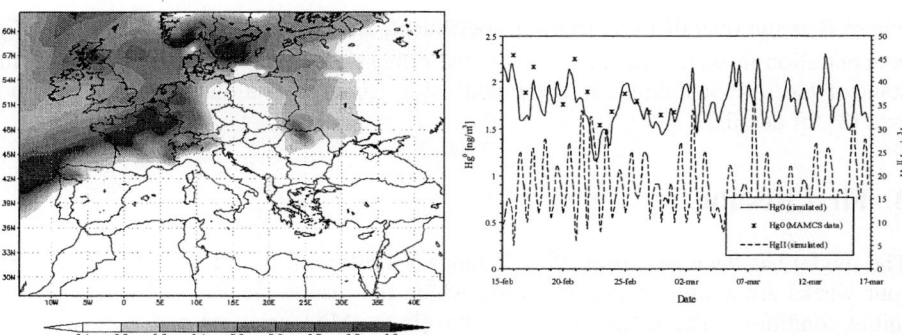

Fig. 6. Cloud cover data from RAMS   **Fig. 7.** Modelled $Hg^0_{(g)}$ and $Hg^{II}_{(g)}$ and MAMCS data

4 Conclusions

Our current scientific understanding of atmospheric mercury cycling is incomplete but can be significantly improved through atmospheric modelling and simulation. Besides, simulation models are currently widely used to inform governmental policy makers about the most effective emission control options to reduce the impact of mercury pollution on human health.

On the other hand atmospheric mercury is subjected to a number of complex physical-chemical processes before its deposition. In addition simulations are required on a large scale to properly account for the Hg long-range transport phenomena. Therefore performing useful simulations of the mercury cycle results in a hard computational task and parallel computing is usually required.

In this work the various modules needed to model Hg cycling have been linked developing an effective parallel simulation framework. First results are consistent with both measurements and our understanding of the atmospherics chemistry of Hg. More tests and longer runs are required to validate the model and to assess the importance of boundary and initial conditions, and fortunately there is a good amount of data now available from a number of atmospheric-mercury monitoring stations and measurement campaigns.

References

1. Petersen, G., et al. A comprehensive Eulerian modelling framework for airborne mercury species: model development and application in Europe. *Atmospheric Environment*, 35-17 (2001), 3063-3074
2. Berg, T., et al. Atmospheric mercury species in the European Arctic: measurements and modelling. *Atmospheric Environment*, 35-14 (2001), 2569-2582
3. ATMET: ATmospheric Meteor. and Environmental Technologies: *http://www.atmet.com*
4. Walko, R.L., et al. Coupled atmosphere-biophysics-hydrology models for environmental modeling. *J. Appl. Meteor.*, 39 (2000), 931-944
5. Pacyna, E., Pacyna, J.M. and Pirrone, N. Atmospheric Mercury Emissions in Europe from Anthropogenic Sources. *Atmospheric Environment*, 35 (2001), 2987-2996.
6. Wild, O., X. Zhu, and M.J. Prather, Fast-J: Accurate simulation of in- and below-cloud photolysis in tropospheric chemical models, *J. Atmos. Chem.*, 37 (2000), 245-282
7. V. Damian, A. Sandu, M. Damian, F. Potra, and G.R. Carmichael: ``The Kinetic PreProcessor KPP -- A Software Environment for Solving Chemical Kinetics'', Computers and Chemical Engineering, Vol. 26:11 (2002), 1567-1579.
8. Dynamics of Mercury Pollution on Regional and Global Scales. Editors: N. Pirrone, K.R. Mahaffey. *Kluwer Academic/Plenum Publishers*, 2004 (in press)
9. Petersen, G., Iverfeldt, A., Munthe, J. Atmospheric mercury species over central and northern Europe. *Atmospheric Environment* 29:1 (1995), 47-67.
10. Wesely, M.L., Hicks, B.B., "Some factors that affect the deposition of sulfur dioxide and similar gases on vegetation" *Journ. of the Air Poll. Control Association* 1977, 1110-1116
11. Slinn, S.A. and W.G.N. Slinn. Prediction for particle deposition on natural waters, *Atmospheric Environment*, 14 (1980), 1013-1016
12. Lindberg, S.E. et al. Atmosphere-surface exchange of mercury in a forest: results of modeling and gradient approaches. *Journ. of Geophys. Research* 97 (1992), 2519-2528
13. Xu, X. et al. Formulation of bi-directional air-surface exchange of elemental mercury. *Atmospheric Environment*, 33:27 (1999), 4345-4355
14. Lindberg, S.E. Forests and the global biogeochemical cycle of mercury: the importance of understanding air/vegetation exchange processes. In: Baeyens, W., et al. (Ed.), *Global and Regional Mercury Cycles: Sources, Fluxes, and Mass Balances*. Kluwer (1996), 359-380.
15. Carpi, A., Lindberg, S.E. Application of a Teflon dynamic flux chamber for quantifyng soil mercury flux: tests and results over background soil. *Atmos. Env.*, 32:5 (1998), 873-882
16. Munthe, J. et al. Distribution of atmospheric mercury species in Northern Europe: final results from the MOE project. *Atmospheric Environment*, Vol. 37 (2003)
17. Pirrone, N. et. al. Dynamic Processes of Mercury Over the Mediterranean Region. *Atmospheric Environment*, 37 (2003)

A Cognitive Perspective for Choosing Groupware Tools and Elicitation Techniques in Virtual Teams

Gabriela N. Aranda[1], Aurora Vizcaíno[2], Alejandra Cechich[1], and Mario Piattini[2]

[1] Universidad Nacional del Comahue, Departamento de Ciencias de la Computación,
Buenos Aires 1400, 8300 Neuquén, Argentina
{garanda, acechich}@uncoma.edu.ar
[2] Universidad de Castilla-La Mancha, Departamento de Informática,
Paseo de la Universidad 4, 13071 Ciudad Real, España
{Aurora.Vizcaino, Mario.Piattini}@uclm.es

Abstract. Nowadays groupware tools, as well as requirement elicitation techniques, are chosen without a clear strategy that takes into account stakeholders' characteristics. When the chosen technology is not appropriate for all the group members it might affect their participation and the quality of the requirement elicitation process itself.

In order to improve communication, and therefore stakeholders' participation, we propose choosing an appropriate set of groupware tools and elicitation techniques according to stakeholders' preferences. This paper presents a prototype tool that makes a selection based on cognitive techniques.

1 Introduction

Multi-site development is a current matter of study and discussion, since global development is becoming a usual style of software production. On the other hand, it is a fact that during a traditional requirement elicitation process, stakeholders must face many problems that have been detected and analyzed for decades [3, 7, 16]. Moreover, when participants are distributed geographically new problems often arise. For example, communication and coordination are more difficult because of differences in culture, timetable, language, etc. [2, 6].

CSCW and Cognitive Informatics are two areas of research that try to minimize the impact of these problems. On the one hand, CSCW (Computer-Supported Cooperative Work), addresses research into experimental systems and the nature of organizations [12], taking into account human behaviour and the technical support people need to work as a group in a more productive way. On the other hand, Cognitive Informatics [5, 21] is an interdisciplinary research area that tackles the common root problems of modern informatics, computation, software engineering, artificial intelligence (AI), neural psychology, and cognitive science. This research area, initiated few years ago, focuses on the nature of human processing mechanisms, especially respect to information acquisition, memory, categorization, retrieve, generation, representation, and communication [21].

Cognitive science is related to computer science in many research areas (artificial intelligence, expert systems, knowledge engineering, etc). Cognitive informatics principles of software engineering encompass some basic characteristics of the software development process that include the difficulty of establishing and stabilizing requirements, changeability or malleability of software, etc. [22].

Since our main goal is to enhance interpersonal communication in geographically distributed teams, we are particularly interested in some techniques from the field of psychology called Learning Style Models (LSM). LSM classify people according to the way they perceive and process information, and analyse relationships between students and instructors. Considering that during requirement elicitation a person acts like student and instructor alternatively; we propose using LSM as a base for improving the requirements elicitation process. In doing so, we propose choosing a set of groupware tools and requirement elicitación techniques that support not only the communication itself but also the stakeholders' preferences.

In the following two sections we present some basic concepts about groupware tools and learning style models. In section four, we describe a model that supports stakeholders' personal preferences in geographically distributed processes, and a prototype tool that uses the previous model to automate the selection process. Finally, we present some related works and conclusions.

2 What Is Groupware?

Generally speaking, groupware is software to enable communication between cooperating people who work on a common task. It may include different communication technologies, from simple plain-text chat to advanced videoconferencing [11]. To avoid ambiguities we will refer to every simple piece of communication technology as a groupware tool, and to the systems that combine them as groupware packages.

The most common groupware tools used during multi-site developments are e-mails, newsgroups, mailing lists, forums, electronic notice boards, shared whiteboards, document sharing, chat, instant messaging, and videoconferencing. [6, 11, 13].

At first glance, groupware tools can be divided into *synchronous* and *asynchronous*; whether the users have to work at the same time or not [14]. Examples of synchronous tools are chat and videoconferencing, while e-mails and document sharing are examples of asynchronous. A second classification can be made according to the way in which they show the information: based primarily on images, figures, diagrams, etc. (shared whiteboards, videoconferencing) or based on words (chat, instant messaging, e-mails, forums, etc.).

Virtual teams choose a combination of groupware tools according to their possibilities or the kind of activities they are carrying out. They can choose between using a groupware package (that offers a combination of tools) and using individual tools in an ad-hoc way. Respect to using synchronous or asynchronous tools in group work, some authors note that both types of communication are important. In the one hand, asynchronous collaboration allows team members to construct requirements individually and contribute to the collective activity of the group for a later discussion. On the other hand, real time collaboration and discussions give the stakeholders the chance of getting instant feedback [13]. However, according to the classification of styles presented previously, people would have preferences for one or the other, in the same way that some people would prefer working with tools based on visual or verbal characteristics.

3 Learning Style Models

A learning process involves two steps: *reception* and *processing* of information. During the first step, people receive external information –which is observable through senses– and internal information –which emerges from introspection–, then they select a part to process and ignore the rest. Processing involves memorization or reasoning (inductive or deductive), reflection or action, and introspection or interaction with others [8, 9].

Learning Style Models (LMS) classify people according to a set of behavioural characteristics pertaining to the ways they receive and process information and this classification is used to improve the way people learn a given task.

These models have been discussed in the context of analyzing relationships between instructors and students. We take advantage of this model and discussions by adapting their application to a virtual team that deals with a distributed requirement elicitation process. To do so, we consider an analogy between stakeholders and roles in LSM since during the elicitation process everybody "learns" from others [17], so that stakeholders play the role of student or instructor alternatively, depending on the moment or the task they are carrying out.

After analyzing five LSM in [17] we found out that every item in the other models was included in the model proposed by Felder-Silverman [9], so that we may build a complete reference framework choosing this as a foundation.

The Felder-Silverman (F-S) Model classifies people into four categories, each of them further decomposed into two subcategories. Characteristics of each subcategory are:

- *Sensing* (concrete, practical, oriented toward facts and procedures) or *Intuitive* (conceptual, innovative, oriented toward theories and meanings);
- *Visual* (visual representations of presented material – pictures, diagrams, flow charts) or *Verbal* (written and spoken explanations);
- *Active* (working by trying things out, working with others) or *Reflective* (thinking things through, working alone);
- *Sequential* (linear, orderly, learn in small incremental steps) or *Global* (holistic, systems thinkers, learn in large leaps).

People are classified by a multiple-choice test that gives them a rank for each category. Depending on the circumstances people may fit into one category or the other, being, for instance, "sometimes" active and "sometimes" reflective, so preference for one category is measured as *strong*, *moderate*, or *mild*. Only when there is a strong preference, people can be catalogued as a member of a certain group.

The test, proposed by Barbara Soloman and Richard Felder, is available at http://www.engr.ncsu.edu/learningstyles/ilsweb.html.

4 A Personal Preference-Based Model to Enhance Distributed Elicitation

A first step, before proposing a methodology for supporting distributed elicitation, is determining the aspects that have to be considered and the way in which they relate to each other. With the aim of recommending a set of suitable groupware tools and elicitation techniques during a particular elicitation process, we have defined a model, which is depicted in Figure 1, whose primary concepts are described in Table 1.

Table 1. Main concepts of our personal preference-based model

Concept	Description
Virtual Team	Virtual team [19] virtual community [10], distributed group [15] are terms used to refer to a group of people who work together on a project. Their main characteristic is that people are distributed over many sites, and they use information technology to communicate and coordinate efforts. In our model the common project or task carried out is the requirement elicitation process, which is the process of "extract and inventory the requirements from a combination of human stakeholders" [20].
Stakeholder	Typical stakeholders are users (those who will operate the system), customers (those who have commissioned the system), system developers, etc. [20] Each person in a virtual team is supposed to play (at least) one *Role* during the elicitation process, and, as it is a person, he or she has some *Personal Characteristics* that tell us about his or her preferences when he/she perceives and process information.
Groupware Tools	As we have mentioned before, groupware is software to enable communication. According to the way in which they show the information, groupware tools have different *Representation Modes* and different *Interaction Modes*.
Elicitation Techniques	Elicitation is fundamentally a human activity where communication plays a transcendental role [20]. The election of elicitation techniques plays a very important role in distributed teams. Since face-to-face interaction is not possible, techniques have to be adapted to be used in combination with groupware. Some techniques that seem to be adaptable to the distributed elicitation process are question and answer methods, customer interviews, brainstorming, idea reduction, storyboards, prototyping, questionnaires, and use cases [15]. Like groupware tools, elicitation techniques have different *Representation Modes*.
Representation Mode	The way groupware tools and elicitation techniques present the information. For instance, it can be based on images or based on words.
Interaction Mode	The way people interact with others depends on the characteristics of the groupware tools. For instance, interaction can be synchronous or asynchronous.
Personal Characteristics	It is information about personal characteristics and preferences of stakeholders. For instance the result of the classification of Felder-Soloman test is information about his or her preferences when perceiving and processing information.
Role	It represents information about the role that stakeholders play during the requirement elicitation process. Examples of *roles* are end-user, client, analyst, project manager, etc.

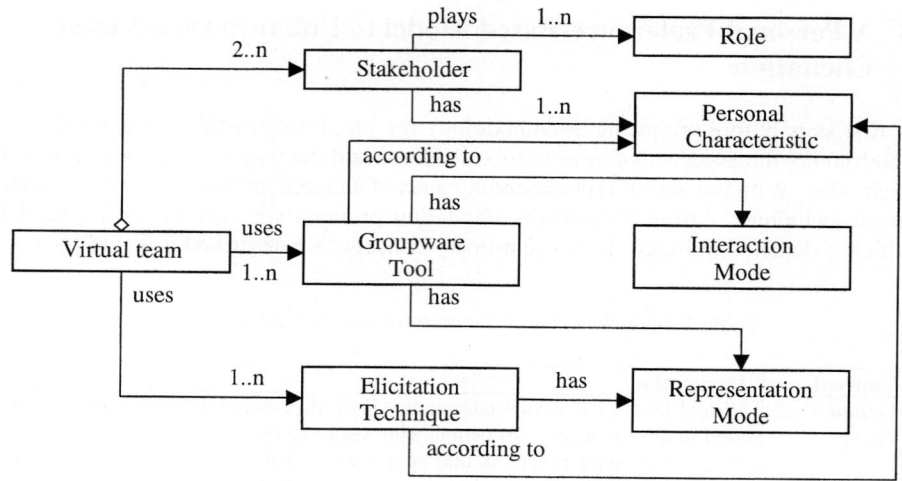

Fig. 1. A model to support personal preferences in a virtual community

Relationships between previous concepts can be expressed generally as:

- A *Virtual Team* represents a group of *Stakeholders* that work cooperatively on a common task (which in our case is the Requirement Elicitation Process).
- *Stakeholders* play *Roles* that imply rights and responsibilities associated to their job. In our case the roles involved in the elicitation process are: users, clients, managers, analysts, project managers, etc.
- *Stakeholders* communicate with each other using some *Groupware Tools* and build different models of a problem using a set of *Elicitation Techniques*.
- *Groupware Tools*, as well as *Elicitation Techniques*, are supposed to be chosen according to the stakeholders' *Personal Characteristics*, in order to make them feel comfortable and improve their performance.
- Each *Groupware Tool* has a *Representation Mode* (verbal or visual) and an *Interaction Mode* (synchronous or asynchronous) which are important in deciding the suitability for a stakeholder's personal preferences.
- In a similar way, each *Elicitation Technique* has a predominant *Representation Mode* (verbal, visual, or a possible good combination of both) that we will take into account to suggest their use or non-use.

4.1 A Cognitive Approach to Choose Groupware Tools

In order to support personal preferences, in [17], we have proposed a classification of the most commonly used groupware tools focusing on Visual/Verbal and Active/Reflective categories of the F-S model. The classification is based on the description and the strategies suggested by Felder and Silverman for each subcategory.

Figure 2 shows the results of such classifications. The sign "++" is used to indicate those groupware tools which are more suitable for people with a strong preference for a given subcategory. The sign "+" indicates that a groupware tool would be mildly preferred by a stakeholder with those characteristics. Finally, the sign "-" suggests that a particular groupware tool would be "not suitable" for that subcategory.

		Visual	Verbal	Active	Reflective
Asynchronous Tools	E-mails	+	++	-	++
	Mailing lists, Newsgroups	-	++	-	++
	Asynch. shared whiteboards	++	-	-	++
	Forums	-	++	-	++
Synchronous Tools	Instant messaging	+	++	++	-
	Synch. shared whiteboards	++	-	++	-
	Chat	-	++	++	-
	Videoconferencing	++	++	++	-

Fig. 2. Characterization of groupware tools based on F-S model

To choose a set of groupware tools for a given group of stakeholders we have suggested representing the information we know about each participant in a two-way matrix [17]. By doing so, we can have a view of stakeholders' preferences in general and, according to the quadrant that contains more instances, choose those groupware tools that adapt to most people in the group. Figure 3 shows an example of such a matrix.

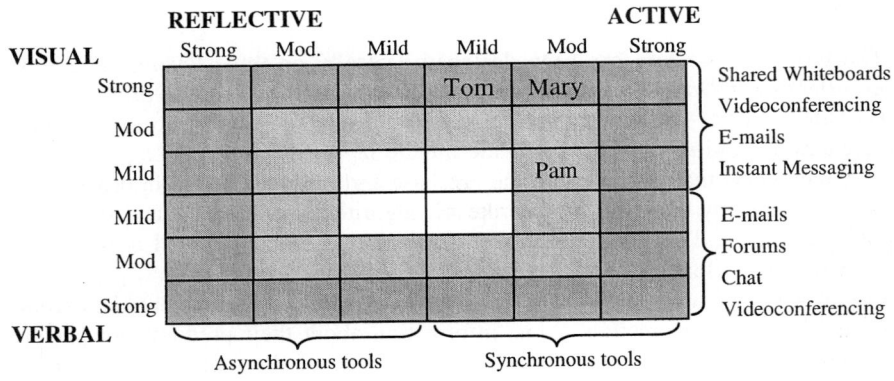

Fig. 3. Choosing a set of groupware tools according to F-S Model categories

In order to obtain a set of rules that tell us which groupware tool is more suitable according to the stakeholder's cognitive style, we have presented a model based on fuzzy logic and fuzzy sets [1]. The model takes four inputs (X_1, X_2, X_3, X_4), which are the preferences for each category of the F-S Model, and an output variable (Y) that is the preference of a stakeholder for one of a given set of groupware tools.

For each input variable we have defined a domain using the adverbs (and their correspondent abbreviations): Very (*V*), Moderate (*M*) and Slight (*S*), which correspond to *strong*, *moderate* and *mild*, respectively, in the F-S model. We have changed their names to avoid confusion with respect to the use of the first letter, so that the definition domain for the category Reflective-Active, for example, would be: Very reflective (VRe), Moderately reflective (MRe), Slightly reflective (SRe), Slightly active (SAc), Moderately active (MAc), Very active (VAc).

Using a machine learning algorithm it is possible to obtain rules such as: *if X_1 is VAc and X_3 is VVi then y is IM;* which is interpreted as: *"If a user has a strong preference for the Active subcategory and a strong preference for the Visual subcategory, the tool that this person would prefer is Instant Messaging"*

In a similar way it is possible to find a suitable set of elicitation techniques according to the preferences for each category of the F-S model.

4.2 Automation of the Selection Process

With the aim of finding a set of groupware tools and elicitation techniques that are suitable for a given group of stakeholders, we have designed a prototype tool – based on the model previously explained – that do it in an automatic way. Its mechanism can be simply explained as follows:

1. Stakeholders are asked to fill in a multiple-choice test so as to know their preferences. This information is maintained throughout the cooperative process.
2. Once a group of stakeholders is defined, our tool analyses their personal preferences using the sets of rules previously generated.
3. As a result, the tool returns the most suitable groupware tools and elicitation techniques for that group of people.

The tool's architecture has been designed basically on three layers: (1) a lower layer –*Persistent Data*– keeps the information concerning personal preferences of stakeholders, rules of suitability preferences-groupware tools and rules of suitability preferences-elicitation techniques; (2) the middle layer –*Application Logic*– contains those components that interact with the database and interface layers in order to find information and, by applying the appropriate algorithms, analyses it and produces a suitable answer; and (3) the upper layer –*User Interface*– contains all those components with which users of the tool interact.

Figure 4 shows a screen of our prototype tool where three stakeholders (Mary, Tom and Pam) are interacting. Some information about their predominant personal characteristics is shown on the upper right hand side of the screen. On the bottom there are two lists of suggested groupware tools and elicitation techniques that would be most suitable for them.

Our tool is currently under test and validation by categorizing people with different profiles and from different organizations. First experiences would indicate that results might be considered to guide elicitation; however further validation is needed to reach more conclusive results.

A Cognitive Perspective for Choosing Groupware Tools and Elicitation Techniques 1071

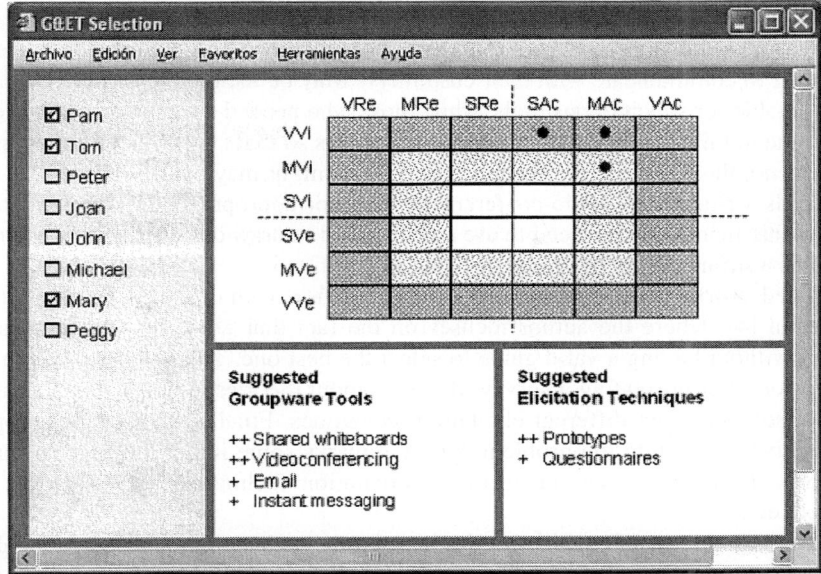

Fig. 4. An interface that shows the suggestions for a particular group of stakeholders

5 Related Work

There are some previous works about analysis of groupware tools and effectiveness in distributed teams. For instance in [6] it is described a case study of a real multi-site organization, where stakeholders use a mix of synchronous and asynchronous tools (teleconferencing, document sharing, email.) to interact. The authors collected data from inspecting documents, observed requirements meetings, and performed semi-structured interviews with stakeholders. Some problems detected for stakeholders are: (1) lack of informal or face-to-face communication; (2) difficulty in sharing drawings on a whiteboard during spontaneous discussions.

A second example is an exploratory empirical study about effectiveness of requirement engineering in a distributed setting, presented in [15]. Students from different graduate Software Engineering courses played the role of customers or engineers in separate groups, using a previously selected set of groupware tools (audio-conferencing, chat, email). They were restricted to do just four planned audio-conferencing meetings but the use of other technologies was unrestricted. Participants playing the role of software engineers wrote a Software Requirements Specification (SRS) that was analyzed with a set of metrics to assess document quality. Results concluded: (1) students who played the role of software engineers chose the elicitation techniques according to their previous experience; (2) groups producing high quality SRS were those that had only used the synchronous tools and did not need to use asynchronous elicitation methods.

Results in both cases have interesting points to be analyzed from a cognitive perspective. Why did students who wrote the highest quality SRS documents not need to use email to communicate with their customers? May be their personal characteristics were suitable for synchronous tools, while those who needed email interaction needed more time to think and prepare questions or answers so that synchronous communication was not the best way of communication for them; or may be they needed "to see" the words written, and audio-conferencing was not appropriate. In addition, stakeholders that mentioned the need to use a whiteboard to draw during discussions in [6], indicates a strong preference for visual subcategory.

Related work concerning selection of elicitation techniques is found, for instance, in [4], where the author focuses on the fact that elicitation techniques are chosen without having a valid guide to select the best one. The author comes to this conclusion after presenting a survey of works about theories, empirical analysis and comparisons between different elicitation techniques. Finally, he does not propose a definitive solution to the problem, but he remarks some important aspects for us, like how to measure the adequacy of elicitation techniques, how metrics are defined, etc.

Relative to the use of cognitive styles in software engineering there is a work about a mechanism for software inspection teams construction [18]. The paper describes an experiment that aims to prove that heterogeneous software inspection teams have better performance than homogeneous ones. Heterogeneity concept is analyzed according to the cognitive style of participants. To classify the possible participants they use the MBTI (Myers-Briggs Type Indicator) which is a tool similar to the F-S model we have presented so far. A difference between both approaches is that they focus on choosing people, according to their personal characteristics, to form the best group of inspectors, while our approach looks for the best technologies and methodologies according to personal preferences of a given group of stakeholders.

6 Conclusions

Global or multi-site software development seems to be more common every day. Organisations have adopted a decentralised, team-based, distributed structure where members communicate through groupware tools.

As the quality of requirements depends on the selection of appropriate technology, we think that by improving the communication during the elicitation process, the elicitation process itself will be also improved. Having this in mind, we have proposed a model and its supporting tool to recommend the more suitable elicitation techniques and groupware tools according to the stakeholders' learning preferences.

An aspect that needs further discussion is the possibility of solving conflicts when stakeholders' preferences seem to be opposite. We are working on that restriction. Additionally, we are using this tool in academic and industrial environments, to evaluate its effectiveness in real situations.

References

1. Aranda, G., Cechich, A., Vizcaíno, A., and Castro-Schez, J.J. Using fuzzy sets to analyse personal preferences on groupware tools. In X Congreso Argentino de Ciencias de la Computación, CACIC 2004. San Justo, Argentina, (2004) 549-560.
2. Boland, D. and Fitzgerald, B. Transitioning from a Co-Located to a Globally-Distributed Software Development Team : A Case Study at Analog Devices Inc. In 3rd. International Workshop on Global Software Development. Edinburgh, Scotland, (2004) 4-7.
3. Brooks, F.P., No Silver Bullet: Essence and accidents of Software Engineering. IEEE Computer, 4(20), (1987) 10-19.
4. Carrizo Moreno, D. Selección de Técnicas de Educción de Requisitos: Una Revisión Conjunta de la Ingeniería de Software y la Ingeniería del Conocimiento. In IV Jornadas Iberoamericanas de Ingeniería de Software e Ingeniería del Conocimiento, JIISIC 2004. Madrid, Spain, (2004) 159-174.
5. Chiew, V. and Wang, Y. From Cognitive Psychology to Cognitive Informatics. In Second IEEE International Conference on Cognitive Informatics, ICCI'03. London, UK, (2003) 114-120.
6. Damian, D. and Zowghi, D. The impact of stakeholders geographical distribution on managing requirements in a multi-site organization. In IEEE Joint International Conference on Requirements Engineering, RE'02. Essen, Germany, (2002) 319-328.
7. Davis, A., Software Requirements: Objects, Functions and States. Upper Saddle River. New Jersey: Prentice Hall, (1993).
8. Felder, R., Matters of Styles. ASEE Prism, 6(4), (1996) 18-23.
9. Felder, R. and Silverman, L., Learning and Teaching Styles in Engineering Education. Engineering Education, 78(7), (1988) 674-681.
10. Geib, M., Braun, C., Kolbe, L., and Brenner, W. Measuring the Utilization of Collaboration Technology for Knowledge Development and Exchange in Virtual Communities. In 37th Hawaii International Conference on System Sciences, HICSS-38. Big Island, Hawaii, (2004) 1-10.
11. Gralla, P., How Intranets Work. Emeryville, California: Ziff-Davis Press, (1996).
12. Grudin, J., Computer-Supported Cooperative Work: History and Focus. IEEE Computer, 27(5), (1994) 19-26.
13. Herlea, D. and Greenberg, S. Using a Groupware Space for Distributed Requirements Engineering. In 7th IEEE Int'l Workshop on Coordinating Distributed Software Development Projects. Stanford, California, USA, (1998) 57-62.
14. Johansen, R., Groupware: Computer Support for Business Teams, ed. Ed. T.F. Press. New York and London, (1988).
15. Lloyd, W., Rosson, M.B., and Arthur, J. Effectiveness of Elicitation Techniques in Distributed Requirements Engineering. In 10th Anniversary IEEE Joint International Conference on Requirements Engineering, RE'02. Essen, Germany, (2002) 311-318.
16. Loucopoulos, P. and Karakostas, V., System Requirements Engineering. International series in Software Engineering, ed. Mc Graw-Hill. New York, NY, USA, (1995).
17. Martin, A., Martinez, C., Martinez, N., Aranda, G., and Cechich, A. Classifying Groupware Tools to Improve Communication in Geographically Distributed Elicitation. In IX Congreso Argentino de Ciencias de la Computación, CACIC 2003. La Plata, Argentina, (2003) 942-953.
18. Miller, J. and Yin, Z., A Cognitive-Based Mechanism for Constructing Software Inspection Teams. IEEE Transactions on Software Engineering, 30(11), (2004) 811-825.

19. Peters, L. The Virtual Environment: The "How-to" of Studying Collaboration and Performance of Geographically Dispersed Teams. In Twelfth IEEE International Workshops on Enabling Technologies: Infrastructure for Collaborative Enterprises, WETICE'03. Linz, Austria, (2003) 137-141.
20. SWEBOK, Guide to the Software Engineering Body of Knowledge, ed. Software Engineering Coordinating Committee (IEEE-CS y ACM), (2004).
21. Wang, Y. On Cognitive Informatics. In First IEEE International Conference on Cognitive Informatics, ICCI'02. Calgary, Alberta, Canada, (2002) 34-42.
22. Wang, Y. On the Cognitive Informatics Foundations of Software Engineering. In Third IEEE International Conference on Cognitive Informatics, ICCI'04, (2004).

A Fast Method for Determination of Solvent-Exposed Atoms and Its Possible Applications for Implicit Solvent Models

Anna Shumilina

Institute of Mathematics, Technical University of Berlin,
Strasse des 17. Juni 136, 10623 Berlin, Germany

Abstract. A new approach to identify solvent-exposed atoms in a protein and determine approximate locations of contacted water molecules is presented. The idea of the method is to generate for each atom of the protein molecule a surface grid consisting of twelve uniformly distributed points. These points describe potential locations of water molecules closely packed around an isolated atom. The grid surface coordinates are updated before each energy evaluation step by random rotation of the grid around the atom center. Each grid point is then checked with regard to its accessibility by water and the corresponding hydration status is assigned to it. Thus the information about the hydration degree of each protein atom and the positions of surrounding water molecules becomes available for energy evaluation procedure.

Possible applications of the described method for implicit solvent models accounting for hydrophobic effect, electrical and van der Waals constituents of free energy of solvation, media-dependent electrical permittivity as well as hydrogen bonding with water are outlined.

1 Introduction

Interactions with environment, in particular, with surrounding solvent, essentially determine energetically preferable conformations of macromolecules, in particular, proteins. Typically water plays the role of solvent. The polarity of its molecules and their propensity to form hydrogen bonds conditions the features of these interactions. Hydrogen bonds are directional, therefore apolar groups tend to avoid contact with water. Their relative inability to form hydrogen bonds forces water molecules to reorient, causing an entropy decrease [1]. Polar groups, on the other hand, win from the contact with water in electrical free energy. Consequently, the solute molecule tends to accept a conformation that minimizes the surface area of apolar groups, often living charged groups exposed. Hydrophobic forces, in particular, play a leading role on the early stages of protein folding, when the molten globule is created.

Another significant effect arising from interactions with water is dielectric screening of charged atoms. Water has much higher electrical permittivity then that of the hydrophobic core of proteins [1–3]. Hence, in the case of an implicit solvent representation for an estimation of the strength of electrostatic interactions it is important to know which atoms are located at the surface and which are buried into the solute interior.

As proposed first by Lee and Richards [4], a solvent accessible surface is conventionally defined as the surface that is described by the center of a probe sphere of a radius of 1.4 Å, approximating a water molecule, rolled over the van der Waals envelope of the solute. A number of methods for calculation of solvent accessible surface area was already introduced.

Lee and Richards [4] estimated static accessibility by sectioning the molecule structure by a set of parallel planes and summation the approximate areas of segments confined between planes. Surface areas of segments they calculated basing on the arc lengths of atom sections after elimination of the intersecting parts.

Shrake and Rupley [5] used a set of 92 fixed test points that were nearly uniformly distributed over the solvated van der Waals sphere to determine the exposure of an atom. Points were checked for an occlusion by test atoms by comparison of the ratio of the solvated sphere radius of the test atom to the distance from the center of the test atom to the test point.

Connolly [6] utilized the definition of molecular surface, introduced by Richards [7] as a part of the van der Waals surface accessible to a probe sphere, and presented a computer algorithm for its analytical calculation by subdividing the surface into a set of pieces of spheres and tori.

Richmond [8] proposed an analytical approach for exact calculation of solvent accessible surface area (SASA) providing an expression for the surface area exterior to an arbitrary number of overlapping spheres.

Futamura et al. [9] presented a Monte Carlo algorithm for computing the SASA and corresponding error bounds. They also suggested sequential algorithms and their parallelizations to reduce computational time for spherical intersection checking. These algorithms can be also used with other methods for SASA computation.

Talking about protein folding simulations one should not forget, however, about computational costs. Despite the constant improvement of the methods, exact computations of SASAs as well as their good approximation is computationally very demanding task due to a large number of atoms to be processed.

Wodak and Janin [10] proposed probabilistic analytical expression for a fast approximation to the accessible surface area and its partial derivatives relative to the distances between atoms. Their approach is based on the assumption of randomly distributed atoms which are not allowed to penetrate each other. Clearly, this condition does not hold for covalently bound atoms. Hasel et al. [11] and further Cavallo et al. [12] introduced some modifications of this method, which imply separate treatment of covalently bound and non-bound atoms. These procedures involve some adjustable parameters optimized on a set of specifically chosen molecules. Although the predictions of SASAs given by probabilistic approach were reported to be in a reasonably good agreement with SASAs calculated using geometrical algorithms, it remains to be questionable whether one can expect reliable results in folding simulations of complex proteins.

The above-mentioned approaches were developed on the basis of the observation of a linear correlation between solvation energies and SASAs. However, since the number of water molecules that can directly contact an atom is discrete, it can be of advantage to estimate this number instead of more exact calculation or estimation of the solvent accessible surface area. Thus, it is not unreasonable to assume that an effect of the interaction of one water molecule with a cavity on solute surface remains essentially the same after doubling the cavity SASA, if the size of the cavity is still not sufficient to accommodate more molecules. The idea to estimate the number of contacted water molecules for each atom is laid as a basis for the below described method.

2 Description of the Method

2.1 Geometrical Motivation

One can place around a sphere S of radius R at maximum 12 spheres of the same radius. This becomes obvious if we consider an icosahedron, inscribed in a sphere with the radius $2R$ around the center of the sphere S. The length of the ridge of this icosahedron is approximately 1.05 times larger than the distance from a vertex to the icosahedron center, equal to $2R$, which means that the small spheres of radius R placed at the vertices connected by a ridge almost touch each other (Fig. 1).

The conventional probe radius r_p approximately corresponds to an overage van der Waals radius of atoms constituting typical organic molecules, in particular, proteins. Therefore, taking into account an approximate nature of the definition of solvent accessible surface, we assume that 12 water molecules at maximum can be placed around each atom, neglecting for simplicity the difference of van der Waals radii.

Let r_i be the van der Waals radius of atom i. Let us further denote $r_{s_i} = r_i + r_p$ as a hydration radius of the atom i, and the sphere of radius r_{s_i} around an atom center as a hydration sphere of the atom i. To determine, which of 12 potential positions of water molecules are in fact accessible, a grid of 12 uniformly distributed points is generated on the surface of the atom hydration sphere. The location of each grid point is described by its surface coordinates (ϕ, θ) defined as in Fig. 2.

2.2 Grid Generation

Using the fact that the grid points correspond to the vertices of the inscribed icosahedron (Fig. 3), we initialize the grid on the stage of creation by the following easily derivable coordinates: $(0, 0)$, $(0, \theta_a)$, $\left(\frac{2\pi}{5}, \theta_a\right)$, $\left(-\frac{2\pi}{5}, \theta_a\right)$, $\left(\frac{4\pi}{5}, \theta_a\right)$, $\left(-\frac{4\pi}{5}, \theta_a\right)$, $\left(\frac{\pi}{5}, \pi - \theta_a\right)$, $\left(-\frac{\pi}{5}, \pi - \theta_a\right)$, $\left(\frac{3\pi}{5}, \pi - \theta_a\right)$, $\left(-\frac{3\pi}{5}, \pi - \theta_a\right)$, $(\pi, \pi - \theta_a)$, $(0, \pi)$, where $\theta_a = \arccos\left(1 - \frac{4}{5+\sqrt{5}}\right)$.

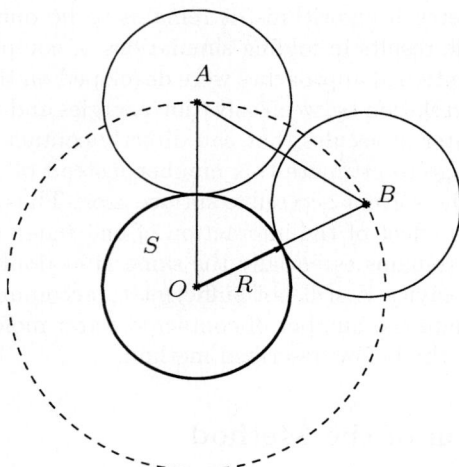

Fig. 1. A cross section through the center of the sphere S of radius R and two vertices A and B of the discussed icosahedron

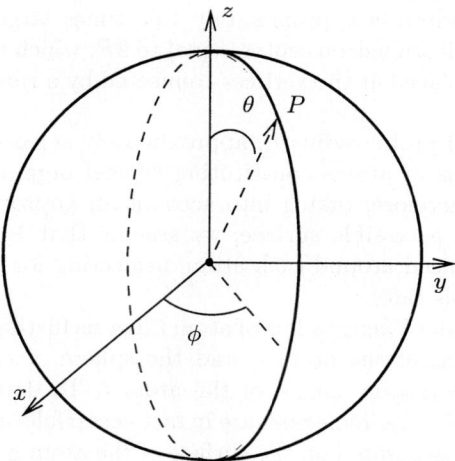

Fig. 2. Surface coordinates (ϕ, θ) of location of the point P on the surface of the atom sphere and their relation to the cartesian coordinates

At the beginning of each energy evaluation procedure the grid is randomly rotated around the atom center. In this way all the surface falls into consideration with an equal probability and the effect of discretization is somewhat smoothed due to averaging in course of simulation. In order to compute the

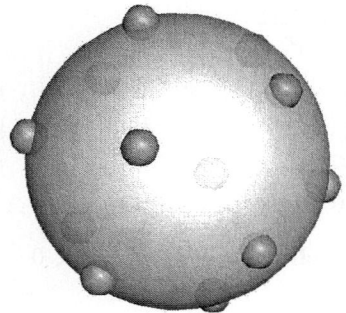

Fig. 3. Atom solvation grid

new point positions, three rotation angles θ_r, ψ_r, and ϕ_r are generated as uniformly distributed random numbers from the intervals $[0, \pi]$, $[0, \pi]$ and $[0, 2\pi]$ respectively. The new coordinates of the k-th vertex are calculated according to (1),(2):

$$\theta_k = \arccos\left(s_1 \sin\theta_k \sin(\psi_r - \phi_k) + c_1 \cos\theta_k\right), \quad (1)$$

$$\phi_k = \begin{cases} 0 & \text{if } \theta_k = 0, \\ \arctan\frac{f_1}{f_2} + \frac{\pi}{2}\left(1 - \text{sign}\frac{f_2}{\sin\theta_k}\right) & \text{otherwise}, \end{cases} \quad (2)$$

where $c_1, c_2, s_1, s_2, f_1, f_2$ are intermediate variables that can be computed using (3)-(5).

$$c_1 = \cos\theta_r, \quad c_2 = \cos\phi_r, \quad s_1 = \sin\theta_r, \quad s_2 = \sin\phi_r, \quad (3)$$

$$f_1 = \sin\theta_k \left(c_2 \cos(\phi_k - \psi_r) + c_1 s_2 \sin(\phi_k - \psi_r)\right) + s_1 s_2 \cos\theta_k, \quad (4)$$

$$f_2 = \sin\theta_k \left(-s_2 \cos(\phi_k - \psi_r) + c_1 c_2 \sin(\phi_k - \psi_r)\right) + s_1 c_2 \cos\theta_k. \quad (5)$$

The new coordinates are then used as the positions of the solvation grid points.

2.3 Accessibility Check

Initially each grid point obtains a hydration status 1, i.e. considered to be exposed to solvent. Then for each pair of atoms with intersecting hydration spheres the hydration status of the points laying within the intersection is set to 0, denoting that these positions are not accessible for water molecules.

To check which grid points of the atom i lay within the intersection, we first describe the intersection cone through the surface coordinates $(\phi_{I_i}, \theta_{I_i})$ of the point I_i and the angle ψ_{I_i} (Fig. 4). The point I_i lays on the intersection of the axis connecting atom centers O_i and O_j with the hydration sphere of the atom i. ψ_{I_i} is the angle between $O_i I_i$ and the line, connecting O_i with a point on the intersection circle. The coordinates $(\phi_{I_i}, \theta_{I_i})$ and the angle ψ_{I_i} are calculated according to (6)-(8):

$$\theta_{I_i} = \arccos \frac{z_{r_{ij}}}{|r_{ij}|}, \tag{6}$$

$$\phi_{I_i} = \begin{cases} 0, & \text{if } x_{r_{ij}} = 0,\ y_{r_{ij}} = 0, \\ \frac{\pi}{2}, & \text{if } x_{r_{ij}} = 0,\ y_{r_{ij}} > 0, \\ -\frac{\pi}{2}, & \text{if } x_{r_{ij}} = 0,\ y_{r_{ij}} < 0, \\ \arctan \frac{y_{r_{ij}}}{x_{r_{ij}}}, & \text{if } x_{r_{ij}} > 0, \\ \pi + \arctan \frac{y_{r_{ij}}}{x_{r_{ij}}}, & \text{if } x_{r_{ij}} < 0. \end{cases} \tag{7}$$

$$\psi_{I_i} = \arccos \left(\frac{\frac{r_{s_i}^2 - r_{s_j}^2}{|r_{ij}|} + |r_{ij}|}{2 r_{s_i}} \right). \tag{8}$$

where $x_{r_{ij}}, y_{r_{ij}}$ and $z_{r_{ij}}$ denote the first, the second and the third component of cartesian coordinates of the vector r_{ij} pointing from the center of the atom i to the center of the atom j.

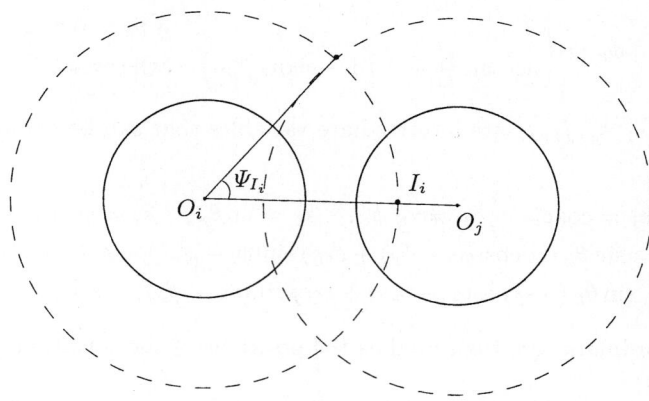

Fig. 4. Two atoms with intersecting hydration spheres

Then for each grid point $P_k = (\phi_k, \theta_k)$ of the atom i the following test is performed: if the unequality (9) is valid, then the point lays in intersection, and reciprocally for the grid points of the atom j.

$$\sin \theta_k \sin \theta_{I_i} \cos(\phi_k - \phi_{I_i}) + \cos \theta_k \cos \theta_{I_i} > \cos \psi_{I_i}. \tag{9}$$

After all atoms pairs in the list[1] are checked, only those grid points that correspond to accessible positions of water molecules have the hydration status 1. The outcome of this procedure can be then used for a computation of the atom hydration degree and subsequent estimation of the solvation energy.

3 Discussion

A new method for determination of solvent-exposed atoms and their hydration degree is proposed. It is based on an estimation of the average number of directly contacted water molecules surrounding solute.

The algorithm was implemented in C++ as a part of a new simulation software, which is currently under development. Figure 5 shows the result of an application of the program for minimization of a small peptide. Atoms are represented by their van der Waals spheres. Small dark points on the atom spheres denote potential places of contact with water molecules. A virtual position of the center of water molecule can be obtained by extending the segment connecting the atom center with a grid point by a distance equal to the probe radius. Light points lie in cavities with no direct contact to solvent.

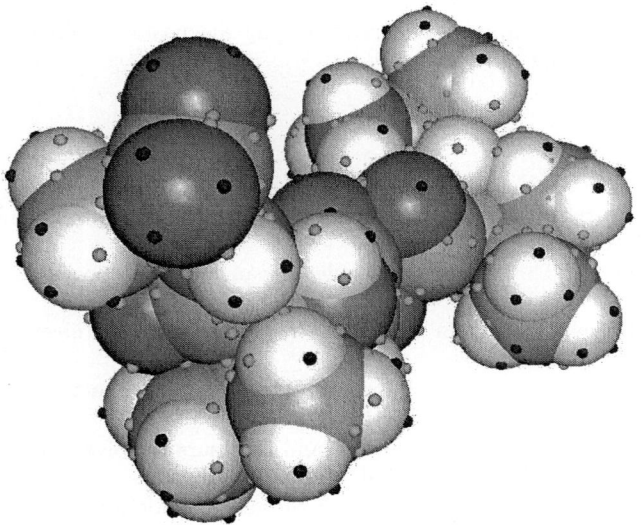

Fig. 5. A small peptide with its solvation grid. Hydrated points are dark-colored and the grid points laying in cavities are light-colored

An implicit solvent model is developed that utilizes the number of hydrated points to evaluate the strength of hydrophobic effect, van der Waals interaction with solvent and electrical energy of solvation depending on atom types.

Furthermore, the information about possible positions of the surrounding water molecules allowed for modeling hydrogen bonding with water implicitely.

---

[1] The list does not have to contain all the macromolecule atoms. One can try to eliminate unnecessary checks, see, for example, [9].

It is incorporated into the model by usage of surface coordinates of hydrated points for an estimation of the propensity of polar groups to build hydrogen bonds with water in a current macromolecule conformation.

The computed hydration degree of the atoms can be used to account for media-dependent electrical permittivity. In the current model, higher electrical permittivity is used for evaluation of the electrostatic interactions between hydrated atoms.

The complete model and its implementation as a simulation software for protein folding still needs some further refinement and determination of the optimal parameter sets. The results of this development will be reported elsewhere.

Acknowledgment

I would like to thank Prof. Dr. A. Unterreiter for an opportunity to work on this topic, organizational support and encouraging discussions, which helped to set a basis for this model.

References

1. A. V. Finkelstein, O. B. Ptizin: Protein Physics. Russian edition: University Book House, Moscow (2002)
2. K. Hamaguchi: The Protein Molecule. Japan Scientific Societies Press, Tokyo (1992)
3. M. J. E. Sternberg: Protein Structure Prediction. Oxford University Press, New York (1996)
4. B. Lee, F. M. Richards: The Interpretation of Protein Structures: Estimation of Static Stability. J. Mol. Biol. **55** (1971) 379-400
5. A. Shrake, J. A. Rupley: Environment and Exposure to Solvent of Protein Atoms. Lysozyme and Insulin. J. Mol. Biol. **79** (1973) 351-371
6. M. L. Connolly: Analytical Molecular Surface Calculation. J. Appl. Cryst. **16** (1983) 548-558
7. F. M. Richards: Areas, Volumes, Packing and Protein Structure. Ann. Rev. Biophys. Bioeng. **6** (1977) 151-176
8. T. J. Richmond: Solvent Accessible Surface Area and Excluded Volume in Proteins. J. Mol. Biol. **178** (1984) 63-89
9. N. Futamura, S. Aluru, D. Ranjan, B. Hariharan: Efficient Parallel Algorithms for Solvent Accessible Surface Area of Proteins. IEEE Trans. on Paral. and Distr. Sys. **13** (2002) 544-555
10. S. J. Wodak, J. Janin: Analytical approximation to the accessible surface area of proteins. Proc. Natl. Acad. Sci. USA **77** (1980) 1736-1740
11. W. Hasel, T. F. Hendrickson, W. C. Still: A rapid approximation to the solvent accessible surface areas of atoms. Tetrahedron Comput. Methodol. **1** (1988) 103-116
12. L. Cavallo, J. Kleinjung, F. Fraternali: POPS: a fast algorithm for solvent accessible surface areas at atomic and residue level. Nucleic Acids Res. **31** (2003) 3364-3366

Thermal Rate Coefficients for the $N + N_2$ Reaction: Quasiclassical, Semiclassical and Quantum Calculations

Noelia Faginas Lago[1,2], Antonio Laganà[1],
Ernesto Garcia[2], and X. Gimenez[3]

[1] Dipartimento di Chimica, Universitá di Perugia,
Via Elce di Sotto, 8, 06123 Perugia, Italy
{noelia, lag}@dyn.unipg.it
[2] Departamento de Quimica Fisica,
Universidad del Pais Vasco, Vitoria, Spain
qfpgapae@vc.ehu.es
[3] Departamento de Quimica Fisica,
Universidad de Barcelona, Barcelona, Spain
xgimenez@ub.edu

Abstract. The thermal rate constants for the $N + N_2$ gas-phase reaction have been calculated using the quasiclassical semiclassical and reduced dimensionality quantum methods. The improvement of semiclassical over quasiclassical calculations (with respect to quantum ones) is significant.

1 Introduction

Recently there has been renewed interest in the study of the

$$N(^4S_u) + N_2(^1\Sigma_g^+, \nu) \rightarrow N(^4S_u) + N_2(^1\Sigma_g^+, \nu') \tag{1}$$

reaction because of intensive activity in modeling the reentry of spacecrafts landing on solar planets [1, 2, 3, 4, 5] whose atmosphere contains a large amount of Nitrogen. In fact, to perform a realistic modeling of spacecraft reentry extended matrices of detailed rate coefficient values related to the involved processes are needed. In spite of that, not only little experimental information [6] is available on the dependence from the temperature of the rate coefficient of reaction (1), but also theoretical information is quite scarce. In particular the only potential energy surface (PES) avalaible from the literature is that of ref. [7] that is a LEPS-type surface having a collinear transition state 36 kcal/mol higher in energy than the asymptotes. More recently, a PES made of an exponential factor multiplied by an expansion in terms of Legendre polynomials has been fitted to 3326 ab initio points calculated using a coupled cluster singles and doubles method with perturbation-correction of triples has been worked out [8]. On this PES (WSHDSP) three dimensional zero total angular momentum calculations of the reactive probabilities have been performed. Unfortunately, we have been unable to obtain a copy of the WSHDSP PES. For this reason we are carrying out afresh extended ab initio calculations and we are planning a new fit of the calculated values. In the meantime, our simulations to develop new computational methods and to improve our understanding of reaction (1) are still based on the LEPS surface. After all, most of the previous work has been already performed using the LEPS surface [9] (including the recent zero total angular momentum quantum calculations of reactive and non reactive probabilities [1]). The specific goal of this paper is the comparison of the values of the rate coefficients of reaction (1) obtained using different approaches in order to establish what level of confidence can be attributed to the usual simulations based on quasiclassical techniques.

2 Theoretical and Computational Approaches

Calculations were performed using quasiclassical, semiclassical and quantum approaches. In this section related formulations of the calculated quantities are outlined.

2.1 The Quasiclassical Calculations

For reaction (1) the Quasiclassical (QCT) estimates of the vibrational state (ν) to state (ν') rate coefficient $k_{\nu,\nu'}(T)$ at temperature T are obtained using the relationship

$$k_{\nu,\nu'}(T) = \frac{\sum_j g(2j+1)e^{-\varepsilon_j/kT}}{(k_B^3 T^3 \pi \mu/8)^{1/2} Q_R} \int_0^\infty dE_{tr} E_{tr} e^{E_{tr}/kT} \sigma_{\nu j,\nu'}(E_{tr}) \quad (2)$$

where g is the degeneracy factor (that is 2 for even and 1 for odd rotational levels), μ is the reduced mass of the reactant atom-diatom system, k_B is the Boltzmann's constant, Q_R is the diatom rotational partition function, E_{tr} is the traslational energy, ε_j is the energy of the jth rotational state, and $\sigma_{\nu j,\nu'}$ is the state to rotationally averaged vibrational state cross section derived from the fully detailed state (νj) to state $(\nu' j')$ reactive cross section $\sigma_{\nu j,\nu' j'}$ summed over the product rotational states j'. The quasiclassical state to state cross section is defined as a five dimensional integral. When using a Monte Carlo technique the integral is usually approximated as

$$\sigma_{\nu j,\nu' j'} = \frac{\pi b_{max}^2}{M} \sum_{i=1}^M f_{\nu j,\nu' j'}(\xi_1, \xi_2, \xi_3, \xi_4, \xi_5) \quad (3)$$

where M is the number of events (trajectories starting with a given quintet of initial values of the parameters ξ) considered for the Monte Carlo integration, b_{max} is the maximun value of the impact parameter leading to reactive encounters and $f_{\nu j, \nu' j'}(\xi_1, \xi_2, \xi_3, \xi_4, \xi_5)$ is a Boolean function whose value is 1 only when, after integrating the related trajectory, the final outcome can be assigned to the $\nu' j'$ quantum state of the products. To calculate the state-to-state rate coefficients values reported in this paper, a prototype grid set up was used [10] in order to be able to run concurrently batches of trajectories of the order of 10^6 using the computational machinery of the molecular dynamics simulator of ref. [11].

2.2 The Quantum Reduced Dimensionality

As already mentioned, full quantum calculations of the state-to-state rate coefficients are being calculated using the time dependent approach [1]. In spite of the experimental use of the above mentioned prototype computing grid infrastructure, the calculations are taking very long because of the huge demand of computing time from individual fixed total angular momentum calculations. For this reason the comparison is made here with the results of the more agile reduced dimensionality Reactive Infinite Order Sudden (RIOS) quantum code [12]. In this case, the degeneracy averaged detailed cross section $\sigma_{\nu j, \nu'}$ needed to evaluate rate coefficients of equation 2 is approximated in terms of the ground rotational state cross section $\sigma_{\nu j=0, \nu'}$ values using the relationship $\sigma_{\nu j, \nu'}(E_{tr}) = \sigma_{\nu j=0, \nu'}(E_{tr} - \varepsilon_{\nu j})$ where $\varepsilon_{\nu j}$ is the energy of the νjth vibrational state. The ground rotational state cross section is calculated from the RIOS **S** matrix elements $S_{\nu l, \nu'}(\Theta; E_{tr})$ using the relationsship

$$\sigma_{\nu j, \nu'}(E_{tr}) = \frac{\pi}{k_\nu^2} \sum_l (2l+1) \int_{-1}^{1} |S_{\nu l, \nu'}(\Theta; E_{tr})|^2 d\cos\Theta \quad (4)$$

where $k_\nu^2 = 2\mu(E - \varepsilon_\nu)$, E is the total energy and ε_ν is the energy of the vibrational state ν. The **S** matrix is evaluated by integrating the coupled differential equations obtained from the Infinite Order Sudden formulation of the Schrödinger equation for atom diatom systems [12]. Also the computational machinery of the RIOS approach has been structured to exploit the advantage of distributing the fixed Θ and E_{tr} integration of the scattering matrix equations to concurrently run on the grid [13].

2.3 The Semiclassical Calculation

Semiclassical (SC) calculations were carried out using the Initial Value Representation (IVR) method of W. H. Miller and coworkers [14]. The method reconducts the evaluation of the vibrational state to state rate coefficients to the calculation of the flux-flux correlation function.

$$k_{\nu\nu'}(T) = \frac{1}{\hbar Q_{tr}(T) Q_{int}(T)} \int_0^\infty dt\, C_{ff}(t) \quad (5)$$

In eq. 5 $Q_{tr}(T)$ and $Q_{int}(T)$ are, the partition functions for the relative translational motion of the atom diatom system and the internal motion of the target molecule, respectively. $C_{ff}(t)$ is the flux-flux correlation function with t being the time variable. For computational purposes the integral of eq. 5 is factored as

$$\int_0^\infty C_{ff}(t)dt = C_{ff}(0) \int_0^\infty dt R_{ff}(t) \qquad (6)$$

with $R_{ff}(t) = C_{ff}(t)/C_{ff}(0)$. This factoring allows the calculation to be split into that of the "static factor" $C_{ff}(0)$ (for which use of imaginary-time path integral techniques is made) and that of the "dynamical" factor $R_{ff}(t)$ (for which a combined use of the semiclassical initial value representation and path integrals technique is made). Also the computational machinery of this SC-IVR approach has been structured to exploit the advantage of distributing the calculations concurrently on the grid.

3 The Theoretical Estimate of the $N + N_2$ Rate Coefficients

The three methods were applied to the

$$N_a + N_b N_c \rightarrow \begin{cases} N_a N_b + N_c \\ N_a N_c + N_b \end{cases}$$

(exchange) reactive processes to calculate the above mentioned thermal rate coefficients.

3.1 Vibrational State to State Rate Coefficents

A first extended campaign of calculations was carried out using the quasiclassical method at T=500 K and T=1000 K. The values of $k_{\nu\nu'}(T)$ calculated at T=500 K and T=1000 K are given in the left hand side panel of Table 1 and 2, respectively.

In Table 1 the column for $\nu \leq 10$ is not shown since related values are smaller than 10^{-15}. At $\nu = 15$ the vibrational state to state reactivity begins to be appreciable and the reactive process increasingly populates higher vibrational states of the products to reach a maximum at $\nu' = 15$. At $\nu = 20$ the vibrational state to state reactivity is even larger are the maximum of the (product vibrational) distribution is located at $\nu' = 19$. This confirms the tendency already singled out for the $H + H_2$ reaction and for which the $\nu \rightarrow \nu - 1$ (one less vibrational number) reactive transitions are the most efficient ones. Reactive vibrational deexcitations ($\nu' < \nu$) become even more efficient when temperature is increased to 1000 K. In this case, however, (see Table 2) the

Table 1. Comparison of QCT and quantum RIOS state to state rate coefficients at 500 K

$\nu'\backslash\nu$	QCT T=500 15	QCT T=500 20	RIOS T=500 15	RIOS T=500 20
0	.235(-14)	.115(-13)	.458(-15)	.250(-14)
1	.101(-14)	.632(-14)	.211(-15)	.283(-14)
2	.119(-14)	.734(-14)	.314(-15)	.188(-14)
3	.920(-15)	.475(-14)	.266(-15)	.210(-14)
4	.964(-15)	.616(-14)	.354(-15)	.248(-14)
5	.642(-15)	.140(-13)	.379(-15)	.260(-14)
6	.236(-14)	.120(-13)	.462(-15)	.310(-14)
7	.624(-15)	.103(-13)	.520(-15)	.357(-14)
8	.246(-14)	.119(-13)	.675(-15)	.409(-14)
9	..232(-14)	.111(-13)	.821(-15)	.506(-14)
10	.765(-15)	.128(-13)	.103(-14)	.582(-14)
11	.359(-14)	.193(-13)	.132(-14)	.722(-14)
12	.610(-14)	.233(-13)	.163(-14)	.819(-14)
13	.417(-14)	.253(-13)	.234(-14)	.109(-13)
14	.586(-14)	.278(-13)	.340(-14)	.125(-13)
15	.655(-14)	.282(-13)	.382(-14)	.157(-13)
16		.373(-13)		.204(-13)
17		.443(-13)		.249(-13)
18		.401(-13)		.333(-13)
19		.718(-13)		.449(-13)
20		.517(-13)		.455(-13)

$\nu \to \nu - 1$ rule is not fulfilled showing that the assumption popular among scientists modeling V-V (vibration to vibration) and T-V (translation to vibration) energy transfer processes is not valid, in general, and this is a weakness of models based on a single vibrational quantum cascade. At the same time, the reactivity of the system (for corresponding vibrational states of the reactant molecule) increases more than one order of magnitude. Accordingly, several vibrational quantums are released during a reactive collision and reactivity is quite high even at $\nu = 10$.

For comparison, also RIOS results are given in the right hand side of both Table 1 and 2 (in this case the values smaller than 10^{-15} are not shown). The first clear evidence brought by the RIOS results is that vibrational state to state reactive rate coefficients calculated at $\nu = 10$ and T=500 K are also so small that they are not shown. At the same temperature (T=500 K) $\nu = 15$ and $\nu = 20$ vibrational state to state rate coefficients calculated using the RIOS method have about the same magnitude and trend as those calculated using the QCT method (though the former are most often smaller than the latter). Yet, there is, indeed, at least one marked difference. This is concerned with the fact that

Table 2. Comparison of QCT and quantum RIOS state to state rate coefficients at 1000 K

$\nu'\backslash\nu$	QCT T=1000			RIOS T=1000		
	10	15	20	10	15	20
0	.423(-14)	.508(-13)	.191(-12)	.141(-14)	.199(-13)	.416(-13)
1	.271(-14)	.507(-13)	.110(-12)	.326(-14)	.120(-13)	.390(-13)
2	.103(-13)	.390(-13)	.114(-12)	.344(-14)	.148(-13)	.308(-13)
3	.135(-13)	.364(-13)	.724(-13)	.324(-14)	.143(-13)	.342(-13)
4	.542(-14)	.293(-13)	.738(-13)	.393(-14)	.173(-13)	.383(-13)
5	.123(-13)	.591(-13)	.109(-12)	.484(-14)	.191(-13)	.418(-13)
6	.193(-13)	.603(-13)	.138(-12)	.576(-14)	.224(-13)	.482(-13)
7	.782(-14)	.683(-13)	.165(-12)	.757(-14)	.261(-13)	.549(-13)
8	.201(-13)	.708(-13)	.171(-12)	.104(-13)	.316(-13)	.629(-13)
9	.365(-13)	.727(-13)	.142(-12)	.165(-13)	.381(-13)	.757(-13)
10	.427(-13)	.124(-12)	.207(-12)	.242(-13)	.467(-13)	.876(-13)
11	.506(-14)	.146(-12)	.283(-12)	.159(-14)	.586(-13)	.101(-12)
12		.104(-12)	.287(-12)	.100(-15)	.733(-13)	.119(-12)
13		.212(-12)	.321(-12)		.992(-13)	.148(-12)
14		.298(-12)	.303(-12)		.141(-12)	.173(-12)
15		.272(-12)	.460(-12)		.181(-12)	.213(-12)
16		.305(-13)	.662(-12)		.143(-13)	.266(-12)
17		.325(-14)	.609(-12)		.106(-14)	.329(-12)
18			.870(-12)			.429(-12)
19			.112(-11)			.582(-12)
21			.129(-12)			.671(-13)
22			.131(-13)			.606(-14)
23						.585(-15)

quantum results show a clear vibrational adiabaticity since $\nu = 10, 15$ and 20 preferentially populate $\nu' = 10, 15$ and 20, respectively. Similar conclusions can be drawn for values calculated at T=1000 K.

3.2 Thermal Rate Coefficients and Semiclassical Results

On the ground of the analysis previously made, we carried out a calculation of the value of the fully thermalized rate coefficient $k(T)$ estimated by making a weighed sum of the various $k_{\nu\nu'}(T)$ populated at the considered temperature. Our expectation was to obtain almost the same results from both QCT and RIOS approaches. Figure 1 shows indeed that this is not the case, since the QCT $k(T)$ values are invariably larger than RIOS ones. This prompted the question on whether the difference was caused by the lack of quantum effects in QCT calculations or by the lack of coupling among the angular degrees of freedom in the infinite order sudden quantum ones. For this reason we decided to carry out a crossed comparison of the QCT and quantum RIOS vibrational state to state

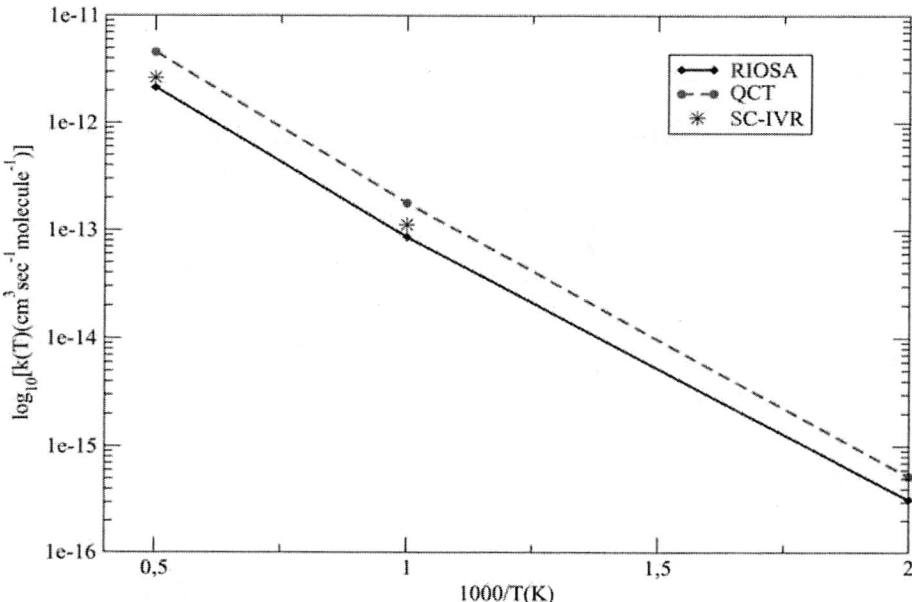

Fig. 1. QCT (red dashed line) and RIOS (solid black line) thermal reactive rate coefficients $k(T)$ plotted as a function of $1000/T$. For comparison also SC-IVR values are plotted as stars

reactive rate coefficients with the semiclassical ones. To this end we plotted in Figures 2 and 3 the value of the vibrational state to state rate coefficients as a function of the difference between the final and initial vibrational numbers $\nu' - \nu = n$.

As apparent from both Figures 2 and 3 a surprising feature of QCT results is the highly structured shape of the obtained distributions (that is likely to be due to some intrinsic weakness of the related computational procedure such as the arbitrary discretization of the asymptotic internal energies, the small number of events at the reactive threshold, etc). On the contrary, the RIOS results, contrary to what usually one expects from quantum results which incorporate all those quantum effects (such as resonances, interferences, tuneling) which are likely to generate structures in the probability curves, have a smooth dependence on n. A clear response of the SC-IVR calculations is that semiclassical estimates of the vibrational state to state rate coefficients are always closer to quantum than to quasiclassical results. This indicates that for this strongly collinearly dominated reaction, a RIOS treatment properly deals with the microscopic nature of the reactive process. This indicates also that the semiclassical approach regains (though in an approximate way) most of the quantum features

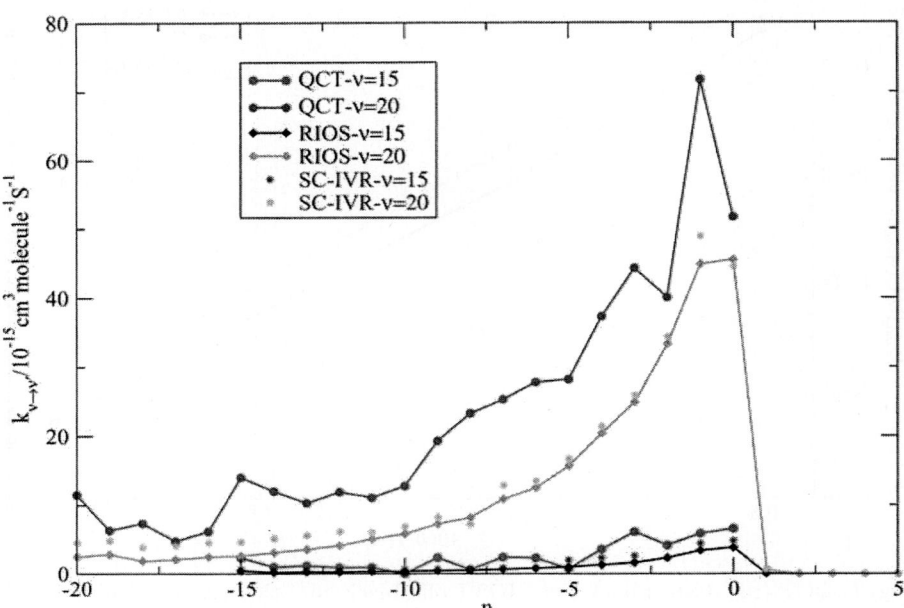

Fig. 2. Vibrational reactive state to state rate coefficients $k_{\nu\nu'}(T)$ calculated at T= 500 K and calculated plotted as a function of $n = \nu' - \nu$ using QCT (circles connected by solid line), RIOS (diamonds connected by a dashed line) and SC-IVR (stars) methods

of the reactive process, and leads to results close to quantum ones. A peculiarity of the semiclassical results is, however, that of being impredictably higher or lower than quantum results (while the QCT ones are always higher) making it difficult to set error boundaries and use the SC-IVR estimates as predictive tools.

4 Conclusions

Extended calculations of vibrational state to state rate coefficients have been performed for the $N + N_2$ reaction to the end of checking the accuracy of a quasiclassical results vis-a-vis the approximate quantum RIOS ones. The systematic deviation of QCT estimates of the thermalized rate coefficients from RIOS ones and the smoother nature of quantum results prompted a crossed comparison with SC-IVR estimates. The semiclassical results were found to be always closer to the quantum ones than quasiclassical. This suggests that the semiclassical treatment captures all the main features of the quantum results in spite of being

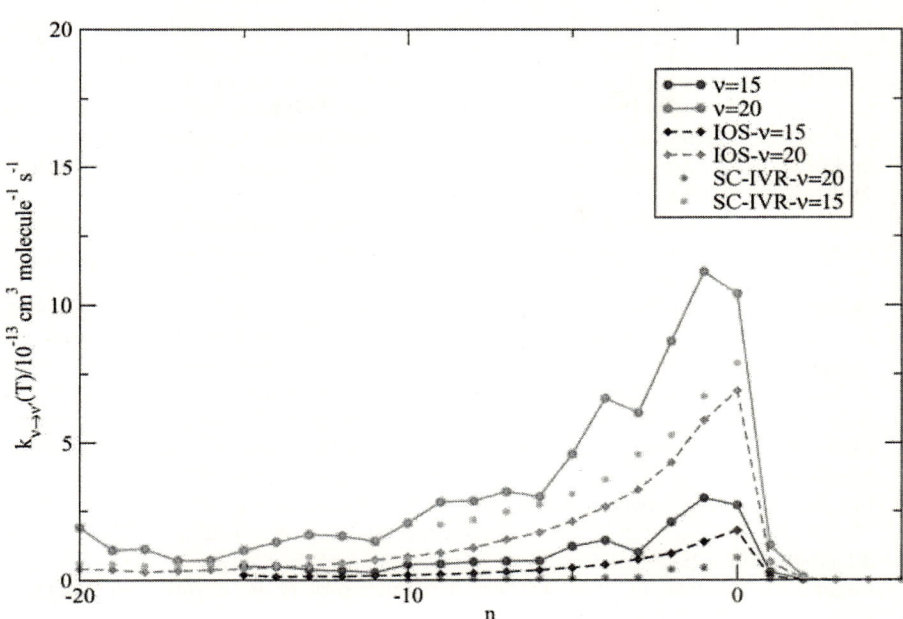

Fig. 3. Vibrational reactive state to state rate coefficients $k_{\nu\nu'}(T)$ calculated at T= 1000 K and calculated plotted as a function of $n = \nu' - \nu$ using QCT (circles connected by solid line), RIOS (diamonds connected by a dashed line) and SC-IVR (stars) methods

unable to provide an indication on whether the deviation from the quantum values is by excess or defect.

Acknowledgments

This work has been supported by the Italian MIUR FIRB Grid.it project (RBNE01KNFP) on High performance Grid Platforms and Tools, and by the MIUR CNR Strategic Project L 499/97-2000 on High performance Distributed Enabling Platforms. Noelia Faginas Lago acknowledges the financial support provided through a postdoctoral fellowship from Basque Government is acknowledged. Thanks are also due to COST in Chemistry and ASI.

References

1. Laganá, A., Pacifici, L., Skouteris, D.: A time dependent study of the nitrogen atom nitrogen molecule reaction. Lecture Notes in Computer Science. **3044** (2004) 357-365.

2. Giordano, D., Maraffa, L.: Proceedings of the AGARD-CP-514 Symposium in Theoretical and Experimental Method in Hypersonic Flows. **26** (1992) 1.
3. (a) Capitelli, M.: Non-equilibrium vibrational kinetics, Ed. Springer-Verlag, Berlin (1986); Capitelli, M., Barsdley, J. N.: Non-equilibrium processes in partially ionized gas. Plenum, New York (1990); (b) Armenise, I., Capitelli, M., Garcia, E., Gorse, C., Laganá, A., Longo, S.: Deactivation dynamics of vibrationally excited nitrogen molecules by nitrogen atoms. Effects on non-equilibrium vibrational distribution and dissociation rates of nitrogen under electrical discharges. Chem. Phys. Letters. **200** (1992) 597-604.
4. (a) Esposito, F., Capitelli, M.: Quasiclassical molecular dynamic calculations of vibrationally and rotationally state selected dissociation cross-section $N+N_2(\nu,j) \rightarrow$ 3N , Chem. Phys. Letters, **302** (1999) 49-54.
5. (a) Balucani, N., Alagia, M., Cartechini, L., Casavecchia, P., Volpi, G. G., Sato, K., Takayanagi, T., Kurosaki, Y.: Cyanomethylene Formation from the Reaction of Excited Nitrogen Atoms with Acetylene: A Crossed Beam and ab Initio Study. J. Am. Chem. Soc. **122(18)** (2000) 4443-4450. (b) Balucani, N., Alagia, M., Cartechini, L., Casavecchia, P., Volpi, G., G.: Observation of Nitrogen-Bearing Organic Molecules from Reactions of Nitrogen Atoms with Hydrocarbons: A Crossed Beam Study of $N(^2D)$ + Ethylene. J. Phys. Chem. A. **104** (2000) 5655-5659.
6. (a) Bar Nan, E., Lifshitz, A.:Kinetics of the Homogeneous Exchange Reaction: $^{14-14}N_2 + ^{15-15}N_2 \rightleftharpoons 2^{14-15}N_2$. Single-Pulse Shock-Tube Studies. J. Chem. Phys. **47** (1969) 2878-2888.
7. Laganá, A., Garcia, E.: Temperature dependence of $N + N_2$ rate coefficients. J. Phys. Chem. **98** (1994) 502-507.
8. Wang, D., Stallcop, J. R., Huo, W. M., Dateo, C. E., Schwenke, D. W., Partridge, H.: Quantal study of the exchange reaction for $N + N_2$ using an ab initio potential energy surface. J. Chem. Phys. **118** (2003) 2186-2189.
9. Laganá, A., Ochoa de Aspuru, G., Garcia, E.: Temperature dependence of quasiclassical and quantum rate coecients for $N + N_2$. Technical report AIAA- (1994) 94-1986.
10. Storchi, L., Manuali, C., Gervasi, O., Vitillaro, G., Laganá, A., Tarantelli, F.: Linear algebra computation benchmarks on a model grid platform. Lecture Notes in Computer Science. **2658** (2003) 297-306.
11. Gervasi, O., Laganá, A.: SIMBEX: a portal for the a priori simulation of crossed beam experiments. Future Generation of Computer systems, **20** (2004) 703-715.
12. Laganá, A., Garcia, E., Gervasi, O.: Improved infinite order sudden cross sections for the $Li + HF$ reaction. J. Chem. Phys. **89** (1988) 7238-7241.
13. Baraglia, V., Ferrini, R., Laforenza, D., Laganá, A. :An optimized task-farm model to integrate reduced dimensionality Schrdinger equations on distributed memory architectures. Future Generation of Computer Systems. **15** (1999) 497-512.
14. Yammamoto, T., Miller, W. H.: Semiclassical calculation of thermal rate constants in full cartesian space: The benchmark reaction $D + H_2 \rightarrow DH + H$. J. Chem. Phys. **118** (2003) 2135-2152.

A Molecular Dynamics Study of Ion Permeability Through Molecular Pores

Leonardo Arteconi and Antonio Laganà

Dipartimento di Chimica, Università di Perugia, Perugia
bodynet@dyn.unipg.it, lag@unipg.it

Abstract. The paper carries out a preliminary analysis of the basic physical and chemical mechanisms of ion mobility in carbon nanotubes taken as models of molecular pores. Then, the selective permeability of monovalent and divalent cations in ionic molecular pores is evaluated by carrying out molecular dynamics calculations for the nanotube model and results are compared with those of a statistical treatment.

1 Introduction

The study of ion transport through molecular pores (hereon called micropores) is essential to several scientific investigations, including biological research since ionic permeability of the micropores determines many properties of cell membranes. For this reason the development of models for ionic transport through micropores has been the concern of neuroscientists in their investigations of the electrical properties of the membrane of excitable cells [1].

The major pathway for ion transport through the membrane of excitable cells is represented by voltage-gated ion channels [2]. These are transmembrane proteins endowed with an aqueous pore that allows the passage of certain ions, while excluding others, a property called selectivity. For example the neuronal action potential, the electrical message propagating throughout the nervous system, involves the sequential opening of two types of voltage-gated ion channels, respectively selective to Na^+ and K^+ ions. It is, therefore, important for physiological studies to single out the structural determinants governing the passage of ions through ion channels, and to identify the mechanism by which they select the various ions.

In this paper we examine the problem of developing possible models for these channels by calculating ionic permeability (σ) through a carbon nanotube (CNT) using molecular dynamics. The emphasis of the paper is on the analysis of the functional representation adopted for the molecular interaction in the dynamical investigation in view of proposing alternative formulations of the potential energy and rationalizing dynamical results. The paper is articulated as follows: in section 2 the carbon nanotube model is described; in section 3 the outcomes of the molecular dynamics investigation of the model are discussed and compared with those of the statistical approach.

2 Carbon Nanotubes as Simple Models for Ionic Channels

Ion channels are hollow proteins with a cavity in their central region around the axis of cylindrical symmetry. Ion channels can be found in nearly all biological membranes and play the role of controlling the flux of ions inside and ouside the cell. They act as gate-keepers and govern functions of key biological and medical importance.

An important function of these channels is to produce electrical signals in the nervous system. They also provide, in nearly every cell of living bodies, control of transport of ions and several other functions among which those interacting with most of the drugs used in clinical medicine.

Channels come in many distinct types because they are designed and built by evolution. Hundreds of types of channels have been already discovered, many other remain to be discovered by experimental and computational molecular science.

A simple and elegant theory of ionic selectivity (inspired by the selectivity of special glasses to bind specific ions [3]) was first proposed by Eisenman and Horn in terms of the difference between the hydration free energy of the ion and the energy of interaction between the ion and a charged binding site within the channel. This theory predicts the existence of eleven selectivity sequences typical of ionic channels. The link between ionic selectivity and the electrostatic interaction of the ion with charged and/or polar groups within the channel has been also investigated in refs. [3] and [4].

Additional studies singled out that a model description of the ionic mobility in a molecular pore has to take into account charge and polar residues [5], the monovalent and divalent [6, 7] as well as the positive and negative [8, 9, 10] nature of the ions. Other studies, instead, suggest that the interaction of ions with charged residues within the selectivity filter is not relevant to ionic selectivity, at least in some ion channels [10]. This better justifies the absence of charge within the carbon nanotube model studied in this paper.

For this purpose the molecular pore can be schematized (see Fig. 1) as a sequence of a selectivity filter, a transbilayer pore and a gate.

To this end, it is of extreme importance to rely on an appropriate description of the relevant molecular interactions. Yet, being impossible for biological systems to construct a potential energy surface (PES) based on rigorous ab initio calculations, the PES is usually built by assembling some realistic few body potential energy terms. In our case this was obtained by using the force field formulation of DL_POLY_2 [11] and the force field parameters of the CHARMM database [12].

We focus here on the selectivity function (the bottom section of Fig. 1) and examine the ionic currents which flow through a simple uniform model channel. Accordingly, we assume this segment to be the flow-determining region of the channel that can be modeled as an uncharged open-end carbon nanotube (CNT). The CNT (see Fig. 2) is therefore made of a regular cylinder of molec-

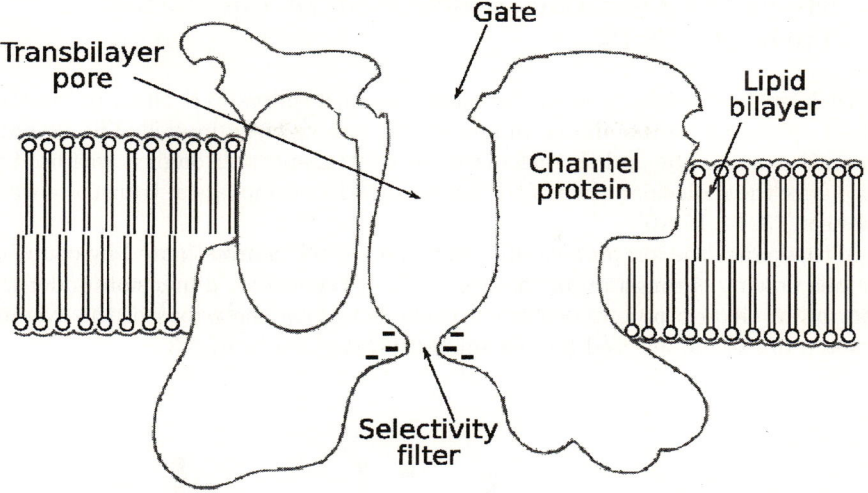

Fig. 1. Simple representation of a generic molecular channel

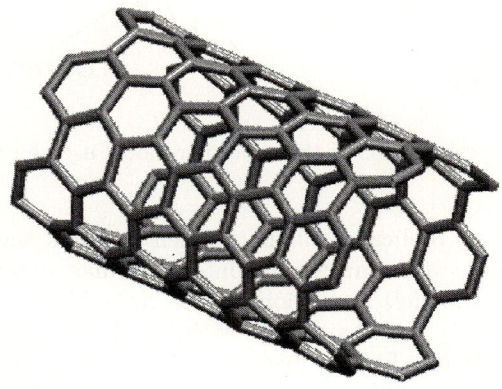

Fig. 2. Carbon Nanotube (CNT) example

ular dimension made of pure benzene rings (pristine nanotube) whose axis of cylindrical symmetry coincides with a cartesian axis (say z) of the molecular dynamics frame.

The channel's response (the permeability σ) will be measured in terms of the number of particles allowed to go through the pore in a molecular dynamics calculation. In this approach the permeability is determined by the competition of the interaction of the ion with the pore (forced by the applied potential difference (ΔV) to flow through the channel) and the solvation energy (trying to keep the ion dressed by the solvent (water) molecules).

3 Molecular Dynamics Calculation of Ion Permeability Through CNTs

CNTs have been already used for modeling micropores and their properties. Indeed, micropore water filling and emptying was recently modelled as an open end carbon nanotube and the calculated water transport through it was used to rationalize the molecular scale of the ion permeation in biological transmembrane channels [13, 14, 15, 16].

In this paper the ion permeability through a CNT is investigated by counting the ions flowing through on uncharged CNT of reasonable dimension under the effect of an electric potential difference applied along the z-axis of the system. A sketch of the model used for the simulation is given in Fig. 3.

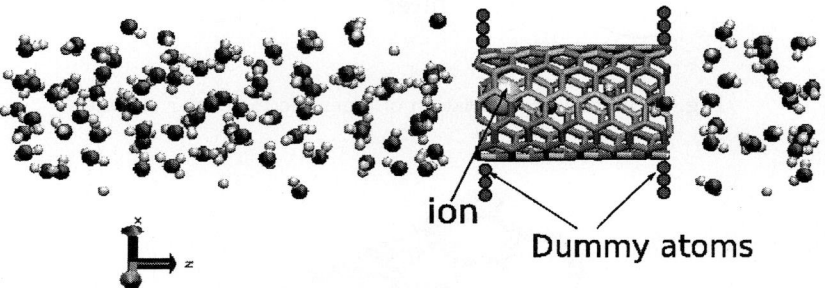

Fig. 3. Representation of the molecular model used for the simulation

The periodically replicated simulation cell contains one nanotube with 144 sp^2 carbon atoms, 109 transferable intermolecular potential 3 point water molecules (TIP3P model) and 1 ion (either Na^+, Mg^{++}, K^+, Ca^{++} or Cs^+). Each CNT is of (6,6) "armchair" type with length and diameter of $\sim 13.4 \mathring{A}$ and $\sim 8.1 \mathring{A}$, respectively (see Fig. 3). Furthermore 32 dummy atoms are used to simulate the lipid bilayer surrounding the pore. This layer has also the function of preventing ion mobility outside the CNT.

A time step of 1fs was chosen to carry out the molecular dynamics simulation at constant temperature (298K) and volume of the system. Particle-mesh Ewald summation [17] is used to calculate electrostatics interaction in the 3D periodic system.

Five simulations were performed for each ion type by roughly doubling each time the applied potential difference ΔV. The permeability (σ) is taken to be proportional to the number (ν) of ions passing through the uncharged CNT in a time interval of 1 ns.

As intuitively expected, ν increases with ΔV. Such an increase is nearly linear with ΔV although for less mobile ions ν seems to level off at high voltages. This trend is more apparent in Fig. 4 where ν is plotted as a function of ΔV.

A Molecular Dynamics Study of Ion Permeability Through Molecular Pores

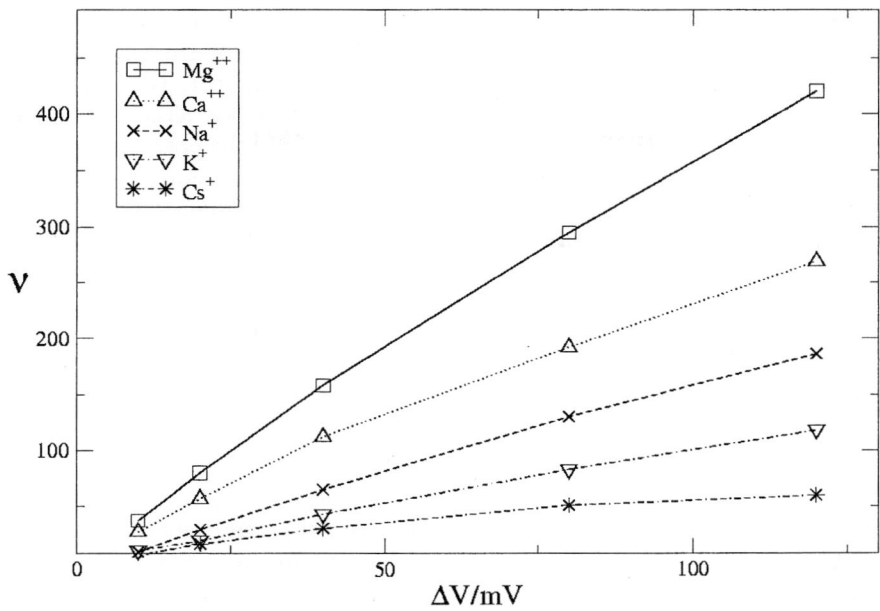

Fig. 4. Number of Mg^{++}, Ca^{++}, Na^+, Cs^+ ion crossing the micropore in 1ns plotted as a function of ΔV. Connecting lines are given only for illustrative purpose

Table 1. Hydration standard hentalpies [1]

Ions	ΔH^0_{hydrat} (kJ/mole)
Mg^{++}	-1990
Ca^{++}	-1659
Na^+	-439
K^+	-355
Cs^+	-297

The value of ν decreases along the series Mg^{++}, Ca^{++}, Na^+, K^+, Cs^+, indicating, as intuitively expected, that doubly charged ions permeate better the CNT. It is interesting to notice that the properties of ν for this simple model (uncharged CNT) reasonably well correlate with the hydration standard hentalpies ΔH^0_{hydrat} (for comparison the value of ΔH^0_{hydrat} taken from ref. [1] for the ions considered are listed in Table 1).

The values given in Table 1 clearly show that there is a sharp difference between the ΔH^0_{hydrat} of monovalent and divalent ions in analogy with what happens for the number of ions crossing the CNT. It can also be pointed out that the ranking of the ions of Table 1 and Figure 4 is the same.

A more quantitative relationship has been worked out [10] between the permeability ratio (σ_A/σ_B for species A and B) and the Gibbs solvation free energy.

In the approach of Laio & Torre, the permeability through the micropore, say of ion A, depends on the Gibbs free energy G_A for undressing the ion of the solvated water molecules in the micropore environment. G_A depends on the radius of the pore and varies with the position of the ion along the channel. Accordingly, the value of G_A (that tends to $\Delta G^0{}_{hydrat}$ for the pore radius tending to infinity) shows repeated regular maxima on its profile along the length of the nanotube (let us call G_A^{max} the largest of them). Therefore, in a transition state teory approach, the ratio of permeability between ion A and B can be formulated as:

$$\frac{\sigma_A}{\sigma_B} \propto exp\left[-\frac{1}{RT}(G_A^{max} - G_B^{max})\right] \quad (1)$$

Using eq. 1, the permeability ratio of Na^+, K^+ and Cs^+ was worked out for different values of the pore radius (see Table 2).

Table 2. Logarithm of the permeability ratio estimated from the eq. 1 (as derived from Fig. 4B of ref. [10]) for different values of the micropore radius with a value of $0.17 Å$ for the r.m.s. of radius fluctuation

radius/$Å$	2.5	3	3.5	4.0
$\log_{10}\frac{\sigma_{Na^+}}{\sigma_{Cs^+}}$	2.0	3.2	3.2	1.0
$\log_{10}\frac{\sigma_{K^+}}{\sigma_{Cs^+}}$	0.3	0.9	1.6	1.2
$\log_{10}\frac{\sigma_{Na^+}}{\sigma_{K^+}}$	1.7	2.3	1.6	-0.2

Table 3. Logarithm of the permeability ratio for various ΔVs obtained from the values of ν calculated for the CNT model

ΔV/mV	10	20	40	80	120
$\log_{10}\frac{\sigma_{Na^+}}{\sigma_{Cs^+}}$	0.18	0.26	0.34	0.40	0.49
$\log_{10}\frac{\sigma_{K^+}}{\sigma_{Cs^+}}$	0.22	0.07	0.16	0.21	0.29
$\log_{10}\frac{\sigma_{Na^+}}{\sigma_{K^+}}$	-0.045	0.18	0.18	0.19	0.20

To compare our dynamical calculations with result of Table 2, we estimated the logarithm of the permeability ratio for different values of ΔV (see Table 3) by exploiting the fact that σ is proportional to ν.

As is apparent from a comparison of Tables 2 and 3 the values calculated using the two approaches differ significantly. For example, if we take as a reference the situation in which the permeability ratio $\sigma_{Na^+}/\sigma_{K^+}$ obtained from dynamics calculations is -0.045 (at $\Delta V = 10$ mV, i.e. the smallest ΔV considered in our dynamical calculations, see column 1 row 3 of Table 3), the corresponding value for the transition state approach at a radius of $4.0 \mathring{A}$ (that well reproduces the radius of the CNT used for the dynamic calculation, see column 4 row 3 of Table 2)is -0.2. The $\sigma_{Na^+}/\sigma_{Cs^+}$ and $\sigma_{K^+}/\sigma_{Cs^+}$ ratios obtained from the two approaches also differ by a factor of 5 clearly indicating that statistical treatments may lead systematically to results appreciably different from those of molecular dynamics.

4 Conclusions

Molecular dynamics calculations have been carried out to evaluate the permeability of molecular pores to cations by taking as a model an open-end carbon nanotube. The calculated permeability of several cations shows an increase with both the potential difference and the charge of the ion in qualitative correlation with the hydration standard hentalpy.

Yet, the more quantitative estimates of the permeability ratio obtained from the hydration Gibbs free energy (this approach is based on a transition state model of the permeability process) appreciably deviate from those obtained from molecular dynamic calculations.

This suggests that before carrying out ion permeability studies for micropores based on thermodynamic or statistic approaches care should be put in assessing the validity of the results versus dynamical calculations.

Acknowledgments

This work has been supported by the Italian MIUR FIRB Grid.it project (RBNE 01KNFP) on High performance Grid Platforms and Tools, and by the MIUR CNR Strategic Project L 499/97-2000 on High performance Distributed Enabling Platforms. Thanks for financial support are also due to ASI and COST in Chemistry. Thanks for very useful discussions are due to Luigi Catacuzzeno.

References

1. Hille, B.: Ionic channels of excitable membranes. Sinaeur Associates Inc. (1984)
2. Hille, B.: Ionic channels: molecular pores of excitable membranes. Harvey Lect **82** (1986) 47–69
3. Eisenman, G., Horn, R.: Ionic selectivity revisited: the role of kinetic and equilibrium processes in ion permeation through channels. J Membr Biol **76** (1983) 197–225
4. Reuter, H., Stevens, C.: Ion conductance and ion selectivity of potassium channels in snail neurones. J Membr Biol **57** (1980) 103–118
5. Imoto, K., Busch, C., Sakmann, B., Mishina, M., Konno, T., Nakai, J., Bujo, H., Mori, Y., Fukuda, K., Numa, S.: Rings of negatively charged amino acids determine the acetylcholine receptor channel conductance. Nature **335** (1988) 645–648

6. Heinemann, S., Terlau, H., Sthmer, W., Imoto, K., Numa, S.: Calcium channel characteristics conferred on the sodium channel by single mutations. Nature **356** (1992) 441–443
7. Kim, M., Morii, T., Sun, L., Imoto, K., Mori, Y.: Structural determinants of ion selectivity in brain calcium channel. FEBS Lett **318** (1993) 145–148
8. Galzi, J., Devillers-Thiry, A., Hussy, N., Bertrand, S., Changeux, J., Bertrand, D.: Mutations in the channel domain of a neuronal nicotinic receptor convert ion selectivity from cationic to anionic. Nature **359** (1992) 500–505
9. Dorman, V., Partenskii, M., Jordan, P.: A semi-microscopic Monte Carlo study of permeation energetics in a gramicidin-like channel: the origin of cation selectivity. Biophys J **70** (1996) 121–134
10. Laio, A., Torre, V.: Physical origin of selectivity in ionic channels of biological membranes. Biophys J **76** (1999) 129–148
11. Smith, W., Forester, T.: Dlpoly 2.0: a general-purpose parallel molecular dynamics simulation package. J Mol Graph **14** (1996) 136–141
12. Brooks, B.R., Bruccoleri, R.E., Olafson, B.D., States, D.J., Swaminathan, S., Karplus, M.: Charmm: A program for macromolecular energy, minimization, and dynamics calculations. J Comp Chem **4** (1983) 187–217
13. Kalra, A., Garde, S., Hummer, G.: Osmotic water transport through carbon nanotube membranes. Proc Natl Acad Sci U S A **100** (2003) 10175–10180
14. Berezhkovskii, A., Hummer, G.: Single-file transport of water molecules through a carbon nanotube. Phys Rev Lett **89** (2002) 064503
15. Mann, D.J., Halls, M.D.: Water alignment and proton conduction inside carbon nanotubes. Phys Rev Lett **90** (2003) 195503
16. Zhu, F., Schulten, K.: Water and proton conduction through carbon nanotubes as models for biological channels. Biophys J **85** (2003) 236–244
17. York, D., Wlodawer, A., Pedersen, L., Darden, T.: Atomic-level accuracy in simulations of large protein crystals. Proc Natl Acad Sci U S A **91** (1994) 8715–8718

Theoretical Investigations of Atmospheric Species Relevant for the Search of High-Energy Density Materials

Marzio Rosi

Istituto di Scienze e Tecnologie Molecolari (ISTM) del CNR,
c/o Dipartimento di Chimica, Università degli Studi di Perugia,
Via Elce di Sotto 8, 06123 Perugia (Italy)
marzio@thch.unipg.it

Abstract. We report in this paper the study, at *ab initio* level, of species important in the chemistry of the atmosphere, which are relevant also for the search of high-energy density materials (HEDM). We will discuss in detail, as examples, the recently discovered metastable species N_2CO and CO_4. All the calculations are carried out by using the DFT hybrid functional B3LYP in conjunction with a triple zeta basis set augmented with polarization and diffuse functions to characterize the potential energy surfaces, while the energetics has been evaluated at CCSD(T) level.

1 Introduction

The great effort put in the last decades in the study of the natural cycles of ozone and carbon dioxide, of the chemistry of air pollution and its effect on the depletion of ozone and on the greenhouse effect caused by combustion processes has shown the great importance of simple, but very reactive, species, as ions, radicals and "exotic" molecules. Being these species in gas phase, it is particularly appropriate for their study the joint use of theoretical methods, at *ab initio* level, with experimental techniques, mass spectrometry specifically. Both these investigation methods improved in the last few years and, for this reason, they are particularly suitable for applications involving the study of the atmosphere. On the one hand, the high performance computing allows nowadays the theoretical investigation of more and more complex species at a great level of accuracy. On the other hand, mass spectrometry, with the development of techniques of neutralization/reionization, can now study neutral species (radicals, unstable molecules) which are of great interest in the atmospheric chemistry [1]. The combined use of accurate theoretical methods and mass spectrometry techniques allowed us to discover recently several new species, both cationic and neutral, which can be important for the chemistry of terrestrial and planetary atmospheres [2, 3, 4, 5, 6, 7]. The neutral species are experimentally detected starting from a charged precursor of appropriate connectivity for the neutralization experiments aimed at the formation of the neutral by a vertical process.

Some of these new species show also the peculiarity of being metastable: these species can dissociate towards the products through very exothermic process; however, the dissociation usually presents an activation energy. This "metastability" suggests that these species should be promising candidates of high energy density materials (HEDM), which can be the next generation of environmentally benign propellants and explosives, especially in relation to spacecraft propulsion [8]. It is worth reminding that a high-energy density material is a species with a high heat of formation, a low molecular weight, a high kinetic stability, which dissociates with a very exothermic process. In this paper we will present, as examples of high-energy density materials, the discovery and the characterization of CO_4 [5] and N_2CO [7].

2 Computational Methods

Density functional theory, using the hybrid [9] B3LYP functional [10, 11], was used in order to optimize the geometry of relevant species and evaluate their vibrational frequencies. Although it is well known that density functional methods using nonhybrid functionals sometimes tend to overestimate bond lengths [12], hybrid functionals as B3LYP usually provide geometrical parameters in excellent agreement with experiment [13]. Single-point energy calculations at the optimized geometries were performed using the coupled-cluster single- and double-excitation method [14] with a perturbational estimate of the triple excitation [CCSD(T)] approach [15], to include extensively correlated contributions [16]. The geometry of some minima of N_2CO were reoptimized also at CCSD(T) level. Transition states were located using the synchronous transit-guided quasi-Newton method of Schlegel and co-workers [17]. The 6-311+G(3df) basis set was used [18] for CO_4 and the 6-311+G(3d) basis set was used [18] for N_2CO. Zero-point energy corrections evaluated at the B3LYP level were added to CCSD(T) energies. For the CCSD(T) optimized geometries, the zero-point energy corrections evaluated at the CCSD(T) level were used. The 0 K total energies of the species of interest were corrected to 298 K by adding translational, rotational, and vibrational contributions. The absolute entropies were calculated by using standard statistical-mechanistic procedures from scaled harmonic frequencies and moments of inertia relative to the B3LYP optimized geometries. All calculations were performed using Gaussian 98 [19] or Gaussian 03 [20].

3 Results and Discussion

3.1 CO_4

The search for new molecules, often metastable and short-lived, made up by atoms of the first periods of the Table, is currently the focus of great interest and active research, that mark a true renaissance of the inorganic chemistry of the main-group elements. Much of the impetus to these studies arises from their great multidisciplinary impact on rapidly developing research fields, in particular the chemistry of the natural and polluted terrestrial atmosphere, the chemistry of planetary atmospheres, and most recently the search for high energy density materials (HEDM), actively developed as environmentally benign propellants and explosives.

In this context, the CO_4 molecule, whose experimental discovery has been recently reported [5], is important on both accounts. Indeed, in addition to its well established and amply documented role in the chemistry of the atmospheres, CO_4 is considered a promising HEDM, because its dissociation into environmentally benign CO_2 and O_2 has been estimated to release as much as 100 kcal mol^{-1}.

Figure 1 shows the stationary points found for the CO_4^+ species, while Table 1 shows their energies. The most stable isomer of CO_4^+ is the C_s $^2A''$ ion **1**, a cluster

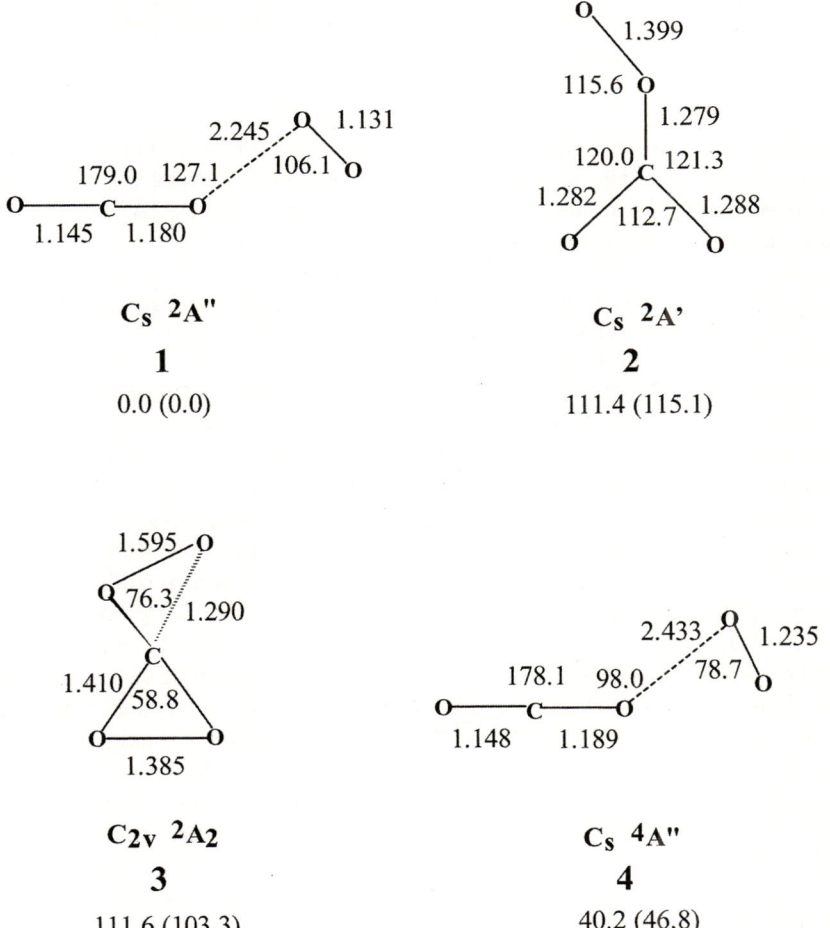

Fig. 1. Optimized geometries and relative energies of the CO_4^+ species computed at the B3LYP and CCSD(T) (in parentheses) level of theory. Bond lengths are in Å, angles in degrees and energies in kcal mol^{-1}

Table 1. Total energies (hartree) of the CO_4^+ and CO_4 species

	1 C_s $^2A"$	2 C_s $^2A'$	3 C_{2v} $^2A_2$	4 C_s $^4A"$
E_{B3LYP}	-338.611738	-338.433771	-338.431779	-338.545491
ZPE^a	0.016678	0.017943	0.016125	0.014207
$E_{CCSD(T)}$	-338.035146	-337.851308	-337.868395	-337.958372

	5 C_{2v} $^1A_1$	6 D_{2d} $^1A_1$	7 C_{2v} $^3B_1$	8 C_s $^3A"$
E_{B3LYP}	-338.933648	-338.870169	-338.831503	-339.040037
ZPE^a	0.018934	0.017212	0.015923	0.015638
$E_{CCSD(T)}$	-338.360531	-338.306524	-338.258494	-338.460164

^{a} Zero point energy.

formed by CO_2 and O_2 units joined by electrostatic forces and characterized by a large separation of the monomers. The C_s $^2A'$ ion **2** and the C_{2v} $^2A_2$ doublet **3** lay at considerably higher energies. Another cluster, the C_s $^4A"$ species **4**, has been identified on the quartet PES.

Passing to neutral CO_4, the reaction between CO and O_3 gives rise to a species of C_{2v} symmetry, the $^1A_1$ molecule **5**, whose optimized geometry is shown in Figure 2. This species shows a singlet ground state and it is more stable than the reactants by 46.9 kcal mol^{-1} at the B3LYP level of calculation (42.2 kcal mol^{-1} at the CCSD(T) level of calculation) at 298 K. At higher energy (38.8 kcal mol^{-1} at B3LYP; 32.9 kcal mol^{-1} at CCSD(T)) we were able to localize on the potential energy surface of CO_4 a singlet state of D_{2d} symmetry (species **6**), while the first triplet state lies at even higher energy (62.3 kcal mol^{-1} at B3LYP and CCSD(T)). The molecule **5** is unstable with respect to dissociation into CO_2 and O_2 in their ground states, by 67.5 kcal mol^{-1} at B3LYP and 62.7 kcal mol^{-1} at CCSD(T) level of calculation. This is a spin forbidden reaction. However, taking into account the experimental separation between O_2 ($X^3\Sigma_g^-$) and O_2 ($a^1\Delta_g$), which is 22.64 kcal mol^{-1} [21], **5** is unstable also with respect to the dissociation into CO_2 in its ground state and O_2 in its first excited singlet state. This reaction, however, implies a barrier of 31.5 kcal mol^{-1} at B3LYP and 36.2 kcal mol^{-1} at CCSD(T). The saddle is not planar, being the dihedral angles OCOO and OOCO 169.9 and 169.6 °. In the triplet surface we were able to localize an adduct of CO_2 and O_2 (species **8**) which, however, is unstable with respect to dissociation, once we include the zero point energy correction. In the triplet surface we were able to localize also a saddle point for the dissociation into CO_2 and O_2; this saddle point is interesting since it can be described as CO^+ and O_3^-.

Our computational results are consistent with those of previous studies as concerns the geometries of the species of interest [22, 23, 24]. In addition, the

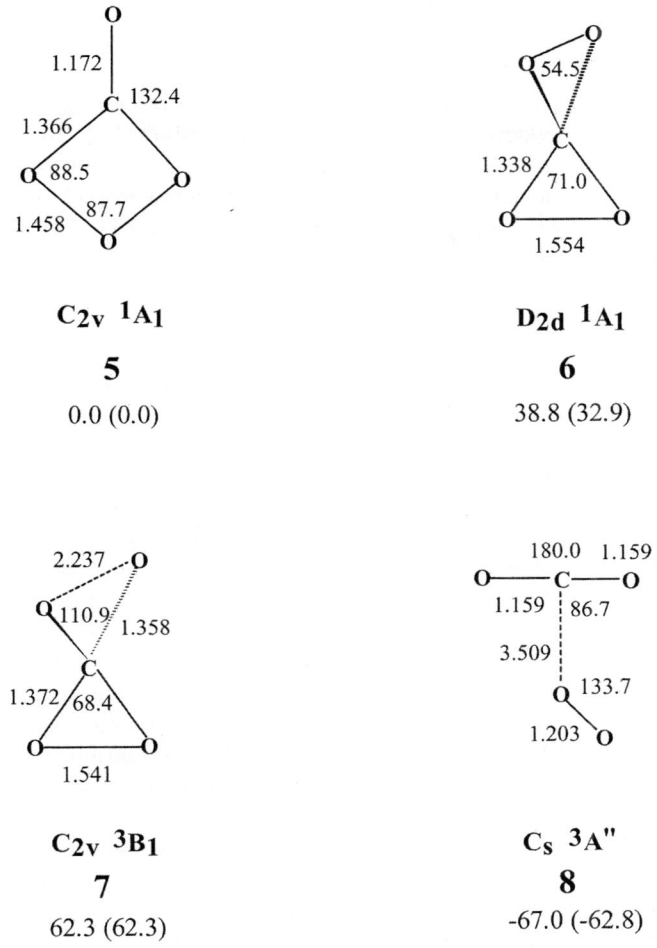

Fig. 2. Optimized geometries and relative energies of the CO_4 species computed at the B3LYP and CCSD(T) (in parentheses) level of theory. Bond lengths are in Å, angles in degrees and energies in kcal mol^{-1}

reported thermochemical stability difference between **5** and **6**, namely 32 kcal mol^{-1} [22, 23, 24], compares well with the 32.9 kcal mol^{-1} difference computed in this work at the CCSD(T) level. Both **5** and **6** are energetically unstable with respect to the lowest adiabatic limit, CO_2 ($X^1\Sigma_g^+$) + O_2 ($a^1\Delta_g$), whereas the C_{2v} $^3B_1$ triplet **7** is correspondingly unstable with respect to dissociation into CO_2 ($X^1\Sigma_g^+$) + O_2 ($X^3\Sigma_g^-$). In view of their lack of thermodynamic stability, the lifetime of the CO_4 species depends exclusively on the existence of a sizable kinetic barrier to dissociation. This restricts the discussion to the D_{2d} $^1A_1$ singlet **6**, since no appreciable kinetic barriers to dissociation of the C_{2v} $^1A_1$ singlet **5** and the C_{2v} $^3B_1$ triplet **7** could be identified.

The study of the CO_4^+ system suggests as a possible charged precursor the C_{2v} $^2A_2$ ion **3**, whose structure appears sufficiently similar to that of **6** to allow survival of the latter following the vertical neutralization process. Clearly **3**, located some 4 eV above the most stable cation **1**, is not expected to be a major component of the CO_4^+ population. Nevertheless, excited O_2^+ and/or CO_2^+ ions could justify the presence of **6** in the mixed CO_4^+ population.

3.2 N_2CO

Also the N_2CO molecule, whose experimental discovery has been recently reported [7], is important both in the chemistry of the atmospheres and as a promising HEDM. Figure 3 shows the geometries and the relative energies of the N_2CO^+ and N_2CO species identified on the PES, while Table 2 reports thir absolute energies. In figure 3 we have reported only the species with connectivity N-N-C-O which seem to be relevant in the experimental detection of N_2CO [7]. However, for the sake of completeness we must say that we have other isomers, both for the neutral and the cation, with different connectivity, not reported here. Let us focus our attention to the species with connectivity N-N-C-O shown in figure 3. According to previous results, the only minimum found on the doublet surface of N_2CO^+ is the trans-planar ion **1**. We could not find any stable T-shaped minimum on the doublet and quartet surfaces, since the doublet ion **2** is unstable towards dissociation once we include the zero point energy correction. On the quartet surface, we found the ions **3** and **4**, higher in energy than ion **1** by 97.3 and 106.8 kcal mol^{-1}, respectively, at the CCSD(T) level. Both the **3** and **4** quartet ions, however, are stable with respect to dissociation towards the spin-allowed asymptotes. As far as N_2CO is concerned, we identified on the singlet PES the diazirinone C_{2v} **7**, that according to previous results is the most stable neutral species [25]. Geometry optimization of the triplet open-chain N_2CO led to the linear D_{2h} species **5**. Previous geometry optimizations [26] at MP2 level however led to a bent structure for this species. For this reason, we decided to optimize the geometry of this state at CCSD(T) level. The CCSD(T) geometry optimization led to the bent C_s species **6**. However, the CCSD(T) energies of **5** and **6** differ only by 0.7 kcal mol^{-1}, both being less stable than diazirinone by about 7 kcal mol^{-1}.

The analysis of the Franck-Condon energies suggests that the experimentally detected neutral should be the triplet open chain $^3A"$ formed probably from the ionic species $^4A'$. This neutral species is stable towards dissociation into the spin-allowed products N_2 ($X^1\Sigma_g^+$) and CO (a$^3\Pi$) by 36 kcal mol^{-1}, whereas it is unstable by 104 kcal mol^{-1} with respect to the spin-forbidden dissociation into

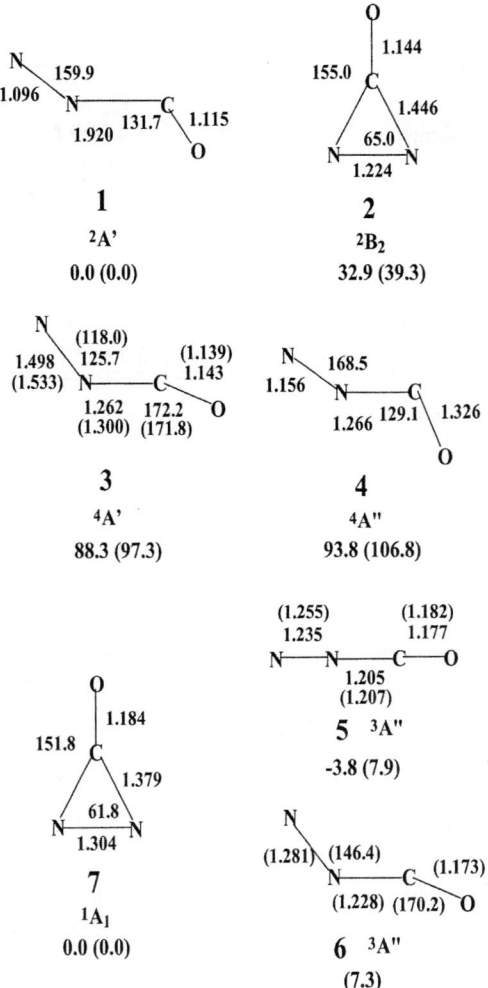

Fig. 3. Optimized geometries and relative energies of the N_2CO and N_2CO^+ species with connectivity N-N-C-O, computed at the B3LYP and CCSD(T) (in parentheses) level of theory. The geometrical parameters in parentheses are optimized at CCSD(T) level. Bond lengths are in Å, angles in degrees and energies in kcal mol^{-1}

N_2 and CO in their ground state [26]. Its "metastability" therefore arises from the possible access to singlet open-chain NNCO states that dissociate without barrier into N_2 ($X^1\Sigma_g^+$) and CO ($X^1\Sigma^+$).

Table 2. Total energies (hartree) of the N_2CO^+ and N_2CO species

	1 $C_s\ ^2A'$	2 $C_{2v}\ ^2B_2$	3 $C_s\ ^4A'$	4 $C_s\ ^4A''$
E_{B3LYP}	-222.446918	-222.393939	-222.303914	-222.295603
ZPE^a	0.012473	0.013428	0.011460	0.011827
$E_{CCSD(T)}$	-221.996981	-221.933812	-221.839666	-221.824877

	5 $C_s\ ^3A''$	6 $C_s\ ^3A''$	7 $C_{2v}\ ^1A_1$
E_{B3LYP}	-222.793007		-222.787278
ZPE^a	0.013206	0.012914	0.014820
$E_{CCSD(T)}$	-222.307373	-222.308752	-222.320371

^{a} Zero point energy.

4 Conclusions

In this paper we have reported the study, at *ab initio* level, of species important in the chemistry of the atmosphere, which are relevant also for the search of high-energy density materials (HEDM). As examples, we have provided a short description of the "metastability" of the recently discovered N_2CO and CO_4 species. The "metastability" of these species arises from the presence of states which show a very exothermic dissociation reaction, but with a consistent activation barrier.

Acknowledgements

This work was supported by the MIUR-FIRB project RBNE01KNFP "Piattaforme abilitanti per griglie computazionali a elevate prestazioni orientate a organizzazioni virtuali scalabili".

References

1. Wesdemiotis, C., McLafferty, F.W.: Neutralization-reionization mass spectrometry (NRMS). Chem. Rev. **87** (1987) 485–500; Schalley, C.A., Hornung, G., Schröder, D., Schwarz, H.: Mass spectrometric apppproaches to the reactivity of transient neutrals. Chem. Soc. Rev. **27** (1998) 91–104
2. Cacace, F., de Petris, G., Rosi, M., Troiani, A.: Formation of O_3^+ upon ionisation of O_2. The role of isomeric O_4^+ complexes. Chem. Eur. J. **8** (2002) 3653–3659; Cacace, F., de Petris, G., Rosi, M., Troiani, A.: Charged and neutral NO_3 isomers from the ionization of NO_x and O_3 mixtures. Chem. Eur. J. **8** (2002) 5684–5693

3. Pepi, F., Ricci, A., D'Arcangelo, G., Di Stefano, M., Rosi, M.: Thionyl fluoride from sulfur hexafluoride corona discharge decomposition. Gas-phase chemistry of [SOF$_2$]H$^+$ ions. J. Phys. Chem. A **106** (2002) 9261–9266; Pepi, F., Ricci, A., Di Stefano, M., Rosi, M.: Sulfur hexafluoride corona discharge decomposition. Gas-phase ion chemistry of SOF$_x^+$ (x=1-3) ions. Chem. Phys. Letters **381** (2003) 168–176
4. de Petris, G., Cartoni, A., Rosi, M., Troiani, A., Angelini, G., Ursini, O.: Isotope exchange in ionised CO$_2$/CO mixtures: the role of asymmetrical C$_2$O$_3^+$ ions. Chem. Eur. J. **10** (2004) 6411–6421
5. Cacace, F., de Petris, G., Rosi, M., Troiani, A.: Carbon tetroxide: theoretically predicted and experimentally detected. Angew. Chem. Int. Ed. **42** (2003) 2985–2990
6. Cacace, F., de Petris, G., Rosi, M., Troiani, A.: Discovery of the new metastable HONF radical. ChemPhysChem. **5** (2004) 503–508
7. de Petris, G., Cacace, F., Cipollini, R., Cartoni, A., Rosi, M., Troiani, A.: Experimental detection of theoretically predicted N$_2$CO. Angew. Chem. Int. Ed. **44** (2005) 462–465
8. Klapötke, T.M., Holl, G.: The greening of explosives and propellants using high-energy nitrogen chemistry. Green Chem. **3** (2001) G75–G77
9. Becke, A.D.: Density-functional Thermochemistry. III. The Role of Exact Exchange. J. Chem. Phys. **98** (1993) 5648–5652
10. Becke, A.D.: Density-functional Exchange-energy Approximation with Correct Asymptotic Behavior. Phys. Rev. A **38** (1988) 3098–3100
11. Lee, C., Young. W., Parr, G.G.: Development of the Colle-Salvetti Correlation-energy Formula into a Functional of the Electron Density. Phys. Rev. B **37** (1988) 785–789
12. Mannfors, B., Koskinen, J.T., Pietilä, L.-O., Ahjopalo, L.: Density functional studies of conformational properties of conjugated systems containing heteroatoms. J. Molec. Struct. (Theoch.) **393** (1997) 39-58
13. Bauschlicher, C.W., Ricca, A., Partridge, H., Langhoff, S.R.: Chemistry by density functional theory. in *Recent Advances in Density Functional Theory* (Ed.: Chong, D.P.), World Scientific Publishing Co., Singapore (1997) Part II, p.165
14. Pople, J.A., Krishnan, R., Schlegel, H.B., Binkley, J.S.: Electron correlation theories and their application to the study of simple reaction potential surfaces. Int. J. Quant. Chem. **14** (1978) 545–560
15. Raghavachari, K., Trucks, G.W., Pople, J.A., Head-Gordon, M.: Size consistent Brueckner theory limited to double substitutions. Chem. Phys. Lett. **157** (1989) 479-483
16. Olsen, J., Jorgensen, P., Koch, H., Balkova, A., Bartlett, R.J.: Full configuration-interaction and state of the art correlation calculations on water in a valence double-zeta basis with polarization functions. J. Chem. Phys. **104** (1996) 8007-8015
17. Peng, C., Ayala, P.J., Schlegel, H.B., Frisch, M.J.: Using redundant internal coordinates to optimize geometries and transition states. J. Comp. Chem. **17** (1996) 49-56
18. Frisch, M.J., Pople, J.A., Binkley, J.S.: Self-consistent Molecular Orbital Methods 25. Supplementary Functions for Gaussian Basis Sets. J. Chem. Phys. **80** (1984) 3265–3269 and references therein
19. Gaussian 98, Revision A.7, Frisch, M.J. *et al.*: Gaussian, Inc., Pittsburgh PA, 1998.
20. Gaussian 03, Revision B.04, Frisch, M.J. *et al.*: Gaussian, Inc., Pittsburgh PA, 2003.

21. Huber, K.P., Herzberg, G.: Constants of Diatomic Molecules. Van Nostrand, New York, 1978
22. Averyanov, A.S., Khait, Yu.G., Puzanov, Yu.V.: Ab initio investigation of the CO_4 and CO_2N_2 molecules as possible high-energy metastable species. J. Mol. Struct. (Theoch.) **367** (1996) 87–95
23. Song, J., Khait, Yu.G., Hoffmann, M.R.: A theoretical study of substituted dioxiranes: difluorodioxirane, fluorofluoroxydioxirane, (fluoroimino)dioxirane, and hydrazodioxirane. J. Phys. Chem. A **103** (1999) 521–526
24. Averyanov, A.S., Khait, Yu.G., Puzanov, Yu.V.: Singlet and triplet states of the CO_3 and CO_4 molecules. J. Mol. Struct. (Theoch.) **459** (1999) 95–102
25. Hochlaf, M., Léonard, C., Ferguson, E.E., Rosmus, P., Reinsch, E.-A., Carter, S., Handy, N.C.: Potential energy function and vibrational states of N_2CO^+. J. Chem. Phys. **111** (1999) 4948–4955
26. Korkin, A.A., Schleyer, P.v.R., Boyd, R.J.: Theoretical study of metastable N_2CO isomers. New candidates for high energy materials? Chem. Phys. Lett. **227** (1994) 312–320. Korkin, A.A., Balkova, A., Bartlett, R.J., Boyd, R.J., Schleyer, P.v.R.: The 28-electron tetraatomic molecules: N_4, CN_2O, BFN_2, C_2O_2, B_2F_2, CBFO, C_2FN, and BNO_2. Challenges for computational and experimental chemistry. J. Phys. Chem. **100** (1996) 5702–5714

ID Face Detection Robust to Color Degradation and Facial Veiling

Dae Sung Kim and Nam Chul Kim

Laboratory for Visual Communications, Department of Electronic Engineering,
Kyungpook National University, Daegu, 702-701 Korea

Abstract. We proposed an efficient method for detecting an identifiable face (ID face) that is not veiled in an input image to establish his or her identity. This method is especially designed to be robust to color degradation and facial veiling, which is composed of two parts: face-like region segmentation and unveiled-face detection. In the face-like region segmentation, measuring horizontal symmetry that is an important property of facial components is introduced to overcome the difficulty of facial region segmentation only by using skin color (SC) under nonuniform illumination causing severe color degradation. As a result, the segmentation leads to extraction of non-SC facial components and their neighbor degraded SC regions as well as undegraded SC regions. The unveiled-face detection is based on analysis of face constellations and statistical averages of facial patterns. The detection especially investigates statistical averages of each facial component pattern and its horizontal symmetry, which leads to detection of a face where all facial components are unveiled. Experimental results for AR and VCL facial databases show that the proposed method yields the improvement of 22.9% in detection rate over a face detection method without consideration of color degradation and facial veiling.

Keywords: ID face detection, color degradation, facial veiling, horizontal symmetry, skin color, eigenface.

1 Introduction

Recently, many researchers have studied face detection which is to decide whether a face exists in an input image or not and where it is if exists. It is a preprocessing essential to face recognition, model-based video coding, content-based image retrieval, and so on [1,2]. In some cases, we not only have to determine the existence and position of a face, we also have to decide whether facial components such as eyes, a nose, and a mouth which are closely related to one's individual identity are veiled or not, that is, whether the face is an identifiable face (ID face) or not. For example, consider a situation that someone wants to draw money from an ATM (automated teller machine). If he or she veils his or her face with sunglasses, a mask, etc., the withdrawal can be rejected detecting the veiled face in an image captured at the situation. Though such an ID face detection may be implemented by conventional face detection methods, complementary measures are required to obtain satisfactory detection performance robust even to color degradation under no good illumination and facial veiling in malice.

Conventional face detection methods can be classified into three categories: feature-based, appearance-based, and hybrid approach [1]. The feature-based approach is based on features such as skin color [2], texture [3], and edge [4] which are often used in image processing area. The appearance-based approach which includes the exampled-based learning method [5], the eigenface method [6], and the neural network method [7] is based on patterns learned from a training set. In typical methods of the feature-based approach, a face is detected by constellation analysis of facial components, which are found by edge features in face-like regions which consist of skin colored pixels specified by a given skin color distribution on a color space such as RGB, YCrCb, HSV, etc.[1]. Under proper illumination these show good performance, but under varying illumination they may not extract skin colored regions effectively [1]. Besides, when they are applied to ID face detection, they may not distinguish facial components from veiling objects which have edge features similar to those of facial components.

The eigenface method classified into the appearance-based approach decides whether each face candidate is a face or not by measuring a distance between the candidate and its approximation with the weighted sum of eigenvectors (eigenfaces), which are obtained from several training faces. This method usually yields a good performance but has high computational complexity due to multiresolution window scanning [1] and may detect partially veiled faces [6] which have to be rejected in ID face detection. In some methods [1, 8] of the hybrid approach, whether each face candidate found by the feature-based approach is a face or not is decided by the appearance-based approach. This approach is considered a good candidate that provides sufficient performance of ID face detection in quasi-real time if it makes provision against color degradation and facial veiling.

We propose an ID face detection method robust to color degradation and facial veiling, which is a sort of the hybrid approach. It is composed of two parts: face-like region segmentation and unveiled-face detection. In the face-like region segmentation, for all blocks partitioned from an input image, their densities of skin color (SC) pixels and densities of horizontal symmetry (HS) pixels are measured. Then, clusters of blocks with dense SC pixels and clusters of blocks with non dense SC pixels, which have any neighbor with dense HS pixels and any neighbor with dense SC pixels, are extracted into face-like regions. The former clusters correspond to undegraded SC regions and the latter to non-SC facial components and their neighbor degraded SC regions.

In the unveiled-face detection, face candidates which are rectangular windows satisfying face-like constellations are first extracted in each face-like region and normalized for the approximation with eigenfaces. Then if the normalized candidate which is closest to its approximation with eigenfaces is far from the approximation, the detection results in failure. Otherwise, if any of facial components in the best candidate is far from its approximation with the corresponding eigen facial component or any of some moments of its horizontal symmetry magnitudes deviates from a given range of the moment, then the detection also results in failure. Otherwise the detection results in success.

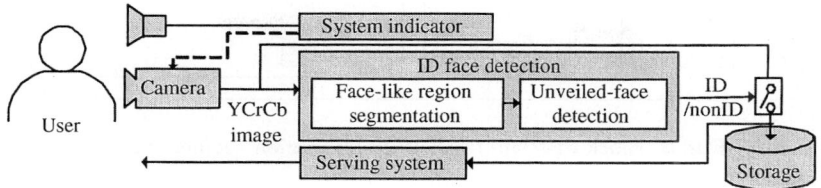

Fig. 1. Block diagram of a serving system with the proposed ID face detection

2 Proposed ID Face Detection

Fig. 1 shows the block diagram of a serving system with the proposed ID face detection. Suppose that when a legal user accesses the system, he follows the system indication that he has to look at the system camera with his unveiled face to receive a service that he desires. In the system, the proposed algorithm which is composed of face-like region segmentation and unveiled-face detection is applied to an YCrCb image of the user which is acquired from the camera immediately after his acceptance of the indication. If an ID face is detected in the input image, the image is stored in the system and the service is provided to the user. Otherwise the service is not provided.

2.1 Face-Like Region Segmentation

The proposed face-like region segmentation is based on the skin-colored facial region segmentation [9] by Chai et al. and the generalized symmetry transform (GST) [11] by Reisfield et al. The former is a sort of morphological process where a processing unit is not a pixel but a quantized density of SC pixels in a block, which leads to being robust to local variations of illumination. However, when colors of a skin region are severely degraded by nonuniform illumination, a facial region may not be properly segmented.

To overcome the difficulty of the SC region segmentation, an estimation of densities of HS pixels is introduced into the proposed segmentation method. The HS density in a block is computed by a variation of the GST which is modified to be robust to noise and to emphasize horizontal symmetry compared to other symmetries. Fig. 2 shows the block diagram of the proposed segmentation which consists of SC density map computation, HS density map computation, face-like block detection, and post-processing.

In the proposed segmentation, the input is a color image of YCrCb 4:1:1. First, in the SC density map computation, whether each chrominance pixel is a SC pixel or not is determined according to whether its pair of Cr and Cb components falls within a limited range R_{Chai} used in Chai's segmentation or not as follows:

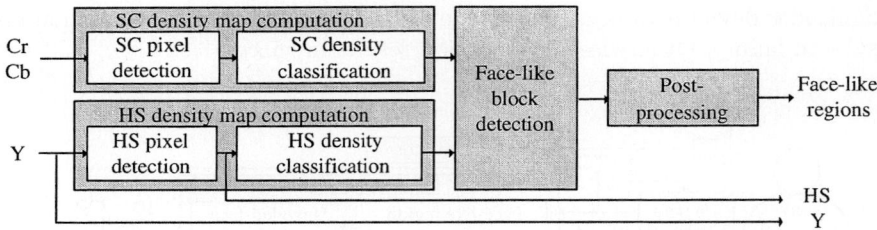

Fig. 2. Block diagram of the face-like region segmentation

$$S(x,y) = \begin{cases} 1, & \text{if } (Cr(x,y), Cb(x,y)) \in R_{\text{Chai}} \\ 0, & \text{otherwise} \end{cases} \quad (1)$$

where (x, y) denotes the position of a chrominance pixel and the value 1 a SC pixel. Next, for each of 4x4 blocks partitioned from a chrominance image, which corresponds to an 8x8 block of a luminance image, its SC density is computed which is defined as the ratio of the number of SC pixels in the block to the number of all pixels in the block. As a result, the SC density map is obtained classifying the SC densities as

$$SM(m,n) = \begin{cases} SC, & \text{if } d_S(m,n) \geq 0.5 \\ NSC, & \text{otherwise} \end{cases} \quad (2)$$

where (m, n) denotes the positional index of a 4x4 block in a chrominance image, $d_S(m, n)$ its SC density, and NSC non-skin color. In Eq. (2), setting the threshold for classification of SC densities as 0.5 which is much smaller than 1 used in Chai's method is for extracting SC blocks as much as possible. Instead, skin colored but non-facial regions or veiled face regions are strictly removed in the unveiled-face detection.

In the HS density map computation, for each pixel of a luminance image, its HS magnitude is first calculated using the variant of the GST as follows:

$$M(p) = \sum_{(q,r) \in \Gamma(p)} P(q,r) e_q e_r \quad (3)$$

where $\Gamma(p)$ denotes the set of pairs of a pixel q and a pixel r such that the pixel p is the midpoint between q and r which are distant within a given range, that is, $\Gamma(p) = \{(q,r) | (q+r)/2 = p \text{ and } \|q-r\| < C_1\}$. The $P(q,r)$ denotes a weighting function associated with the phase of gradient of q and that of r, and e_q and e_r quantities reflecting the magnitude of gradient of q and that of r, respectively.

The quantity e_q is given as

$$e_q = \begin{cases} 1, & \text{if } \|\nabla Y_q\| < C_2 \\ 0, & \text{otherwise} \end{cases} \quad (4)$$

where Y_q denotes the luminance value of q, ∇Y_q the gradient of Y_q, $\|\nabla Y_q\|$ the magnitude of ∇Y_q, and C_2 a threshold. Thresholding the gradient magnitude in

(4) is for removing the influence of noisy pixels with small gradient magnitudes on the symmetry magnitude $M(p)$. The phase weighting function $P(q,r)$ is given as

$$P(q,r) = \begin{cases} 1, \text{ if } \pi/3 < \theta_q < 2\pi/3 \text{ and } -2\pi/3 < \theta_r < -\pi/3 \\ 0, \text{ otherwise} \end{cases} \quad (5)$$

where θ_q and θ_r denote the gradient phase of q and that of r, respectively. The condition for pairs of gradient phases in (5) is for emphasizing the pairs of gradient phases which are horizontally symmetrical with respect to the pixel p.

After obtaining the HS magnitudes, whether each luminance pixel is a HS pixel or not is determined as

$$H(p) = \begin{cases} 1, \text{ if } M(p) \geq C_3 \\ 0, \text{ otherwise} \end{cases} \quad (6)$$

where the value 1 denotes a HS pixel and C_3 a threshold. Fig. 3 shows the comparison between the GST and the method specified in the Eq. (3)-(5). The figure (a) shows the original image, (b) the result of binarizing the symmetry magnitudes of (a) obtained by the GST, and (c) the result by the modified method. We notice that unlike in (b), edges of objects which show horizontal symmetry are mostly extracted in (c).

Next, for each of 8x8 blocks partitioned on the luminance image, which corresponds to a 4x4 block of chrominance images, its HS density is computed which is defined as the ratio of the number of HS pixels in the block to the number of all pixels in the block. As a result, the HS density map is constructed classifying the HS densities as

$$HM(m,n) = \begin{cases} \text{HS}, \text{ if } d_H(m,n) > \frac{1}{8} \\ \text{NHS}, \text{ otherwise} \end{cases} \quad (7)$$

where $d_H(m,n)$ denotes the HS density of an 8x8 block in the luminance image, and NHS non-horizontal symmetry. The selection of a relatively low threshold in Eq. (7) is for extracting blocks which contains even a part of edge boundaries of horizontal symmetry objects.

In the face-like block detection, whether each block is a face-like block or not is determined according to the SC density map and the HS density map as follows:

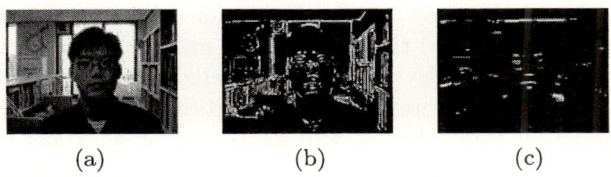

(a)　　　　　　　　(b)　　　　　　　　(c)

Fig. 3. (a) Original image; (b) result of binarizing the symmetry magnitude of (a) obtained by the GST; (c) result by the modified method

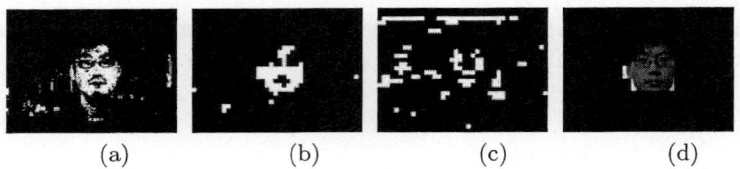

(a)   (b)   (c)   (d)

Fig. 4. Intermediate and final results of the proposed face-like region segmentation for Fig 3(a). (a) The SC pixels; (b) SC density map; (c) HS density map; (d) face-like region

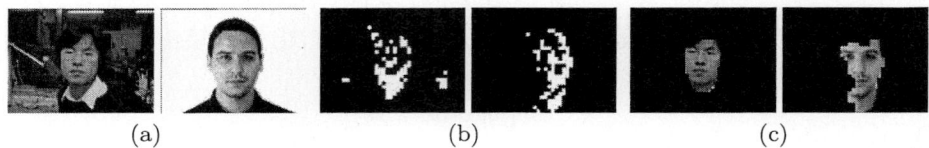

(a)   (b)   (c)

Fig. 5. Results of the proposed face-like segmentation for other two images. (a) Original images; (b) SC density maps; (c) face-like regions

$$FL_B(m,n) = \begin{cases} 1, & \text{if } SM(m,n) = \text{SC} \\ 1, & \text{if } SM(m,n) = \text{NSC and } N_{\text{SC}}(m,n) \geq 1 \text{ and } N_{\text{HS}}(m,n) \geq 1 \\ 0, & \text{otherwise} \end{cases}$$
(8)

where the value 1 means a face-like block and $N_{\text{SC}}(m,n)$ and $N_{\text{HS}}(m,n)$ denote the number of SC blocks and the number of HS blocks among the (m,n)-block and its 8 neighbor blocks, respectively. In Eq. (8), the first condition tells us that a block of high SC density is classified as a face-like block. The second condition means that even though a block is of low SC density it is determined as a face-like block if there are one or more SC blocks and one or more HS blocks among the block and its neighborhood blocks. Finally, in the post-processing, small isolated clusters of face-like blocks or non-face-like blocks are removed and then the remaining clusters of face-like blocks are segmented as face-like regions.

Fig. 4 shows the intermediate and final results of the proposed face-like region segmentation for Fig. 3(a). The figure (a) shows the skin color pixels, (b) the SC density map, (c) the HS density map, and (d) the face-like region. We see that the face-like region in (d) includes even the eyes which are not the skin colored region in (b). Fig. 5 shows SC density maps and the face-like regions obtained by the proposed segmentation for other two images. The figure (a) shows original images, (b) their SC density maps, and (c) their face-like regions. We see that the face-like regions in (c) include even eyes, mouths, or their neighborhoods which are not in the skin colored regions in (b). It is also shown that using horizontal symmetry as well as skin color as a feature of face-like regions overcomes the difficulty of face-like region segmentation only by skin color.

2.2 Unveiled-Face Detection

The unveiled-face detection is based on the face-like constellation analysis [8] and the eigenface method, which consists of face candidate detection, face detection, and facial component detection as shown in Fig. 6. Face candidates which satisfy the face-like constellations are first detected from face-like regions. Next, the detection of a face among the face candidates is accomplished by the eigenface method. Finally, whether facial components of the detected face are veiled or not is decided by investigating each facial component pattern and its horizontal symmetry.

In the face candidate detection, the centroid of each cluster which consists of HS pixels in the face-like regions is first calculated. Next, whether each triple among all the centroids satisfies given face-like constellations or not is determined. The face-like constellations indicate the structural conditions of a face on the distance between the centroids of the eyes, the distance between the midpoint of the two eye centroids and the centroid of the mouth, and the angle between the line passing the eye centroids and the horizontal line. If there is no triple which satisfies the constellations, the ID face detection results in failure. Otherwise, the rectangular window surrounding each triple is extracted as a face candidate. Each face candidate passes through the affine transform [12] for setting it upright, the bilinear interpolation [12] for adjusting its size to the size of eigenfaces, the histogram specification [13] for reducing the effect of nonuniform illumination, and the intensity normalization for forcing it to have unity mean and variance.

In the face detection, each of the normalized face candidates is first approximated as the weighted sum of eigenfaces obtained from facial images in the training set. That is,

$$\widetilde{\mathbf{f}}_i = \sum_{j=1}^{L} \mathbf{u}_j^T \mathbf{f}_i \mathbf{u}_j \qquad (9)$$

where $\mathbf{f}_i$ denotes the ith normalized face candidate, $\mathbf{u}_j$ jth eigenface, and L the number of eigenfaces. Among all the normalized face candidates, the normalized face candidate whose distance between itself and its approximation is minimal is found as

$$\mathbf{f}^* = \arg\min_{\mathbf{f}_i} \left\| \mathbf{f}_i - \widetilde{\mathbf{f}}_i \right\| \qquad (10)$$

If the minimum distance is larger than a given threshold, the ID face detection results in failure. Otherwise, the minimum face candidate is detected as a face.

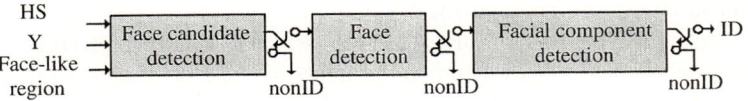

Fig. 6. Block diagram of the unveiled-face detection

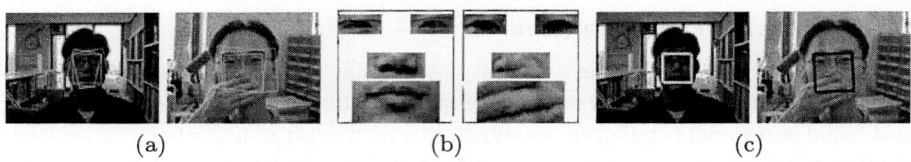

Fig. 7. Intermediate and final results of the unveiled-face detection. (a) several face candidates; (b) facial components regions in the detected faces; (c) an ID face and a nonID face

In the facial component detection, four facial component regions corresponding to a left eye, a right eye, a nose, and a mouth whose sizes and locations are previously fixed are first extracted from the detected face. For each facial component region, the HS magnitudes are computed according to Eq. (3)-(5) and their mean, centroid, and first-order invariant moment [12] are found. If among the three statistics there is one which deviates from its given range, the ID face detection results in failure. Otherwise, for each of passed facial component regions, the distance between the facial component data $\mathbf{f}_c$ and its approximation $\widetilde{\mathbf{f}}_i$ is computed similarly in Eq. (9) and (10). If the distance is larger than its given threshold, the ID face detection results in failure. Otherwise, if all distances for the four facial components are smaller than their thresholds, the ID face detection finally results in success. Fig. 7 shows the intermediate and final results of the unveiled-face detection. The figure (a) shows several face candidates, (b) facial component regions in the detected faces, and (c) an ID face and a nonID face.

3 Experimental Results and Discussions

The AR database (DB) and VCL DB are used to evaluate the performance of the proposed ID face detection method. The former is derived from the AR face DB [10] and composed of 1035 color images of 320x240 pixels. It is classified into 9 subsets of the same size: ID face images under front, left, and right lightings (named IDF, IDL, and IDR), respectively, nonID face images with sunglasses under front, left, and right lightings (named nIDSF, nIDSL, and nIDSR), respectively, and nonID face images with a scarf under front, left, and right lightings (named nIDCF, nIDCL, and nIDCR), respectively. The latter is collected by the authors and contains 280 color images of 320x240 pixels, which is classified into 2 subsets of the same size: ID face images under complex backgrounds and varying lightings (named IDC) and nonID face images veiled with sunglasses, a hat, a mask, or a hand (named nIDV).

The threshold C_1 for $\Gamma(p)$ in Eq. (3) is chosen as $C_1 = 14$ which corresponds to the maximum vertical size of facial components of the DB images, the thresholds C_2 in Eq. (4) and C_3 in Eq. (6) are obtained by the entropy threshold technique [4] which is known to be an efficient edge thresholding, and the gradient ∇Y_q in Eq. (4) is computed by using the Sobel edge operator [12]. A training set of

30 face images outsides the DBs were used to obtain eigenfaces and eigen facial components for the face detection and facial component detection. The threshold for the minimum distance in the face detection is given as the maximum among the minimum distances obtained from the eigenfaces and another test inputs of 30 face images outside the DBs. The threshold for each of the four distances in the facial component detection is given as the maximum among the distances obtained from the corresponding eigen facial components and the test inputs. The ranges for the three moments of four facial component regions are set so that they may include all the corresponding moments obtained from the test inputs.

As a performance measure, ID face detection rate is used which is defined as the ratio of the number of ID and nonID face images correctly detected to the total number of images in a DB. Table 1 shows the ID face detection rate of three detection methods for the AR DB and VCL DB. Method 1 is a modified version of the proposed method without HS density map and facial component detection. This method corresponds to a simple combination of a feature-based approach using constellation analysis and the eigenface method. Method 2 is another version without facial component detection. As shown in Table 1, in case of the AR DB, the proposed method yields 64.9% detection rate gain for the ID face images, 5.8% for the nonID face images over the Method 1 and 31.3% for the nonID face images over the Method 2. As a result, the method yields 25.5% and 20.9% average detection rate gain over the Method 1 and Method 2, respectively. In case of the VCL DB, the proposed method yields 13.2% and 9.6% average detection rate gain over the Method 1 and Method 2, respectively. These results underscore the effectiveness of the face-like region segmentation which leads to complementary extraction of non-SC facial components and their neighbor SC regions degraded under varying illumination and the unveiled-face detection which leads to detection of a face where all facial components are unveiled.

Fig. 8. (a)Examples of results detected by the proposed detection method. (a) detected ID faces; (b) detected nonID faces; (c) false dismissals; (d) false alarms

Fig. 8 shows examples of face images detected by the proposed detection method. In the figure (a), we see that the ID faces are detected regardless of varying illumination conditions, complex backgrounds, a light mustache, or glasses. It is shown in the figure (b) that the nonID face images are detected regardless of veiling with sunglasses, a scarf, a hat, or a hand. The figure (c) and (d) show examples of ID face images detected as nonID faces (false dismissals) and nonID face images detected as ID faces (false alarms) by the proposed method, respectively. In the figures, we see that the false dismissals are caused by heavy mustaches and the false alarms are caused by resemblance between facial components and veiled appearances. As a result, it is found that the proposed method is robust to color degradation and facial veiling. Further work however is required to reduce the false dismissals and alarms like in the figure (c) and (d).

Table 1. ID face detection rate of three detection methods for the AR and VCL DB

	AR DB										
	ID face				nonID face						
Class	IDF	IDL	IDR	Subtotal	nIDSF	nIDSL	nIDSR	nIDCF	nIDCL	nIDCR	Subtotal	Total
Method 1	89.6	2.6	1.7	31.3	91.3	100.0	100.0	58.3	95.7	99.1	90.7	70.9
Method 2	96.5	96.5	95.7	96.2	87.8	69.6	85.2	64.3	41.7	42.6	65.2	75.5
Proposed	96.5	97.4	94.8	96.2	100.0	98.3	96.5	95.7	92.2	96.5	96.5	96.4
	VCL DB										
Class	ID face(IDC)				nonID face(nIDV)							Total
Method 1	85.7				78.6							82.5
Method 2	96.4				75.7							86.1
Proposed	96.4				95.0							95.7

Acknowledgement. This research was supported by the Program for the Training of Graduate Students in Regional Innovation which was conducted by the Ministry of Commerce Industry and Energy of the Korean Government.

References

[1] E. Hielmas, "Face detection: A survey," Computer Vision and Image Understanding, no. 83, pp. 236-274, 2001.
[2] R.-L. Hsu, M. Abdel-Mottaleb, and A. K. Jain, "Face detection in color images," IEEE Trans. Pattern Anal. Machine Intell., vol. 24, no.5, pp. 696-706, May 2002.
[3] C. Garcia and G. Tziritas, "Face detection using quantized skin color regions merging and wavelet packet analysis," IEEE Trans. Multimedia, vol. 1, no. 3, pp. 264-277, Sep. 1999.
[4] J. Fan, D. K. Y. Yau, A. K. Elmagarmid, and W. G. Aref, "Automatic image segmentation by integrating color-edge extraction and seed region growing," IEEE Trans. Image Processing, vol.10, no.10, pp. 1454-1463, Oct. 2001.
[5] K. K. Sung and T. Poggio, "Example-based learning for view-based human face detection," IEEE Trans. Pattern Anal. Machine Intell., vol. 20, no. 1, pp. 39-51, Jan. 1998.
[6] M. Turk and A. Pentland, "Eigenfaces for recognition," Journal of Cognitive Neuroscience, vol. 3, no.1, pp. 71-86, 1999.

[7] J. Zang, Y. Yan and M. Lades, "Face recognition: Eigenface, elastic matching, and neural nets," Proc. IEEE, vol. 85, no. 9, pp. 1423-1435, Sep. 1997.

[8] K. W. Wong, K. M. Lam, and W. C. Siu, "A robust scheme for live detection of human faces in color images," Signal processing: Image communication, vol. 18, no.2, pp. 103-144, Feb. 2003.

[9] D. Chai and K. N. Ngan, "Face segmentation using skin-color map in videophone applications," IEEE Trans. Circuits Syst. Video Technol., vol. 9, no. 4, pp. 551-564, June 1999.

[10] A. M. Martinez and R. Benavente, The AR face database, CVC Tech. Report #24, 1998.

[11] D. Reisfeld, H. Wolfson, and Y. Yeshurun, "Context free attentional operators: The generalized symmetry," Int. J. Comput. Vis., vol.14 no.2, pp.119-130, Mar. 1995.

[12] R. C. Gonzalez and R. E. Woods, Digital Image Processing, Addision-Wesley publiching Company, 1992.

[13] P. J. Phillips and Y. Vardi, "Efficient illumination normalization of facial images," Pattern Recognition Lett. vol. 17, no. 8, pp. 921-927, July 1996.

Detection of Multiple Vehicles in Image Sequences for Driving Assistance System

SangHoon Han[1], EunYoung Ahn[2], and NoYoon Kwak[3]

[1] Dept. of Information Security
Korea National College of Rehabilitation & Welfare,
5-3, JangAn-Dong, Pyung Taek-Si, Gyeong Gi-Do, Rep. of Korea
hansh@baubau.com
[2] Div. of Information and Communication Engineering, Cheonan University,
115 Anseo-Dong Cheonan-City, Chungcheongnam-Do,
330-704, Rep. of Korea
ahnyoung@cheonanac.kr
[3] Div. of Information and Communication Engineering,
Cheonan University,115 Anseo-Dong Cheonan-City,
Chungcheongnam-Do, 330-704, Rep. of Korea
nykwak@cheonan.ac.kr

Abstract. This study suggests a method to detect multiple vehicles, which is important for driving assistance system. In a frame of color image, shadow information and edge elements are used to detect vehicle candidate areas. Detecting the areas of multiple vehicles requires to analyze Estimation of Vehicle (EOV) and Accumulated Similarity Function (ASF) from the vehicle candidate areas that exist in image sequences. Later by evaluating the possibility of vehicles, it determines the vehicle areas. Most studies focus on detecting a single vehicle in front. This study, however, focuses on detecting multiple vehicles even in heavy traffic and frequent change of lanes.

1 Introduction

As vehicles increase, demands are high in the system that assists drivers to drive safely on the road. Accordingly, a variety of researches is being conducted to enhance the safety of driving by installing high-tech electronic communications devices and controllers on the vehicles and the roads. As a part of these researches, there are Advanced Vehicle Highway System (AVHS), Advanced Safety Vehicle (ASV), and Advanced Driver Assistance Systems (ADAS). One of the key technologies in these researches is to detect and track obstacles and vehicles in front of a vehicle that is running. Especially, detecting obstacles and vehicles on its way and warning to the driver will be of great help to safe driving.

Researches using image processing have three methods: a method of using stereo vision and mono vision, a method of using color image, and a method of using gray image. When using stereo vision, most of cases obtain disparity map to detect obstacles [1, 2, 3]. When using mono vision, most of cases use gray image while using motion analyses such as optical flow and entropy [4, 5].

Researches using color images try to detect vehicles from varied color areas [6, 7, 8, 9]. Most of these methods detect a single vehicle and they are applicable to the cases that traffic is not heavy and vehicle areas are larger. This study suggests a method to detect multiple vehicles in front from image sequences taken by a CCD camera even in heavy traffic and complicated markings on the road.

2 System Makeup and Detection of Vehicle Candidate Areas

2.1 System Makeup

When an RGB image comes in as an input from a camera, it converts it into HSV color space. By using saturation and value in the converted color space, a vehicle candidate point is obtained by a preprocessing process to find shadow areas that exist below the vehicle. Then the vehicle candidate areas are gained. The vehicle candidate areas obtained like this, by using the accumulated information of vehicle candidate areas extracted from the current frame and the previous n-time frames through image sequences analysis, determines whether the areas obtained from the current frame are vehicle or not, and then detects its location [10].

2.2 Preprocessing

As a method to find shadow areas of a vehicle that show in the road image, a dichromatic reflection model is used [9]. When viewed the condition of this model with the color vector space, it is expressed as Formula 1 below.

$$C_L = m_d C_d + m_s C_s + C_a \tag{1}$$

Here, C is [r, g, b] color space vector. C_d is diffuse reflection, C_s is specular reflection and C_a is ambient reflection. m_d and m_s are proportional factors due to geometric characteristics, which has a value between 0 and 1.

Supposing that sunshine is white light in the road images, the areas that have sunshines have smaller value of saturation and brightness component is increasing. By using this feature, candidate points that have the possibility of vehicle are extracted from the road images. To extract vehicle candidate points, first by using the difference of the values such as saturation and brightness, a standard value is set as Formula 2. Shown as Formula 3, when there is a big difference between the standard value and the average of basic areas, it must be the pixels that have the high possibility of vehicle [10].

$$F(x, y) = S(x, y) \times 255 - V(x, y) \tag{2}$$

Here, S(x, y) refers to saturation component and V(x, y) refers to brightness component.

$$W(x,y) = k_a \frac{(F(x,y)-\mu_r)^2}{\sigma_r^2}$$

$$R(x,y) = \begin{bmatrix} 1, & W(x,y)>1 & and & F(x,y)>\mu_r \\ 2, & W(x,y)>1 & and & F(x,y)<\mu_r \\ 3, & & else & \end{bmatrix} \quad (3)$$

μ_r means the average of the designated areas and σ_r means the standard deviation of the designated areas. k_a as weight shows how crowded it is to the average and it is a constant that adjusts the distance from the average. When R(x,y) becomes 1, it becomes the vehicle candidate point.

2.3 Detection of Vehicle Candidate Areas

Detection of vehicle candidate areas makes it clear the neighboring areas from the extracted image through the preprocessing process and separate the areas through morphology closing and connection element labeling to eliminate small holes. Going through the processes of merging and dividing the areas from the separated areas, the final vehicle candidate areas are separated.

3 Multiple Vehicles Detection

The candidate areas of vehicle have the mixed areas of vehicle and background and all of the candidate areas are not the vehicle areas. If the areas are vehicle, they should exist in other frames by the analysis of image sequences and maintain similar features. This is based on the fact that a vehicle does not show up and disappear suddenly in a series of road images. Surrounding background areas are hidden by vehicles and disappeared since the vehicle is moving. So the same features cannot be maintained in a number of frames. Therefore, the areas that show stable features in image sequences are determined as vehicle areas. Though the areas are not vehicle areas, but they show similar features in a series with low possibility of vehicle, they can be considered as non-vehicle areas.

To determine the vehicle areas, a ratio that certain areas of each frame can be determined as vehicle is obtained, which is Estimation of Vehicle (EOV). It is a method that does not study vehicle features and it is widely applied to such methods that use image processing. However, since this EOV cannot be the perfect estimation factor to recognize a vehicle, similarity function is applied as well.

This study suggests Accumulated Similarity Function (ASF) that applies similarity between accumulated EOV in multiple frames. For the candidate areas of vehicle obtained from the current frame, in the condition that they are high possibility of vehicle by continued observing information in the previous n-time frames, when similarity is high between frames, the areas are determined as vehicle areas.

Figure 1 illustrates a flow of image sequences analysis. EOV is a value that is obtained from the current frame and the figures in the circles are the values of EOV that are obtained from the current frame. In the figure, the arrow connects to the area that is the highest in Area Matching Probability (AMP) between t frame and t-1 frame. When there is no value on the arrow, it means 1.00 and when a value on the arrow is low, it is the case that it has no corresponding areas. In the rest of areas has low value, but the arrows are omitted. When you see the values between t-n frame and t-3 frame, the number of the areas is different. So the two areas in the t-n frame correspond to one area in the t-3 frame. In this case, the one that has lower AMP is omitted for its consideration object.

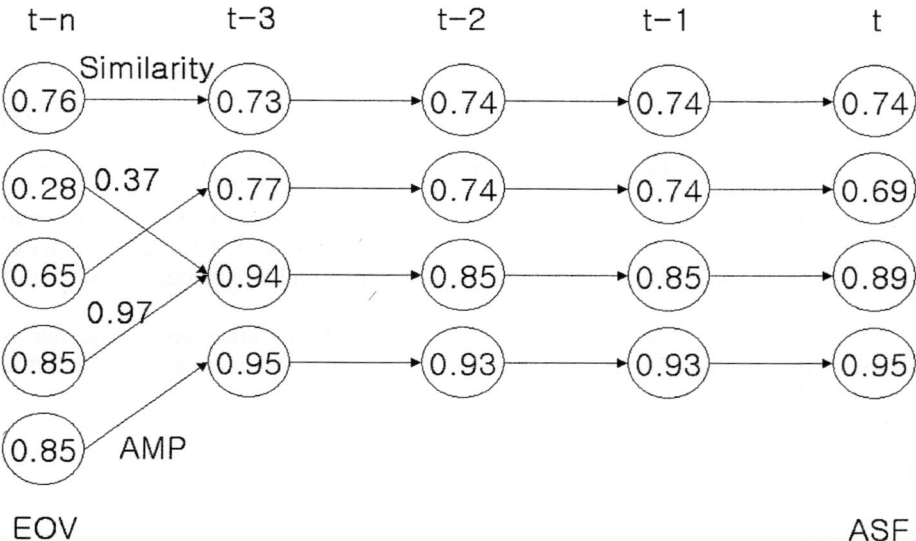

Fig. 1. Suggested image sequences analysis process

ASF is the sum of EOV from the current t frame to the t-n frame and weight of similarity. In the current frame, the vehicle areas and the non-vehicle areas are determined using the ASF value. When using ASF value, it considers only the similarity between the corresponding areas, which decreases processing time.

EOV is a factor that evaluates the possibility of vehicle. It uses basic features of vehicle to determine vehicle areas. In the road images, vehicle features are divided into the following factors: vertical edge existence, horizontal edge existence, symmetry of edge or brightness information, and shape of line segment. Other researches use these features and each factor is mainly used to find out a vehicle area [11].

This study uses the following three features.

① Vehicle areas have horizontal lines, which are similar to the size of the areas.
② To the left and right sides, vertical edges are symmetrical.
③ In the vehicle areas, the edges are symmetrical.

A function for each feature is defined as F_m, which can be expressed by Formula 4.

$$EOV(R_i^t(s)) = \sum_{m=1}^{3} W_m F_m(R_i^t(s)) \quad (4)$$

W_m shows the weight of each feature, $R_i^t(s)$ indicates a segment of i-th that exists in t time. Each weight W is [0.33, 0.33, 0.34].

ASF is an estimation factor that checks whether the possibility of being vehicle maintains high during the previous K frames. As shown in Formula 5, ASF is an accumulated EOV and similarity value from the current frame to the previous K frame divided by K.

$$ASF(R_i^t(s)) = \frac{1}{k} \times \sum_{m=t,\, j \in \max AMP_m}^{m=t-k} F \quad (5)$$

$$F = \omega_1 EOV(R_i^m(s)) + \omega_2 SIM(R_i^{m-1}(s), R_j^m(s))$$

Here, m is frame; i and j are indexes of the corresponding areas. i is the area of m frame and j is the index of the area that AMP is the largest in m-1 frame. SIM shows similarity of the areas that correspond to the two frames and k indicates the number of frame sequences.

If the value of ASF is more than the critical value, the areas are considered as vehicles. If the value of ASF is less than the critical value, the areas are considered as non-vehicle areas.

4 Experimental Results

4.1 Experimental Environment

Image sequences are taken by a USB-type PC camera on the Olympic highway in Seoul and its dimensions is 320 x 240. The camera is mounted in front of driver's seat and its angle is adjusted to capture the front situation in auto focusing mode. No additional devices are used to correct the camera location and maintain its even level. For this experiment, a notebook computer is used whose system has Intel Pentium III 650MHz and RAM 256MB. In Windows 2000, 12000 images are tested.

4.2 Experimental Results

The result of detecting vehicle areas by analyzing the image sequences is shown in Table 1. Seen from the results based on Y axis and Areas, the front vehicles show a better result than that of neighboring vehicles. In general, however, it shows a good detection rate.

Supposing the method based on symmetry, the Table 2 shows the result comparison between EOV applied result and ASF applied result. When reviewed the ratio of determining vehicle areas as non-vehicle areas, the ratios of wrong determination is 8.4% for only EOV applied and 7.6% for ASF applied. It is found that a better result is shown when ASF is applied together with EOV.

Table 1. Vehicle detection result

	Y axis	Areas Width
Front vehicle	98.7%	92.0%
Neighboring vehicle	97.5%	90.5%
Total	98.1%	91.2%

Table 2. Comparison of EOV and ASF results

Method	Vehicle => Non-vehicle	Non-vehicle => Vehicle	Total
ASF	5.7%	1.9%	7.6%
EOV	7.3%	1.1%	8.4%

Figure 2 illustrates the results of detecting vehicle areas through the analysis of image sequences. Even in heavy traffic and complicated markings on the road, the vehicle areas are well detected. A frontal left car in the image (d) of Figure 2 is recognized as non-vehicle areas. It is the case that the area is detected, but it is considered as non-vehicle areas because the vehicle features continue to change whenever new vehicles show up in heavy traffic.

(a) When a vehicle is changing its lane

(b) When vehicles are away

(c) When it is dark

(d) When vehicles are close

Fig. 2. Result of Vehicles detection

5 Conclusions

When determining vehicle areas with multiple images, using accumulated similarity information of neighboring images could have better results than only using the information of vehicles' basic features.

By detecting front vehicles and lanes and calculating relative speed considering the distance from the front vehicle, it can alarm drivers when they drive too fast so that it can help avoid a collision. If this is implemented in real time, this method can be used for the system that informs front situations by calculating the collision possible time with the front vehicle.

This method can be used for a major feature for the system that recognizes vehicles and traffic flow by detecting the flow of vehicles in front. In addition, it can be a foundation that allows to construct a front vehicle recognition system using a low-cost hardware based on image processing on the highways.

References

[1] Stefan Ernst, Christop Stiller, Jens Goldbeck, Christoph Roessig, : Camera Calibration for Lane and Obstacle Detection. Proceedings of the International Conference on Intelligent Transportation Systems (1999) 356-361
[2] Nobuhiro TSUNASHIMA, Masato NAKAJIMA, : Extraction of the Front Vehicle using Projected Disparity Map. Conference Visual Communications and Image Processing '99, California, January (1999) 1297-1304
[3] Ernst Lissel, Peter Andreas, Ralf Bergholz, Hubert Weisser, : From Automatic Distance Regulation to Collision Avoidance. AVEC' 96, International Symposium on Avoidance Vehicle Control (1996) 1367-1378
[4] S. M. Smith and J. M. Brady, "ASSET-2: Real-Time Motion Segmentation and Shape Tracking", IEEE Trans. Pattern Analysis and Machine Intelligence, vol. 17, no. 8 (1995) 814-820
[5] M. Werner, W.v. Seelen, : An Image processing system for assistance. Image and Vision Computing (2000) 367-376
[6] De Micheli, E., Prevete, R.; Piccioli, G., Campani, M., : color cues for traffic scene analysis. Intelligent Vehicles '95 Symposium (1995) 466-471
[7] Betke, M., Haritaoglu, E., Davis, L.S., : Highway Scene Analysis in Hard Real-Time. Intelligent Transportation System (1997) 812-817
[8] B. Heisele, W. Ritter, : Obstacle Detection Based On Color Blob Flow. Proceedings of the Intelligent Vehicles '95 Symposium (1995) 282-286
[9] Steven A. Shafer, : Using color to separate reflection components. COLOR research and application, vol. 10, no. 4 (1985) 210-218
[10] Sanghoon Han, Hyungje Cho, : HSV Color Model Based Front Vehicle Extraction and Lane Detection using Shadow Information. Journal of KOREA MULTIMEDIA SOCIETY, Vol. 5, No. 2 (2002) 176-190
[11] Y. Du, N.P. Papanikolopoulos, : Real-time vehicle following through a novel ymmetry-based approach. Robotics and Automation, Proceedings, Vol. 4 (1997) 3160-3165

A Computational Model of Korean Mental Lexicon[*]

Heui Seok Lim[1], Kichun Nam[2], and Yumi Hwang[2]

[1] Dept. of Software, Hanshin University, Korea
limhs@hs.ac.kr
http://nlp.hs.ac.kr
[2] Dept. of Psychology, Korea University, Korea
kichun{beleco}@korea.ac.kr
http://coglab.korea.ac.kr

Abstract. This paper proposes a computational model of Korean mental lexicon which can explain very basic findings that are observed using the visual Korean lexical decision task. The model is Frequency-based trie which is based on the trie data structure. Alphabets in each node of the Frequency-based trie is sorted in descending order by by their frequency in the corpus. It is the main idea which enables the model to simulate the Korean mental lexicon. There will be evaluations and comparison results with human data: how well the model simulates Korean word processing and how much the simulated results coincide with that of human processing. We also discuss both the strength and weakness of the computational model.

1 Introduction

Computational modeling of cognitive process is an approach to simulate and to understand the central principles involved in human cognitive process mechanism. Some advantages of using computational models in researches on cognitive mechanism in human brain is as follows. First, it gives us detailed explanation of the cognitive process by mimicking results which are already known. Second, it enable us to perform lesion study of human brain function which is impossible with human subjects. Furthermore, it can predict some unknown phenomena which are worthy of investigating with human subjects.

Researches by using the computational models in cognitive language process are rapidly growing. Many computational models are proposed and implemented to explain many important principles involved in human language processing[1], [2], [3]. The processing mechanism of different languages have not only universal characteristics but only very peculiar ones of each language. However, many

---

[*] This Work was Supported by the Korea Ministry of Science & Technology (M10413000008-04N1300-00811), Corresponding Author : Kichun Nam {kichun@korea.ac.kr}.

of the models are related with lexical access and knowledge representations of English human mental lexicon. There rarely have been researches on developing computational models of Korean language processing. This paper proposes a computational model of Korean mental lexicon which can explain and simulate some aspects of Korean lexical access process.

2 Backgrounds

In order to measure the processes that are involved in lexical access, we need a task which guarantees that a lexical entry is indeed accessed. LDT(lexical decision task) is most frequently adopted in the study of lexical access[7]. In a lexical decision experiment, subjects are required to discriminate words from nonwords by pressing one button if the presented stimuli is a word and another button if it is not. The LDT has proven to be a very fruitful method for exploring the nature of the mental lexicon. As main purpose of this paper is building a computational model which can simulate major characteristics of Korean lexical access, we propose a computational model which explain several basic finding that are consistently obtained using visual Korean lexical decision task[4], [5], [6], [7].

1) Frequency effect. The time taken to make a lexical decision response to a word of high frequency in the language is less than that to a word of low frequency in the language.

2) Length effect. The time taken to make a lexical decision response to a long word is less than that to a short word.

3) Word similarity effect. When a legal nonword is sufficiently similar to a word, it is hard to classify as a nonword. While there are many other effects observed using LDT, the above are perhaps the most fundamental in Korean LDT and the target effects which our proposed model try to simulate.

3 Inspiration from Machine Readable Dictionaries

It is very helpful to find out how mental lexicon is organized and how a lexical item is accessed to understand human lexical processing. It is also very important to make a human-oriented efficient machine readable dictionary(MRD) for natural language processing. Efficient memory structure and fast accessibility are ultimate goals in building the MRD. There have been many structures proposed to make efficient MRD such as B-tree structure, hashing, and Trie. We can guess some principles on the mental lexicon through investigation of the previous MRDs' structure and their operating algorithms. A B-tree is an m-way search tree to which we can make easy modification through insertion or deletion of an item whereas maximal search time is increased, proportional to the total number of items in the dictionary. In hashing, we store items in a fixed size table called a hash table. An arithmetic function, f which is called a hashing

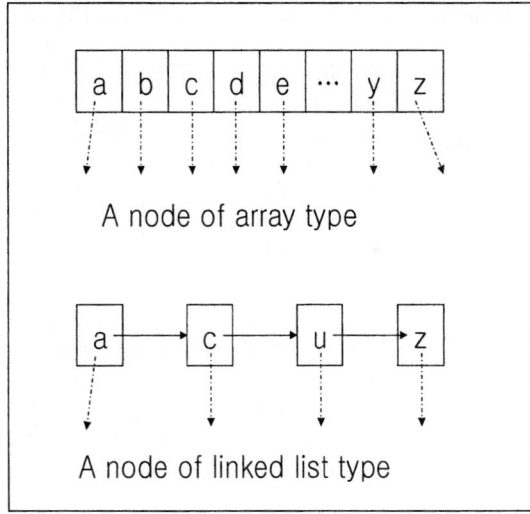

Fig. 1. Two Types of Trie nodes

Table 1. Data Structure of a Node of a FB-Trie

	B-Tree	Hashing	Trie
Freq. Effect	X	X	X
Length effect	X	X	○
Sim. effect	X	X	X

function is used to determine the address of an item when inserting and searching the item. If we use perfect hashing function, we can get information of an item by a direct access regardless of the total number of items in the dictionary. However, we are bothered to construct a new hashing table whenever an item is inserted or deleted. A trie is for storing strings, in which there is one node for every common prefix. The strings are stored in extra leaf nodes. The origin of the name is from the middle section of the word "reTRIEeval", and this origin hints on its usage. The trie data structure is based on two principles: a fixed set of indices and hierarchical indexing. The first requirement is usually met when you can index dictionary items by the alphabet. For example, at the top level we have a 26-element array or a linked list which we call a node from now on. Each of the nodes' elements may point to another 26-element node, and so on. Figure 1 represents structures of a 26-element array and a linked list. One of the advantages of the trie data structure is that its search time depends on only the length of an item, not on the number of stored items. Another advantage is that we can easily get a set of words with the same prefix, which is not easy in other dictionary structures presented in above.

Table 2. Data Structure of a Node of a FB-Trie

```
typedef struct node {
char ch;
int freq;
struct node *next;
struct node *child;
short int isfinal;
} nodetype;
```

Table 2 shows the MRDs' capacities of the characteristics of Korean mental lexicon and lexical recognition. As shown in 2, it is impossible to simulate frequency effect and length effect by using conventional hashing or B-tree index structure. In the trie structure, it can simulate length effect due to the principle of hierarchical indexing though frequency effect and word similarity effect are hard to be simulated.

4 Frequency-Based Trie Model

Our goal of modeling Korean mental lexicon is to build a model which can simulate or reflect phenomena induced from Korean lexical recognition; full-form representation, length effect, and frequency effect. To do that, we decided that the previous trie structure is the most promising. If we make some modification to reflect frequency effect in trie structures, the trie structure can satisfy our requirements. We propose a frequency-based trie(FB-trie) to make the simple trie structure reflect frequency effect. The proposed FB-trie is a modified trie structure which can satisfy the following requirements. 1) A full-form of Korean words is stored in the trie. 2) The structure of a node is linked list structure. 3) The alphabet in a node is a set of Korean phonemes. 4) The alphabet elements in a node are sorted by descending order of frequencies of the alphabets in a Korean corpus.

Whereas in the general trie structure, the alphabets in a node are sorted alphabetically, the 4th requirement is needed to model frequency effect, which means more frequent words are accessed faster by visiting minimal elements and nodes. We tried to model language proficiency by adjusting the size of the corpus in indexing; larger for proficiency of an adult or an expert and smaller for that of an infant or novice. Data structure for a node and an indexing algorithm of FB-trie are presented in table 2 and table 3 respectively.

In table 2, 'ch' is a variable to store a phoneme and 'freq' is a variable to store frequency of the phoneme. Variable 'next' points to the next element in a node in which the current element exists and variable 'child' points to a hierarchically next node. Variable 'isfinal' is a flag to indicate whether a node is a final node of a word or not.

Table 3. Indexing algorithm of FB-trie

1) Get a set of unique words in a corpus and count the frequency of each unique word
2) For each unique word
2.1) Convert the word into sequences of phonemes
2.2) Insert the sequences of phonemes into a trie.
2.3) Increase 'freq' of each phoneme by frequency of the current word
3) For each unique word 4) Traverse all nodes of the trie made in step 2) and sort phoneme elements of each node by descending order of frequency of the phoneme
4) Output the trie made in step 3)

5 Experimental Results of the Proposed FB-Trie

We constructed two FB-tries with two training corpora of size about 12 million words and 7 million words. We also made two tries with both training corpora for the purpose of comparison with FB-tries. Table 2 shows the experiment results of the correlation between frequency and human reaction time and between the length and human reaction time in word recognition. In table 2, Trie1 and Trie2 represent tries trained with 7 million word size corpus and 12 million word size corpus. Simillary, FB-trie1 and FB-trie2 represent FB-tries trained with 7 million word size corpus and 12 million word size corpus. Human result of the second row is from [6].

Table 4. Comparison of Correlations with Human Data

	Corr. of Frequency	Corr. of Length	Corr. of Similarity
Human	**-0.22200**	**0.22700**	**0.25500**
Trie1	-0.03258	0.58241	0.62241
Trie2	-0.06674	0.497034	0.51254
FB-Trie1	-0.15169	0.460327	0.44926
FB-Trie2	**-0.21718**	**0.337082**	**0.28731**

The experiment results were very promising in that correlations of frequency and length with bigger training corpus are more similar to those of human recognition than with the smaller. We aimed to model language ability or proficiency with the size of training corpus. Correlations of frequency of Trie1 and Trie2 are very small while those of FB-trie1 and FB-trie2 are relatively large and very similar to those of Human recognition. The reason of high correlations of length of Trie1 and Trie2 is due to inherent characteristic of trie indexing structure requiring more time for retrieving a longer string. We can see that this characteristic is alleviated in FB-trie resulting in more similar correlation with human word recognition.

6 Conclusion

In this chapter, we introduced some aspects in Korean lexical and morphological processing and a computational model for human mental lexicon. The proposed model is based on trie composed of nodes of linked list type and elements in each node is sorted by frequency of phonemes in corpus. We showed with some experimental results that the model reflects frequency effect and length effect which matter in Korean word recognition. In addition, we found some interesting phenomena when simulating FB-trie. First, words with high frequent neighborhood of syllables are likely to be accessed faster than with less frequent ones. Second, the difference between frequencies of phonemes which is the boundary between morphemes are very high and prominent. We guessed that this would be a clue to acquire a morpheme with many experiences with words with the morpheme. We are preparing a subsequent study on building a computational model which can encompass the neighborhood effect and the morpheme acquisition.

References

1. Bradley, A. D. Lexical representation of derivational relation. In M. Aronoff and M. L. Kean(Eds.), Juncture, 37-55. Cambridge, MA: MIT Press, 1980.
2. Caramazza. A., Laudanna, A., Romani, C. Lexical access and inflectional morphology. Cognition, 28, 207 - 332, 1988.
3. Foster, K. I. Accessing the mental lexicon. In R. J. Wales, E. Walker (Eds.), New approches to language mechanisms, 257-287. Amsterdam : North-Holland, 1976.
4. Jung, J., Lim, H., Nam, K., Morphological Representations of Korean compound Nouns. Journal of Speech and Hearing Disorders, 12, 77-95, 2003.
5. Kim, D., Nam. K., The Structure an Processing of the Korean functional category in Aphasics. Journal of Speech and Hearing Disorders, 12, 21-40, 2003.
6. Nam, K., Seo, K., Choi, K., The word length effect on Hangul word recognition. Korean Journal of Experimental and Cognitive Psychology, 9, 1-18, 1997.
7. Taft, M., Reading and the mental Lexicon. Hillsdale, NJ : Erlbaum, 1991.

A Realistic Human Face Modeling from Photographs by Use of Skin Color and Model Deformation

Kyongpil Min and Junchul Chun

Department of Computer Science, Kyonggi University,
Yui-Dong Suwon, Korea
{cabbi, jcchun}@kyonggi.ac.kr
http://giplab.kyonggi.ac.kr

Abstract. This paper presents a novel approach to produce a realistic 3D face model from multi-view face images. To extract facial region and facial feature points from color image a new nonparametric skin color model that is proposed. Conventionally used parametric skin color models for face detection have lack of robustness for varying lighting conditions and need extra work. To resolve the limitation of current skin color model, we exploit the Hue-Tint chrominance components and represent the skin chrominance distribution as a linear function. Thus, the facial color distribution is simply described as a combination of the maximum and minimum values of Hue and Tint components. Moreover, the minimal facial feature positions detected by the proposed skin model are adjusted by using edge information of the detected facial region along with the proportions of the face. To produce the realistic face model, we adopt log RBF(Radial-Based Function) to deform the generic face model according to the detected facial feature points from face images. The experiments show that the proposed approach efficiently detects facial feature points and produces a realistic 3D face model.

1 Introduction

The requirements of a realistic and feasibly animated facial model have been increased because facial modeling has been an important field of diverse application areas such as virtual character animation for entertainment, 3D avatars in the internet, 3D teleconferencing, and face recognition. Since Frederic I. Parke's[1] the pioneering work in animating face in the early 70's, many significant research efforts have tried to create the realistic facial model. Unfortunately, because of an extremely complex geometric form, countless tiny creases and wrinkles in the face, and subtle variations in color and texture, the realistic facial modeling system has not been developed yet.

Most of the facial modeling techniques rely on measured three-dimensional surface data. Surface measurement techniques fall into two major categories, one based on 3D scan system and the other based on photogrametric techniques. Unfortunately, not everyone has access to 3D scan systems, and an alternative surface measurement approach is through the use of photogrametric approach. The basic idea is to take multiple simultaneous photographs of a face, each from a different point of view.

Parke [2] improves a simple photogrametric method that uses orthogonal views of the face. Pighin et al. [3] have used an approach for creating photo-realistic textured 3D facial models from photographs of a human subject, and for creating smooth transitions of different facial expressions by morphing these different models. V. Blanz [4] proposes an approach that uses only a single photograph and a database of several hundred facial data and R. Enciso et al. [5] uses an approach that is based on computer vision techniques in which different pairs of stereo images are used to create precise geometry. But these approaches require physical markers on the face, extensive human intervention because of correspondence or the need for a huge database of human faces.

Meanwhile, the analysis of facial information has been one of the challenging problems in computer vision field. Especially, the facial region and feature detection is considered a critical work for developing various face recognition and face modeling systems. However, due to variations in illumination, background, and facial expression, the face detection and recognition have complex problems. Many of the works to detect facial region can be broadly classified as feature-based methods and image-based methods [6]. The feature-based methods make explicit use of face knowledge and follow the classical detection methodology in which low-level features are derived prior to knowledge-based analysis. Skin color segmentation can be performed using appropriate skin color thresholds where skin color is modeled through histograms. Skin color model, which are very efficient for detecting face from color images, have some difficulties in robust detection of skin colors in the presence of complex background and light variations [7].

In this paper we propose an automated 3D face modeling from color face images. The process includes a novel approach to detect facial region based on a new HT skin color model and to deform the 3D face model with log BRF. The proposed method shows efficiency for detecting face and facial feature points guided by MPEG-4 and is robust to various lighting conditions and input images. The facial feature points are subsequently used to create a realistic 3D face model. Figure 1 shows the block diagram of the proposed system.

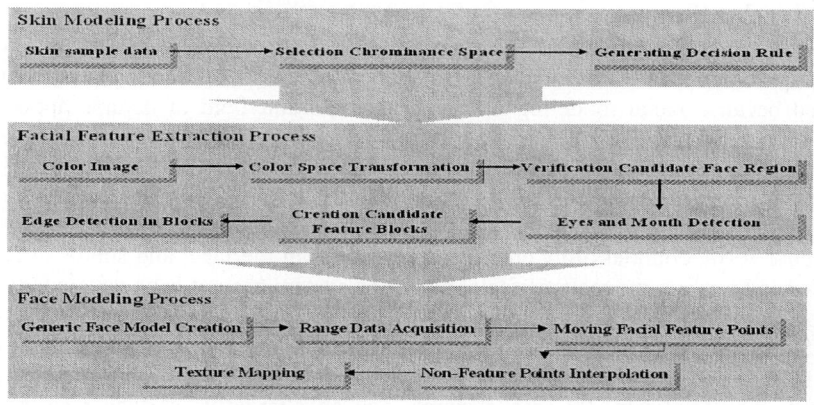

Fig. 1. Overview of the 3D Face Modeling Process

The system consists of three major phases: skin modeling process and facial feature extraction process and Face modeling process. In skin modeling process, we plot distribution histograms in combination of different chrominance information of skin sample data and generate proper skin chrominance model. The skin chrominance distribution by a function adopted in this work is Hue-Tint components. The decision rule to detecting candidate facial region is based on measuring distance between input data and the model to detect candidate face region. In the second phase, we verify candidate face region using feature's relational position information and apply local edge detection to extract facial feature. The third phase is the face modeling process. In this phase, we utilize the generic face model is for 3D reconstruction of input face image. The depth information of 3D face can be derived using 3D range data. The deformation of feature points and non feature points are performed to create a realistic 3D face model. The texture map of the detected facial region is applied to the face model in final stage.

2 Skin Color Modeling

Color information is efficient for identifying skin region. In computer vision, since every color spaces have different properties, color space selection is a very important factor for face detection. We need to consider some criteria for selecting an efficient skin color models: how it separates color information with chrominance and luminance data, a chrominance data can describe complex shaped distributions in a given space, and the amount of overlap between the skin and non-skin distributions in that space.

RGB is one of the most widely used color spaces for image processing. But, because it consists of combination chrominance with luminance data, we must transform from RGB into 2D chrominance space to get robustness to changes in illumination conditions. To date, various color models have been proposed for face detection, but many researches do not provide strict justification of their color models. In order to select proper color components for face detection, we plot 2D distribution of the two chrominance data that have often been used for skin detection. In order to plot distribution, we use skin sample images of 14 Asian that consist of dark and light images. Generally, since YCbCr, HSV, and TSL, have been proposed to achieve better color constancy, we select Cb, Cr components of YCbCr, H(hue), S(saturation) components of HSV, and T(tint), s components (we use small letter s in distinction from S component of HSV) of TSL[8].

Figure 1 shows 15 different chrominance spaces. Because skin sample data has 14 different face regions, the most of chrominance space have broad distribution. But, H-T chrominance distribution is denser and easier to be represented by modeling method than other chrominance distribution. And we calculate amount of the intersection between the normalized skin and non-skin histogram to evaluate the degree of overlap between the skin and non-skin distribution. Table 1 illustrates the value of intersection for each chrominance space.

Table 1. Comparison of amount of the intersection between skin and non-skin region

Color space	Intersection ratio	Color space	Intersection ratio
Cb-Cr	0.1311	Cr-H	0.1331
Cb-H	0.1311	Cr-S	0.1182
Cb-S	0.1218	Cr-T	0.1214
Cb-T	0.1218	Cr-s	0.1318
Cb-s	0.1318	H-S	0.1218
H-s	0.1282	S-s	0.1318
H-T	0.1182	T-s	0.1182
S-T	0.1311		

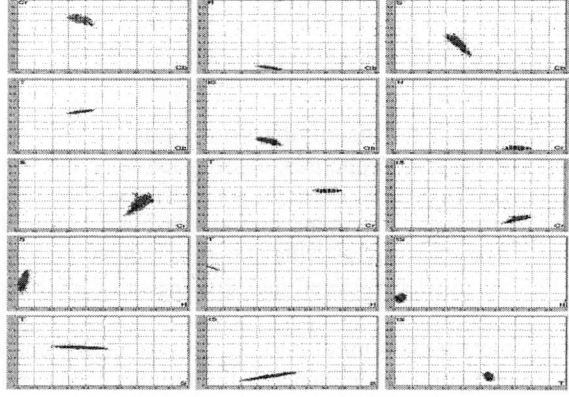

Fig. 2. Histograms in 15 different chrominance spaces of skin sample images. From top to bottom and left to right: Cb-Cr, Cb-H, Cb-S, Cb-T, Cb-s, Cr-H, Cr-S, Cr-T, Cr-s, H-S, H-s, H-T, S-T, S-s and T-s spaces

Once the combination of two chrominance components suitable for skin representation is selected, the function that models skin color distribution should be established. This function has to classify the color distribution into skin field and non-skin field and measures distance of input color value to skin field. In parametric skin modeling method it is common that the skin chrominance distribution is modeled by an elliptical Gaussian joint probability density function. But, because the boundary is defined by training data, in case of skin detection from different training image, a wrong result come out often. Therefore, nonparametric skin modeling methods can be used to evaluate skin color distribution from the training data without deriving an explicit model of the skin color. These methods are fast and independent to the shape of skin distribution.

The chrominance components we select here are Hue and Tint for skin detection. The shape of H-T distribution always looks like a straight line although distribution scatters longer than combinations of other components. We set up two points using the maximum and minimum values of H and T components. The skin chrominance distribution is modeled by a linear function defined as

$$f(h) = \frac{T_{min} - T_{max}}{H_{max} - H_{min}}(h - H_{min}) + T_{max} \quad , H_{min} \leq h \leq H_{max} \quad (1)$$

where h denote the H chrominance value of a pixel with coordinates (i, j), the T_{min}, H_{min} are minimum values of T and H components, the T_{max}, H_{max} represent the maximum values of T and H components. To detect skin region, a decision rule is used

$$d(x, y) = |f(x) - c(x, y)| \quad (2)$$

where $c(x, y)$ represents the measured values of the chrominance (H(i, j), T(i, j)) of a pixel with coordinates (i, j) in an image. If equation (4) is less than threshold λ_s, then we call it a face candidate pixel. Threshold value is obtained by comparison of every pixel over the skin samples with $f(c)$.

3 Facial Feature Extraction

For facial feature analysis, we divide human face in the middle horizontally and can get two areas that include eyes and mouth respectively. We inspect intensity variation to extract facial features in each area. In face candidate region, intensity of eyes, mouth and eyebrows have lower brightness than other region. The positions of eyes and mouth are determined by searching for minima in the topographic relief. We compute the mean value of every row and then search for minima in x-direction at the minima in the resulting y-relief. In figure 4, the eyes positions can be acquired if significant minima in x-direction are detected. We can get the mouth position using the same method.

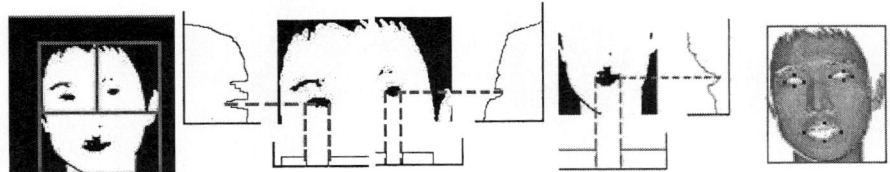

Fig. 4. Face, eyes and mouth position extraction

A feature points are significant factors of human face analysis, like the corner of the mouth or the tip of the nose. We extract minimal feature points that it is needed to face recognition and 3D face modeling. In previous section, we extract eyes and mouth positions, but cannot detect other facial features as the nose, cheekbone, brow ridge, jaw and so on. The standard proportions for the human face can be used to determine its proper feature position and find their orientation [9]. If we adapt this proportion information of a human face, we can detect rough positions of other facial features. But all the face is not composed by these proportions for the human face. We create a proper block that contains rough feature positions sufficiently and apply 4x4 sobel edge detection method to a created block. Figure 5 shows a step for extracting eye position. First and last variation positions in edge histogram are eye positions.

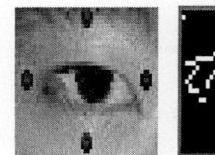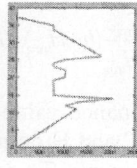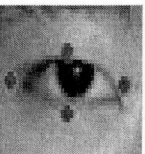

Fig. 5. Eye position extraction using edge detection

4 3D Face Modeling

For 3D face modeling a parametric pseudo-muscle face model developed by [10] is basically adopted. Then we modify the generic model into a 3D individual facial model using feature points detected in color images and fiducial points defined on a generic facial model.. We use one pair of orthogonal views of color images, and acquire a range data from these images using the analysis of fiducial points. This approach measures distance from the origin point to fiducial points and coordinates of these points. Fiducial points, which are extracted in the facial feature detection process, are some basic feature points to create face model. In this experiment, we consider 28 fiducial points. The tip of the nose is origin point in the front view and the rear of the ear is origin point in the side view. But it is difficult to detect the ear. So we set the nose position to origin point in the side view. Coordinates of the points may be stored with the corresponding points. These coordinates have to match with a generic model's fiducial points, which are already defined manually.

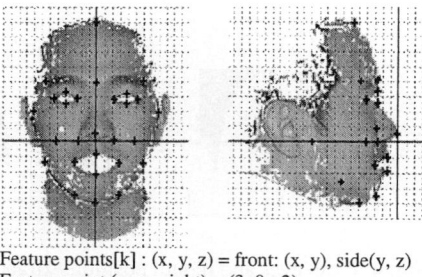

Feature points[k] : (x, y, z) = front: (x, y), side(y, z)
Feature point (nose_right) = (3, 0, -2)
Feature point (nose_top) = (0, 2, 0)
Feature point (lips_top) = (0, -4, -1)
Feature point (eyes_right_right) = (6, 10, -5)
Feature point (head) = (0, 16, -7)

Fig. 6. Fiducial points in a pair of orthogonal views of photographs and the coordinates of these points

We have to consider two fitting process; the one fits estimated fiducial points in generic model to corresponding feature points and the other modify non-fiducial points in generic model using interpolation technique. Interpolation is one way to manipulate flexible surfaces such as those used in facial models and probably the most widely used technique for facial animation. Basically, this method determines intermediate points using the given set of points in R^3. Since we have estimated two corresponding coordinates for feature points in color images and generic model in previous section, a fitting process using estimated fiducial points is conveniently computed. The displacement vector D^g is defined as a vector form the position in the original generic model p_i^{origin} to the corresponding position p_i^g in color images.

$$D_i = p_i - p_i^{origin} \tag{3}$$

The displacement vectors of non-fiducial points can be derived by the interpolation techniques. We use a radial basis function technique to interpolate scattered data. Unlike most other methods, this approach does not require any information about the connectivity of the data points and has mostly been used to solve the function reconstruction problem. An radial function is

$$\phi(D_k) = \phi(\sqrt{(x-x_k)^2 + (y-y_k)^2 + (z-z_k)^2}) \tag{4}$$

We find a smooth vector-valued funcion $f(p)$ fitted to the known data $D_i=f(p_i)$, from which we can compute $D_j=f(p_j)$ at vertices j, D_j is the displacement for every unconstrained vertex [10].

$$F(x,y,z) = \sum_i c_i \phi(D_i) \tag{5}$$

To determine the coefficients c_i, we solve the set of linear equations $D_i=f(p_i)$. Observe that the x, y, and z components of the c_i vectors can be computed independently using the set of equations

$$D_i = \sum_j c_i \phi(D_{ij}), \quad D_{ij} = \sqrt{(x_i-x_j)^2 + (y_i-y_j)^2 + (z_i-z_j)^2} \tag{6}$$

We use $\phi(D) = e^{-D/64}$ as radial function.

5 Experimental Results

To evaluate the efficiency of the proposed color model to detect facial region, we have applied the model to various color images that are exposed to varying light conditions. For the comparison of the skin color detection results, we both apply the proposed model and decision technique and currently used elliptical Gaussian joint probability density function for other color spaces. Figure 7 shows detected rough feature positions by H-T chrominance and detected feature positions by edge detection.

From the results of all of those cases illustrated in figure 7, we can prove the proposed skin color model efficiently detects skin color region rather than previously introduced skin color models. Subsequently, we apply our facial feature detection technique to 3D face modeling process. The figure 8 shows the the deformed face model and generated 3D human face model from photographs.

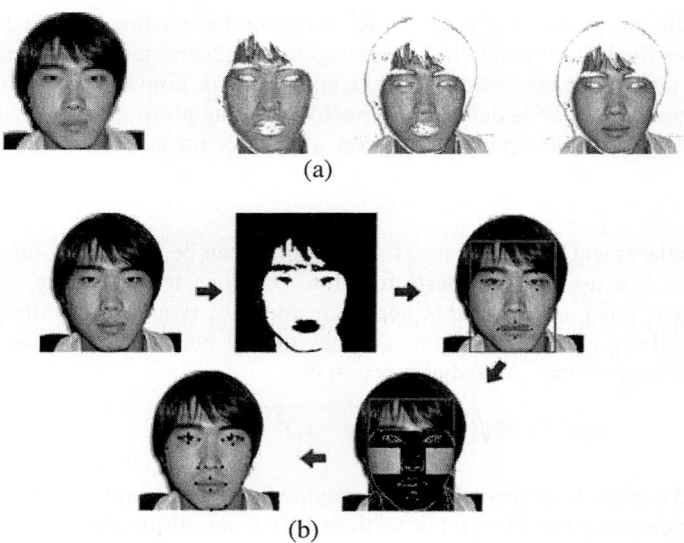

Fig. 7. (a) Skin region detection results (From left to right: Original image, results by H-T, Cb-Cr, T-s chrominance spaces) (b) Feature points detection results

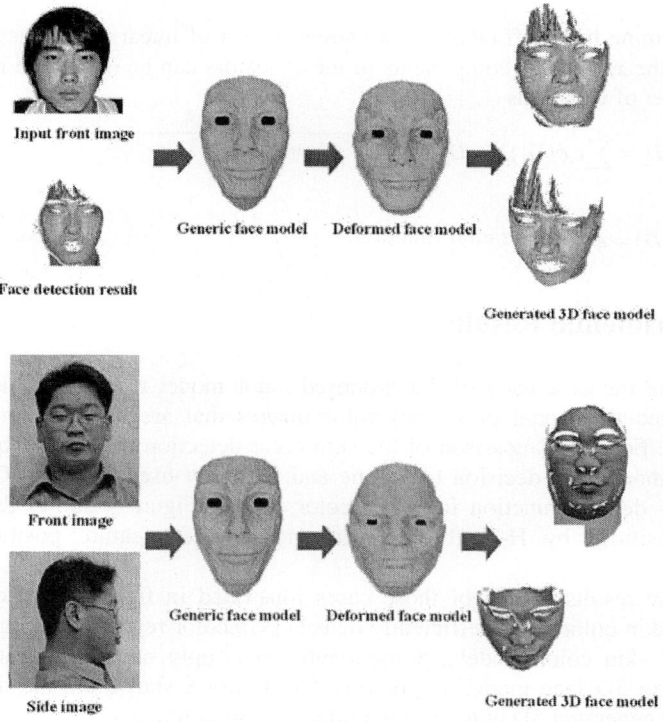

Fig. 8. Generated 3D face model

6 Concluding Remark

In this paper, we have introduced a unique method to detect face and facial feature points based on Hue-Tint color model. The proposed method projects color image on Tint-Hue chrominance space and detects skin region by using nonparametric skin model that is defined straight line equation. The facial modeling is processed by scattered data interpolation using radial basis function and texture mapping. Since there is difficulty to detect corresponding facial feature points from images, the most of facial modeling method require human intervention. However, based on the proposed approach, we can automate the overall 3D face modeling procedure. From experimental results, we can prove that the proposed skin color model is suitable to characterize human faces under various circumstances that are different lighting conditions and complex background. The deteted facial features and their exact poistions are used for generating a realistic 3D human face model from 2D images.

References

1. 1.F. I. Parke, *Computer generated animation of faces. Master's thesis*, University of Utah, Salt Lake City, UT, June 1972.
2. F. I. Parke, *A parameteric model for human faces*. PhD thesis, University of Utah, salt Lake City, UT, December 1974.
3. F. Pighin, J. Hecker, D. Lischinski, R. Szeliski, and D. H. Salesin, Synthesizing realistic facial expressions from photographs. *In Computer Graphics SIGGRAPH'98 Proceedings*, 1998, 231-242.
4. Volker Blanz, Thomas Vetter, A morphable model for the synthesis of 3D faces, *In Computer Grphics, SIGGRAPH'99 Conference Proceedings,* 1999, 187-194.
5. Richard Lengagne, Pascal Fua, Olivier Monga, 3D Face modeling from stereo and differential constraints.
6. E. Hjelmas, Boon Kee Low, Face detection: A Survey, Computer Vision and Image Understanding, 2001, 236-274
7. D. Maio and Maltoni, "Real-time face location on gray-scale static images," Pattern Recognition, vol. 33, no. 9, 2000, 1525-1539
8. Jean-christophe Terrillon, Shigeru Akamatsu, Comparative performance of different chrominance spaces for color segmentation and detection of human faces in complex scene images. In Proceedings of the 12th Conference on Vision Interface, 1999, 180-18
9. Bill Fleming, Darris Dobbs, *Animating facial features and expressions*, Jenifer L. Niles, MA:Charles River Media, 1999
10. Frederic I. Parke, Keith Waters, *Computer Facial Animation*, Wellesley, MA: A K Peters, 1996.

An Optimal and Dynamic Monitoring Interval for Grid Resource Information System

Angela Song-Ie Noh[1], Eui-Nam Huh[1], Ji-Yeun Sung[1], and Pill-Woo Lee[2]

[1] Division of Information & Communication Engineering,
Seoul Women's university,
Kongnung 2-dong, Nowon-gu, Seoul, Korea
{angela,huh,iris@swu.ac.kr}
[2] Korea Institute of Science and Technology Information,
Taejon, Korea
pwlee@kisti.re.kr

Abstract. Grid technology uses geographically distributed resources from multiple domains. For that reason, resource monitoring services or tools will run on various kinds of systems to find static resource information and dynamic resource information, such as architecture vendor, OS name and version, MIPS rate, memory size, CPU capacity, disk size, NIC information, CPU usage, network usage (bandwidth, latency), and memory usage, etcs. Thus monitoring itself may cause the system overhead. This paper proposes the optimal monitoring interval to reduce the cost of monitoring services and the dynamic monitoring interval to measure the monitoring events accurately. By simulating and implementing those two factors, we find out that unnecessary system overhead is significantly reduced and accuracy of events is satisfied.

1 Introduction

Grid computing is a form of distributed computing that involves coordinating and sharing computing, application, data, storage, or network resources across dynamic and geographically dispersed organizations [1]. Grid environment integrates computing resources in various kinds of applications, storage devices, and research devices. Eventually, Grid technology promises to solve the complex computational problems that lots of scientific researchers and business face to.

Grid resource management system composed of three big components manages a huge amount of diverse resources is managed by resource brokering services, information services, and resource manager services. Resource brokering services request the allocation of one or more resources for a specific purpose and the scheduling of tasks on the appropriate resources. Information services collect and offer the information about locality and status of resources. Moreover, resource manager services manipulate each resource and application.

We can infer that monitoring resources comprised of data collections and diagnoses of resource status in order to manage applications, networks and hosts detected with an error message or the system overload [2][3]. This monitoring must be done in real-time because the resources of clusters and nodes are dynamically changed. However, if a job is monitored in real-time with an intrusive approach, the system efficiency will be degraded. That is why an efficient monitoring cycle is important. For example, if we monitor in a short cycle even though there is no big changes in resource status, this will cause an increasing overhead by unnecessary monitoring. Otherwise, the resource management system will use the inaccurate information in its management. In this study, we analyze the changes in the resource status and propose a model for a monitoring cycle that minimizes the system cost by considering the operating loss and the system overhead.

This paper is organized as follows. In section 2, we present background needed for the monitoring interval problem. Section 3 and 4 introduce the dynamic monitoring interval and optimal monitoring interval. Section 5 shows the experiments and result of those theories and we will conclude in section 6.

2 Grid Monitoring Architecture

In this section, we will look over the Grid monitoring architecture (GMA). In GMA, the basic unit of timed monitoring data is called an event. The component that makes the event data available is called a producer, and a component that requests or accepts event data is called a consumer. A directory service is used to restore the event data in producer, which is currently available, and to contact for consumer to request it. These components and its architecture are shown in Fig.1 [4][5].

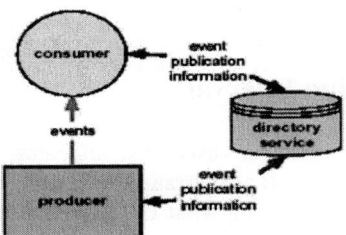

Fig. 1. Grid Monitoring Architecture Components

In detail, a producer accesses to directory service and notifies its existence by LDAP (Lightweight Directory Access Protocol). It can also have various kinds of producer interface to send event data to a consumer. A consumer can use directory service to request an event data from a producer and it has its own interface to receive an event data from a producer. A directory service allows the discovery of existence and properties of resources. Directory service allows users to add new entries by component, '*Add*', to record the changes in entries by '*Update*' component, and to delete entries by '*Remove*'. '*Search*' component helps users to query resources by name and attributes such as types, availabilities, or loads.

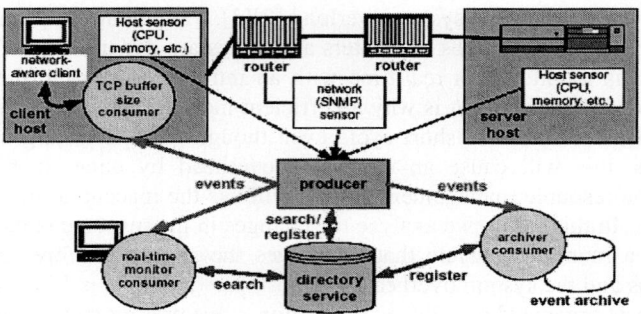

Fig. 2. Grid Monitoring Architecture

Fig.2 shows an example of the Grid monitoring architecture designed by NASA (National Aeronautics and Space Administration). Each event data is collected through each host and gives this information to producers. Producers register their information to directory service and network-aware clients, optimize the TCP buffer size with shielded data and transfer their events.

Current Grid monitoring systems do not even solve the fundamental problem of resource overhead, but just get more accurate resource information and devise a counterplan for an error. In Grid environment, periodic monitoring is essential to offer the correct information to users. However, in reality, periodic monitoring will bring out frequent overhead. Therefore, in next section, we propose the dynamic monitoring interval and the periodic monitoring system in order to resolve the periodic monitoring overhead.

3 Dynamic Monitoring Interval

Grid resource information manager discovers, allocates, and negotiates the use of network-accessible capabilities. It also arranges and utilizes this for its use and monitors its state description for checking what Grid resource information manager does. When the resource management system discovers the resource information, this system puts it into the list and monitors its system repeatedly. These monitored resources are composed of S/W and H/W information, dynamic system status information, and the decision to select proper resources is made by resource management system. The dynamic information, such as CPU usage, memory usage, and network status, is an important element in allocation and manipulation by resource managers.

Resource status and other monitoring components depend on CPU changes not only in distribute computing program, but also in Grid program. Therefore, if we observe the changes in CPU usage, we can approximately predict the changes of the other resources. For example, assume there are almost no changes in CPU usage, then there will be no big changes in other resource status. If there is a huge change in CPU usage, there will be many changes in other resource status. In other words, Grid monitoring events change dependent to CPU changes, so if there is a huge change in CPU usage, other resources should be monitored immediately. What is more, if we just observe the amount of CPU usage, there will be a frequent monitoring in high performance system using CPU. In this case, unnecessary overhead will be on the increase. To prevent this problem, this study will prove how efficient the monitoring by changes in CPU usage is.

If the CPU usage changes dynamically, it means the resource status has a strong possibility of changing widely. On that account, we should regulate the monitoring interval; so let the resource information service attain the updated resource status information. Namely, if the CPU usage is currently high, but begins to fall to almost zero, it means the resource status has been also changed. Therefore, the monitoring should start from in this point and the monitoring interval should be longer because the loss of the data was decreased. The change of the CPU usage was calculated on the basis of the difference between the predicted CPU usage and the current CPU usage. At this moment, if the difference is huge, we consider this as a big change in system, so control the monitoring interval [6].

In short, this method is comparing the predicted CPU usage and the current CPU usage. The current CPU usage is expressed as a percentage. e(k) is the result of subtracting current CPU usage from 100%, total CPU usage. In this study, we consider not only about current status, but also about previous status. In other words, we consider about dynamic threshold that uses PID (Proportional Integrated Differential) controller [7]. We predict u(k) as the required amount of resources for next allocation to reach 100% of CPU.

$$u(k) = (k_p \times e(k) + k_i \times \sum_{j=1}^{k} e(j) + k_d \times (e(k) - e(k-1)) \quad (1)$$

Here, each PID parameter is set as $k_p = -0.5$, $k_i = 0.125$, $k_d = -0.125$. Precisely, u(k) is expressed by CPU idle value, e(k) in PID controller. In addition, |u(k)-u(k-1)| shows how much of CPU usage will be changed from the current CPU information.

$$|u(k) - u(k-1)|$$
$$= (k_p \times e(k) + k_i \times \sum_{j=1}^{k} e(j) + k_d \times (e(k) - e(k-1))$$
$$- (k_p \times e(k-1) + k_i \times \sum_{j=1}^{k} e(j) + k_d \times (e(k-1) - e(k-2))$$
$$= k_p \times (e(k) - e(k-1)) + k_i (\sum_{j=1}^{k} e(j) - \sum_{j=1}^{k-1} e(j))$$
$$\cong k_i \times e(k) \quad (2)$$

If the system is steady state, e(k), e(k-1), and e(k-2) are identical. Therefore, we can derive |u(k)-u(k-1)| by employing control value, u(k). In short, we can simply compare |u(k)-u(k-1)| and $k_i * e(k)$. When both two values equal to each other, or |u(k)-u(k-1)| is smaller than $k_i * e(k)$, it means the current CPU usage has not been changed dynamically comparing with the previous stage. However, when the situation is reversed, then it means that the system changes are significant. Therefore, it is time to change the monitoring interval. This is our approach to control the monitoring interval dynamically.

4 Optimal Monitoring Interval

In Grid environment, the resource management system continuously asks resource status information through monitoring system, so the monitoring system is always updated when the new information comes. Therefore, in previous section, we proposed to control the monitoring interval by the observation of CPU usage changes. In this section, we propose to change monitoring interval by intention depending on the condition of the system, so we deserve the precise information to the monitoring system and prevent wasting unnecessary monitoring cost and the overhead. There are three variables to check efficient monitoring system. First variable is the cost for the overhead for monitoring to deliver accurate resource information to users. Second variable is the loss ratio caused by having inaccurate resource status information and the last variable is resource status changing rate. Those three variables will be applied to formulate the monitoring interval.

We define the changing possibility of resource status during a monitoring interval as p. p is dependent to the condition of the system, which relies on the number of users and processes. To get precise condition of the system, we have to check real-time monitoring data. However, this will cause the overhead by running both real-time monitoring system and the other with the controlled monitoring interval. Instead, the administrator considers the number of current users and processes, and sets initial value for p. The overhead by using system resources during a monitoring interval will be defined as M. The cost of overhead is calculated by the sum of the usage of CPU, memory, and network, so it will be shown by percentage. The loss ratio is caused by uncompleted jobs because of the changed resource status information and it will be defined as L. Lastly, the optimal monitoring interval will be set as t. The changing possibility, p, is set up by the administrator, and the cost of overhead relies on each system. Therefore, in short, the optimal monitoring interval is dependent to the value, L.

With those four variables, we formulate the optimal monitoring interval. As the first step, the monitoring interval is set to t, and the changing possibility of resources during t time is set to p.

$$\sum_{i=0}^{t} P_i = p \sum_{i=0}^{t} i = \frac{pt(t+1)}{2} \qquad (3)$$

Next, the cost of loss caused by changed resource status during the monitoring interval equals to the loss ratio during t hours and it can be shown like this.

$$L \frac{pt(t+1)}{2} \qquad (4)$$

If we add the cost of the overhead, it indicates the total cost for whole monitoring system during the monitoring interval. In short, it is the total cost during the monitoring interval.

$$M + L \frac{pt(t+1)}{2} \qquad (5)$$

This formula shows the cost of monitoring per period, t.

$$c = \frac{M + L\frac{pt(t+1)}{2}}{t} \tag{6}$$

To get the optimal monitoring interval, t_{opt}, we should minimize the cost of monitoring per an hour. With differentiation, the cost of monitoring can be minimized like this.

$$\frac{\partial C}{\partial t} = \frac{Lp}{2} - \frac{M}{t^2} = 0, t_{opt} = \sqrt{\frac{2M}{LP}} \tag{7}$$

If we use Formula (7), we can simply calculate the optimal monitoring interval by the loss ratio. In other words, if the changing rate is small, so the loss ratio, L, becomes small, then the optimal monitoring interval will be longer.

5 Experiments and Results

We made three experiments based on the theory of section 3 and 4. First one is about the optimal monitoring interval, and the second one is finding out when to start monitoring by measuring CPU status changing rate. The last experiment shows the previous experiments are working efficiently in real Grid system.

In this first experiment, we will find out the optimal monitoring interval, t_{opt}, using previously mentioned variables. Suppose multimedia data is resource. The ratio of lost packets is expressed as L. The total cost for monitoring, M, is calculated by the average of the usage rate of CPU, memory and network. The changing possibility of resource information is set to p, and we will find out the optimal monitoring interval, t_{opt}, with this value, p, the loss ratio, L, and the overhead, M.

Table 1. Optimal Monitoring

\multicolumn{3}{c}{L = 20 %}			L = 40 %			L = 80 %	
P	M(%)	t_{opt}	p	M(%)	t_{opt}	p	M(%)	t_{opt}
0.2	20	3.1623	0.2	20	2.2361	0.2	20	1.5811
0.2	50	5	0.2	50	3.5355	0.2	50	2.5
0.2	80	6.3246	0.2	80	4.4721	0.2	80	3.1623
0.5	20	2	0.5	20	1.4142	0.5	20	1
0.5	50	3.1623	0.5	50	2.2361	0.5	50	1.5811
0.5	80	4	0.5	80	2.8284	0.5	80	2
0.8	20	1.5811	0.8	20	1.118	0.8	20	0.7906
0.8	50	2.5	0.8	50	1.7678	0.8	50	1.25
0.8	80	3.1623	0.8	80	2.2361	0.8	80	1.5811

The first experiment shows the result as shown in Table 1. The bigger the changing possibility of system is, the shorter the monitoring interval should be. This result is observed when the loss ratio and the overhead by monitoring are fixed. Furthermore, we find out that the monitoring interval for the same resource should be shorter when the rate of the loss is bigger. To simply this fact, we changed this data into a graph. This is the case when the overhead, M, is 20% and the changing possibilities of resources are 0.2, 0.5, and 0.8.

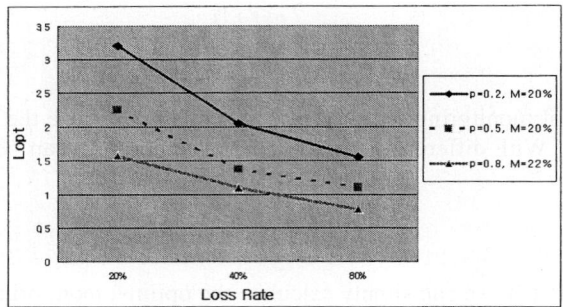

Fig. 3. Overhead by monitoring is 20%

When the possibility, p, increases from 0.2 to 0.8, it signifies the resource status is changing quite a lot. Hence, the monitoring interval, t_{opt}, should be shorter. Moreover, if p is fixed by 0.5, and the loss ratio is increasing from 20% to 80%, then the monitoring interval, t_{opt}, is also getting shorter.

We will prove the necessity of monitoring by checking the changed CPU amount and resource status. We divided into two situations, - when the changing rate of the system is smooth and dynamic.

Table 2. Smooth changing rate of the system

		Smooth monitor			
CPU usage	e(k)	u(k)	u(k-1)	|u(k)-u(k-1)|	ki*e(k)
53	47	117.75	128.75	11	5.875
57	43	96.875	102.125	5.25	5.375
56	44	101.825	96.875	4.95	5.5
54	46	111.875	101.625	10.25	5.75
53	47	117.375	111.875	5.5	5.875
52	48	122.875	117.375	5.5	6
51	49	128.5	122.875	5.625	6.125
61	39	78.125	82.875	4.75	4.875
60	40	82.375	78.125	4.25	5
59	41	87	82.375	4.625	5.125
• 58	42	91.75	87	4.75	5.25
• 53	47	118.875	91.75	27.125	5.875
54	46	112.25	116.875	4.625	5.75
55	45	107	112.25	5.25	5.625
56	44	101.875	107	5.125	5
• 55	45	106.75	101.875	4.875	4.875
• 60	40	83.125	106.75	23.625	5
61	39	78.125	83.125	5	4.875
60	40	82.375	78.125	4.25	5
61	39	78.125	82.375	4.25	5.625
60	40	82.375	78.125	4.25	5.75
55	45	106.25	82.375	23.875	5.625
54	46	112	106.25	5.75	5.75
55	45	107	112	5	5.625
66	44	101.875	107	5.125	5.5
55	45	106.75	101.875	4.875	5.625
66	34	57.25	55.125	2.125	4.25
65	35	61.125	57.25	3.875	4.375
66	34	57.5	61.125	3.625	4.25

First experiment is the case when the system changed smoothly as shown in Table 2 and Fig.4. Even though the resource status has changed a little, it still means it has changed. Therefore, we have to change the monitoring interval, applying this rule.

When the change of the resources is big as the interval between '*' marks in Table 2, |u(k)=u(k-1)| is larger than ki * e(k), so the system controlled its monitoring interval. As a result, the change of the resources has been decreased than before.

An Optimal and Dynamic Monitoring Interval for Grid Resource Information System

The second environment is the system that has a dynamic changing possibility shown in Fig. 5. For example, when the resource information has changed widely as 5 to 20, if we hastily conclude and changed the monitoring interval shorter, the system can be overhead. In this case, we should carefully see the rate of the CPU usage regularly and check the big change and the small change. After seeing that, we can conclude that the resource status has been changed. When the value, |u(k)-u(k-1)|, is smaller than the value, $k_i * e(k)$, the change of the resource status is small, so we should reset the monitoring interval longer. For instance, the availability of resource information has increased from 46% to 60%, then the value, |u(k)-u(k-1)|, will be larger than $k_i * e(k)$ is. This will be simplified at the Fig. 5.

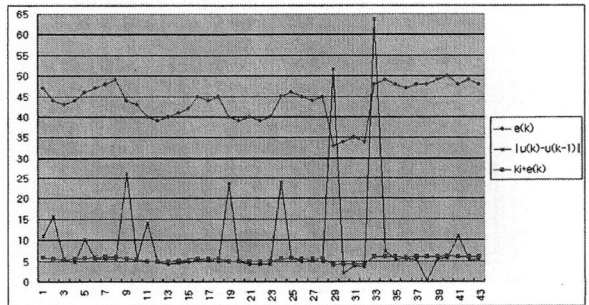

Fig. 4. Having Smooth changing Possibility

Even the value, |u(k)-u(k-1)|, is very large, if much bigger change comes, then the edge will be much higher. At this moment, |u(k)-u(k-1)| will be larger than the value, $k_i * e(k)$, so we will control the monitoring interval into much shorter one. The proposed model is also applied to the CPU intensive military applications that are real-time task sets running on very dynamic environment. We initially put the constant threshold by 5% of CPU to detect the system changes. In this case, we observed many unnecessary triggers. However, when our model is applied to those systems, 20% of triggers have been decreased. In short, we checked our theory for the monitoring interval really works.

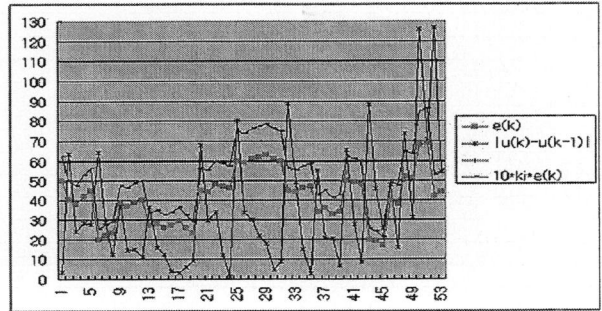

Fig. 5. Having Dynamic changing possibility

Lastly, we compare the monitoring system we suggested and the Cluster, which has the same environment as Grid environment. This cluster is Pentium III CPU 820 MHz, Memory 256 MB, HDD 10GB. This system monitoring itself in a second precisely shows the current resource status. Therefore, you can see how compatible this method is for controlling the monitoring interval.

In this experiment as shown in Fig. 6, we test the changing rate of optimal interval as the rate of the loss changes. When the rate of the loss is larger, we regulate the monitoring interval shorter and make the rate of the loss is getting smaller. In addition, when the rate of the predicted loss is smaller than calculated data, then the monitoring system doesn't check itself and prevents its overhead. The interval with no graph means the changing rate of the system will not effect to the system, so the monitoring system, itself, has reduced its overhead by resources.

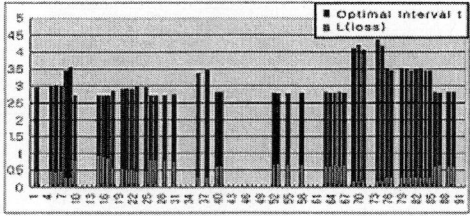

Fig. 6. Optimal Interval, t_{opt}, by the rate of the loss, L

Last experiment is about the comparison of the monitoring data in each second by the rate of the loss, L, and the status of real resources. Those two make the similar shape of graphs, and it shows the optimal monitoring system truly reduces the overhead by monitoring and gets precise Grid resource status information. Therefore, it proves this system is much efficient than the system that has the regular monitoring interval.

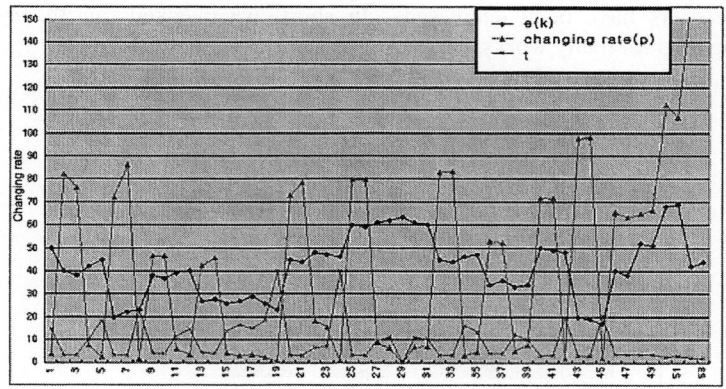

Fig. 7. The change of monitoring interval, t, by the rate of resource status

Unfortunately, this system has a defect that it cannot grasp the resource status in dynamic changes in a short period of time. However, the Grid system does not shows the dynamic changes in short time when transporting resources, but shows the smooth changes. Therefore, this theory can be applied to real Grid monitoring system.

6 Conclusion

In Grid environment, the monitoring process should be preceded efficiently before the resource management system analyzes the resource information. This study infers the resource status by changing rate of CPU usage is collected accurately and non-intrusively. When we notice the big changes in resources, we start the monitoring for resources so to reduce the overhead by wrong prediction and changed resource status. This theory is based on current Grid monitoring system. In this study, we add an effective module, the optimal monitoring system, to existing monitoring system. This study can be easily applied into current Grid monitoring system without changing any source code or application.

References

[1] I. Foster, "The Grid: A New Infrastructure for 21st Century Science." Physics Today, 55(2):42-47 2002.
[2] A. Chervenak, E. Deelman, I. Foster, L. Guy, W. Hoschek, A. Iamnitchi, C. Kesselman, P. Kunst, M. Ripeanu, B, Schwartzkopf, H, Stock-inger, K. Stockinger, B. Tierney, "Giggle: A Framework for Constructing Scalable Replica Location Services." Proceedings of Supercomputing 2002 (SC2002), November 2002.
[3] Tierney, B., B. Crowley, D. Gunter, M. Holding, J. Lee, M. Thompson "A Monitoring Sensor Management System for Grid Environments" Proceedings of the IEEE High Performance Distributed Computing conference (HPDC-9), August 2000, LBNL-45260.
[4] B. Tierney, R. Aydt, D. Gunter, W. Smith, M. Swany, V. Taylor, R. Wolski, "A Grid Monitoring Architecture", Global Grid Forum Performance Working Group, March 2000
[5] TopoMon: A Monitoring Tool for Grid Network Topology, Proceedings of ICCS 2002.
[6] Monitoring event, "Discovery and Monitoring Event Description (DAMED-WG)", http://www- didc.lbl.gov/ damed/
[7] PID Tutorial, http://www.engin.umich.edu/group /ctm/PID/PID.html

Real Time Face Detection and Recognition System Using Haar-Like Feature/HMM in Ubiquitous Network Environments

Kicheon Hong, Jihong Min, Wonchan Lee, and Jungchul Kim

Dept. of Information and Telecommunications Engineering,
The University of Suwon, South Korea
{kchong, cklove77, wondolcp, chaos11}@suwon.ac.kr
http://misl.suwon.ac.kr

Abstract. In this paper, a real time face detection and recognition system is introduced for applications or services in Ubiquitous network environments. The system is realized based on a Haar-like features algorithm and a Hidden Markov model (HMM) algorithm using communications between a WPS(Wearable Personal Station) 350MHz development board and a Pentium III 800 MHz main server communicating with each other by a Bluetooth wireless communication method. Through experimentation, the system identifies faces with 96% accuracy for 480 images of 48 different people and shows successful interaction between the WPS board and the main server for intelligent face recognition service in Ubiquitous network environments.

1 Introduction

Face recognition is a technique of identifying a person by comparing the person's still image or moving picture with a given database of face images. Face recognition acquires biometric information from a person to be verified in a less intrusive manner than other biometric recognition techniques, such as fingerprint recognition, without requiring the person to directly contact a recognition system with part of his or her body. However, the face is subjected to various changes and is more sensitive to the surroundings than other parts of the body. Thus, recognition rates of face recognition systems are relatively lower than recognition rates of other biometric recognition systems. Therefore, this paper suggests a Haar-like features-based algorithm [1] and a HMM algorithm method as a way to efficiently extract a face image and enhance face recognition rates. A detailed description of a face recognition system based on Haar-like features and the HMM approach will be presented below. In addition to these algorithms, a real-time face recognition system is realized based on the assumption that a WPS 350MHz application development board [2] and a Pentium III computer are used as a portable device and a main server, respectively, for intelligent face recognition service in ubiquitous network environments.

2   Face Detection Method Using Haar-Like Features

Haar-like features and the AdaBoost learning algorithm are two leading face detection algorithms. Haar-like features are widely used in face searching and numerous prototypes have been trained to accurately represent human faces through the AdaBoost learning algorithm [3]. In a Haar-like feature approach, feature values are obtained by summing up the values of pixels in each region of a face image and weighting and then summing up the regional sums, instead of directly using the values of the pixels of the face image. Fig. 1 illustrates various prototypes of Haar-like features. Referring to Fig. 1, the locations of Haar-like features may vary in a window according to the characteristics of a face. Therefore, the Haar-like features may have various values depending on the type of image to be recognized.

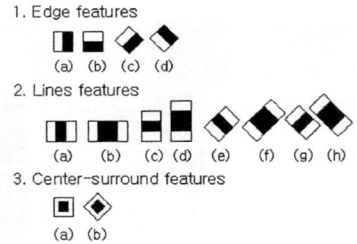

Fig. 1. Prototypes of Haar-like features

The AdaBoost algorithm, which is simple but efficient compared to other boost algorithms, is used for more precisely differentiating a facial area from a non-facial area and boosting face recognition rates. In the AdaBoost learning algorithm, locations of prototypes of Haar-like features can render more detailed facial characteristics especially at a higher stage of classification. Haar-like features obtained using the AdaBoost learning algorithm are classified in stages, as shown in Fig. 2.

The classification of Haar-like features is carried out in stages in order to realize a more robust face recognition algorithm. Image data of a face image is received before the classification and a window is laid over a facial area of the face image to obtain Haar-like features of the facial area. Some of the Haar-like

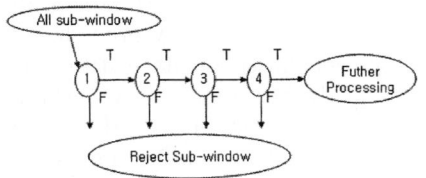

Fig. 2. Stage Classifiers Using AdaBoost

features are selected using the AdaBoost learning algorithm, and the selected Haar-like features are classified and then stored. In a first stage of classification, the smallest number of Haar-like features, i.e., 9 Haar-like features, are classified. In a 25th stage of classification, a total of 200 Haar-like features are classified. Data obtained by the first through 25th stages of classification is stored as a text file and is used as a hidden cascade in face recognition.

In this paper, the size of the window is set to 24 × 24. In addition, the number of Haar-like features processed at a higher stage of classification is larger than the number of Haar-like features processed at a lower stage of classification, and prototypes of Haar-like features at a higher stage of classification can render more detailed facial characteristics than at a lower stage of classification [4]. A total of 25 stages of classification are performed, and 200 Haar-like features are generated at the 25th stage of classification.

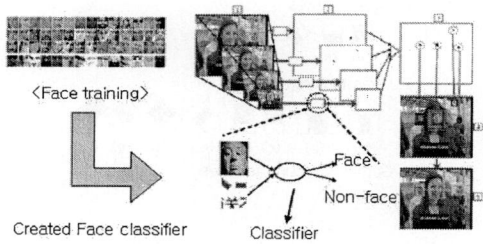

Fig. 3. Block diagram of Haar Face Detection

Fig. 3 illustrates a block diagram of a face detection method using Haar-like features. Referring to Fig. 3, a face area of an image of one frame is determined using a trained group of face images while gradually decreasing the size of the image in a pyramid manner. A plurality of candidate areas for the face area of the image are generated in the process of recovering the size of the image, and an average of the candidate areas is output.

3 Face Recognition Method Using HMM Algorithm

The HMM algorithm is based on the Markov assumption that the probability of an event at a given moment of time depends only on the given moment of time, rather than any previous moment of time. Markov states are hidden but may be observable only through other probability processes. According to the HMM algorithm, a face is recognized based on its features. Prominent features of a face include the hair, forehead, eyes, nose, and mouth. These facial features are modeled using a one-dimensional (1D) HMM. Markov states are dependent on this modeling process and are applied to a face image in the order from the top to the bottom of the face image. The face image is segmented in consideration of such elements of a face as the eyes, nose, ears, and mouth, whose locations

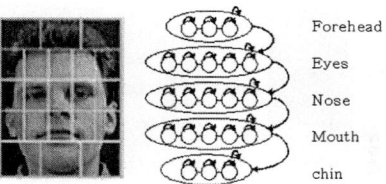

Fig. 4. 2D HMM Used for Face Recognition

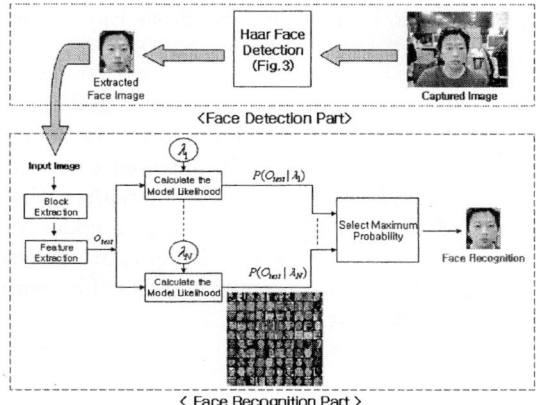

Fig. 5. Block diagram of Face Detection and Recognition System

in the face are fixed. The 1D HMM achieves a face recognition rate of about 85% [5], [6], [7], [8]. The 1D HMM can be extended to a pseudo two-dimensional (2D) HMM [9]. The pseudo 2D HMM is generated through the expansion of each block of a face image. Referring to Fig. 4, a super state is divided into a number of states.

The probabilities of the states are calculated, and then the probability of the super states is calculated based on the probabilities of the states. In this manner, subsequent super states are processed. Fig. 5 shows a block diagram of a face detection and recognition system.

4 Realization of System

A real-time face recognition system may be realized based on the assumption that the WPS and a Pentium III computer are used as a portable device and a server, respectively, according to the following two scenarios. In a first scenario, the WPS only captures an image rendering a person's face and the surroundings, compresses and stores the captured image using a JPEG codec, and transmits the compressed image to a server, i.e., a Pentium III computer through a Bluetooth wireless communication method. The server automatically extracts a face area using Haar-like features and searches a database for a face image that is a match

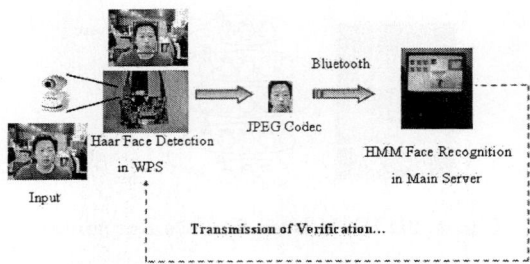

Fig. 6. Scenario Diagram in Ubiquitous Environments

for the extracted face area using an HMM algorithm. Finally, it identifies a person to which the searched face image belongs, and transmits the name of the person to the WPS. In a second scenario, the WPS automatically extracts a face area from a captured image using Haar-like features. Thereafter, the WPS compresses and stores the extracted face image using a JPEG codec and transmits the compressed face image to the server using a Bluetooth wireless communication method as shown in Fig. 6 in ubiquitous environments. The remaining operations are exactly the same as in the first scenario.

4.1 Photo of System

Fig. 7 shows the WPS designed with a size of 9 cm × 5.5 cm and a wireless communication part between the WPS and the main server. Fig. 8 shows a real-time face recognition system designed for virtual experimentation using real-time face recognition processes based on communications between the WPS and the main server.

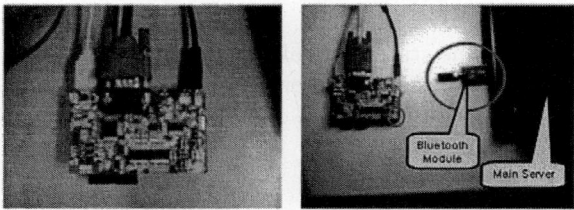

Fig. 7. WPS development board(left) and wireless communication parts of WPS and Main Server(right)

5 Simulation Results

5.1 Case of no Scenarios

Experiments were carried out using the main server without any communication path. The main server was assumed to be a system for performing both

Fig. 8. Realization of Face Recognition System

face detection and recognition. In the experiments, 48 face images of 8 people (6 face images per person) and a trained database of 240 face images of 40 people (6 face images per person) were used. Face areas were extracted from the respective face images using Haar-like features and then stored as BMP files, as shown in Fig. 5. In the experiments, the 48 face images of 8 people were taken care of separately from the 240 face images of 40 people to show that face areas can be automatically extracted not only from still pictures but also from moving pictures. Table 1 shows trained location information and critical values of prototypes of Haar-like features in a 24 × 24 window.

Table 1. Classification of Trained Database of Face Images

Stage Numbers	extern const char*FaceCascade[]
Stage 0	" 9 1 2 6 4 12 9 0 -1 6 7 12 3 0 3 haar_y3 -0.031512 0 -1 2.087538 -2.217210 1 2 6 4 12 7 0 -1\n" "10 4 4 7 0 3 haar_x3 0.012396 0 -1 -1.863394 1.327205 1 2 3 9 18 9 0 -1 3 12 18 3 0 3\n"......" haar_y2 0.005974 0 -1 -0.859092 0.852556 -5.042550\n"
Stage 1	" 16 1 2 6 6 12 6 0 -1 6 8 12 2 0 3 haar_y3 -0.021110 0 -1 1.243565 -1.571301 1 2\n" " 6 4 12 7 0 -1 10 4 4 7 0 3 haar_x3 0.020356 0 -1 -1.620478 1.181776 1 2 1 8 19 12 0 -1\n"
. .	. .

Fig. 9 shows that the size of the forehead in face images of those who failed to be verified is larger than in face images of those who were successfully verified. Because of a larger forehead, the eyes in the face images of those who failed to be verified deviate from expected locations when segmenting the corresponding face images in a 2D HMM-based training process. Table 2 shows the results of testing a face recognition system.

5.2 Case of Two Scenarios

The performance of a real-time face recognition system (hereinafter referred to as a first system) realized based on the first scenario was compared with the

Fig. 9. Face Identification Results

Table 2. Test Results

Number of Experiments	Total Number of Participants in Exeriments	Number of People Who Were Successfully Verified	Number of People Who Failed to be Verified	Verification Success Rate	Verification Failure Rate
10	48	46	2	95.83%	4.17%

Table 3. Experimental Results of First and Second Scenarios

Scenarios / Results	First Scenario	Second Scenario
Number of Experiments	10	10
Total Number of Participants in Experiments	48	48
Number of People Who Were Successfully Verified	46	45
Numerber of People Who Failed to be Verified	2	3
Verification Success Rate	95.83%	93.75%
Verification Failure Rate	4.17%	6.25%
WPS's Face Detecting Time	×	About 4.1 seconds
Total of Data Communication and Main Server Processing Time	About 2.1 seconds	About 0.7 seconds

performance of a real-time face recognition system (hereinafter referred to as a second system) realized based on the second scenario through experimentation using a database of 48 face images of 48 people. The time taken for the development board of the first or second system to obtain the name of each person to be verified from the server of the first or second system was measured. Table 3, the second system in the second scenario has a higher verification failure rate than the first system in the first scenario because a communication error occurred in the process of transmitting face images from the WPS to the server. Table 3 also shows that the first system can complete face verification about 2.7 seconds faster than the second system. Even though the total processing time in the second scenario is longer than the time taken in the first scenario, the second system is very useful for reducing the transmission bandwidth by splitting the face detection and recognition operations into two separate parts. In addition to it, the system is also efficient for a face recognition system that operates on a large data base requiring high computational load and a large memory size not suitable for WPS in ubiquitous environments.

6 Conclusions and Discussion

The structure, operation, and performance of a face verification system using Haar-like features and an HMM algorithm in ubiquitous network environments have been described above.The face verification system is expected to be very useful because it can perform face recognition processes not only on still images but also on moving pictures even in real time. The face verification system can verify faces very quickly using Haar-like features and can be easily applied to moving pictures because of its short face verification time. In addition, since the face verification system applies face area data extracted from a face image to an HMM, it is possible to quickly perform face recognition on video image data in real time. Face recognition rates of the face verification system can be boosted by HMM using more detailed information on the location of the eyes and the mouth and the skin color of faces to be verified. Moreover, the face verification system can be separated into two parts and be realized based on Bluetooth wireless communications between a portable device (e.g., Motorolas i.MX21 application development board) and a server (e.g., a computer) for ubiquitous network environments.

References

1. P.Viola, M.J.Jones, "Robust real-time object detection", Technical Report Series, Compaq Cambridge research Laboratory, CRL 2001/01,Feb. 2001.
2. ETRI Technical Report, "WPS Test Bed specification" , 2004.
3. Sunghoon Park, Jaeho Lee, Whoiyul Kim, "Face Recognition Using Haar-like feature/LDA", Workshop on Image Processing and Image Understanding '04 (IPIU 2004), Jan. 2004.
4. Alexander Kuranov, Rainer Lienhart, and Vadim Pisarevsky. An Empirical Analysis of Boosting Algorithms for Rapid Objects With an Extended Set of Haar-like Features. Intel Technical Report MRL-TR, July02-01, 2002.
5. Ziad M.Hafed and Martin D. Levine, "Face Recognition using the discrete cosine transform," International Journal of Computer Vision, Vol. 43, No.3,pp.167-188, July 2001.
6. F. Samaria and S. Young, "HMM based architecture for face identification," Image and Computer Vision, vol. 12, pp.537-543, October 1994.
7. A. V. Nefian and M. H. hayes, "A Hidden Markov Model for face recognition," in ICASSP 98, vol. 5, pp. 2721-2724, 1998.
8. A. V. nefian and M. H. hayes, "Face detection and recognition using Hidden Markov Models," in International Conference on Image Processing, 1998.
9. S. Kuo and O. Agazzi, "Keyword spotting in poorly printed documents using pseudo 2-D Hidden Markov Models, "IEEE Transactions on Patten Analysis and Machine Intelligence, vol. 16, pp. 842-848, August 1994.

A Hybrid Network Model for Intrusion Detection Based on Session Patterns and Rate of False Errors

Se-Yul Lee[1], Yong-Soo Kim[2], and Woongjae Lee[3]

[1] Department of Computer Science, Chungwoon University,
San29 Namjang-Ri, Hongseong-Eup, Hongseong-Gun, Chungnam, 350-701 Korea
Pirate@cwunet.ac.kr
[2] Division of Computer Engineering, Daejeon University,
96-3 Yongun-Dong, Dong-Gu, Daejeon, 300-716 Korea
kystj@dju.ac.kr
[3] Division of Information & Communication Engineering, Seoul Women's University,
9126 KongNeung-Dong, Nowon-Gu, Seoul, Korea 139-774
wjlee@swu.ac.kr

Abstract. Nowadays, computer network systems play an increasingly important role in our society. They have become the target of a wide array of malicious attacks that can turn into actual intrusions. This is the reason why computer security has become an essential concern for network administrators. Intrusions can wreak havoc on LANs. And the time and cost to repair the damage can grow to extreme proportions. Instead of using passive measures to fix and patch security holes, it is more effective to adopt proactive measures against intrusions. Recently, several IDS have been proposed and they are based on various technologies. However, these techniques, which have been used in many systems, are useful only for detecting the existing patterns of intrusion. It can not detect new patterns of intrusion. Therefore, it is necessary to develop new technology of IDS that can find new pattern of intrusion. In this paper, we propose a hybrid network model for IDS based on reducing risk of false negative errors and false positive errors that can detect intrusion in the forms of the denial of service and probe attack detection method by measuring the resource capacities. The "IDS Evaluation Data Set" made by MIT was used for the performance evaluation.

1 Introduction

Nowadays, computer network systems play an increasingly important role in our society. They have become the targets of a wide array of malicious attack that invariably turn into actual intrusions. This is the reason why computer security has become an essential concern for network administrators.

Intrusions often can wreak havoc on LANs. And the time and cost to repair the damage can grow to extreme proportions. Instead of using passive measures to fix and patch security holes, it is more effective to adopt proactive measures against intrusions.

In addition to the well-established security measures such as data encryption, message integrity, user authentication, user authorization, intrusion detection techniques can be viewed as additional safeguard for computer networks.

A basic premise for intrusion detection is that a distinct evidence of legitimate activities and intrusions will be manifest in the audit data. Because of the amount of data which consists of records and system features, efficient and intelligent data analysis tools are required to discover the effect of intrusions on systems. IDS techniques in many systems are useful only against existing patterns of intrusion and can not detect new patterns of intrusions. Therefore it is necessary to develop IDS that find new patterns of intrusions.

Many researchers discussed the problems of current intrusion detection systems. The followings are the limitations of current IDS.

- Mainly misuse detection systems based on rule-based expert system: vulnerable to intruders who use new patterns of behavior.
- The acquisition of these rules is a tedious and error-prone process (lack of flexibility and maintainability).
- Lack of predictive capability: can not predict the possibility of attack.
- Lack of automatic machine learning capability.
- High rate of false alarm or missing alarm.
- Difficult for organizations to apply their security policies to IDS.

Current rule-based systems suffer from an inability to detect attacker's scenarios that may occur over an extended period of time. Rule-based systems also lack flexibility in the rule to audit. Slight variations in a sequence of attacks can affect the intrusion detection. While increasing the level of abstraction of the rules does provide a partial solution to this weakness, it also reduces the granularity of the intrusion detection system [1, 2].

A false positive is that IDS sensor misinterprets one or more normal packets or activities as an attack. It can degrade the effectiveness of the systems in protection by invoking unnecessary countermeasures. IDS operators spend too much time on distinguishing events that require immediate attention from events that have low priority or are normal events in a particular environment.

Most of IDS sensors show less than a 5% rate of false positives. A false negative occurs when an attacker is misclassified as a normal user. It is difficult to distinguish between intruder and normal users and to predict all possible false negatives and false positives due to the enormous variety and complexity of today's network. IDS operators rely on experience to identify and resolve unexpected false errors.

In this paper, a false negative error is denoted by "F_n error". And a false positive error is denoted by "F_p error". Table 1 describes the result about the F_n error and F_p error.

Table 1. F_n error and F_p error

Predicted / Actually	Predicted non-attack	Predicted attack
Actually non-attack	Correct	F_p error
Actually attack	F_n error	Correct

The purpose of IDS is to distinguish between attackers and normal users. It is difficult to remove all possible errors due to the enormous variety and complexity of today's networks. The success of IDS can be characterized by both false alarm rates and detection efficiency. Our main objective is to improve IDS accuracy by reducing false alarms F_n by detecting new attacks. In open network environment, intrusion detection is rapidly improved by reducing F_n more than F_p.

We propose Hybrid Probe Detection Algorithm (HPDA) based on reducing risk rate of F_n and F_p by the detection of DoS attack using resource capacities and packet analysis. A DoS attack typically appears as a probe and syn flooding attack. The syn flooding attack takes advantage of the vulnerability of three-way handshake between the stations of TCP, which is connection-oriented transmission service [3-5, 7, 8].

The HPDA, which utilizes the F_n, F_p, and session patterns, captures and analyzes parsing packet information to detect syn flooding attack. Using the result of decision module analysis, which uses the session patterns and fuzzy cognitive maps, the decision module measures the risk of F_n and F_p [6, 7]. Section 2 explains details of KDD'99 Data Set and probe detection algorithm. Section 3 shows the result of performance evaluation.

2 Hybrid Probe Detection Algorithm

The current IDS have contributed to identifying attacks using previous patterns. But they have difficulty in identifying attacks which have new patterns or do not have any pattern. Previous studies have utilized a rule-based approach such as USTAT, NADIR and W&S [7, 8]. Their systems lack flexibility in the rule to audit. Slight variations in a sequence of attacks can affect the intrusion detection systems. While increasing the level of abstraction of the rules does provide a partial solution, it also reduces the granularity of the intrusion detection system. The hybrid probe detection algorithm is a network based detection model that uses network data to analyze packet information. We used the KDD'99 Data Set on the test-bed network. It consists of several files that account for the various connections detected on a host. The intrusion logs were generated in 1998 as part of the DARPA Intrusion Detection Evaluation Program. The raw TCP dump data of nine weeks were split into a seven week training data set and the remaining logs were used as test data. Since the KDD'99 Data Set was created to be used in a contest, the test data did not have any target labels. Only the contest organizers had the label corresponding to each test record, so that they could determine which participating team obtained the best intrusion detection results. Given that our intrusion detection project was carried out without a second party that would confirm our results, we needed to know the labels of all data sets. This allowed us to evaluate our own models. This is why we divide the KDD'99 training data set into our custom training set and validation set. TCP syn flooding attacks come from abnormal packets, so detection of abnormal packets is the same as the detection of syn flooding attack on TCP networks.

The HPDA model consists of network based probe detection model and monitoring tool (**Fig. 1, 2**) [5-8]. The HPDA adopts the problem solving methodology which uses previous problem solving situations to solve new problems. The model does preprocessing by packet analysis module and packet capture module. The packet capture module captures and controls packet. The packet capture module does real-time capturing and packet filtering by using the monitoring tool of Detector4win Ver. 1.2. In the packet filtering process, packets are stored according to the features which distinguish between normal packets and abnormal packets. The packet analysis module stores data and analyzes half-open state. After storing packets, the packets, which are extracted by audit record rules in the packet analysis module, are sent to the detection module.

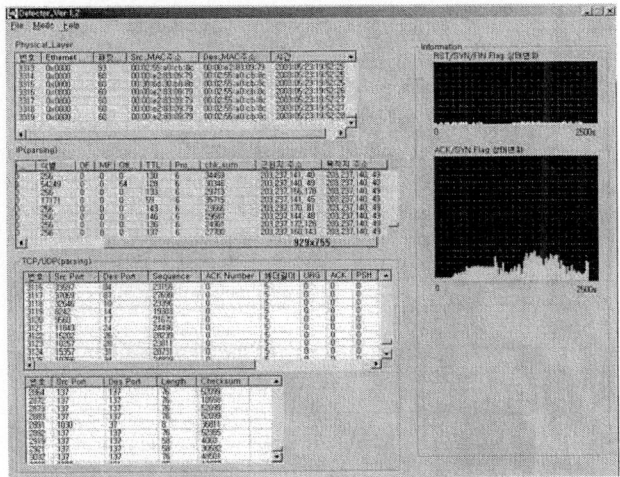

Fig. 1. Monitoring tool

The input and the output of HPDA detection module, namely STEP 1, is traffic and alert, respectively. The traffic is an audit packet and the alert is generated when an intrusion is detected. The HPDA detection module consists of session classifier, pattern extractor, and pattern comparator.

The session classifier takes packet of the traffic and checks whether or not the source is the same as the destination. There is a buffer for the specific session to be stored. And, if the next packet is arrived, it is stored in the correspond buffer. If all packets of the corresponding buffer are collected, all packets of the corresponding buffer are output as one session. The output session becomes an input to the pattern extractor or pattern comparator according to action mode. The action mode consists of learning mode and pre-detection mode.

The output session from the session classifier is sent to the pattern extractor in the learning mode and to the pattern comparator in the pre-detection mode. Fig. 3 is the block diagram of the STEP 1.

The pattern extractor collects the sessions, which have the same destination, and extract common pattern. Each pattern consists of two features. The first feature is a head part which appears in common sessions, which have the same destination, when sessions are arranged by size of packets using the time sequence. The second feature is the minimum length of the sessions which have the same destination. The length of session is the number of packets of a session.

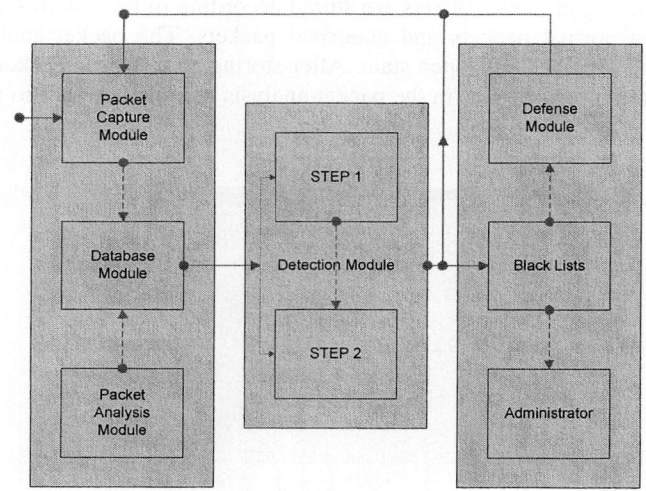

Fig. 2. Architecture of the HPDA

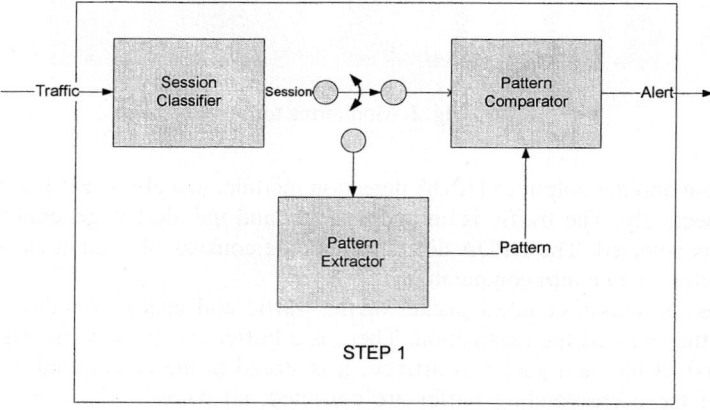

Fig. 3. A Block Diagram of STEP 1

The pattern comparator compares probe packets with the rule-based pattern. If the probe packets and the rule-based pattern do not correspond, the pattern comparator considers the probe packets as the abnormal session and generates an alert signal. Thus, the pattern comparator receives a session and the rule-based pattern as an input. From the input session the data size and the length of session are extracted. If there is a mismatch in one of two features, the pattern comparator considers a session as the abnormal session.

What we must consider for the pattern extraction is whether we extract the pattern continuously or we extract the pattern periodically. We generally call the former the real-time pattern extraction and the latter the off-line pattern extraction. The real-time pattern extraction is better than off-line pattern extraction in the viewpoint of updating the recently changed pattern. But, it is difficult to update the pattern when probes occur. For the pattern, if possible, normal traffic becomes a rule-based pattern. Otherwise, an abnormal traffic sometimes becomes a rule-based pattern. And an abnormal intrusion traffic is considered as an normal traffic. It is called false negative error.

The HPDA uses detection module, namely STEP 2, to compensate the false negative error by using fuzzy cognitive maps. The detection module of HPDA is intelligent and uses causal knowledge reason utilizing variable events which hold mutual dependences. For example, because CPU capacity increases when syn packet increases, the weight of a node, W_{ik}, has the value of range from 0 to 1. The total weighted value of a node depends on path between nodes and iteration number. It is expressed as the following equation.

$$N_k(t_{n+1}) = \sum_{i=1}^{n} W_{ik}(t_n) N_i(t_n)$$

$N_k(t_n)$: the value of the node k at the iteration number t_n
t_n : iteration number
$W_{ik}(t_n)$: weight between the node i and the node k at the iteration number t_n

On the above equation, the sign of weight between the node i and the node k depends on the effect from the source node to the destination node. The value of a weight is the degree of effect in Path Analysis which is calculated using Quantitative Micro Software's Eview Ver. 3.1.

3 Performance Evaluation

The best false error rates (Table 2) are the results of simulation for connection records of DoS attack in 2 weeks. The true positive rate (T_p) of our result is 97.064%. In the KDD'99 competition, Bernhard's true positive rate (T_p) is 97.1%. We compared Bernhard's true positive rate with that of HPDA and found that the result of HPDA is as good as Bernard's result.

Table 2. Best false error rates

Data Set \ Detector	Wenke Lee	Dr. Bernhard	HPDA
DoS	79.90%	97.10%	97.06%
Probe	97.00%	83.30%	99.10%

From the result of test-bed and the aspect of resource capacity, Fig. 3 and Fig. 4 show how attack counts affect to the vulnerability of hardware capacity when the number of counts was increased from 0 to 70,000. In Fig. 3 and Fig. 4, we set the bandwidth on the range from 40% to 60% for the hardware capacity limitation as the bandwidth of the probe detection. And the above bandwidth is the deadline in the probe attack.

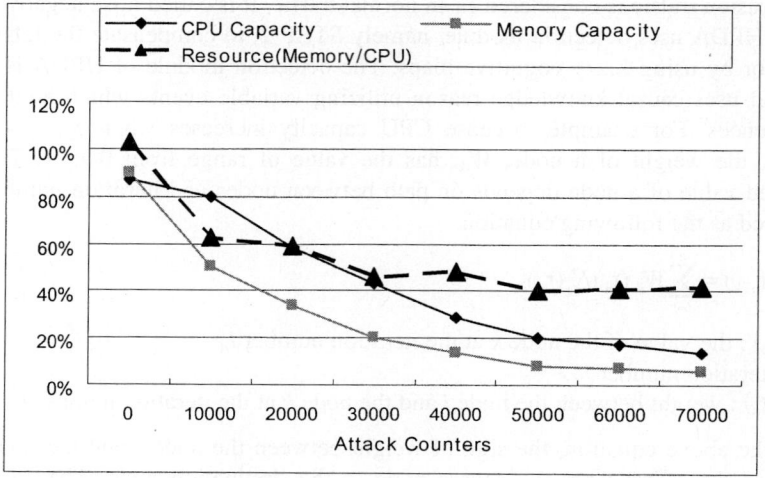

Fig. 3. Attack counters vs. hardware resource capacity on test-bed network

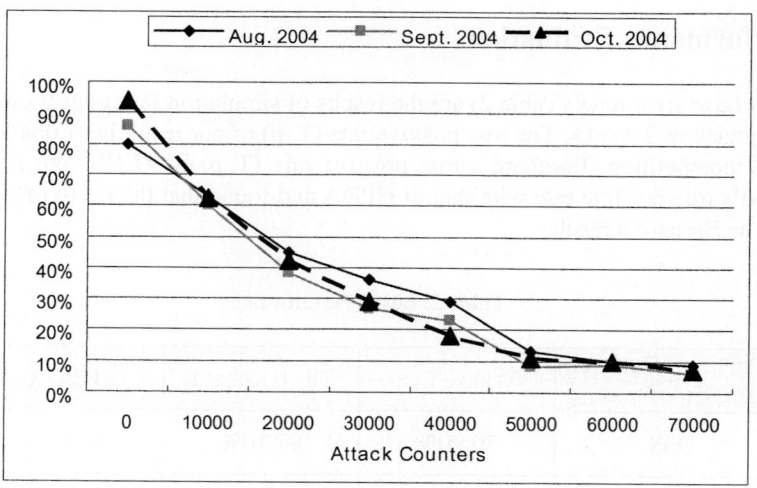

Fig. 4. Attack counters vs. hardware resource capacity on real-time network

4 Conclusions

This paper presented a hybrid probe detection model, namely HPDA, which utilizes the session patterns and rate of false errors. For the performance evaluation, the "KDD'99 Competition Data Set" made by MIT Lincoln Labs was used. The result shows that the probe detection rates of the HPDA for the KDD'99 Data Set were over 97 percentages in average. Though other probe detection systems always have an emphasis on detection rate, we considered the aspect of cost ratio between false positive error and false negative error for information security.

References

1. A. Siraj, S. M Bridges, and R. B. Vaughn, "Fuzzy cognitive maps for decision support in an intelligent intrusion detection system," IFSA World Congress and 20th NAFIPS International Conference, Vol. 4, pp. 2165-2170, 2001.
2. H. S. Lee, Y. H. Im, "Adaptive Intrusion Detection System Based on SVM and Clustering," Journal of Fuzzy Logic and Intelligent Systems, Vol. 13, No. 2, pp. 237-242, 2003.
3. D. J. Joo, The Design and Analysis of Intrusion Detection Systems using Data Mining, Ph. D. Dissertation, KAIST, 2003.
4. C. L. Schuba, I. V. Krsul, M. G. Khun, E. H. Spaford, A. Sundram, and D. Zamboni, "Analysis of a denial of service attack on tcp," IEEE Symposium on security and Privacy, 1997.
5. S. Y. Lee, Y. S. Kim, "Design and Analysis of Probe Detection Systems for TCP Networks," International Journal of Advanced Computational Intelligence & Intelligent Informatics, Vol. 8, No. 4, pp. 369-372, 2004.
6. W. Lee, S. J. Stolfo, "A Framework for Constructing Features and Models for Intrusion Detection Systems," In Proceedings of the 5th ACM SIGKDD International Conference on Knowledge Discovery and Data Mining, 1999.
7. S. Y. Lee, An Adaptive Probe Detection Model using Fuzzy Cognitive Maps, Ph. D. Dissertation, Daejeon University, 2003.
8. S. J. Park, A Probe Detection Model using the Analysis of the Session Patterns on the Internet Service, Ph. D. Dissertation, Daejeon University, 2003.

Energy-Efficiency Method for Cluster-Based Sensor Networks

Kyung-Won Nam[1], Jun Hwang[2], Cheol-Min Park[1],
and Young-Chan Kim[1]

[1] Chung-Ang Univ., Bobst-Hall 5 floor, System Software Lab,
Dept. of Computer Science & Engineering,
Seoul, 221, 156-756, Republic of Korea
{expert911, raphael66}@sslab.cse.cau.ac.kr
[2] Seoul Woman's Univ., Dept. of Information & Media,
Seoul, 126, 139-774, Republic of Korea
hjun@swu.ac.kr

Abstract. Wireless sensor networks have recently emerged as an important computing platform. The sensors in this system are power-limited and have limited computing resources. Therefore sensors' energy has to be managed wisely in order to maximize the lifetime of the network. There have been many studies on considering sensors' energy. Among these, clustering sensor nodes is more efficient and adaptive approach in sensor networks. Based on this approach, we introduce a method that prolongs the network lifetime. The method is that a cluster-head with less-energy-constraint, distributes energy load evenly among its members based on their energy usage by sub-clustering. The simulation result shows that the proposed method in this paper has achieved better performance than other clustering method in the lifetime of the network.

Keywords: Wireless sensor networks, network lifetime, clustering, energy-efficiency method.

1 Introduction

Wireless sensor networks have recently emerged as an important computing platform. These sensors are deployed in the area of interest to sense the environment and to monitor events. For example, in combat field surveillance and disaster management.

Because these devices are cheap and small, it is possible to achieve high quality. However, they are power-limited and have limited computing resources. Sensors are typically disposable and expected to last until their energy drains[1]. Therefore, energy efficient communication protocols are required in order to manage sensor's energy wisely and prolong the network lifetime.

The existing routing models are may be grouped into one of the following three models[2] and here is briefly summarized.

One-hop Model. As shown in Figure1, This model is very simple. Every node in the network directly transmits to the base station. Because sensors' transmission range is limited, sensors which are far from base station cannot reach it.

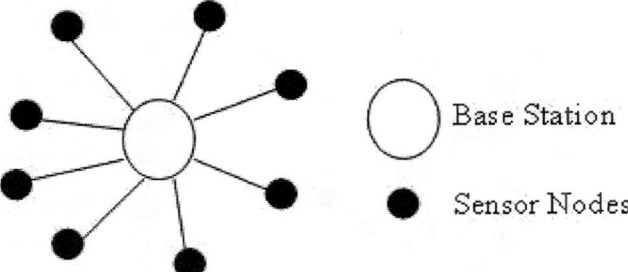

Fig. 1. One Hop Model

Moreover, if the network is dense, collisions that degrade the network efficiency may occur. Therefore this model isn't feasible for wireless sensor networks.

Multi-hop Model. As shown in Figure2, the information travels from source to destination by hop from one node to another until it arrives at the destination in this model. In a dense network, this model has high latency. Moreover, self-induced black hole effect is another drawback. Nodes that are closer to the base station would have to act as intermediaries to all traffic being sent to the base station. As they handle all the traffic, they will die first creating a black hole around the base station for incoming traffic.

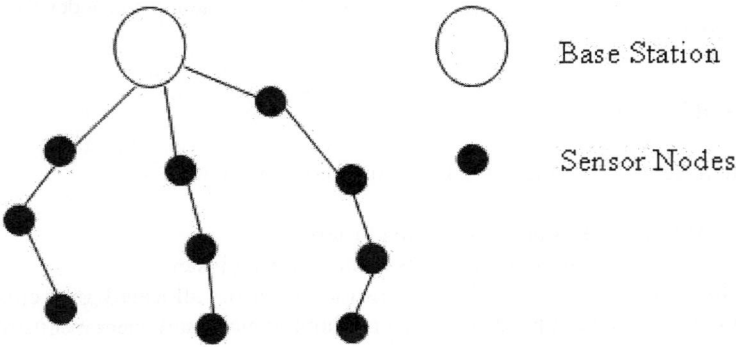

Fig. 2. Multi Hop Model

Cluster-based Model. In this model, nodes are grouped into clusters with a cluster head that has the responsibility of routing from the cluster to the other cluster head or base stations. While the nodes that are located far from the base station will consume more energy and therefore die sooner in one hop model, nodes' energy is conserved. Moreover, the latency is less than that in multi hop model. Many clustering methods have been propped in [3], [4], [5]. Although there are many possible clustering models, we consider the clustering model in Figure3.

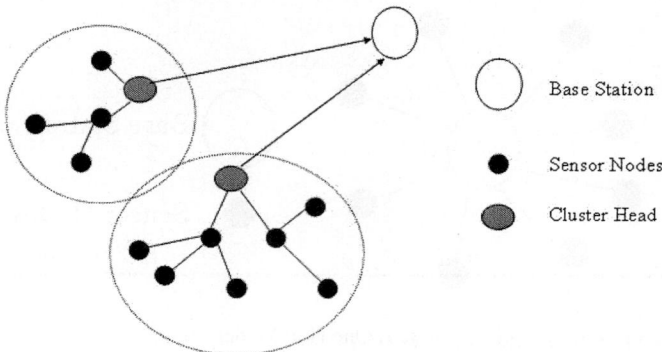

Fig. 3. Cluster-based Model

Cluster-based model is believed to be more feasible for wireless sensor networks than other approach. Based on this model, we introduce a method that prolongs the network lifetime. In this model, it can be seen from the fact that nodes that are closer to the cluster-head act as routers for other nodes' data in addition to sensing the environment and may die sooner. Therefore, in this paper, we introduce a method that a cluster-head with less-energy-constraint, distributes energy load evenly among its members based on their energy usage by sub-clustering.

This paper is organized as follows: Section 2 describes the network model of our approach. Then in Section 3, sub-clustering protocol is presented. *Agent node* election can be found in Section 4. Section 5 presents the simulation results. Finally Section 6 concludes the paper and discusses future works.

2 Network Model

As shown in figure 3, we make the following assumption.

- the sensor network is organized into clusters.
- the cluster-head with high-energy is deployed in a cluster.
- the cluster-head knows the ID and location of sensors allocated to its cluster.
- all nodes excluding the cluster-heads are homogenous and energy-constrained
- each node in a cluster can directly communicate with its cluster-head.

It is desirable to make cluster-head with high-energy decide routing because sensors' energy is limited and much consumed during updating routing table.

2.1 Radio Model[6]

The radio model is based on [6] and its characteristics are presented in Table1.
To transmit a k-bit message a distance d, the radio expends:

$$E_{Tx}(k,d) = E_{Tx\text{-}elec}(k) + E_{Tx\text{-}amp}(k,d)$$
$$= E_{elec} * k + \in_{amp} * k * d^2$$

Table 1. Radio characteristics

Operation	Energy Dissipated
Transmitter Electronics($E_{Tx\text{-}elec}$) Receiver Electronics($E_{Rx\text{-}elec}$) ($E_{Tx\text{-}elec} = E_{Rx\text{-}elec} = E_{elec}$)	50 nJ/bit
Transmit Amplifer($\in_{amp}$)	100 pJ/bit/m²

and to receive this message, the radio expends:

$$E_{Rx}(k) = E_{Rx\text{-}elec}(k) = E_{elec} * k$$

3 Sub-clustering Protocol

As mentioned above, in a simple clustering, it can be seen from the fact that nodes that are closer to the cluster-head act as routers for other nodes' data in addition to sensing the environment and may die sooner. Thus, we introduce a sub-clustering method. The method is that nodes within a same clusters are organized into other clusters as shown in Figure 4.

Each nodes i maintains a unique identification, ID(i), a cluster identification to which i belongs, CID(i), and its remaining battery power, CRP(i)[7]. We add a sub-cluster identification to which i belongs, SCID(i), which is used for sub-clustering. After the sensor network is organized into clusters, cluster-heads inform its member node of SCID(i). Nodes that have same SCID(i) is organized into same sub-cluster.

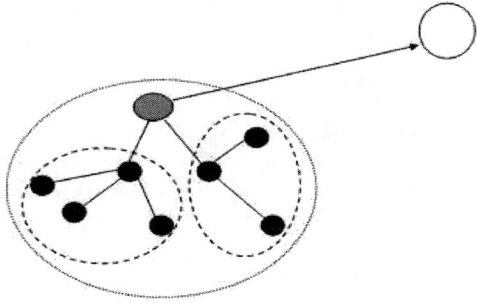

Fig. 4. Sub-clustering the network

4 *Agent Node* Election

After the network is organized into sub-cluster, the cluster-head elects *agent-node*, which act as router for other nodes' data within sub-cluster. The *agent-node* aggregates data sensed by other nodes within sub-cluster and send it to its cluster-head.

Because the *agent-node* consumes more energy than other nodes within sub-cluster, the cluster-head elects another node with the most residual power *agent-node* in next cluster cycle. This information is stored in CRP(i). That is, when a cluster cycle is over, nodes are informed about their newly elected *agent node*. Nodes within sub-cluster transmit to their newly elected *agent node*.

Figure5 and Figure6 illustrate this process.

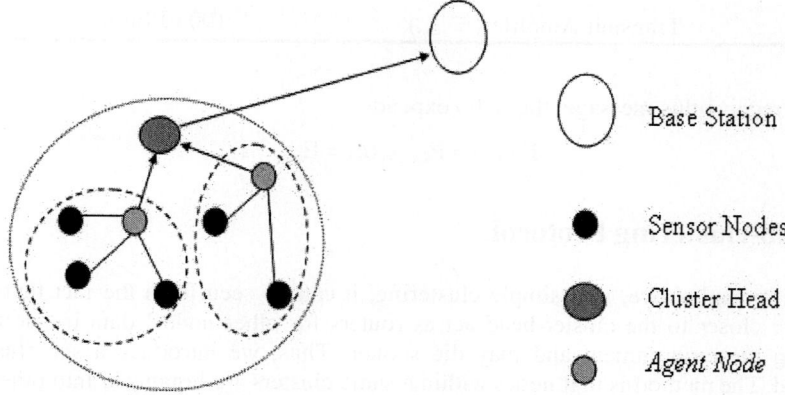

Fig. 5. *Agent Node* election based on their energy usage

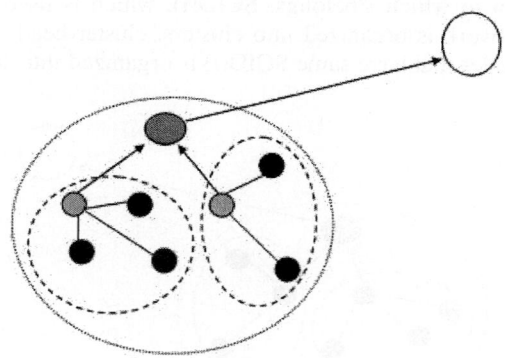

Fig. 6. Newly assigned *Agent Node*

5 Simulation Results

For the simulation, we use the NS-2 network simulator. The environment consists of 120 nodes distributed randomly in a 100m X 100m area. Each node is equipped with the initial energy of 5 joules.

We compare our approach with the conventional clustering based on network life time and total energy dissipated in system.

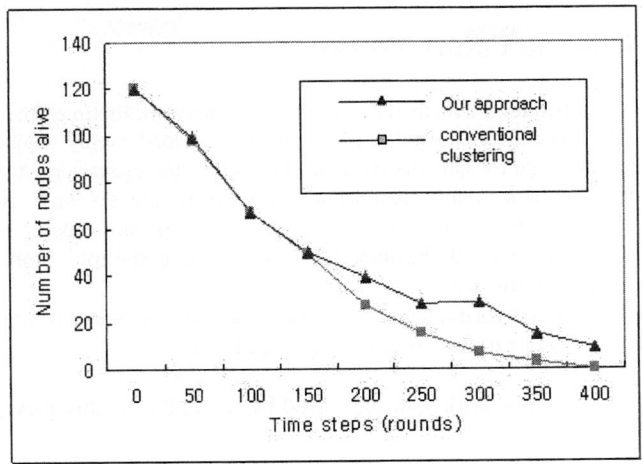

Fig. 7. Comparison between our approach and conventional clustering without *agent node*

Figure 7 shows the performance comparison for number of nodes alive. At initial stage, our approach and conventional clustering without *agent node* produced similar result. However, after about the 150 time steps, it can be seen that our approach slightly achieved better performance in number of nodes alive.

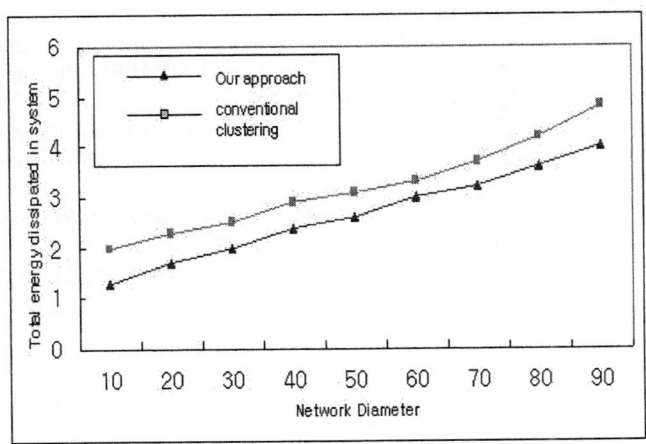

Fig. 8. Comparison between our approach and conventional clustering

Figure 8 shows the performance comparison for total energy dissipated in system. From the initial stage, our approach achieved better performance than conventional clustering. Consequently, we can see that our approach prolong the network life.

6 Conclusions and Future Works

In this paper, we introduced a method that prolongs the network lifetime. The method is that a cluster-head with less-energy-constraint, distributes energy load evenly among its members based on their energy usage by sub-clustering. In sub-cluster, the *agent node* aggregates data sensed by other nodes within sub-cluster and send it to its cluster-head. In next cluster cycle, another node with the most residual energy within sub-cluster is elected the *agent node* (by its cluster-head) because the *agent node* consumes more energy than other nodes within sub-cluster.

We will study the performance of our approach under various conditions. For example, we will consider nodes' mobility and fault.

Acknowledgements. This work was supported by the ITRI of Chung-Ang University in 2005.

References

1. Gaurav Gupta, Mohamed Younis, "Falut-Tolerant Clustering of Wireless Sensor Networks", IEEE, pp.1579-1584, 2003.
2. Jamil Ibriq, Imad Mahgoud , "Cluster-Based Routing in Wireless Sensor Networks: Issues and Challenges", SPECTS'04, pp. 759-766, 2004.
3. K. Dasgupta, K. Kalpakis, and P. Namjoshi. "An efficient clustering-based heuristic for data gathering and aggregation in sensor networks", In Proceedings of the IEEE Wireless Communications and Networking Conference, 2003
4. W. Heinzelman, J. Kulik, and H. Balakrishnan, "Adaptive protocols for information dissemination in WSNs", In Proceeding of the Fifth Annual ACM/IEEE International Conference on Mobile Computing and Networking, 1999.
5. H. Luo, F. Ye, J. Cheng, S. LU, and L. Zhang. TTDD: a two-tier data dissemination model for large scale WSNs. In Proceedings of the 8th Annual International Conference on Mobile computing and networking, 2002.
6. W. B. Heinzelman, "Application-Specific Protocol Architectures for Wireless Newworks", Ph.D. Thesis, Massachusetts Institute of Technology, June 2000
7. Jain-Shing Liu, Chun-Hung Richard Lin, "Power-Effiency Clustering Method with Power-Limit Constraint for Sensor Networks.

A Study on an Efficient Sign Recognition Algorithm for a Ubiquitous Traffic System on DSP

Jong Woo Kim, Kwang Hoon Jung,
and Chung Chin Hyun

Department of Information and Control Engineering,
Kwangwoon University, 447-1, Wolgye-dong,
Nowon-gu, Seoul, 139-701, Korea
chung@kw.ac.kr

Abstract. This paper presents an efficient Ubiquitous computing algorithm to detect a Korean traffic sign board using adaptive segmentation by estimating an inequality and proceeding the geometrical morphology, and matching with an adaptive local affine transform. Our approach shows significant performance with efficient detection about the traffic sign board and robust recognition of object sign. Also, detection with adaptive segmentation and recognition with adaptive local affine transformation by using TMS320C6711 DSP Vision Board are set to highlight the advantages of our algorithm.

1 Introduction

The goal of this paper is to present an efficient Ubiquitous computing algorithm of detection of Korean traffic sign by using adaptive segmentation with estimate of inequality and geometrical morphology, and recognition the traffic sign by using skeleton pattern matching with adaptive local affine transform. The segmentation of background and traffic sign board includes in section 2.1. Adaptive segmentation begins by making the Estimate of inequality between background model [1] and input image. And geometrical morphology [2] proceeded to express the object of interest and obtained a extracted edge map. Also, projection in x, y axis is proceeded to extract only a sign image which the traffic sign essentially providing. In section 2.2, we deal with a obtainment of skeleton [3] of sign. In section 3, we carried out a computation of pattern matching rate by using iterative application of adaptive LAT (Local Affine Transform [4]) with adaptive window. The adaptive LAT is an iterative technique for gradually deforming a mask binary image with successive local affine transform operations so as to yield the best pattern matching to input binary images. In section 4.1, we deal with experimental Ubiquitous computing environments. In that section, we deal with formation of system hardware: acquire image, image processing and image output. And we deal about video display [5, 6, 7] and video capture

[5, 6, 7] of DSP Vision Board. Our approach shows significant performance advantages with efficient detection about the traffic sign and robust recognition of object sign. Also, detection with adaptive Segmentation and recognition with adaptive local affine transformation made to highlight the advantages of our algorithm.

2 Traffic Sign Detection

2.1 Segmentation Background and Traffic Sign Board

Object detection is one of the initial steps and activity recognition system [8]. The target images of adaptive segmentation is compared and filtered with the background image and an estimate of inequality is computed for the entire distribution of edge density and directional values. The edge-extraction employs Kirsch Edge Operator [9]. The estimate of inequality measures the difference between the input image and the background as Fig. 1(a) and Fig. 1(b). Edge density represents the ratio of the edges in a window centered around pixel x, y. The difference between the edge densities is given by:

$$\triangle E_d = E_d(x,y)|_{image} - E_d(x,y)|_{ref\ Image} \qquad (1)$$

If this difference between the edge densities is greater than a particular threshold, then the pixel passes the edge density filter. The Fig. 1(c) shows the result of conventional background subtraction and Fig. 1(d) shows the result of estimate of inequality algorithm.

Moreover, the geometrical morphology algorithm uses the gradient distribution of the neighboring pixels. On the adaptive geometrical erosion, a pixel is not declared edge pixel if it is randomly distributed noise pixel by the adaptive geometrical erosion. And on the basis adaptive geometrical dilation, a pixel will not be regenerated if the probability [10] of finding edges around it is less then a particular threshold.

And the object completion and filling is done by dividing the image into quadrants, and the quadrants further into sub-quadrants. Fig. 2(a) shows the

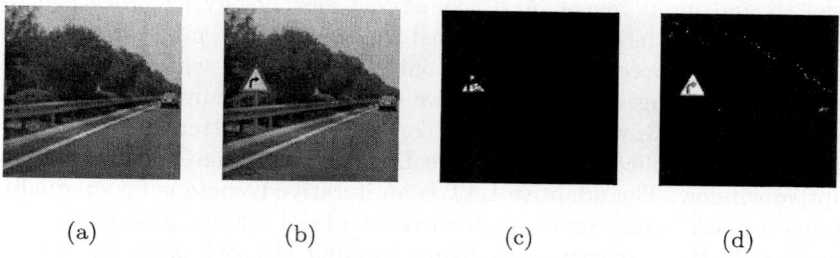

(a)   (b)   (c)   (d)

Fig. 1. (a)Input model (b)The selected optimal background model

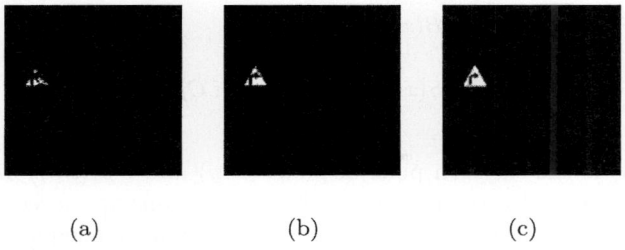

Fig. 2. (a)Result of adaptive geometrical erosion (b)Result of adaptive geometrical dilation (c)Result of the adaptive object completion and filling

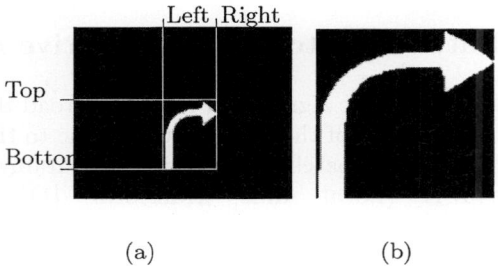

Fig. 3. (a)Compute size of image (b)"Right Corner"

result of the adaptive geometrical erosion and Fig. 2(b) shows the dilation. And Fig. 2(c) shows the result of the object completing and filling. For the final step for traffic Sign Detection, projection in x, y axis proceeded to extract only a sign image which the traffic sign essentially providing.

2.2 Skeleton Algorithm

A skeleton is presumed to represent the shape of the object in a relatively small number of pixels, all of which are structural and therefore necessary. In line images the skeleton conveys all of the information found in the original, wherein lies the value of the skeleton. Our skeletonization algorithm is based on a repeated stripping away of layers of pixels until no more layers can be removed. We need two geometrical preconditions as follows.

Precondition

1. A object pixel needs neighboring pixels more than two and less than six. And end or inner points of skeleton were not to be erased.
2. The object pixels which are connecting both sections were not to be erased.

Due to the geometrical precondition, proposed skeleton algorithm of this paper making strong provisions for keeping the skeleton connected.

$$S_1(x,y) \leftarrow \text{if } Cp_1(x,y) \begin{cases} B(x,y) & \text{false} \\ S(x,y) & \text{true, then if } Cp_2(x,y) \begin{cases} B(x,y) & \text{false} \\ S(x,y) & \text{true.} \end{cases} \end{cases} \quad (2)$$

where $S_1(x,y)$ is the skeleton pixel of point (x,y) and $Cp_1(x,y)$ and $Cp_2(x,y)$ is the first and second precondition. Each preconditions are based on geometrical specification. If these geometrical specifications are inadequate to the each precondition, then the object point does not change. After the first precondition applied over all correspond boundaries, proceed the second precondition on the boundary same as first precondition.

3 Skeleton Pattern Matching by Adaptive LAT

Almost all of the sign of traffic sign board informing road directions to driver through the arrowhead shapes of the sign. However, due to the irregular shapes of skeleton pattern, pattern matching procedure inevitably needed by using iterative adaptive LAT (Local Affine Transform) [11] when it proceed recognition.

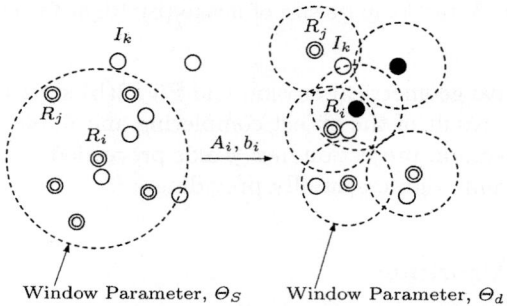

Fig. 4. Local affine transformation

To begin with the adaptive local affine transform, $7 \times 9 (9 \times 7)$ size reference pattern masks $(R\{i\})$ [4,12] defined as Table. 1. Those $R\{i\}$ are unimpaired arrowhead skeleton points. Hence, when it find out the arrowhead input patterns $(I\{k\})$ [4] from the skeleton image, since the positions of eight points of $7 \times 9 (9 \times 7)$ size $I\{k\}$ are exactly same as $R\{i\}$, we process the LAT on that $I\{k\}$ points with size $7 \times 9 (9 \times 7)$. As it turned out, the $7 \times 9 (9 \times 7)$ size arrowhead input pattern found as Table. 1, moreover, the $I\{k\}$ has observed as impaired skeleton points.

Also, the key idea of adaptive LAT is an iterative technique for gradually deforming a mask binary image with successive local affine transform operations so as to yield the best pattern matching to input binary images. Furthermore,

Table 1. Reference pattern mask and input patterns

Reference	Input	Reference	Input

adaptive LAT by supplement of preconditions only for application of arrowhead pattern matching with LAT developed.

Precondition

1. If the position of a reference pattern point (R_i) and a input pattern point (I_k) is equal, that point should erase as it seems to be matched point.
2. Eventually, when the same distance detected between two nearest input pattern points from the reference points, the LAT on that reference points skipped.

By the way, if we using the searching window on the R_i or R_j points which are consisting in near boundaries of mask, then some strayed out region will be occur itself from the arrowhead pattern. As one idea for resolving this problem, the adaptive searching window defined as below.

As Fig. 5(b), the distances computed from point R_i to the end of window and pattern mask in four directions (up, down, left and right). As result of above procedure, the window size(7×7) of rest directional fragment will be preserve

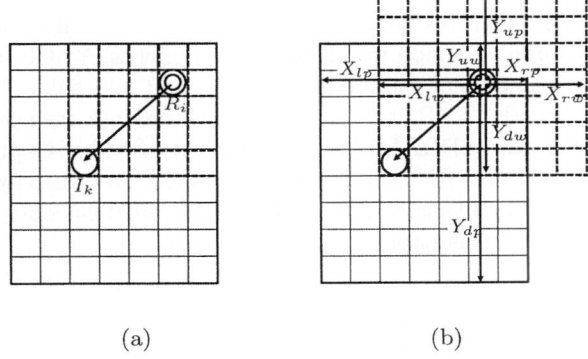

Fig. 5. (a)Compute window space (b)Adapted searching window

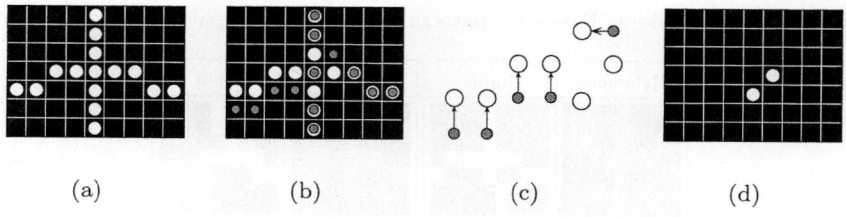

Fig. 6. (a)Reference pattern (white dots) (b)Input pattern (gray dots) (c)LAT (d)Erase matched point

as it is, simultaneously only the fragment of strayed away from pattern region will be modify.

Eventually the matching rate computation presented. As we mentioned above, we using the 7×9 or 9×7 size arrowhead skeleton pattern mask. Also, there are fifteen points of skeleton in the mask, in terms of position of eight points of $R\{i\}$ are exactly same as $I\{k\}$, we proceed the LAT on that $I\{k\}$ points with size $7 \times 9(9 \times 7)$ with assuming that there are arrowhead pattern in the $I\{k\}$ mask. Then, we continue accumulate matching rate of surplus points of $I\{k\}$ which does not matched with $R\{i\}$ by adaptive LAT. If there are unmatched point in $I\{k\}$ although we had use the 7×7 size adaptive window, then leave them alone as "Dissimilar point".

4 Experiments

4.1 System Construction

The DSP Vision Board [5, 6, 7] has been developed as a platform for development and demonstration of image / video processing applications on TMS320C 6711 DSP. The IDK is based on the floating point C6711 DSP may also be useful to developers using this platform to develop other algorithms for image, video, graphics processing. We summarized this concept in Fig. 7.

There are three separates buffers used to implement the triple buffering scheme and each buffer is made up of three components, Y, Cb, and Cr. Also there are separate component buffers for the even and odd fields. The captured image data inputs to the TVP5022 decoder and these data transfer to the user buffers of Video Capture SDRAM. And the data from Video Capture SDRAM transfer to the User buffers of Video Capture SDRAM. And display the final result images through Video Display SDRAM and TVP3026 Encoder after applied the unique algorithm of this paper.

4.2 Results of the Algorithm

This section presents results of sign detection and recognition when the whole algorithms applied. And six kinds of traffic sign board used here as input images. Furthermore, we implemented present algorithm with an CCD Camera, IDK

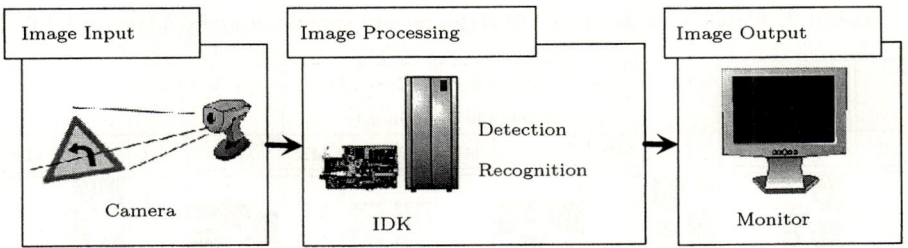

Fig. 7. Construction of system hardware

(Imaging Developer's Kit - TMS320C6711 DSK, Frame Grabber) board and CCS (Code Composer Studio) from TI. 300 times of experiments using present algorithm with six traffic sign board was run to detect and recognition of sign.

As Table. 2, the proposed algorithm results competent outcomes for detection and recognition of sign on each steps. Moreover, on account of the adaptive LAT performs the high matching rates, the results reveals that the Adaptive LAT is a suitable algorithm to estimate matching rate of sign. The matching rates of six kinds of sign of 10 times experiments using Adaptive LAT shows in Table. 3.

In turn, we estimated the matching rates with absence of adaptive windows. That is to say, we perform the LAT without adaptive windows on the arrowhead input pattern only with 3×3 mask along with unconcern about boundaries. As shown in Table. 4, the average matching rates of absence of adaptive windows are quite low.

Fig. 8(a) shows the distribution rates of adaptive LAT without adaptive windows. On the other hand, Fig. 8(b) shows the distribution rates of adaptive LAT with adaptive windows. As Fig. 8(a) shows, while outcomes of adaptive LAT without adaptive window shows the average matching rate is 73.52% of its matching rates, the average matching rate of adaptive LAT with adaptive window is 98.97% as Fig. 8(b). By defining, "Set A" is the 100 times experiments with six kinds of traffic sign board. Moreover, the "Set B" and "Set C" are same as "Set A".

Table 2. Result images of each steps

Table 3. Matching Rates (MR) of skeleton pattern matching by adaptive LAT

T	Left Corner		Right Corner		Right-left winding path		Left-right winding path		No U-turn		No across	
	I{k}	MR	I{k}	MR	I{k}	MR	I{k}	MR	I{k}	MR	I{k}	MR
1th		93%		87%		100%		93%		100%		93%
2th		93%		100%		100%		93%		100%		100%
3th		100%		93%		100%		93%		100%		93%
4th		100%		87%		100%		93%		100%		80%
5th		100%		100%		100%		100%		100%		87%
6th		100%		93%		100%		93%		100%		93%
7th		93%		100%		100%		100%		100%		80%
8th		93%		100%		100%		93%		87%		87%
9th		100%		100%		87%		73%		93%		93%
10th		100%		100%		93%		87%		93%		87%

Table 4. Comparison of matching rates of presence adaptive window

	Average Matching Rate	Average Error Rates
Without Adaptive Windows	73.52%	26.48%
With Adaptive Windows	98.97%	1.03%

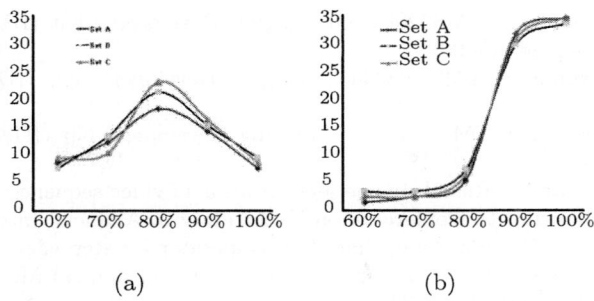

Fig. 8. (a)Without adaptive windows (b)With adaptive windows

5 Conclusion

Our approach shows significant performance with efficient detection about the traffic sign board and robust recognition of object sign. Also, detection with adaptive segmentation and recognition with adaptive local affine transformation are made to highlight the advantages of our algorithm. Moreover, an iterative technique for gradually deforming a mask of binary image with successive adaptive local affine transform operates so as to yield the best pattern matching to input binary arrowhead patterns. Furthermore, the preconditions of obtain skeleton leads to construct precise connected skeleton points. Also the preconditions of application of adaptive LAT leads to best pattern matching to arrowhead patterns. As it turned out, as Fig. 8(a) shows, while outcomes of adaptive LAT without adaptive window shows the average matching rate is 73.52% of its matching rates, the average matching rate of adaptive LAT with adaptive window is 98.97% as Fig. 8(b) on basis 300 times experiments. We expect that the result of this paper can be contributed to develop enhanced traffic sign detection and recognition system development. We are planning to further develop this algorithm to make it more trustful by undergoing the field test for result verification.

References

1. K. Toyama, J. Krumm, B. Brummit and B. Meyers Wallflower: Principles and practice of background maintenance. IEEE International Conference on Computer Vision (1999)
2. M. Raffay Hamid, Aijaz Baloch and Ahmed Bilal Nauman Zaffar: Object segmentation using feature based conditional morphology. Proceedings of the 12th IEEE International Conference on Image Analysis and Processing (2003)
3. Rafael C. Gonzalez and Richard E. Woods: Digital Image Processing Second Edition. Prentice Hall, New Jersey (2002)
4. Turu Wakahara: Shape matching using LAT and its application to handwritten numeral recognition. IEEE Transactions on Pattern Analysis and Machine Intelligence **16** (June 1994) P.618–P.629

5. Texas Instrument: TMS320C6000 Imaging Developer's Kit (IDK) video device driver user's guide (2000)
6. Texas Instrument: TMS320C6000 Imaging Developer's Kit (IDK) user's guide (2001)
7. Texas Instrument: TMS320C6000 Imaging Developer's Kit (IDK) programmer's guide (2001)
8. N. Friedman and S. Russel: Image segmentation in video sequences: A probabilistic approach. Thirteenth Conference on Uncertainty in Artificial Intelligence (1997)
9. Shen J. and S. Casten: An optimal linear operator for step edge detection. Computer Vision , Graphics, and Image Processing : Graphical Models and Understanding **54** (1992) 112 – 133
10. Henry Stark and John W. Woods: Probability and Random Processes with Applications to Signal Processing Third Edition. Prentice Hall, New Jersey (2002)
11. S. Christy and R. Horaud: Euclidean shape and motion from multiple perspective views by affine iteration. IEEE Transactions on Pattern Analysis and Machine Intelligence **18** (November 1996) pp. 1098–1104
12. P. A. Beardsley, A. Zisserman and D. W. Murray: Navigation using affine structure and motion. European Conference on Computer Vision **LNCS 800/801** (1994) pp. 85–96

Real-Time Implementation of Face Detection for a Ubiquitous Computing

Jin Ok Kim[1] and Jin Soo Kim[2]

[1] Faculty of Multimedia, Daegu Haany University,
290, Yugok-dong, Gyeongsan-si,
Gyeongsangbuk-do, 712-715, Korea
bit@dhu.ac.kr
[2] Department of Information and Control Engineering,
Kwangwoon University, 447-1, Wolgye-dong,
Nowon-gu, Seoul, 139-701, Korea
chung@kw.ac.kr

Abstract. Human face detection is the most important process in applications such as video surveillance, human computer interface, face recognition, and image database management. Face detection algorithms have primary factors that decrease a detection ratio : variation by lighting effect, location and rotation, distance of object, complex background. Due to variations in illumination, background, visual angle and facial expressions, the problem of machine face detection is complex. Algorithms were discussed in several papers about face detection and face recognition. But we know that implementation of these algorithm is not easy. We propose a face detection algorithm for color images in the presence of varying lighting conditions as well as complex background. We use the YC_bC_r color space since it is widely used in video compression standard and multimedia streaming services. Our method detects skin regions over the entire image, and then generates face candidate based on the spatial arrangement of these skin patches. The algorithm constructs eye, mouth, nose, and boundary maps for verifying each face candidate.

1 Introduction

Human activity is a major concern in a wide variety of applications such as video surveillance, human computer interface, face recognition, and face image database management. And machine face recognition is a research field of fast increasing interest. Although a lot of work has already been done, a robust extraction of facial regions and features out of complex scenes is still a problem [1]. In the first step of face detection, the localization of facial regions and the detection of facial features, e.g. eyes and mouth, is necessary. Detecting face is a crucial step in these identification applications. Most face detection algorithms assume that the face location is known. Similarly, face tracking algorithms often assume the initial face location is known. Note that face detection can be viewed as a two-class (face versus non-face) classification problem. Therefore, some techniques developed for face detection (e.g., holistic / template approaches, feature based approaches, and their combination) have also been used to detect faces, but they are computationally very demanding and cannot handle large variations in face images. Various approaches to face detection are discussed in [2, 3, 4, 5, 6]. For recent surveys on face detection, see [3, 6]. These approaches utilize techniques such as principal component analysis, neural networks, machine learning, information theory, geometrical modeling, (deformable) template matching, Hough transform, motion extraction, and color analysis. We propose a face detection algorithm that is able to handle a compensation technique. We construct skin color model in YC_bC_r color space and suggestion the lighting compensation algorithm for variative light condition. also we get the facial candidate region using pixel connectivity by morphological operation. At last we determine the eye, mouth feature point in color image.

2 Face Detection System

Face detection is the most important process in applications such as video surveillance, human computer interface, face recognition, and image database management. Face detection algorithms have primary factors that decrease a detection ratio: variation by lighting effect, location and rotation, distance of object, complex background. Due to variations in illumination, background, visual angle and facial expressions, the problem of machine face detection is complex.

We propose a face detection algorithm for color images in the presence of varying lighting conditions as well as complex background. We use the YC_bC_r color space since it is widely used in video compression standard and multimedia streaming services. Our method detects skin regions over the entire image, and then generate face candidate based on the spatial arrangement of these skin patches. The algorithm constructs eye, mouth, nose, and boundary maps for verifying each face candidate.

An overview of our face detection algorithm is depicted in Fig. 1, which contains two major modules: 1) face segmentation for finding face candidates and 2) facial feature extraction for verifying detected face candidates. Our approach for face localization is based on the observation that human faces are characterized by their oval shape and skin color, also in the case of varying light conditions. Therefore, we locate face-like regions on the base of shape and color information. We employ the YC_bC_r color space by using the RGB to YC_bC_r transformation. The hypotheses for faces are verified by searching for facial features in side the facial regions. We extract facial features based on the observation that eyes and mouth differ from the rest of the face in chrominance because of their conflictive response to C_b, C_r.

This algorithm reduces the error ratio by using lighting compensation process that over exposure. Also compensated RGB image transformed by YC_bC_r color model, then we present skin color model make use luma-independent C_bC_r model.

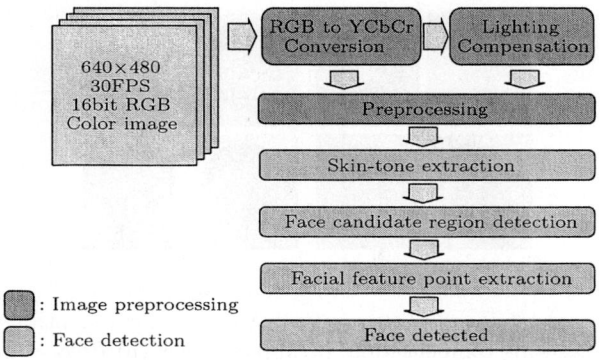

Fig. 1. Face detection algorithm

2.1 Lighting Compensation Algorithm

Skin-tone color depends on the lighting conditions. We introduce a lighting compensation technique that use "reference white" to normalize the color appearance. We regard pixels with to top 5 percents of the luminance values in the image as the reference white only if the number of these pixels in sufficiently large [1].

Figure 2 demonstrates an example of our lighting compensation method. Note that the various lighting condition image in Fig. 2(a) has been removed, as shown in Fig. 2(b). Note that the variations in skin color from different facial groups, reflection characteristics of human skin and its surrounding objects (including clothing), and camera characteristics will all affect the appearance of skin color and hence the performance of face detection. Therefore, if models of the lighting source and cameras are available, additional lighting correction should be made to remove color bias.

3 Contrast Limited Adaptive Histogram Equalization

CLAHE (Contrast Limited Adaptive Histogram Equalization) seems a good algorithm to obtain a good looking image directly from a raw image, without window and level adjustment. This is one possibility to automatically display an image without user intervention. Further investigation of this approach is necessary.

CLAHE was originally developed for medical imaging and has proven to be successful for enhancement of low-contrast images such as portal films.

The CLAHE algorithm partitions the images into contextual regions and applies the histogram equalization to each one. This evens out the distribution of used grey values and thus makes hidden features of the image more visible. The full grey spectrum is used to express the image.

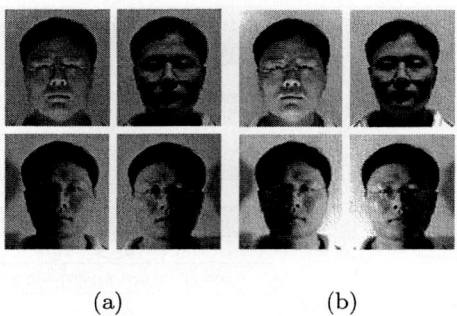

(a)   (b)

Fig. 2. Lighting compensation: (a) Input image; (b) Precessed image

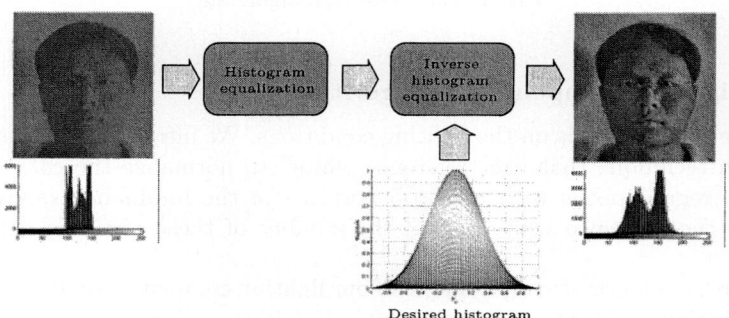

Fig. 3. CLAHE algorithm

4 Skin Color Modeling and Extraction

We demonstrate the luma dependency of skin-tone color in different color spaces in Fig. 5, based on skin patches collected from IMDB [7] in the Intelligent Multimedia Laboratory image database. These pixels from an elongated cluster that shrinks at high and low luma in the YC_bC_r space, shown in Fig. 5(b), 5(c). Detecting skin tone based on the cluster of training samples in the C_bC_r subspace, show in Fig. 5(a). And Fig. 6 show extracting result.

4.1 Construct Facial Features

The facial coordinate system based on Bookstein's [8] is used to describe the geometric shape of a face. Two categories of coordinates are contained and used to indicate the location and shape of each component respectively. The main coordinate system is used to describe the centers of brows, eyes, mouth, and nose [9]. Its origin is set to the center of left eye, and the distance of the left and right eyes is set to unity. For a face almost frontal, the distance between two eyes can be approximated with their horizontal distance. Then the coordinate of the right eye is $(1, y_{re})$.

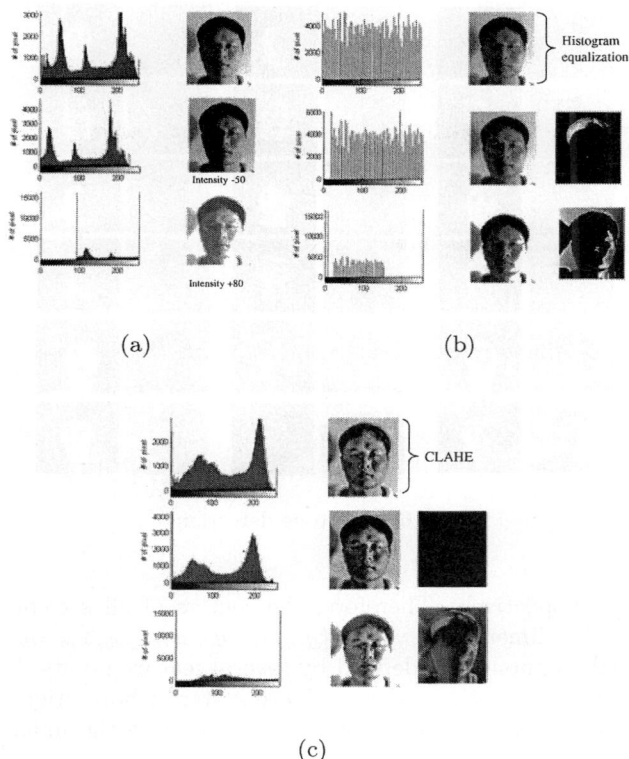

Fig. 4. Enhanced image: (a) Original image; (b) Histogram equalization; (c) CLAHE

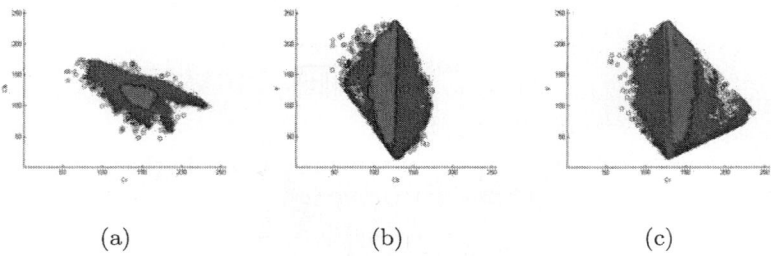

(a) (b) (c)

Fig. 5. The YC_bC_r color space and the skin tone color model (red dots represent skin color samples): (a) A 2D projection in the $CbCr$ subspace; (b) A 2D projection in the YCb subspace; (c) A 2D projection in the YCr subspace

4.2 Feature Vector Construction

Suppose the screen coordinates of the centers of left and right eyes are (x_{le}, y_{le}) and (x_{re}, y_{re}), then their corresponding facial coordinates are $(0,0)$

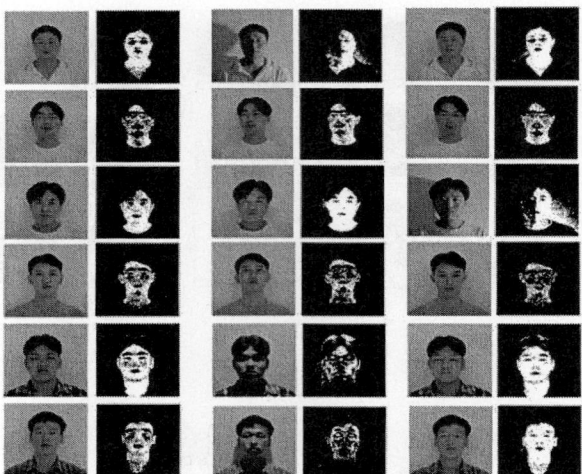

Fig. 6. Skin tone detection

and $(1, \frac{y_{re}-y_{le}}{x_{re}-x_{le}})$ respectively. Therefore, the centers of all six components can be described by a 9 dimensional vector $(y_{re}, x_{lb}, y_{lb}, x_{rb}, y_{rb}, x_m, y_m, x_n, y_n)$. The shape of a facial component is defined by several feature points. For each component, we define a componential coordinate system whose origin is set to the componential center and whose unity is set to equal to the unity of the main coordinate.

– Determine of feature vector [10]

$$\|u\|\|v\|cos\theta = u \cdot v$$

$$cos\theta = \frac{u \cdot v}{\|u\|\|v\|}$$

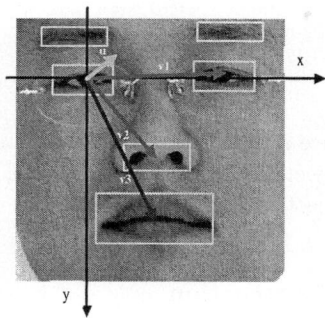

Fig. 7. Facial feature coordinate

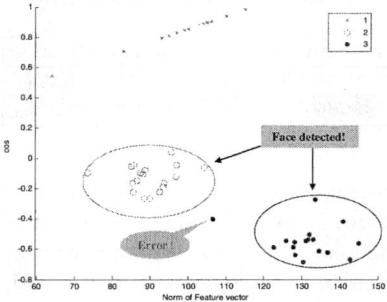

Fig. 8. Distribution of facial feature vector

- The length (or norm) of v (feature vector) is the nonnegative scalar $\|v\|$ defined by
$$\|v\| = \sqrt{v \cdot v} = \sqrt{v_1^2 + v_2^2 + \cdots + v_n^2},$$
and
$$\|v\|^2 = v \cdot v$$

In previous sections, we have shown that the feature of a point can be described by its facial coordinates and responses in mapped image. Our search for features, we might try to capitalize on the observation that mouth is typically more far from right eye then nose. Now we have ten features for classifying face- the angle x_1 and the length x_2 [11] and geometric position of face components $x_3 - x_{10}$. We realize to a point or feature vector x in a two-dimensional feature space, where

$$\mathbf{x} = \begin{bmatrix} x_1 \\ x_2 \end{bmatrix}$$

And distribution of facial feature vector show Fig. 8.

5 Experimental Result

In our experiments, we have used the POSTECH face database [7], which consists of 1802 frontal images of 106 person (17 images each). Images are 256×256 pixels in full-color images. The test platform is a P4/2.4GHz computer with 512MB RAM under Windows 2K.

We realized the face region detection process using YC_bC_r color space. We use response to a chrominance component in order to find eye and mouth. In a general, images like digital photos have problems (e.g., complex background, variation of lighting condition). Thus it is difficult process to determine skin- tone's special features and find location of eye and mouth. Nevertheless we can

Table 1. Feature extraction generalization rates

	left eye	right eye	nose	mouth	left brow	right brow
female	83.5%	84.1%	84.0%	82.0%	80.2%	80.2%
male	84.2%	84.5%	83.5%	82.4%	83.2%	83.2%

make to efficiency algorithm that robustness to variation lighting condition to use chrominance component in YC_bC_r color space. Also we can remove a fragment regions by using morphological process and connected component labeling operation. We find eye and mouth location use vertical, horizontal projection. This method is useful and show that operation speed is fast.

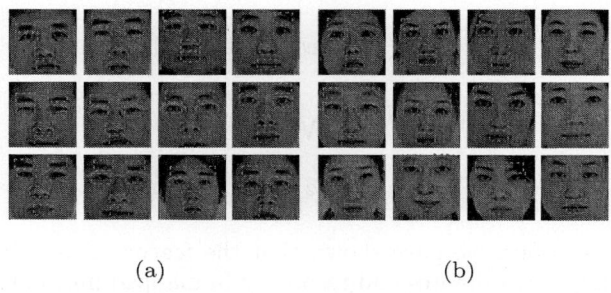

(a) (b)

Fig. 9. Some face and facial component extraction: (a) Male (b) Female

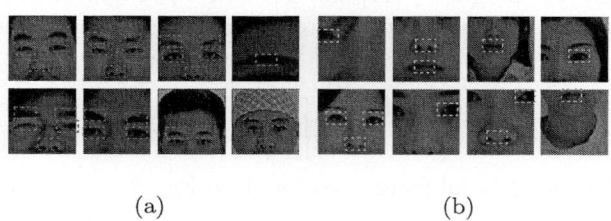

(a) (b)

Fig. 10. Worst case of facial component extraction: (a) Male (b) Female

Figure 9 demonstrates that our face detection algorithm can successfully detect facial candidate region. Fig. 10 shows the worst results for subjects with some facial variations (e.g., rotation or lighting variation).

The face detection result on the POSTECH image database [7] are presented in Table 1. The POSTECH image database contains 951 images, each of size 255 × 255 pixels. Lighting conditions (including overhead light and side lights) change from one image to another.

6 Conclusion and Future Work

In this paper, a new feature extraction method for real-time face detection system is proposed. We have represented in this paper a face detection system by using chrominance component of skin tone. Performance improvement of this method is demonstrated through our own experiments. Our face detection method detects skin regions over the entire image, and then generates face candidates based on the spatial spatial arrangement of these skin patches. Our algorithm constructs eye, mouth maps for detecting the eyes, mouth, and face region. Detection results on several photo collections have been shown Fig. 9 in experiment result. Especially, the result of several experiments in real life show that the system works well and is applicable to real-time tasks. This level of performance is achieved through a careful system design of both software and hardware, and tells about the possibility of various applications. It is a future work to make the training stage faster or to make code optimize for efficiency vectors calculation and face recognition. Also hardware integration may be considered for faster system.

References

1. Hsu, R.L., Abdel-Mottaleb, M.: Face detection in color images. IEEE Pattern Analysis and Machine Intelligence **24** (2002) 696–706
2. Feraud, R., Bernier, O., Viallet, J.E., Collobert, M.: A fast and accurate face detection based on neural network. IEEE Trans. Pattern Analysis and Machine Intelligence **23** (2001) 42–53
3. Hjelmas, E., Low, B.: Face detection : A survey. Computer Vision and Image Understanding **83** (2001) 236–274
4. Maio, D., Maltoni, D.: Real-time face location on gray-scale static images. Pattern Recognition **33** (1999) 1525–1539
5. Pantic, M., Rothkrantz, L.: Automatic analysis of facial expressions : The state of the art. IEEE Trans. Pattern Analysis and Machine Intelligence **22** (1996) 1424–1445
6. Yang, M.H., Kreigman, D.J., Ahuja, N.: Detecting faces in images : A survey. Pattern Analysis and Machine Intelligence **24** (2002) 34–58
7. IMDB. Intelligent Multimedia Laboratory, POSTECH, KOREA. (2001)
8. Bookstein, F.L.: A statistical method for biological shape comparison. J. Theor. Biol. **107** (1984) 475–520
9. Dihua Xi, Igor T. Podolak, S.W.L.: Facial component extraction and face recognition with support vector machines. IEEE International Conference on Automatic Face and Gesture Recognition (FGR) (2002)
10. Lay, D.C.: Linear Algebra And Its Applications 2nd Ed. Addison-wesley (1999)
11. Duda, R.O., Hart, P.E., Stork, D.G.: Pattern Classification 2nd Ed. John Wiley & sons, New York (2001)

On Optimizing Feature Vectors from Efficient Iris Region Normalization for a Ubiquitous Computing

Bong Jo Joung[1] and Woongjae Lee[2]

[1] Department of Information and Control Engineering, Kwangwoon University,
447-1, Wolgye-dong, Nowon-gu, Seoul, 139-701, Korea
chung@kw.ac.kr
[2] Division of Information and Communication Eng., Seoul Women's University,
126, Kongnung2-dong, Nowon-gu, Seoul, 139-774, Korea
wjlee@swu.ac.kr

Abstract. Iris patterns are believed to be an important class of biometrics suitable for subject verification and identification applications. An efficient approach for iris recognition is presented in this paper. An efficient iris region normalization consists of a doubly polar coordinate and noise region exclude. And then a Haar wavelet transform is used to extract features from iris region of normalized. From this evaluation, we obtain iris code of small size and very high recognition rate. This effort is intended to enable a human authentication in small embedded systems, such as an integrated circuit card (smart cards).

1 Introduction

The recent advances of information technology and the increasing requirement for security have resulted in a rapid development of intelligent personal identification based on biometrics. Biometrics is known as a way of using physiological or behavioral characteristics as measuring means. Some physiological or behavioral characteristics are so unique to each individual that they can be used to prove the person's identity through automated system. Today, biometric recognition is a common and reliable way to authenticate the identity of a living person based on physiological or behavioral characteristics. A physiological characteristic[1] is relatively stable physical characteristics, such as fingerprint, iris pattern, facial feature, hand silhouette, etc. The iris begins to form in the third month of gestation and the structures creating its pattern are largely complete by the eighth month, then does not change after two or three years. Recently, Daugman [2][3] developed the feature extraction based on 2D Gabor filter. He obtained 2048 bits iris coding by coarsely quantizing the phase information according to complex-valued coefficients of 1024 wavelets, chose a separate point between same match and different match. His research work has been the mathematic basis of most commercial iris recognition systems. But, the system of Daugman concentrated on ensuring that repeated image captures produced irises on the same location within the image, had the same resolution, and were glare-free under fixed illumination. These constraints may restrict to apply it in practical experiences. Wildes [4] proposed a prototype system based on automated iris recognition, which registered iris image to a stored model, filtered with four resolution levels and exploited spatial correlations and Fisher liner discrimination for pattern matching. This system is very computationally demanding. Boles[5] implemented a feature extraction algorithm via zero-crossing representation of the dyadic wavelet transform. It is tolerant to illumination variation, but only feature extraction and matching algorithm are considered.

In this paper, we propose an iris region to be normalized and haar wavelet transform using to extract features from iris region.

2 Image Preprocessing

2.1 Iris Localization

Iris is circular and much darker than the neighboring sclera, the iris region can be easily detected in the input image. Both the inner boundary and the outer boundary of a typical iris is usually not co-centric. The outer and inner boundaries are detected using equation (1) method and the center coordinates and radii of the outer and inner boundaries are obtained. We can see that the iris can be exactly localized using this technique. These detection operations[2][3] are accomplished by integro-differential operators the form

$$\max(r, x_0, y_0) \left[G_\sigma(r) * \frac{\partial}{\partial r} \oint_{r, x_0, y_0} \frac{I(x, y)}{2\pi r} ds \right] \qquad (1)$$

where contour integration parametrized for size and location coordinates r, x_0, y_0 at a scale of analysis σ set by some blurring function $G_\sigma(r)$ is performed over the image data array $I(x, y)$. The result of this optimization search is the determination of the circle parameters r, x_0, y_0 which best fit the inner and outer boundaries of the iris.

2.2 Iris Region Normalization

The iris region is normalized as shown in Fig. 3(b) using Fig. 2 and equation (2) by the obtained the center coordinates and radii of the outer and inner boundaries. Fig. 2 shows both the inner boundary and the outer boundary of a typical iris is usually not co-centric and besides iris region is change due to variableness of light or illumination. The localized iris region is transformed into doubly polar coordination system in an efficient way so as to facilitate the next process, the feature extraction process.

Fig. 2 shows the process of converting the Cartesian coordinate system into the doubly polar coordinate system for the iris region. Fig. 2(a) Here, **p** and **q** mean the center of pupil and iris. The center of the iris and pupil is not located in same place and therefore **first step** iris data must extract in identical angle

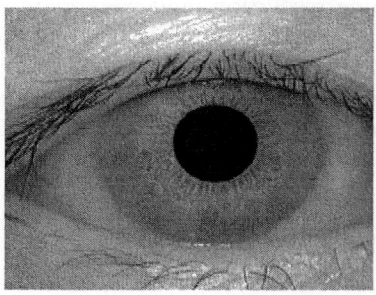
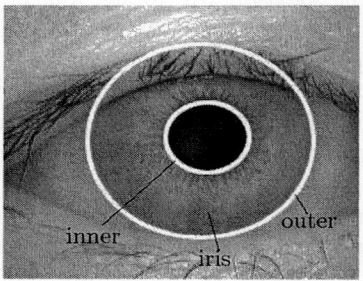

(a) Original iris image (b) Detected inner and outer boundaries of iris

Fig. 1. Iris localization

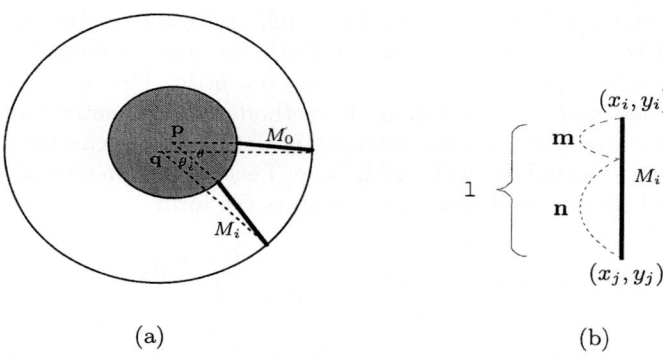

(a) (b)

Fig. 2. Illustration of the method of iris data sampling.(a) sampling direction (b) sampling interval

($\theta = \theta_i$). The **second step** we extract an iris data in the direction like M_i and M_0. Fig. 2(b) Here, l mean r of polar coordinate (r, θ). The length of l is changed due to iris region is change and therefore the iris data extract to the rate of the m and n.

$$\left[\frac{(m \times x_j) + (n \times x_i)}{m+n}, \frac{(m \times y_j) + (n \times y_i)}{m+n}\right] = \begin{cases} (x_l, y_l) & \text{for } 0 < m \leq \frac{1}{2} \\ (\text{ Reject }) & \text{for } m > \frac{1}{2} \end{cases} \quad (2)$$

Fig. 3 shows extracted a part **U** region of iris region by Equation (2). It is because other region takes influence of the eyelid or eyebrow. Also and Fig. 3(d) shows intensity of this part region(noise region) with the feature which the change is not extreme. Extracted features from this part region(noise region) have low matching rate and fall the Recognition rate. Therefore this part region(noise region) is exclusion.

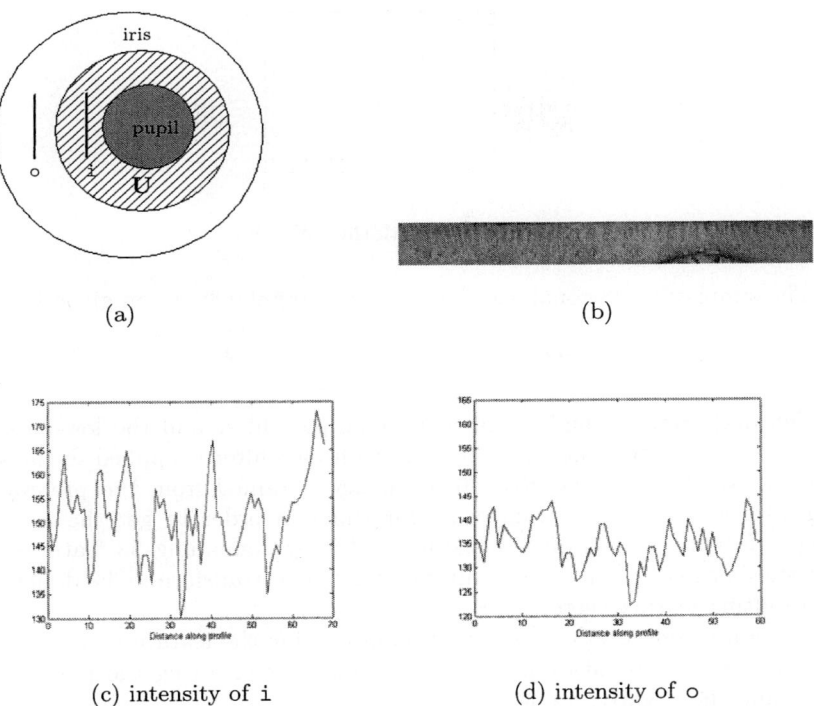

Fig. 3. Illustration of the method of iris region normalized. (a)Noise region and interested region (b)Iris region normalized by a doubly polar coordinate and noise region exclusion

Fig. 3(c) shows intensity of this part **U** region with the feature which the change is extreme. Therefore Fig. 3(b) shows this part **U** region(interested region) is used.

3 Iris Feature Extraction by Haar Wavelet Transform

In this paper, a wavelet transform is used to extract features from iris region[6][7][8]. Any particular local features of a signal can be identified from the scale and position of the wavelets in which it is decomposed[9]. Wavelets are a powerful tool for presenting local features of a signal. When the size and shape of a wavelet are exactly the same as a section of the signal, the wavelet transform gives a maximum absolute value, a property which can be used to detect transients in a signal. Thus the wavelet transform can be regarded as a procedure for comparing the similarity of the signal and the chosen wavelet. we use Haar wavelet illustrated in Fig. 4 and equation(3) as a basis function[10].

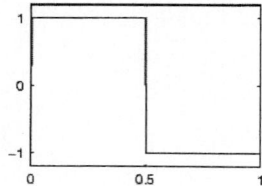

Fig. 4. Haar Mother Wavelet

The simplest orthogonal wavelet is the rectangular function given by

$$w(t) = \begin{cases} -1 & \text{for} \quad 0 \leq t < \frac{1}{2} \\ +1 & \text{for} \quad \frac{1}{2} \leq t < 1 \end{cases} \quad (3)$$

Fig. 5(a) Here, H and L mean the high-pass filter and the low-pass filter, respectively, and HH indicates that the high-pass filter is applied to the signals of both axes[11]. For the 450×30 iris image obtained from the preprocessing stage, we apply wavelet transform four times in order to get the 29×2 sub-images. Finally, we organize a feature vector by combining 58 features in the HH sub-image of the high-pass filter of the fourth transform. The dimension of the resulting feature vector is 58.

To reduce computational time for manipulating the feature vector, we quantize each real value into binary value by simply converting the positive value into 1 and the negative value into 0.

(a) Subband form of wavelet transform

(b) Subband image of wavelet transform

Fig. 5. Iris subband image and form

$$W(t) = \begin{cases} 1 & \text{if} & w(t) \geq 0 \\ 0 & \text{if} & w(t) < 0 \end{cases} \qquad (4)$$

Equation(4)made iris codes consist of 1 and 0. Using Hamming Distance(HD)[12] such as the Equation(5) and compare a two iris codes of 58bit to be inputed.

$$HD = \frac{1}{N} \sum_{t=1}^{N} A_t(XOR)B_t \qquad (5)$$

4 Experiment Results

For this experiment, we use 7 data per person from 20 persons. In order to determine a threshold separating False Reject Rate(FRR) and False Accept Rate(FAR), we using a Hamming distance.

Fig. 6 shows the distribution of Hamming distances computed between 420 pairs of different images of the same iris. In the figure, x-axis and y-axis indicate the multiply HD by 100 and the count of date.

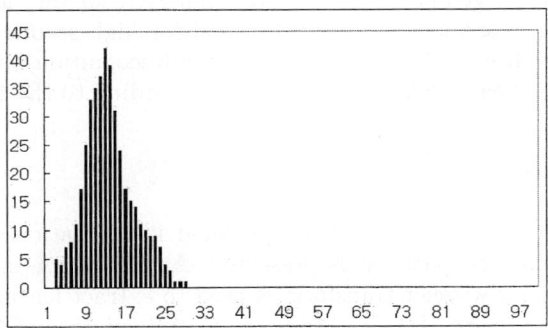

Fig. 6. Hamming Distances for Authentics

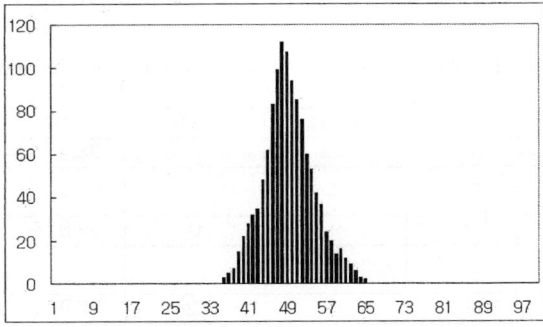

Fig. 7. Hamming Distances for Imposters

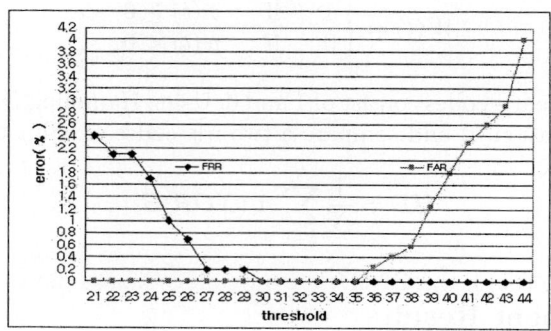

Fig. 8. Result in verification with Hamming distance

Fig. 7 shows the distribution of Hamming distances computed between 1211 pairs of different images of the different iris. In the figure, x-axis and y-axis indicate the multiply *HD* by 100 and the count of date.

It can be seen in Fig. 8, that Equal Error Rate(EER), the cross point between the FAR and the FRR curves, achieves a 0%. But what is more important, obtained both the range of cross point and a null FAR for very low rates of False Rejection, which means this system is optimal for high security environments. When we use the threshold of 32, we can get the Recognition rate(RR) of about 99.8%. Table. 1 indicates FAR, FRR and RR according to the threshold.

5 Conclusion

In this paper, an efficient method for personal identification and verification by means of human iris patterns is presented. An iris region is proposed to be normalized and haar wavelet transform is used to extract features from the iris region. With these methods, we obtain iris feature vectors of 58 bits. Table 2 with only 58bits shows that we could present an iris pattern without any negative influence and maintain an high rates of recognition.

Table 1 and Table 2 show that the system can achieve high rates of security.

Table 1. FAR, FRR and RR according to the Threshold

threshold	FRR	FAR	RR
29	0.2	0	98.4 %
⋮	⋮	⋮	⋮
32	0	0	99.8 %
⋮	⋮	⋮	⋮
36	0	0.25	97.9 %

Table 2. Comparing the proposed method and Gabor transform

	Proposed method	Gabor transform
Code Size	58 bit	256 byte
RR	99.8 %	99.6 %

Acknowledgements

We would like to thank the CASIA for providing the iris database.

References

1. Ashbourn, D.M.: Biometrics: Advanced identify verification: The complete guide. Springer (2000)
2. Daugman, J.G.: High confidence visual recognition of persons by a test of statistical independence. IEEE Trans. Pattern Analysis and Machine Intelligence **15** (1993) 1148–1161
3. Daugman, J.G.: Recognizing persons by their iris patterns. Cambridge University (1997)
4. Wildes, R.P.: Iris recognition: An emerging biometric technology. Proceedings of the IEEE **85** (1997) 1348–1363
5. Boles, W.W., Boashash, B.: A human identification technique using images of the iris and wavelet transform. IEEE Trans. on Signal Processing **46** (1998) 1185–1188
6. Young, R.K.: Wavelet and signal processing. Kluwer Academic Publisher (1992)
7. Rioul, O., Vetterli, M.: Wavelet and signal processing. IEEE Signal Processing Magazine (1981) 14–38
8. Strang, G., Nguyen, T.: Wavelet and filter banks. Wesley-Cambridge Press (1996)
9. Gonzalez, R.C., Woods, R.E.: Digital image processing second edition. Addison Wesley (2002)
10. Sonka, M., Hlavac, V., Boyle, R.: Image processing, analysis, and machine vision second edition. International Thomson (1999)
11. Lim, S., Lee, K., Byeon, O., Kim, T.: Efficient iris recognition through improvement of feature vector and classifier. ETRI Journal **23** (2001)
12. Daugman, J.G.: High confidence recognition of persons by rapid video analysis of iris texture. European Convention on Security and Detection (1995)

On the Face Detection with Adaptive Template Matching and Cascaded Object Detection for Ubiquitous Computing Environment

Chun Young Chang[1] and Jun Hwang[2]

[1] Department of Information and Control Engineering,
Kwangwoon University, 447-1, Wolgye-dong,
Nowon-gu, Seoul, 139-701, Korea
chung@kw.ac.kr
[2] Division of Information and Communication Eng.,
Seoul Women's University, 126, Kongnung2-dong,
Nowon-gu, Seoul, 139-774, Korea
hjun@swu.ac.kr

Abstract. This paper presents the template matching and efficient cascaded object detection. The proposed template matching method is superior to previous face detection. Furthermore, the proposed cascade method has some merits to the face changes. Thus, we can detect the object effectively and this can inevitably lead to the Ubiquitous Computing Environment. We also expand the more detection algorithms through this method.

1 Introduction

Traditionally, computer vision systems have been used in specific tasks such as performing tedious and repetitive visual tasks of assembly line inspection[1]. Current trend in computer vision is moving toward generalized vision applications. For example, face recognition and video coding techniques etc.

Many of the current face recognition techniques assume the availability of frontal faces of similar sizes[1].

Fig. 1 pictures are typical test images used in face classification research.The background in Fig. 1 images is necessary for face classification techniques[1]. However, with the face of Fig. 2 could exist in a complex background and in many different positions.

Most face recognitions are achieved practically in such condition. Consequently, We must consider such condition for efficient face recognition. As Well, we must consider efficient recognition about various objects. However, in computer vision is not easy. The solution to the problem involves segmentation, extraction, and verification of faces and possibly facial features from an uncontrolled background[1]. To be concrete, the problem of such object recognition uses a object image of two dimension normally to the input and use the output

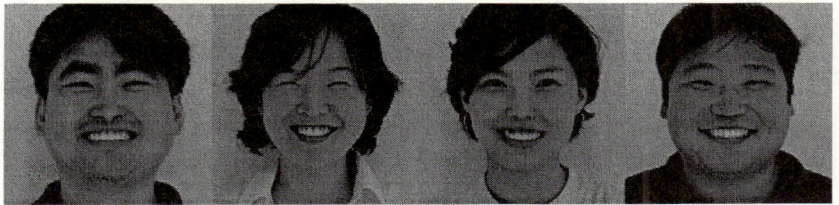

Fig. 1. Typical training images for face recognition

Fig. 2. A realistic face detection scenario

to assort who that image is. For example, a face image of the three dimension is to be reflected to the two dimension. Accordingly, the information deeply, magnitude, rotation etc. has the loss of important many information in the recognition.The recognition comes to be difficult basically as the pattern due to the complication which the object has and illumination,background and the complication of the environment etc.

The object recognitions of comprehensive concept must accomplish the course to find the location of the object in a two dimension image of the random above all. Afterwards, preprocessing course of the back of a noise removal is performed. And also, Normalization course of a object image comes to be continuously performed which set the size of the object or location with inside the image's size to want to the location.

2 Detection Algorithm

The active shape models can express which it appears in the Table 1. They can distinguish generally to three types . The first type uses a generic active contour called snakes, first introduced by Kass et al. in 1987[2]. Deformable Templates were then introduced by Yuille et al.[3] to better the performance of snakes. Cootes et al.[4] later proposed the use of a new generic flexible model which they termed smart snakes and PDM to provide an efficient interpretation of the human face.

Table 1. Object detection divided into approaches

Approach	Method	Representative work
Feature-based	Low-level	Edge: Grouping of edges
		Gray-level (texture): Space Gray-Level Dependence matrix(SGLD)
		Color: Mixture of Gaussian
		Motion: Second order temporal edge operator
		Multiple Features: Integration of skin color, size, shape
	Feature analysis	Feature Searching: anthropometrics measures
		Constellation analysis: gradient-type operator
	Active shape	Snake: term sensitive to the image gradient
		Deformable template: integral energy
		Point distribution models (PDMs):flexible model

Approach	Method	Representative work
Appearance-based	Eigenface	Eigenvector decomposition and clustering
	Distribution-based	Gaussian distribution and multiplayer perception
	Neural networks	Ensemble of neural networks and arbitration schemes
	Support Vector Machine(SVM)	SVM with polynomial kernels
	Naive Bayes Classifier	Joint statistics of local appearance and position
	Hidden Markov Model(HMM)	Higher order statistics with HMM
	Information-Theoretical App.	Kullback relative information

3 Template Matching for Object Detection

The template matching belongs to wide criteria which can regard to the "feature-centric". Minute explanation about this explains in a next section.

The template matching finds similar image pattern in the image inside to check to be given beforehand when the image was given. At this time, template is the kind of a model image.We overlap small template of the image at the starting point on left corner which compare a template image with the part of the overlapping image to check. This comparison standard amount can choose so that it is suitable according to the purpose.After store comparison standard amount to be calculated, We shift again a one pixel to left which does the template. And we compare again a template image with the part of the overlapping image to check.

The template matching is important to well select the comparison standard amount. The determine the comparison standard amount has some kind of the subject to consider. It must be insensible at an image noise and at intensity variation. It must have also small computation quantity.

The current standard of template matching is based on computed in Fast Fourier Transform(FFT)[5]. This can be extended to shift of template by a suitable sampling of the template[5]. We can use generally two kinds method. The first method is MAD(Mean Absolute Difference).

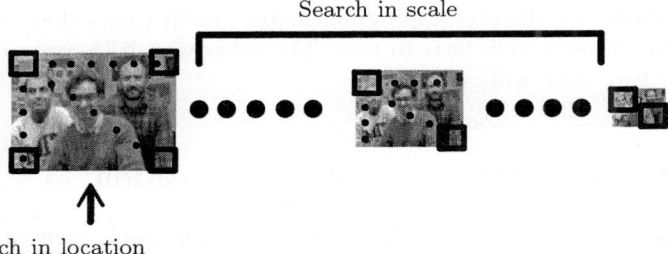

Search in scale

↑
Search in location

Fig. 3. Exhaustive search object detection

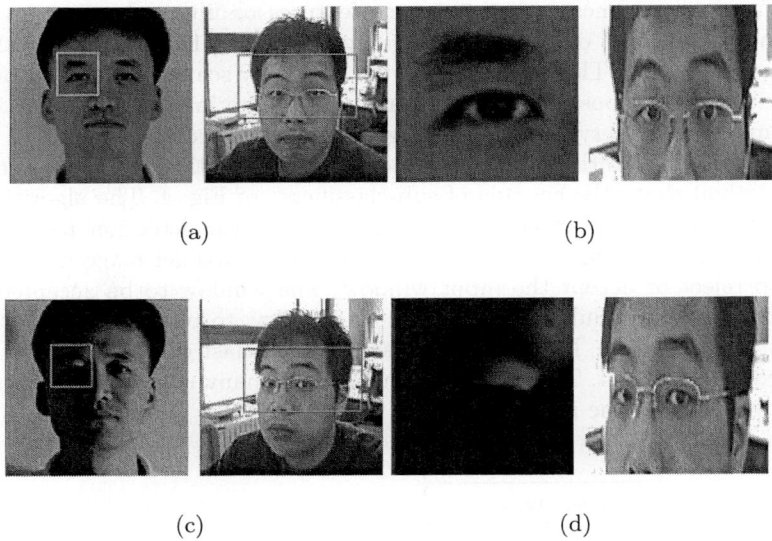

(a) (b)

(c) (d)

Fig. 4. (a)The image to set the template(The comparison standard amount). (b)Template image.(c)The image to search.(d)The image to compare a template image with the part of the overlapping image to check

$$MAD = \frac{1}{MN}\sum_{i=0}^{M}\sum_{j=0}^{N} \mid T(x_i, y_i) - I(x_i, y_i) \mid \qquad (1)$$

where M and N are width and length of template image, $T(x_i, y_i)$ is template image, $I(x_i, y_i)$ is the overlapping image to check. The second method is MSE(Mean Square Error).

$$MSE = \frac{1}{MN}\sum_{i=0}^{M}\sum_{j=0}^{N}[T(x_i, y_i) - I(x_i, y_i)]^2 \qquad (2)$$

If template and the overlapping image are similar each other, MAD or MSE will become computation near to zero. The other way, if they are different each other, the two value will grow bigger.

4 Cascaded Method for Object Detection

The cascaded method for object detection approach uses a novel organization of the first cascade stage called "feature-centric" like the Templates.

One of the point of detection is coping with variation in object size and location. There are general two approach methods for this. The first is "Invariant" methods. These attempt to use features or filters that are invariant to geometry[6][7][8][9] or photometric properties[10][11][12]. Another method is "exhaustive-search". This method finds the object by scanning classifier over an exhaustive range of possible locations and scales in an image[13].But the defect of this method has very time consuming to find the object to want.

The method to do the supplementation the defect of the "exhaustive search" is the method that "the cascade of sub-classifiers" of Fig. 4. The algorithm for constructing a cascade of classifiers achieves increased detection performance while radically reducing time consuming. Each sub-classifier stage makes a decision to reject or accept the input window. The window to be accepted goes to a next stage(next Sub-Classifier) and the window to be rejected goes to the Classify as non-object. The window to remain in the last goes via such course is classified as the object. This is designed to remove many non-object windows to the computation of the minimum.

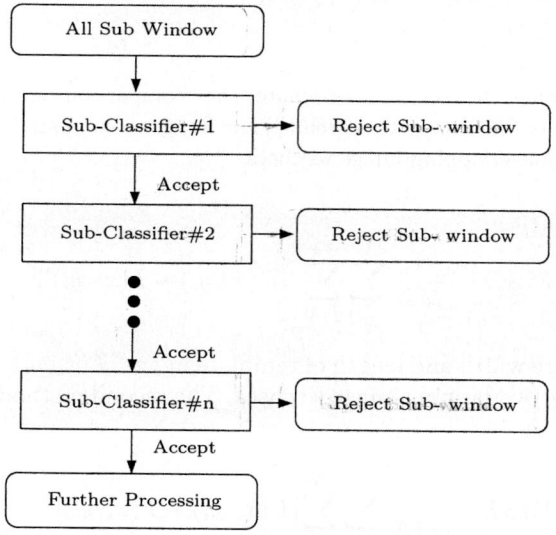

Fig. 5. Schematic depiction of a the detection cascade

The idea of using cascade-like methods has existed for several decades, and, in particular, was used widely in the automatic target recognition techniques of the 1970s and 80s[14].

Each sub-classifier is represented as a semi-naïve Bayes classifier[15].

$$H(X_1,\ldots,X_n) = \log\frac{P(S_1|\omega_1)}{P(S_1|\omega_2)} + \log\frac{P(S_2|\omega_1)}{P(S_2|\omega_2)} + \ldots + \log\frac{P(S_m|\omega_1)}{P(S_m|\omega_2)} > \lambda$$

$$S_1,\ldots,S_m \subset \{X_1,\ldots,X_n\} \tag{3}$$

where $X_1,\ldots,X_n$ are the input variables within the classification window, $S_1,\ldots,S_r$ are subsets of these variables, and ω_1 and ω_2 indicate the two classes. If ω_1 is the face and ω_2 is the non-face, the classifier chooses class ω_1. Otherwise, it chooses class ω_2(if $f(X_1,\ldots,X_n) < \lambda$).

Each stage in the cascade reduces the false positive rate and decreases the detection rate.In most classifiers including more features will achieve higher detection rates and lower false positive rates. At the classifiers with more features require more time to compute[16].

(a)

(b)

Fig. 6. (a)The image which the face of the many people existent. (b)The image by the cascade method.

Figure 6 of the downside is the experiment result by Ming-Hsuan Yang in Honda Research Institute Honda Research Institute Mountain View, California, USA[17].

5 Experiments

This chapter describes about the experiment environment and result. The face image data and object data did not put the limit in the lighting for a face detection. Also, we did not consider to wear glasses and the size of face area. The data used to be holding at the laboratory and on MIT-CMU test set. We experimented with the gray scale image of 256×256 size on an pentium4, 1.7GHz processor.

5.1 Template Matching

Like formula (1),(2), if the template of (M×N) size and the image to check of R×C size are given, the number to compare of the overlapping image happens (R-M)×(C-N) times. If size of template is 100×100 and size of image to check is 640×640, the number of the overlapping image happens 540×380 times. This is not little the number to overlapping. According to, it happens time complexity. If the time complexity is high, the computation time to similar pattern takes long.

We trained for finding the part of the face to want from image. The experiments was processed to two kinds. The first, the image used male of 56 persons and female of 51 persons including expression of four kinds. The template set in the sequence of eye, nose, mouth and ear. The second, it applies the template matching at the layer when the face of many people exists.

5.2 Cascaded Method

The complete face detection cascade has 38 stages with over 6000 features[16]. On a difficult data, containing 507 faces and 75 million sub-windows, faces are detected using of 10 feature evaluations per sub-window. This system is about 15 times faster than an implementation of the detection system constructed by Rowley et al.[16][18].

This experiments was processed also to two kinds as the template matching.

5.3 The Result of the First Experiment

The detection probability of the cascade method is superior generally but the template matching method fells off remarkably from the detection probability of the nose and eye. Because, this regards the open mouth of the appearance to smile wrong as the eye and is the experiment result of the case to wear glasses.

Table 2. Comparison of template matching and cascade

	Eye	Nose	Mouth	Ear
Template Matching	78.75 %	98.7 %	56.4 %	87.7 %
Cascade	85.65 %	95.7 %	82.4 %	92.7 %

Table 3. False positives

False Positives	50	100	150	200
Template Matching	73.2 %	73.8 %	72.4 %	71.8 %
Cascade	75.3 %	76.42 %	77.31 %	78.3 %

5.4 The Result of the Second Experiment

This experiment's detection probability is lower than the first experiment. we can see high cascade method's detection probability as the false positive come to be high.

6 Conclusion

This paper describes the template matching and cascaded object detection for efficient face detection. The template matching method is superior to previous face detection. Since the template matching and cascade method has an advantage to find the object better, it can find the object to correspond completely. Therefore, the method can detect the many faces mixed with different objects better and can detect the various expressions of face, provided that the cascade method can maximize the face detection probability. We expect that the result of this paper can be contributed to develop face detection methods and face recognition methods.

References

1. Low, B.K.: Face detection (a survey) (2001)
2. Kass, M., Witkin, A., Terzopoulos, D.: Snakes: active contour models. on Computer Vision, London (1987) in Proc. of 1st Int Conf.
3. Yuille, A.L., Hallinan, P.W., Cohen, D.S.: Feature extraction from faces using deformable templates. Int.J Comput Vision8 (1992) pp. 99–111
4. Cootes, T.F., Taylor, C.J.: Active shape models-smart snakes. in Proc. of British Machine Vision Conference (1992) pp. 266–275
5. Fredriksson, K., Ukkonen, E.: Faster template matching without FFT. IEEE (2001)
6. Fergus, R., Perona, P., Zisserman, A.: Object class recognition by unsupervised scale-invariant learning. CVPR (2003)
7. Forsyth, D.A.: Invariant descriptors for 3d recognition and pose,. PAMI (1991)

8. Wood, J.: Invariant pattern recognition: A review. Pattern Recognition (1996) pp. 1–17
9. Zisserman, A.: 3D object recognition using invariance. Artificial Intelligence 78(1-2):239-288 (1995)
10. Chen, H., Belhumeur, P., Jacobs, D.: In search of illumination invariants. CVPR (2000) pp. 254–261
11. Nagao, K., Grimson, W.E.L.: Using photometric invariants for 3D object recognigion,. CVIU (1998) pp. 74–93
12. Slater, D., Healey, G.: The illumination-invariant recognition of 3D objects using local color invariants. PAMI (1996) pp. 206–210
13. Schneiderman, H.: Feature-centric evaluation for efficient cascaded object detection. (2004) Robotics Institute Carnegie Mellon University Pittsburgh.
14. Bhanu, B.: Automatic target recognition: a state of the art survey. IEEE Trans. (1986) pp. 364–379 on aerospace and Electronic Systems.
15. Kononenko, I.: Semi-naïve bayesian classifier. Sixth European Working Session on Learning. (1991) pp. 206–219
16. Viola, P., Jones, M.: Rapid object detection using a boosted cascade of simple features. IEEE (2001)
17. Yang, M.H.: Recent advances in face detection. (2004) pp. 82
18. Rowley, H., Baluja, S., Kanade, T.: Neural network-based face detection. In IEEE Patt. Anal. Mach. Intell. **vol. 20** (1998) pp. 22–38

On Improvement for Normalizing Iris Region for a Ubiquitous Computing

Bong Jo Joung[1], Chin Hyun Chung[1], Key Seo Lee[1],
Wha Young Yim[1], and Sang Hyo Lee[1]

Department of Information and Control Engineering,
Kwangwoon University, 447-1, Wolgye-dong,
Nowon-gu, Seoul, 139-701, Korea
chung@kw.ac.kr

Abstract. Iris patterns are believed to be an important class of biometrics suitable for subject verification and identification applications. An efficient approach for iris recognition through an iris region normalization is presented in this paper. An efficient iris region normalization consists of a doubly polar coordinate and noise region exclude. From this evaluation, we obtain iris code of small size and very high recognition rate. This effort is intended to enable a human authentication in small embedded systems, such as an integrated circuit card (smart cards).

1 Introduction

The recent advances of information technology and the increasing requirement for security have resulted in a rapid development of intelligent personal identification based on biometrics. Biometrics is known as a way of using physiological or behavioral characteristics as measuring means. Some physiological or behavioral characteristics are so unique to each individual that they can be used to prove the person's identity through automated system. Today, biometric recognition is a common and reliable way to authenticate the identity of a living person based on physiological or behavioral characteristics. A physiological characteristic [1] is relatively stable physical characteristics, such as fingerprint, iris pattern, facial feature, hand silhouette, etc. The iris begins to form in the third month of gestation and the structures creating its pattern are largely complete by the eighth month, then does not change after two or three years. Recently, Daugman [2][3] developed the feature extraction based on 2D Gabor filter. He obtained 2048 bits iris coding by coarsely quantizing the phase information according to complex-valued coefficients of 1024 wavelets, chose a separate point between same match and different match. His research work has been the mathematic basis of most commercial iris recognition systems. But, the system of Daugman using whole region which the noise is included. Extracted features from this whole region have low matching rate and fall the Recognition rate. Also Gabor filter extract many iris feature codes. Wildes [4] proposed a prototype system based on automated iris recognition, which registered iris image to a stored model, filtered with four resolution levels and exploited spatial correlations and Fisher liner discrimination for pattern matching. This system both did not consider an iris normalizing and very computationally demanding. Boles [5] implemented a feature extraction algorithm via zero-crossing representation of the dyadic wavelet transform. It is tolerant to illumination variation, but only feature extraction and matching algorithm are considered.

In this paper, we propose an iris region to be efficient normalized. With these methods, we obtaining iris feature vectors which the size small and high matching Rate.

2 Iris Localization

Iris is circular and much darker than the neighboring sclera, the iris region can be easily detected in the input image. Both the inner boundary and the outer boundary of a typical iris is usually not co-centric. The outer and inner boundaries are detected using equation (1) method and the center coordinates and radii of the outer and inner boundaries are obtained. We can see that the iris can be exactly localized using this technique. These detection operations [2][3] are accomplished by integro-differential operators the form

$$\max(r, x_0, y_0) \left[G_\sigma(r) * \frac{\partial}{\partial r} \oint_{r, x_0, y_0} \frac{I(x, y)}{2\pi r} ds \right] \quad (1)$$

where contour integration parametrized for size and location coordinates r, x_0, y_0 at a scale of analysis σ set by some blurring function $G_\sigma(r)$ is performed over the image data array $I(x, y)$. The result of this optimization search is the determination of the circle parameters r, x_0, y_0 which best fit the inner and outer boundaries of the iris.

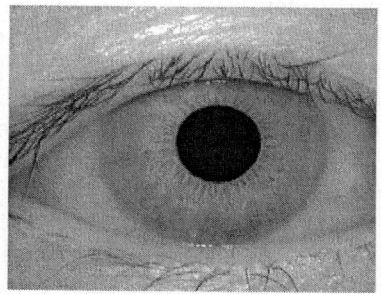
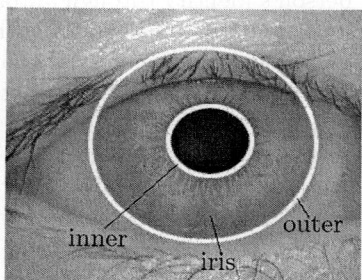

(a) riginal iris image

(b) Detected inner and outer boundaries of iris

Fig. 1. Iris localization

3 Iris Region Normalization

The iris region is general normalized with the process of converting the cartesian coordinate system into the polar coordinate system using sampling direction and sampling interval as shown in Fig. 2 by the obtained the pupil coordinates and radii of the inner boundaries. Fig. 2(a) Here, **p** mean the center of pupil.

But this method did not consider the iris coordinate and radii. Also iris region is change due to variableness of light or illumination. Fig. 2(b) shows the size of iris region has different area like M_0 and M_i. As the result Fig. 2(b) shows the case to loss data with M_i happens.

Fig. 3(a) Here, **p** and **q** mean the center of pupil and iris. Fig. 3 shows both the inner boundary and the outer boundary of a typical iris is usually not co-centric and besides iris region is change due to variableness of light or illumination.

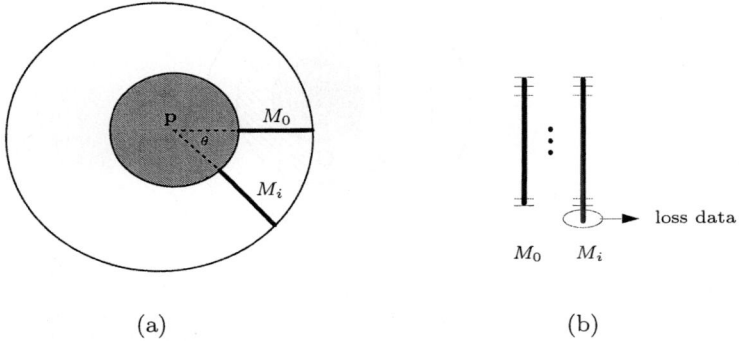

Fig. 2. Illustration of the general method of iris data sampling. (a) sampling direction (b) sampling interval

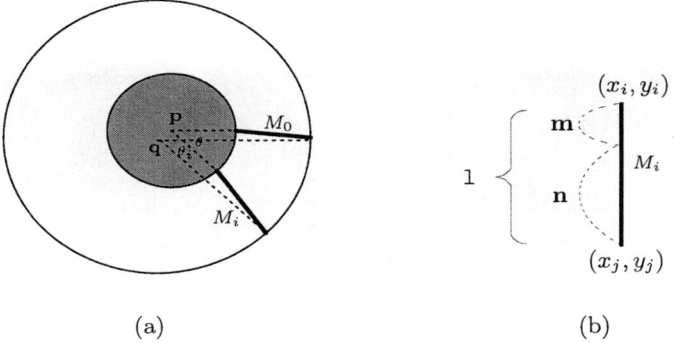

Fig. 3. Illustration of the efficient method of iris data sampling. (a) sampling direction (b) sampling interval

Fig. 4. Iris region efficient normalized by a doubly polar coordinate

Fig. 3 shows the process of converting the cartesian coordinate system into the doubly polar coordinate system for the iris region. The center of the iris and pupil is not located in same place and therefore **first step** iris data must extract in identical angle($\theta = \theta_i$). The **second step** we extract an iris data in the direction like M_i and M_0. Fig. 3(b) Here, l mean r of polar coordinate (r, θ). The length of l is changed due to iris region is change and therefore the iris

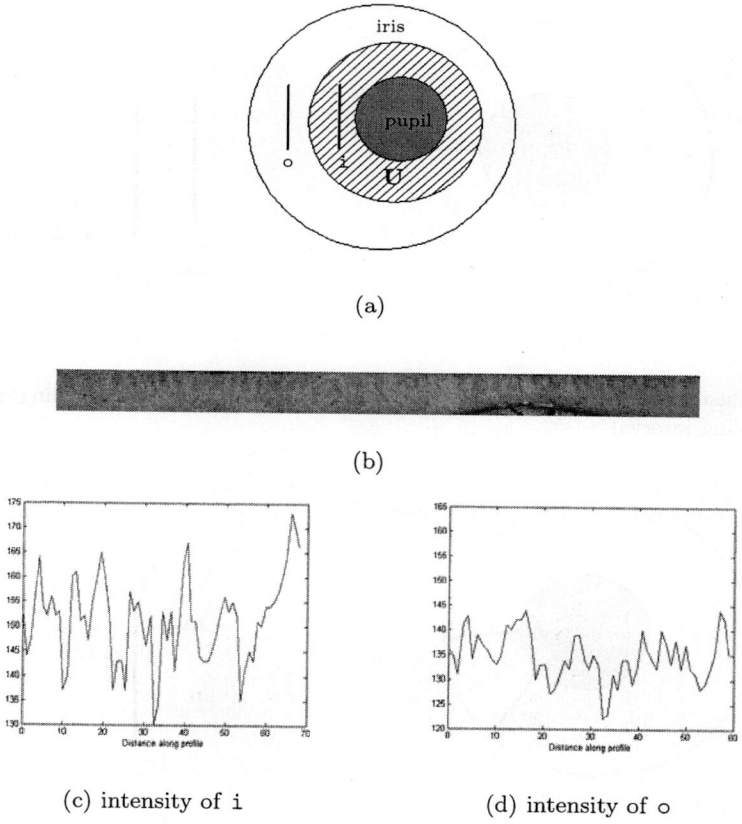

Fig. 5. Illustration of the efficient method of iris region normalized. (a)Noise region and interested region (b)Iris region efficient normalized by a doubly polar coordinate and noise region exclusion

data extract to the rate of the **m** and **n**. The iris region is efficient normalized as shown in Fig. 4 using Fig. 3 and equation (2).

$$\left[\frac{(m \times x_j) + (n \times x_i)}{m+n}, \frac{(m \times y_j) + (n \times y_i)}{m+n}\right] = \begin{cases} (x_l, y_l) & \text{for } 0 < m \leq \frac{1}{2} \\ (\text{Reject}) & \text{for } m > \frac{1}{2} \end{cases} \quad (2)$$

Fig. 5 shows extracted a part **U** region of iris region by Equation (2). It is because other region takes influence of the eyelid or eyebrow. Also and Fig. 5(d) shows intensity of this part region(noise region) with the feature which the change is not extreme. Extracted features from this part region(noise region) have low matching rate and fall the Recognition rate. Therefore this part region(noise region) is exclusion.

Fig. 5(c) shows intensity of this part **U** region with the feature which the change is extreme. Therefore Fig. 5(b) shows this part **U** region(interested region) is used.

4 Experiment Results

For this experiment, we use 2 different data images of the same iris as shown in Fig. 6. The iris region of data images is normalized as shown in Fig. 7,8,9 using each normalizing method. In order to obtain a Matching Rate(MR) of the each Figures, Haar wavelet transform [6][7][8] and Hamming Distance(HD)[9] is used. Haar wavelet transform is using to extract features from each images. With these methods, we obtaining iris feature vectors. Using Hamming Distance(HD) such as the Equation(3) and compare a two iris feature vectors.

$$HD = \frac{1}{N}\sum_{t=1}^{N} A_t(XOR)B_t \quad (3)$$

Table. 1 shows Matching Rate of each normalizing method express the fact that the normalize went well. Therefore most good normalizing method is the proposed method. Proposed method has matching rate of 86 % and code size of 58 bit.

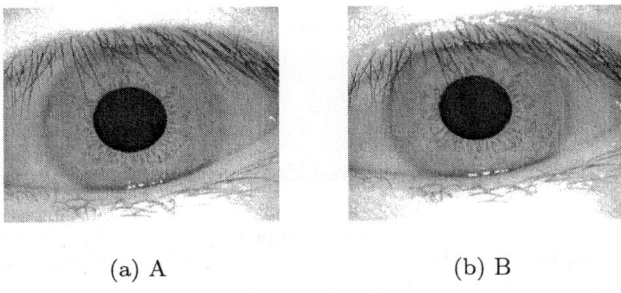

(a) A (b) B

Fig. 6. Different data images of the same iris

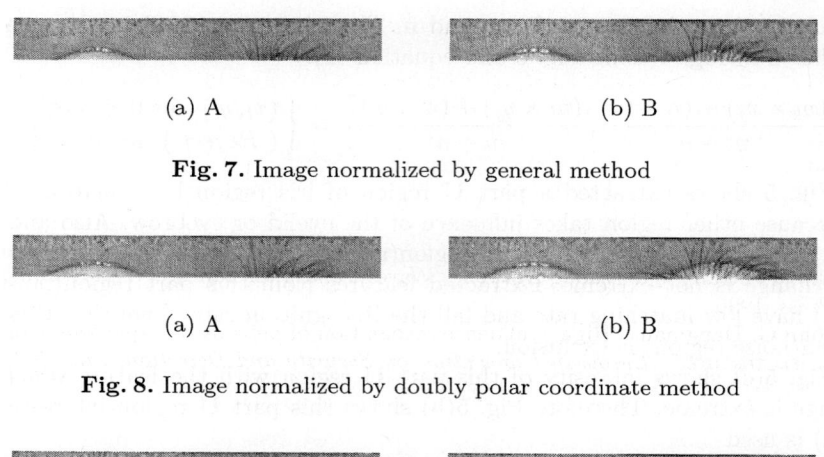

(a) A (b) B

Fig. 7. Image normalized by general method

(a) A (b) B

Fig. 8. Image normalized by doubly polar coordinate method

(a) A (b) B

Fig. 9. Image normalized by proposed method

Table 1. Comparing the Proposed method and General, Doubly polar coordinate method

	Matching Rate	Code Size
General method (Fig. 7)	69.8 %	116 bits
Doubly polar coordinate method (Fig. 8)	74 %	116 bits
Proposed method (Fig. 9)	86 %	58 bits

5 Conclusion

In this paper, an efficient method for iris region normalizing is presented. Wavelet transform is used to extract features from iris region normalized. With these methods, we could present iris feature vectors without any negative influence and maintain the high rates of recognition. The results show that the proposed method can achieve high rates of security.

References

1. D. M. Ashbourn, "Biometrics: Advanced identify verification: The complete guide," *Springer*, 2000.
2. John G. Daugman, "High confidence visual recognition of persons by a test of statistical independence," *IEEE Trans. Pattern Analysis and Machine Intelligence*, vol. 15, no. 11, pp. 1148–1161, November 1993.
3. John G. Daugman, "Recognizing persons by their iris patterns," *Cambridge University*, 1997.

4. R. P. Wildes, "Iris recognition: An emerging biometric technology," *Proceedings of the IEEE*, vol. 85, no. 9, pp. 1348–1363, 1997.
5. W. W. Boles and B. Boashash, "A human identification technique using images of the iris and wavelet transform," *IEEE Trans. on Signal Processing*, vol. 46, no. 4, pp. 1185–1188, 1998.
6. Randy K. Young, "Wavelet and signal processing," *Kluwer Academic Publisher*, 1992.
7. O. Rioul and M. Vetterli, "Wavelet and signal processing," *IEEE Signal Processing Magazine*, pp. 14–38, October 1981.
8. Gilbert Strang and Truong Nguyen, "Wavelet and filter banks," *Wesley-Cambridge Press*, 1996.
9. John G. Daugman, "High confidence recognition of persons by rapid video analysis of iris texture," *European Convention on Security and Detection*, , no. 408, May 1995.

Author Index

Abawajy, J.H.   III-60, IV-1272
Ahiska, S. Sebnem   IV-301
Åhlander, Krister   I-657
Ahmad, Uzair   II-1045
Ahn, Beumjun   IV-448
Ahn, Byeong Seok   IV-360
Ahn, Chang-Beom   I-166
Ahn, EunYoung   I-1122
Ahn, Hyo Cheol   IV-916
Ahn, Jaewoo   I-223
Ahn, Joung Chul   II-741
Ahn, Kwang-Il   IV-662
Ahn, Seongjin   I-137, I-242, I-398, II-676, II-696, II-848, IV-1036
Ahn, Yonghak   II-732
Akbari, Mohammad K.   IV-1262
Akyol, Derya Eren   IV-596
Alcaide, Almudena   III-729, IV-1309
Alexander, Phillip J.   IV-1180
Ali, A.   II-1045
Alkassar, Ammar   II-634
Aloisio, Giovanni   III-1
Alonso, Olga Marroquin   II-1156
Altas, Irfan   III-463
An, Sunshin   I-261
Anikeenko, A.V.   I-816
Anton, François   I-669, I-683
Aquino, Adélia J.A.   I-1004
Araújo, Madalena M.   IV-632
Aranda, Gabriela N.   I-1064
Arteconi, Leonardo   I-1093
Aylett, Ruth   IV-30

Bacak, Goksen   III-522
Baciu, George   I-737
Bae, Hae-Young   IV-812
Bae, Hanseok   I-388
Bae, Hyeon   IV-1075, IV-1085
Bae, Hyerim   III-1259
Bae, Ihn Han   II-169
Bae, Jongho   IV-232
Bae, Kyoung Yul   I-204
Baek, Dong-Hyun   IV-222
Baek, Jang Hyun   IV-528

Baek, Jang-Mi   III-964
Baek, Jun-Geol   IV-148
Baek, Sunkyoung   I-37
Baig, Meerja Humayun   I-806
Baik, MaengSoon   III-89, IV-936
Baker, Robert G.V.   III-143
Bang, Young-Cheol   IV-989
Bannai, Hideo   III-349
Bao, Hujun   III-215
Bao, XiaoMing   II-1167
Barbatti, Mario   I-1004
Barco, Raquel   IV-958
Barlow, Jesse   IV-843
Baumgartner, Robert   II-988
Bayhan, G. Mirac   IV-596
Bekker, Henk   IV-397
Bénédet, Vincent   I-838
Bernholdt, D.E.   III-29
Bernholt, Thorsten   I-697
Bertazzon, Stefania   III-118, III-152
Bhat, M.S.   IV-548
Bhatia, Davinder   IV-1190
Bhattacharyya, Chiranjib   IV-548
Bierbaum, Aron   III-1119
Borruso, Giuseppe   III-126
Bozer, Yavuz A.   IV-437
Braad, Eelco P.   IV-397
Brucker, Peter   IV-182
Brunstrom, Anna   IV-1331
Burns, John   II-1254
Burrage, Kevin   II-1245
Byun, Sang-Yong   III-788
Byun, Yung-Cheol   III-788

Caballero-Gil, Pino   III-719
Cafaro, Massimo   III-1
Cai, Guoyin   III-173
Çakar, Tarık   IV-1241
Caminati, Walther   I-1046
Castro, Julio César Hernández   IV-1292
Catanzani, Riccardo   I-921
Cattani, Carlo   III-604
Cechich, Alejandra   I-1064
Cha, ByungRae   II-254

Cha, Jae-Sang   II-332, II-341, II-373,
    II-411, II-429, II-449, IV-1319
Cha, Jeon-Hee   I-11
Cha, Jung-Eun   III-896
Cha, Kyungup   III-1269
Chae, Jongwoo   II-147
Chae, Kijoon   I-591
Chae, Oksam   II-732, IV-20
Chae, Soo-young   II-458
Chan, Choong Wah   II-657
Chanchio K.   III-29
Chang, Chun Young   I-1204
Chang, Dong Shang   IV-577
Chang, Elizabeth   II-1125
Chang, Hangbae   IV-128
Chang, Hung-Yi   IV-1007
Chang, Jae Sik   IV-999
Chang, Jae-Woo   I-77
Chang, Ok-Bae   III-758, III-836,
    III-878, III-945
Chang, Pei-Chann   IV-172, IV-417
Chang, Soo Ho   I-46
Chau, Rowena   II-956
Che, Ming   III-284
Che, Yinghui   III-225
Chen, M.L.   III-29
Chen, Chia-Ho   IV-417
Chen, Chun   IV-826
Chen, J.C.   IV-333
Chen, Jianwei   IV-519
Chen, Jinwen   II-1217
Chen, Ling   III-338
Chen, Shih-Chang   IV-1017
Chen, Taiyi   I-967
Chen, Tse-Shih   I-19
Chen, Tung-Shou   IV-1007
Chen, Weidong   I-865
Chen, Wen   II-806
Chen, Yefang   IV-1
Chen, Yun-Shiow   IV-172
Chen, Zhiping P.   IV-733
Cheng, Xiangguo   IV-1046
Cheon, Seong-Pyo   IV-1075
Chernetsov, Nikita   III-133
Chi, Jeong Hee   II-977
Ching, Wai-Ki   IV-342, IV-843
Cho, Byung Rae   IV-212
Cho, Cheol-Hyung   I-707, III-993
Cho, Chiwoon   III-1297
Cho, Dongyoung   I-232

Cho, Eun-Sook   III-778, III-868
Cho, Hyeon Seob   II-832
Cho, Kyung Dal   II-474
Cho, Miyoung   I-37
Cho, Nam Wook   III-1297
Cho, Seokhyang   I-498
Cho, Sok-Pal   II-781, II-896
Cho, Sung-Keun   IV-48
Cho, SungHo   I-204
Cho, Wanhyun   IV-867
Cho, Yongju   III-1289
Cho, Yongsun   I-1
Cho, Yongyun   II-1008
Cho, Yookun   II-353
Cho, You-Ze   I-378
Cho, Youngsong   I-707, I-716, III-993
Cho, YoungTak   IV-20
Choi, In Seon   II-889
Choi, DeaWoo   IV-103
Choi, Deokjai   I-195
Choi, Dong-seong   IV-62
Choi, Eunmi   II-187, III-858
Choi, Gyunghyun   IV-261
Choi, Hee-Chul   III-938
Choi, Honzong   III-1289
Choi, Hyang-Chang   II-82
Choi, Hyung Jo   II-1207
Choi, Ilhoon   III-1229
Choi, Jaemin   II-567
Choi, Jaeyoung   II-1008, II-1018, IV-10
Choi, Jong Hwa   IV-86
Choi, Jong-In   III-1148
Choi, Jonghyoun   I-271
Choi, Kee-Hyun   III-99
Choi, Kun Myon   I-448
Choi, Mi-Sook   III-778
Choi, Sang-soo   II-458
Choi, Sang-Yule   II-341, II-429, IV-1319
Choi, Sung-ja   II-215
Choi, SungJin   III-89, IV-936
Choi, Wonwoo   I-137
Choi, WoongChul   IV-1231
Choi, Yeon-Sung   II-71
Choi, YoungSik   I-186
Chong, Kil To   II-1207, II-1293
Chong, Kiwon   I-1
Choo, Hyunseung   I-291, I-448, I-468,
    I-529, I-540, IV-989
Choudhary, Alok Kumar   IV-680
Chow, K.P.   III-651

Chow, Sherman S.M.   III-651
Choy, Yoon-Chu   I-847
Chua, Eng-Huat   II-1167
Chun, Junchul   I-1135
Chun, Kilsoo   II-381
Chun, Kwang Ho   II-749
Chun, Kwang-ho   II-723
Chung, Chin Hyun   I-638, I-1213
Chung, Hyun-Sook   III-788
Chung, Jinwook   I-137
Chung, Kwangsue   IV-1231
Chung, Min Young   I-348, I-448, I-529
Chung, Mokdong   II-147
Chung, Tae-sun   IV-72
Chung, Tae-Woong   IV-836
Chung, Tai-Myung   I-146, I-468
Chung, YoonJung   II-92, II-274
Chung, Youn-Ky   III-769
Chunyan, Yu   I-875, I-974
Cornejo, Oscar   IV-712
Corradini, Flavio   II-1264
Costantini, Alessandro   I-1046
Cotrina, Josep   II-527, II-624
Couloigner, Isabelle   III-181
Croce, Federico Della   IV-202
Cruz-Neira, Carolina   III-1070, III-1119
Cui, Kebin   I-214
Cui, Shi   II-657

Das, Amitabha   I-994
Das, Sandip   I-827
Dashora, Yogesh   IV-680
Datta, Amitava   I-87, II-686, III-206
Da-xin, Liu   IV-753
Debels, Dieter   IV-378
de Frutos Escrig, David   II-1156
Deris, M. Mat   III-60
Dévai, Frank   I-726
Dew, Robert   III-49
Díez, Luis   IV-958
Dillon, Tharam S.   II-914, II-1125
Ding, Jintai   II-595
Ding, Yongsheng   III-69
Djemame, Karim   IV-1282
Doboga, Flavia   III-563
Dol'nikov, Vladimir   III-628
Dongyi, Ye   I-875, I-974
Du, Tianbao   I-1040
Duan, Pu   II-657

Dumas, Laurent   IV-948
Duong, Doan Dai   II-1066

Emiris, Ioannis   I-683
Enzi, Christian   II-988
Eong, Gu-Beom   II-42
Epicoco, Italo   III-1
Ercan, M. Fikret   III-445
Ergenc, Tanil   III-463
Escoffier, Bruno   IV-192, IV-202
Espírito-Santo, Isabel A.C.P.   IV-632
Estévez-Tapiador, Juan M.   IV-1292, IV-1309
Eun, He-Jue   II-10

Faudot, Dominique   I-838
Feng, Jieqing   III-1023
Fernandes, Edite M.G.P   IV-488, IV-632
Fernandez, Marcel   II-527, II-624
Ferreira, Eugenio C.   IV-632
Fiore, Sandro   III-1
Fong, Simon   II-1106
For, Wei-Khing   II-1167
Frank, A.O.   II-1018
Froeklich, Johannes   I-905, I-938
Fu, Haoying   IV-843
Fúster-Sabater, Amparo   III-719

Gaglio, Salvatore   III-39
Gálvez, Akemi   III-472, III-482, III-502
Gao, Chaohui   I-1040
Gao, Lei   III-69
Garcia, Ernesto   I-1083
Gardner, William   II-1125
Gatani, Luca   III-39
Gaur, Daya Ram   IV-670
Gavrilova, M.L.   I-816
Gavrilova, Marina L.   I-748
Geist A.   III-29
Gerardo, Bobby   II-71, II-205
Gervasi, Osvaldo   I-905, I-921, I-938
Ghelmez, Mihaela   III-563
Ghinea, G.   II-1018
Ghose, Debasish   IV-548
Gil, JoonMin   IV-936
Gimenez, Xavi   I-1083
Goetschalckx, Marc   IV-322
Goff, Raal   I-87
Goh, Dion Hoe-Lian   II-1177
Goh, John   IV-1203

Goh, Li Ping   IV-906
Goi, Bok-Min   I-488, IV-1065
Gold, Christopher M.   I-737
Goldengorin, Boris   IV-397
Goscinski, Andrzej   III-49
Goswami, Partha P.   I-827
Gower, Jason   II-595
Grinnemo, Karl-Johan   IV-1331
Großschädl, Johann   II-665
Grząślewicz, Ryszard   II-517
Gu, Mi Sug   II-966
Guan, Xiucui   IV-161
Guan, Yanning   III-173
Guo, Heqing   III-691, IV-1028
Guo, X.C.   IV-1040
Gupta, Pankaj   IV-1190

Ha, JaeCheol   II-245
Haji, Mohammed   IV-1282
Han, Chang Hee   IV-222, IV-360
Han, In-sung   II-904
Han, Joohyun   II-1008
Han, Jung-Soo   III-748, III-886
Han, Kyuho   I-261
Han, SangHoon   I-1122
Han, Young-Ju   I-146
Harding, Jenny A.   IV-680
Hartling, Patrick   III-1070, III-1119
He, Ping   III-338
He, Qi   III-691, IV-1028
He, Yuanjun   III-1099
Hedgecock, Ian M.   I-1054
Heng, Swee-Huay   II-603
Henze, Nicola   II-988
Heo, Hoon   IV-20
Herbert, Vincent   IV-948
Hernández, Julio C.   IV-1301
Herrlich, Marc   II-988
Herzog, Marcus   II-988
Higuchi, Tomoyuki   III-381, III-389
Hirose, Osamu   III-349
Hong, Changho   IV-138
Hong, Choong Seon   I-195, I-339
Hong, Chun Pyo   I-508
Hong, Helen   IV-1111
Hong, In-Sik   III-964
Hong, Jung-Hun   II-1
Hong, Jungman   IV-642
Hong, Kicheon   I-1154
Hong, Kiwon   I-195

Hong, Maria   I-242, IV-1036
Hong, Seok Hoo   II-1076
Hong, Xianlong   IV-896
Hou, Jia   II-749
Hsieh, Min-Chi   II-1055
Hsieh, Ying-Jiun   IV-437
Hsu, Ching-Hsien   IV-1017
Hu, Bingcheng   II-1274
Hu, Guofei   I-758
Hu, Xiaohua   III-374
Hu, Xiaoyan   III-235
Hu, Yifeng   I-985
Hu, Yincui   III-173
Hua, Wei   III-215
Huang, Zhong   III-374
Huettmann, Falk   III-133, III-152
Huh, Eui-Nam   I-311, I-628, I-1144
Hui, Lucas C.K.   III-651
Hung, Terence   I-769, IV-906
Hur, Nam-Young   II-341, IV-1319
Hur, Sun   II-714, IV-606
Huynh, Trong Thua   I-339
Hwang, Chong-Sun   III-89, IV-936, IV-1169
Hwang, Gi Yean   II-749
Hwang, Ha Jin   II-304, III-798
Hwang, Hyun-Suk   II-127
Hwang, Jae-Jeong   II-205
Hwang, Jeong Hee   II-925, II-966
Hwang, Jun   I-1170, I-1204
Hwang, Seok-Hyung   II-1
Hwang, Suk-Hyung   III-827, III-938
Hwang, Sun-Myung   II-21, III-846
Hwang, Yoo Mi   I-1129
Hwang, Young Ju   I-619
Hwang, Yumi   I-1129
Hyun, Chang-Moon   III-927
Hyun, Chung Chin   I-1177
Hyuncheol, Kim   II-676

Iglesias, Andrés   III-502, III-472, III-482, III-492, III-547, III-1157
Im, Chae-Tae   I-368
Im, Dong-Ju   II-420, II-474
Imoto, Seiya   III-349, III-389
In, Hoh Peter   II-274
Inceoglu, Mustafa Murat   III-538, IV-56
Iordache, Dan   III-614
Ipanaqué, Ruben   III-492
Iqbal, Mahrin   II-1045

Iqbal, Mudeem II-1045
Izquierdo, Antonio III-729, IV-1309

Jackson, Steven Glenn III-512
Jahwan, Koo II-696
Jalili-Kharaajoo, Mahdi I-1030
Jang, Dong-Sik IV-743
Jang, Injoo II-102, II-111
Jang, Jongsu I-609
Jang, Sehoon I-569
Jang, Sung Man II-754
Jansen, A.P.J. I-1020
Javadi, Bahman IV-1262
Jeon, Hoseong I-529
Jeon, Hyong-Bae IV-538
Jeon, Nam Joo IV-86
Jeong, Bongju IV-566
Jeong, Chang Sung I-601
Jeong, Eun-Hee II-322, II-585
Jeong, Eunjoo I-118
Jeong, Gu-Beom II-42
Jeong, Hwa-Young I-928
Jeong, In-Jae IV-222, IV-312
Jeong, JaeYong II-353
Jeong, Jong-Youl I-311
Jeong, Jongpil I-291
Jeong, Kugsang I-195
Jeong, KwangChul I-540
Jeong, Seung-Ju IV-566
Ji, Joon-Yong III-1139
Ji, Junfeng III-1167
Ji, Yong Gu III-1249
Jia, Zhaoqing III-10
Jiang, Chaojun I-1040
Jiang, Xinhua H. IV-733
Jiao, Xiangmin IV-1180
Jin, Biao IV-1102
Jin, Bo III-299
Jin, Guiyue IV-1095
Jin, Jing III-416
Jin, YoungTaek III-846
Jin, Zhou III-435
Jo, Geun-Sik IV-1131
Jo, Hea Suk I-519
Joo, Inhak II-1136
Joung, Bong Jo I-1196, I-1213
Ju, Hak Soo II-381
Ju, Jaeyoung III-1259
Jun, Woochun IV-48
Jung, Changho II-537

Jung, Ho-Sung II-332
Jung, Hoe Sang III-1177
Jung, Hye-Jung III-739
Jung, Jason J. IV-1131
Jung, Jin Chul I-252
Jung, Jung Woo IV-467
Jung, KeeChul IV-999
Jung, Kwang Hoon I-1177
Jung, SM. II-1028

Kang, Euisun I-242
Kang, HeeJo II-420, II-483
Kang, Kyung Hwan IV-350
Kang, Kyung-Woo I-29
Kang, MunSu I-186
Kang, Oh-Hyung II-195, II-284, II-295
Kang, Seo-Il II-177
Kang, Suk-Hoon I-320
Kang, Yeon-hee II-215
Kang, Yu-Kyung III-938
Karsak, E. Ertugrul IV-301
Kasprzak, Andrzej IV-772
Kemp, Ray II-1187
Khachoyan, Avet A. IV-1012
Khorsandi, Siavash IV-1262
Kiani, Saad Liaquat II-1096
Kim, B.S. II-1028
Kim, Byunggi I-118
Kim, Byung Wan III-1306
Kim, Chang Han II-647
Kim, Chang Hoon I-508
Kim, Chang Ouk IV-148
Kim, Chang-Hun III-1080, III-1129, III-1139, III-1148
Kim, Chang-Min I-176, III-817, IV-38
Kim, Chang-Soo II-127
Kim, Chul-Hong III-896
Kim, Chulyeon IV-261
Kim, Dae Hee II-1284
Kim, Dae Sung I-1111
Kim, Dae Youb II-381
Kim, Daegeun II-1035
Kim, Deok-Soo I-707, I-716, III-993, III-1060, IV-652
Kim, D.K. II-1028
Kim, Do-Hyeon I-378
Kim, Do-Hyung II-401
Kim, Dong-Soon III-938
Kim, Donghyun IV-877
Kim, Dongkeun I-857

Kim, Dongkyun   I-388
Kim, Dongsoo   III-1249
Kim, Donguk   I-716, III-993
Kim, Dounguk   I-707
Kim, Eun Ju   I-127
Kim, Eun Suk   IV-558
Kim, Eun Yi   IV-999
Kim, Eunah   I-591
Kim, Gi-Hong   II-771
Kim, Gui-Jung   III-748, III-886
Kim, Guk-Boh   II-42
Kim, Gye-Young   I-11
Kim, Gyoung-Bae   IV-812
Kim, Hae Geun   II-295
Kim, Hae-Sun   II-157
Kim, Haeng-Kon   II-1, II-52, II-62,
    II-137, III-769, III-906, III-916
Kim, Hak-Keun   I-847
Kim, Hang Joon   IV-999
Kim, Hee Sook   II-483, II-798
Kim, Hong-Gee   III-827
Kim, Hong-jin   II-781, II-896
Kim, HongSoo   III-89
Kim, Hoontae   III-1249
Kim, Howon   II-1146
Kim, Hwa-Joong   IV-538, IV-722
Kim, Hwankoo   II-245
Kim, HyoungJoong   IV-269
Kim, Hyun Cheol   I-281
Kim, Hyun-Ah   I-427, III-426, IV-38
Kim, Hyun-Ki   IV-887
Kim, Hyuncheol   I-137, II-676
Kim, Hyung Jin   II-789, II-880
Kim, InJung   II-92, II-274
Kim, Jae-Gon   IV-280, IV-322
Kim, Jae-Sung   II-401
Kim, Jae-Yearn   IV-662
Kim, Jae-Yeon   IV-743
Kim, Jang-Sub   I-348
Kim, Jee-In   I-886
Kim, Jeom-Goo   II-762
Kim, Jeong Ah   III-846
Kim, Jeong Kee   II-714
Kim, Jin Ok   I-638, I-1187
Kim, Jin Soo   I-638, I-1187
Kim, Jin-Geol   IV-782
Kim, Jin-Mook   II-904
Kim, Jin-Sung   II-31, II-567
Kim, Jong-Boo   II-341, IV-1319
Kim, Jong Hwa   III-1033

Kim, Jong-Nam   I-67
Kim, Jongsung   II-567
Kim, Jong-Woo   I-1177, II-127
Kim, Ju-Yeon   II-127
Kim, Jun-Gyu   IV-538
Kim, Jung-Min   III-788
Kim, Jungchul   I-1154
Kim, Juwan   I-857
Kim, Kap Sik   III-798
Kim, Kibum   IV-566
Kim, KiJoo   I-186
Kim, Kwan-Joong   I-118
Kim, Kwang-Baek   IV-1075
Kim, Kwang-Hoon   I-176, III-817, IV-38
Kim, Kwang-Ki   III-806
Kim, Kyung-kyu   IV-128
Kim, Mihui   I-591
Kim, Mijeong   II-1136
Kim, Minsoo   II-225, II-1136, III-1249,
    III-1259
Kim, Misun   I-550, I-559
Kim, Miyoung   I-550, I-559
Kim, Moonseong   IV-989
Kim, Myoung Soo   III-916
Kim, Myuhng-Joo   I-156
Kim, Myung Ho   I-223
Kim, Myung Won   I-127
Kim, Myung-Joon   IV-812
Kim, Nam Chul   I-1111
Kim, Pankoo   I-37
Kim, Sang Ho   II-977, IV-79
Kim, Sang-Bok   I-628
Kim, Sangjin   IV-877
Kim, Sangkyun   III-1229, III-1239,
    IV-122
Kim, Seungjoo   I-498, II-1146
Kim, Soo Dong   I-46, I-57
Kim, Soo-Kyun   III-1080, III-1129,
    III-1139
Kim, Soung Won   III-916
Kim, S.R.   II-1028
Kim, Sung Jin   II-1076
Kim, Sung Jo   III-79
Kim, Sung Ki   I-252
Kim, Sung-il   IV-62
Kim, Sung-Ryul   I-359
Kim, Sungshin   IV-1075, IV-1085
Kim, Tae Hoon   IV-509
Kim, Tae Joong   III-1279
Kim, Tae-Eun   II-474

Kim, Taeho   IV-280
Kim, Taewan   II-863
Kim, Tai-Hoon   II-341, II-429, II-468, II-491, IV-1319
Kim, Ungmo   II-936
Kim, Won-sik   IV-62
Kim, Wooju   III-1289, IV-103
Kim, Y.H.   III-1089
Kim, Yon Tae   IV-1085
Kim, Yong-Kah   IV-858
Kim, Yong-Soo   I-320, I-1162
Kim, Yong-Sung   II-10, II-31, III-954
Kim, Yongtae   II-647
Kim, Young Jin   IV-212, IV-232
Kim, Young-Chan   I-1170
Kim, Young-Chul   IV-10
Kim, Young-Shin   I-311
Kim, Young-Tak   II-157
Kim, Youngchul   I-107
Ko, Eun-Jung   III-945
Ko, Hoon   II-442
Ko, Jaeseon   II-205
Ko, S.L.   III-1089
Koh, Jae Young   II-741
Komijan, Alireza Rashidi   IV-388
Kong, Jung-Shik   IV-782
Kong, Ki-Sik   II-1225, IV-1169
Koo, Jahwan   II-696, II-848
Koo, Yun-Mo   III-1187
Koszalka, Leszek   IV-692
Kravchenko, Svetlana A.   IV-182
Kriesell, Matthias   II-988
Krishnamurti, Ramesh   IV-670
Kuo, Yi Chun   IV-577
Kurosawa, Kaoru   II-603
Kutyłowski, Jarosław   II-517
Kutyłowski, Mirosław   II-517
Kwag, Sujin   I-418
Kwak, Byeong Heui   IV-48
Kwak, Kyungsup   II-373, II-429
Kwak, NoYoon   I-1122
Kwon, Dong-Hee   I-368
Kwon, Gihwon   III-973
Kwon, Hyuck Moo   IV-212, IV-232
Kwon, Jungkyu   II-147
Kwon, Ki-Ryong   II-557
Kwon, Ki-Ryoung   III-1209
Kwon, Oh Hyun   II-137
Kwon, Soo-Tae   IV-624
Kwon, Soonhak   I-508

Kwon, Taekyoung   I-577, I-584
Kwon, Yong-Moo   I-913

La, Hyun Jung   I-46
Lægreid, Astrid   III-327
Laganà, Antonio   I-905, I-921, I-938, I-1046, I-1083, I-1093
Lago, Noelia Faginas   I-1083
Lai, Edison   II-1106
Lai, K.K.   IV-250
Lamarque, Loïc   I-838
Lázaro, Pedro   IV-958
Ledoux, Hugo   I-737
Lee, Bo-Hee   IV-782
Lee, Bong-Hwan   I-320
Lee, Byoungcheon   II-245
Lee, Byung Ki   IV-350
Lee, Byung-Gook   III-1209
Lee, Byung-Kwan   II-322, II-585
Lee, Chang-Mog   III-758
Lee, Chong Hyun   II-373, II-411, II-429, II-449
Lee, Chun-Liang   IV-1007
Lee, Dong Chun   II-714, II-741, II-762, II-889, II-896
Lee, Dong Hoon   I-619, II-381
Lee, Dong-Ho   IV-538, IV-722
Lee, DongWoo   I-232
Lee, Eun-Ser   II-363, II-483
Lee, Eung Jae   II-998
Lee, Eung Young   III-1279
Lee, Eunkyu   II-1136
Lee, Eunseok   I-291
Lee, Gang-soo   II-215, II-458
Lee, Geuk   II-754
Lee, Gi-Sung   II-839
Lee, Hakjoo   III-1269
Lee, Ho Woo   IV-509
Lee, Hong Joo   III-1239, IV-113, IV-122
Lee, Hoonjung   IV-877
Lee, Hyewon K.   I-97, I-118
Lee, Hyoung-Gon   III-1219
Lee, Hyun Chan   III-993
Lee, HyunChan   III-1060
Lee, Hyung-Hyo   II-82
Lee, Hyung-Woo   II-391, II-401, IV-62
Lee, Im-Yeong   II-117, II-177
Lee, Insup   I-156
Lee, Jae-deuk   II-420
Lee, Jaeho   III-1060

Lee, Jae-Wan II-71, II-205, II-474
Lee, Jee-Hyong IV-1149
Lee, Jeongheon IV-20
Lee, Jeongjin IV-1111
Lee, Jeoung-Gwen IV-1055
Lee, Ji-Hyen III-878
Lee, Ji-Hyun III-836
Lee, Jongchan I-107
Lee, Jong chan II-781
Lee, Jong Hee II-856
Lee, Jong-Hyouk I-146, I-468
Lee, Joon-Jae III-1209
Lee, Joong-Jae I-11
Lee, Joungho II-111
Lee, Ju-Il I-427
Lee, Jun I-886
Lee, Jun-Won III-426
Lee, Jung III-1080, III-1129, III-1139
Lee, Jung-Bae III-938
Lee, Jung-Hoon I-176
Lee, Jungmin IV-1231
Lee, Jungwoo IV-96
Lee, Kang-Won I-378
Lee, Keon-Myung IV-1149
Lee, Keun Kwang II-474, II-420
Lee, Keun Wang II-798, II-832, II-856
Lee, Key Seo I-1213
Lee, Ki Dong IV-1095
Lee, Ki-Kwang IV-427
Lee, Kwang Hyoung II-798
Lee, Kwangsoo II-537
Lee, Kyunghye I-408
Lee, Malrey II-71, II-363, II-420, II-474, II-483
Lee, Man-Hee IV-743
Lee, Mi-Kyung II-31
Lee, Min Koo IV-212, IV-232
Lee, Moon Ho II-749
Lee, Mun-Kyu II-314
Lee, Myung-jin IV-62
Lee, Myungeun IV-867
Lee, Myungho IV-72
Lee, NamHoon II-274
Lee, Pill-Woo I-1144
Lee, S.Y. II-1045
Lee, Sang Ho II-1076
Lee, Sang Hyo I-1213
Lee, Sangsun I-418
Lee, Sang Won III-1279
Lee, Sang-Hyuk IV-1085
Lee, Sang-Young II-762, III-945
Lee, Sangjin II-537, II-567
Lee, SangKeun II-1225
Lee, Sangsoo II-816
Lee, Se-Yul I-320, I-1162
Lee, SeongHoon I-232
Lee, Seoung Soo III-1033, IV-652
Lee, Seung-Yeon I-628, II-332
Lee, Seung-Yong II-225
Lee, Seung-youn II-468, II-491, II-499
Lee, SiHun IV-1149
Lee, SooBeom II-789, II-880
Lee, Su Mi I-619
Lee, Suk-Hwan II-557
Lee, Sungchang I-540
Lee, Sunghwan IV-96
Lee, SungKyu IV-103
Lee, Sungyoung II-1096, II-1106, II-1115
Lee, Sunhun IV-1231
Lee, Suwon II-420, II-474
Lee, Tae Dong I-601
Lee, Tae-Jin I-448
Lee, Taek II-274
Lee, TaiSik II-880
Lee, Tong-Yee III-1043, III-1050
Lee, Wonchan I-1154
Lee, Woojin I-1
Lee, Woongjae I-1162, I-1196
Lee, Yi-Shiun III-309
Lee, Yong-Koo II-1115
Lee, Yonghwan II-187, III-858
Lee, Yongjae II-863
Lee, Young Hae IV-467
Lee, Young Hoon IV-350
Lee, Young Keun II-420, II-474
Lee, YoungGyo II-92
Lee, YoungKyun II-880
Lee, Yue-Shi II-1055
Lee, Yung-Hyeon II-762
Leem, Choon Seong III-1269, III-1289, III-1306, IV-79, IV-86, IV-113
Lei, Feiyu II-806
Leon, V. Jorge IV-312
Leung, Stephen C.H. IV-250
Lezzi, Daniele III-1
Li, Huaqing IV-1140
Li, Jin-Tao II-547
Li, JuanZi IV-1222
Li, Kuan-Ching IV-1017

Li, Li   III-190
Li, Minglu   III-10
Li, Peng   III-292
Li, Sheng   III-1167
Li, Tsai-Yen   I-957
Li, Weishi   I-769, IV-906
Li, Xiao-Li   III-318
Li, Xiaotu   III-416
Li, Xiaowei   III-266
Li, Yanda   II-1217
Li, Yun   III-374
Li, Zhanhuai   I-214
Li, Zhuowei   I-994
Liao, Mao-Yung   I-957
Liang, Xiaohui   III-225
Liang, Y.C.   I-1040
Lim, Cheol-Su   III-1080, III-1148
Lim, Ee-Peng   II-1177
Lim, Heui Seok   I-1129
Lim, Hyung-Jin   I-146
Lim, In-Taek   I-438
Lim, Jongin   II-381, II-537, II-567, II-647
Lim, Jong In   I-619
Lim, Jongtae   IV-138
Lim, Myoung-seob   II-723
Lim, Seungkil   IV-642
Lim, Si-Yeong   IV-606
Lim, Soon-Bum   I-847
Lim, YoungHwan   I-242, IV-1036
Lim, Younghwan   I-398, II-676, II-848
Lin, Huaizhong   IV-826
Lin, Jenn-Rong   IV-499
Lin, Manshan   III-691, IV-1028
Lin, Ping-Hsien   III-1050
Lindskog, Stefan   IV-1331
Lischka, Hans   I-1004
Liu, Bin   II-508
Liu, Dongquan   IV-968
Liu, Fenlin   II-508
Liu, Jiming   II-1274
Liu, Jingmei   IV-1046
Liu, Joseph K.   II-614
Liu, Ming   IV-1102
Liu, Mingzhe   II-1187
Liu, Xuehui   III-1167
Liu, Yue   III-266
Lopez, Javier   III-681
Lu, Chung-Dar   III-299
Lu, Dongming   I-865, I-985

Lu, Jiahui   I-1040
Lu, Xiaolin   III-256
Luengo, Francisco   III-1157
Luo, Lijuan   IV-896
Luo, Xiangyang   II-508
Luo, Ying   III-173
Luo, Yingwei   I-301, II-822

Ma, Fanyuan   II-1086
Ma, Liang   III-292
Ma, Lizhuang   I-776
Mackay, Troy D.   III-143
Mał afiejski, Michal   I-647
Manera, Jaime   IV-1301
Mani, Venkataraman   IV-269
Manzanares, Antonio Izquierdo   IV-1292
Mao, Zhihong   I-776
Maris, Assimo   I-1046
Markowski, Marcin   IV-772
Márquez, Joaquín Torres   IV-1292
Martoyan, Gagik A.   I-1012
Medvedev, N.N.   I-816
Meng, Qingfan   I-1040
Merelli, Emanuela   II-1264
Miao, Lanfang   I-758
Miao, Yongwei   III-1023
Michelot, Christian   IV-712
Mielikäinen, Taneli   IV-1251
Mijangos, Eugenio   IV-477
Million, D.L.   III-29
Min, Byoung Joon   I-252
Min, Byoung-Muk   II-896
Min, Dugki   II-187, III-858
Min, Hyun Gi   I-57
Min, Jihong   I-1154
Min, Kyongpil   I-1135
Min, Seung-hyun   II-723
Min, Sung-Hwan   IV-458
Minasyan, Seyran H.   I-1012
Minghui, Wu   I-875, I-974
Minhas, Mahmood R.   IV-587
Mirto, Maria   III-1
Miyano, Satoru   III-349
Mnaouer, Adel Ben   IV-1212
Mo, Jianzhong   I-967
Mocavero, Silvia   III-1
Moon, Hyeonjoon   I-584
Moon, Kiyoung   I-609

Morarescu, Cristian   III-556, III-563
Moreland, Terry   IV-1120
Morillo, Pedro   III-1119
Moriya, Kentaro   IV-978
Mourrain, Bernard   I-683
Mun, Ki-Young   I-311
Mun, Young-Song   I-97, I-118, I-242, I-271, I-398, I-408, I-459, I-550, I-559, I-569, I-628, II-676, II-848, IV-1036
Murat, Cécile   IV-202
Muyl, Frédérique   IV-948

Nait-Sidi-Moh, Ahmed   IV-792
Nakamura, Yasuaki   III-1013
Nam, Junghyun   I-498
Nam, Kichun   I-1129
Nam, Kyung-Won   I-1170
Nandy, Subhas C.   I-827
Nariai, Naoki   III-349
Nasir, Uzma   II-1045
Nassis, Vicky   II-914
Ng, Michael Kwok   IV-843
Ng, See-Kiong   II-1167, III-318
Nicolay, Thomas   II-634
Nie, Weifang   III-284, III-292, III-416
Nikolova, Mila   IV-843
Ninulescu, Valerică   III-635, III-643
Nodera, Takashi   IV-978
Noël, Alfred G.   III-512
Noh, Angela Song-Ie   I-1144
Noh, Bong-Nam   II-82, II-225
Noh, Hye-Min   III-836, III-878, III-945
Noh, Seung J.   IV-615
Nozick, Linda K.   IV-499
Nugraheni, Cecilia E.   III-453

Offermans, W.K.   I-1020
Ogiela, Lidia   IV-852
Ogiela, Marek R.   IV-852
Oh, Am-Sok   II-322, 585
Oh, Heekuck   IV-877
Oh, Nam-Ho   II-401
Oh, Sei-Chang   II-816
Oh, Seoung-Jun   I-166
Oh, Sun-Jin   II-169
Oh, Sung-Kwun   IV-858, IV-887
Ok, MinHwan   II-1035
Olmes, Zhanna   I-448
Omar M.   III-60

Ong, Eng Teo   I-769
Onyeahialam, Anthonia   III-152

Padgett, James   IV-1282
Páez, Antonio   III-162
Paik, Juryon   II-936
Pan, Hailang   I-896
Pan, Xuezeng   I-329, II-704
Pan, Yi   III-338
Pan, Yunhe   I-865
Pan, Zhigeng   II-946, III-190, III-245
Pandey, R.B   II-1197
Pang, Mingyong   III-245
Park, Myong-soon   II-1035
Park, Bongjoo   II-245
Park, Byoung-Jun   IV-887
Park, Byungchul   I-468
Park, Chan Yong   II-1284
Park, Chankwon   III-1219
Park, Cheol-Min   I-1170
Park, Choon-Sik   II-225
Park, Daehee   IV-858
Park, Dea-Woo   II-235
Park, DaeHyuck   IV-1036
Park, DongGook   II-245
Park, Eung-Ki   II-225
Park, Gyung-Leen   I-478
Park, Hayoung   II-442
Park, Hee Jun   IV-122
Park, Hee-Dong   I-378
Park, Hee-Un   II-117
Park, Heejun   III-1316
Park, Jesang   I-418
Park, Jin-Woo   I-913, III-1219
Park, Jonghyun   IV-867
Park, Joon Young   III-993, III-1060
Park, Joowon   II-789
Park, KwangJin   II-1225
Park, Kyeongmo   II-264
Park, KyungWoo   II-254
Park, Mi-Og   II-235
Park, Namje   I-609, II-1146
Park, Sachoun   III-973
Park, Sang-Min   IV-652
Park, Sang-Sung   IV-743
Park, Sangjoon   I-107, I-118
Park, Seon Hee   II-1284
Park, Seoung Kyu   III-1306
Park, Si Hyung   III-1033
Park, Soonyoung   IV-867

Park, Sung Hee II-1284
Park, Sung-gi II-127
Park, Sung-Ho I-11
Park, Sungjun I-886
Park, Sung-Seok II-127
Park, Woojin I-261
Park, Yongsu II-353
Park, Youngho II-647
Park, Yunsun IV-148
Parlos, A.G II-1293
Paschos, Vangelis Th. IV-192, IV-202
Pedrycz, Witold IV-887
Pei, Bingzhen III-10
Peng, Jiming IV-290
Peng, Qunsheng I-758, III-1023
Penubarthi, Chaitanya I-156
Pérez, María S. III-109
Phan, Raphael C.-W. I-488, III-661, IV-1065
Piattini, Mario I-1064
Pietkiewicz, Wojciech II-517
Ping, Lingdi I-329, II-704
Pirani, Fernando I-1046
Pirrone, Nicola I-1054
Podoleanu, Adrian III-556
Ponce, Eva IV-1301
Porschen, Stefan I-796
Prasanna, H.M. IV-548
Przewoźniczek, Michał IV-802
Pusca, Stefan III-563, III-569, III-614

Qi, Feihu IV-1140
Qing, Sihan III-711
Qu, Na III-225

Rahayu, Wenny II-914, II-925
Rajugan, R., II-914, II-1125
Ramadan, Omar IV-926
Rasheed, Faraz II-1115
Ravantti, Janne IV-1251
Raza, Syed Arshad I-806
Re, Giuseppe Lo III-39
Ren, Lifeng IV-30
Rhee, Seung Hyong IV-1231
Rhee, Seung-Hyun III-1259
Rhew, Sung Yul I-57
Riaz, Maria II-1096
Ribagorda, Arturo III-729
Riganelli, Antonio I-905, I-921, I-938
Rim, Suk-Chul IV-615

Rob, Seok-Beom IV-858
Rocha, Ana Maria A.C. IV-488
Roh, Sung-Ju IV-1169
Rohe, Markus II-634
Roman, Rodrigo III-681
Rosi, Marzio I-1101
Ruskin, Heather J. II-1254
Ryou, Hwang-bin II-904
Ryu, Keun Ho II-925, II-977
Ryu, Han-Kyu I-378
Ryu, Joonghyun III-993
Ryu, Joung Woo I-127
Ryu, Keun Ho II-966
Ryu, Keun Hos II-998
Ryu, Seonggeun I-398
Ryu, Yeonseung IV-72

Sadjadi, Seyed Jafar IV-388
Sætre, Rune III-327
Sait, Sadiq M. IV-587
Salzer, Reiner I-938
Sánchez, Alberto III-109
Sarfraz, Muhammad I-806
Saxena, Amitabh III-672
Seo, Dae-Hee II-117
Seo, Jae Young IV-528
Seo, Jae-Hyun II-82, II-225, II-254
Seo, Jeong-Yeon IV-652
Seo, Jung-Taek II-225
Seo, Kwang-Kyu IV-448, IV-458
Seo, Kyung-Sik IV-836
Seo, Young-Jun I-928
Seong, Myoung-ho II-723
Seongjin, Ahn II-676, II-696
Serif, T. II-1018
Shang, Yanfeng IV-1102
Shao, Min-Hua III-701
Shehzad, Anjum II-1096, II-1106
Shen, Lianguan G. III-1003
Shen, Yonghang IV-1159
Sheng, Yu I-985
Shi, Lie III-190
Shi, Lei I-896
Shi, Xifan IV-1159
Shim, Bo-Yeon III-806
Shim, Donghee I-232
Shim, Young-Chul I-427, III-426
Shimizu, Mayumi III-1013
Shin, Byeong-Seok III-1177, III-1187
Shin, Chungsoo I-459

Shin, Dong-Ryeol   I-348, III-99
Shin, Ho-Jin   III-99
Shin, Ho-Jun   III-806
Shin, Hyo Young   II-741
Shin, Hyoun Gyu   IV-86
Shin, Hyun-Ho   II-157
Shin, In-Hye   I-478
Shin, Kitae   III-1219
Shin, Myong-Chul   II-332, II-341, II-499, IV-1319
Shin, Seong-Yoon   II-195, II-284
Shin, Yeong Gil   IV-1111
Shin, Yongtae   II-442
Shu, Jiwu   IV-762
Shumilina, Anna   I-1075
Sicker, Douglas C.   IV-528
Siddiqi, Mohammad Umar   III-661
Sierra, José M.   IV-1301, IV-1309
Sim, Jeong Seop   II-1284
Sim, Terence   III-1197
Simeonidis, Minas   III-569
Sinclair, Brett   III-49
Singh, Sanjeet   IV-1190
Sivakumar, K.C.   IV-1341
Skworcow, Piotr   IV-692
Smith, Kate A.   II-956
So, Yeon-hee   IV-62
Soares, João L.C.   IV-488
Soh, Ben   III-672
Soh, Jin   II-754
Sohn, Bangyong   II-442
Sohn, Chae-Bong   I-166
Sohn, Hong-Gyoo   II-771
Sohn, Sungwon   I-609
Sokolov, B.V.   IV-407
Solimannejad, Mohammad   I-1004
Son, Bongsoo   II-789, II-816, II-863
Song, Hoseong   III-1259
Song, Hui   II-1086
Song, Il-Yeol   III-402
Song, MoonBae   II-1225
Song, Teuk-Seob   I-847
Song, Yeong-Sun   II-771
Song, Young-Jae   I-928, III-886
Soriano, Miguel   II-527, II-624
Sourin, Alexei   III-983
Sourina, Olga   IV-968
Srinivas   IV-680
Steigedal, Tonje Stroemmen   III-327
Sterian, Andreea   III-585, III-643

Sterian, Andreea-Rodica   III-635
Sterian, Rodica   III-592, III-598
Strelkov, Nikolay   III-621, III-628
Su, Hua   II-1293
Suh, Young-Joo   I-368
Sun, Dong Guk   III-79
Sun, Jizhou   III-284, III-292, III-416, III-435
Sun, Weitao   IV-762
Sung, Jaechul   II-567
Sung, Ji-Yeon   I-1144
Suresh, Sundaram   IV-269
Swarna, J. Mercy   IV-1341

Ta, Duong Nguyen Binh   I-947
Tadeusiewicz, Ryszard   IV-852
Tae, Kang Soo   I-478
Tai, Allen H.   IV-342
Tamada, Yoshinori   III-349
Tan, Chew Lim   III-1197
Tan, Kenneth Chih Jeng   IV-1120
Tan, Soon-Heng   III-318
Tan, Wuzheng   I-776
Tang, Jiakui   III-173
Tang, Jie   IV-1222
Tang, Sheng   II-547
Tang, Yuchun   III-299
Taniar, David   IV-1203
Tao, Pai-Cheng   I-957
Tavadyan, Levon A.   I-1012
Techapichetvanich, Kesaraporn   III-206
Teillaud, Monique   I-683
Teng, Lirong   I-1040
Thuy, Le Thi Thu   II-1066
Tian, Haishan   III-1099
Tian, Tianhai   II-1245
Tillich, Stefan   II-665
Ting, Ching-Jung   IV-417
Tiwari, Manoj Kumar   IV-680
Toi, Yutaka   IV-1055
Toma, Alexandru   III-556, III-569
Toma, Cristian   III-556, III-592, III-598
Toma, Ghiocel   III-563, III-569, III-576, III-585, III-614
Toma, Theodora   III-556, III-569
Tomaschewski, Kai   II-988
Torres, Joaquin   III-729
Trunfio, Giuseppe A.   I-1054
Turnquist, Mark A.   IV-499
Tveit, Amund   III-327

Author Index

Ufuktepe, Ünal   III-522, III-529
Umakant, J.   IV-548
Urbina, Ruben T.   III-547

Vanhoucke, Mario   IV-378
Velardo, Fernando Rosa   II-1156
Varella, E.   I-938
Vita, Marco   II-1264
Vizcaíno, Aurora   I-1064

Wack, Maxime   IV-792
Wahala, Kristiina   I-938
Walkowiak, Krzysztof   IV-802
Wan, Zheng   I-329, II-704
Wang, Chen   II-1086
Wang, Chengfeng   I-748
Wang, Chuanpeng   III-225
Wang, Gi-Nam   IV-702
Wang, Guilin   III-701, III-711
Wang, Hao   III-691, IV-1028
Wang, Hei-Chia   III-309
Wang, Hui-Mei   IV-172
Wang, Jianqin   III-173
Wang, Jiening   III-284
Wang, K.J.   IV-333
Wang, Lei   IV-733
Wang, Pi-Chung   IV-1007
Wang, Ruili   II-1187
Wang, Shaoyu   IV-1140
Wang, Shu   IV-1
Wang, S.M.   IV-333
Wang, Weinong   II-806
Wang, Xiaolin   II-822
Wang, Xinmei   IV-1046
Wang, Xiuhui   III-215
Wang, Yanguang   III-173
Wang, Yongtian   III-266
Weber, Irene   III-299
Wee, H.M.   IV-333
Wee, Hyun-Wook   III-938, IV-333
Wei, Sun   IV-753
Weng, Dongdong   III-266
Wenjun, Wang   I-301
Wille, Volker   IV-958
Wirt, Kai   II-577
Won, Chung In   I-707
Won, Dongho   I-498, I-609, II-92, II-1146
Won, Hyung Jun   III-1259
Won, Jae-Kang   III-817

Wong, Duncan S.   II-614
Woo, Gyun   I-29
Woo, Seon-Mi   III-954
Woo, Sinam   I-261
Wu, C.G.   I-1040
Wu, Chaolin   III-173
Wu, Enhua   III-1167
Wu, Hulin   IV-519
Wu, Yong   III-1099
Wu, Yue   IV-250
Wu, Zhiping   II-595

Xia, Yu   IV-290
Xiaolin, Wang   I-301
Xinpeng, Lin   I-301
Xiong, Guomin   II-822
Xirouchakis, Paul   IV-538, IV-722
Xu, Bing   II-946
Xu, Dan   III-274
Xu, Guilin   IV-30
Xu, Jie   I-758
Xu, Qing   III-292
Xu, Shuhong   I-769, IV-906
Xu, Xiaohua   III-338
Xu, Zhuoqun   II-822
Xue, Yong   III-173

Yamaguchi, Rui   III-381
Yamamoto, Osami   I-786
Yamashita, Satoru   III-381
Yan, Chung-Ren   III-1043
Yan, Dayuan   III-266
Yan, Hong   III-357
Yang, Byounghak   IV-241
Yang, Chao-Tung   IV-1017
Yang, Ching-Nung   I-19
Yang, Dong Jin   II-647
Yang, Hae-Sool   II-1, II-52, III-739, III-827, III-938
Yang, Hongwei   II-946
Yang, Jie   III-416
Yang, KwonWoo   III-89
Yang, Tao   III-266
Yang, X.S.   III-1109
Yang, Xin   I-896, IV-1102
Yang, Yoo-Kil   III-1129
Yantır, Ahmet   III-529
Yao, Xin   II-1217
Yates, Paul   I-938
Yazici, Ali   III-463

Ye, Dingfeng II-595
Ye, Lu III-190, IV-30
Yeh, Chung-Hsing II-956
Yen, Show-Jane II-1055
Yi, Yong-Hoon II-82
Yim, Wha Young I-1213
Yin, Jianfei III-691, IV-1028
Yin, Ming II-1177
Yingwei, Luo I-301
Yiu, S.M. III-651
Yoo, Cheol-Jung III-758, III-836, III-878, III-945
Yoo, Chun-Sik II-31, III-954
Yoo, Hun-Woo IV-458, IV-743
Yoo, Hyeong Seon II-102, II-111
Yoo, Jin Ah II-889
Yoo, Seung Hwan I-252
Yoo, Seung-Jae II-870
Yoo, Sun K. II-1028
Yoon, Chang-Dae II-332, II-373, II-429
Yoon, Mi-sun IV-62
Yoon, Yeo Bong III-19
Yoshida, Ryo III-389
You, L.H. III-197
You, Jinyuan III-10
You, Peng-Sheng IV-368
Youn, Chan-Hyun I-320
Youn, Hee Yong I-519, II-936, III-19, IV-916, IV-1149
Youn, Hyunsang I-291
Youn, Ju-In II-10
Younghwan, Lim II-676
Youngsong, Mun II-676
Yu, Eun Jung III-1306
Yu, HeonChang III-89, IV-936
Yu, Jiangying III-225
Yu, Sang-Jun I-166
Yuan, Qingshu I-865, I-985
Yun, HY. II-1028
Yun, Sung-Hyun II-391, II-401, IV-62
Yun, Won Young IV-558
Yunhe, Pan I-875, I-974
Yusupov, R.M. IV-407

Zantidis, Dimitri III-672
Zaychik, E.M. IV-407
Zhai, Jia IV-702
Zhai, Qi III-284, III-435
Zhang, Changshui II-1217
Zhang, Fuyan III-245
Zhang, Jian J. III-197, III-1003, III-1109
Zhang, Jianzhong IV-161
Zhang, Jiawan III-292, III-416, III-435
Zhang, Jin-Ting IV-519
Zhang, Jun III-691, IV-1028
Zhang, Kuo IV-1222
Zhang, Mingmin II-946, III-190, III-245
Zhang, Mingming IV-1, IV-30
Zhang, Qiaoping III-181
Zhang, Qiong I-967
Zhang, Shen II-686
Zhang, Ya-Ping III-274
Zhang, Yan-Qing III-299
Zhang, Yi III-435
Zhang, Yong-Dong II-547
Zhang, Yu III-1197
Zhang, Yang I-214
Zhang, Yuanliang II-1207
Zhao, Jane II-1235
Zhao, Qinping III-235
Zhao, Weizhong IV-1159
Zhao, Yang III-274
Zhao, Yiming IV-1
Zheng, Jin Jin III-1003
Zheng, Weimin IV-762
Zheng, Zengwei IV-826
Zhong, Shaobo III-173
Zhou, Hanbin IV-896
Zhou, Hong Jun III-1003
Zhou, Jianying III-681, III-701
Zhou, Qiang IV-896
Zhou, Suiping I-947
Zhou, Xiaohua III-402
Zhu, Jiejie IV-30
Zhu, Ming IV-1102
Zhuoqun, Xu I-301
Żyliński, Paweł I-647

Lecture Notes in Computer Science

For information about Vols. 1–3400
please contact your bookseller or Springer

Vol. 3525: A.E. Abdallah, C.B. Jones, J.W. Sanders (Eds.), Communicating Sequential Processes. XIV, 321 pages. 2005.

Vol. 3517: H.S. Baird, D.P. Lopresti (Eds.), Human Interactive Proofs. IX, 143 pages. 2005.

Vol. 3516: V.S. Sunderam, G.D.v. Albada, P.M.A. Sloot, J.J. Dongarra (Eds.), Computational Science – ICCS 2005, Part III. LXIII, 1143 pages. 2005.

Vol. 3515: V.S. Sunderam, G.D.v. Albada, P.M.A. Sloot, J.J. Dongarra (Eds.), Computational Science – ICCS 2005, Part II. LXIII, 1101 pages. 2005.

Vol. 3514: V.S. Sunderam, G.D.v. Albada, P.M.A. Sloot, J.J. Dongarra (Eds.), Computational Science – ICCS 2005, Part I. LXIII, 1089 pages. 2005.

Vol. 3510: T. Braun, G. Carle, Y. Koucheryavy, V. Tsaousidis (Eds.), Wired/Wireless Internet Communications. XIV, 366 pages. 2005.

Vol. 3508: P. Bresciani, P. Giorgini, B. Henderson-Sellers, G. Low, M. Winikoff (Eds.), Agent-Oriented Information Systems II. X, 227 pages. 2005. (Subseries LNAI).

Vol. 3503: S.E. Nikoletseas (Ed.), Experimental and Efficient Algorithms. XV, 624 pages. 2005.

Vol. 3502: F. Khendek, R. Dssouli (Eds.), Testing of Communicating Systems. X, 381 pages. 2005.

Vol. 3501: B. Kégl, G. Lapalme (Eds.), Advances in Artificial Intelligence. XV, 458 pages. 2005. (Subseries LNAI).

Vol. 3500: S. Miyano, J. Mesirov, S. Kasif, S. Istrail, P. Pevzner, M. Waterman (Eds.), Research in Computational Molecular Biology. XVII, 632 pages. 2005. (Subseries LNBI).

Vol. 3498: J. Wang, X. Liao, Z. Yi (Eds.), Advances in Neural Networks – ISNN 2005, Part III. L, 1077 pages. 2005.

Vol. 3497: J. Wang, X. Liao, Z. Yi (Eds.), Advances in Neural Networks – ISNN 2005, Part II. L, 947 pages. 2005.

Vol. 3496: J. Wang, X. Liao, Z. Yi (Eds.), Advances in Neural Networks – ISNN 2005, Part II. L, 1055 pages. 2005.

Vol. 3495: P. Kantor, G. Muresan, F. Roberts, D.D. Zeng, F.-Y. Wang, H. Chen, R.C. Merkle (Eds.), Intelligence and Security Informatics. XVIII, 674 pages. 2005.

Vol. 3494: R. Cramer (Ed.), Advances in Cryptology – EUROCRYPT 2005. XIV, 576 pages. 2005.

Vol. 3492: P. Blache, E. Stabler, J. Busquets, R. Moot (Eds.), Logical Aspects of Computational Linguistics. X, 363 pages. 2005. (Subseries LNAI).

Vol. 3489: G.T. Heineman, I. Crnkovic, H.W. Schmidt, J.A. Stafford, C. Szyperski, K. Wallnau (Eds.), Component-Based Software Engineering. XI, 358 pages. 2005.

Vol. 3488: M.-S. Hacid, N.V. Murray, Z.W. Raś, S. Tsumoto (Eds.), Foundations of Intelligent Systems. XIII, 700 pages. 2005. (Subseries LNAI).

Vol. 3486: T. Helleseth, D. Sarwate, H.-Y. Song, K. Yang (Eds.), Sequences and Their Applications - SETA 2004. XII, 451 pages. 2005.

Vol. 3483: O. Gervasi, M.L. Gavrilova, V. Kumar, A. Laganà, H.P. Lee, Y. Mun, D. Taniar, C.J.K. Tan (Eds.), Computational Science and Its Applications – ICCSA 2005, Part IV. LXV, 1362 pages. 2005.

Vol. 3482: O. Gervasi, M.L. Gavrilova, V. Kumar, A. Laganà, H.P. Lee, Y. Mun, D. Taniar, C.J.K. Tan (Eds.), Computational Science and Its Applications – ICCSA 2005, Part III. LXV, 1340 pages. 2005.

Vol. 3481: O. Gervasi, M.L. Gavrilova, V. Kumar, A. Laganà, H.P. Lee, Y. Mun, D. Taniar, C.J.K. Tan (Eds.), Computational Science and Its Applications – ICCSA 2005, Part II. LV, 1316 pages. 2005.

Vol. 3480: O. Gervasi, M.L. Gavrilova, V. Kumar, A. Laganà, H.P. Lee, Y. Mun, D. Taniar, C.J.K. Tan (Eds.), Computational Science and Its Applications – ICCSA 2005, Part I. LXV, 1234 pages. 2005.

Vol. 3479: T. Strang, C. Linnhoff-Popien (Eds.), Location- and Context-Awareness. XII, 378 pages. 2005.

Vol. 3477: P. Herrmann, V. Issarny, S. Shiu (Eds.), Trust Management. XII, 426 pages. 2005.

Vol. 3475: N. Guelfi (Ed.), Rapid Integration of Software Engineering Techniques. X, 145 pages. 2005.

Vol. 3468: H.W. Gellersen, R. Want, A. Schmidt (Eds.), Pervasive Computing. XIII, 347 pages. 2005.

Vol. 3467: J. Giesl (Ed.), Term Rewriting and Applications. XIII, 517 pages. 2005.

Vol. 3465: M. Bernardo, A. Bogliolo (Eds.), Formal Methods for Mobile Computing. VII, 271 pages. 2005.

Vol. 3463: M. Dal Cin, M. Kaâniche, A. Pataricza (Eds.), Dependable Computing - EDCC 2005. XVI, 472 pages. 2005.

Vol. 3462: R. Boutaba, K. Almeroth, R. Puigjaner, S. Shen, J.P. Black (Eds.), NETWORKING 2005. XXX, 1483 pages. 2005.

Vol. 3461: P. Urzyczyn (Ed.), Typed Lambda Calculi and Applications. XI, 433 pages. 2005.

Vol. 3460: Ö. Babaoglu, M. Jelasity, A. Montresor, C. Fetzer, S. Leonardi, A. van Moorsel, M. van Steen (Eds.), Self-star Properties in Complex Information Systems. IX, 447 pages. 2005.

Vol. 3459: R. Kimmel, N.A. Sochen, J. Weickert (Eds.), Scale Space and PDE Methods in Computer Vision. XI, 634 pages. 2005.

Vol. 3458: P. Herrero, M.S. Pérez, V. Robles (Eds.), Scientific Applications of Grid Computing. X, 208 pages. 2005.

Vol. 3456: H. Rust, Operational Semantics for Timed Systems. XII, 223 pages. 2005.

Vol. 3455: H. Treharne, S. King, M. Henson, S. Schneider (Eds.), ZB 2005: Formal Specification and Development in Z and B. XV, 493 pages. 2005.

Vol. 3454: J.-M. Jacquet, G.P. Picco (Eds.), Coordination Models and Languages. X, 299 pages. 2005.

Vol. 3453: L. Zhou, B.C. Ooi, X. Meng (Eds.), Database Systems for Advanced Applications. XXVII, 929 pages. 2005.

Vol. 3452: F. Baader, A. Voronkov (Eds.), Logic for Programming, Artificial Intelligence, and Reasoning. XI, 562 pages. 2005. (Subseries LNAI).

Vol. 3450: D. Hutter, M. Ullmann (Eds.), Security in Pervasive Computing. XI, 239 pages. 2005.

Vol. 3449: F. Rothlauf, J. Branke, S. Cagnoni, D.W. Corne, R. Drechsler, Y. Jin, P. Machado, E. Marchiori, J. Romero, G.D. Smith, G. Squillero (Eds.), Applications of Evolutionary Computing. XX, 631 pages. 2005.

Vol. 3448: G.R. Raidl, J. Gottlieb (Eds.), Evolutionary Computation in Combinatorial Optimization. XI, 271 pages. 2005.

Vol. 3447: M. Keijzer, A. Tettamanzi, P. Collet, J.v. Hemert, M. Tomassini (Eds.), Genetic Programming. XIII, 382 pages. 2005.

Vol. 3444: M. Sagiv (Ed.), Programming Languages and Systems. XIII, 439 pages. 2005.

Vol. 3443: R. Bodik (Ed.), Compiler Construction. XI, 305 pages. 2005.

Vol. 3442: M. Cerioli (Ed.), Fundamental Approaches to Software Engineering. XIII, 373 pages. 2005.

Vol. 3441: V. Sassone (Ed.), Foundations of Software Science and Computational Structures. XVIII, 521 pages. 2005.

Vol. 3440: N. Halbwachs, L.D. Zuck (Eds.), Tools and Algorithms for the Construction and Analysis of Systems. XVII, 588 pages. 2005.

Vol. 3439: R.H. Deng, F. Bao, H. Pang, J. Zhou (Eds.), Information Security Practice and Experience. XII, 424 pages. 2005.

Vol. 3437: T. Gschwind, C. Mascolo (Eds.), Software Engineering and Middleware. X, 245 pages. 2005.

Vol. 3436: B. Bouyssounouse, J. Sifakis (Eds.), Embedded Systems Design. XV, 492 pages. 2005.

Vol. 3434: L. Brun, M. Vento (Eds.), Graph-Based Representations in Pattern Recognition. XII, 384 pages. 2005.

Vol. 3433: S. Bhalla (Ed.), Databases in Networked Information Systems. VII, 319 pages. 2005.

Vol. 3432: M. Beigl, P. Lukowicz (Eds.), Systems Aspects in Organic and Pervasive Computing - ARCS 2005. X, 265 pages. 2005.

Vol. 3431: C. Dovrolis (Ed.), Passive and Active Network Measurement. XII, 374 pages. 2005.

Vol. 3429: E. Andres, G. Damiand, P. Lienhardt (Eds.), Discrete Geometry for Computer Imagery. X, 428 pages. 2005.

Vol. 3428: Y.-J. Kwon, A. Bouju, C. Claramunt (Eds.), Web and Wireless Geographical Information Systems. XII, 255 pages. 2005.

Vol. 3427: G. Kotsis, O. Spaniol (Eds.), Wireless Systems and Mobility in Next Generation Internet. VIII, 249 pages. 2005.

Vol. 3423: J.L. Fiadeiro, P.D. Mosses, F. Orejas (Eds.), Recent Trends in Algebraic Development Techniques. VIII, 271 pages. 2005.

Vol. 3422: R.T. Mittermeir (Ed.), From Computer Literacy to Informatics Fundamentals. X, 203 pages. 2005.

Vol. 3421: P. Lorenz, P. Dini (Eds.), Networking - ICN 2005, Part II. XXXV, 1153 pages. 2005.

Vol. 3420: P. Lorenz, P. Dini (Eds.), Networking - ICN 2005, Part I. XXXV, 933 pages. 2005.

Vol. 3419: B. Faltings, A. Petcu, F. Fages, F. Rossi (Eds.), Constraint Satisfaction and Constraint Logic Programming. X, 217 pages. 2005. (Subseries LNAI).

Vol. 3418: U. Brandes, T. Erlebach (Eds.), Network Analysis. XII, 471 pages. 2005.

Vol. 3416: M. Böhlen, J. Gamper, W. Polasek, M.A. Wimmer (Eds.), E-Government: Towards Electronic Democracy. XIII, 311 pages. 2005. (Subseries LNAI).

Vol. 3415: P. Davidsson, B. Logan, K. Takadama (Eds.), Multi-Agent and Multi-Agent-Based Simulation. X, 265 pages. 2005. (Subseries LNAI).

Vol. 3414: M. Morari, L. Thiele (Eds.), Hybrid Systems: Computation and Control. XII, 684 pages. 2005.

Vol. 3412: X. Franch, D. Port (Eds.), COTS-Based Software Systems. XVI, 312 pages. 2005.

Vol. 3411: S.H. Myaeng, M. Zhou, K.-F. Wong, H.-J. Zhang (Eds.), Information Retrieval Technology. XIII, 337 pages. 2005.

Vol. 3410: C.A. Coello Coello, A. Hernández Aguirre, E. Zitzler (Eds.), Evolutionary Multi-Criterion Optimization. XVI, 912 pages. 2005.

Vol. 3409: N. Guelfi, G. Reggio, A. Romanovsky (Eds.), Scientific Engineering of Distributed Java Applications. X, 127 pages. 2005.

Vol. 3408: D.E. Losada, J.M. Fernández-Luna (Eds.), Advances in Information Retrieval. XVII, 572 pages. 2005.

Vol. 3407: Z. Liu, K. Araki (Eds.), Theoretical Aspects of Computing - ICTAC 2004. XIV, 562 pages. 2005.

Vol. 3406: A. Gelbukh (Ed.), Computational Linguistics and Intelligent Text Processing. XVII, 829 pages. 2005.

Vol. 3404: V. Diekert, B. Durand (Eds.), STACS 2005. XVI, 706 pages. 2005.

Vol. 3403: B. Ganter, R. Godin (Eds.), Formal Concept Analysis. XI, 419 pages. 2005. (Subseries LNAI).

Vol. 3402: M. Daydé, J.J. Dongarra, V. Hernández, J.M.L.M. Palma (Eds.), High Performance Computing for Computational Science - VECPAR 2004. XI, 732 pages. 2005.

Vol. 3401: Z. Li, L.G. Vulkov, J. Waśniewski (Eds.), Numerical Analysis and Its Applications. XIII, 630 pages. 2005.